W9-CTM-419

NINTH EDITION

Automotive Mechanics

WILLIAM H. CROUSE ♦ DONALD L. ANGLIN

GREGG DIVISION
McGRAW-HILL BOOK COMPANY

New York Atlanta Dallas St. Louis San Francisco Auckland Bogotá Guatemala Hamburg Johannesburg Lisbon London Madrid Mexico Montreal New Delhi Panama Paris San Juan São Paulo Singapore Sydney Tokyo Toronto

ABOUT THE AUTHORS

♦ William H. Crouse

Behind William H. Crouse's clear technical writing is a background of sound mechanical engineering training as well as a variety of practical industrial experience. After finishing high school, he spent a year working in a tin-plate mill. Summers, while still in school, he worked in General Motors plants, and for three years he worked in the Delco-Remy division shops. Later he became director of field education in the Delco-Remy Division of General Motors Corporation, which gave him an opportunity to develop and use his writing talent in the preparation of service bulletins and educational literature.

Mr. Crouse became editor of technical education books for the McGraw-Hill Book Company and was the first editor-in-chief of McGraw-Hill's *Encyclopedia of Science and Technology*. He has contributed numerous articles to automotive and engineering magazines and has written many outstanding books.

William H. Crouse's outstanding work in the automotive field has earned for him membership in the Society of Automotive Engineers and in the American Society of Engineering Education.

♦ Donald L. Anglin

Trained in the automotive and diesel service field, Donald L. Anglin has worked both as a mechanic and as a service manager. He has taught automotive courses and has also worked as curriculum supervisor and school administrator. Interested in all types of vehicle performance, he has served as a racing-car mechanic and as a consultant to truck fleets on maintenance problems.

Currently he devotes full time to technical writing teaching, and visiting automotive instructors and service shops. Together with William H. Crouse he has co-authored magazine articles on automotive education and many automotive books published by McGraw-Hill.

Donald L. Anglin is a Certified General Automotive Mechanic, a Certified General Truck Mechanic, and holds many other licenses and certificates in automotive education service and related areas. His work in the automotive service field has earned for him membership in the American Society of Mechanical Engineers and the Society of Automotive Engineers. In addition, he is a member of the Board of Trustees of the National Automotive History Collection.

Sponsoring Editor: D. Eugene Gilmore
Editing Supervisor: Larry Goldberg
Design and Art Supervisor/Text Designer: Nancy Axelrod
Production Supervisor: Laurence Charnow

Cover Designer: David Archambault
Part Opening Illustration: Edward Butler
Technical Studio: Fine Line, Inc.

Library of Congress Cataloging in Publication Data

Crouse, William Harry, [Date]
Automotive mechanics.

Includes index.
1. Automobiles. 2. Automobiles—Maintenance and repair. 3. Automobiles—Maintenance and repair—Examinations, questions, etc. I. Anglin, Donald L. II. Title.
TI205.C86 1984 629.28'722 84-4388
ISBN 0-07-014860-0

AUTOMOTIVE MECHANICS, NINTH EDITION

2 3 4 5 6 7 8 9 0 DOWDOW 8 9 1 0 9 8 7 6 5

ISBN 0-07-014860-0

CONTENTS

PREFACE

This is the ninth edition of *Automotive Mechanics.* In the five years since the publication of the last edition, tremendous changes have taken place in the automotive industry. Automotive vehicles imported from Japan and Europe have made a major impact on the domestic market in the United States. They now account for about 25 percent of all automotive vehicles sold in this country.

To counteract this flood of imports, American manufacturers have brought out new lines of small, highly driveable, fuel-efficient cars and trucks. Both the imported and the domestic vehicles have many innovations and feature many technological advances. All these changes have resulted in a major revision of *Automotive Mechanics.*

This new edition covers new suspension systems, including MacPherson front and rear suspensions; rack-and-pinion steering; electronic control of fuel and ignition systems, including fuel metering and spark advance; new gasoline fuel-injection systems; new developments in emission control; transversely mounted engines; transaxles; four-wheel drive and transfer cases; new instrument-panel indicating devices; engine self-diagnostic systems; "talking" cars; and much more.

This edition continues the feature of including the metric equivalents of all United States Customary System measurements.

In addition to the new developments covered in this edition, the text has a comprehensive section on automotive safety in the shop and on the hand and power tools students use in the shop.

The text has been largely rewritten to simplify explanations, shorten sentences, and improve readability. There are objectives at the beginning of each chapter so that the student and teacher both know what the learning outcome should be. There are multiple-choice tests at the end of each chapter.

The ancillary materials developed for use with *Automotive Mechanics* have also been revised. These include a study guide, a shop workbook, a testbook, and an instructor's planning guide.

All these materials and the text provide the instructor with a flexible teaching package that should fit any type of teaching situation. The instructor's planning guide explains how the various materials can be used, either alone or with others, to meet any teaching requirement. It also contains the answers to the multiple-choice tests in the text.

The *Automotive Mechanics* program is correlated with the recommendations of the Motor Vehicle Manufacturers Association—American Vocational Association Industry Planning Council. The program also incorporates guidelines for automotive mechanics certification, state plans for vocational education, and recommendations for automotive trade preapprenticeship and apprenticeship training. The program is flexible: it will fit classroom instruction, shop activities, individual instructions, or "how-to" courses for hobbyists and consumers.

This edition of *Automotive Mechanics* covers the subjects recommended by the American National Standards Institute in their detailed standards booklet "American National Standard for Training of Automotive Mechanics for Passenger Cars, Recreational Vehicles, and Light Trucks."

In addition, *Automotive Mechanics* covers in depth the subjects tested by the National Institute for Automotive Service Excellence (ASE). These tests are used for certifying general automotive mechanics and other automotive-service technicians working in specific areas of specialization under the ASE voluntary mechanic testing and certification program.

During the planning and preparation of the ninth edition of *Automotive Mechanics,* the authors and publisher had the advice and assistance of many people—educators, researchers, artists, editors, and automotive industry service specialists. Special thanks must go to Frank C. Derato, Westhill High School, Stamford, Conn.; John Flaherty, Sandhills Community College, Carthage, N.C.; and John Steck; who took time from their busy schedules to attend the master planning session for the new edition. Sincere thanks also go to A. William Palmer, East Los Angeles College, for his careful and thoughtful review of the final manuscript.

William H. Crouse
Donald L. Anglin

TO THE STUDENT

Automotive Mechanics was designed with you, the student, in mind. It was put together to help you learn, in the quickest and easiest way possible, all about automobiles. By the time you have finished studying this book, and have done the related shopwork, you will be ready to enter the world of automotive servicing as a wage earner.

Various special materials have been developed to make learning easier. These materials include a study guide, a shop workbook, and a testbook. Each of these was prepared with one thought in mind—to help you learn all about automobiles.

Your major job now is to study this book and to do your shopwork. Studying is usually hard for everyone. It takes will power to sit down and read this technical material. But if you follow the suggestions listed below, you will find studying much easier.

1. The first thing to do when you pick up your textbook to study your assignment is to turn the pages one by one. Look at the pictures. Read the numbered section headings. Study each section heading carefully. This will give you an idea of what the assignment is about.
2. If you are starting a new chapter in the textbook, read the objective that tells you what you should be able to do after you have studied the chapter. This emphasizes what you should learn from the chapter.
3. Read the first section in your assignment. Then read the first section again slowly and carefully so that you make sure you understand it.
4. Continue studying the pages assigned to you. Read each section carefully.
5. If you come to a sentence that you don't understand, read it aloud. Think about it. If this does not help, write the sentence on a piece of paper. When you have a chance, ask your instructor to explain the sentence.
6. Don't hesitate to admit that something puzzles you. Everybody gets stuck once in a while. Your instructor is there to help you understand.
7. Don't worry about not getting everything the first time you read it. Most good students read and reread their lessons several times.

ACKNOWLEDGMENTS

Special thanks are owed to the following organizations for information and illustrations they supplied: AC Spark Plug Division of General Motors Corporation; American Honda Motor Company, Inc.; American Motors Corporation; Ammco Tools Inc.; ATW; Automotive Division of Applied Power Inc.; Bendix Corporation; B. F. Goodrich Company; Black & Decker, Inc.; BMW of North America, Inc.; Buick Motor Division of General Motors Corporation; Cadillac Motor Car Division of General Motors Corporation; Champion Spark Plug Company; Chevrolet Motor Division of General Motors Corporation; Chicago Pneumatic Tool Company; Chrysler Corporation; Clayton Manufacturing Company; Clevite Division of Gould, Inc.; CR Industries; Delco Moraine Division of General Motors Corporation; Delco-Remy Division of General Motors Corporation; Dow Corning Corporation; Eaton Corporation; Federal-Mogul Corporation; Fiat Motors of North America, Inc.; Firestone Tire & Rubber Company; Ford Motor Company; Ford Motor Company of Germany; General Motors Corporation; Goodyear Tire & Rubber Company; Graymills Corporation; Guaranteed Parts Company; Hamilton Test Systems Subsidiary of United Technologies; Harrison Radiator Division of General Motors Corporation; Hastings Manufacturing Company; Hennessy Industries, Inc.; Hunckler Products, Inc.; Hunter Engineering Company; Inland Manufacturing Company; International Harvester Company; Jenny Division of Homestead Industries; Kwik-Way Manufacturing Company; Lempco Industries, Inc.; Lincoln St. Louis Division of McNeil Corporation; Lisle Corporation; The L. S. Starret Company; Mazda Motors of America, Inc.; McCord Replacement Products Division of McCord Corporation; McQuay-Norris, Inc.; Mercedes-Benz of North America, Inc.; Mobil Oil Corporation; Monroe Auto Equipment Company; Moog Automotive, Inc.; Motor Vehicle Manufacturers Association; Nissan Motor Company, Ltd.; Oldsmobile Division of General Motors Corporation; Owatonna Tool Company; Pontiac Division of General Motors Corporation; Proto Tool Company; Renault USA Inc.; Robert Bosch Corporation; Rochester Products Division of General Motors Corporation; Rockwell International Corporation; Rubber Manufacturers Association; Saab-Scania of America, Inc.; Schwitzer Division of Wallace-Murray Corporation; Sealed Power Corporation; Sioux Tools, Inc.; Snap-on Tools Corporation; Subaru of America, Inc.; Sun Electric Corporation; Sunnen Products Company; Texaco, Inc.; Texas Instruments, Incorporated; Tire Industry Safety Council; Toyota Motor Sales, Inc.; TRW, Inc.; Volkswagen of America, Inc.; Wagner Division of McGraw-Edison Company; Warner Electric Brake and Clutch Company; and Weaver Manufacturing Company.

William H. Crouse
Donald L. Anglin

PART 1

Part 1 of *Automotive Mechanics* discusses automotive-shop safety: how to work safely and protect yourself and others from harm. It also explains how to use and take care of the various tools you will use when you go out into the shop. This part of the book has seven chapters:

Automotive-Shop Safety and Tools

CHAPTER 1
The Automotive-Service Business

After studying this chapter, you should be able to:

1. List the places where automotive service is performed.
2. Explain what the person who hires you expects you to know.
3. List the major domestic manufacturers of automobiles sold in the United States.
4. Describe how you will learn about automobiles in this course.

♦ 1-1 AUTOMOTIVE SERVICE

Today about 160 million automobiles, trucks, and buses operate on the streets and highways of the United States (Fig. 1-1). There are millions more off-the-road vehicles such as tractors, power mowers, motorcycles, dune buggies, and snowmobiles. More than a million men and women work to keep all this equipment going. Here are some of the places where trained automotive mechanics and technicians work:

- Dealerships, where cars and trucks are sold and serviced
- Independent garages, which service all makes and types of vehicles
- Service stations, where vehicles get fuel, oil, and related products and services
- Tire and battery dealers
- Service facilities set up by chains such as Sears Roebuck
- Specialty shops, which handle wheel alignment, transmissions, body repair, and tuneup work
- Fleet garages, where fleet operators, truck lines, and bus lines have their own service shops
- Parts stores, where automotive parts are sold

♦ 1-2 GETTING READY FOR A JOB IN AUTOMOTIVE SERVICE

The first step toward getting a job in the automotive-service business is to learn the fundamentals. By this, we mean finding out how all the components of the automobile work; how they are constructed; how they work; and what to do when they don't work.

You are taking that important first step right now. You are starting to learn all about automobiles. This automotive course you are taking, this book you are reading, and your related shopwork will give you the fundamentals.

When someone hires you as an automotive mechanic or automotive technician, that person expects you to have a basic knowledge of automobiles. Just stop for a moment and turn back to the contents at the beginning of this book. Look at the chapter headings. You will see that the book covers engines, fuel systems, lubricating systems, cooling systems, electrical equipment, and so on. This is what you will be studying in this course. And you will be applying what you learn from the book and in the classroom in the shop where you will put your knowledge to work. In most shops, you

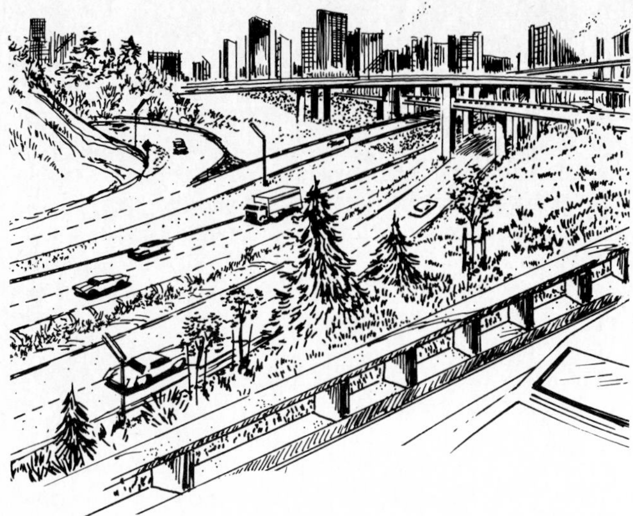

Fig. 1-1 About 160 million automotive vehicles operate on the streets and highways of the United States.

must have your own hand tools to get a job as a mechanic. Tools are discussed in later chapters.

Every chapter you complete, every shop job you master, takes you that much closer to getting a good job in the automotive-service business.

♦ 1-3 THE AUTOMOTIVE INDUSTRY

Before we begin studying how automobiles operate and how to service them, let us look at how automobiles are made. There are four major companies (domestic manufacturers) in the United States making automobiles:

- ♦ American Motors Corporation
- ♦ Chrysler Corporation
- ♦ Ford Motor Company
- ♦ General Motors Corporation

In addition, some foreign-car makers have built manufacturing plants in the United States; Volkswagen, or example. Also, there are many foreign companies that ship motor vehicles into this country. These include Honda, Toyota, Datsun, BMW, Mercedes, and others.

Let us follow the production of a new-model automobile from its start as an idea to the complete car. It takes three or more years to translate a new design into the complete new car. Detailed drawings of every part have to be made. Machinery has to be set up to make the parts. Material has to be ordered ahead of time. The assembly lines have to be set up and machinery installed. People who will work on the machines and assembly lines have to be trained in their operation.

Meantime, the factory service engineers are producing the shop manuals that explain how to test and service all the components in the car. When new cars come out, service engineers organize schools to train mechanics in the new servicing procedures required on the new cars.

All this is a very complicated process that involves hundreds of thousands of people. When you see a new car, think of how many millions of hours that men and women spent to think up, design, and make the car.

♦ 1-4 OPPORTUNITIES IN THE AUTOMOTIVE-SERVICE BUSINESS

Automotive service offers many excellent opportunities for men and women who know automobiles. You are making a fine start toward learning about automobiles by studying this book. The better jobs require a knowledge of how the automobile works and how to service and maintain it.

What are some of these job opportunities?

The efficient automotive mechanic is a highly respected, well-paid worker who earns as much as men and women in other skilled professions. The mechanic's services are always in demand. And there is a shortage of good automotive mechanics.

The job of automotive mechanic can be a stepping-stone to greater things. The automotive mechanic or technician can become the service manager or parts manager of an automotive shop or new-car agency. The automotive mechanic might someday want to open his or her own business. Most independent garages, specialty shops, and service stations are owned by automotive mechanics who have moved up the ladder.

Many automotive salespeople, manufacturers' representatives, and even top executives in the factories were once automotive mechanics. The opportunities are almost unlimited. Men and women who know automobiles and have the right personality and knowledge will achieve success.

All successful men and women have one thing in common, the ability to study and work hard. Success for you lies in the same direction, and you can succeed in the same way. Get everything you can from this book and your courses.

♦ 1-5 THE SERVICE STATION

There are more than 158,000 service stations in the United States, varying from the small two- or three-pump station adjacent to the country store to the huge complexes you see along the interstate highways. The services they offer are just as varied. Some are simply "filling" stations, where you get gas and oil. Others are equipped to do a complete tuneup. Some have mechanics who can take care of many of the services a car might need—balance wheels, align wheels, repair brake systems, make minor engine repairs, and the like.

♦ 1-6 AUTOMOTIVE DEALERS

There are about 25,000 new-car dealers in the United States (Fig. 1-2). These are franchised dealers. By this, we mean that the dealer has a contract—a written agreement—with the vehicle manufacturer, spelling out the dealer-manufacturer agreement. The manufacturer supplies the dealer with vehicles at wholesale prices.

The dealer, in turn, is obligated to keep all customers satisfied with the vehicles sold and with the service work done in the service shop. The dealer must make sure that all new cars are properly prepared for delivery to customers. Service work on cars driven into the service department by customers must be good so that customers are satisfied. At the same time, the work must make a profit for the department. If trouble develops in a new car, it is the dealer's responsibility to see that it is corrected under the manufacturer's warranty agreement. This agreement is a warranty that new cars will be fixed at little or no cost to the car owner. Under the warranty agreement, the cost of parts and labor is paid by the manufacturer.

Fig. 1-2 Automotive dealers not only sell new and used cars, but they also have a well-equipped service shop.

A dealership is a complicated business. But it offers many opportunities for the person who wants to get ahead in the automotive-service business. The service mechanic or technician can become a specialist in engines, brakes, electrical systems, suspension and steering systems, or some other area. Another step up the ladder would be service or sales manager and possibly, at last, an automotive dealer.

♦ 1-7 INDEPENDENT GARAGES

There are an estimated 177,000 independent and franchised auto repair shops in the United States (Fig. 1-3). These vary from the small shop with only a few workers to the big, all-purpose garage with a large staff of automotive mechanics, parts people, and related personnel. The fact that there are so many independent garages means that many car owners do not go back to the dealer for service. These owners, for one reason or another, prefer to go to an independent garage. Also, many cars are second- or third-owner, bought from a used-car dealer not connected with a new-car dealership. These cars are more likely to end up at an independent garage when service is required.

The independent garage, if large, offers a complete line of automotive service, from tuneups, alignments, and body work to complete overhaul of engines, transmissions, differentials, and other major car components. As a result, the automotive mechanics in these garages may work on a great variety of jobs. Here again, the best mechanics rise to the top. Mechanics can become department managers, service managers, and garage managers or owners.

Fig. 1-3 Independent garages handle all types of automobile service work.

♦ 1-8 SPECIALTY SHOPS

A great variety of specialty shops provide backup service for the automotive-service industry. They include shops that do major machine work such as brake-drum or disk turning, crankshaft and camshaft grinding, brake-shoe relining, and the like. Some shops specialize in engine rebuilding, or "remanufacturing." Also, there are such places as the AAMCO franchised automotive-transmission shops, which deal basically with one component. Another example is the Midas Muffler shops, specializing in replacing mufflers. Some service shops specialize in repairing or rebuilding small components such as alternators, carburetors, and distributors. Some specialty shops are set up to handle body repair and repainting or interior upholstery work.

There are many opportunities in these shops for the person who likes to stay in one place and do one kind of job, rather than hop from one job to another as in a service station. Here again, the person who knows the job can rise to a better job—shop supervisor, manager, owner.

♦ 1-9 FLEET GARAGES

Small fleet operators with only a few trucks, buses, or other vehicles usually have their service work done at an independent garage equipped to handle their heavy vehicles. However, the larger fleet owner usually sets up a garage to handle service and maintenance of the vehicles in the fleet. Often, the work is done on a preset schedule that calls for periodic checks and servicing of various components. Some items are checked every time the truck or bus goes out on the road. Other items are checked less frequently. The procedures are called *driver maintenance* and *preventive maintenance*. The equipment is *maintained* in good working condition to *prevent* failures. A break-down on the highway could result in a loss of thousands of dollars. This would happen if the truck that broke down was carrying frozen food which thawed and spoiled. Likewise, a bus breaking down in some remote place could result in many angry passengers and lost future ridership.

Here again, there are many opportunities for the man or woman who knows the vehicles and can handle the service jobs.

♦ 1-10 PARTS DEALER

The parts dealer operates much like a retail store. The dealer keeps thousands of different automotive-service parts in stock. Also, the dealer has contact with a network of distributors and suppliers who can get needed parts quickly. Service stations and garages needing parts order them and the parts dealer delivers them. Often, the garage sends someone in to the parts dealer to pick up parts needed in a hurry.

Working in a parts store is similar to working in any other retail store. You must know the merchandise. This makes being an automotive mechanic very helpful, because the merchandise is all automotive parts. You must find the proper parts numbers in the parts books, know where the parts are located, keep an inventory of what is in stock, and order more parts from the suppliers when the stock gets low. For example, a customer might want a set of main bearings for an engine. First you find the part number in the parts catalog (Fig. 1-4). Then you locate the bearings and bring them to the counter. Parts dealers have parts books or catalogs that

Fig. 1-4 Automotive-service parts books can be very large.

list and sometimes picture the various parts. These parts books are very large, sometimes 2 feet [0.61 meter] thick.

Today, many manufacturers have replaced the parts books with a microfiche system. Each fiche is a small sheet of microfilm that may have on it hundreds of pages reduced to almost microscopic size (Fig. 1-5). To find a part number on a microfiche, you pick out the correct fiche from the file and put the fiche into the microfiche reader. Then, by operating a control, the fiche is shifted until the reader shows the enlarged page on a screen.

♦ 1-11 DEPARTMENT, ACCESSORY, AND AUTOMOTIVE-SUPPLY STORES

Many department stores, chains such as Sears and specialty stores offer automotive service. They sell all sorts of accessories and necessities from tires and batteries to complete rebuilt engines. Many have complete servicing facilities. There are many job opportunities in these stores, ranging from actual service work to selling automotive parts and accessories. The range of service offered by these stores is large. Therefore, there is a great variety of jobs, with many opportunities to advance to more responsible positions.

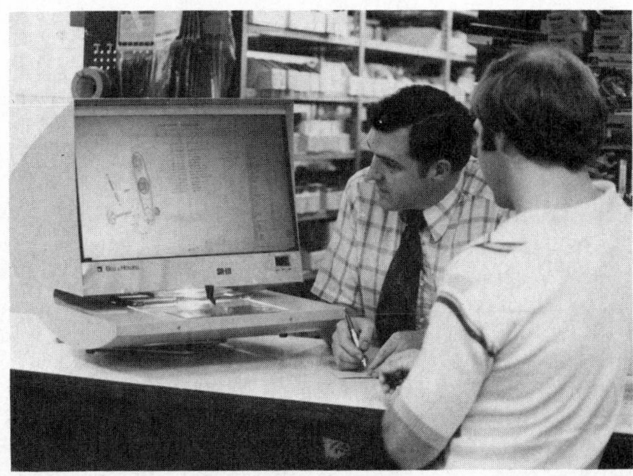

Fig. 1-5 A microfiche can contain many pages with thousands of parts listed, reduced to almost microscopic size.

♦ 1-12 THIS IS HOW YOU WILL LEARN ABOUT AUTOMOBILES

There are two parts to the job of learning about automobiles and how to service them. One part is to study this book just as you are now doing, and to do your classroom work. The other part is to go out into the shop and work on the automotive components you have been studying. For example, when you come to the chapter in the book on engine-valve service, you will first study the chapter. Then you will go out into the shop and do the jobs you were reading about.

You may be using the *Workbook for Automotive Mechanics*. It tells you, step-by-step, how to do each service job you are assigned.

Another aid that some schools use is the *Study Guide for Automotive Mechanics*, which gives you quick reviews of all the chapters in the *Automotive Mechanics* textbook. For example, suppose you have just studied the chapter in this book on valve service. There is a lot to learn in that chapter. The study guide helps you review the chapter. It goes over the chapter, item by item, emphasizing the key points you should know. This review helps you to learn the essentials of valve service.

♦ 1-13 KEEPING A NOTEBOOK

Keeping a notebook is a valuable part of your training to become an expert automotive mechanic. Start it now, at the beginning of your study of this textbook. Your notebook will help you in many ways. It will be a record of your progress in your studies. It will become a storehouse of valuable information you will refer to time after time. It will help you learn. And it will help you organize your training program to do you the most good. Do not overlook this valuable way of becoming the automotive expert you want to be. Keep a notebook!

Here's how to keep your notebook. Get yourself a large 8½ by 11 inch [216 by 279 millimeter (mm)] ring binder and a set of notebook dividers or index tabs. Organize your notebook into 10 parts, just as *Automotive Mechanics* is divided into 10 parts. These parts are:

1. Automotive-shop Safety and Tools
2. Automotive Engines
3. Automotive Engine Systems
4. Automotive Electrical and Electronic Equipment
5. Automotive Emission Controls
6. Engine Trouble Diagnosis and Tuneup
7. Automotive Engine Service
8. Automotive Power Trains
9. Automotive Suspension, Steering, and Brakes
10. Automotive Heating and Air Conditioning

When you study a lesson in the textbook, have your notebook open in front of you. Start with a fresh notebook page at the beginning of the lesson. Write the lesson number or textbook page numbers and date at the top of the page. As you read your lesson, jot down the highlights.

You may not want to carry your big notebook around with you while you work on a car. You can put a small scratch pad or a few 3 by 5 inch [76.2 by 127 mm] cards in your pocket.[1] You will probably need this paper anyway to write down part

[1]Note that in this text, metric or United States Customary equivalents are shown in brackets [ ].

numbers, measurements, and specifications. And be sure you always have a pencil with you in the shop. Keep your big notebook in a drawer or on the bench. After completing a job, jot down on the pad or cards the points covered or special problems. These notes are merely reminders. Redo the notes on separate pages in your big notebook in the evening.

You can also make sketches in your notebook. These could be wiring or hose diagrams, fuel circuits, sketches of parts, or anything that is more easily drawn than described in words.

You can save articles and illustrations you find in the technical and hot-rod magazines. For example, you might find a description of how a new fuel-injection system works. You could clip this, punch the pages to fit your notebook, and insert it in the proper section.

Other important material you should save in your notebook includes the instructions you find with service parts. For example, a set of piston rings will include an instruction sheet explaining how to install the rings. Save this sheet. Cement or tape it onto a sheet of paper and file it in your notebook.

Your notebook will become one of your most valuable possessions—a permanent record of how you made your way to the top, to become an expert automotive mechanic.

So, *keep a notebook!*

♦ REVIEW QUESTIONS ♦

Select the *one* correct, best, or most probable answer to each question. You can find the answers in the section indicated at the end of each question.

1. In the automotive shop, you may work on (♦1-2)
   a. engines
   b. fuel systems
   c. electrical systems
   d. all of the above
2. The contents at the beginning of the book gives you (♦1-2)
   a. a list of the jobs you must do
   b. a list of the chapter titles
   c. a list of the places you will work
   d. none of the above
3. From the time that a new car starts as an idea until it is rolling off the assembly line takes about (♦1-3)
   a. six months
   b. a year
   c. three years or more
   d. ten years or more
4. The service facility that most likely performs warranty work is (♦1-6)
   a. a new-car dealer
   b. an independent garage
   c. a service station
   d. a department store
5. The two major parts to the job of learning about automobiles and how to service them are (♦1-12)
   a. studying the book and doing classroom work
   b. studying the book and doing shopwork
   c. reading car manuals and doing shopwork
   d. studying shop manuals and following the teacher's instructions

CHAPTER 2
Safety in the Shop

After studying this chapter, you should be able to:

1. Discuss shop layouts.
2. Discuss shop safety.
3. List shop hazards due to various causes such as faulty working habits, working conditions, equipment defects, and incorrect use of hand tools.
4. Discuss fire prevention in the shop, including various types of fire extinguishers and how to use them.
5. List the safety rules.

♦ 2-1 SHOPWORK AND SAFETY

Shopwork is varied and interesting. You will learn to do many automotive jobs in the shop—adjusting wheel alignment, checking engine performance, checking charging systems, adjusting engine valves, and many other jobs. All of these jobs can be done easily and safely if you follow the safety rules.

Safety means protecting yourself and others from possible danger and injury. You do not want to get hurt, and you do not want to hurt others. But you could hurt yourself or others if you become careless and thoughtless.

This chapter describes the rules you should follow to protect yourself from harm. When everybody obeys the rules, the shop is a much safer place to work in than your home! Many more people are hurt in the home than in the shop. So—always *think safety.*

♦ 2-2 SAFETY IS YOUR JOB

Yes, safety *is* your job. In the shop, you are "safe" when you protect your eyes, your fingers, your hands—all of yourself—from danger. And, just as important, when you look out for the safety of the people you work with.

The rules of safety are listed and discussed in the next few pages. Follow the rules for your protection, and for the protection of others around you.

♦ 2-3 SHOP LAYOUTS

The term *shop layout* means the locations of workbenches, car lifts, machine tools, and so on (Fig. 2-1). Shop layouts vary. The first thing you should do in a shop is find out where everything is located. This includes the different machine tools and the workbenches, car lifts, and work areas. Many shops have painted lines on the floor to mark off work areas. These lines guide customers and workers away from danger zones where machines are being operated. The lines also remind workers to keep their tools and equipment inside work-area lines.

Many shops have warning signs posted around machinery. Some of these signs may have the name OSHA, which stands for Occupational Safety and Health Administration. This is the federal agency with the responsibility of studying and correcting conditions and equipment that present hazards to workers (see ♦(2-4). The signs are posted to remind everyone about safety, and about how to use machines safely. Follow the posted instructions at all times. The most common cause of accidents in the shop is failure to follow instructions.

Fig. 2-1 A typical automotive-shop layout.

♦ 2-4 SHOP HAZARDS TO WATCH OUT FOR

Federal laws have been enacted which are designed to ensure healthful and safe working conditions. The federally established National Institute for Occupational Safety and Health (NIOSH) makes studies of shop working conditions and reports on potential hazards that should be corrected. Further, the law requires that the shops with such hazards must eliminate them. Hazards found are sometimes the fault of management, and sometimes the fault of the workers. In the following three sections, we discuss hazards that might be considered as due to working habits, hazards due to faulty or improperly used shop equipment, and hazards due to faulty or improperly used hand tools.

♦ 2-5 HAZARDS DUE TO FAULTY WORKING HABITS OR CONDITIONS

Here are some of the major hazards that might be due to working habits of the employees or to the general working conditions:

1. Smoking while handling dangerous materials such as gasoline or solvents (Fig. 2-2). This can result in a major fire or explosion.

Fig. 2-2 Do not smoke or have open flames around combustibles such as gasoline or solvents.

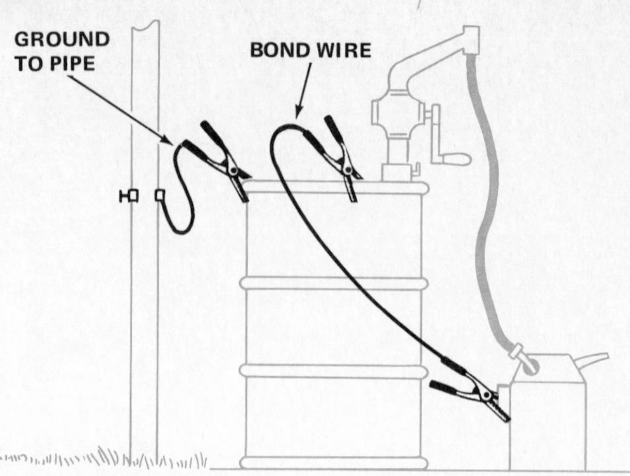

Fig. 2-3 Setup for pumping a flammable liquid from a large container into a small one.

2. Careless or incorrect handling of paint, thinners, solvents, or other flammable fluids. Figure 2-3 shows the correct arrangement for pumping a flammable fluid from a large container into a small one. Note the bond and ground wires. Without these, a spark might jump from the nozzle to the small container. This could cause an explosion and fire.
3. Blocking exits. Areas around exit doors and passageways leading to exits must be kept free of all obstructions. If you wanted to get out in an emergency—as, for example, when a fire or explosion occurred—a blocked exit could mean serious injury or even death.

♦ 2-6 HAZARDS DUE TO EQUIPMENT DEFECTS OR MISUSE

Here are the most common hazards in the shop due to equipment that is faulty or which is improperly used:

1. Incorrect safety guarding of moving machinery. For example, fans should have adequate guards, as shown in Fig 2-4. Air compressors should have proper guards over the belt and pulley (Fig. 2-5).
2. Misuse of flexible electric cords, cords that are worn or frayed, or cords that have been improperly spliced. Flexible cord should not be run through holes in the wall or tacked onto the wall. Any of these could cause a fire or could electrocute someone.
3. Compressed-gas cylinders improperly stored or misused. These cylinders should never be stored near radiators or other sources of heat. They should never be kept in unventilated enclosures such as lockers or closets. There should be at least 20 feet [6.1 meters (m)] between stored oxygen and stored acetylene cylinders. Cylinders should not stand free but should be secured with a chain or lashing. Cylinders should never be used as supports or as rollers to move an object. Such treatment could cause the cylinder to explode.
4. Hand-held electric tools not properly grounded. All such tools must have a separate ground lead (Fig. 2-6), or be double insulated to guard against shock.
5. Hydraulic car lifts improperly used. Passengers should not remain in the car when it is lifted. All doors, hood,

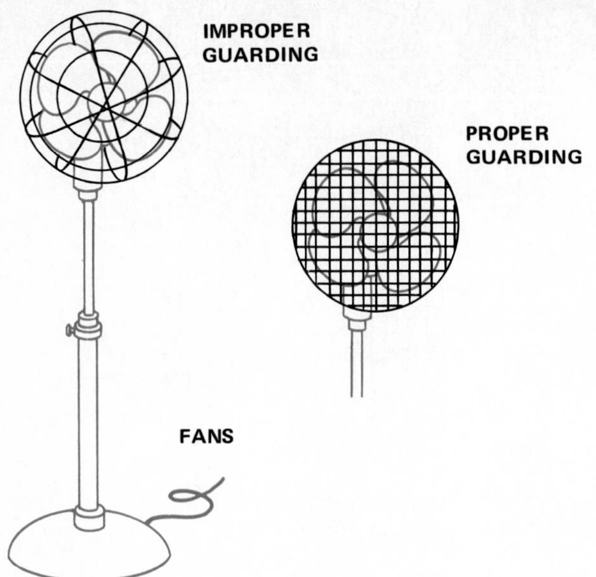

Fig. 2-4 Fans, properly and improperly guarded.

and trunk lid should be closed. Otherwise, they could be damaged as the car is lifted. If the lift has a mechanical locking device, it should be engaged before you go under the lift. Do not use a lift that is not working properly and jumps or jerks when raised, settles slowly when it should not, rises or settles too slowly, blows oil out of the exhaust line, or leaks oil at the packing gland.

6. Using a wheel-and-tire balancer which does not have the hood in place. OSHA regulations call for all wheel balancers of the spinner type to have hoods (Fig. 2-7). These hoods protect the workers in case a stone or other object is thrown from the spinning tire tread.
7. Letting tester leads fall into the engine fan when the engine is running. This, at the least, can ruin the leads. At worst, the fan can throw out leads, topple the tester, and injure the mechanic.
8. Many of the engines that are mounted transversely at the front of the car (crosswise) have electric engine fans. These can run even though the engine is turned off. As a safety measure, when working around this type of engine and fan, disconnect the electric lead to the fan. Otherwise, if the engine is hot, the fan might start unexpectedly and hurt you.
9. Leaving a running machine unattended. Whenever you are using a power tool, and have to leave it for a moment, turn it off! If you leave it running, someone might come along and, not realizing it is running, be injured.
10. Playing with fire extinguishers. There have been cases where people thought it was fun to play with the fire extinguishers. But then someone got hurt from slipping and falling down, or injured his or her eyes from the liquid or foam. Also, this leaves the extinguishers empty, so they are useless if a fire should break out.

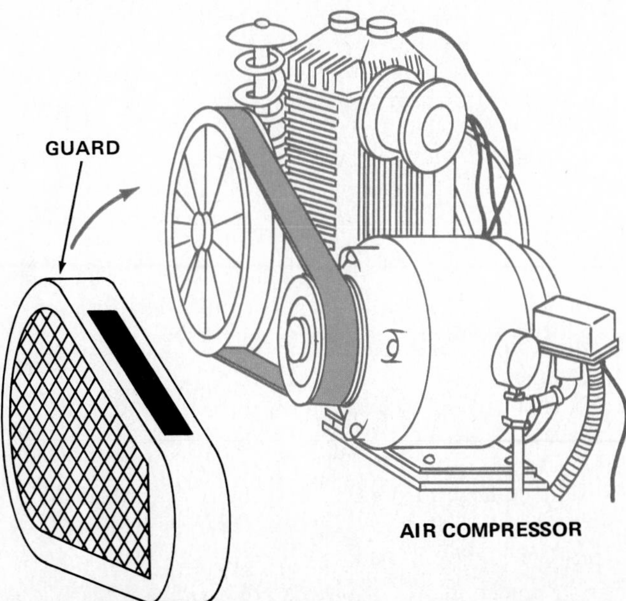

Fig. 2-5 Belts and pulleys on shop equipment should always be protected with guards.

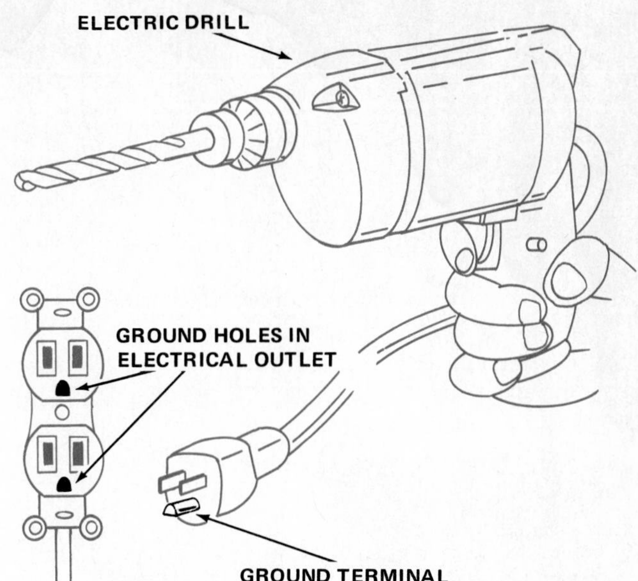

Fig. 2-6 Electric drill with three-wire cord. The third wire and terminal are to ground the electric-drill motor.

♦ 2-7 HAND-TOOL HAZARDS TO WATCH OUT FOR

Hand tools should be kept clean and in good condition. Greasy and oily tools are hard to hold and use. Always wipe them before trying to use them. Do not use a hardened hammer or punch on a hardened surface. Hardened steel is brittle and can shatter from heavy blows. Slivers may fly out and enter the hand, or worse, the eye. Hammers with broken or cracked handles, chisels and punches with mushroomed heads, or broken or bent wrenches are other tool hazards that should be avoided.

Never use a tool that is in poor condition or that does not fit the job. Avoid hand-tool hazards so you won't get hurt.

♦ 2-8 FIRE PREVENTION

Gasoline is used so much in the shop that people forget it is very dangerous if not handled properly. A spark or lighted match in a closed place filled with gasoline vapor can cause an explosion. Even the spark from a light switch can set off an explosion. So you must always be careful with gasoline. Here are some tips.

Fig. 2-7 An electronic wheel balancer used to check tire and wheel balance. The hood protects the operator from stones and other objects that might be thrown from the tire tread. The hood pivots down to a horizontal position before the wheel is spun. *(Hunter Engineering Company)*

There will be gasoline vapors around, if gasoline is spilled or a fuel line is leaking. You should keep the shop doors open or keep the ventilating system going. Wipe up the spilled gasoline at once, and put the rags outside to dry. Never smoke or light cigarettes around gasoline. When you work on a leaky

Fig. 2-8 Always store gasoline and all flammable liquids in approved safety containers. *(ATW)*

Fig. 2-9 Safety container for the storage of oily rags.

fuel line, carburetor, or fuel pump, catch the leaking gasoline in a container or with rags. Put the soaked rags outside to dry. Fix the leak as quickly as possible. And don't make sparks around the car, for example, by connecting a trouble light to the battery.

Store gasoline in an approved safety container (Fig. 2-8). Never, *never* store gasoline in a glass jug. The jug could break and could cause an explosion and fire.

Oily rags can also be a source of fire. They can catch fire without a spark or flame. Oily rags and waste should be put into a special safety container where they can do no harm (Fig. 2-9).

♦ 2-9 FIRE EXTINGUISHERS

Note the location of the fire extinguishers in the shop. Make sure you know how to use them. Figure 2-10 is a chart showing different types of fires and the types of fire extinguisher to use for each type. The quicker you begin to fight a fire, the easier it is to control. But you have to use the right kind of fire extinguisher, and use it correctly. The chart explains this. Talk over any questions with your instructor.

♦ 2-10 THE SAFETY RULES

Some people say, "Accidents will happen!" But safety experts do not agree. They say, "Accidents are caused; they are caused by careless actions, by inattention to the job, by using damaged or incorrect tools." Fewer accidents occur in a shop that is neat and clean.

To help prevent accidents from happening, follow these simple rules:

1. Work quietly and give the job you are doing your full attention.
2. Keep your tools and equipment under control.
3. Keep jack handles out of the way. Stand the creeper against the wall when it is not in use.
4. Never indulge in horseplay or other foolish activities. You could cause someone to get seriously hurt.
5. Don't put sharp objects, such as screwdrivers, in your pocket. You could cut yourself or get stabbed. Or you could damage the upholstery in a car.
6. Make sure your clothes are right for the job. Dangling sleeves or ties can get caught in machinery and cause

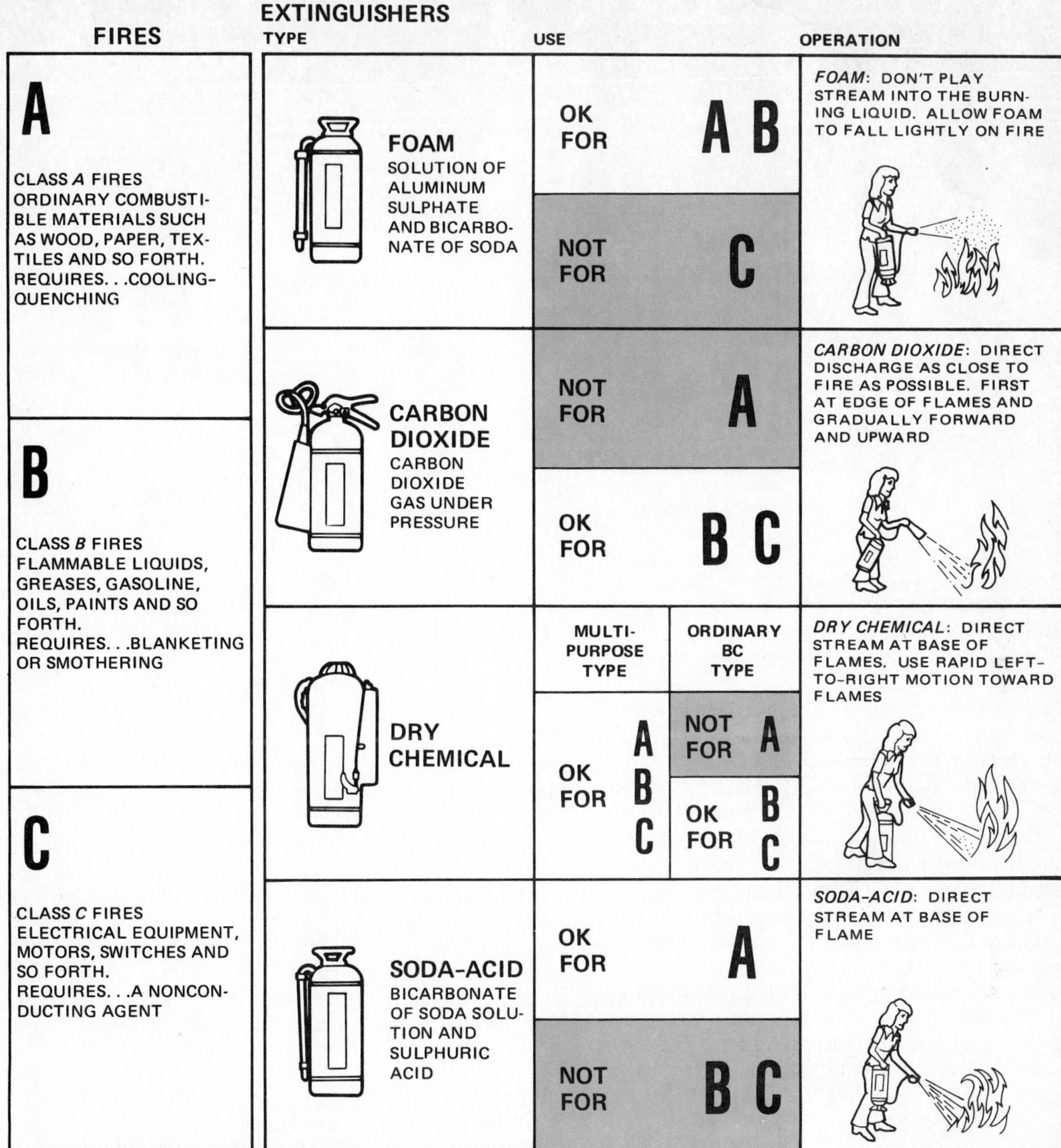

Fig. 2-10 Chart showing types of fire extinguishers and the classification of fires. *(Ford Motor Company)*

serious injuries. Do not wear sandals or open-toe shoes. Wear full leather shoes with nonskid rubber heels and soles. Steel-toe safety shoes are best for shopwork. Keep long hair out of machinery by wearing a cap.

7. Do not wear rings, bracelets, or watches when working around moving machinery or electrical equipment. Jewelry can catch in moving machinery with very serious results. Also, if a ring or bracelet should accidentally create a short circuit of the car battery, the metal of the ring or bracelet could become white hot in an instant. This would produce serious burns.
8. Wipe excess oil and grease off your hands and tools so that you can get a good grip on tools or parts.
9. If you spill oil, grease, or any liquid on the floor, clean it up so that no one will slip and fall.
10. Never use compressed air to blow dirt from your clothes. Never point a compressed-air hose at another person. Flying particles could put out an eye.
11. Always wear safety glasses, safety goggles, or a face shield when there are particles flying about. Always wear eye protection when using a grinding wheel (Fig. 2-11).
12. Watch out for sparks flying from a grinding wheel or welding equipment. The sparks can set your clothes on fire.
13. To protect your eyes, wear goggles when using chemicals, such as solvents. If you get a chemical in your eyes, wash them with water at once (Fig. 2-12). Then see the school nurse or a doctor as soon as possible.
14. When using a car jack, make sure it is centered so that it won't slip. *Never* jack up a car while someone is working

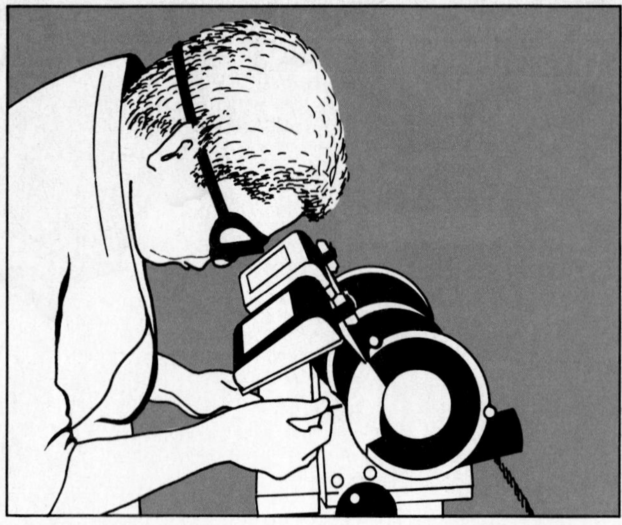

Fig. 2-11 Always wear safety glasses, safety goggles, or a face shield when using a grinder or machine that can throw chips or sparks.

under it! People have been killed when the jack slipped and the car fell on them! Always use safety stands or supports, properly placed, when going under a car (Fig. 2-13).

15. Always use the right tool for the job. The wrong tool could damage the part being worked on and could cause you to get hurt.
16. Keep your hands away from the engine fan and belt when the engine is running. You could be badly cut or even lose fingers if your hand got caught in the fan or fan belt.
17. Do not stand directly in line with the engine fan when the engine is running. Some fans, especially the flex fans, have been known to throw off a blade when spinning. Anyone standing in line with the fan could get seriously hurt if hit by the blade.
18. Many transverse engines have electric fans which can run even if the engine is off, if the engine is hot. Disconnect the lead to the fan so there is no danger while you are working around an engine with an electric fan.

Fig. 2-12 If solvent or some other chemical splashes in your eyes, immediately wash them out with water.

Fig. 2-13 Safety stands should be properly placed to support the car before you work under it.

♦ CAUTION ♦ Never run an engine in a closed garage that does not have a ventilating system. The exhaust gases contain carbon monoxide. Carbon monoxide is a colorless, odorless, tasteless, poisonous gas that can kill you! In a closed one-car garage, enough carbon monoxide to kill you can collect in only 3 minutes.

♦ 2-11 USING POWER-DRIVEN EQUIPMENT

A lot of power-driven equipment is used in the automobile shop. The instructions for using any equipment should be studied carefully before the equipment is operated. Hands and clothes should be kept away from moving machinery. Keep hands out of the way when using any cutting device, such as a drum lathe. Do not attempt to feel the finish while the machine is in operation. There may be slivers of metal that will cut your hands. When using grinding equipment, keep hands away from rotating parts. Do not try to feel the finish with the machine in operation. Sometimes you will work on a device with compressed springs, such as a clutch or valves. Use great care to prevent the springs from slipping and jumping loose. If this happens, the spring may take off at high speed and hurt someone.

Never attempt to adjust or oil moving machinery unless the instructions tell you that this should be done.

♦ 2-12 WHAT TO DO IN EMERGENCIES

If there is an accident and someone gets hurt, notify your instructor at once. The instructor will know what to do—give first aid, phone for the school nurse, a doctor, or an ambulance. Be very careful in giving first aid. You must know what you are doing. Trying first aid on an injured person can do more harm than good if it is done wrong. For example, a serious back injury could be made worse if the injured person is moved improperly. However, quick mouth-to-mouth resuscitation may save the life of a person who has suffered an electric shock. Talk to your instructor if you have any questions about this.

♦ 2-13 DRIVING CARS IN THE SHOP

Cars have to be moved in the shop. They must be brought in for service, and may have to be moved from one work area to another. When the job is finished, they have to be moved out of the work area. Be careful when you drive a car in the shop. Make sure the way is clear. Make sure no one is under a nearby car. Someone might suddenly stick out an arm or a leg. Make sure there are no tools on the ground.

When you take a car out for a road test, *fasten your seat belt*, even though you're going only a short distance.

♦ CAUTION ♦ Always fasten your safety belt in a moving car, whether you are the driver or a passenger. Seat belts save lives; your seat belt could save yours. Buckle up for safety!

♦ 2-14 TOW-TRUCK OPERATION

Driving a tow truck and doing emergency work at the scene of a collision takes experience. Each wreck is a special problem. Here are some recommendations for working with tow trucks:

1. Make sure the fire extinguisher is properly serviced, in good working condition, and mounted securely on the truck.
2. Do not exceed the unit's maximum hoisting capacity.
3. Make sure the truck floodlights are in good working condition.
4. The control mechanism for the hoist should be inspected periodically to make sure it is in good operating condition, and that the cable, hooks, drum, and other parts are okay.

♦ REVIEW QUESTIONS ♦

Select the *one* correct, best, or most probable answer to each question. You can find the answers in the section indicated at the end of each question.

1. Carrying screwdrivers or other tools in your back pocket can (♦2-10)
   a. injure you and damage upholstery
   b. keep tools clean
   c. keep tools handy
   d. protect the tools
2. Long loose sleeves around moving machinery could (♦2-10)
   a. protect your arms
   b. protect the machinery
   c. get caught in the machinery and injure you
   d. keep you clean
3. A compressed-air hose must never be (♦2-10)
   a. used in the shop
   b. pointed at someone else
   c. left connected overnight
   d. kept on a reel
4. Horseplay in the shop is (♦2-10)
   a. permitted
   b. seldom permitted
   c. frequently permitted
   d. never permitted
5. An engine running in a closed room is dangerous because the exhaust gas contains (♦2-10)
   a. water
   b. blue smoke
   c. air
   d. carbon monoxide
6. Before you get under a car to work on it, make sure that (♦2-10)
   a. the jack handle is down
   b. the car is supported on safety stands
   c. you use a creeper
   d. the tires are inflated
7. Tools may slip out of your hands if your hands are covered with (♦2-7)
   a. gloves
   b. sand
   c. grease or oil
   d. protective cream
8. When working in the automotive shop, your attention should be directed to (♦2-10)
   a. the clock
   b. the weather outside
   c. any new car that has been brought into the shop
   d. the job you are doing
9. Oil-soaked rags should be (♦2-8)
   a. washed in the sink
   b. stored in a special safety container
   c. piled in a corner out of the way
   d. used to mop up the floor
10. Fewer accidents occur in a shop that is (♦2-10)
    a. cluttered
    b. neat and clean
    c. overcrowded
    d. poorly lighted

CHAPTER 3
Shopwork and Shop Manuals

After studying this chapter, you should be able to:

1. List the six steps in automotive service and explain each.
2. List the various sources of automotive-servicing information.

♦ 3-1 THE SIX STEPS IN AUTOMOTIVE SERVICE

Servicing jobs vary from simple to difficult. But no job requires more than six steps. These are:

1. Measuring
2. Disassembling
3. Machining
4. Installing new or serviced parts
5. Reassembling
6. Adjusting

Some jobs require fewer steps. Let's look at these six basic steps. They are performed by using hand tools, measuring tools, and power tools. Tools and their uses are described in following chapters.

Note There is a difference between "reinstall" and "replace." When you *reinstall* a part or component, you are putting back the *same part or component* that you removed. But when you *replace* a part or component, you have discarded the old part or component and are installing a new one. These are the definitions used in the automotive-service business and in this book.

1. Measuring Before you can work on a car, you must find out what is wrong with it. You often begin by measuring (Fig. 3-1).

Linear measurements are the most common kind of measurement. They are measurements you take in a straight line. For example, you might measure an opening or a diameter. Using the familiar United States Customary System (USCS), you take measurements in inches or fractions of an inch. Using the metric system, you take measurements in millimeters, centimeters, or meters. All imported and many domestic vehicles are measured with the metric system. Chapter 4 describes the USC and metric systems.

There are other ways to measure. Sometimes the measuring is done by listening—as when you listen to a running engine. When you check the oil in an engine, you measure its level in the crankcase. You use test instruments to measure battery conditions. When you check engine vacuum or compression, you measure engine performance. The results of your measurements tell you what sort of job you have to do.

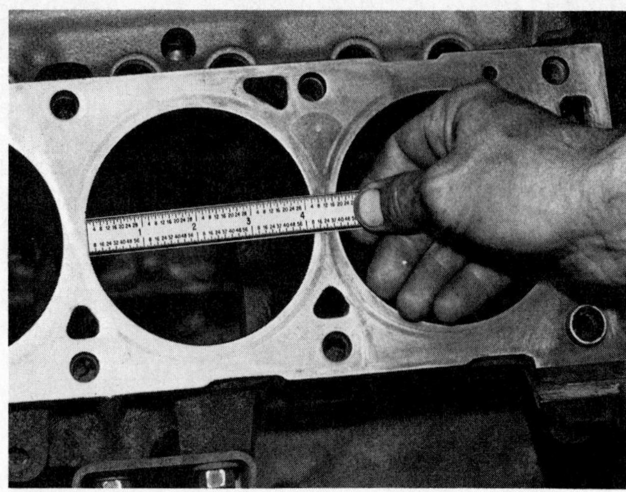

Fig. 3-1 Measuring the diameter of an engine cylinder with a steel rule. *(ATW)*

2. Disassembling Sometimes the measurements show that there is trouble. You then have to disassemble, or "take apart," the component to get at the trouble. Suppose your measurements show that the valves are not doing their job. You then have to take some parts of the engine off, to get to the valves and repair them (Fig. 3-2).

Disassembly is also called *teardown*. For example, you disassemble, or tear down, an engine. However, you do it carefully, part by part.

3. Machining Sometimes you have to remove metal from a part. Using a machine to remove metal is called *machining*. Suppose you find valve trouble. This could require machining, or "grinding," the valves and valve seats. Or you might find that the engine cylinders require machining. Special machines are required to do these jobs.

4. Installing New Parts You might find that some parts are so worn that they must be thrown away. Then new parts must be installed in their place. For example, engine bearings sometimes wear out, and new ones must be installed in place of the old ones. Even new parts may require machining to make them fit.

5. Reassembling After a repair, you may have to put some parts back together. This is called *reassembly*. You put the parts back together to make a complete assembly.

6. Adjusting As an automobile is operated, parts normally wear. This requires adjustments from time to time. Also, adjustments may be required after a service job. For example, after grinding the valves, you put everything back together. Then you measure the valve action. If it is not right, adjustments must be made.

♦ 3-2 SPECIFICATIONS

You will hear the word "specifications," or "specs," quite often in your shopwork. The specs give you the right measurements for the cars you work on. The car manufacturer sets the specs. You find the specs in the manufacturer's shop manual. These specs include valve setting, ignition timing, piston clearances, piston-ring clearances, and hundreds of other measurements. These terms are explained in later chapters.

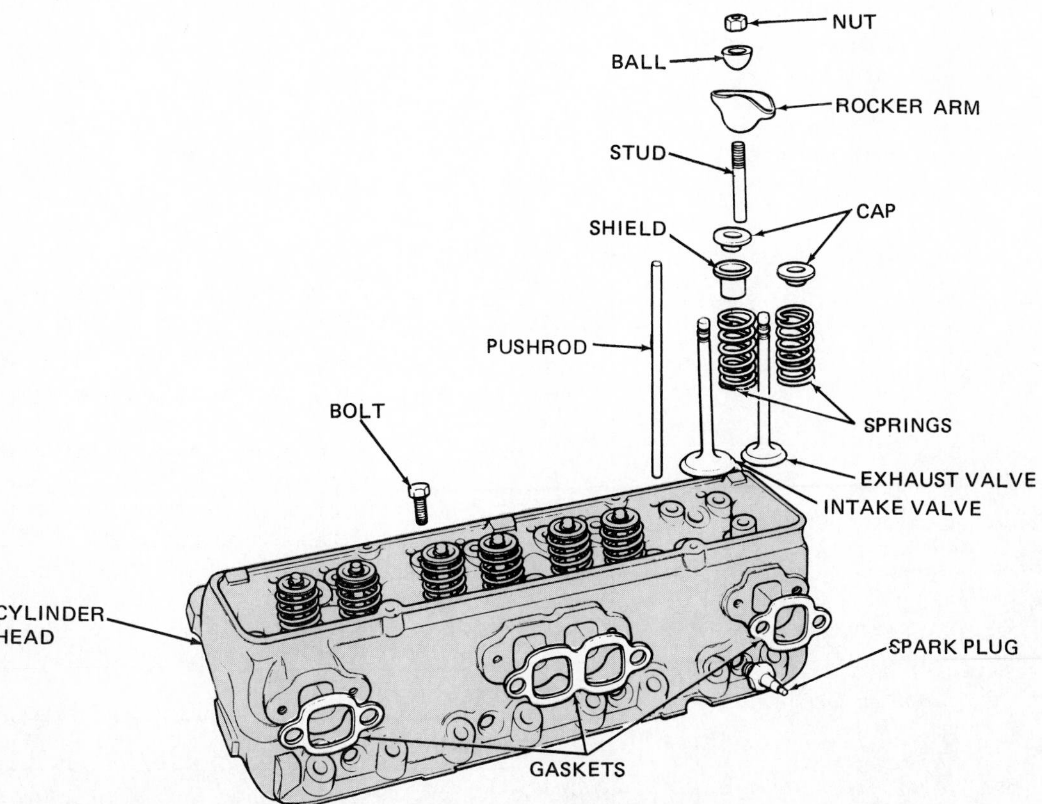

Fig. 3-2 Cylinder head from a V-8 engine, showing the valves for one cylinder removed. *(Chevrolet Motor Division of General Motors Corporation)*

♦ 3-3 MANUFACTURERS' SERVICE MANUALS

Each year, every car manufacturer issues a service manual covering the cars manufactured that year (Fig. 3-3). Each manual covers all service procedures, provides the specs, names the tools needed, and explains how to do all service jobs on the car models produced that year.

You should take a careful look at any manufacturer's service manual you get your hands on. Note how it is arranged in groups or sections. Figure 3-4 shows the section index for a recent Chevrolet service manual. Each section is further divided into descriptions of specific components. These explain how to find and repair troubles with those components, and list the special tools required to service them. Get acquainted with car manufacturers' service manuals. You will be using them when you work in the shop. Practice using these manuals. Look up "Engine," "Transmission," and so on in a manual. Note how each section is divided into specific service jobs.

♦ 3-4 FLAT RATE

Two ways that automotive mechanics are paid are by fixed income and by flat rate (or "piecework"). With a fixed income, the mechanic is paid so much per hour or per week. When the pay is by flat rate, the mechanic earns so much per job. *Flat rate* tells the number of hours and minutes it normally takes to do a service job. Car makers and independent companies print flat-rate manuals. These are also called *time-labor standards.* They list every service job and the amount of time it takes to do the job. For example, one flat-rate manual says that it takes 10 hours to install new piston rings in a certain engine. If the labor charge is $30.00 per hour, the total labor charge for the job is $300.00.

Many shops divide the labor charge with the mechanic. With the right tools, the mechanic can often do service jobs in less than the flat-rate time. Suppose a mechanic does a 10-hour job in 8 hours, for example. The mechanic would still get paid for 10 hours even though the job actually took less time.

♦ 3-5 OTHER USEFUL PUBLICATIONS

Many automotive-service magazines are published. They contain information on servicing cars and articles on specific service jobs. They often have tips on how to make hard jobs a little easier.

Fig. 3-3 Manufacturers' service manuals supply detailed servicing information on their latest models. *(ATW)*

Testing-equipment manufacturers, parts makers, and tool and service-equipment manufacturers publish manuals on how to install their parts or use their equipment. These publications can be very helpful.

TABLE OF CONTENTS	
SECTION NAME	SECTION NO.
GENERAL INFORMATION	
General Information	**0A**
Maintenance and Lubrication	**0B**
HEATING AND AIR CONDITIONING	
Heating and Ventilation	**1A**
Air Conditioning	**1B**
Air Conditioning Compressor	**1C**
FRAME AND BUMPERS	
Frame and Body Mounts	**2A**
Bumpers	**2B**
Sheet Metal	**2C**
STEERING, SUSPENSION, WHEELS AND TIRES	
Diagnosis	**3**
Front End Alignment	**3A**
Manual Steering Gear	**3B1**
Power Steering Rack & Pinion and Pump	**3B2**
Steering Wheels and Columns	**3B3**
Front Suspension	**3C**
Rear Suspension	**3D**
Wheels and Tires	**3E**
BRAKES	**5**
ENGINE	
Diagnosis	**6**
L4 Engine	**6A1**
V6 Engine	**6A2**
Diesel Engine	**6A7**
Engine Cooling	**6B**
Engine Fuel	**6C**
E2SE Carburetor	**6C3**
Diesel Fuel Injection	**6C5**
Engine Electrical	**6D**
Emission Control Systems	**6E1**
EFI (Electronic Fuel Injection)	**6E2**
Engine Exhaust	**6F**
Vacuum Pumps	**6H**
TRANSAXLE	
Automatic	**7A**
Manual	**7B**
Clutch	**7C**
Drive Axle	**7D**
CHASSIS ELECTRICAL	
Electrical Troubleshooting	**8A**
Chassis Electrical	**8B**
Instrument Panel and Gages	**8C**
ACCESSORIES	**9**
METRIC INFORMATION	**10**
BODY	

Fig. 3-4 Section index from a recent Chevrolet service manual.

♦ 3-6 PAPERWORK

You cannot run any kind of business without paperwork, and the automotive-service business is no exception. From the time that a car drives in for service, until the time that the bill for service is paid, there is paperwork to be done. Actually, paperwork begins well before the car arrives. Let's see how it works.

1. The Parts Department The parts department has to have on hand all fast-moving parts that might be needed to service cars. These parts include spark plugs, fan belts, antifreeze, engine oil, brake fluid, distributor contact points, ignition coils, gaskets, and screws and bolts. All these parts and supplies have to be ordered by the parts manager. The parts manager writes out orders for the things needed to stock the department. The parts manager also has to keep a record, or inventory, of the parts in stock. That way, more parts or supplies can be ordered when the inventory runs low. All this record keeping and ordering is paperwork.

2. The Service Department The phone rings and the service manager or an assistant answers. Someone has car trouble. The caller wants to bring a car in for service. The service manager consults the work schedule (often called the *dispatch sheet*), and tells the caller that the department can handle the car at ten the next morning. The service manager notes this down on the schedule.

When the car arrives the next morning, it is met by the service manager or a service-order writer, who is usually called a *service writer* or a *service advisor*. The driver explains the problem. The service writer writes the customer complaint on the repair order, together with labor instructions and notes to the mechanic. The repair order accompanies the car as it is "dispatched" or sent to the mechanic assigned to do the service job. The mechanic follows the labor instructions on the repair order, doing whatever is necessary to repair the car. If new parts are needed, the mechanic gets them from the parts department.

To get parts from some parts departments, the mechanic presents the mechanic's copy of the repair order to the parts counter attendant. The attendant issues the parts and writes the price of each part on the repair order. When the job is completed, the service writer or service manager writes, on the repair order, the flat-rate time or the time taken by the mechanic to do the job. To pick up the car, the customer pays the price of the parts, plus the price of the mechanic's labor charge, and any taxes required by law.

3. Billing Everything is added up on the service order to get the total cost of the repair. If the job is to be paid for in cash, the driver pays the bill when taking delivery of the car. If credit has been established, the car owner is sent a bill which should be paid after so many days. Even here the paperwork does not end. When the payment comes in, it has to be credited against the car owner's account. And, if the car owner is slow to pay, letters have to go out asking for payment. If the car owner refuses to pay, it may be necessary to take legal action. All this is paperwork.

4. Other Paperwork The owner or manager of an automotive-service business also keeps books of various kinds. These include periodic P&L (profit and loss) statements, employee records, general inventories, and local, state, and federal tax records.

You now have a brief review of the paperwork involved in the automotive-service business. Whatever type of job you take in the business will involve some paperwork. You will learn exactly how to handle your paperwork when you go out into the shop.

♦ REVIEW QUESTIONS ♦

Select the *one* correct, best, or most probable answer to each question. You can find the answers in the section indicated at the end of each question.

1. The six steps in automotive service include (♦3-1)
   a. measuring and disassembling
   b. machining and installing new parts
   c. reassembling and adjusting
   d. all of the above
2. To "reinstall" means to (♦3-1)
   a. repair a disassembled part
   b. put back a component or part you have removed
   c. reassemble a component
   d. remove and replace a part
3. In shop talk, to "replace" means to (♦3-1)
   a. put back a part that was removed
   b. discard the old part and install a new one
   c. reassemble a component
   d. readjust a part that has been replaced
4. Tools used in the shop include (♦3-1)
   a. hand tools and power tools
   b. measuring tools and cutting tools
   c. hand tools and cutting tools
   d. all of the above
5. Flat rate is (♦3-4)
   a. the rate in dollars required to do a job
   b. the speed with which you can do a job when working flat out
   c. the time it normally takes to do a service job
   d. all of the above

CHAPTER 4
Measuring Systems and Measuring Tools

After you have studied this chapter, you should be able to:

1. Explain why the United States is changing to the metric system.
2. Discuss the basic differences between the metric system and the USC system.
3. Describe the relationship among linear measurements, weight, and volume in the metric system.
4. Explain how to convert measurements in one system into the other (metric into USCS or USCS into metric).
5. Describe the basic automotive-shop measuring tools (thickness gauges, micrometers, dial indicators, vernier calipers, and small-hole gauges) and demonstrate how to use them.

♦ 4-1 MEASUREMENTS

In the automotive shop, you will take many different kinds of measurements. Sometimes this means you measure the engine power output, or engine vacuum, or alternator output. Sometimes it means measuring angles in the front-suspension system. Most often, it means measuring length, diameter, or clearance. These are called *linear* measurements (measurements in a *line*). For example, in engine work you often measure the diameter, or bore, of engine cylinders (Fig. 4-1).

You make these measurements either in inches or in centimeters and millimeters. You will use either the United States Customary System (USCS) or the metric system. Figure 4-2 compares the two systems. This shows a rule, or steel scale, marked off on the upper edge with centimeters and millimeters (metric system) and on the lower edge with inches and fractions of inches (USC system).

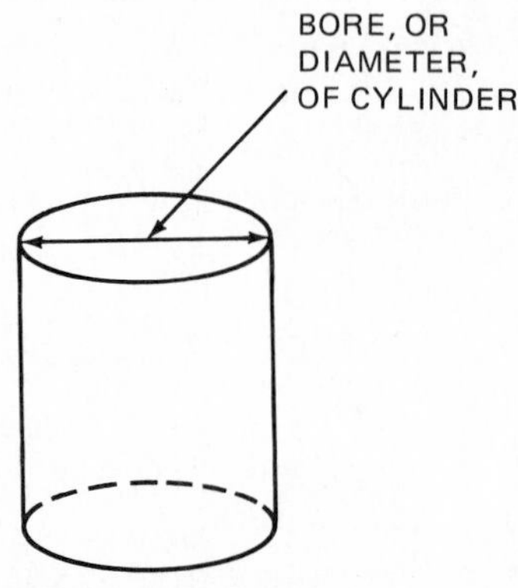

Fig. 4-1 The bore, or diameter, of a cylinder is the distance across it.

♦ 4-2 USC AND METRIC SYSTEMS

In the United States, we have grown up with the USC system and its inches, feet, miles, pints, quarts, gallons, and so on. We all know what is meant when we say that a piston measures 3 inches in diameter. In the metric system, this translates into 76.2 millimeters (mm). One inch equals 25.4 mm.

All imported cars and many cars manufactured in the

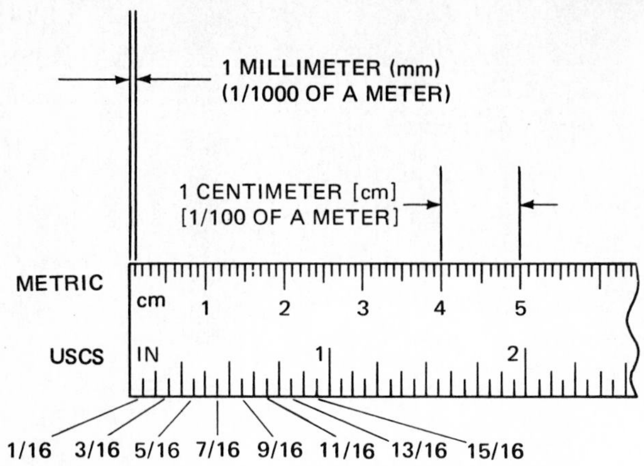

Fig. 4-2 Rule, or steel scale, marked in both the metric and the USC systems.

United States are now made with all measurements in the metric system. For example, the diameter of the pistons in one engine is given as 79.9 mm, which is 3.157 inches.

Note In this book, we give the USCS measurement first, followed by the metric equivalent in brackets, as 1 inch [25.4 mm].

♦ 4-3 REASON FOR GOING METRIC

The metric system is used by every other major country in the world. The system is based on multiples of 10, just like our money system. Ten cents equals one dime and ten dimes equals one dollar. In the same way, 10 mm equals 1 centimeter (cm), 10 cm equals 1 decimeter (dm), and 10 dm equals 1 meter (m) (Fig. 4-3). And 1000 m equals 1 kilometer (km). Kilo- means a thousand. Compare this with our USC system:

12 inches = 1 foot
3 feet = 1 yard, or 36 inches
1760 yards = 1 mile, or 5280 feet, or 63,360 inches

10 millimeters (mm)	= 1 centimeter (cm)
10 centimeters	= 1 decimeter (dm) = 100 millimeters
10 decimeters	= 1 meter (m) = 1000 millimeters
10 meters	= 1 dekameter (dam)
10 dekameters	= 1 hectometer (hm) = 100 meters
10 hectometers	= 1 kilometer (km) = 1000 meters

Fig. 4-3 Complete list of metric linear measurements. You convert from one unit of measurement to another by moving the decimal point.

♦ 4-4 USING THE USC SYSTEM

When you make small measurements in the USC system, you must deal with small fractions of an inch (Fig. 4-2), such as 1/4, 1/8, 1/16, 1/32, and 1/64. But even these are not small enough when we take many measurements on automobiles. These require that we work in thousandths and ten-thousandths of an inch. For example, 1/64 inch is 0.0156 inch. A bearing clearance might be given as 0.002 inch, which is two-thousandths of an inch. To convert common fractions into decimal fractions, you need a table of decimal equivalents (Fig. 4-4). But in the metric system, all you have to do is move the decimal point (Fig. 4-2).

♦ 4-5 WEIGHT AND VOLUME IN THE METRIC SYSTEM

The metric unit of mass, or weight, is the *gram* (g). It is the weight of 1 cubic centimeter (cc) of water at its temperature of maximum density. A cubic centimeter is a cube that measures 1 cm [1/100 m] on a side (Fig. 4-5). One thousand grams, or 1 kg, is equal to 1000 cc (Fig. 4-6).

The metric unit of volume, or liquid measurement, is the *liter* (L). It is the volume of a cube that measures 10 cm on a side [or 1000 cc (Fig. 4-6)]. This is the same measurement for weight or mass. One liter of water weighs 1 kg [1000 g].

Inches			Inches			Inches		
Fraction	Decimal	mm	Fraction	Decimal	mm	Fraction	Decimal	mm
1/64	0.0156	0.3969	23/64	0.3594	9.1281	11/16	0.6875	17.4625
1/32	0.0312	0.7938	3/8	0.3750	9.5250	45/64	0.7031	17.8594
3/64	0.0469	1.1906	25/64	0.3906	9.9219	23/32	0.7188	18.2562
1/16	0.0625	1.5875	13/32	0.4062	10.3188	47/64	0.7344	18.6531
5/64	0.0781	1.9844	27/64	0.4219	10.7156	3/4	0.7500	19.0500
3/32	0.0938	2.3812	7/16	0.4375	11.1125	49/64	0.7656	19.4469
7/64	0.1094	2.7781	29/64	0.4531	11.5094	25/32	0.7812	19.8438
1/8	0.1250	3.1750	15/32	0.4688	11.9062	51/64	0.7969	20.2406
9/64	0.1406	3.5719	31/64	0.4844	12.3031	13/16	0.8125	20.6375
5/32	0.1562	3.9688	1/2	0.5000	12.7000	53/64	0.8281	21.0344
11/64	0.1719	4.3656	33/64	0.5156	13.0969	27/32	0.8438	21.4312
3/16	0.1875	4.7625	17/32	0.5312	13.4938	55/64	0.8594	21.8281
13/64	0.2031	5.1594	35/64	0.5469	13.8906	7/8	0.8750	22.2250
7/32	0.2188	5.5562	9/16	0.5625	14.2875	57/64	0.8906	22.6219
15/64	0.2344	5.9531	37/64	0.5781	14.6844	29/32	0.9062	23.0188
1/4	0.2500	6.3500	19/32	0.5938	15.0812	59/64	0.9219	23.4156
17/64	0.2656	6.7469	39/64	0.6094	15.4781	15/16	0.9375	23.8125
9/32	0.2812	7.1438	5/8	0.6250	15.8750	61/64	0.9531	24.2094
19/64	0.2969	7.5406	41/64	0.6406	16.2719	31/32	0.9688	24.6062
5/16	0.3125	7.9375	21/32	0.6562	16.6688	63/64	0.9844	25.0031
21/64	0.3281	8.3344	43/64	0.6719	17.0656	1	1.0000	25.4000
11/32	0.3438	8.7312						

Fig. 4-4 Table for changing common fractions (in inches) to their decimal or metric equivalent.

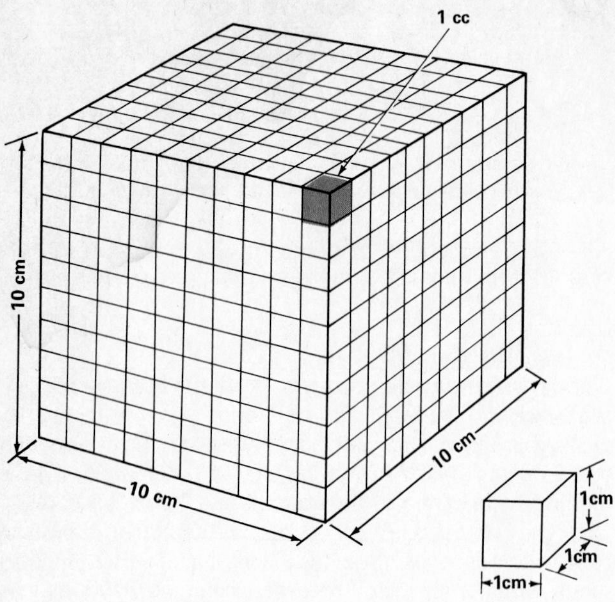

Fig. 4-5 A cubic centimeter is a cube measuring 1 cm on a side. A liter is 1000 cc. Since the gram is the weight of 1 cc of water, a liter weighs 1000 g, or 1 kg.

	Length	
1 in (inch)	= 25.4 mm (millimeters)	= 2.54 cm (centimeters)
1 ft (foot)	= 304.8 mm	= 0.3048 m (meter)
1 mi (mile)	= 1.609 km (kilometers)	= 1609 m
1 mm	= 0.039 in	= 0.1 cm
1 cm	= 0.390 in	= 10 mm
1 m	= 3.28 ft	= 39.37 in
1 km	= 0.62 mi	= 3281 ft
	Capacity and Volume	
1 cu in (cubic inch)	= 16.39 cc (cubic centimeters)	= 0.016 L (liter)
1 fl oz (fluid ounce)	= 29.57 cc	= 29.57 mL (milliliter)
1 qt (quart)	= 32 fl oz	= 0.946 L
1 gal (gallon)	= 4 qt	= 3.78 L
1 cc	= 0.061 cu in	= 1 mL
1 L	= 61.02 cu in	= 1.057 qt
	Mass and Weight	
1 oz (ounce)	= 0.0625 lb (pound)	= 28.35 g (grams)
1 lb	= 16 oz	= 454 g
1 ton	= 2000 lb	= 908 kg (kilograms)
1 g	= 0.035 oz	= 1000 mg (milligrams)
1 kg	= 2.2 lb	= 35.2 oz

Here are the metric units that you will work with most often.

	Length	
1 km	= 1000 m	= 100,000 cm
1 m	= 100 cm	= 1000 mm (millimeters)
	Capacity and Volume	
1 kL (kiloliter)	= 1000 liters	= 100,000 cL (centiliters)
	= 1000 cc	= 1000 mL (milliliters)
	Mass and Weight	
1 kg	= 1000 g	= 100,000 cg (centigrams)

Fig. 4-6 Table to convert metric to USCS units, or USCS units to metric.

♦ 4-6 WEIGHT AND VOLUME IN THE USC SYSTEM

In the USC system, there is no direct relationship among linear measurements, weight, and volume.

Length

12 inches = 1 foot
3 feet = 1 yard
5280 feet = 1 mile

Also:

Liquid or volume

16 fluid ounces (fl oz) = 1 pint (pt)
2 pints = 1 quart (qt)
4 quarts = 1 gallon (gal)

Weight

16 ounces (oz) = 1 pound (lb)
2000 pounds = 1 ton

When you compare the two systems, the metric system is basically simpler and easier to use.

♦ 4-7 CONVERTING USCS TO METRIC

You may have to convert USCS measurements to metric, and metric to USCS. The table in Fig. 4-6 will help you do this. You will find additional conversion tables at the end of the book.

♦ 4-8 RULES

The simplest tool for measuring linear distances is the steel scale, or rule. Figure 4-2 shows a steel scale marked off in both metric and USC (inch) systems.

♦ 4-9 THICKNESS GAUGES

Thickness gauges, or "feeler" gauges (Fig. 4-7), are strips or blades of metal of various thicknesses. Many thickness gauges are dual-dimensioned. For example, the 3 and 0.08 mm on the first blade in Fig. 4-7 means it is 0.003 inch [or 0.08 mm] thick.

Some thickness gauges are stepped. The tip is thinner than the rest of the blade. For example, the top blade in Fig. 4-8 is 0.004 inch [0.10 mm] thick at the tip and 0.006 inch [0.15 mm] along the rest of the blade.

Thickness gauges are used to measure small distances such as the clearance between two parts. For example, Fig. 4-9 shows a thickness gauge being used to check the clearance

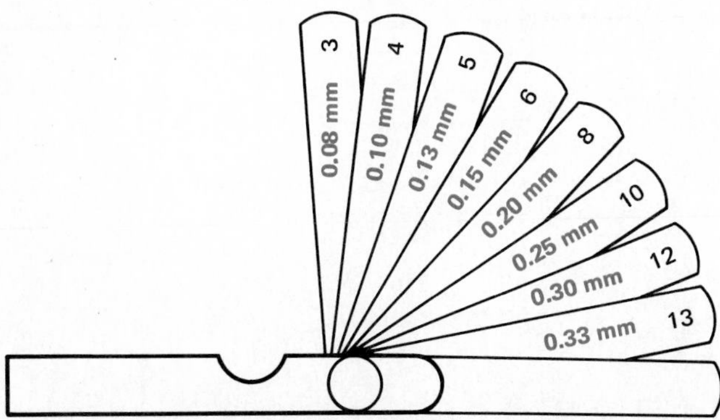

Fig. 4-7 Set of thickness gauges.

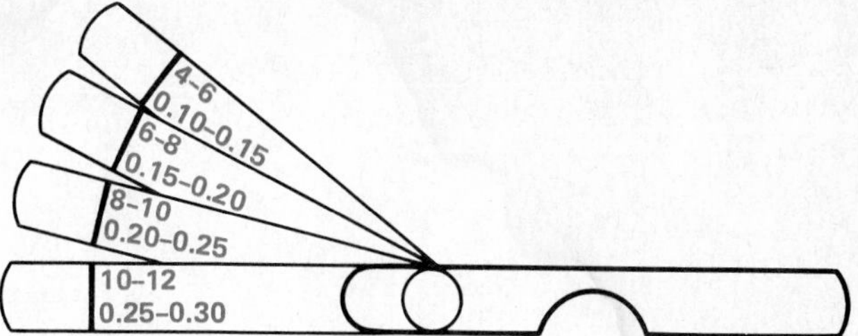

Fig. 4-8 Set of stepped thickness gauges.

between the rocker arm and valve stem in an engine. The clearance is adjusted by turning the adjusting screw.

Stepped thickness gauges are often used where close adjustments are required. For example, suppose the specifications call for a valve clearance of 0.005 inch [0.13 mm]. You would use a 0.004 to 0.006 inch [0.10 to 0.15 mm] stepped thickness gauge (Fig. 4-9). You would adjust the clearance so that the 0.004 inch [0.10 mm] end would fit but the thicker part would not. Stepped thickness gauges are often called *go–no–go* gauges.

♦ 4-10 WIRE GAUGES

Wire gauges are made of round wire. They are used to measure spark-plug gaps and other openings.

♦ 4-11 MICROMETERS

The *micrometer* is a precision measuring tool that can measure thicknesses in thousandths or ten-thousandths of an inch (USC system), or in hundredths of a millimeter (metric system). There are two types, outside micrometers (Fig. 4-10) and inside micrometers (Fig. 4-11). In the shop, micrometers are called "mikes."

A micrometer is accurately calibrated at the time of manufacture. To ensure accurate measuring, check the calibration of the micrometer before each use. Outside micrometers are checked by cleaning the contact faces. Then close them against each other or against a standard test gauge. This gauge is usually included in a micrometer set. Inside mi-

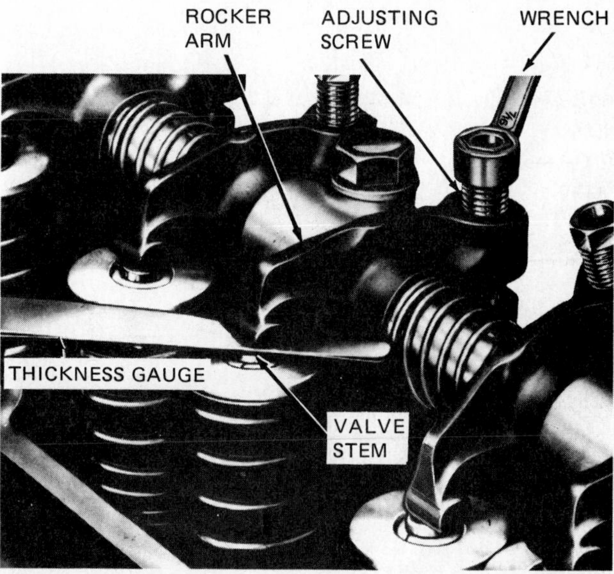

Fig. 4-9 Using a thickness gauge to check the clearance between the rocker arm and valve stem in an engine. *(Ford Motor Company)*

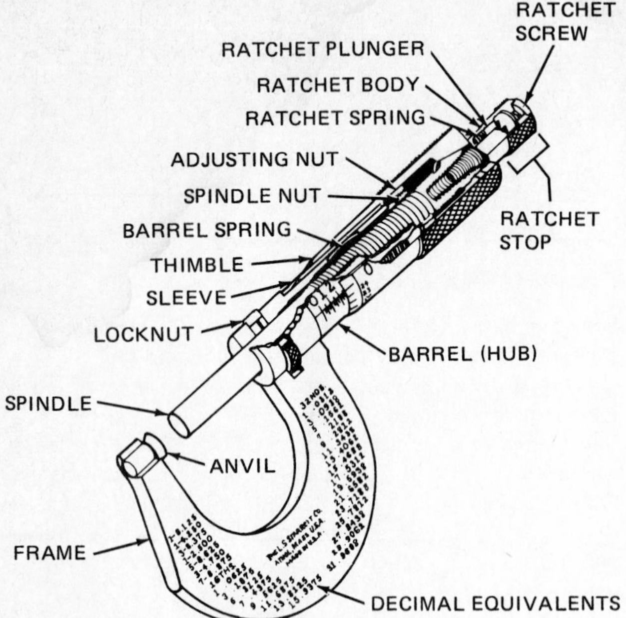

Fig. 4-10 Principle parts of an outside micrometer. *(The L. S. Starrett Company)*

crometers should be assembled. Then the overall length should be checked using an outside micrometer of known accuracy.

♦ 4-12 OUTSIDE MICROMETER

The outside micrometer (Fig. 4-10) has a frame and a movable spindle. The spindle is moved toward or away from the anvil by turning the thimble. The thimble has screw threads that cause the thimble to move when it is turned.

Careful The mike is an expensive measuring tool that must be used with care! After using it, wipe it clean with a clean shop towel and return it to its case or rack. Never drop the mike on the bench or treat it roughly; this could ruin its precision.

Figure 4-12 shows you how to use the outside mike to measure the diameter of a rod. Hold the rod against the anvil and turn the ratchet stop of the thimble until the spindle touches the rod. The ratchet stop on the end of the thimble keeps you from applying excessive force on the mike. When

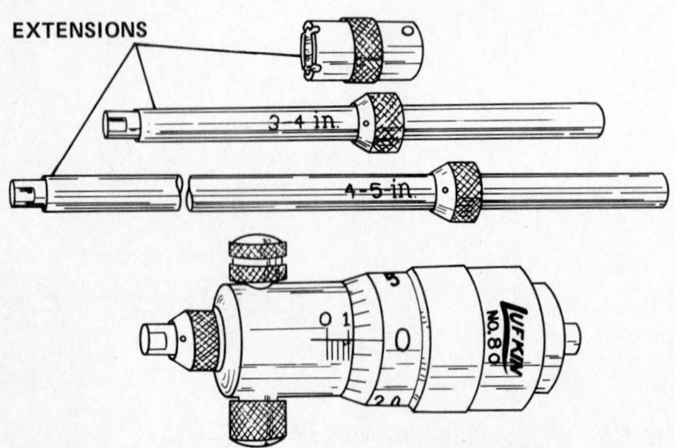

Fig. 4-11 Inside micrometer with extensions. *(Lufkin)*

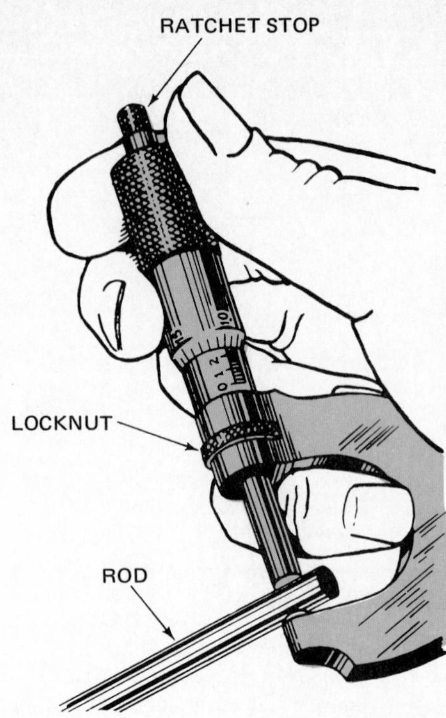

Fig. 4-12 Using an outside micrometer to measure the diameter of a rod.

the spindle touches the rod, the ratchet stop slips and clicks so no further force can be applied.

♦ CAUTION ♦ Never try to measure a part that is moving, such as a rotating shaft. The mike might jam on the shaft, be whirled around, and break. Then you could be seriously injured by the mike or flying particles.

♦ 4-13 READING THE MICROMETER

If you are using a USC system mike, you are reading decimal fractions in thousandths or ten-thousandths of an inch. To read the mike, you must look at both the revolution line on the hub and the thimble position (Fig. 4-13). Every revolution of the thimble moves it one marking on the revolution line. Each marking means twenty-five thousandths (0.025) of an inch. The markings on the thimble run from 0 to 24. There are 25 markings on the thimble. When the thimble is turned from one mark to the next, the spindle has moved exactly one-thousandth (0.001) of an inch. When the thimble is turned one complete revolution, it has moved the spindle 25 markings, or 0.025 inch. This is the distance between markings on the hub.

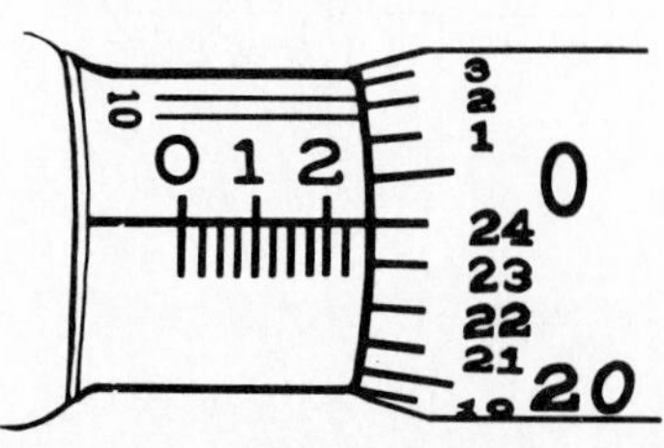

Fig. 4-13 Hub and thimble markings on a USCS micrometer which measures in thousandths of an inch.

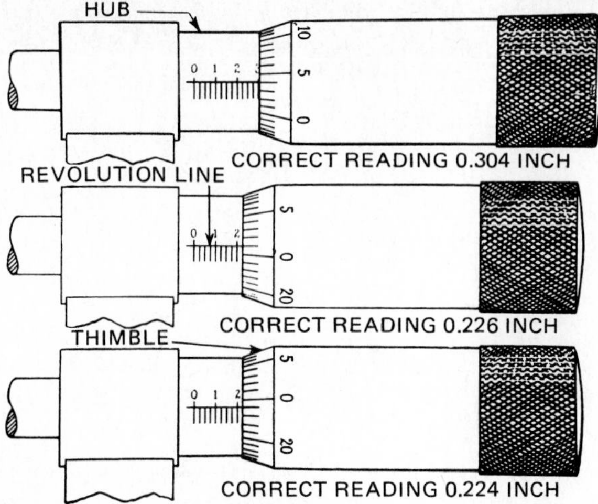

Fig. 4-14 Reading the mike.

To read the mike setting shown in Fig. 4-13, first notice that the "2" and one of the in-between markings are exposed on the hub. This means the distance between the spindle and anvil is 0.2 inch plus 0.025 inch, or 0.225 inch—plus something more. That something more is the amount the thimble has been turned away from the 0.225 reading on the hub. This is 24, or 0.024 inch. Add 0.024 to the 0.225 to get the total reading, which is 0.249 inch. If you want to convert this into a common fraction, refer to the decimal equivalent chart (Fig. 4-4).

Other mike readings are shown in Fig. 4-14.

♦ 4-14 METRIC MICROMETERS

With the metric mike, the measurements are read directly from the hub line and thimble in millimeters and hundredths of a millimeter (0.01 mm). Figure 4-15 shows a metric mike with the reading 11.45 mm. The hub is marked off in millimeters above the line and half millimeters below the line. The thimble has been backed off to expose the 10 (10 mm) marking and one of the upper markings (1.0 mm). This makes 11 mm, to which must be added the thimble marking 45 (0.45 mm). Therefore the reading is 11.45 mm.

Some metric mikes read directly in millimeters (Fig. 4-16). A counter inside the mike shows directly the measurement to which the mike is adjusted (7.00 mm in Fig. 4-18, for example). These mikes are called *digital* metric mikes.

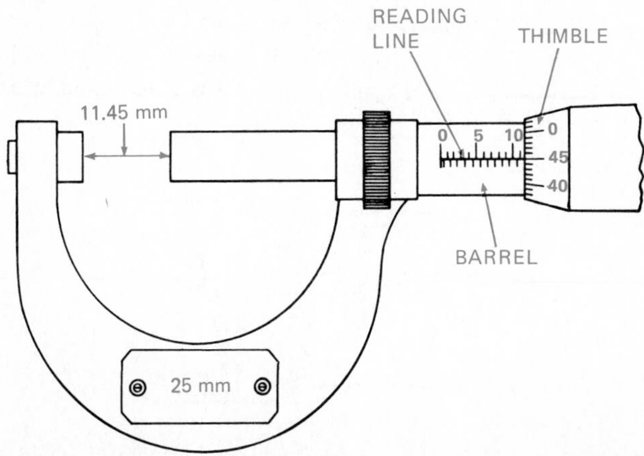

Fig. 4-15 Metric micrometer with a reading of 11.45 mm. *(Volkswagen of America, Inc.)*

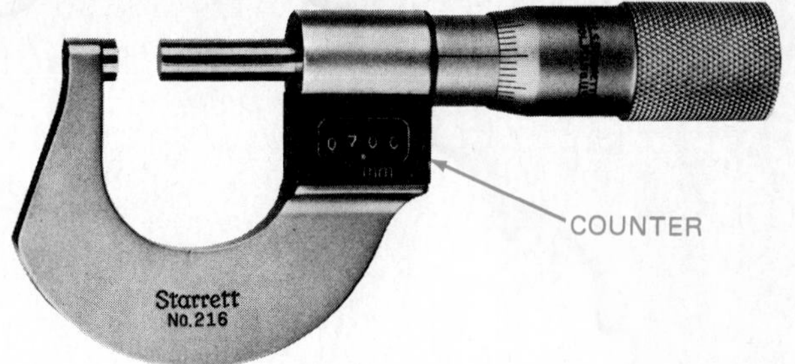

Fig. 4-16 Digital metric micrometer with a reading of 7.00 mm. *(The L. S. Starrett Company)*

♦ 4-15 INSIDE MICROMETERS

Figure 4-11 shows an inside micrometer with extensions. If the inside measurement to be made is greater than the basic mike can measure, an extension can be added. Figure 4-17 shows an inside mike being used to measure the diameter, or bore, of an engine cylinder.

An outside mike and a telescope gauge can be used to measure inside diameters as shown in Fig. 4-18. The telescope gauge is adjusted to the diameter. Then the outside mike is used to measure the gauge setting.

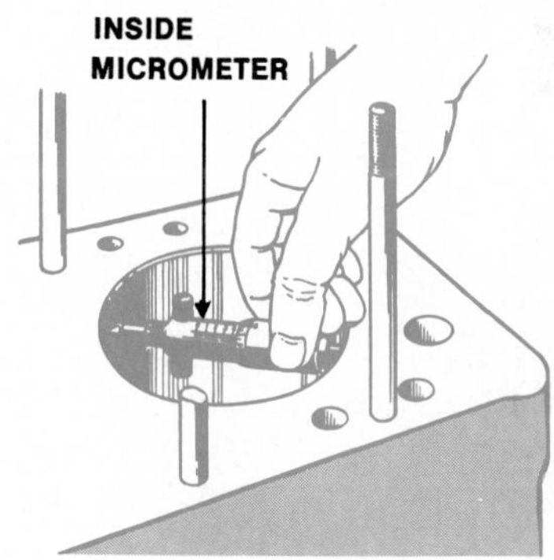

Fig. 4-17 Using an inside micrometer to measure the diameter, or bore, of an engine cylinder.

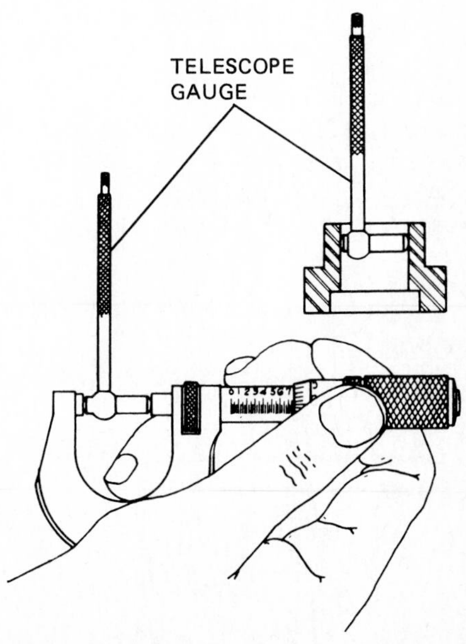

Fig. 4-18 Using a telescope gauge and an outside micrometer to measure the diameter of a small cylinder. *(The L. S. Starrett Company)*

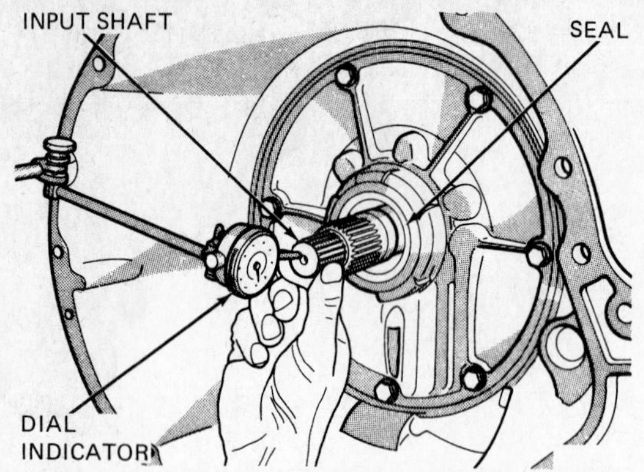

Fig. 4-19 Checking the end play of the input shaft of an automatic transmission with a dial indicator. *(Chrysler Corporation)*

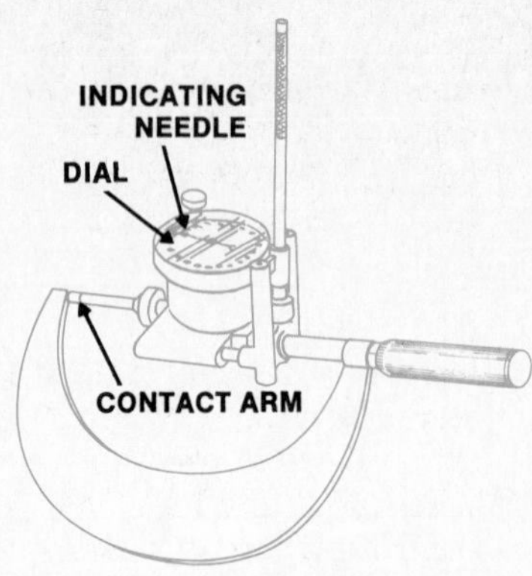

Fig. 4-21 A dial indicator and micrometer can be used to measure the diameter, or bore, of an engine cylinder. After the reading is taken with the dial indicator, the dial is set to zero. The reading is then measured with the micrometer. *(Pontiac Motor Division of General Motors Corporation)*

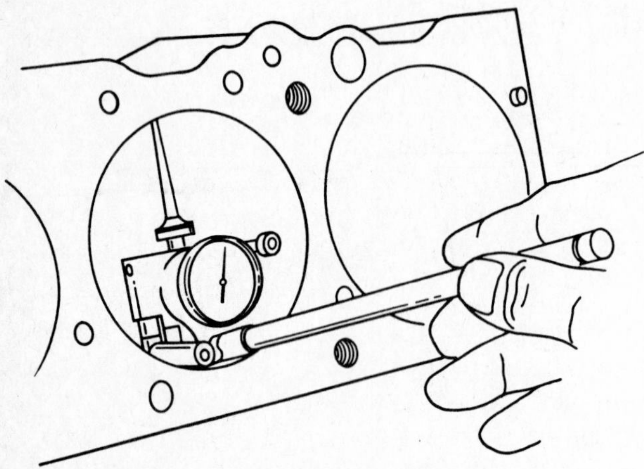

Fig. 4-20 Using a dial indicator to measure the taper wear in an engine cylinder. *(Pontiac Motor Division of General Motors Corporation)*

♦ 4-16 MACHINE-TOOL MICROMETER ADJUSTMENTS

Many machine tools in the automotive shop have micrometer adjustments. The machines have knobs or dials with markings similar to those on mikes. This enables the operator to set the machine precisely.

♦ 4-17 DIAL INDICATORS

The *dial indicator* is a gauge that uses a dial face and a needle to register measurements (Figs. 4-19 to 4-21). It has a movable plunger or contact arm. As the plunger or arm is moved, the needle rotates on the dial face to indicate the distance in thousandths of an inch (or in hundredths of a millimeter). The dial indicator can be used to measure end play in shafts or gears (Fig. 4-19). Also, it can be used to measure taper in engine cylinders (Fig. 4-20). When the dial indicator is moved up and down in the cylinder, movement of the needle will show the amount of taper.

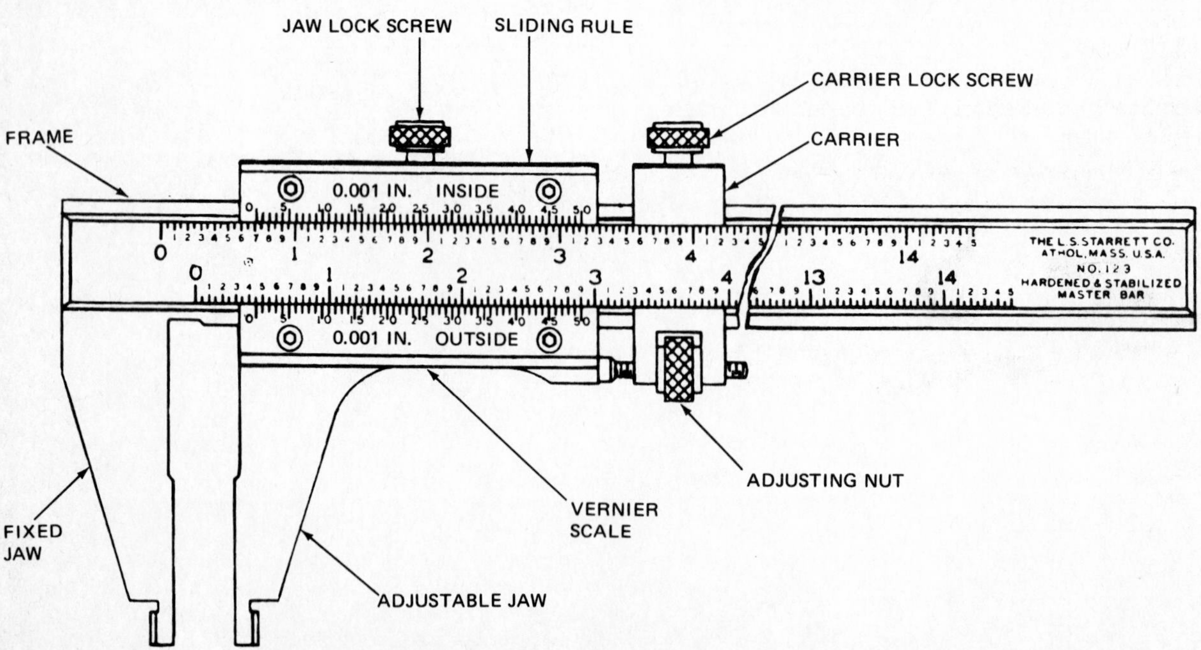

Fig. 4-22 Vernier caliper. *(The L. S. Starrett Company)*

The dial indicator and an outside micrometer can be used to measure the diameter, or bore, of the cylinder. First, the position of the needle is noted when the dial indicator is in the cylinder. The dial is set to zero and the reading is measured with the mike as shown in Fig. 4-21.

♦ 4-18 VERNIER CALIPER

For some mechanics, the vernier caliper is more convenient than the mike. The vernier caliper shown in Fig. 4-22 can take both inside and outside measurements to within thousandths of an inch. To read it, add the number of inches on the fixed scale on the frame to the number of tenths that are seen between the last inch marking and the zero on the vernier scale. Then add the number of 0.025-inch markings seen between the last tenth reading and the zero on the vernier scale. Finally, read the number of lines from zero on the reverse scale to the point where the line on the vernier scale exactly meets a line on the fixed scale. Each of these lines represents 0.001 inch. There is also a metric vernier caliper.

♦ 4-19 DEPTH GAUGE

The depth gauge is a type of micrometer. It is used to measure the depth of a hole.

♦ 4-20 SMALL-HOLE GAUGE

The small-hole gauge is used to measure the diameter of small holes such as the valve-guide bore in a cylinder head. The gauge is adjusted until the split ball slides in the hole with a slight drag. Then the distance between the two sides of the split ball is measured with an outside micrometer.

♦ REVIEW QUESTIONS ♦

Select the *one* correct, best, or most probable answer to each question. You can find the answers in the section indicated at the end of each question.

1. Most nations in the world use the metric system because (♦4-3)
   a. it is based on multiples of 10
   b. most other countries use it
   c. it is less complex than the USC system
   d. all of the above
2. The bore of an engine cylinder is 3.400 inches. In the metric system the bore is (♦4-7)
   a. 86.36 m (meters)
   b. 86.36 cm (centimeters)
   c. 86.36 mm (millimeters)
   d. 86.36 km (kilometers)
3. A fuel tank holds 21 gallons. How many liters would it hold? (♦4-7)
   a. 7.9 L (liters)
   b. 79 L
   c. 79 cc (cubic centimeters)
   d. 79 kL (kiloliters)
4. The contact-point gap for an imported car is given as 0.45 mm. What is this in the USC system? (♦4-7)
   a. 0.010 inch
   b. 0.015 inch
   c. 0.018 inch
   d. 0.024 inch
5. Connecting-rod-bearing clearance for an imported car is given as 0.051 mm. What is this in the USC system? (♦4-7)
   a. 0.0002 inch
   b. 0.002 inch
   c. 0.020 inch
   d. 0.200 inch
6. A dual-dimensioned thickness gauge (♦4-9)
   a. is marked in inches and millimeters
   b. has a pair of blades
   c. is marked in inches and fractions of inches
   d. has a blade of two thicknesses
7. A thickness-gauge blade which has two thicknesses is often called a (♦4-9)
   a. wire gauge
   b. micrometer gauge
   c. go–no-go gauge
   d. double-thickness gauge
8. The two basic types of micrometers are (♦4-11)
   a. direct and indirect
   b. parallel and perpendicular
   c. inside and outside
   d. upside and downside
9. The USCS micrometer reads in (♦4-13)
   a. thousandths of an inch
   b. centimeters
   c. thousandths of millimeters
   d. meters
10. The metric micrometer reads in (♦4-14)
    a. thousandths of an inch
    b. centimeters
    c. hundredths of a millimeter
    d. meters

CHAPTER 5
Automotive Fasteners

After studying this chapter, you should be able to:

1. List and explain the four ways that screw threads are described.
2. Explain how bolt heads are marked to indicate their strength.
3. Discuss six types of screw and bolt heads and the tools used with them.
4. Describe four ways that nuts and bolts can be prevented from loosening.
5. Discuss snap rings and their uses.
6. Explain the purpose of keys and splines.
7. Describe thread inserts and how they are installed.

♦ 5-1 FASTENERS

Fasteners are the parts that hold the automobile together. Examples are screws, nuts and bolts, studs (Fig. 5-1), rivets, snap rings, and cotter pins. Most types of fasteners can be removed so that the assembly can be taken apart. There are also permanent ways to fasten things together, such as welding and soldering. For example, the car body is made by welding metal parts and panels together.

♦ 5-2 USCS SCREW THREADS

Screws, bolts, and studs have external (outside) threads. Nuts and threaded holes for screws, bolts, and studs have internal (inside) threads. Threads, or *screw threads,* are described in four ways:

- By size (Fig. 5-2)
- By threads per inch or pitch (Fig. 5-2)
- By thread series (coarseness or fineness of thread—Fig. 5-3)
- By thread class (quality of finish and fit)

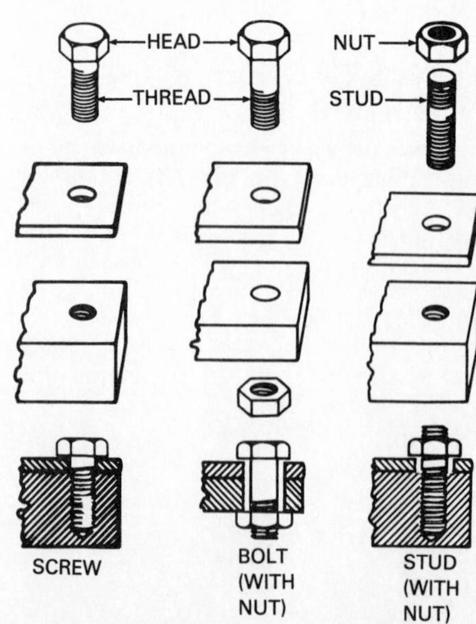

Fig. 5-1 Screw, bolt, and stud. Top shows the attaching parts separated but aligned for assembly. Bottom shows the parts together, in sectional view.

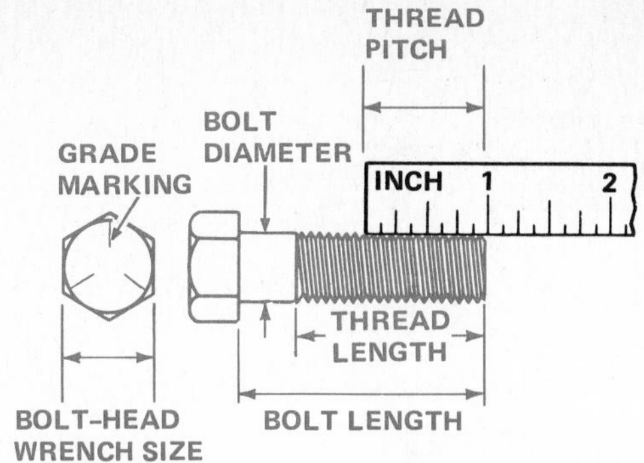

Fig. 5-2 Various measurements of a bolt. *(ATW)*

Figure 5-4 is a table of USCS screw-thread sizes and pitches. When you are doing a service job, you must use the correct screw, bolt, or nut. A ¼-inch screw, for example, can have 20, 28, or 32 threads per inch. You cannot use a 20-thread (coarse) screw in a 32-thread (extra-fine) hole.

Figure 5-2 shows one way to determine pitch (threads per inch). You count how many threads there are to an inch. Figure 5-5 shows another way; using a thread gauge. You find the blade that fits the threads. Blades are marked with the pitch, or number of threads per inch.

♦ 5-3 OTHER THREAD DESIGNATIONS

Other thread designations are series and class.

1. Thread Series The three series in the USC system are coarse, fine, and extra fine (as listed in Fig. 5-4). The difference is in the number of threads per inch.

2. Thread Class This designates the closeness of fit. Class 1 has the loosest fit. It is easily removed and installed, even if the threads are dirty and somewhat battered. Class 2 is a tighter fit. Class 3 has a very close fit and is used where maximum strength and holding power are required.

3. A and B Designations External threads, as on screws and bolts, are *A threads.* Internal threads, as in holes and nuts, are *B threads.*

4. Complete Thread Designation A thread is designated by size, pitch, series, and class. For example, a ¼-20 UNC-2A bolt is ¼ inch in diameter and has coarse threads of 20 threads per inch. The thread is an external, class 2 thread.

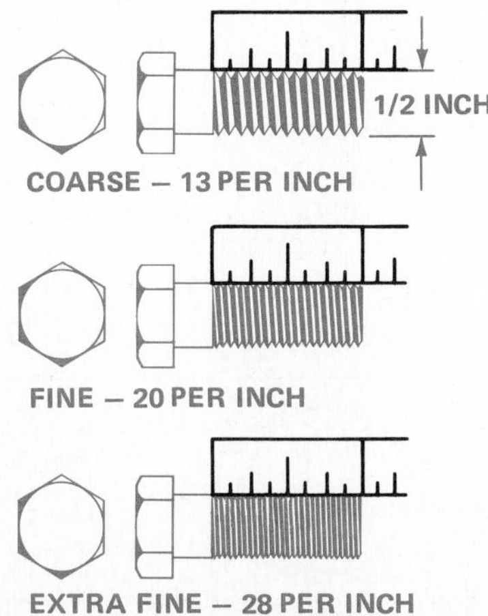

Fig. 5-3 Thread series on a ½-inch bolt.

Size	Diameter (Decimal)	Threads per Inch: Coarse (UNC or NC)	Fine (UNF or NF)	Extra-Fine (UNEF or NEF)
0	0.0600	. . .	80	
1	0.0730	64	72	
2	0.0860	56	64	
3	0.0990	48	56	
4	0.1120	40	48	
5	0.1250	40	44	
6	0.1380	32	40	
8	0.1640	32	36	
10	0.1900	24	32	
12	0.2160	24	28	32
¼	0.2500	20	28	32
5/16	0.3125	18	24	32
⅜	0.3750	16	24	32
7/16	0.4375	14	20	28
½	0.5000	13	20	28
9/16	0.5625	12	18	24
⅝	0.6250	11	18	24
¾	0.7500	10	16	20
⅞	0.8750	9	14	20

Fig. 5-4 Thread series for various sizes of screw, bolt, stud, and nut.

Note For screws, bolts, nuts, and threaded holes to match, they must match as to size, pitch, series, and class. Using the wrong screw or nut could be disastrous. Even if you could force a fit, it would be weak and would probably fail.

5. Length The length of a bolt or screw is the length from the bottom of the head to the end of the threads. However, the threads do not always extend the entire length of the bolt or screw.

♦ 5-4 METRIC SCREW THREADS

Metric bolts, screws, and threads are measured in millimeters. Bolt sizes and threads are different from those in the USC system. Because of this, an automotive mechanic working on both domestic and imported cars needs two sets of fasteners and two sets of wrenches. Figure 5-6 shows the way bolts are designated in the metric system. The number on the head of the bolt indicates the class of bolt, or fastener strength. The length of the bolt is designated in millimeters.

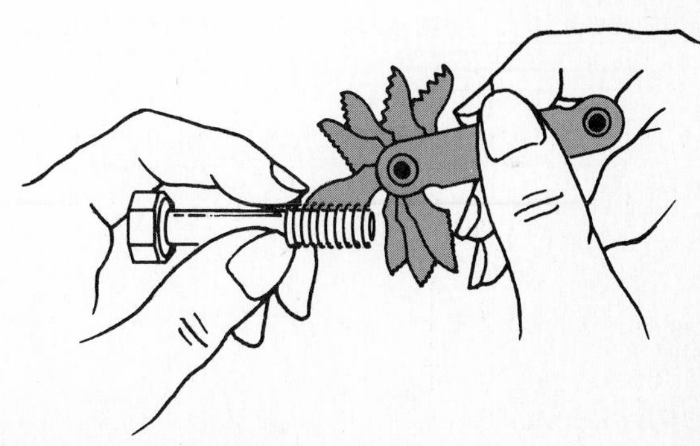

Fig. 5-5 Using a thread gauge.

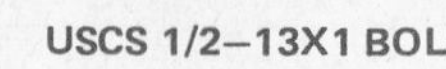

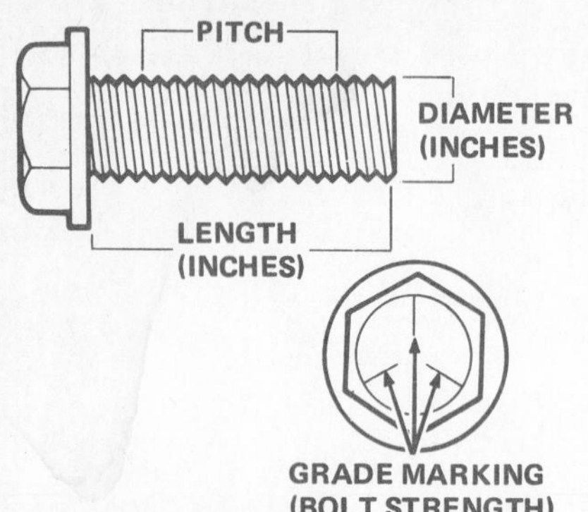

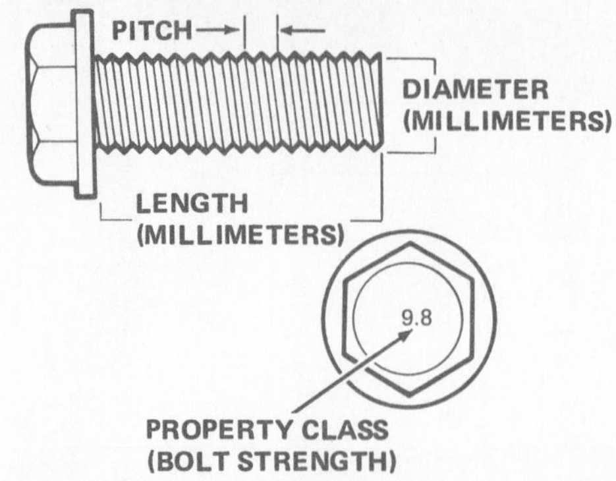

INCH					
½	—	13	X	1
Thread major diameter in inches		Number of threads per inch		Length of bolt in inches

METRIC				
M12	—	1.75	X	25
Thread major diameter in millimeters		Distance between threads in millimeters		Length of bolt in millimeters

Fig. 5-6 Basic measurements of a USCS bolt compared with a metric bolt. *(Ford Motor Company)*

The basic major diameter of the thread is also designated in millimeters.

Thread pitch is measured differently in the metric system (Fig. 5-6). It is the distance between individual threads. Typically, pitch may run from 1 to 2 mm as the diameter of the threads increases. A bolt with a basic thread diameter of 6 mm has a pitch of 1 mm. A bolt with a thread diameter of 16 mm has a pitch of 2 mm.

♦ 5-5 BOLT AND SCREW STRENGTH

Bolts and screws are made from materials having different strengths. Figure 5-7 shows how USCS screw and bolt heads are marked to indicate their strength. The minimum tensile strength is the pull, in pounds, that a round rod with a cross section of 1 square inch can stand before it tears apart or breaks. The higher-strength bolts and screws are more expensive. They are used only where the added strength is required.

♦ 5-6 SCREW AND BOLT HEADS

A great variety of screw and bolt heads have been used. Figure 5-8 shows several, along with the screwdrivers and wrenches required to install and remove them. Most bolts have six-sided, or hex (hexagonal), heads.

USAGES IN VEHICLES	SOME USED	MUCH USED	FOR SPECIAL EQUIPMENT	CRITICAL POINTS	COMPETITION MAXIMUM REQUIREMENTS
TYPICAL APPLICATIONS	FENDERS	CLUTCH HOUSINGS	HEAD BOLTS	BEARING CAPS	RACE CARS
MINIMUM TENSILE STRENGTH, PSI	64,000	105,000	133,000	150,000	160,000
MATERIAL	LOW-CARBON STEEL	MEDIUM-CARBON STEEL	MEDIUM-CARBON STEEL	ALLOY STEEL	SPECIAL ALLOY STEEL
QUALITY	INDETERMINATE	MINIMUM COMMERCIAL	MEDIUM COMMERCIAL	BEST COMMERCIAL	BEST QUALITY
HEAD MARKINGS					

Fig. 5-7 Meaning of USCS bolt-head markings.

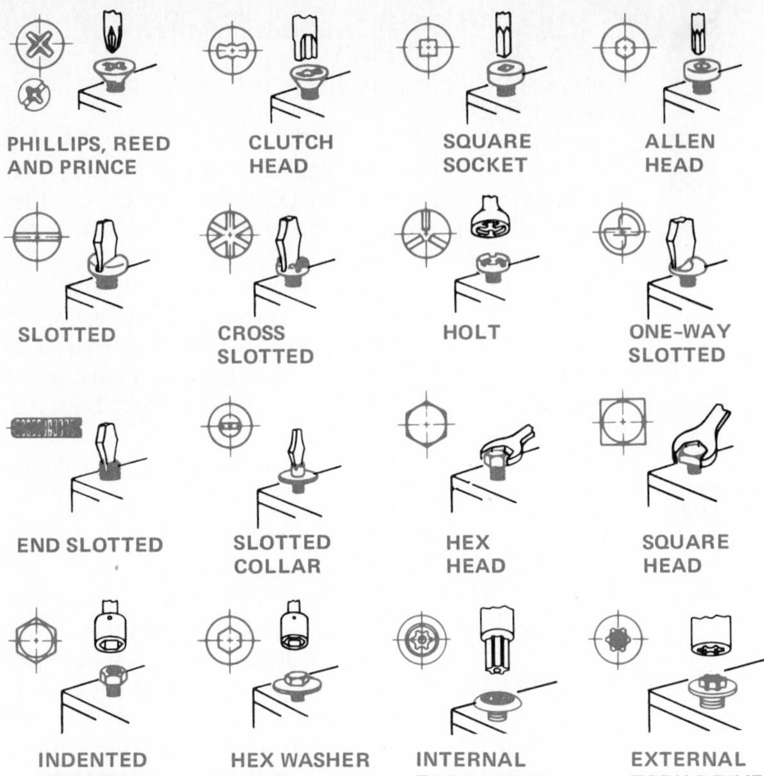

Fig. 5-8 Screwdrivers and wrenches required to drive various types of screws and bolts *(ATW)*

♦ 5-7 NUTS

The most common nut used in the automotive shop is the hex nut (Fig. 5-9B). The slotted hex and the castle nut are for use with a cotter pin (Fig. 5-10). After the nut is tightened, the pin is inserted through a hole in the bolt and two slots in the nut. Then the two ends of the pin are bent around the nut, as shown in Fig. 5-10. This prevents the nut from loosening.

Another way to lock a nut on a bolt is to use a second nut, tightened against the first. The second nut is called a *jam nut.*

♦ 5-8 LOCK WASHERS

A common way to lock an ordinary nut or bolt in place is to use a lock washer (Fig. 5-11) under the nut or bolt head. The sharp edges of the lock washer bite into the metal and keep the bolt or nut from turning.

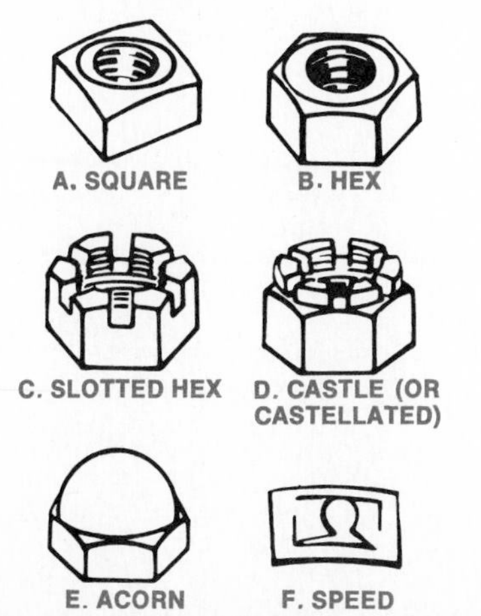

Fig. 5-9 Several common nuts.

Fig. 5-10 A cotter pin before installation (at top) and after installation (at bottom).

♦ 5-9 PREVAILING-TORQUE NUTS AND BOLTS

Many nuts and bolts have an interference fit on the threads. They do not need cotter pins or lock washers to prevent them from loosening. The interference fit prevents the nut or bolt from loosening after it is tightened. This type of fastener is called a *prevailing-torque fastener.*

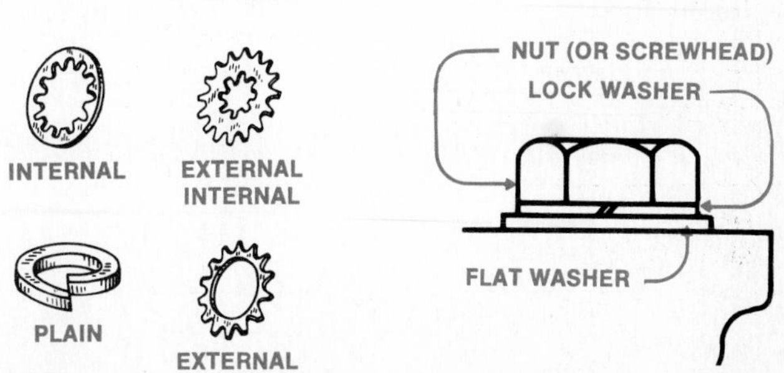

Fig. 5-11 Lock washers (left), and a plain lock washer installed between a flat washer and a nut or bolt (right).

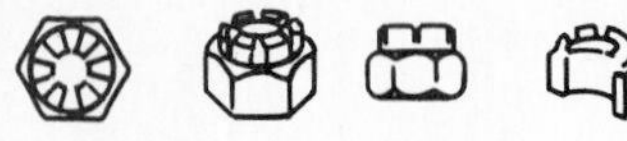

Fig. 5-12 Prevailing-torque nuts. *(General Motors Corporation)*

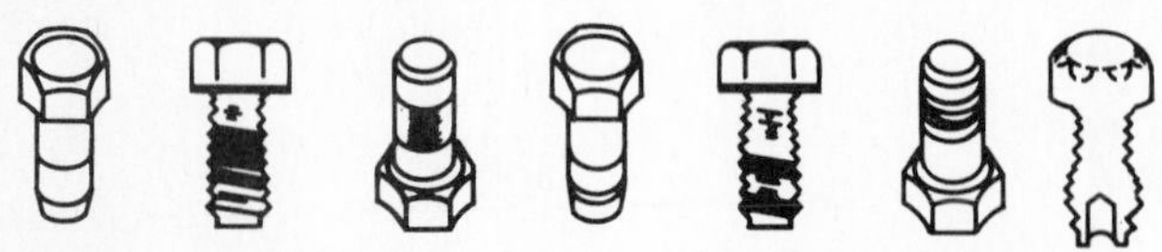

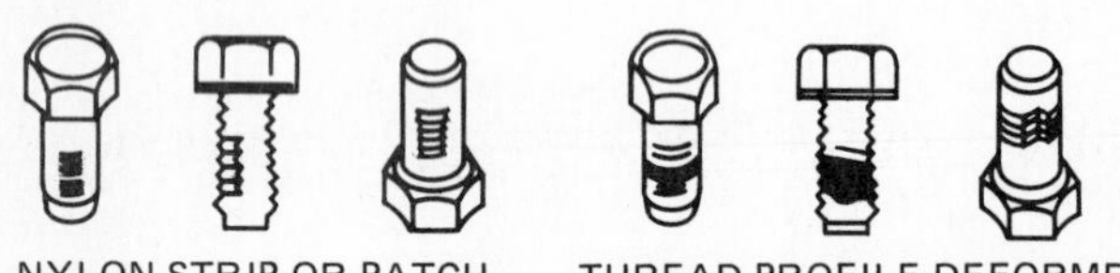

Fig. 5-13 Prevailing-torque bolts. *(General Motors Corporation)*

In nuts (Fig. 5-12), interference is usually achieved by distorting the top of all-metal nuts. Other nuts may achieve interference by distorting at the middle of a hex flat, by a nylon patch on the threads, by a nylon washer insert at the top of the nut, or by a nylon insert through the nut.

In bolts (Fig. 5-13), the interference is achieved by distorting some of the threads, by applying a nylon patch or strip, or by adhesive coatings on the threads.

♦ 5-10 REUSE OF PREVAILING-TORQUE FASTENERS

Clean, unrusted prevailing-torque bolts and nuts may be reused as follows:

1. Clean dirt or other foreign material from nut or bolt.
2. Inspect the bolt or nut for cracks, elongation, or other signs of overtightening or other misuse. If the bolt or nut looks damaged, discard it.
3. Lightly lubricate the threads and assemble the parts.
4. Start the bolt or nut by hand. Then use a torque wrench to tighten the bolt or nut. (Torque wrenches are described in ♦6-9). Before the fastener seats, it develops a prevailing torque as shown in Fig. 5-14. If the fastener turns too easily, replace it.
5. Tighten the fastener to the torque specified in the manufacturer's service manual.

♦ 5-11 ANTISEIZE COMPOUND

When a steel bolt is turned into an aluminum part such as a cylinder head or block, the bolt threads should first be coated with an *antiseize compound.* This prevents corrosion, seizing, galling, and pitting. Without such a compound, the steel bolt tends to lock, or seize, in the aluminum threads. Then, when the bolt is removed, the aluminum threads will be seriously damaged. But if the bolt was coated with antiseize compound, the bolt may be removed without damaging the aluminum threads.

♦ 5-12 THREAD INSERTS

Damaged or worn threads in a cylinder block, cylinder head, or other part can often be repaired with a thread insert (Fig. 5-15). In the installation of one type of thread insert, first the old threads are drilled out. Next the hole is tapped with the special tap that comes with the repair package. Then the thread insert is screwed into the tapped hole to bring the hole back to its original thread size.

PREVAILING TORQUE FOR METRIC FASTENERS

	Millimeters	6 & 6.3	8	10	12	14	16	20
Nuts and all-metal bolts	N-m	0.4	0.8	1.4	2.2	3.0	4.2	7.0
	in-lb	4.0	7.0	12	18	25	35	57
Adhesive- or nylon-coated bolts	N-m	0.4	0.6	1.2	1.6	2.4	3.4	5.6
	in-lb	4.0	5.0	10	14	20	28	46

PREVAILING TORQUE FOR USCS FASTENERS

	Inch	1/4	5/16	3/8	7/16	1/2	9/16	5/8	3/4
Nuts and all-metal bolts	N-m	0.4	0.6	1.4	1.8	2.4	3.2	4.2	6.2
	in-lb	4.0	5.0	12	15	20	27	35	51
Adhesive- or nylon-coated bolts	N-m	0.4	0.6	1.0	1.4	1.8	2.6	3.4	5.2
	in-lb	4.0	5.0	9.0	12	15	22	28	43

Fig. 5-14 Torque specifications for prevailing-torque nuts and bolts. *(General Motors Corporation)*

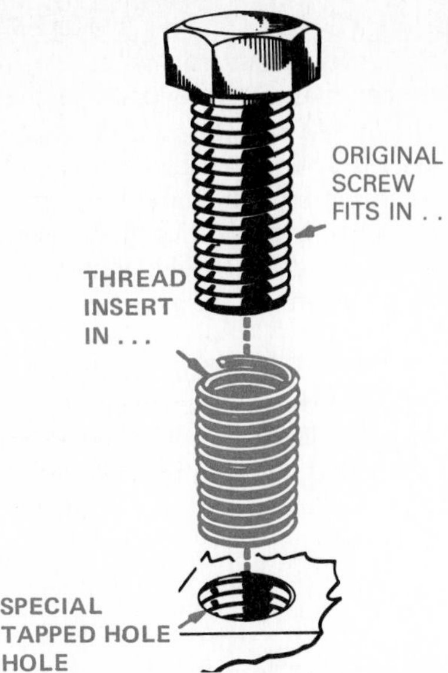

Fig. 5-15 Installing a thread insert in a tapped hole.

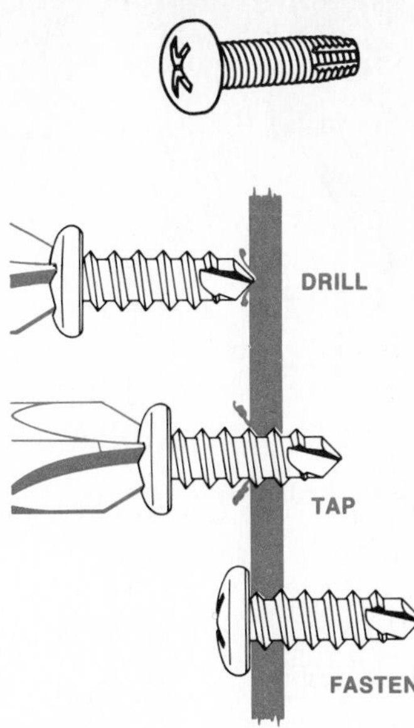

Fig. 5-17 Self-tapping screws. The lower one drills and then taps the hole.

♦ 5-13 SETSCREWS

The setscrew is a special type of screw (Fig. 5-16). Its purpose is to secure a collar or gear on a shaft. When the screw is tightened, it "sets" the collar or gear into place.

♦ 5-14 SELF-TAPPING SCREWS

These cut their own threads when turned into drilled holes (Fig. 5-17). One type also drills the hole.

♦ 5-15 SNAP RINGS

There are two types of snap rings—external and internal (Fig. 5-18). External snap rings are used on shafts to prevent gears or collars from sliding on the shaft. Internal snap rings are used in housings to keep shafts or other parts in position. Some snap rings have holes which make it easier to install and remove them. Figure 5-19 shows how they are used. Some snap rings have tangs instead of holes.

♦ 5-16 KEYS AND SPLINES

Keys and splines lock gears, pulleys, and hubs to shafts so that they will rotate with the shaft. Figure 5-20 shows a typical key installation. The wedge-shaped key fits into slots (*keyways*) in the shaft and hub.

Splines are external and internal teeth cut in both the shaft and the installed part (Fig. 5-21). This is the same as having a number of keys. The shaft and installed part must turn together. In many machines, the splines fit loosely so that the installed part is free to slide along the shaft but still must turn with the shaft. Splines may be straight or curved.

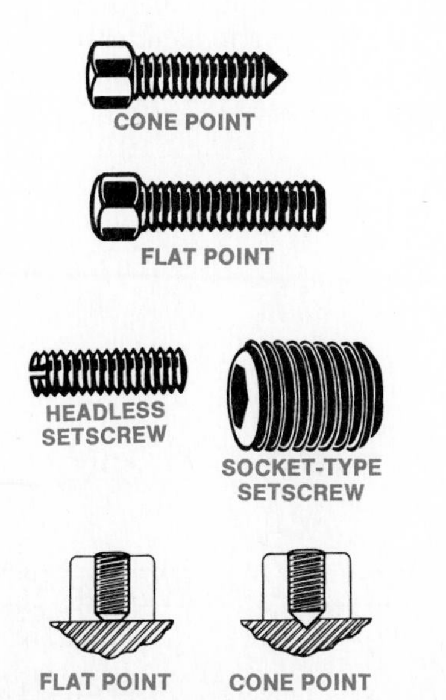

Fig. 5-16 Types of setscrew points.

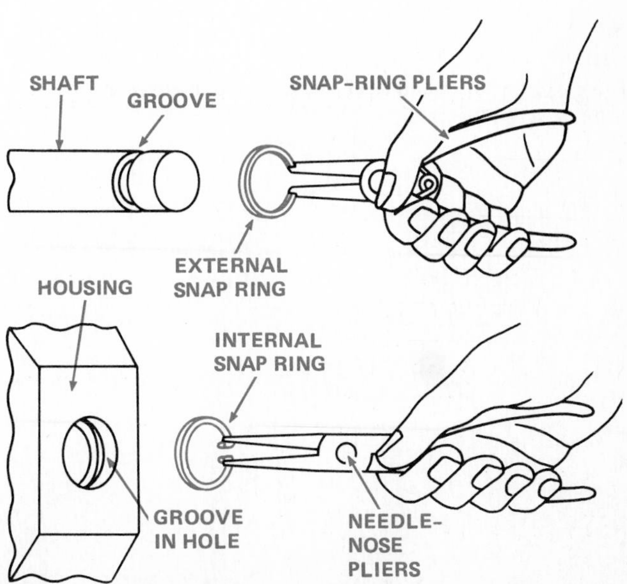

Fig. 5-18 Internal and external snap rings. *(ATW)*

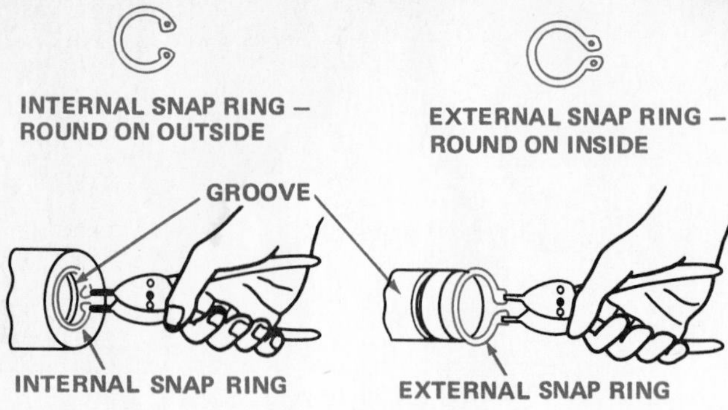

Fig. 5-19 Installing internal and external Truarc retaining rings. *(ATW)*

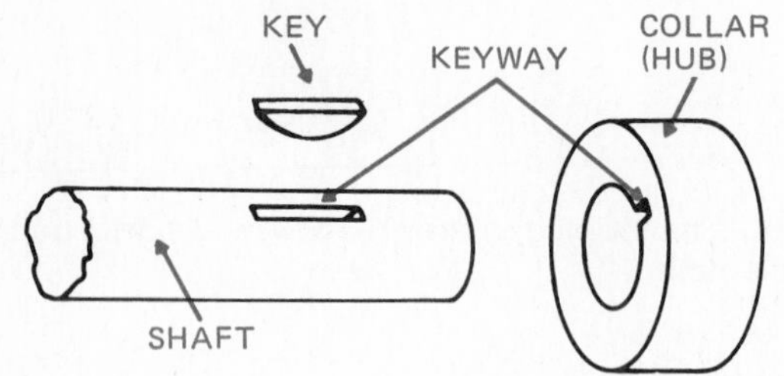

Fig. 5-20 A key locks parts together by fitting into slots called *keyways*.

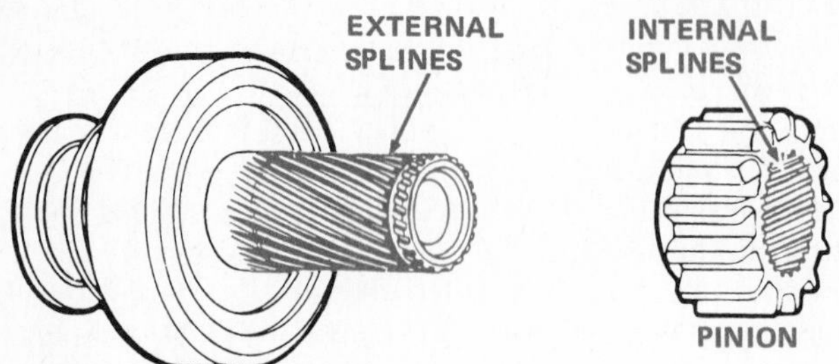

Fig. 5-21 Internal and external splines.

♦ 5-17 RIVETS

Rivets (Fig. 5-22) are metal pins used to fasten parts together. One end of the rivet has a head. After a rivet is in place, a driver, or hammer and rivet set, is used to form a head on the other end of the rivet.

♦ 5-18 BLIND RIVETS

Blind rivets can be installed in blind holes (Fig. 5-23). These are holes where the end of the rivet cannot be reached to form a head. Figure 5-23 shows how to install a blind rivet.

♦ 5-19 GASKETS

A *gasket* is a thin layer of soft material such as paper, cork, rubber, or synthetic material. It is placed between two flat surfaces to make a tight seal. An example is the gasket between the cylinder head and cylinder block (Fig. 5-24). When the gasket is squeezed by tightening the fasteners, the soft material fills any small irregularities in the mating surfaces. This prevents any leakage of fluid, vacuum, or pressure through the joint. Sometimes the gasket is used as a shim to take up space.

Many different types of gaskets are used in the automobile, particularly in the engine. In the engine, they seal joints between the oil pan, manifolds, and water pump and the cylinder head and block (Fig. 5-24). A complete engine-overhaul gasket set for an in-line six-cylinder engine is shown in Fig. 5-25.

♦ 5-20 FORMED-IN-PLACE GASKETS

These gaskets are formed by a bead of plastic gasket or sealant material (Fig. 5-26). This material can be used instead of preformed gaskets in many places in the automobile. It is used under valve covers, thermostat housings, water pumps, and differential covers. When you use a tube of the gasket material, follow the instructions on the tube.

There are two types of plastic gasket material, *aerobic* and *anaerobic*. Aerobic material is also known as *self-curing* or *room-temperature-vulcanizing* (RTV) silicone rubber. At room temperature it "vulcanizes," or cures. When applied (Fig. 5-26), it forms a skin in 15 minutes after exposure to the air. The parts can be brought together as soon as it is applied to the surfaces to be sealed. The sealant forms a good gasket. RTV is normally used on surfaces that flex, such as a valve cover.

Anaerobic material cures only in the absence of air. This means it hardens when squeezed between two surfaces. It will not cure if exposed to air. It is used as an adhesive, a sealer, and a locking cement. It is frequently used on bolts, nuts, screws, and some bushings. To remove a screw that is locked in place with anaerobic material, you must first soften the material. Heat the area around the screw with an electric soldering iron. Never use a torch or open flame. Sometimes applying fresh anaerobic material will soften the material locking the screw in place.

RIVETS BEFORE INSTALLATION

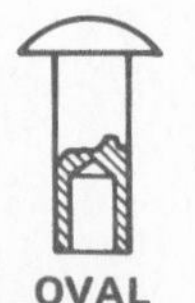

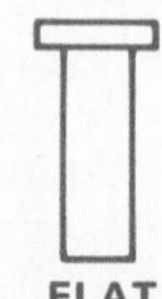

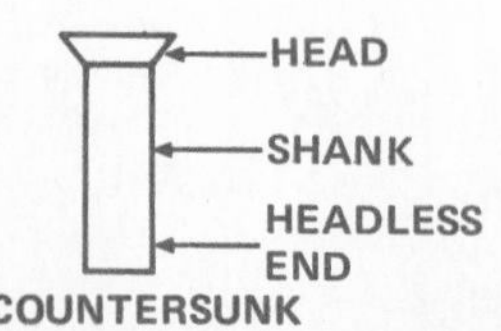

RIVETS AFTER INSTALLATION

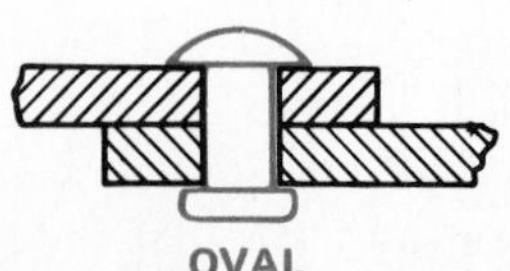

Fig. 5-22 Rivets before installation (left) and after installation (right).

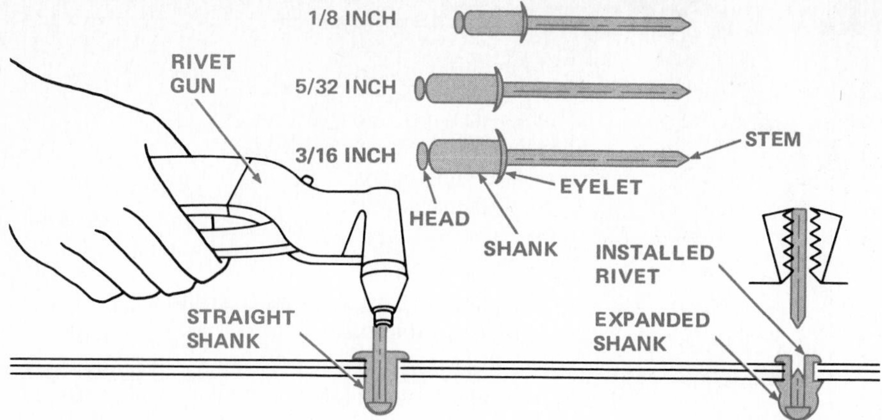

Fig. 5-23 Installing a blind rivet. *(ATW)*

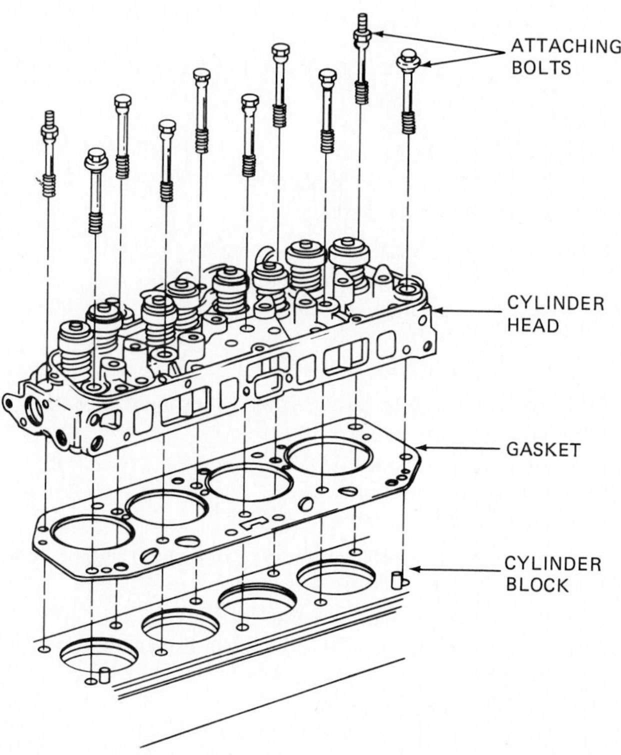

Fig. 5-24 A gasket is placed between the cylinder head and the cylinder block to seal the joint. *(Chevrolet Motor Division of General Motors Corporation)*

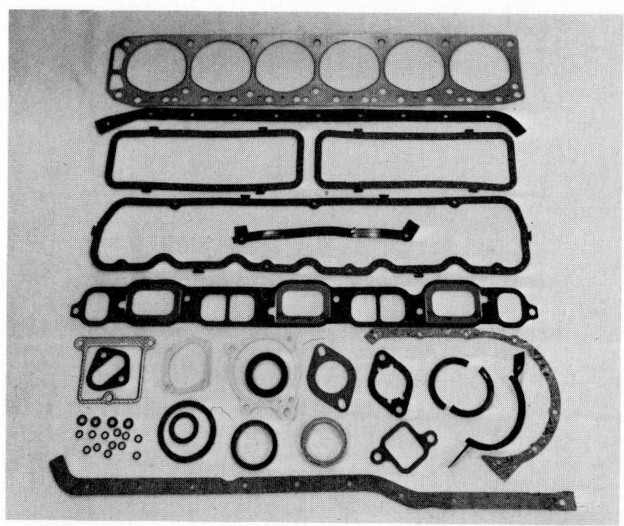

Fig. 5-25 Engine-overhaul gasket set for a six-cylinder engine, showing all gaskets and seals used in the engine. *(McCord Replacement Products Division of McCord Corporation)*

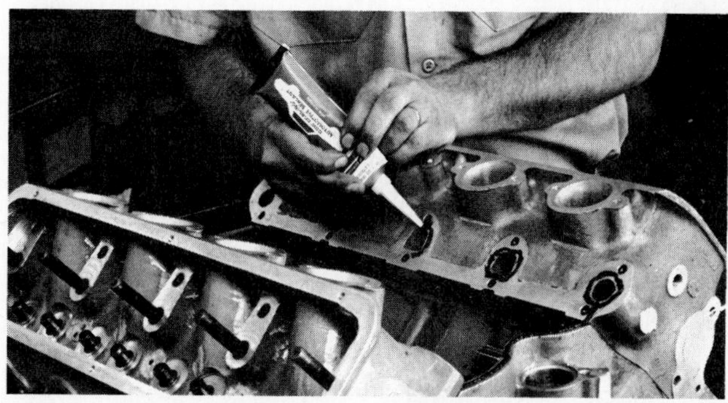

Fig. 5-26 Using a tube of RTV silicone rubber to make a formed-in-place gasket. *(Dow Corning Corporation)*

♦ REVIEW QUESTIONS ♦

Select the *one* correct, best, or most probable answer to each question. You can find the answers in the section indicated at the end of each question.

1. Pitch is the (♦5-2)
   a. depth of the threads
   b. number of threads per inch
   c. thread series
   d. thread class
2. The three thread series in the USC system are (♦5-3)
   a. 20, 28, and 32 threads per inch
   b. coarse, find, and superfine
   c. coarse, fine, and extra fine
   d. classes 1, 2, and 3
3. Thread class designates (♦5-3)
   a. the closeness of fit
   b. number of threads per inch
   c. thread size
   d. thread pitch
4. The A and B designations indicate (♦5-3)
   a. whether threads are coarse or fine
   b. the size of the threads
   c. whether threads are internal or external
   d. the thread series
5. The most commonly used bolts have (♦5-6)
   a. hexagonal heads
   b. six-sided heads
   c. hex heads
   d. all of the above
6. The purpose of the cotter pin is to (♦5-7)
   a. prevent the nut from loosening
   b. fasten the cotter securely
   c. prevent the cotter from dropping off
   d. keep the splines from loosening
7. The slotted hex and the castle nut are for use with (♦5-7)
   a. lock washers
   b. snap rings
   c. cotter pins
   d. bolt heads
8. The two types of snap rings are (♦5-15)
   a. threaded and press fit
   b. internal and external
   c. compression and oil-control
   d. split and solid
9. Thread inserts are used to (♦5-12)
   a. repair damaged internal threads
   b. repair damaged external threads
   c. replace faulty studs
   d. repair damaged nuts
10. Keys and splines (♦5-16)
    a. lock nuts to bolts
    b. lock gears, pulleys, or hubs to shafts
    c. fasten studs into place
    d. take the place of safety wire
11. A gasket is used to (♦5-19)
    a. seal in fuel, oil, or coolant
    b. seal out dirt, water, and air
    c. take up space
    d. all of the above
12. Gaskets that are made by squeezing material from a tube are called (♦5-20)
    a. precut gaskets
    b. preshaped gaskets
    c. framed-by-the-part gaskets
    d. formed-in-place gaskets
13. An aerobic material is also known as (♦5-20)
    a. RTV (room-temperature-vulcanizing)
    b. self-curing silicone rubber
    c. both a and b
    d. neither a nor b
14. To make a gasket on a valve cover, you should use (♦5-20)
    a. RTV silicone rubber
    b. anaerobic sealant
    c. antiseize compound
    d. none of the above
15. To loosen a screw held with anaerobic sealant, you should (♦5-20)
    a. hit the screw with a hammer
    b. heat the sealant with an electric soldering gun
    c. spray the sealant with oil
    d. soak the area with water

CHAPTER 6
Shop Hand Tools

After studying this chapter, you should be able to:

1. List the basic hand tools used in the automotive shop.
2. Describe the various kinds of screwdrivers.
3. Explain the differences among the open-end wrench, the box wrench, and the combination wrench.
4. Demonstrate the correct way to use a hammer.
5. Explain when to use the various types of pliers.
6. Explain the purpose of pullers and how to use them.
7. List the cutting tools used in the automotive shop.
8. Demonstrate the use of various cutting tools.

♦ 6-1 HAND TOOLS AND POWER TOOLS

You will use two main types of tools in the shop. One type is known as *hand* tools because your hand supplies the energy to operate them. The other type is called *machine,* or *power,* tools. Electricity, compressed air, or hydraulic pressure operate these tools. Power tools are described in Chap. 7.

Hand tools include screwdrivers, hammers, pliers, wrenches, and pullers. Figure 6-1 shows a complete set of mechanic's hand tools.

♦ 6-2 SCREWDRIVERS

The screwdriver is used to drive, or turn, screws. The most common type has a single flat blade for driving screws with slotted heads (Fig. 6-2). There are also the Phillips-head and Reed-and-Prince screwdrivers (Fig. 6-2). Other screwdriver types are shown in Fig. 5-8. Always use the correct screwdriver. When using a single-blade screwdriver, for example, select the one with a tip that properly fits the slot in the screw.

♦ 6-3 HAMMERS

The ball-peen hammer (Fig. 6-3) is the one you will use the most in the shop. It should be gripped on the end of the handle. When you swing the hammer, the face should strike the object squarely, and not at an angle. Rawhide, plastic-tip, brass, and rubber hammers are used to strike easily marred surfaces.

♦ 6-4 WRENCHES

Wrenches are used to turn screws and nuts and bolts with hexagonal heads. "Hexagonal" means six-sided (Fig. 5-2). A variety of wrenches are used in the shop. To work on both domestic and imported cars, you will need two sets of wrenches. One set is for USCS nuts and bolts. The other set is for metric nuts and bolts. A ⅜-inch to 1-inch USCS set and a 6-mm to 19-mm metric set will handle most jobs. Chapter 4 explains the USC and the metric systems. There are three types of wrenches, open end, box, and combination (Fig. 6-4).

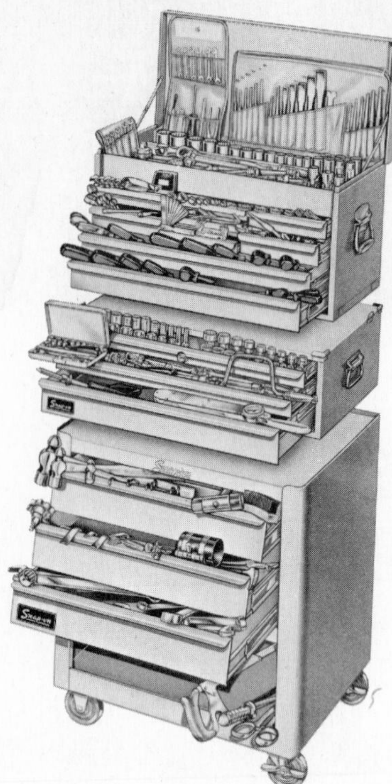

Fig. 6-1 A complete set of mechanic's hand tools. *(Snap-On Tool Corporation)*

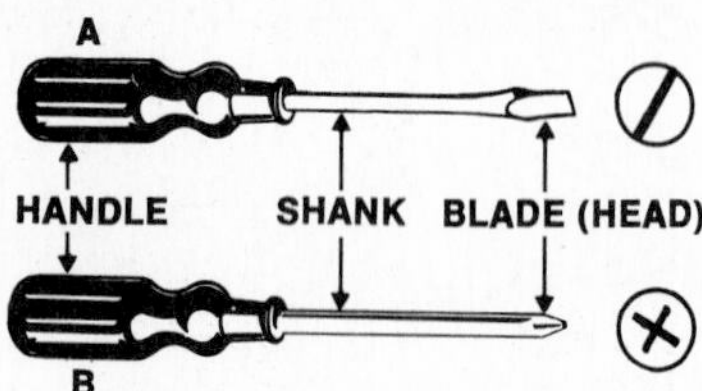

Fig. 6-2 (A) A typical slotted-head screwdriver. (B) A Phillips or Reed-and-Prince screwdriver.

♦ 6-5 OPEN-END WRENCHES

These wrenches are the simplest to use. The opening should be the right size to fit the nut or bolt (Fig. 6-5). If the wrench opening is too large, it could round off the corners of the hex. This makes the use of the proper wrench more difficult. Open-end wrenches usually have a different size on each end.

♦ 6-6 BOX WRENCH

The box wrench (Fig. 6-4) does the same job as the open-end wrench. However, the opening for the nut or bolt head surrounds, or "boxes," the nut or bolt head. Box wrenches can be used in very tight places because the wrench head is thin. The most common box wrench has 12 notches, or "points," in the head. Having 12 points means that a nut or bolt can be turned even if the wrench can swing only 30 degrees. Six-point box wrenches are also made. They hold better on a nut or bolt but are of limited use in close spaces. Box wrenches have a different size on each end.

♦ 6-7 COMBINATION WRENCH

The combination wrench has a box on one end and an open-end on the other (Fig. 6-4). Both ends are the same size. The box wrench is more convenient for final tightening or breaking loose of a nut or bolt because it will not slip off. But the box must be lifted completely free after each swing. The open-end wrench is more likely to slip off. Therefore it is less convenient for final tightening or loosening of a nut or bolt. However, once the nut or bolt is loose, it can be turned faster with the open-end wrench. The combination wrench lets the mechanic use one end to first break loose a nut or bolt and then the other end to turn it rapidly.

Flare-nut wrenches are a special type of combination wrench. The open end has thicker jaws than those of other wrenches of the same size. This helps prevent the wrench from slipping and rounding off the points on soft metal fittings, which are used to connect tubing. The other end has a six-point box end with one of the flats cut out. This opening usually is large enough to allow the box end to slip over the tubing.

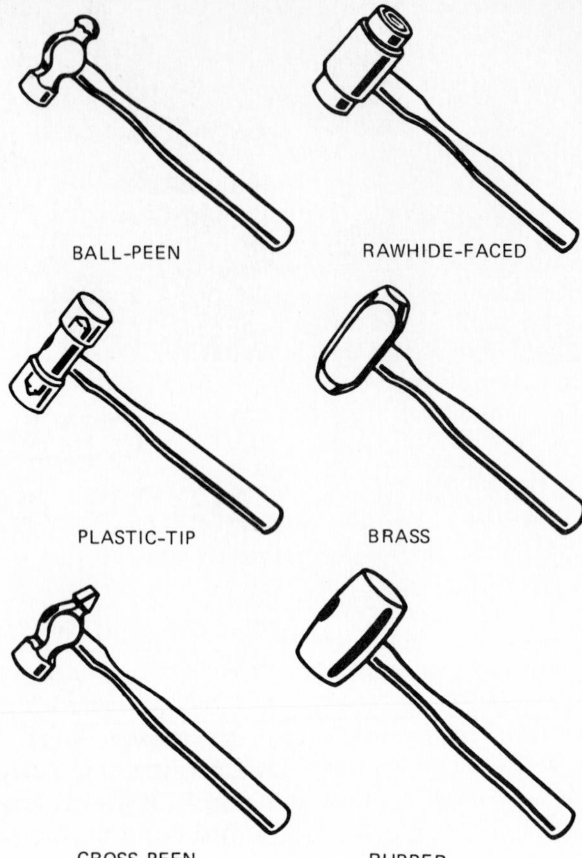

Fig. 6-3 Various hammers used in the shop.

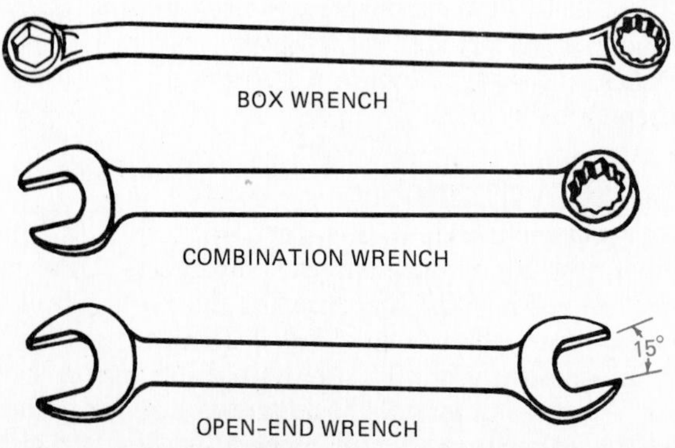

Fig. 6-4 Types of wrenches.

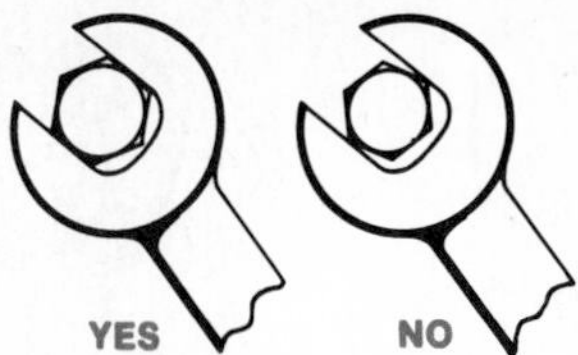

Fig. 6-5 The wrench should fit the nut or bolt head.

♦ 6-8 SOCKET WRENCHES

Socket wrench sets (Fig. 6-6) are the most widely used tools in the shop. They are the same as box wrenches, except that the head, or socket, is detachable. You make up the socket wrench that you need from the set in your toolbox.

First you select the handle or driver—nut spinner, ratchet, speed handle, or others shown in Fig. 6-6. Then select the socket that will fit the bolt head or nut. There are several kinds and sizes of sockets. The 12-point socket is most common (Fig. 6-7). Attach the socket to the driver (Fig. 6-8). If you need an extension, or a universal joint, attach it to the driver first, before attaching the socket.

A set such as shown in Fig. 6-6 is convenient for several reasons. You can pick out immediately which part you need. Also, you can check at a glance to see if anything is missing.

The drive end of the socket, which snaps onto the driver (Fig. 6-8), is square and always sized in fractions of an inch. The most common socket sets have ¼-inch, ⅜-inch, or ½-inch drive.

Note To work on both domestic and imported cars, you will need two socket sets, one for USCS and the other for metric measurements.

Fig. 6-7 Various types of sockets. *(ATW)*

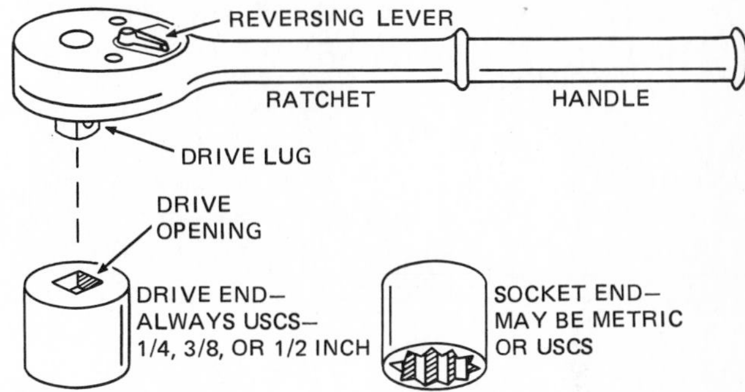

Fig. 6-8 Attaching the socket to the ratchet handle. *(ATW)*

The reason sockets have different numbers of points is the application. The 12-point socket allows you to turn a bolt or nut in tight spots with a swing of only 30 degrees. But if a nut

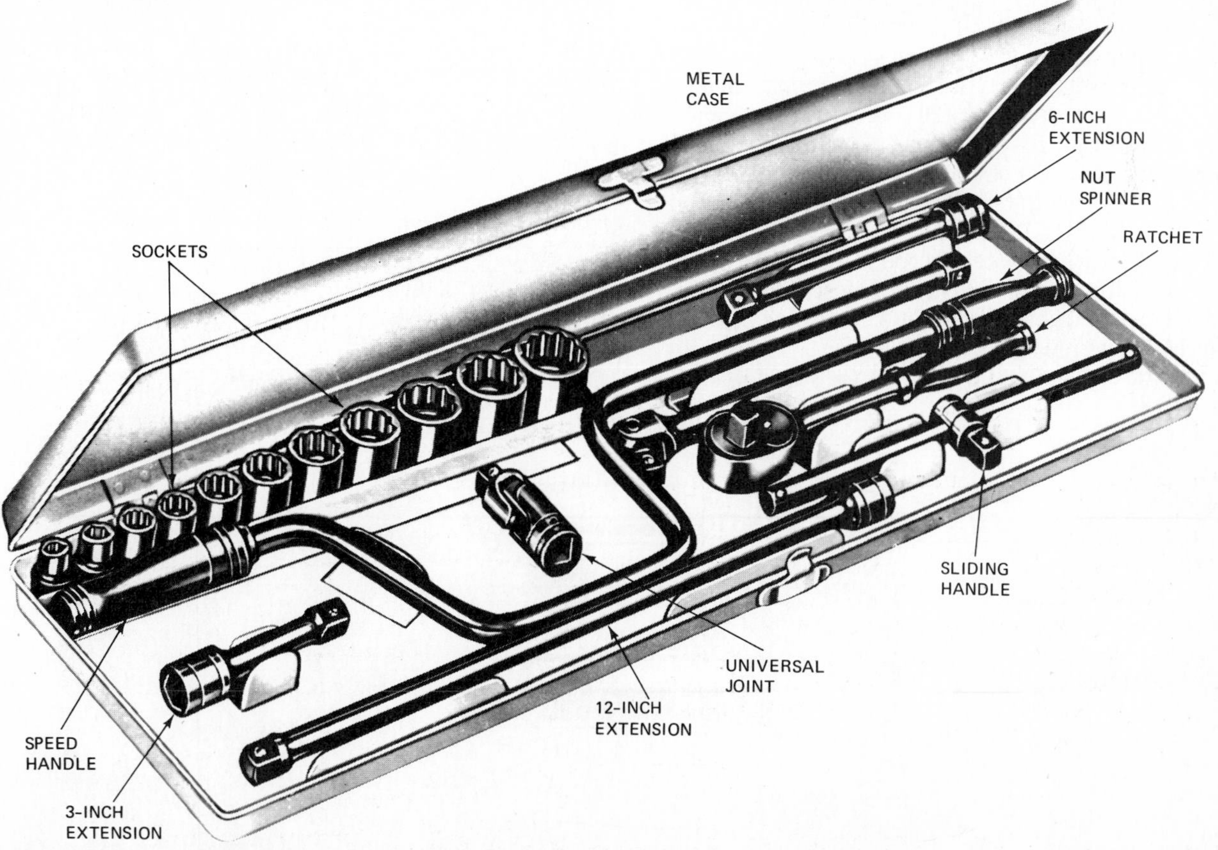

Fig. 6-6 Set of sockets with handles, extensions, and universal joint. *(Snap-On Tools Corporation)*

or bolt has rounded corners, the 12-point socket may slip off. The six-point socket is used for these. The eight-point socket is used for turning square heads, such as on pipe plugs.

Deep sockets are used for nuts on studs or bolts that are too long for the standard socket to reach. Spark plugs are removed and installed with a special spark-plug socket. This is a six-point deep socket with a rubber insert which holds the plug for removal after it is loosened.

Several handles are used. The ratchet handle has a mechanism that permits motion in one direction only (Fig. 6-8). It releases in one direction but catches in the other. You select the direction by flipping the reversing lever. You can quickly tighten or loosen a nut or bolt by selecting which direction the ratchet will catch.

Extensions of various lengths can be used between the driver and the socket. These enable you to work on hard-to-reach bolts or nuts.

The universal joint is used if you must turn a nut or bolt while holding the driver at an angle. Adaptors are available which enable you to use a driver with a socket having a different drive opening. Also, various kinds of screwdrivers, Torx-drives, and Allen wrenches can be fitted to the drivers.

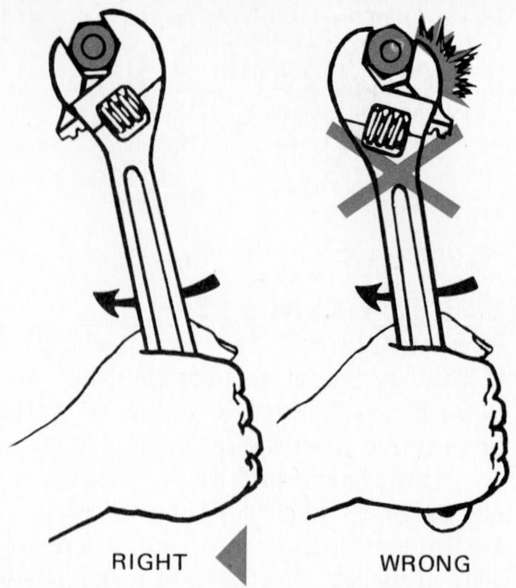

Fig. 6-10 How to use an adjustable wrench. *(Ford Motor Company)*

♦ 6-9 TORQUE WRENCHES

Nuts and bolts must be tightened properly. If they are not tight enough, they could loosen and something could fall apart. If the nut or bolt is tightened too much, threads might be stripped or the nut or bolt strained to the breaking point. Then an accident could result.

Torque wrenches are used to ensure proper tightening (Fig. 6-9). Torque wrenches have indicators which show the amount of torque being applied. The manufacturer's shop manual gives you the proper tightening specifications. For example, the specs might call for tightening a bolt to "20 lb-ft." This means that you apply a 20 pound pull at a distance of 1 foot from the bolt. To use a torque wrench, you snap on the proper socket and pull on the wrench handle. When the torque wrench indicates the specified torque, the bolt is properly tightened.

Most torque wrenches indicate in pound-feet (lb-ft). Some indicate in pound-inches (lb-in). These are used where bolts and nuts must be tightened to close tolerances. Twelve pound-inches equals one pound-foot.

In the metric system, torque wrenches are scaled in kilogram-meters (kg-m), kilogram-centimeters (kg-cm), and newton-meters (N-m). Newton-meters is the preferred metric unit, although others are still used in manufacturer's specifications. To convert pound-feet to kilogram-meters, multiply by 0.138. To convert to newton-meters, multiply pound-feet by 1.35.

Threads must be clean and in good condition. Dirty or damaged threads put a drag on the threads as the bolt or nut is tightened. This gives a false reading and can result in insufficient tightening.

♦ 6-10 ADJUSTABLE WRENCH

This wrench has jaws that can be adjusted to fit nuts and bolt heads of various sizes. Figure 6-10 shows the way to use it. Be sure the jaws are tightened on the nut or bolt head.

♦ 6-11 PLIERS

Pliers are a special type of adjustable wrench. The two legs move on a pivot so that items of various sizes can be gripped. Figure 6-11 shows several. There are two basic types, gripping pliers and cutting pliers.

Channellock pliers have wide-opening jaws and long han-

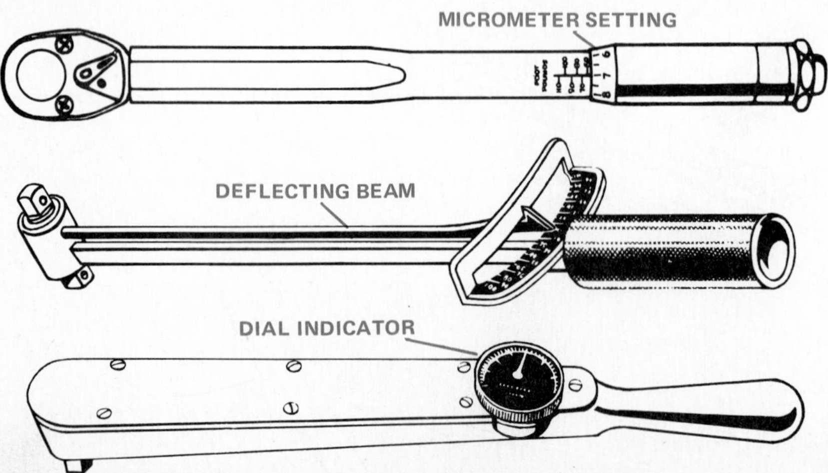

Fig. 6-9 Torque wrenches. *(General Motors Corporation)*

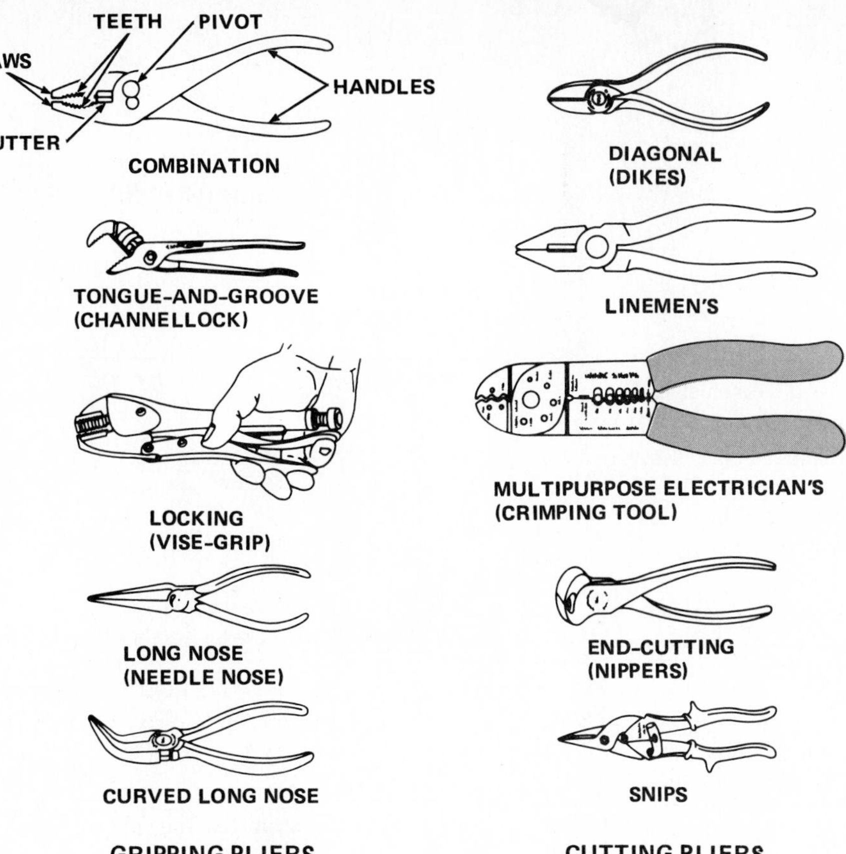

Fig. 6-11 Various types of gripping pliers and cutting pliers. *(ATW)*

dles. They have a "tongue-and-groove" or "groove-and-land" design. The tongues or lands are on one jaw, and the grooves on the other. Changing the relative positions of the grooves and lands changes the distance the jaws can open. The design permits the jaws to be parallel to each other at any setting.

Vise-Grip pliers (Fig. 6-11) have locking jaws that make them useful as pliers, wrenches, clamps, or small vises. The jaws can be locked by turning a screw in the end of the handle to adjust the size of the opening. Then closing the handles locks the jaws into place. They are released by pulling the release lever.

Careful Never use gripping pliers on hardened steel surfaces. This dulls the pliers teeth. Never use pliers on nuts or bolt heads. They may slip and round off the edges of the hex. Then a wrench or socket will no longer fit the nut or bolt head properly.

Cutting pliers (Fig. 6-11) are made in a variety of types. They have a variety of uses.

Figure 6-12 shows pliers that have been adapted, by drilling holes in the jaws, so that they can be used to handle wire-spring hose clamps.

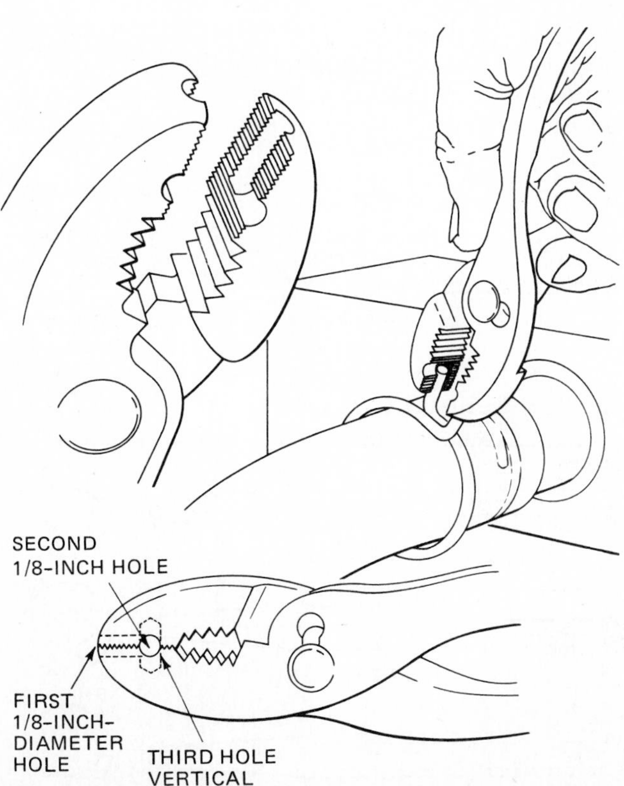

Fig. 6-12 How to adapt pliers so they can handle wire-spring hose clamps.

♦ 6-12 REMOVING BROKEN STUDS

If a stud or bolt breaks off, the broken part must be removed. If the break is above the surface, you may be able to file flats on two sides. Then use a wrench to back the broken part out. Or a slot can be cut so that a screwdriver can be used. Penetrating oil can be applied and left for a while to help loosen the part.

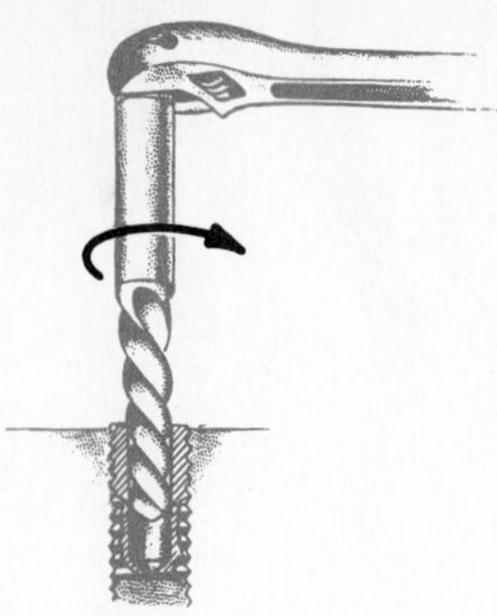

Fig. 6-13 Using a stud extractor to remove a broken stud.

If the break is below the surface, use a bolt or stud extractor. One type is shown in Fig. 6-13. First, center punch the broken part and then drill it. Use a drill that makes a hole almost as large as the small diameter of the threads. Then use an extractor of the right size and turn it so the broken part is backed out.

♦ 6-13 PULLERS

Pullers (Fig. 6-14) are used to remove gears and hubs from shafts, bushings from blind holes, and cylinder liners from engine blocks. A puller set has many pieces that can be fitted together to form the proper combination for any pulling job. There are three basic types, pressure screw, slide hammer, and combination. Figures 6-15 and 6-16 show pullers in use.

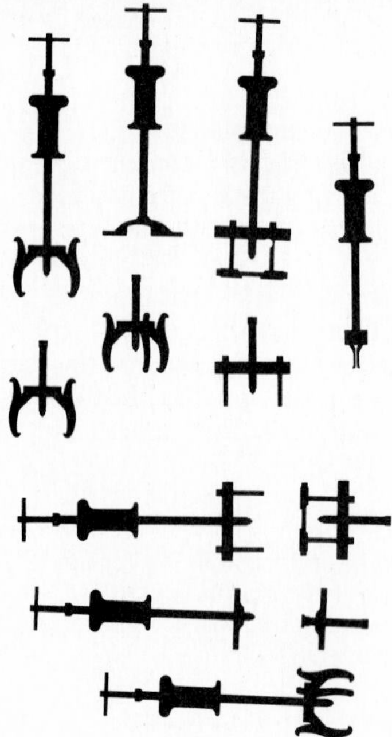

Fig. 6-14 Various assembled pullers. *(Hastings Manufacturing Company)*

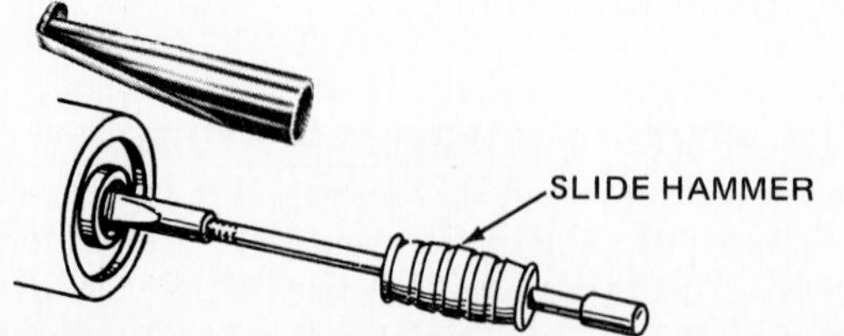

Fig. 6-15 Slide-hammer puller being used to remove a grease retainer. *(Proto Tool Company)*

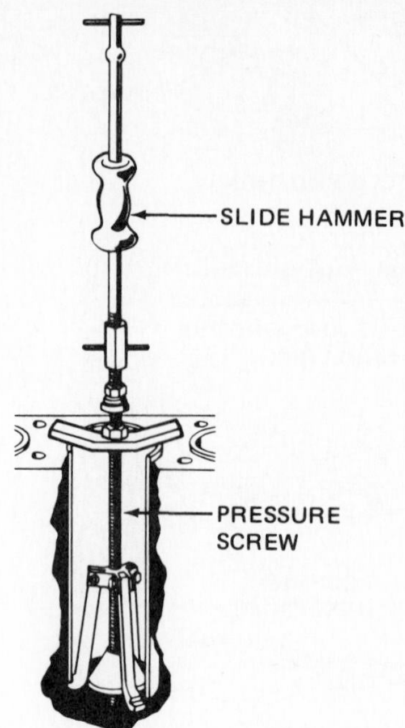

Fig. 6-16 Combination puller being used to remove a sleeve from an engine cylinder. *(Owatonna Tool Company)*

♦ 6-14 CUTTING TOOLS

Cutting tools are used to remove metal. They include chisels, hacksaws, files, punches, drills, and taps and dies. You will learn how to use all of these in the shop. This part of the chapter describes these tools and how to use them.

♦ 6-15 CHISELS

Chisels are made in various sizes and shapes (Fig. 6-17). The chisel has a single cutting edge. It is driven with a hammer to cut metal (Fig. 6-18). Hold the chisel in one hand and the hammer in the other. Strike the end of the chisel squarely with the hammer.

You can use a tool holder, or a pair of pliers, to hold the

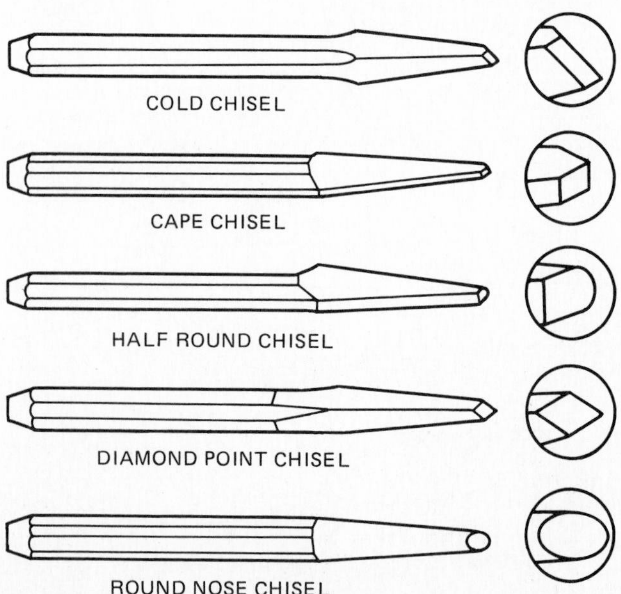

Fig. 6-17 Various types of chisels.

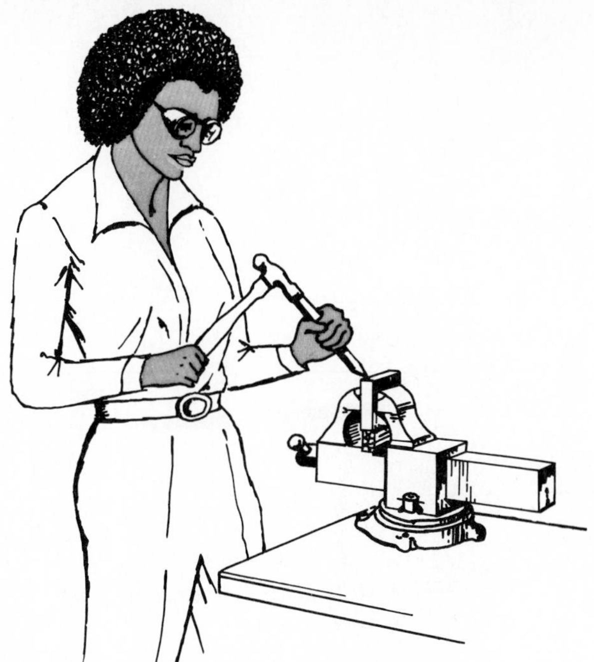

Fig. 6-18 How to use a chisel and hammer.

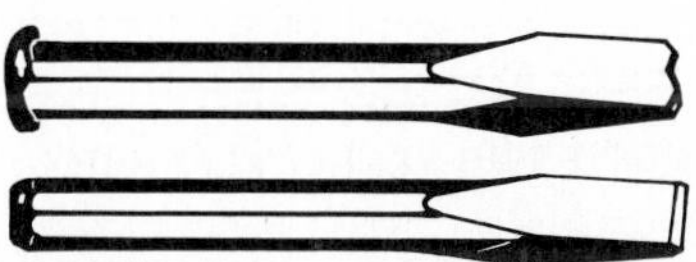

Fig. 6-19 At top, a chisel with a mushroom head and chipped cutting edge. These should be ground off, as shown at the bottom.

chisel. This protects your hand in case the hammer does not strike squarely and slips off the end of the chisel.

After some use, the cutting edge may chip and get dull. The head may "mushroom" (Fig. 6-19). The cutting edge and head should be ground on a grinding wheel as shown in Fig. 6-20. Chapter 7 describes the grinding wheel.

♦ CAUTION ♦ Always wear safety goggles or a face shield to protect your eyes when you use a chisel or the grinding wheel. Never use a chisel with a mushroomed head. A piece may break off when you strike the head and hit you, giving you a bad cut. Keep the chisel sharp. It will do a better job.

♦ 6-16 HACKSAW

The hacksaw (Fig. 6-21) has a replaceable steel blade which has a series of sharp teeth. The teeth act like tiny chisels. As the blade is pushed over a piece of metal, the teeth cut fine shavings, or filings, off the metal. The forward strokes, which do the cutting, should be smooth and steady, not jerky. On the back stroke, lift the blade slightly so that it does not drag. This would dull the teeth.

When using a hacksaw, select a blade with the proper number of teeth per inch for the job. The teeth must be close enough so that at least two teeth will be cutting metal at the same time. If the teeth are too fine and too close together, they will get clogged and stop cutting.

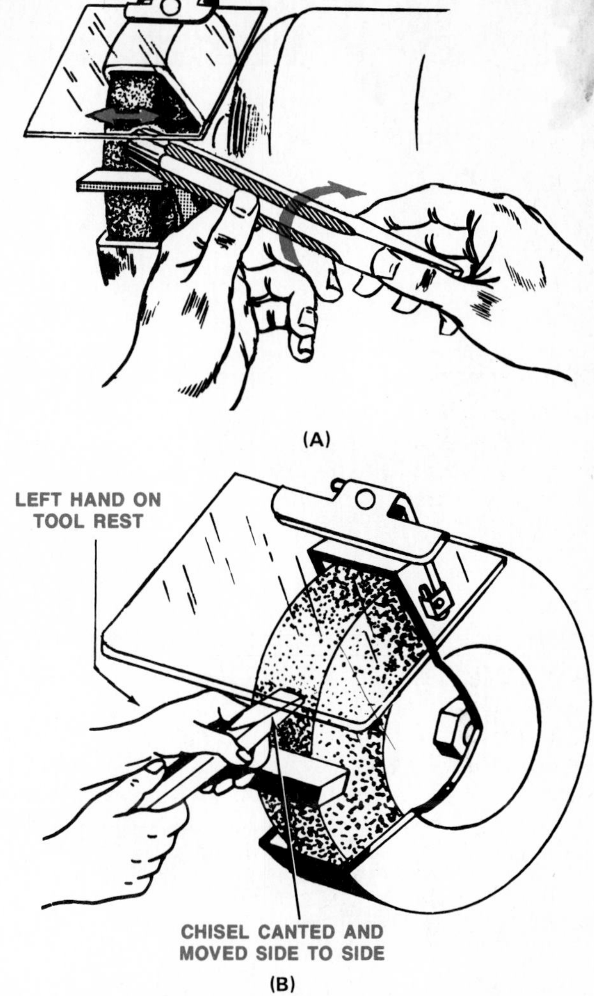

Fig. 6-20 (A) Grinding the mushroom from the head of a chisel. (B) Grinding the cutting edge of a chisel.

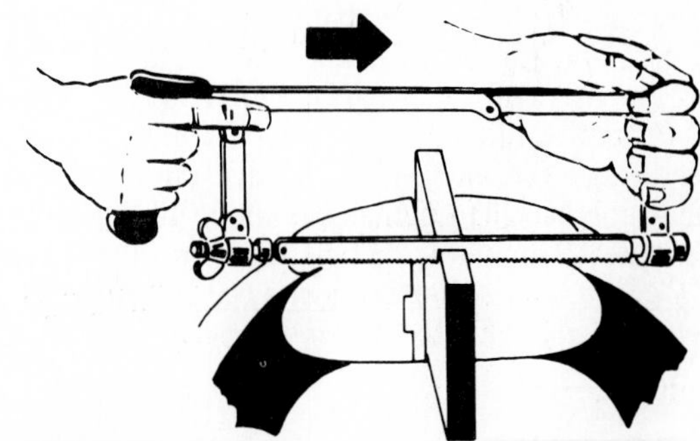

Fig. 6-21 Using a hacksaw.

♦ 6-17 FILES

Files are cutting tools with a large number of cutting edges, or teeth. A typical file with the various parts named is shown in Fig. 6-22. Files have many uses, so there are hundreds of types, with various kinds of cut. "Cut" refers to the grooves, or cuts, that form the file teeth. The most commonly used file is the flat file. When the cuts are far apart, so that there are only a few per inch, the file is called a *rough* or *coarse-cut* file. When they are close together, the file is a *smooth* or *dead-smooth* file.

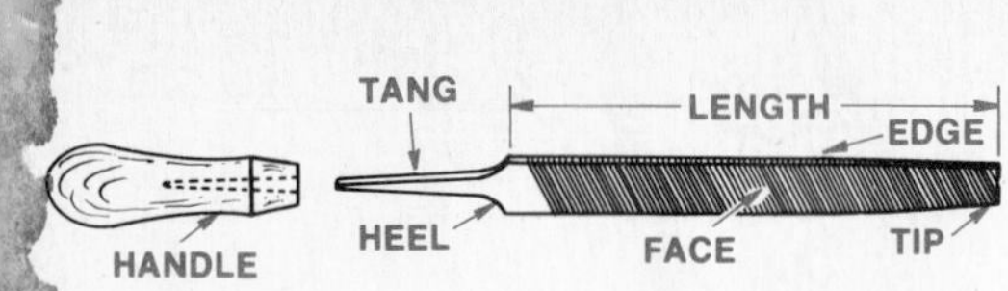

Fig. 6-22 A typical file with the parts named.

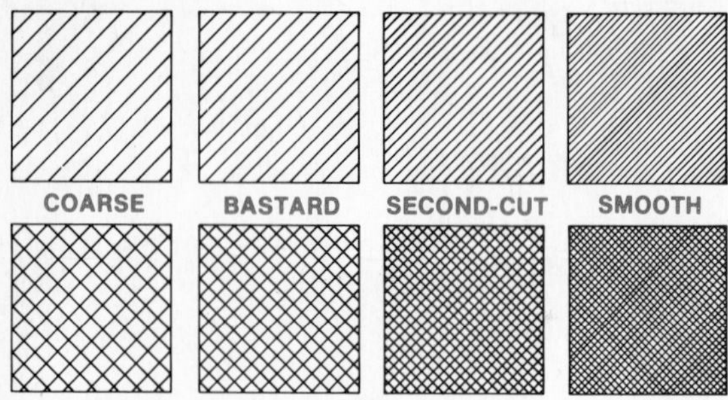

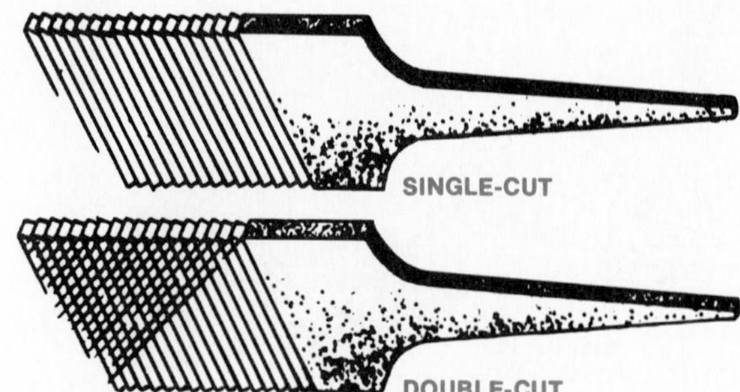

Fig. 6-23 Various file cuts.

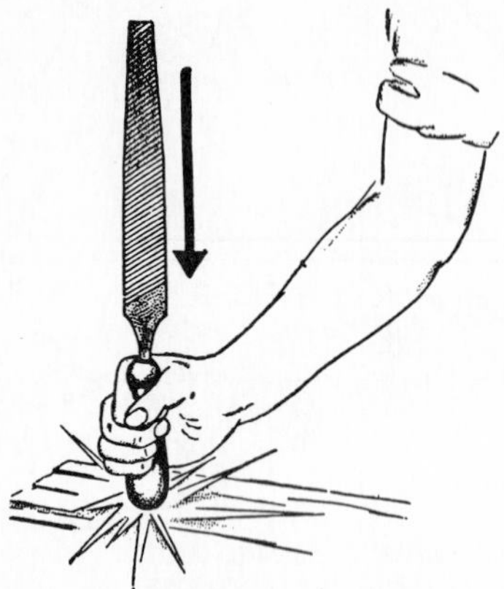
Fig. 6-24 To tighten a file tang in a handle, tap the end of the handle on the bench.

Figure 6-23 shows various file cuts. Files may have a single cut, which leaves a series of sharp blades. Or they may be double cut, which means that a second cut was made at an angle to the first cut (lower part of Fig. 6-23). This leaves a series of sharp teeth.

When using a file, always use a handle. Tap the end of the handle on the bench to tighten the file in the handle (Fig. 6-24). File teeth should be kept clean by brushing as needed with a *file card*. It has short, stiff wire bristles which remove chips and dirt from file teeth.

♦ 6-18 PUNCHES

Punches are used to knock out rivets and pins and to align parts for assembly (Fig. 6-25). They are also used to mark where holes are to be drilled. To remove rivets, two kinds of punches are used, starting and pin (Fig. 6-26). First, the rivet head is ground off or cut off with a chisel. Then the starting punch is used to break the rivet loose. The pin punch is then used to drive the rivet out.

The center punch can be used to mark the point where a hole is to be drilled. This gives the drill bit a place to start so that it stays centered (Fig. 6-27). The center punch is also used to mark parts before they are disassembled. Then, on reassembly, the punch marks can be lined up so that the parts go back in the right positions.

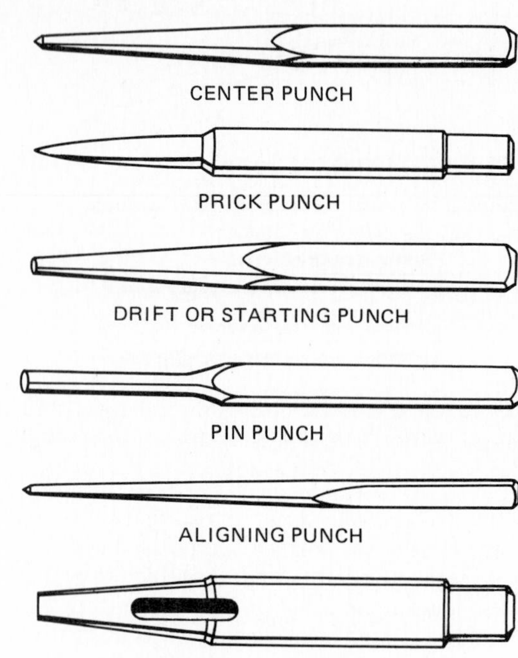

Fig. 6-25 Various types of punches.

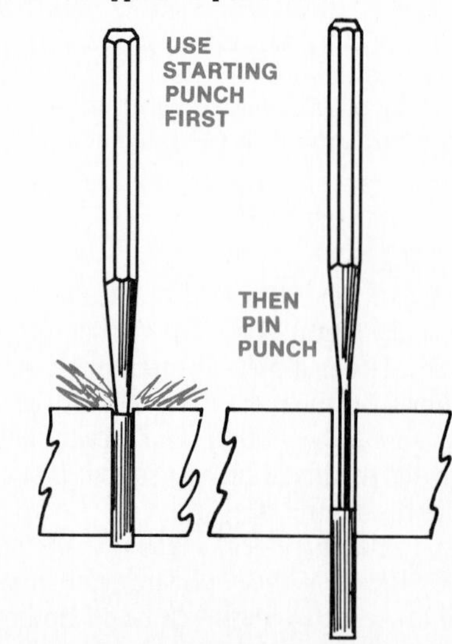

Fig. 6-26 Using starting and pin punches.

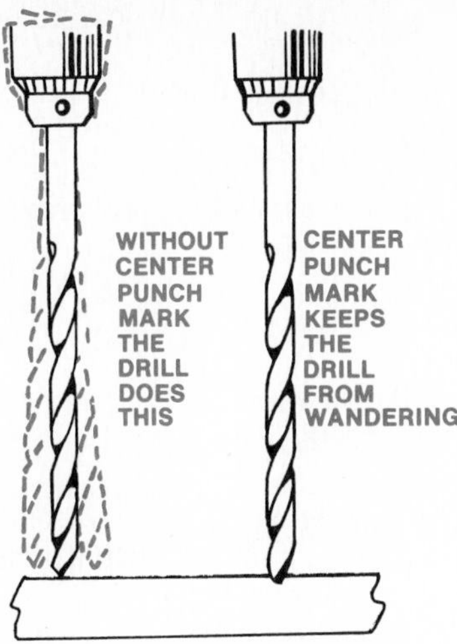

Fig. 6-27 Center-punching a hole location will keep the drill from wandering.

♦ 6-19 DRILL BITS

Drill bits, or *twist drills,* are tools for making holes. The material to be drilled determines the type of drill bit that must be used. The drill bit is a round bar with curved grooves and a point (Fig. 6-28). The point is shaped to provide cutting edges. The grooves allow the chips that are cut to pass away from the working surface. For most shopwork, electric drills are used to drive the drill bits. Drill motors may be portable or permanently assembled into a drill press (Fig. 6-29).

♦ 6-20 TAPS AND DIES

Taps and dies cut inside and outside threads (Figs. 6-30 and 6-31). Taps cut inside threads and are made in several styles. The taper tap is used to thread a hole completely through a piece of metal. The plug tap is used to thread a hole only partway. The bottoming tap is used for a hole that does not go all the way through the metal.

Dies cut outside threads. Dies are held in *diestocks* (Fig. 6-31). The rod end should be chamfered, or beveled, on the end so that the die starts easily. The basic procedure is the same for both taps and dies. Lubricant should be applied to start with. Then every two turns the tap or die should be backed off, and more lubricant applied.

♦ 6-21 APPRENTICE'S TOOLBOX

If you want to build up your toolbox, buy a tool every week or so. Get tools as you need them. Start out with a good quality toolbox with a top tray. Your instructor can advise you on what kind of toolbox to get and the tools to put in it. On the following page is a suggested list of tools that you should acquire while you are studying automotive mechanics.

Fig. 6-29 A drill press. The electric drill is a permanent part of the assembly.

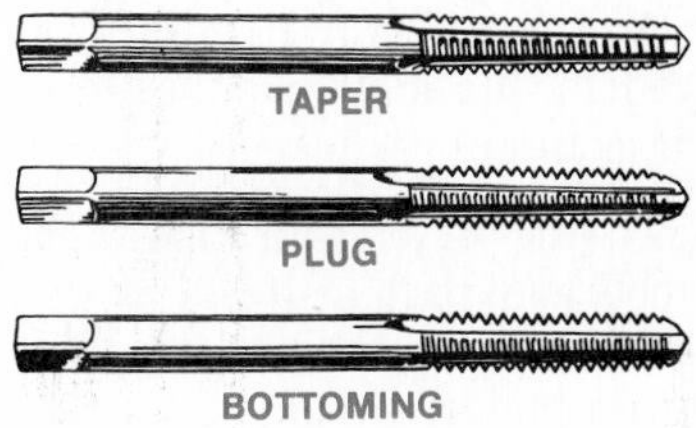

Fig. 6-30 Hand taps.

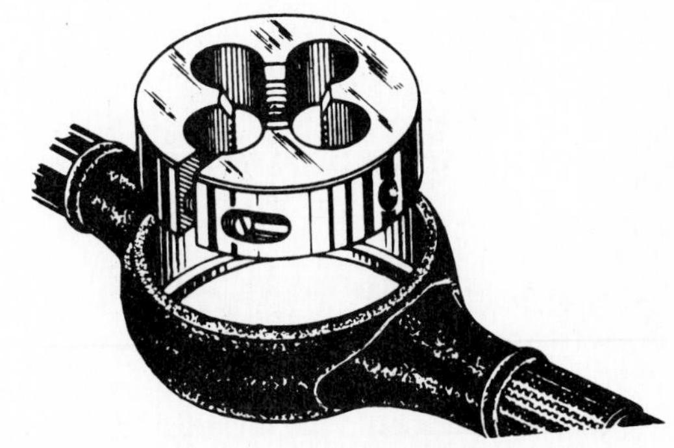

Fig. 6-31 Diestock with die above it, ready to be put into place.

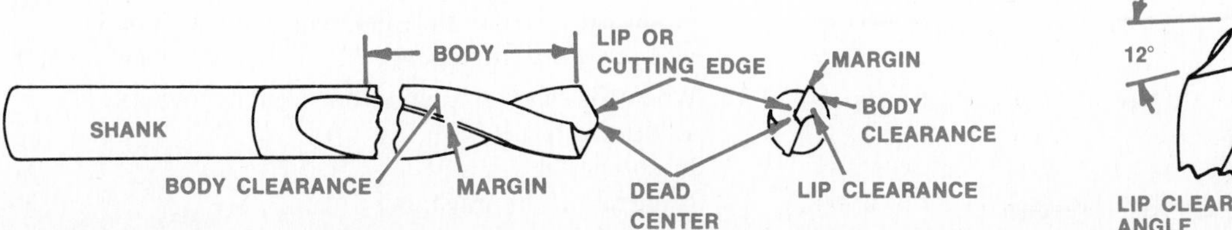

Fig. 6-28 The parts of a twist drill.

Suggested List of Tools

- ♦ 16-ounce [0.45-kg] ball-peen hammer.
- ♦ Plastic hammer.
- ♦ Five-piece screwdriver set (two Phillips).
- ♦ Standard pliers.
- ♦ Needle-nose pliers.
- ♦ Channellock pliers.
- ♦ Diagonal pliers.
- ♦ Spark-plug ratchet (3/8-inch drive with break-over handle).
- ♦ Spark-plug sockets (5/8 and 13/16 inch)
- ♦ Set of combination wrenches, 1/4 through 7/8 inch.
- ♦ Set of metric combination wrenches, 6 mm thru 22 mm.
- ♦ Flashlight (one that works, with new batteries).
- ♦ Short jumper wires. (You can make these, and more as needed.)
- ♦ 12-volt continuity light.
- ♦ Tape measure.
- ♦ 6-inch [152.4-mm] scale (metric on one side, inches on the other).
- ♦ Screw starter with magnet on one end.
- ♦ Tire-pressure gauge.
- ♦ Thickness-gauge set (can be go–no-go type). Buy combination thickness gauges, marked with both inches and millimeters.
- ♦ Spark-plug gauge set (round wire).
- ♦ Point file.
- ♦ Center punch.
- ♦ Set of 3/8-through 11/16-inch flare nut wrenches.
- ♦ 1/2-inch drive socket set.
- ♦ Ratchet.
- ♦ Speed handle.
- ♦ Breaker bar.
- ♦ 4-inch [100-mm] extension.
- ♦ 8-inch [200-mm] extension.
- ♦ U-joint.
- ♦ Set of sockets, 3/8 through 7/8 inch (12 point).
- ♦ Set of metric sockets, 6 mm through 22 mm (12 point).

Optional

- ♦ Vise-Grip pliers.
- ♦ Insulated pliers.
- ♦ Small adjustable wrench.
- ♦ Large adjustable wrench.

♦ REVIEW QUESTIONS ♦

Select the *one* correct, best, or most probable answer to each question. You can find the answers in the section indicated at the end of each question.

1. The two basic types of tools used in the shop are (♦6-1)
   a. power tools and machine tools
   b. hand tools and power tools
   c. screwdrivers and wrenches
   d. hand tools and cutting tools
2. The most common screwdriver used in the shop is the (♦6-2)
   a. double blade
   b. Phillips head
   c. flat blade
   d. hex head
3. The hammer used most often in the shop is the (♦6-3)
   a. ball-peen hammer
   b. plastic-tipped hammer
   c. brass hammer
   d. rawhide hammer
4. To work on both domestic and imported cars, you will need (♦6-4)
   a. a USCS wrench set
   b. a metric wrench set
   c. both a and b
   d. neither a nor b
5. The main difference between the box wrench and the socket is that (♦6-8)
   a. the socket is larger
   b. they are both combination wrenches
   c. the socket does not have an attached handle
   d. the box wrench has exchangeable handles
6. The most common box wrench has (♦6-6)
   a. 6 notches, or points
   b. 8 notches, or points
   c. 12 notches, or points
   d. 16 notches, or points
7. The combination wrench (♦6-7)
   a. has boxes at both ends
   b. is open at both ends
   c. has replaceable sockets
   d. has a box at one end and is open at the other
8. To tighten nuts or bolts accurately, use (♦6-9)
   a. a tension wrench
   b. a torque wrench
   c. two box wrenches
   d. an open-end wrench
9. The simplest wrench to use is the (♦6-5)
   a. open-end wrench
   b. box wrench
   c. socket
   d. adjustable wrench
10. The three basic types of pullers are (♦6-13)
    a. Channellock, screw, and Vise-Grip
    b. pressure screw, pusher, and twister
    c. pressure screw, slide hammer, and combination
    d. adjustable, interchangeable socket, and combination
11. When the head of a chisel mushrooms, (♦6-15)
    a. discard it
    b. sharpen it
    c. grind off the mushroom
    d. hammer it flat

12. You can use a tool holder to hold the chisel and (♦6-15)
    a. protect the hammer
    b. protect your hand
    c. prevent the chisel head from mushrooming
    d. avoid dulling the chisel
13. The blade should be put into the hacksaw so that it cuts on (♦6-16)
    a. the forward stroke
    b. the return stroke
    c. both strokes
    d. none of the above
14. The purpose of wearing safety goggles when using a chisel or a grinding wheel is to (♦6-16)
    a. sharpen your vision so that you can see what you are doing
    b. focus your eyes on the job
    c. protect your eyes from flying chips or sparks
    d. obey the shop rules
15. The most commonly used file is the (♦6-17)
    a. flat file
    b. oval file
    c. round file
    d. cut file
16. When a file has two cuts made at an angle with each other, the file is a (♦6-17)
    a. smooth file
    b. two-cut file
    c. double-cut file
    d. coarse-cut file
17. To knock out rivets and pins, you need a (♦6-18)
    a. pin punch and an aligning punch
    b. starting punch and a pin punch
    c. center punch and a pin punch
    d. drill punch and center punch
18. To cut threads in a hole, use a (♦6-20)
    a. die
    b. stud extractor
    c. threaded insert
    d. tap
19. To cut threads on a rod, use a (♦6-20)
    a. die
    b. stud extractor
    c. threaded insert
    d. tap
20. To remove a bolt that has broken off flush with the cylinder block, use (♦6-12)
    a. a file and wrench
    b. a drill and stud extractor
    c. a hacksaw and screwdriver
    d. any of the above

CHAPTER 7
Shop Equipment and Power Tools

After studying this chapter, you should be able to:

1. List the basic shop equipment and power tools found in the typical automotive shop.
2. Describe the use of each power tool in the automotive shop.
3. Demonstrate the use of the power tools in the shop.

♦ 7-1 POWER TOOLS AND SHOP EQUIPMENT

This chapter is divided into four parts, electric tools, pneumatic tools, hydraulic tools, and cleaning equipment. Only the basic power tools are described. Specialized power tools, such as those used to work on engines, are covered in later chapters, where their operation and use are explained.

♦ 7-2 TYPES OF POWER TOOLS

There are three types of power tools: electric, pneumatic, and hydraulic. *Electric tools,* such as drills, valve grinders, cylinder hones, and grinders, use electric motors. The word *pneumatic* means "of or pertaining to air." Pneumatic tools are tools operated by compressed air. They include air hammers, impact wrenches, air hoses, air ratchets, and air jacks.

The word *hydraulic* means "of or pertaining to a fluid or liquid." Hydraulic tools are tools that work because of pressure on a liquid. Hydraulic tools used in the automotive shop include car lifts, floor jacks, portable cranes, and hydraulic presses.

♦ 7-3 BENCH VISE

One of the most common shop tools is the bench vise (Fig. 7-1). It is used to hold a part that is being sawed, filed, chiseled, or otherwise worked on. When the handle is turned, a screw moves the movable jaw toward or away from the stationary jaw. To protect the parts clamped in the vise, caps of soft metal are placed over the steel jaws of the vise. These are usually called "soft jaws."

ELECTRIC TOOLS

♦ 7-4 ELECTRIC DRILL

The electric drill (Fig. 7-2) has an electric motor that drives a chuck. The chuck has jaws that can be opened and then

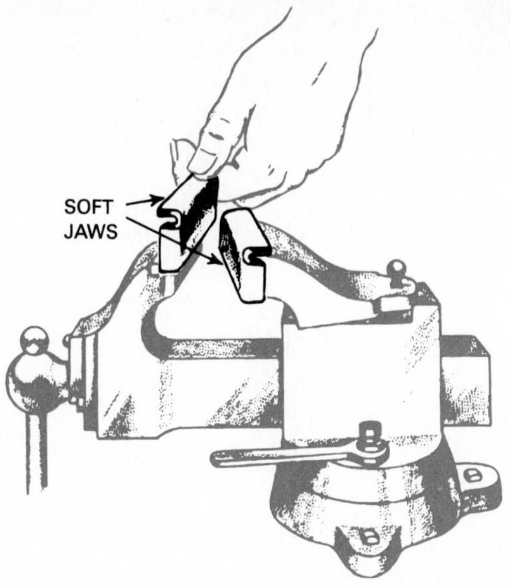

Fig. 7-1 A bench vise, showing soft jaws being put into place in the vise jaws.

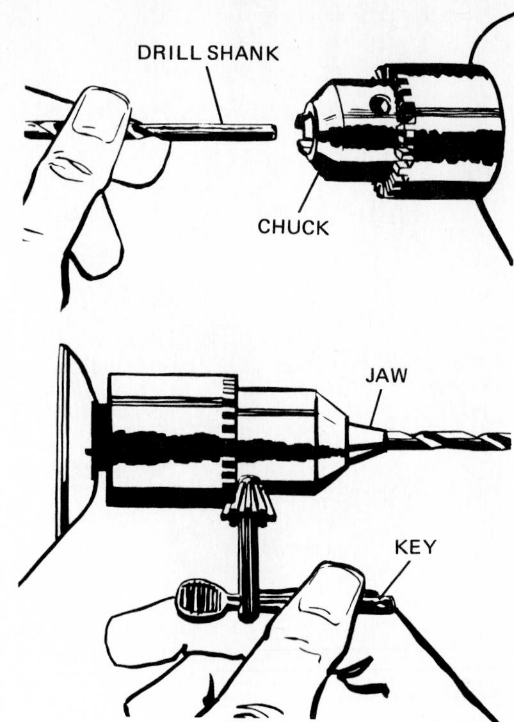

Fig. 7-3 Installing a twist drill in a drill chuck.

closed to grip a drill bit (Fig. 7-3). A special chuck key is required to operate the jaws. Drill bits are described in ♦6-19. Here are some cautions to observe when using an electric drill:

1. Make sure the drill is properly grounded (Fig. 7-4) with the third blade on the plug, or that the drill is double-insulated.
2. Do not drag the drill around by its cord. Do not kink the cord or allow anyone to step on it or run a machine over it. Any of these things can damage the insulation and cause someone to get a dangerous electric shock.
3. Keep your hands and clothes away from the moving bit.
4. Keep a firm grip on the drill and be ready to shut it off if it jams. When a bit is about to break through metal, it sometimes jams. Be prepared for this.
5. If the drill bit jams, do not try to break it free by turning it on and off. This can damage the drill. Instead, pull it back to free it.
6. When you are through with the drill, disconnect it, wipe it clean and put it away in a safe place. Oil it occasionally if it requires oiling. Some drills have preoiled bearings and require no oiling.

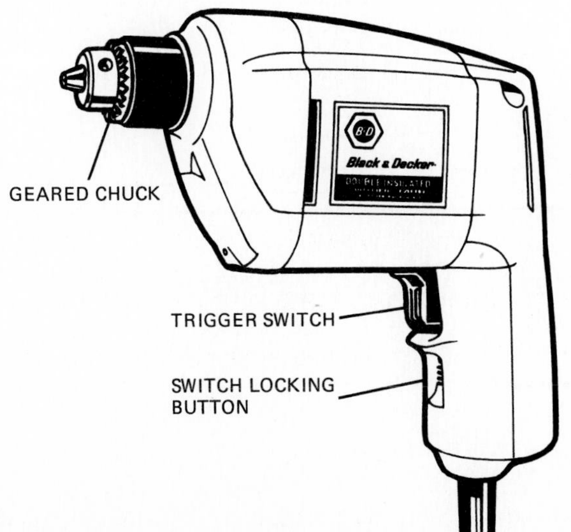

Fig. 7-2 Electric drill. (*The Black & Decker Manufacturing Company*)

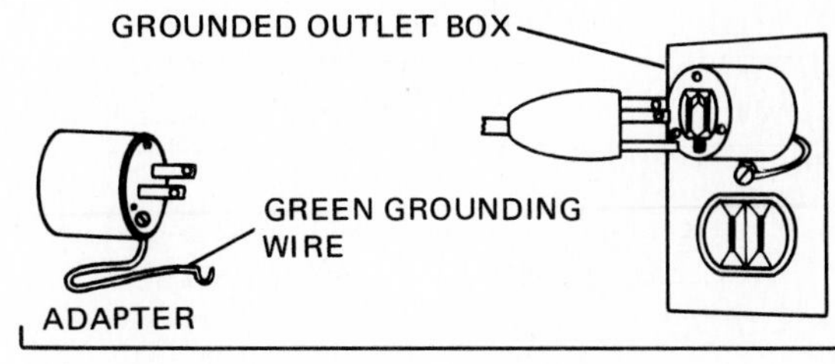

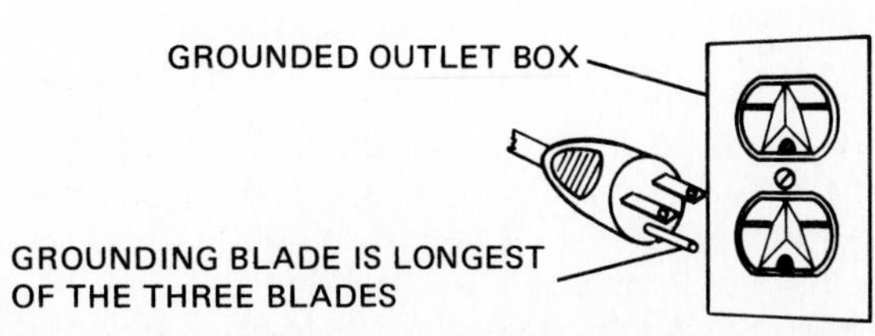

Fig. 7-4 Various methods of grounding electric tools. (*The Black & Decker Manufacturing Company*)

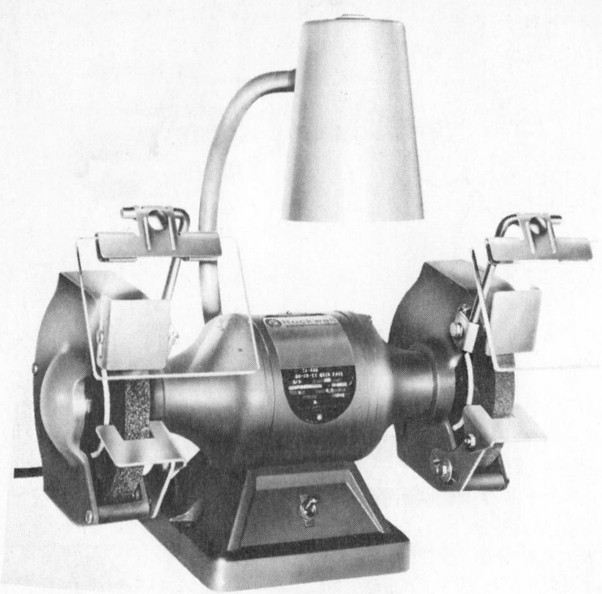

Fig. 7-5 Bench grinder. (*Rockwell International*)

♦ 7-5 GRINDING WHEELS

The grinder (Fig. 7-5) can be either bench-mounted or installed on a pedestal. It may have a grinding wheel and a wire wheel, or two grinding wheels as shown in Fig. 7-5. Then one wheel is coarse, the other fine. Figure 6-20 shows a grinder in use. Observe the following cautions when using a grinding wheel:

1. Do not hammer on the wheel or apply excessive force against it.
2. Be alert for the danger of flying sparks from the wheel when metal is being ground. These sparks are hot and can set fire to your clothing or burn you.
3. Never try to adjust the tool rest while the grinder is running. Shut the grinder off and wait until the wheels stop moving.
4. Do not touch the rotating wheel. It can take skin and flesh off on contact.
5. Always wear eye protection, even if the wheel is equipped with an eye shield.
6. Adjust the light so that you can see clearly what you are doing.
7. When grinding a tool, do not overheat it. This will draw the temper so the tool will get soft and not hold an edge. To prevent overheating, dip the tool in water repeatedly during the grinding operation.

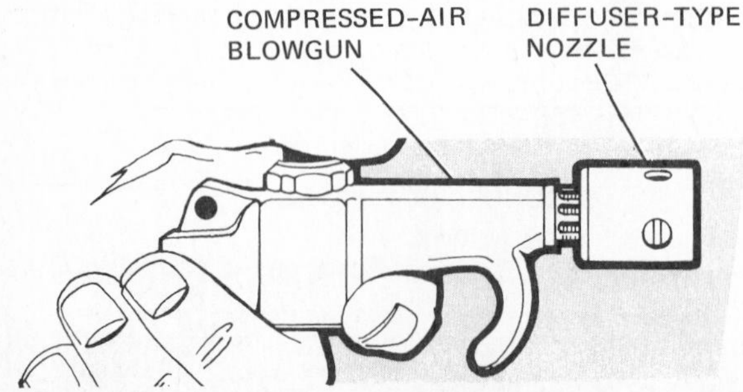

Fig. 7-6 Compressed-air blowgun, with approved diffuser-type nozzle. The diffuser reduces the air pressure to a safer value. (*Ford Motor Company*)

♦ 7-6 VACUUM CLEANER

Many shops have vacuum cleaners for cleaning the floor and also for cleaning car interiors after service. This is especially important in a body shop, where dust is created during auto-body repair and refinishing. The shop vacuum cleaner can be used to remove asbestos dust from wheel-brake mechanisms during brake jobs. Asbestos dust is harmful to breathe as explained in the chapter on brake service.

PNEUMATIC TOOLS

♦ 7-7 CAUTIONS TO OBSERVE WHEN USING PNEUMATIC TOOLS

Pneumatic tools operate on compressed air supplied by the shop air system. A compressed-air system is needed to pressurize tires, to blow-dry parts that have been washed in solvent, and to operate pneumatic tools. Here are cautions to observe when using pneumatic tools:

1. Shops have air nozzles—often called *blowguns*—for blowing parts dry and clean (Fig. 7-6). They are often equipped with a device to reduce the pressure to a safe maximum. Never use the blowgun to blow dust off your clothes. Dirt particles are driven at high speed by the air. They can penetrate the flesh or the eyes. If high-pressure air is directed toward an open wound, it can send air into the bloodstream. This could possibly cause death.
2. Never point the blowgun at anyone.
3. Never look into the air outlet of any pneumatic tool.
4. Never operate air impact chisels without a chisel installed. This can damage the tool.

♦ 7-8 AIR CHISEL

The air chisel, or air hammer (Fig. 7-7), uses reciprocating motion to drive a cutting or hammering tool. Figure 7-7 shows an air hammer driving a chisel to cut off a nut that has frozen to a stud. A few strokes will split the nut so that it can be removed. Air hammers can be used with a variety of tools—cutters, chisels, punches—to do many jobs.

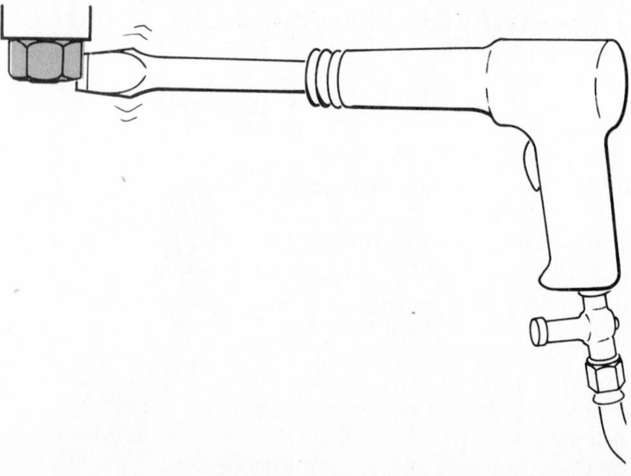

Fig. 7-7 Using an air hammer. (*Chicago Pneumatic Tool Company*)

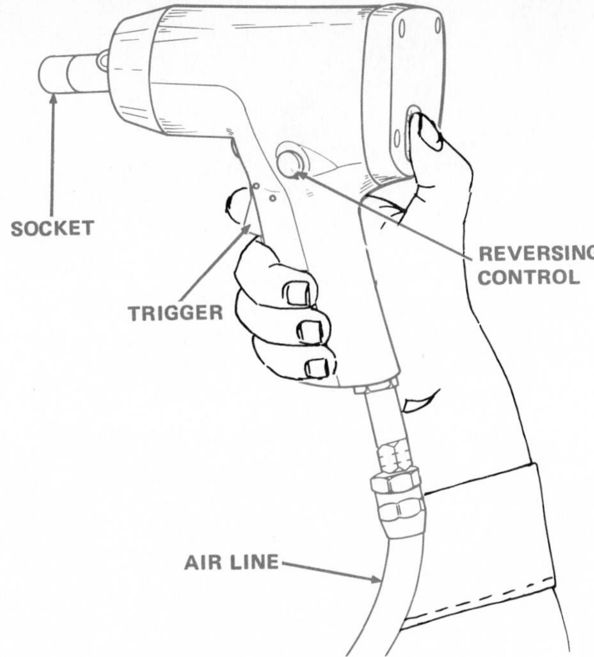

Fig. 7-8 Using an air impact wrench.

♦ 7-9 AIR IMPACT WRENCHES

Air impact wrenches (Fig. 7-8) use a pounding or impact force to loosen or tighten nuts or bolts. They do the job very rapidly and are widely used in the shop. The direction of rotation can be changed by operating a reversing control.

Note Some impact wrenches are electric. They have an electric motor.

Here are some rules for using an impact wrench:

1. Always use *impact sockets* of the correct size.
2. Hold the wrench so the socket fits squarely on the nut or bolt. Apply a slight forward force to hold the socket in place.
3. Once a nut or bolt tightens, never impact it beyond an additional one-half turn. Continued impacting can strip threads or break the bolt.
4. For accurate tightening of a bolt or nut, always use a torque wrench (♦6-9).
5. Soak rusty nuts or bolts with penetrating oil before impacting them.

♦ 7-10 AIR DRILL

Air drills do the same job as electric drills (♦7-4). Figure 7-9 shows the hook-up for an air drill. One advantage of the air drill is that it is lighter than a comparable electric drill. Also, repeated stalling or overloading does not damage or overheat the air drill.

♦ 7-11 AIR RATCHET

Most air ratchets have a ⅜-inch drive. They can use the sockets and attachments from a standard socket set. Air ratchets do not apply as much force as impact wrenches. Therefore, standard sockets can be used.

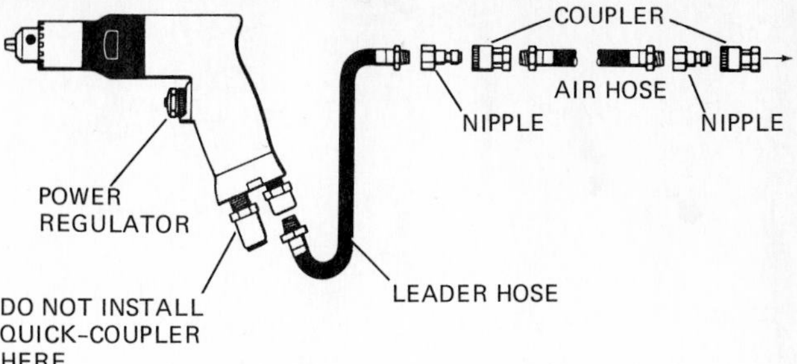

Fig. 7-9 Air drill with hose and couplings. (*Chicago Pneumatic Tool Company*)

♦ 7-12 PNEUMATIC FLOOR JACK

Many types of pneumatic jacks are used in the shop. The pneumatic floor jack, or bumper-type end lift (Fig. 7-10), uses compressed air to jack up one end of the car. When the air valve is operated, compressed air flows into the jack cylinder, causing the ram to extend and raise the vehicle. When the lever position is reversed, the air is exhausted so that the vehicle settles back to the floor.

Fig. 7-10 Bumper-type end lift, or air jack.

♦ CAUTION ♦ Never go under a vehicle that is supported only by a jack. Always use safety stands placed under the vehicle to support it safely before working under a vehicle (Fig. 7-11). A jack could slip or release, allowing the vehicle to come down on you. You could be seriously injured or even killed.

♦ 7-13 CARE OF AIR TOOLS

Air tools should be given the same care as any other power tool. They should not be dragged around by their hoses, dropped on the floor, or otherwise abused. When not in use, they should be disconnected from the air line and put away in their designated storage place. Every day, before using an air tool, apply three or four squirts of flushing oil to the air inlet (Fig. 7-12). Then connect and operate the tool. This flushes

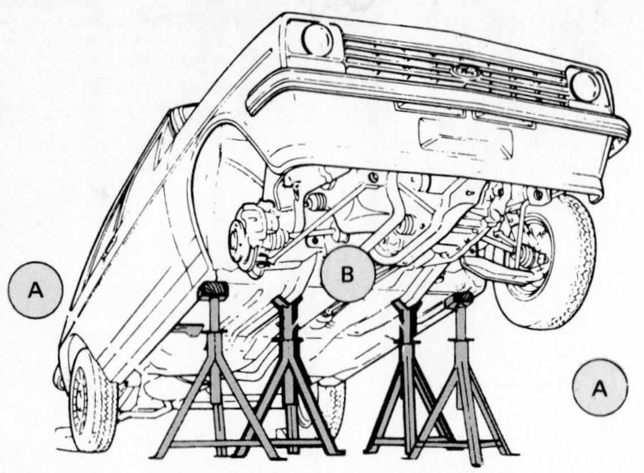

Fig. 7-11 Always place safety stands under a car at the proper lift points (A or B) before going under a car. (*Ford Motor Company*)

out any dirt or moisture and lubricates moving parts of the air tools.

♦ CAUTION ♦ Do not flush the air tool around an open flame. The mist coming out is flammable. Always point the tool air exhaust away from your body.

HYDRAULIC TOOLS

♦ 7-14 CAR LIFTS

Car lifts are either single- or double-post (Figs. 7-13 and 7-14). Both have lift pads that must be positioned under the designated lift points of the car frame. There is also the drive-on type of above-ground lift which has parallel tracks onto which the car is driven. This type is convenient for such jobs as draining oil and working on transmissions, drive lines, and differentials.

The car lift is operated by an air or electric motor that drives a hydraulic pump. The pump sends liquid under pressure into the lift cylinder or cylinders. This raises the car.

♦ CAUTION ♦ Never allow passengers to remain in a car that is being lifted. Make sure all doors, the hood, and the trunk lid are closed. Otherwise, they could get ripped off. If the lift has a mechanical lock, make sure it is engaged before you go under the car. If the lift is not working properly—jerks or jumps when raised, leaks oil, blows oil out the exhaust line—don't use it.

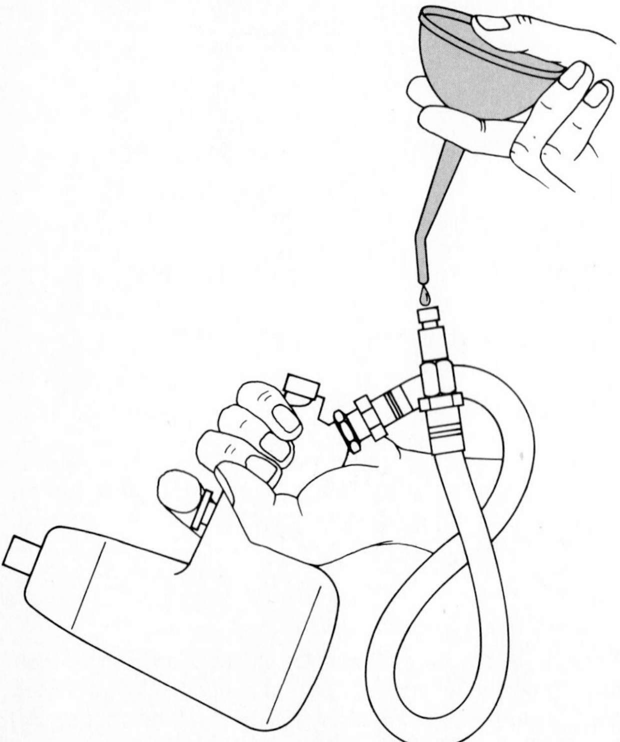

Fig. 7-12 Squirting a few drops of flushing oil into an impact-wrench air line. (*Chicago Pneumatic Tool Company*)

Fig. 7-13 Single-post air lift. The four lifting arms can be moved so the lift pads are positioned directly under the lift points on vehicles. (*Lincoln St. Louis Division of McNeil Corporation*)

Fig. 7-14 Two-post lift. The post under the front wheels can be moved forward or backward to accommodate cars of different lengths. (*Lincoln St. Louis Division of McNeil Corporation*)

Fig. 7-15 Hydraulic floor jack. (*Blackhawk Division of Applied Power, Inc.*)

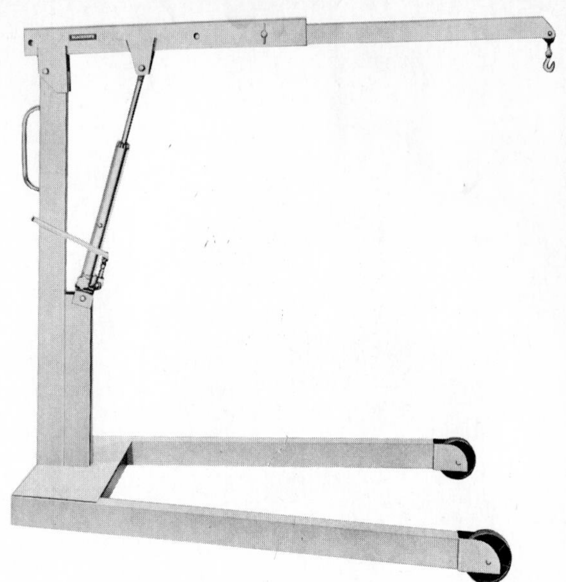

Fig. 7-16 Portable crane. (*Blackhawk Division of Applied Power, Inc.*)

♦ 7-15 HYDRAULIC FLOOR JACK

The hydraulic floor jack (Fig. 7-15) is operated by pumping the handle. This raises the lifting saddle. A lever on the handle releases the pressure so that the saddle and load will settle back down. When using the jack, be sure the saddle is firmly placed under the proper lift point on the car. If the saddle should slip, the car could drop and be damaged.

♦ CAUTION ♦ Never work under a car that is supported only by a jack. See the detailed Caution in ♦7-12 about properly supporting a car (Fig. 7-11).

♦ 7-16 PORTABLE CRANE

The portable crane (Fig. 7-16) is used for such jobs as lifting the engine out of the car. It is operated hydraulically by a hand pump. It can also be used for other jobs where a heavy object must be lifted and transported.

♦ CAUTION ♦ When using the portable crane or other lifting equipment, stand clear of the part that is being lifted. That way, you will not be injured if the part should slip or the lift should topple over. Never work on a part that is suspended in the air. Instead, lower it onto your workbench or onto a support stand before working on it.

♦ 7-17 HYDRAULIC PRESS

The hydraulic press (Fig. 7-17) applies force on bent parts to straighten them. It can also do such jobs as press bushings in or out and press out rivets. A manual press, operated by hand, is the *arbor press* (Fig. 7-18). It also applies force, but less than the hydraulic press.

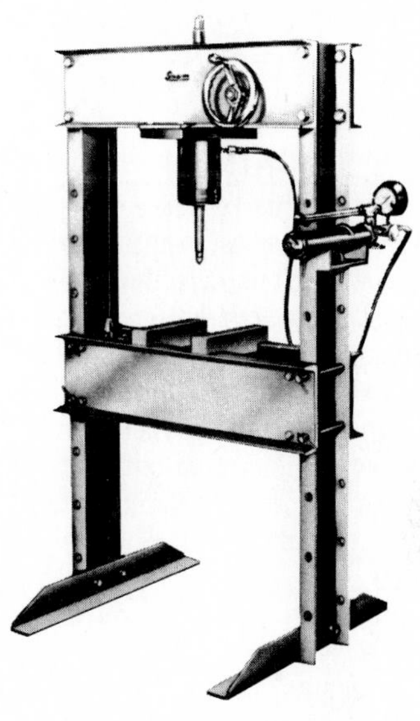

Fig. 7-17 Hydraulic press. (*Snap-on Tools Corporation*)

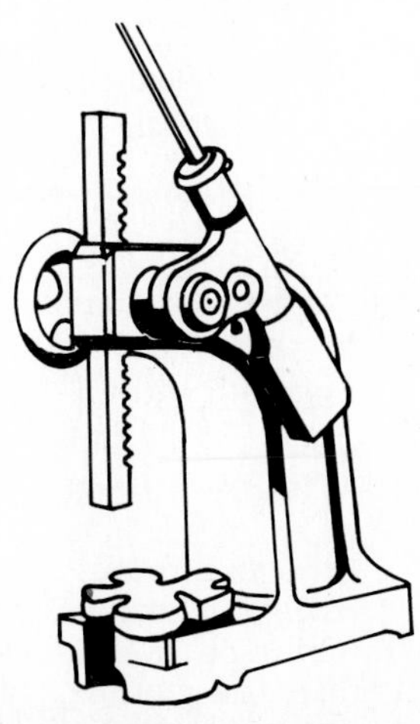

Fig. 7-18 A hand-operated arbor press.

Fig. 7-19 Solvent tank. (*Gray Mills Corporation*)

CLEANING EQUIPMENT

♦ 7-18 PARTS CLEANERS

Proper cleaning of the parts that are being serviced is an important first step. The most experienced mechanic cannot tell whether a dirty part is defective. Also, without proper cleaning, the most expert overhaul job can be ruined by dirt left in during reassembly. Parts cleaning is so important that many professional mechanics will not let anyone else clean parts. It is during the cleaning operation that clues to the trouble often show up. Many automotive-service shops and machine shops have a solvent tank, a hot tank or steam cleaner, and a glass-bead cleaner. These are described in the following sections.

The putty knife or scraper is the most common cleaning tool in the shop. Next is the solvent tank (Fig. 7-19). Solvent tanks are made in many sizes and types. Even the cutoff end of a barrel has been used to hold solvent. Most commercial models have shelves and flexible hoses to accommodate a variety of parts. Also, they have filters and sediment trays to keep the solvent clean.

Solvent is one type of cold liquid cleaner. Another type is the widely used *carburetor cleaner*.

♦ 7-19 HOT TANKS

Hot tanks, or boil tanks (Fig. 7-20), are used to clean larger parts, such as cylinder heads. The solution used is usually caustic soda or lye mixed with water. The solution is heated by a gas or electric heater. To aid in the cleaning process, some models use compressed air to agitate the solution.

♦ CAUTION ♦ Extreme care must be used when working around the hot tank. The tank is filled with a boiling caustic solution. Parts must be lowered into and removed from the tank slowly to avoid splashing the solution. A face shield and protective clothing should be worn when working around a hot tank.

Fig. 7-20 Hot tank with parts basket and electric lift. (*Simmons Parts Company, Inc.*)

♦ 7-20 STEAM CLEANER

Figure 7-21 shows one model of steam cleaner. A kerosene burner heats water to high-pressure steam. A spray of this steam, with soap, is used to remove grease and dirt from parts, or to steam-clean engines or the underside of a vehicle.

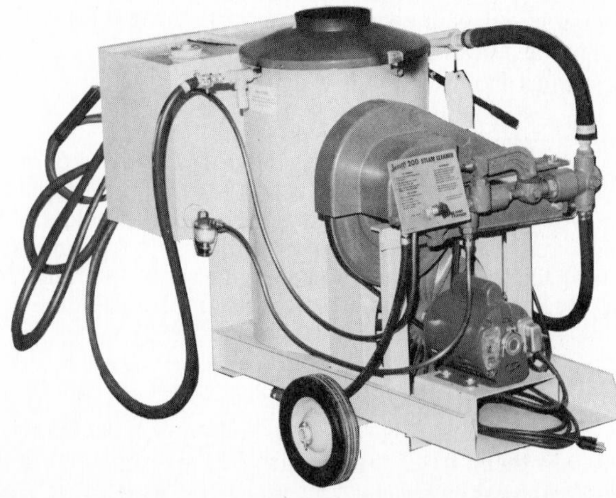

Fig. 7-21 Steam cleaner. (*Jenny Division of Homestead Industries*)

♦ 7-21 GLASS-BEAD CLEANERS

This type of cleaner (Fig. 7-22) uses compressed air to blow a stream of glass beads or other abrasives at the part to be cleaned. The glass beads scrub dirt, grease, and grime off the object. This type of cleaner is preferred in many shops because it is cleaner and safer to use than a liquid chemical cleaner. However, the glass-bead cleaner will not clean out internal passages in metal parts, such as the fuel passages in a carburetor.

♦ 7-22 OTHER SHOP EQUIPMENT

Specialized shop equipment such as valve grinders, cylinder hones, wheel balancers, and electrical testers are covered in later chapters.

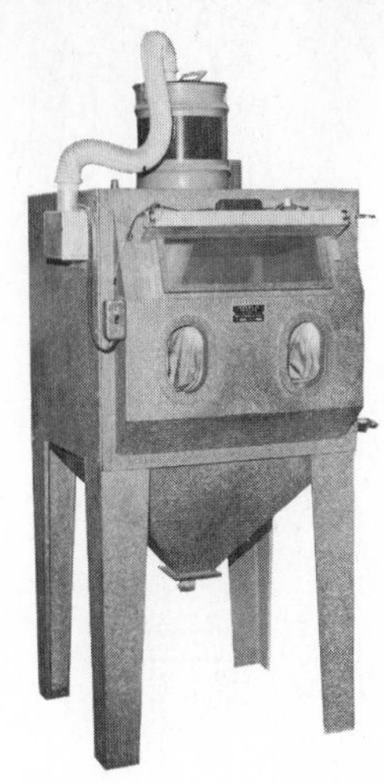

Fig. 7-22 Glass-bead cleaner. (*Inland Manufacturing Company*)

♦ REVIEW QUESTIONS ♦

Select the *one* correct, best, or most probable answer to each question. You can find the answers in the section indicated at the end of each question.

1. The three types of power tools are (♦7-2)
   a. pneumatic, air, and electric
   b. pneumatic, hydraulic, and electric
   c. hydraulic, brake, and electric
   d. electric, mechanical, and hydraulic
2. Pneumatic tools include (♦7-2)
   a. air hammers, impact wrenches, and air jacks
   b. air hoses, air ratchets, and air jacks
   c. air hammers, air ratchets, and air hoses
   d. all of the above
3. Hydraulic tools include (♦7-2)
   a. car lifts, floor jacks, impact wrenches, and air hammers
   b. car lifts, portable cranes, air ratchets, and air jacks
   c. car lifts, floor jacks, portable cranes, and hydraulic presses
   d. floor jacks, hydraulic presses, air hammers, and ratchets
4. When using a grinding wheel, (♦7-5)
   a. always wear safety goggles
   b. never adjust the tool rest when the grinder is running
   c. never touch the wheel when it is rotating
   d. all of the above
5. Mechanic A says it is all right to go under a car supported by a floor jack provided the saddle is properly located. Mechanic B says you should never go under a car unless it is supported by safety stands. Who is right? (♦7-12)
   a. mechanic A
   b. mechanic B
   c. neither A nor B
   d. both A and B
6. When lifting a car, (♦7-14)
   a. make sure everyone is out of the car
   b. make sure all doors, the hood, and the trunk lid are closed
   c. engage the mechanical lock, if available
   d. all of the above
7. The portable crane is used to (♦7-16)
   a. lift the engine out of the car
   b. lift the car to change oil
   c. suspend parts so they can be worked on
   d. apply force to straighten parts
8. Three pieces of cleaning equipment used in the shop are (♦7-18)
   a. solvent tank, cold tank, and hot tank
   b. solvent tank, hot tank, and glass cleaner
   c. solvent tank, hot tank, and glass-bead cleaner
   d. hot tank, cold tank, and steam cleaner
9. Pneumatic tools are operated by (♦7-2)
   a. hydraulic pressure
   b. electricity
   c. air pressure
   d. fluid pressure
10. When using an air nozzle, (♦7-7)
    a. never point it at anyone
    b. never blow dust off your clothes
    c. do not direct the airstream against your face or hands
    d. all of the above

PART 2

This part of *Automotive Mechanics* describes the construction and operation of automotive engines. Later in the book we describe the engine systems that are necessary to make the engine run. These include the fuel, lubricating, cooling, and electrical systems. This part of the book—Automotive Engines—has six chapters:

Chapter 8 Engine Fundamentals
Chapter 9 Piston-Engine Operation
Chapter 10 Engine Types
Chapter 11 Engine Construction
Chapter 12 Valves and Valve Trains
Chapter 13 Engine Measurements and Performance

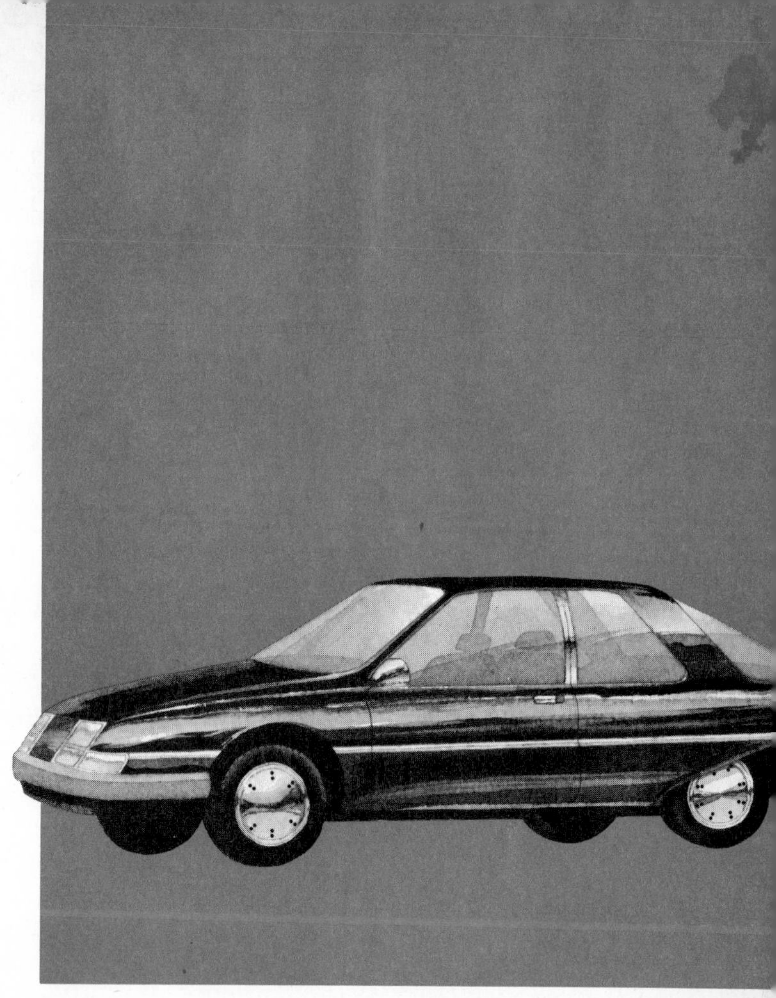

Automotive Engines

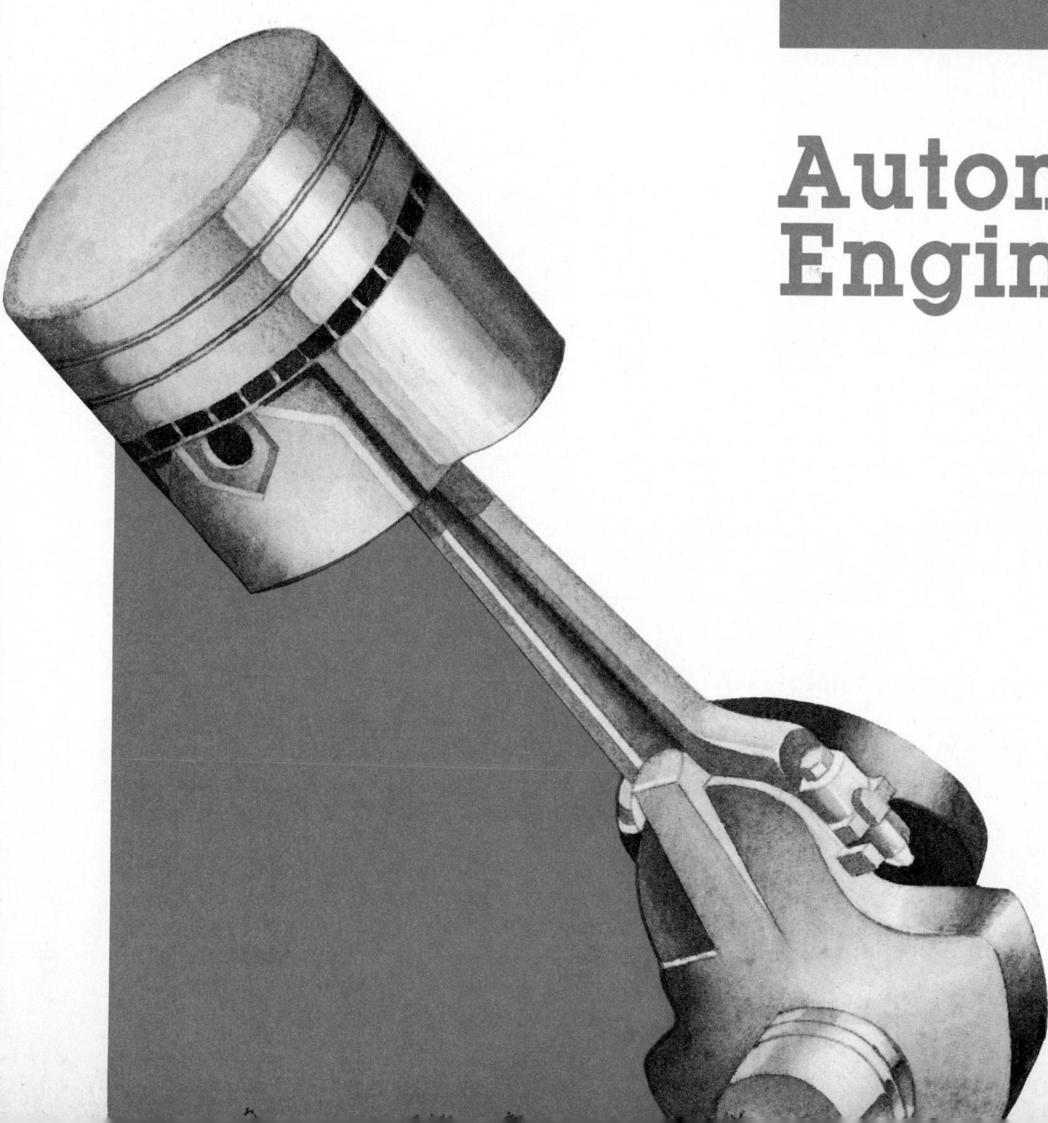

CHAPTER 8
Engine Fundamentals

After studying this chapter, you should be able to:

1. Explain what happens during combustion.
2. Discuss combustion of fuel in the engine.
3. Explain why combustion is not ideal, or complete, in the engine.
4. Describe the two rules—increase of pressure with increasing temperature, and increase of temperature with increasing pressure—and explain how they work in the engine.

♦ 8-1 ATOMS

Atoms are the basic "building blocks" of the world. They are extremely tiny. There are about 100 billion billion atoms in a single drop of water! There are about a hundred varieties of atoms. Each has a special name: gold, silver, iron, aluminum, oxygen, hydrogen, and so on. The copper in a penny is made up of billions upon billions of one kind of atom. About 20 percent of our air is a gas called oxygen. Any substance made up of only one kind of atom is called an *element.*

♦ 8-2 CHEMICAL REACTIONS

Atoms of different elements can connect up with each other. The process is called a *chemical reaction.* Chemical reactions go on all around us—all the time. The burning of fuel in an automobile engine is a chemical reaction. We normally refer to this chemical reaction as *combustion.*

♦ 8-3 COMBUSTION

Automotive fuels are made up mostly of two elements, *hydrogen* and *carbon,* which have the chemical symbols H and C. It is therefore called a *hydrocarbon* (HC). During complete combustion in the engine, these two elements unite with a third element, the gas oxygen (O). Oxygen forms about 20 percent of the air we breathe. Each oxygen atom connects up with two hydrogen atoms (Fig. 8-1). Each carbon atom connects up with two oxygen atoms. The combining of hydrogen and oxygen produces H_2O, which is just plain water. The combining of carbon and oxygen produces the gas *carbon dioxide* (CO_2).

Note Only a few atoms are shown in Fig. 8-1. However, there are billions of atoms uniting every second in the engine when it is running.

During the combustion of gasoline in the engine, the burning gases get very hot. Their temperature may go as high as 6000°F [3319°C]. This high temperature produces the pressure that makes the engine run and produce power.

Note The symbol for *degree* is °. For example, 6000 degrees is usually written as 6000°.

With ideal, or perfect, combustion, all of the hydrogen and all of the carbon in the gasoline would combine with oxygen to form harmless H_2O (water) and CO_2 (carbon dioxide). However, in the engine, we do not get ideal combustion. Instead, some of the gasoline (HC) does not burn. Also, some

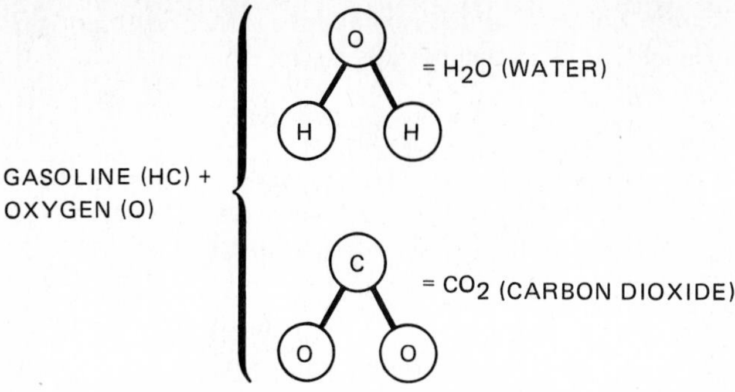

Fig. 8-1 When gasoline burns (combines with oxygen), the result is water and carbon dioxide.

only partly burns, producing *carbon monoxide* (CO) instead of CO_2. The unburned gasoline (HC) and partly burned gasoline (CO) cause pollution of the air as they exit through the tail pipe with the exhaust gases. This is the reason that modern cars are equipped with antipollution devices (called *emission controls*). These devices reduce the amount of HC and CO coming from the engine. They are covered later.

♦ 8-4 EXPANSION OF SOLIDS WITH HEAT

Any solid, such as the metal in an engine piston, expands and gets larger as its temperature increases. However, the piston must be free to move up and down in the cylinder, even if it gets very hot (Fig. 8-2). The piston is designed so that it does not expand too much. If it did, it would stick in the cylinder and the engine would be damaged.

♦ 8-5 EXPANSION OF FLUIDS WITH HEAT

Fluids are substances that can flow. Examples are water and air. Both expand when heated. One cubic foot of water [0.0283 m^3] at 32°F [0°C] becomes 1.01 cubic foot [0.0286 m^3] when heated to 100°F [37.8°C]. One cubic foot of air [0.0283 m^3] at 32°F [0°C] expands to 1.14 cubic feet [0.0323 m^3] when heated to 100°F [37.8°C] without a change of pressure.

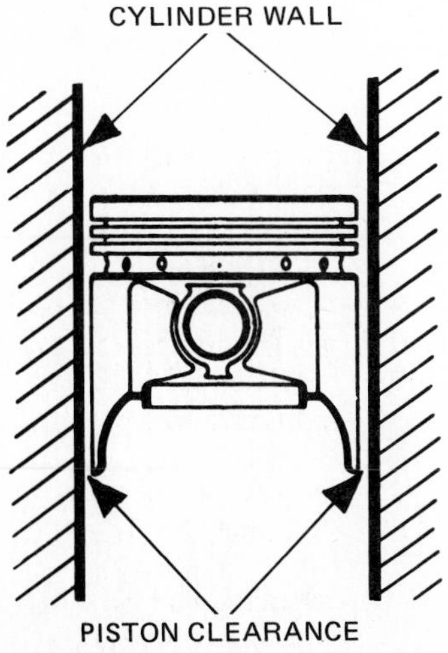

Fig. 8-2 The piston is designed so that even though it gets hot and tends to expand, it will not expand enough to stick in the engine cylinder.

Note If the air is heated in a closed container so its volume cannot increase, the pressure will go up.

♦ 8-6 INCREASE OF PRESSURE WITH TEMPERATURE

Here is what happens when air is heated in a closed container that will not allow it to expand. If we start with a cubic foot of air at 32°F and heat it to 100°F (from 0 to 37.8°C), the pressure will go up about 10 percent. If we started with a pressure of 15 pounds per square inch (psi) [103 kilopascals (kPa)], the pressure would go up to 17 psi [117 kPa].

This is the first rule: *Pressure increases with increasing temperature.*

♦ 8-7 INCREASE OF TEMPERATURE WITH PRESSURE

The rule works the other way, too: *Temperature increases with increasing pressure.* This is the second rule.

Both rules are at work in the engine cylinders. First, a mixture of air and gasoline vapor is taken into the cylinder. Then the piston moves up and compresses this mixture. Compressing the mixture—increasing the pressure on the mixture—makes it hot (temperature increases with increasing pressure). Next, the compressed mixture is ignited, or set on fire. It burns, producing a very high temperature. The high temperature causes high pressure (pressure increases with increasing temperature). The high pressure pushes the piston down. This motion is carried to the car wheels so that they turn and the car moves.

♦ 8-8 THE THERMOMETER

The thermometer (Fig. 8-3) is a familiar example of expansion with heat. It is just a hollow glass tube partly filled with a liquid, such as mercury. As the temperature increases, the mercury expands. Part of it is forced up the tube. The higher the temperature, the more the mercury expands, and the far-

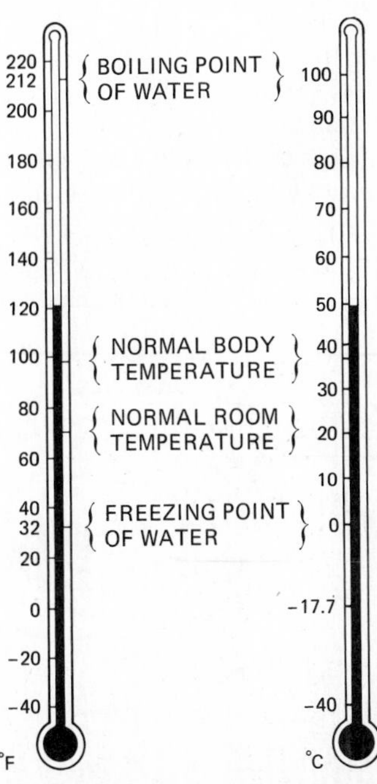

Fig. 8-3 Fahrenheit and Celsius thermometers.

ther up the tube the mercury rises. The tube is marked off in degrees to indicate the temperature.

♦ 8-9 THE THERMOSTAT

Different metals expand different amounts when they are heated. For example, aluminum expands about twice as much as iron. Such differences in expansion are used in many thermostats. One type is shown in Fig. 8-4. It is a coil made of two strips of different metals, such as brass and steel, welded together. When the coil is heated, one metal expands more than the other. This causes the coil to wind up or unwind. The motion of the coil can be used to control fluids or electricity. It is used in this manner in several places in the automobile.

♦ 8-10 GRAVITY

Gravity is the attractive force between the earth and all other objects. When we release a stone from our hand, it falls to earth. When a car is driven up a hill, part of the engine power is used to raise the car against gravity. Likewise, a car coasts down a hill with the engine not running, because gravity pulls downward on the car.

Gravitational attraction is usually measured in terms of weight. We put an object on a scale and note that it weighs 10 pounds [4.5 kg]. What we mean is that the earth registers that much pull on it. Gravitational attraction gives any object its weight. Astronauts out in space, where the earth's gravitational attraction is zero, are weightless. They float freely around in the interior of their spaceships.

♦ 8-11 ATMOSPHERIC PRESSURE

The air is also pulled toward the earth by gravity. At sea level and average temperature, a cubic foot of air weighs about 0.08 pound, or about 1.25 ounces [0.035 kg]. This seems like very little. But the blanket of air surrounding the earth—our atmosphere—is many miles thick. It is like thousands of cubic feet of air piled one on top of another, all adding their weight. The total weight, or downward push, is about 15 psi [103 kPa]. Since the human body has a total surface area of several square feet, the weight of the air pressing on your skin amounts to several tons!

The pressure does not crush you because pressures inside your body balance this atmospheric pressure.

Atmospheric pressure also affects the amount of power that an engine can produce. At sea level, an engine gets more air (due to atmospheric pressure) and produces more power. In Denver, Colorado, which is a mile above sea level, an engine gets less air and produces less power.

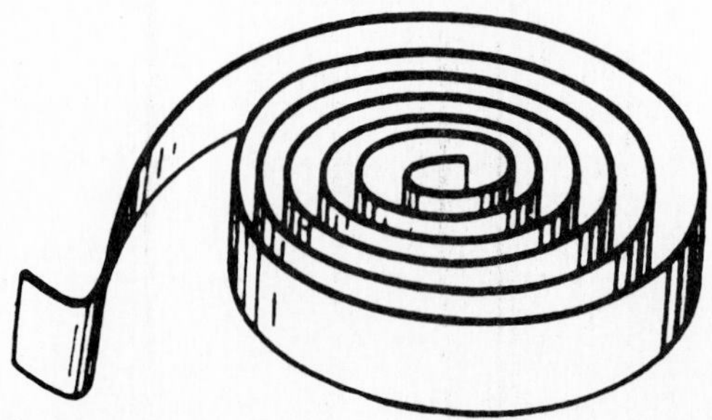

Fig. 8-4 Coil-type thermostat. The thermostat winds up or unwinds as the temperature changes. This motion can be used to operate a control.

♦ 8-12 VACUUM

A *vacuum* is the absence of air or any other matter. Astronauts on their way to the moon soon pass through our atmosphere and into a vast region of empty space. This is a vacuum.

♦ 8-13 PRODUCING A VACUUM

There are many ways to produce a vacuum here on earth, in a small volume. The automobile engine, as it operates, produces a partial vacuum in the engine cylinders. Atmospheric pressure then pushes the air-fuel mixture into the cylinders. This is how the engine gets the air-fuel mixture it needs to run.

1. Barometer The barometer is another device that works by vacuum. You can make a barometer by filling a long tube with mercury (Fig. 8-5). Then close the end with your finger. Next, turn the tube upside down, and put the end into a dish of mercury. Now open the end. Some of the mercury will run out of the tube. This will leave the upper end of the tube empty (a vacuum). The barometer is used to measure atmospheric pressure. Changes in atmospheric pressure indicate weather conditions. If the atmospheric pressure drops, the mercury "falls," and a storm is probably on the way.

2. Vacuum Gauge Vacuum gauges are used in automotive service. It is really a pressure gauge and is often referred to as an *absolute pressure gauge.* It measures the pressure in a closed space and compares it with atmospheric pressure. If the pressure in the closed space is lower, the vacuum gauge records so many "inches of mercury." Figure 8-6 shows a vacuum gauge used to check automotive engines. When the vacuum gauge is connected, it measures the amount of vacuum in the engine. If the reading is low or unsteady, then there is engine trouble.

The vacuum gauge contains a bellows, or diaphragm, linked to a needle on the dial face. When vacuum is applied,

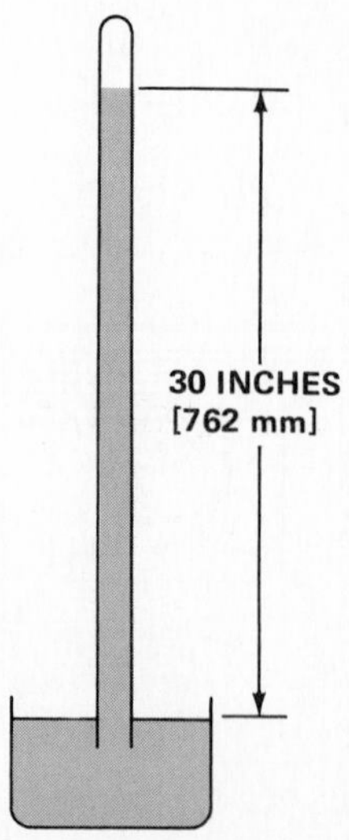

Fig. 8-5 In a barometer, the mercury will stand in the tube about 30 inches (762 mm) above the surface of the mercury in the dish when the atmospheric pressure is 15 psi (103 kPa).

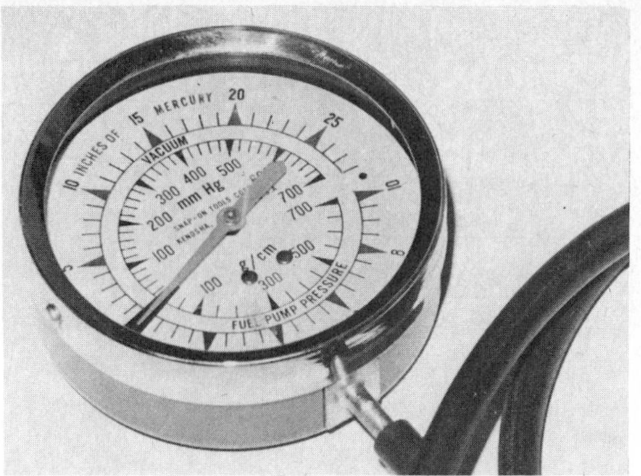

Fig. 8-6 A vacuum gauge. (*Snap-on Tools Corporation*)

the bellows or diaphragm moves. This causes the needle to move, to show the amount of vacuum.

♦ 8-14 HUMIDITY

Almost all air has some water vapor (evaporated water) in it. When the air is carrying a good deal of water vapor, the air is said to be *humid.* This means it has high humidity. You find humid air around bodies of water. Air over deserts has low humidity. It has very little water vapor. Humidity is measured in terms of percentages. Zero percent humidity means the air has no water vapor. Hundred percent humidity means the air is holding all the water vapor it can hold. A reading of 50 percent humidity means that the air is holding half as much water vapor as it could.

Humidity affects engines. An engine develops less power in hot, dry conditions than in cool, moist conditions. The cool air is denser (molecules are closer together) so that more air enters the engine. This means more power. At the same time, the moisture displaces some of the oxygen from the air. This richens the mixture, which makes the engine run more smoothly.

♦ 8-15 ATMOSPHERIC FACTORS AFFECTING COMBUSTION

Changes in temperature, atmospheric pressure, and humidity affect combustion in the engine. They affect the way the fuel burns and the power output of the engine. Accurate testing of engines requires that all readings be corrected to account for temperature, atmospheric pressure, and humidity.

♦ REVIEW QUESTIONS ♦

Select the *one* correct, best, or most probable answer to each question. You can find the answers in the section indicated at the end of each question.

1. During combustion of gasoline in the engine, (♦8-3)
   a. oxygen unites with carbon to form carbon dioxide
   b. hydrogen unites with oxygen to form water
   c. some carbon unites with oxygen to form carbon monoxide
   d. all of the above
2. Which of these statements is correct? (♦8-6 and 8-7)
   a. temperature increases with increasing pressure
   b. pressure increases with increasing temperature
   c. both a and b
   d. neither a nor b
3. During combustion of gasoline in the engine, (♦8-3)
   a. some of the gasoline does not burn
   b. some gasoline only partly burns, producing CO
   c. polluting gases exit from the tail pipe
   d. all of the above
4. If we start with a cubic foot of air in a closed container at 32°F, and heat it to 100°F [from 0 to 37.8°C], the pressure will go up about (♦8-6)
   a. 10 percent
   b. 1 percent
   c. 37 percent
   d. 15 percent
5. As the piston moves up in the cylinder, compressing the air-fuel mixture, (♦8-7)
   a. pressure goes up
   b. temperature goes up
   c. the mixture gets hotter
   d. all of the above
6. The device used to indicate temperature is called the (♦8-8)
   a. thermostat
   b. thermometer
   c. gravity
   d. pressure
7. The device that can provide control as temperature changes is the (♦8-9)
   a. thermostat
   b. thermometer
   c. gravity
   d. vacuum
8. Atmospheric pressure is produced by (♦8-11)
   a. vacuum
   b. gravity
   c. temperature
   d. barometer
9. Vacuum is an absence of (♦8-12)
   a. air or other matter
   b. gravity
   c. temperature
   d. humidity
10. Combustion in the engine is affected by (♦8-15)
    a. atmospheric pressure
    b. changes in air temperature
    c. humidity
    d. all of the above

CHAPTER 9
Piston-Engine Operation

After studying this chapter, you should be able to:

1. Explain the basic differences between spark-ignition and diesel engines.
2. Discuss the two types of piston rings and the jobs they do.
3. Describe how the crankshaft and connecting rod change rotary motion to reciprocating motion.
4. Explain the purpose of the valves.
5. Discuss the four piston strokes of a spark-ignition engine and what happens in the engine during each stroke.
6. Describe the four piston strokes of a compression-ignition engine and what happens in the engine during each stroke.
7. Explain how a pushrod valve train works.
8. Discuss the flywheel and its function.

♦ 9-1 INTERNAL-COMBUSTION ENGINES

Automotive engines are called *internal-combustion* (IC) *engines* because the fuel that runs them is burned internally, or inside the engines. There are two types, reciprocating and rotary (Fig. 9-1). *Reciprocating* means moving up and down, or back and forth. Almost all automotive engines are of the reciprocating type. In these engines, pistons move up and down, or reciprocate, in cylinders (Fig. 9-2). This type of engine is called a *piston engine.*

Rotary engines have rotors that spin, or rotate. The only such engine now used in automobiles is the Wankel engine. It is described in a following chapter.

♦ 9-2 TWO KINDS OF PISTON ENGINES

There are two kinds of piston engines—spark-ignition and compression-ignition (diesel). The differences between the two are:

- The type of fuel used
- The way the fuel gets into the engine cylinders
- The way the fuel is ignited

The *spark-ignition engine* uses a highly volatile fuel which turns to vapor easily, such as gasoline or gasohol. The fuel is mixed with air *before* it enters the engine cylinders. The fuel turns into a vapor and mixes with the air to form a combustible air-fuel mixture. This mixture then enters the cylinders and is compressed. Next, an electric spark produced by the ignition system sets fire to, or ignites, the compressed air-fuel mixture.

In the *compression-ignition,* or *diesel, engine,* the fuel is mixed with the air *after* the air enters the engine cylinder. Air alone is taken into the cylinders of the diesel engine. The air is then compressed as the piston moves up. The air is compressed so much that its temperature goes up to 1000°F [538°C] or higher. Then the diesel-engine fuel—a light oil—

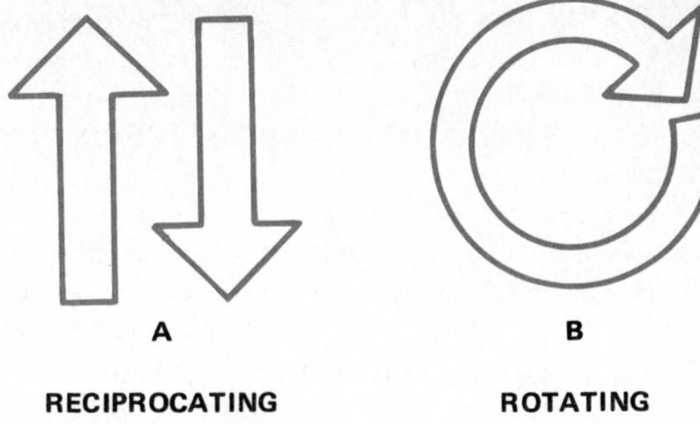

Fig. 9-1 Reciprocating motion is up-and-down or back-and-forth motion as contrasted with rotary (rotating) motion.

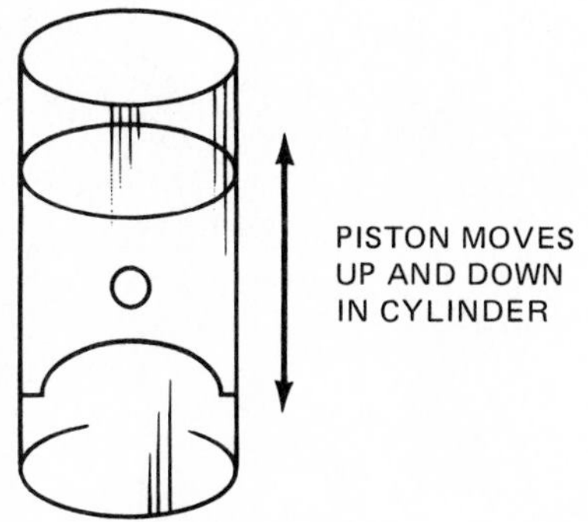

Fig. 9-2 In automotive engines using pistons, the pistons move up and down in the cylinders.

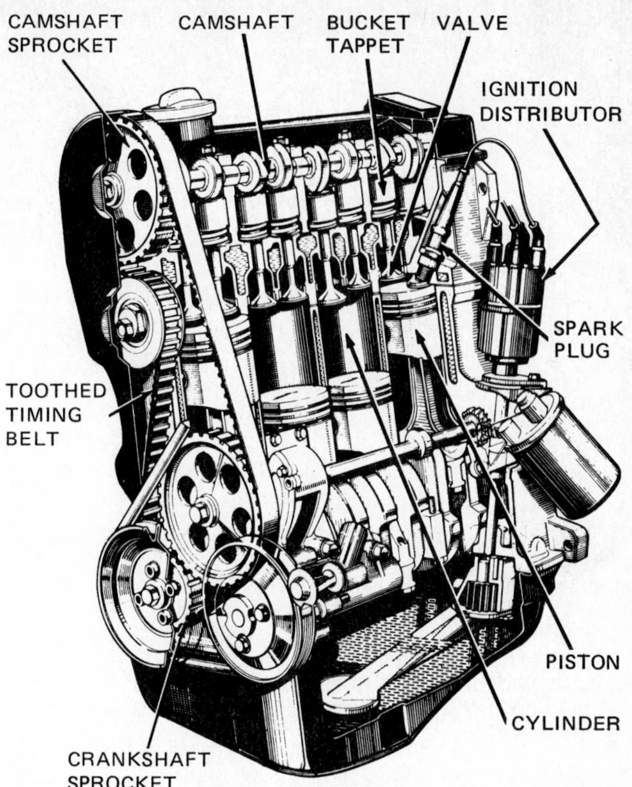

Fig. 9-3 A four-cylinder in-line, overhead-camshaft spark-ignition engine. (*Chrysler Corporation*)

is injected (sprayed) into the engine cylinder. The hot air, or *heat of compression,* ignites the fuel. This is why the diesel engine is called a compression-ignition engine.

♦ 9-3 ENGINE CONSTRUCTION

Both the spark-ignition and the compression-ignition engines are very much alike. They both have pistons that move up and down in cylinders (Fig. 9-2). Most automotive engines have four, five, six, or eight cylinders. Since the same actions take place in each cylinder, let us study one cylinder to see what makes the engine run.

Figure 9-3 shows a four-cylinder spark-ignition engine, sliced in two so that the pistons and cylinders can be seen. Each cylinder is about 4 inches [102 mm] in diameter. The cylinder is a long, round air pocket, somewhat like a tin can with the bottom cut out. Each cylinder has a piston which is slightly smaller in size than the cylinder. The piston (Fig. 9-4) is a metal plug that slides up and down in the cylinder.

Note The example below is a spark-ignition engine. Diesel-engine operation is described later.

In Fig. 9-5A, the piston is shown at its lower limit of travel. The space above it is filled with a mixture of air and gasoline vapor. Figure 9-5B shows the piston at the upper limit of its travel. In moving up, it has compressed the air-fuel mixture. Figure 9-5C shows a spark igniting the air-fuel mixture. The high pressure that results has pushed the piston down. It is this downward movement that produces the power from the engine.

As the engine operates, the piston continues to move up and down. The downward movements of the piston that produce the power continue as long as the engine runs.

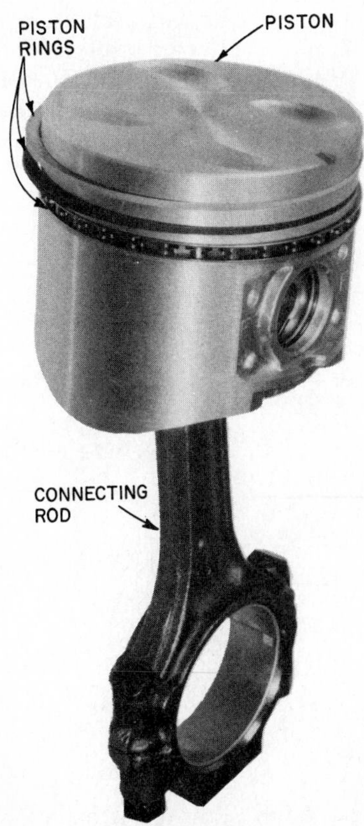

Fig. 9-4 Piston with piston rings and connecting rod. (*Chrysler Corporation*)

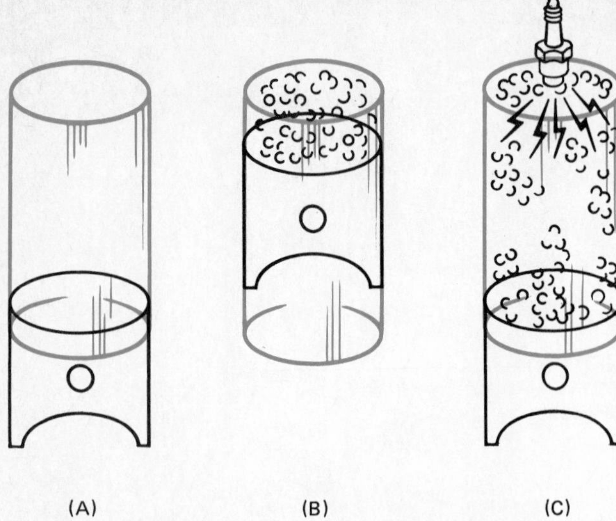

Fig. 9-5 Three views showing the actions in an engine cylinder. (A) The piston is a metal plug that fits snugly into the cylinder. (B) When the piston is pushed up into the cylinder, air-fuel mixture is trapped and compressed. (C) When the compressed air-fuel mixture is ignited by a spark at the spark plug, the high pressure pushes the piston down in the cylinder.

Figure 9-6 shows the two limiting positions of the piston—top dead center (TDC) and bottom dead center (BDC). A piston stroke occurs when the piston moves from TDC to BDC, or from BDC to TDC.

♦ 9-4 PISTONS AND PISTON RINGS

Figure 9-4 shows an engine piston. Pistons are usually made of aluminum alloy, which is a mixture of aluminum and other metals. Automotive pistons weigh about 1 pound [0.454 kg]. They are a sliding fit in the cylinders. This means the pistons can slide up and down in the cylinders. Therefore the pistons are slightly smaller in diameter than the cylinders.

Because the pistons are slightly smaller, there is a gap, or clearance, between the piston and the cylinder. This gap must be filled. Otherwise, some of the compressed air-fuel mixture would leak out through the clearance. Also, when combustion took place, much of the high-pressure gas would leak out. These leaks would greatly reduce the efficiency of the engine. Much of the power would be lost.

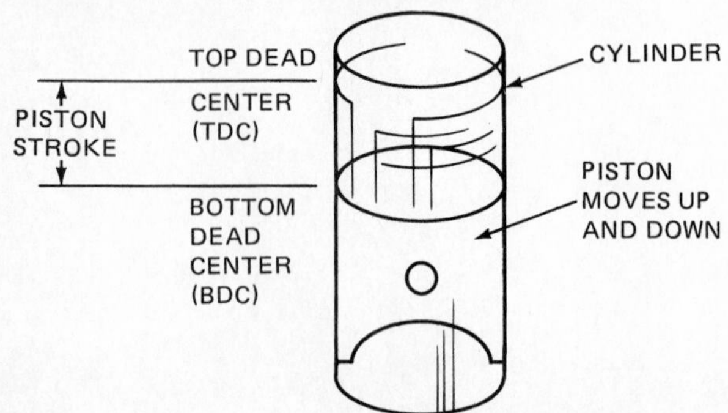

Fig. 9-6 The two limiting positions of the piston—top dead center (TDC) and bottom dead center (BDC). A piston stroke occurs when the piston moves from BDC to TDC, or from TDC to BDC.

To prevent this, *piston rings* are installed on the pistons (Fig. 9-7). The piston rings are split at one point so they can be expanded slightly and slipped over the piston and into grooves cut in the piston. Piston rings are of two types and they do two jobs:

1. Form a sliding seal between the piston and the cylinder wall. These are called *compression rings.*
2. Scrape off most of the oil that is splashed on the wall so that it does not get up into the combustion chamber where it would burn. These are called *oil-control rings.*

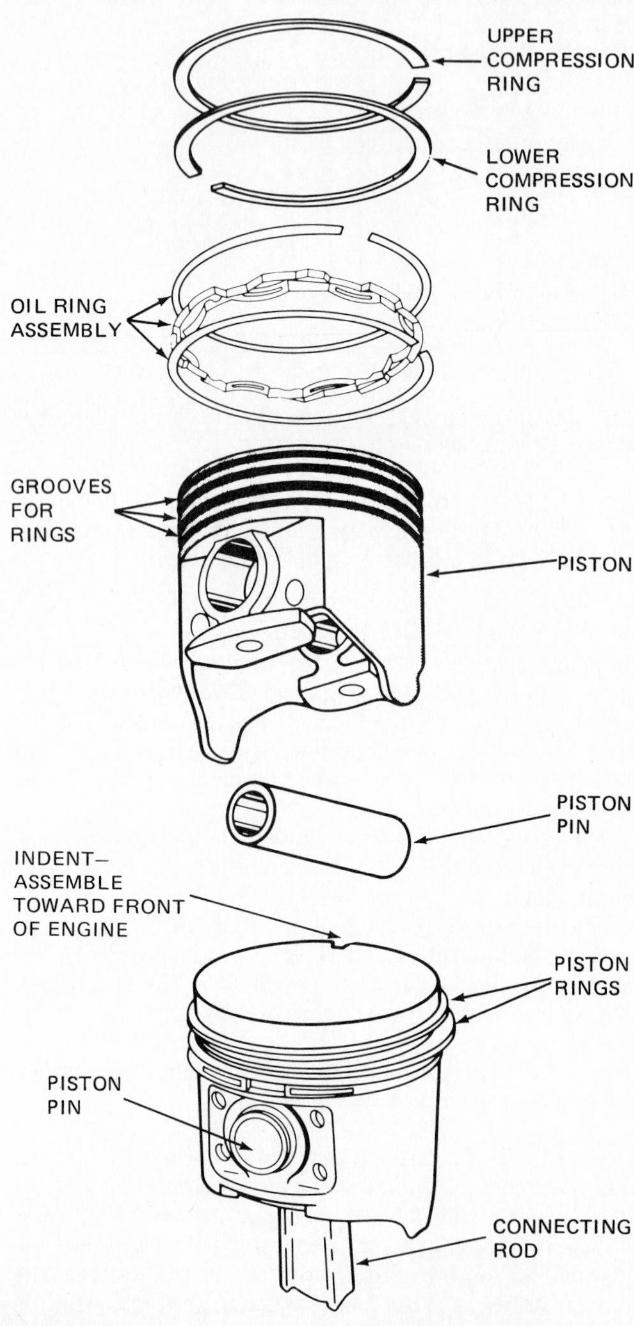

Fig. 9-7 Piston and piston rings. Top, rings separated and above the piston. Bottom, piston rings installed in grooves in the piston. Piston is attached to the connecting rod by the piston pin. Only the upper part of the connecting rod is shown.

♦ 9-5 RECIPROCATING TO ROTARY MOTION

Forcing a piston out of a cylinder just once is not enough to make a car move. The piston must remain in the cylinder and move up and down rapidly. Then this up-and-down, or reciprocating, motion must be turned into rotary motion to rotate the car wheels. The connecting rod and crankshaft do the job of changing the reciprocating motion of the piston into rotary motion.

The piston pin and connecting rod connect the piston to a crankpin on the crankshaft (Figs. 9-8 and 9-9). The purpose of this is to change the reciprocating, or up-and-down, motion of the piston into rotary motion of a shaft. The rotary motion is carried through shafts and gears to the car wheels. Then the wheels rotate and the car moves.

The *crankpin* is an offset part of the crankshaft, as shown in Fig. 9-9. When the shaft rotates, the crankpin swings in a circle around the shaft (Fig. 9-10). The big end of the connecting rod is connected to the crankpin by a rod cap and rod-cap bolts (Fig. 9-8). Two bearing halves are positioned between the rod and cap and the crankpin. They allow the crankpin to rotate easily within the rod and cap. The small end of the connecting rod is connected to the piston by the piston pin.

As the piston moves up and down, the crankshaft rotates. Figure 9-11 shows the action. As the piston starts moving down (A in Fig. 9-11), the connecting rod tilts to one side so that the big end can follow the circular path of the crankpin. As shown in the sequence, the connecting rod tilts first in one direction and then in the other. As the piston moves up and down and the crankshaft rotates, the reciprocating motion of the piston is converted into rotary motion.

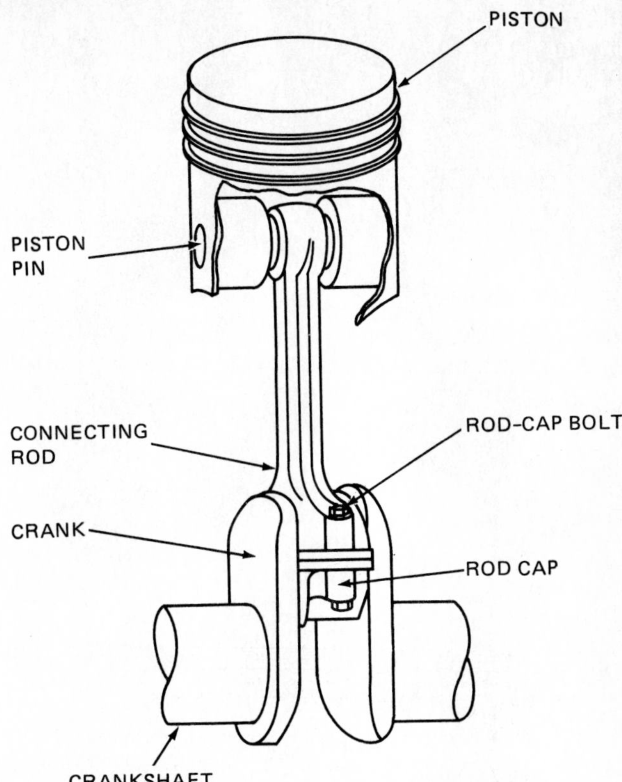

Fig. 9-9 Piston-and-connecting-rod assembly attached to a crankpin on a crankshaft. The piston is partly cut away to show how the piston pin attaches it to the connecting rod.

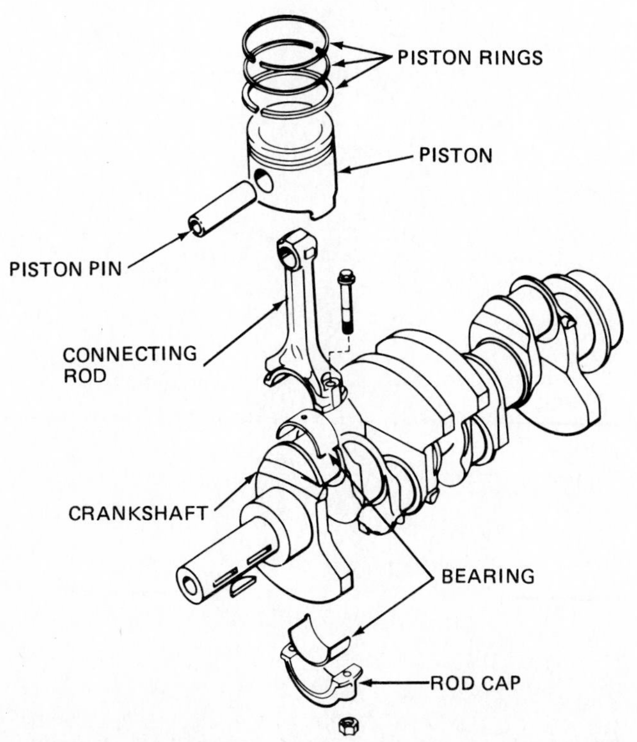

Fig. 9-8 Crankshaft with one piston-and-connecting-rod assembly, showing how the piston is attached, through the connecting rod, to the crankpin on the crankshaft.

Fig. 9-10 As the crankshaft rotates, the crankpin swings in a circle around it.

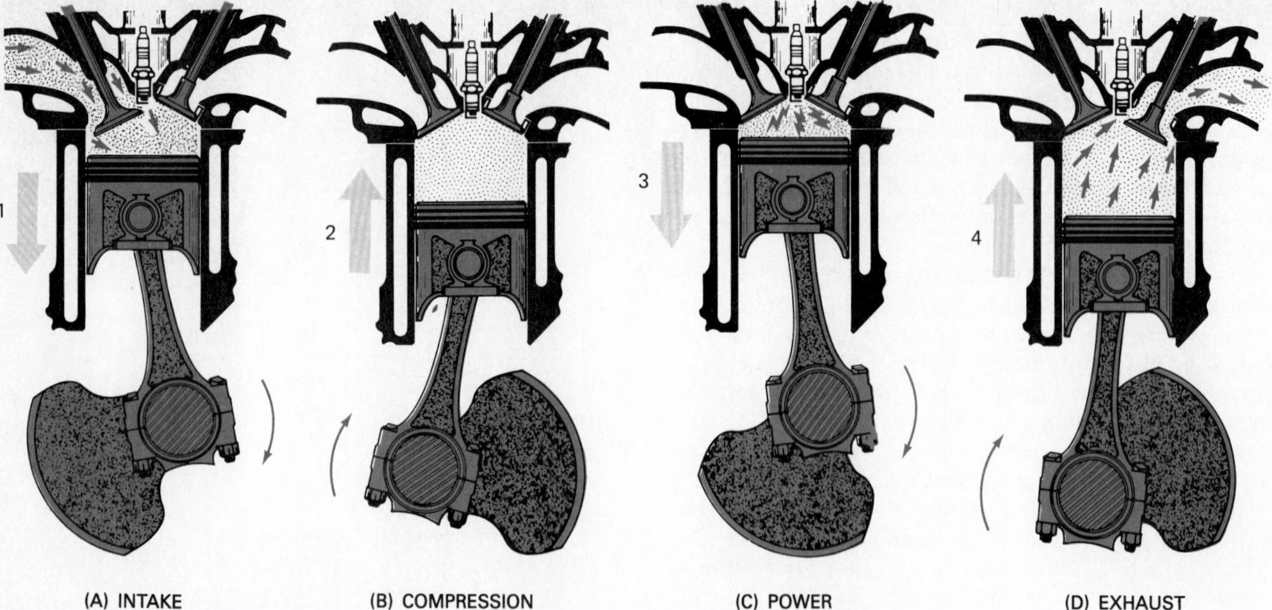

Fig. 9-11 (A) Intake stroke. The intake valve (at left) has opened. The piston is moving downward, allowing a mixture of air and gasoline vapor to enter the cylinder. (B) Compression stroke. The intake valve has closed. The piston is moving up, compressing the mixture. (C) Power stroke. The ignition system has delivered a spark to the spark plug that ignites the compressed mixture. As the mixture burns, high pressure is created, pushing the piston down. (D) Exhaust stroke. The exhaust valve has opened. The piston is moving upward, as the burned exhaust gases escape from the cylinder.

♦ 9-6 THE VALVES

Most engines have two holes, or *ports,* in the enclosed upper end of the cylinder. One of the two ports is the *intake port.* It allows the mixture of gasoline vapor and air to enter the cylinder. The other is the *exhaust port.* It allows the burned gases to exhaust from, or leave, the cylinder.

These two ports are open only part of the time. The rest of the time they are closed off by the intake and exhaust valves

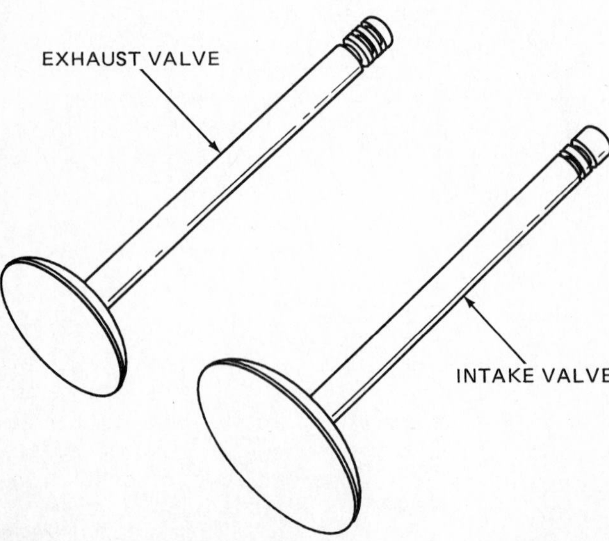

Fig. 9-12 Intake and exhaust valves for one cylinder of an engine. Note that the intake valve is larger.

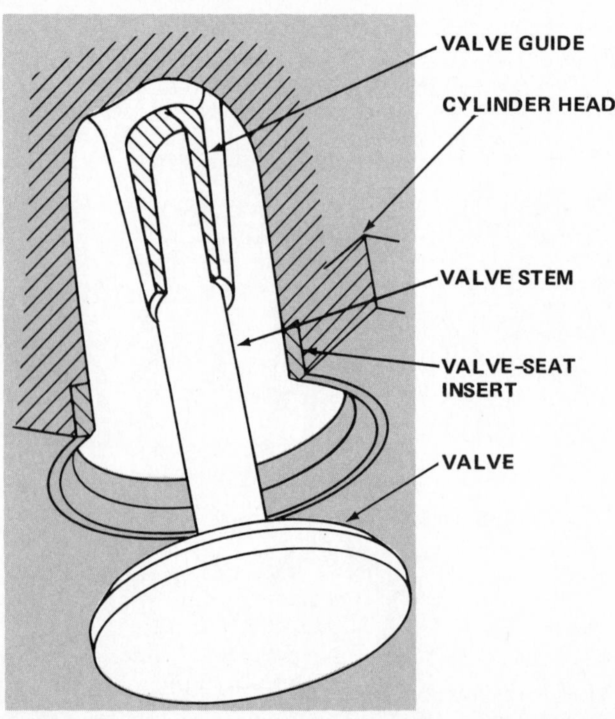

Fig. 9-13 Valve and valve seat in a cylinder head. Some engines have valve-seat inserts, especially for the exhaust valves.

(Fig. 9-12). The valves are metal plugs that fit the round holes, or ports (Fig. 9-13). When a valve moves up into its port (or *seats*), it seals off the port. When the valve is pushed down off its port, the port is open.

The valves are operated by several engine parts that make up the *valve train.* The valve train causes the valves to open (move down off their seats in the ports) and close (move up in their ports, and seat to seal off the ports). Valve trains are described later.

♦ 9-7 ENGINE OPERATION

The actions in the spark-ignition engine can be divided into four parts. (Diesel-engine operation is described later.) Each part consists of a piston stroke (Fig. 9-6). A piston stroke is the movement of a piston in a cylinder from top to bottom, or from bottom to top. The top position is called *top dead center* (TDC). The bottom position is called *bottom dead center* (BDC).

The complete cycle of events in the engine cylinder requires four piston strokes—intake, compression, power, and exhaust. This requires two revolutions of the crankshaft. The four strokes make the engine a *four-stroke-cycle* engine. This is usually shortened to "four-cycle" engine.

Note The word *cycle* means a series of events that repeat themselves. The cycle of the seasons (spring, summer, fall, and winter) is an example. The cycle of the four strokes (intake, compression, power, exhaust) in the four-stroke-cycle (or four-cycle) engine is another.

♦ 9-8 INTAKE STROKE

During the intake stroke (Fig. 9-11A), the piston is moving down. The intake valve is open (has been pushed down off its seat in the intake port). A mixture of fuel vapor and air enters the cylinder. This mixture is supplied by the fuel system, which is described later.

Note The reason the air-fuel mixture enters the cylinder is that when the piston moves down, it leaves above it a *vacuum.* A vacuum is a space with nothing in it. Atmospheric pressure forces the air-fuel mixture into the cylinder to fill this vacuum.

As the piston passes through BDC, the intake valve closes (moves up onto its seat). Now, the upper end of the cylinder is sealed off.

♦ 9-9 COMPRESSION STROKE

After the piston passes through BDC, it starts moving up (Fig. 9-11B). The air-fuel mixture is compressed, or squeezed into a smaller space. The mixture is compressed into one-eighth of its original volume—or less. This is like taking a quart of air and compressing it into about half a cup (Fig. 9-14). The amount the mixture is compressed is called the *compression ratio.* If the mixture is compressed to one-eighth of its original volume, then the compression ratio is 8 to 1 (written 8:1). Compression ratio is discussed later.

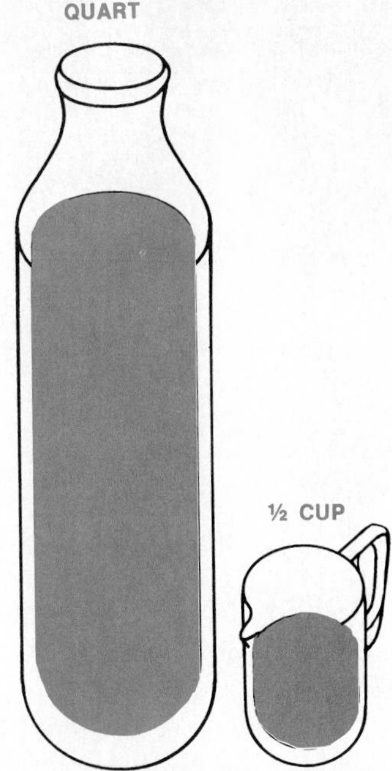

Fig. 9-14 With a compression ratio of 8:1, the air-fuel mixture is compressed to one-eighth of its original volume. This is like compressing a quart of air into half a cup. There are eight half-cups in a quart.

♦ 9-10 POWER STROKE

As the piston nears TDC on the compression stroke, an electric spark is produced in the cylinder. The spark comes from the ignition system and takes place at the spark plug. The heat of the spark ignites, or sets fire to, the compressed air-fuel mixture. By the time the piston has moved up over TDC and has started down (Fig. 9-11C), combustion is well advanced. Momentary temperatures in the burning mixture may reach as high as 6000°F [3316°C]. The very high temperature results in a very high pressure. This is an application of the rule in ♦8-6; *Pressure increases with increasing temperature.*

The pressure goes up to as much as 600 psi [41,369 kPa]. This means the pressure pushing down on the piston may be as much as 4000 pounds [1814 kg].

The high pressure pushes the piston down. The downward push is carried through the connecting rod to the crankpin on the crankshaft. The crankshaft is forced to turn. The rotary motion is carried through shafts and gears to the car wheels. The wheels turn and the car moves.

♦ 9-11 EXHAUST STROKE

As the piston approaches BDC on the power stroke, the exhaust valve opens (is pushed down off its seat). Now, as the piston moves up again after BDC, the burned gases escape through the exhaust port (Fig. 9-11D). Then, as the piston nears TDC, the intake valve opens. Now, as the piston passes through TDC and starts down again, the exhaust valve closes. Another intake stroke takes place (Fig. 9-11A) and the whole cycle—intake, compression, power, and exhaust—is repeated. This goes on continuously in all engine cylinders, while the engine is running.

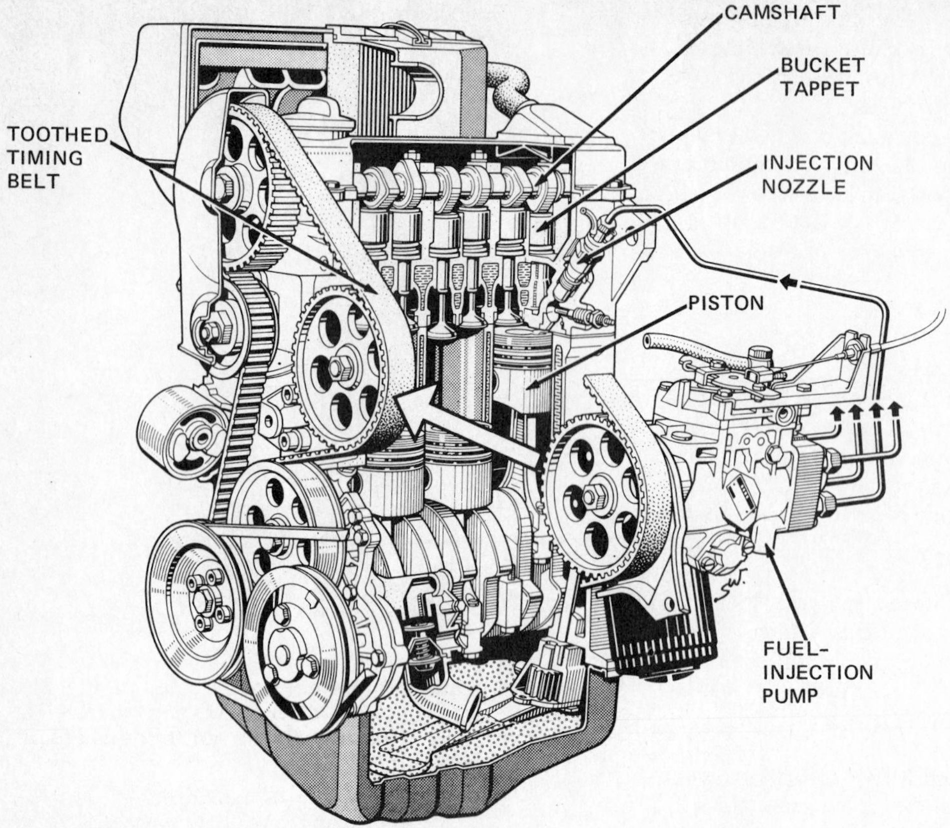

Fig. 9-15 A four-cylinder overhead-camshaft diesel engine. The basic construction and valve train of this engine are about the same as for the spark-ignition engine shown in Fig. 9-3.

♦ 9-12 VALVE ACTION

There are usually two valves per cylinder. The valves are opened and closed by the action of the valve train. One of the simplest valve trains is shown in Figs. 9-3 and 9-15. It has a camshaft (Fig. 9-16) mounted on top of the cylinder head. The camshaft is driven (rotated) from the crankshaft by a belt with teeth on its inside. The teeth match the teeth on the sprockets on the crankshaft and camshaft.

The camshaft (Fig. 9-16) has two cams for each cylinder, one for the intake valve and one for the exhaust valve. The *cam* is a round collar with a high spot, or lobe (Fig. 9-17). Note that the cams are directly above the valve stems (Fig. 9-15).

The top of the valve stem is covered with a small hollow cylinder called a "bucket" valve tappet. It is called a bucket valve tappet because it is shaped like an upside-down bucket.

When the camshaft and cam rotate, the lobe comes around and pushes the valve tappet, and valve stem, down. The valve opens. When the lobe passes out from the valve tappet, the valve spring pulls the valve up so that the valve closes.

Engines with the camshaft on the cylinder head are called *overhead-camshaft* (OHC) *engines*. They are more responsive than other types of valve train. This is because there are fewer parts in the valve train.

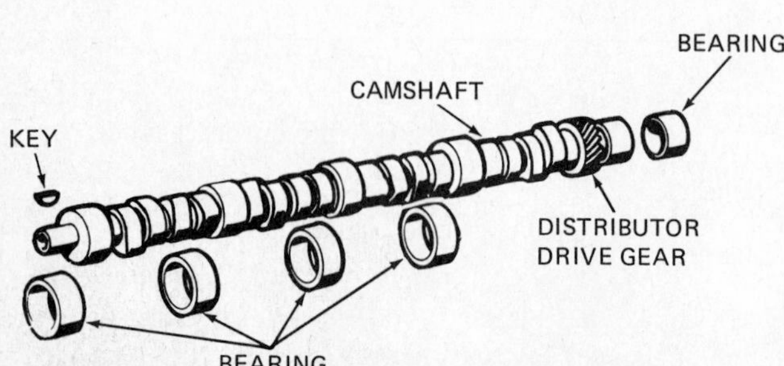

Fig. 9-16 Camshaft and bearings for a V-8 engine.

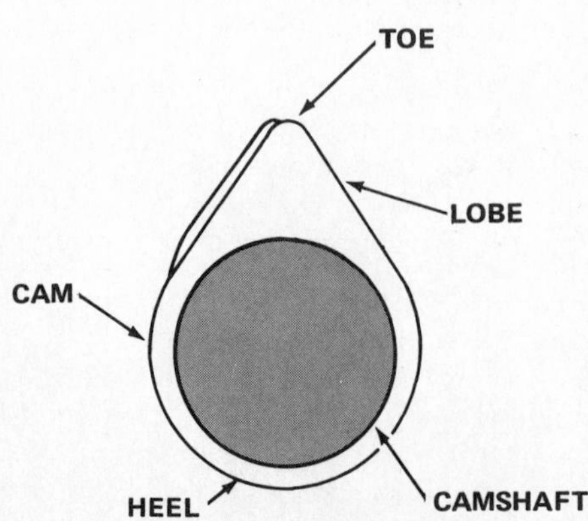

Fig. 9-17 Cam on a camshaft. Note the locations of the toe and heel.

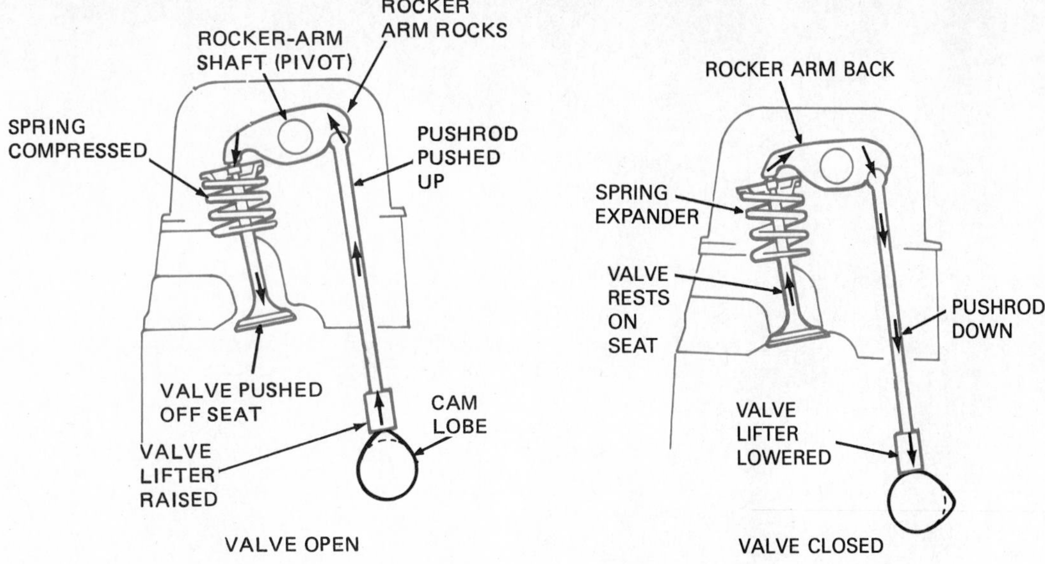

Fig. 9-18 Operation of the valve train in a pushrod engine.

♦ 9-13 PUSHROD VALVE TRAIN

There are several other valve-train arrangements. The most popular arrangement for many years uses a pushrod. For this reason, the engine is often called a *pushrod engine.* This arrangement has the camshaft in the cylinder block. Figure 9-18 is a simplified drawing of the valve train. The way the valve train looks in an engine is shown in Fig. 9-19. The valve train for each valve includes a cam on the camshaft, a valve lifter (also called a *valve tappet*), a pushrod, a rocker arm, and a valve spring (Fig. 9-19). Each part is briefly described below.

1. Cam on camshaft. Camshaft and cam action is described in ♦9-12 and illustrated in Figs. 9-16 and 9-17.
2. The valve lifter is a round cylinder that rides on the cam. The cam rotates under it.
3. The pushrod is a long rod that goes from the valve lifter up to one end of the rocker arm.
4. The rocker arm is pivoted at about its middle. The pushrod pushes against one end of the rocker arm. The other end of the rocker arm rests on the valve stem.
5. The valve spring is a coil spring that fits between the cylinder head and a retainer on the stem end of the valve.

Figure 9-18 shows how the system works. The camshaft is driven by a chain and sprockets (or gears) from the crankshaft. When the camshaft and cams rotate, the cam lobes come up under the valve lifters. When a lobe moves up under a valve lifter, the lobe causes the lifter to be pushed upward. The upward push is carried by the pushrod to the rocker arm. The rocker arm is pivoted at its center. When one end of the rocker arm is pushed up by the pushrod, the other end moves down. This pushes down on the valve stem and causes it to move down, opening the port.

When the cam lobe moves out from under the valve lifter, the valve spring pulls the valve back up onto its seat. At the same time, the valve stem pushes up on the rocker arm, forcing it to rock back. This pushes the pushrod and the valve lifter down.

Valve-train arrangements are described later in the book.

♦ 9-14 MULTIPLE-CYLINDER ENGINES

A single-cylinder four-cycle engine has only one power impulse for every two crankshaft revolutions. The engine cylinder delivers power only one-fourth of the running time. During the other three strokes—intake, compression, and exhaust—the engine is not delivering power. For a more even flow of power, most automotive engines have four or more cylinders. The cylinders fire one after another to produce a more steady flow of power.

♦ 9-15 FLYWHEEL AND DRIVE PLATE

The power impulses in an automotive engine follow each other to provide a fairly even flow of power. To further smooth out power flow, engines have flywheels or drive plates (on automatic transmissions). Figure 9-20 shows a crankshaft and flywheel, with other related parts. The *flywheel* is a large wheel bolted to the end of the crankshaft. The flywheel smooths out the flow of power by resisting any change in its speed of rotation.

The flywheel also serves as part of the clutch on cars with manual transmissions. Also, the flywheel has teeth on its outer rim. A small pinion gear on the starting motor meshes with these teeth when the starting motor is operated. This spins the flywheel and the crankshaft so the engine starts.

On vehicles using automatic transmissions, a drive plate, or "flex plate," takes the place of the flywheel (see Fig. 11-2). The automatic transmission has a doughnut-shaped device called a *torque converter.* It is attached to the drive plate and the combination serves the same purpose as the flywheel. Automatic transmissions are covered in a later chapter.

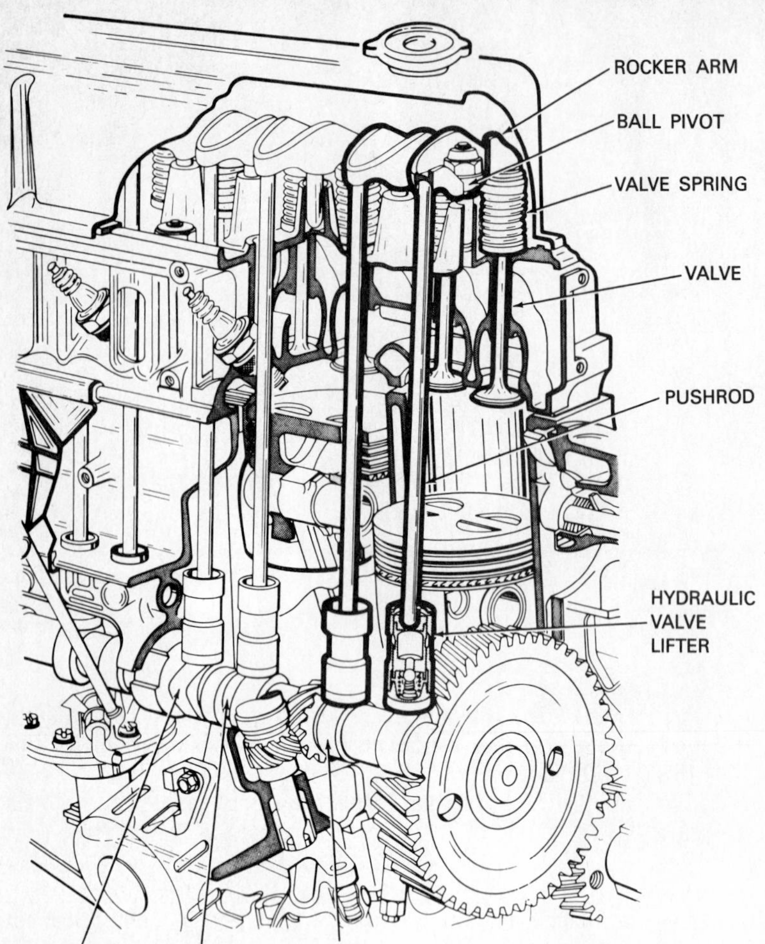

Fig. 9-19 Spark-ignition engine cut away to show valve-train parts. This is a pushrod engine. The camshaft is driven by gears from the crankshaft.

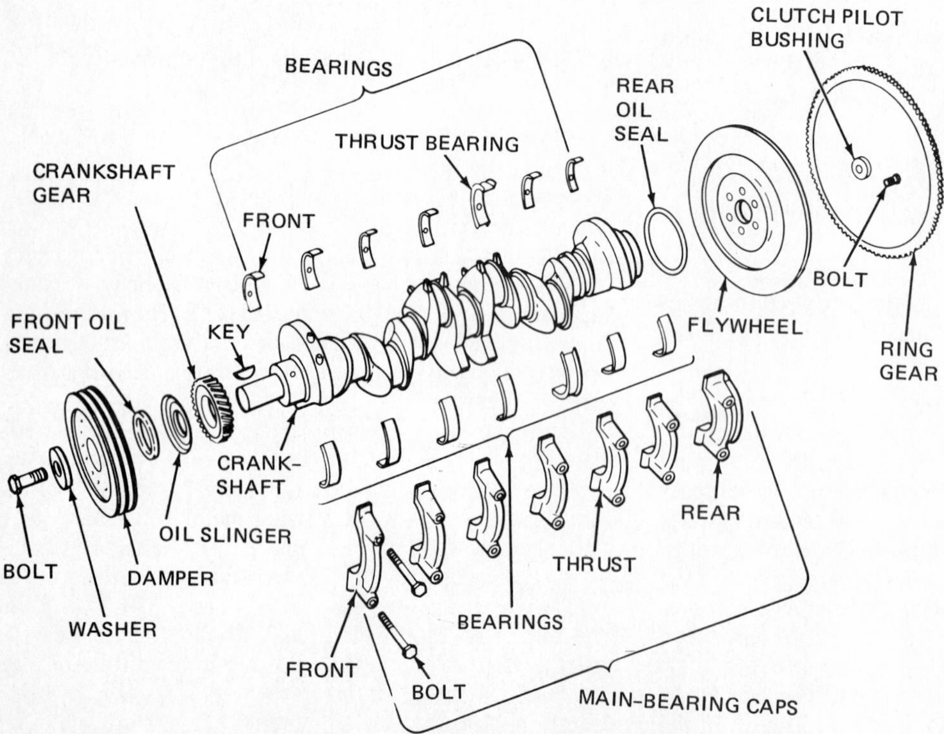

Fig. 9-20 Crankshaft and related parts for an in-line six-cylinder engine. The flywheel is to the right. (*Ford Motor Company*)

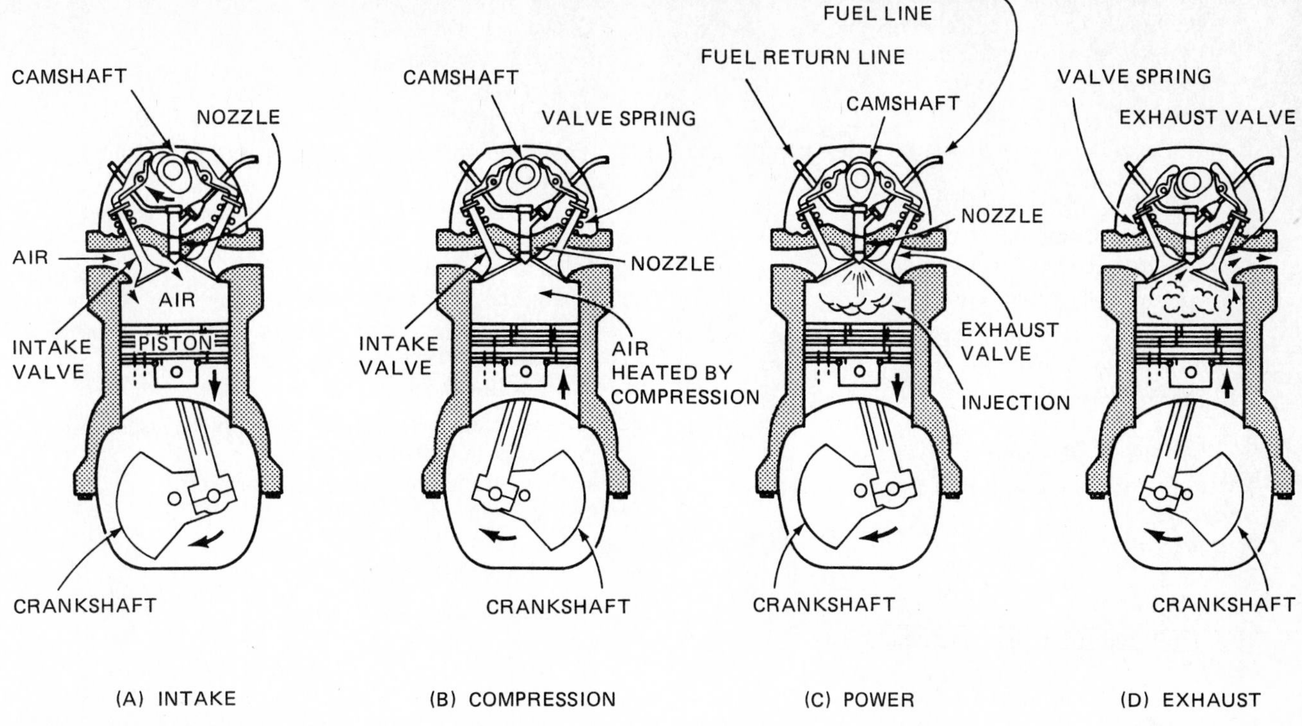

Fig. 9-21 The four piston strokes in a four-cycle diesel engine. (*General Motors Corporation*)

♦ 9-16 DIESEL ENGINES

There is a second way to ignite the fuel—by compression ignition. This is the system used in diesel engines. When air is compressed, it gets very hot. The more it is compressed, the hotter it gets. Temperature increases with increasing pressure (♦8-7). In the diesel engine (Fig. 9-15), the air is compressed to as little as 1/22 of its original volume. This is a compression ratio of 22:1. (Compression ratio is described in ♦9-9.) When air is compressed this much, its temperature goes as high as 1000°F [538°C]. As the piston nears TDC on the compression stroke, the fuel system sprays diesel fuel (a light oil) into the combustion chamber. The high air temperature ignites the fuel and the power stroke follows.

Figure 9-21 shows the four strokes in the diesel engine. They are very similar to those for the spark-ignition-engine intake, compression, power, and exhaust strokes. The major differences are:

1. The diesel engine compresses air alone.
2. The fuel is sprayed into the combustion chamber as the piston nears TDC on the compression stroke.
3. The temperature of the air ignites the fuel.

Diesel engines and diesel-engine fuel systems are described later in the book.

♦ REVIEW QUESTIONS ♦

Select the *one* correct, best, or most probable answer to each question. You can find the answers in the section indicated at the end of each question.

1. The two basic types of internal-combustion engines are the (♦9-1)
   a. piston and reciprocating
   b. reciprocating and rotary
   c. reciprocating and pushrod
   d. rotary and spark-ignition
2. The two basic types of piston engines are the (♦9-2)
   a. rotary and reciprocating
   b. pushrod and reciprocating
   c. spark-ignition and compression-ignition
   d. gasoline and gasohol
3. Mechanic A says that, in the engine, the air temperature increases with increasing pressure. Mechanic B says that pressure increases with temperature. Who is right? (♦9-10)
   a. mechanic A
   b. mechanic B
   c. both A and B
   d. neither A nor B
4. Engine power is produced by the (♦9-10)
   a. rotation of the crankshaft
   b. combustion pressures pushing on pistons
   c. up-and-down movement of pistons
   d. valve action

5. The two kinds of piston rings are (♦9-4)
   a. compression and oil-control rings
   b. sliding-seal and compression rings
   c. oil-scraper and oil-control rings
   d. pressure and sealing rings
6. To change reciprocating motion to rotary motion, the engine has (♦9-5)
   a. a crankshaft and a camshaft
   b. pistons and connecting rods
   c. connecting rods and a crankshaft
   d. a camshaft and rocker arms
7. The two valves that engines use are the (♦9-6)
   a. inlet and outlet valves
   b. plug and port valves
   c. port valves
   d. intake and exhaust valves
8. A piston stroke is movement of the piston (♦9-7)
   a. from TDC to BDC
   b. from BDC to TDC
   c. both a and b
   d. neither a nor b
9. The four piston strokes follow each other in the order of (♦9-7)
   a. intake, compression, power, exhaust
   b. intake, power, exhaust, compression
   c. compression, power, intake, exhaust
   d. exhaust, compression, intake, power
10. The valve train of a pushrod engine includes a cam and a (♦9-13)
   a. valve lifter, bucket tappet, and rocker arm
   b. valve spring, pushrod and bucket tappet
   c. valve lifter, pushrod, rocker arm, and valve spring
   d. connecting rod, camshaft gear, and pushrod
11. OHC means (♦9-12)
   a. overhead camshaft
   b. overhead crankshaft
   c. overhead cylinder head
   d. off-center camshaft
12. A basic difference between the spark-ignition engine and the diesel engine is (♦9-2)
   a. the diesel engine compresses air alone instead of an air-fuel mixture
   b. air temperature ignites the fuel in the diesel engine
   c. the fuel is sprayed into the combustion chamber in the diesel engine as the piston nears TDC on the compression stroke
   d. all of the above
13. The overhead-camshaft engine is a more responsive engine because (♦9-12)
   a. there are fewer parts in the valve train
   b. the engine is smaller
   c. the pistons are lighter
   d. it is a spark-ignition engine
14. The engine flywheel (♦9-15)
   a. smooths out the flow of power
   b. serves as part of the clutch on cars with manual transmissions
   c. has teeth that mesh with the starting-motor drive pinion
   d. all of the above
15. The basic principle of the diesel engine is that (♦9-2)
   a. when air is compressed, it gets very hot
   b. the electronic ignition system is very simple
   c. it has fewer moving parts
   d. it is easier to start

CHAPTER 10
Engine Types

After studying this chapter, you should be able to:

1. List the various ways in which engines are classified and explain what each classification means.
2. Define *variable-displacement engine.*
3. Explain what firing order means and why it is important in the design and operation of an engine.
4. Describe the basic differences between spark-ignition and compression-ignition engines.

♦ 10-1 ENGINE CLASSIFICATIONS

Automotive engines are classified in many different ways. All automotive engines are of the internal-combustion type (combustion takes place *inside* the engine). Internal-combustion engines can be classified according to:

1. Number of cylinders
2. Arrangement of cylinders
3. Arrangement of valves
4. Type of cooling
5. Number of strokes per cycle (two or four)
6. Type of fuel used
7. Method of ignition (spark or compression)
8. Firing order
9. Reciprocating or rotary (Wankel and turbine)

♦ 10-2 NUMBER AND ARRANGEMENT OF CYLINDERS

American passenger-car engines have four, six, or eight cylinders. Foreign-made cars offer a greater variety, including three, four, five, six, eight, and twelve cylinders. Engines with four, five, six, and eight cylinders are described and illustrated in the next few sections. Cylinders can be arranged in several ways:

1. In a row (in line)
2. In two rows or banks set at an angle (V type)
3. In two rows opposing each other (flat, or pancake)
4. Like spokes on a wheel (radial airplane type)

Only the first three have been used for automotive engines. Figure 10-1 shows various cylinder arrangements.

♦ 10-3 FOUR-CYLINDER ENGINES

The cylinders of a four-cylinder engine can be arranged in any of three ways: in line, V, or opposed. In the V type, the cylinders are in two banks, or rows, of two cylinders each. The two rows are set at an angle to each other. In the opposed type, the cylinders are in two banks of two cylinders each, set opposite each other.

The four-cylinder engine has become increasingly popular in recent years. A basic reason is the trend toward small, lightweight, fuel-efficient cars. This trend has been caused by the oil shortage and government regulations regarding car size and fuel-mileage requirements. These regulations step up the miles-per-gallon requirements, year by year. For example, the requirements for 1980 were that all cars produced by a manufacturer must average 20 miles per gallon. In 1985, the average must be 27.5 miles per gallon.

The only way these requirements can be met is for the automotive industry to use smaller engines, to "down-size" their big cars, and to bring out new generations of small cars. A four-cylinder engine in a small car can give 40 miles per gallon [14 km/L] or more on the highway. The large cars, with eight-cylinder engines, can often do no better than 12 to 15 miles per gallon [4 to 5 km/L]. Another factor to be considered is the pollution problem. Emission standards set by federal, state, and local governments (discussed in a later chapter) require controls on cars that tend to make fuel mileage worse. Therefore, the small, lightweight car with a small engine appears to be the car of the future.

The small four-cylinder engine does not give the car the acceleration it has with a bigger engine. However, *turbocharging* the engine can improve this. Turbocharging (♦15-14) forces more air into the engine cylinders so that more fuel can be burned. As a result, more power is produced. There are various types of four-cylinder engines.

1. In-Line Engines Figure 10-2 shows a four-cylinder spark-ignition in-line engine, partly cut away so that the internal construction can be seen. Figure 9-15 shows a four-cylinder compression-ignition (diesel) engine. The four-cylinder engine is used widely in the modern, small, fuel-efficient cars. Note that the engines in Figs. 9-15 and 10-2 have the camshaft on the cylinder head. They are OHC engines.

Many four-cylinder in-line and V-type engines mount transversely in the car (Fig. 10-3). With this arrangement, the engine drives the front wheels.

Figure 10-4 shows a four-cylinder diesel engine. It has a single overhead camshaft and uses bucket tappets set directly on top of the valve stems. Bucket tappets (Figs. 9-3 and 9-15) are described in ♦9-12. A toothed belt drives both the camshaft and the fuel-injection pump.

2. V-4 Engines The V-4 engine has two rows of two cylinders each, set at an angle, or a V, to each other. The crankshaft has only two cranks. Connecting rods from opposing cylinders in the two rows are attached to the same crankpin. Therefore, each crankpin has two connecting rods attached to it. Figure 10-5 shows the internal moving parts of a V-4 engine. This type of engine is difficult to balance with counterweights on the crankshaft. The engine in Fig. 10-5 is balanced by a balance shaft that turns in a direction opposite to the crankshaft.

3. Flat-Four-Cylinder Engine One of the early flat-fours is the Volkswagen engine, used in the "Beetle." The four cylinders are arranged in two opposing rows of two cylinders each. This engine is mounted at the rear of the car and drives the rear wheels. The engine is air-cooled. The cylinders have flat metal rings, or fins, surrounding them. These fins provide large surfaces from which heat can dissipate so the engine will not overheat.

Another flat-four is shown in Fig. 10-6. This engine is liquid-cooled. It is mounted at the front of the car and drives the front wheels. A transaxle is assembled directly to the engine.

Engine cooling systems—air cooling and liquid cooling—are described in a later chapter.

♦ 10-4 FIVE-CYLINDER ENGINES

Some five-cylinder automotive engines are being built. Mercedes produces a five-cylinder diesel engine. Volkswagen has a five-cylinder in-line spark-ignition engine (Fig. 10-7) for a front-drive car. Note the axle coming out the side of the transmission. The engine is mounted ahead of the front-wheel drive axles.

♦ 10-5 SIX-CYLINDER IN-LINE ENGINES

Most six-cylinder engines are in line ("straight six"), although there are V-6 and flat-six engines. This compares to the four-cylinder engines, which can also be in line, V type, or flat.

Figure 10-8 shows a six-cylinder spark-ignition in-line engine. The valves are overhead and the camshaft is in the cylinder block. The crankshaft is supported by seven main bearings. This engine is also known as a "slant six" because the cylinders are slanted to one side. This makes additional room vertically so the hood line can be lowered. This engine is made with either a cast-iron or an aluminum-alloy cylinder block.

Other six-in-line engines have the cylinders vertical, and not slanted as in Fig. 10-8. Most of these engines use pushrods and have the camshaft in the cylinder block. One manufacturer (Porsche) produces a flat-six, air-cooled, 200-hp engine for their 911 Carrera.

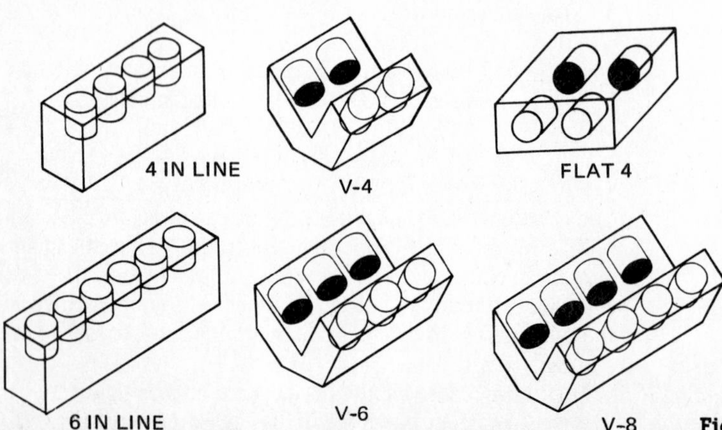

Fig. 10-1 Various ways of arranging engine cylinders.

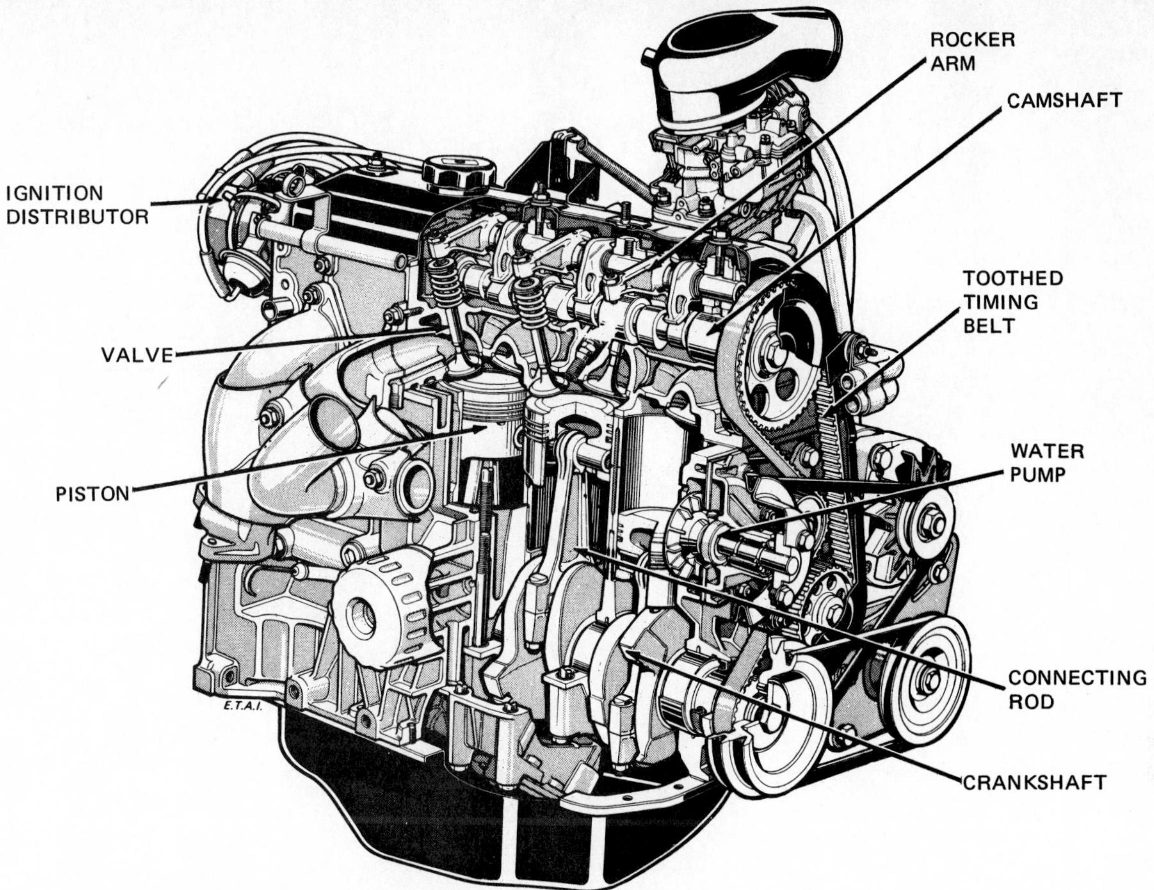

Fig. 10-2 A four-cylinder spark-ignition engine with overhead camshaft. The toothed belt that drives the camshaft and the ignition distributor are direct-coupled to the ends of the camshaft. (*Renault USA, Inc.*)

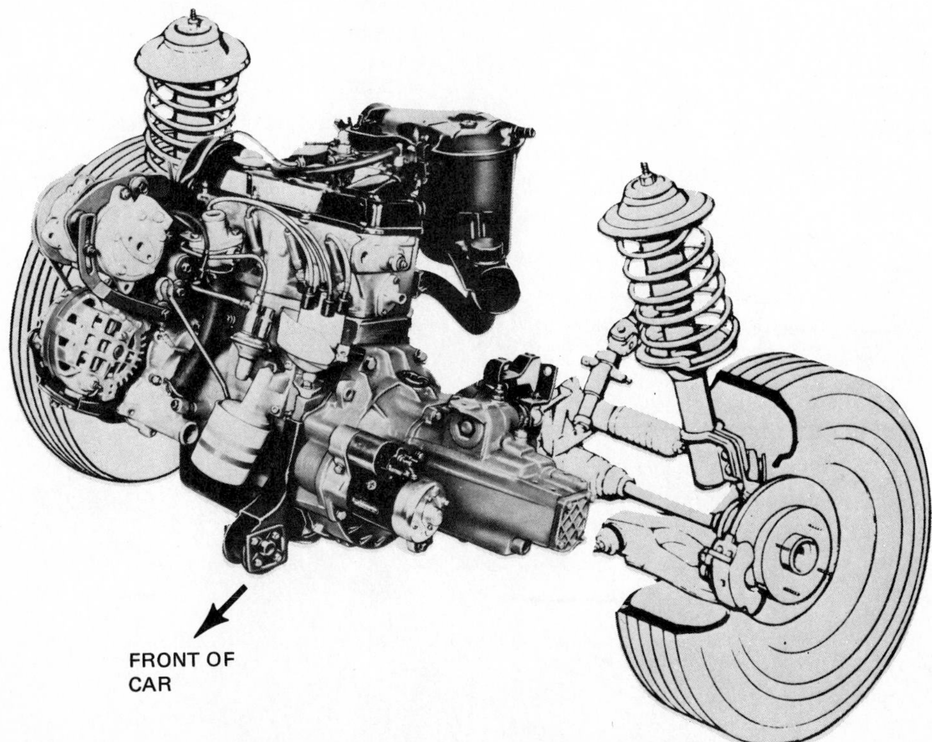

Fig. 10-3 Mounting and drive arrangement for a transversely mounted four-cylinder OHC engine. (*Chrysler Corporation*)

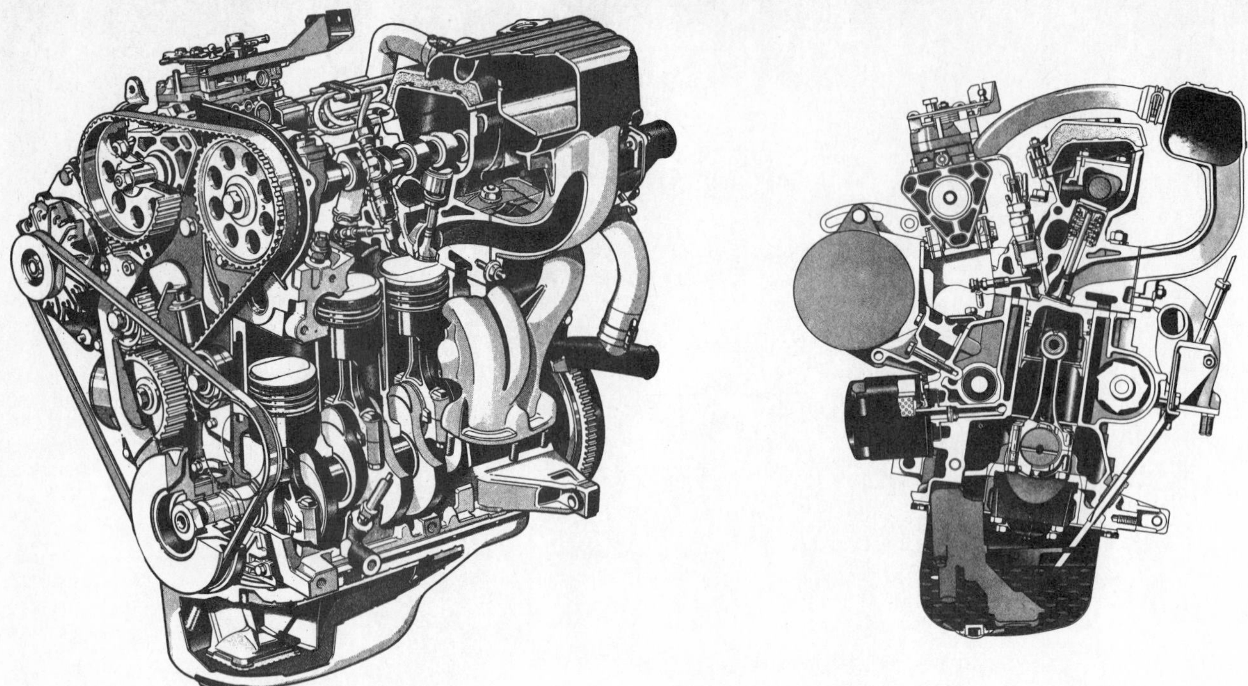

Fig. 10-4 A four-cylinder OHC diesel engine. The bottom end (except for the pistons) has been adapted from a spark-ignition engine. The pistons are contoured and provide a high compression ratio of 20:1. (*Fiat Motors of North America, Inc.*)

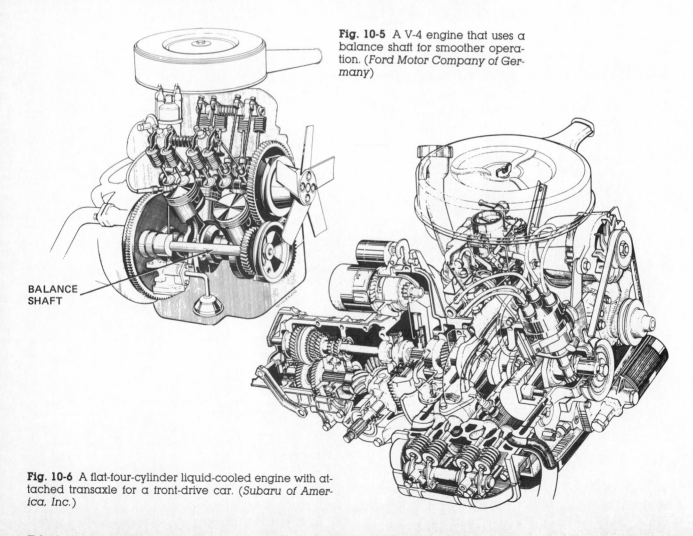

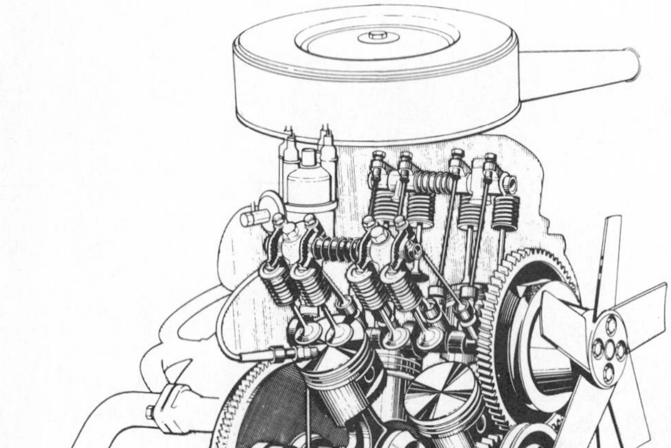

Fig. 10-5 A V-4 engine that uses a balance shaft for smoother operation. (*Ford Motor Company of Germany*)

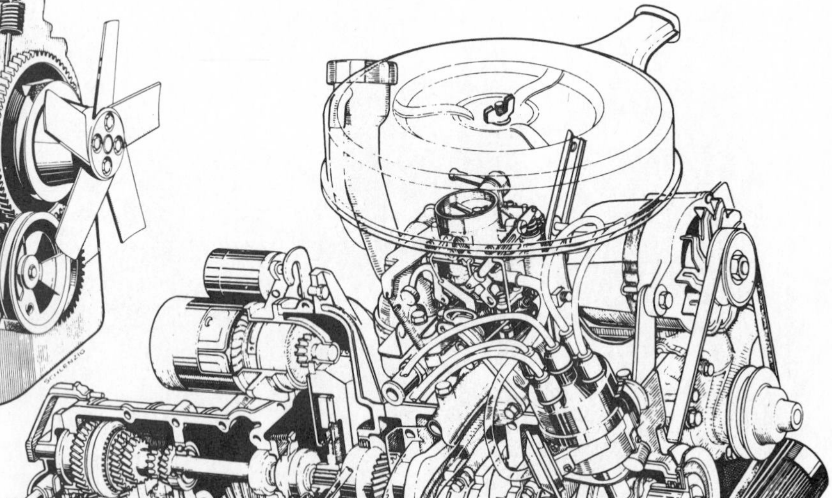

BALANCE SHAFT

Fig. 10-6 A flat-four-cylinder liquid-cooled engine with attached transaxle for a front-drive car. (*Subaru of America, Inc.*)

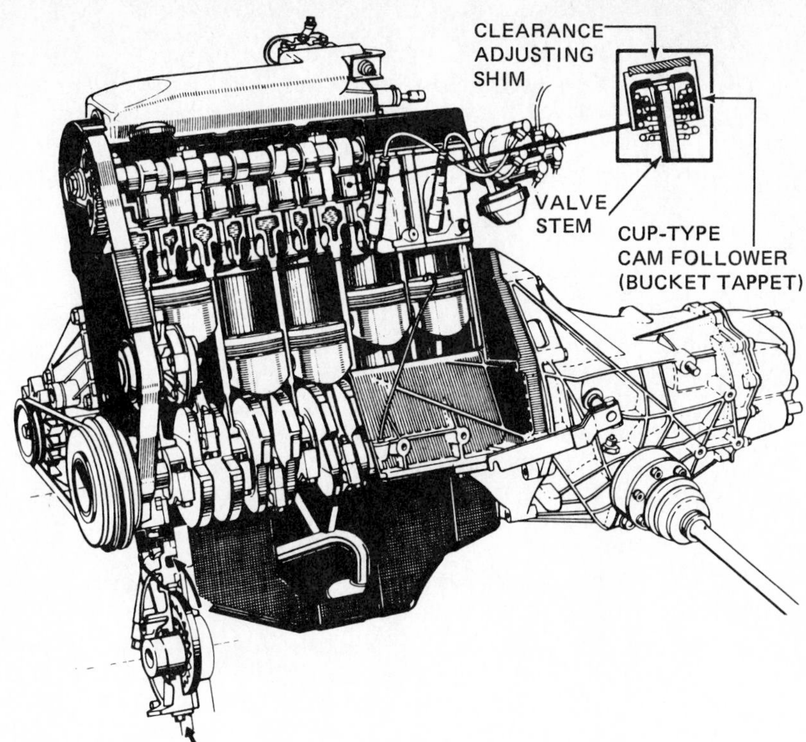

Fig. 10-7 A five-cylinder in-line engine with overhead camshaft for a front-drive car. The small insert shows a bucket tappet. (*Volkswagen of America, Inc.*)

Fig. 10-8 A slant-six in-line, overhead-valve pushrod engine. The cylinders are slanted to one side to permit a lower hood line. (*Chrysler Corporation*)

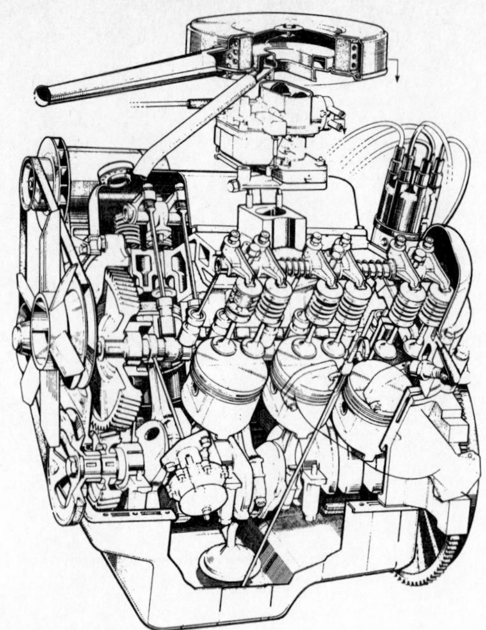

Fig. 10-9 A V-6 overhead-valve engine. (*Ford Motor Company of Germany*)

♦ 10-6 V-6 ENGINES

A V-6 engine has two rows of three-cylinders each, set at an angle to form a V (Fig 10-9). Figure 10-10 shows a V-6 engine mounted transversely between the front wheels. It also shows the transmission and axles to the front wheels. This is a spark-ignition engine, with the camshaft in the cylinder block.

♦ 10-7 V-8 ENGINES

In the V-8 engine (Fig. 10-11), the cylinders are arranged in two rows, or banks, set at an angle to each to form a V. This engine is like two four-cylinder in-line engines mounted on a single crankcase and using one crankshaft. The crankshaft has four cranks. Connecting rods from opposing cylinders in the two banks are attached to a single crankpin. Two rods are attached to each crankpin. Two pistons are connected to each crankpin. The crankshaft usually is supported by five main bearings.

The engine shown in Fig. 10-11 has the camshaft in the cylinder block. It is a pushrod engine. Some of the high-performance V-8 engines have the camshafts in the cylinder head. This arrangement requires either two or four camshafts. Overhead camshaft engines are discussed later.

FRONT

Fig. 10-10 A transversely mounted V-6 pushrod engine. (*Chevrolet Motor Division of General Motors Corporation*)

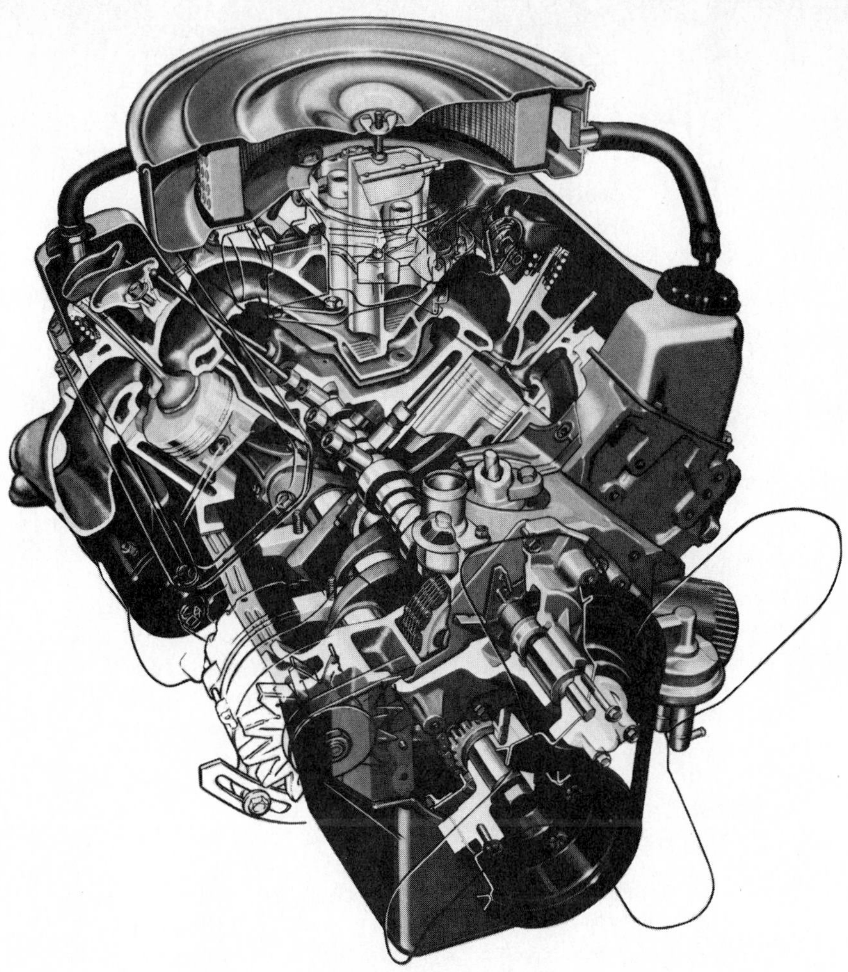

Fig. 10-11 A spark-ignition pushrod V-8 engine. (*Ford Motor Company*)

♦ 10-8 VARIABLE-DISPLACEMENT V-8 ENGINES

This engine (Fig. 10-12) has electronic controls that selectively cut out two or four cylinders at a time. The number of cylinders the controls cut out depends on the power requirements. When the engine is idling, or cruising at a steady speed on a level highway, the controls cut out four of the cylinders. Only four cylinders are needed to provide sufficient power for these operating conditions.

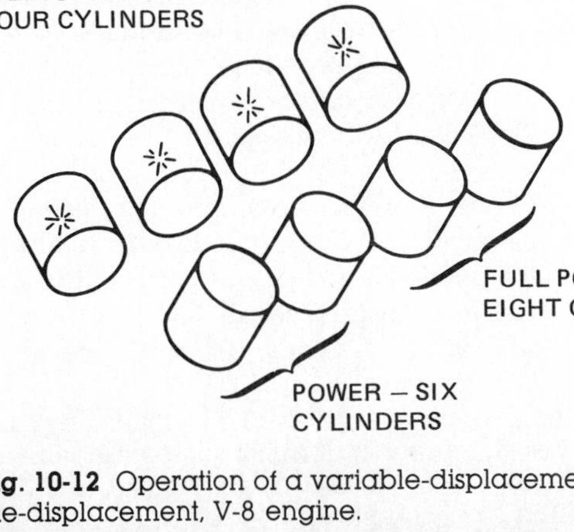

IDLING AND CRUISING — FOUR CYLINDERS

FULL POWER — EIGHT CYLINDERS

POWER — SIX CYLINDERS

Fig. 10-12 Operation of a variable-displacement, or multiple-displacement, V-8 engine.

However, when the car encounters a hill, additional power is needed and the electronic controls put additional cylinders to work. The same thing happens when the driver "steps on the gas" to increase car speed, as, for example, passing another car. Additional cylinders go to work. The number of cylinders that go back to work depends on the amount of additional power that is needed. This arrangement saves fuel without sacrifice of performance, according to the manufacturer.

♦ 10-9 TWELVE- AND SIXTEEN-CYLINDER ENGINES

Twelve- and sixteen-cylinder engines have been used in passenger cars, buses, trucks, and industrial plants. The cylinders are mostly in two banks (V type or pancake type). Sometimes they are in three banks (W type) or four banks (X type). The pancake engine is similar to a V engine, but the two rows are flat and opposing. The cylinders work to the same crankshaft. The only passenger cars now being made with a twelve-cylinder engine are the Ferrari, the Jaguar, and the Maserati.

♦ 10-10 FIRING ORDER

The *firing order* is the order in which the cylinders deliver their power strokes. This is a built-in part of the engine de-

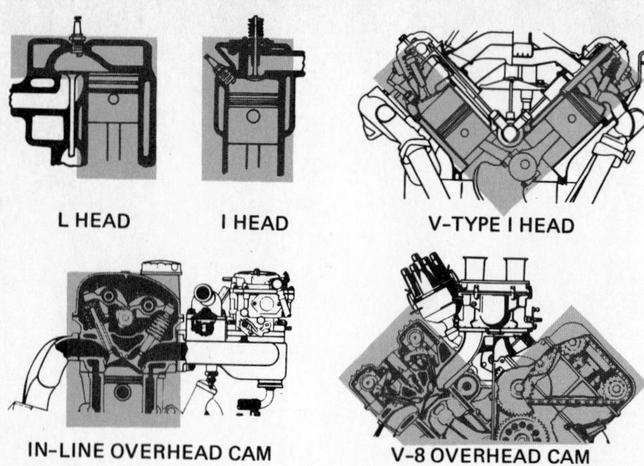

Fig. 10-13 Various valve arrangements used in automotive engines.

sign. The strokes are divided along the crankshaft so that a well-distributed pattern results. This minimizes the strain on the crankshaft. Two cylinders firing at one end of the crankshaft one after the other is usually avoided. In in-line engines, the cylinders are numbered from front to back. The two firing orders used in four-cylinder engines are 1–3–4–2 and 1–2–4–3.

Two firing orders are possible in six-cylinder in-line engines, 1–5–3–6–2–4 or 1–4–2–6–3–5. All modern six-cylinder in-line engines use 1–5–3–6–2–4.

The firing order of V-6 and V-8 engines is more complicated. This is because the cylinders are numbered in different ways by different manufacturers. For example, some V-6 cars have the left bank numbered from front to back, 1, 3, and 5. The right bank is numbered 2, 4, and 6. On other V-6 engines, the right bank is numbered 1, 2, and 3. The left bank is numbered 4, 5, and 6.

When working on a V-type engine, refer to the manufacturer's service manual to find out how the cylinders are numbered and the firing order. On many engines, the firing order is cast into the intake manifold.

♦ 10-11 VALVE ARRANGEMENTS

The intake and exhaust valves can be arranged in various ways in the cylinder head or block (Fig. 10-13). The valve-in-head (I-head) design is almost universally used today for automotive engines.

1. L-Head Engines Many years ago, most automobile engines used the L-head arrangement shown in Fig. 10-13. In the L-head engine, the combustion chamber and cylinder forms an inverted L. The intake and exhaust valves are located side by side in the cylinder block, and to one side of the cylinder. This is a relatively simple and dependable arrangement. However, the design has two drawbacks. First, it cannot achieve high compression ratios. *Compression ratio* is a measure of how much the air-fuel mixture is compressed. High-compression engines produce more power, as explained later. Second, the L-head design causes greater pollution. Its exhaust gases contain too much unburned and partly burned fuel (HC and CO). The reason is that the combustion chamber surfaces are large and relatively cool. This prevents combustion of the layers of air-fuel mixture close to these surfaces. Automobile manufacturers switched to overhead-valve engines many years ago.

You will still see small L-head engines, however. They are used in power mowers and other equipment where light weight and simplicity are important.

2. Pushrod Camshaft-in-Block Engines In the overhead-valve pushrod engine, the valves are in the cylinder head (Fig. 10-13). In in-line engines, the valves are usually in a single row (Fig. 10-8). In V-type engines, the valves may be in a single row in each bank, or in a double row in each bank. When the camshaft is in the cylinder block (pushrod type), only one camshaft is needed to operate all valves in a V-type engine. The valve lifters, pushrods, and rocker arms carry the motion from the cams to the valves. This action is described in ♦9-13 and illustrated in Fig. 9-18.

Overhead-valve engines are used in all modern automobiles. These engines have higher compression ratios and are more efficient that L-head engines. In an overhead-valve engine, the clearance volume can be smaller. *Clearance volume* is the volume in the combustion chamber, above the piston, with the piston at TDC. Having the valves directly above the piston permits a smaller clearance volume. In some I-head engines, there are pockets in the piston heads. The valves can move into these pockets when the valves are open with the piston at TDC.

3. Overhead Camshafts (Figs. 10-2 and 10-4) Pushrod engines use pushrods and rocker arms to operate the valves. However, the pushrods and rocker arms impose some inertia that affects valve action. The pushrods and rocker arms flex, or bend, slightly before they open the valve. This slows the valve action. At lower speeds, this does not matter. As speed increases, the flexing also increases. This causes an increasing lag in valve action which tends to limit top engine speed. However, with the overhead camshaft, the cams work directly on the rocker arms or cam followers. This gives quicker valve response so that higher engine speeds are possible. When the camshaft is overhead (OHC), the engine may use only one camshaft (in-line engines). However, some in-line engines use two camshafts. Some V-type OHC engines use one camshaft for each bank. Other, high-performance engines may use two camshafts for each bank, or four camshafts. The single-overhead-camshaft engine (one camshaft in each cylinder head) is called an SOHC engine. The double-overhead-camshaft engine is a DOHC engine. These different engines are described later.

4. Additional Valves Some engines have more than two valves per cylinder. Figure 10-14 shows a four-valve arrangement. Each cylinder has two intake valves and two exhaust valves. This is a four-cylinder engine that uses 16 valves. Two overhead camshafts are required. They are driven by a duplex chain as shown in Fig. 10-15. Bucket tappets are used (♦9-12). The additional valves improve engine performance. They provide additional ports for the air-fuel mixture to enter the cylinder, and for exhaust gases to leave. Uniform distribution of air-fuel mixture to all four cylinders is achieved by a fuel-injection system.

Note Some engines use a small third valve in connection with the combustion-chamber design. In one arrangement, the small third valve admits a very rich air-fuel mixture. This is described in a later chapter.

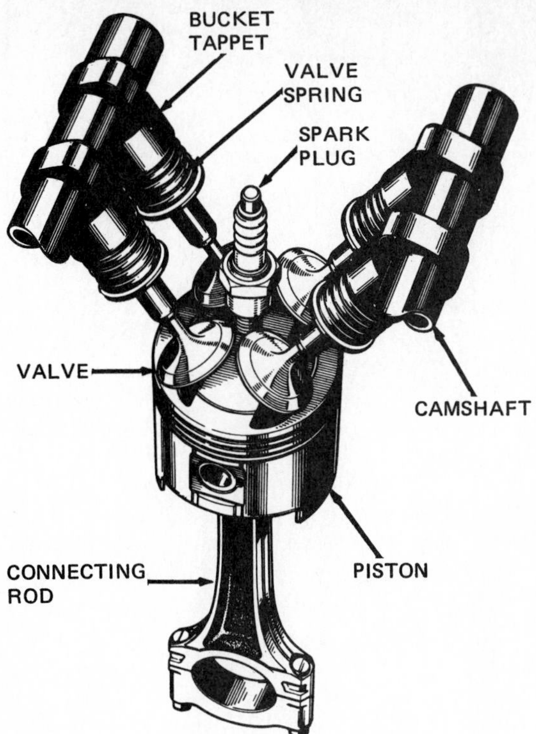

Fig. 10-14 Valve and camshaft arrangement for a four-cylinder engine using four valves per cylinder (two intake and two exhaust) and two camshafts. This engine uses bucket tappets. (*Nissan Motor Corporation*)

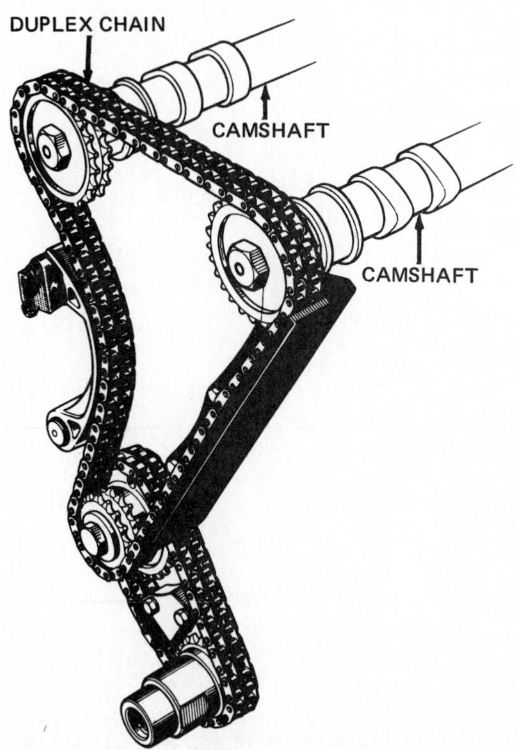

Fig. 10-15 Two overhead camshafts driven by a duplex chain. (*Nissan Motor Corporation*)

♦ 10-12 CLASSIFICATION BY COOLING

Engines can be liquid-cooled or air-cooled. Most present-day automobile engines are liquid-cooled. The Corvair, Volkswagen, and some other early automobile engines were air-cooled. Also, the small one- and two-cylinder engines on power mowers and other garden equipment are air-cooled. In air-cooled engines, the cylinders are usually separate. They have metal fins which provide a large surface. This permits engine heat to be carried away from the cylinders. Air-cooled automotive engines have shrouds which direct the airflow around the cylinders for cooling.

Liquid-cooled engines use a liquid to take heat from the engine. The liquid, called the *coolant*, is water mixed with antifreeze. These engines have water jackets surrounding the cylinders and combustion chambers (Fig. 10-16). A later chapter describes liquid-cooling systems.

♦ 10-13 CLASSIFICATION BY CYCLES

Engines can be either two-stroke-cycle or four-stroke-cycle. The four-stroke-cycle engine (usually called a four-cycle engine) was discussed in ♦9-7 to 9-11. In it, the complete cycle requires four piston strokes (intake, compression, power, and exhaust). In the two-stroke-cycle, or two-cycle, engine, the intake and compression strokes, and the power and exhaust strokes, are combined. This permits the engine to produce a power stroke every two piston strokes, or every crankshaft rotation.

The two-cycle engine is simple in construction, having no valve train. But it is a relatively inefficient and "dirty" engine. It produces more pollutants than a comparable four-cycle engine. Also, the power strokes of the two-cycle engine are less powerful than the power strokes in a four-cycle engine. This is because the power stroke and exhaust stroke are combined in the two-cycle engine. In the four-cycle engine, the power stroke is separate from the other three piston strokes.

There are no two-cycle automotive engines sold in this country today. However, small two-cycle engines are widely used in power lawn mowers and similar equipment. Many motorcycles, snowmobiles, and similar machines use two-cycle engines. They have the advantage of operating at high rpm, thereby providing a high power-to-weight ratio.

♦ 10-14 CLASSIFICATION BY FUEL

Internal-combustion engines can be classified according to the type of fuel they use. Automotive spark-ignition engines, in general, use gasoline or gasohol. Some bus and truck engines use liquefied petroleum gas (LPG). These are usually gasoline engines adapted for LPG. Diesel engines use diesel fuel oil. Chapter 15 describes these fuels in detail.

♦ 10-15 DIESEL ENGINES

The diesel engine is a compression-ignition engine (♦9-16). Air alone is highly compressed and becomes very hot. As the piston nears the end of the compression stroke, the fuel is injected, or sprayed, into the combustion chamber. The heat of compression ignites the fuel so that the power stroke follows.

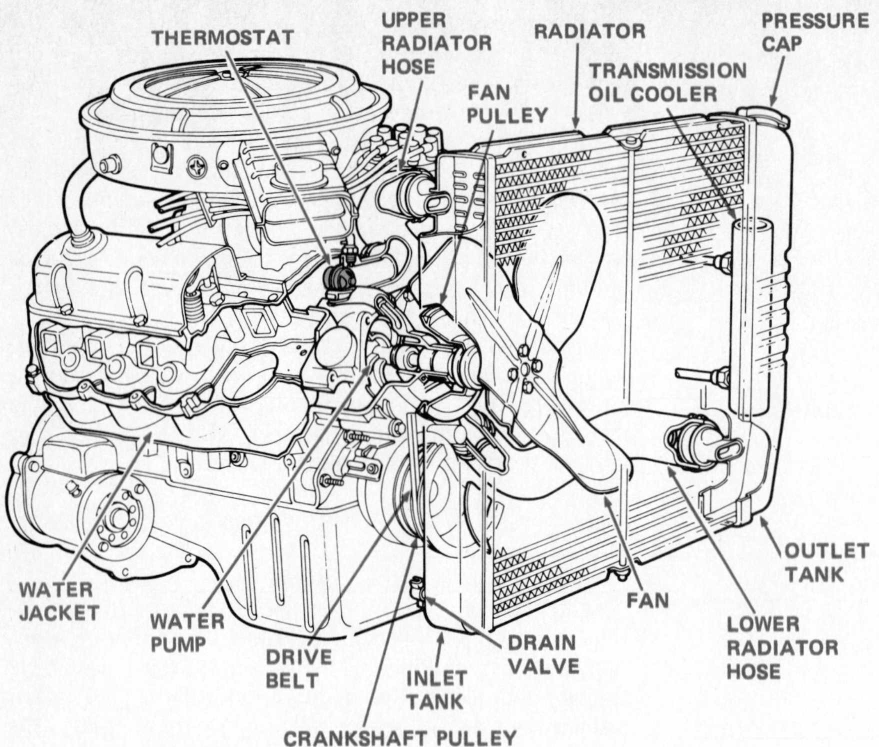

Fig. 10-16 Components in a V-8 engine liquid-cooling system. (*Ford Motor Company*)

♦ 10-16 DIESEL-ENGINE APPLICATIONS

Diesel engines have been made in many sizes and outputs, from 1.5 to 140,000 horsepower (hp) [1.1 to 104,440 kilowatts (kW)]. They are used in passenger cars, trucks, buses, farm and construction machinery, ships, electric power plants, and other mobile and stationary applications.

Diesel engines have been used in passenger cars for many years. Until recently no automobile manufacturer in the United States was producing diesels. For many years, the major manufacturer of passenger cars with diesel engines was Mercedes-Benz. It has produced more than half a million diesel-powered cars since 1950 (Fig. 10-17). Many of these have

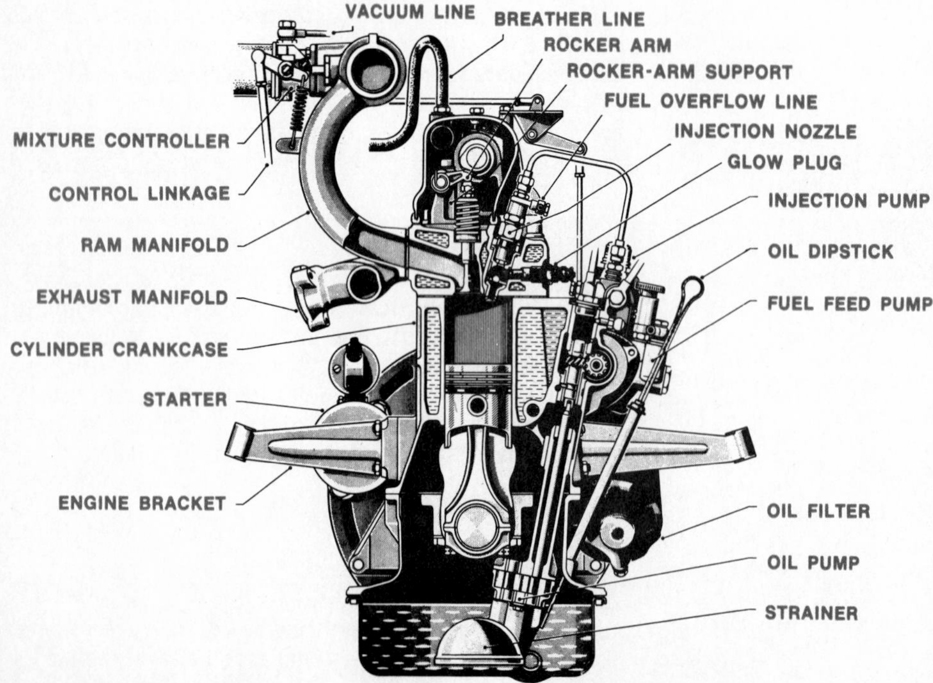

Fig. 10-17 A four-cylinder automotive diesel engine. (*Mercedes-Benz of North America, Inc.*)

been imported into the United States. Mercedes-Benz now makes a five-cylinder diesel engine. Peugeot is another European automobile manufacturer producing diesel-powered cars.

Volkswagen and General Motors build diesel-powered cars. Figure 9-15 shows a four-cylinder diesel engine produced by Volkswagen. Figure 10-18 shows a V-8 automotive diesel engine from General Motors. Other automotive companies are also producing diesel engines. There are several reasons for this increased emphasis on diesels. The diesel is a relatively fuel-efficient engine, especially under part load. You can get more miles per gallon with a diesel. The diesel engine has no electric ignition system. However, there are disadvantages. It requires more cranking power to get it started and needs a heavier starting motor and battery. It is noisier than a comparable spark-ignition engine. Also, it lacks the acceleration of a well-tuned spark-ignition engine. It is a heavier engine because of the high internal pressures that develop in the diesel-engine combustion chambers. Therefore the diesel engine weighs more per horsepower than the spark-ignition engine.

Despite these drawbacks, some experts believe that the diesel engine has a good future as a passenger-car engine. It is already the standard engine for heavy-duty trucks, buses, and off-the-road construction machinery. American manufacturers, including General Motors, Caterpillar, Cummins, International, and Mack, have been building diesel engines for these applications for years.

♦ 10-17 ROTARY ENGINES

Until now, we have been describing reciprocating engines. These engines have pistons that move up and down, or reciprocate, in cylinders. Another type of engine has no pistons. Instead, it has a rotor that is spun by the burning of the fuel in the engine. Two general types of rotary engines are the gas *turbine* and the *Wankel*.

♦ 10-18 GAS TURBINES

Gas turbines have been used in some buses and trucks. But many engineers believe we will never see them as the main power plant for cars. Figure 10-19 shows a gas-turbine engine. It has two sections: the gasifier section, where the fuel is burned, and the power section, where the power from the burning fuel is produced. The turbine can use gasoline, kerosene, or oil for fuel.

The compressor in the gasifier section has an air-intake rotor with a series of blades on it. When the air-intake rotor spins, it acts as an air pump and supplies the burner with high-pressure air. In the burner, fuel is sprayed into the compressed air. The fuel burns, and the burned gases then flow, at still higher pressures (because of the combustion process), through the blades of the gasifier section. This spins the turbine rotor. The gasifier turbine rotor is mounted on the same shaft as the air-intake rotor. Therefore, when the turbine rotor spins, the air-intake rotor spins, supplying the gasifier section with compressed air.

The regenerator core is a rotating heat exchanger. It takes heat from the hot exhaust gases and carries it back to the incoming air. This preheats the air. This does two things. It improves the efficiency of the turbine by utilizing some of the heat in the exhaust gases. It also reduces the temperature of the exhaust gas.

The power of the spinning turbine rotor is carried through shafts and gears to the vehicle's drive wheels.

♦ 10-19 WANKEL ENGINE

The Wankel engine (Fig. 10-20) has a rotor that spins in an oval chamber shaped like a fat figure 8. It is often called a *rotary-combustion* (RC) engine because the combustion chambers rotate in a somewhat circular pattern. The engine rotor has three lobes. The rotor rotates in an eccentric pattern. The lobes remain in contact with the oval housing to

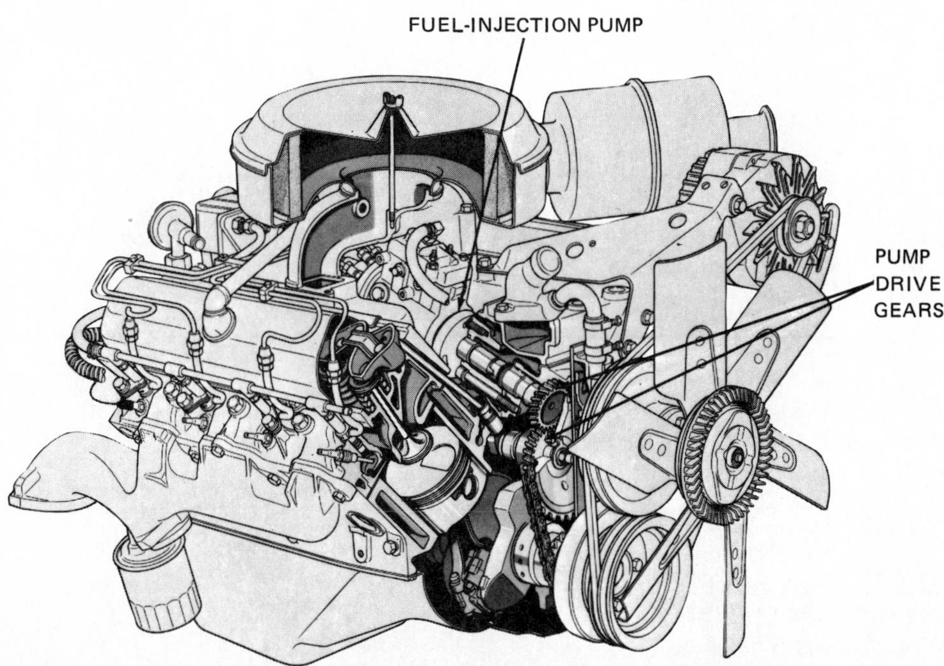

Fig. 10-18 A V-8 diesel engine for passenger cars. (*General Motors Corporation*)

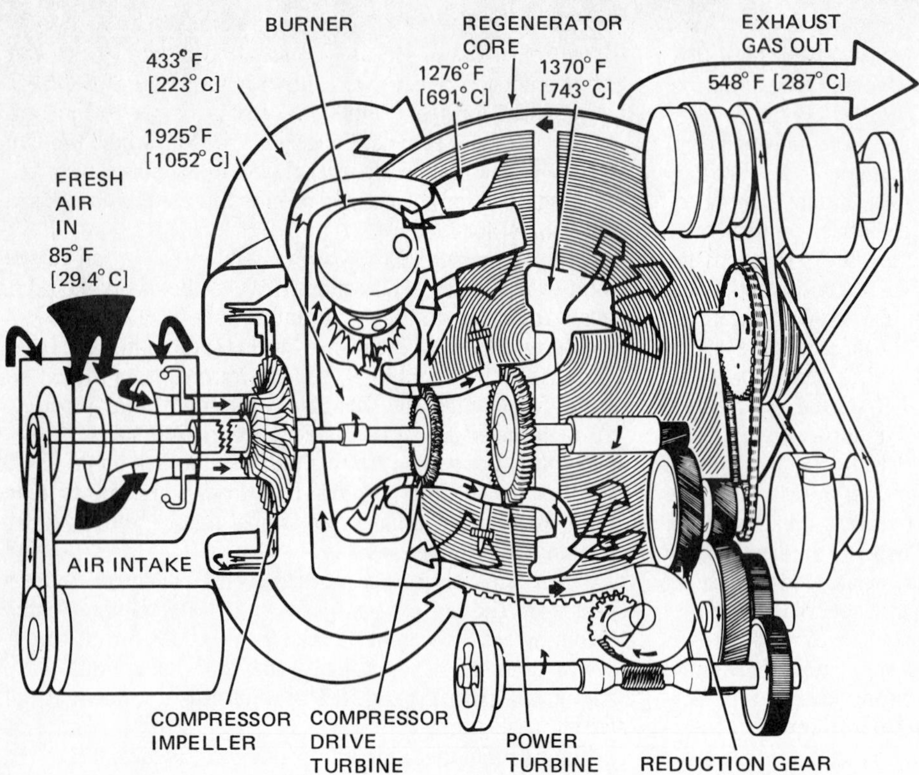

Fig. 10-19 A gas turbine, showing airflow through the engine. (*Chrysler Corporation*)

form a tight seal. This seal compares with the seal formed by the piston rings against the cylinder wall in a reciprocating engine.

The four actions—intake, compression, power, and exhaust—are going on at the same time around the rotor while the engine is running. Figure 10-21 shows how the engine works. The rotor lobes A, B, and C seal tightly against the side of the oval housing. The rotor has recesses in its three faces between the lobes. The dashed lines on the rotors in Fig. 10-21 show the locations of the recesses and how deep they are. It is in these recesses that combustion actually starts. The spaces between the rotor lobes are where intake, compression, power, and exhaust take place.

Let us follow the rotor around as it goes through a complete cycle—intake, compression, power, and exhaust. At I (upper left), lobe A has passed the intake port, and the air-fuel mixture is starting to enter ①. As the rotor moves around, at II (upper right), the space between lobes A and C increases ②. This motion produces a vacuum, which causes the air-fuel mixture to enter. This action compares with the intake stroke of the piston in the reciprocating engine. At III (lower right), the air-fuel mixture continues to enter as the space between lobes A and C continues to increase ③. Then lobe C starts to move past the intake port, as shown in IV (lower left). Further rotor movement carries lobe C past the intake port, so the air-fuel mixture is sealed between lobes A and C ④.

To understand what happens to the air-fuel mixture, let us go back to I (upper left) again. Here the air-fuel mixture has been trapped between lobes A and B ⑤. Further rotation of the rotor decreases the space between lobes A and B. By the time the rotor reaches the position shown in III, the space ⑦ is at a minimum. This action is the same as the piston reaching TDC on the compression stroke in the reciprocating engine. Now the spark plug fires and ignites the compressed mixture. Pressure is exerted on the side of the rotor and this forces the rotor to move around. See IV (lower left). This action is the same as the power stroke in the reciprocating engine.

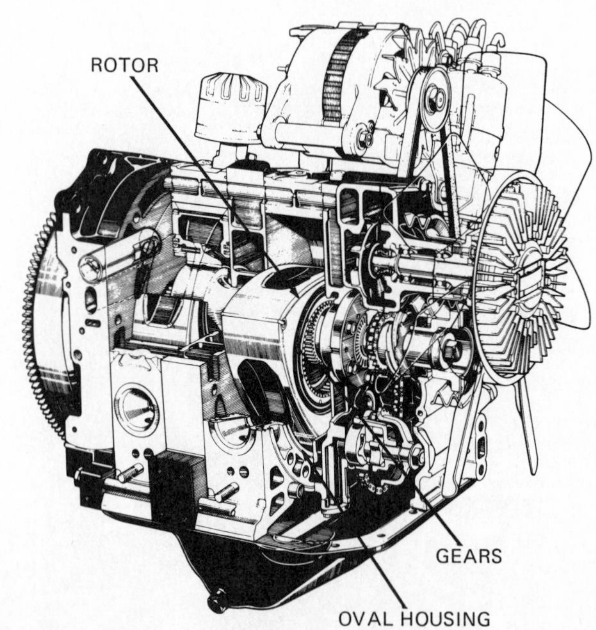

Fig. 10-20 A two-rotor Wankel engine. (*Mazda Motors of America, Inc.*)

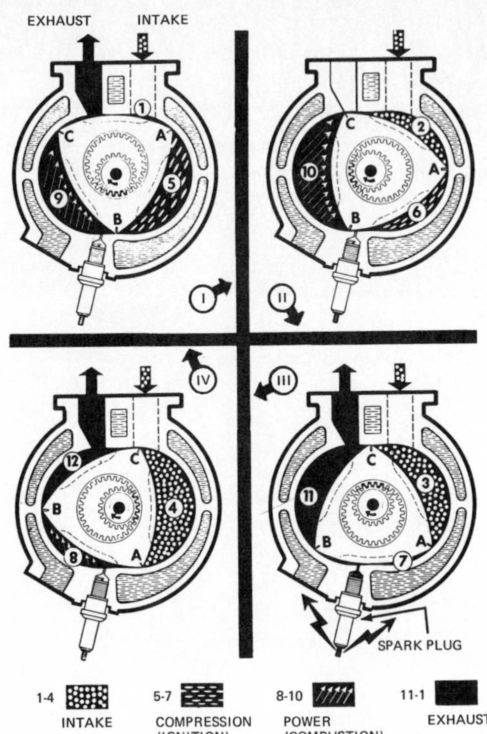

Fig. 10-21 Actions in a Wankel engine during one complete revolution of the rotor.

At IV, the high pressure of the burned air-fuel mixture ⑧ forces the rotor around to position I again. Continued expansion of the burned gases continues to rotate the rotor until the leading lobe passes the exhaust port. Then the burned gases begin to exhaust from between the lobes, as shown by ⑪ and ⑫ in III and IV. As the rotor continues to rotate, the space between the lobes decreases and the gases are exhausted. This action is the same as the exhaust stroke of the piston in the reciprocating engine.

Following the exhaust stroke, the leading lobe passes the intake port, and the whole cycle is repeated. Note that there are three lobes and three spaces between the lobes. This means that there are three complete cycles of intake, compression, power, and exhaust going on at the same time. The engine is delivering power almost continuously. In a way, the engine is equivalent to a three-cylinder piston engine. A two-rotor Wankel engine would then be equivalent to a six-cylinder piston engine.

♦ REVIEW QUESTIONS ♦

Select the *one* correct, best, or most probable answer to each question. You can find the answers in the section indicated at the end of each question.

1. The three basic cylinder arrangements for automotive engines are (♦10-2)
   a. flat, radial, and V
   b. in a row, in line, and opposed
   c. in line, V, and opposed
   d. V, double row, and opposed
2. The power from a small engine can be increased without increasing engine weight by (♦10-3)
   a. making cylinders larger
   b. turbocharging
   c. increasing piston stroke
   d. adding more cylinders
3. Compared to an engine with the camshaft in the block, the overhead-camshaft engine has (♦10-11)
   a. more parts
   b. fewer parts
   c. lower rpm
   d. none of the above
4. When the piston is at TDC, the volume above the piston in the combustion chamber is the (♦10-11)
   a. clearance volume
   b. compression ratio
   c. volumetric efficiency
   d. none of the above
5. In the multiple-displacement V-8 engine, electronic controls can cut out (♦10-8)
   a. one, two, three, or four cylinders
   b. two, three, or four cylinders
   c. two or four cylinders
   d. all the cylinders
6. The firing order is the (♦10-10)
   a. order in which the cylinders are numbered
   b. order in which the cylinders deliver their power strokes
   c. standard arrangement which can be changed by changing the crankshaft
   d. order in which the pistons are arranged
7. The pushrod engine (♦10-11)
   a. uses pushrods
   b. has an overhead camshaft
   c. has an L-type valve arrangement
   d. uses bucket tappets
8. In the diesel engine, the fuel is ignited by (♦10-15)
   a. an electric spark
   b. the heat of compression
   c. the hot exhaust gas
   d. none of the above
9. Two types of rotary engines are the (♦10-17)
   a. V type and in line
   b. Wankel and diesel
   c. Wankel and gas turbine
   d. Wankel and OHC
10. The Wankel engine has a (♦10-19)
    a. rotor that spins eccentrically in an oval chamber
    b. rotor made up of a compressor and a turbine
    c. gasifier section and a power section
    d. spark plug for each lobe of the rotor

CHAPTER 11
Engine Construction

After studying this chapter, you should be able to:

1. Discuss cylinder blocks and how they are made.
2. Discuss cylinder heads and how they are made.
3. Describe three types of cylinder heads for gasoline engines.
4. Explain the purpose and function of intake and exhaust manifolds, vibration dampers, and flywheels.
5. Discuss crankshafts and how they are made.
6. Describe engine bearings, their function, and how they are lubricated.
7. Identify and name the parts of disassembled engines.
8. Discuss the construction, purpose, operation, and lubrication of the connecting rods.
9. Discuss the construction, purpose, operation, and lubrication of pistons and piston rings.
10. Explain the purpose of piston-expansion control and how it is achieved.

♦ 11-1 THE ENGINE

In the spark-ignition engine, a mixture of air and fuel vapor is delivered by the fuel system to the engine cylinders. There it is compressed, ignited by the spark plug, and burned. The high pressure resulting from this combustion pushes the piston down on the power stroke so the crankshaft turns and the engine produces power.

In the compression-ignition, or diesel, engine, air alone is compressed. Near the end of the compression stroke, fuel is injected into the combustion chamber. The high temperature of the compressed air ignites the fuel and the power stroke follows.

Now, in this and the following chapter, the various parts of the engine are described in detail.

♦ 11-2 CYLINDER BLOCK

The cylinder block is the foundation of the engine. Everything else is put inside of or attached to the block. Figure 11-1 shows the cylinder block for a V-8 spark-ignition engine, with the major parts that go into the block. The pan is plastic or metal shaped to fit on the bottom of the block. It holds a reservoir of lubricating oil to lubricate the engine. The oil pump for this engine is attached to the bottom of the block.

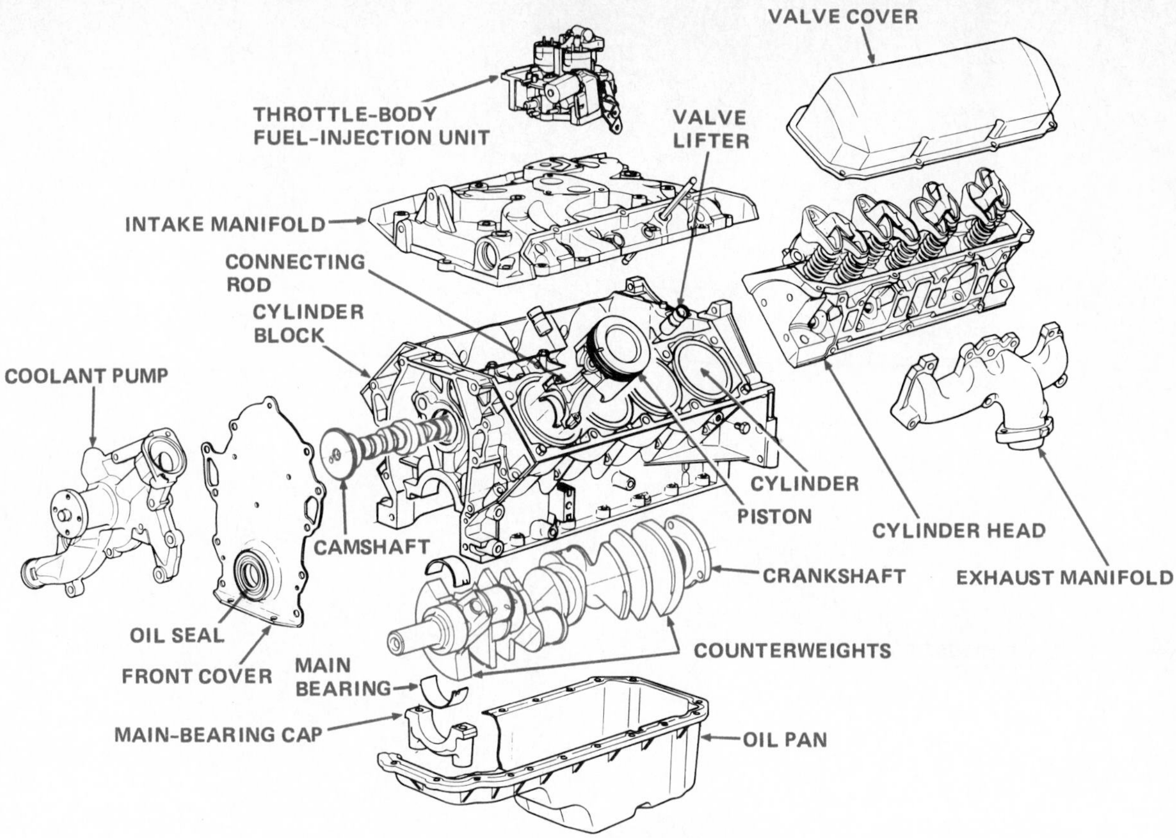

Fig. 11-1 Major components that are placed in and attached to a V-8 engine cylinder block. *(Cadillac Motor Car Division of General Motors Corporation)*

The oil pickup pipe extends down into the reservoir of lubricating oil in the oil pan.

Figure 11-2 shows the cylinder block for a four-cylinder in-line, camshaft-in-block spark-ignition engine, with the major parts that go into the block. The oil pump for this engine also mounts under the block (Fig. 11-3).

Most cylinder blocks are cast in one piece from gray iron or iron mixed with other metals, such as nickel or chromium. These added metals give the cylinder block greater resistance against wear as well as greater strength. Some cylinder blocks are cast from an aluminum alloy. They are described later in the chapter.

The cylinder block is a complicated casting (Fig. 11-4). It must have the large holes for the cylinder bores, and also water jackets and coolant passages, boltholes, and core clean-out holes. The *water jackets* are the spaces between the cylinder bores and the outer shell of the block. Coolant (water mixed with antifreeze) flows through these spaces to pick up heat and take it away from the engine. Cooling systems are described later.

The *core clean-out holes* allow the cores that formed the water jackets to be removed. These cores are made of sand. They are put in place before the hot metal is poured into the cylinder-block mold. After the metal has hardened, these sand cores are broken up and removed through the clean-out holes, leaving the water jackets. These holes are then sealed with plugs. The plugs are called by several names—*block core plugs* (see item 18 in Fig. 11-2), *freeze plugs,* and *expansion plugs.* The reason for the last two names is that if the coolant in the block starts to freeze and expand, sometimes the coolant can push the plugs out before it expands enough to crack the block.

Spark-ignition cylinder blocks and compression-ignition cylinder blocks are very much alike. However, the diesel-engine block is heavier and stronger. It must be stronger because the compression ratios and internal pressures in the combustion chambers of diesel engines are higher.

♦ 11-3 MACHINING THE BLOCK

After the cores are removed, the block is machined. Figures 11-1 and 11-2 show the many finished surfaces, passages, bores, and tapped holes in the block. The machining operations include:

1. Drilling holes for attachment of various parts
2. Machining the cylinders
3. Boring the camshaft-bearing holes (camshaft-in-block type)
4. Smoothing the surfaces to which parts are attached
5. Drilling oil passages
6. Boring the valve-lifter bores (pushrod engines)
7. Cleaning out coolant passages

♦ 11-4 PARTS ATTACHED TO AND INSTALLED IN THE BLOCK

The cylinder head is mounted on top of the block, and the oil pan is attached to the bottom of the block. The crankshaft is hung from the bottom of the block by the crankshaft main bearings and bearing caps. The piston-and-rod assemblies are

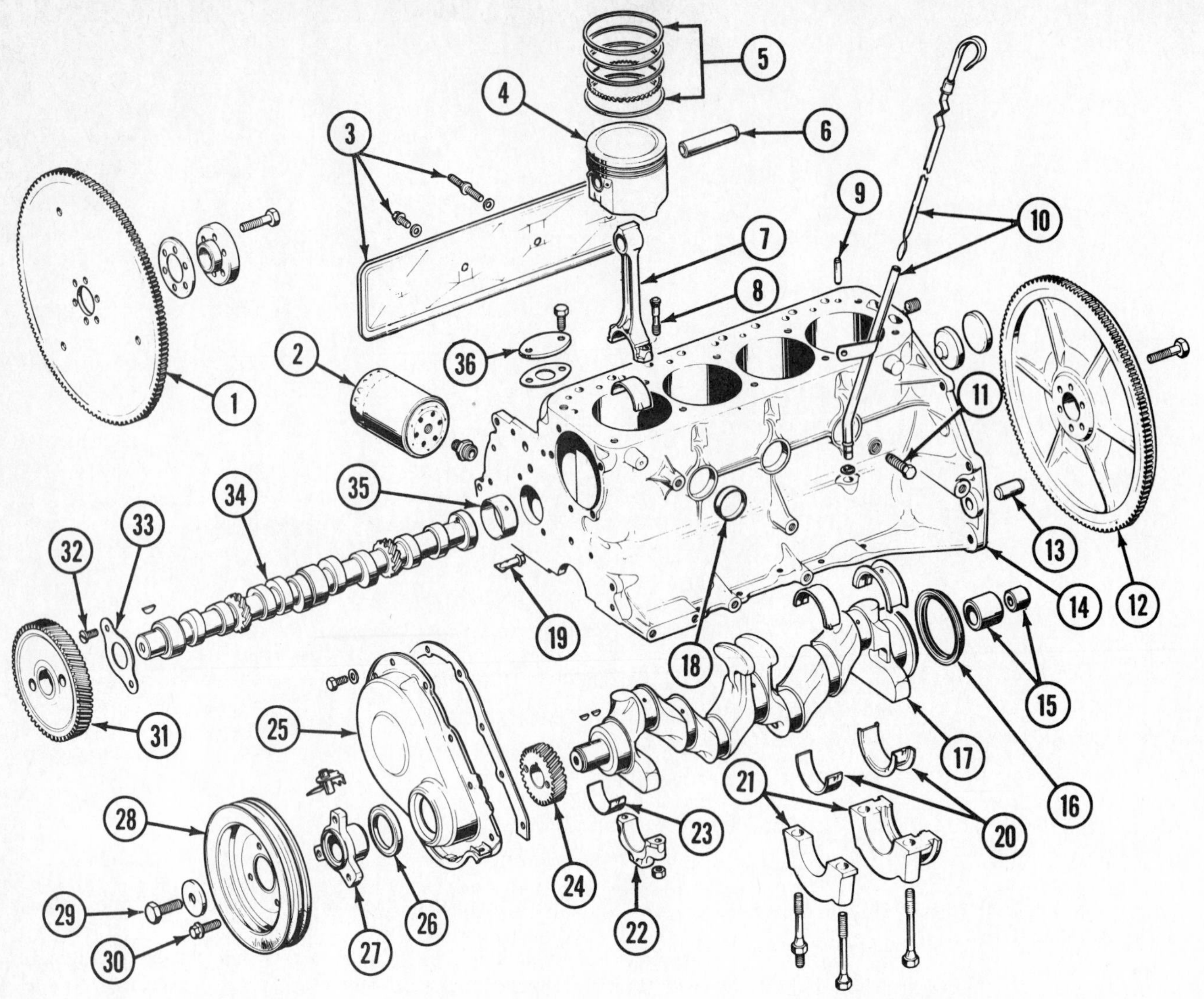

1. DRIVE PLATE AND RING GEAR (AUTOMATIC TRANSMISSION)
2. OIL FILTER
3. PUSHROD COVER AND BOLTS
4. PISTON
5. PISTOR RINGS
6. PISTON PINS
7. CONNECTING ROD
8. CONNECTING-ROD BOLT
9. DOWEL
10. OIL LEVEL INDICATOR AND TUBE
11. BLOCK DRAIN
12. FLYWHEEL AND RING GEAR (MANUAL TRANSMISSION)
13. DOWEL
14. CYLINDER BLOCK
15. PILOT AND/OR CONVERTER BUSHING
16. REAR OIL SEAL
17. CRANKSHAFT
18. BLOCK CORE PLUG
19. TIMING GEAR OIL NOZZLE
20. MAIN BEARINGS
21. MAIN-BEARING CAPS
22. CONNECTING-ROD BEARING CAP
23. CONNECTING-ROD BEARING
24. CRANKSHAFT GEAR
25. TIMING-GEAR COVER
26. TIMING-GEAR-COVER OIL SEAL
27. CRANKSHAFT PULLEY HUB
28. CRANKSHAFT PULLEY
29. CRANKSHAFT PULLEY HUB BOLT
30. CRANKSHAFT PULLEY BOLT
31. CAMSHAFT GEAR
32. CAMSHAFT THRUST PLATE SCREW
33. CAMSHAFT THRUST PLATE
34. CAMSHAFT
35. CAMSHAFT BEARING
36. OIL-PUMP DRIVE-SHAFT RETAINER PLATE, GASKET, AND BOLT

Fig. 11-2 The various major parts that are attached to or go into the cylinder block of a four-cylinder engine which has the camshaft in the block. *(American Motors Corporation)*

installed from the top of the block. Then the lower ends of the connecting rods are attached to the crankshaft cranks with the rod bearings and bearing caps.

When the major parts shown in Figs. 11-1 and 11-2 are installed in the cylinder block, the assembly is called a *short block*. This is a service item that can be purchased from many manufacturers. Sometimes a service problem can best be solved by buying a short block. The rest of the parts from the old engine, such as the cylinder head, can then be installed on the new short block.

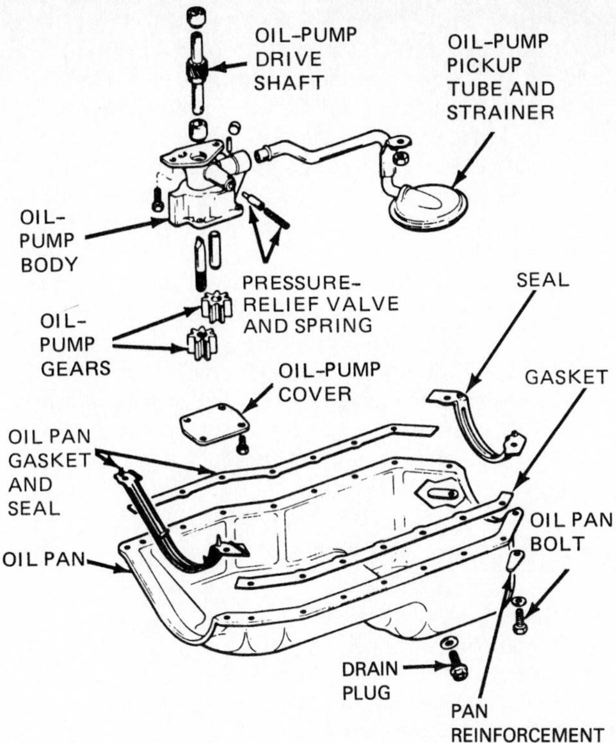

Fig. 11-3 The oil pan and pump for the engine block shown in Fig. 11-2. *(American Motors Corporation)*

♦ 11-5 OIL PAN

The oil pan (Fig. 11-3) is attached to the bottom of the cylinder block. The bottom of the cylinder block, plus the oil pan, form the *crankcase.* They enclose, or *encase,* the cranks on the crankshaft. The oil pan holds from about 4 to 9 quarts [3.8 to 8.5 L] of oil, depending on the engine design. Preformed or formed-in-place gaskets (♦5-19 and 5-20) are used to seal the joint between the cylinder block and the oil pan. The gaskets prevent loss of oil from the oil pan.

When the engine is running, the oil pump sends oil from the oil pan up to the moving engine parts. The engine lubricating system is described later.

♦ 11-6 ALUMINUM CYLINDER BLOCK

Some engines have aluminum cylinder blocks. Aluminum is a light metal, weighing much less than cast iron. Aluminum also conducts heat more rapidly than cast iron. Lightness and heat conductivity are the two main reasons that aluminum has been used for cylinder blocks. However, aluminum is too soft to use as cylinder-wall material. It would wear very rapidly. Therefore, aluminum cylinder blocks must have cast-iron cylinder liners, or be cast from a special aluminum alloy that has silicon particles in it (♦11-7). Cylinder liners are sleeves that are either cast into the cylinder block or installed later.

In the cast-in type, the cylinder liners are installed in the mold, and the aluminum is poured around them. They then become a permanent part of the cylinder block.

Two kinds of liner, dry and wet, can be installed later. Dry liners are pressed into the cylinder block. They touch the cylinder bore along their full length. Wet liners touch the cylinder block only at the top and bottom. The rest of the liner touches only the coolant. Wet liners are removable and can be replaced if they become worn or damaged. This method is also used in some truck diesel engines.

♦ 11-7 SLEEVELESS ALUMINUM CYLINDER BLOCKS

These cylinder blocks, used by Mercedes-Benz, Porsche, and others, are cast from aluminum which has silicon particles in it. Silicon is a very hard material. After the cylinder block is cast, the cylinders are honed. They are then treated with a chemical that etches, or eats away, the surface aluminum. This leaves only the silicon particles exposed (Fig. 11-5). The pistons and piston rings slide on the silicon with minimum wear.

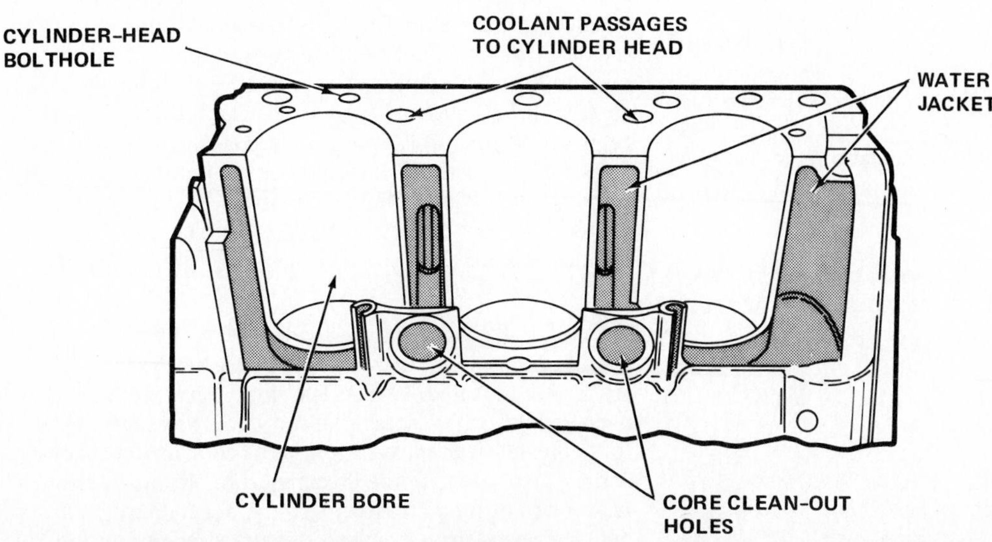

Fig. 11-4 One bank of a V-6 engine block partly cut away, showing the internal construction. *(General Motors Corporation)*

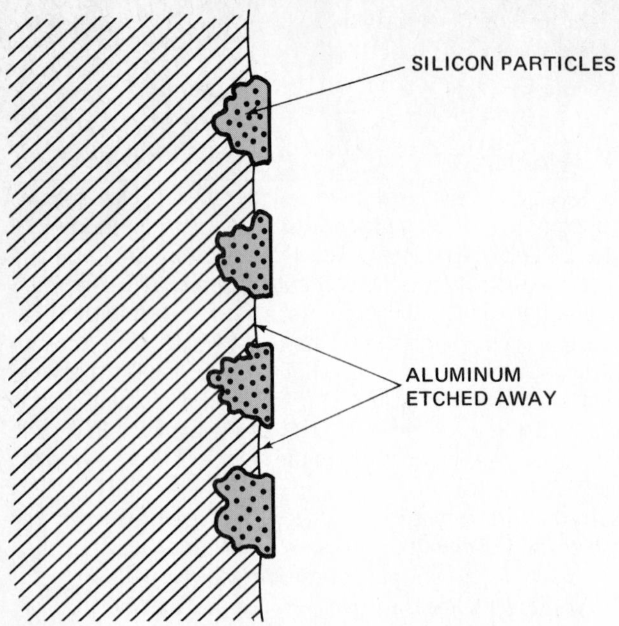

Fig. 11-5 In the finishing process, the aluminum is etched away, leaving only the hard particles of silicon.

♦ 11-8 AIR-COOLED ENGINES

In air-cooled engines, the cylinders are separated and have fins. The *fins* are metal rings which aid in conducting heat away from the cylinders. In the assembled air-cooled engine, the cylinders are installed on the crankcase. Then the cylinder head is installed on top of the cylinders.

♦ 11-9 CYLINDER HEAD

The cylinder head encloses one end of the engine cylinders and forms the upper end of the combustion chambers. The piston head and piston rings form the lower end. Figures 11-6 and 11-7 show disassembled cylinder heads with the related parts that are attached to them. One of these is for a camshaft-in-block engine (Fig. 11-6). The other is for an OHC engine (Fig. 11-7).

Cylinder heads are cast in one piece from gray iron, iron alloy (iron mixed with other metals), or aluminum alloy. Most cylinder heads are of cast iron. The cylinder head includes water jackets and passages from the valve ports to the openings in the manifolds (intake and exhaust). The casting method uses sand molds and is similar to the casting process for cylinder blocks. Many machining operations are required to convert the rough castings into finished cylinder heads, including:

1. Drilling and tapping holes for the spark plugs (or injection nozzles in diesel engines), for bolts and studs to attach the manifolds, and for other attached parts
2. Drilling coolant and oil-passage holes
3. Smoothing mounting surfaces
4. Drilling valve-guide holes
5. Finishing valve seats

Figure 11-6 shows the cylinder head for a four-cylinder in-line, camshaft-in-block spark-ignition engine. This is the head for the block shown in Fig. 11-2. The engine uses eight valves (two for each cylinder), eight pushrods, and eight rocker arms. The rocker arms are supported on ball studs. Figure 9-19 shows how this type of rocker arm is attached.

Figure 11-7 shows a cylinder head for a four-cylinder in-line, OHC spark-ignition engine, with related parts that are assembled to the cylinder head. The engine uses eight valves and springs. This engine uses no pushrods. Instead, one end of the rocker arm rides on a cam on the camshaft. When a cam lobe comes up under the end of the rocker arm, the arm pivots on the rocker-arm shaft. This causes the other end of the rocker arm to push down on the valve stem. This moves the valve down to the open position. When the lobe passes out from under the end of the rocker arm, the valve spring pulls the valve up to the closed position.

The camshaft is mounted on the cylinder head with bearing caps. It is driven by a timing chain from a crankshaft sprocket. OHC engines are discussed in detail in a later chapter.

♦ 11-10 SWIRL-TYPE COMBUSTION CHAMBER

The cylinder head forms the top of the combustion chamber. Basic combustion-chamber shapes are shown in Fig. 11-8. Some cylinder heads have additional features that help promote good combustion. For example, the cylinder head shown in Fig. 11-9 has a turbulence-generating pot (TGP). The purpose of this pot is to produce high turbulence, or swirl, of the air-fuel mixture during combustion. During compression, part of the air-fuel mixture is forced into the pot. On ignition, the mixture in the pot starts burning first and streams out at high velocity. This helps spread the flame rapidly so that better combustion results.

Another engine design that uses a high-swirl intake port and a masked intake-valve seat to produce high swirl is shown in Fig. 11-10. The shape of the valve is, in effect, changed by the off-center location of the seat in the port (due to the mask). This arrangement gives the incoming air-fuel mixture a high swirl as it enters. This helps promote complete combustion. Several other high-swirl designs have been developed. The purpose of all such designs is to improve combustion.

An engine that uses a second intake valve to produce high swirl is shown in Fig. 11-11. The second intake valve is much smaller than the main intake valve. The small valve admits only air from the upper part of the carburetor. Both valves open together. The main valve admits air-fuel mixture. The air from the second valve port is not restricted by the carburetor venturi or throttle valve. Therefore it flows in at high speed (a "jet" stream) into the combustion chamber. This sets up a high swirl of the air-fuel mixture as it is compressed. Combustion is improved, with more complete burning of the air-fuel mixture.

♦ 11-11 PRECOMBUSTION CHAMBER

In this arrangement, there is a pot with a small valve and the spark plug in it (Fig. 11-12). The small valve works with the primary intake valve. When the two valves open, air-fuel mixture enters. A lean mixture enters the combustion chamber. A rich mixture enters the precombustion chamber. Ignition takes place in the rich mixture in the precombustion chamber. Then the burning air-fuel mixture streams out into the main combustion chamber. This produces good turbulence and rapid burning of the mixture.

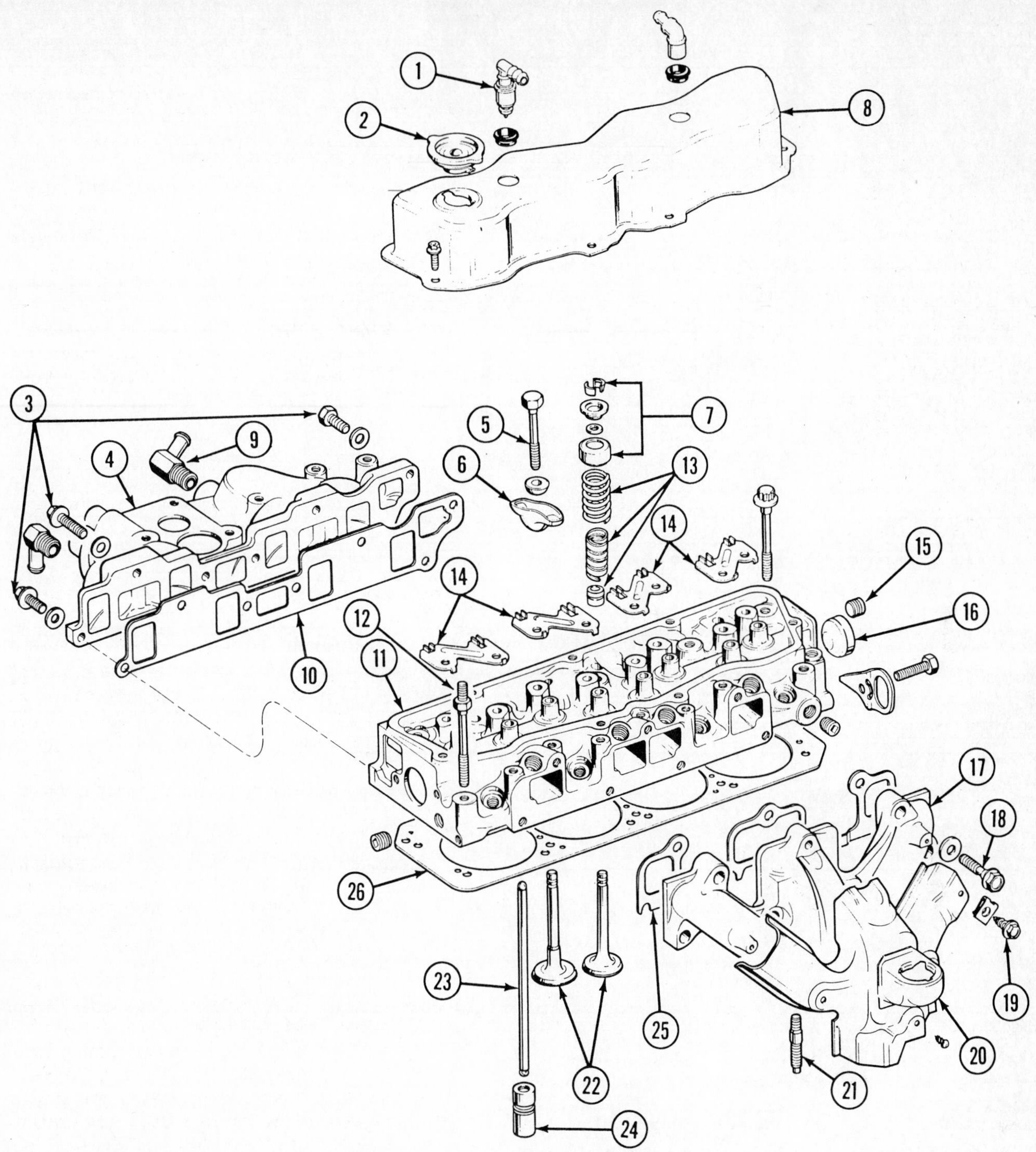

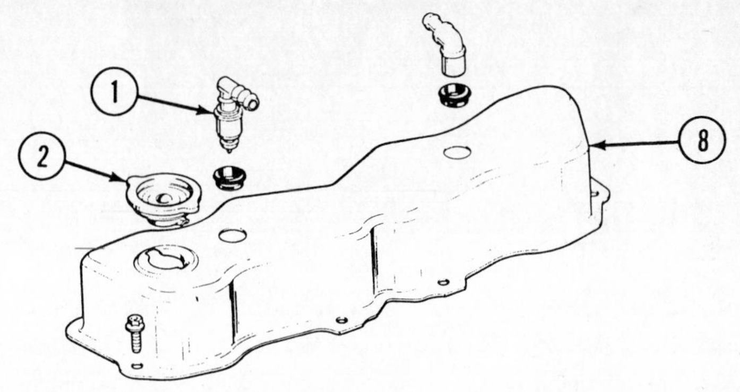

1. PCV VALVE
2. OIL FILLER CAP
3. INTAKE MANIFOLD ATTACHING BOLTS
4. INTAKE MANIFOLD
5. ROCKER-ARM CAPSCREW
6. ROCKER ARM
7. VALVE SPRING RETAINER ASSEMBLY
8. CYLINDER-HEAD COVER (VALVE COVER)
9. COOLANT HOSE FITTING
10. INTAKE MANIFOLD GASKET
11. CYLINDER HEAD
12. CYLINDER-HEAD STUD BOLT
13. VALVE SPRING AND OIL DEFLECTOR
14. PUSHROD GUIDE
15. CYLINDER-HEAD PLUG
16. CYLINDER-HEAD CORE PLUG
17. EXHAUST MANIFOLD
18. EXHAUST MANIFOLD BOLT
19. OIL-LEVEL-INDICATOR TUBE ATTACHING SCREW
20. EXHAUST MANIFOLD HEAT SHROUD (HEAT SHIELD)
21. EXHAUST MANIFOLD TO EXHAUST PIPE STUD
22. VALVES
23. PUSHROD
24. TAPPET
25. EXHAUST MANIFOLD GASKET
26. CYLINDER-HEAD GASKET

Fig. 11-6 Cylinder head for a four-cylinder camshaft-in-block engine, showing the major parts that are installed in and attached to the cylinder head. *(American Motors Corporation)*

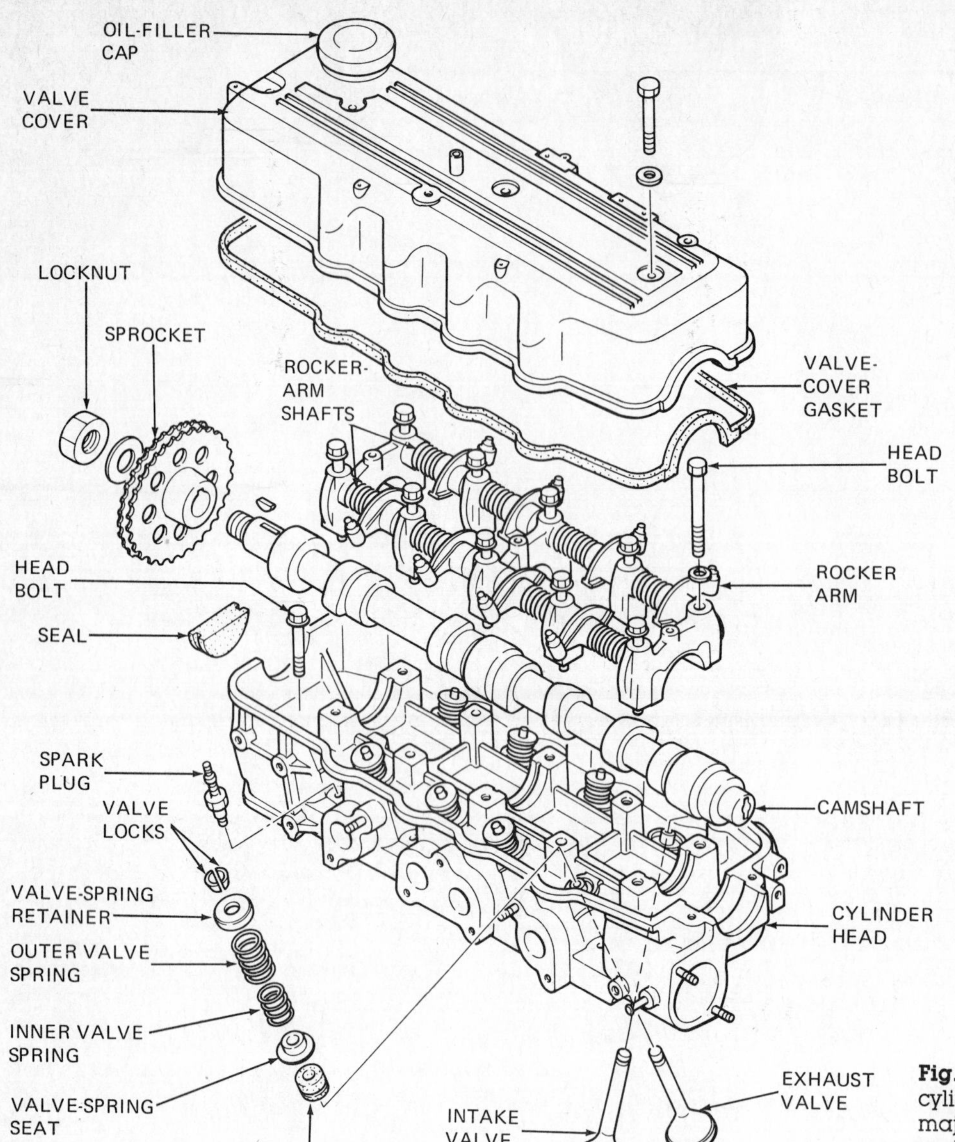

Fig. 11-7 Cylinder head for a four-cylinder OHC engine, showing the major parts that are attached to and installed in the cylinder head. *(Mazda Motors of America, Inc.)*

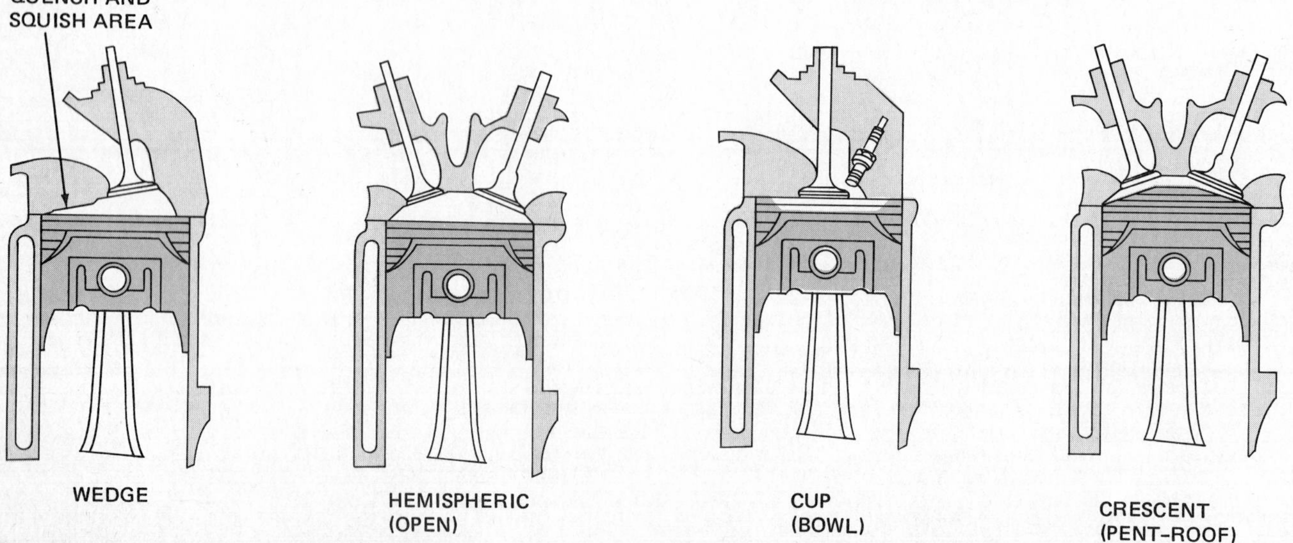

Fig. 11-8 Shapes of combustion chambers. *(ATW)*

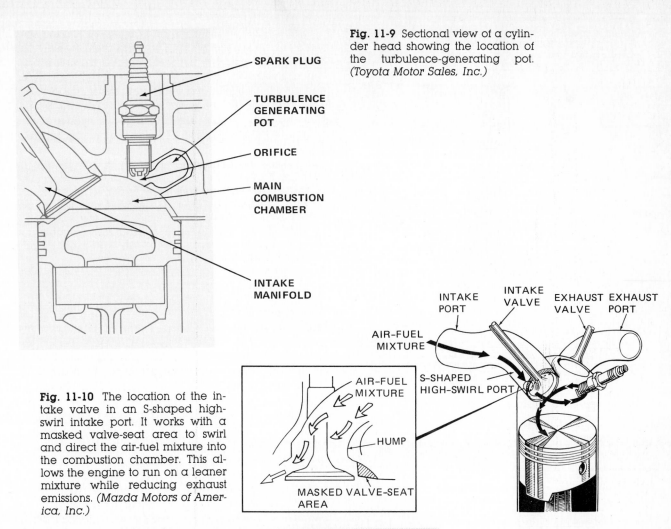

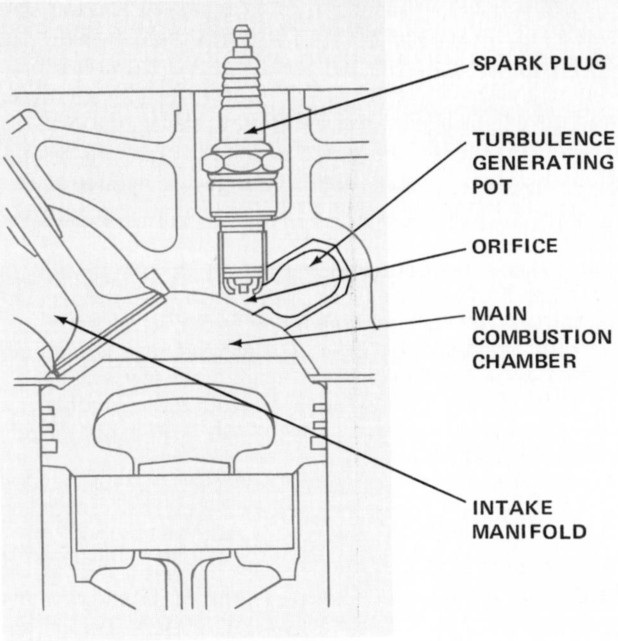

Fig. 11-9 Sectional view of a cylinder head showing the location of the turbulence-generating pot. *(Toyota Motor Sales, Inc.)*

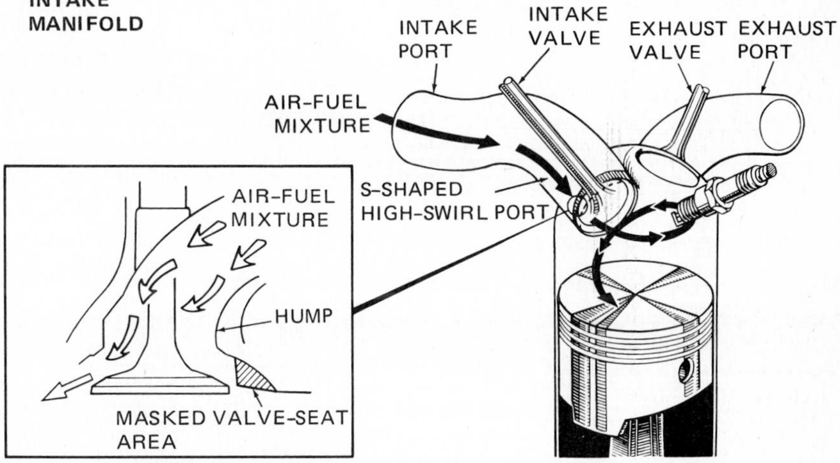

Fig. 11-10 The location of the intake valve in an S-shaped high-swirl intake port. It works with a masked valve-seat area to swirl and direct the air-fuel mixture into the combustion chamber. This allows the engine to run on a leaner mixture while reducing exhaust emissions. *(Mazda Motors of America, Inc.)*

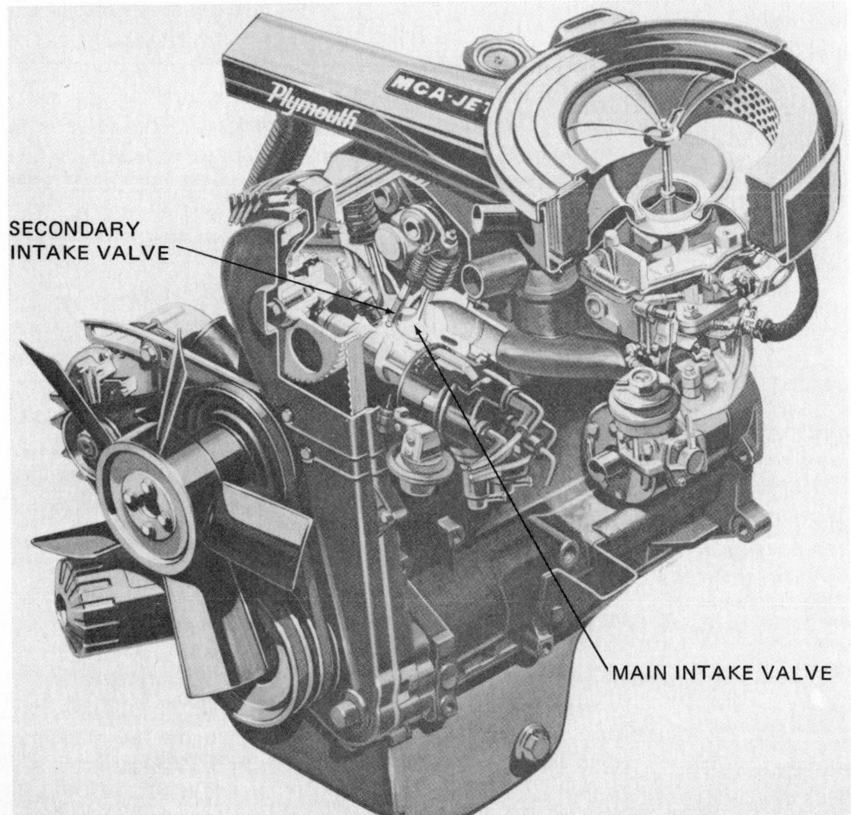

Fig. 11-11 A four-cylinder OHC engine using two intake valves per cylinder. One valve admits air-fuel mixture and the other admits air only. *(Chrysler Corporation)*

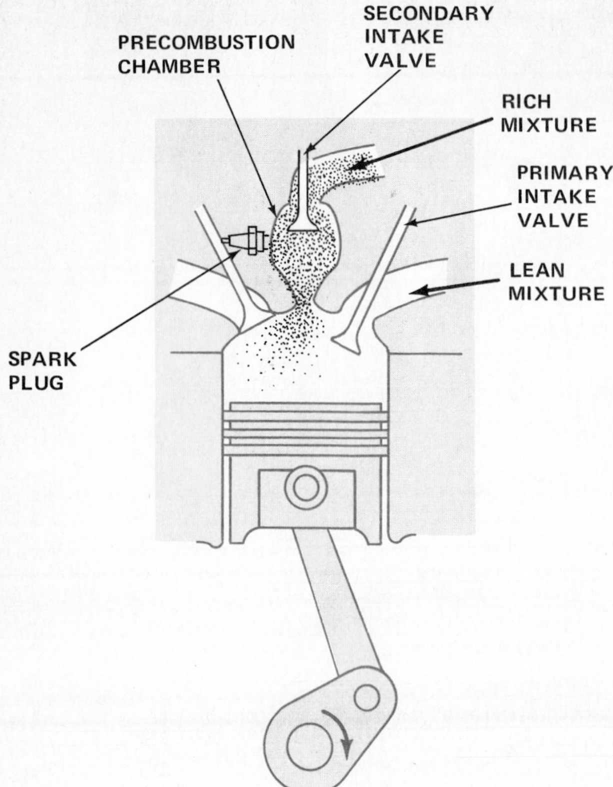

Fig. 11-12 Precombustion chamber in a cylinder head with an added intake valve. The spark plug is located in the precombustion chamber so ignition starts there. *(Chrysler Corporation)*

♦ 11-12 DIESEL-ENGINE CYLINDER HEAD

The diesel engine ignites the fuel in the combustion chamber by the heat of compression. As the air is compressed into the combustion chamber, the air becomes very hot. When the fuel is sprayed in, it is ignited by this high temperature. In cold weather, the cold head and cylinder block may take heat away so fast that the air does not get hot enough to ignite the fuel. Then glow plugs (Fig. 11-13) are needed. Each glow plug is a small electric heater.

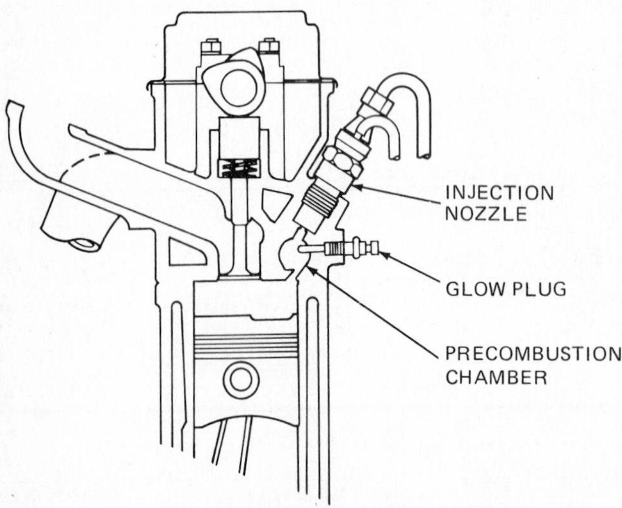

Fig. 11-13 A diesel-engine precombustion chamber, showing the locations of the glow plug and the fuel-injection nozzle. *(Volkswagen of America, Inc.)*

To start a cold diesel engine, first the driver turns the ignition switch to RUN (not START). This turns the glow plugs on. After a few seconds, they add enough heat to the precombustion chambers to ignite fuel. When that temperature is reached, a light signals the driver to start the engine. Now the driver turns the ignition switch to START to crank the engine. After the engine begins to run, the key is released and it returns to the RUN position. The glow plugs then turn off.

Note The above arrangement, with a precombustion chamber is used in diesel engines for passenger cars. Diesel engines for trucks and other heavy-duty vehicles usually have *direct injection*. The fuel is injected directly into the engine cylinders. This utilizes the fuel more efficiently. But it is noisier and the engine runs somewhat more roughly. For this reason, direct injection is not widely used in passenger-car diesel engines.

♦ 11-13 EXHAUST MANIFOLD

The *exhaust manifold* is a set of tubes that carry the exhaust gases from the engine cylinders. Figure 11-14 shows a disassembled cylinder head for a six-cylinder spark-ignition, in-line camshaft-in-block engine. The two manifolds, exhaust and intake, attach to the same side of the head. On some engines, the two manifolds are on opposite sides of the head (Fig. 11-6).

In in-line engines, only one exhaust manifold is required. But in V-type engines, two are used, one for each bank of cylinders, attached to the outsides of the heads. Figure 11-15 shows the exhaust system for a V-type engine. The two exhaust manifolds are connected by a crossover pipe. The exhaust gases then pass through the catalytic converter. It converts most of the HC and CO remaining in the exhaust gases into water and CO_2. Then the gases pass through the muffler, which reduces the exhaust noise, and the resonator, which further reduces the noise. Exhaust systems and catalytic converters are described later in the book.

Some exhaust manifolds for in-line engines have a *heat-control valve* (Fig. 11-16). The purpose of this valve is to provide quick heating of the air-fuel mixture when the engine is cold. This improves fuel vaporization, provides better cold-engine performance, and reduces harmful emissions. The heat-control valve is covered later.

♦ 11-14 INTAKE MANIFOLD

The *intake manifold* is a series of tubes that carry air, and in most systems, fuel vapor, to the engine cylinders. Figure 11-14 shows an intake manifold for a six-cylinder spark-ignition in-line engine which uses a carburetor. The *carburetor* is a mixing device that mixes fuel vapor with the air going to the cylinders to form a combustible mixture. It mounts on top of the intake manifold on a mounting pad.

Other spark-ignition engines do not have a carburetor. Instead, a fuel-injection system is used. This system has injection valves that inject, or spray, fuel into the intake manifold. This system, and carburetors, are covered in later chapters.

The intake manifold for diesel engines is an assembly of tubes that carry air from the air cleaner to the engine cylinders. In these engines, the fuel is injected into the cylinders.

On V-type engines, the intake manifold sits between the two banks of cylinders (Figs. 10-10, 10-11, and 10-18). It supplies air, or air-fuel mixture, to both cylinder banks.

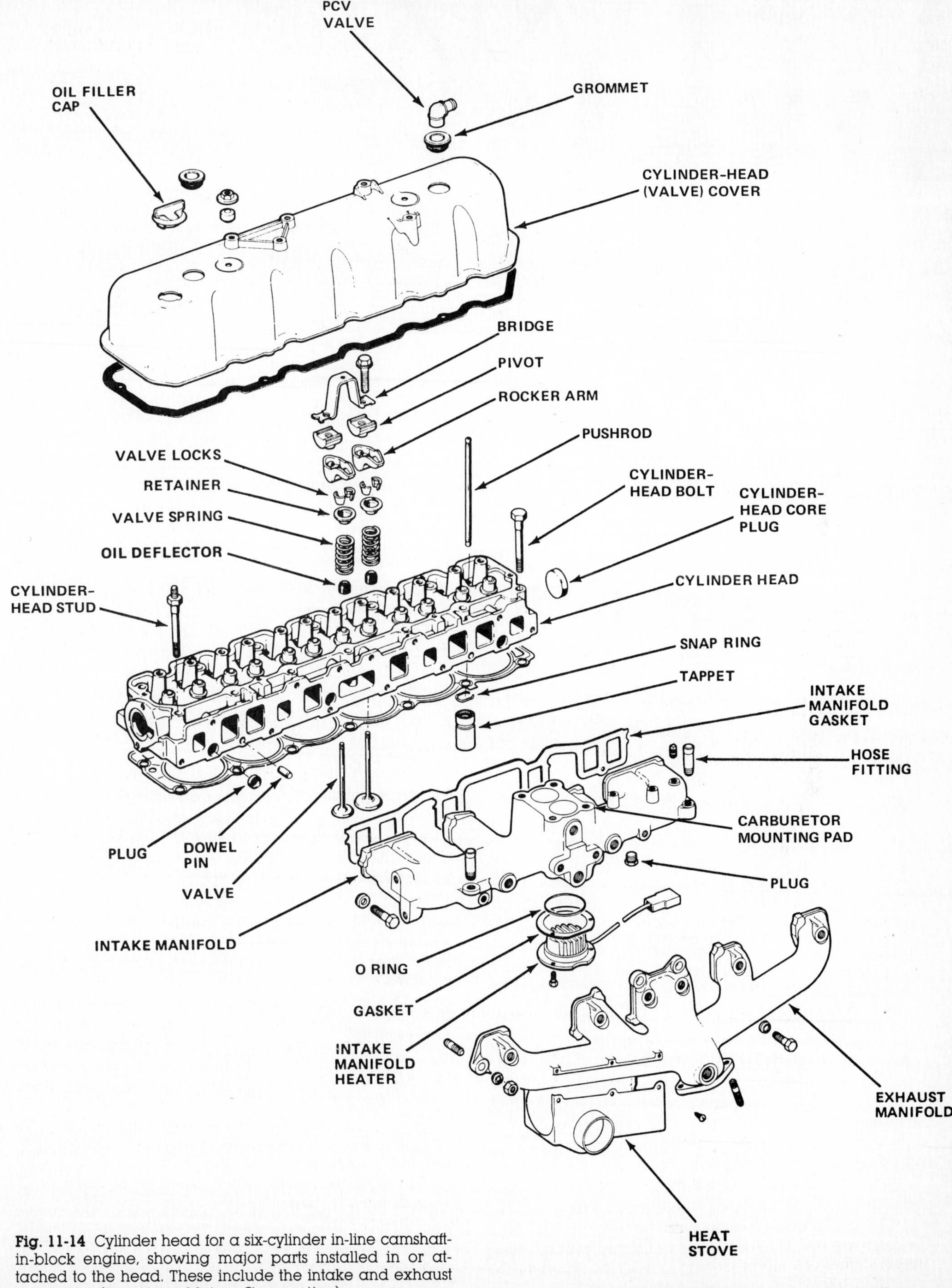

Fig. 11-14 Cylinder head for a six-cylinder in-line camshaft-in-block engine, showing major parts installed in or attached to the head. These include the intake and exhaust manifolds. *(American Motors Corporation)*

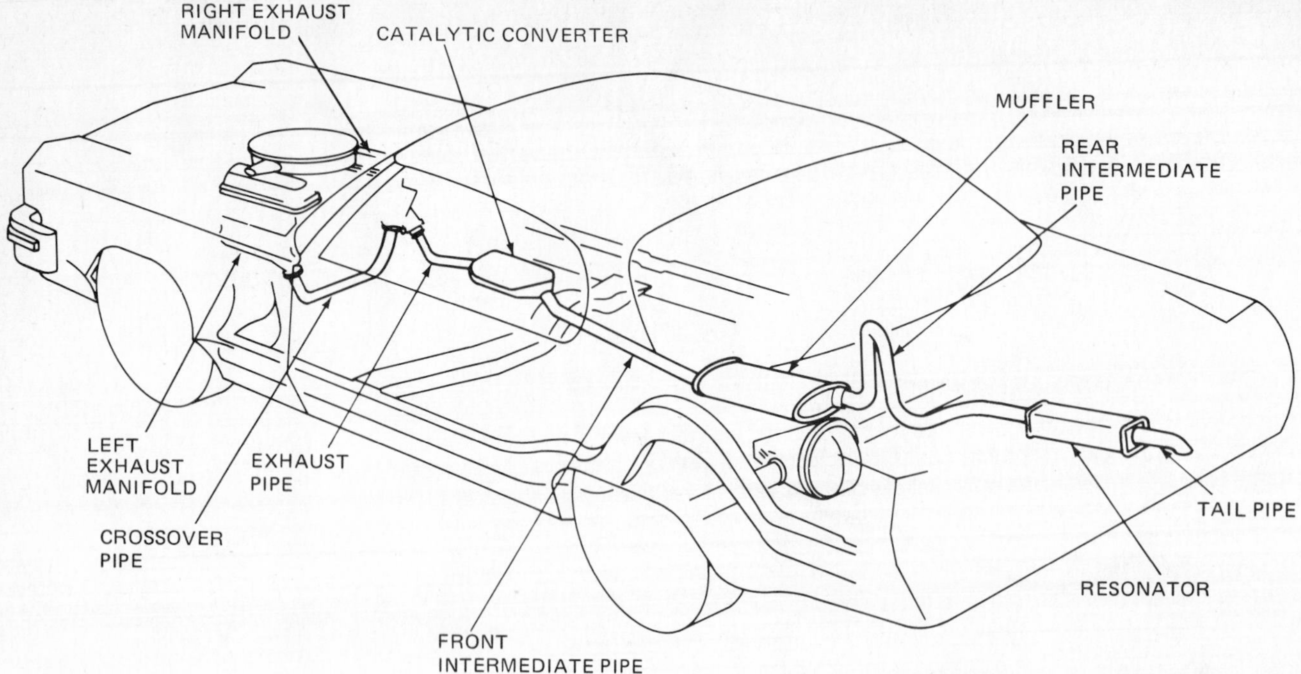

Fig. 11-15 Single exhaust system for a V-8 engine that uses a catalytic converter. *(Oldsmobile Division of General Motors Corporation)*

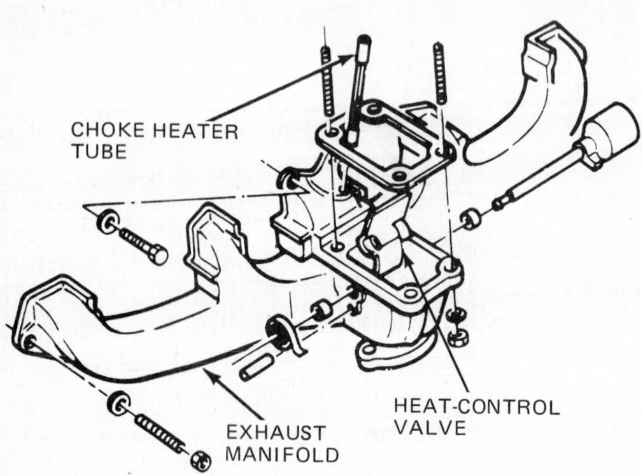

Fig. 11-16 Exhaust manifold for a six-cylinder in-line engine with heat-control valve disassembled.

♦ 11-15 CRANKSHAFT

The *crankshaft* is a strong one-piece casting or forging of heat-treated alloy steel (Figs. 11-1 and 11-2). The crankshaft must be strong enough to take the downward push of the pistons during the power strokes without excessive twisting. In addition, the crankshaft must be carefully balanced to eliminate undue vibration resulting from the weight of the offset cranks.

To provide balance, crankshafts have counterweights opposite the cranks. Crankshafts have drilled oil passages (Fig. 11-17) through which oil flows from the main to the connecting-rod bearings. Many V-6 engines have *splayed* crankpins. The word "splay" means to spread out. As applied to the crankpin, it means they are split into two parts. Each rod has its own crankpin (Fig. 11-18). Moving the crankpins apart this way reduces any out-of-balance condition.

The rear end of the crankshaft has the flywheel or drive plate attached to it (♦9-15). The front end of the crankshaft has three devices attached to it. These are the gear or sprocket that drives the camshaft, the vibration damper (♦11-16), and the drive-belt pulleys.

♦ 11-16 VIBRATION DAMPER

The power impulses tend to set up a twisting vibration in the crankshaft. When a piston moves down on its power stroke, it thrusts, through the connecting rod, against a crankpin with a force that may exceed 2 tons. This force tends to twist the crankpin ahead of the rest of the crankshaft. Then, as the force against the crankpin recedes, it tends to untwist, or move back into its original relationship with the rest of the crankshaft. This twist-untwist action, repeated with every power impulse, tends to set up an oscillating motion in the crankshaft. This is called *torsional vibration.* If it were not controlled, it could cause the crankshaft to break at certain speeds. To control torsional vibration, devices which are called *vibration dampers,* or *harmonic balancers,* are used. These dampers are usually mounted on the front end of the crankshaft and the drive-belt pulleys are incorporated into them.

A typical damper is made in two parts, a small inertia ring or damper flywheel and the pulley. They are bonded to each other by a rubber insert about ¼-inch [6-mm] thick (Fig. 11-19). The damper is mounted to the front end of the crankshaft. As the crankshaft speeds up or slows down, the damper flywheel has a dragging effect. This effect, which slightly flexes the rubber insert, tends to hold the pulley and crankshaft to a constant speed. This tends to check the twist-untwist action, or torsional vibration, of the crankshaft.

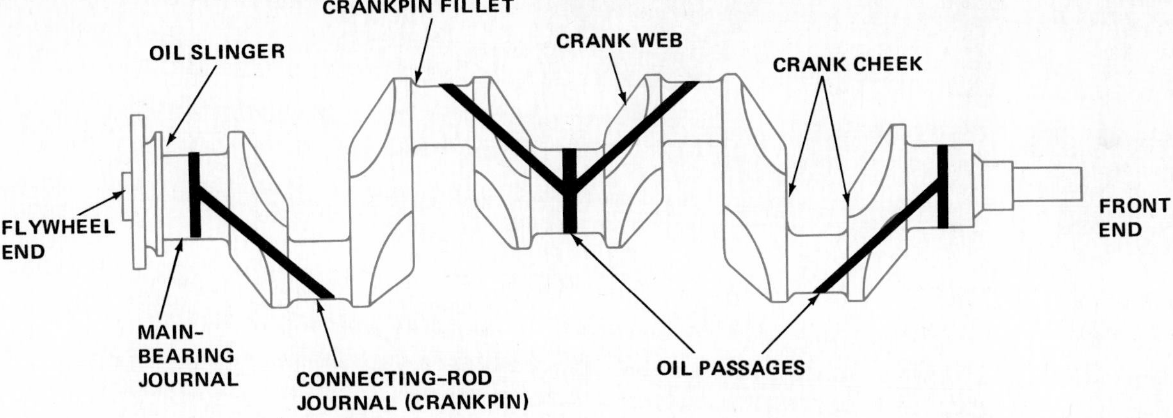

Fig. 11-17 The crankshaft has oilholes drilled through it to carry lubricating oil from the main bearings to the connecting-rod bearings. *(Johnson Bronze Company)*

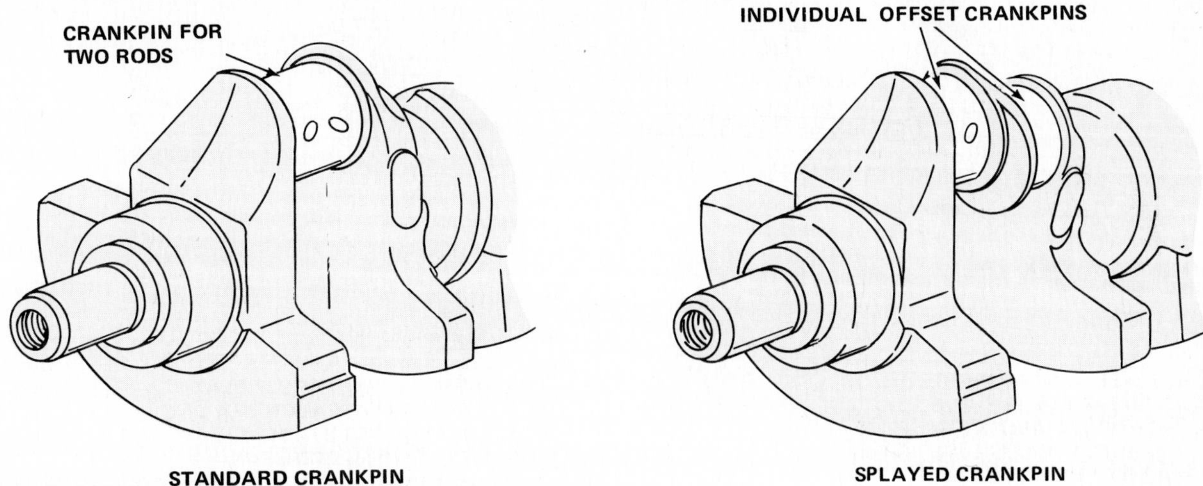

Fig. 11-18 Standard crankshaft for a V-6 engine (left) compared with a V-6 crankshaft with splayed crankpins (right). *(Buick Motor Division of General Motors Corporation)*

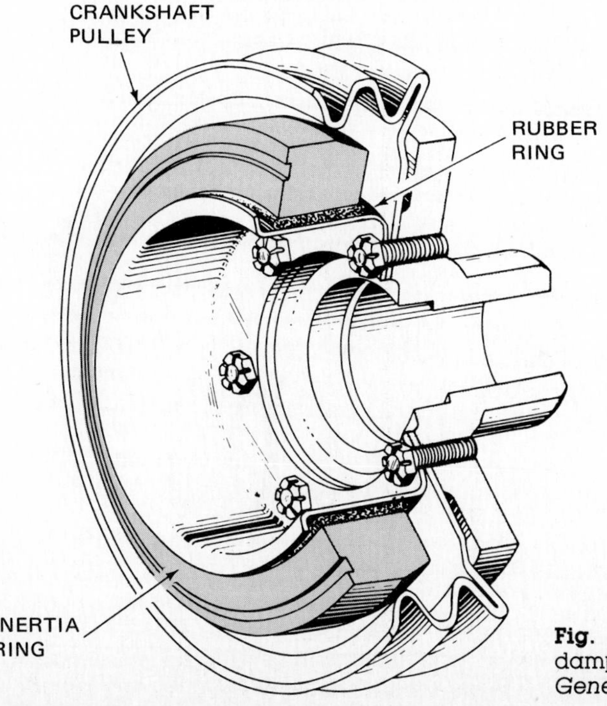

Fig. 11-19 A torsional-vibration damper. *(Pontiac Motor Division of General Motors Corporation)*

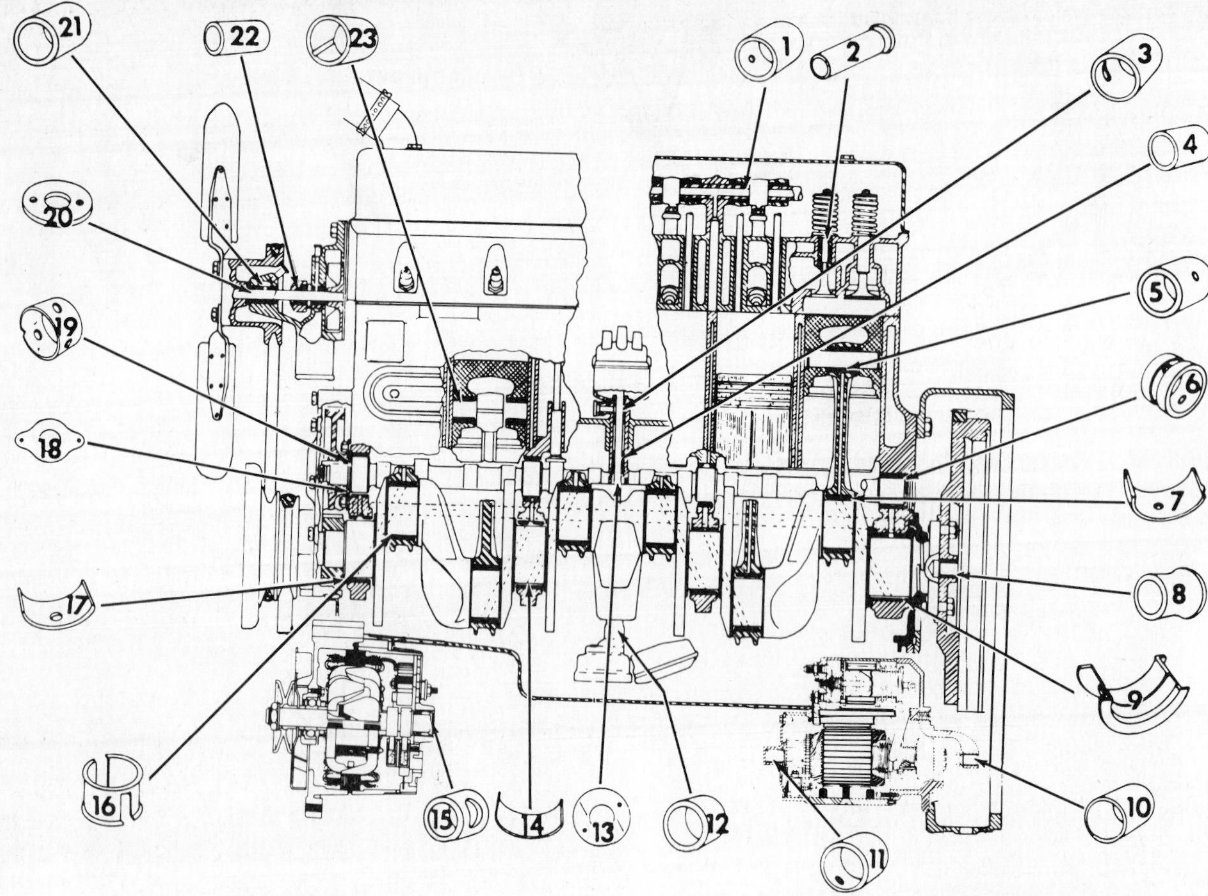

1. ROCKER-ARM BUSHING
2. VALVE-GUIDE BUSHING
3. DISTRIBUTOR BUSHING, UPPER
4. DISTRIBUTOR BUSHING, LOWER
5. PISTON-PIN BUSHING
6. CAMSHAFT BUSHINGS
7. CONNECTING-ROD BEARING
8. CLUTCH PILOT BUSHING
9. CRANKSHAFT THRUST BEARING
10. STARTING-MOTOR BUSHING, DRIVE END
11. STARTING-MOTOR BUSHING, COMMUTATOR END
12. OIL-PUMP BUSHING
13. DISTRIBUTOR THRUST PLATE
14. INTERMEDIATE MAIN BEARING
15. ALTERNATOR BEARING
16. CONNECTING-ROD BEARING, FLOATING TYPE
17. FRONT MAIN BEARING
18. CAMSHAFT THRUST PLATE
19. CAMSHAFT BUSHING
20. FAN THRUST PLATE
21. WATER-PUMP BUSHING, FRONT
22. WATER-PUMP BUSHING, REAR
23. PISTON-PIN BUSHING

Fig. 11-20 Bearings and bushings used in a typical pushrod spark-ignition engine. *(Johnson Bronze Company)*

♦ 11-17 ENGINE BEARINGS

Bearings are placed in the engine wherever there is rotary motion between engine parts (Fig. 11-20). These engine bearings are called *sleeve bearings* because they are shaped like sleeves that fit around the rotating shaft. The part of the shaft that rotates in the bearing is called a *journal.* Connecting-rod and crankshaft (also called *main*) bearings are of the split, or half, type (Figs. 11-1, 11-2, and 11-21). The upper half of a main bearing is installed in the counterbore in the cylinder block. The lower half is held in place by the bearing cap (Fig. 11-2). The upper half of a connecting-rod big-end (or crankpin) bearing is installed in the rod. The lower half is placed in the rod cap (item 22 in Fig. 11-2).

The typical bearing half is made up of a steel or bronze back, with up to five linings of bearing material. The bearing material is soft. Therefore, the bearing wears, and not the more expensive engine part. Then, the bearing, and not the engine part, can be replaced when it has worn too much.

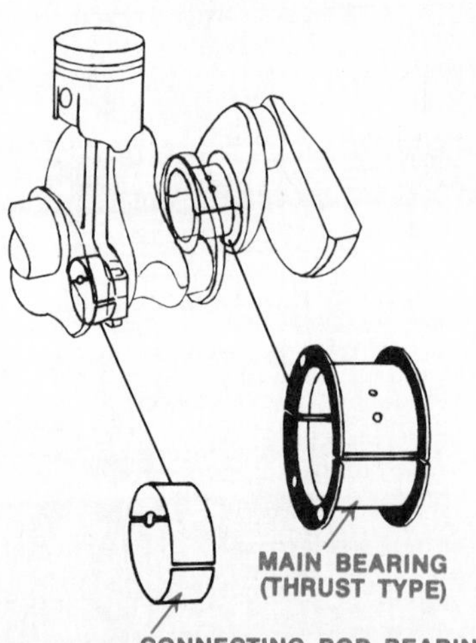

Fig. 11-21 A thrust-type main bearing and a connecting-rod bearing, showing their positions on the crankshaft.

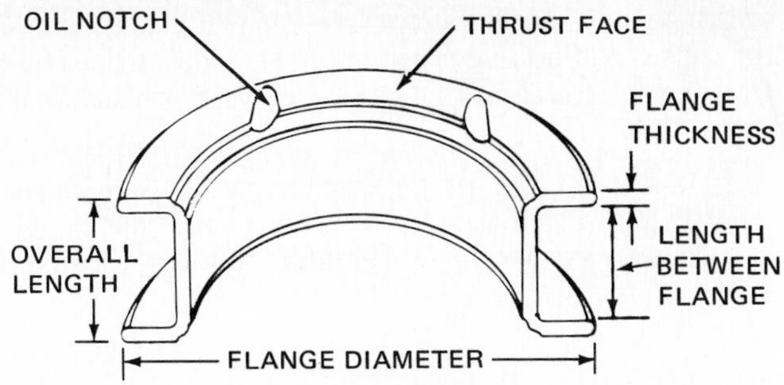

Fig. 11-22 A typical crankshaft thrust bearing half. *(McQuay-Norris Manufacturing Company)*

♦ 11-18 THRUST BEARING

The crankshaft has to be kept from moving back and forth in the block. To prevent back-and-forth movement, one of the main bearings is a thrust, or end-thrust, bearing. This bearing has flanges on its two sides, as shown in Figs. 11-21 and 11-22. In the engine, flanges on the crankshaft fit close to the flanges on the thrust bearing. If the crankshaft tends to shift forward or backward, the crankshaft flange comes up against the thrust-bearing flange. This prevents endwise movement.

♦ 11-19 BEARING LUBRICATION

Oil from the engine oil pump flows onto the bearing surfaces. The rotating shaft journals are supported on layers of oil. The journal must be smaller than the bearing (Fig. 11-23) so that there is a clearance (called *oil clearance*) between the two. In the engine, oil moves through this clearance.

The lubricating system feeds oil to the main bearings. It enters through the oilholes (Fig. 11-24), and the rotating journals carry it around to all parts of the bearings. The oil works its way to the outer edges of the main bearings. From there, it is thrown off and drops back into the oil pan. The oil thrown off helps lubricate other engine parts, such as the cylinder walls, pistons, and piston rings.

The connecting-rod bearings are lubricated through the oilholes drilled in the crankshaft (Fig. 11-17). As the oil moves across the faces of the bearings, it also helps to cool them. The oil is relatively cool as it leaves the oil pan. It picks

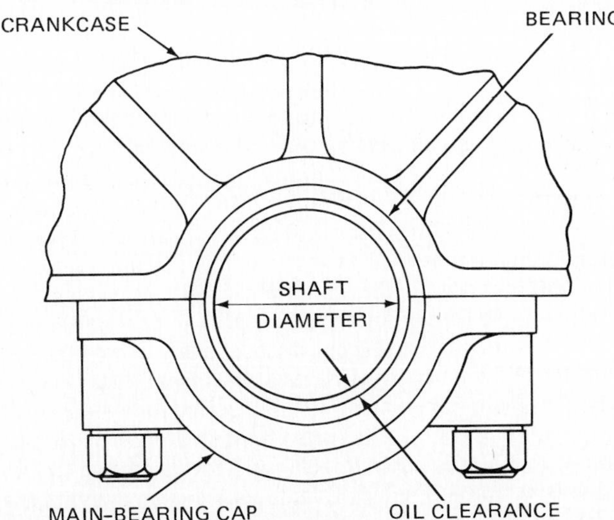

Fig. 11-23 Oil clearance (exaggerated) between the main bearing and the crankshaft journal. *(Clevite)*

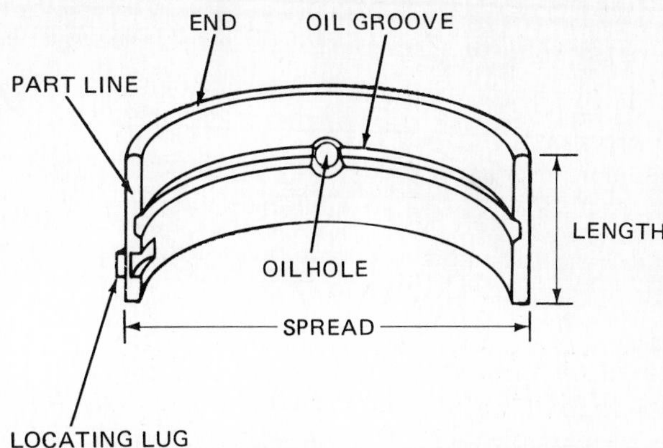

Fig. 11-24 Typical sleeve-type bearing half, with its parts named. *(McQuay-Norris Manufacturing Company)*

up heat in its passage through the bearings. This heat is carried down to the oil pan and released to the air around the oil pan. The oil also flushes and cleans the bearings. It flushes out particles of grit and dirt from the bearings. The particles are carried back to the oil pan by the oil. They then settle to the bottom of the oil pan, or are removed from the oil by the oil screen or filter.

♦ 11-20 BEARING OIL CLEARANCES

The greater the oil clearance (Fig. 11-23), the faster oil flows through the bearing. Proper clearance varies with different engines, but 0.0015 inch [0.037 mm] is a typical clearance. As the clearance becomes greater (owing to bearing wear, for example), the amount of oil flowing through and being thrown off increases. With a 0.003-inch [0.076-mm] clearance (only twice 0.0015 inch) [0.037 mm], the oil throwoff increases as much as five times. A 0.006-inch [0.152-mm] clearance allows 25 times as much oil to flow through and be thrown off.

As bearings wear, more and more oil is thrown onto the cylinder walls. The piston rings cannot handle so much oil. Part of it works up into the combustion chambers, where it burns and forms carbon. Carbon deposits in the combustion chambers reduce engine power and cause other engine troubles as explained in a later chapter.

Excessive oil clearances can also cause some bearings to fail from oil starvation. An oil pump can deliver only a certain amount of oil. If the oil clearances are excessive, most of the oil will pass through the nearest bearings. There won't be enough for the more distant bearings. Then these will probably fail from lack of oil. An engine with excessive bearing oil clearances usually has low oil pressure: The oil pump cannot build up normal pressure because of the large oil clearances in the bearings.

If bearing oil clearances are too small, there will be metal-to-metal contact between the bearing and the shaft journal. Very rapid wear and quick failure will result. Also, there will not be enough oil throwoff to lubricate cylinder walls, pistons, and rings.

♦ 11-21 BEARING REQUIREMENTS

Bearings must be able to do other things besides carry loads. Some of these are listed below.

1. Load-Carrying Capacity Modern engines are

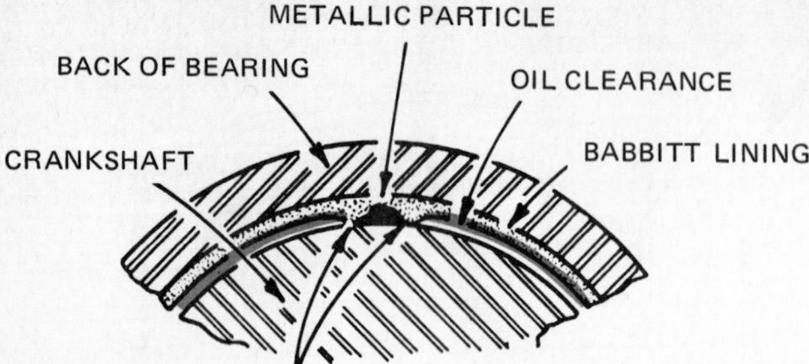
METALLIC PARTICLE
BACK OF BEARING
OIL CLEARANCE
CRANKSHAFT
BABBITT LINING
BABBITT DISPLACED BY PARTICLE AND RAISED UP AROUND IT, GREATLY REDUCING OR DESTROYING THE OIL CLEARANCE LOCALLY.

Fig. 11-25 Effect of a metallic particle that is embedded in the bearing material. *(Federal-Mogul Corporation)*

lighter and more powerful. They have higher compression ratios which impose greater bearing loads. Only a few years ago, bearing loads were around 1600 to 1800 psi [11,032 to 12,411 kPa]. Today, connecting-rod bearings carry loads of up to 6000 psi [41,369 kPa].

2. Fatigue Resistance When a piece of metal is bent back and forth, over and over, it hardens and finally breaks. This is called *fatigue failure.* You have probably done this with a piece of wire or sheet metal. Bearings are subject to such loads and must withstand them without failing from fatigue.

3. Embeddability This term refers to the ability of a bearing to permit foreign particles to embed in it. Dirt and dust particles enter the engine despite the air cleaner and oil filter. Some of them work onto the bearings and are not flushed away by the oil. A bearing protects itself by letting such particles sink into, or embed in, the bearing lining material. If the bearing were too hard to allow this, the particles would lie on the surface. They would scratch the shaft journal and probably gouge out the bearing. This would cause overheating and rapid bearing failure. Therefore, the bearing material must be soft enough for adequate embeddability.

4. Conformability This is associated with embeddability. It is the ability of the bearing material to conform to variations in shaft alignment and journal shape. For example, suppose that a shaft journal is slightly tapered. The bearing under the larger diameter will be more heavily loaded. If the bearing material has high conformability, it will "flow" slightly, from the heavily loaded areas to the lightly loaded areas. This slight flow evens the load on the bearing. A similar action takes place when foreign particles embed in the bearing. As they embed, they displace bearing material, producing local high spots (Fig. 11-25). However, with high conformability, the material flows away from the high spots. This prevents local heavy loading that could cause bearing failure.

5. Corrosion resistance The by-products of combustion may form corrosive substances, harmful to some metals. Bearing materials must be resistant to corrosion. Unleaded gasoline, required on cars using catalytic converters, changes the chemistry of the engine oil. Catalytic converters, discussed in later chapters, are installed in the exhaust systems to reduce the pollutants coming out the tail pipe. The unleaded gasoline, in changing the chemistry of the oil, tends to increase bearing corrosion. Therefore, the composition of engine bearings has been changed. For example, instead of the copper-lead bearings used for years, some engines now have aluminum-lead bearings. These appear to withstand corrosion better.

6. Wear Rate The bearing material must be so hard and tough that it will not wear too fast. At the same time, it must be soft enough to permit good embeddability and conformability.

♦ 11-22 ENGINE MOUNTS

Engines are usually attached to the engine frame in three places. Some engines have four attachments, or *mounts.* The mounts must not only provide flexible supports for the engine, they must also absorb engine noises and vibration. Otherwise, these would be carried up to the driver and passengers through the car frame and body. Also, vibration would be carried to the radiator, electric relays, and electronic controls and they could be damaged.

The engine mounts are somewhat flexible to permit the car frame or body to twist, as it does during normal operation, without putting any strain on the engine block. Figures 11-26 and 11-27 show typical engine mounts. Figure 11-28 shows two of the three mounts for a transverse engine. The third mount is attached between the side of the engine and a reinforced section of the body. Engine mounts have pads of rubber or similar material which are flexible to absorb noise and vibration.

♦ 11-23 CONNECTING ROD

The connecting rod (Fig. 11-29) is attached at one end to a crankpin on the crankshaft. It is attached at the other end to a piston, through a piston pin (sometimes called a *wrist pin*). The connecting rod must be very strong and rigid, and as light as possible. The connecting rod carries the power thrusts from the piston to the crankpin. At the same time, the rod is in eccentric, or off-center, motion (Fig. 9-11). To minimize vibration and bearing loads, the rod must be light in weight.

The crankpin end of the rod (the big end) is attached to the crankpin by the rod cap and bolts (Fig. 11-29). A split-type bearing is installed between the crankpin and the rod and rod cap.

The piston end of the rod (the small end) is attached to the piston by means of a piston pin. The pin passes through bearing surfaces in both the piston and the connecting rod. There are five ways to connect the rod and piston with the piston pin (Fig. 11-30). The most common method is to press-fit the pin to the connecting rod (Fig. 11-30D). The press fit is tight enough to keep the pin from moving out of position. When an aluminum piston is used, as shown in Fig. 11-30D, the piston pin rests directly on the aluminum of the piston, with no intervening bushing. Another method is to provide the rod with a bushing so that the pin can float freely in both the rod bushing and the bearing surfaces in the piston (Fig. 11-30A). The pin is kept from moving out and scoring the cylinder wall by a pair of lock rings. The rings fit into undercuts or grooves in the piston. The other three arrangements, shown in Fig. 11-30B, C, and E, use lock bolts to lock the pin to either the rod or piston. These methods are seldom used today.

To provide connecting-rod big-end bearing lubrication, oil travels the following path: The oil pump sends oil to oil lines

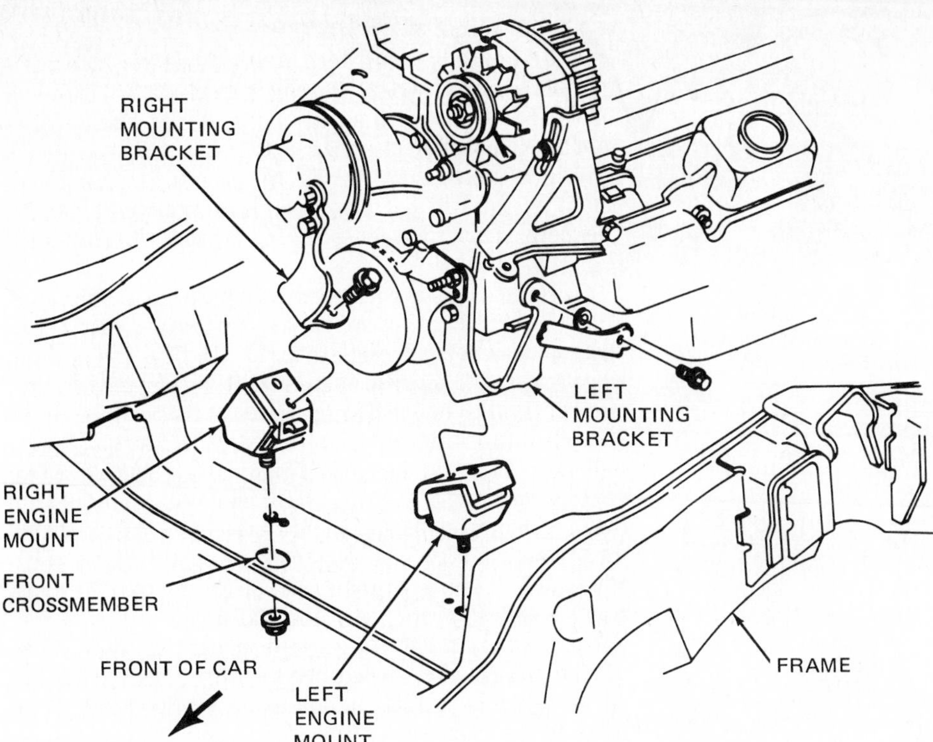

Fig. 11-26 Front engine mounts that insulate the engine from the frame. *(Cadillac Motor Car Division of General Motors Corporation)*

or galleries in the cylinder block. From these oil lines, the oil moves to the main bearings. From the main bearings, oil travels through oil passages drilled in the crankshaft (Fig. 11-17). Finally, the oil moves through the connecting-rod-bearing oilholes onto the bearing surfaces.

On many engines, the piston pins are lubricated by oil scraped from the cylinder walls. As discussed later in this chapter, the piston rings scrape excess oil from the cylinder walls. Some of this oil flows through slots or holes in the piston to lubricate the piston pin.

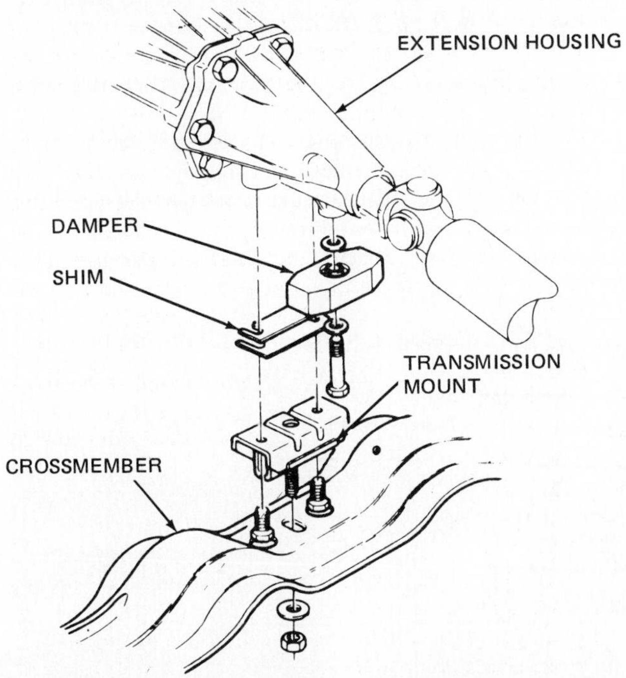

Fig. 11-27 Location of the rear engine mount, or transmission mount, between the extension housing of the transmission and the cross member of the frame. *(Buick Motor Division of General Motors Corporation)*

Some connecting rods have oil passages drilled from the connecting-rod big end to the piston-pin bushing. The oil gets to the piston-pin bushing in the connecting rod by the following path: From the oil pump, oil travels through oil lines in the block, and through oil passages in the crankshaft, to the connecting-rod big-end bearings. The oil then moves through the oil passage in the connecting rod to the piston-pin bushing. This arrangement has been used in engines that have free-floating pins in the connecting-rod bushings.

Fig. 11-28 On the transverse engine, the weight is supported by a mount at each end of the engine. *(Chrysler Corporation)*

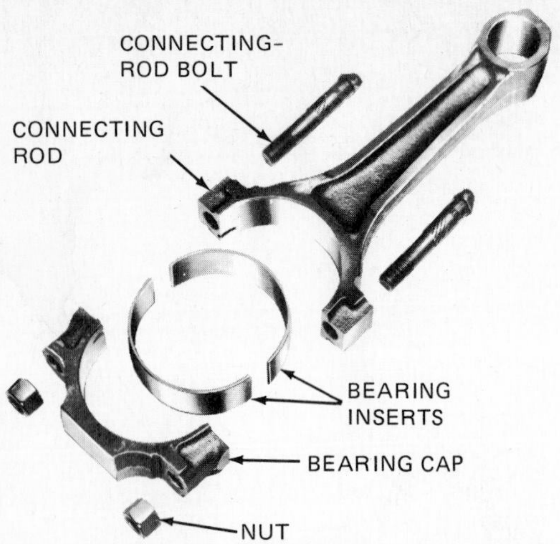

Fig. 11-29 A disassembled connecting-rod assembly, showing the bearing and cap. *(Cadillac Motor Car Division of General Motors Corporation)*

On many V-type engines, cylinder walls and piston pins are lubricated by oil jets from opposing connecting rods. Each rod has a groove or hole that lines up with an oil-passage hole in the crank journal every crankshaft revolution. When this happens, a jet of oil spurts into the opposing cylinder in the other cylinder bank.

Rod caps and rods are carefully matched during original assembly. After a cap is installed on a rod, the bearing bore is machined. That means the cap will fit that rod and only that rod. Caps must not be interchanged between rods during service jobs. If a cap goes on the wrong rod, the assembly will not fit the crank on the crankshaft. If a cap is lost or damaged, then a new or reconditioned cap-rod assembly is required.

♦ 11-24 PISTONS AND PISTON RINGS

The *piston* (Fig. 9-7) is a cylindrical plug that moves up and down in the engine cylinder. It has piston rings to provide a good seal between the cylinder wall and piston. Although the piston appears to be a simple part, it is actually quite a complex design. The piston works with the piston rings to contain the compression and combustion pressures within the cylinder. But before we discuss pistons, let us examine piston rings.

♦ 11-25 PISTON RINGS

A good seal must be maintained between the piston and cylinder wall to prevent blowby. *Blowby* is the name that describes the escape of unburned air-fuel mixture and burned gases from the combustion chamber, past the pistons, and into the crankcase. These gases "blow by" the pistons. The piston cannot fit in the cylinder closely enough to prevent blowby. Therefore, piston rings must be used to provide the necessary seal.

The rings are installed in grooves in the piston, as shown in Fig. 9-7. Actually, there are two types of rings, compression rings and oil-control rings (Fig. 11-31). The compression rings seal in the air-fuel mixture as it is compressed. They also seal in the combustion pressures as the mixture burns. The oil-control rings scrape excessive oil from the cylinder wall and return it to the oil pan. Figure 11-32 shows a typical set of piston rings. The rings have joints (they are split) so they can be expanded and slipped over the piston head and into grooves in the piston. Rings for automotive engines usually have butt joints. In some heavy-duty engines, the joints may be angled, lapped, or of the sealed type.

The rings are slightly larger in diameter than they will be when in the cylinder. Then, when the rings are installed, they are compressed so that the joints are nearly closed. Compressing the rings puts them in tension. Then they press tightly against the cylinder wall.

The distance between the ends of the ring when it is compressed in the cylinder is called the *ring gap* (Fig. 11-33). Figure 11-33 also shows other important ring and ring-groove measurements. Some of these you will check when you service pistons and rings.

♦ 11-26 COMPRESSION RINGS

Compression rings are usually made of cast iron. This material has good wearing qualities. It also provides good ring tension, or force, on the cylinder walls. Some heavy-duty rings are made of *ductile iron.* These rings are more flexible than cast iron. Ductile-iron rings are widely used in racing, diesel, turbocharged, and other high-performance engines.

Compression rings, besides reducing blowby, also help control oil. On the intake stroke, the compression rings scrape all but a fine film of oil from the cylinder walls. Then, on the exhaust stroke, the rings are moving upward and tend to "skate" over the oil film.

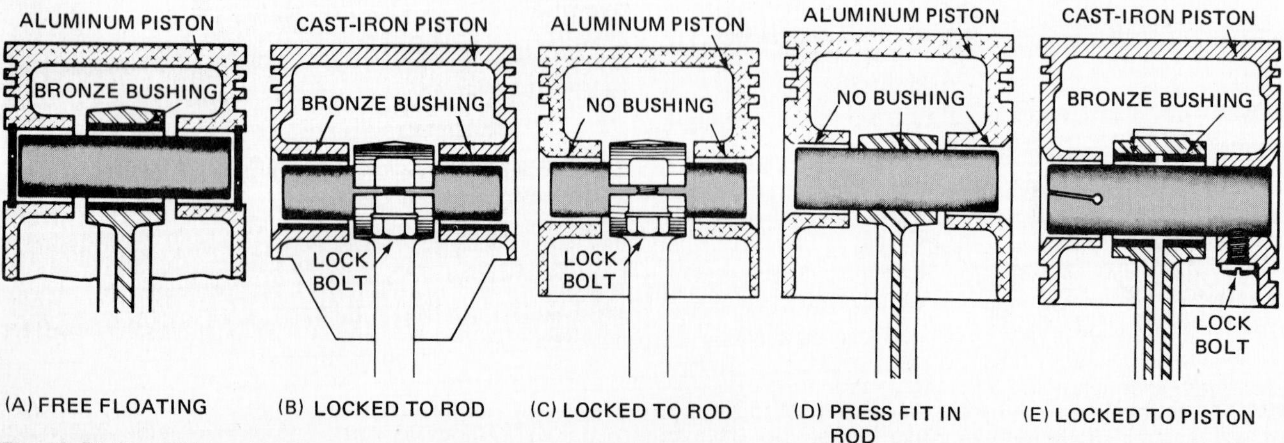

Fig. 11-30 Five piston-pin arrangements. *(Sunnen Products Company)*

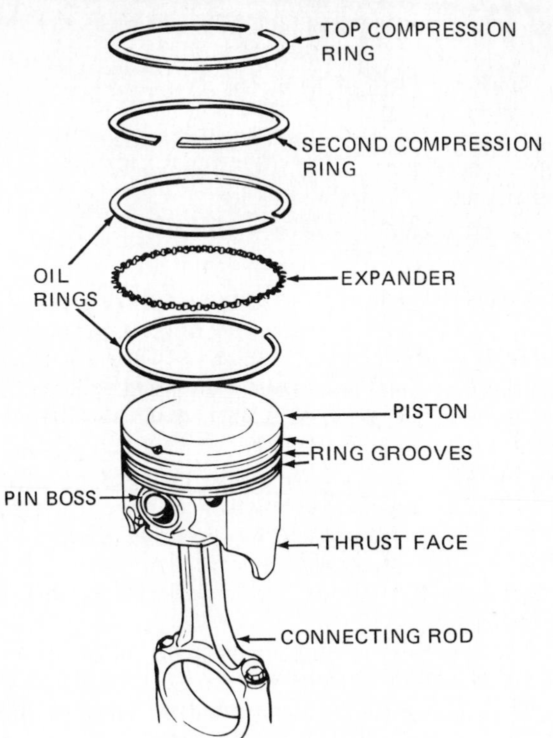

Fig. 11-31 Piston and connecting-rod assembly. Piston rings are shown above the piston. *(Buick Motor Division of General Motors Corporation)*

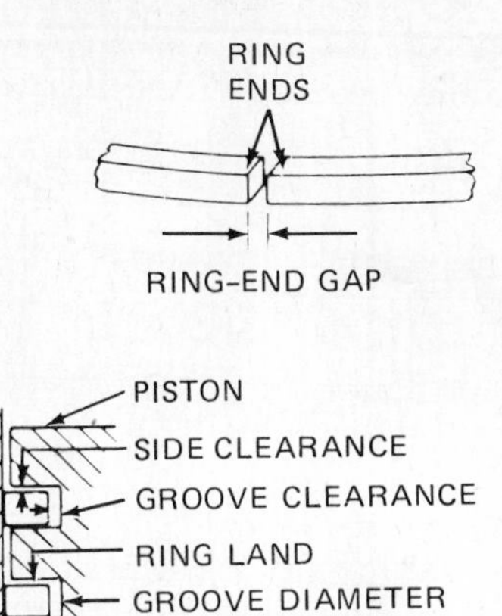

Fig. 11-33 Top, the ring end gap in a piston ring with a butt joint. Bottom, clearances between the installed ring and the ring groove in the piston. *(Sealed Power Corporation)*

♦ 11-27 COMPRESSION-RING SHAPES

Compression rings are made in a variety of shapes. The reverse twist and the barrel face are two of the latest which have certain advantages over earlier rings.

Figure 11-34 shows the action of the reverse-twist ring as compared to a positive-twist ring (an earlier-type ring). At wide-open throttle, when combustion pressures are the greatest, both are forced down flat against the lower sides of the grooves by these pressures. At part throttle, they do not flatten as much. During the intake stroke, vacuum above the piston tries to draw oil up the cylinder wall, past the piston rings. The reverse-twist ring (left in Fig. 11-34) prevents this. That is because the twist pushes the lower side of the ring

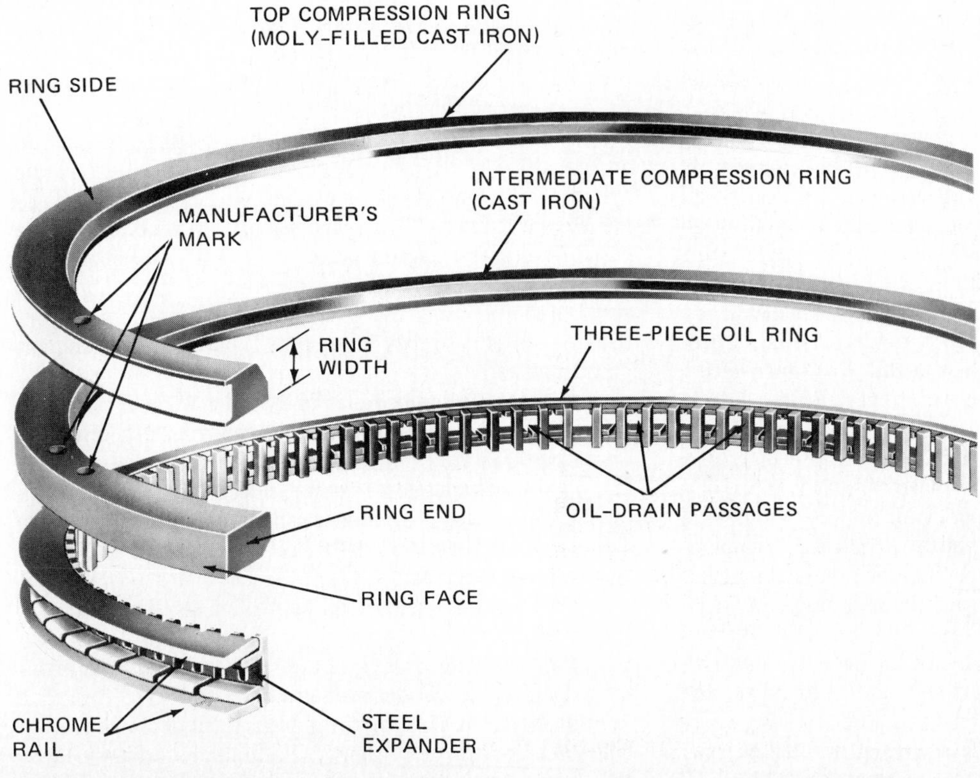

Fig. 11-32 A typical set of piston rings. *(TRW, Inc.)*

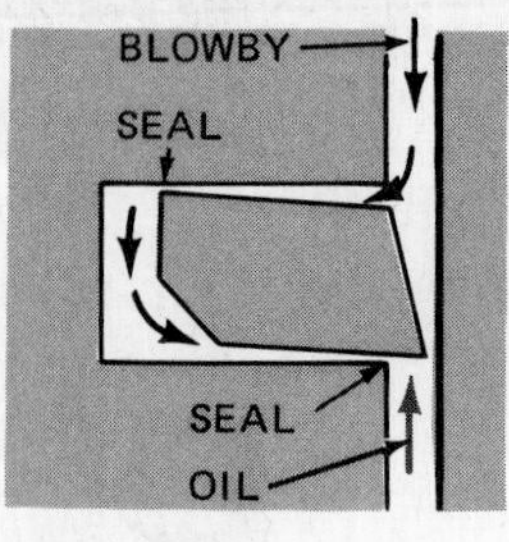

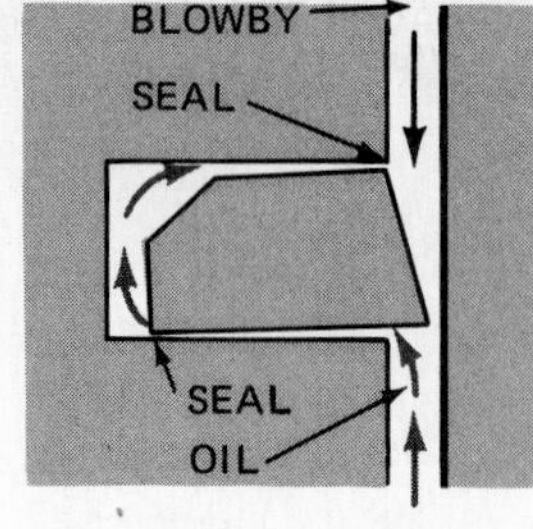

Fig. 11-34 Action of a taper-face compression ring with (left) reverse twist and (right) positive twist. *(Chevrolet Motor Division of General Motors Corporation)*

against the edge of the ring groove to produce a seal. The positive-twist ring (right in Fig. 11-34) is twisted the other way so that some oil can work through under the ring.

The barrel-face ring has a slightly rounded face. The curvature is so slight you cannot see it with the naked eye. The ring has only a narrow line of contact with the cylinder wall so the pressure against the wall is relatively high. This ring rocks slightly as it changes direction at the top of the cylinder (end of compression and exhaust strokes). But the line of contact is not lost, so the force of the ring on the cylinder wall is maintained. This provides good compression control. At the same time, the rocking action reduces the formation of the ring ridge at the top of the cylinder. An explanation of how the ring ridge is formed is given in ♦11-32.

♦ 11-28 COMPRESSION-RING COATINGS

Coatings of various substances have been applied to compression rings to help wear-in and prevent rapid wear of rings and cylinder walls. "Wear-in" means that when new, the rings and cylinder wall have slight irregularities so the fit is not perfect. However, after a time, these irregularities are worn away so the fit is improved. Relatively soft materials such as phosphate, graphite, and iron oxide, which wear rapidly, help the wear-in. Also, they hold some oil so that improved lubrication results.

Another factor to consider in selecting compression-ring coatings is the type of wear the rings will meet in service. At one time, the most common type of ring and cylinder-wall wear was caused by abrasives that worked their way through the air filter and got in the engine oil. These abrasives—fine dust particles—would deposit on the cylinder walls and cause abrasive wear of the walls and rings. Where abrasive wear is a problem, compression rings are chromium-plated. Chromium (or "chrome") is a very hard metal. It is finished to a very smooth surface so the rings cause very little cylinder-wall wear.

Abrasive wear is less important today because of greatly improved air-filtering systems. Now scuff wear is more of a problem. In the modern engine, cylinder walls, rings, and pistons operate at higher temperatures. These high temperatures increase the possibility that the rubbing metal surfaces will form local hot spots that reach the melting point. Then small-area welds can develop. These welds are most likely to occur only when the piston is momentarily at rest at TDC. The slightest movement of the piston and rings breaks the welds, leaving rough, scuffed spots and surfaces. Rings, pistons, and cylinder walls will be scratched and failure may soon occur.

To combat scuff wear, high-temperature ring coatings such as molybdenum are used. Iron, from which the compression rings are made, melts at about 2250°F [1233°C]. Chromium melts at about 3450°F [1898°C]. However, molybdenum has to be heated to 4800°F [2449°C] before it melts. Therefore, compression rings coated with molybdenum can be worked at higher temperatures without danger of scuff wear.

♦ 11-29 NEED FOR TWO RINGS

Two compression rings are used to reduce the pressure on each one. At the start of the power stroke, pressures in the combustion chamber may go as high as 1000 psi [6895 kPa]. Pressure in the crankcase is about atmospheric. Therefore, the pressure difference could be about 1000 psi. It would be difficult for a single compression ring to hold this much pressure. However, with the second ring, the pressure is, in effect, divided between the two. This does more than reduce blowby, or loss of pressure past the upper ring. It also reduces the load on the upper ring, so it does not press quite as hard on the cylinder wall. Ring friction and cylinder and ring wear are reduced.

The two compression rings are different. Compression and combustion pressures get behind the top compression ring and push it out against the cylinder wall and down against the lower side of the ring groove. This improves the sealing. However, the pressure reaching the lower compression ring is often not great enough to produce this seal. Therefore, this lower compression ring is often of the twist type to improve sealing. In some installations, an expander or inner tension ring is installed behind the lower compression ring to improve its sealing force against the cylinder wall.

♦ 11-30 OIL-CONTROL RINGS

The oil-control ring or rings prevent oil from working up into the combustion chamber. Oil throwoff from the main and rod bearings lubricates the cylinder walls, pistons, and rings. Some connecting rods have an oil-spit hole which spits oil onto the cylinder wall every time it lines up with the oilhole in the crankpin. Most of the time, more oil is thrown onto the cylinder walls than is needed. Most of it must be scraped off and returned to the oil pan. If too much oil is left on the cylinder walls, it works its way up into the combustion chamber, where it burns. This increases oil consumption, so the engine requires extra oil quite often. Also, the burned oil fouls the spark plugs, hampers the action of the compression rings, and increases the possibility of engine detonation. (Detonation is described later in the chapter diagnosing engine troubles.)

The oil that is scraped off the cylinder walls does several things. It carries away particles of carbon, dust, and dirt. These particles are then removed by the oil screen or filter. The oil also provides some cooling effect. In addition, the oil on the rings helps provide a seal between the rings and the cylinder wall. Therefore, as the oil circulates, it lubricates, cleans, cools, and seals.

There are four types of oil-control rings shown in Fig. 11-35. One-piece oil-control rings are usually installed with an expander spring in back of them. Then they have greater force against the cylinder wall and improved oil control. The multiple-piece oil-control rings have expander springs that help force the rails out against the cylinder wall. Figure 11-36 shows how this works. The rails are forced upward, down-

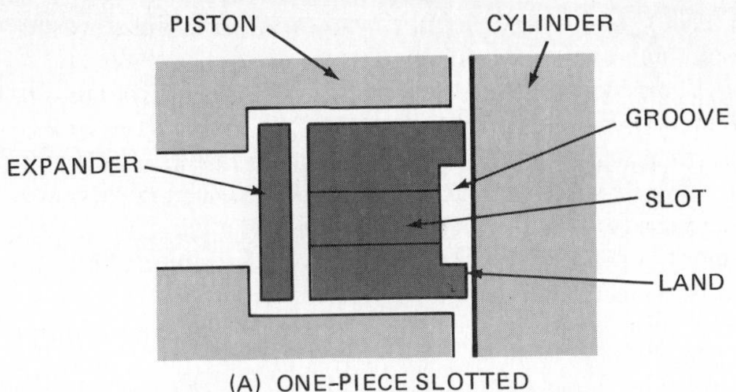

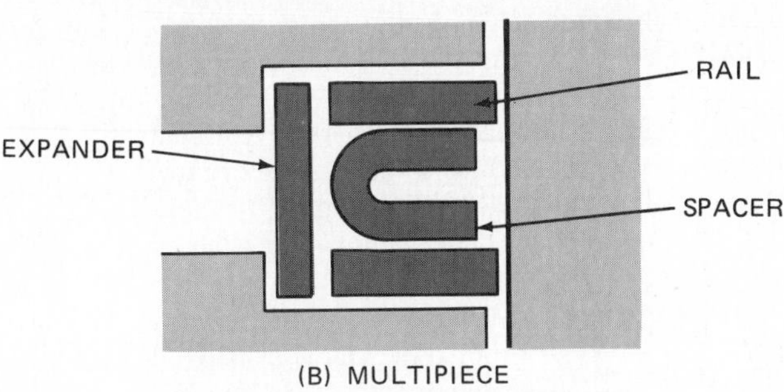

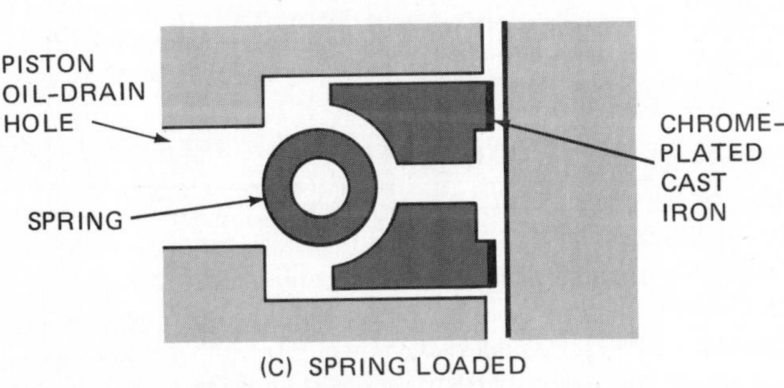

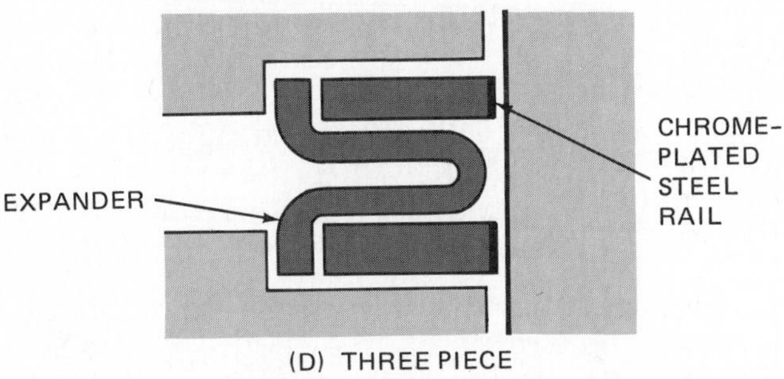

Fig. 11-35 Various types of oil rings. *(Perfect Circle Division of Dana Corporation)*

ward, and outward to provide good sealing on the cylinder walls and also on the sides of the ring groove.

Ring grooves for oil-control rings have drain holes or slots which allow the oil the rings scrape off the cylinder walls to drain to the inside of the piston. There, the oil drops off and falls down into the oil pan. Some lubricates the piston pin, as previously explained.

♦ 11-31 EFFECT OF SPEED ON OIL CONTROL

As engine speed increases, the oil-control rings have a harder time controlling the oil and preventing excessive amounts

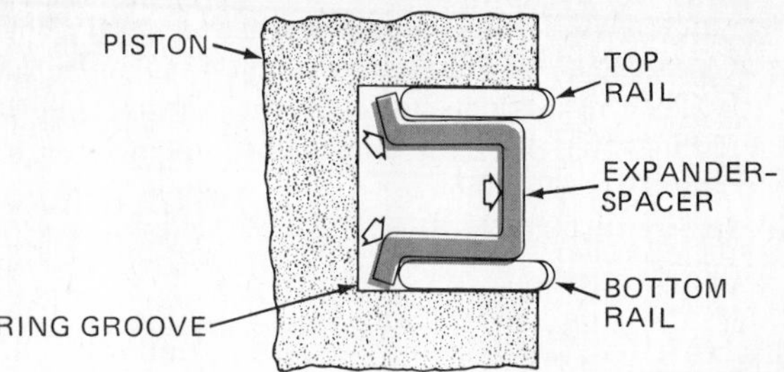

Fig. 11-36 Action of the expander spacer. As shown by the arrows, it forces the rails out against the cylinder wall and up and down against the sides of the ring groove. *(Perfect Circle Corporation)*

from passing them. There are several reasons for this. The engine and engine oil are hotter. Hot oil is thinner and can pass the rings more easily. More oil is pumped at high speed, so that more oil is thrown onto the cylinder walls. This means the oil-control rings have a harder job to do. And they have less time to do it. Therefore, at high speed, more oil gets past the rings and is burned in the combustion chamber. This greatly increases oil consumption. An engine may use two or three times as much oil at high speed as at low speed. Much, but not all, of this is due to the reduced effectiveness of the oil-control rings at high speeds.

♦ 11-32 EFFECT OF ENGINE WEAR ON OIL CONTROL

As engine miles pile up, parts wear. The cylinder walls wear unevenly. They wear more at the top, where the combustion pressures are the greatest. The higher the combustion pressure, the harder the rings are pressed against the cylinder walls. The maximum pressure and wear at the top of the cylinder wall leaves a ring ridge, just above the upper limit of ring travel (Fig. 11-37).

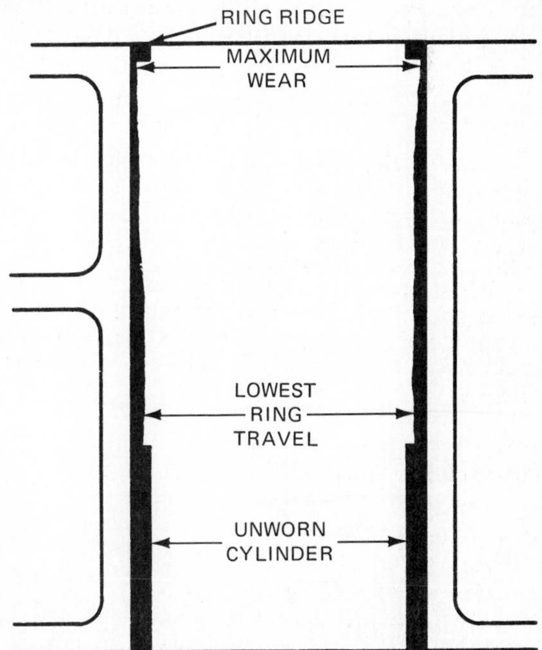

Fig. 11-37 Wear of a cylinder due to movement of the rings on the wall. The amount of taper is the difference between the cylinder diameter at the point of maximum wear and the diameter of the unworn part of the cylinder.

The uneven wear from top to bottom of the ring-travel area makes it harder for the rings to maintain good contact with the wall. The rings have to change in size as they move from one limiting position to the other. The result is they do a poorer job of scraping the oil off the cylinder wall. More oil gets up into the combustion chamber, where the oil is burned.

♦ 11-33 OIL CONSUMPTION

The amount of oil that an engine uses depends on engine speed and engine wear. All engines burn some oil. A new engine operated at moderate speed would probably not require any extra oil between oil changes. An old engine operated at high speed could require a quart of oil every few hundred miles.

♦ 11-34 REPLACEMENT RINGS

As cylinder walls wear, power is lost, and oil is burned in the combustion chambers. There comes a time when the engine is losing so much power and is burning so much oil that repair is required. This means that the cylinder head must come off so that the cylinder bores can be checked to determine their condition. Mainly, you should be interested in the amount of *taper*. This is the difference in diameters at the upper and lower limits of ring travel (Fig. 11-37). Depending on the amount of taper, you can install a set of new standard rings, install a set of high-tension rings (*severe* rings), or service the cylinders by boring them. These procedures are covered in a later chapter on engine service.

♦ 11-35 PISTONS

Pistons undergo great stress. During the power stroke, up to 4000 pounds [1814 kg] is suddenly applied to the piston head. This happens 30 to 40 times a second to each piston at highway speed. Temperatures above the piston head reach 4000°F [2204°C]. Today, pistons are designed to take these stresses for thousands of miles.

The piston must be strong. But it must be light, too, to reduce inertia loads on the bearings. Inertia is a property of all material things. Any object in motion resists any effort to change its speed or direction of travel. However, the piston is constantly changing speed and reversing directions as it moves up and down. How this affects bearing wear is described in the servicing chapters.

Pistons are made as light as possible. Because aluminum is a light metal, modern automotive engines use aluminum alloy pistons (Fig. 11-38) of the full-slipper type. For automotive engines, they vary from about 3 to 4 inches [76 to 102 mm] in diameter and weigh about 1 pound [454 g]. All pistons in an engine must be of the same weight. Otherwise, the engine would be unbalanced. Then it would vibrate excessively. To simplify service, when oversize replacement pistons are required (after honing or boring the cylinder), the new pistons weigh the same as the original pistons. Therefore, engine balance is not upset if only part of the pistons are replaced.

During manufacture, pistons are often *plated*. They are given a thin coat of tin or other material. This plating helps prevent scuffing when the engine is first started and during break-in. Scuffing is a type of wear that occurs when parts slide against each other without lubrication. It shows up as pits and grooves in the mating surfaces.

♦ 11-36 PISTON SHAPES

Pistons have been made in many shapes. The old-style small-bore, long-stroke engines used full-skirt pistons. Then, as lower hood lines and short-stroke engines became popular, the slipper piston came into use (Fig. 11-38). The cutaway skirt on the slipper piston saved weight. Also, it made room for the counterweights on the crankshaft (Fig. 11-39). In addition, the slipper-piston design helped control heat expansion of the piston (♦11-38).

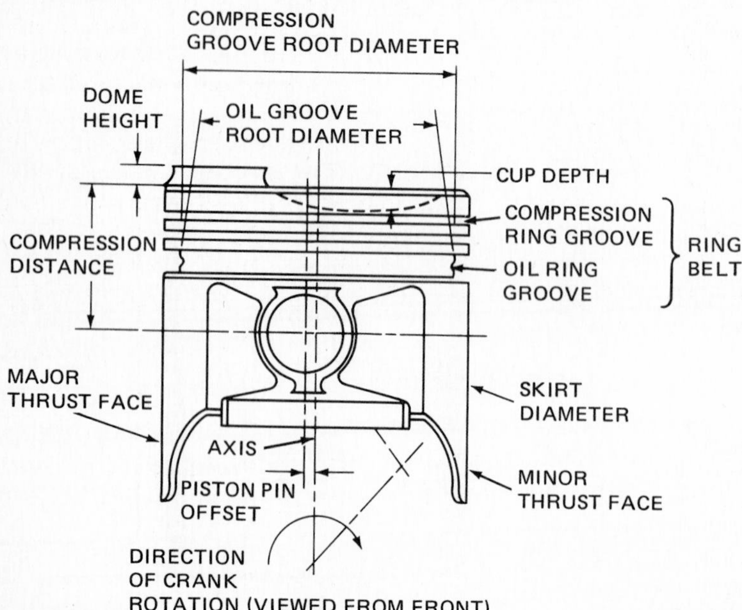

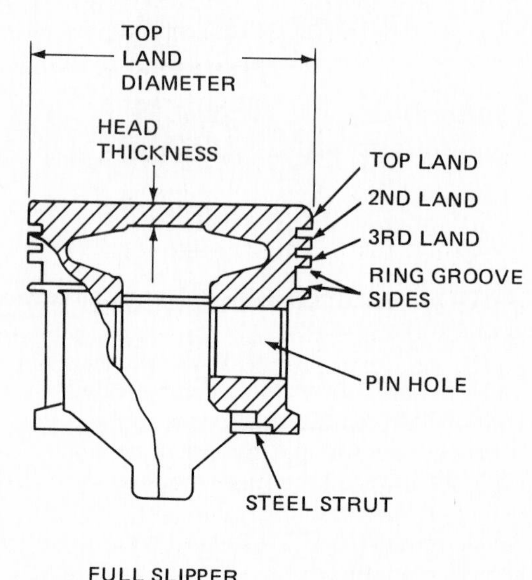

Fig. 11-38 Slipper piston with parts named.

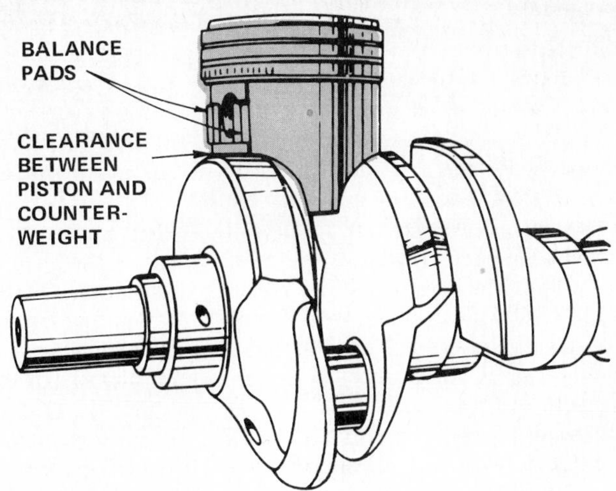

Fig. 11-39 Slipper piston and connecting rod assembled to the crankshaft. Note the small clearance between the piston and the counterweights on the crankshaft. *(Chevrolet Motor Division of General Motors Corporation)*

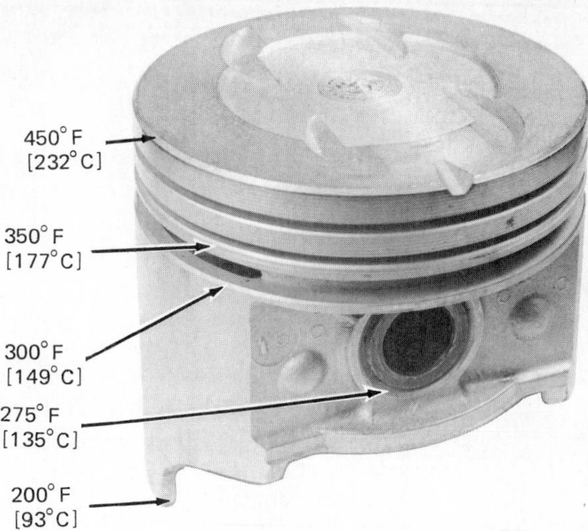

Fig. 11-40 Typical operating temperatures of various parts of a piston. *(Ford Motor Company)*

♦ 11-37 PISTON CLEARANCE

Piston clearance is the distance between the cylinder wall and the skirt, or lower part, of the piston. It is also called *skirt clearance.* Correct clearance varies with different engines. Racing engines, for example, have greater piston clearance. They must have enough clearance to allow adequate amounts of oil to get up to the upper compression ring.

Clearance is usually between 0.001 and 0.004 inch [0.025 and 0.102 mm]. In operation, this clearance is filled with oil so that the piston and rings move on films of oil.

If the clearance is too small, there will be loss of power from excessive friction, severe wear, and possible seizure of the piston in the cylinder. Seizure would cause complete engine failure. If clearances are too large, *piston slap* will result. Piston slap is caused by the sudden tilting of the piston in the cylinder as the piston starts down on the power stroke. The piston shifts from one side of the cylinder to the other, with enough force to produce a noise. Usually, piston slap is a problem only in older engines with worn cylinder walls and worn or collapsed piston skirts. Any of these can produce excessive clearance.

Pistons run many degrees hotter than cylinder walls and therefore expand more. This expansion must be controlled in order to avoid loss of piston clearance. Such loss could lead to serious engine trouble. There is even more of a problem with aluminum pistons. This is because aluminum expands more rapidly than iron with increasing temperature.

♦ 11-38 EXPANSION CONTROL IN PISTONS

Aluminum expands with temperature more than the cast iron of the cylinder block. This expansion must be controlled. Otherwise, piston clearance would be lost, pistons would stick, and the engine would "freeze." One way to do this is to control the amount of heat that reaches the various parts of the piston. The piston head should run hot, but not too hot. If it is not hot enough, the chilling effect of the cool metal will partly quench combustion. The layer of air-fuel mixture next to this metal will not burn. Unburned fuel will escape from the engine, increasing atmospheric pollution and reducing engine efficiency. If the metal is too hot, it can cause surface ignition, or preignition. This can result in misfiring and damage to the engine as explained in a later chapter.

The skirt of the piston should be comparatively cool to prevent excessive expansion. Figure 11-40 shows a typical operating-temperature pattern of a piston. However, the piston must not expand too much.

Many pistons are made so that they have a slightly oval shape when cold. These pistons are called *cam-ground* pistons (Fig. 11-41). They are finish-ground on a machine that uses a cam to move the piston toward and away from the grinding wheel as the piston is revolved. When a cam-ground piston warms up, it becomes round. Its area of contact with the cylinder wall therefore increases (Fig. 11-42). "Contact" here does not mean actual metal-to-metal contact. There must be clearance between the piston and cylinder wall. What is meant is that, when cold, the oval shape of the piston permits normal clearance in only a small area (with excessive clearance elsewhere). But as the piston warms up, this area of normal clearance increases. The head of the piston expands uniformly in all directions. But the stiff piston-pin bosses (the round sections that support the piston pins) are more effective in transmitting this outward thrust. Therefore, these bosses move outward, causing the piston to assume a round shape.

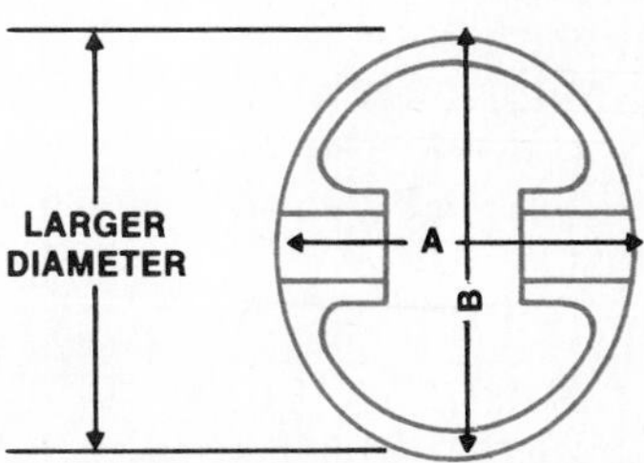

Fig. 11-41 Cam-ground piston viewed from the bottom. When the piston is cold, its diameter at A (through the piston-pin holes) may be 0.002 to 0.003 inch [0.05 to 0.08 mm] smaller than at B. *(Chrysler Corporation)*

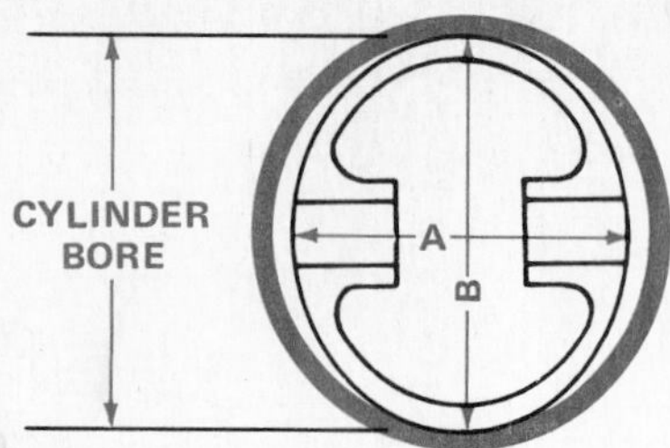

Fig. 11-42 As the cam-ground piston warms up, the expansion of the skirt distorts the piston from an elliptical to a round shape. This increases the area of normal clearance between the piston and the cylinder wall.

Another method of controlling piston expansion is to use bands or belts cast into the piston (Fig. 11-43). These cause the outward thrust of the expanding piston head to be carried more toward the piston-pin bosses than toward the thrust faces. The effect is similar to that in cam-ground pistons.

♦ 11-39 PISTON HEAD SHAPE

The simplest piston head is the flat head (Fig. 11-44). However, increasing compression ratios have made it necessary to reduce the clearance volume (the volume above the piston at TDC). But the valves must have room to open without striking the piston head. Notches or hollows in the piston head can be used to provide adequate valve clearance when the piston is at TDC. Also, some piston heads have a cup, or are dished, to improve the turbulence, or swirling, of the air-fuel mixture (Fig. 11-44). Such turbulence improves combustion.

♦ 11-40 PISTON-PIN OFFSET

In some engines, the piston pin is offset from the centerline of the piston toward the major thrust face. This is the face that bears most heavily against the cylinder wall during the power stroke (Fig. 11-45, left). If the pin is centered, the minor thrust face remains in contact with the cylinder wall until the end of the compression stroke. Then, as the power stroke starts, the rod angle changes from left to right (in Fig. 11-45, left). This causes a sudden shift of the side thrust on the piston, from the minor thrust face to the major thrust face. If there is too much clearance, piston slap will result.

However, if the piston pin is offset (Fig. 11-45, right), the combustion pressure causes the piston to tilt as the piston nears TDC. The lower end of the major thrust face makes first contact with the cylinder wall. Then, after the piston passes TDC and the reversal of side thrust occurs, full major-thrust-face contact is made. There is less chance for piston slap to occur.

The tilting action occurs because there is more combustion pressure on the right side of the piston (which measures $R + O$, or piston radius plus offset) than on the left side of the piston (which measures $R - O$).

♦ 11-41 RING-GROOVE FORTIFICATION

The compression rings continually move up and down in the ring grooves. At the beginning of the power stroke, as the piston goes up to TDC and starts down again, the top compression ring is forced down hard against the lower side of the ring groove. The increasing combustion pressures produce this action. Then, during the intake stroke, the vacuum in the cylinder causes the top compression ring to move up and into contact with the upper side of the ring groove. Therefore, the ring repeatedly strikes the upper and lower sides of the ring groove.

In high-performance engines, these repeated impacts can cause rapid ring-groove wear. The top ring groove is most critical. It receives the greater part of the combustion pressures. To combat this wear, pistons for some high-performance engines have top-ring-groove fortification. The fortifica-

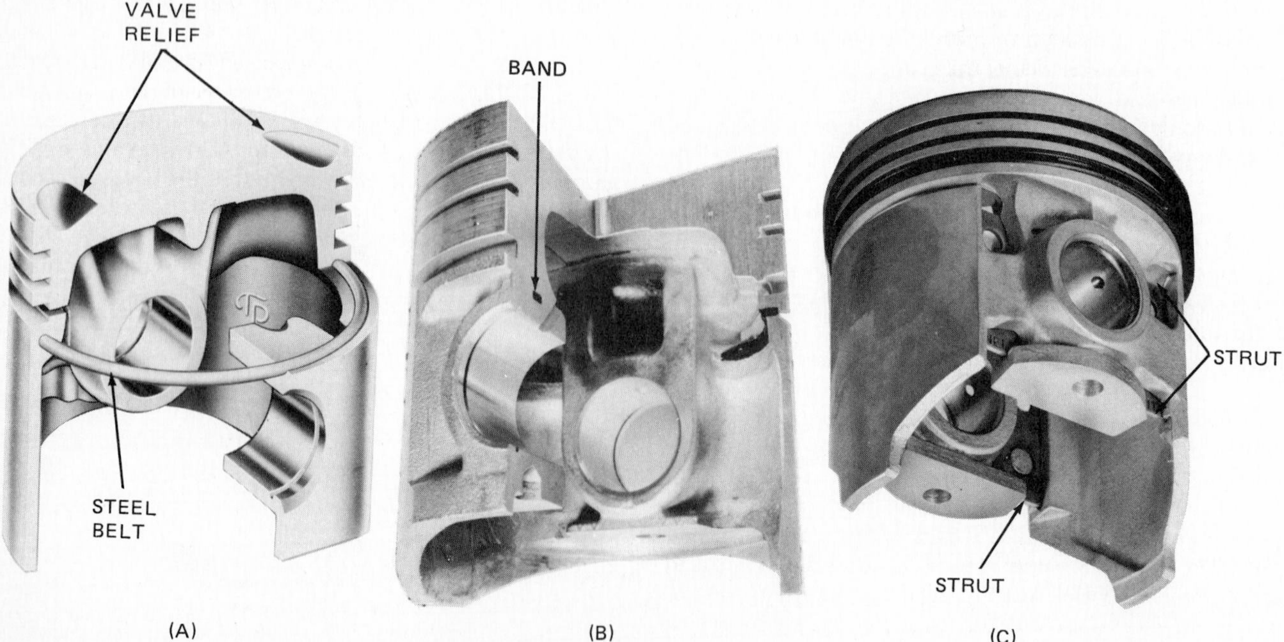

Fig. 11-43 Three methods of piston-expansion control. (A) Cast-in steel belt. (B) Cast-in band. (C) Cast-in strut. *(TRW, Inc.)*

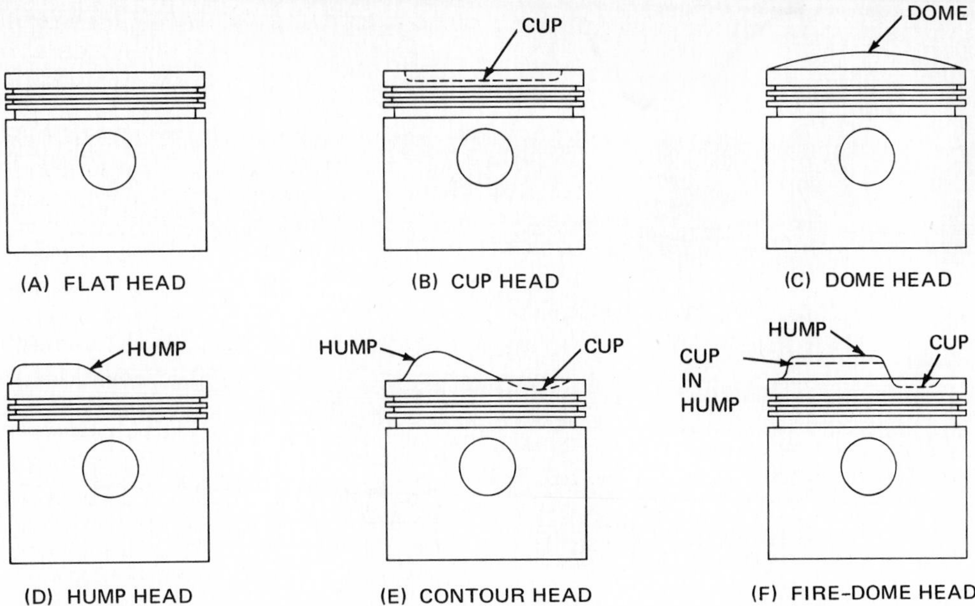

Fig. 11-44 Shapes of piston heads. *(TRW, Inc.)*

tion consists of a ring of cast iron or nickel-iron alloy which is cast into the piston (Fig. 11-46). For cast pistons in medium-duty engines, the inserts are stamped from steel sheets. In forged pistons, inserts cannot be used. Instead, if ring-groove fortification is required, the groove area is sprayed with molten metal having the proper wear resistance. Then, the groove is machined to the proper dimensions.

♦ 11-42 HIGH-PERFORMANCE PISTONS

Aluminum pistons can be either cast or forged. Cast pistons are made by pouring molten aluminum into molds. Forged pistons are "hammered out" from slugs of aluminum alloy. The alloy, subjected to high forging pressure, flows, or extrudes, into dies to form pistons. Both must be heat treated. The forged piston is denser, forming a better heat path.

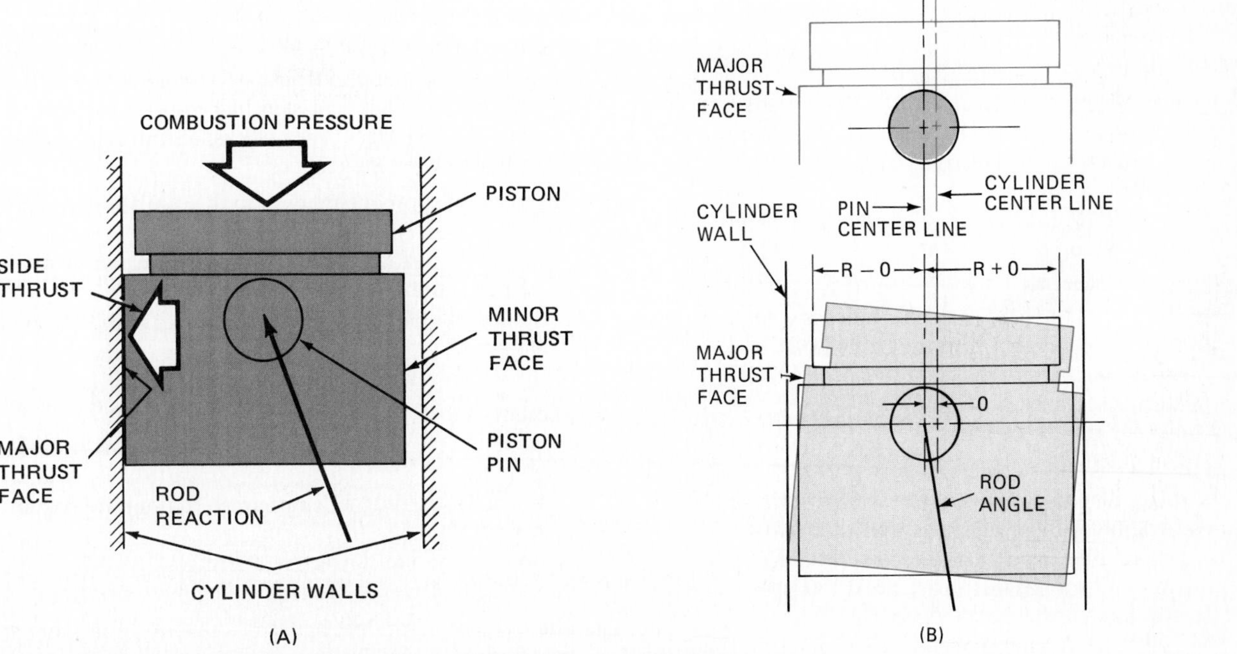

Fig. 11-45 (A) As the combustion pressure is applied to the piston head, and the connecting-rod angle changes from left to right, side thrust on the piston causes it to shift abruptly toward the major thrust face. (B) If the piston pin is offset toward the major thrust face, combustion pressure will cause the piston to tilt to the right, to reduce piston slap. *R* is the radius of the piston; *O* is the offset of the piston. *(Bohn Aluminum and Brass Company, Division of Universal American Corporation)*

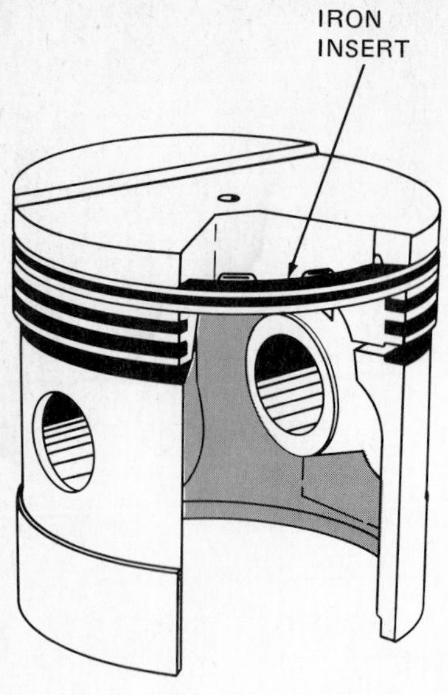

Fig. 11-46 Left, a piston with a cast-in groove insert. Right, a piston with an insert for the upper side of the groove. *(TRW, Inc.; Sealed Power Corporation)*

♦ REVIEW QUESTIONS ♦

Select the *one* correct, best, or most probable answer to each question. You can find the answers in the section indicated at the end of each question.

1. The basic part of the engine, which everything else is attached to or assembled into, is the (♦11-2)
   a. cylinder head
   b. crankshaft
   c. cylinder block
   d. oil pan
2. Mechanic A says many engines have the camshaft in the block. Mechanic B says many engines have the camshaft on the head. Who is right? (♦11-2)
   a. mechanic A
   b. mechanic B
   c. neither A nor B
   d. both A and B
3. The purpose of the core clean-out holes is to permit removal of the cores that formed the (♦11-2)
   a. cylinder bores
   b. water jackets
   c. holes for the freeze plugs
   d. valve guides
4. The major difference between the spark-ignition and the diesel-engine cylinder blocks is that the (♦11-2)
   a. spark-ignition block is more complicated
   b. spark-ignition block is heavier and stronger
   c. diesel block is heavier and stronger
   d. diesel block is more complicated
5. A short block includes the (♦11-4)
   a. cylinder block
   b. crankshaft
   c. pistons and rods
   d. all of the above
6. The upper end of the combustion chamber is formed by the cylinder head. The lower end is formed by the (♦11-9)
   a. piston and rings
   b. connecting rods and crankshaft
   c. valves and valve stems
   d. cams and camshaft
7. The two basic types of cylinder heads for spark-ignition engines are for the (♦11-9)
   a. valve-in-head and overhead-valve engines
   b. camshaft-in-block and camshaft-in-head engines
   c. crankshaft-in-block and crankshaft-in-head engines
   d. in-line and V-type engines
8. The purpose of producing swirl in the combustion chamber is to (♦11-10)
   a. allow the burned gases to exhaust more freely
   b. permit more air-fuel mixture to enter
   c. spread the flame more rapidly so that combustion is improved
   d. allow the valves to open more evenly
9. The purpose of the glow plugs is to (♦11-12)
   a. preheat the precombustion chambers for easier starting
   b. warm the passenger compartment to improve passenger comfort
   c. preheat the fuel for easier starting
   d. warn the driver if the engine begins to overheat
10. The purpose of the heat-control valve in some exhaust manifolds is to (♦11-13)
    a. preheat the air so that the engine will start easier
    b. improve fuel vaporization for better cold-engine performance
    c. help in starting a diesel engine
    d. all of the above

11. Of the five piston-pin arrangements, the most common is to (♦11-23)
    a. lock the pin to the rod with a bolt
    b. lock the pin to the piston with a bolt
    c. free-float the pin in both the piston and rod
    d. press-fit the pin to the rod
12. The name that describes the escape of unburned air-fuel mixture and burned gases from the combustion chamber, past the pistons, and into the crankcase is (♦11-25)
    a. blow-up
    b. blow-past
    c. blowby
    d. blow-off
13. The two basic types of piston rings are (♦11-25)
    a. compression and oil-control
    b. compression and combustion
    c. combustion and exhaust
    d. scuffed and scored
14. Two of the latest compression-ring shapes are the (♦11-27)
    a. twist and reverse twist
    b. barrel face and flat
    c. reverse twist and barrel face
    d. positive twist and reverse twist
15. Scuff wear is caused by (♦11-28)
    a. abrasive wear
    b. momentary welds at TDC
    c. momentary welds at BDC
    d. excessive oil in combustion chamber
16. The purpose of the expander spring in the multiple-piece oil-control ring is to (♦11-30)
    a. force the two rails upward and downward
    b. force the two rails outward
    c. improve oil control
    d. all of the above
17. Two factors that increase oil consumption are (♦11-33)
    a. type of ring and engine speed
    b. engine speed and cylinder-wall wear
    c. thin oil and engine temperature
    d. engine speed and temperature
18. Modern engines use aluminum pistons of the (♦11-35)
    a. full-slipper type
    b. full-skirt type
    c. no-skirt type
    d. semiskirt type
19. The purpose of piston-pin offset is to (♦11-40)
    a. center the piston in the cylinder
    b. reduce piston slap
    c. prevent piston slip
    d. improve piston-pin oiling
20. The purpose of cam-grinding pistons is to (♦11-38)
    a. improve expansion control
    b. improve the camming action of the valve train
    c. increase the oval shape as the piston warms up
    d. stiffen the piston skirt

CHAPTER 12
Valves and Valve Trains

After studying this chapter, you should be able to:

1. Discuss the purpose, construction, and operation of engine valves.
2. Describe the advantages, construction, and operation of overhead-camshaft engines.
3. Explain what valve timing means and how valves are timed for optimum economy and performance.
4. Explain the purpose of hydraulic valve lifters and how they work.
5. Discuss the purpose of valve rotators and how they work.
6. Identify and name the parts in valve trains and explain what they are for.

♦ 12-1 TYPES OF VALVE TRAINS

The two basic types of valve trains used in automotive engines are:

1. Camshaft in the cylinder block, as shown in Figs. 9-19 and 10-8. This type of valve train uses pushrods and the engine is called a *pushrod engine.* The valve train includes a lifter, pushrod, rocker arm, and valve spring. Figure 9-18 shows its operation. How the pushrod valve train works is described in ♦9-13.
2. Camshaft on the cylinder head, as shown in Figs. 10-2 and 11-7. These engines are known as *overhead-camshaft* (OHC) *engines.*

Note Camshafts are described in ♦9-12 and 9-13. Figure 9-16 shows a camshaft.

The OHC engine uses various means of carrying the cam-lobe action to the valve. The simplest arrangement is shown in Fig. 9-15. It uses a bucket tappet. The operation of the bucket-tappet valve train is shown in Fig. 12-1. The cam lobe is directly above the bucket tappet. As the lobe comes around, it pushes the tappet, and the valve, down. The valve opens. As the lobe passes the tappet, the spring pulls the valve closed.

A second arrangement, using a rocker arm, is shown in Fig. 12-2. As the cam lobe comes up under the valve lifter, it causes the rocker arm to rock. This pushes down on the valve stem so that the valve moves down and opens. When the lobe passes out from under the valve lifter, the valve spring pulls the valve closed. The rocker arm rocks back, pushing the valve lifter down on the cam. Other OHC arrangements are described later in the chapter.

♦ 12-2 DRIVING THE CAMSHAFT

The camshaft is driven by gears (Fig. 12-3), by sprockets and chain (Figs. 12-4 and 12-5), or by sprockets and toothed belt (Fig. 12-6). The crankshaft must turn two times to turn the camshaft once. There are four piston strokes to a complete

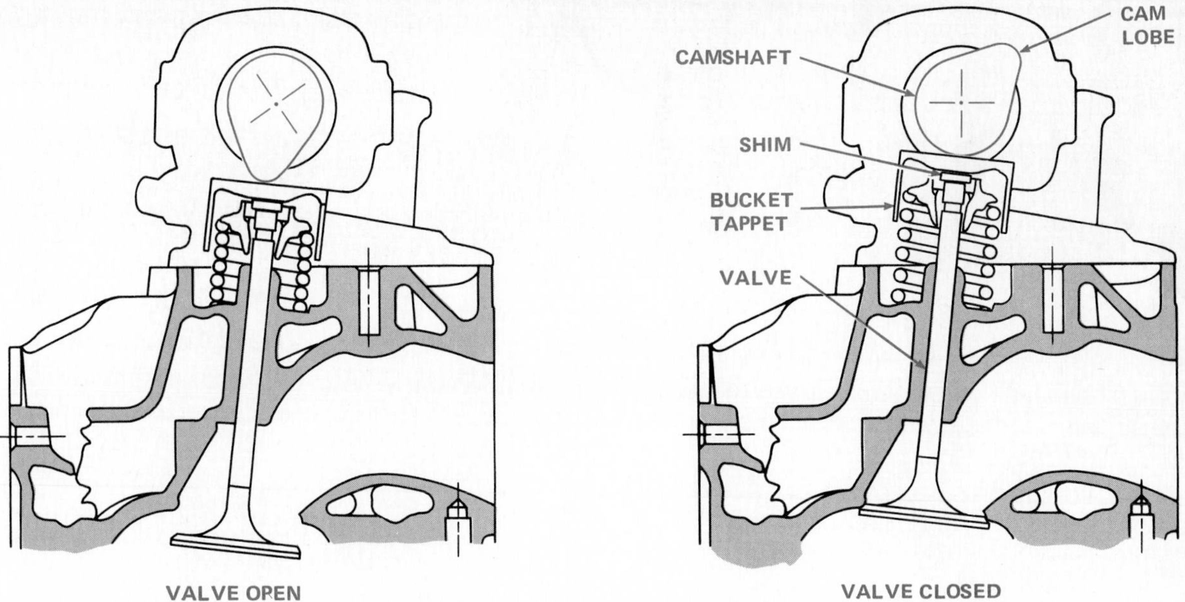

Fig. 12-1 OHC-engine valve train using a bucket tappet. The cam works directly on the tappet.

cycle of actions in the engine. This requires two crankshaft revolutions. Each valve must open only once during a complete cycle. Since each valve opens every camshaft revolution, the camshaft must rotate only once while the crankshaft rotates twice. This 1:2 gear ratio is achieved by making the camshaft gear or sprocket twice as large as the crankshaft gear or sprocket.

The gears are called *timing gears*. The chain is called the *timing chain*. The toothed belt is called the *timing belt*. The reason for this is that the gears, chain and sprockets, or belt and sprockets "time" the opening and closing of the valves.

The camshaft is mounted either in the cylinder block (camshaft in block) or on the cylinder head (OHC). In either position, the camshaft is supported by bushings, or sleeve bearings.

Many camshafts have a spiral gear to drive the distributor and oil pump, and an eccentric to drive the fuel pump (Fig. 12-7).

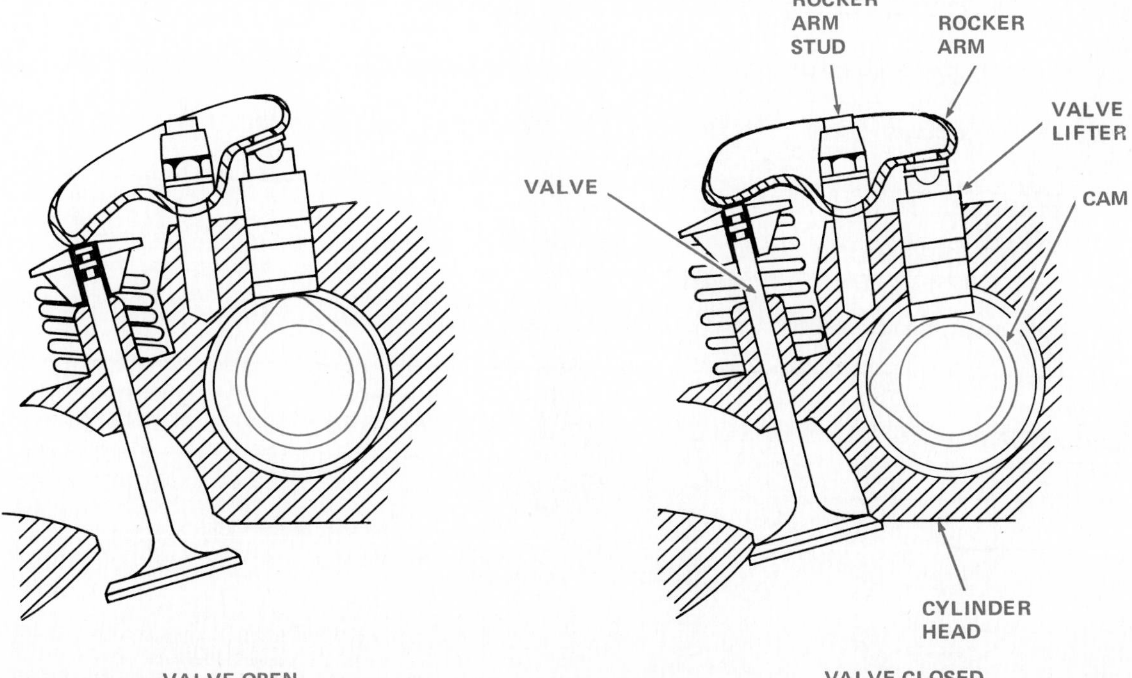

Fig. 12-2 OHC-engine valve train using a rocker arm between the valve lifter and valve stem.

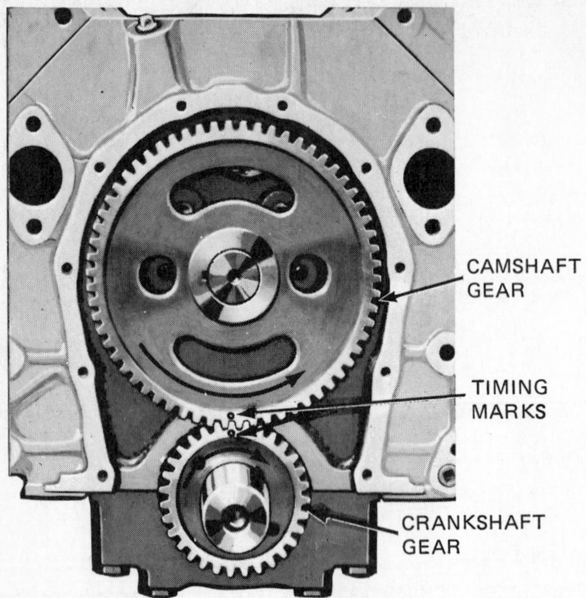

Fig. 12-3 Crankshaft and camshaft gears for a V-type engine. Note the timing marks on the gears. *(Chevrolet Motor Division of General Motors Corporation)*

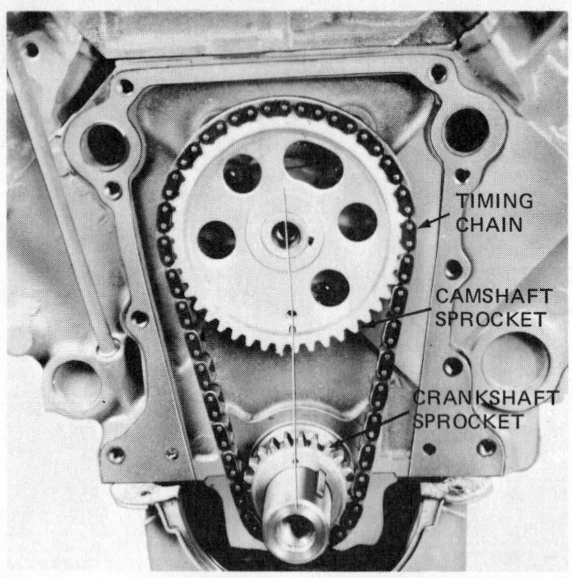

Fig. 12-4 Crankshaft and camshaft sprockets with chain drive for a V-type engine. *(Chrysler Corporation)*

♦ 12-3 VALVES

Each cylinder has two valves, an intake valve and an exhaust valve (Fig. 12-8). Some high-performance engines have four valves per cylinder—two intake valves and two exhaust valves (Fig. 10-14). Usually, the intake valve is larger than the exhaust valve. The reason is that when the intake valve is opened, the only force moving air-fuel mixture into the cylinder is atmospheric pressure. When the exhaust valve is opened, the piston is moving up, and there is a high pressure driving the exhaust gases out. Therefore, the intake port

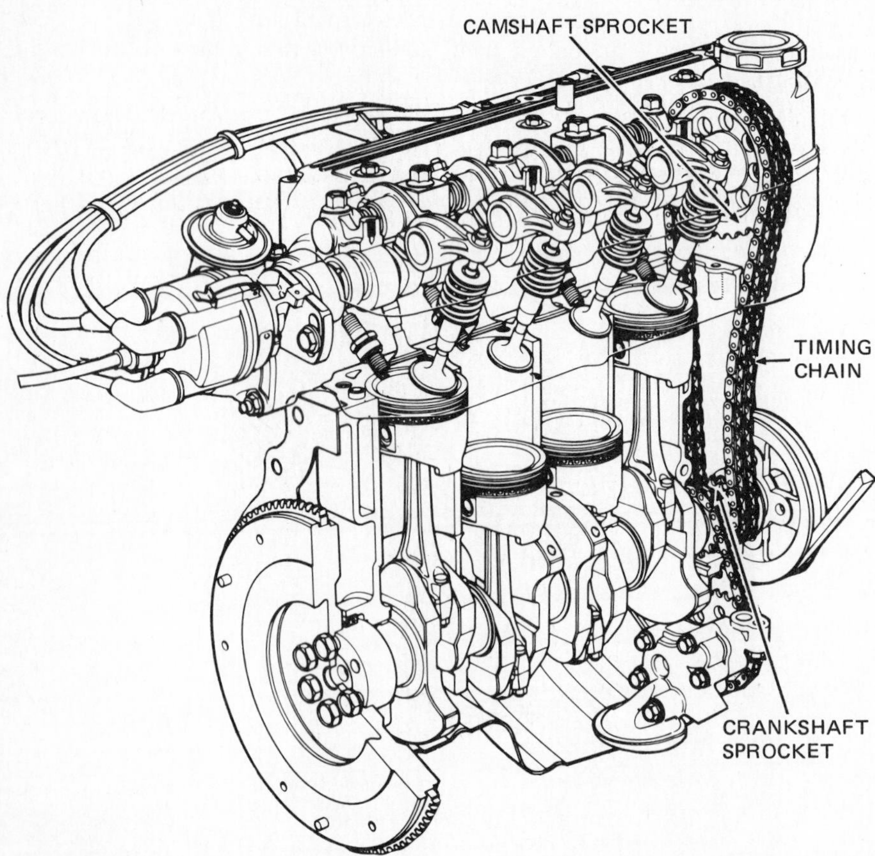

Fig. 12-5 An OHC engine using a timing chain to drive the camshaft. *(Mazda Motors of America, Inc.)*

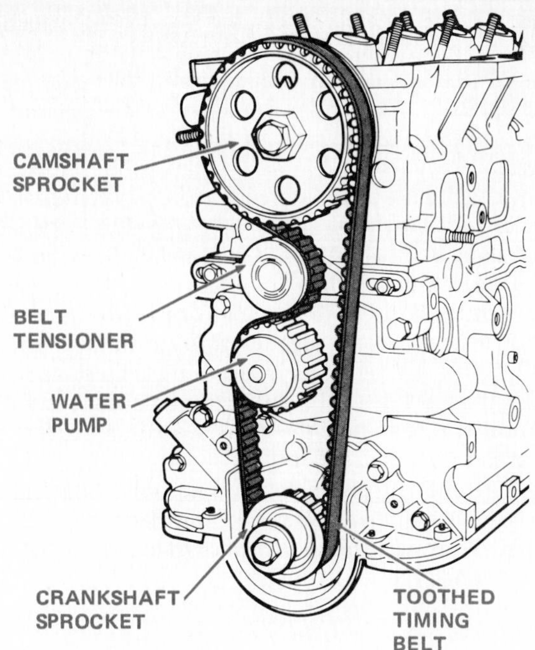

Fig. 12-6 Front end of an OHC engine using a toothed belt and sprockets to drive the camshaft. A belt tensioner is used to prevent the belt from jumping time. *(Ford Motor Company)*

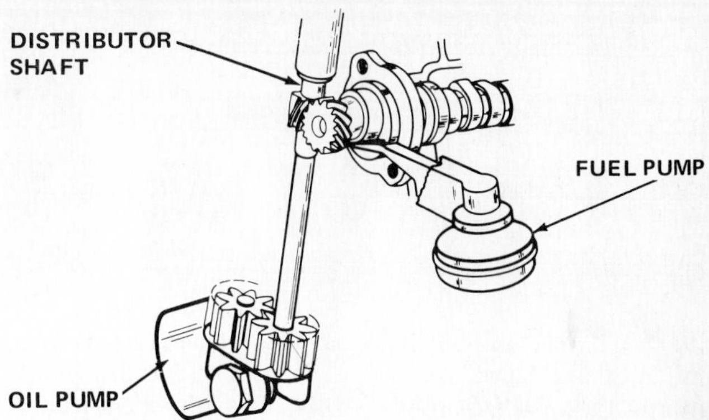

Fig. 12-7 Oil-pump, distributor, and fuel-pump drives. The gear on the camshaft drives the oil pump and distributor. The eccentric on the camshaft drives the fuel pump. *(Buick Motor Division of General Motors Corporation)*

must be larger to allow enough air-fuel mixture to enter.

Various types of valves have been used in the past; among them are the sliding-sleeve and rotary types. But the valve in general use today is the mushroom, or poppet, valve (Fig. 12-9). When the valve is closed, it is held on the valve seat by the valve spring.

Modern engines using low-lead or no-lead gasoline work the valves harder. The reason for removing the lead from gasoline is that excessive amounts of lead can pollute the atmosphere. This is discussed in later chapters. The fact that there is little or no lead in gasoline creates a potential wear problem for valves and valve seats. Lead in gasoline forms a fine coating on the valve faces and seats. This coating acts as a lubricant. Without the lead, the valve faces and seats lack lubricant and can wear rather rapidly. For this reason, many engines have valves with special coatings on their faces. These coatings reduce wear. For severe service, the valve faces may be made of stellite, a very hard metal.

In addition, some valves for modern, high-performance engines have a chrome-plated stem and a hard-alloy tip that has been welded onto the stem. This reduces wear on these two critical areas (Fig. 12-9). Also, some valves are made with a hollow stem. This reduces the valve weight so it has less inertia. Engine power and responsiveness are increased.

♦ 12-4 VALVE COOLING

The intake valve runs relatively cool, since it passes only the air-fuel mixture. But the exhaust valve must pass the very hot exhaust gases. The exhaust valve may actually become red-hot in operation, at temperatures well above 1000°F

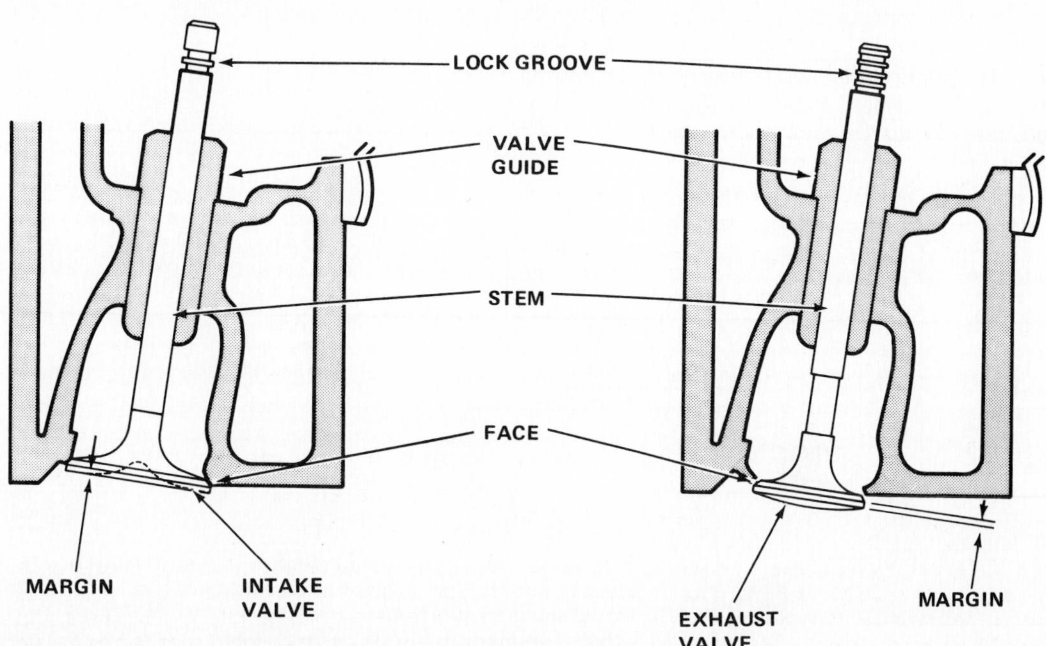

Fig. 12-8 Typical engine valves. *(Chrysler Corporation)*

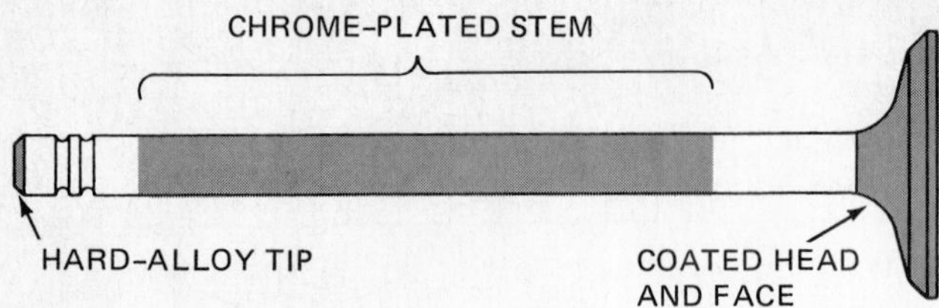

Fig. 12-9 A valve, showing final steps in manufacture. *(ATW)*

[537.8°C]. Figure 12-10 shows a typical temperature pattern for an exhaust valve. The valve stem is coolest, and the part nearest the valve face is next coolest. The valve stem passes heat to the valve guide, which helps to keep the valve stem cool. Likewise, the valve face passes heat to the valve seat. This helps to keep the valve face cool.

The valve seat and valve guide are cooled by the engine cooling system. The cylinder head is carefully designed to permit good coolant circulation through the water jackets around the seat and guide. Some I-head engines have nozzles that force coolant circulation around the valve seats. In some engines, deflectors in the head improve coolant circulation around the valve seats.

To assist in valve cooling, some heavy-duty engines have an exhaust valve with a fat hollow stem partly filled with the metal sodium. Sodium melts at 208°F [97.8°C]. When the engine is operating, the sodium is a liquid. The liquid is thrown up and down in the stem by the valve movement. This circulation takes heat from the valve head and carries it down to the stem, which is cooler. The action keeps the head cooler for longer valve life. However, sodium-cooled valves are seldom used in automotive engines.

♦ CAUTION ♦ Sodium is a highly reactive element. If a piece of sodium is dropped into water, it will burst into flame with explosive violence. If it gets on your skin, it will cause deep and serious burns. Never cut into a sodium-cooled valve stem! The sodium could explode and burn you.

♦ 12-5 VALVE SEAT

The exhaust-valve seat is heated to extremely high temperatures by the exhaust gases. The cylinder-head material is adequate to take these temperatures under normal operating conditions. But where the engine is put to severe service, as for example in trucking, special provisions must be made.

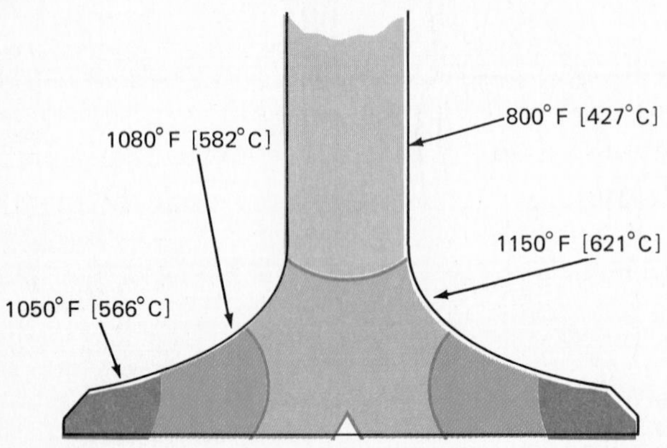

Fig. 12-10 Temperatures in an exhaust valve. *(Eaton Corporation)*

One is to harden the valve seats by the electric induction-hardening process (Fig. 12-11). The other is to install seat inserts. These are special heat-resistant steel-alloy insert rings (Fig. 9-13). These rings can withstand the higher temperatures without undue wear. If they do wear, they can be replaced.

Many engines use an interference angle between the valves and valve seats. As shown at the top of Fig. 12-11, the interference angle is usually attained by grinding the valve at an angle 1 degree flatter than the seat angle. This produces greater seating force at the outer edge of the valve seat. Therefore, the valve-seat edge tends to cut through any deposits that have formed. This produces a good seal. As shown at the bottom of Fig. 12-11, an interference angle is not always recommended where the valve is faced with stellite and the seat is induction hardened.

The difference in angles between the valve face and the valve seat gradually disappears as the valve face and seat wear. The contact between the two changes from line contact (interference angle) to area contact.

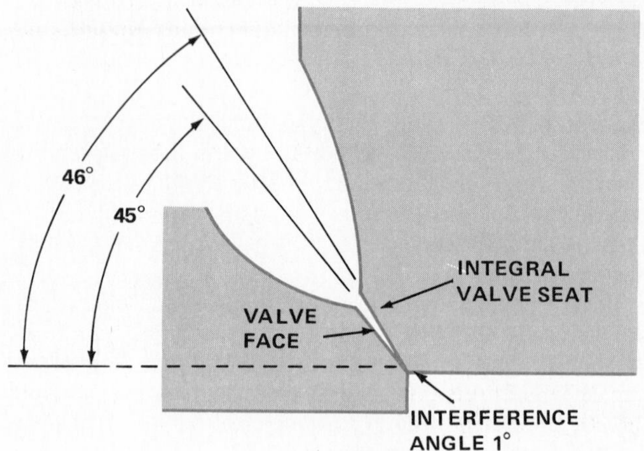

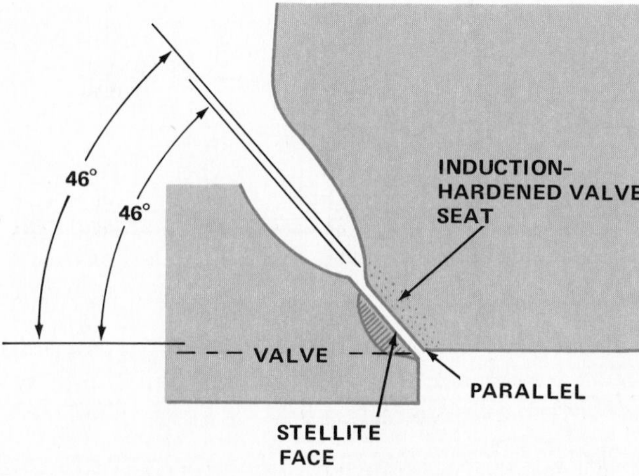

Fig. 12-11 Valve and valve-seat angles. Top, interference angle. Bottom, an induction-hardened valve seat. The valve shown is stellite-faced. Stellite is resistant to heat and wear. The faces of the valve and valve seat (bottom) are parallel. *(Chevrolet Motor Division of General Motors Corporation)*

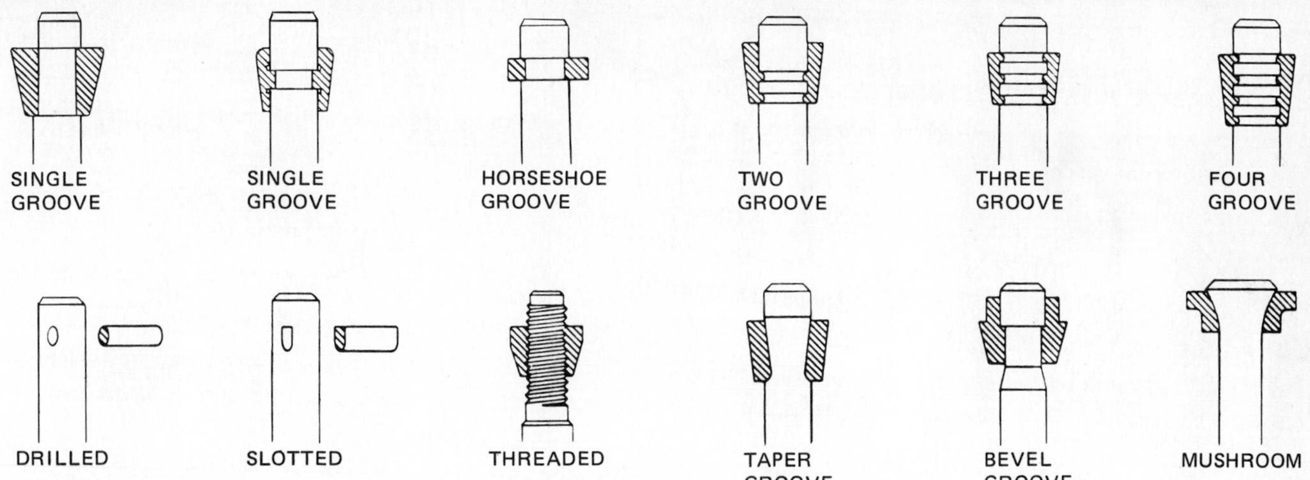

Fig. 12-12 Various types of valve-spring-retainer locks (also called *valve keepers*).

♦ 12-6 VALVE-SPRING ATTACHMENT

In I-head engines, one end of the valve spring presses against the cylinder head. The other end is attached to the end of the valve stem with a spring retainer and a retainer lock, or keeper (Fig. 12-12). When installed, the spring is compressed with the retainer above it. Then the retainer lock is installed in the groove in the valve stem. When the spring is released, the retainer presses against the lock, holding it in place in the valve stem. The tension spring puts a heavy pull on the valve stem. This keeps the valve seated tightly on the valve seat—except while the valve train holds the valve open.

♦ 12-7 ROCKER ARMS

There are several different types of rocker arms (Fig. 12-13). Some rocker arms have a means of adjustment. The purpose is to provide a minimum valve-tappet clearance, or gap, in the valve train. On valve trains with hydraulic valve lifters, this is not important because the hydraulic valve lifter automatically takes care of any clearance (♦12-14). Some rocker arms have an adjusting screw. It can be turned in or out to adjust the valve-tappet clearance. The stud- or pedestal-mounted rocker-arm valve train is adjusted by turning the stud nut or attaching bolt.

Valve-tappet clearance should be kept to a minimum to reduce noise and wear from parts hitting together when the valves are opened. The clearance should be large enough to assure complete closing of the valves. If the adjustment were made to give no clearance when the engine is cold, trouble would result. When the engine warmed up, the valve-train parts would expand enough to prevent the valves from closing completely. Hot combustion gases would flow between the valve face and valve seat. This would result in burned valves and seats and an expensive engine repair job.

Overhead-cam engines may also use rocker arms (Figs. 10-2, 11-7, and 12-2). In some engines, the rocker arms are mounted on shafts. In others, the rocker arm floats (Fig. 12-14). One end of the rocker arm rests on the hydraulic valve-lash adjuster. This device takes up any clearance that occurs. The other end of the rocker arm rests on the valve stem. The cam on the camshaft rides on a pad on the rocker arm. When the cam lobe moves around above the rocker arm, it pushes down on the rocker arm. The rocker arm pivots on the valve-lash adjuster so the valve-stem end is pushed down. This opens the valve. When the cam lobe moves out from above the rocker arm, the valve spring closes the valve. Meantime, the valve-lash adjuster takes up any lash, or clearance. The operation of the hydraulic valve lifter is described in ♦12-14.

♦ 12-8 VALVE STEM OIL SEALS

There is always considerable oil on top of the cylinder head. Oil is needed to lubricate the valve stems, rocker arms, and pushrods (where used). Oil must be prevented from getting past the valve stems and into the combustion chamber. The oil would burn, leaving carbon deposits. Valves and piston rings would not work properly. Compression ratio could go so high that detonation would occur. Spark plugs would foul and misfire.

To prevent all this from happening, valve stems are protected from oil by either an oil shield or an oil seal (Figs. 12-15 and 12-16). The shield or seal allows just enough oil to get on the stem to provide proper stem lubrication.

♦ 12-9 VALVE ROTATION

If the exhaust valve rotates as it opens, there is less chance of valve-stem deposits causing the valve to stick. In addition, valve rotation results in more even valve-head temperature. Some parts of the valve seat may be hotter than others. Hot spots may develop. If the same part of the valve face continues to seat on the hot spot, a hot spot develops on the valve face. The hot spot on the valve face wears or burns away faster. But if the exhaust valve rotates, no one part is always subjected to the higher temperature. Therefore, longer exhaust-valve life results.

In the typical engine design, the rocker arm is slightly offset from the centerline of the valve. Every time the valve is opened, there is an off-center push on the valve stem. This tends to rotate it. There are also special *valve rotators* that are part of some valve trains. These are of two types, free and positive.

♦ 12-10 FREE-TYPE VALVE ROTATOR

Figure 12-17 shows a free-type valve rotator. In this design, the spring-retainer lock has been replaced by two parts: a

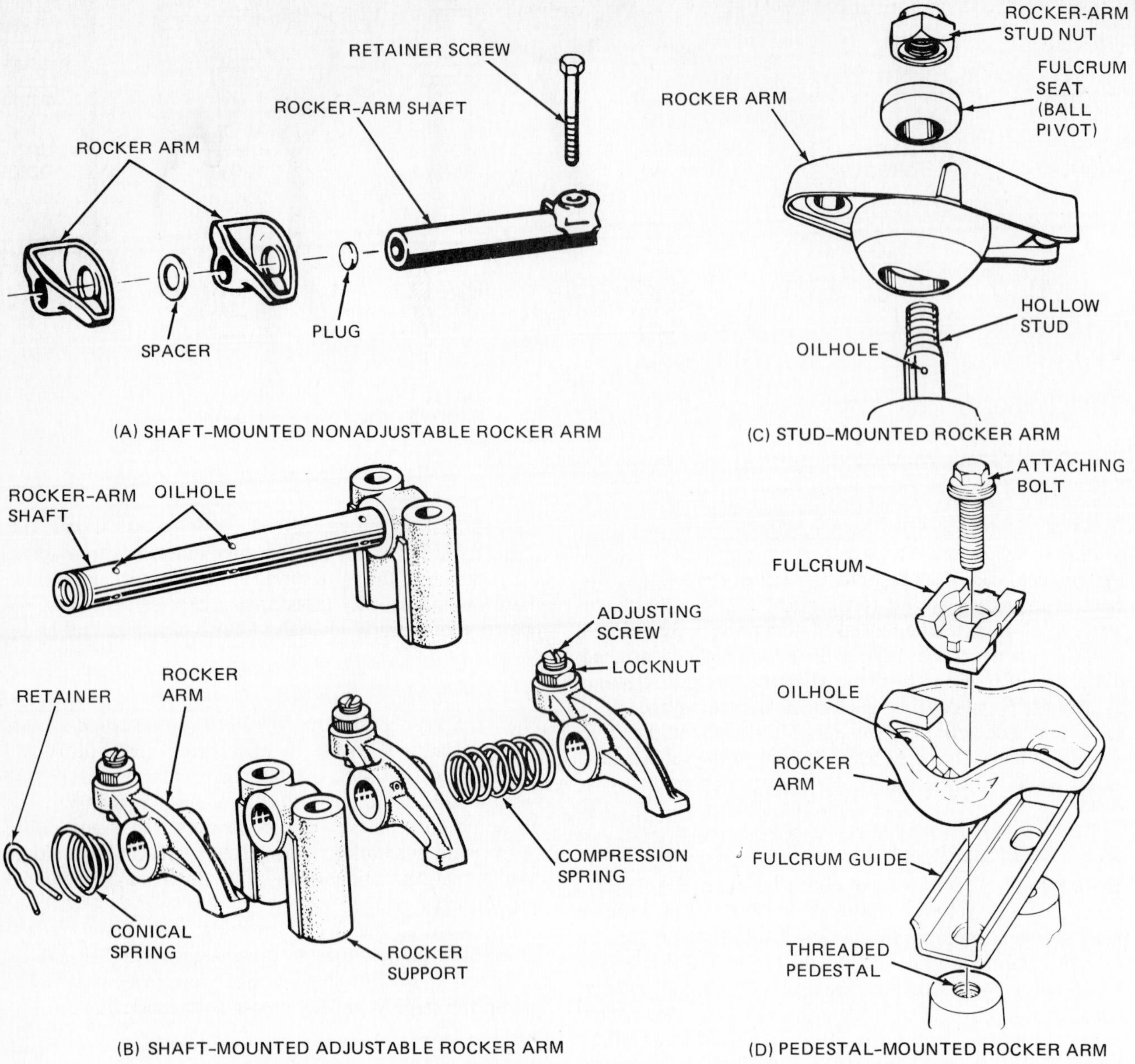

RETAINER SCREW
ROCKER-ARM SHAFT
ROCKER ARM
SPACER
PLUG

(A) SHAFT-MOUNTED NONADJUSTABLE ROCKER ARM

ROCKER-ARM STUD NUT
FULCRUM SEAT (BALL PIVOT)
ROCKER ARM
HOLLOW STUD
OILHOLE

(C) STUD-MOUNTED ROCKER ARM

ROCKER-ARM SHAFT
OILHOLE
ADJUSTING SCREW
LOCKNUT
RETAINER
ROCKER ARM
COMPRESSION SPRING
CONICAL SPRING
ROCKER SUPPORT

(B) SHAFT-MOUNTED ADJUSTABLE ROCKER ARM

ATTACHING BOLT
FULCRUM
OILHOLE
ROCKER ARM
FULCRUM GUIDE
THREADED PEDESTAL

(D) PEDESTAL-MOUNTED ROCKER ARM

Fig. 12-13 Different types of rocker arms used in overhead-valve engines.

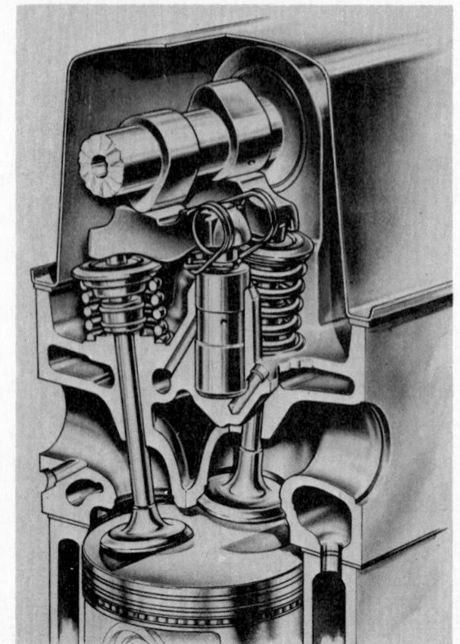

Fig. 12-14 A cylinder head and valve train using an overhead camshaft, showing the valve train for one valve. The rocker arm is located between the valve stem and the automatic valve-lash adjuster or hydraulic valve lifter. The center of the rocker arm rides on the cam. *(Ford Motor Company)*

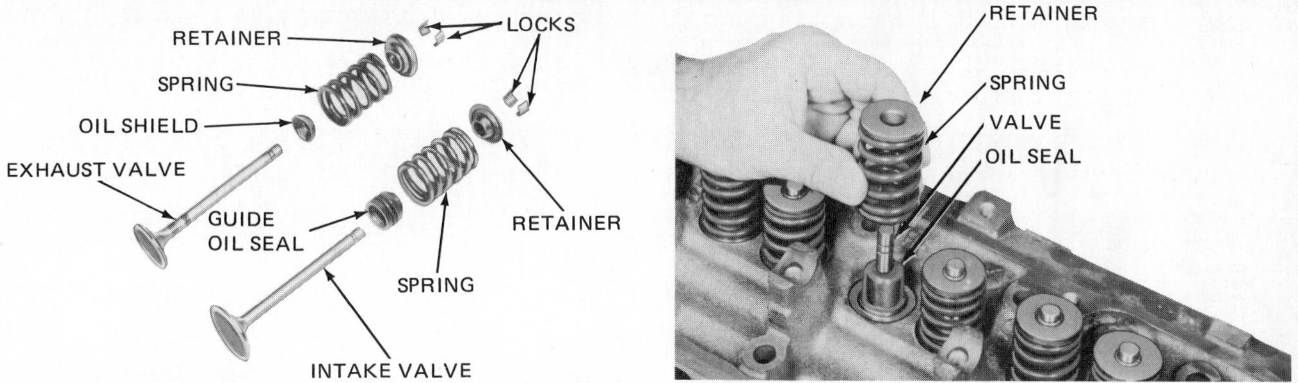

Fig. 12-15 Disassembled and assembled views of a valve-and-spring assembly with oil seal and oil shield. *(Chrysler Corporation)*

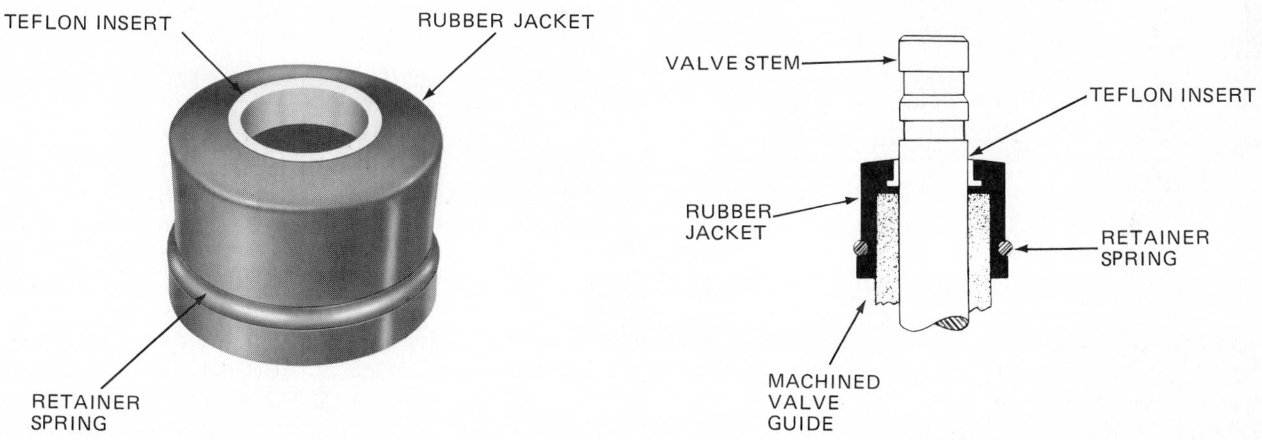

Fig. 12-16 Oil seal for intake-valve stem (to left). Right, how the oil seal fits around the valve stem and valve guide. *(Perfect Circle Division of Dana Corporation)*

washer-type split lock and a tip cup. As the valve lifter moves up, the rocker arm presses down against the tip cup. The tip cup then carries the motion to the lock and valve retainer. The valve retainer is pushed down, thereby taking up the valve-spring force. Then the bottom of the tip cup moves down against the end of the valve stem so that the valve is opened. The spring pressure is taken off the valve stem, which leaves the valve free to rotate. Engine vibration causes valve rotation.

♦ 12-11 POSITIVE VALVE ROTATOR

The positive valve rotator turns the valve, not by engine vibration, but by a positive push. Figure 12-18 shows how the positive valve rotator fits into the valve train. It takes the place of the spring retainer. The valve rotator applies a rotating force on the valve stem each time the valve opens. A seating collar is spun over the outer lip of the spring retainer. The valve spring rests on the seating collar. The collar encloses a flexible washer placed below a series of spring-loaded balls. The tops of the grooves (races) are inclined. When the lifter is raised, the rocker arm lifts the valve and applies in-

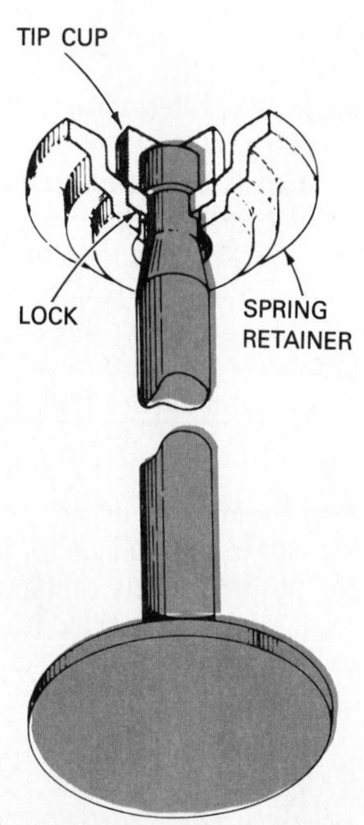

Fig. 12-17 Construction of a free-type valve rotator. *(TRW, Inc.)*

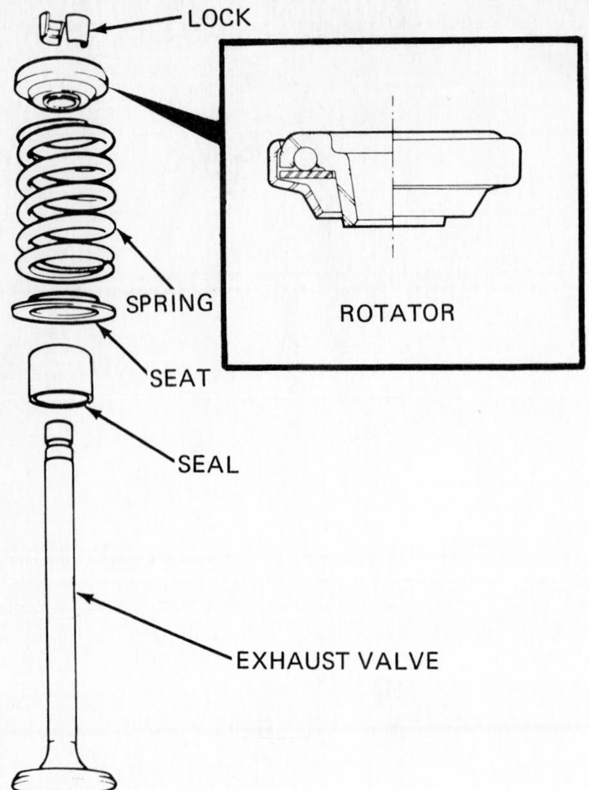

Fig. 12-18 Positive valve rotator on exhaust valve. *(American Motors Corporation)*

creasing force on the seating collar. This flattens the flexible washer so that the washer applies the spring load on the balls. As the balls receive this load, they roll up the inclined races. This causes the retainer to turn a few degrees, which turns the valve a few degrees. When the valve closes, the spring force is reduced. The balls return to their original positions, ready for the next valve motion.

♦ 12-12 VALVE LIFTERS

There are two types of valve lifters (also called *valve tappets*), the solid or mechanical lifter and the hydraulic lifter (♦12-14). The solid lifter is just a cylinder placed between the cam (on the camshaft) and the pushrod. The valve lifter is rotated in much the same way that the valve is rotated by the rocker arm. The face of the lifter is offset slightly from the center of the cam. This rotation of the valve lifter prevents sludge from accumulating in the lifter bore in the cylinder block. At the same time, the lifter rotates the pushrod. This keeps clean the pushrod bearing surfaces with the lifter and rocker arm.

A rough or worn cam lobe can often be located by seeing which pushrod is not turning, or not turning at the same speed as the other pushrods. This rotation distributes the wear from the cam over the face of the lifter. The hydraulic valve lifter has an internal construction that reduces noise and valve-train clearance.

♦ 12-13 ROLLER TAPPETS

The valve lifters described above are *flat tappets.* The face that is in contact with the cam is flat or slightly convex. Some engines, such as diesels and racing engines, use *roller tappets,* or *roller lifters* (Fig. 12-19). These have a hardened steel roller on the end. The roller rolls over the cam instead of sliding over it. This reduces the friction. The roller lifter must not rotate because this would cause the roller to turn sideways. The lifters have flat guide surfaces, and guides are used to hold the lifters in position so they do not turn (Fig. 12-20).

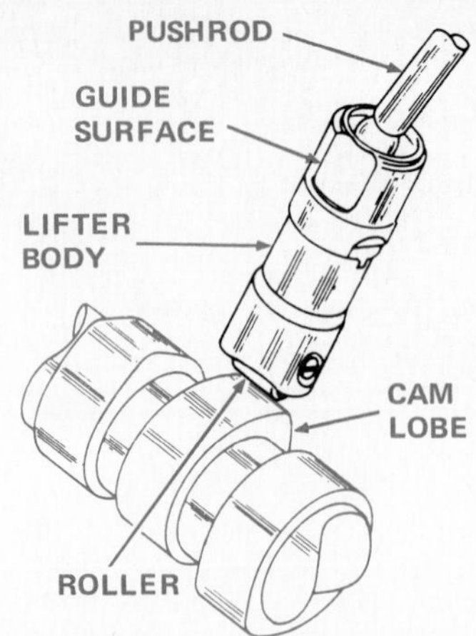

Fig. 12-19 A roller-type valve lifter, or roller tappet, has rolling contact with the cam lobe instead of sliding contact. *(Oldsmobile Division of General Motors Corporation)*

♦ 12-14 HYDRAULIC VALVE LIFTERS

The hydraulic valve lifter is used in many engines. It is very quiet because it assures zero tappet clearance (or valve lash). Also, it usually requires no adjustment in normal service. Variations due to temperature changes or to wear are taken care of hydraulically.

Figure 12-21 shows the details of a hydraulic valve lifter. Oil is fed into the valve lifter from the oil pump, through an oil gallery that runs the length of the engine. When the valve is closed, oil from the pump is forced into the valve lifter

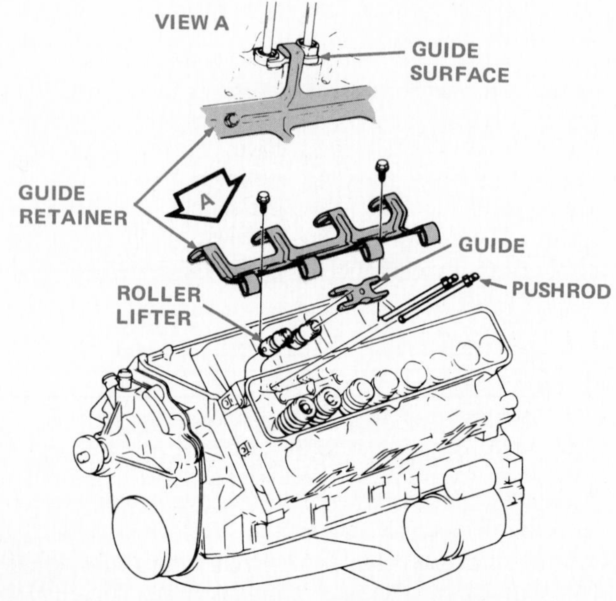

Fig. 12-20 Guides and guide retainers are used with roller lifters to prevent them from turning in the lifter bores. *(Oldsmobile Division of General Motors Corporation)*

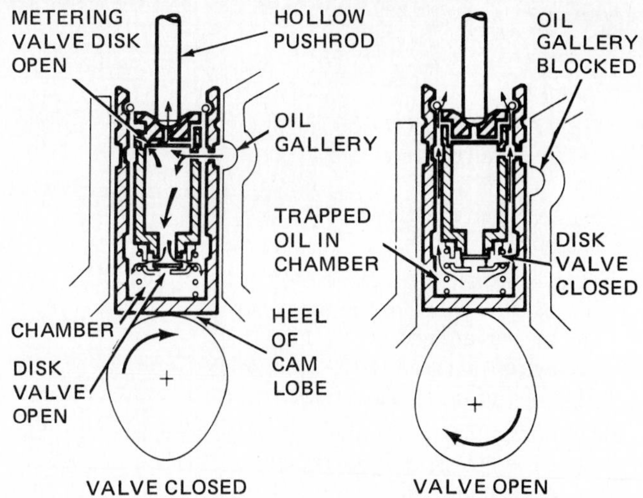

Fig. 12-21 Hydraulic valve lifter with valve closed and open. *(American Motors Corporation)*

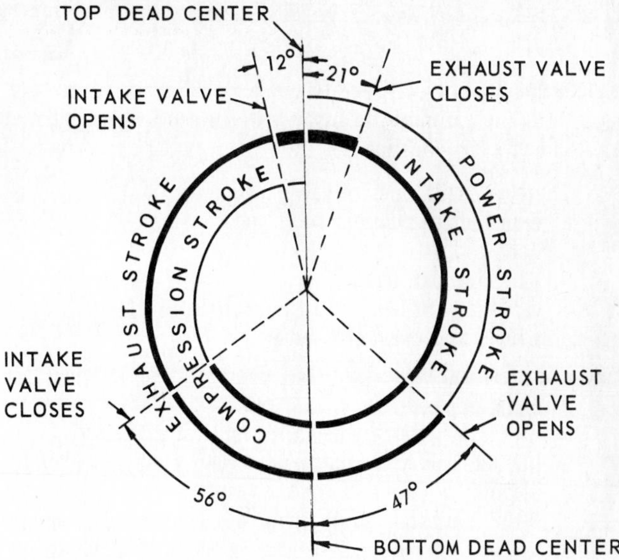

Fig. 12-22 Intake- and exhaust-valve timing. The complete cycle of events is shown as a 720 degree spiral, which represents two complete crankshaft revolutions. Timing of valves differs for different engines.

through oilholes in the lifter body and plunger. The oil forces the ball-check valve in the plunger to open. Oil then passes the ball-check valve and enters the space under the plunger. The plunger is forced upward until it touches the valve pushrod. This takes up any clearance in the system.

When the cam lobe moves around under the lifter body, the lifter is raised. Since there is no clearance, there is no tappet noise. The raising of the lifter and the opening of the valve suddenly increases the pressure in the body chamber under the plunger. This causes the ball-check valve to close. Oil is therefore trapped in the chamber. Because liquids such as oil are not compressible, the lifter acts like a simple one-piece lifter. It moves up as an assembly and causes the valve to open. Then, when the valve closes, the lifter moves down, and the pressure on the plunger is relieved. If any oil has been lost from the chamber under the plunger, oil from the engine oil pump causes the ball-check valve to open. Engine oil can then refill the chamber.

♦ 12-15 VALVE TIMING

Figure 12-22 shows a typical valve-timing diagram. In this diagram, the exhaust valve starts to open at 47 degrees before BDC on the power stroke. It stays open until 21 degrees after TDC on the intake stroke. This gives more time for the exhaust gases to leave the cylinder. By the time the piston reaches 47 degrees before BDC on the power stroke, the combustion pressures have dropped considerably. Little power is lost by giving the exhaust gases this extra time to leave the cylinder.

In a similar manner, the intake valve starts to open at 12 degrees before TDC and remains open for 56 degrees past BDC after the intake stroke. This gives additional time for air-fuel mixture to flow into the cylinder. The delivery of adequate amounts of air-fuel mixture to the engine cylinders is a critical item in engine operation. Actually, the cylinders are never quite "filled up" when the intake valve closes.

The exhaust valve closes 21 degrees after the intake valve opens in Fig. 12-22. This provides an overlap of 33 degrees during which both valves are open at the same time. Most automotive engines have valve overlap during the end of the exhaust stroke and the beginning of the intake stroke. Then the exhaust gases are moving rapidly from the cylinder into the exhaust port. Holding the exhaust valve open well past TDC after the exhaust stroke gives the gases more time to leave. At the same time, starting to open the intake valve before TDC on the exhaust stroke gives the incoming air-fuel mixture a "head-start" toward entering the cylinder.

Timing of the valves is controlled by the shape of the lobe on the cam and the relationship between the gears or sprockets and chain on the camshaft and crankshaft. Changing the relationship between the driving and driven gears or sprockets changes the timing at which the valves open and close. For example, suppose the timing chain is worn and this allows the chain to "jump time." Then the chain slips a tooth and this causes the camshaft to fall behind that one tooth. The valves would then open and close later. Now, for example, the valve action has been moved back 15 degrees. The exhaust valve will open at 32 degrees before BDC on the power stroke. It will close at 36 degrees after TDC on the exhaust stroke (in the example shown in Fig. 12-20). The intake-valve actions would likewise be moved back. These valve-action delays would reduce engine performance and cause engine overheating. The gears or sprockets are marked so that they can be properly aligned on assembly (Fig. 12-3).

♦ REVIEW QUESTIONS ♦

Select the *one* correct, best, or most probable answer to each question. You can find the answers in the section indicated at the end of each question.

1. The two basic types of valve trains are (♦12-1)
   a. camshaft in cylinder head and camshaft above cylinder head
   b. I head and L head
   c. camshaft on head and camshaft in block
   d. I head and overhead valve
2. The pushrod valve train has five basic parts. These are cam, (♦12-1)
   a. lifter, pushrod, rocker arm, and valve spring
   b. bucket tappet, adjustment screw, spring, and valve spring
   c. lifter, pedestal, adjustment screw, and valve spring
   d. lifter, pushrod, hydraulic adjuster, and valve spring
3. The camshaft is driven by sprockets and chain or toothed belt, or by (♦12-2)
   a. the distributor shaft
   b. an oil-pump gear
   c. timing gears
   d. timing belt
4. The purpose of the gear on some camshafts is to (♦12-2)
   a. drive the gear train
   b. time the gears
   c. drive the fan belt
   d. drive the distributor and oil pump
5. Mechanic A says that all OHC engines use rocker arms. Mechanic B says some use bucket tappets and no rocker arms. Who is right? (♦12-1)
   a. mechanic A
   b. mechanic B
   c. both A and B
   d. neither A nor B
6. Valve overlap is the number of degrees of camshaft rotation during which (♦12-15)
   a. both valves are closed
   b. both valves are open
   c. each valve is closed
   d. none of the above
7. The camshaft turns at (♦12-2)
   a. half the speed of the crankshaft
   b. the same speed as the crankshaft
   c. twice the speed of the crankshaft
   d. none of the above
8. In normal operation, the part of the exhaust valve that gets the hottest is the (♦12-4)
   a. valve stem
   b. valve seating face
   c. edge of the margin
   d. center of the head
9. When a valve-face angle is ground 1 degree less than the seat angle, it is called (♦12-5)
   a. sloppy workmanship
   b. completion angle
   c. interference angle
   d. bearing interference
10. As the hydraulic valve lifter moves up, opening the valve, the disk or ball valve is (♦12-14)
   a. opening
   b. closing
   c. open
   d. closed

CHAPTER 13
Engine Measurements and Performance

After studying this chapter, you should be able to:

1. Define *work, energy, power,* and *torque* and explain how they are measured and their importance to engine performance.
2. Define *friction* and list and explain the three kinds of friction.
3. Define *bore, stroke, piston displacement, compression,* and *volumetric efficiency.*

♦ 13-1 WORK

Work is the moving of an object against an opposing force. The object is moved by a push, a pull, or a lift. For example, when a weight is lifted, it is moved upward against the pull of gravity. Work is done on the weight (Fig. 13-1).

Work is measured in terms of distance and force. If a 5-pound weight is lifted off the ground 5 feet, the work done on the weight is 25 foot-pounds (ft-lb), or 5 feet times 5 pounds (Fig. 13-1). *Distance times force equals work.*

In the metric system, work can be measured in meter-kilograms (mkg) or in joules (J). For example, lifting a 5-kg weight [11 pounds] a distance of 1 m [3.28 feet] requires 5 mkg of work [36.08 ft-lb]. The joule is a combination unit which includes both distance and weight. One foot-pound is equal to 1.356 J.

♦ 13-2 ENERGY

Energy is the ability, or capacity, to do work. When work is done on an object, energy is stored in that object. When work is done on a weight by lifting it (Fig. 13-1), energy is stored in it. In Fig. 13-1, 25 ft-lb of work has been done on the weight. The weight can do that much work if it is released. When an automobile is accelerating, work is being done on it. The engine must work to produce car movement. Energy is being stored in the car.

♦ 13-3 POWER

Work can be done slowly, or it can be done rapidly. The rate at which work is done is measured in terms of power. A machine that can do a great deal of work in a short time is called a *high-powered* machine. *Power* is the rate, or speed, at which work is done.

♦ 13-4 TORQUE

Torque is a twisting or turning force. You apply torque to the top of a screw-top jar when you loosen it (Fig. 13-2). You apply torque to the steering wheel when you steer a car around a turn. The engine applies torque to the wheels to make them rotate.

Torque must not be confused with power. Torque is turning force which *may or may not result in motion.* Power is the rate at which work is done. This means that something must be moving.

Torque is measured in pound-feet (lb-ft, not to be confused with ft-lb of work) or, in the metric system, in newton-meters (N-m). For example, a 20-pound push on a 1½-foot crank produces 30-pound-feet torque (Fig. 13-3). You would be applying this torque to the crank regardless of whether the crank was turning. The torque is applied as long as you push on the crank.

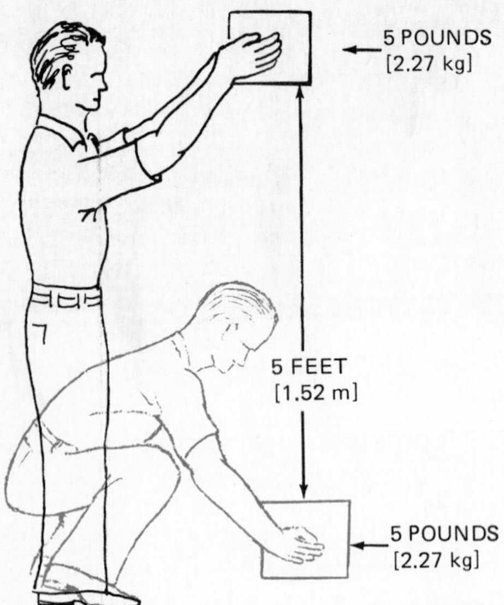

Fig. 13-1 When a weight is lifted, work has been done on it.

♦ 13-5 HORSEPOWER

A horsepower (hp) is the power of one horse, or a measure of the rate at which a horse can work. A 10-hp engine, for example, can do the work of 10 horses.

A horsepower is 33,000 ft-lb of work per minute. In Fig. 13-4, the horse walks 165 feet in 1 minute, lifting the 200-pound weight. The amount of work involved is 33,000 ft-lb (165 feet × 200 pounds). The time is 1 minute. If the horse did this work in 2 minutes, then it would be only "half" working; it would be putting out only ½ hp. One formula for horsepower is

$$\text{hp} = \frac{\text{ft-lb per minute}}{33{,}000} = \frac{L \times W}{33{,}000 \times t}$$

where hp = horsepower
L = length, in feet, through which W is exerted
W = force, in pounds, exerted through distance L
t = time, in minutes, required to move W through L

In the metric system, power output from an engine is measured in kilowatts (kW), which is an electrical term. It is the amount of electricity the engine can produce if it were used to drive an electric generator. One horsepower is equal to 0.746 kW, and 1 kW is equal to 1.34 hp.

A second formula for horsepower is

$$\text{hp} = \frac{\text{torque} \times \text{rpm}}{5252}$$

It is more commonly used because modern dynamometers (described in ♦13-13) measure engine performance in rpm (revolutions per minute), torque, and horsepower. This makes the second formula easier to work with. Torque is defined in ♦13-4.

Fig. 13-2 Torque, or twisting force, must be applied to loosen and remove the top from a screw-top jar.

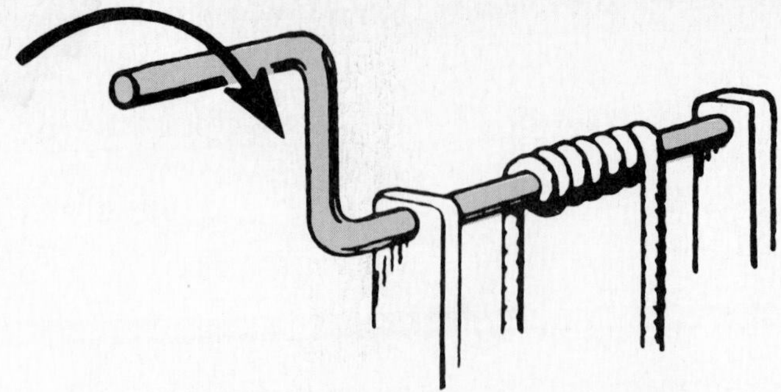

Fig. 13-3 Torque is measured in pound-feet (lb-ft) or in newton-meters (N-m). It is calculated by multiplying the push by the crank offset, or the distance of the push from the rotating shaft.

♦ 13-6 INERTIA

Inertia is a property of all material objects. It causes them to resist any change of speed or direction of travel. A motionless object tends to remain motionless. A moving object tends to keep moving at the same speed and in the same direction.

When the automobile is standing still, its inertia must be overcome by applying power to make it move. To increase its speed, more power must be applied. To decrease its speed, the brakes must be applied. The brakes must overcome the car's inertia to slow it down. Also, when the car goes around a curve, its inertia tends to keep it moving in a straight line. The tires on the road must overcome this tendency. Otherwise, the inertia of the car will send it into a skid.

♦ 13-7 FRICTION

Friction is resistance to motion between two objects in contact with each other. If you put a book on a table and then pushed the book, you would find that it took a certain force to

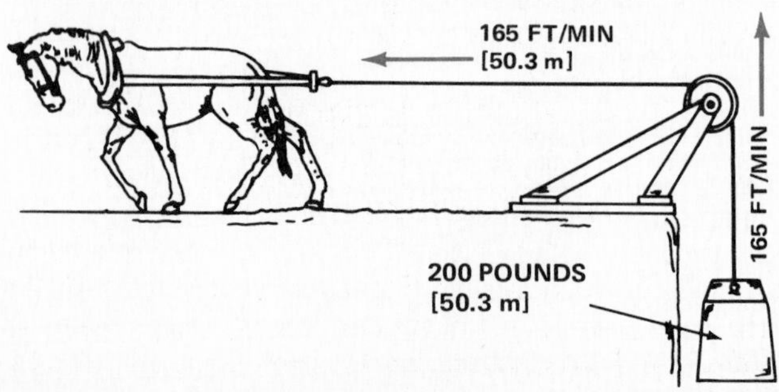

Fig. 13-4 One horse can do 33,000 ft-lb of work a minute.

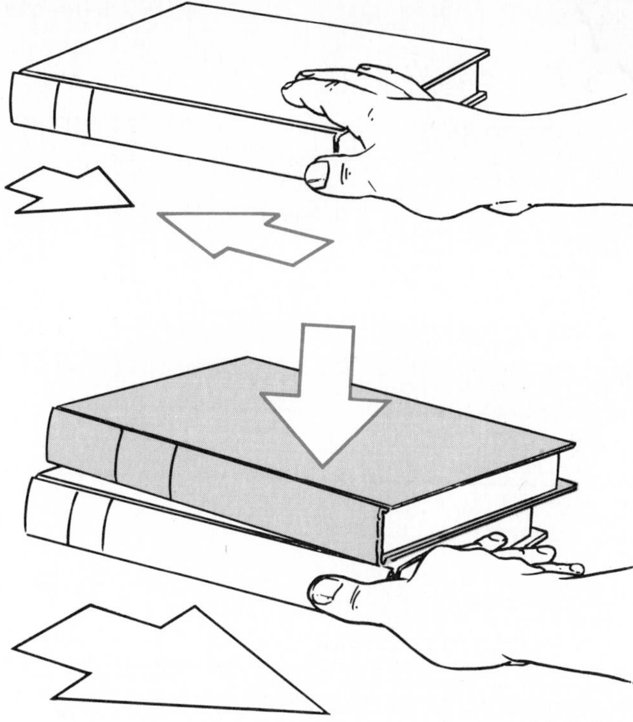

Fig. 13-5 Top, friction resists the push on the book. Bottom, increasing the weight, or load, increases the friction.

move it (Fig. 13-5). If you put a second book on top of the first book, you would find that you had to push harder to move the two books on the table top (Fig. 13-5). Therefore, friction, or resistance to motion, increases with the load. The higher the load, the greater the friction. There are three kinds of friction: dry, greasy, and viscous.

1. Dry Friction This is the resistance to motion between two dry objects, for example, a board being dragged across a floor.

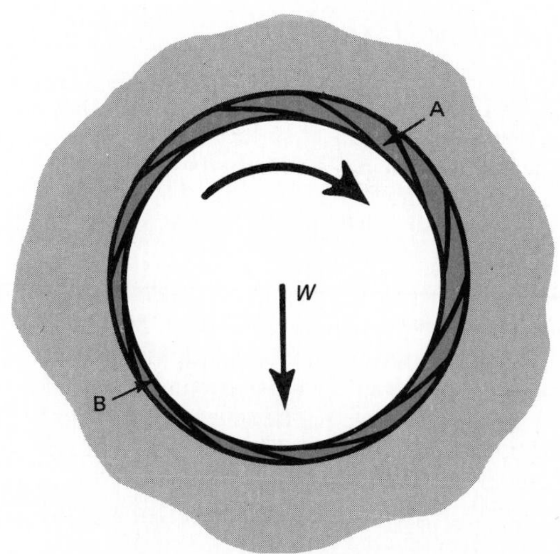

Fig. 13-6 Shaft rotation causes layers of clinging oil to be dragged around with it. The oil moves from the wide clearance A and is wedged into the narrow clearance B, thereby supporting the shaft weight W on an oil film. The clearances are shown exaggerated.

2. Greasy Friction This is the friction between two objects thinly coated with oil or grease. In an automobile engine, greasy friction may occur in an engine on first starting. Most of the lubricating oil may have drained away from the bearing surfaces and from the cylinder walls and piston rings. When the engine is started, only the small amount of oil remaining on these surfaces protects them from undue wear. However, the lubricating system quickly supplies additional oil. But before this happens, greasy friction exists on the moving surfaces. The lubrication between the surfaces where greasy friction exists is not enough to prevent wear. This is why initial starting and warm-up of the engine is hardest on the engine and wears it the most.

3. Viscous Friction *Viscosity* is a term that refers to the tendency of liquids, such as oil, to resist flowing. A heavy oil is more viscous than a light oil and flows more slowly. (It has a higher viscosity, or higher resistance to flowing.) *Viscous friction* is the friction, or resistance to motion, between layers of liquid. In an oiled engine bearing, layers of oil adhere to the bearing and shaft surfaces. Layers of oil clinging to the shaft are carried around by the rotating shaft. They wedge between the shaft and the bearing (Fig. 13-6). The wedging action lifts the shaft so that the oil supports the weight, or load. Now, since the shaft is supported ("floats") on layers of oil, there is no metal-to-metal contact. However, the layers of oil must move over each other. Some energy is needed to make them do so. The resistance to motion between these oil layers is viscous friction.

4. Bushings and Bearings In the engine, as in almost all machinery, the moving parts are lubricated with oil. Therefore the surfaces that move against each other are protected against dry friction. These surfaces are of special materials, specially prepared. The cylinder walls, for example, against which the pistons and piston rings slide, are of smooth gray iron or other metal with good wearing qualities. The cylinder walls in some small engines are chrome-plated to improve their resistance to wear. The piston rings are also made of material that gives long life. Shafts are supported by bushings or bearings. Three types of bearing surfaces in engines are shown in Fig. 13-7.

♦ 13-8 BORE AND STROKE

The size of an engine cylinder is given by its bore and stroke (Fig. 13-8). The *bore* is the diameter of the cylinder. The *stroke* is the distance the piston travels from BDC to TDC. The bore is always stated first. For example, in a 4- by 3½-inch cylinder, the diameter, or bore, is 4 inches [101.6 mm] and the stroke is 3½ inches [88.9 mm]. These measurements are used to figure the piston displacement (♦13-9).

Before about 1955, most engines were built with a long stroke and smaller bore, such as a 3 by 4 engine. Later, engines were designed with a shorter stroke and large bore. For example, a 350-hp Chevrolet engine had a 4-inch [101.6 mm] bore and a 3¼-inch [82.5 mm] stroke. Such engines are called "oversquare." A "square" engine has a bore and stroke of equal lengths.

There are several reasons for the use of the oversquare engine. With the shorter piston stroke, there is less friction loss (♦13-15) and shorter piston-ring travel (which means less wear). Also, the shorter stroke reduces the loads on the engine bearings. In addition, the shorter stroke permits a re-

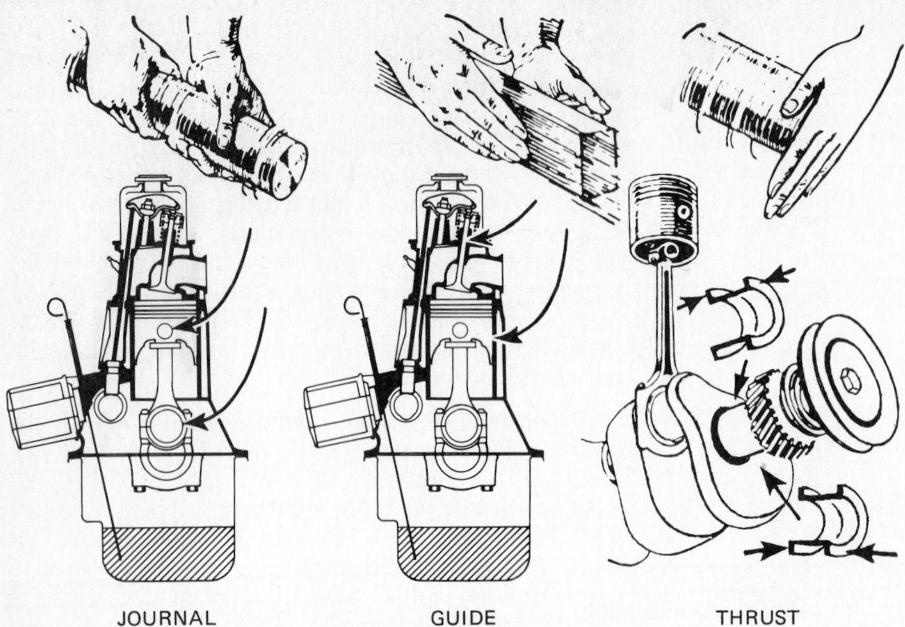

Fig. 13-7 Three types of friction-bearing surfaces in an automobile engine.

duction of engine height. Therefore, the car can have a lower hood line.

Despite the advantages of the shorter-stroke, oversquare engine, concern for atmospheric pollution has forced automotive manufacturers to lengthen the stroke on some engines. For example, one model of the Ford six-cylinder engine now has a stroke of 3.91 inches [99.3 mm]. An earlier engine of the same design had a stroke of 3.13 inches [79.5 mm]. The longer stroke provides more burning time for better combustion, so fewer pollutants are emitted.

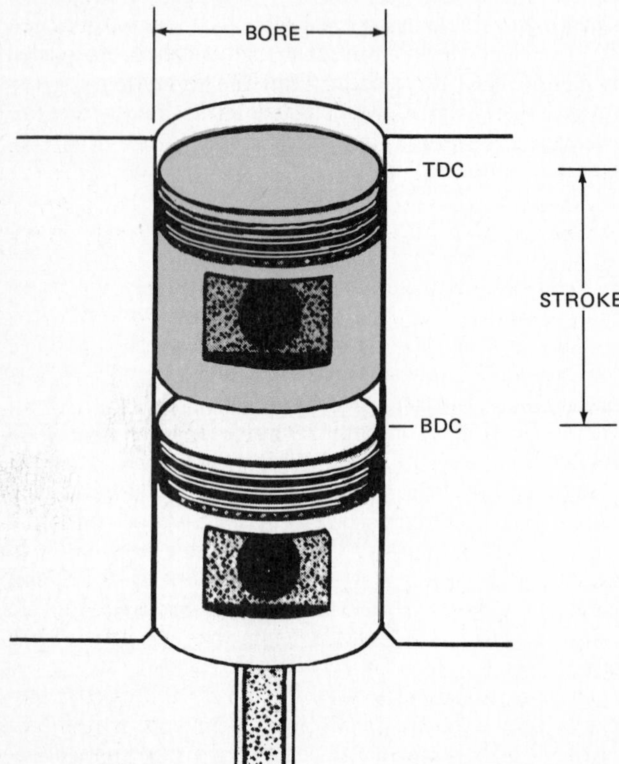

Fig. 13-8 Bore and stroke of an engine cylinder.

♦ 13-9 PISTON DISPLACEMENT

Piston displacement is the volume that the piston displaces, or "sweeps out," as it moves from BDC to TDC. The piston displacement of a 4- by 3½-inch [101.6 by 88.9 mm] cylinder, for example, is the volume of a cylinder 4 inches in diameter and 3½ inches long. To find the piston displacement of this engine, use the formula

$$\frac{\pi \times D^2 \times L}{4} = \frac{3.1416 \times 4^2 \times 3\frac{1}{2}}{4} = \frac{3.1416 \times 16 \times 3\frac{1}{2}}{4} = 43.98 \text{ in.}^3$$

If the engine has eight cylinders, the total displacement is 43.98 times 8, or 351.84 cubic inches.

In the metric system, displacement is given in cubic centimeters (cc). Therefore, a 200-cubic-inch (in^3) displacement would be 3280 cc in metric measurements. Since 1000 cc equals 1 L, 3280 cc is 3.28 L. One cubic inch equals 16.39 cc.

In competitive racing, displacement limitations are set. At the Indianapolis 500—the "Indy 500"—the maximum allowable displacement for a race was set at 305.1 cubic inches for nonsupercharged engines. In many races, the displacement is given in terms of liters. Therefore, the Indy-500 specification (305.1 cubic inches) is 5 L. (One liter is 61.02 cubic inches, so 305.1 divided by 61.02 is 5.)

The Wankel engine does not have pistons, so you cannot figure piston displacement on the Wankel. But you can figure the displacement the rotor produces as the volume in the combustion chamber goes from maximum to minimum (Fig. 13-9). For example, suppose the volume is reduced 490 cc as it goes from maximum to minimum (Fig. 13-9). This is the displacement in one of the three chambers of the rotor. Instead of being referred to as *piston displacement,* the volume is called *single-chamber capacity.*

♦ 13-10 COMPRESSION RATIO

The *compression ratio* of an engine is a measure of how much the air-fuel mixture is compressed in an engine cylinder. It is calculated by dividing the air volume in one cylinder

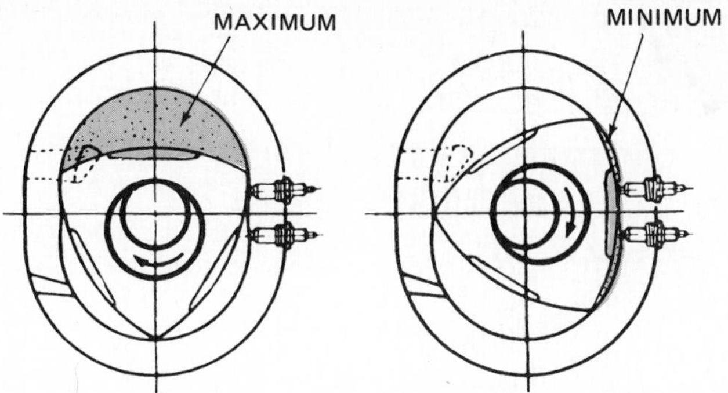

Fig. 13-9 Single-chamber capacity of a Wankel engine.

with the piston at BDC by the air volume with the piston at TDC (Fig. 13-10).

The air volume with the piston at TDC is called the *clearance volume.* It is the clearance that remains above the piston at TDC.

For example, the engine of one car has a cylinder volume of 42.35 cubic inches [694 cc] at BDC (A in Fig. 13-10). It has a clearance volume of 4.45 cubic inches [73 cc] (B in Fig. 13-10). The compression ratio, therefore, is 42.35 divided by 4.45 [694 ÷ 73], or 9.5:1. This means that during the compression stroke, the air-fuel mixture is compressed from a volume of 42.35 cubic inches [694 cc] to 4.45 cubic inches [73 cc] or to 1/9.5 of its original volume.

♦ 13-11 INCREASING COMPRESSION RATIO

Up until recently, the compression ratios of automotive engines were gradually increasing year by year. This increase offers several advantages. The power and economy of an engine increases as the compression ratio goes up (within limits). This does not require an increase in engine size or weight. An engine with a higher compression ratio "squeezes" the air-fuel mixture harder (compresses it more). This causes the air-fuel mixture to produce more power on the power stroke. The reason is that a higher compression ratio causes a higher pressure at the end of the compression stroke. This means higher pressures which exert a greater force on the piston during the power stroke. The burning gases also expand to a greater volume. The result is more push on the piston for a longer part of the power stroke. Therefore, more power is obtained from each power stroke.

Because of these advantages, compression ratios of engines increased year after year. In 1955, the average compression ratio was less than 8:1. By 1969, the ratio had increased to 9.5:1. However, after that, the average compression ratio began to drop. Recently, it was down to 8.2:1.

The basic reason for the drop was concern for atmospheric pollution. The engine produces pollutants as it runs (♦8-3). To reduce these pollutants, antipollution, or emission control devices, are installed on cars. This required removal of lead (tetraethyl lead) from gasoline because these control devices are damaged by lead.

Formerly, lead had been added to gasoline to prevent *ping,* or detonation. Detonation occurs if the compressed air-fuel mixture ignites spontaneously after the timed spark. This condition is discussed in a later chapter. For now, detonation can be summarized as follows.

The higher the compression ratio, the more the air-fuel mixture is compressed on the compression stroke. The more

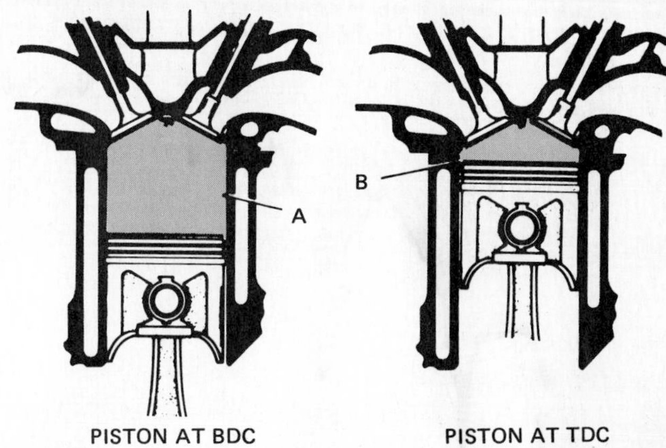

Fig. 13-10 Compression ratio is the volume in a cylinder with the piston at BDC divided by its volume with the piston at TDC, or A divided by B.

the mixture is compressed, the higher its temperature.

If the temperature of the compressed mixture goes high enough, it will ignite from the high temperature. Formerly, lead in the gasoline controlled the problem. But with lead removed, compression ratios had to be dropped.

A second factor to be considered is the effect of carbon buildup in the engine cylinders. Carbon can accumulate as a result of incomplete fuel combustion. The carbon takes up some of the clearance volume. This increases the actual compression ratio. In a severe case of carbon accumulation, the compression ratio will be greatly increased (Fig. 13-11).

♦ 13-12 VOLUMETRIC EFFICIENCY

The amount of air-fuel mixture taken into the cylinder on the intake stroke is a measure of the engine's volumetric efficiency. If the mixture were drawn into the cylinder very slowly, a full measure could get in. But the mixture must pass rapidly through narrow openings and bends in the carburetor and intake manifold. In addition, the mixture is heated (from engine heat) and therefore expands. The rapid movement and heating reduce the amount of mixture that can get into the cylinder. A full charge of air-fuel mixture cannot enter, because the time is too short and because the air becomes heated.

Volumetric efficiency is the ratio of the amount of air-fuel mixture that actually enters the cylinder to the amount that could possibly enter. For example, a certain cylinder has an air volume (A in Fig. 13-10) of 47 cubic inches [770 cc]. If the cylinder were allowed to completely "fill up," it would take in 0.034 ounce [0.964 g] of air. However, suppose that the engine is running at a high speed, so only 0.027 ounce [0.765 g] of air can enter during each intake stroke. This means that the volumetric efficiency is only about 80 percent (0.027 is 80 percent of 0.034). Actually, 80 percent is a good volumetric efficiency for an engine running at fairly high speed. The volumetric efficiency of some engines may drop to as low as 50 percent at high speeds. This is another way of saying that the cylinders are only "half-filled" at high speeds.

This is one reason why engine speed and output cannot increase without limit. At higher speed, the engine has a harder time "breathing," or drawing in air. It is "starved" for air and cannot produce any further increase in power output.

To improve volumetric efficiency, intake valves can be

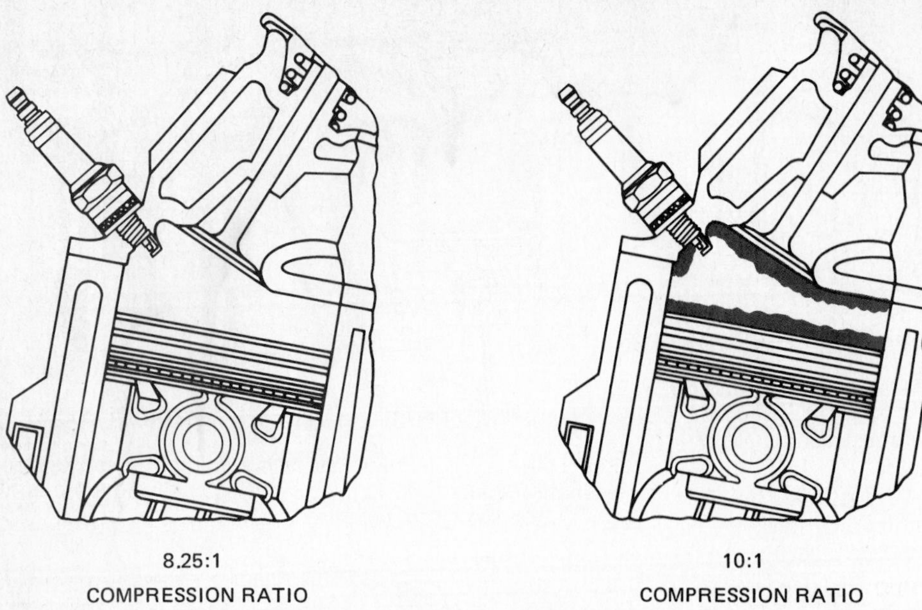

Fig. 13-11 Carbon deposits in the combustion chamber can raise the compression ratio from 8.25:1 (left) to as high as 10:1 (right). *(Chrysler Corporation)*

made larger. In addition, the number of valves per cylinder can be increased (Fig. 10-14). Also, valve lift can be increased by making the cam lobes on the cams larger so that the valve opens wider. However, when this is done, there is danger of the piston head striking the valve head. Unless the piston design takes this into account, serious engine damage could result (♦11-39).

Volumetric efficiency can also be increased by making the intake-manifold passages wider, and as straight and short as possible. Also, the smoothness of the inside surfaces of the intake manifolds is important. Rough surfaces slow down the flow of air-fuel mixture. Another way to improve volumetric efficiency is to use carburetors with extra air passages (called "barrels"). These open at high speed to improve engine breathing. A later chapter describes carburetors in detail. In addition, turbochargers can be used to increase volumetric efficiency (♦15-14). All these changes help produce more power at higher speeds.

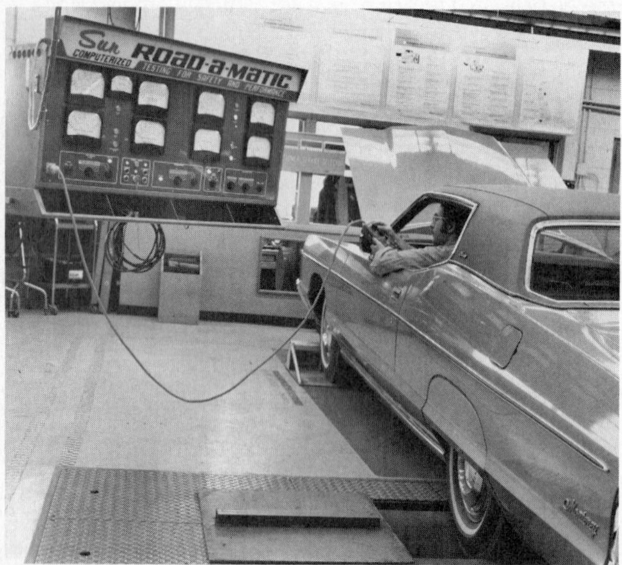

Fig. 13-12 Automobile in place on a chassis dynamometer. The drive wheels turn the dynamometer rollers, which measure the power available at the wheels. *(Sun Electric Corporation)*

♦ 13-13 BRAKE HORSEPOWER

Engine power output is measured in terms of *brake horsepower* (bhp). The name comes from the braking device that is used to hold engine speed down while horsepower is measured. When an engine is rated at 300 hp [224 kW], for example, it is really brake horsepower that is meant. This is the amount of usable power the engine can produce at a certain speed at wide-open throttle.

The usual way to rate an engine is with a *dynamometer* (Fig. 13-12). This device has a power absorber, such as an electric generator or a water brake, which can put different loads on the engine. Therefore, the dynamometer can measure the amount of horsepower the engine can develop under various operating conditions.

Some dynamometers are used to test engines that have been removed from cars. However, the dynamometer used in the service shop checks the engine *in* the car. This type of unit is called a *chassis dynamometer.* On these, the drive wheels of the car are placed on rollers. Then, the engine drives the wheels, and the wheels drive the rollers. The rollers can be loaded varying amounts so that engine power can be measured. The use of the chassis dynamometer is becoming more common in the automotive-servicing field. It can give a very quick report on engine conditions (by measuring power at various speeds and loads). This type of dynamometer is also used to test and adjust automatic transmissions in the shop. No road testing is necessary.

♦ 13-14 INDICATED HORSEPOWER

Indicated horsepower (ihp) is the power that the engine develops inside the combustion chambers during the combustion process. A special device is required to measure ihp. It

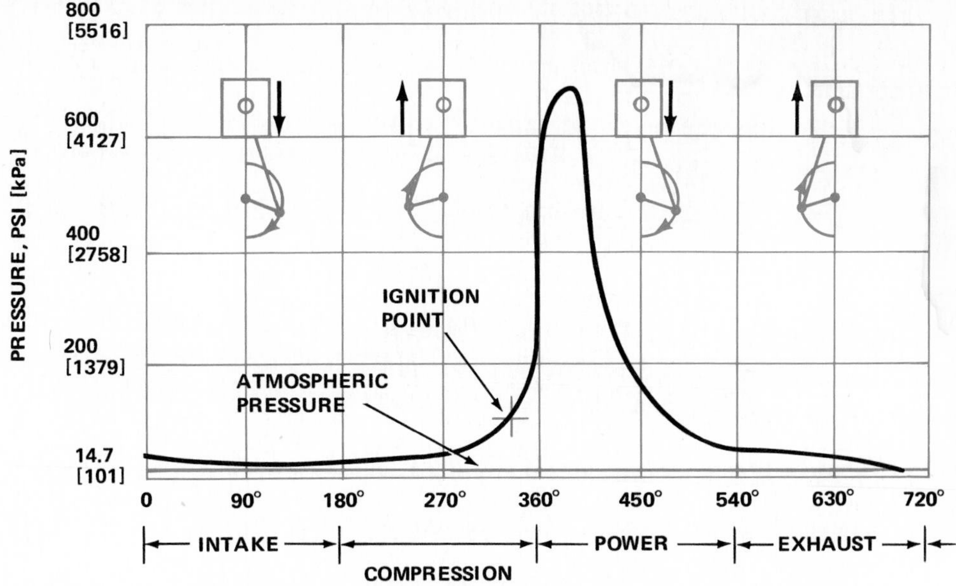

Fig. 13-13 Pressures in an engine cylinder during the four piston strokes. The four strokes require two crankshaft revolutions (360 degrees each), for a total of 720 degrees of rotation. This curve is for one particular engine operating at one definite speed and throttle opening. Changing the speed and throttle opening would change the power curve for this engine.

measures the pressures in the engine cylinders (Fig. 13-13). The four small drawings show the four piston strokes. The curve shows the pressures in the cylinder during these four strokes. These pressures are used to figure ihp. Indicated horsepower is always higher than bhp. This is because some of the power developed in the engine cylinders is used to overcome the engine's internal friction.

♦ 13-15 FRICTION HORSEPOWER

Friction horsepower (fhp) is the power required to overcome the friction of the moving parts in the engine. One of the major causes of friction loss (or fhp) is piston-ring friction. Under some conditions, the friction of the rings moving on the cylinder walls accounts for 75 percent of all friction losses in the engine. This points up one advantage of the short-stroke, oversquare engine. With a short stroke, the piston rings do not have as far to move. Therefore, ring friction is lower. Figure 13-14 shows a curve of friction horsepower for one engine operating under certain specified conditions.

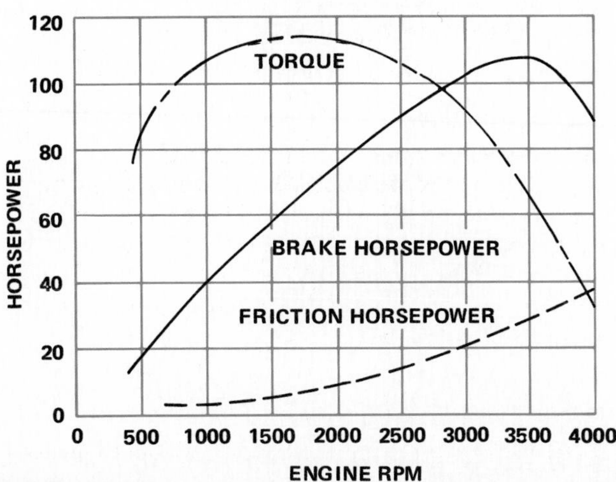

Fig. 13-14 Torque-bhp-fhp curves for an engine.

♦ 13-16 RELATING BHP, IHP, and FHP

Brake horsepower is the power delivered, ihp is the power developed in the engine, and fhp is the power lost due to friction. The relationship among the three is

$$\text{bhp} = \text{ihp} - \text{fhp}$$

The horsepower delivered by the engine (bhp) is equal to the horsepower developed (ihp) minus the power lost due to friction (fhp).

♦ 13-17 ENGINE TORQUE

Torque is turning force. When the piston is moving down on the power stroke, it applies torque to the engine crankshaft (through the connecting rod). The harder the push on the piston, the greater the torque applied. Therefore, the higher the combustion pressures, the greater the amount of torque.

The dynamometer is normally used to check engine torque. Torque can be measured at the same time as horsepower on the dynamometer.

♦ 13-18 BRAKE HORSEPOWER VERSUS TORQUE

The torque that an engine can develop changes with engine speed (Fig. 13-14). During intermediate speeds, volumetric efficiency is high. There is sufficient time for the cylinders to

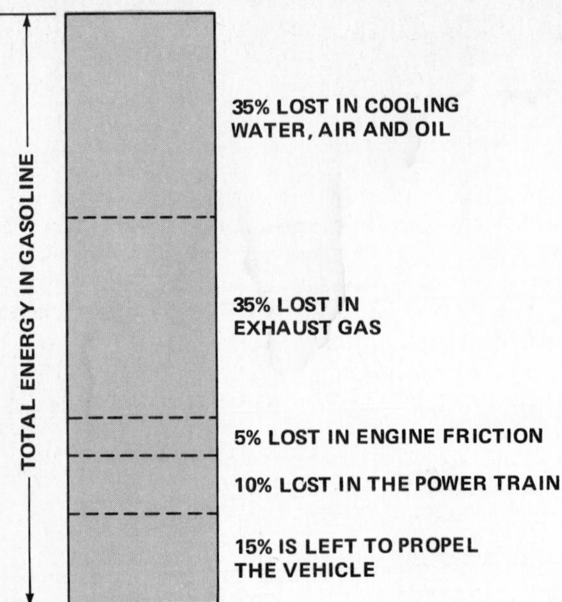

Fig. 13-15 Energy lost from cylinders to wheels.

become fairly well "filled up." This means that with a fairly full charge of air-fuel mixture, higher combustion pressures will develop. With higher combustion pressures, the engine torque is higher.

But, at higher speed, volumetric efficiency drops off. There is not enough time for the cylinders to become filled up with air-fuel mixture. Since there is less air-fuel mixture to burn, the combustion pressures are not as high. There is less push on the pistons. Therefore, engine torque is lower. Figure 13-14 shows how the torque drops off as engine speed increases.

The bhp curve of an engine is different from the torque curve. Figure 13-14 compares the bhp of the same engine for which the torque curve is shown. It starts low at low speed and increases until a high engine speed is reached. Then, at still higher engine speeds, bhp drops off.

The drop-off of bhp is due to reduced torque at higher speed and to increased fhp at the higher speed. Figure 13-14 compares the curves of torque, bhp, and fhp for an engine.

Note that the curves in Fig. 13-14 are for one particular engine only. Different engines have different torque, bhp, and fhp curves. Peaks may be at higher or lower speeds, and the relationships may not be as shown in the curves in Fig. 13-14.

♦ 13-19 ENGINE EFFICIENCY

The term *efficiency* compares the effort exerted and the results obtained. For engines, efficiency is the relation between the power delivered and the power that could be obtained if the engine operated without any power loss. Engine efficiency can be computed in two ways, as *mechanical* efficiency and as *thermal* efficiency.

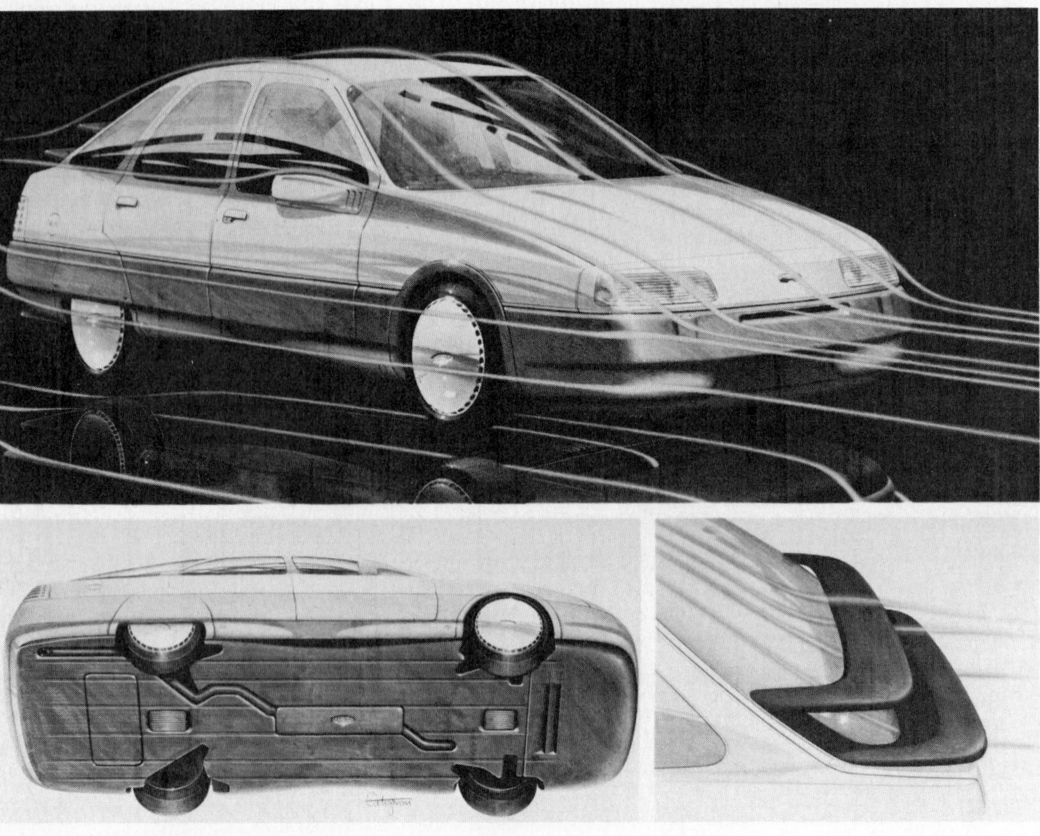

Fig. 13-16 Streamlining a car makes it more "slippery" so that it has less wind resistance. The white lines illustrate how the air moves up, over, and around the car. The bottom view shows how a cover is applied so that the bottom of the car presents minimum air resistance. *(Ford Motor Company)*

1. Mechanical Efficiency This is the relationship between bhp and ihp. It is

$$\text{Mechanical efficiency} = \frac{\text{bhp}}{\text{ihp}}$$

EXAMPLE At a certain speed, the bhp of an engine is 116, and its ihp is 135. Therefore mechanical efficiency is bhp/ihp = 116/135 = 0.86, or 86 percent. This means that 86 percent of the power developed in the cylinders is delivered by the engine. The remaining 14 percent, or 19 hp (14.17 kW), is consumed as fhp.

2. Thermal Efficiency *Thermal* means "of or related to heat." The thermal efficiency of an engine is the relation between the power produced and the energy in the fuel burned to produce this power.

Some of the heat produced by combustion is carried away by the engine cooling system. Some heat is lost in the exhaust gases. They are hot when they leave the cylinder. These are heat (thermal) losses that reduce the thermal efficiency of the engine. They do not add to the power output of the engine. The remainder of the heat is used by the engine to develop power. Because a large quantity of heat is lost during engine operation, thermal efficiencies may be as low as 20 percent. They are seldom higher than 25 percent.

♦ 13-20 OVERALL EFFICIENCY

The gasoline enters the engine with a certain energy content, a certain ability to do work. At every step in the process, from the burning of the gasoline in the cylinders to the rotation of the car wheels, energy is lost. Figure 13-15 illustrates these losses for one engine and car during one test run. As little as 15 percent of the energy in the gasoline remains to move the car. This energy is used to overcome rolling resistance, air resistance, and power-train resistance and to accelerate the car.

1. Rolling Resistance This results from irregularities in the road over which the wheels ride. It is also a result of the flexing of the tires as they turn under the car. Rolling resistance is less with radial tires as explained in a later chapter on tires.

2. Air Resistance Air resistance is the resistance of the air to the passage of the car body through it. As car speed increases, so does the air resistance. At 90 miles per hour (mph) [145 km/h], as much as 75 percent of the power that reaches the wheels is used up in overcoming air resistance, or the drag of the air on the car. Even at lower speeds the power required is significant. It can reduce fuel economy by two or more miles per gallon. Then the car goes fewer miles on a gallon of fuel. This loss of mileage is due to the drag of the air.

To improve fuel economy, modern automobiles are being made more "slippery" so that they slide through the air more easily. The process is called *streamlining.*

Figure 13-16 shows a new car design that has a very low resistance to air. The white lines are designed to show how the air slides up and around the car. The underbody is completely enclosed.

Air resistance is measured as the *drag coefficient.* The lower the drag coefficient, the less fuel it takes to move the car through air. The car shown in Fig. 13-16 has a drag coefficient of 0.22. This is about 50 percent less than the ratings of most new cars. It can go about 2.5 miles farther on a gallon of fuel than other cars of comparable size built today.

3. Power Train Resistance Some engine power is lost between the engine and the drive wheels. One reason is because of friction (♦13-7) between moving parts in the power train, which includes the clutch, transmission, and drive axle.

4. Acceleration Power is required to increase car speed. The power applied to accelerate the car overcomes the inertia of the car. Energy in the form of speed is stored in the car.

♦ REVIEW QUESTIONS ♦

Select the *one* correct, best, or most probable answer to each question. You can find the answers in the section indicated at the end of each question.

1. Distance times force equals (♦13-1)
   a. torque
   b. work
   c. power
   d. horsepower
2. The rate, or speed, at which work is done is (♦13-3)
   a. torque
   b. work
   c. power
   d. horsepower
3. The ability, or capacity, to do work is (♦13-2)
   a. torque
   b. energy
   c. power
   d. horsepower
4. The application of a twisting force, with or without motion, is (♦13-4)
   a. torque
   b. energy
   c. work
   d. power
5. The size of an engine cylinder is referred to in terms of its (♦13-8)
   a. diameter and bore
   b. displacement and efficiency
   c. bore and stroke
   d. bore and length

PART 3

Part 3 of *Automotive Mechanics* describes the construction, operation, and servicing of automotive engine systems. These include fuel systems (both carbureted and fuel-injected), lubrication systems, and cooling systems. Ignition systems are covered in Part 4, Automotive Electrical and Electronic Equipment. There are nine chapters in Part 3:

Automotive Engine Systems

CHAPTER 14
Automotive Engine Fuels

After studying this chapter, you should be able to:

1. Discuss the purpose of the additives used in gasoline.
2. Explain volatility and octane ratings and why they are important in gasoline.
3. Describe knocking, detonation, and preignition and explain what causes them and how they can be prevented.
4. Discuss the composition and special characteristics of diesel-engine fuel.
5. Explain what gasohol is.

♦ 14-1 AUTOMOTIVE ENGINE FUELS

The most common fuel for automotive engines is gasoline. It is used in spark-ignition engines. Some drivers use gasohol, which in a mixture of alcohol and gasoline. Other vehicles with spark-ignition engines use liquefied petroleum gas (LPG). These vehicles require a special fuel system. Diesel engines use a light oil called *diesel fuel*. All of these fuels are discussed in this chapter.

♦ 14-2 GASOLINE

Gasoline is a hydrocarbon (abbreviated HC), made up largely of hydrogen and carbon compounds. In ♦8-3 there is an explanation of what happens in the engine when gasoline is burned. The resulting exhaust gas contains water vapor, carbon dioxide, and the pollutants carbon monoxide (partly burned gasoline) and hydrocarbons (unburned gasoline). Later chapters describe the antipollution devices used to reduce these pollutants.

♦ CAUTION ♦ Gasoline is a very dangerous liquid if it is not handled properly. Vapor rising from an open pan of gasoline could ignite from a spark, a flame, or a lighted cigarette. The result could be an explosion or fire. Gasoline from a leaky fuel pump could be ignited by a hot engine or any electric spark.

Gasoline is often referred to as "gas," which can be confusing. The sort of gas that you burn in a gas stove or to heat a house is actually a vapor or gas that is delivered through gas lines or pipes. So there is a gas that is a *gas* and "gas" which is a slang expression for the liquid fuel gasoline.

♦ 14-3 SOURCE OF GASOLINE

Gasoline is made from crude oil by a refining process that also produces engine lubricating oil, diesel fuel oil, and other products. During the refining process, several additives are put into the gasoline to improve its characteristics. Good gasoline should have:

1. Proper volatility, which determines how quickly gasoline vaporizes
2. Resistance to spark knock, or detonation

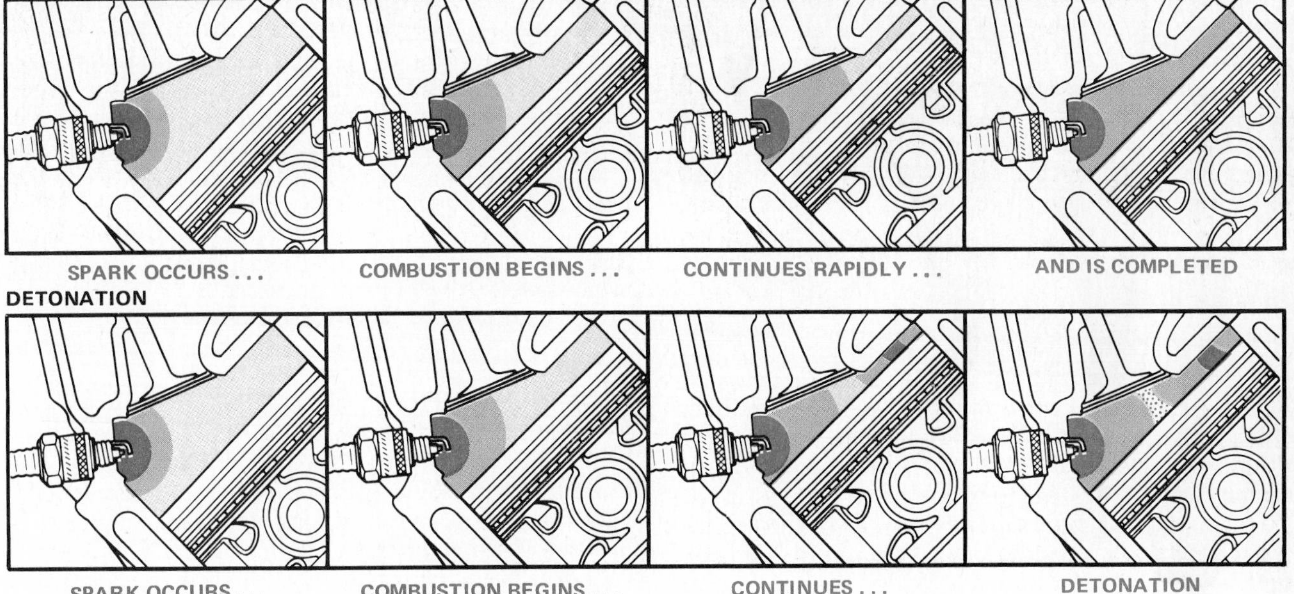

Fig. 14-1 Normal combustion without detonation is shown in the top row. The fuel charge burns smoothly from beginning to end, providing an even, powerful force to the piston. Detonation is shown in the bottom row. The last part of the fuel explodes, or burns, almost instantly, to produce detonation, sometimes called *spark knock*. *(Champion Spark Plug Company)*

3. Oxidation inhibitors, which prevent formation of gum in the fuel system
4. Antirust agents, which prevent rusting of metal parts in the fuel system
5. Anti-icers, which retard carburetor icing and fuel-line freezing
6. Detergents, which help keep the carburetor clean
7. Dye for identification

Let's talk about volatility and antiknock value first.

♦ 14-4 VOLATILITY

After gasoline is mixed with air in the carburetor, the gasoline must vaporize quickly, before it enters the engine cylinders. If the gasoline is slow to vaporize, tiny drops of liquid gasoline will enter the cylinders. Because these drops do not burn, some of the fuel is wasted. It goes out the tail pipe and helps create atmospheric pollution. Also, the gasoline drops tend to wash the lubricating oil off the cylinder walls. This increases the wear on the cylinder walls, piston rings, and pistons.

The ease with which gasoline vaporizes is called its *volatility*. A high-volatility gasoline vaporizes very quickly. A low-volatility gasoline vaporizes slowly. A good gasoline should have just the right volatility for the climate in which the gasoline is to be used. If the gasoline is too volatile, it will vaporize in the fuel system. The result will be a condition called *vapor lock*. It prevents the flow of gasoline to the carburetor. Vapor lock causes the engine to stall from lack of fuel.

The volatility of gasoline is seasonally adjusted by the refiner. This tends to increase the chances of vapor lock occurring under certain conditions. For example, suppose the fuel tank of a car is filled with a gasoline blended for winter driving. This gasoline has a high volatility. If the weather suddenly becomes unseasonably hot, the gasoline will vaporize in the fuel system. Then the car may stall because of vapor lock.

♦ 14-5 ANTIKNOCK VALUE

Spark knock also is called *detonation*. If you have ever been in a car that had detonation, you have heard the sound. The engine pings, usually under light load. It sounds like someone is tapping on the cylinder walls with a hammer. Look at Fig. 14-1. The horizontal row at the top of the figure shows what happens during normal combustion. The fuel charge—the mixture of air and fuel—starts burning as soon as the spark occurs at the spark plug. The flame sweeps smoothly and evenly across the combustion chamber. Now look at the horizontal row at the bottom. The spark starts combustion in the same way. However, before the flame can reach the far side of the combustion chamber, the last part of the charge explodes. The result is a very quick increase in pressure. This is detonation, and it gives off a pinging sound.

Detonation can ruin an engine. The heavy shocks on the piston put a great strain on the engine parts. Continued detonation can cause pistons to chip and parts to break. So detonation must be avoided.

Gasoline refiners have various ways to make gasoline that does not detonate easily. A gasoline that detonates easily is called a *low-octane* gasoline. A gasoline that resists detonation is called a *high-octane* gasoline.

♦ 14-6 INCREASING THE OCTANE RATING

One way to increase the octane rating is to change the refining process. Another way is to add a small amount of tetraethyl lead, also known as "lead," "ethyl," or "tel." This additive tends to prevent the last part of the fuel charge from detonating. However, there are two problems with using tetraethyl lead.

One problem is that when gasoline containing lead is burned, some of it gets into the air. Lead is a poison. Breathing air containing lead can cause lead poisoning. Lead poisoning can cause illness and death. The other problem is that the lead keeps exhaust emission controls from working properly. These are two reasons why gasoline without lead is now required for most new cars.

♦ 14-7 TWO KINDS OF GASOLINE

One of the emission controls is called a *catalytic converter*. (It is discussed in a later chapter on emission controls.) The exhaust gases from the engine flow through this device. It reduces the amount of unburned gasoline vapor (HC) and carbon monoxide (CO) in the exhaust gases. However, it will stop doing its job if leaded gasoline is used in the engine. That is the reason service stations have pumps labeled "No Lead" or "Unleaded." These pumps sell gasoline without lead for cars with catalytic converters. Other pumps may sell leaded gasoline.

♦ 14-8 COMPRESSION RATIOS AND DETONATION

Over the years the compression ratios of automobile engines have gone up. The reason is that higher compression ratios give engines more power. Compression ratio is the amount that the air-fuel mixture is compressed on the compression stroke. The more the air-fuel mixture is compressed, the higher the compression ratio is.

But a high compression ratio can cause a problem. It increases the temperature of the air-fuel mixture. The higher heat of compression may cause the remaining air-fuel mixture to explode before normal combustion is completed. This is detonation. If the air-fuel mixture explodes before the spark occurs at the spark plug, this condition is called *preignition*. The compression ratio must be kept low enough to make sure that the fuel charge will not ignite prematurely from the heat of compression. Also, the higher combustion temperatures cause increased amounts of another exhaust-gas pollutant, nitrogen oxides (NO_x).

♦ 14-9 REDUCING COMPRESSION RATIOS

Increasing the combustion temperature increases the formation of NO_x. To have cleaner air, it is desirable to reduce the combustion temperature. One method is to reduce compression ratios to slightly less than 9:1 (for spark-ignition engines). A lower compression ratio means lower combustion temperatures. Then less NO_x is formed during the combustion process. This has also reduced engine performance and fuel economy. Other methods of lowering combustion temperature are also used. These are discussed in later chapters on emission controls.

♦ 14-10 GASOHOL

Gasohol is usually 10 percent ethyl alcohol and 90 percent unleaded gasoline. Ethyl alcohol can be made from sugar, grain, and other organic substances. Gasohol can be used in engines without any change in the fuel system. With a shortage of crude oil, many people believe that gasohol will help ease the demand for oil and the gasoline made from oil.

If more than 10 percent ethyl alcohol is added, the carburetor or fuel-injection system must be reworked to provide a richer mixture. For straight ethyl alcohol, the air-fuel ratio would have to be about 9:1. This is much richer than the usual 14.5:1 for gasoline, or 10 percent gasohol.

Note Air-fuel ratio is the ratio, by weight, of the air and fuel in the mixture. A 14.5:1 ratio, for example, means that there is 14.5 times more air, by weight, than fuel.

♦ 14-11 METHYL ALCOHOL

Methyl alcohol is also known as "methanol" and "wood alcohol." It can be used straight or blended with gasoline, in about the same way as gasohol. However, to use straight methyl alcohol requires reworking of the carburetor or fuel-injection system to provide an air-fuel ratio of 6.4:1. Methanol can be manufactured from coal, oil shale, wood, manure, garbage, and many other substances. It has been used for years as a fuel for race cars. It is the only clean and readily storable liquid fuel that we know how to make from coal.

However, there are disadvantages to using methanol. It is highly toxic (poisonous) and corrosive. It attacks aluminum, solder, plastics, and other materials. Also, it attracts water. If water gets into a blend, the gasoline and alcohol will separate. Then the engine stalls because the fuel-delivery system is calibrated to use only the blend.

♦ 14-12 LIQUEFIED PETROLEUM GAS

Propane is a type of liquefied petroleum gas (LPG) made from crude oil. When put under pressure, propane turns to liquid. When the pressure is released, it turns back into a gas. Some car manufacturers offer a propane fuel system as an option (Fig. 14-2).

One advantage to using propane is that it has an octane rating of over 100. This means the engine compression ratio can be raised for greater power and efficiency. Also, propane burns cleanly. Less engine wear results. No fuel pump and little emission control equipment is required.

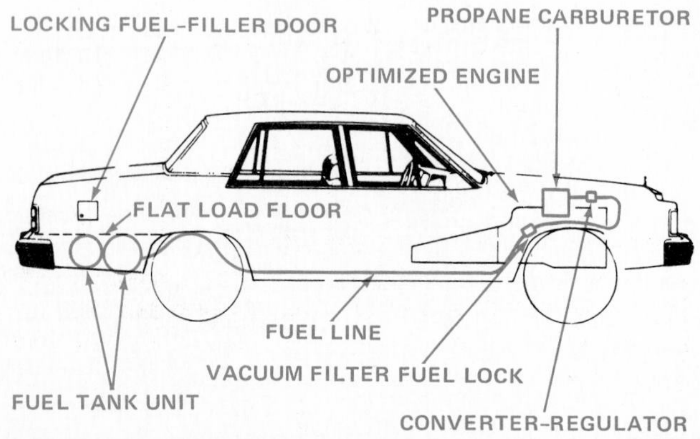

Fig. 14-2 A propane fuel system on a new car. *(Ford Motor Company)*

However, the fuel system requires a pressurized fuel tank and a converter-regulator. It allows the propane to turn to gas before it enters the special carburetor.

♦ 14-13 DIESEL-ENGINE FUEL

Diesel engines use diesel fuel oil. The fuel oil is sprayed, or injected, into the engine cylinders toward the end of the compression strokes (Fig. 9-21). Heat of compression ignites the fuel oil, and the power stroke follows.

Diesel fuel is made from crude oil by the same refining process that produces gasoline. Diesel fuel is a light oil with the proper viscosity, volatility, and cetane number for use as a fuel.

1. Volatility This is a measure of how easily a liquid evaporates. Diesel fuel has a low volatility. It boils at a temperature of 700°F [371°C] or less. There are two grades for automotive diesel engines, number 1 diesel and number 2 diesel. Number 1 diesel is more volatile. It is used where temperatures are very low. Number 2 diesel is recommended for most driving conditions. It is used in most automotive diesel engines.

2. Viscosity This refers to the ease with which a liquid flows. The lower the viscosity, the more easily the liquid flows. Diesel fuel must have a relatively low viscosity. It must flow through the fuel-system lines and spray into the engine cylinders with little resistance. If the oil has high viscosity, it will not break up into fine particles. Then it will not burn rapidly enough. Engine performance will be poor. But if the viscosity is too low, the fuel oil will not lubricate the moving parts in the fuel pump and fuel injectors. Damage may result.

Number 2 diesel has the right viscosity for most driving conditions. Number 1 diesel has lower viscosity so it will flow and spray properly at low temperatures.

♦ 14-14 CETANE NUMBER

The *cetane number* of diesel fuel refers to the ease with which the fuel ignites. With a high cetane number, the fuel ignites easily (or at a relatively low temperature). The lower the cetane number, the higher the temperature needed to ignite the fuel.

Also, the low-cetane-number fuel takes a little longer to ignite. This is called *ignition lag*. During this slight delay, the fuel tends to accumulate in the cylinder. Then, when ignition does occur, all the accumulated fuel ignites at once. The pressure goes up suddenly and a combustion knock results. This is similar to detonation in a spark-ignition engine (♦14-5).

High-cetane fuel ignites as soon as it enters the cylinder. There is no accumulation of fuel. The result is a smooth pressure rise, so no combustion knock takes place.

♦ 14-15 NEED FOR CLEAN DIESEL FUEL

Diesel fuel must be clean and free of water. Even almost invisible dirt particles can clog the injection nozzles and cause poor engine performance. Water can rust internal fuel-pump and injection-nozzle parts. Many cars and trucks equipped with diesel engines have an antiwater system which signals the driver by a warning light when excessive water accumulates in the fuel tank. The water can then be removed from the tank. This is covered in a later chapter.

At temperatures below 20°F [−6.7°C], wax forms in diesel fuel. The wax may clog the fuel filters and injection nozzles. One method of preventing problems with wax is to blend number 1 diesel fuel (kerosene) with the number 2 diesel fuel normally used. Another recommendation is to switch to number 1 fuel in temperatures below 20°F [−6.7°C]. Fuel heaters (♦19-15) are also used to heat diesel fuel. They keep the fuel above the temperature at which the wax forms.

♦ REVIEW QUESTIONS ♦

Select the *one* correct, best, or most probable answer to each question. You can find the answers in the section indicated at the end of each question.

1. The exhaust gas contains (♦14-2)
   a. gasoline vapor and water vapor
   b. carbon dioxide and gasoline vapor
   c. water vapor and carbon monoxide
   d. all of the above
2. The ease with which gasoline vaporizes is called its (♦14-4)
   a. oxidation
   b. octane number
   c. volatility
   d. cetane number
3. When the last part of the air-fuel mixture—the end gas—explodes before being ignited by the flame traveling from the spark plug, the result is (♦14-5)
   a. detonation
   b. preignition
   c. stalling
   d. vaporization
4. A gasoline that detonates easily is called (♦14-5)
   a. high-octane gasoline
   b. low-octane gasoline
   c. unleaded gasoline
   d. blended fuel
5. When the air-fuel mixture ignites before the spark takes place at the spark plug, the condition is called (♦14-8)
   a. detonation
   b. ignition
   c. preignition
   d. rumble

CHAPTER 15
Automotive Fuel and Exhaust Systems

After studying this chapter, you should be able to:

1. List and describe the purpose, construction, and operation of all components in the carbureted fuel system except the carburetor.
2. Explain the basic differences between the carbureted fuel system and the fuel-injection system.
3. Explain the purpose and operation of the turbocharger.
4. Identify and explain the operation and construction of the various fuel-system components on the car.

♦ 15-1 FUEL SYSTEMS FOR GASOLINE ENGINES

This chapter covers all the components in the fuel systems using carburetors except the carburetor itself. It is covered in the following chapter. The fuel tank, fuel lines, fuel pump, air cleaner, and exhaust system are described in following sections.

For many years, the gasoline engines used in automobiles have had carburetors in their fuel systems. The carburetor is a mixing device which mixes air and gasoline in the proper proportions to produce a combustible mixture. Figure 15-1 is a simplified view of carburetor action. The combustible mixture flows from the carburetor to the engine cylinders.

A second type of fuel system is used on many cars with spark-ignition engines. This is the fuel-injection system (Fig. 15-2). In this system, the carburetor is replaced by a throttle body whose only purpose is to control the amount of air entering the intake manifold. Most fuel-injection systems include a fuel pump, fuel lines, an electronic control unit, and one or more fuel-injection valves. The *injection valve* is a nozzle with a small hole through which fuel is sprayed on a signal from the electronic control unit.

In some fuel-injection systems, there is an injection valve for each cylinder (Figs. 15-2 and 15-3). At the proper moment before the piston reaches TDC on the exhaust stroke and the intake stroke begins, each injection valve sprays a metered amount of fuel into the intake manifold. The fuel is sprayed into the air before it reaches the intake valve (Fig. 15-3). Then, when the intake valve opens, the air-fuel mixture flows into the engine cylinder.

In other fuel-injection systems, the injection valve is located in the throttle body (Fig. 15-4). This is known as a *throttle-body fuel-injection* (TBFI) system or a throttle-body injection (TBI) system. If the throttle body has only one throttle valve, there is one injection valve. For throttle bodies with two throttle valves, there are two injection valves. These and other gasoline fuel-injection systems are covered in a later chapter.

♦ 15-2 PURPOSE OF THE FUEL SYSTEM

Both the carbureted fuel system and the fuel-injection system have the same job. That is to supply a combustible mixture of air and fuel to the engine. The fuel system must change the

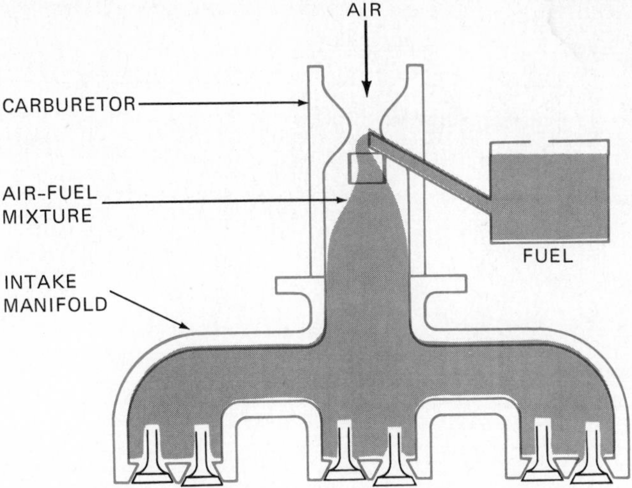

Fig. 15-1 Simplified view of a carbureted fuel system.

proportions of air and fuel for different operating conditions. When the engine is cold, for example, the mixture must be rich (have a high proportion of fuel). This is because the fuel does not vaporize readily at low temperatures. Extra fuel must be added to the mixture so that enough will vaporize to form a combustible mixture.

♦ 15-3 CARBURETED-FUEL-SYSTEM COMPONENTS

The carbureted fuel system consists of the fuel tank, fuel pump, fuel filter, carburetor, intake manifold, and fuel lines. The fuel lines are tubes connecting the tank, fuel pump, and carburetor (Fig. 15-5). Most of these components are the same in both the carbureted and the fuel-injection systems. Each component is described in following sections.

♦ 15-4 FUEL TANK

The fuel tank (Fig. 15-6) is normally located at the rear of the vehicle. It is usually made of sheet metal or plastic. It is attached to the frame or body. The filler opening of the tank is closed by a cap. The fuel line is attached to a pickup tube which is usually part of the tank fuel-gauge assembly. In most tanks, there is a strainer on the end of the fuel-pickup tube.

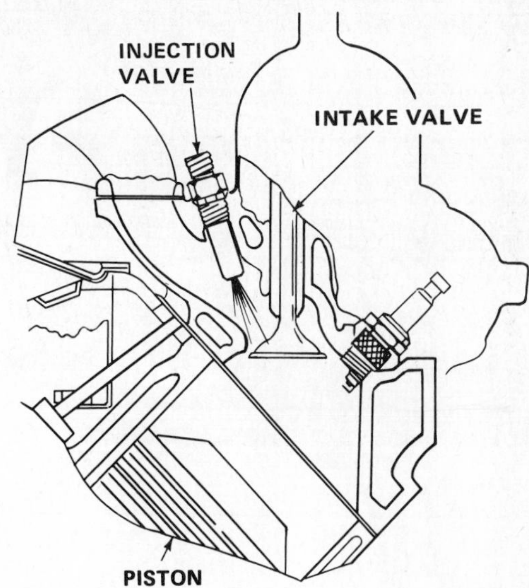

Fig. 15-3 Simplified view showing the method of injecting fuel into the intake manifold at the intake-valve port. This is port fuel injection.

Fuel tanks in older cars have a vent pipe to allow air to escape when the tank is being filled.

Vaporized fuel (HC) escaping from the fuel tank through the vent pipe contributes to air pollution. Therefore, cars manufactured since 1970 have been equipped with an evaporative emission control system. In this system, the fuel-tank vent is connected to a charcoal canister. It holds the vapor and prevents its escape into the air.

♦ 15-5 FUEL FILTERS AND SCREENS

Fuel systems have filters and screens to prevent dirt in the fuel from entering the fuel pump or carburetor. Dirt could prevent normal operation of these units and cause poor engine performance. Filters may be a separate unit connected into the fuel line between the tank and the fuel pump, or between the fuel pump and the carburetor (Fig. 15-7), or in or on the carburetor itself. Figure 15-8 shows the type that is in the carburetor. The screw threads enter a tapped hole in the carburetor. The fuel line fits on the opposite end of the filter. Figure 15-9 shows an in-line fuel filter that has a magnet to pick up metal particles in the fuel.

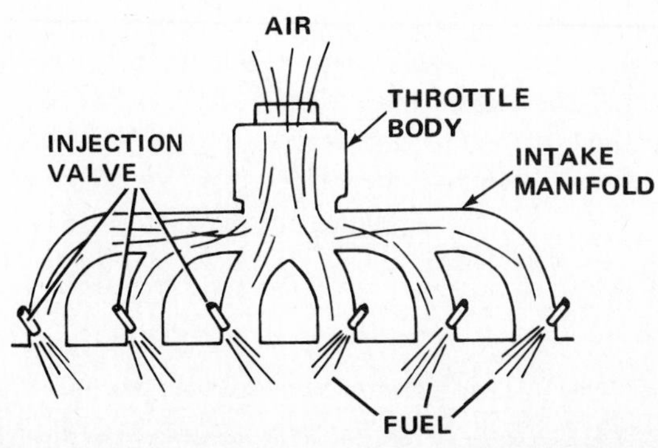

Fig. 15-2 Simplified view of a port or multiple-point fuel-injection system.

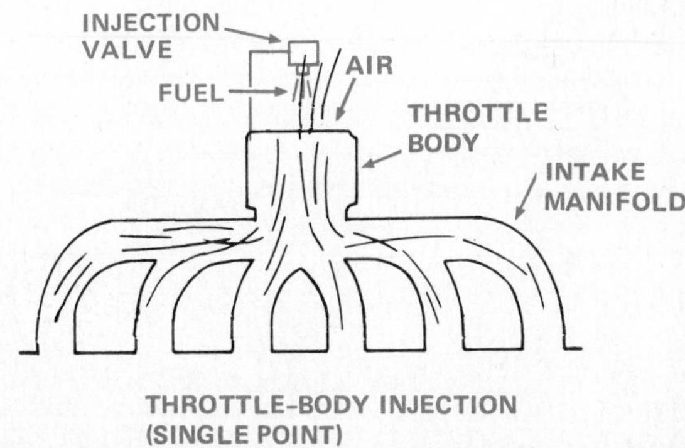

Fig. 15-4 The injection valve sprays fuel directly into the air passing through the throttle body.

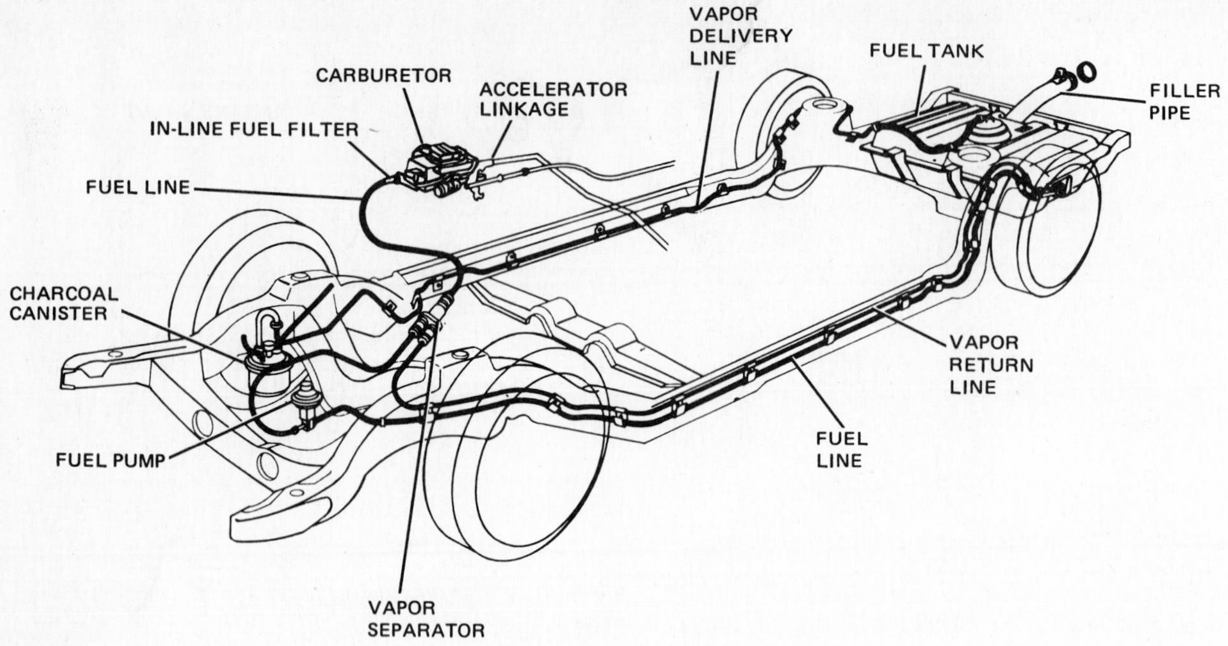

Fig. 15-5 Fuel system for a car with a V-8 engine. *(Ford Motor Company)*

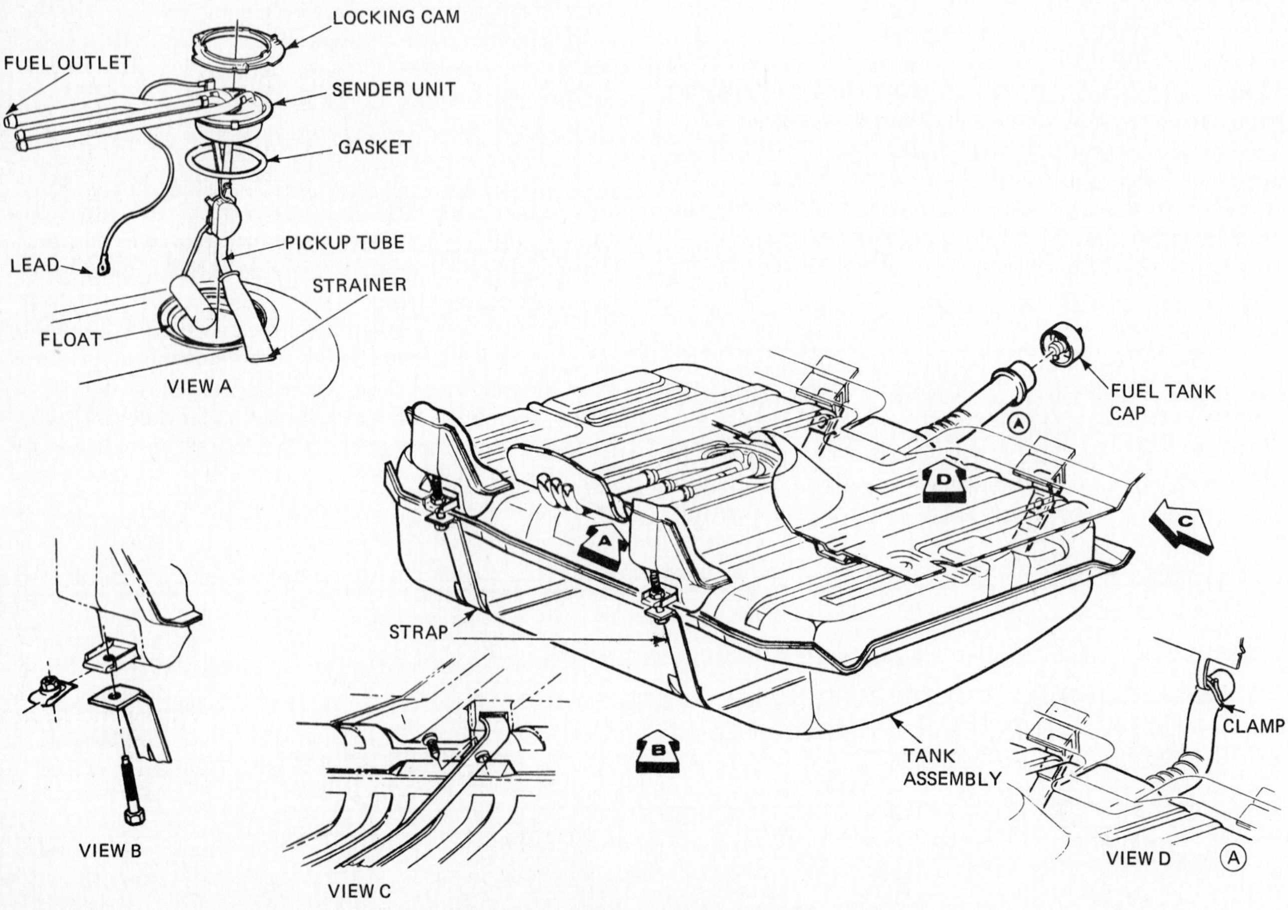

LOCKING CAM
FUEL OUTLET
SENDER UNIT
GASKET
PICKUP TUBE
LEAD
STRAINER
FLOAT
VIEW A
FUEL TANK
CAP
STRAP
CLAMP
TANK
ASSEMBLY
VIEW B
VIEW C
VIEW D

Fig. 15-6 Fuel tank, showing the mounting arrangement. View A shows the fuel pickup tube and sender unit for the fuel gauge. *(Buick Motor Division of General Motors Corporation)*

Fig. 15-7 In-line fuel-filter installation on a six-cylinder engine. *(Chrysler Corporation)*

♦ 15-6 FUEL GAUGES

There are two types of fuel gauges, *magnetic* and *thermostatic*. Each of these gauges has a tank unit and an instrument-panel unit.

1. Magnetic (Fig. 15-10) The tank unit in this fuel gauge contains a sliding contact. The contact slides back and forth on a resistor as the float moves up and down in the fuel tank. This changes the amount of electric resistance the tank unit offers. As the tank empties, the float drops and the sliding contact moves to reduce the resistance. The instrument panel unit contains two coils, as shown in Fig. 15-10. When the ignition switch is turned on, current from the battery flows through the two coils. This produces a magnetic field that acts on the armature, to which the pointer is attached. When the resistance of the tank unit is high (tank filled and float up), the current through the empty coil also flows through the full coil. Therefore, the armature is pulled to the right, so that the pointer is on the full side of the dial. But when the tank begins to empty, the resistance of the tank unit drops. Then more of the current flowing through the empty coil passes through the tank unit. Since less is flowing through the full coil, its magnetic field is weaker. As a result, the empty coil pulls the armature toward it. The pointer swings around toward the empty side of the dial.

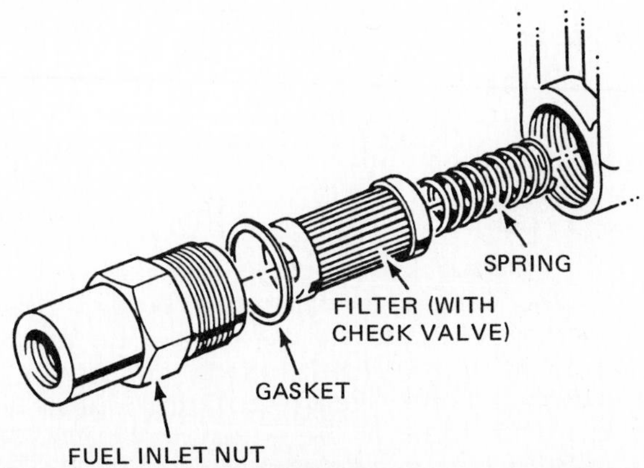

Fig. 15-8 Fuel filter that fits in the fuel inlet of the carburetor. *(Buick Motor Division of General Motors Corporation)*

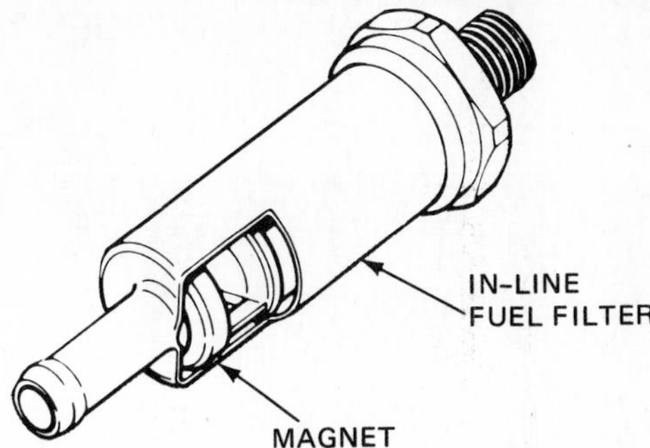

Fig. 15-9 Carburetor fuel filter which contains a magnet. *(Ford Motor Company)*

2. Thermostatic Figure 15-11 is the wiring circuit of a thermostatic fuel gauge. It has a fuel-tank unit much like the magnetic system. The tank unit has a float and a sliding contact that moves on a resistor. Current flows from the battery through the resistance in the tank unit. When the fuel is low in the tank, most of the resistance is in the circuit. Very little current can flow. When the tank is filled, the float moves up, and the sliding contact cuts most of the resistance out of the circuit. Now more current flows. As it flows through the heater coil in the fuel gauge, the current heats the thermostat. The thermostat blade bends because of the heat. This moves the needle to the right, toward the full mark.

3. Low-level-fuel indicator This indicator uses a fuel-tank unit similar to that shown in Fig. 15-10. As the indicator needle on the fuel gauge moves toward empty, the reduced resistance allows increasing current flow through the low-level-fuel indicator system. When this current flow is great enough, indicating a nearly empty fuel tank, the system turns on the low-fuel-level light on the instrument panel. In some late-model cars, the electronic system gives a spoken warning of low fuel level.

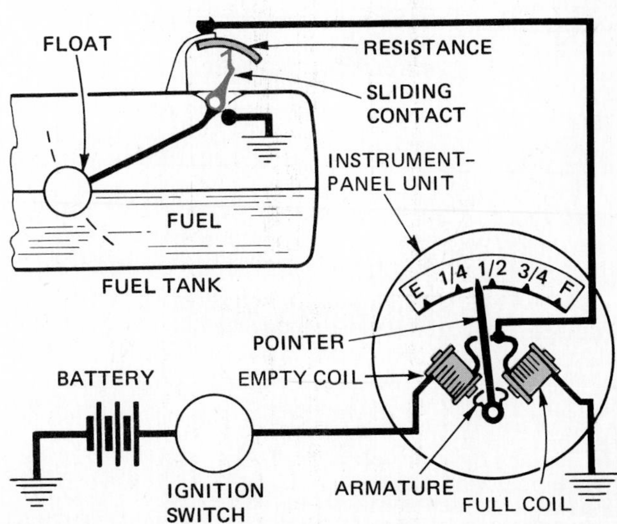

Fig. 15-10 Schematic wiring diagram for a magnetic fuel-gauge system.

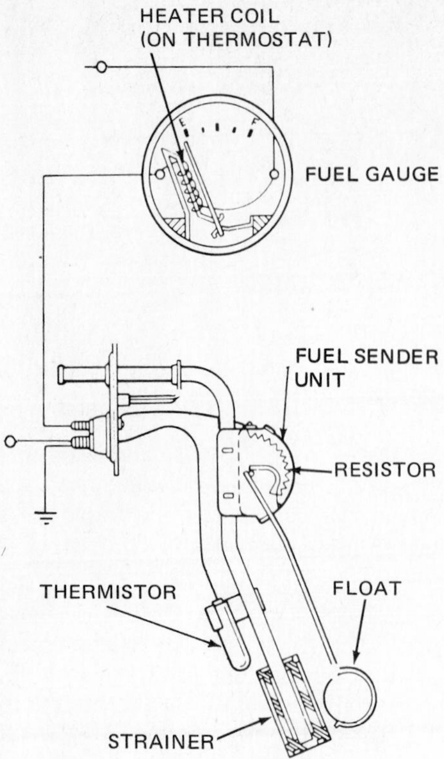

Fig. 15-11 Schematic wiring diagram for a thermostatic fuel-gauge system. It uses a variable-resistance tank unit and a thermostatic instrument-panel unit (fuel gauge). *(Ford Motor Company)*

♦ 15-7 MILES-TO-EMPTY FUEL INDICATOR

This system shows the driver how much further the car can be driven before it runs out of fuel. The indicator is usually located just under the fuel gauge on the instrument panel (Fig. 15-12). When the button is pressed, the *miles to empty* will display for a few seconds. (If the system is metric, the readout would be in *kilometers to empty*.) If there is less than 50 miles of fuel remaining, the mileage reading will remain on. Figure 15-13 shows the components of the system.

Fig. 15-12 Fuel gauge and miles-to-empty indicator. *(Ford Motor Company)*

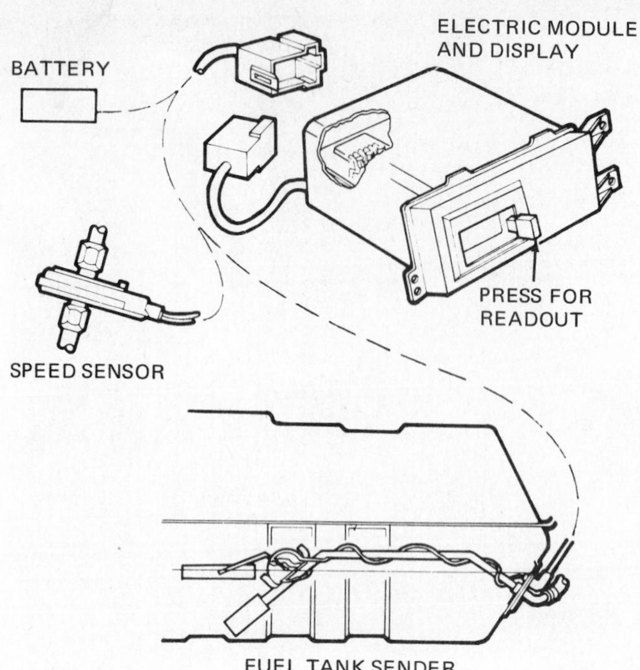

Fig. 15-13 Components in the miles-to-empty indicator system. *(Ford Motor Company)*

The system continuously adjusts for changes in driving conditions and the way the car is used. If the car is driven around town, where fuel economy is poor, the system will adjust to this mileage. If the car is generally used on the highway, where fuel economy is better, the system will adjust to this type of driving. The system measures car speed, distance traveled, fuel level in the tank, and the rate at which fuel is being pumped from the fuel tank.

♦ 15-8 FUEL PUMPS

The fuel system uses a fuel pump to deliver fuel from the tank to the carburetor. There are two types of fuel pump, *mechanical* and *electric*. Electric fuel pumps are discussed in ♦15-10. Mechanical fuel pumps are operated by an eccentric (an off-center section) on the engine camshaft (Fig. 12-7), as explained below. The mechanical fuel pump is mounted on the side of the cylinder block on in-line engines (Fig. 15-14). In some V-type engines, the pump is mounted between the two cylinder banks. Most V-type engines mount the fuel pump on the side, at the front of the engine (Fig. 15-15).

The mechanical fuel pump has a rocker arm whose end rests on the camshaft eccentric. Many V-type engines use a pushrod from the eccentric to the rocker arm (Fig. 15-15).

As the camshaft rotates, the eccentric rocks the rocker arm back and forth. The inner end of the rocker arm is linked to a flexible diaphragm. The diaphragm is clamped between the upper and lower pump housings (Fig. 15-16). There is a spring over the diaphragm that keeps tension on it. As the rocker arm rocks, it pulls the diaphragm up and then releases it. The spring then forces the diaphragm down. The diaphragm moves up and down as the rocker arm rocks.

This diaphragm movement produces partial vacuums and pressures in the space below the diaphragm. When the diaphragm moves up, a partial vacuum is produced. Then, atmospheric pressure, acting on the fuel in the tank, forces fuel through the fuel line and into the pump. The inlet valve in

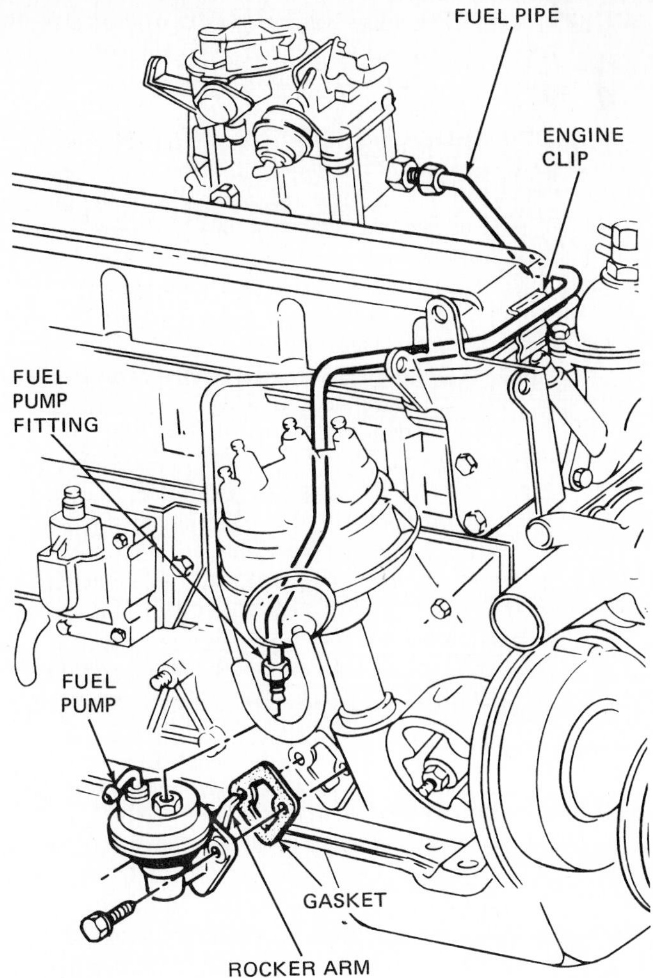

Fig. 15-14 Installation of a fuel pump on an in-line engine. *(Chevrolet Motor Division of General Motors Corporation)*

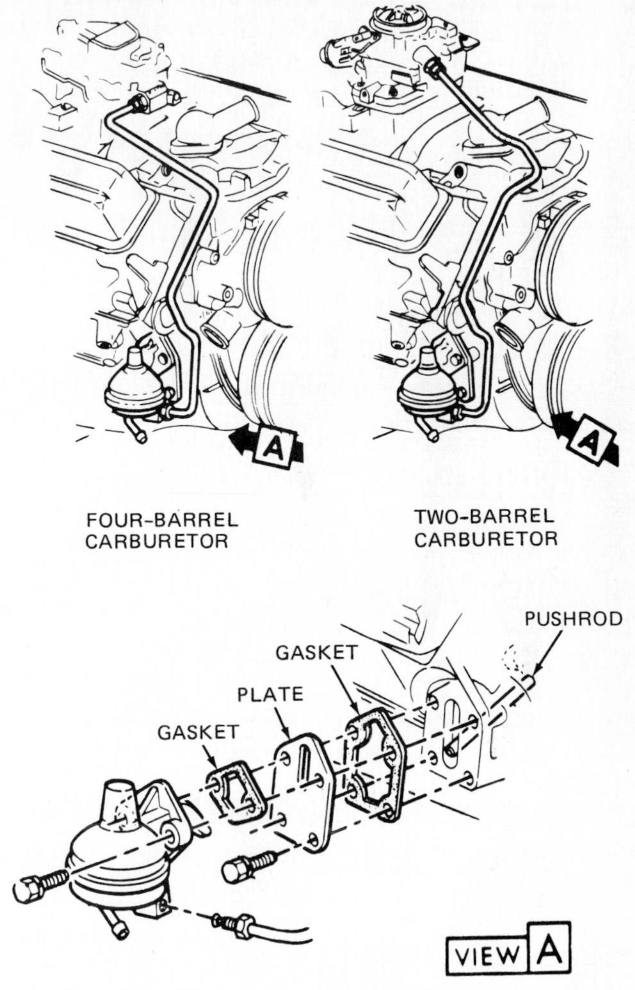

Fig. 15-15 Mounting of the fuel pump on a V-type engine with a pushrod between the fuel-pump rocker arm and the camshaft eccentric. *(Chevrolet Motor Division of General Motors Corporation)*

the pump opens to admit fuel, as shown by the arrows in Fig. 15-16.

When the diaphragm is released by the rocker arm, the spring forces the diaphragm downward. This produces pressure in the space under the diaphragm. The pressure closes the inlet valve and opens the outlet valve. Now fuel is forced from the fuel pump through the fuel line to the carburetor, as shown by the arrows in Fig. 15-17.

The actions in the pump as the eccentric rotates are shown in Figs. 15-16 and 15-17. The fuel from the fuel pump enters the carburetor past a needle valve in the float bowl. If the bowl is full, the needle valve closes so that no fuel can enter. When this happens, the fuel pump cannot deliver fuel to the carburetor. Then the rocker arm continues to rock. However, the diaphragm remains at or near its upper limit of travel. Its spring cannot force the diaphragm downward as long as the carburetor float bowl will not accept fuel. However, as the carburetor uses up fuel, the needle valve opens to admit fuel to the float bowl. Now the diaphragm can move down (on the rocker-arm return stroke) to force fuel into the float bowl.

Most mechanical fuel pumps used on cars today (Figs. 15-14 and 15-15) are serviced by complete replacement. They are crimped together when assembled at the factory and cannot be disassembled. No service parts are available for these pumps. However, some fuel pumps have been put together with screws. These can be repaired if parts are available.

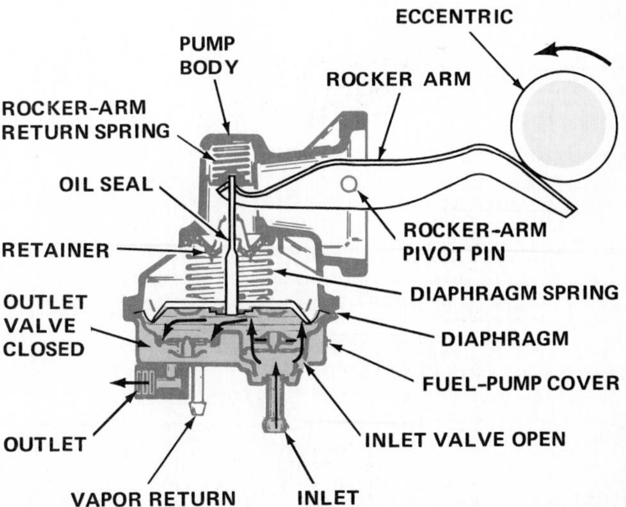

Fig. 15-16 When the eccentric rotates so as to push the rocker arm down, the arm pulls the diaphragm up. The inlet valve opens to admit fuel into the space under the diaphragm.

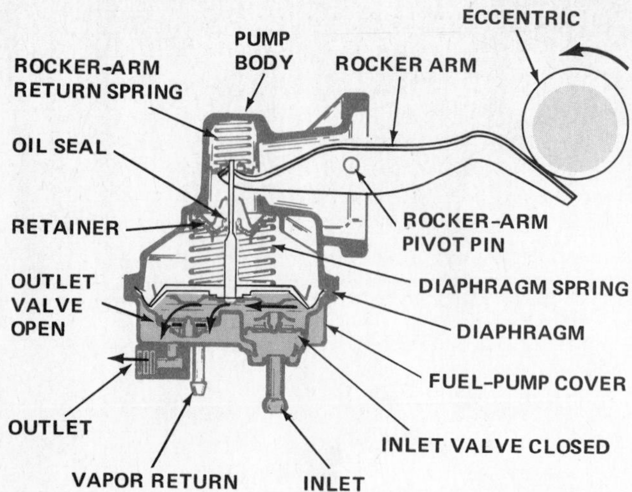

Fig. 15-17 When the eccentric rotates, the rocker arm moves up under it. This releases the diaphragm so it can move down, pushed down by the diaphragm spring. The pressure produced under the diaphragm then closes the inlet valve and opens the outlet valve. Now, fuel flows to the carburetor.

♦ 15-9 VAPOR-RETURN LINE

Some cars have a vapor-return line running from the fuel pump or the fuel filter to the fuel tank (Fig. 15-5). Figure 15-16 shows the connection at the fuel pump (to lower left) for the vapor-return line. The purpose of this line is to return to the fuel tank any vapor that forms in the fuel pump. The fuel pump can handle liquid only. Pumping stops if vapor forms in the fuel pump. With air conditioning, under-the-hood temperatures are likely to be higher. The air-conditioning condenser delivers more heat under the hood. Also, during idle, the engine cooling system is not very efficient. This allows under-the-hood temperatures to increase. The higher temperatures tend to cause vapor to form in the fuel pump.

Vapor can form in the fuel pump because the pump alternately produces vacuum and pressure. During the vacuum phase, the boiling, or vaporizing, temperature of the fuel is lowered. The lower the pressure, the lower the temperature at which any liquid vaporizes. For example, some gasolines with high volatility will boil at 188°F [87°C] at sea level, which has an atmospheric pressure of 14.7 psi [101 kPa]. But at 16,000 feet [4877 m] above sea level, where the pressure is about 7 psi [48 kPa], this gasoline would boil at 164°F [73°C].

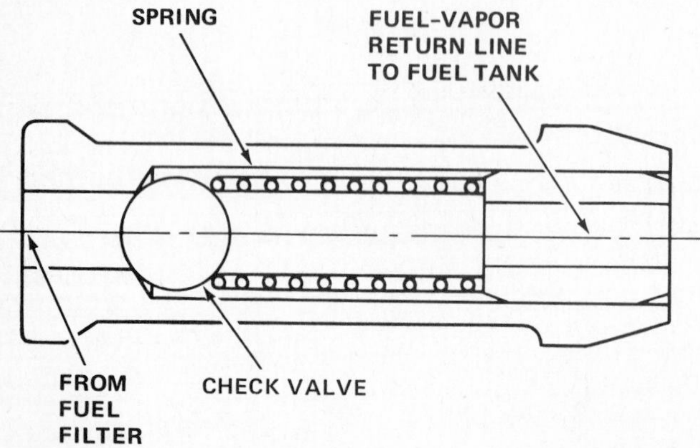

Fig. 15-18 In-line check valve which is located between the fuel filter and the fuel tank. *(American Motors Corporation)*

The combination of increased temperature and lower pressure or partial vacuum in the fuel pump can cause fuel to vaporize. This produces *vapor lock*, a condition that prevents normal delivery of fuel to the carburetor. The engine stalls.

The vapor-return line is connected to a special outlet in the fuel pump. This allows any vapor to return to the fuel tank. The vapor-return line also permits excess fuel being pumped by the fuel pump to return to the tank. This excess fuel, in constant circulation, helps keep the fuel pump cool. Therefore, it prevents vapor from forming.

Some cars have a vapor separator connected between the fuel pump and the carburetor (Fig. 15-5). It consists of a sealed can, a filter screen, an inlet and outlet fitting, and a metered orifice, or outlet, for the return line to the fuel tank. Any fuel vapor that the fuel pump produces enters the vapor separator (as bubbles) along with fuel. These bubbles of vapor rise to the top of the vapor separator. The vapor then is forced, by fuel-pump pressure, to pass through the fuel-return line and back to the fuel tank. In the tank, the vapor condenses back into liquid fuel.

Some vapor-return lines have an in-line check valve (Fig. 15-18). It prevents fuel from feeding back to the carburetor from the fuel tank through the vapor-return line. If fuel does attempt to feed back in this way, the pressure of the fuel forces the check ball to seat. This blocks the line. In normal operation, the pressure of the fuel vapor from the fuel pump unseats the ball and allows the fuel vapor to flow to the fuel tank.

♦ 15-10 ELECTRIC FUEL PUMPS

Electric fuel pumps have some advantages over mechanical fuel pumps. Fuel delivery starts as soon as the ignition is turned on. The pump can deliver more fuel than the engine will require even under maximum operating conditions. The engine will never be fuel starved. Therefore, electric fuel pumps are used in many high-performance, heavy-duty and fuel-injected vehicles.

There are two general types of electric fuel pumps, the in-line and the fuel-tank-mounted type. The in-line fuel pump can be operated either by an electric motor or by an electromagnetic solenoid (Fig. 15-19). The latter has a flexible metal bellows that is operated by an electromagnet. The electromagnet is connected to the battery by turning on the ignition switch. Then the electromagnet pulls down the armature and extends the bellows. This produces a vacuum in the bellows. Fuel from the fuel tank enters the bellows through the inlet valve. Then, as the armature reaches its lower limit of travel, it opens a set of contact points. This disconnects the electromagnet from the battery. The return spring pushes the armature up and collapses the bellows. This forces fuel from the bellows through the outlet valve and to the carburetor. As the armature reaches the upper limit of its travel, it closes the contacts. The electromagnet is again energized and again pulls the armature down. These actions are repeated as long as the ignition is turned on.

The in-tank electric fuel pump has an electric motor that drives an impeller. When the impeller spins, it sends fuel through the outlet pipe to the carburetor or fuel-injection system. An advantage of having the fuel pump in the fuel tank is that there is pressure on the fuel all the way from the tank to the engine. Therefore, no vapor lock can occur. The type

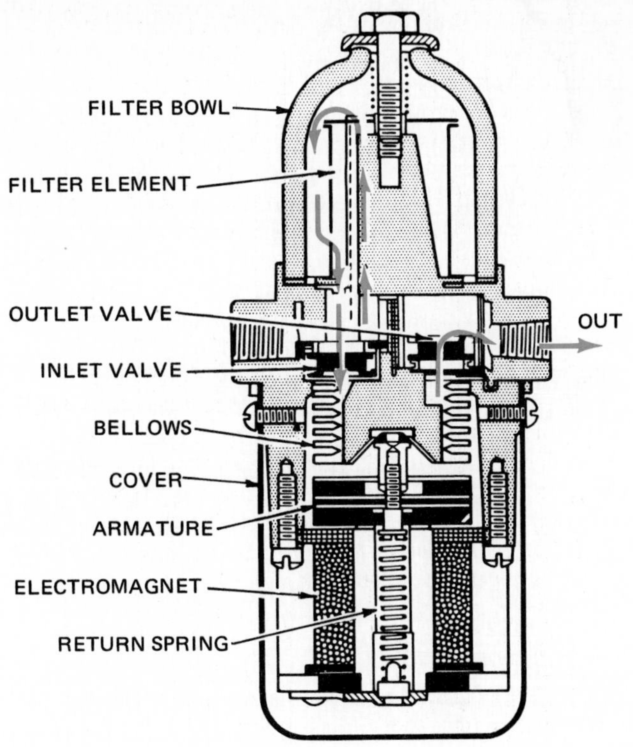

Fig. 15-19 Sectional view of an electric fuel pump.

mounted in the fuel line (mechanical or electric) may allow vapor lock to occur under some conditions (♦14-4).

Figure 15-20 shows the wiring system for an electric in-tank fuel pump. The wiring circuit connects the fuel pump to the battery through contacts in the starting-motor relay. This allows the fuel pump to start delivering fuel as soon as the starting motor begins to crank the engine. After the engine starts, the circuit is maintained through the oil-pressure switch. This arrangement shuts off the fuel pump whenever the engine stops or the oil pressure drops.

Some wiring systems for electric fuel pumps include an *inertia switch*. If the car rolls over, the inertia switch opens. This shuts off the fuel pump so that no more fuel flows, reducing the possibility of fire.

♦ 15-11 AIR CLEANERS

The fuel system mixes air and fuel to produce a combustible mixture. A large volume of air passes through the carburetor and engine—as much as 100,000 cubic feet [2832 m^3] of air every 1000 car miles [1609 km]. Air always contains a lot of floating dust and grit. The grit and dust could cause serious damage if they entered the engine. To prevent this, an air cleaner is mounted on the air entrance of the carburetor or fuel-injection system (Fig. 15-21).

All air entering the engine through the carburetor must first pass through the air cleaner. The air cleaner contains a ring of filter material (fine-mesh metal threads or ribbons, pleated paper (Fig. 15-22), cellulose fiber, or polyurethane). The air must pass through this material. It provides a fine maze that traps most of the airborne particles. Some air cleaners have an oil bath. This is a reservoir of oil which the incoming air flows past. The moving air picks up particles of oil and carries them up into the filter. There the oil washes any dust back down into the oil reservoir. The oiliness of the

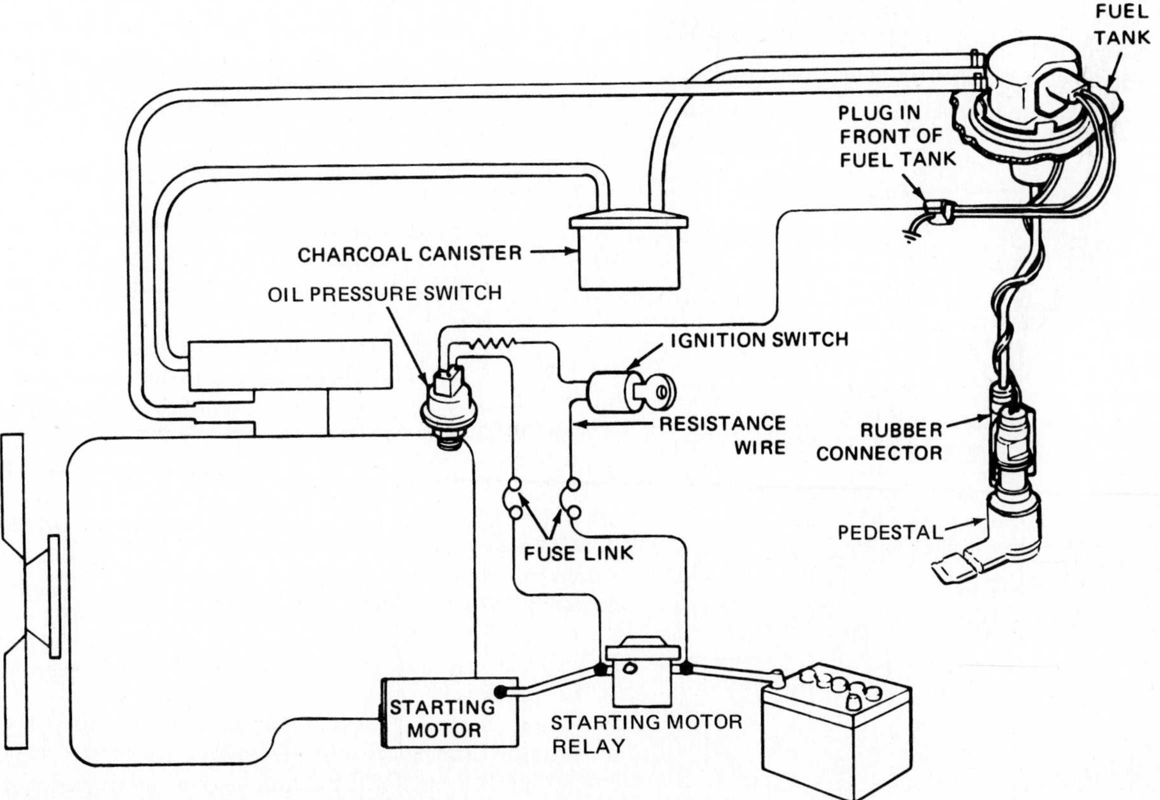

Fig. 15-20 Electric fuel-pump wiring diagram for the type of fuel pump that is located in the fuel tank. *(Ford Motor Company)*

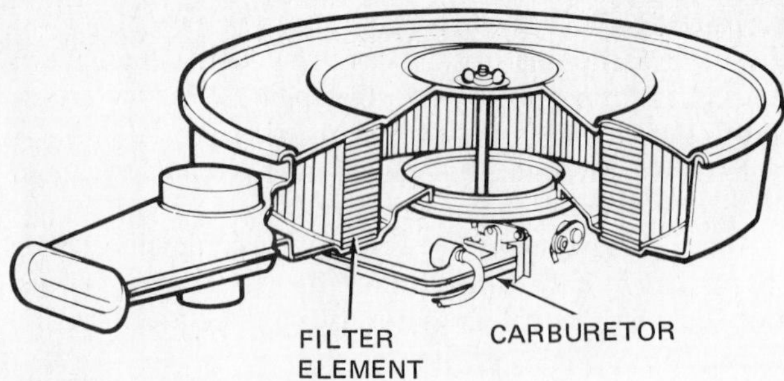

Fig. 15-21 Typical air cleaner, partly cut away to show the filter element. *(Chrysler Corporation)*

filter material improves the filtering action.

The air cleaner also muffles the noise of the intake of air through the carburetor or fuel-injection system, manifold, and valve ports. This noise would be very noticeable if it were not for the air cleaner. In addition, the air cleaner acts as a flame arrester in case the engine backfires through the intake manifold. Backfiring may occur if the air-fuel mixture is ignited in the cylinder before the intake valve closes. Then there is a momentary flashback through the air intake. The air cleaner prevents this flame from escaping and igniting gasoline fumes outside the engine.

Most automotive vehicles today are equipped with an emission control system which uses a thermostatic air cleaner. It improves cold-engine operation by sending heated air to the air cleaner while the engine is cold. The system is covered in a later chapter on emission controls.

♦ 15-12 EXHAUST SYSTEM

The exhaust system includes the exhaust manifold, exhaust pipe, catalytic converter, muffler, and tail pipe (Fig. 15-23). Some V-type engines have a crossover pipe to connect their two exhaust manifolds. Other cars with V-type engines use two separate exhaust systems (dual system), one for each cylinder bank. This improves the "breathing" ability of the engine, allowing it to exhaust more freely. This may slightly increase engine power.

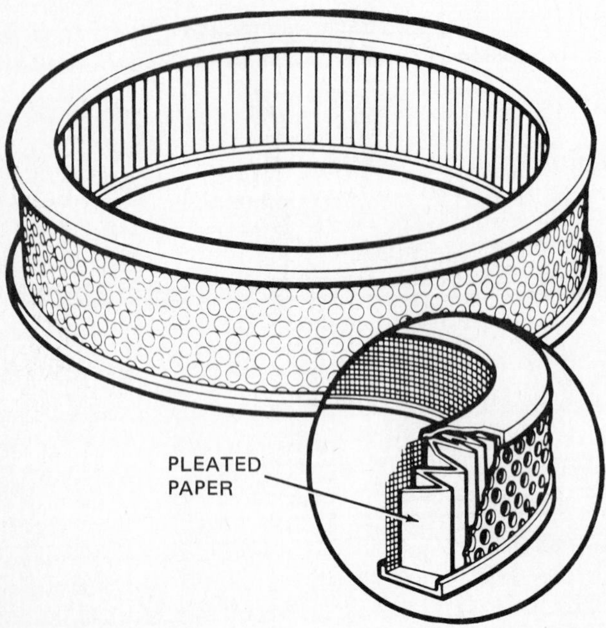

Fig. 15-22 Construction of a pleated-paper type of filter element. *(Chrysler Corporation)*

Exhaust systems on modern cars have *catalytic converters*. One arrangement is shown in Fig. 15-23. The catalytic converter converts unburned fuel (HC) and partly burned fuel (CO) into water (H_2O) and carbon dioxide (CO_2). These devices and how they work are described in a later chapter.

Some exhaust manifolds, especially on older cars, have heat-control valves. These valves close when the engine is cold. This directs exhaust heat to the intake manifold. The heat helps vaporize the ingoing gasoline, which improves cold-engine operation.

Some engines have exhaust manifolds equipped with air-injection systems. The system includes an air pump and a series of injection tubes in the exhaust manifold. In operation, the air pump sends a flow of air into the exhaust manifold opposite the exhaust valves. This extra air helps to burn any HC or CO still left in the exhaust gases. The air-injection system is covered in a later chapter.

♦ 15-13 MUFFLER

The muffler (Fig. 15-24) contains a series of holes, passages, and resonance chambers. These openings act to absorb and damp out the high-pressure surges introduced into the exhaust system as the exhaust valves open. This quiets the exhaust. Some exhaust systems do not use a muffler. Instead, the exhaust pipe has a series of scientifically shaped restrictions. These also act to damp out the exhaust noises without unduly restricting the flow of exhaust gases.

To further reduce exhaust noise, some exhaust pipes are made of laminated pipe. Two-ply laminated pipe consists of one pipe inside of another. A three-ply laminated pipe includes a layer of plastic sandwiched between the two metal layers. Either combination damps out exhaust-pipe ring, which can occur on some exhaust systems.

♦ 15-14 SUPERCHARGERS AND TURBOCHARGERS

Many engines use a special air pump to deliver a "supercharge" of air or air-fuel mixture to the engine. This increases the power output of the engine. In the spark-ignition engine, more air-fuel mixture enters the cylinders during the intake stroke. This means higher pressures during the power stroke and higher engine power output. In the diesel engine, more air in the cylinders allows more fuel to be injected into the cylinders. Greater engine power results.

There are two types of engine supercharging devices. Both use a rotary air pump, or compressor. The major difference is in the way the compressor is driven. One type, usually called a *supercharger*, has the compressor driven by a belt from the engine crankshaft. The other type, called a *turbocharger*, has the compressor driven by the exhaust gas from the engine. This is the type most commonly installed on car engines.

The turbocharger (Fig. 15-25) has two basic parts, an air pump (or compressor) and a turbine. Exhaust gas flows through the turbine and spins the turbine rotor. The other end of the shaft on which the turbine is mounted supports the compressor rotor. When the turbine spins, the compressor rotor spins, producing pressurized air or pressurized air-fuel mixture.

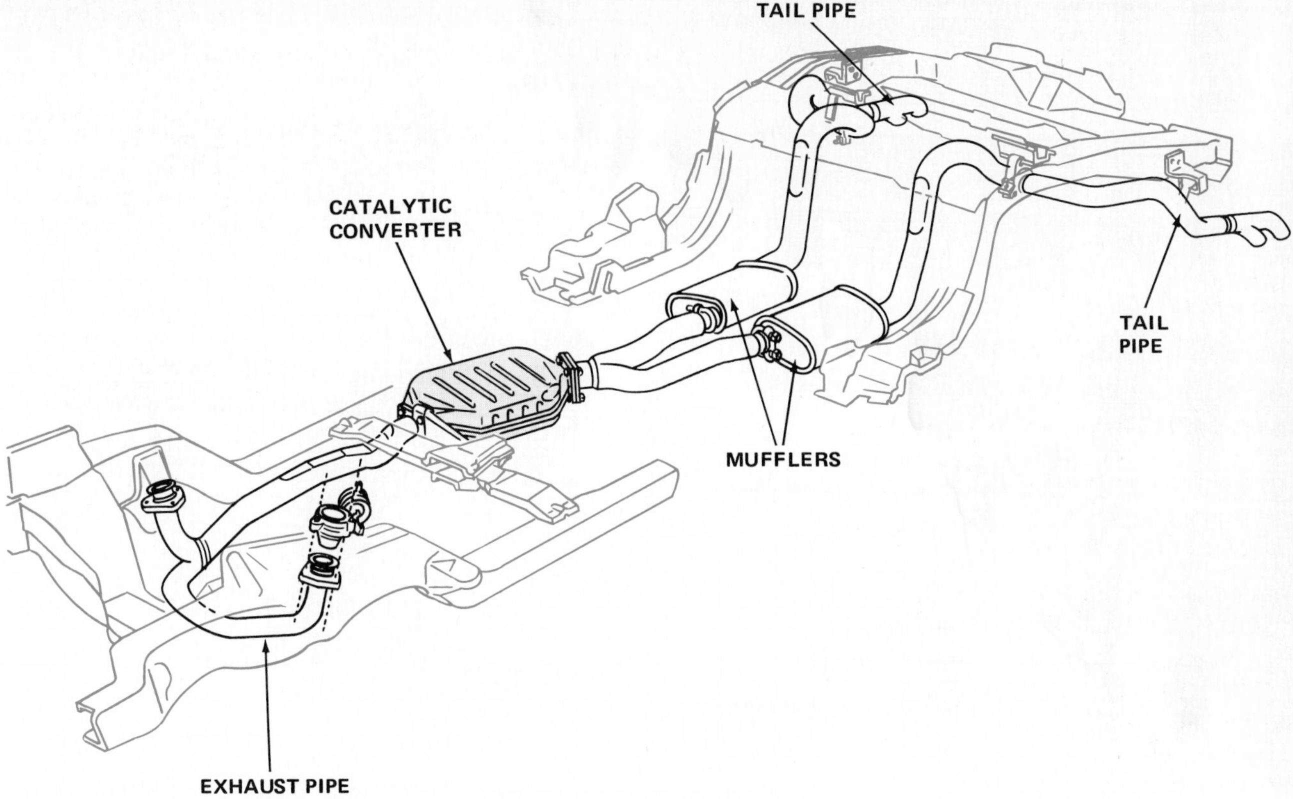

Fig. 15-23 Dual exhaust system with a single catalytic converter and twin mufflers and tail pipes. *(American Motors Corporation)*

In some systems, the compressor produces pressurized air which is sent to the carburetor (Figs. 15-26 and 15-27). In other systems, the compressor is placed between the carburetor and the intake manifold (Fig. 15-28) and delivers compressed air-fuel mixture to the intake manifold.

With either arrangement, additional air-fuel mixture enters the engine cylinders when the turbocharger is operating in *boost*. This means extra power is produced. Engine power may be increased 30 percent or more. However, the turbocharger provides boost only part of the time. This is when the engine is called upon to deliver maximum power, such as when passing on the highway or going up a hill. During highway driving, the turbocharger on a spark-ignition engine provides boost less than 10 percent of the time. In city traffic, it operates very seldom, if at all.

Figure 15-27 shows a modern turbocharger. The *wastegate* (item 9) prevents the turbocharger from overcharging the engine. The wastegate opens if the intake-manifold pressure

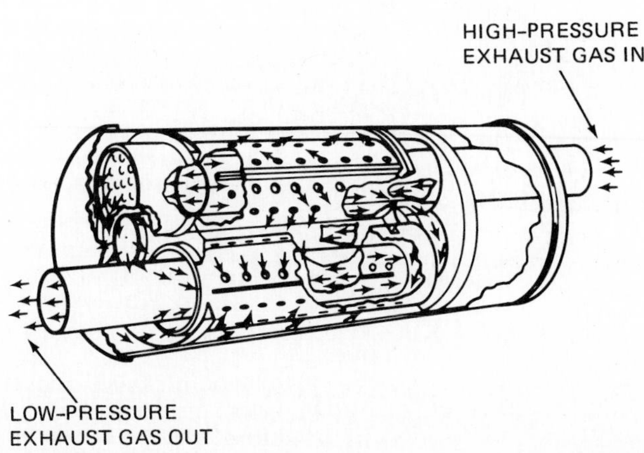

Fig. 15-24 Exhaust muffler in cutaway view. The arrows show the path of exhaust-gas flow through the muffler. *(Chevrolet Motor Division of General Motors Corporation).*

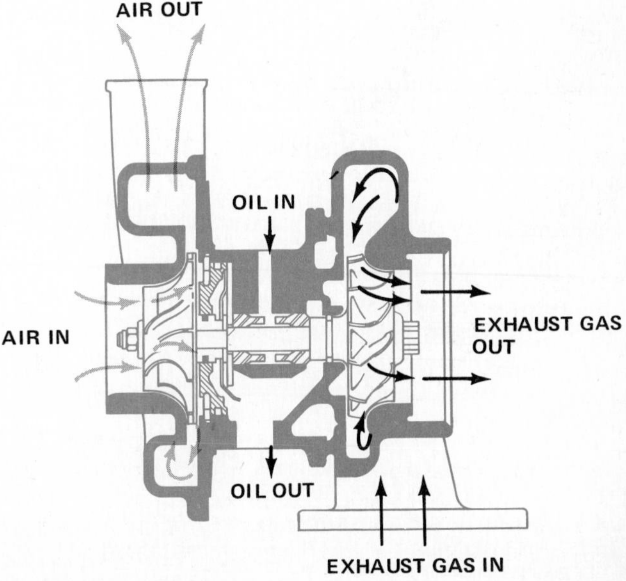

Fig. 15-25 Operation of a turbocharger. *(Schwitzer Division, Wallace-Murry Corporation)*

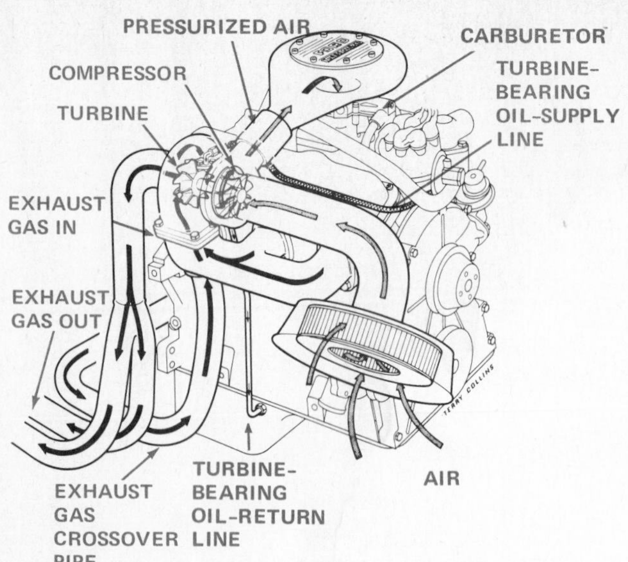

Fig. 15-26 Turbocharged version of the Ford 182-cubic-inch (3-L) V-6 engine. *(Ford Motor Company)*

goes too high. If this should happen, the engine would receive air-fuel mixture at excessive pressures. The result would be detonation and engine damage. The compression pressures would go so high that detonation of the compressed air-fuel mixture in the cylinders would occur (♦14-5 and 14-8). To prevent this, the wastegate actuator senses the pressure of the air or air-fuel mixture coming from the compressor. If it goes too high, the actuator opens the wastegate.

AIR AT ATMOSPHERIC PRESSURE
PRESSURIZED CHARGE AIR
PRESSURIZED AND RECOOLED CHARGE AIR
EXHAUST GASES

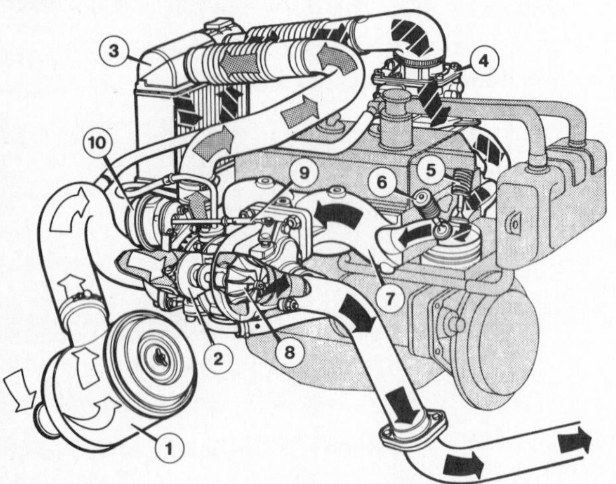

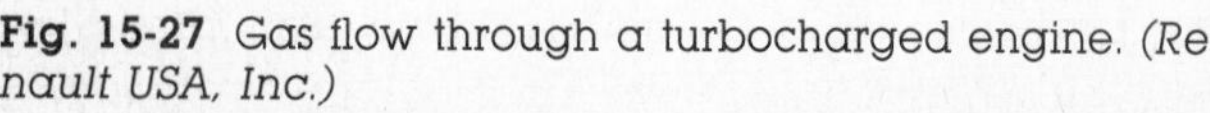

1. AIR FILTER
2. COMPRESSOR ROTOR
3. CHARGE AIR INTERCOOLER
4. PRESSURIZED CARBURETOR
5. INTAKE VALVE
6. EXHAUST VALVE
7. EXHAUST MANIFOLD
8. TURBINE
9. WASTEGATE
10. WASTEGATE ACTUATOR (PRESSURE SENSOR)

Fig. 15-27 Gas flow through a turbocharged engine. *(Renault USA, Inc.)*

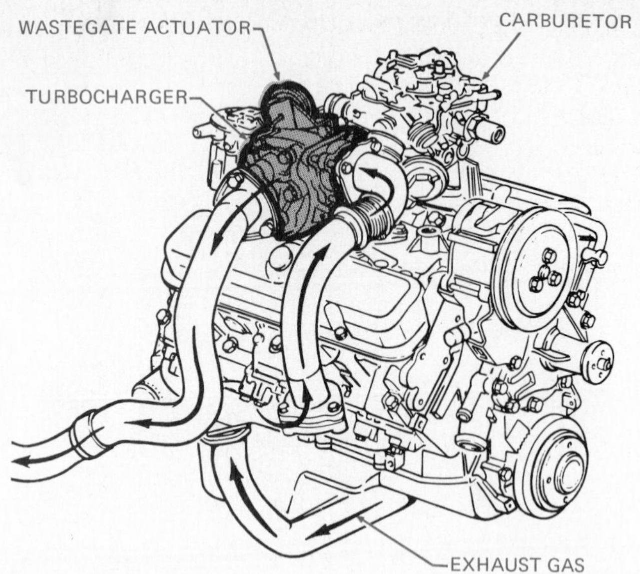

Fig. 15-28 Turbocharger installation on a V-type engine. The wastegate prevents excessive boost pressure which could damage the engine. *(Pontiac Motor Division of General Motors Corporation)*

Then part of the exhaust gas bypasses the turbine. This reduces turbine and compressor speed so that the air or air-fuel-mixture pressure is reduced. Figure 15-29 shows the connections between the turbocharger wastegate-boost actuator and the wastegate-bypass valve.

Another safeguard against detonation, used on many turbocharged engines, is an *electronic spark control* (ESC). This system uses a detonation sensor to signal an electronic controller. When detonation begins, the controller signals the ignition system to retard the spark. This system is covered in a later chapter on ignition systems.

In the design shown in Fig. 15-27, note that the air from the compressor is passed through a radiator, called the *charge-air intercooler*. This cools the air before it enters the

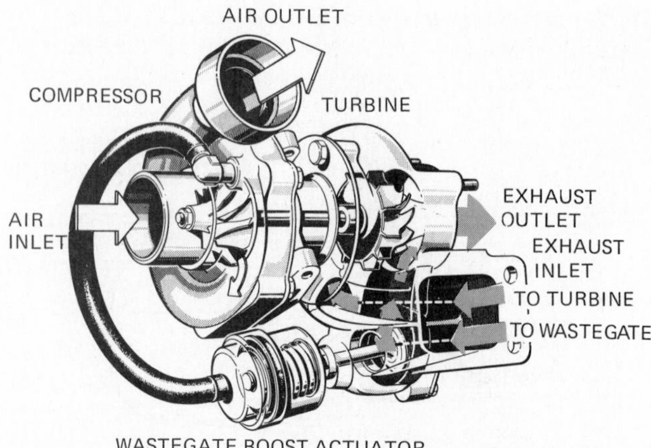

Fig. 15-29 This turbocharger installation produces about 75 kPa (11 psi) boost, resulting in maximum manifold pressure of 171 kPa (24.6 psi). The compressor and turbine spin at about 100,000 rpm at highway cruising speed. The unit adds about 15 kg (30 lb) to the weight of the engine, but boosts engine power by 43 percent. *(Mercedes-Benz of North America, Inc.)*

carburetor and engine. The air-fuel mixture therefore enters the engine at a lower temperature. This reduces the tendency for detonation to occur.

The rotors in the turbocharger may turn at very high speeds of more than 100,000 revolutions per minute (rpm). This requires that the shaft bearings have adequate lubrication. Therefore, engine oil circulates through the turbocharger. Some manufacturers recommend more frequent oil changes for engines with turbochargers. Because of the high speeds, only clean lubricating oil must reach the bearings.

Note Engines designed for racing and using turbochargers show dramatic boosts of horsepower. For example, the Offenhauser (or "Offy") four-cylinder engine that has been so successful in the Indianapolis 500 is a relatively small engine. But with turbocharging, it can produce 600 to 800 hp (448 to 597 kW).

♦ REVIEW QUESTIONS ♦

Select the *one* correct, best, or most probable answer to each question. You can find the answers in the section indicated at the end of each question.

1. The two types of fuel systems for spark-ignition engines are (♦15-1)
   a. diesel and gasoline
   b. carbureted and fuel-injected
   c. gasoline and gasohol
   d. LPG and alcohol
2. The two types of fuel gauges are (♦15-6)
   a. thermostatic and magnetic
   b. electrical and mechanical
   c. pressure and vacuum
   d. none of the above
3. The two types of fuel pumps are (♦15-8)
   a. mechanical and electric
   b. motorized and solenoid-operated
   c. unit replacement and serviceable by disassembly
   d. none of the above
4. The purpose of the vapor-return line is to return vapor from the (♦15-9)
   a. fuel pump to the carburetor
   b. carburetor to the fuel pump
   c. fuel pump to the fuel tank
   d. exhaust manifold to the carburetor
5. The two types of electric fuel pumps are (♦15-10)
   a. in line and in tank
   b. in tank and mechanical
   c. diaphragm and bellows
   d. motorized and mechanical
6. The in-tank fuel-pump wiring circuit is completed through the oil-pressure switch. The purpose of this is to (♦15-10)
   a. prevent excessive oil pressure
   b. maintain correct fuel pressure
   c. shut off the fuel pump when the engine stops and oil pressure drops
   d. shut off the oil pressure when the fuel pump stops
7. The purpose of the inertia switch in the in-tank fuel-pump wiring circuit is to (♦15-10)
   a. maintain the inertia of the fuel pump to prevent fuel starvation of the engine
   b. shut off the fuel pump if the car rolls over
   c. keep inertia from stalling the engine
   d. prevent inertia from shutting off the fuel pump
8. A major difference between superchargers and turbochargers is that the (♦15-14)
   a. supercharger is driven by a belt from the engine crankshaft
   b. turbocharger is driven by the force of the exhaust gas
   c. both a and b
   d. neither a nor b
9. The purpose of the wastegate is to (♦15-14)
   a. waste some of the engine power to prevent detonation
   b. allow some exhaust gas to bypass the turbine and prevent detonation
   c. prevent excessive fuel consumption during low-speed operation
   d. prevent excessive power output during part-throttle operation
10. The rotors in the turbocharger may rotate at speeds of (♦15-14)
    a. 1000 rpm
    b. 10,000 rpm
    c. 100,000 rpm
    d. 1,000,000 rpm

CHAPTER 16
Automotive Carburetors

After studying this chapter, you should be able to:

1. Discuss carburetors and explain how the basic fixed-venturi carburetor works.
2. Explain the difference between a fixed-venturi and a variable-venturi carburetor.
3. List and describe the operation of the six systems in a fixed-venturi carburetor.
4. Explain how a variable-venturi carburetor works.
5. Open the hoods of cars and identify the various visible carburetor components.
6. Examine disassembled carburetors and identify the major parts.

♦ 16-1 CARBURETOR TYPES

There are two basic types of carburetors, fixed venturi and variable venturi (VV). The venturi is the restricted area in the carburetor air passage through which the air must flow. As explained later, this restriction produces a partial vacuum. The vacuum causes a fuel nozzle to discharge gasoline into the air passing through (Fig. 16-1). The gasoline mixes with the air to produce the combustible mixture the engine needs to run.

Most carburetors installed in cars made in the United States are of the fixed-venturi type. Many foreign cars and some domestic cars use the variable-venturi carburetor. This type of carburetor is described later in the chapter.

♦ 16-2 CARBURETION

Carburetion is the mixing of the gasoline fuel with air to get a combustible mixture. The function of the carburetor is to supply a combustible mixture of varying degrees of richness to suit engine operating conditions. The mixture must be rich (have a higher percentage of fuel) for starting, acceleration, and high-speed operation. A less rich (leaner) mixture is desirable at intermediate speed with a warm engine. The carburetor has several systems through which air-fuel mixture flows during different operating conditions. These systems produce the varying mixtures required for the different operating conditions.

♦ 16-3 VAPORIZATION

When a liquid changes to a vapor, it is said to evaporate. Water placed in an open pan will evaporate. The water changes from a liquid to a vapor. Wet clothes hung on a line dry: the water in the clothes turns to vapor. When the clothes are spread out, they dry more rapidly than when they are bunched together. This illustrates an important fact about evaporation. The greater the surface exposed, the more rapidly evaporation takes place. Water in a tall glass takes longer to evaporate than water in a shallow pan. Much more area is exposed in the pan.

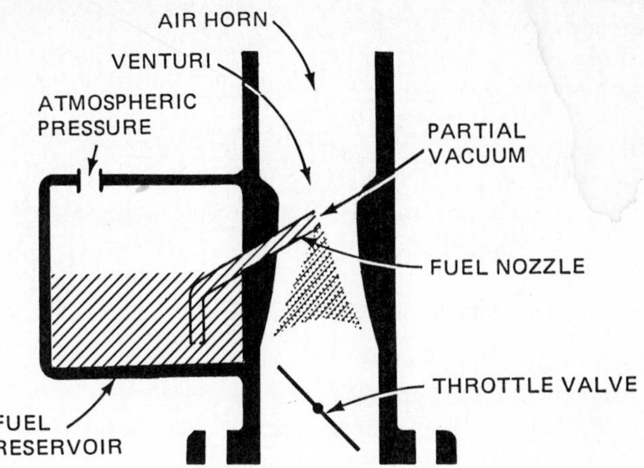

Fig. 16-1 Basic carburetor, consisting of an air horn, a fuel nozzle, a throttle valve, and a fuel reservoir.

♦ 16-4 ATOMIZATION

To produce very quick vaporization of the liquid gasoline, it is sprayed into the air passing through the carburetor. Spraying the liquid turns it into many fine droplets. This effect is called *atomization* because the liquid is broken up into small droplets (but not actually into atoms, as the name implies). Each droplet is exposed to air on all sides so that it vaporizes very quickly. Therefore, during normal running of the engine, the fuel sprayed into the air passing through the carburetor turns to vapor, or vaporizes, almost instantly.

♦ 16-5 CARBURETOR FUNDAMENTALS (FIXED VENTURI)

A simple fixed-venturi carburetor can be made from a round cylinder with a constricted section, a fuel nozzle, and a round disk, or valve (Fig. 16-1). The round cylinder is called the *air horn*, the constricted section the *venturi*, and the valve the *throttle valve*. The throttle valve can be tilted more or less to control the air flow (Fig. 16-2). In the horizontal position, the throttle valve shuts off, or *throttles*, the airflow through the air horn. When the throttle valve is turned away from this position, air can flow through the air horn.

Fig. 16-2 Throttle valve in the air horn of a carburetor. When the throttle valve is closed, as shown, little air can pass through. But when the throttle valve is opened, as shown dashed, there is little throttling effect.

♦ 16-6 VENTURI EFFECT

The engine is, in a sense, a vacuum pump. As the pistons move down on the intake strokes, a partial vacuum is produced in the cylinders. A partial vacuum is any pressure less than atmospheric pressure. Atmospheric pressure pushes air, or air-fuel mixture, into the cylinders to fill the vacuum. (See ♦8-13, Producing a Vacuum, and ♦9-8, Intake Stroke.)

As the air flows toward the engine cylinders, it must first pass through the carburetor. A *venturi* (Fig. 16-1) is located in the air passage through the carburetor. As the air flows through the venturi, a partial vacuum is produced in it. The venturi restricts the flow of air so that the air pressure in the venturi is reduced. The air particles before the venturi are at atmospheric pressure (normal air pressure, or particles close together.) But as they move through the venturi, they speed up and spread out (pressure drops, or a partial vacuum develops).

The fuel nozzle (in Fig. 16-1) is located in the venturi. Atmospheric pressure is pushing down on the fuel in the float bowl. Since there is a partial vacuum around the venturi end of the fuel nozzle (the pressure is lower), atmospheric pressure pushes fuel up through the nozzle and into the air flowing through the venturi. The fuel sprays out, or atomizes, and quickly turns to vapor.

♦ 16-7 THROTTLE-VALVE ACTION

The throttle valve is a round disk below the venturi and fuel nozzle in the carburetor (Fig. 16-2). The air horn is the round cylinder through which air flows on its way to the engine cylinders. The air picks up a charge of fuel vapor while passing through the venturi. The throttle valve can be tilted more or less to allow more or less air-fuel mixture to flow through. If it is tilted as shown by the dashed lines in Fig. 16-2, more air can flow. More air flowing through the venturi increases the venturi vacuum. This causes more fuel to flow from the nozzle. If the throttle valve is tilted toward the closed position (shown solid in Fig. 16-2), less air will flow. The vacuum in the venturi will be less. Therefore less fuel will discharge into the passing air.

This arrangement provides a relatively constant air-fuel ratio from open to closed throttle. When the driver pushes down on the accelerator pedal, the throttle valve opens and the engine begins to deliver more power. If the driver releases the accelerator pedal, the throttle valve closes. Less air-fuel mixture flows to the engine. Engine power drops off.

Under the hood of a car with a carburetor, the accelerator pedal is connected by cable or linkage to the carburetor. The cable or linkage not only controls the position of the throttle valve or valves in the carburetor, but it can do other jobs. For example, in many cars the position of the accelerator pedal controls the upshift and downshift points of the automatic transmission under certain conditions.

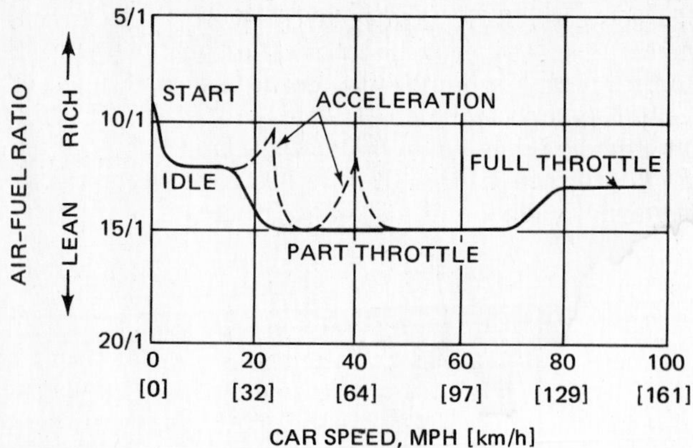

Fig. 16-3 Graph of air-fuel ratios for different car speeds. The graph is typical. Car speeds at which the various ratios are obtained may vary with different cars and engines. Also, there may be some variations in the ratios.

♦ 16-8 AIR-FUEL-RATIO REQUIREMENTS

The fuel system must vary the air-fuel ratio to suit different operating requirements. The mixture must be rich (have a high proportion of fuel) for starting. It must be leaner (have a lower proportion of fuel) for part-throttle medium-speed operation. Figure 16-3 is a graph showing typical air-fuel ratios as related to various car speeds. Ratios, and the speeds at which they are obtained, vary with different cars. In the example shown in Fig. 16-3, a rich mixture of about 9:1 (9 pounds [4 kg] of air for each pound [0.45 kg] of fuel) is supplied for starting. Then, during idle, the mixture leans out to about 12:1. At medium speeds, the mixture further leans out to about 15:1 or leaner. Some engines run on mixtures as lean as 20:1. But at higher speeds, with a wide-open throttle, the mixture is enriched to about 13:1. Opening the throttle for acceleration at any speed causes a momentary enrichment of the mixture. This results from special carburetor systems which are described later. Two examples are shown in Fig. 16-3 (at about 23 mph [37 km/h] and at 40 mph [64 km/h]).

You might think that the engine itself demands varying air-fuel ratios for different operating conditions. This is not quite true. For example, the mixture must be very rich for starting because fuel vaporizes slowly under starting conditions. The engine and carburetor are cold, the air speed is low, and much of the fuel does not vaporize. Therefore, an extra amount of fuel must be delivered by the carburetor so that enough will vaporize for starting. Likewise, sudden opening of the throttle for acceleration allows a sudden inrush of air. Extra fuel must quickly enter (the mixture must be enriched). This is because only part of the fuel vaporizes and mixes with the ingoing air to provide the proper proportions of air and fuel in the engine.

The following sections describe the various systems in carburetors that supply the air-fuel mixture required for different operating conditions.

In many late-model cars, the air-fuel ratio is controlled electronically. The electronic control systems are explained in the chapter on emission controls.

♦ 16-9 CARBURETOR SYSTEMS

The systems (or *circuits* as they are sometimes called) in the carburetor are:

1. Float system
2. Idle system
3. Main-metering system
4. Power system
5. Accelerator-pump system
6. Choke system

These systems are discussed in detail in following sections.

♦ 16-10 FLOAT SYSTEM

The float system includes the float bowl and a float and needle-valve arrangement. The float and the needle valve maintain a constant level of fuel in the float bowl. If the level is too high, then too much fuel will feed from the fuel nozzle. If it is too low, too little fuel will feed. In either event, poor engine performance will result. Figure 16-4 shows the basic float system. If fuel enters the float bowl faster than it is withdrawn, the fuel level rises. This causes the float to move up and push the needle valve into the valve seat. This shuts off the fuel inlet so that no fuel can enter. Then, if the fuel level drops, the float moves down and releases the needle so that the fuel inlet is opened. Now fuel can enter. In actual operation, the fuel is kept at an almost constant level. The float tends to hold the needle valve partly closed so that the incoming fuel just balances the fuel being withdrawn.

Figure 16-5 shows a carburetor with a dual float assembly. The carburetor has a float bowl that partly surrounds the carburetor air horn. The two floats are attached by a U-shaped lever and operate a single needle valve. Some carburetors have an auxiliary fuel valve and inlet. During heavy-load or high-speed operation, fuel may be withdrawn from the float bowl faster than it can enter through the main fuel inlet. If this happens, the fuel level drops. The end of the float lever presses against the auxiliary value, pushing it upward. This opens the auxiliary fuel inlet so that additional fuel can enter.

♦ 16-11 FLOAT-BOWL VENTS

The float bowls of carburetors are vented into the carburetor air horn at a point above the choke valve (top of Fig. 16-5, upper left in Fig. 16-6). The purpose of the vent is to help equalize the effects of a clogged air cleaner. For example, suppose the air cleaner has become clogged with dirt. The passage of air through it is then restricted. As a result, a partial vacuum develops in the carburetor air horn. Therefore, a greater-than-normal vacuum is applied to the fuel

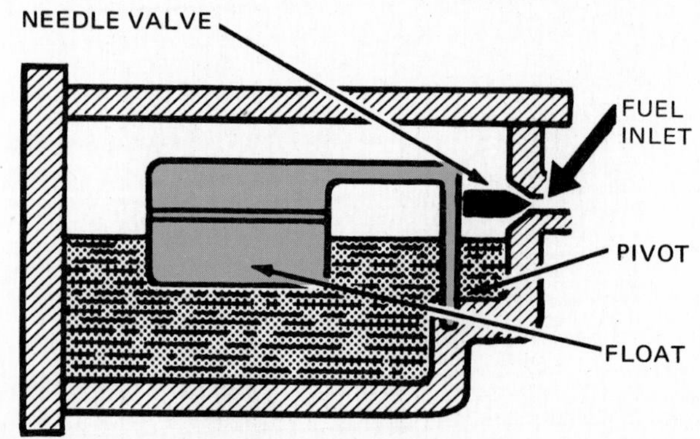

Fig. 16-4 A basic carburetor float system.

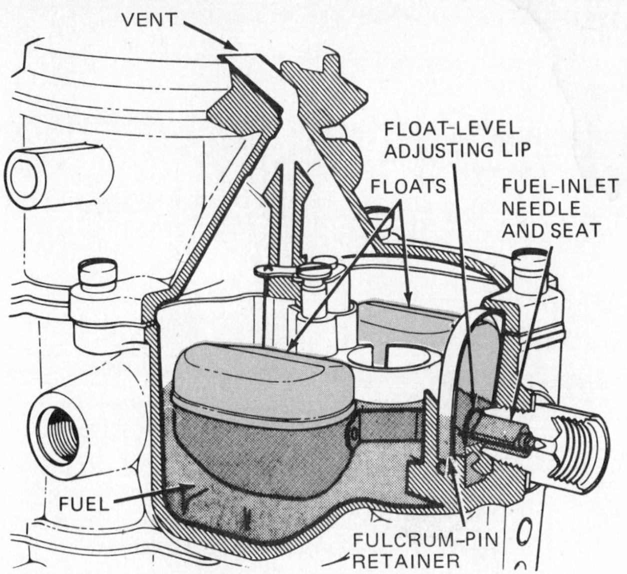

Fig. 16-5 The float system in a carburetor. *(Chrysler Corporation)*

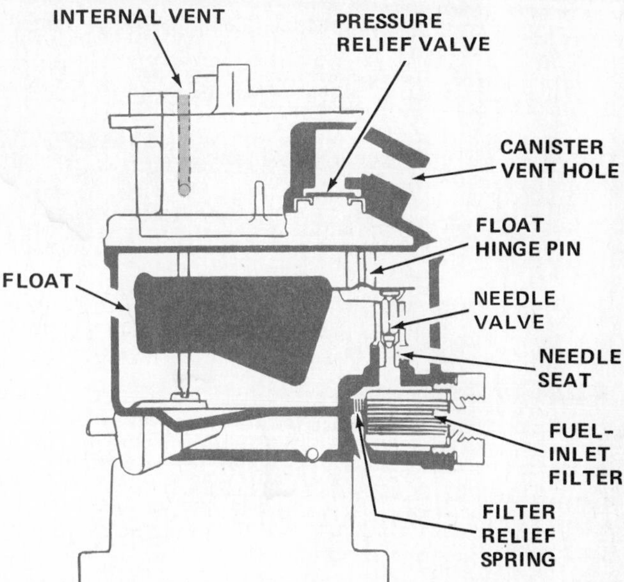

Fig. 16-6 Float system with two vents, one internal and the other to the charcoal canister. *(Chevrolet Motor Division of General Motors Corporation)*

nozzle (since this vacuum is added to the venturi vacuum). However, the partial vacuum resulting from the clogged air cleaner is also applied to the float bowl (through the vent). Therefore, the only force that pushes fuel from the fuel nozzle is the air pressure in the air cleaner. This is less than atmospheric pressure. The vent makes up for the effect of a clogged air cleaner. If the float bowl were vented to the atmosphere, then the force would be atmospheric pressure. This would produce a greater fuel flow from the fuel nozzle. Then the mixture would be too rich.

The carburetor which has the bowl vented into the carburetor air horn is called a *balanced carburetor*. Almost all carburetors used today are of this type.

The float bowl has another vent, shown to the upper right in Fig. 16-6. This vent is connected by a tube to the charcoal canister which is part of the fuel-vapor-recovery system. It is covered in the chapter on emission controls. In the carburetor in Fig. 16-6, the float bowl has a pressure-relief valve. The valve opens when vapor pressure increases in the float bowl. This allows the fuel vapor to flow to the charcoal canister. In other carburetors, the vent to the charcoal canister has a valve operated by the accelerator-pump lever. The valve is opened when the engine is idling or when it has been turned off.

In some carburetors, the vent valve to the charcoal canister is controlled by a solenoid. This provides positive action. When the ignition is turned on, the solenoid is connected to the battery. This pulls the valve down to block off the vent to the charcoal canister. The float bowl is vented to the top of the air horn. When the ignition is turned off, the solenoid is disconnected from the battery and the vent valve is released. It now moves up to open the vent to the charcoal canister. At the same time, the vent to the air horn is closed.

♦ 16-12 HOT-IDLE COMPENSATOR VALVE

The internal vent could be a problem during idling or low-speed operation, especially during hot weather. Fuel vapor from the float bowl can pass through the internal vent in sufficient amounts to upset the air-fuel ratio. The fuel vapor adds to the normal air-fuel mixture. Then the mixture becomes too rich. To take care of this, some carburetors have a hot-idle compensator valve. This valve is operated by a thermostatic blade. When temperatures reach a preset value, the blade bends enough to open the valve port. Now additional air can flow through the auxiliary air passage. This additional air bypasses the idle system. It leans out the mixture enough to make up for the added fuel vapor coming from the float bowl.

♦ 16-13 IDLE SYSTEM

When the throttle valve is closed or only slightly opened, only a small amount of air can pass through the air horn. The air speed is low, and very little vacuum develops in the venturi. This means that the fuel nozzle does not discharge fuel. Therefore, the carburetor must have another system to supply fuel when the throttle is closed or slightly opened.

This system, called the *idle system*, is shown in operation in Fig. 16-7. It includes passages through which air and fuel can flow. The air passage is called the *air bleed*. With the throttle valve closed as shown in Fig. 16-7, there is a high vacuum below the throttle valve from the intake manifold. Atmospheric pressure pushes air and fuel through the passages. They mix and flow past the tapered point of the idle-mixture screw. The mixture has a high proportion of fuel (is very rich). However, the mixture leans out slightly as it mixes with the small amount of air that gets past the closed throttle valve. But the final mixture is still rich enough for good idling. In some carburetors, the richness is adjusted by turning the idle-mixture screw in or out. This permits less or more air-fuel mixture to flow past the screw.

Note Many carburetors have limiting caps on the idle-mixture screws. On others the idle-mixture screws are recessed in the throttle body and sealed with hardened-steel plugs. The limiting caps permit some limited adjustment of

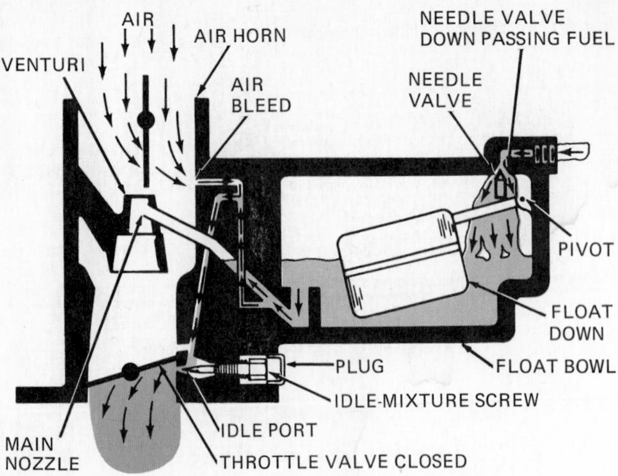

Fig. 16-7 Idle system in a carburetor. The throttle valve is closed so that only a small amount of air can get past it. All fuel is being discharged past the idle-mixture screw.

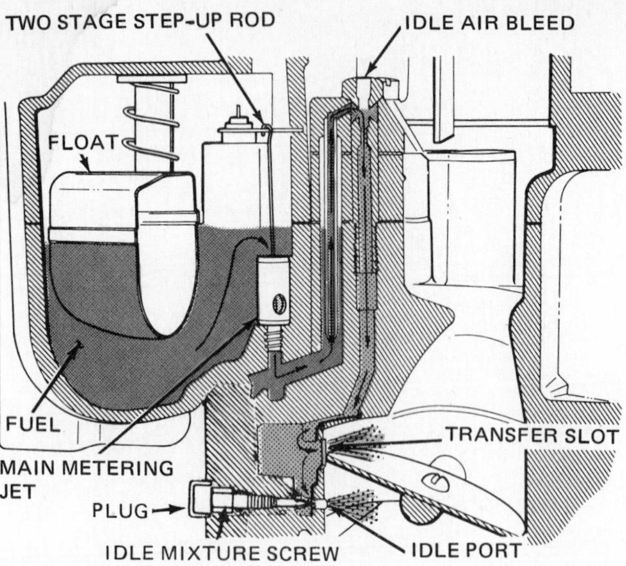

Fig. 16-9 Low-speed system in operation in a carburetor. *(Chrysler Corporation)*

the idle mixture. The steel plugs discourage tampering with the idle-mixture screws. The only way the plugs can be removed is by disassembling the carburetor.

Some carburetors have an extra feature in the idle system to provide some enrichment of the idle mixture during cold start-up. This enrichment is in addition to the choke action. The arrangement includes a *restrictor valve* in the air bleed to the idle system, which is described later. This restrictor valve is controlled by a vacuum diaphragm. When vacuum is applied to the vacuum diaphragm, the restrictor valve partly closes off the airflow. The air loss in the idle system then causes additional fuel to discharge from the idle system. This improves cold idle and combats any tendency for the engine to stall. The vacuum diaphragm gets its vacuum through a thermal switch that is mounted in the cooling system. When the coolant is cold, the thermal switch opens. This allows intake-manifold vacuum to operate the vacuum diaphragm. Therefore, the idle air-fuel mixture is enriched. When the engine warms up, the thermal switch closes to shut off the vacuum to the vacuum diaphragm. Now, normal hot-engine idle results, with normal hot-engine air-fuel mixture discharging from the idle system.

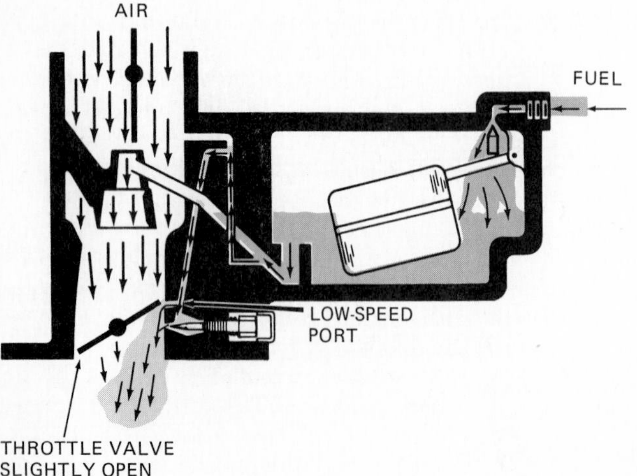

Fig. 16-8 Low-speed operation. The throttle valve is slightly open, and fuel is being discharged through the low-speed port and through the idle port.

Note Many late-model fuel systems have an electronic control system that provides accurate control of the carburetor idle system. It can change the richness of the idle mixture to hold exhaust-gas pollutants to a minimum. The system is covered later (♦16-22 to 16-24).

♦ 16-14 LOW-SPEED OPERATION

When the throttle valve is opened slightly, as shown in Fig. 16-8, the edge of the throttle valve moves past the low-speed port (called the *transfer slot* in Fig. 16-9) in the side of the air horn. This port is a vertical slot or a series of small holes, one above the other. Therefore, additional fuel is fed into the intake manifold through the low-speed port. This fuel mixes with the additional air moving past the slightly opened throttle valve. The additional fuel provides sufficient mixture richness for part-throttle low-speed operation.

Some air bleeds around the throttle valve through the low-speed port when the edge of the throttle valve is only partway past this port. This air improves the atomization of the fuel coming from the low-speed port.

♦ 16-15 OTHER IDLE SYSTEMS

There are many varieties of idle systems in addition to those shown here. In some two-barrel carburetors, each barrel has its own idle system. In many four-barrel carburetors, only the primary barrels have idle systems, as explained later.

♦ 16-16 MAIN-METERING SYSTEM

Suppose the throttle valve is opened enough so that its edge moves well past the low-speed port. Now there is little differ-

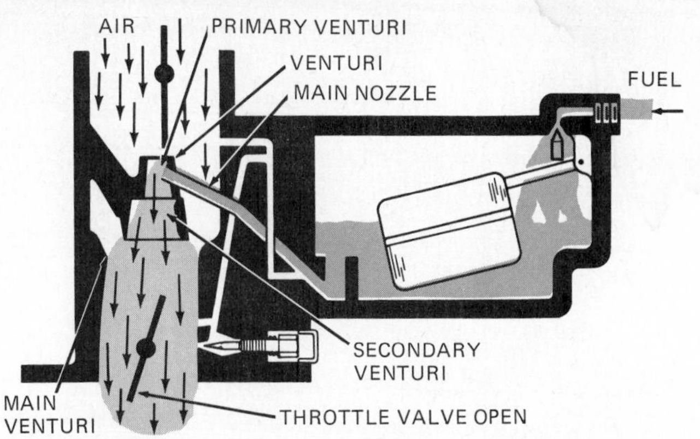

Fig. 16-10 Main-metering system in the carburetor. The throttle valve is open, and fuel is being discharged through the high-speed, or main, nozzle.

ence in vacuum between the upper and lower parts of the air horn. Therefore, little air-fuel mixture discharges from the low-speed port. However, under this condition, enough air moves through the air horn to produce a vacuum in the venturi. As a result, the fuel nozzle centered in the venturi (called the *main nozzle* or the *high-speed nozzle*) begins to discharge fuel (as explained in ♦16-6). The main nozzle supplies the fuel during operation with the throttle partly to fully opened. Figure 16-10 shows this action. The system from the float bowl to the main nozzle is called the *main-metering system*.

The wider the throttle is opened and the faster the air flows through the air horn, the greater the vacuum in the venturi. This means that additional fuel will be discharged from the main nozzle (because of the greater vacuum). As a result, a nearly constant air-fuel ratio is maintained by the main-metering system from part- to wide-open throttle. However, the maximum amount of fuel that can flow is controlled by the size of the fuel passage through the main-metering jet.

Note Many late-model fuel systems have electronic control systems that provide accurate control of the carburetor main-metering system. They constantly readjust the mixture richness from the main-metering system to hold exhaust-gas pollutants to a minimum. The systems are covered later (♦16-22 to 16-24).

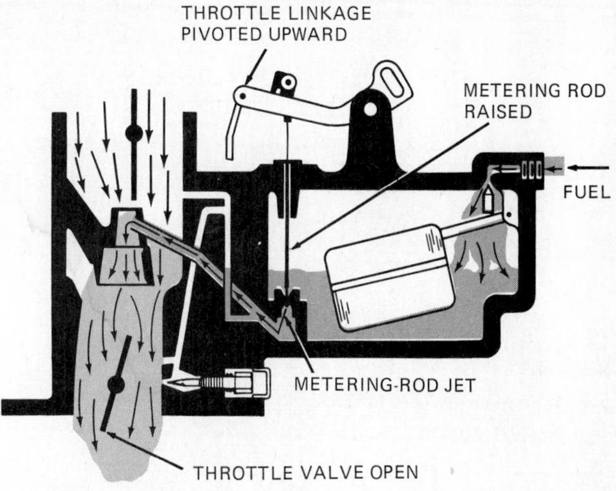

Fig. 16-12 Mechanically operated power system. When the throttle valve is opened, as shown, the metering rod is raised so that the smaller diameter of the rod clears the jet. This allows additional fuel to flow.

♦ 16-17 POWER SYSTEM

For high-speed full-power wide-open-throttle operation, the air-fuel mixture must be enriched. Additional devices are incorporated in the carburetor to provide this enriched mixture during high-speed full-power operation. They are operated mechanically or by intake-manifold vacuum.

♦ 16-18 MECHANICALLY OPERATED POWER SYSTEM

This system includes a metering-rod jet (an accurately calibrated orifice, or opening) and a metering rod with two or more steps of different diameters (Fig. 16-11). The metering rod is attached to the throttle linkage (Fig. 16-12). When the throttle is opened, the metering rod is lifted. But when the throttle is partly closed, the larger diameter of the metering

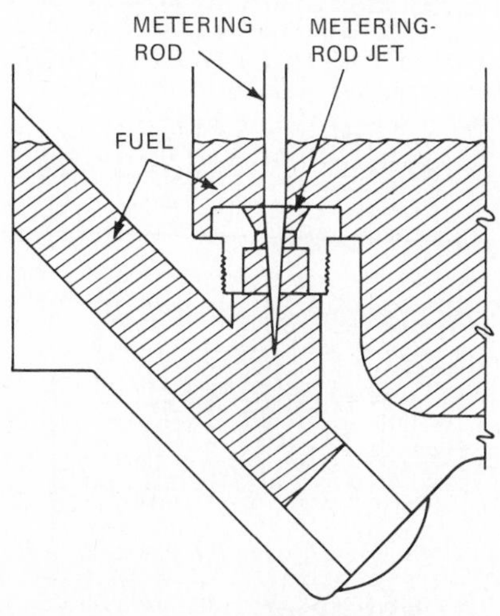

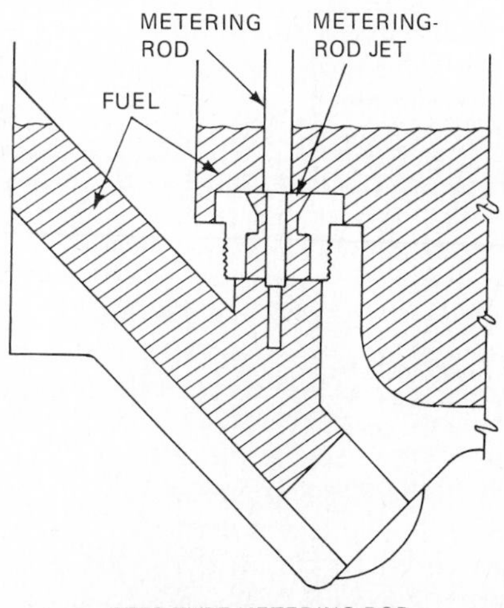

Fig. 16-11 Two types of metering rods that control fuel flow through the metering-rod jet into the power system.

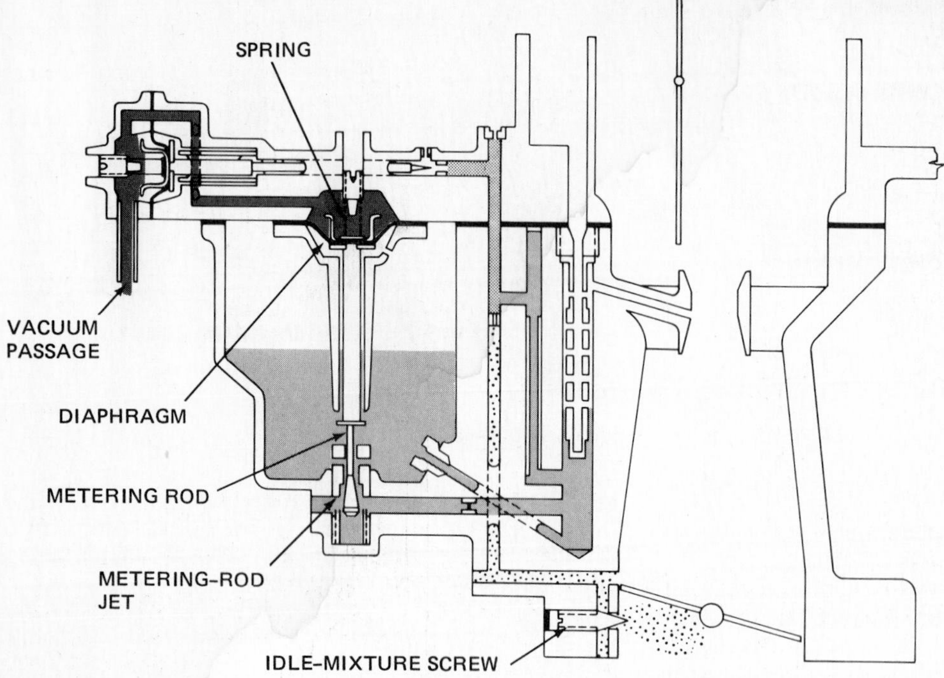

Fig. 16-13 A carburetor with a metering rod controlled by a vacuum-operated diaphragm. *(Pontiac Motor Division of General Motors Corporation)*

rod is in the metering-rod jet. This partly restricts fuel flow to the main nozzle. However, enough fuel does flow for normal part-throttle operation. When the throttle is opened wide, the rod is lifted enough to cause the smaller diameter of the rod to move up into the metering-rod jet. Now, the jet is less restricted, and more fuel can flow. The main nozzle is therefore supplied with more fuel, and the resulting air-fuel mixture is richer.

♦ 16-19 VACUUM-OPERATED POWER SYSTEM

This system is operated by intake-manifold vacuum. It includes a vacuum piston or diaphragm linked to a metering rod similar to the one shown in Fig. 16-11.

A carburetor using a spring-loaded diaphragm to control the position of the metering rod is shown in Fig. 16-13. When the throttle valve is opened so that intake-manifold vacuum is reduced, the spring pushes the diaphragm down. This lowers the metering rod so that its smaller diameter opens the jet, allowing more fuel to flow. Some carburetors use a spring-loaded piston instead of a diaphragm to control the metering rod.

♦ 16-20 COMBINATION POWER SYSTEMS

In some carburetors, a combination power system is used. It is operated both mechanically and by vacuum from the intake manifold. In one such carburetor, the metering rod is linked to a vacuum diaphragm as well as to the throttle linkage. Movement of the throttle to "full open" lifts the metering rod to enrich the mixture. Also, loss of intake-manifold vacuum (as during acceleration or hill climbing) causes the vacuum-diaphragm spring to raise the metering rod for an enriched mixture.

♦ 16-21 AIR-FUEL RATIOS WITH DIFFERENT SYSTEMS

Figure 16-14 shows the air-fuel ratios with the different carburetor systems in operation. This is a typical curve only. Actual air-fuel ratios may be different for different carburetors and different operating conditions. Note that the idle system supplies a very rich mixture to start with. Then, as engine speed increases, the mixture leans out. Between about 25 and 40 mph [40 and 64 km/h], when the throttle valve is only partly opened, both the idle and main-metering systems are supplying air-fuel mixture. Next (in Fig. 16-14) the main-metering system takes over completely. It continues to supply air-fuel mixture until a relatively high speed is reached. Then the power system comes into operation (ear-

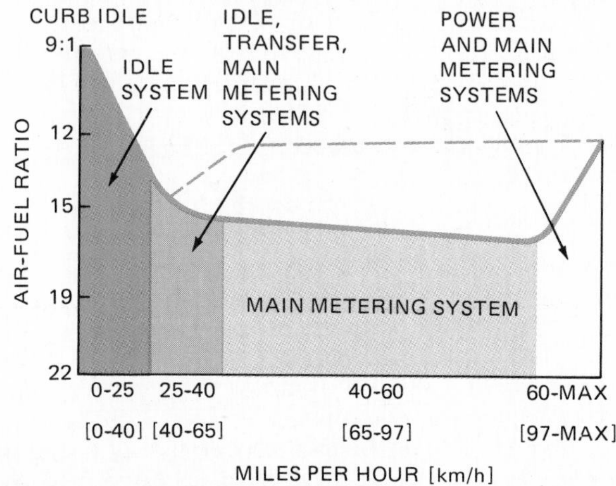

Fig. 16-14 Air-fuel ratios with different carburetor systems operating at different speeds. *(Chevrolet Motor Division of General Motors Corporation)*

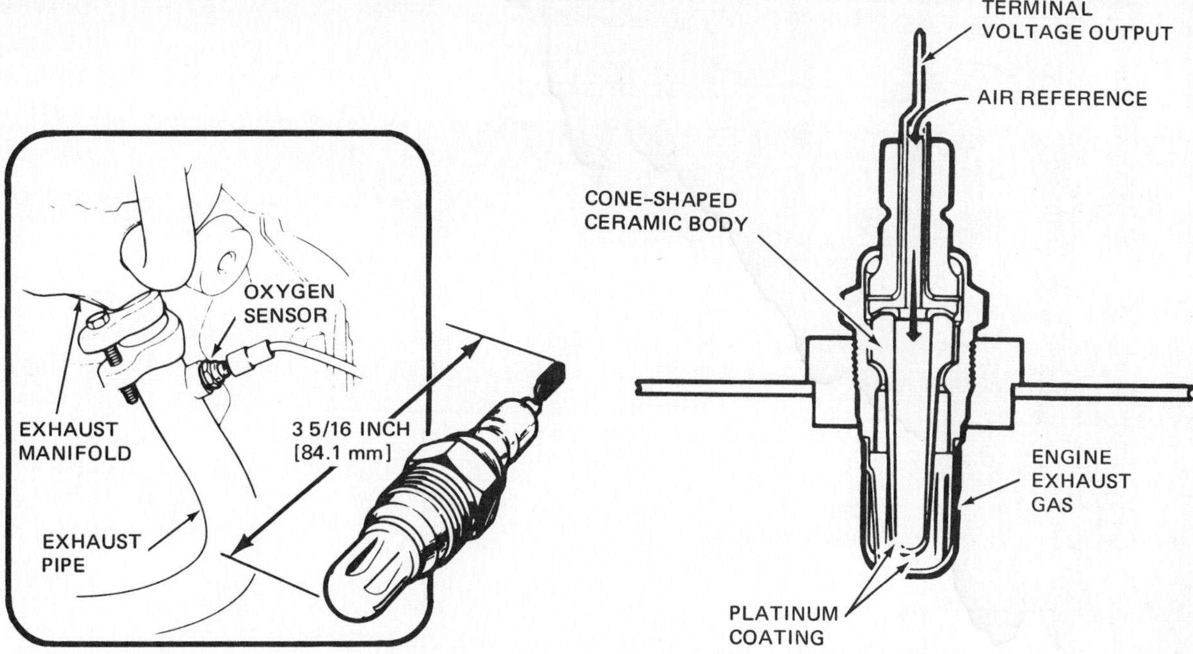

Fig. 16-15 Construction and installation of an oxygen sensor. *(AC Spark Plug Division of General Motors Corporation)*

lier, if the throttle valve is opened wide at lower speeds). Now the air-fuel mixture gets richer as speed increases.

♦ 16-22 ELECTRONIC CONTROL OF AIR-FUEL RATIO

The idle, low-speed, and power systems described in previous sections supply varying air-fuel ratios to meet different engine operating conditions. These systems depend on mechanical controls which often are not as accurate as the emission laws require. Inaccurate fuel metering may cause excessive HC and CO in the exhaust gas, which is usually due to an excessively rich mixture. When the mixture is too lean, excessive NO_x may appear in the exhaust gas and the engine may stall.

Modern electronic systems control the air-fuel ratio of the mixture flowing from the carburetor. These systems monitor the amount of oxygen in the exhaust gases. They use an oxygen sensor which is installed in the exhaust pipe (Fig. 16-15). If the exhaust gas is low in oxygen, the air-fuel mixture is too rich. It is using up most of the oxygen in the air-fuel mixture. If the exhaust gas is too high in oxygen, the air-fuel mixture is too lean. There is not enough fuel to use up the proper amount of oxygen in the air-fuel mixture.

The oxygen sensor is tied in with other components of the system, as explained in following sections. As the oxygen sensor reports the amount of oxygen in the exhaust gas, the system makes the adjustments required to correct the air-fuel ratio. If there is too much oxygen in the exhaust gas, the system signals the carburetor to enrich the mixture. If there is too little oxygen in the exhaust gas, the system signals the carburetor to lean out the mixture.

Two systems that electronically control the air-fuel ratio are described in following sections. These are the General Motors Computer Command Control (CCC) system, and the Ford Feedback-Carburetor Electronic Engine Control (FCEEC) system.

♦ 16-23 GENERAL MOTORS COMPUTER COMMAND CONTROL SYSTEM

This system (Fig. 16-16) monitors up to 15 engine-vehicle operating conditions. The information received by the electronic control module (ECM) then controls up to nine engine-related systems, as shown in Fig. 16-17.

The General Motors CCC system (Fig. 16-16) includes three subsystems:

1. Air injection system
2. Dual catalytic converter
3. Electronic feedback carburetor

The system includes an oxygen sensor (Fig. 16-15) and a coolant- (or engine-) temperature sensor (upper left in Fig. 16-16). The air-fuel ratio delivered by the carburetor is controlled by a mixture-control solenoid valve instead of a vacuum feedback diaphragm as in other carburetors. Figures 16-18 and 16-19 show the locations of the mixture-control solenoid valve.

Figure 16-20 is a sectional view of the solenoid. When the solenoid is energized, it pulls the plunger down. This pushes the valve down into the opening. The valve blocks off the opening and stops fuel flow through it. However, fuel is still being fed past the lean-mixture screw (which is factory adjusted). Figures 16-18 and 16-19 show how the fuel flowing past the lean-mixture screw can feed the idle or the main-metering system. However, this fuel is not sufficient to produce good engine operation. It takes additional fuel from the mixture-control solenoid valve.

The mixture-control solenoid valve, in operation, enriches the mixture. It supplies varying amounts of additional fuel according to the operating requirements. The solenoid cycles ON-OFF 10 times a second. The amount of additional fuel the valve supplies depends on what percentage of the time the valve is in the up position. If the solenoid holds the valve down longer each ON time, less fuel will flow. But if the solenoid stays off longer each OFF time, more fuel will flow.

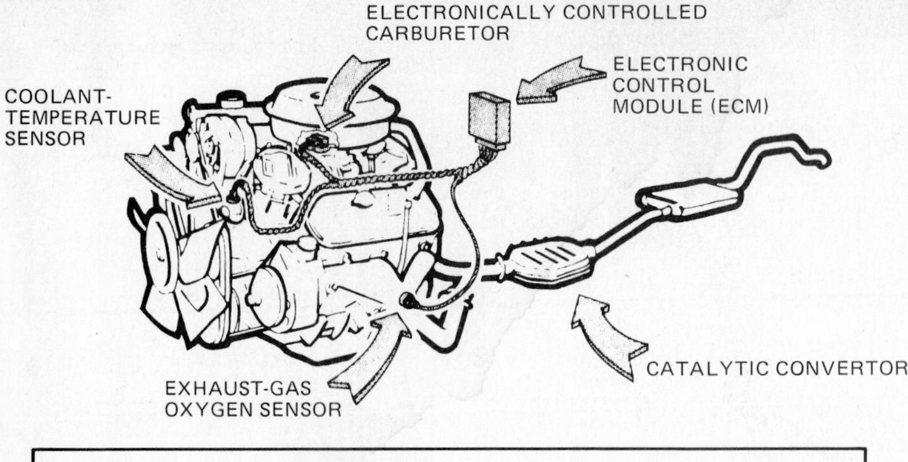

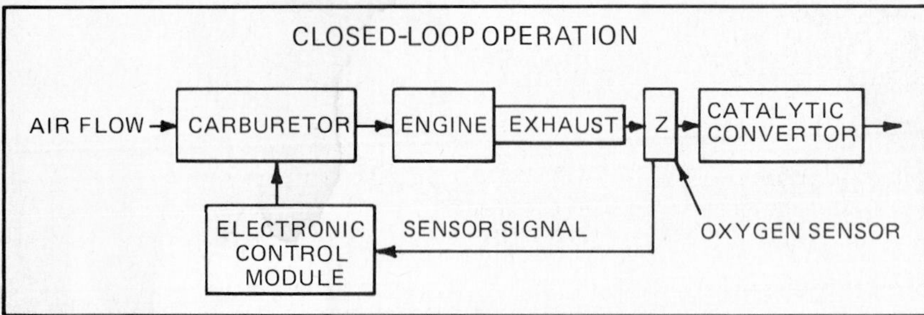

Fig. 16-16 General Motors CCC system of electronic engine control. *(Rochester Products Division of General Motors Corporation)*

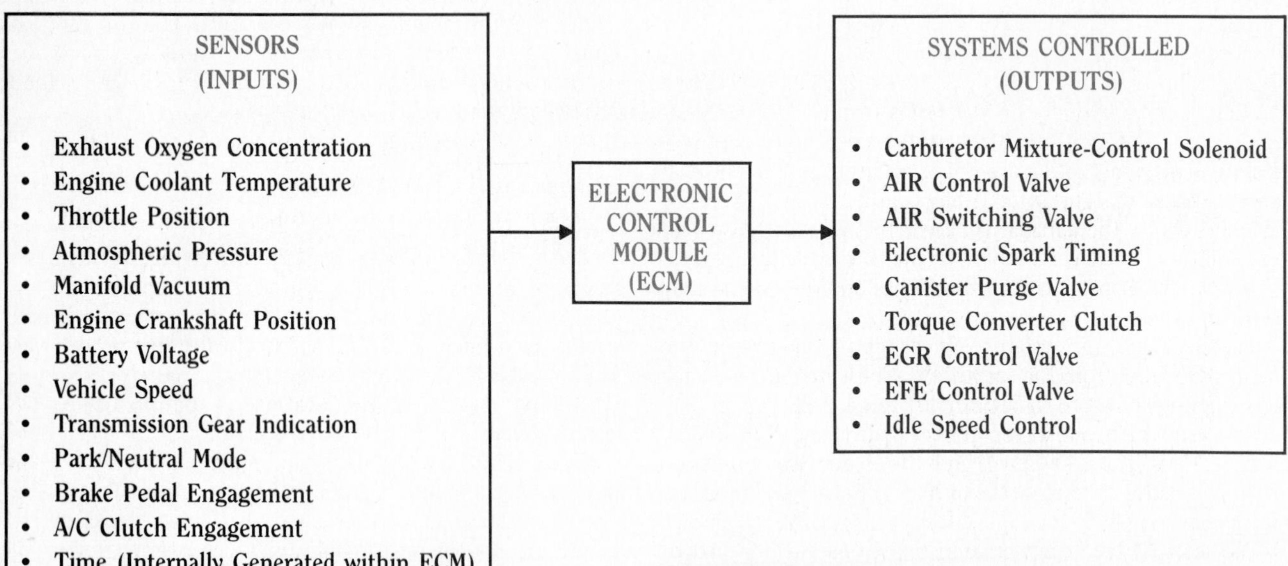

SENSORS (INPUTS)
• Exhaust Oxygen Concentration
• Engine Coolant Temperature
• Throttle Position
• Atmospheric Pressure
• Manifold Vacuum
• Engine Crankshaft Position
• Battery Voltage
• Vehicle Speed
• Transmission Gear Indication
• Park/Neutral Mode
• Brake Pedal Engagement
• A/C Clutch Engagement
• Time (Internally Generated within ECM)
• Cold Start Program Modifier Condition
• Engine Detonation
ELECTRONIC CONTROL MODULE (ECM)
SYSTEMS CONTROLLED (OUTPUTS)
• Carburetor Mixture-Control Solenoid
• AIR Control Valve
• AIR Switching Valve
• Electronic Spark Timing
• Canister Purge Valve
• Torque Converter Clutch
• EGR Control Valve
• EFE Control Valve
• Idle Speed Control

Fig. 16-17 Left, sensors that send information to the ECM. Right, systems and devices controlled by the ECM. All features are not used on all engines. *(Chevrolet Motor Division of General Motors Corporation)*

The electronic-control module receives signals from the various sensors (Fig. 16-17). Then the ECM "decides" how long to hold the solenoid valve closed each ON-OFF cycle. This determines how much more fuel is added to the basic flow past the lean-mixture screw.

If the oxygen sensor reports that the oxygen content of the exhaust gas is low, indicating a rich mixture, the ECM will hold the valve closed longer each ON-OFF cycle. This reduces the fuel flow from the mixture-control solenoid valve, leaning out the mixture. If the oxygen sensor reports that the oxygen content of the exhaust gas is high, this indicates a lean mixture. Then the ECM will hold the valve open longer each ON-OFF cycle. This increases the fuel flow from the mixture-control solenoid valve, enriching the mixture.

The system also controls the flow of air from the air pump. In Fig. 16-21, the engine is cold. Air from the air pump is flowing to the exhaust manifold. This is the *air-injection system.* Its purpose is to supply additional oxygen to burn up CO

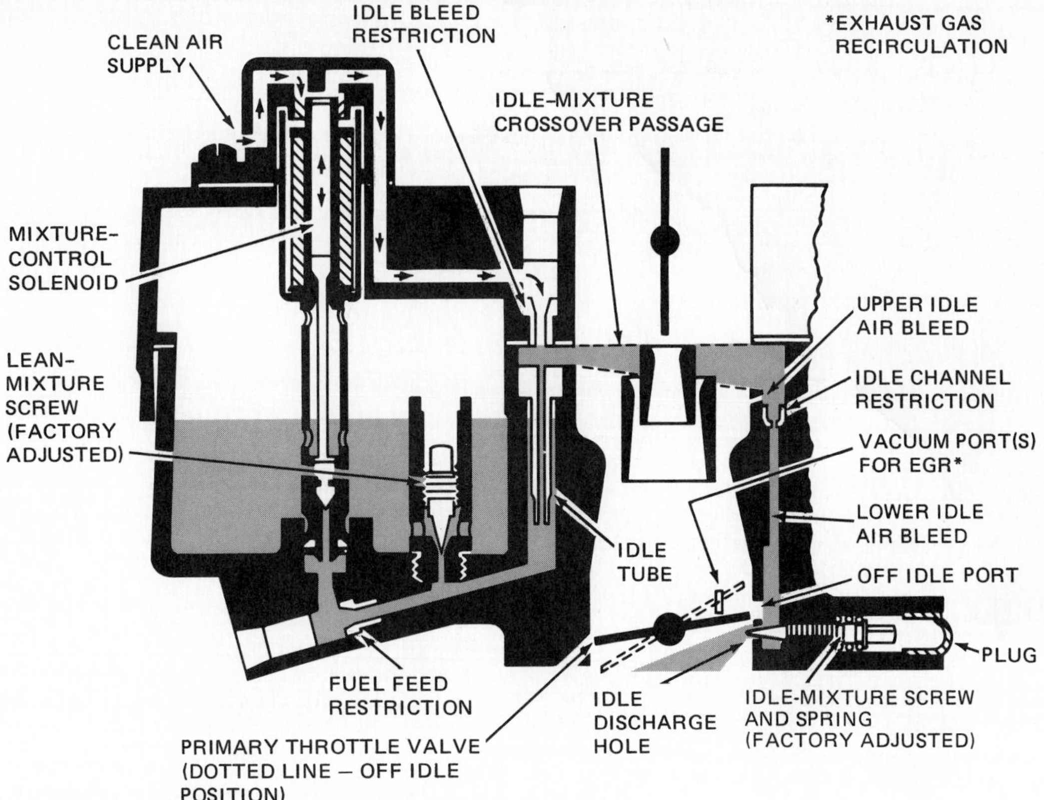

Fig. 16-18 Idle system of the CCC carburetor, which includes a mixture-control solenoid. *(Chevrolet Motor Division of General Motors Corporation)*

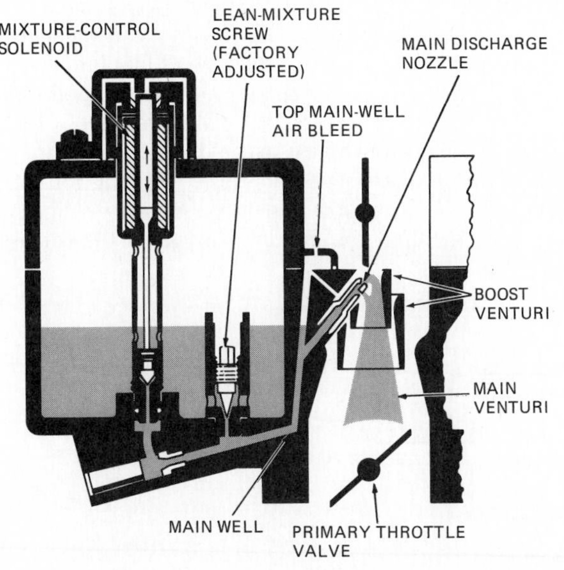

Fig. 16-19 Main-metering system of the CCC carburetor, which contains a mixture-control solenoid. *(Chevrolet Motor Division of General Motors Corporation)*

and HC in the exhaust gases. During the cold-engine operation, the air-fuel mixture is rich. Therefore, the exhaust gases are rich with HC and CO.

In Fig. 16-22, the engine is warmed up. Now the air from the air pump flows to the catalytic converter. This supplies additional oxygen to the converter to help it reduce the amount of CO and HC in the exhaust gases. The air-injection system and catalytic converter are described in a later chapter on emission control systems.

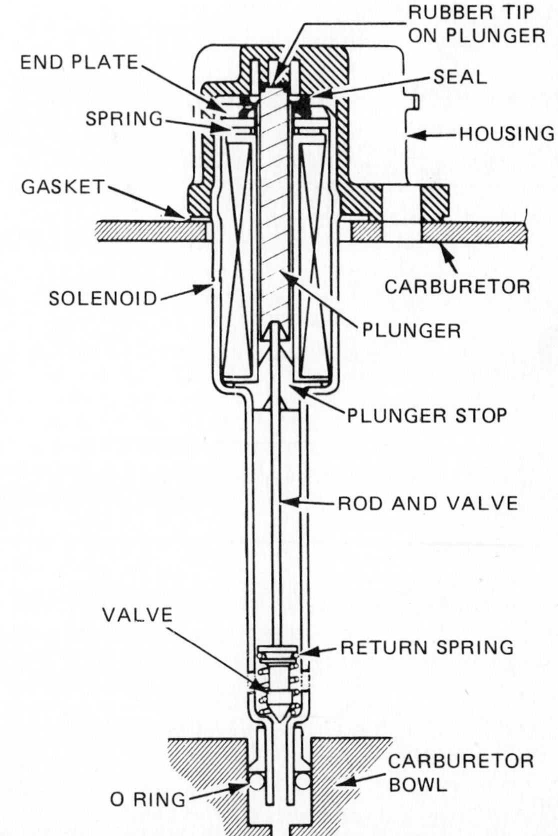

Fig. 16-20 Mixture-control solenoid valve. *(Chevrolet Motor Division of General Motors Corporation)*

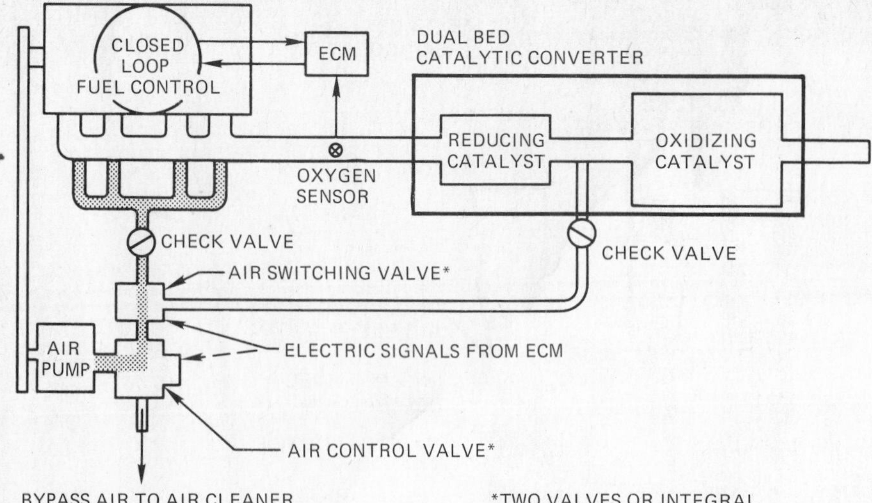

Fig. 16-21 Operation of the CCC system when the engine is cold. The system (called the *air-management system*) is sending air from the air pump to the exhaust manifold. *(Chevrolet Motor Division of General Motors Corporation)*

♦ 16-24 FORD FEEDBACK-CARBURETOR ELECTRONIC ENGINE CONTROL SYSTEM

The Ford FCEEC system for electronically controlling the air-fuel mixture also controls the *thermactor* (air-pump) system and the dual-catalytic-converter action. The thermactor-system air pump blows fresh air into the exhaust manifold. This helps burn up any HC and CO in the exhaust gases. The catalytic converter handles HC, CO, and NO_x. Both the thermactor system and the catalytic converter are described in a later chapter on emission control systems.

The Ford FCEEC system (Figs. 16-23 and 16-24) consists of three subsystems:

1. Thermactor, or air-injection, system
2. Dual catalytic converter
3. Electronic feedback carburetor

The electronic feedback carburetor is similar to the General Motors unit (Figs. 16-18 and 16-19). However, the Ford carburetor uses a feedback diaphragm to control the main-metering system (Fig. 16-25). The system uses an oxygen sensor in the exhaust pipe and an engine-temperature switch.

If the oxygen sensor reports that the oxygen content of the exhaust gas is low, this indicates a rich mixture. Then the ECM signals the vacuum-switch system to send more vacuum to the feedback diaphragm. This causes the diaphragm to lift the metering rod to restrict the flow of fuel into the main-metering system. If the oxygen sensor reports that the oxygen in the exhaust gas is excessive, then the mixture is too lean. Now, the vacuum-switch system reduces the vacuum to the diaphragm. It relaxes and allows the metering rod to move down. More fuel can flow to enrich the mixture. The system constantly monitors the amount of oxygen in the exhaust gas.

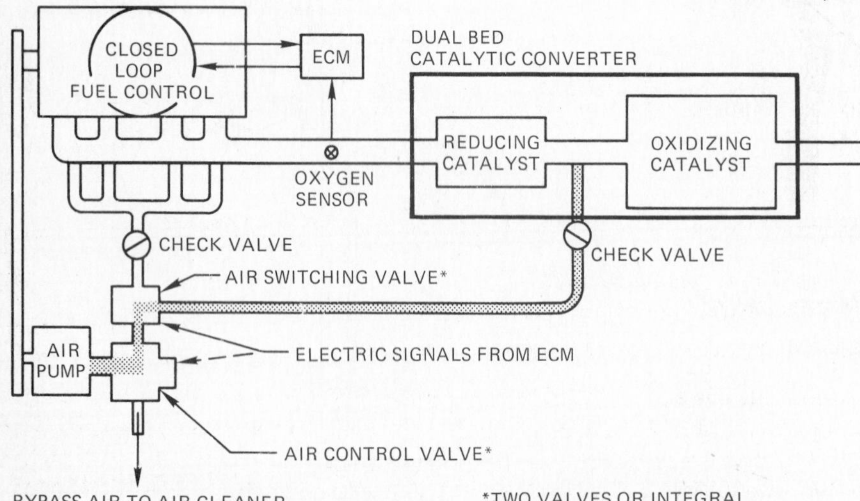

Fig. 16-22 Operation of the CCC system when the engine is at normal operating temperature. The air-management system is sending air from the air pump to the catalytic converter. *(Chevrolet Motor Division of General Motors Corporation)*

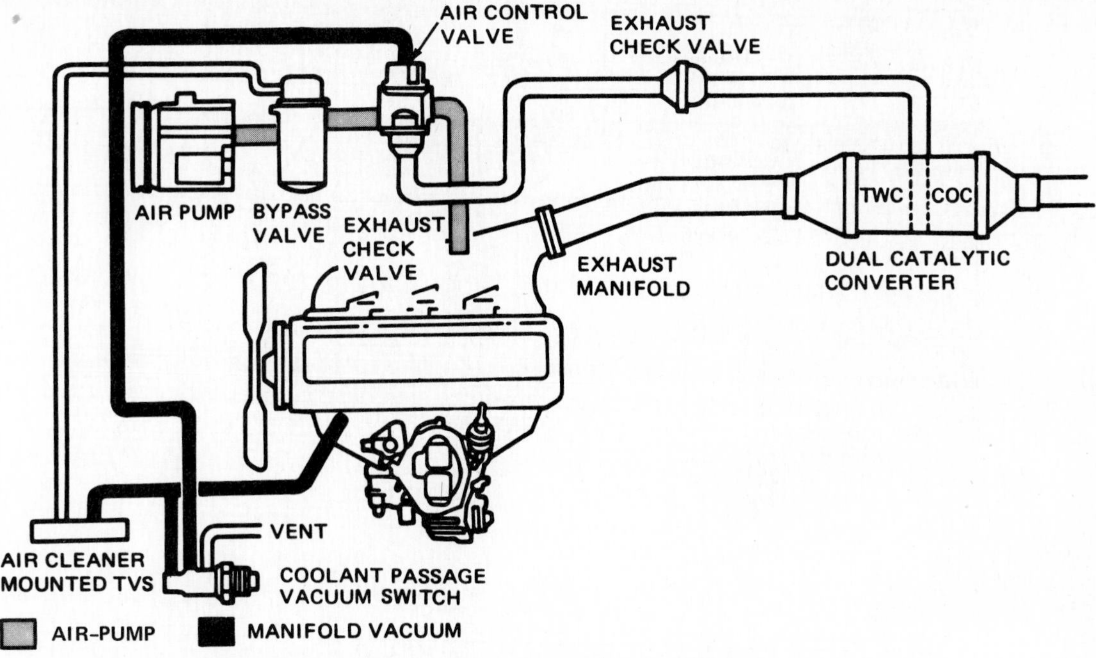

Fig. 16-23 Air-injection system for the electronic engine control system, with dual catalytic converter. With the engine cold, air-pump air flows to the exhaust manifold. *(Ford Motor Company)*

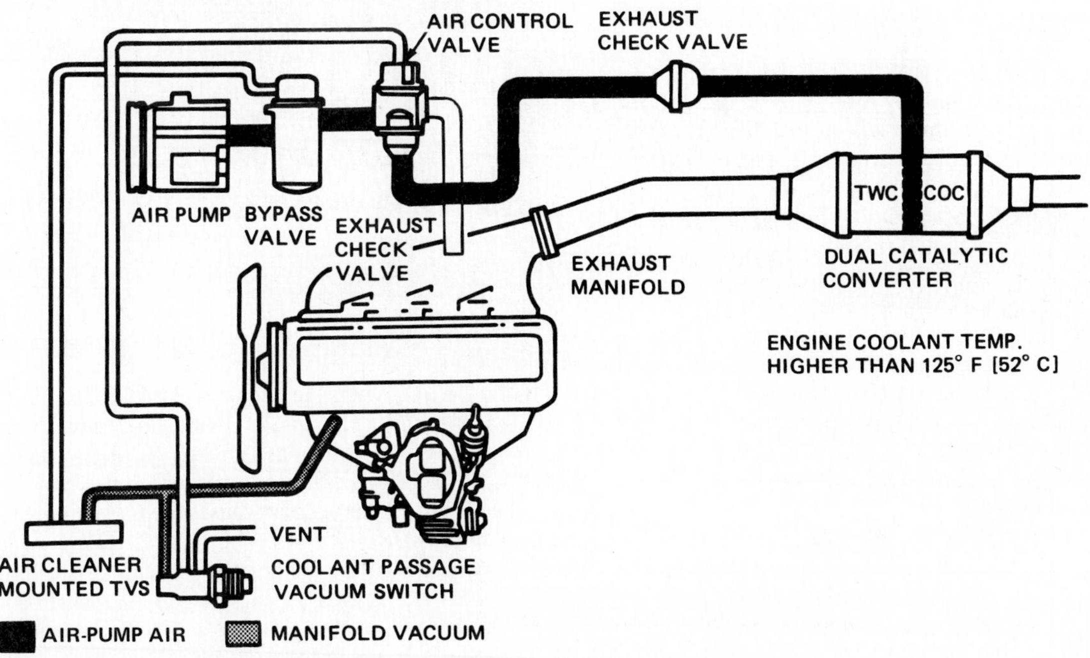

Fig. 16-24 Air-injection system with the engine at normal operating temperature. Air from the air pump is flowing to the dual catalytic converter. *(Ford Motor Company)*

Adjustments are made automatically as necessary to maintain the desired air-fuel ratio.

The system operates in two modes (Figs. 16-23 and 16-24). When the engine is cold (Fig. 16-23), the air-control valve sends the air from the air pump into the exhaust manifold. During cold-engine operation, the air-fuel mixture is rich. Therefore, the exhaust gas is rich with HC and CO. The fresh air entering the exhaust manifold helps to burn up these pollutants.

When the engine warms up, the system switches to the hot-engine mode (Fig. 16-24). The coolant-passage vacuum switch, sensing that the engine has warmed up, shuts off the vacuum to the air-control valve. Now, air from the air pump is sent to the catalytic converter. The oxygen in the air helps the converter to change the pollutants into harmless gases. Catalytic-converter action is described in a later chapter. Also discussed is the necessity for additional air to make the catalyst work.

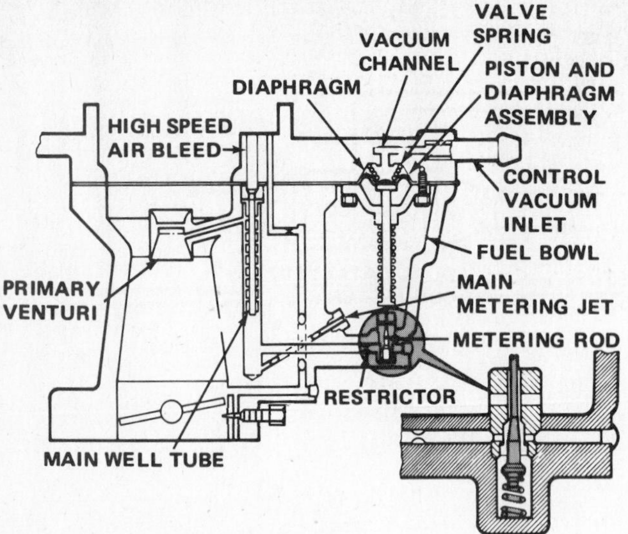

Fig. 16-25 Feedback carburetor for an electronic engine control system, using a diaphragm for control of the main-metering system. *(Ford Motor Company)*

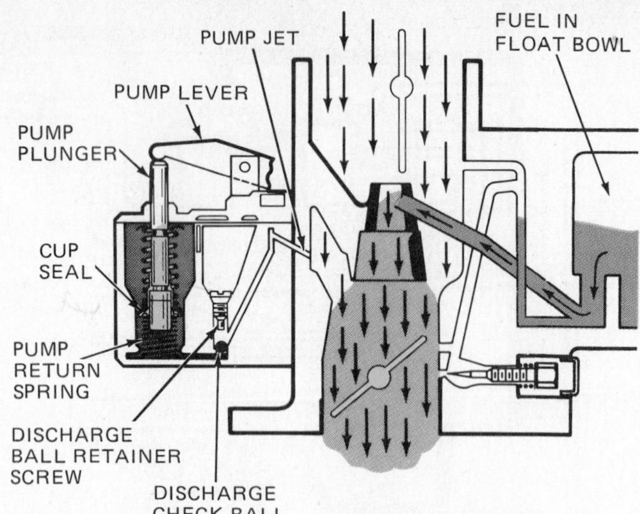

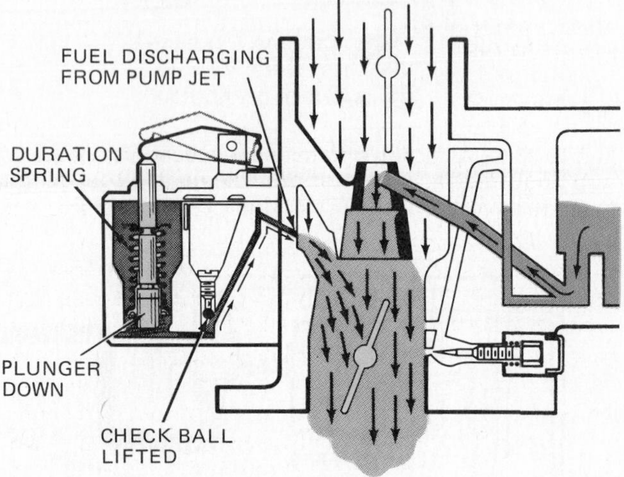

Fig. 16-26 Operation of the accelerator-pump system.

♦ 16-25 ACCELERATOR-PUMP SYSTEM

For acceleration, the carburetor must deliver additional fuel (♦16-8). Rapid opening of the throttle allows a sudden inrush of air. This causes a sudden demand for additional fuel. Carburetors have accelerator-pump systems to provide this extra fuel. Figure 16-26 shows one type. It includes a pump plunger which is forced downward by a pump lever that is linked to the throttle. When the throttle is opened, the pump lever pushes the pump plunger down. This forces fuel to flow through the accelerator-pump system and out the pump jet (Fig. 16-26). This fuel enters the air passing through the carburetor to supply the additional fuel needed.

The action carries through a plunger (or "duration") spring (Fig. 16-26). When the throttle valve is opened, the spring is compressed. This applies a force against the plunger which pushes it down. The spring maintains this force all the time that the throttle is held open until the pump plunger is all the way down, as shown in Fig. 16-26. This arrangement allows the accelerator-pump system to discharge fuel for several seconds, or until the power system can take over. The result is smooth acceleration.

One type of accelerator-pump system uses a diaphragm instead of a plunger. When the throttle is opened, linkage pushes the diaphragm into a fuel-filled pump chamber. This forces additional fuel from the pump chamber through the accelerator-pump system and out the pump jet.

♦ 16-26 CHOKE SYSTEM

When the engine is being started, the carburetor must deliver a very rich mixture to the intake manifold. With the engine and carburetor cold, only part of the fuel vaporizes. Therefore, extra fuel must be delivered. Then, enough evaporates to make a combustible mixture that permits the engine to start.

During cranking, air speed through the carburetor air horn is very low. Vacuum from the venturi action and vacuum below the throttle valve would be insufficient to produce adequate fuel flow for starting. To produce enough fuel flow during cranking, the carburetor has a *choke* (Fig. 16-27). The choke is a round valve, shaped like the throttle valve, located in the top of the air horn. It is controlled mechanically or by an automatic device. When the choke is closed, it is almost horizontal, as shown in Fig. 16-27. Only a small amount of air can get past it. The valve has "choked off" the airflow. Then, when the engine is cranked, a fairly high vacuum develops in the air horn. This vacuum causes the main nozzle to discharge a heavy stream of fuel. The quantity delivered is sufficient to produce the air-fuel mixture needed for starting the engine.

As soon as the engine starts, the speed increases from a cranking speed of around 250 to 300 rpm to over 600 rpm. Now more air and a slightly leaner mixture are required. One method of getting more air into the engine as soon as it starts is to mount the choke valve off center on its shaft in the air horn. Then, by adding a spring to the choke linkage, the vacuum produced by the running engine causes the valve to partly open against the spring force. More air can then flow through. Another arrangement includes a small spring-loaded section in the valve. This section opens to admit the additional air.

Mechanically controlled chokes are operated by a pull rod on the dash. The pull rod is linked to the choke valve. When the rod is pulled out, the choke valve is closed. However, the driver must remember to push the control rod into the

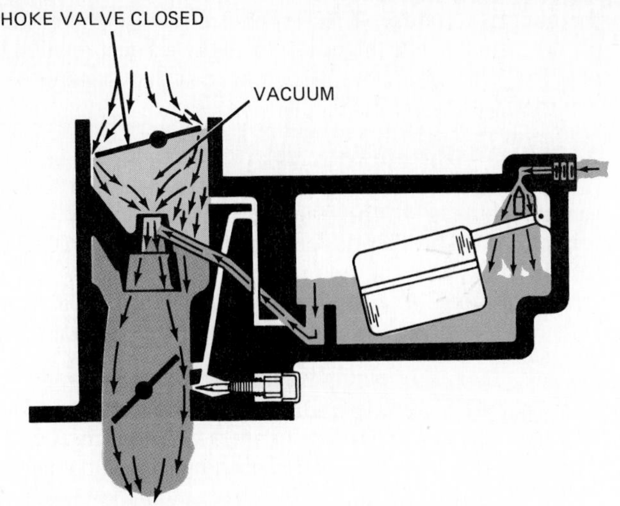

Fig. 16-27 With the choke closed, intake-manifold vacuum is introduced into the carburetor air horn. This causes the main nozzle to discharge fuel.

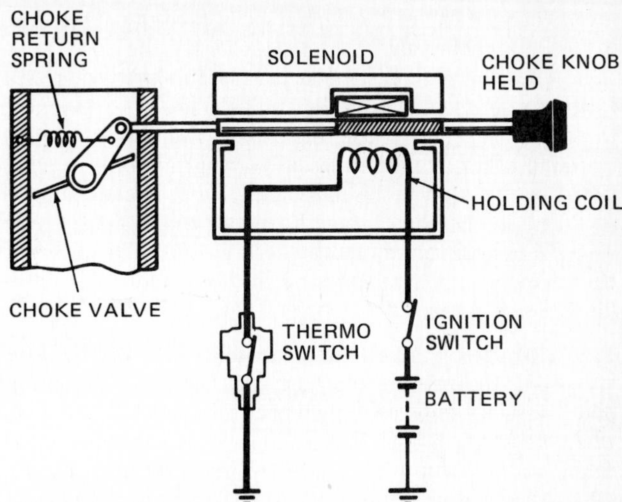

Fig. 16-28 Manual choke with automatic return. When the engine warms up, the thermostatic (thermo) switch opens, disconnecting the solenoid from the battery. This allows the choke spring to open the choke valve. *(Toyota Motor Sales Company, Ltd.)*

dechoked position as soon as the engine begins to warm up. If not, the carburetor will continue to supply a very rich mixture to the engine. This excessively rich air-fuel mixture will cause poor engine performance, high exhaust emissions, fouled spark plugs, poor fuel economy, and many other problems.

To prevent the problems caused by improper driver operation of the choke, carburetors have *automatic chokes*. These devices close the choke valve when the engine is cold and gradually open it as the engine warms up. The automatic choke devices are all similar, although they vary in detail. Most operate on exhaust manifold temperature and intake manifold vacuum.

♦ 16-27 AUTOMATIC CHOKES

1. Manual Choke with Automatic Return This system is a partial answer to the problem of leaving the choke valve in the closed position (Fig. 16-28). When the ignition switch is turned on, with the engine cold, the solenoid is connected to the battery. Now, when the choke knob is pulled out, it will be held in the choking position by the solenoid. The solenoid overcomes the pull of the choke return spring. However, when the engine warms up, the thermo switch will open. This disconnects the solenoid from the battery. The choke return spring now pulls the choke valve to the open, or dechoked, position.

2. Automatic Choke The above semiautomatic arrangement (Fig. 16-28) is not fast enough for most vehicles. It could leave the choke valve in the choked position too long. This would cause excessive pollutants in the exhaust gases. Most modern vehicles have an automatic choke that works on manifold temperature and intake-manifold vacuum. On some cars the choke has an electric heating element. There are also other arrangements as explained below. The purpose of these arrangements is to get fast choke opening after the engine has started. This minimizes the time that the exhaust gases contain excessive pollutants.

Figure 16-29 shows a carburetor with an automatic choke that works on heat from the exhaust manifold and vacuum from the intake manifold. It includes a thermostatic spring and a vacuum piston, both linked to the choke valve. The thermostatic spring is made up of two different metal strips welded together and formed into a spiral. Owing to differences in the expansion rates of the two metals, the thermostatic spring winds up or unwinds with changing tempera-

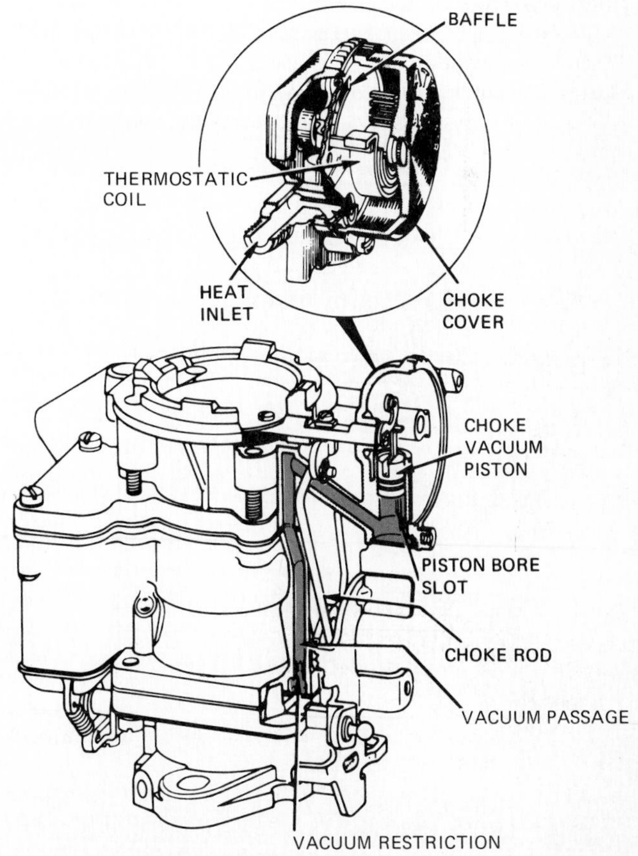

Fig. 16-29 Automatic-choke system on a carburetor. *(American Motors Corporation)*

ture. When the engine is cold, the spring is wound up enough to close the choke valve and spring-load it in the closed position. When the engine is cranked, a rich mixture is delivered to the intake manifold. As the engine starts, air movement through the air horn causes the choke valve to open slightly (working against the thermostatic-spring tension). In addition, the vacuum piston is pulled outward by intake-manifold vacuum. This produces some further opening of the choke valve. The choke valve is now properly positioned to cause the carburetor to supply the proper rich mixture for cold-engine idling.

3. Automatic-Choke Action When the throttle valve is opened, the mixture must be enriched. The accelerator pump provides some extra fuel, but still more fuel is needed when the engine is cold. This additional fuel is secured by the action of the vacuum piston. When the throttle is opened, intake-manifold vacuum is lost. The vacuum piston releases and is pulled inward by the thermostatic-spring tension. The choke valve therefore moves toward the "closed" position and causes the mixture to be enriched. During the first few moments of engine operation, the choke valve is controlled by the vacuum piston.

4. Automatic-Choke Action during Warm-Up The thermostatic spring begins to release as the engine warms up. The thermostatic spring is in a housing that is heated through a tube connected to the exhaust manifold. As the engine warms up, the spring unwinds. This causes the choke valve to move toward the "opened" position. When operating temperature is reached, the thermostatic spring has unwound enough to fully open the choke valve. No further choking takes place.

When the engine is stopped and cools, the thermostatic spring again winds up. This closes the choke valve and spring-loads it in the "closed" position.

Figure 16-29 shows a carburetor and the construction of the automatic choke. The vacuum passage to the vacuum piston is shown, but the heat tube to the exhaust manifold is not. The heat tube sends heat from the exhaust manifold to the thermostatic-spring housing, as shown in Fig. 16-30.

5. Thermostat in Manifold In many engines, the thermostat is located in a well in the intake manifold. There it can quickly react to the manifold heat as the engine starts (Fig. 16-31). The thermostat is connected by a link to the carburetor. Some carburetors using this arrangement have vacuum pistons. Others have vacuum diaphragms. Both work with the thermostat as previously described to control the choke-valve position during warm-up.

6. Coolant-Operated Choke Some carburetors use heat from the engine coolant to operate the thermostat. The thermostat housing has a passage through which the coolant flows. The action is similar to other automatic chokes described above.

7. Vacuum Diaphragm Instead of a vacuum piston, many automatic chokes use a vacuum-operated diaphragm (called a *vacuum-break* or *vacuum-kick diaphragm*) as shown in Fig. 16-31. The operation is quite similar. However, the diaphragm provides more force to break the choke valve loose if it gets stuck. The linkage from the diaphragm to the choke-valve lever rides freely in a slot in the lever. During certain phases of warm-up operation, the changing vacuum causes the linkage to ride to the end of the slot in the choke lever and move the choke valve. For example, when first starting, vacuum develops in the intake manifold. This vacuum acts on the vacuum-break diaphragm. The linkage rides to the end of the slot with enough force to "break" the choke valve away from the full-choke position. This partly opens the choke so that an overrich mixture is prevented. The vacuum-break diaphragm exerts a more positive and stronger force than a choke piston, such as shown in Fig. 16-29.

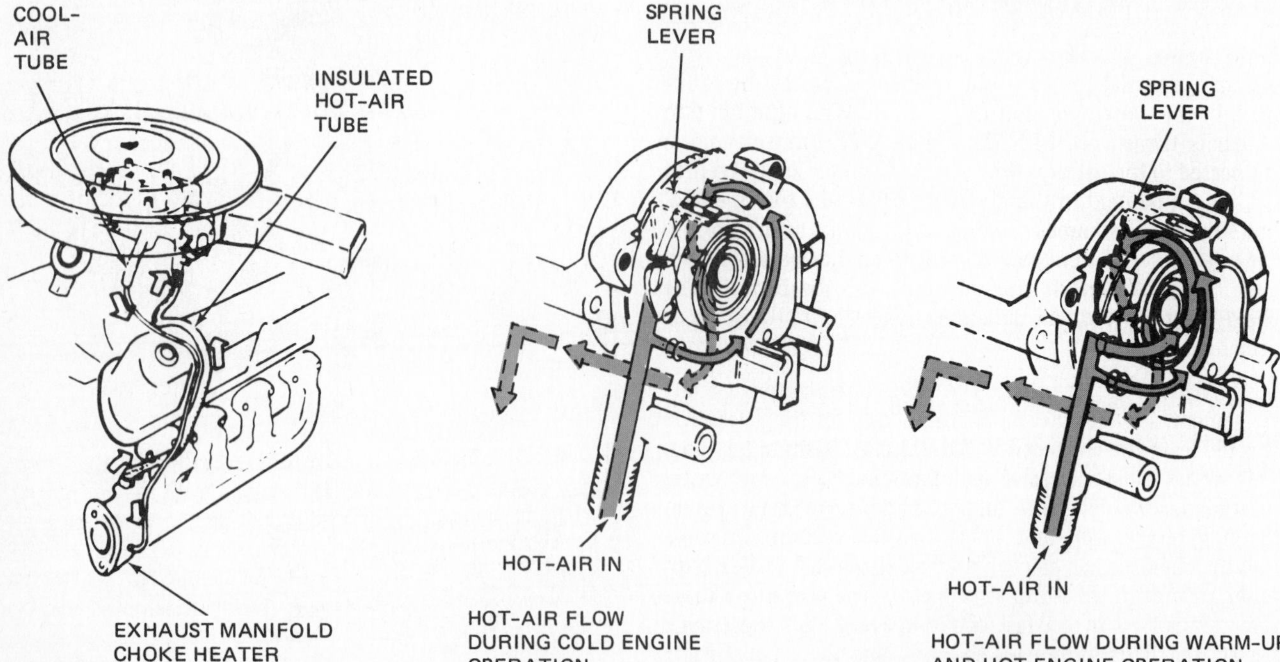

Fig. 16-30 Flow of hot air through the automatic choke. *(Ford Motor Company)*

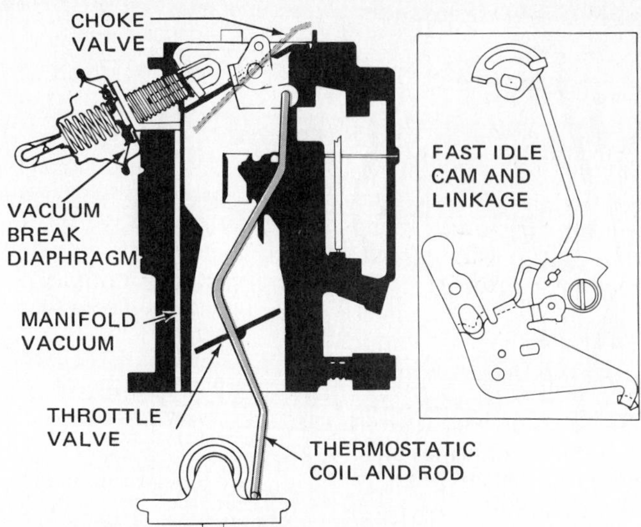

Fig. 16-31 Choke system with the thermostat located in a well in the intake manifold. The vacuum-break diaphragm opens the choke valve. *(Chevrolet Motor Division of General Motors Corporation)*

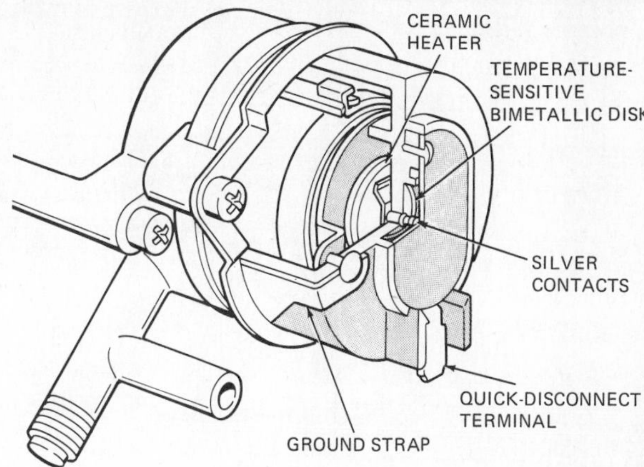

Fig. 16-32 An electric-assist choke. At low temperatures, the ceramic heater turns on, adding heat to the choke so that it opens more quickly. *(Ford Motor Company)*

8. Electric Choke Many cars have electric automatic chokes. This type of choke includes an electric heating element (Fig. 16-32). The purpose of the heater is to assure faster choke opening. This helps reduce emissions from the engine. Emissions (HC and CO) are relatively high during the early stages of engine warm-up. At low temperatures, the electric heater adds to the heat coming from the intake manifold. This reduces choke-opening time to as short as 1 ½ minutes or less.

Some automatic chokes do not rely on heat from the engine to actuate the thermostatic coil. Instead, they operate only on electricity and intake-manifold vacuum. The electric heating element is calibrated to produce as quick a choke opening as possible without causing poor driveability. This quick choke opening reduces the time that a rich mixture is going to the engine cylinders. Therefore, there will be excessive HC and CO in the exhaust gas for a shorter time.

♦ 16-28 MANIFOLD HEAT-CONTROL VALVE

During initial warm-up of a cold engine, the fuel vaporizes very poorly. This is because the fuel and the entering air are cold. To improve fuel vaporization and cold-engine operation, some engines have a *manifold heat-control valve*. This device improves fuel vaporization. When the engine is cold, the heat-control valve quickly adds heat from the exhaust gases to the intake manifold. The valve is built into the exhaust and intake manifolds. There are two basic arrangements, one for in-line engines and the other for V-type engines.

Note Many engines have a heated-air system with a thermostatic air cleaner. This system also adds heat from the exhaust gases to the ingoing air-fuel mixture. For this reason, engines with the heated-air system usually do not have a heat-control valve. The heated-air system is covered in a later chapter on emission controls.

1. In-Line Engines In these engines, the exhaust manifold is located under the intake manifold. At a central point, there is an opening from the exhaust manifold into a chamber, or oven, surrounding the intake manifold (Fig. 16-33). A butterfly valve is placed in this opening (see Fig. 11-16). When the valve is turned one way, the opening is closed off. The position of the valve is controlled by a thermostat. When the engine is cold, the thermostatic spring winds up and moves the valve to the "closed" position (top in Fig. 16-33). Now, when the engine is started, the hot exhaust gases pass through the opening and circulate through the oven around the intake manifold. Heat from the exhaust gas quickly warms the intake manifold and helps the fuel to vaporize. Therefore, cold-engine operation is improved. As the engine warms up, the thermostatic spring unwinds and the valve moves to the "opened" position (bottom in Fig. 16-33). Now, the exhaust gases pass directly into the exhaust pipe.

2. V-Type Engines In V-type engines, the intake manifold is placed between the two banks of cylinders. It has a special passage (Fig. 16-34) through which exhaust gases can pass. One of the exhaust manifolds has a thermostatically controlled valve that closes when the engine is cold. This causes exhaust gases to pass from that exhaust manifold through the special passage in the intake manifold. The exhaust gases then enter the other exhaust manifold. Heat from the exhaust gases passing through heats the air-fuel mixture in the intake manifold for improved cold-engine operation. As the engine warms up, the thermostatically controlled valve opens. Then, the exhaust gases from both exhaust manifolds pass directly into the exhaust pipes.

♦ 16-29 EARLY FUEL EVAPORATION (EFE) SYSTEM

In 1975, a vacuum-controlled manifold heat-control valve was introduced. It uses a vacuum motor instead of a thermostat to control the heat-control valve (Fig. 16-35). The system gets its name—early fuel evaporation (EFE) system—from its quick action. The vacuum to operate the vacuum motor comes from the intake manifold through a thermal vacuum switch.

The EFE valve does the same job as the thermostatically controlled valve described in ♦16-28. However, the EFE valve does the job much faster. This reduces the time that heat is going into the intake manifold, and reduces the chances of

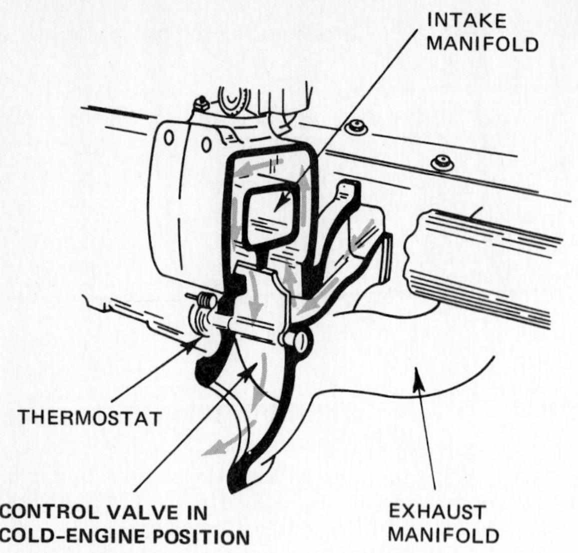

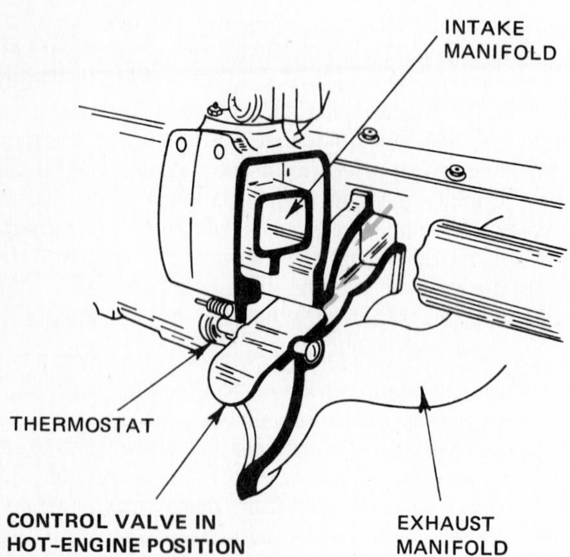

Fig. 16-33 Intake and exhaust manifolds of an in-line engine that uses a manifold heat-control valve. At top, the heat-control valve is in the cold-engine (heat on) position. It is directing hot exhaust gases up and around the intake manifold, as shown by the arrows. At bottom, the valve is in the hot-engine (heat off) position. Flow of hot gases up around the intake manifold is blocked off. *(Ford Motor Company)*

overheating the ingoing air-fuel mixture. Overheating the air-fuel mixture will reduce engine power. Expansion of the mixture due to heat reduces the amount of mixture entering the engine.

When the engine is cold, the thermal vacuum switch admits vacuum to the EFE vacuum motor. This causes the heat-control valve to take the cold-engine position (top in Fig. 16-33). The intake manifold is heated by hot exhaust gases passing up around the intake manifold. When the engine warms up, the thermal vacuum switch closes. With no vacuum on the EFE vacuum motor, the heat-control valve swings to the hot-engine position (bottom, Fig. 16-33).

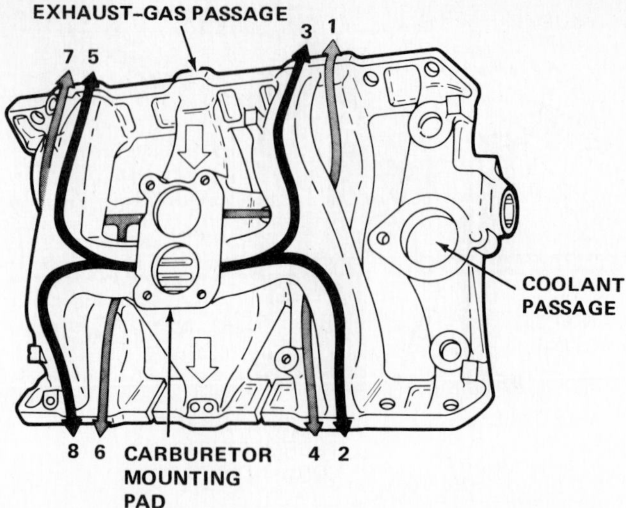

Fig. 16-34 Intake manifold for a V-8 engine. The arrows show the flow of air-fuel mixture from the two barrels of the carburetor to the eight cylinders in the engine. The central passage connects the two exhaust manifolds. Exhaust gas flows through this crossover passage during engine warm-up. *(Pontiac Motor Division of General Motors Corporation)*

♦ 16-30 INTAKE-MANIFOLD ELECTRIC HEATER

Some engines use an electric intake-manifold heater. Its location is shown in Fig. 11-14. Figure 16-36 shows how it is mounted on the intake manifold. Operation of the heater is controlled by coolant temperature. When it and the engine are cold, the electric-heater relay is closed. Then, when the ignition is turned on, the electric-heater relay is connected to the battery. The relay closes its points, which connects the manifold electric heater to the battery. The heater immediately begins to warm the intake manifold. This improves fuel vaporization and cold-engine operation. As the engine and coolant warm up, extra heating is no longer needed. The coolant-temperature sensor signals the electric-heater relay. It opens its points so the electric heater is disconnected from the battery.

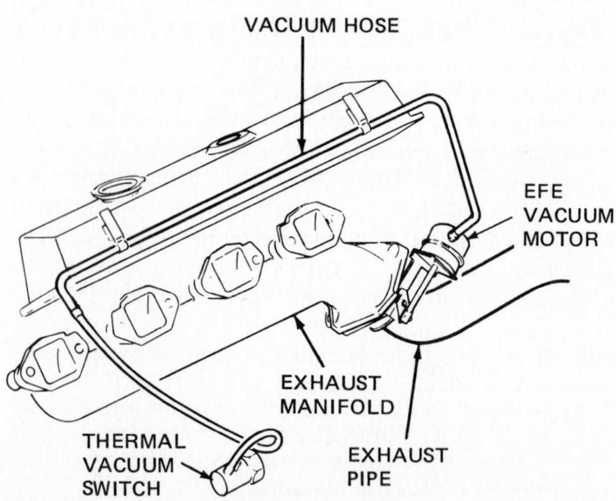

Fig. 16-35 Location of the vacuum motor and thermal-vacuum switch (TVS) in an EFE system. *(Cadillac Motor Car Division of General Motors Corporation)*

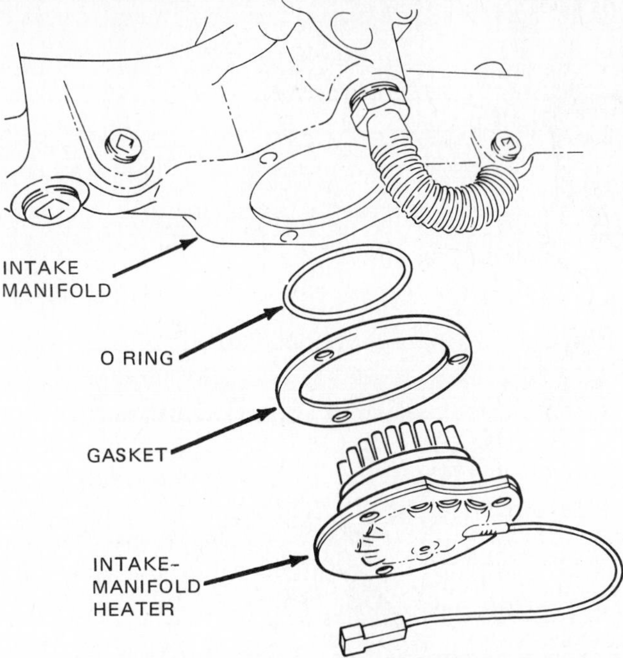

Fig. 16-36 An intake-manifold heater. When the engine is cold, the heater adds heat to the intake manifold. This improves cold-engine operation. *(American Motors Corporation)*

♦ 16-31 ANTI-ICING

When fuel is sprayed into the air passing through the air horn, the fuel evaporates, or turns to vapor. During evaporation, the fuel takes heat from the surrounding air and metal parts. This is the same effect you get if you put a few drops of water on your hand. Your hand feels cold. If you blow on your hand, causing the water to evaporate faster, your hand will feel even colder. The faster the evaporation takes heat from your hand, the cooler your hand feels.

In the carburetor, spraying and evaporation of the fuel "rob" the surrounding air and carburetor of heat. Under certain conditions, the surrounding metal parts are so cooled that moisture in the air condenses and then freezes on the metal parts. The ice can build up sufficiently, if conditions are right, to cause the engine to stall. This is most apt to occur during the warm-up period following the first start-up of the day. It happens more with air temperatures in the range of 40 to 60°F [4.4 to 15.6°C] and fairly humid air.

To prevent such icing, many carburetors have a special anti-icing system. During the warm-up period, the manifold heat-control valve sends hot exhaust gases from one exhaust manifold to the other. Part of this hot exhaust gas circulates around the carburetor idle ports and near the throttle-valve shaft. This adds enough heat to prevent icing. Another system has coolant passages in the carburetor. A small amount of engine coolant passes through a special manifold in the carburetor throttle body. This adds enough heat to the carburetor to prevent icing.

♦ 16-32 FAST IDLE

When the engine is cold, slight throttle-valve opening must be maintained. This is needed to make the engine idle faster. Otherwise, the slow idle and cold engine might cause the engine to stall. With fast idle, enough air-fuel mixture gets through, and air speeds are great enough, to produce adequate vaporization and a sufficiently rich mixture. Fast idle is obtained by a fast-idle cam linked to the choke valve (Fig. 16-37). When the engine is cold, the automatic choke holds the choke valve closed. In this position, the linkage has revolved the fast-idle cam so that the adjusting screw rests on the high point of the cam. The adjusting screw prevents the throttle valve from moving to the "fully closed" position. The throttle valve is held partly open for fast idle. As the engine warms up, the choke valve opens. This rotates the fast-idle cam so that the high point moves from under the adjusting screw. The throttle valve closes for normal hot-engine slow idle.

♦ 16-33 AIR-BLEED AND ANTISIPHON PASSAGES

In all the systems of the carburetor except the accelerator-pump system, there are small openings to permit air to enter, or *bleed* into, the system. This produces some premixing of the air and fuel so that better atomization and vaporization result. It also helps maintain a more uniform air-fuel ratio. At higher speeds, a larger amount of fuel tends to discharge from the main nozzle. But at the same time the faster fuel movement through the high-speed system causes more air to bleed into it. Therefore, the air-bleed holes tend to equalize the air-fuel ratio.

Air-bleed passages are also sometimes called *antisiphon* passages. They act as air vents to prevent the siphoning of fuel from the float bowl at intermediate engine speeds.

If air-bleed passages become plugged, they may cause the float bowl to be emptied after the engine shuts off. When the engine is shut off, the intake manifold cools down and a slight vacuum forms as a result. With open air bleeds, air can move through the bleeds to satisfy the vacuum. If air bleeds are plugged, then the vacuum will cause the float bowl to empty through the idle system.

♦ 16-34 SPECIAL CARBURETOR DEVICES

Carburetors have other devices to improve driveability and fuel economy and reduce air pollution. Some of these are described in later chapters on emission controls. Others are described below.

1. Idle Solenoid Some engines have a tendency to continue to run after the ignition switch is turned off. This is called "run-on" or *dieseling*. The engine runs like a diesel engine, without electric ignition. Dieseling can be caused by high engine-cylinder temperatures, particles of carbon that are hot, or hot spark-plug electrodes. Enough air-fuel mixture can leak around the throttle valve and through the idle system to maintain the dieseling. To prevent this, many carburetors are equipped with an antidieseling solenoid or idle solenoid. Figure 16-38 shows one arrangement which controls the closing of the throttle valve. When the engine is running, the solenoid is connected to the battery. This causes the solenoid plunger to extend. The plunger serves as the idle stop and prevents complete closing of the throttle. Therefore, normal hot-idle operation results when the driver releases the accelerator pedal. However, when the engine is turned off, the solenoid is disconnected from the battery. Now, the solenoid plunger retracts. The throttle closes almost completely, shutting off all airflow. The engine stops running.

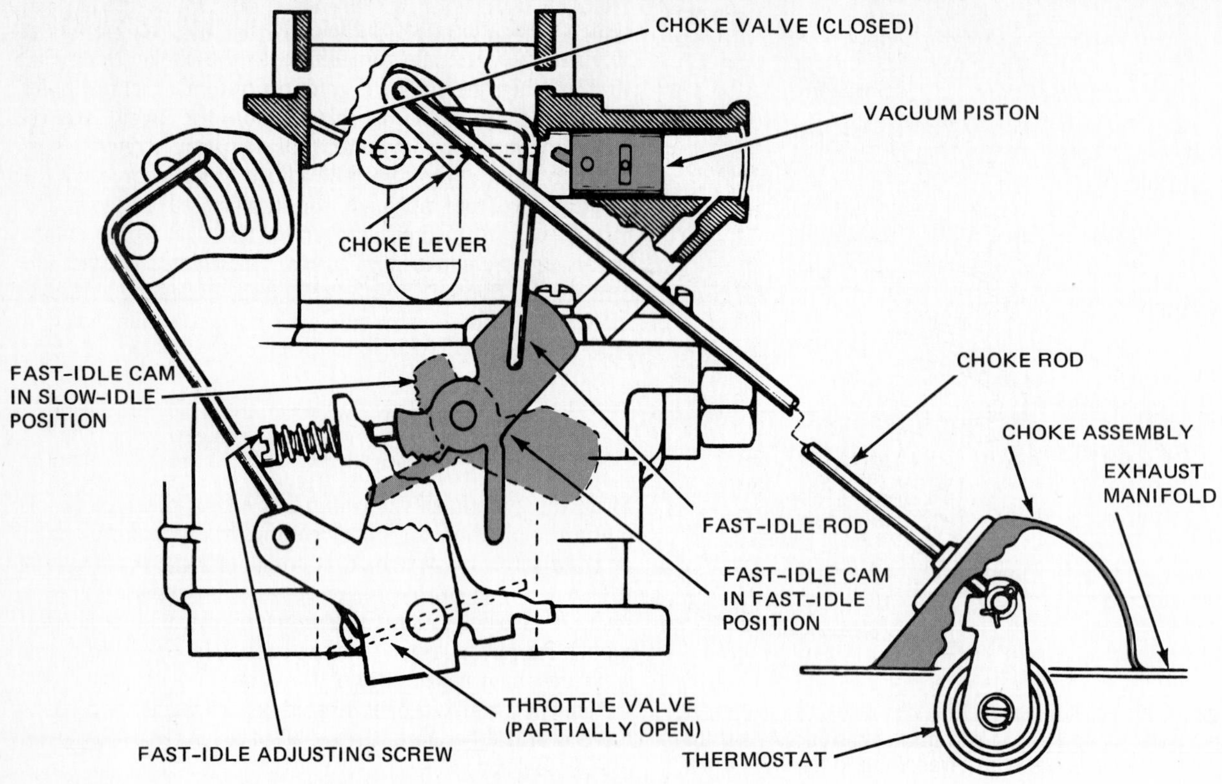

Fig. 16-37 Vacuum and thermostatically operated choke with the thermostat located in the exhaust manifold. Note the two positions of the fast-idle cam. *(Chrysler Corporation)*

A second arrangement uses the solenoid to shut off the fuel flow in the idle system when the engine is turned off. During normal operation, the solenoid pulls in its plunger so normal fuel flow can continue through the idle system. However, when the engine is shut off, the solenoid releases its plunger. This blocks the idle system so no fuel can flow through it. Therefore the engine stops.

2. Electric Speedup Solenoid Many cars with air conditioning have an *electric speedup solenoid*. It looks like the idle solenoid described above. However, the speedup solenoid increases the engine idle speed when the air conditioner is turned on. This helps prevent engine stalling at idle.

3. Throttle-Return Check If the throttle closes too fast after the driver releases the accelerator pedal, the air-fuel mixture can be momentarily excessively enriched. This is because the fuel nozzle continues to dribble fuel for a moment even though the airflow is largely shut off. The idle system will also momentarily feed a rich mixture. This is due to the high vacuum that results when the engine is running fairly fast with the throttle closed. This very rich mixture can cause the engine to stumble, or hesitate, because it will not burn properly. Also, the high level of HC and CO in the exhaust, because of poor combustion, may damage the catalytic converter. To prevent this, many carburetors are equipped with a throttle-return check. It slows down the closing of the throttle enough so that the momentary excessive richness is prevented.

Note Carburetors equipped with the antidieseling solenoid, described in item 1 on page 165, do not require the throttle-return check.

4. Electric Kickdown Switches on Some Cars Equipped with Automatic Transmissions These switches provide an electrical means to downshift the transmission into a lower gear when the throttle is opened wide.

5. Governors to Control or Limit Top Engine Speed The use of governors is largely confined to heavy-duty vehicles. They prevent overspeeding and rapid wear of the engine. One type directly controls the throttle valve. It tends to close the valve as rated engine speed is reached. Another type has a throttle plate between the carburetor throttle valve and the intake manifold. The throttle plate moves toward the closed position as rated speed is reached. This prevents delivery of additional amounts of air-fuel mixture and any further increase in engine speed.

6. Vacuum-Operated Devices There is a vacuum in the intake manifold when the engine is running. This vacuum is used by other devices on the engine and in the car. These uses are listed below and described later.

a. Ignition-Distributor Vacuum-Advance Mechanism This advances the spark during part-throttle operation. In electronic spark-control systems, the vacuum is carried to an electronic control unit (ECU). The ECU "reads" the vacuum and decides when and how much to advance the spark.

b. Positive Crankcase Ventilating System This is a system for ventilating the crankcase without polluting the atmosphere.

c. Fuel-Vapor Emission Control System This system traps gasoline vapor from the fuel tank and carburetor float bowl.

d. Heated-Air System This system provides rapid heating of the air entering the carburetor when the engine is cold. This improves cold-engine operation.

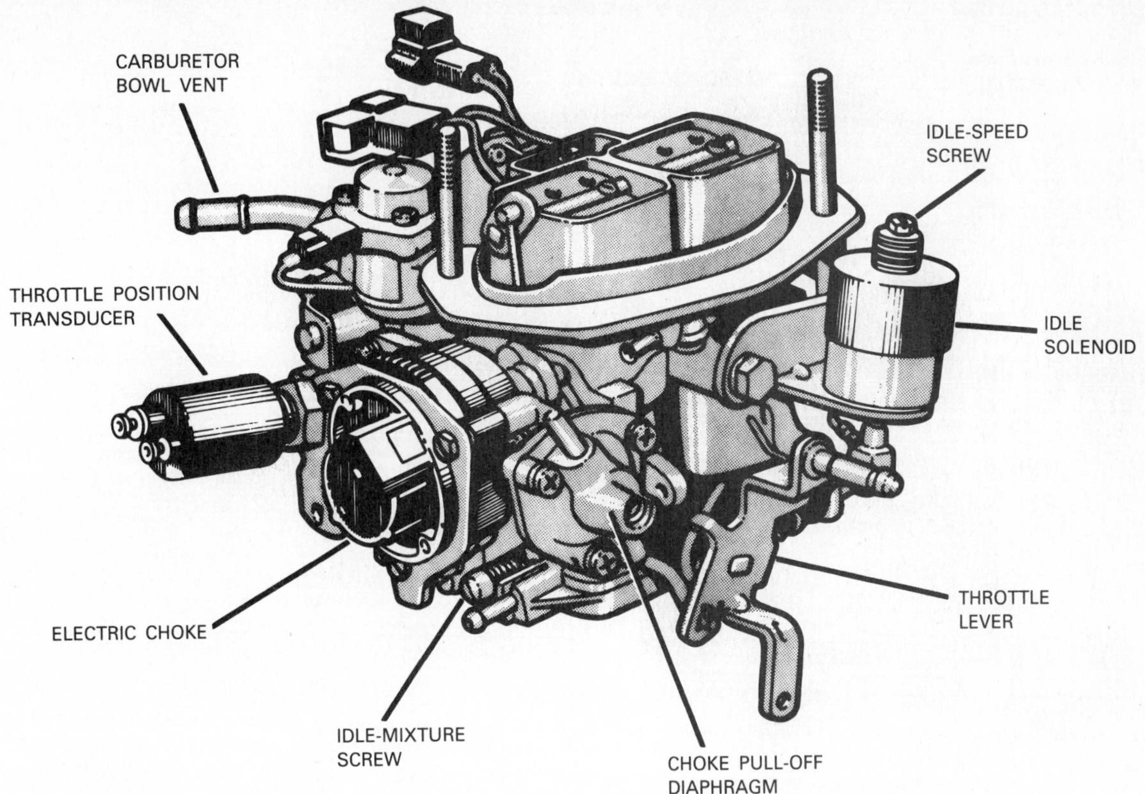

Fig. 16-38 Carburetor with an idle, or antidieseling, solenoid. *(Buick Motor Division of General Motors Corporation)*

e. Exhaust-Gas Recirculation System The exhaust-gas recirculation (EGR) system introduces some exhaust gas into the air-fuel mixture going into the engine cylinders. This reduces the formation of one of the atmospheric pollutants (nitrogen oxides) during the combustion process.

f. Vacuum Motors for Air Conditioners Many air conditioners have vacuum motors for operating the air-conditioner doors.

g. Vacuum for Vacuum-Operated Power Brakes Diesel engines do not develop vacuum in the intake manifold. Therefore, to operate the various vacuum devices on a diesel vehicle, a separate vacuum pump is installed.

♦ 16-35 TWO-BARREL AND FOUR-BARREL CARBURETORS

Carburetors with more than a single barrel are used on many engines. Many carburetors have two barrels (dual carburetors) and others have four barrels (quad carburetors). The purpose of the additional barrels is to improve engine "breathing," particularly at high speeds. The extra barrels permit more air and fuel to enter the engine. If air were the only consideration, then a single large-diameter barrel could be used. But with only a single large barrel, venturi action would be poor. Proper air-fuel ratios would be hard to achieve under varying operating conditions.

1. Two-Barrel Carburetors The two-barrel carburetor is essentially two single-barrel carburetors in a single assembly (Fig 16-39). The second barrel is used in two different ways, according to the carburetor design. In one arrangement, each barrel handles the air-fuel requirements of half the engine cylinders. For example, Fig. 16-34 shows the air-fuel delivery pattern in a V-8 engine. One carburetor barrel supplies cylinders 2, 3, 5, and 8. The other barrel supplies cylinders 1, 4, 6, and 7. The arrows indicate the pattern in Fig. 16-34. Each barrel has a complete set of systems. The throttle valves are fastened to a single throttle shaft so that both valves open and close together.

The second design is most commonly used on four-cylinder in-line engines. In this arrangement, the secondary barrel comes into operation only after the primary throttle valve has opened about 45 degrees. When the primary throttle valve moves past the 45 degree opening, linkage to the secondary throttle valve starts to open it. The secondary barrel then comes into operation and starts to supply additional air-fuel mixture to the four cylinders. This action increases the supply of air-fuel mixture to the engine for improved medium- to high-speed performance.

2. Four-Barrel Carburetor The four-barrel carburetor (Fig. 16-40) consists essentially of two two-barrel carburetors combined in a single assembly. The carburetor has four barrels and four main nozzles. Therefore, it is often called a *Quadrajet,* or *quad carburetor.* One pair of barrels makes up the primary side, the other pair the secondary side (Fig. 16-40). Under most operating conditions, the primary side alone takes care of engine requirements. However, when the throttle is moved toward the "wide-open" position for acceleration or full-power operation, the secondary side comes into operation. It supplies additional amounts of air-fuel mixture. Volumetric efficiency (♦13-12) is higher, and the engine produces more horsepower.

Two ways to control the secondary-barrel action are by mechanical linkage from the primary throttle shaft or by a

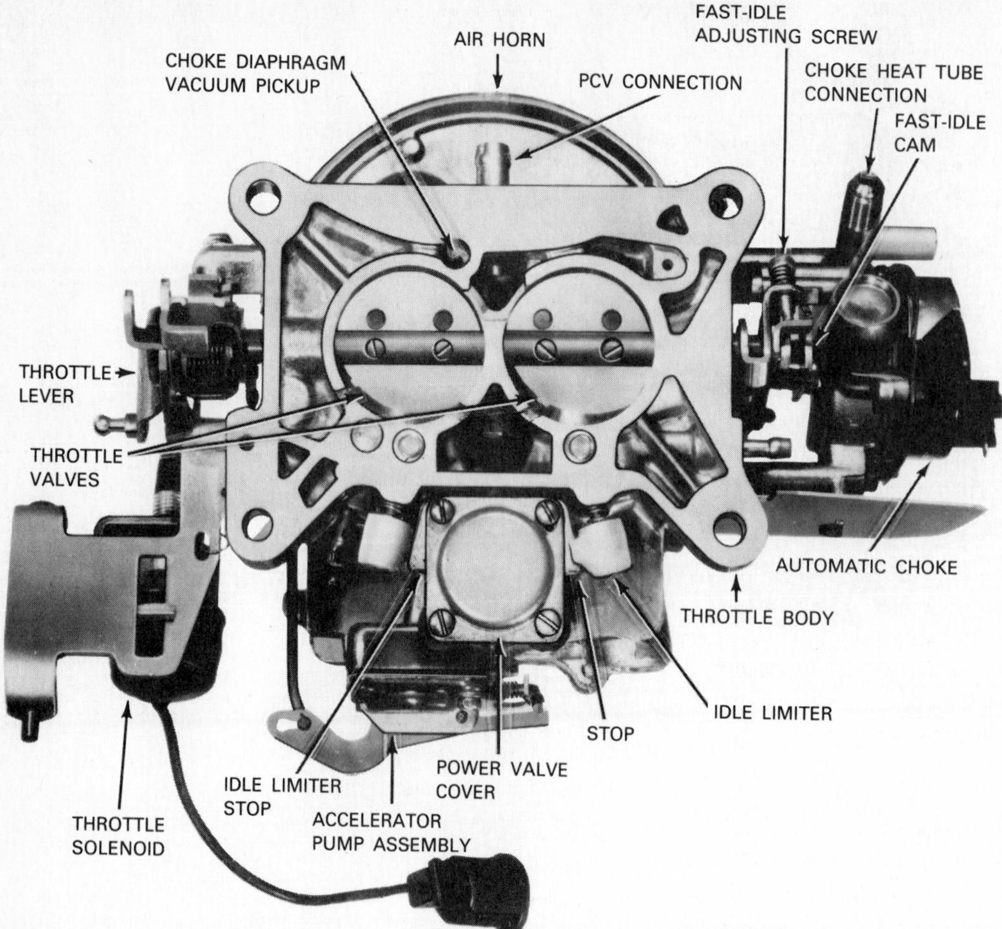

Fig. 16-39 A two-barrel carburetor, showing the locations of the two throttle valves.

vacuum device. During part throttle, only the primary throttle valves are open. In the mechanical system, when the throttle is opened wide for additional power, the linkage between the primary and secondary throttle valves causes the secondary throttle valves to open. The secondary barrels now supply air-fuel mixture for full-power engine operation.

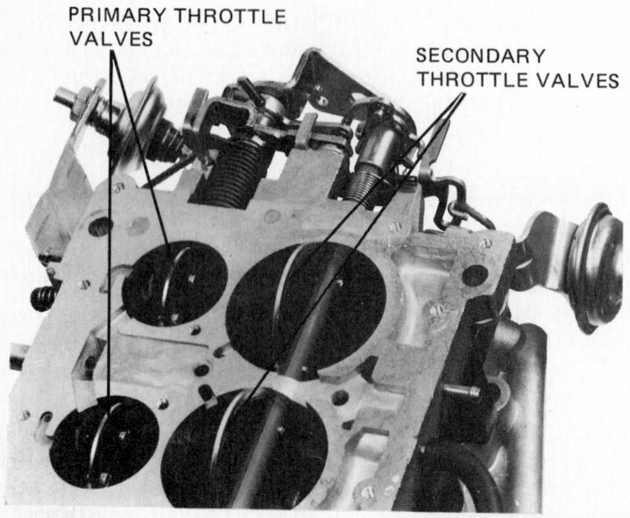

In the vacuum-controlled system, there is a vacuum-operated diaphragm. The vacuum is picked up from one of the primary-barrel venturis (venturi vacuum). As air speed through the primary barrels increases, so does venturi vacuum. When the vacuum reaches a predetermined amount—indicating a rather high engine rpm—the vacuum actuates the diaphragm. This opens the secondary throttle valves so the secondary barrels begin to supply air-fuel mixture.

Fig. 16-40 A four-barrel, or quad, carburetor, showing the locations of the four throttle valves. The small throttle valves are on the primary side.

♦ 16-36 FACTORY-ADJUSTED PART THROTTLE (APT)

To provide a more accurate factory adjustment of the fuel flow during part-throttle operation, some carburetors have an additional metering rod and a fixed metering-rod jet. The metering rod is adjusted at the factory by turning the adjustment screw. If it is turned to lift the metering rod, more fuel can flow through the metering-rod jet. If the screw is turned to lower the metering rod, less fuel can flow. Adjusting the metering rod at the factory compensates for any slight differences in fuel flow resulting from manufacturing tolerances.

♦ 16-37 ALTITUDE COMPENSATION

Some carburetors with APT also have an automatic altitude adjustment. This is an *aneroid* surrounding the APT metering rod. An aneroid is a sealed bellows which is sensitive to changes in atmospheric pressure. When pressure is increased, the aneroid is squeezed so it shortens. When pressure is reduced, the aneroid expands. These actions lower or raise the APT metering rod as atmospheric pressure changes.

For example, when a car is driven up a mountain, the car encounters reduced atmospheric pressure. Without any compensating device, the air-fuel mixture would become enriched. This happens because less air enters the carburetor (air pressure is lower). To compensate for this, the aneroid expands as a result of the reduced pressure. This action lowers the APT metering rod so that less fuel can flow. As a result, a more even air-fuel ratio is maintained.

♦ 16-38 MULTIPLE CARBURETORS

To achieve still better engine "breathing," some high-performance engines are equipped with more than one carburetor. The additional carburetors supply more air-fuel mixture to improve high-speed, full-power engine performance. Two carburetors mounted on an engine are called *dual carburetors.* Three carburetors mounted on an engine are called *triple carburetors* or sometimes "tri-power." The ultimate in this is to equip each engine cylinder with its own carburetor. This arrangement is commonly used on motorcycles and race cars.

♦ 16-39 VARIABLE-VENTURI (VV) CARBURETORS

All the carburetors described so far are fixed-venturi units. The size and shape of the venturi does not change. However, there is another design of carburetor in which the size of the venturi can change. These are called *variable-venturi,* or VV, carburetors. Many foreign cars, and some cars made in the United States, have VV carburetors. The size of the venturi varies as operating conditions change.

Two types are the round-piston type and the rectangular venturi-valve type. The round-piston type has been used on some foreign cars and motorcycles for years. The rectangular venturi-valve type is a relatively new development from Ford. Both types have float systems similar to those used in the fixed-venturi carburetors.

♦ 16-40 ROUND-PISTON VV CARBURETOR

Figures 16-41 and 16-42 show exterior and sectional views of one model round-piston VV carburetor. The carburetor can be disassembled as shown in Fig. 16-43. The piston is an assembly of two basic parts: the outer two-diameter piston and the inner oil-damper reservoir. The piston moves up and down in the piston chamber in response to the amount of vacuum between the piston and throttle valve. When the piston moves down, it reduces the size of the venturi. The venturi is formed by the end of the piston and the throttle body. Figure 16-44 shows how downward movement of the piston reduces the size of the venturi.

Movement of the piston also moves the tapered needle. The needle is fastened to the bottom of the piston. When the piston moves downward, the needle moves down into the fuel jet. This reduces the operating size of the jet opening and therefore the amount of fuel that can flow. At the same time, the downward movement of the piston reduces the size of the venturi, which limits the amount of air that can flow through it. This provides a balanced condition. As the size of the venturi changes, the size of the jet opening also changes. Therefore, the proper proportions of air and fuel are maintained. The air-fuel ratio stays about the same.

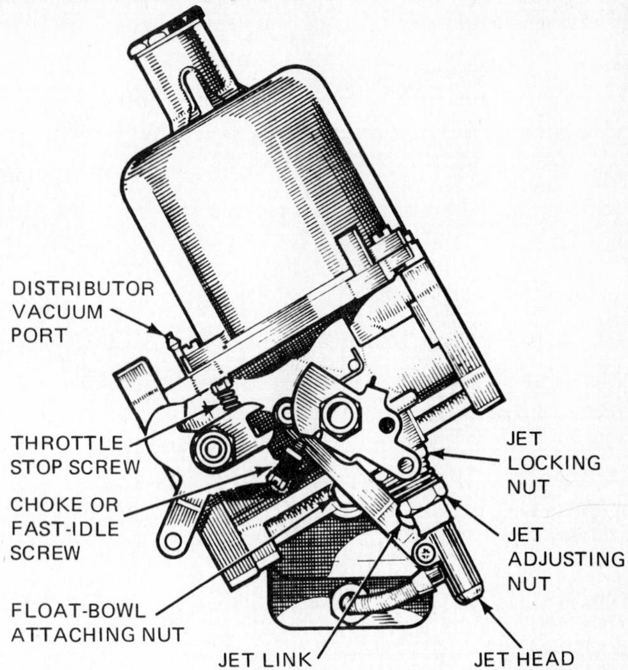

Fig. 16-41 A round-piston VV carburetor. *(British Leyland Motors Incorporated)*

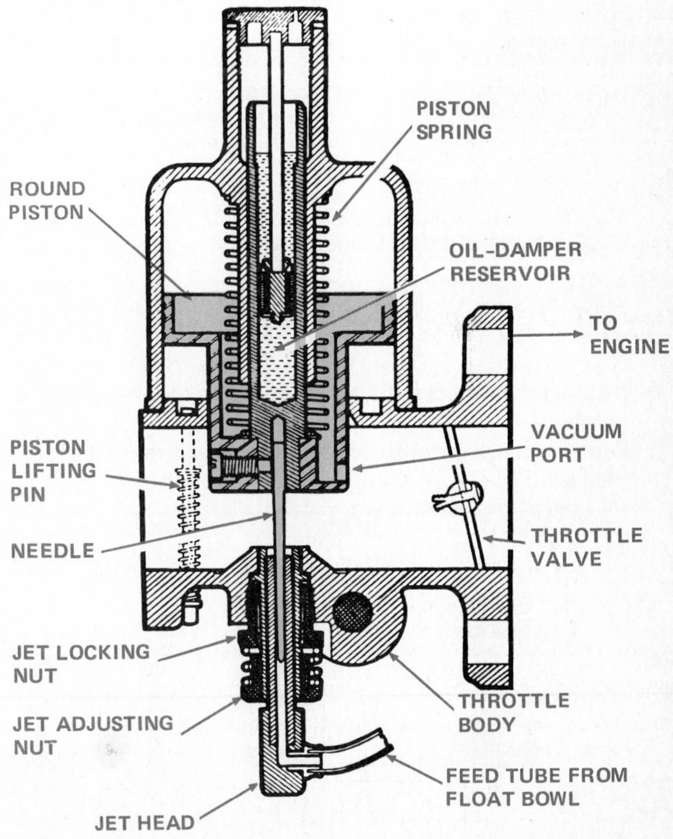

Fig. 16-42 Sectional view of a round-piston VV carburetor. *(British Leyland Motors Incorporated)*

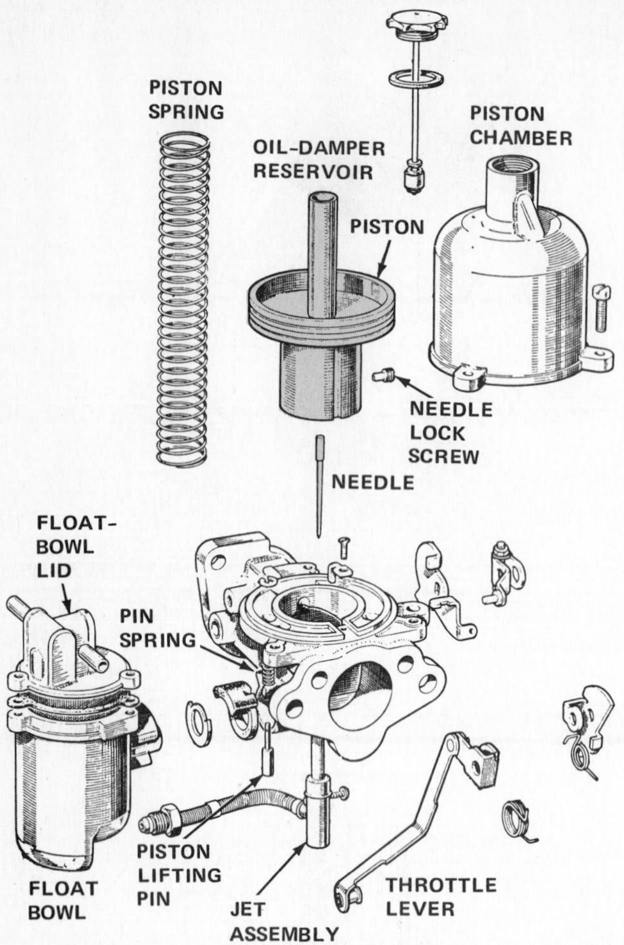

Fig. 16-43 A disassembled round-piston VV carburetor. *(British Leyland Motors Incorporated)*

The piston is raised or lowered in response to the movement of the throttle valve (Fig. 16-42). When the throttle valve is closed to the idling position, intake-manifold vacuum is cut off from the throttle body. The piston spring pushes the piston down to its lowest position. A small amount of air flows around the throttle valve and through the venturi. It produces just enough vacuum at the venturi to cause the fuel jet to deliver enough fuel for idling.

When the throttle is opened, intake-manifold vacuum enters the throttle body. This vacuum draws air from the space above the piston, acting through the vacuum port in the lower part of the piston. The piston is raised by the vacuum, partly compressing the piston spring. As the piston moves up, more air can pass through the venturi. At the same time, the needle moves up in the jet. This increases the effective size of the jet. More fuel flows and mixes with the air passing through. The piston and tapered needle work together to provide the proper air-fuel mixture for the engine operating condition. The taper on the needle is of critical importance in balancing the air-fuel ratio so that it will be uniform through the entire range of throttle opening. The pin taper permits additional fuel to flow at full throttle so that the mixture is enriched for acceleration and full-power operation.

The oil-damper reservoir that is part of the piston acts like a tiny shock absorber. It prevents excessive movements of the piston as the throttle valve is moved and vacuum conditions change. Without this shock-absorber action, the piston could bounce up and down. This would cause erratic carburetor action and poor engine performance.

♦ 16-41 FORD RECTANGULAR VV CARBURETOR

This carburetor (Fig. 16-45) is Ford's version of the VV carburetor, which it introduced on some 1978 cars. Externally, this carburetor is similar in appearance to the fixed-venturi carburetor. Internally, however, it is quite different. It has two air horns, or "throats." In each throat there is a rectangular-shaped piston, or venturi valve which slides back and forth across the throat (Fig. 16-46). This changes the size of the opening (the venturi) above the throttle valve. The two venturi valves are connected together. Their positions are controlled by intake-manifold vacuum and throttle position. Each is connected to a tapered needle or metering rod which is positioned in a fuel jet.

This carburetor has all the main elements of the round-piston VV carburetor discussed in the previous section. The major difference is in the shapes and locations of the elements. The carburetor has several special systems including the float system, main-metering system, cold-cranking enrichment system, cold-running enrichment system, accelerator-pump system, and others. This carburetor has all the spe-

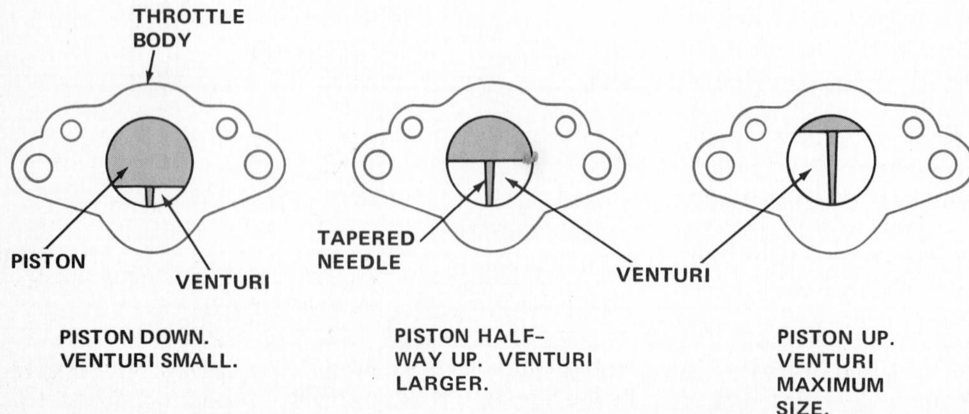

Fig. 16-44 Looking into the throttle body to see how the up-and-down movement of the piston changes the size of the venturi.

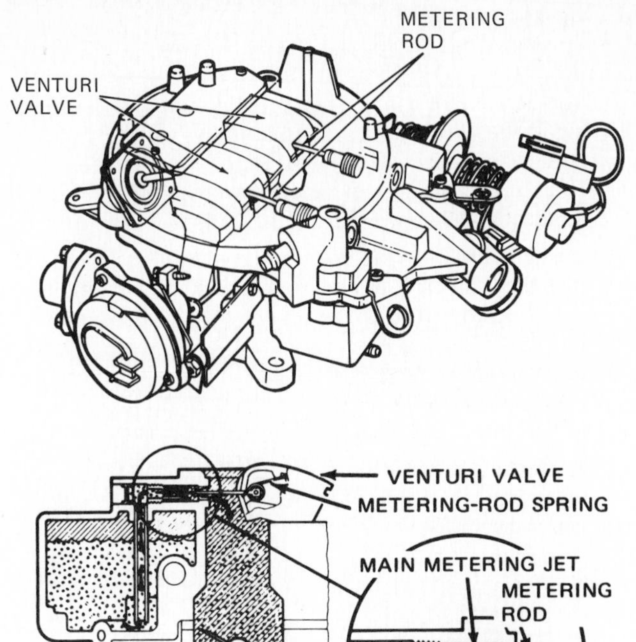

Fig. 16-45 A two-barrel Ford VV carburetor using rectangular-shaped venturi valves. *(Ford Motor Company)*

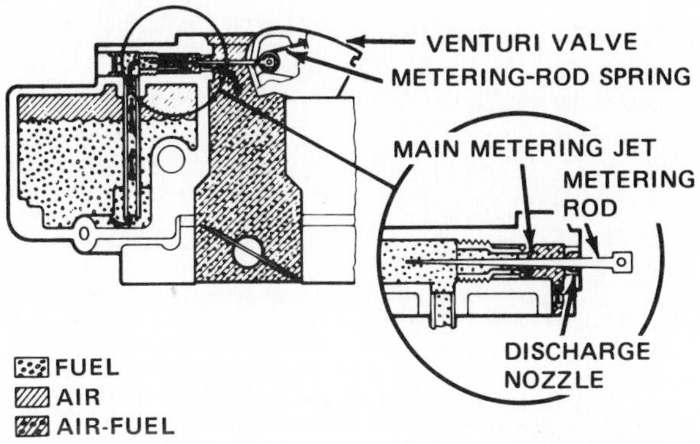

Fig. 16-46 Main-metering system of the Ford VV carburetor. *(Ford Motor Company)*

cial controls built into it that the fixed-venturi carburetors have. However, this VV carburetor does not need a separate idle system, choke valve, and enrichment-valve system.

1. Vacuum Control The position of the venturi valves and therefore the tapered metering rods is controlled by vacuum. The vacuum control includes a spring-loaded vacuum diaphragm which is connected by a rod to the venturi valve. When the throttle is opened, intake-manifold vacuum can work on the vacuum diaphragm. The vacuum causes the venturi valve to move back to increase the venturi opening. More air can flow through. At the same time, the tapered metering rod is raised in the fuel jet.

2. Float System This system is similar to the float system used in fixed-venturi carburetors and works in the same way.

3. Venturi-Valve Limiter At wide-open throttle, under some conditions the control vacuum will not be strong enough to open the venturi valve fully. Then, the valve is opened by the venturi-valve limiter lever on the throttle shaft.

4. Cold-Cranking Enrichment Systems This system provides extra fuel when a cold engine is being cranked. It includes a solenoid valve that is normally closed. Fuel to this valve is controlled by a thermostatic-blade valve that is closed above 75°F [24°C]. When the ignition key is turned to START, the engine is cranked. At the same time, the solenoid valve is opened. Now, fuel can flow if the thermostatic-blade valve is also opened. This provides extra fuel for starting.

5. Cold-Running Enrichment System After a cold engine starts, it must be supplied with a rich mixture for a short time until it warms up. In the fixed-venturi carburetor, this is taken care of by the choke system. In the Ford VV carburetor, it is taken care of by a cold-running enrichment system that includes a bimetal spring heated from the exhaust gas. The bimetal spring controls a vacuum regulator and a cold-running enrichment rod. After a cold start, the bimetal spring has raised the cold-running enrichment rod to allow additional fuel to flow. At the same time, the vacuum regulator cuts off part of the vacuum to the vacuum-control unit. Now, intake-manifold vacuum overrides the control and determines the position of the venturi valve. The combination produces an extra-rich mixture for cold-engine operation. As the engine warms up, the bimetal spring relaxes and allows the cold-running enrichment rod to seat. This cuts off the flow of additional fuel. At the same time, the vacuum regulator releases the vacuum to the vacuum-control unit. Then, normal hot-engine operation results.

6. Fast-Idle Cam The fast-idle cam is tied in with the cold-running enrichment system. The position of the cam is controlled by the bimetal spring, just as in other carburetors. When the engine is cold, the cam is positioned to provide a fast idle. There is one unique feature of the system, however. During cold starts, a special lever is inserted between the fast-idle cam and fast-idle lever. This provides additional throttle-plate opening for starting. When the engine starts, manifold vacuum is applied to a separate vacuum diaphragm (called the *high-cam-speed positioner,* or HCSP). This causes the lever to be withdrawn after the first throttle movement.

7. Other Special Systems The Ford VV carburetor also has other systems to meet varying operating conditions. The accelerator-pump system is similar to those used in fixed-venturi carburetors. Also included is a hot-idle compensator similar to those used in fixed-venturi carburetors. Some of these VV carburetors also include automatic altitude compensation. These systems are similar in construction and operation to those used in some fixed-venturi carburetors.

♦ REVIEW QUESTIONS ♦

Select the *one* correct, best, or most probable answer to each question. You can find the answers in the section indicated at the end of each question.

1. All of these statements are true about the idle-stop (anti-dieseling) solenoid *except* (♦16-34)
   a. it controls engine idle speed
   b. it can cause dieseling (after-run) when not adjusted correctly
   c. it opens the carburetor throttle plates when the ignition is on
   d. it controls engine fast idle during warm-up
2. The portion of the carburetor air horn that reduces pressure to cause fuel to flow is called the (♦16-5)
   a. throttle body
   b. air bleed
   c. venturi
   d. fuel nozzle
3. The throttle valve (♦16-5)
   a. when closed, allows little or no air to flow through the air horn
   b. when open, allows air to flow freely through the air horn
   c. both of the above
   d. none of the above
4. The reason that a richer mixture must be delivered when first starting a cold engine is that (♦16-26)
   a. this allows a higher cranking speed
   b. only part of the gasoline will vaporize when cold
   c. the thick engine oil must be thinned out
   d. none of the above
5. An air-fuel ratio of 12:1 means that the mixture has (♦16-8)
   a. 12 pounds of gasoline to 1 pound of air
   b. 12 pounds of air to 1 pound of gasoline
   c. 1 gallon of gasoline to 12 gallons of air
   d. 12 gallons of air to 1 gallon of gasoline
6. The purpose of the hot-idle compensator is to (♦16-12)
   a. supply additional air for idling when the engine is hot
   b. slow down engine speed
   c. prevent hot engine from running when the ignition is shut off
   d. increase idle speed while the engine is cold
7. When the engine is hot and running at 600 rpm, the gasoline is supplied by the (♦16-13)
   a. idle system
   b. low-speed system
   c. choke system
   d. main-metering system
8. The operating mechanism of the vacuum-operated metering rod includes either a (♦16-19)
   a. vacuum piston or pump
   b. diaphragm or pump
   c. vacuum piston or diaphragm
   d. none of the above
9. The accelerator pump operates (♦16-25)
   a. all the time the engine is running
   b. during initial throttle opening
   c. automatically, when vacuum drops
   d. during wide-open throttle operation
10. In the main-metering system, the maximum amount of fuel that can flow during normal driving conditions is controlled by (♦16-16)
    a. main-metering jets
    b. main air bleeds
    c. the choke valve
    d. the inlet check ball
11. Mechanic A says the automatic choke is opened by manifold vacuum. Mechanic B says the automatic choke is closed by spring force. Who is right? (♦16-27)
    a. A only
    b. B only
    c. both A and B
    d. neither A nor B
12. The APT carburetor has (♦16-36)
    a. an extra throttle valve
    b. an extra barrel
    c. an extra metering rod adjusted at the factory
    d. all of the above
13. In the VV carburetor, venturi size is varied by the movement of (♦16-39)
    a. the throttle
    b. the piston or venturi valve
    c. the choke valve
    d. none of the above
14. In the Ford VV carburetor, the position of the venturi valves and tapered metering rods is controlled by (♦16-41)
    a. intake-manifold vacuum
    b. linkage to the throttle valve
    c. fast-idle cam
    d. slow-idle cam
15. The exhaust-manifold heat-riser valve is stuck in the open position. Mechanic A says this can cause poor gas mileage. Mechanic B says this can cause the intake-manifold vacuum to be lower than normal. Who is right? (♦16-28)
    a. A only
    b. B only
    c. both A and B
    d. neither A nor B

CHAPTER 17
Carbureted-Fuel-System Service

After studying this chapter, and with proper instruction and equipment, you should be able to:

1. List and describe the various troubles that might occur in fuel systems using carburetors and explain the possible causes of each trouble.
2. Describe the cautions to observe in fuel-system work.
3. Check and service the carburetor air cleaner.
4. Remove and install a fuel pump.
5. Remove, service, and install various carburetors.
6. Adjust the carburetor idle mixture by using the lean-best-idle method and by using propane.

♦ 17-1 DEFINING TERMS

Various terms are used to describe engine troubles (Fig. 17-1). These troubles can be caused by conditions in the fuel system. They can also be caused by conditions in the other engine systems, or the engine itself.

1. Detonation This is a secondary explosion that occurs after the spark at the spark plug. It produces excessively rapid burning of the air-fuel mixture and a pinging or knocking sound. Usually worse under acceleration. Can damage the engine if severe.

2. Dieseling The engine continues to run after the ignition is turned off. Runs unevenly and may knock. Ignition is caused by hot spots in the combustion chambers, such as carbon, exhaust valve, and spark plugs. Supported by air and fuel leaks past the throttle valve and through the idle system.

3. Hesitation Momentary lack of response as the throttle valve is opened. Can occur at all car speeds. Most severe when first starting out. May cause engine to stall.

4. Spark Knock See Detonation.

5. Miss Failure of a cylinder to fire. Causes steady pulsation or jerking. Usually more noticeable when the load increases. Not normally noticeable at higher engine speeds. Exhaust can have a steady spitting sound at low speeds.

6. Roll Engine speed varies under steady throttle or cruise. Feels as if the car speeds up and slows down with no change in throttle-valve position.

7. Preignition Ignition before the spark occurs, due to hot spots in the combustion chamber. Can cause a rumbling sound and rough engine operation.

8. Rough Idle Engine runs roughly at idle, sometimes severe enough to cause the car to shake.

9. Sag See Hesitation.

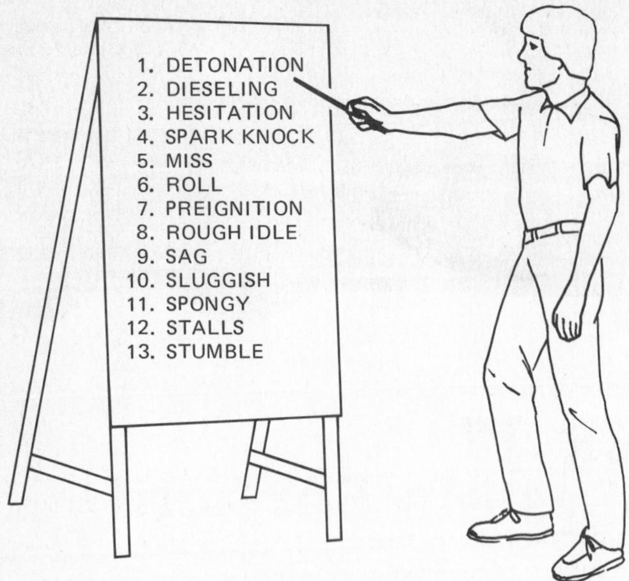

Fig. 17-1 The basic engine troubles that might be caused by conditions in the fuel system.

10. Sluggish Engine power limited under load or when climbing hills. Slow on acceleration. Loses too much speed going up hills.

11. Spongy Less than anticipated response from engine when throttle is opened. Little or no speed increase when throttle is opened a small amount. Further throttle opening will finally get some response.

12. Stalls Engine quits running. This may occur when idling or when car is moving, as for example when the throttle is suddenly opened. See Stumble.

13. Stumble Opening of the throttle causes the engine power to drop off. Can even cause engine to stall. Like hesitation but more severe. Engine will recover if throttle is closed and then tapped a couple of times.

These troubles can be caused by conditions in the fuel system. They can also be caused by conditions in other engine systems or in the engine itself.

♦ 17-2 ANALYZING CARBURETED-FUEL-SYSTEM TROUBLES

Fuel-system troubles usually show up as faulty engine operation, or what is called car "driveability." A car that runs well is "driveable." A car with such problems as poor acceleration, hard starting, missing, loss of power, stumble, hesitation, and stalling has poor driveability. All engine or driveability problems caused by the fuel system, ignition system, cooling system, or troubles in the engine itself are discussed in detail in a later chapter.

The following sections discuss briefly driveability problems that could be due to troubles in the fuel system. The conditions discussed here could also be caused by troubles outside the fuel system.

Many late-model cars have built-in electronic diagnostic systems. When a fault occurs, the system records the fault in its memory. Either a CHECK ENGINE light comes on, or the electronic system gives the driver verbal warning that something is wrong. Then, when the mechanic checks the system, all that is necessary is to activate the system. One or more code numbers then appear on the instrument panel, called up from the memory. The code numbers tell the mechanic what the trouble is. This system is covered further in following chapters on ignition-system and engine service.

♦ 17-3 ENGINE CRANKS NORMALLY BUT WILL NOT START

This could be due to ignition problems, lack of fuel, underchoking or overchoking, or the failure of the carburetor to deliver air-fuel mixture normally. Make sure that there is gasoline in the fuel tank and in the carburetor.

Many chokes have an electric heater element. If you wait too long between turning on the ignition key and trying to start, the choke may open. Then underchoking occurs. The remedy is to turn the ignition off. Wait long enough for the choke to cool so that the choke valve closes. Then turn on the ignition and start the engine without any delay.

♦ CAUTION ♦ Do not crank the engine with the air cleaner off. The engine could backfire up through the open carburetor and cause a fire, or you could get your face badly burned.

You can check for fuel in the carburetor by pushing the throttle to the floor several times. This should cause the accelerator pump to squirt fuel into the carburetor air horn. Watch for this with the air cleaner off (engine not running). If fuel squirts out, you can assume there is fuel in the float bowl. Reinstall the air cleaner. Try starting now, after having primed the engine with the accelerator pump. If the engine starts and runs briefly, the carburetor is probably at fault. If the engine does not start, there may be other trouble, such as a clogged fuel-tank strainer, a defective fuel pump, a stopped-up fuel-tank vent or cap, or the wrong tank cap.

♦ 17-4 ENGINE STARTS, SPEEDS UP MOMENTARILY, THEN DIES

This could be due to the choke vacuum-break (vacuum-kick) setting being too wide. When the vacuum is applied, the vacuum break kicks the choke too far open. It could also be due to a defect in the exhaust-gas recirculation (EGR) system. The EGR system is allowing exhaust gas to reenter the engine cylinders when the engine is cold. This can upset the combustion process so the engine dies. The trouble could be a stuck EGR valve, or some abnormal condition in the EGR control system. Other conditions that could cause this trouble include a fast idle set too low, a defective fuel pump not delivering enough fuel, or a low fuel level in the float bowl.

♦ 17-5 ENGINE RUNS BUT MISSES

This can be caused by defective ignition, spark plugs, valve action, piston rings, leaking intake manifold or gasket, or a vacuum line disconnected or split. In the fuel system, it could be caused by too lean or too rich a mixture. This could indicate a problem in the carburetor, fuel pump, flex line, or fuel tank.

♦ 17-6 ENGINE LACKS POWER

This complaint is best checked with a dynamometer and oscilloscope, as explained later. If the problem is poor acceleration, check to see if the throttle valve opens fully. On four-barrel carburetors make sure that the secondary throttle valves are opening properly. Also make sure the accelerator-pump system is working. If the complaint is lack of power cold, it could be due to a defective choke or the manifold heat-control or EGR valve stuck open. Lack of power hot could mean that the choke or manifold heat-control valve is stuck closed. A defective fuel pump or clogged fuel filter could starve the engine so it does not get enough fuel. The result is loss of power under almost any condition, particularly when full-power performance is called for. Many other conditions outside of the fuel system can cause lack of power. For example, automatic transmission problems can cause the car to lack acceleration. Diagnosis of these conditions is explained in a later chapter.

♦ 17-7 ENGINE STALLS COLD OR AS IT WARMS UP

This could be due to a choke opening too quickly or to a manifold heat-control valve that is stuck open. It could also indicate that not enough fuel is getting to or through the carburetor because of fuel-pump or carburetor troubles. There are several other possibilities in other engine components.

♦ 17-8 COLD-ENGINE DRIVEABILITY PROBLEMS

Newer cars are more likely to have problems after the first start of a cold engine. This is because of emission controls and close carburetor settings. No one worried in years gone by if an engine ran rich during and after warm-up. Today, however, such richness is not allowed. Lean carburetor settings and fast-acting chokes prevent it. The lean settings and fast-acting chokes can cause cold-running problems if everything is not properly adjusted. Here are some possible troubles and their causes:

1. Engine stalls when transmission is shifted into gear. This could be due to improper choke vacuum-break setting, fast-idle setting, or incorrect ignition timing.
2. Engine stalls, hesitates, or sags during first mile when throttle is opened. This could be due to improper choke vacuum-break setting, electric choke kicking off too fast, EGR system releasing exhaust gas to the intake manifold, low float-level adjustment, defective accelerator pump, or no distributor vacuum advance.
3. Engine stalls, hesitates, or sags after first warm-up mile. This could be due to everything mentioned in ♦17-7. It could also be due to a defective fuel pump that is not delivering enough fuel for sustained intermediate speeds.

♦ 17-9 WARM-ENGINE DRIVEABILITY PROBLEMS

Many warm-engine driveability problems cause loss of engine power (♦17-6). Other problems include hesitation, stumble, or sag. These can result from incorrect ignition timing, defective spark-advance operation, a defective accelerator pump, or the heated-air-system inlet door in the air-cleaner snorkel being stuck. The conditions described in ♦17-6 may also cause warm-engine driveability problems.

♦ 17-10 SURGE

Another driveability problem that sometimes occurs is surge. While you are driving along at a steady speed, the engine suddenly rolls—increasing and then decreasing its power output. This can result from a vacuum leak (one of the hoses to the carburetor loose, or carburetor mounting loose), incorrect spark advance or faulty mechanism, a defective PCV valve (stuck in high-flow position), lack of sufficient fuel to engine (low float level, defective fuel pump, clogged fuel filter), or the heated-air-system inlet door being stuck.

♦ 17-11 DIESELING, OR RUN-ON

This is a condition where the engine continues to run after the ignition is turned off. It could be due to a high idle-speed adjustment or to the idle-stop solenoid not allowing the throttle valve to close completely when the ignition is turned off. As a result, enough air-fuel mixture gets through the carburetor to allow the engine to continue to run. Ignition is from hot spots in the combustion chamber.

♦ 17-12 EXCESSIVE FUEL CONSUMPTION

This can be due to many causes—driving habits, high speed, short runs, choke partly closed, high carburetor float level, worn float-bowl needle valve or seat, worn jets in the carburetor, internal carburetor leaks, external gasoline leaks, or stuck metering rod or accelerator pump. Engine troubles such as defective rings or valve action, excessive friction in the drive line, low tire pressure, and an improperly operating automatic transmission can also increase fuel consumption.

♦ 17-13 CARBURETOR TROUBLES

Engine troubles can come from many other causes besides problems in the fuel system and carburetor. The chapter on engine trouble diagnosis lists and describes engine troubles and their possible causes. Troubles caused by conditions inside the carburetor itself are listed and described below.

1. Excessive fuel consumption can result from:
   a. A high float level, or a leaky or saturated float
   b. A sticking or dirty float needle valve
   c. Worn jets or nozzles
   d. A stuck metering rod or power piston
   e. Idle too rich or too fast
   f. A stuck accelerator-pump check valve
   g. A leaky carburetor
   h. A dirty air cleaner
2. Lack of engine power, acceleration, or high-speed performance can result from:
   a. The power step-up on the metering rod not clearing the jet
   b. Dirt or gum clogging the fuel nozzle or jets
   c. A stuck power piston or valve
   d. A low float level
   e. A dirty air filter
   f. The choke stuck or not operating
   g. Air leaks into the manifold
   h. The throttle valve not fully opening
   i. A rich mixture, due to causes listed under item 1, above
3. Poor idle can result from a leaky vacuum hose, stuck PCV valve, or retarded timing. Also, it could be due to an

incorrectly adjusted idle mixture or speed, a clogged idle system, or any of the causes listed under item 2, above.

4. Failure of the engine to start unless primed could be due to no gasoline in the fuel tank or carburetor, the wrong tank cap, or a plugged tank or cap vent. The latter causes a vacuum to develop in the tank, which prevents delivery of fuel to the carburetor. Holes in the fuel-pump flex line will allow air leakage, which prevents fuel delivery. In addition, consider carburetor jets or lines clogged, a defective choke, a clogged fuel filter, or air leaks into the manifold.
5. Hard starting with the engine warm could be due to an inoperative choke.
6. Slow engine warm-up could be due to a defective choke or manifold heat-control valve.
7. A smoky, black exhaust is due to an over-rich mixture. Carburetor conditions that could cause this are listed in item 1, above.
8. If the engine stalls as it warms up, this could be due to a defective choke.
9. If the engine stalls after a period of high-speed driving, this could be due to a malfunctioning antipercolator.
10. If the engine backfires, this could be due to an excessively rich or lean mixture. If the noise is in the exhaust system, it is usually caused by an excessively rich mixture in the exhaust gas. This results from a defective air-injection-system antibackfire valve. Lean mixtures usually cause a pop-back through the intake manifold to the carburetor.
11. If the engine runs but misses, the most likely cause is a vacuum leak. It may be caused by a vacuum hose or intake-manifold leak. In addition, it could be that the proper amount and ratio of air-fuel mixture are not reaching the engine. This might be due to clogged or worn carburetor jets or to an incorrect fuel level in the float bowl.

Some of the above conditions can be corrected by external adjustments on some carburetors. Others require removal of the carburetor from the engine so that it can be disassembled, repaired, and reassembled. Sections at the end of the chapter describe briefly carburetor adjustment and servicing procedures.

♦ 17-14 CARBURETOR QUICK CHECKS

Several quick checks can be made to indicate how the carburetor is working. More accurate analysis requires test instruments such as an exhaust-gas analyzer and an intake-manifold vacuum gauge, as explained in a later chapter.

1. Float-Level Adjustment With the engine warmed up and running at idle speed, remove the air cleaner. Carefully note the condition of the high-speed nozzle. If the nozzle tip is wet or is dripping gasoline, the float level probably is too high. This could cause a continuous discharge of fuel from the nozzle, even at idle.

2. Idle System If the engine does not idle smoothly after it is warmed up, the idle system could be at fault. Slowly open the throttle until the engine is running at about 3000 rpm. If the speed does not increase evenly and the engine runs roughly through this speed range, the idle or main-metering system is probably defective.

3. Accelerator-Pump System With the air cleaner off and the *engine not running*, open the throttle suddenly. See if the accelerator-pump system discharges a squirt of fuel into the air horn. The flow should continue for a few seconds after the throttle valve reaches the wide-open position.

4. Main-Metering System With the engine warmed up and running at 2000 rpm, slowly cover part of the air horn with a piece of stiff cardboard. The engine should speed up slightly, since this causes a normally operating main-metering system to discharge more fuel.

♦ CAUTION ♦ Do not use your hand to cover the air horn.

♦ 17-15 CAUTIONS IN FUEL-SYSTEM SERVICE

The following cautions should be carefully observed during fuel-system work:

1. A trace of dirt in a carburetor or fuel pump can cause fuel-system and engine trouble. Be careful about dirt when repairing these units. Your hands, the workbench, and the tools should be clean.
2. Gasoline vapor is very explosive.Wipe up spilled gasoline at once, and put the cloths outside to dry. Never smoke or bring an open flame near gasoline! Reread the Caution at the end of ♦14-2.
3. When using the solvent tank, be careful not to splash solvent in your eyes. Do not add gasoline to the solvent tank. This increases the danger of fire. Dump any gasoline from the carburetor or fuel pump in a container before cleaning them in the solvent.
4. Gasoline may be under pressure in the fuel lines. Whenever you disconnect a fuel line, especially between the fuel pump and carburetor, have a cloth ready to soak up any gasoline that spurts or leaks out. Wipe up all gasoline and put the cloth outside in a safe place to dry. The procedure of loosening fuel-line connections is described later.
5. When drying parts with the air hose, handle the hose with care, as noted in ♦2-10. Also observe the other safety cautions listed in Chap. 2.

♦ CAUTION ♦ Do not operate the engine with the air cleaner off unless it is necessary during a test. If the engine backfires with the air cleaner off, flames could ignite fuel vapors in the engine compartment.

♦ 17-16 AIR-CLEANER SERVICE

Air-cleaner service recommendations vary with different automotive manufacturers. Always check the manufacturer's shop manual for specifications. Here are sample recommendations and service procedures.

1. Paper Element The paper filter element (Fig. 17-2) in the air cleaner should be replaced periodically. One recommendation is to replace it every 30,000 miles [48,000 km]. It should be replaced more often if the car is driven in dusty or sandy areas.

Remove the cap or cover from the air cleaner and lift out the paper filter element. Examine it for oil. If the element has oil on it for more than half its circumference, check the

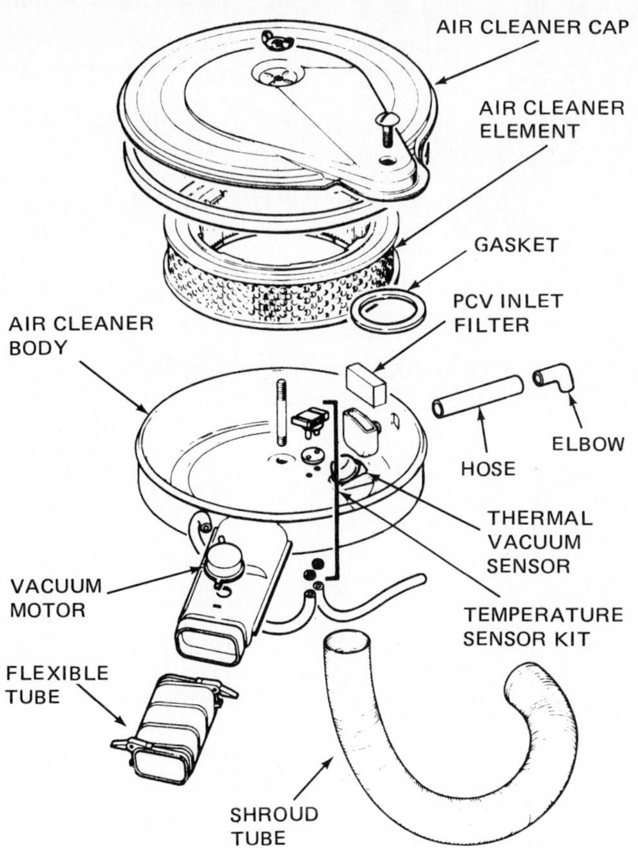

Fig. 17-2 Disassembled carburetor air cleaner and related parts for an in-line engine. *(American Motors Corporation)*

crankcase ventilating system. It is carrying oil up to the air cleaner.

Before replacing the filter element, clean all dust and oil out of the air-cleaner body. Make sure that the plastic rings on both sides of the element are smooth and seal properly at the top and bottom.

Note Chevrolet recommends inspection of the element for dust leaks after the first 15,000 miles (24,000 km) of operation. Look for dust in the cleaner housing on the inside of the element. If dust is found, replace the element. If no dust is found, rotate the element one-half turn, and reinstall it.

Note Some manufacturers recommend cleaning paper filter elements with compressed air at periodic intervals, as shown in Fig. 17-3. The air is blown from the inside out. If the element is damaged or very dirty, replace it. Figure 17-3 shows how to clean a paper filter element for an automotive diesel engine.

2. Polyurethane Element Some air cleaners have a plastic wraparound, or polyurethane element, outside the paper element (Fig. 17-4). When servicing this air cleaner, remove the polyurethane element. Inspect it carefully for rips or other damage. If it is serviceable, wash it in an approved filter cleaner. Then squeeze it gently to remove the excess cleaner.

Careful Some solvents may ruin the polyurethane element. Do not wring it out or shake or swing it. This could tear it. Instead, fold the element and squeeze it gently.

Clean the cover and other parts of the air-cleaner assembly. Then dip the cleaned element in SAE 10W-30 engine oil.

Fig. 17-3 Cleaning paper filter element by blowing air from the inside. *(Toyota Motor Sales Company, Ltd.)*

Squeeze out the excess. Install the element and its support in the air cleaner. Be sure that the element is not folded or creased. It must seal all the way around. Use a new gasket when installing the air cleaner on the carburetor.

3. Oil-Bath Air Cleaner Figure 17-5 shows an oil-bath air cleaner. After removing the filter element, clean it by sloshing it up an down in clean solvent. Dry it with compressed air. Dump the dirty oil from the cleaner body, wash the body with solvent, and dry it. Refill the body to the full mark with SAE 10W-30 engine oil. Reinstall the filter element and air-cleaner body on the carburetor.

Air filters that are not installed directly on the carburetor must have airtight connections through flexible hose. The hose must have no tears or punctures that could leak unfiltered air into the carburetor.

♦ 17-17 THERMOSTATICALLY CONTROLLED AIR CLEANER

Figure 17-6 shows the components of a heated-air system (HAS) that uses a thermostatic air cleaner. This system and its servicing procedures are discussed in a later chapter on emission controls.

♦ 17-18 AUTOMATIC-CHOKE SERVICE

Automatic chokes may require cleaning to eliminate sticking. Some chokes have an adjustable cover that can be turned one way or the other to enrich or lean out the mixture. Other types of choke are adjusted by bending a linkage rod.

A thermometer is needed to check the operation of the electric automatic choke. Specifications vary, so look up the testing procedure in the shop manual for the car you are checking. A typical procedure for a carburetor-mounted choke is to tape a bulb thermometer to the choke housing, and then start the engine. Note the temperature at which the choke opens and the length of time required.

♦ 17-19 FUEL-GAUGE SERVICE

Fuel gauges require very little in the way of service. Defects in either the instrument-panel unit or the tank unit usually require replacement of the unit. If you run into a fuel-gauge problem, test the system with a fuel-gauge tester, if available. Another method is to substitute a good tank unit for the one on the car. Connect the good tank unit between the dash unit and ground. Move the float arm up and down to see if the movement registers on the dash unit. The thermostatic fuel

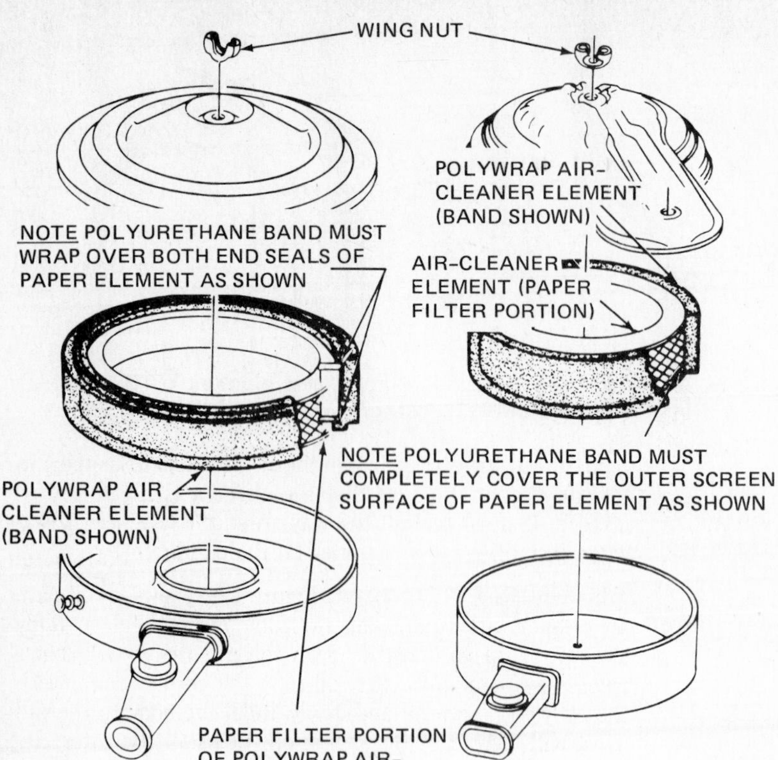

Fig. 17-4 Air cleaners which use a polyurethane band, showing locations of the band. *(Chevrolet Motor Division of General Motors Corporation)*

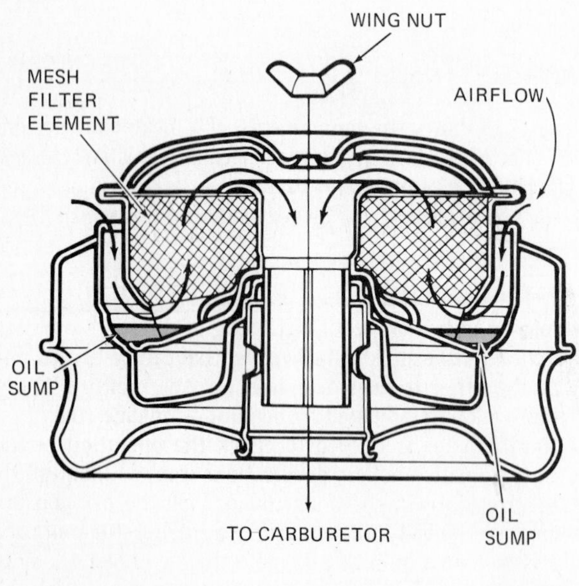

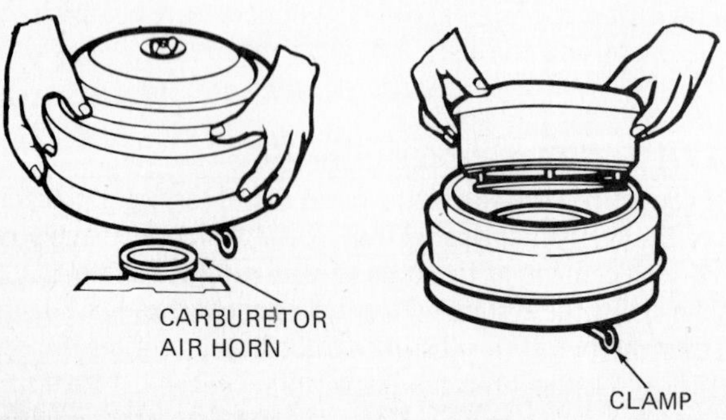

Fig. 17-5 An oil-bath air cleaner. *(Chrysler Corporation)*

gauge (thermostatic blade in dash unit) takes a few seconds to register. If the dash unit works normally, the tank unit on the car is defective. If the dash unit does not work, it is defective.

♦ 17-20 FUEL-LINE CONNECTIONS

When you remove and install fuel pumps and carburetors, you must disconnect and reconnect fuel lines. Fuel-line connections, or couplings, take various forms (Fig. 17-7). When loosening a coupling of the type having two nuts, use two flare-nut wrenches to avoid damaging the line (Fig. 17-8). When using two wrenches, hold one steady while you turn the other, as shown in Fig. 17-8.

♦ CAUTION ♦ The gasoline in the fuel line may be under pressure. Be ready with a cloth to soak up gasoline that spurts or leaks out. Put the cloth outside to dry.

♦ 17-21 FUEL-PUMP AND FILTER SERVICE

Fuel-pump pressure and capacity can be checked with a fuel-pump tester. Low pump pressure causes fuel starvation and poor engine performance. High pressure will cause an over-rich mixture, excessive fuel consumption, and such troubles as fouled spark plugs, rings, and valves (from excessive carbon deposits). Fuel-pump testers are connected into the fuel line from the pump. They measure either the pressure that the pump can develop or the amount of fuel the pump can deliver during a timed interval. The vacuum that the fuel pump can develop should also be checked. This is done by connecting the tester to the vacuum side of the pump. Defective fuel pumps are usually discarded.

Fuel filters (Fig. 17-9) require no service except periodic checks to make sure that they are not clogged, and replace-

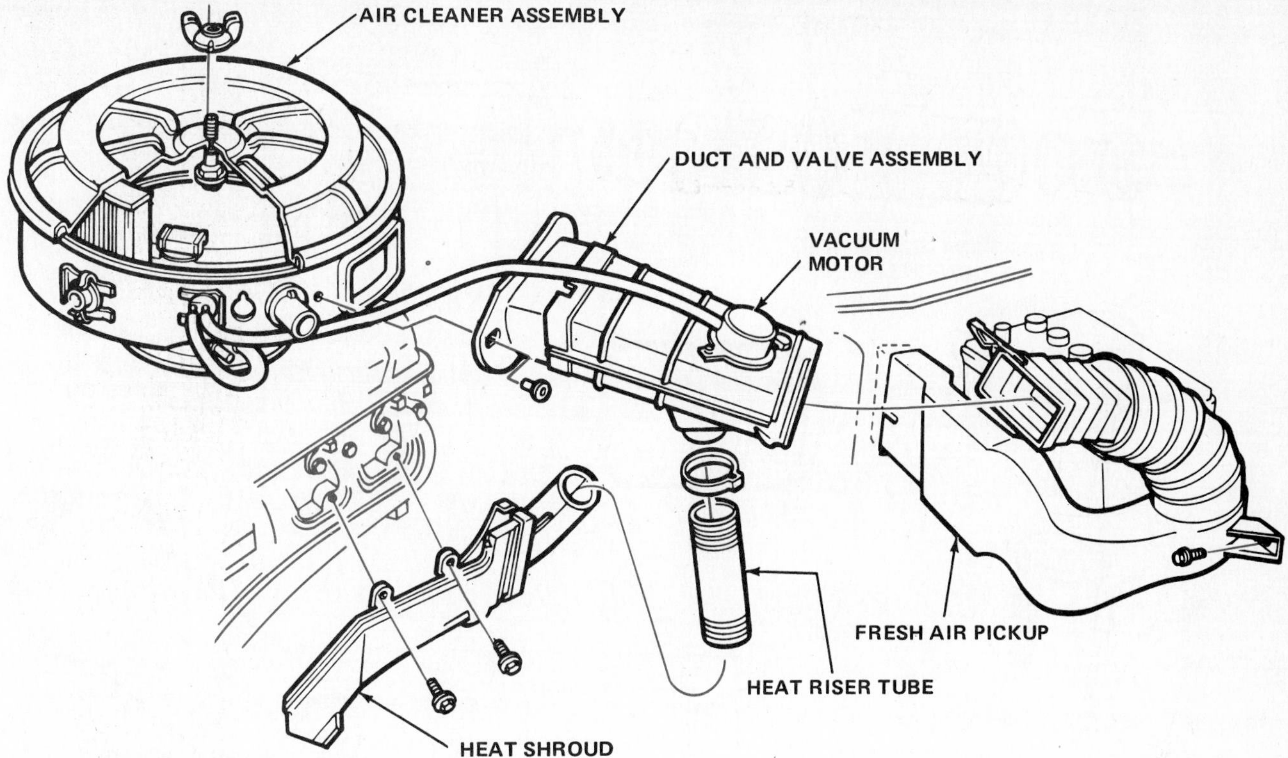

Fig. 17-6 Air cleaner and duct system for a V-type engine with a heated-air system. *(Ford Motor Company)*

ment of the filter element or cleaning of the filter, according to type. On some cars, the filter is part of the fuel pump and can be removed so that the element can be replaced. Another type is the carburetor-mounted filter. The filter is replaced by unclamping and detaching the fuel hose from the filter. Then, the filter can be unscrewed from the carburetor and replaced.

♦ 17-22 FUEL-PUMP TROUBLES

Fuel-system troubles that might be caused by the fuel pump are discussed below.

1. Insufficient Fuel Delivery This could result from low pump pressure, which in turn could be due to any of the following:

a. Broken, worn-out, or cracked diaphragm
b. Improperly operating fuel-pump valves
c. Broken or damaged rocker arm
d. Clogged pump-filter screen or filter
e. Leakage of air into sediment bowl because of loose bowl or worn gasket

These are all causes of insufficient fuel delivery due to conditions within the pump. In addition, there are conditions outside the pump that could prevent delivery of normal

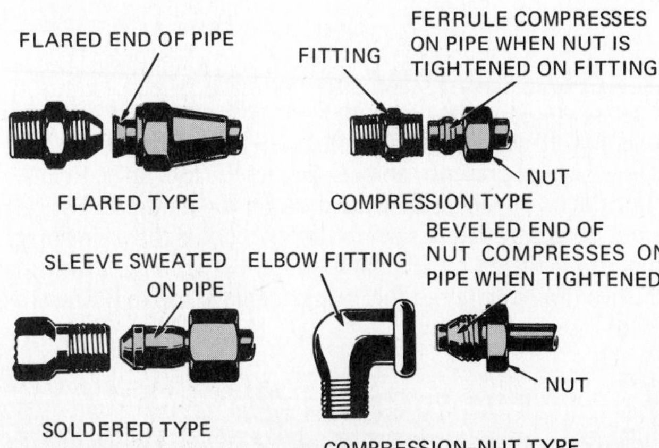

Fig. 17-7 Types of fuel-line couplings, or connectors.

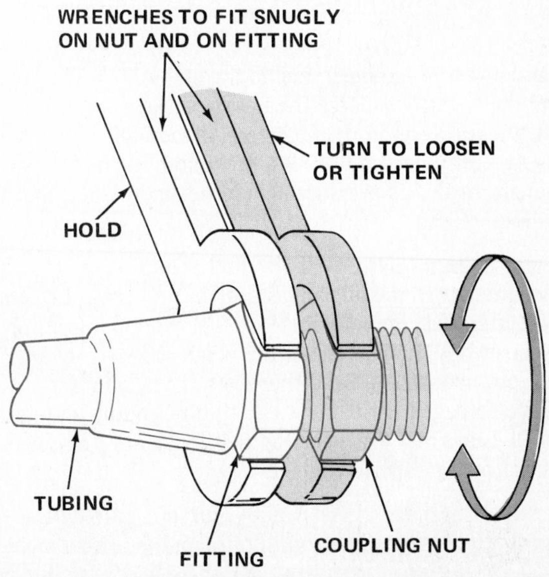

Fig. 17-8 Using two wrenches to loosen or tighten coupling nuts to avoid twisting and damaging the line.

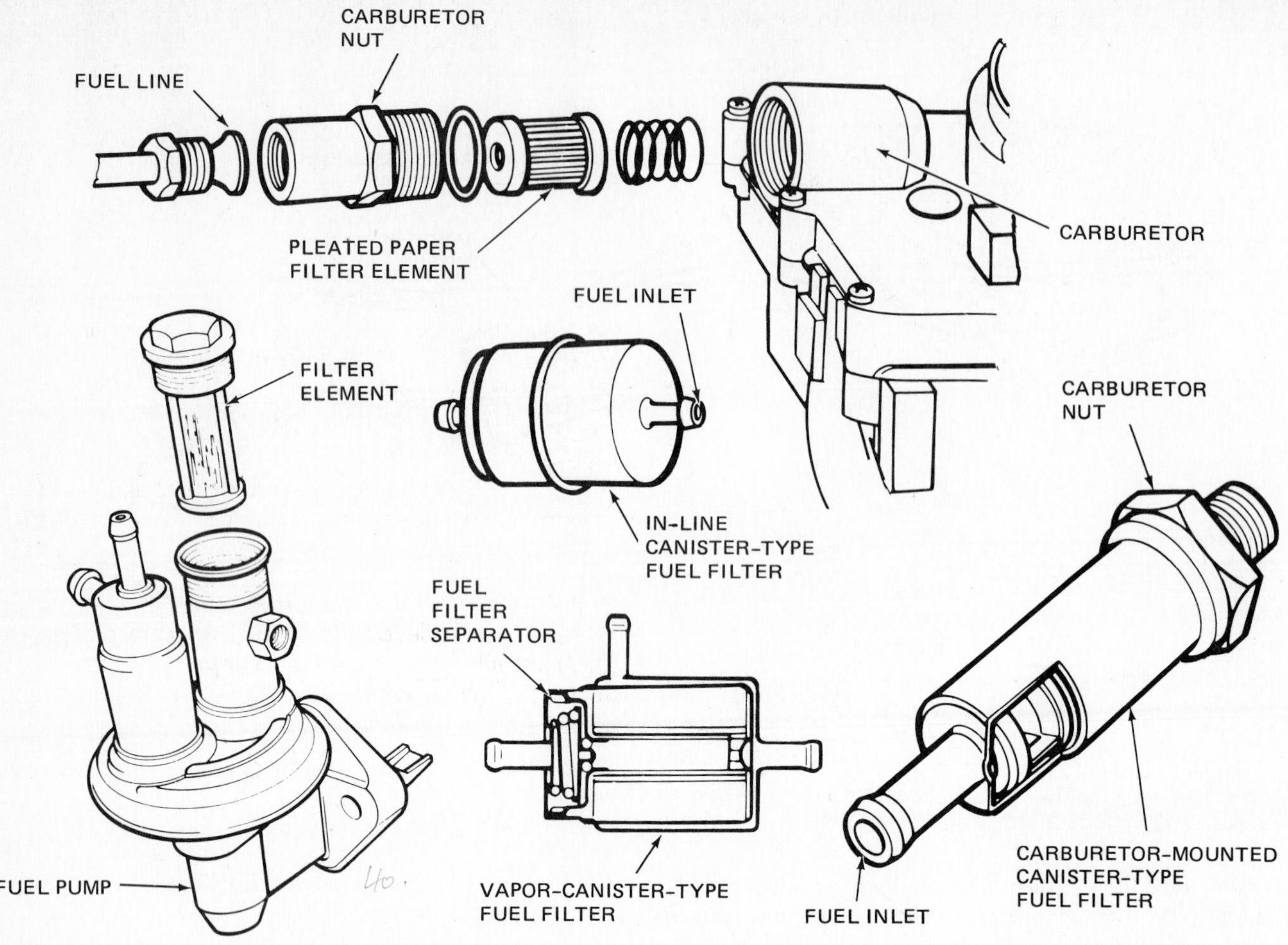

Fig. 17-9 Various types of fuel filters. *(Chrysler Corporation)*

amounts of fuel. These include such things as a clogged fuel-tank-cap vent, clogged fuel line or filter, air leaks into the fuel line, and vapor lock. In the carburetor, an incorrect float level, a clogged inlet screen, or a malfunctioning inlet needle valve would prevent delivery of adequate amounts of fuel to the float bowl.

2. Excessive Pump Pressure High pump pressure causes delivery of too much fuel to the carburetor. The excessive pressure tends to force the needle valve off its seat. As a result, the fuel level in the float bowl is too high. This would be a very unusual condition. If a fuel pump has been operating satisfactorily, its pressure should not increase in normal operation.

3. Fuel-Pump Leaks If the fuel pump leaks, gasoline can accumulate in the oil pan faster than the crankcase ventilating system can remove it. Most fuel pumps are crimped together on assembly. If the crimp is not properly made, the diaphragm may leak around the edges. The diaphragm itself could be damaged. In either case, the fuel pump should be replaced. Leaks may also occur at fuel-line connections which are loose or improperly coupled.

4. Fuel-Pump Noises A noisy pump is usually the result of worn or broken parts within the pump. These include a weak or broken rocker-arm spring, a worn or broken rocker-arm pin or rocker arm, or a broken diaphragm spring. In addition, a loose fuel pump or a scored rocker arm or cam on the camshaft may cause noise. Fuel-pump noise may sound similar to engine valve-tappet noise. Its frequency is the same as camshaft speed. If the noise is bad enough, it can actually be "felt" by gripping the fuel pump firmly with your hand. The noise may also be isolated by using a mechanic's stethoscope.

♦ CAUTION ♦ Don't do this if reaching into the fuel pump puts your hand too close to the moving fan or belts. Careful listening will usually tell you whether the noise is coming from the fuel pump. Tappet noise is usually heard all along the engine.

♦ 17-23 FUEL-PUMP REMOVAL

Before removing the fuel pump, wipe off any dirt or grease so that it will not get into the engine. Then take off the heat shield (where present), and disconnect the fuel lines. Remove the attaching nuts or bolts, and lift off the pump. If it sticks, work it gently from side to side or pry lightly under its mounting flange or attaching studs. On an engine using a pushrod to operate the fuel pump, remove the rod. Examine it for wear or sticking.

♦ 17-24 FUEL-PUMP SERVICE

Most fuel pumps are assembled by crimping and cannot be disassembled. If defective, another fuel pump must be installed.

♦ 17-25 FUEL-PUMP INSTALLATION

Make sure fuel-line connections are clean and in good condition. Connect the fuel lines to the pump before attaching the pump to the engine (see ♦17-20). Then place a new gasket on the studs of the fuel-pump mounting or over the opening in the crankcase. The mounting surface of the engine should be clean. Insert the rocker arm of the fuel pump into the opening. Make sure that the arm goes on the proper side of the camshaft (or that it is centered over the pushrod). It may be difficult to get the holes in the fuel-pump flange to align with the holes in the block or timing cover. Then crank the engine until the low side of the camshaft eccentric is under the fuel-pump rocker arm. Now the pump can be installed without forcing or prying it into place. Attach it with bolts or nuts. Check the pump operation, as explained in ♦17-21.

♦ 17-26 CARBURETOR ADJUSTMENTS

At one time, there were several adjustments that could be made on carburetors. Now the requirements of the Clean Air Act regarding automotive emissions have eliminated most adjustments. Today, the only carburetor adjustment recommended for most cars is to adjust the idle speed. The idle mixture is preset at the factory, and limiter caps, dowels, or plugs are installed to prevent tampering. The adjustment procedure is shown in a tuneup decal (vehicle emission control information label) in the engine compartment (Fig. 17-10). The decal procedure must be followed. If the carburetor requires disassembly for servicing, the limiter caps, dowels, or plugs must be removed so that the idle mixture can be readjusted. Then new limiter caps, dowels, or plugs must be installed.

In 1975, the procedure of setting the idle mixture was changed for some cars. This procedure required the use of propane. It is described in the following section (♦17-27). Setting the idle speed on older cars is described below.

♦ CAUTION ♦ When the engine is running, do not put your hands near the pulleys, belts, or fan. Do not stand in direct line of the fan. Never wear loose clothing that could catch in moving parts. On engines with electric cooling fans, disconnect the fan so that it does not run.

Adjusting the Idle Speed (Older Car)

1. Disconnect the fuel-tank hose from the charcoal canister.
2. Disconnect the vacuum hose to the distributor. Plug the hose leading to the carburetor.
3. Make sure distributor contact-point dwell and ignition timing are correct.
4. Adjust idle speed by the means provided. In earlier models, this was a screw in the throttle linkage at the carburetor. In later models equipped with idle-speed solenoids, the adjustment screw is in the solenoid. Regardless of the location of the adjustment screw, use a tachometer to measure engine speed. Make the adjustment to get the specified idle speed.
5. Reconnect the distributor vacuum hose and the fuel-tank hose.

Setting Idle Mixture This adjustment is permissible only if the carburetor has required major service. A typical procedure is:

1. With limiter caps off, turn mixture screws in until they lightly touch the seats. Then back them off two full turns.
2. Adjust idle speed as described above.
3. Connect an exhaust-gas analyzer with a CO meter to the exhaust system. Adjust the idle-mixture screws to get a satisfactory idle at the specified rpm with the CO reading at or below the specified allowable maximum. Engine should be running at normal idle with automatic transmission in D (drive) or manual transmission in neutral.
4. After setting the idle mixture, recheck the idle speed. If everything is within specifications, install new limiter caps on the idle-mixture screws.

♦ 17-27 ADJUSTING IDLE MIXTURE WITH PROPANE

The catalytic converter cleans up the exhaust gas by converting the HC and CO into harmless gases. It also makes it practically impossible to adjust the idle mixture by reading the CO content of the exhaust gas. Therefore, an adjustment procedure using propane to artificially enrich the mixture during

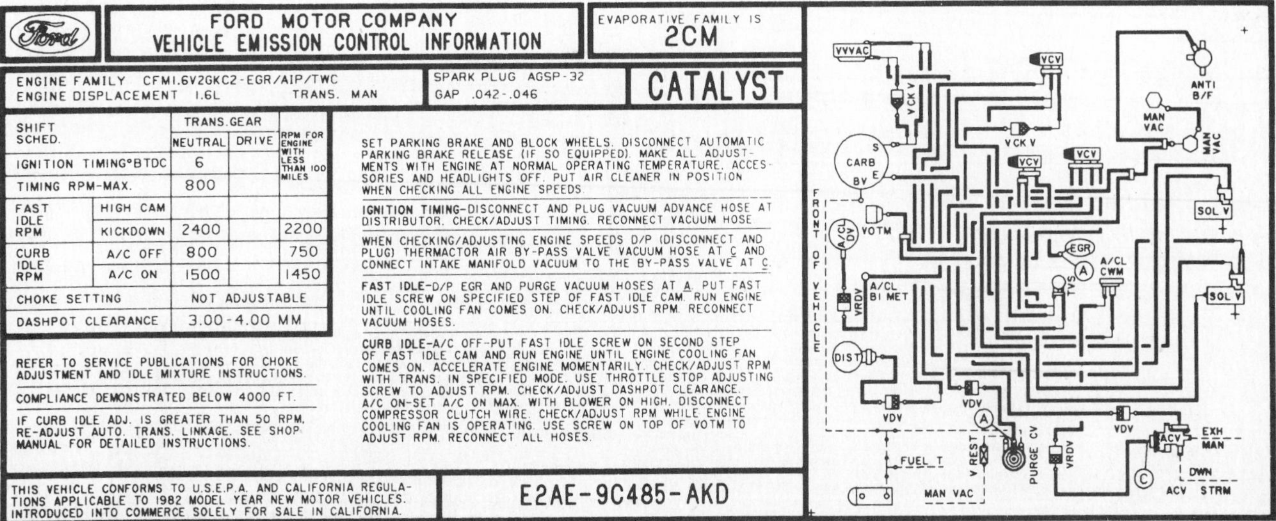

FORD MOTOR COMPANY
VEHICLE EMISSION CONTROL INFORMATION

EVAPORATIVE FAMILY IS
2CM

ENGINE FAMILY CFM1.6V2GKC2-EGR/AIP/TWC
ENGINE DISPLACEMENT 1.6L TRANS. MAN

SPARK PLUG AGSP-32
GAP .042-.046

CATALYST

SHIFT SCHED.		TRANS. GEAR NEUTRAL	TRANS. GEAR DRIVE	RPM FOR ENGINE WITH LESS THAN 100 MILES
IGNITION TIMING°BTDC		6		
TIMING RPM-MAX.		800		
FAST IDLE RPM	HIGH CAM			
	KICKDOWN	2400		2200
CURB IDLE RPM	A/C OFF	800		750
	A/C ON	1500		1450
CHOKE SETTING		NOT ADJUSTABLE		
DASHPOT CLEARANCE		3.00-4.00 MM		

REFER TO SERVICE PUBLICATIONS FOR CHOKE ADJUSTMENT AND IDLE MIXTURE INSTRUCTIONS.

COMPLIANCE DEMONSTRATED BELOW 4000 FT.

IF CURB IDLE ADJ. IS GREATER THAN 50 RPM, RE-ADJUST AUTO. TRANS. LINKAGE. SEE SHOP MANUAL FOR DETAILED INSTRUCTIONS.

SET PARKING BRAKE AND BLOCK WHEELS. DISCONNECT AUTOMATIC PARKING BRAKE RELEASE (IF SO EQUIPPED). MAKE ALL ADJUSTMENTS WITH ENGINE AT NORMAL OPERATING TEMPERATURE, ACCESSORIES AND HEADLIGHTS OFF. PUT AIR CLEANER IN POSITION WHEN CHECKING ALL ENGINE SPEEDS.

IGNITION TIMING—DISCONNECT AND PLUG VACUUM ADVANCE HOSE AT DISTRIBUTOR. CHECK/ADJUST TIMING. RECONNECT VACUUM HOSE.

WHEN CHECKING/ADJUSTING ENGINE SPEEDS D/P (DISCONNECT AND PLUG) THERMACTOR AIR BY-PASS VALVE VACUUM HOSE AT C AND CONNECT INTAKE MANIFOLD VACUUM TO THE BY-PASS VALVE AT C.

FAST IDLE—D/P EGR AND PURGE VACUUM HOSES AT A. PUT FAST IDLE SCREW ON SPECIFIED STEP OF FAST IDLE CAM. RUN ENGINE UNTIL COOLING FAN COMES ON. CHECK/ADJUST RPM. RECONNECT VACUUM HOSES.

CURB IDLE—A/C OFF—PUT FAST IDLE SCREW ON SECOND STEP OF FAST IDLE CAM AND RUN ENGINE UNTIL ENGINE COOLING FAN COMES ON. ACCELERATE ENGINE MOMENTARILY. CHECK/ADJUST RPM WITH TRANS. IN SPECIFIED MODE. USE THROTTLE STOP ADJUSTING SCREW TO ADJUST RPM. CHECK/ADJUST DASHPOT CLEARANCE. A/C ON—SET A/C ON MAX. WITH BLOWER ON HIGH. DISCONNECT COMPRESSOR CLUTCH WIRE. CHECK/ADJUST RPM WHILE ENGINE COOLING FAN IS OPERATING. USE SCREW ON TOP OF VOTM TO ADJUST RPM. RECONNECT ALL HOSES.

THIS VEHICLE CONFORMS TO U.S.E.P.A. AND CALIFORNIA REGULATIONS APPLICABLE TO 1982 MODEL YEAR NEW MOTOR VEHICLES. INTRODUCED INTO COMMERCE SOLELY FOR SALE IN CALIFORNIA.

E2AE-9C485-AKD

Fig. 17-10 Vehicle emission control information decal. *(Ford Motor Company)*

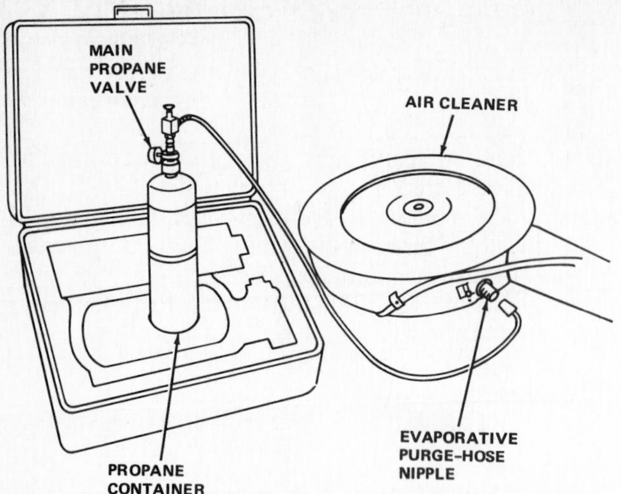

Fig. 17-11 Setup for adjusting idle mixture with propane. *(Ford Motor Company)*

the adjustment was developed. The procedure is called the *propane enrichment adjustment procedure.* The engine rpm is used to determine the proper setting. Figure 17-11 shows the setup for doing the job.

The emission control information label in the engine compartment gives the steps for setting the mixture using propane (Fig. 17-12). The label explains how to disconnect and plug hoses before the adjustment is made. Manufacturers' service manuals also explain how to use propane to make the adjustment.

Note Making any adjustment contrary to the steps in the manufacturers' procedure may violate federal, state, or local laws.

♦ 17-28 LEAN-DROP IDLE-MIXTURE ADJUSTMENT

The idle mixture on vehicles built in 1981 and later is *not* adjusted by using propane (♦17-27). These cars use the *lean-drop,* or *idle-drop, adjusting procedure.* This procedure is usually performed only if the idle-mixture screws were removed or tampered with during carburetor overhaul. A typical procedure follows:

1. After the carburetor is installed on the engine, and all vacuum and electric connections made, start the engine and warm it to normal operating temperature.
2. Connect an accurate tachometer with an expanded 400 to 800 or 0 to1000 rpm scale.
3. Set the parking brake firmly. Block the drive wheels if the vehicle has automatic transmission. Position the selector lever in NEUTRAL (manual transmission) or DRIVE (automatic transmission). If the vehicle has an automatic transmission and an automatic brake release, disconnect it. This is required so that the parking brake will remain set in DRIVE. *Do not accelerate the engine!*
4. Set idle speed to specifications.
5. Turn idle-mixture screw (or screws) in (leaner) until there is a noticeable loss of speed. Turn the idle-mixture screw (or screws) out (richer) until the highest rpm is obtained. Do not turn the screw (or screws) any further than the point at which the highest rpm is first obtained.
6. As a final adjustment, turn the screw or screws in (leaner) to obtain the specified drop in idle rpm. This is known as *lean best idle.* Turn the screw or screws in by small amounts until the specified idle drop is obtained. This specification may vary. A typical specification for idle drop is 50 rpm. For example, the idle speed would drop from 800 rpm to 750 rpm. Refer to the manual for the vehicle you are working on for the proper specification.
7. Reinstall the plugs or dowels.

♦ 17-29 CARBURETOR REMOVAL

To remove a carburetor, first disconnect the battery ground terminal and wires from any electrical devices on the carburetor. This prevents sparks that might ignite any spilled gasoline. Then disconnect the air and vacuum lines and take off the air cleaner. Then disconnect the throttle and choke linkages. Disconnect the hot-air tube to the choke (if present). Disconnect the fuel line and the distributor vacuum-advance line from the carburetor. Use two wrenches, as necessary, to avoid damage to the lines or couplings.

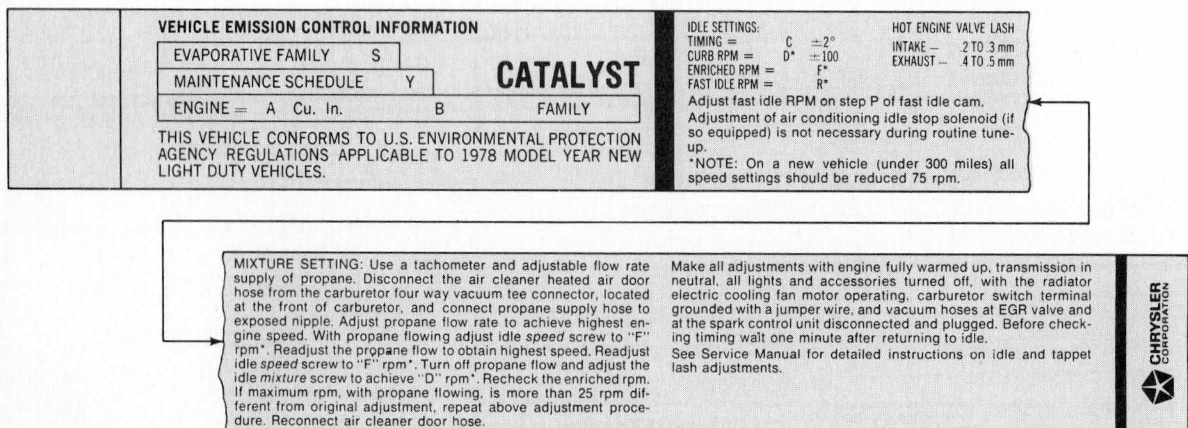

Fig. 17-12 Typical engine compartment decal for a car that requires idle-mixture adjustment using propane. *(Chrysler Corporation)*

♦ **CAUTION** ♦ Note Caution 4 in ♦17-15 about the gasoline being under pressure in the line between the fuel pump and carburetor.

Take off the carburetor-attaching nuts or bolts, and lift off the carburetor. Try to avoid jarring the carburetor. It may have accumulations of dirt in the float bowl. Rough treatment may stir up this dirt and cause it to get into carburetor jets and systems.

After the carburetor is off, it should be put in a clean place where dirt and dust cannot get into openings.

As soon as the carburetor is removed, cover the openings in the manifold with masking tape. This prevents engine damage from loose parts dropped into the manifold. Parts dropped into the manifold could end up in the combustion chambers where they could cause serious damage.

♦ 17-30 CARBURETOR OVERHAUL PROCEDURES

Disassembly and reassembly procedures on carburetors vary according to design. The manufacturer's recommendations should be carefully followed. The time required to overhaul a carburetor varies from approximately ¾ to 2 hours, according to type. A few special carburetor tools may be required. Gauges needed to measure float clearance, float centering, float height, and choke clearance are usually included in the carburetor overhaul kit. Some carburetors may require the use of dial indicators, micrometers, and plug gauges to check calibration.

Complete carburetor overhaul kits are supplied for many carburetors. These kits contain instructions and all parts (jets, gaskets) required to overhaul the carburetor and restore it to manufacturer's specifications.

♦ **CAUTION** ♦ When removing and handling a carburetor, avoid spilling gasoline. The carburetor float bowl will have gasoline in it, so keep the carburetor upright. Gasoline is extremely flammable. Any spilled gasoline should be wiped up immediately. Put gasoline-soaked towels outside to dry.

When overhauling a carburetor, get the overhaul kit supplied for that carburetor. Follow the instruction sheet included in the kit. Do not use drills or wires to clean out fuel passages or bleed holes in the carburetor or throttle body. This can enlarge the openings and upset the carburetor calibration. Instead, clean out the openings with a chemical cleaner and blow them out with compressed air.

Safety cautions for servicing carburetors include the following:

1. Do not splash carburetor cleaner in your eyes. It can seriously harm them. Wear goggles to protect your eyes. Wash your eyes thoroughly if you get chemicals in them (Fig. 2-12). See a doctor without delay.
2. Use the air hose with care. Wear goggles. See ♦7-7 on the dangers of misusing compressed air.
3. Always be aware of the fact that gasoline and other solvents are highly flammable. Never smoke or have open flames around these liquids.

♦ 17-31 CARBURETOR INSTALLATION

Use a new gasket to assure a good seal between the carburetor and the mounting pad. Put the carburetor into position on the intake manifold and attach it with nuts or bolts. Connect the fuel line and the distributor vacuum-advance line to the carburetor, using two wrenches if necessary to avoid damage to the lines or couplings. Connect wires to switches and other electric controls (where present). Make idle-speed, idle-mixture, and other adjustments. Install the air cleaner. Many mechanics recommend test driving the car after carburetor adjustments are made.

♦ REVIEW QUESTIONS ♦

Select the *one* correct, best, or most probable answer to each question. You can find the answers in the section indicated at the end of each question.

1. If the engine cranks normally but does not start, the trouble could be (♦17-3)
   a. an ignition problem
   b. overchoking
   c. lack of fuel
   d. all of the above
2. If the engine starts, speeds up, and then dies, the trouble could be (♦17-4)
   a. defective spark plugs
   b. stuck EGR valve
   c. defective ignition coil
   d. low battery
3. The best way to check a complaint that the engine lacks power is with (♦17-6)
   a. a road test
   b. a fuel-pump tester
   c. a dynamometer and oscilloscope
   d. an exhaust-gas analyzer
4. Dieseling, or run-on, can be caused by (♦17-11)
   a. incorrectly adjusted idle-stop solenoid
   b. throttle-valve stuck closed
   c. EGR valve stuck open
   d. defective air pump
5. A vehicle with poor driveability might have (♦17-2)
   a. hard starting
   b. poor acceleration
   c. stalling
   d. all of the above

CHAPTER 18
Gasoline Fuel-Injection Systems: Operation and Service

After studying this chapter, you should be able to:

1. Define *injection valve*.
2. Describe the operation of electronic fuel-injection systems.
3. Explain how the mechanical fuel-injection system works.
4. Locate the components of gasoline fuel-injection systems on various cars.
5. Explain the difference between timed injection and continuous injection.
6. Describe the construction and operation of a throttle-body injection system.
7. Diagnose troubles in various gasoline fuel-injection systems.

♦ 18-1 FUEL INJECTION

The engine must have a continuous supply of combustible air-fuel mixture to run. In most engines, a carburetor (Fig. 16-1) mounted on the intake manifold supplies the mixture. This mixture then flows to the engine cylinders when the intake valves open. The amount of mixture that enters the intake manifold is controlled by the position of the throttle valves. Their position (from fully closed to wide open) determines the amount of air (actually, air-fuel mixture) that can flow through into the intake manifold.

An engine equipped with gasoline fuel injection has a throttle body mounted on the intake manifold, instead of the complete carburetor. The throttle body is very similar to the carburetor throttle body for one- or two-barrel carburetors. It has either one or two throttle valves. As in the carburetor, the position of the throttle valves determines the amount of air that flows through into the intake manifold.

In the gasoline fuel-injection system, the fuel is injected into the intake manifold through fuel-injection valves. There are two basic arrangements, *port injection*, and *throttle-body injection* (TBI). These arrangements are also called *multiple-point* and *single-point injection*. In port injection, there is an injection valve for each engine cylinder (Fig. 18-1). Each injection valve is positioned in the intake port near the intake valve (Fig. 18-2). In a TBI system, an injection valve is positioned slightly above each throat of the throttle body (Fig. 18-3). The injection valve sprays fuel into the air just before it passes the throttle valve and enters the intake manifold (Fig. 15-4).

♦ 18-2 TIMED AND CONTINUOUS INJECTION

Another way to classify fuel-injection systems is by the injection action. Depending on the system, fuel sprays from the injection valves either in pulses (timed injection) or continu-

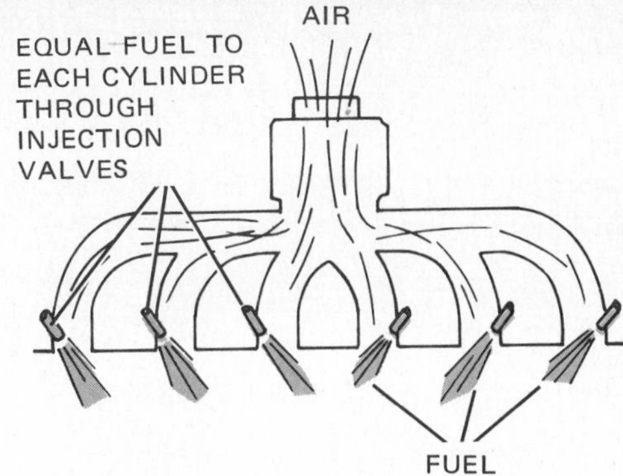

Fig. 18-1 Port injection. An injection valve in each intake port sprays fuel into the intake air.

ously. Port injection and throttle-body systems may be either pulsed or continuous. In continuous injection systems, the fuel sprays continuously from the injection valves. With either system, the amount of fuel injected varies with engine speed and power demands. Later sections in the chapter describe the operation and control of several types of gasoline fuel-injection systems.

♦ 18-3 ADVANTAGES OF PORT FUEL-INJECTION

Regardless of whether the port injection system is pulsed or continuous, it eliminates several intake-manifold distribution problems. One of the most difficult problems in a carbureted system is to get the same amount and richness of air-fuel mixture to each cylinder. The problem is that the intake manifold acts as a sorting device, sending a richer air-fuel mixture to the end cylinders. Figure 18-4 shows how this happens. The air flows readily around corners and through variously shaped passages. However, the fuel, because it is heavier, is unable to travel as easily around the bends in the intake manifold. As a result, some of the fuel particles continue to move to the end of the intake manifold, accumulating or puddling there. This enriches the mixture going to the end cylinders. The center cylinders, closest to the carburetor, get the leanest mixture.

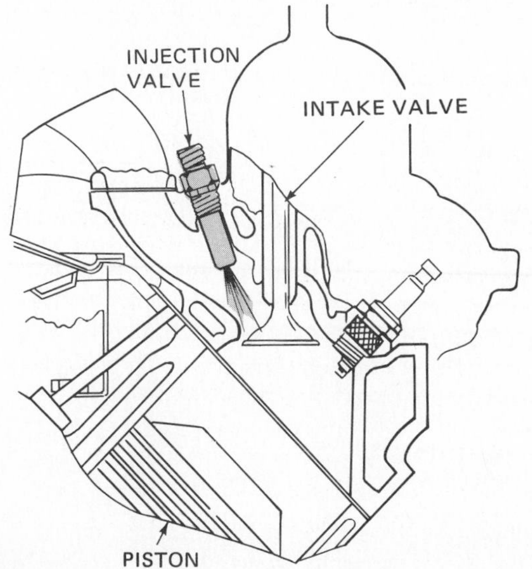

Fig. 18-2 The injection valve sprays fuel into the intake air just before it passes the intake valve and enters the cylinder.

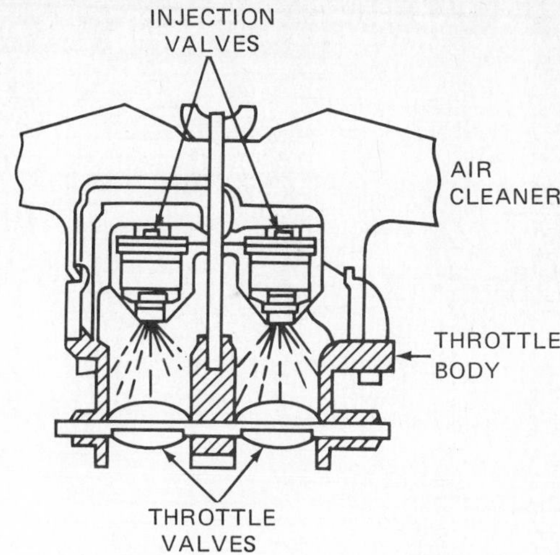

Fig. 18-3 The throttle body for the TBI system is similar to the carburetor throttle body. Throttle valves in both control the amount of air entering the intake manifold.

The port fuel-injection system solves this problem because the same amount of fuel is injected at each intake-valve port. Each cylinder gets the same amount of air-fuel mixture of the same mixture richness.

Another advantage of the port fuel-injection system is that the intake manifold can be designed for the most efficient flow of air only. It does not have to handle fuel. Also, because only a throttle body is used, instead of the complete carburetor, the hood height of the car can be lowered.

With fuel injection, the fuel mixture requires no extra heating during warm-up. No manifold heat-control valve or heated-air system is required. Throttle response is faster because the fuel is under pressure at the injection valves at all times. An electric fuel pump supplies the pressure. The carburetor must depend on differences in air pressure (or vacuum) as the force that causes the fuel to feed into the air passing through.

♦ 18-4 ELECTRONIC FUEL INJECTION

The development of solid-state electronic devices such as diodes and transistors made electronic fuel injection possible.

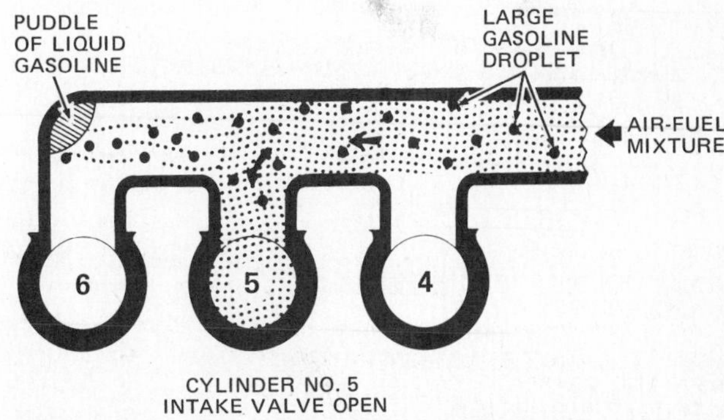

Fig. 18-4 Distribution pattern in an intake manifold. The fuel particles tend to continue to the end of the manifold. This enriches the mixture going into the end cylinders. *(Chevrolet Motor Division of General Motors Corporation)*

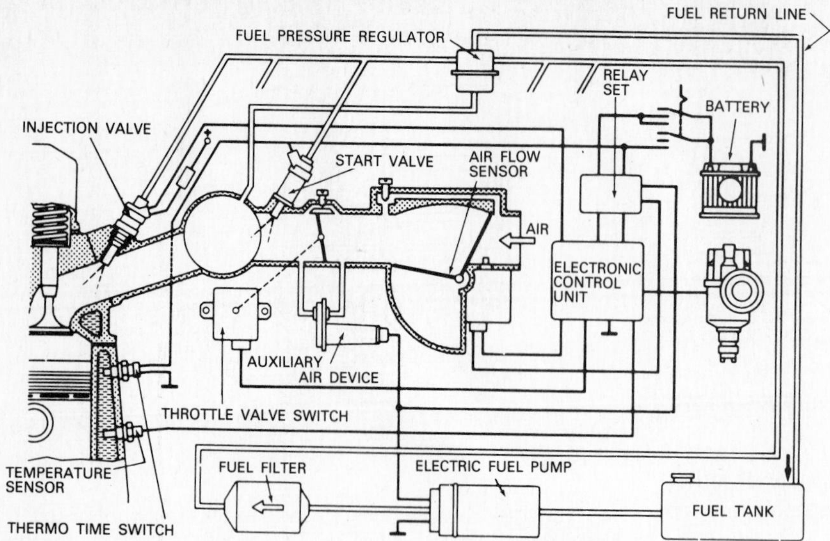

Fig. 18-5 Bosch L-type electronic fuel-injection system. *(Robert Bosch Corporation)*

As early as 1968, some systems began to appear on automobiles. Volkswagen began installing an electronic fuel-injection system that year. The system was developed by the Robert Bosch Corporation. In 1975, a version of the system appeared on some Cadillacs (♦18-5). Other companies also adopted variations of the system. This was a port injection system, known as the *Bosch D-type* electronic fuel-injection system. Later, Bosch developed the L-type system (Fig. 18-5). This was also a port injection system. It was introduced by Porsche and Volkswagen in 1974. In this system, fuel metering is controlled primarily by engine speed and the amount of air that actually enters the engine. This is called *air-mass metering*, or *airflow metering*. The airflow sensor in the air intake measures the amount of air that enters.

Electronic throttle-body injection was first used by Cadillac and Ford on some 1980 cars (Fig. 18-6). In this system, one or two injection valves are mounted in the throttle body. In operation, they spray fuel into the air passing through. These systems are covered in later sections.

Modern electronic fuel-injection systems include an oxygen sensor (Fig. 16-15). The device measures the amount of oxygen in the exhaust gas and sends this information to the electronic control unit (♦16-22). If there is too much oxygen, the mixture is too lean. If there is too little, the mixture is too rich. In either case, the ECU adjusts the air-fuel ratio by changing the amount of fuel injected.

♦ 18-5 CADILLAC 1975–1980 ELECTRONIC PORT FUEL INJECTION

The electronic fuel-injection system used by Cadillac from 1975 to 1980 is shown in Fig. 18-7. It is a port injection system using a separate fuel-injection valve for each cylinder (Fig. 18-8). The eight injection valves are connected to a fuel rail. They are divided into two groups of four injection valves each. Each group of injection valves is alternately turned on by the ECU. The injection valves are turned on once for each two revolutions of the crankshaft.

Figure 18-9 is a block diagram showing (on the left) the sensors that send information to the ECU. With this and

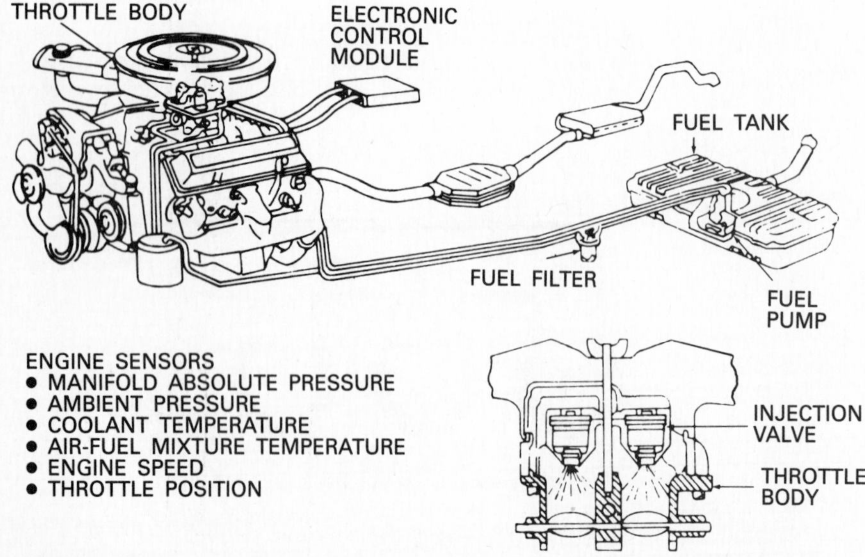

Fig. 18-6 The digital electronic fuel-injection (DEFI) system of throttle-body injection. *(General Motors Corporation)*

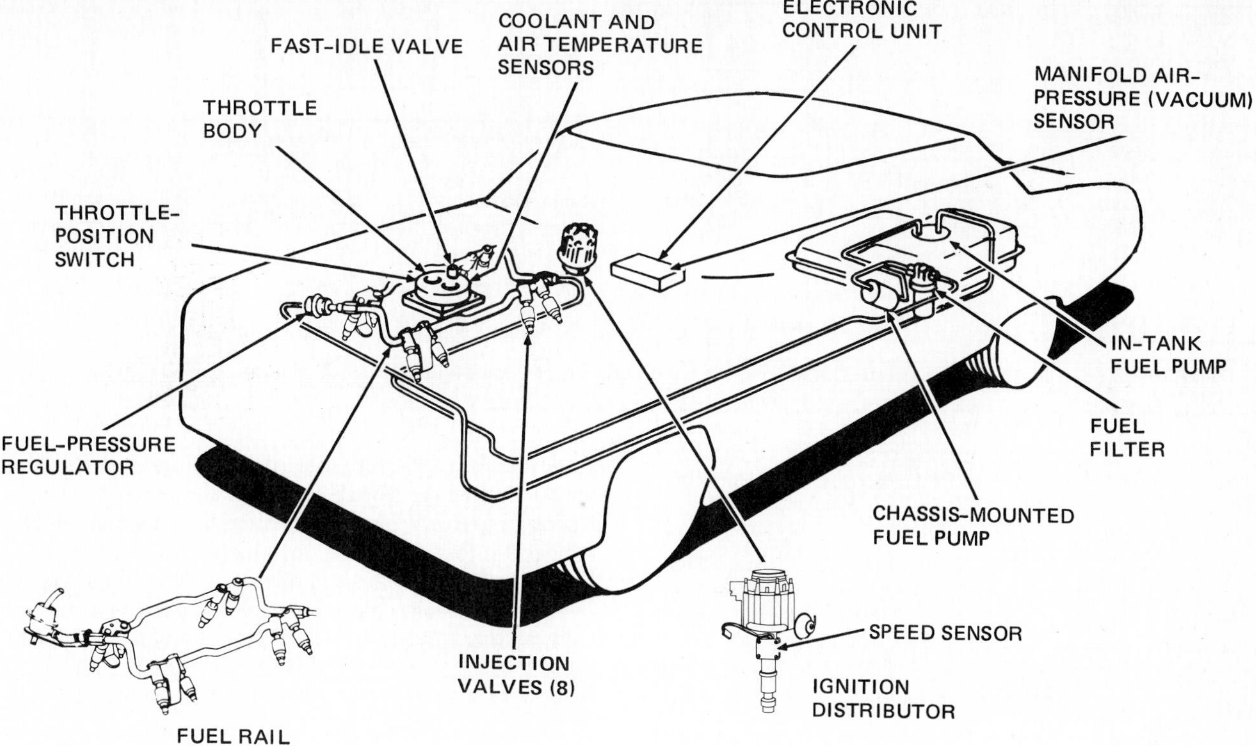

Fig. 18-7 Components of the Cadillac port-type electronic fuel-injection system. *(Cadillac Motor Car Division of General Motors Corporation)*

other information, the ECU computes the amount of fuel that the engine needs. Then the ECU sends signals to the injection valves and other parts of the system (right in Fig. 18-9).

The ECU is a preprogrammed computer installed above the glove box within the passenger compartment. It converts the input information from the sensors into an electric pulse. The length of the pulse opens the injection valves for the proper duration at the proper time. The ECU cannot be adjusted or serviced. When a malfunction is traced to the ECU, it must be replaced. Complete diagnosis of the ECU requires a special tester.

♦ 18-6 FUEL-DELIVERY SYSTEM

Figure 18-8 shows the fuel-delivery system for the Cadillac port injection system. The fuel pump delivers fuel through the fuel filter to the injection valves. Each injection valve is operated by a small electric solenoid (Fig. 18-10). When the

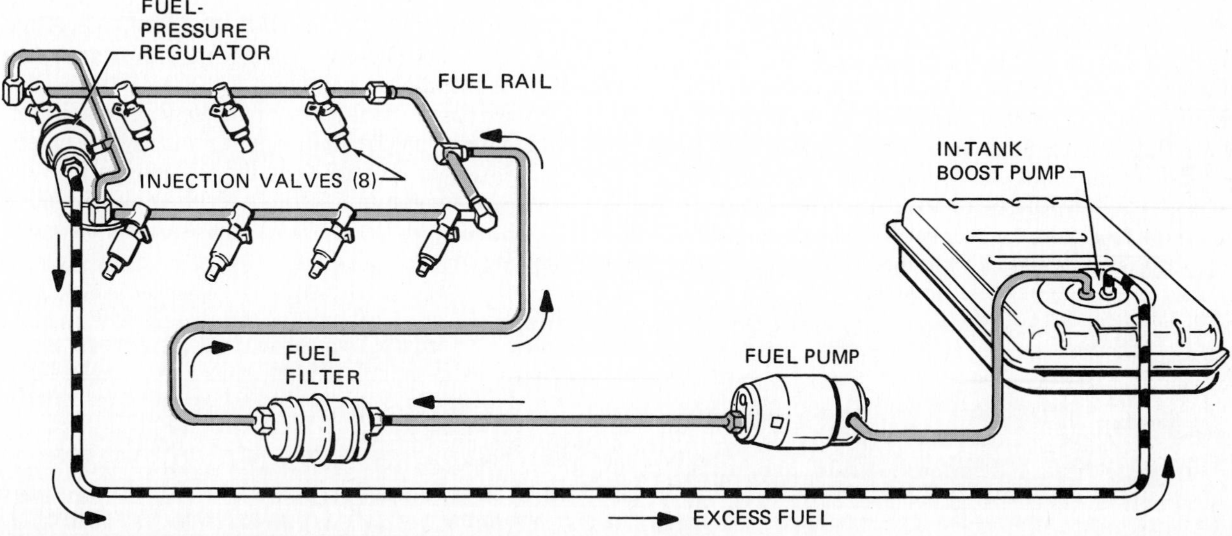

Fig. 18-8 Fuel system used with the Cadillac port EFI system. *(Cadillac Motor Car Division of General Motors Corporation)*

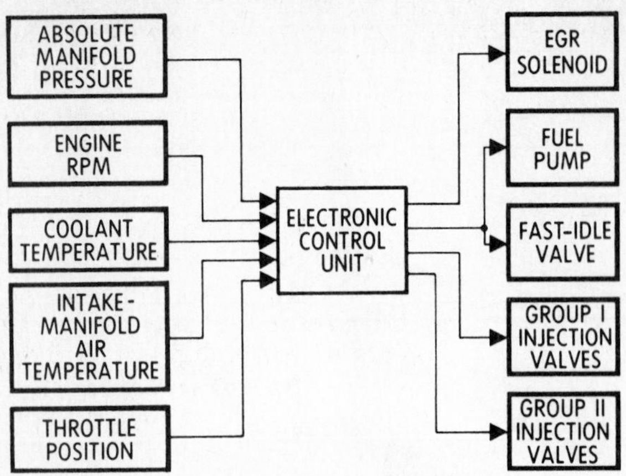

Fig. 18-9 Block diagram showing the sensors (left) that provide information to the ECU *(Cadillac Motor Car Division of General Motors Corporation)*

solenoid is connected to the battery through the ECU, the solenoid pulls the injection-valve needle back. This allows fuel to spray from the valve. The ECU holds the valve open the proper length of time to meet operating conditions. For example, if the car is accelerating, the valve will be held open longer so that the engine gets the added fuel it needs.

The timing of the injection valves—when they open—is determined by the engine-speed sensor. It is located in the ignition distributor (Fig. 18-7). It includes two reed switches and two magnets. Every time a magnet passes a reed switch, the switch closes. This signals the ECU to actuate one set of injection valves. When the other switch is closed, it actuates the other set of injection valves.

The valves are not actuated individually, but in groups. There are two groups of injection valves. In a four-cylinder engine, there are two groups of two valves each. In a six-cylinder engine, there are two groups of three valves each. In an eight-cylinder engine, there are two groups of four valves each. For example, Fig. 18-11 shows the injection-valve grouping for a six-cylinder engine. Valves for cylinders 1, 3, and 5 have opened at the same time and are injecting fuel into the intake manifold. Next, these valves close and the valves for cylinders 2, 4, and 6 open and inject fuel.

Figure 18-12 shows the timing chart for this engine. The individual intake valves open at varying times (crankshaft degrees) after injection. For example, look at the top line, which is for number 1 cylinder. Injection takes place at 300 degrees of crankshaft rotation. Almost 60 degrees later (near 360 degrees), the number 1 intake valve opens and the intake stroke starts. Cylinder 5 is next in the firing order. Its intake valve opens near 480 degrees, or about 180 degrees after injection. The intake valve for number 3 opens about 300 degrees of crankshaft rotation after injection. During these

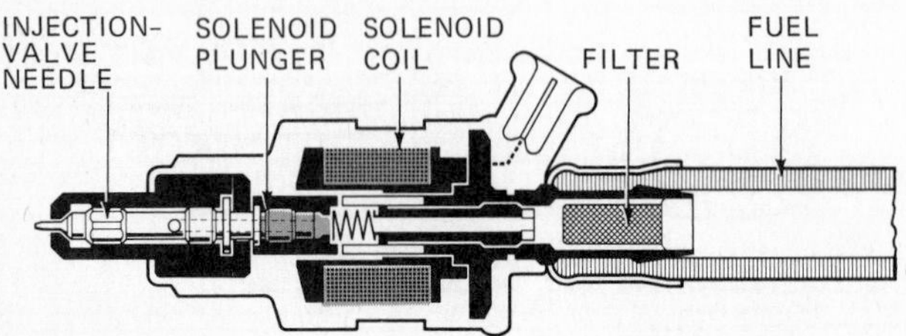

Fig. 18-10 Construction of the solenoid-operated injection valve. *(Robert Bosch Corporation)*

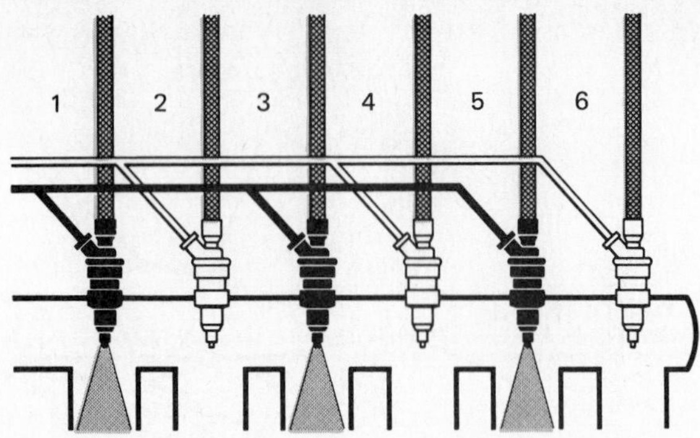

Fig. 18-11 Injection-valve grouping. *(Robert Bosch Corporation)*

varying intervals between fuel injection and intake-valve opening, the fuel is "stored" in the intake ports.

Having only two groups of injection valves simplifies the system. Little loss of engine performance results from this momentary storage of air-fuel mixture in the intake ports. The whole action takes place in a small fraction of a second. At highway speed, the time between injection and opening of the intake valve averages only about 0.01 second.

♦ 18-7 FUEL METERING BY MEASURING AIRFLOW

In the Bosch L-type electronic fuel injection (Fig. 18-5), fuel metering is controlled primarily by engine speed and by measuring the intake airflow. The injection valves all open and close at the same time. They are actuated twice for every camshaft revolution. This allows the injection pulses to be triggered directly by the ignition distributor. No separate speed sensor is needed.

The amount of intake air is measured by the airflow sensor as the air passes through it (Fig. 18-5). A pivoted flap is placed in the air passage. Attached to the flap is a small spring and a voltage sensor. As airflow increases, the spring allows the flap to move according to the amount of air passing through. As the flap moves, the voltage sensor signals the ECU how much air is passing through. This information is then used by the ECU to determine the length of time that the injection valves will be held open. Then the injection valves will deliver the exact amount of fuel needed to provide the proper air-fuel ratio for the operating condition.

A 1984 version of the L-type system uses hot-wire induction instead of the air flap to monitor air intake. The wire is stretched across the air intake and the air tends to keep the wire cool. The amount of electricity required to maintain a specified temperature in the wire is a measure of how much air is flowing in.

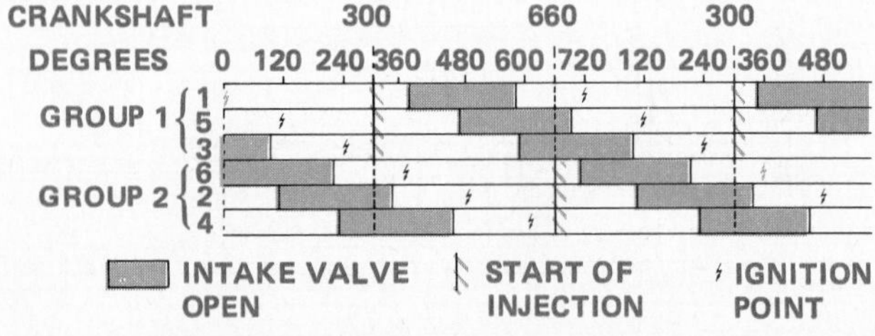

Fig. 18-12 Injection timing chart for a six-cylinder engine. *(Robert Bosch Corporation)*

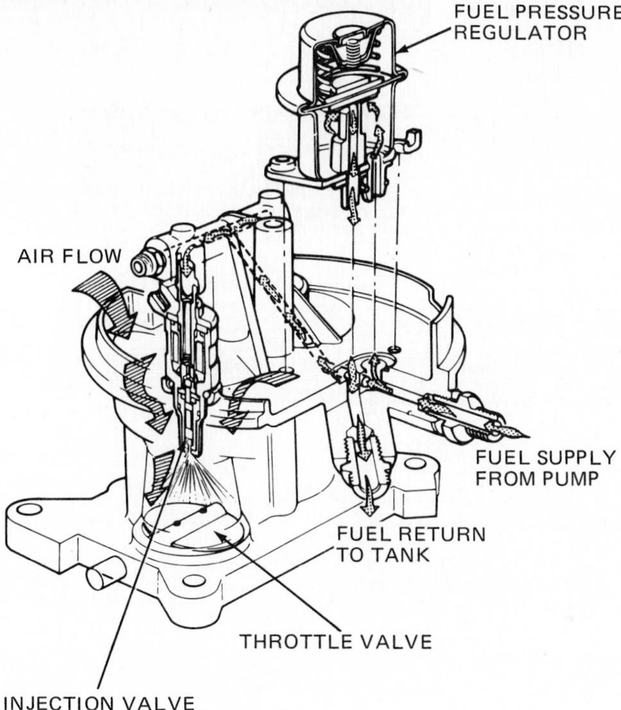

Fig. 18-13 Throttle body and fuel-injection valve in the Ford electronic fuel-injection system. *(Ford Motor Company)*

Buick and AC divisions of General Motors have developed a mass airflow sensor similar to the hot-wire induction system. This airflow sensor uses a heated film. The heated film consists of a nickle grid coated with a high-temperature material. The system feeds current to the film to maintain its temperature at about 166°F [75°C] above the temperature of the incoming air. If it requires more current to maintain this temperature, more air is entering. The entering air subtracts heat from the film. The more air flowing, the more heat is subtracted, and the more current required to maintain the temperature. The amount of current required is therefore a measure of the amount of air entering.

♦ 18-8 FORD THROTTLE-BODY INJECTION

In 1980, Ford began installing throttle-body fuel-injection systems on some cars. The system uses two injection valves, located in the throttle body on the intake manifold (Fig. 18-13). The amount of fuel discharged from the injection valves is controlled by Ford's electronic engine control (EEC) system. This is similar to the Ford EEC system described in ♦16-24. However, this system uses throttle-body fuel injection instead of a feedback carburetor.

Figure 18-14 shows the locations of the components in the Ford throttle-body fuel-injection system. Fuel to the injection valves is supplied by a high-pressure electric fuel pump inside the fuel tank. A primary fuel filter is located in the fuel line under the passenger compartment. A smaller, secondary fuel filter is mounted in the engine compartment. A fuel-pressure regulator on the throttle body maintains the fuel pressure to the injection valves at exactly 39 psi [269 kPa]. Any excess fuel not needed by the engine is returned to the fuel tank through a fuel-return line.

The constant high pressure at the injection valves provides a fine atomized spray of fuel when the injection valves open (Fig. 18-13). They are mounted vertically in the throttle body above the throttle valves so that the fuel is sprayed directly into the airstream.

Sensors monitor engine operating conditions and send this information to a microprocessor, which is a small computer (Fig. 18-14). It continuously calculates how long the injection valves should be open. The frequency of injection is constant at four pulses per engine-crankshaft revolution, two for each injection valve. Fuel volume is controlled by how long each injection valve is open.

Figure 18-15 shows the complete electronic fuel-injection system and EEC III installed on an engine. During starting and warm-up, the EEC III system provides extra fuel on sig-

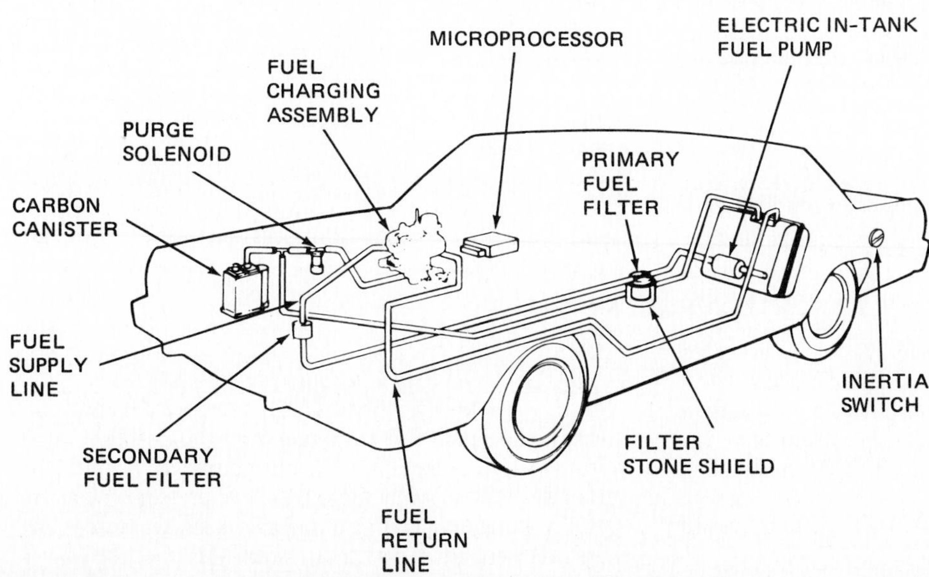

Fig. 18-14 Location of the components in the Ford electronic throttle-body fuel-injection system. *(Ford Motor Company)*

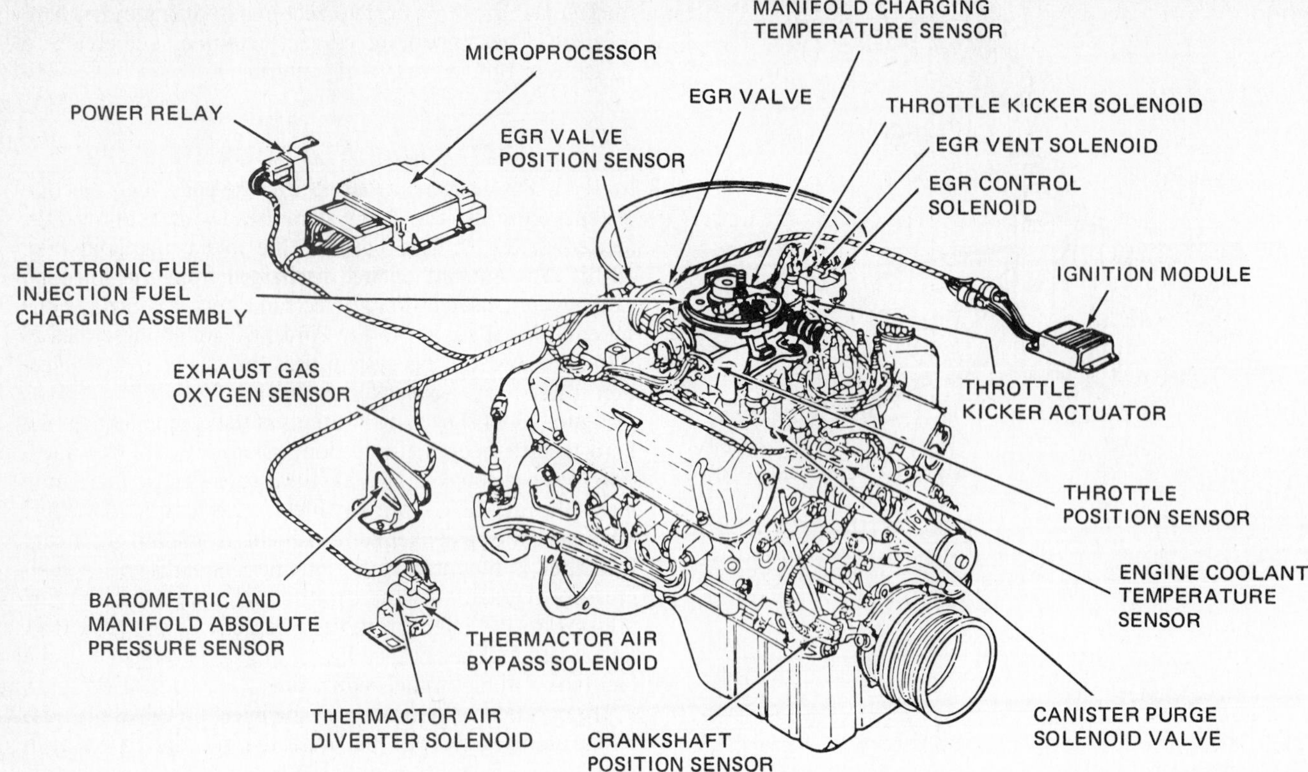

Fig. 18-15 EEC III system, which includes electronic throttle-body fuel injection, installed on an engine. *(Ford Motor Company)*

nal from a bimetal electric switch on the throttle body.

The system includes an exhaust-gas oxygen sensor (Fig. 18-15) which reports to the microprocessor (the ECU) the amount of oxygen in the exhaust. This device (♦16-22) indicates the richness or leanness of the air-fuel mixture. If it is not right, the ECU adjusts the amount of fuel being injected so that the correct air-fuel mixture is achieved.

The system also includes an *inertia switch* (Fig. 18-14). If the car is involved in a collision, contacts in the inertia switch open and stop the electric fuel pump. The inertia switch must be reset by pressing both buttons on it at the same time before the engine can be started. The inertia switch is located in the luggage compartment near the left wheel well on some cars.

♦ 18-9 GM DIGITAL FUEL INJECTION

This system is similar to the TBI system described in ♦18-8. Fuel flow from the injection valves is regulated by an electronic control module (ECM). The ECM monitors operating conditions so that the system will supply the correct air-fuel mixture for good driveability and exhaust emission control. The system uses an oxygen sensor (♦16-22). Fuel is delivered to the injection valves at a constant pressure. The amount of fuel the injection valves deliver is determined by the length of time that the valves are held open. This is controlled by the ECM, calculated from information supplied by the sensors.

♦ 18-10 GM CROSS-FIRE INJECTION

Some General Motors V-type engines have two separate throttle bodies with injection valves. They are controlled by the ECM, as in other TBI systems. Both throttle bodies are mounted on the intake manifold. The intake-manifold passages "cross over" to feed the cylinders in the opposite bank.

♦ 18-11 CONTINUOUS INJECTION SYSTEM

Figure 18-16 shows a continuous injection system (CIS). This is basically a mechanical system. It does not use electronics to time and meter fuel flow. Instead, fuel sprays continuously from the injection valves as long as the engine runs.

The amount of fuel injected is controlled by an airflow sensor plate which continuously measures the amount of air flowing into the intake manifold. As airflow increases, it lifts the sensor plate higher. This causes a lever to lift the control plunger on the fuel distributor. The action increases the amount of fuel that flows to the injection valves. The proper air-fuel ratio results. To meet emission standards, late-model cars have an oxygen sensor and ECU. However, fuel metering is still primarily a mechanical process performed by the airflow sensor plate.

♦ 18-12 CHRYSLER CONTINUOUS-FLOW ELECTRONIC FUEL INJECTION

This system (Fig. 18-17) feeds fuel from the injection valves continuously. The amount of fuel being sprayed from the injection valves is varied by varying the fuel pressure. When the engine is idling or operating at low speed, the pressure is low. Therefore, very little fuel sprays from the injection valves. But when power demands increase, as during intermediate- and high-speed operation, the pressure on the fuel is increased. Now more fuel sprays from the injection valves.

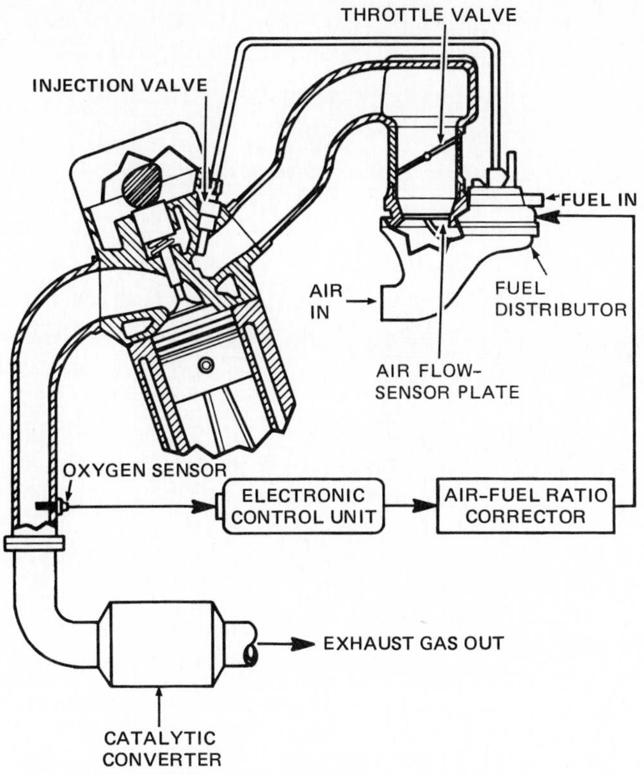

Fig. 18-16 A mechanical CIS which has an oxygen sensor. *(Saab-Scania of America, Inc.)*

FUEL-INJECTION SERVICE

♦ 18-13 ENGINE PERFORMANCE DIAGNOSIS

Diagnostic procedures are supplied in different forms by different manufacturers. For example, Bosch supplies a trouble-diagnosis chart for each type of fuel-injection system (Fig. 18-18). General Motors supplies performance-diagnosis procedures. These list various troubles and the possible causes of each. The opening of the solenoid-operated injection valve can be checked with an oscilloscope (described later in the text).

♦ 18-14 BUILT-IN ELECTRONIC SELF-DIAGNOSTIC SYSTEMS

Many late-model cars have built-in or "on-board" electronic self-diagnostic systems. When a fault occurs, the system stores a number code for the fault in its memory. Either a CHECK ENGINE light comes on or the electronic system gives the driver verbal warning that something is wrong. Then, when the mechanic activates the diagnostic system, the number code for the fault appears on the instrument panel. Referring to the manufacturer's service manual, the mechanic finds the code number which identifies the cause of the fault. This system is covered further in following chapters on ignition-system and engine service.

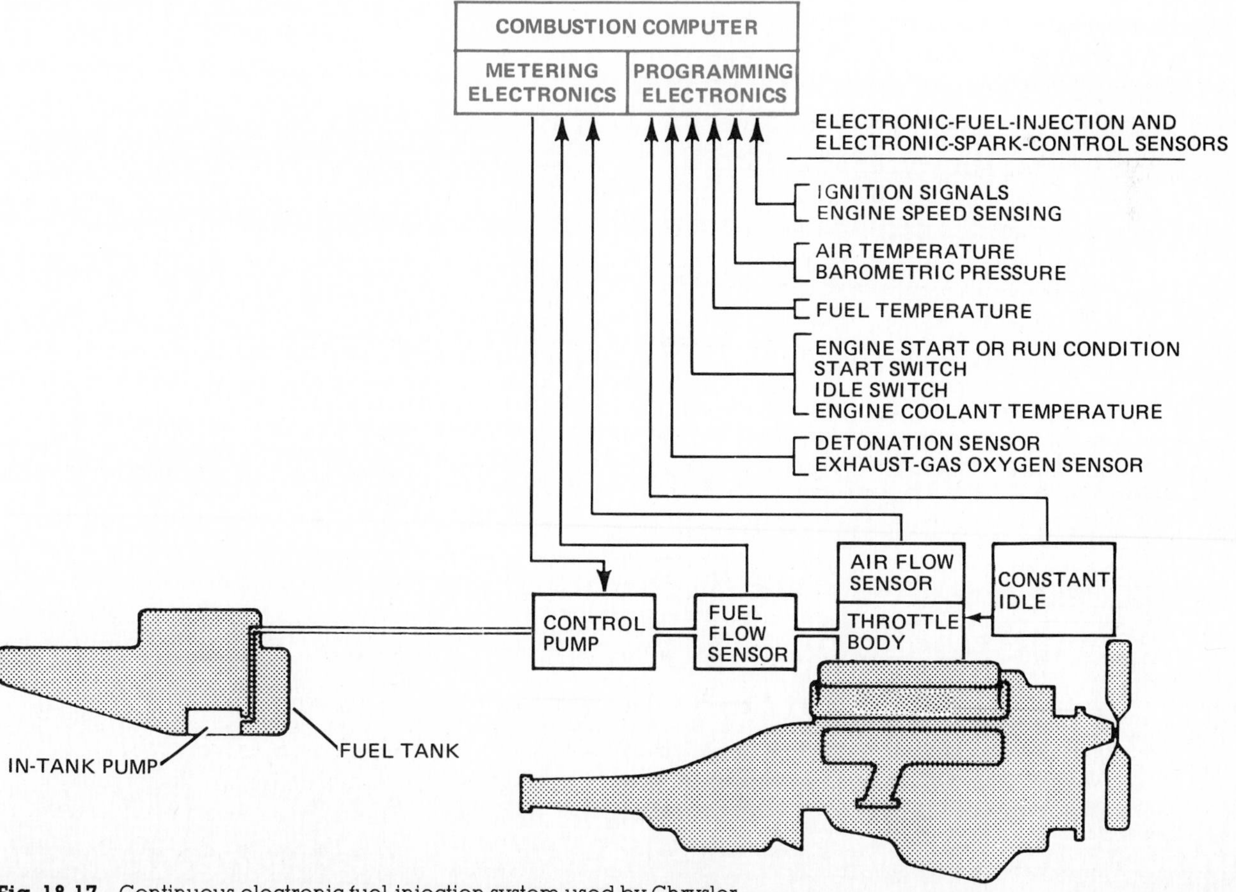

Fig. 18-17 Continuous electronic fuel-injection system used by Chrysler in some cars. The system includes oxygen and detonation sensors. *(Chrysler Corporation)*

Possible cause	Engine does not start in cold condition	Engine does not start in warm condition	Engine starts poorly in cold condition	Engine starts poorly in warm condition	Irregular idle during warm-up	Irregular idle with warm engine	Engine backfires into intake manifold	Engine backfires into exhaust system	Engine misses when driven on road	Driving performance unsatisfactory	Engine runs on	Fuel consumption too high	CO concentration at idle too high	CO concentration at idle too low	Idle speed cannot be adjusted	Check or correction
Electric fuel pump not operating	●	●														Check pump fuse, pump relay, and pump
Loose contact at electric fuel pump	●	●							●							Check pump wiring
"Cold" control pressure outside tolerance	●		●		●											Test pressure
"Warm" control pressure too high		●		●		●	●			●				●		Test pressure
"Warm" control pressure too low		●		●		●		●		●		●	●			Test pressure
Auxiliary air valve does not close				●	●	●						●			●	Check valve for correct function
Auxiliary air valve does not open	●		●		●											Check valve for correct function
Cold start valve does not open	●		●													Check cold start valve
Cold start valve leaking	●		●	●	●	●		●		●	●	●	●			Check cold start valve
Primary system pressure out of tolerance									●							Test pressure and adjust with shims
Air flow sensor plate stop incorrectly set	●	●	●	●												Check and reset
Sensor plate and/or plunger not moving freely	●	●	●	●		●				●	●		●			Check for free movement
Leaks in air intake system (false air)	●	●	●	●	●	●	●			●				●		Check air system for leaks
Fuel system leakage	●	●	●	●	●	●		●	●			●	●			Inspect fuel system for leaks
Injector leaking, opening pressure low		●		●	●	●					●					Check injectors on tester
Idle mixture too rich		●		●		●		●		●		●	●			Check and adjust CO level
Idle mixture too lean		●		●		●	●			●				●		Check and adjust CO level
Throttle butterfly does not open completely										●						Check butterfly and stops in throttle venturi
Thermo-time switch defective	●		●													Test for resistance readings vs. temperature

Fig. 18-18 Trouble-diagnosis chart for the Bosch K-type fuel-injection system. *(Robert Bosch Corporation)*

♦ 18-15 SERVICING GASOLINE FUEL-INJECTION SYSTEMS

When a car equipped with any type of gasoline fuel-injection system has a problem, first be certain that the ignition system or the engine itself is not causing it. In troubleshooting, the fuel-injection system usually is the last place to look when an engine is not running properly. Basically, any fuel-injection system has the same job as the carburetor it replaces—to supply the cylinders with the proper mixture of air and fuel. How the fuel-injection system does this depends on its design.

Although the fuel-injection systems in use today were designed by Bendix and Bosch, some car manufacturers make their own parts and have further adapted the systems to particular engines. Therefore, not all installations of the same type of fuel-injection system are identical in operation or appearance. For this reason, you should always have the manufacturer's service manual when you troubleshoot or service a fuel-injection system. When you have determined that the ECU has failed, replace it.

♦ 18-16 SERVICING BOSCH K-TYPE CONTINUOUS INJECTION SYSTEM

This system has been installed on cars built by Audi, BMW, Mercedes-Benz, Porsche, Saab, Volkswagen, and Volvo. Before attempting to check out or service this system using tools or test instruments, always make a complete and thorough visual inspection. First, unplug the leads from the warm-up regulator and auxiliary-air device. This prevents the heating elements in these parts from overheating. Now, you are ready to begin the visual inspection.

The two major problems to look for when inspecting a gasoline fuel-injection system are fuel leaks and air leaks. Either of these can cause driveability problems, and possibly engine damage and excessive emissions. A fuel leak is a fire hazard.

1. Fuel Leaks Check carefully that there are no fuel leaks from any fuel-line connection, from around the electric fuel pump, or from the fuel accumulator. Heavy layers of dirt may make fuel leaks and their source difficult to find. New seals should be installed in any leaking connections, and any defective hoses should be replaced.

Disconnect the fuel lines from the fuel distributor. Use a flare-nut wrench to disconnect the control pressure connection. Use of regular open-end wrenches on these fittings may damage the adjacent fuel lines.

Some hoses are equipped with screw-type joints. On these, the complete fuel-hose assembly, including the nipples and ring connectors, must be replaced. The complete hose assembly is available through the car dealer's parts department.

Clean all fuel-line connections thoroughly before they are opened. Always install new seals when the connections are restored. When installing the mixture-control unit, tighten the mounting screws evenly. When detaching or tightening the tubing at the injection valves, use a wrench to hold the hexagonal part stationary.

2. Air Leaks Air leaks must not occur in the air-intake system between the mixture-control unit and the engine. Air leaks into the intake manifold cause an excessively lean air-fuel mixture to be delivered to the cylinders. This is because air drawn in through the leak is not metered by the airflow sensor.

Air leaks can occur at several places. These include the connection between the mixture-control unit and the intake manifold, the seal at the flange of the start valve, all hose connections at the intake manifold and at the auxiliary-air device, the seal ring at the injection valves, and the intake-manifold support at the cylinder head.

To test the air-intake system for leaks, remove one air hose from the auxiliary-air regulator. Pressurize this hose with air while holding the throttle open. No air should leak out. Use soapy water on all air-hose and manifold connections to detect leaks. Replace any defective hoses or seals.

Figure 18-18 is a trouble-diagnosis chart for the Bosch K-type continuous fuel-injection system. The chart lists the most frequent complaints and their possible causes and the checks or corrections to be made. Many of the corrections are self-explanatory and can be performed quickly and easily. However, others require you to use testers. Then you must refer to the manufacturer's service manual for the car you are servicing.

♦ 18-17 SERVICING BOSCH L-TYPE ELECTRONIC FUEL INJECTION

This system is found on various models of BMW, Opel, Porsche, Renault, and Volkswagen. It can be checked with a voltmeter, ohmmeter, pressure gauge, and basic hand tools by following the steps in the manufacturer's service manual. As in the D type, the electrical measurements are made at the end of the wiring harness after it is removed from the ECU. However, some tests may be made at the terminals on the individual components. As with other electronic fuel-injection systems, most of the components cannot be repaired. Any component found to be defective must be replaced. A trouble-diagnosis chart for the L-type system is shown in Fig. 18-19.

Listed below are six service cautions that must be followed when working on cars equipped with Bosch L-type system. This is to prevent damage to electronic devices on the car if an improper procedure is used.

1. Never jump the battery to start the car.
2. Never disconnect the cables from the battery with the engine running.
3. Always disconnect the cable from the negative (grounded) terminal of the battery before charging it.
4. Never remove or attach the wiring-harness plug to the ECU with the ignition on.
5. When cranking the engine to check compression, unplug the red cable from the battery to the relays.
6. Before testing the L-type system, make sure that the ignition timing and dwell are within specifications, and that the spark plugs are firing properly.

♦ 18-18 SERVICING DIGITAL ELECTRONIC FUEL INJECTION

In this system (Fig. 18-20), the ECM performs certain diagnostic and backup-system functions. The ECM detects certain system and component problems and identifies them through a coded digital readout. On some cars, diagnostic codes appear as a digital display on the instrument panel (Fig. 18-21). Other cars have only a CHECK ENGINE light (Fig. 18-22).

When the ECM detects an improper sensor signal, indicat-

Engine cranks but does not start	Engine starts but then dies	Rough or unstable idle	Idle speed incorrect	CO value incorrect	Erratic running	Engine misses when driving	Fuel consumption too high	No maximum power	POSSIBLE CAUSE	CHECK OR CORRECTION
●	●	●	●		●	●	●	●	Defect in ignition system	Check battery, distributor, plugs, coil, and timing
●	●	●	●		●	●	●	●	Mechanical defect in engine	Check compression, valve adjustment, and oil pressure
●	●	●	●	●	●	●		●	Leaks in air intake system (false air)	Check all hoses and connections; eliminate leaks
●	●	●						●	Blockage in fuel system	Check fuel tank, filter, and lines for free flow
●									Relay defective; wire to injector open	Test relay; check wiring harness
●									Fuel pump not operating	Check pump fuse, pump relay, and pump
●	●	●		●	●		●	●	Fuel system pressure incorrect	Check pressure regulator
●									Cold start valve not operating	Test for spray, check wiring and thermo-time switch
●	●	●	●	●	●		●		Cold start valve leaking	Check valve for leakage
●									Thermo-time switch defective	Test for resistance reading vs. temperature
●	●	●	●						Auxiliary air valve not operating correctly	Must be open with cold engine; closed with warm
●	●	●		●			●		Temperature sensor defective	Test for 2–3 kΩ at 68°F [20°C]
●	●	●	●	●	●		●	●	Air flow meter defective	Check pump contacts; test flap for free movement
		●	●					●	Throttle butterfly does not completely close or open	Readjust throttle stops
		●						●	Throttle valve switch defective	Check with ohmmeter and adjust
	●	●	●	●					Idle speed incorrectly adjusted	Adjust idle speed with bypass screw
		●		●	●	●		●	Defective injection valve	Check valves individually for spray
	●	●	●	●	●		●		CO concentration incorrectly set	Readjust CO with screw on air flow meter
●	●	●			●	●			Loose connection in wiring harness or system ground	Check and clean all connections
●	●	●				●		●	Control unit defective	Use known good unit to confirm defect

Fig. 18-19 Trouble-diagnosis chart for the Bosch L-type electronic fuel-injection system. *(Robert Bosch Corporation)*

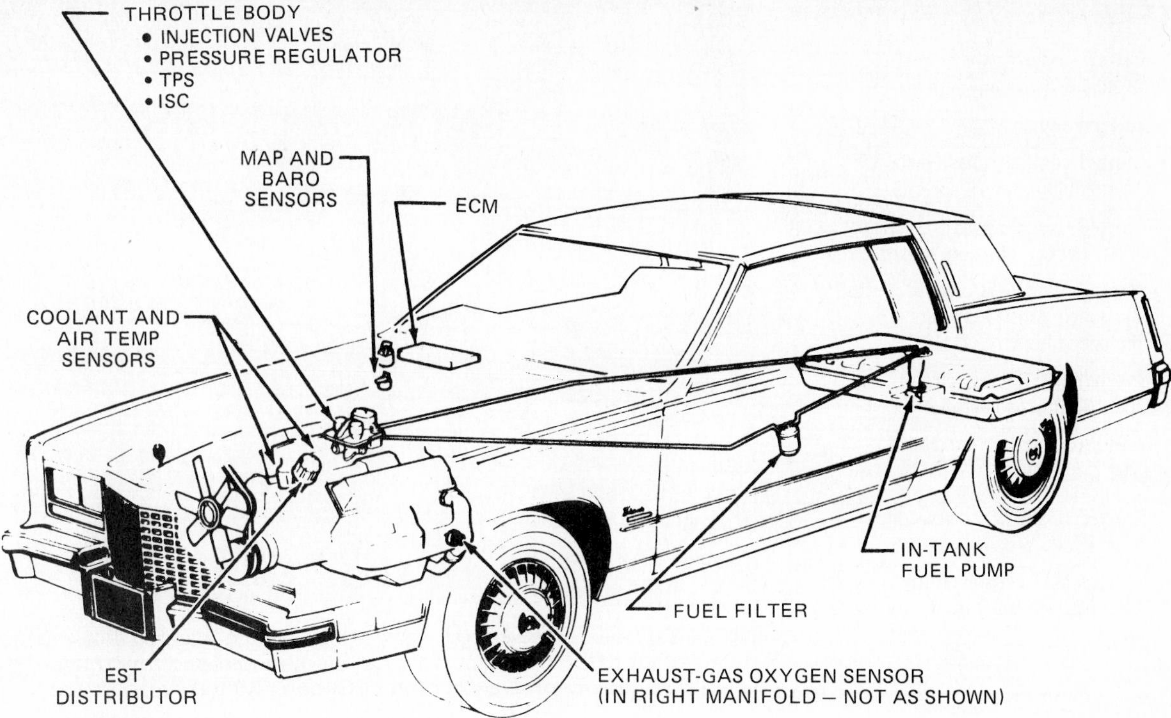

Fig. 18-20 Location of the components in the digital fuel-injection (DFI) system. *(Rochester Products Division of General Motors Corporation)*

ing a failure, it takes over the job of the defective sensor. Substitute values from its stored memory replace the missing information from the sensor. If the ECM itself has failed, an analog backup circuit takes over. This allows the car to be driven, but with severely reduced performance. The ability to determine the cause of its own trouble makes this a self-diagnosing system.

An amber dash-mounted CHECK ENGINE light informs the driver that the ECM has detected a system malfunction (Fig. 18-22). The condition may be related to the various sensors or to the ECM itself. If the fault clears up, the light resets automatically. However, the ECM stores the trouble code associated with the failure until the diagnostic system is cleared by the mechanic.

To read out the stored failure codes on a car with a digital display, the mechanic depresses the OFF and WARMER buttons on the ECM panel (Fig. 18-21). Then the code "88" will appear. This indicates the start of the diagnostic readout. Trouble codes stored in the ECM as a result of troubles that have occurred will now display, beginning with the lowest-numbered code. Figure 18-22 shows how the CHECK ENGINE light flashes the trouble code.

In addition, the ECM is programmed to monitor the operation of several switches on the car. Figure 18-23 shows the codes for trouble with the sensors and ECM. Diagnosis charts in the manufacturer's service manual provide the steps to pinpoint and repair the defects.

88
COOLER
WARMER
ELECTRONIC CLIMATE CONTROL
OFF

Fig. 18-21 Digital trouble-code display ("88") on the instrument panel. *(Cadillac Motor Car Division of General Motors Corporation)*

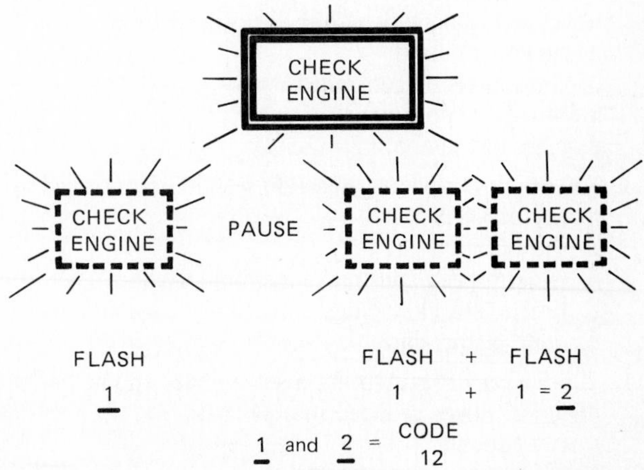

Fig. 18-22 Operation of the CHECK ENGINE light used with the GM computer command control (CCC) system. By grounding a terminal located under the instrument panel, the CHECK ENGINE light will start to flash. These flashes indicate a code (or codes). For example, a code 12 consists of one flash followed by a pause and then two more flashes. *(Rochester Products Division of General Motors Corporation)*

Trouble Code	Circuit Affected
00	All diagnostics complete
12	No tach signal
14	Shorted coolant sensor circuit
15	Open coolant sensor circuit
21	Shorted throttle position sensor circuit
22	Open throttle position sensor circuit
28	Shorted idle speed control circuit
29	Idle speed control circuit
30	Idle speed control circuit
31	Shorted MAP sensor circuit
32	Open MAP sensor circuit
33	MAP/BARO sensor correlation
34	MAP hose
35	Shorted BARO sensor circuit
36	Open BARO sensor circuit
37	Shorted MAT sensor circuit
38	Open MAT sensor circuit
55 56	ECM

Fig. 18-23 Trouble-code chart, showing the circuits affected. Refer to the manufacturer's service manual for further diagnosis and repair information. *(Cadillac Motor Car Division of General Motors Corporation)*

♦REVIEW QUESTIONS♦

Select the *one* correct, best, or most probable answer to each question. You can find the answers in the section indicated at the end of each question.

1. In the gasoline fuel-injection system, the gasoline is sprayed into the air (♦18-1)
   a. in the combustion chambers
   b. before it reaches the intake valve
   c. after it passes the intake valve
   d. all of the above
2. Gasoline fuel-injection systems can be classified in two ways, according to whether they are (♦18-2)
   a. timed or pulsed
   b. continuous or controlled
   c. timed or continuous
   d. none of the above
3. The Bosch D-type fuel-injection system is controlled primarily by (♦18-4)
   a. the ECU
   b. engine speed and intake-manifold vacuum
   c. timers and glow plugs
   d. none of the above
4. The typical EFI system for a six-cylinder engine has the injection valves grouped into (♦18-5)
   a. two groups
   b. three groups
   c. six groups
   d. none of the above
5. The typical gasoline fuel-injection system for an eight-cylinder engine has the injection valves grouped into (♦18-5)
   a. two groups
   b. three groups
   c. four groups
   d. none of the above
6. The amount of fuel injected by the timed injection system depends on (♦18-6)
   a. when the injection valves open
   b. how far the injection valves open
   c. how long the injection valves stay open
   d. all of the above
7. The four factors that control the length of time the fuel-injection valves stay open are throttle position and (♦18-5)
   a. intake-manifold vacuum, air temperature, and coolant temperature
   b. intake-manifold temperature, manifold vacuum, and coolant temperature
   c. amount of oxygen in the exhaust, intake-manifold vacuum, and coolant temperature
   d. none of the above

8. In the mechanical continuous injection system, the amount of fuel being injected is controlled by (♦18-11)
   a. electronic solenoids
   b. the amount of air flowing into the engine
   c. the amount of oxygen in the exhaust
   d. all of the above
9. The gasoline fuel-injection system used by Cadillac from 1975 to 1980 has (♦18-5)
   a. one injection valve for each cylinder
   b. one injection valve for each two cylinders
   c. one injection valve for each bank of cylinders
   d. one injection valve for the engine
10. Throttle-body injection has (♦18-8)
    a. one injection valve for each cylinder
    b. one or two injection valves for the engine
    c. one injection valve for each spark plug
    d. none of the above
11. In the digital fuel-injection system, the injectors are mounted (♦18-9)
    a. in the valve port
    b. in the intake manifold
    c. above the throttle valves
    d. below the throttle valves
12. In the Bosch K-type continuous fuel-injection system, air leaks into the intake manifold cause (♦18-16)
    a. a rich mixture
    b. a lean mixture
    c. no change in the mixture
    d. the airflow sensor to compensate for the leaks
13. The Cadillac electronic fuel-injection system used on 1975 to 1980 models is most similar to (♦18-5)
    a. the Bosch type D
    b. the Bosch type L
    c. the Bosch type K
    d. digital electronic fuel injection
14. When the electronic control unit (ECU) has failed, the normal procedure is to (♦18-17)
    a. replace the ECU
    b. adjust the ECU
    c. rewire the ECU
    d. send the ECU to a rebuilder
15. The opening of the solenoid-operated injection valve can be checked with (♦18-13)
    a. a pressure gauge
    b. a dynamometer
    c. a voltmeter
    d. an oscilloscope

CHAPTER 19
Diesel Fuel-Injection Systems: Operation and Service

After studying this chapter, you should be able to:

1. Explain the difference in operation between diesel (compression-ignition) and gasoline (spark-ignition) engines.
2. Describe the two basic types of fuel-injection pumps used on automotive diesel engines.
3. Locate the components of diesel-engine fuel-injection systems on cars and explain how each component works.
4. Explain how to replace and time a fuel-injection pump.
5. Diagnose and service various automotive diesel-engine fuel-injection systems.

♦ 19-1 DIESEL ENGINES

Diesel engines have been used for many years in trucks, buses, and off-the-road equipment. The diesel engine has a reputation for being a reliable heavy-duty power plant. In recent years, automotive manufacturers have developed diesel engines designed especially for use in passenger cars.

The operation of diesel engines is described in ♦9-16. Air alone is taken into the engine cylinder during the intake stroke. On the compression stroke, the air is highly compressed and reaches temperatures as high as 1000°F [538°C]. Toward the end of the compression stroke, diesel fuel is sprayed into the combustion chamber. The high temperature of the air ignites the fuel. It burns and the power stroke follows. Then the exhaust stroke occurs, followed by the next intake stroke. Figure 19-1 shows the sequence.

♦ 19-2 COMPARING THE PISTON STROKES

A comparison of the piston strokes of the compression-ignition (diesel) and the gasoline (spark-ignition) engines follows.

1. Intake Stroke The diesel engine is taking in air alone. There is no throttle valve to impede the flow of air into the cylinder. In the spark-ignition engine, a mixture of air and fuel is taken into the cylinder during the intake stroke.

2. Compression Stroke In the diesel engine, air alone is compressed during the compression stroke. In the gasoline engine, a mixture of air and fuel is compressed.

3. Power Stroke In the diesel engine, ignition of the fuel as it is sprayed (injected) into the combustion chamber is produced by the compressed and superheated air. In the gasoline engine, ignition of the compressed air-fuel mixture is produced by a spark from the spark plug.

4. Exhaust Stroke The exhaust stroke is essentially the same for both engines. The exhaust valve opens and the burned gases flow out of the cylinder as the piston moves up toward TDC.

To sum up, the diesel engine:

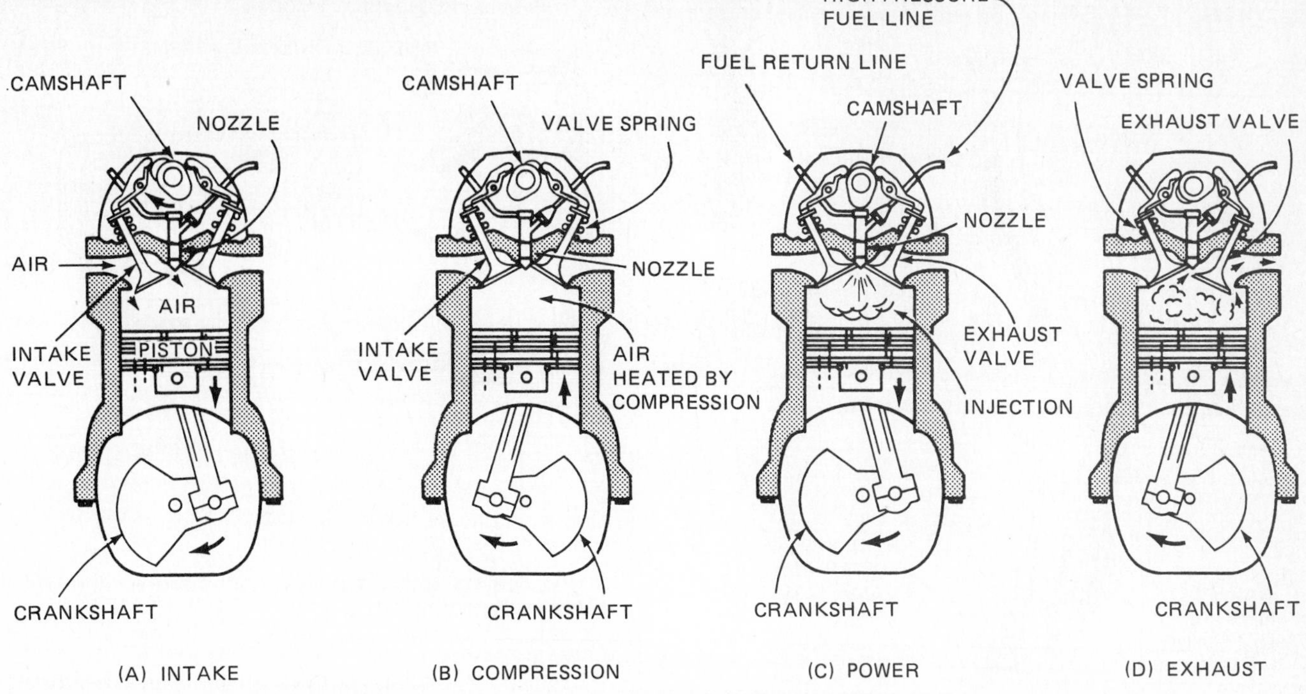

Fig. 19-1 The four piston strokes in a four-stroke-cycle diesel engine.

1. Has no throttle valve to restrict air flow into the engine (except for some Mercedes engines, described in ♦19-10).
2. Compresses only air on the compression stroke.
3. Has a much higher compression ratio.
4. Does not have an electric ignition system. Instead, heat of compression ignites the fuel as it is sprayed into the cylinders.
5. Engine power and speed are controlled only by the amount of fuel sprayed into the cylinders. For more power, more fuel is injected. For less power, less fuel is injected.
6. Has glow plugs which make it easier to start a cold engine (♦19-12).

♦ 19-3 DIESEL FUEL-SYSTEM REQUIREMENTS

The diesel fuel system must:

1. Deliver the exactly right amount of fuel to meet the operating requirements.
2. Change the timing of fuel delivery as engine speed changes. As engine speed increases, the fuel delivery must start earlier. This compares with the advance of the spark in the gasoline engine as engine speed increases. The purpose is the same, to get the ignition of the fuel started before the piston reaches TDC. Without an advance, the piston would be over TDC and starting down on the power stroke before ignition was well started. The piston movement would keep ahead of the pressure rise so that most of the power in the fuel would be wasted.
3. Deliver the fuel to the cylinders under high pressure. The pressure in the cylinder at the end of the compression stroke is more than 500 psi [3447 kPa]. The fuel must be under pressure much higher than this for it to be sprayed into the compressed air.

♦ 19-4 NEED FOR CLEAN DIESEL FUEL

Diesel fuel must be clean and free of water (♦14-15). Even almost invisible dirt particles can clog the injection nozzles and cause poor engine performance. Also, water can rust interior fuel-pump and injection-nozzle parts. Many cars and trucks with diesel engines have a *water-detection system*.

♦ 19-5 WATER IN DIESEL FUEL

Diesel fuel oil will absorb water freely. Water in the diesel fuel system can damage the fuel-injection system and cause it to fail. Many vehicles have a water detector located in the fuel tank (Fig. 19-2). This device is mounted on the tank fuel-pickup tube. When from 1 to 2½ gallons [3.8 to 9.5 L] of water has collected in the bottom of the fuel tank, the detector circuit is completed (through the water). This turns on an instrument-panel light which warns the driver WATER IN FUEL.

When this light comes on, the water should be removed with a pump or by syphoning. The pump or syphon should be hooked up to the fuel-return hose. In many cars, this connection can be made above the rear axle or under the hood near the fuel pump. Syphoning should continue until fuel starts to come out. Then you have removed all the water.

Careful Remove the fuel-tank cap during this operation. Then reinstall it when the job is done.

The tank fuel pickup shown in Fig. 19-2 also includes a check valve. It will permit fuel to pass if the tank filter becomes plugged with wax during cold weather. This plugging can occur at temperatures below 20°F [−6°C] if other than number 1 diesel fuel is used. When the plugging occurs, the last 4 gallons [15 L] of fuel will not be used due to the location of the check valve. When using number 2 fuel, keep the fuel level in the tank above the ¼ mark to prevent running out of fuel.

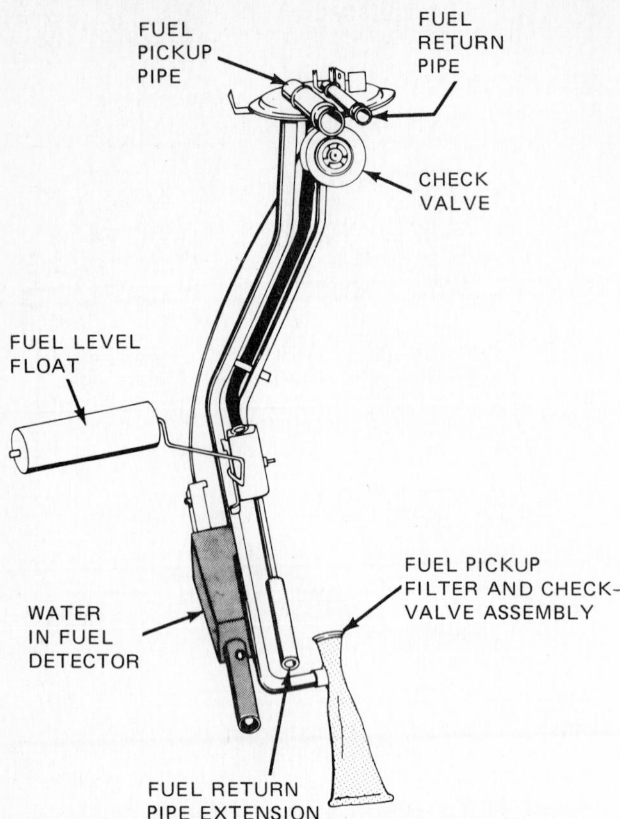

Fig. 19-2 Diesel fuel gauge with water-in-fuel detector, mounted on the bottom of the fuel-pickup tube. *(Buick Motor Division of General Motors Corporation)*

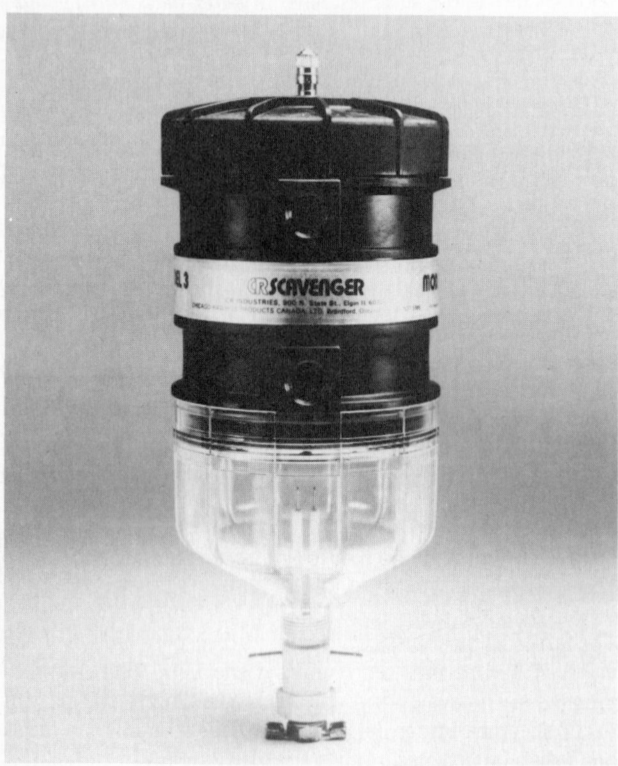

Fig. 19-3 Diesel fuel-water separator. Water separates from the fuel and collects in the bottom of the separator. The water can then be drained from the separator by opening the valve at the bottom. *(CR Industries)*

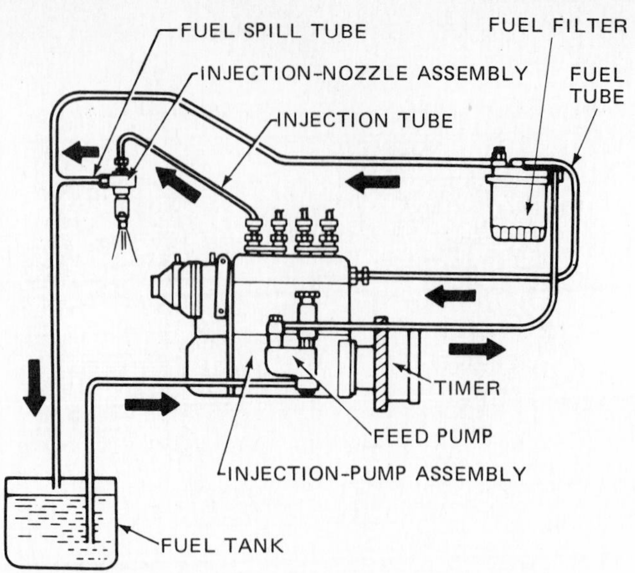

Fig. 19-4 Fuel system for a four-cylinder diesel engine. *(Nissan Motor Company, Ltd.)*

On vehicles without a water-detection sensor, a *fuel-water separator* (Fig. 19-3) can be installed. The fuel passes through the separator. Any water in the fuel settles to the bottom of the separator. Then the water can be drained out as necessary.

♦ 19-6 TWO DIESEL-ENGINE FUEL SYSTEMS

There are two basic types of systems used to feed the fuel to the engine. In one, a centrally located pump pressurizes the fuel, meters it, times it, and delivers it to the cylinders through tubes.

In the other system, the fuel is sent to fuel injectors under a relatively low pressure. The injectors have cam-operated plungers (like the valves) which are adjustable. At the proper instant the cams operate the plungers which force fuel at high pressure into the cylinders.

The centrally located pump system is used on a majority of diesel engines. It can be divided into two types, the cam-operated in-line plunger type and the rotary-distributor type.

♦ 19-7 CAM-OPERATED IN-LINE PLUNGER PUMP

This pump is used on some four- and six-cylinder automotive diesel engines, and on many larger diesel engines in trucks and tractors. The pump has a *barrel-and-plunger assembly* for each cylinder. For example, the in-line plunger pump for a four-cylinder engine has four separate barrel-and-plunger assemblies. The top of each barrel is connected by a separate injection tube to an injection nozzle in each engine cylinder (Fig. 19-4). Most plunger pumps are gear-driven from the engine crankshaft (Fig. 19-5). Figure 19-6 shows the complete fuel-injection system using a plunger pump installed on a six-cylinder engine.

Figure 19-7 shows the construction of a plunger pump. A fuel gallery in the pump keeps the space between the end of the plunger and the delivery valve filled with fuel. When the cam lobe comes up under the plunger, the plunger is raised.

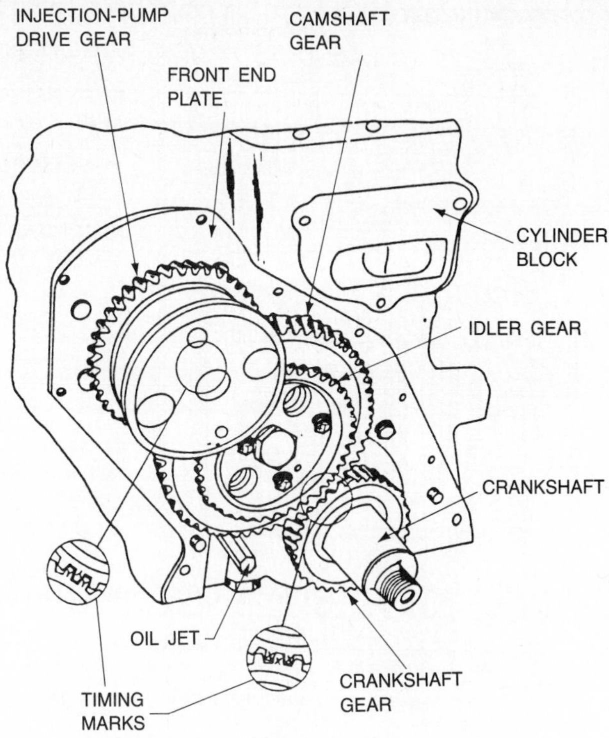

Fig. 19-5 Drive arrangement for an in-line plunger-type fuel-injection pump. *(International Harvester Company)*

This forces the fuel out of the barrel at high pressure. The fuel flows through the delivery valve to the injection nozzle in the engine cylinder (Fig. 19-6).

The diesel-engine injection pump has a speed-advance mechanism that advances the time of injection as engine speed increases. The speed advance supplies the fuel earlier in the compression stroke (toward the end of the stroke). This gives the fuel enough time to ignite and begin burning so that it provides an even pressure rise against the piston head. Without this advance, the piston would be up over the top (TDC) and moving down before the fuel was well ignited. The result would be that the piston would be moving down away from the pressure rise. A weak power stroke would result and fuel would be wasted. The advance is controlled by a governor that is built into the pump (♦19-10).

The amount of fuel injected is varied by varying the effective length of the plunger strokes. When less engine power is needed, the effective length of the plunger strokes is short. Relatively small amounts of fuel are delivered. When more power is needed, the effective length of the plunger strokes is increased. More fuel is delivered. The actual length of plunger travel is always the same. However, the quantity of fuel injected is changed by turning the pump plungers. The top of each plunger has an inclined *helix* machined into it (Fig. 19-8). As the plunger is turned, its effective stroke is varied. The plungers are turned by movement of the control rod. The control rod is connected to the accelerator pedal through the *governor* and linkage. Movement of the accelerator pedal is converted into a corresponding control-rod travel. This rotates the plungers.

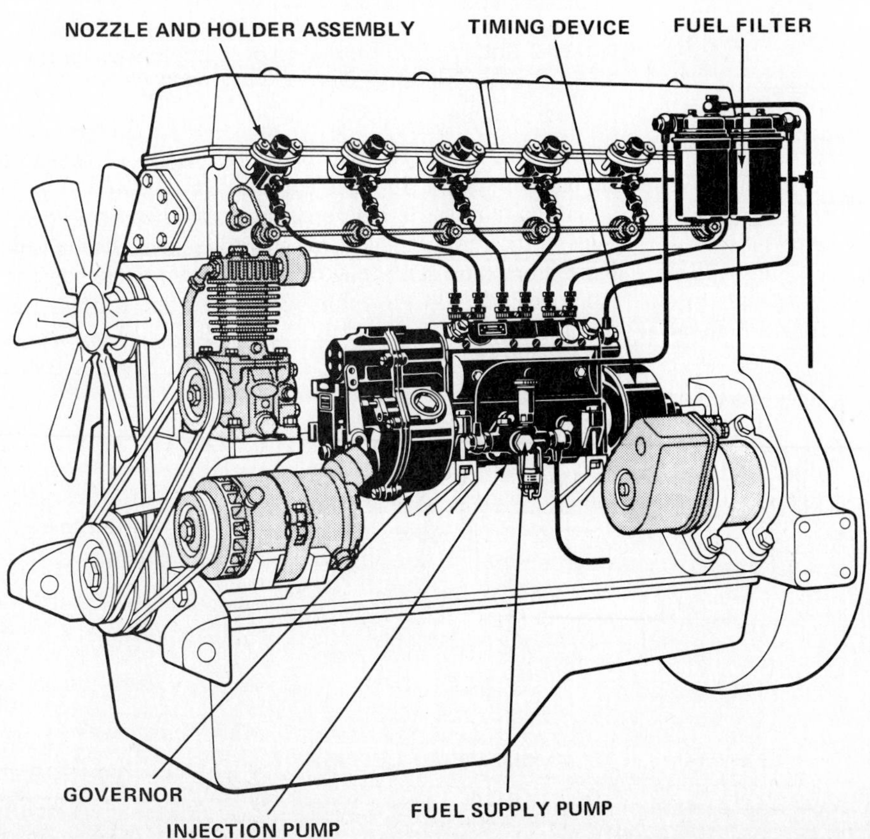

Fig. 19-6 Fuel system for a six-cylinder diesel engine using an in-line fuel-injection pump. *(Robert Bosch Corporation)*

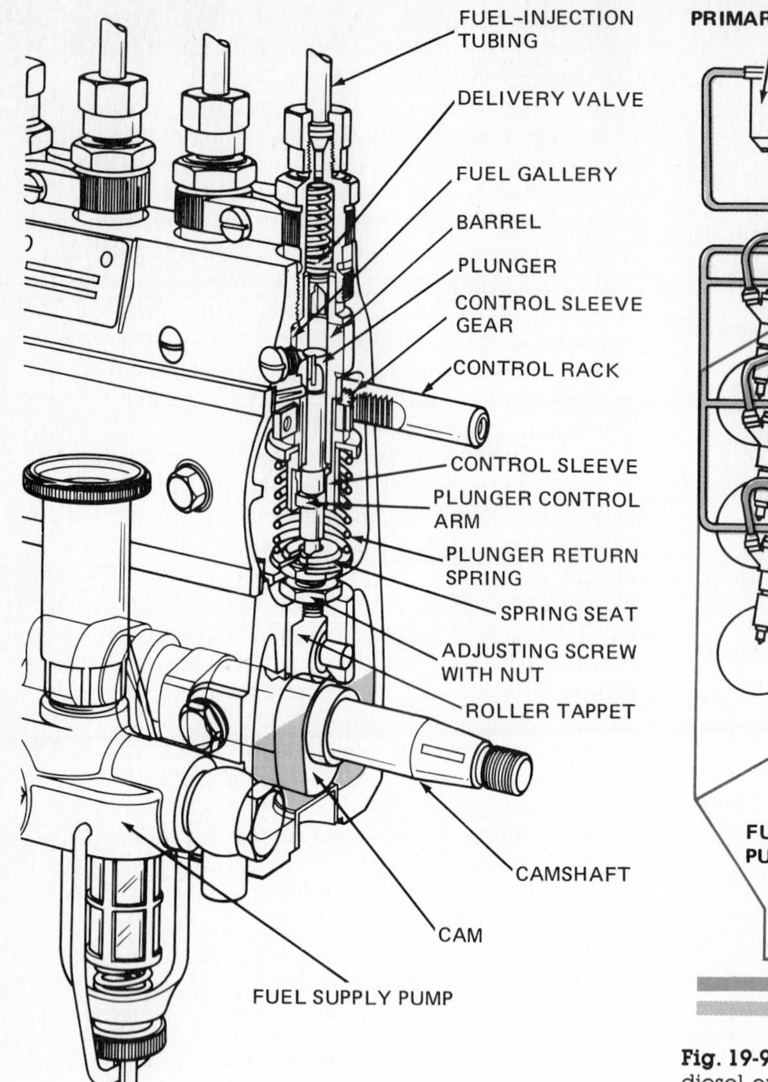

Fig. 19-7 Construction of an in-line plunger-type fuel-injection pump. *(Robert Bosch Corporation)*

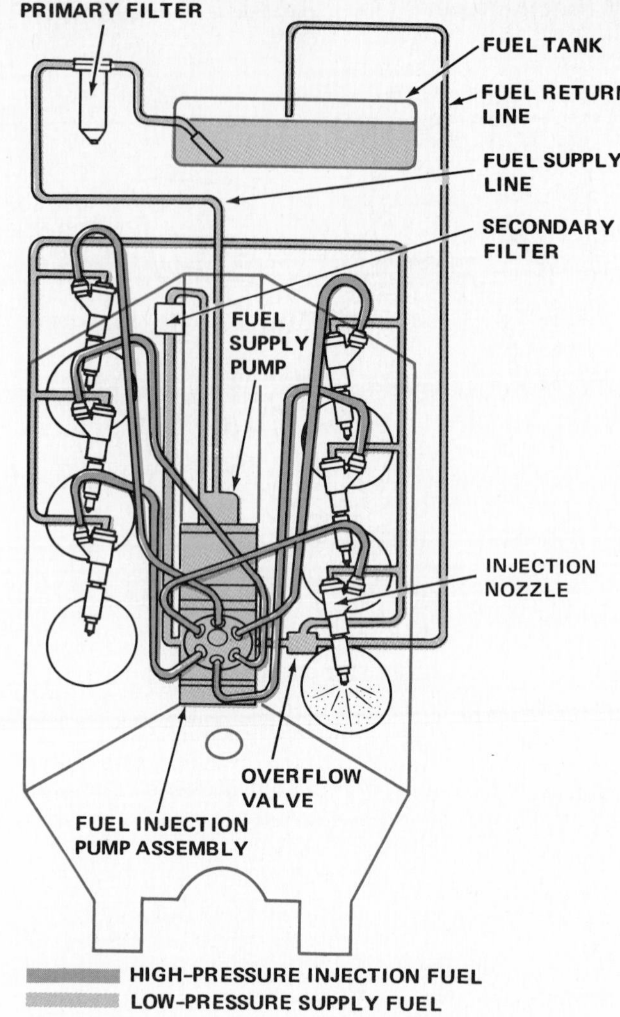

Fig. 19-9 Rotary-distributor injection-pump system for a V-6 diesel engine. *(General Motors Corporation)*

♦ 19-8 ROTARY-DISTRIBUTOR PUMP

This pump compares, in some ways, to the ignition distributor used with spark-ignition (gasoline) engines. Figure 19-9 shows the fuel system for a V-6 diesel engine that uses this pump. Figure 19-10 is a partial cutaway view of a V-8 diesel engine using a rotary-distributor pump. Figure 19-11 shows the pump removed from the engine.

The ignition distributor sends sparks to the engine spark plugs as the rotor inside the distributor rotates. The rotary-distributor pump sends fuel to the injection nozzles in the engine cylinders as the rotor inside the pump rotates. The pump mounts on the engine (Figs. 19-9 and 19-10) and is

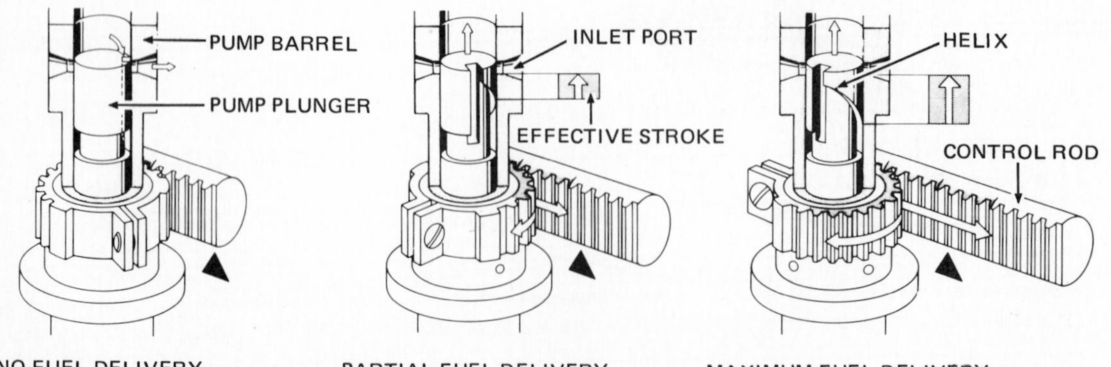

Fig. 19-8 Movement of the toothed control rod turns the pump plungers to vary the amount of fuel injected. *(Robert Bosch Corporation)*

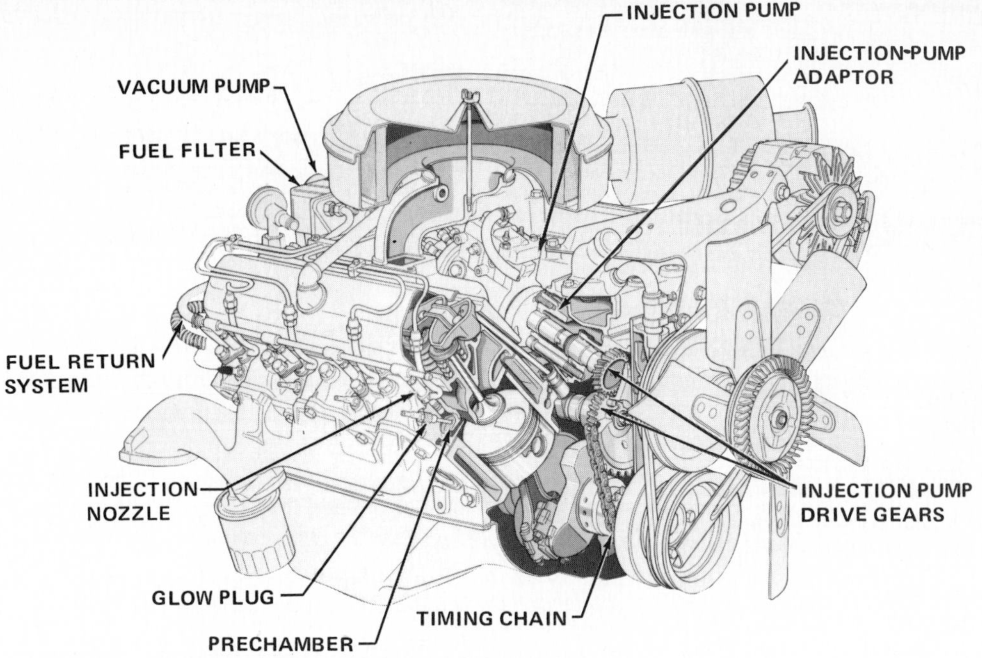

Fig. 19-10 V-8 diesel engine using a rotary-distributor injection pump, partly cut away to show the pump drive arrangement. *(Oldsmobile Division of General Motors Corporation)*

driven from the camshaft by a pair of bevel gears. This causes the pump to turn at one-half crankshaft speed.

Figures 19-12 and 19-13 show how the pump works. The pump has a rotor (Fig. 19-14) that includes a pair of cam rollers and plungers. These rollers roll on an internal cam (lobes on inside, Fig. 19-15). As the rotor rotates, the rollers roll on the inner surface of the cam. They move in and out as they roll over the cam lobes. When they move out, they cause the plungers to move out. This increases the size of the internal chamber. Fuel flows into the chamber.

Then when the plungers are pushed in (rollers meet cam lobes), they apply high pressure to the fuel in the chamber. It is forced out through an opening in the rotor.

This opening indexes with stationary openings in the outer shell of the pump. There are the same number of these openings as there are cylinders in the engine. Each opening is connected by an injection pipe to a nozzle in a cylinder. As the rotor opening indexes with the stationary openings, high-pressure surges of fuel are sent to the cylinder nozzles in the engine firing order.

♦ 19-9 ROTARY-DISTRIBUTOR-PUMP CONTROLS

The rotary-distributor pump has two mechanisms to control the timing and the amount of fuel to be injected. The timing device is connected to the internal cam ring. As engine speed increases, the cam ring is moved ahead. This causes the two plungers to move out and in earlier, advancing the timing of injection (Fig. 19-15). At the same time, the internal gover-

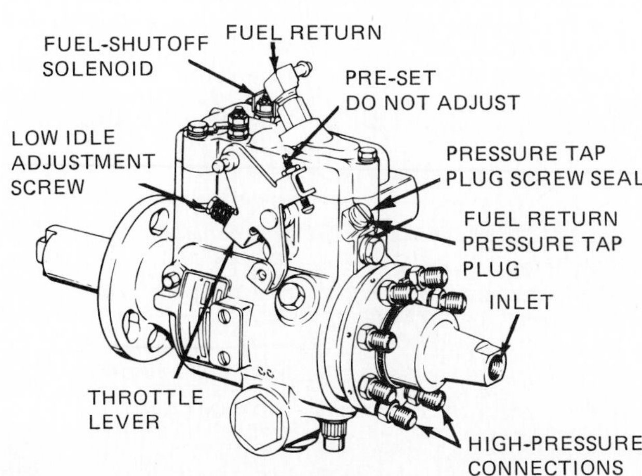

Fig. 19-11 External view of a distributor-type fuel-injection pump. *(Chevrolet Motor Division of General Motors Corporation)*

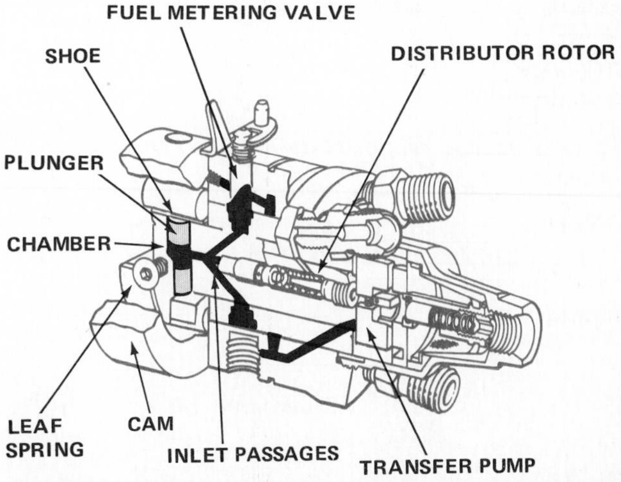

Fig. 19-12 Charging cycle in the distributor pump. The two plungers move apart to cause fuel to enter the chamber. *(Oldsmobile Division of General Motors Corporation)*

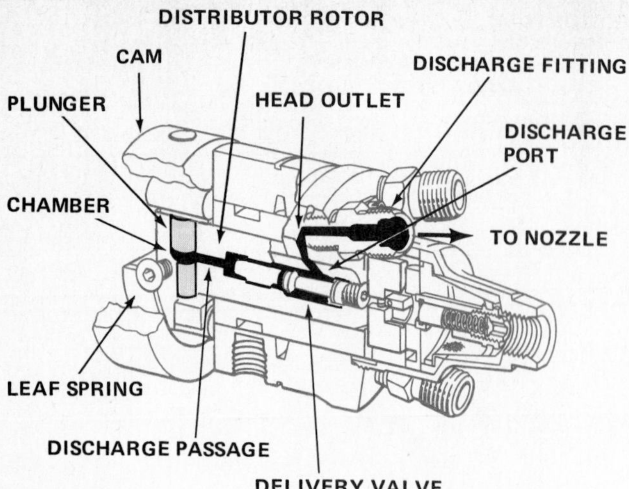

Fig. 19-13 Dischage cycle. The plungers are moving together, forcing fuel from the chamber, past the delivery valve, and through the discharge port. The rotor has turned so that the port in the rotor aligns with the next discharge port. This port is connected through a fuel line to the nozzle in the cylinder in which the piston is approaching TDC on the compression stroke. *(Oldsmobile Division of General Motors Corporation)*

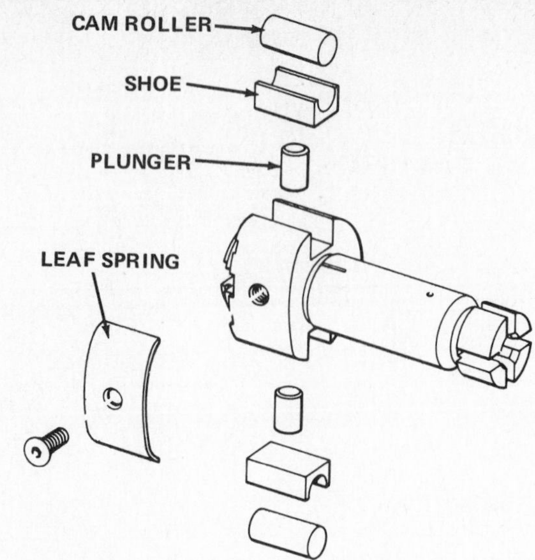

Fig. 19-14 Parts in the injection-pump rotor. *(Chevrolet Motor Division of General Motors Corporation)*

nor regulates the amount of fuel being fed to the engine cylinders.

The governor and the throttle pedal in the driver's compartment are linked to a fuel-metering valve. Movement of the throttle affects the action of the governor. Together, they control the action of the fuel-metering valve. With a light throttle (low speed, low engine power), the governor reduces the amount of fuel being delivered to the engine cylinders. As the throttle is opened for increased engine power, the governor increases the amount of fuel being delivered. The throttle linkage and the governor work together to supply the engine with the correct amount of fuel needed.

♦ 19-10 GOVERNORS FOR FUEL-INJECTION PUMPS

The fuel-injection pump on a diesel engine always has a governor. This device provides automatic control of fuel delivery to meet operating conditions. Actually, it is the governor that directly controls the amount of fuel injected into the engine. The accelerator-pedal movement only changes the setting of the governor.

Without a governor, a diesel engine can stall at low speeds or run so fast it will self-destruct. Automotive diesel engines may use either a mechanical (centrifugal) or a pneumatic governor. Both are variable-speed governors.

Note Constant-speed governors are used for engines that must run at a constant speed, as for example a diesel engine driving an electric generator.

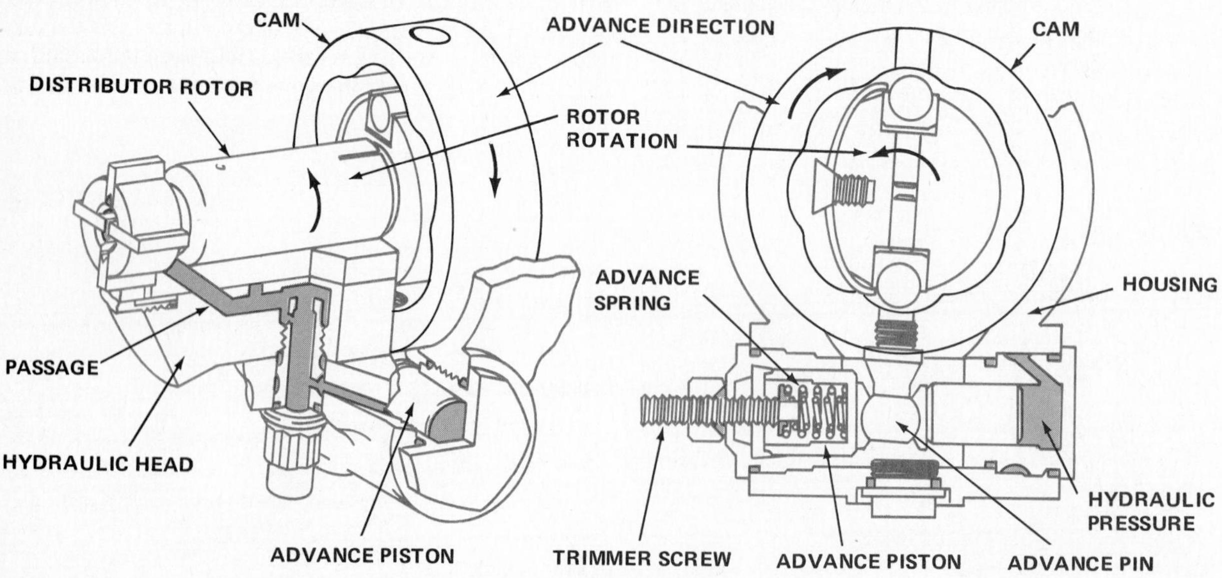

Fig. 19-15 Two views of the automatic advance system, which provides automatic advance of injection timing. Hydraulic pressure, which increases with speed, moves the cam ahead. *(Oldsmobile Division of General Motors Corporation)*

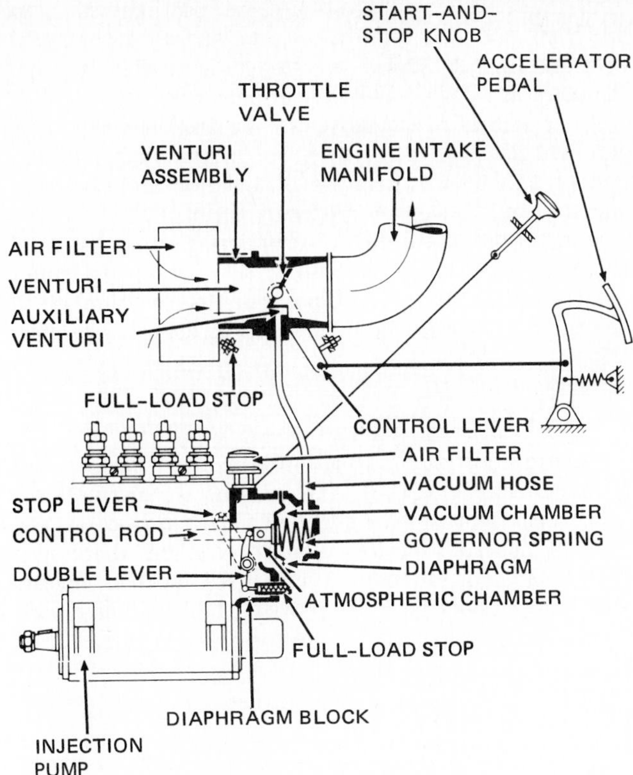

Fig. 19-16 When a pneumatic governor is used, a throttle valve is installed in the intake-air passage into the intake manifold. *(Robert Bosch Corporation)*

A mechanical governor has flyweights that spin with the camshaft. The faster they spin, the further out they move. This acts on the control rod to turn the plungers and adjust the fuel delivery.

Figure 19-16 shows the layout of an engine with a pneumatic governor. Note that there is a throttle valve in the intake manifold. The purpose of this throttle valve is to provide a vacuum signal to the governor. It is not used to control airflow into the engine.

A hose connects from a venturi section in the intake manifold to the vacuum chamber in the diaphragm assembly. The control rod is attached to the vacuum-chamber diaphragm. Any change in vacuum influences the governor. As the throttle opens and closes, the varying vacuum causes the diaphragm and control rod to move. This rotates the plungers to vary the amount of fuel delivered.

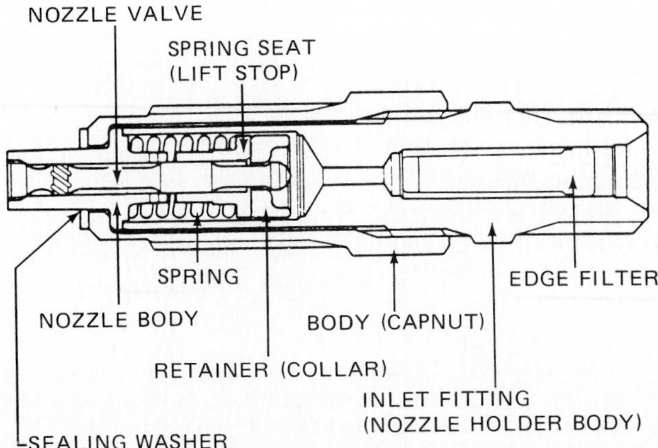

Fig. 19-17 Construction of a diesel-engine injection nozzle. *(Chevrolet Motor Division of General Motors Corporation)*

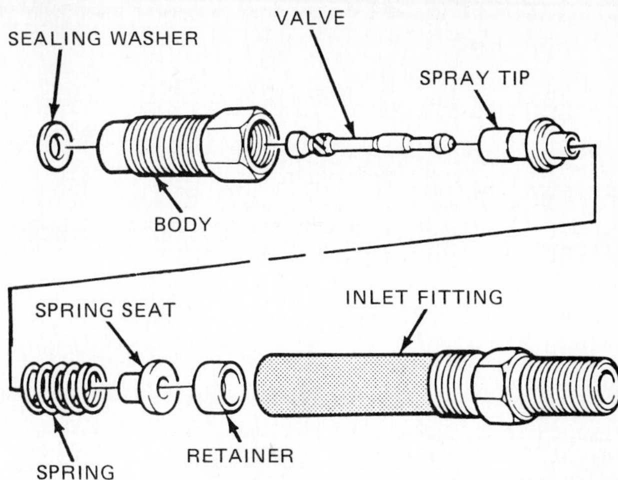

Fig. 19-18 Disassembled diesel-engine injection nozzle. *(Chevrolet Motor Division of General Motors Corporation)*

Some diesel engines are turbocharged (♦15-14). These engines have the governor connected by a tube or hose to the intake manifold. This allows the governor to match fuel delivery with the pressure (the amount of air) in the intake manifold. Air-fuel ratios in a diesel engine range from about 100:1 at idle to about 20:1 under full load. The governor keeps the air-fuel ratio within these limits. An air-fuel ratio richer than 20:1 can cause unacceptable amounts of exhaust smoke.

♦ 19-11 INJECTION NOZZLE

The injection nozzle (Figs. 19-17 and 19-18) has a spring-loaded valve. The valve is held closed by the spring until high fuel pressure is applied, through the injection pipe, to the nozzle. When this happens, the pressure forces the valve off its seat. Then fuel sprays out (is injected) into the engine combustion chamber. The instant that the pressure is relieved, the spring pulls the valve back onto its seat. Then fuel stops spraying from the nozzle.

♦ 19-12 GLOW PLUGS

For easy starting, especially in cold weather, most diesel engines use glow plugs. The glow plugs have electric heating elements that become very hot when connected to the battery. Figure 19-19 shows the location of a plug in a cylinder. Note that it is in a precombustion chamber, close to the fuel-injection nozzle. The precombustion chamber is where the fuel is injected and where the combustion starts. After combustion begins, the burning air-fuel mixture streams out of the precombustion chamber and into the main combustion chamber. There, it mixes with the combustion-chamber air and combustion is completed. In a diesel engine, there is always an excess of air so that combustion of fuel can be relatively complete.

When the engine is cold, and the air temperature is low, the glow plugs are turned on to put some heat into the precombustion chambers. This greatly improves starting because the fuel is sprayed into air that has been preheated by the glow plugs.

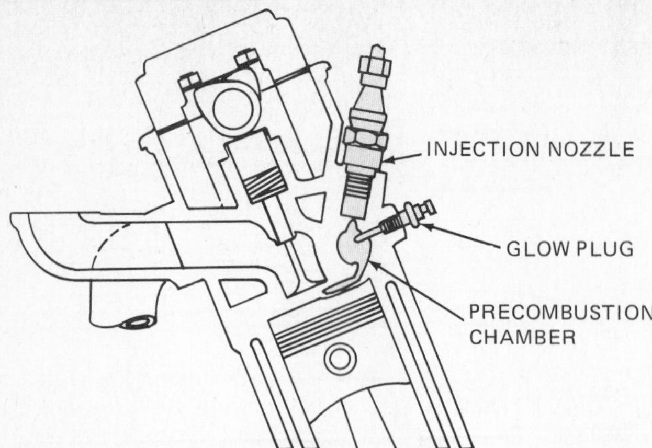

Fig. 19-19 Location of precombustion chamber, glow plug, and injection nozzle in a diesel engine.

♦ 19-13 STARTING INSTRUCTIONS

On some cars, the glow plugs can be turned on by the driver if the driver thinks the engine needs them for starting. On other cars, the glow plugs are turned on automatically during the starting procedure.

Careful Never use starting fluids in an attempt to make starting easier. Some starting fluids can cause serious engine damage.

As an example, here is the starting procedure for the V-type engines used in many General Motors vehicles. The instrument panel has two special lights, WAIT and START. To start the engine:

1. Put the transmission lever in PARK or N. PARK is preferred.
2. Turn the ignition switch (Fig. 19-20) to RUN, not START. Do not turn it to START just yet. When you turn the switch to RUN, an amber WAIT light comes on (if engine is cold). This tells you that the glow plugs are on, heating the engine precombustion chambers.

After the precombustion chambers have been sufficiently heated (usually only a few seconds depending on the temperature), the WAIT light will go out and the START light will come on. Then, push the accelerator pedal halfway down to the floor and hold it there. Turn the ignition switch to START. Normally, the engine will start in only a few seconds. If it does not start in 15 seconds, release the ignition switch. If the WAIT light comes on again, leave the ignition switch in the RUN position until the WAIT light goes off and the START light comes on. Now, try the starting procedure again.

Pumping the accelerator before or during cranking will not aid in starting. The fuel-injection system has no accelerator pump to force additional fuel into the air passing through on its way to the combustion chambers, as does the gasoline-engine fuel system.

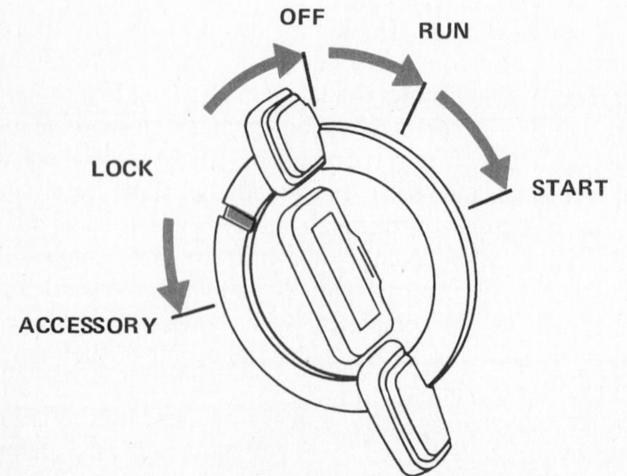

Fig. 19-20 Ignition-switch positions. *(Oldsmobile Division of General Motors Corporation)*

♦ 19-14 OTHER STARTING INSTRUCTIONS

The starting instructions above are for the V-type General Motors diesel engine used in many General Motor vehicles. Other manufacturers have slightly different instructions and different indicating devices. For example, the Volkswagen Rabbit with a diesel engine uses a single light to indicate glow-plug action. The ignition switch has only three positions—OFF, ON, and START. Here is the recommended starting procedure:

1. Temperatures above 32°F [0°C], engine cold—turn ignition switch ON. The glow-plug light will come on and remain on as long as the glow plugs are heating. As soon as the light goes off, the plugs have heated enough and you are ready to start. Turn the ignition switch to START. Do not depress the throttle pedal.
2. Temperatures below 32°F [0°C], engine cold—turn ignition switch to ON. After the glow-plug light goes off, push the throttle all the way open. Pull out the cold-start knob under the instrument panel. This provides extra fuel to the cylinders. Turn the switch to START. Two minutes after starting, push the cold-start knob in.
3. Starting a warm engine—do not depress the throttle. Do not use the glow plugs. Turn the ignition switch past ON to START immediately.

Do not accelerate the engine excessively immediately after starting. Wait for the oil pressure to build up in the lubricating system. Do not operate the starting motor longer than 30 seconds. If the engine does not start, turn ignition switch to OFF. Wait for about 30 seconds. Then pre-glow again and after the glow-plug light goes off, try another start.

♦ 19-15 COOLANT AND FUEL HEATERS

For very cold weather operation, where temperatures get down to zero or below [−18°C or under], coolant and fuel heaters are often used to assist in starting. One type of coolant heater has an electrical element that works off house current (115 volts). It is located in the engine block and has a special electric cord that is plugged into a regular electric outlet.

♦ CAUTION ♦ Make sure to plug into a three-prong outlet. The third prong is the ground. Its use is essential to protect you from electric shock. If the electric cord connected to the heater is not long enough, do not use an ordinary extension cord with only two prongs. This is not heavy enough and does not have the ground wire in it.

ENGINE BLOCK HEATER USAGE			
Oil	32° to 0°F 0°to −18°C	0° to −10°F −18° to −23°C	Below −10°F Below −23°C
30W	Two hours minimum	Eight hours or overnight	Oil not recommended
15W-40	Not required	Two hours minimum	Eight hours or overnight
10W-30	Oil not recommended	Not required	Eight hours or overnight

Fig. 19-21 Chart showing proper usage of the engine-block heater. *(Oldsmobile Division of General Motors Corporation)*

The length of time that the coolant heater should be plugged in depends on the type of oil used and the temperature (Fig. 19-21). The notation OIL NOT RECOMMENDED means that you should not use the oil indicated for the temperature shown.

Careful Do not use "starting aids" such as ether, gasoline, or similar materials in the air intake. These so-called aids can actually delay starting. They can also damage the engine.

Many cars with diesel engines are equipped with a *cold-weather option*. It consists of an engine coolant heater and an in-line fuel heater. This is a metal heating strip wound around the fuel pipe. When the ignition switch is turned on, the heater operates if the fuel temperature is so low that wax may form. Warming the fuel reduces the possibility that wax will plug an engine-mounted fuel filter.

♦ 19-16 VACUUM PUMP

Because there is no throttle valve or venturi in the airstream, there is no vacuum source that can be tapped in the diesel-engine fuel system. Therefore, the diesel engine requires a vacuum pump to provide the vacuum to operate various devices such as power brakes and air-conditioner vacuum doors. On the General Motors automotive diesel engines, the vacuum pump is located at the back of the engine (Fig. 19-10).

♦ 19-17 DIESEL-ENGINE FUEL SYSTEM

The fuel-injection system is so closely tied in with the diesel engine that it is difficult to separate the servicing of the fuel system from the servicing of the engine. For this reason, this chapter describes only the servicing of the fuel-injection nozzles and connecting pipes. Servicing of the injection pump is not covered. This procedure requires special tools and training. If an injection pump is faulty, it usually is exchanged for a new or rebuilt unit.

FUEL-INJECTION SERVICE

♦ 19-18 DIESEL FUEL-SYSTEM TROUBLE DIAGNOSIS

Many diesel-engine troubles can be traced to the fuel system. The chart on the following page lists various engine troubles that could be caused by the fuel system, along with checks or corrections.

♦ 19-19 ENGINE CRANKS NORMALLY BUT WILL NOT START

If the fuel is dirty, it may have clogged the system. This requires cleaning of the pumps, fuel lines, and nozzles. The wrong fuel can also cause failure to start. Use the recommended fuel.

Note Make sure the injection-pump timing mark aligns with the mark on the adaptor. This is to check that the timing is correct.

To check fuel nozzles for fuel flow, loosen the injection line at a nozzle. Do not disconnect it. Wipe the connection dry. Crank for 5 seconds. Fuel should flow from the loose connection. If it does not, check the fuel solenoid and the fuel-supply line to the injection pump. Disconnect the line at the fuel inlet at the injection pump. Connect a hose from this line to a metal container. Crank the engine. If no fuel flows, the trouble is in the fuel-supply line or the fuel-supply pump. If fuel flows, the trouble is in the injection-pump fuel filter or the pump itself. Replace the filter first. If this does not cure the problem, replace the injection pump.

To check for a plugged fuel-return line, disconnect the line at the injection pump and connect this line to a metal container. Connect a hose from the injection pump connection to the metal container. Crank the engine. If it starts and runs, the trouble is in the fuel-return line.

♦ 19-20 ENGINE STARTS BUT STALLS ON IDLE

This could be caused by any of the troubles discussed in ♦19-19. These include incorrect or dirty fuel, limited fuel to nozzles or injection pump, a restricted fuel return, incorrect pump timing, or defects in the injection pump. Also, stalling on idle after starting could be caused by an incorrectly set idle, or by low fuel in the tank.

♦ 19-21 ROUGH IDLE, NO ABNORMAL NOISE OR SMOKE

First check if the low-idle speed is correctly set. Then look for injection-line leaks. Wipe off the lines and run the engine. If there are leaks, tighten connections or replace lines as necessary to eliminate them. Next, check for a restricted fuel-return system (♦19-19). Disconnect the fuel-return line to see if the engine runs normally.

To check for a defective nozzle, start the engine and then loosen the injection-line fitting at each nozzle in turn. This relieves the pressure and prevents normal injection-nozzle action. Be careful to avoid spraying of fuel onto hot engine parts. When a nozzle that is good is prevented from operating in this way, the rhythm of the engine will change. The engine will run more roughly. If you find a nozzle that does not change the idle when partly disconnected, then that nozzle was not working and should be replaced.

To check for internal fuel leaks at the fuel nozzle, disconnect the fuel-return system from the nozzles on one bank at a time. With the engine running, note the fuel seepage at the nozzles. Replace any nozzle with excessive fuel leakage.

A rough idle can also be caused by a fuel-supply problem. The fuel-supply pump, line, and fuel filter should be checked. In addition, rough idle can also be caused by dirty fuel or the wrong fuel.

DIESEL FUEL-INJECTION-SYSTEM TROUBLE-DIAGNOSIS CHART*

Complaint	Possible Cause	Check or Correction
1. Engine cranks normally but will not start (♦19-19)	a. Incorrect or dirty fuel	Flush system—use correct fuel
	b. No fuel to nozzles or injection pump	Check for fuel to nozzles
	c. Plugged fuel return	Check return, clean
	d. Pump timing off	Retime
	e. Inoperative glow plugs, incorrect starting procedure, or internal engine problems	
2. Engine starts but stalls on idle (♦19-20)	a. Fuel low in tank	Fill tank
	b. Incorrect or dirty fuel	Flush system—use correct fuel
	c. Limited fuel to nozzles or injection pump	Check for fuel to nozzles and to pump
	d. Restricted fuel return	Check return, clean
	e. Idle incorrectly set	Reset idle
	f. Pump timing off	Retime
	g. Injection-pump trouble	Install new pump
	h. Internal engine problems	
3. Rough idle, no abnormal noise or smoke (♦19-21)	a. Low idle incorrect	Adjust
	b. Injection line leaks	Fix leaks
	c. Restricted fuel return	Clear
	d. Nozzle trouble	Check, repair or replace
	e. Fuel-supply-pump problem	Check, replace if necessary
	f. Uneven fuel distribution to nozzles	Selectively replace nozzles until condition clears up
	g. Incorrect or dirty fuel	Flush system—use correct fuel
4. Rough idle with abnormal noise and smoke (♦19-22)	a. Injection-pump timing off	Retime
	b. Nozzle trouble	Check in sequence to find defective nozzle
5. Idle okay but misfires as throttle opens (♦19-23)	a. Plugged fuel filter	Replace filter
	b. Injection-pump timing off	Retime
	c. Incorrect or dirty fuel	Flush system—use correct fuel
6. Loss of power (♦19-24)	a. Incorrect or dirty fuel	Flush system—use correct fuel
	b. Restricted fuel return	Clear
	c. Plugged fuel-tank vent	Clean
	d. Restricted fuel supply	Check fuel lines, fuel-supply pump, injection pump
	e. Plugged fuel filter	Replace filter
	f. Plugged nozzles	Selectively test nozzles, replace as necessary
	g. Internal engine problems, loss of compression, compression leaks	
7. Noise—"rap" from one or more cylinders (♦19-25)	a. Air in fuel system	Check for cause and correct
	b. Gasoline in fuel system	Replace fuel
	c. Air in high-pressure line	Bleed system
	d. Nozzle sticking open or with low opening pressure	Replace defective nozzle
	e. Engine problems	
8. Combustion noise with excessive black smoke (♦19-26)	a. Timing off	Reset
	b. Injection-pump trouble	Replace pump
	c. Nozzle sticking open	Clean or replace
	d. Internal engine problems	

*See ♦19-19 to 19-26 for explanations of the trouble causes and corrections listed.

♦ 19-22 ROUGH IDLE WITH ABNORMAL NOISE AND SMOKE

First check the injection-pump timing. Then disable the nozzles, one at a time, to check their operation. Do this by loosening the injection-line connection at the nozzle (♦19-21). When you find a nozzle that does not change the noise or smoke when it is disabled, that is the bad nozzle. It should be replaced.

♦ 19-23 IDLES OK BUT MISFIRES AS THROTTLE OPENS

This can be caused by a plugged fuel filter, incorrect injection-pump timing, or incorrect or dirty fuel. A plugged fuel filter should be replaced. If the condition is caused by incorrect injection-pump timing, reset the timing. Incorrect or dirty fuel should be flushed out. Then the tank should be filled with the correct fuel.

♦ 19-24 LOSS OF POWER

This is a general complaint that could result from many conditions in the engine systems or engine as well as outside the engine. For example, dragging brakes, excessive resistance in the power train, or underinflated tires can produce an impression of low power. In the fuel system, loss of power could result from incorrect or dirty fuel, restricted fuel return, a restricted fuel supply, or plugged nozzles. Previous sections describe how to check for these conditions. Another possible cause is a plugged fuel-tank vent. This would prevent normal fuel flow to the injection pump and engine so that the engine could not produce full power. Also, if the engine overheats, it will lose power.

♦ 19-25 NOISE (RAP) FROM ONE OR MORE CYLINDERS

One possible cause of this condition could be air in the fuel system. Air could cause a very uneven flow of fuel to the nozzles. The air expands and contracts with changing pressure, causing too much or too little fuel to feed. To correct the problem, loosen the injection line at each nozzle to allow the air to bleed from the system. Another cause of the noise or rap could be a nozzle sticking open or having a very low opening pressure. Loosening the injection lines at each nozzle in turn will locate a defective nozzle.

♦ 19-26 COMBUSTION NOISE WITH EXCESSIVE BLACK SMOKE

This could be caused by incorrect injection-pump timing, or injection-pump troubles. The cylinders receive too much fuel, or fuel at the wrong time. It could also be caused by an injection nozzle sticking open and by internal engine problems.

♦ 19-27 DIESEL FUEL-SYSTEM SERVICE PROCEDURES

After a trouble is located, it must be corrected. Specific servicing operations on the diesel-engine fuel-injection system are covered below. These include service to the fuel-injection lines, fuel filter, injection nozzles, and injection and fuel-supply pumps.

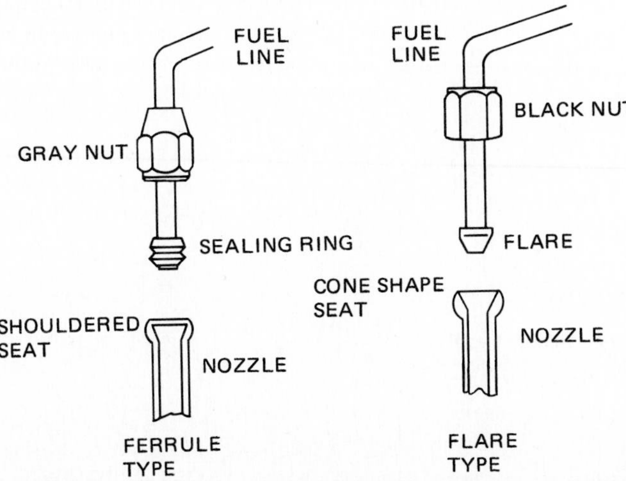

Fig. 19-22 Two types of fittings used to connect the fuel line to the injection nozzle. *(Oldsmobile Division of General Motors Corporation)*

♦ 19-28 INJECTION LINES AND FITTINGS

Injection lines in diesel fuel-injection systems are also called *high-pressure fuel lines, injection tubes,* and *injection pipes* (Fig. 19-4). They carry the fuel, under high pressure, from the injection pump to the injection nozzles in the cylinders. The lines must withstand pressures of several thousand pounds per square inch, and must be noncorrosive. When an injection line requires replacement, always install a line that is recommended or approved by the engine manufacturer.

Figure 19-22 shows two types of fittings used to connect the fuel lines to the injection nozzles. When disconnecting a line of the type shown in Fig. 17-8, use two wrenches. Hold the fitting with one wrench and turn the coupling nut. If you try to loosen the nut without using the backup wrench, you can twist the line and damage it. It is not necessary to use the backup wrench when disconnecting the lines from the fuel-injection pump.

Whenever lines are disconnected, the lines, nozzles, and pump fittings must be capped. This will prevent dirt from entering the fuel system. Cleanliness is very important when working on the fuel system. A dirt particle so tiny as to be almost invisible to your eye can clog an injection nozzle.

New injection lines are preformed. Care should be used in installing them to avoid twisting or bending them out of shape. If the line that is to be replaced is under other lines, you may have to remove these upper lines. Figure 19-6 shows the basic parts of a diesel fuel-injection system on a six-cylinder engine. Note how the injection lines are formed.

♦ 19-29 FUEL FILTER

The fuel for a diesel engine must be clean and free of contaminants and water (♦19-4 and 19-5). Tiny, almost invisible specks of dirt can clog the fuel nozzles or damage the injection-pump parts. Therefore, the fuel filter is an essential part of the system. It works like the filter used in engine lubricating systems. The filter contains a cartridge of filtering material (special pleated paper or fiber mat) through which the fuel must pass. The filter traps any particles and keeps them from entering the fuel system. The filter should be replaced at the interval recommended by the manufacturer.

♦ 19-30 INJECTION NOZZLE

Injection nozzles should not be removed unless there is evidence that they require servicing or replacement. Usual indications of trouble include:

- ♦ One or more cylinders knocking
- ♦ Loss of power
- ♦ Smoky black exhaust
- ♦ Engine overheating
- ♦ Excessive fuel consumption

One way to check for a faulty injection nozzle is to run the engine at fast idle. Then loosen the connector at each nozzle in turn, one at a time. Wrap a cloth around the connection before you loosen it to keep fuel from spurting out. If loosening a connector causes the engine speed to drop off, the nozzle is probably working okay. If the engine speed remains the same, then the nozzle is probably not performing properly. It could be clogged so that no fuel flows through. Or the holes could be partly clogged so that the spray is inadequate or does not have the required pattern.

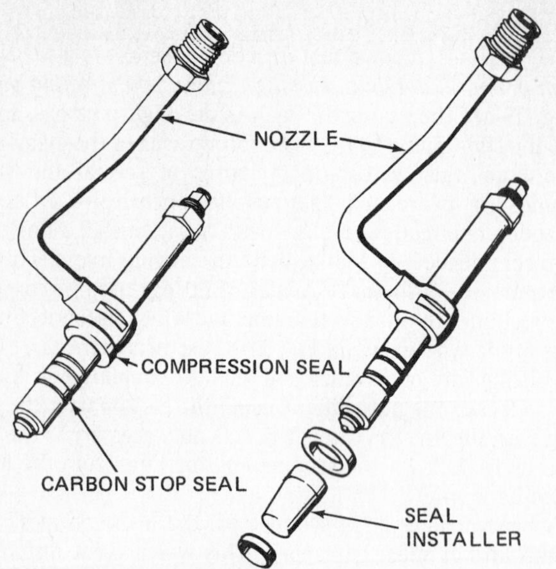

Fig. 19-23 Injection nozzles with seals. *(Chevrolet Motor Division of General Motors Corporation)*

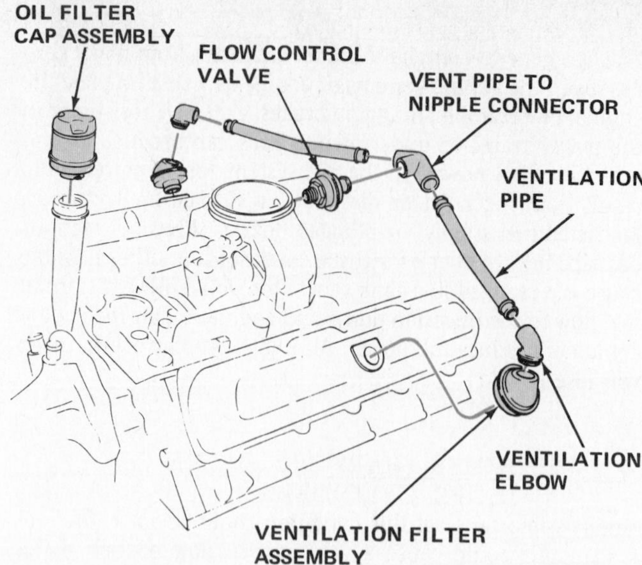

Fig. 19-25 Ventilation system for diesel-engine crankcase. *(Chevrolet Motor Division of General Motors Corporation)*

A variety of nozzles are used, but all have a check valve. The check valve opens when spray pressure is applied so that the fuel can flow through. When the pressure drops, the check valve closes to shut off the flow rapidly and completely. To remove an injection nozzle, first remove the fuel-return-line clamps and return line. Then remove the nozzle hold-down clamp and spacer or other connector arrangement. Remove the nozzle. Cap the nozzle inlet line and the tip of the nozzle. Figure 19-23 shows one installation arrangement which includes a compression seal and a carbon stop seal.

Some manufacturers recommend a spray test of the detached nozzle (Fig. 19-24). This requires a special hydraulic pump which has a pressure gauge. You attach the nozzle to the pump and work the pump. The fuel should spray when the pump pressure reaches the specified value. When the pressure is released, the spray should stop abruptly and the nozzle should not drip.

♦ CAUTION ♦ Direct the spray from the nozzle into a suitable container. Do not allow the spray to hit your skin. The pressure is high enough to force fuel oil through the skin. You can be seriously injured because the oil could cause an infection.

If the nozzle does not work properly, it can be disassembled and cleaned. Some manufacturers recommend replacing the nozzle. If you do disassemble a nozzle, avoid damaging the tip or enlarging the holes.

♦ 19-31 INJECTION-PUMP REPLACEMENT

Figure 19-6 shows the various parts that must be removed to remove the in-line plunger type of fuel-injection pump. First, take off the injection pipes. Use two wrenches on the couplings, as explained in ♦17-20. Then remove other parts, as necessary, until you can remove the injection pump. Installation is usually the reverse of removal.

To replace the distributor type of injection pump (as used on General Motors V-type engines), first do the following:

1. Remove air cleaner, filters, and pipes from valve covers and air crossover (Fig. 19-25). Cap intake manifold with special screened covers.
2. Disconnect the throttle rod and return spring (Fig. 19-26).
3. Remove bell crank and throttle cable (from intake-manifold brackets), and push the cable away from the engine.
4. Remove lines to fuel filter and then remove filter and bracket. Disconnect the fuel line at the fuel pump. Disconnect the fuel-return line from the injection pump.
5. Use two wrenches (Fig. 17-8) and disconnect injection lines from the fuel nozzles. Use the special tool to remove the nuts attaching the injection pump. Now, remove the pump and cap all open lines and nozzles.

To install the pump, remove the protective caps. Line up the offset tang on the pump drive shaft with the pump-driven gear and install the pump (Fig. 19-27). Attach it with the nuts

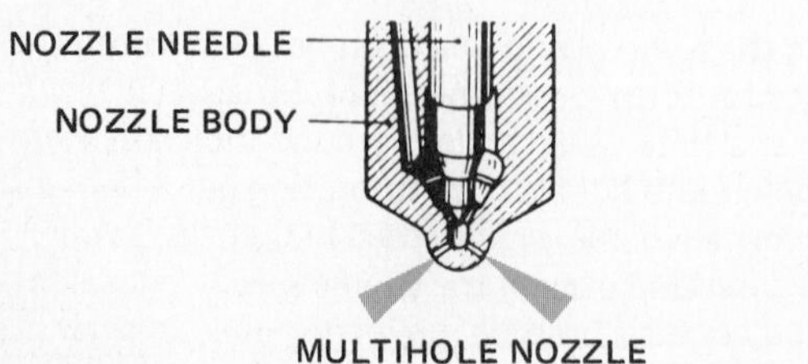

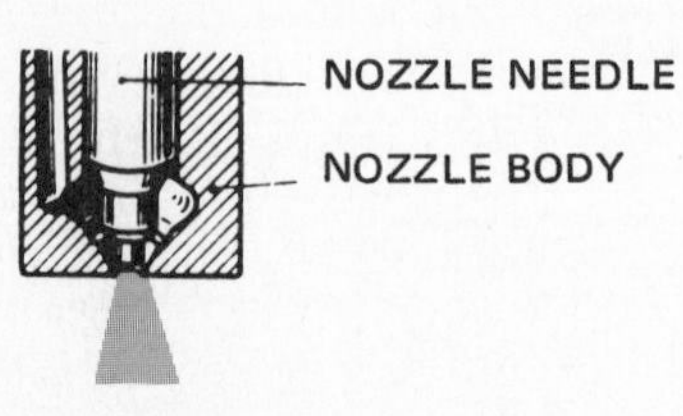

Fig. 19-24 Correct spray patterns for a multihole nozzle and for a pintle (single-hole) nozzle. *(Robert Bosch Corporation)*

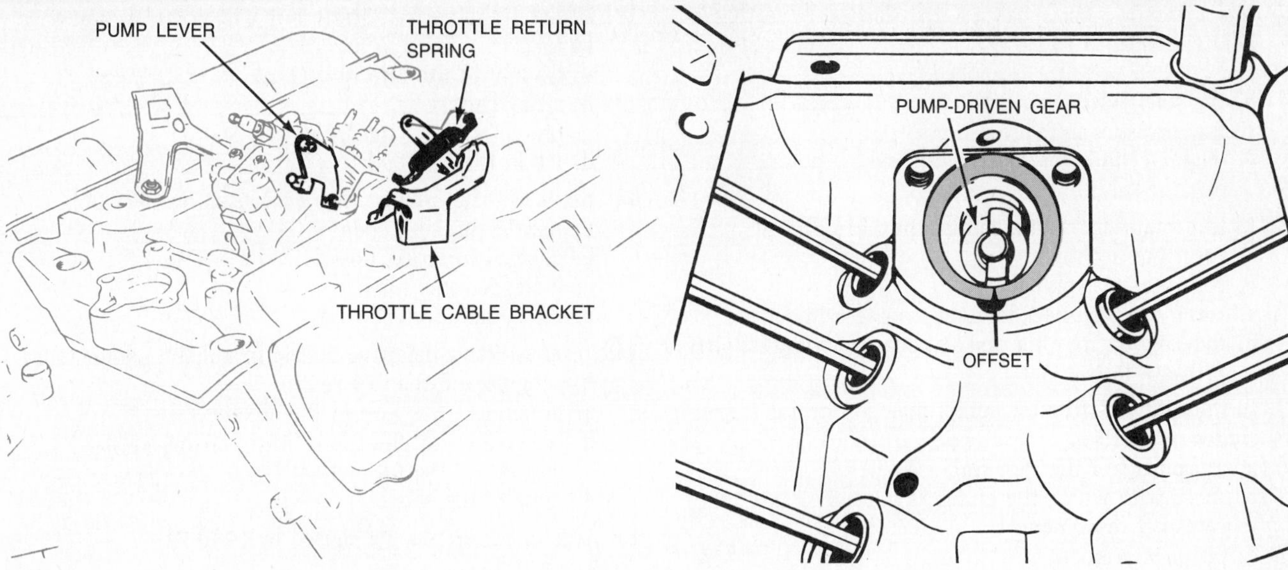

Fig. 19-26 Location of throttle-return spring. *(Chevrolet Motor Division of General Motors Corporation)*

Fig. 19-27 Location of offset tang on the pump-driven gear. *(Chevrolet Motor Division of General Motors Corporation)*

and lock washers, lightly run down on the studs. Connect the injection lines to the nozzles, using two wrenches (Fig. 17-8).

Align the mark on the injection pump with the line on the adapter and tighten the attaching nuts. Use a ¾-inch end wrench on the boss at the front of the injection pump to aid in rotating the pump to align the marks.

Install the fuel line from the fuel pump to the fuel filter. Adjust the throttle rod. Install bell crank and throttle-return spring.

Start the engine and check for fuel leaks. Tighten connections as necessary. Then install the air crossover and air cleaner.

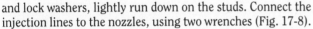

♦ REVIEW QUESTIONS ♦

Select the *one* correct, best, or most probable answer to each question. You can find the answers in the section indicated at the end of each question.

1. In the diesel engine, the fuel is injected into the engine cylinders (♦19-1)
   a. at the end of the power stroke
   b. at the beginning of the power stroke
   c. toward the end of the compression stroke
   d. at the end of the exhaust stroke
2. Which one of the following is correct? (♦19-2)
   a. the diesel engine has a throttle valve to restrict airflow
   b. the power and speed of the diesel engine are controlled, in part, by the spark advance.
   c. the diesel engine has a lower compression ratio than the spark-ignition engine
   d. the diesel engine compresses only air on the compression stroke
3. In the diesel-engine fuel system, the (♦19-3)
   a. timing of fuel delivery changes with engine speed
   b. amount of fuel delivered changes with operating requirements
   c. fuel is delivered to the injection nozzles at high pressure
   d. all of the above
4. The two types of centrally located fuel-injection pumps are (♦19-6)
   a. cam-operated plunger type and rotary-distributor type
   b. cam-operated distributor type and rotary-plunger type
   c. rotary cam and plunger-distributor type
   d. remote pressure and cam pressure
5. The cam-operated in-line plunger pump has (♦19-7)
   a. a plunger for the pump
   b. two plungers for each cylinder in the engine
   c. a rotary-distributor system
   d. a barrel-and-plunger assembly for each engine cylinder
6. In the cam-operated in-line plunger pump, the amount of fuel delivered depends on the (♦19-7)
   a. effective length of the plunger strokes
   b. power demands on the engine
   c. operating conditions
   d. all of the above

7. The injection nozzle has a spring-loaded valve that is held closed until (♦19-11)
   a. high fuel pressure is applied
   b. the solenoid is energized
   c. fuel pressure is relieved
   d. engine-cylinder pressure is relieved by opening of the exhaust valve
8. In the rotary-distributor pump, the (♦19-8)
   a. rotor has a plunger
   b. rotor has two plungers
   c. rotor has a plunger for each engine cylinder
   d. indexing of the rotor and stationary openings applies pressure to the fuel
9. In the rotary-distributor pump, injection timing is controlled by (♦19-9)
   a. movement of the cam ring
   b. speed with which the plungers move
   c. camshaft movement
   d. piston movement
10. The purpose of the glow plugs is to (♦19-12)
    a. provide light so that the driver can see to insert the ignition key
    b. improve engine performance after the engine has warmed up
    c. make it easier to start a cold diesel engine
    d. light up the side markers so that other drivers can see the car
11. When a diesel engine cranks normally but will not start, the cause may be (♦19-19)
    a. water in the fuel system
    b. ice in the fuel system
    c. both a and b
    d. neither a nor b
12. A diesel engine that starts but will not idle may have (♦19-20)
    a. no fuel to the injection pump
    b. a low battery
    c. one or more clogged fuel nozzles
    d. ice in the fuel tank
13. Smoke in the exhaust may be caused by (♦19-26)
    a. nozzles sticking open
    b. clogged injection lines
    c. a clogged fuel filter
    d. dirty fuel
14. To check for a defective nozzle in a diesel engine, start the engine and then (♦19-30)
    a. disconnect the wire to the solenoid
    b. loosen the injection-line fitting at the nozzle
    c. disconnect the fuel-return line
    d. none of the above
15. Loss of power may be caused by (♦19-24)
    a. dirty fuel
    b. restricted fuel-return line
    c. plugged nozzles
    d. all of the above

CHAPTER 20
Engine Lubricating Systems: Operation and Service

After studying this chapter, you should be able to:

1. Explain the operation of the engine lubricating system.
2. Describe six jobs performed by the engine oil.
3. Discuss the properties of a good lubricating oil and the reasons for the additives that are put into it.
4. Explain the service ratings of lubricating oil.
5. Discuss possible troubles in the lubricating system and their causes and corrections.
6. Change the engine oil and oil filter.
7. Service the engine oil pan and oil pump.

♦ 20-1 PURPOSE OF THE LUBRICATING SYSTEM

The engine lubricating system supplies lubricating oil to all engine moving parts. Figure 20-1 shows the system for a four-cylinder OHC spark-ignition engine. The oil pump picks up oil from the oil pan. Then the oil is sent up through the oil lines to the main bearings that support the crankshaft. This lubricates the main bearings. Oil from the main bearings feeds through oilholes drilled in the crankshaft to the rod bearings. Oil feeds through an oil line up to the cylinder head. There, the oil flows through an oil gallery to lubricate the camshaft bearings and the valve-train parts. The pistons, piston rings, and piston pins are lubricated by oil thrown off the connecting-rod bearings. This oil lands on the cylinder walls to lubricate the pistons, rings, and pins.

The oil circulates to all the parts shown by the arrows in Fig. 20-1. Then the oil drops back down into the oil pan. Chapter 11 describes the various engine components that make up the lubricating system. Figure 11-3 shows the engine oil pan and oil pump. The oilholes in the crankshaft are shown in Fig. 11-17. Figures 11-21 to 11-24 show various bearings and bushings that are lubricated by the lubricating system.

Figure 20-2 shows the lubricating system for a six-cylinder camshaft-in-block, pushrod spark-ignition engine. In this engine, the lubricating oil flows up to the rocker arms and valve stems through hollow pushrods. The main bearings, connecting-rod bearings, pistons, piston rings, and piston pins are lubricated in the same way as previously described.

On many V-type engines, the connecting rods have oil-spit holes or grooves that align with the crankpin-journal oilholes once each revolution. When this happens, a spurt of oil sprays onto the cylinder walls of the opposite cylinders in the other cylinder bank (Fig. 20-3).

♦ 20-2 PURPOSE OF LUBRICATING OIL

The lubricating oil performs several jobs in the engine:

1. The oil lubricates moving parts to minimize wear. Clearances between moving parts (the bearing and rotating journal, for example) are filled with oil (♦11-19 to 11-21).

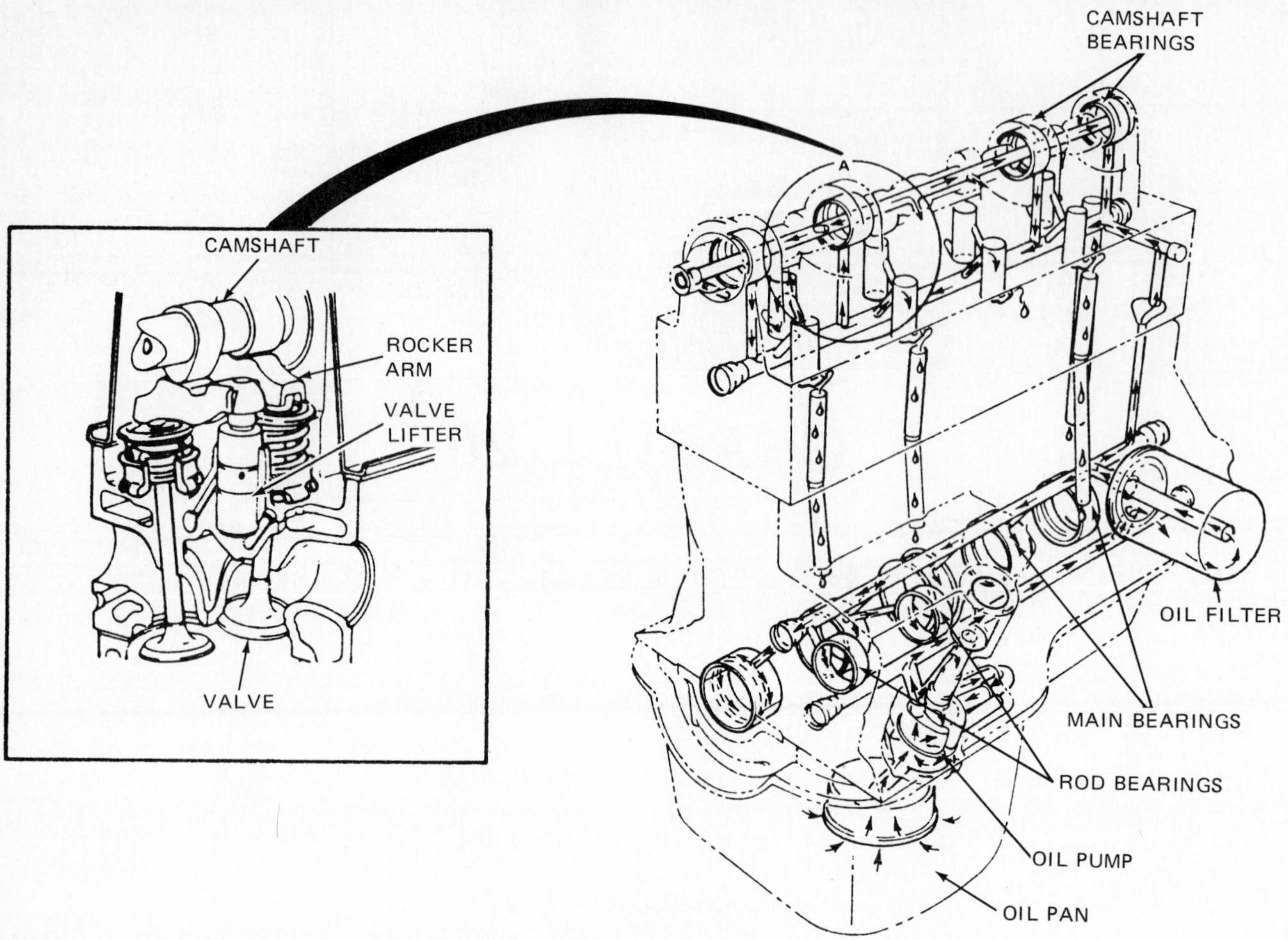

Fig. 20-1 Lubricating system for a four-cylinder OHC spark-ignition engine. The insert to the left shows details of the valve train. *(Ford Motor Company)*

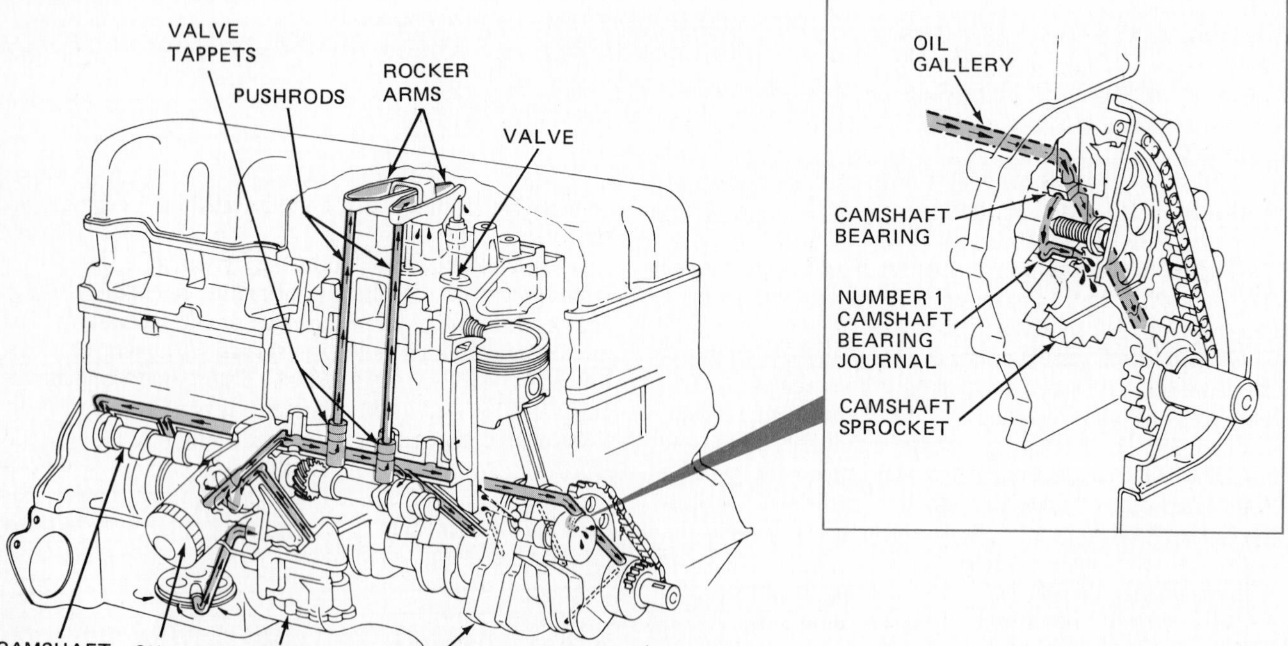

Fig. 20-2 Lubricating system for a six-cylinder camshaft-in-block spark-ignition engine. The insert to the right shows how the camshaft sprockets and chain are lubricated. *(American Motors Corporation)*

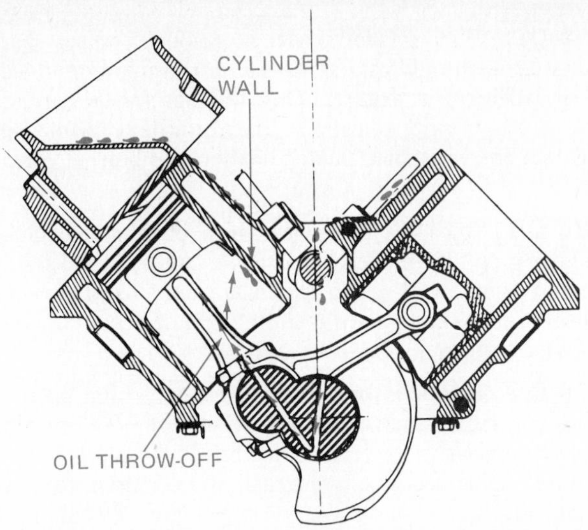

Fig. 20-3 On many V-type engines, cylinder walls are lubricated by oil thrown from connecting-rod bearings of opposing cylinders. *(Chevrolet Motor Division of General Motors Corporation)*

The parts move on the layers of oil to help prevent excessive wear of the parts.

2. Parts moving on layers of oil require a minimum of power to make them move. The oil minimizes power losses in the engine.
3. As the oil moves through the engine, the oil picks up heat. Then, when the hot oil drops down into the cooler oil pan, the oil gives up some of this heat. Therefore, the oil serves as a cooling agent.
4. The clearances between bearings and rotating journals are filled with oil. When heavy loads are suddenly imposed on the bearings, the oil helps to cushion the load. This reduces bearing wear.
5. The oil helps form a gastight seal between piston rings and cylinder walls. Piston rings and the effect that the lubricating oil has on their performance are described in ♦11-24 to 11-34. The oil reduces blowby in addition to lubricating the piston and rings.
6. The oil acts as a cleaning agent. As it circulates through the engine, the oil picks up particles of metal and carbon and carries them back down into the oil pan. Larger particles fall to the bottom of the oil pan. Smaller particles are filtered out by the oil filter.

♦ 20-3 PROPERTIES OF LUBRICATING OIL

The properties that an engine lubricating oil should have include the following:

1. Viscosity Correct viscosity for the operating conditions. Viscosity is a measure of an oil's resistance to flow. A low-viscosity oil is thin and flows easily. A high-viscosity oil is thicker. Therefore, it flows more slowly.

Engine oil should have the proper viscosity so that it flows easily to all moving engine parts. But the oil must not be too thin. Low viscosity reduces the ability of the oil to stay in place between moving engine parts. If the oil is too thin (has a very low viscosity), it will be forced out from between the moving parts. Then rapid wear will result.

If the oil is too thick (has high viscosity), it will flow very slowly to engine parts, especially when the oil and engine are cold. This can cause rapid wear of engine parts. The engine will be operating with insufficient oil when first starting. Also, in cold weather, high-viscosity oil might be so thick that it would prevent normal starting. An excessive amount of cranking power might be required to crank the engine.

A single-viscosity oil (such as SAE 20W, described later) gets thinner as it gets hot.

2. Viscosity Index This is a measure of how much the viscosity of an oil changes with temperature. A single-viscosity oil could be too thick (highly viscous) at low temperatures and very thin at high engine operating temperatures. To take care of this, viscosity-index improvers are added to engine oil so the viscosity stays more nearly the same, hot or cold.

3. Viscosity Numbers There are several grades of single-viscosity oils. They are rated for winter and for other than winter. Winter-grade oils are supplied in three grades, SAE 5W, SAE 10W, and SAE 20W. The "SAE" stands for the Society of Automotive Engineers, which developed the grading system. The "W" stands for winter. For other than winter use, the grades are SAE20, SAE30, SAE40, and SAE50. The higher the number, the higher the viscosity (thickness) of the oil.

4. Multiple-Viscosity Oil Many engine oils have a viscosity-index improver (described in 3, above) added that allows the viscosity of the oil to remain relatively the same, hot or cold. For example, a multiple-viscosity oil may be graded SAE 10W-30. This means that the oil is the same as SAE 10W when cold and SAE 30 when hot. This oil works satisfactorily in engines running in a wide range of outside temperatures. Manufacturers recommend the use of multiple-viscosity oil in many automotive engines.

5. Resistance to Carbon Formation and Oil Oxidation Oil is refined, and chemicals are added, to combat carbon formation and oxidation. These can occur at the high temperatures inside the engine. The additives minimize them.

6. Corrosion and Rust Inhibitors Additives are put in the oil to help the oil fight corrosion and rust in the engine. These additives displace water from metal surfaces so that oil coats them. They also neutralize acids.

7. Foaming Resistance As oil is churned up in the crankcase by the rotating of the crankshaft, the oil tends to foam or *aerate*. This can reduce the lubricating effectiveness of the oil because it has air in it. Also, the foaming can cause the oil to overflow and pass up through the crankcase ventilating system to the intake manifold and air cleaner.

8. Detergent-Dispersants These additives act somewhat like ordinary soap. They loosen and detach particles of carbon and grit from engine parts and carry them down to the oil pan as the oil circulates.

9. Extreme-Pressure Resistance Additives put in engine oils improve the resistance to penetration at critical points in the engine. Modern engines subject the oil to very high pressures in the bearings and valve trains. The extreme-pressure additives react chemically with the metal surfaces. The result is a very strong, slippery film that supplements the oil by providing protection during extreme pressure.

10. "Improved" Oils Some lubricating oils are called "improved" because special additives are used to modify certain characteristics of the oil. There are two types of modifiers. One type is a chemical that dissolves completely in the oil. The other type uses powdered graphite or molybdenum ("moly"), which is held in suspension in the oil. It is claimed that improved oils reduce friction in the engine.

11. Synthetic Oils These oils are made by chemical processes and do not necessarily come from petroleum. Some oil manufacturers claim these synthetics have superior lubricating properties. There are several types. The most commonly used now is made from organic acids and alcohols (from plants). A second type is produced from coal and crude oil. Even though some tests indicate these oils have some superior qualities, no automotive manufacturer uses synthetic oil in new engines.

♦ 20-4 SLUDGE FORMATION

Sludge is a thick, creamy, black substance that sometimes forms in the crankcase. It clogs oil screens and oil lines, preventing oil circulation, so the engine can fail from oil starvation.

1. How Sludge Forms Water collects in the crankcase in two ways. First, water is formed as the product of combustion. Second, the crankcase ventilating system carries air through the crankcase. The air usually has moisture in it. This moisture condenses on cold engine parts. The crankshaft acts like a giant eggbeater, whipping the water and oil up into the thick, black substance called sludge. The black comes from dirt and carbon.

2. Why Sludge Forms If a car is driven for long distances each time it is started, water in the crankcase quickly evaporates. The crankcase ventilating system then removes the water vapor. Therefore, no sludge will form. However, if the engine is operated when cold most of the time, then sludge will form. For example, the home-to-shop-to-home sort of driving is sludge-forming. When a car is used for short-trip start-and-stop driving, the engine never has a chance to warm up enough to get rid of the water. The water remains and forms sludge.

3. Preventing Sludge To prevent sludge, the car must be driven long enough for the engine to heat up and get rid of the water in the crankcase. This means trips of 12 miles [19 km] or longer in winter (but shorter in summer). If trips of this length are impractical, then the oil must be changed frequently. During cold weather, it takes longer for the engine to warm up. Therefore, in cold weather, the trips must be longer or oil must be changed more often to prevent sludge formation.

♦ 20-5 SERVICE RATINGS OF OIL

Oil is rated by viscosity number (♦20-3) and also by its *service designation.* This is the kind of service for which the oil is best suited. There are six service ratings for spark-ignition-engine lubricating oils, SA, SB, SC, SD, SE, and SF. There are four service ratings for compression-ignition-engine lubricating oils, CA, CB, CC, and CD.

1. SA Oil This oil is for utility gasoline and diesel engines operating under mild conditions, so protection by additives is not required. This oil may have pour-point and foam depressants.

2. SB Oil This oil is for service in gasoline engines operated under such mild conditions that only minimum protection by additives is required. Oils designed for this service have been used since the 1930s. They provide only antiscuff capability and resistance to oil oxidation and bearing corrosion.

3. SC Oil This oil is for service typical of gasoline engines in the 1964 to 1967 models of passenger cars and trucks. It is intended primarily for use in passenger cars. This oil provides control of high- and low-temperature engine deposits, wear, rust, and corrosion.

4. SD Oil This oil is for service typical of gasoline engines in passenger cars and trucks beginning with 1968 models. This oil provides more protection from high- and low-temperature engine deposits, wear, rust, and corrosion than do the SC oils.

5. SE Oil This oil is for service typical of gasoline engines in passenger cars and some trucks beginning with 1972 (and some 1971) models. This oil provides more protection against oil oxidation, high-temperature engine deposits, rust, and corrosion than do oils with the SC and SD ratings.

6. SF Oil SF oil is for service typical of gasoline engines in passenger cars and some trucks beginning with 1981 (and some 1980) models. This oil provides more protection than do SE oils against sludge, varnish, wear, oil-screen plugging, and engine deposits.

Notice that this is an open-end series. When the car manufacturers and oil producers see the need for other types of oil, they can bring out SG and SH service-rated oils. SA and SB oils are not recommended for use in automobile engines. These are nondetergent oils. Detergent oils are required in modern automotive engines.

Diesel-engine oils must have different properties than oils for gasoline engines. The CA, CB, CC, and CD ratings indicate oils for increasingly severe diesel-engine operation. For example, CA oil is for light-duty service. CD oil is for severe-duty service typical of tubocharged high-output diesel engines operating on fuel oil with a high sulfur content. Oil for use in automotive diesel engines should be marked SF/CC or SF/CD.

Note The viscosity grade and the service rating of an oil are different. A high-viscosity oil is not necessarily a "heavy-duty" oil. Viscosity grade refers to the thickness of the oil. Thickness is not a measure of heavy-duty quality. An oil has two ratings, viscosity and service. Therefore, an SAE 30 oil may be an SC, SD, or SE oil (for automotive engines). Likewise, an oil of any other viscosity grade can have any one of the service ratings.

♦ 20-6 OIL PUMPS

The two general types of oil pumps used in automotive-engine lubricating systems are shown in Figs. 20-4 and 20-5. The *gear-type pump* uses a pair of meshing gears. As the gears rotate, the spaces between the gear teeth are filled with oil from the oil inlet. Then, as the teeth mesh, the oil is forced out through the oil outlet. The *rotor-type pump* uses an inner rotor and an outer rotor. The inner rotor is driven and causes the outer rotor to turn with it. As this happens, the spaces between the rotor lobes become filled with oil.

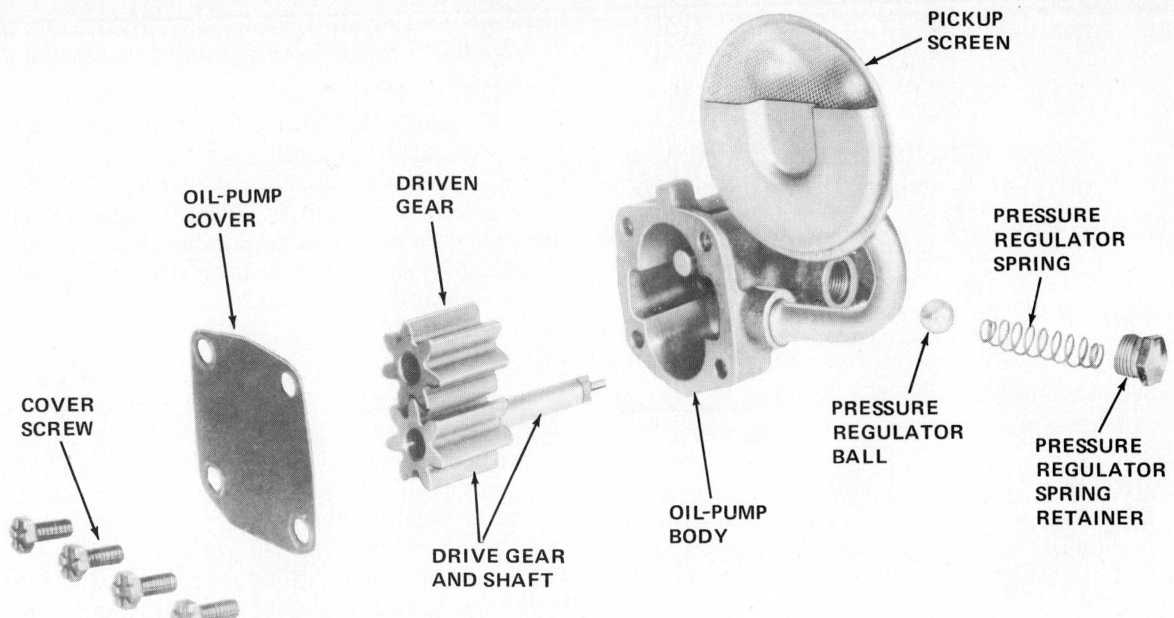

Fig. 20-4 Disassembled gear-type oil pump. *(Pontiac Motor Division of General Motors Corporation)*

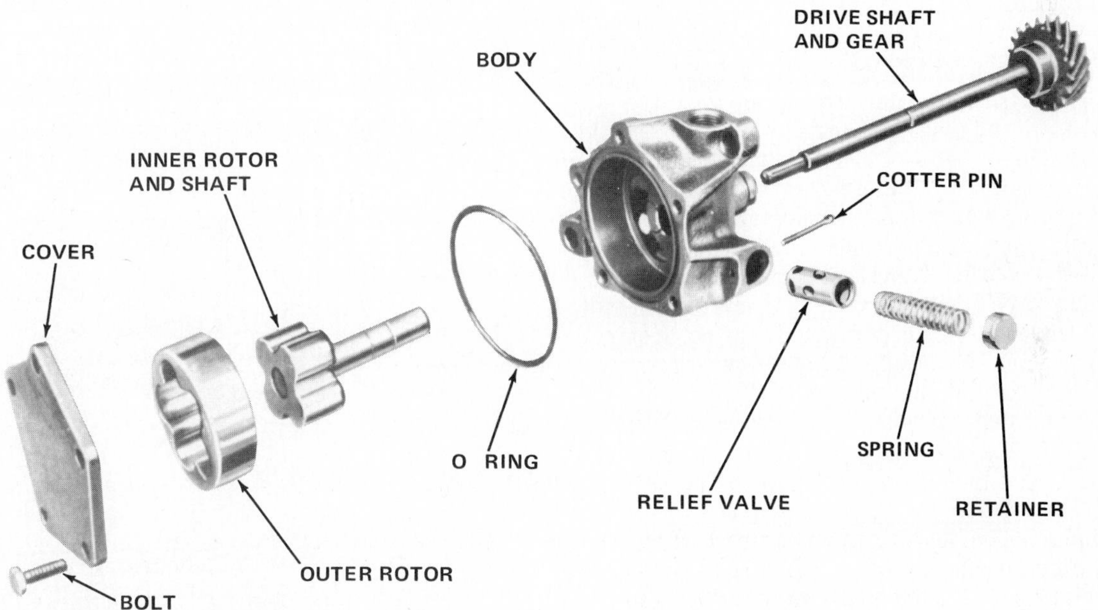

Fig. 20-5 Disassembled rotor-type oil pump. *(Chrysler Corporation)*

When the lobes of the inner rotor move into the spaces in the outer rotor, oil is squeezed out through the outlet.

Oil pumps are usually driven from the engine camshaft, by the same spiral gear that drives the ignition distributor (Fig. 20-6). The oil intake for the oil pump is attached to a float in many engines. This floating intake takes oil only from the top of the oil in the oil pan. Since dirt particles sink, the top oil is cleanest.

♦ 20-7 RELIEF VALVE

To keep the oil pump from building up too much pressure, a relief valve is included in the lubricating system. The valve is a spring-loaded ball (Fig. 20-4) or a spring-loaded plunger (Fig. 20-5). When the pressure reaches the preset value, the ball or plunger is moved against its spring. This opens a port through which oil can flow back to the oil pan. Enough oil flows past the relief valve to prevent excessive pressure. The oil pump can normally deliver much more oil than the engine requires. This is a safety factor that assures delivery of enough oil under extreme operating conditions.

♦ 20-8 OIL COOLER

Some engine lubricating systems have oil coolers. Oil coolers are used on almost all automotive air-cooled engines. One

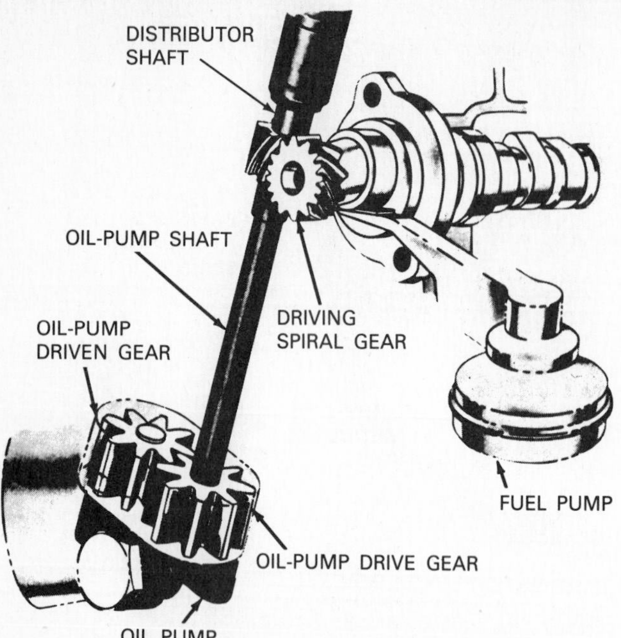

Fig. 20-6 Oil-pump, distributor, and fuel-pump drives. The oil pump is the gear type. A gear on the end of the camshaft drives the distributor. An extension of the distributor shaft drives the oil pump. The fuel pump is driven by an eccentric on the camshaft. *(Buick Motor Division of General Motors Corporation)*

type consists of a small radiator, mounted on the side of the engine block. Oil and coolant circulate through the radiator. The coolant comes from the engine cooling system. As the coolant circulates through the oil cooler, the coolant picks up heat. The heat is carried to the radiator, where the heat is passed on to the cooler air passing through the radiator. This process helps to cool the oil and keep it at a workable temperature. Another type of oil cooler uses a small section of the cooling system radiator. An extra radiator is not required. The oil cooler used in air-cooled engines consists of a small radiator much like the radiators used in liquid-cooling systems. Cooling-system radiators are described in Chap. 21.

♦ 20-9 OIL FILTERS

All automotive engine lubricating systems have an oil filter (Fig. 20-7). Some or all of the oil from the oil pump circulates through this filter. In the filter, an element of filtering material traps particles of foreign matter. Therefore, the filter helps to keep the oil clean and to prevent particles from entering the engine.

Oil filters are of two types. Those which filter part of the oil from the oil pump are called *bypass filters.* Those which filter all the oil in circulation through the system are called *full-flow filters.* The full-flow filter includes a spring-loaded bypass valve. It protects the engine against oil starvation if the filter becomes clogged. When this happens, the valve is opened by increased pressure from the pump trying to push oil through. With the valve opened, oil bypasses the filter. Therefore, the engine is assured of sufficient oil.

♦ 20-10 OIL-PRESSURE INDICATORS

The oil-pressure indicator tells the driver what the oil pressure is in the engine. This gives warning if something in the lubrication system prevents delivery of oil to vital parts. Oil-pressure indicators are of two general types: indicator light and electric gauge.

1. Indicator Light This system, found in most cars, uses an indicator light on the instrument panel. The light is connected in series with the ignition switch, a pressure switch on the engine, and the battery. When the ignition switch is turned to start the engine, the pressure switch is closed. This connects the indicator light to the battery so the light comes on. As soon as the engine starts, oil pressure builds up and opens the pressure switch. This disconnects the indicator light so it goes off.

2. Electric Gauge Figure 20-8 shows an electric oil-pressure gauge system. The engine unit has a diaphragm connected to a sliding contact. The diaphragm has engine oil pressure applied to it. As the oil pressure increases, the diaphragm is pushed in. This causes the sliding contact to move along the resistance (upper left in Fig. 20-8). This decreases the amount of current that can flow through this resistance to ground. As a result, the right coil gets more current and becomes stronger. This pulls the pointer around to the right to indicate the increased oil pressure.

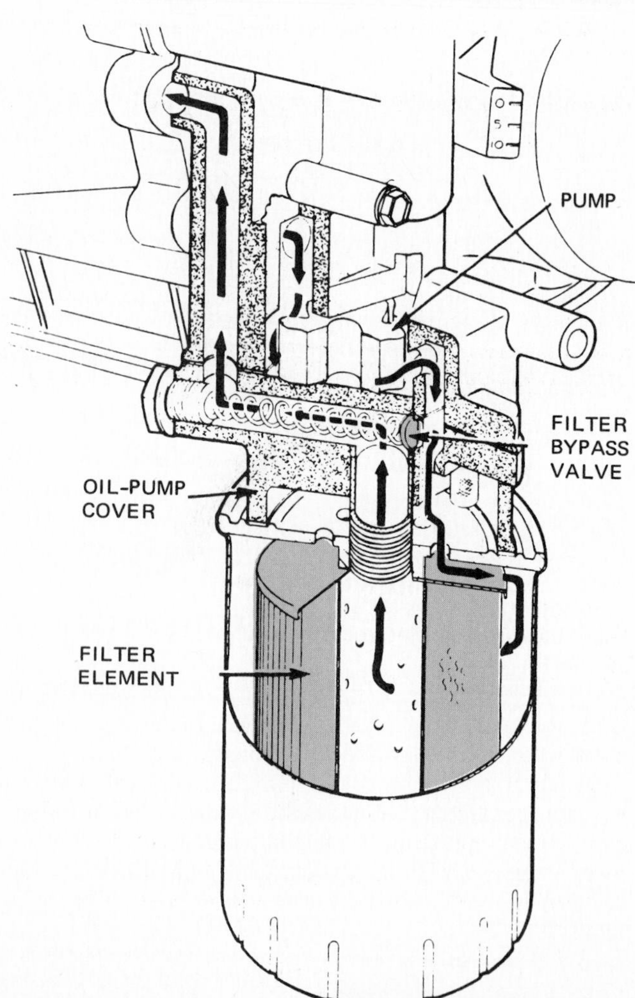

Fig. 20-7 A full-flow oil filter with bypass valve. *(Buick Motor Division of General Motors Corporation)*

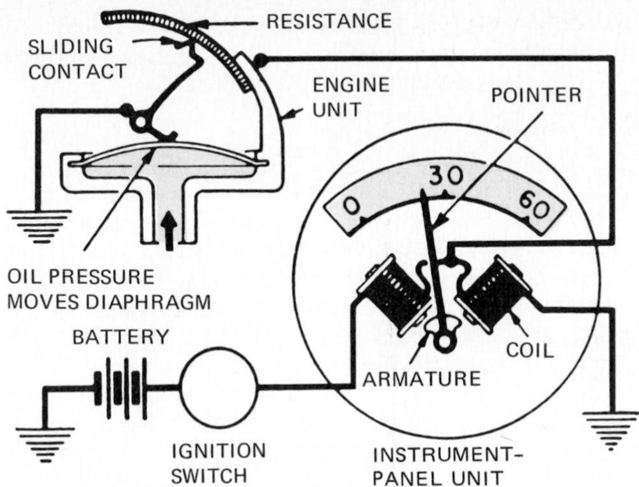

Fig. 20-8 Electric circuit of an oil-pressure gauge.

♦ 20-11 OIL-LEVEL INDICATORS

To determine the level of the oil in the oil pan, an oil-level indicator, or "dipstick," is used (Fig. 20-9). The dipstick is placed so that it sticks down into the oil. The oil level is determined by withdrawing the dipstick and noting how high the oil rises on the dipstick. In the positive crankcase ventilating (PCV) system, the dipstick tube is sealed at the top when the dipstick is in place. This keeps unfiltered air from entering the crankcase and crankcase gases from escaping.

♦ 20-12 CRANKCASE VENTILATION

Air must circulate through the crankcase when the engine is running. This removes water and liquid gasoline that appear in the crankcase when the engine is cold. Also, it removes blowby gases from the crankcase. Unless the water, liquid gasoline, and blowby gases are removed from the crankcase, sludge and acids will form. Sludge can clog oil lines and starve the lubricating system. Acids corrode metal parts. These effects can damage the engine. A later chapter discusses crankcase ventilating systems in detail.

♦ 20-13 OTHER AUTOMOTIVE LUBRICANTS

The automobile needs many other lubricants besides engine oil. There are chassis lubricants, automatic-transmission fluids, steering-gear lubricants, and more. The various lubricants and fluids used in automotive vehicles are briefly described below. In your automotive shopwork, you will probably use all of these lubricants and fluids.

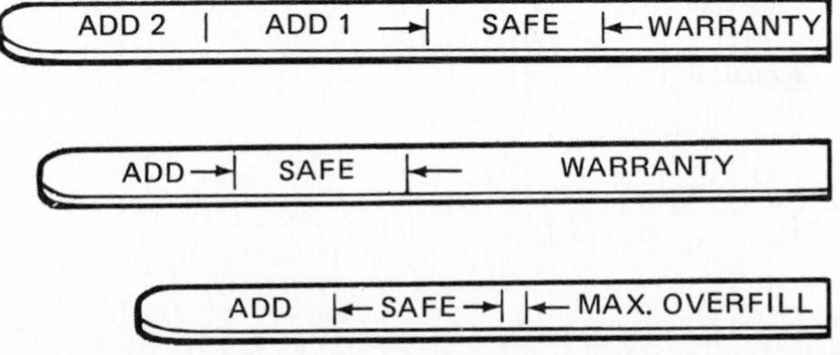

Fig. 20-9 Typical markings on engine oil-level dipsticks. *(Chrysler Corporation)*

1. Grease This is a fluid lubricant such as mineral oil mixed with a thickening agent to make it a semisolid, or plastic. The thickening agent may be a metallic soap or a nonsoap substance such as clay. The soaps commonly used are lithium, calcium, sodium, aluminum, and barium. Each of these, alone or in combination, gives the grease special characteristics. Aluminum gives the grease good adhesion. Sodium gives the grease a thick, fibrous appearance. A number of additives are also mixed in to improve the performance of the grease.

Among the characteristics a good grease must have are consistency, stability, oxidation resistance, ability to protect against friction, wear, and corrosion, and feedability (ability to flow through dispensing equipment).

2. Fields of Use Greases are commonly classified according to their use, as follows:

a. Wheel-Bearing Greases These greases are resistant to high temperatures and the separating effect of centrifugal force.

b. Universal-Joint Greases These greases are compounded to stay in place as the universal joints spin and flex.

c. Chassis Greases These are greases that can be applied with grease guns through fittings. They have the characteristics needed to keep them in place on the moving chassis surfaces without separating or losing lubricating effect.

d. Extended-Lubrication-Interval (ELI) Chassis Greases These are greases with the composition, structure, consistency, life, and antiwear and anticorrosion characteristics for use in "lifetime" applications. These include suspension, drive-line, and steering systems having sealed joints, prepacked during manufacture or assembly. They normally do not need relubrication for long intervals.

e. Multipurpose Greases These greases are compounded to meet the performance requirements for chassis grease, wheel-bearing grease, universal-joint grease, and other automotive uses such as fifth-wheel service.

Some ELI greases are good for multipurpose uses.

f. Extreme-Pressure (EP) Greases These greases are suitable for high-load-carrying applications. Some have a surface-active additive that gives antiwear or antiseize properties beyond those of other greases. "Surface-active" means the agent bonds to metal surfaces to form a barrier that protects the surface if the normal lubricant film is pierced.

g. Other Greases There are other special greases. Brake grease is specified for the moving parts in the drum-brake mechanisms. Distributor breaker-cam grease is specified for the cam in ignition distributors. Speedometer-cable lubricant is another special lubricant.

3. Automatic-Transmission Fluid There are different types of automatic-transmission fluid. Some automatic transmissions have special lubrication requirements. Use the automatic-transmission fluid recommended by the manufacturer.

4. Power-Steering Fluid This is a special fluid that meets the needs of the power-steering unit.

5. Other Fluids There are several other fluids used in automobiles, including antifreeze (ethylene glycol) and brake fluid. In service work, a variety of fluids are used to clean or loosen parts. Carburetor cleaner is one example. The manifold heat-control-valve solvent is another. It is used to loosen the valve if it gets stuck.

Lubricant and fluid makers supply the specific substances needed for each service operation. Automotive manufacturers' service manuals contain lubrication charts. These charts indicate the type of lubricant or fluid to use at every place needing lubrication and the intervals at which these services should be performed.

LUBRICATING-SYSTEM SERVICE

♦ 20-14 LUBRICATING-SYSTEM TROUBLE DIAGNOSIS

The two most common complaints about the lubricating system are:

1. The engine is using too much oil.
2. The indicator light is erratic (or the oil-pressure gauge shows low pressure).

The causes of excessive oil consumption are described in ♦20-15. Erratic indicator-light or oil-pressure-gauge action is discussed in ♦20-16.

♦ 20-15 OIL CONSUMPTION

Oil is lost from the engine in three ways: by burning in the combustion chambers, by leakage in liquid form, and by passing out of the crankcase in the form of a mist. Two main factors affect oil consumption. These are engine speed and engine wear.

High speed produces high temperature. This, in turn, lowers the viscosity of the oil. Now it can more readily work past the piston rings into the combustion chamber, where it is burned. In addition, the high speed exerts a centrifugal effect on oil feeding through the crankshaft to the connecting-rod journals. Therefore, more oil is fed to the bearings and thrown on the cylinder walls. Also, high speeds can cause "ring shimmy," or "ring float." With this condition, the oil-control rings cannot function effectively. Crankcase ventilation causes more air to pass through the crankcase quickly. This causes oil to be lost as mist.

As engine parts wear, oil consumption increases. Worn bearings tend to throw more oil onto the cylinder walls. Tapered and worn cylinder walls prevent normal oil-control-ring action. The rings cannot change shape rapidly enough to conform with the worn cylinder walls as the rings move up and down. Therefore, more oil gets into the combustion chamber, where it burns and fouls spark plugs, valves, rings, and pistons. Carbon formation worsens the condition, since it further reduces the effectiveness of the oil-control rings. Where cylinder-wall wear is not excessive, special oil-control rings can be used to reduce oil consumption. They improve the wiping action so that less oil can move past the rings. After cylinder walls have worn beyond a certain amount, the cylinders must be machined and new rings installed to reduce oil consumption.

Worn intake-valve guides also increase oil consumption. Oil leaks past the valve stems and is pulled into the combustion chamber along with the air-fuel mixture every time the intake valves open. Worn exhaust-valve guides can also cause high oil consumption. Then the oil is burned as the hot exhaust gases hit it when the exhaust valve opens. Installation of new valve guides, reaming of guides and installation of valves with oversize stems, or installation of valve-stem seals will reduce oil consumption from these causes.

♦ 20-16 ERRATIC INDICATOR-LIGHT OR OIL-PRESSURE-GAUGE ACTION

If the light comes on part of the time, or the oil-pressure guage sometimes shows low oil pressure, either there is not enough oil in the crankcase or else the oil pickup is not consistently picking up oil. This could be due to the oil pickup falling off or having been pushed up by a bent oil pan. The oil pan can be bent by hitting a curb or other object.

If the light stays on all the time, or the pressure gauge consistently reads low, then the first check is to see whether the engine oil is low. If it is at normal height, check the oil-pressure sending unit. Remove it and install a pressure gauge to check for pressure with the engine running. If the pressure is okay, then the trouble is a defective sending unit. If the pressure is low, there is other trouble—probably in the engine itself. Trained mechanics can tell by listening whether the oil pressure is low. The clatter of the hydraulic valve lifters and the bearings running without oil warns them that the oil pressure is low. However, an engine operating without oil pressure will be quickly damaged.

Causes of low oil pressure include:

1. A weak relief-valve spring
2. A worn oil pump
3. A broken or cracked oil line
4. Obstructions in the oil lines
5. Insufficient or excessively thin oil
6. Worn bearings that can pass more oil than the oil pump is capable of delivering
7. A defective oil-pressure indicator

Excessive oil pressure may result from:

1. A stuck relief valve
2. An excessively strong valve spring
3. A clogged oil line
4. Excessively heavy oil

♦ 20-17 LUBRICATING-SYSTEM SERVICE

There are certain lubricating-system jobs that are done when the engine is repaired. For example, the oil pan is removed and cleaned during such engine-overhaul jobs as replacing bearings or rings. When the crankshaft is removed, clean out the oil passages in the crankshaft. Also, the oil passages in the cylinder block should be cleaned out. These jobs are described in later chapters on engine service. Sections that follow describe such lubricating-system service jobs as:

1. Checking oil level
2. Changing oil
3. Changing the oil filter
4. Servicing the oil-pressure relief valve
5. Servicing the oil pump and oil-pressure indicators.

♦ 20-18 CHECKING OIL LEVEL

Most engines use a dipstick as the oil-level gauge (Fig. 20-9). To use the dipstick, withdraw it from the crankcase to determine the oil level in the crankcase. The gauge should be withdrawn, wiped clean, reinserted, and again withdrawn so that the oil level on the gauge can be seen. The gauge is usually marked to indicate the proper oil level. The appearance of the oil should be noted to see whether it is dirty, thin, or thick. A few drops of oil can be placed between the thumb and fingers and rubbed to detect dirt. If the oil level is low, oil should be added to the crankcase. If the oil is thin or dirty, it should be drained, and clean oil added.

The car should be on a level surface. If it is on a slope, you will get a false reading.

If the engine has just been shut off, wait a few minutes for the oil to drain back down into the oil pan before checking the oil level.

♦ 20-19 OIL CHANGES

From the day that fresh oil is put into the engine crankcase, it begins to lose its effectiveness as an engine lubricant. This gradual loss of effectiveness is largely a result of the depletion or "wearing out" of the additives. For example, the antioxidant additive becomes used up, allowing gum and varnish to form. The corrosion and rust inhibitors are gradually depleted, allowing corrosion and rust formation to begin.

In addition, during engine operation, carbon tends to form in the combustion chamber. Some of this carbon gets into the oil. Also, the air that enters the engine air cleaner carries a certain amount of dust. Even though the air filter is operating efficiently, it will not remove all the dust. Also, the engine releases fine metal particles as it wears.

All these substances tend to circulate with the oil. As the mileage increases, the oil accumulates more and more of these contaminants. Even though the engine has an oil filter, some of these contaminants remain in the oil. Finally, the oil is so loaded with contaminants that it is not safe to use. Unless the oil is drained and clean oil is put in, engine wear will increase rapidly.

Different automotive manufacturers have different recommendations on how often the engine oil should be changed. For example, Fig. 20-10 lists the recommendations for late-model Chevrolets. Specifications vary for different types of vehicle operation. The oil filter should be changed the first time the oil is changed, and then with every other oil change. For the specifications for other automotive vehicles, see the automotive manufacturers' service manuals.

The viscosity of the oil you put into an engine depends on the temperature range in which the vehicle is expected to operate. Figure 20-11 shows the Chevrolet recommendations.

♦ 20-20 CHANGING ENGINE OIL

To change the engine oil, raise the car on a lift. Place the oil drain pan in position, and remove the drain plug from the oil pan (Fig. 20-12). After the oil is drained, install the plug and lower the car. Then add the recommended amount and grade of oil to the engine. Start the engine and check for leaks.

Note Always put on a new lubrication sticker with the car mileage and date when the oil was changed.

♦ 20-21 OIL-FILTER SERVICE

The oil-filter element should be serviced or replaced regularly. Many manufacturers recommend servicing the oil filter with the first oil change. After that, the oil filter should be serviced with *every other* oil change. Some filters have replaceable elements. On these, a cover is removed, the old element taken out, the inside of the housing cleaned, and the new filter element installed. Then the cover is reinstalled.

On most engines with full-flow oil filters, the filter element and container are replaced as a unit. For example, on the type shown in Fig. 20-13, the old filter is unscrewed and a new filter installed by hand. A drip pan should be placed under the old filter as it is removed, to catch any oil that runs out. With the old filter off, the recess and sealing face of the filter bracket should be wiped with a clean shop towel. Then, the sealing gasket of the new filter should be coated with oil. Finally, the new filter should be hand-tightened until the gasket comes up against the bracket face. It should then be hand-tightened another half turn. After installation, the engine should be operated at fast idle to check for leaks. Check the oil level in the crankcase and add oil if necessary.

Careful Engine oil should be changed before the new filter is installed. A new filter should always start out with new oil.

After a filter element or filter is replaced, the date and the mileage should be marked on the lubrication or maintenance sticker. Then, after the specified time or mileage, the driver and technician will know that filter service is due again.

♦ 20-22 OIL-PAN SERVICE

Whenever the pan is removed for engine or oil-pump service, the pan, oil screen, and oil pump should be cleaned. Before installing the oil pan, be sure that all the old gasket material has been scraped from the gasket surfaces on the pan and

Type of Use	Change Interval
• Operating in dusty areas. • Towing a trailer. • Idling for extended periods and/or low-speed operation such as police, taxi, or door-to-door delivery service. • Operating when outside temperatures remain below freezing and when most trips are less than 4 miles [6.4 km].	• Change engine oil and filter every 3000 miles [4800 km] or 3 months, whichever comes first.
• Operating on a daily basis, as a general rule, for several miles and when none of the above conditions apply.	• Change engine oil every 7500 miles [12,000 km] or 12 months, whichever comes first. Change engine oil filter at first oil change, then every other oil change if mileage determines when you change oil. If time determines change intervals, change the filter with each oil change.

Fig. 20-10 Chart showing recommended oil-change intervals for different types of vehicle use. *(Chevrolet Motor Division of General Motors Corporation)*

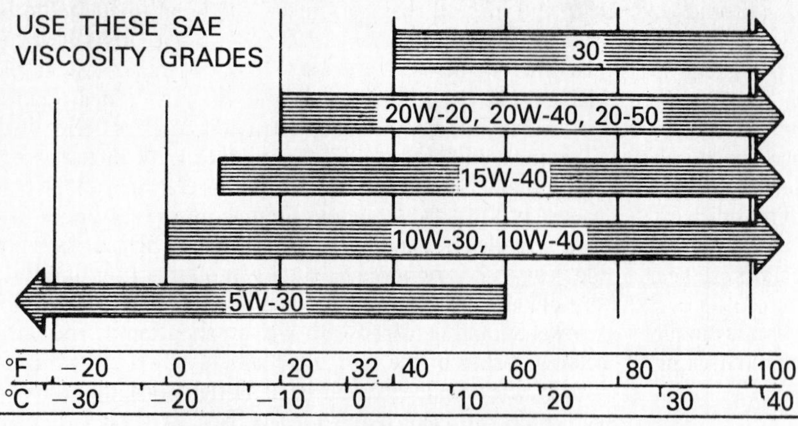

Fig. 20-11 Recommendations for proper engine lubricating oils to use for different temperature ranges in which the vehicle will operate. *(Chevrolet Motor Division of General Motors Corporation)*

block. On a sheet-metal pan, check the gasket surfaces for flatness. Many times, the screws have been overtightened, raising the metal around the screw holes. This will prevent the new gasket from sealing. Tap the holes lightly with a hammer until the gasket surfaces are flat.

Some engines use preformed gaskets (♦5-19). Others use formed-in-place gaskets of RTV silicone rubber (♦5-20). If using a preformed gasket, be sure the gasket is installed right side up, and that the boltholes in the gasket and pan align. Install the oil pan and tighten the bolts to the proper torque.

Fig. 20-12 Draining oil in preparation for adding fresh engine oil. *(Mobil Oil Company)*

♦ 20-23 OIL-PUMP SERVICE

Oil pumps require little service in normal operation. If a pump is badly worn, it will not maintain oil pressure. Remove it for repair or replacement. Refer to the manufacturer's shop manual for details of removal and installation, or replacement.

♦ 20-24 RELIEF-VALVE SERVICE

Relief valves are not usually adjustable. However, springs of different tension may be installed to change the regulating pressure. This is not usually recommended because a spring of the proper tension was originally installed on the engine. Any change of pressure is usually brought about by some defect that requires correction. For example, badly worn bearings may pass so much oil that the oil pump cannot maintain normal pressure in the lines. Installing a stronger spring in the relief valve will not increase oil pressure. The relief valve does not operate when there is low oil pressure.

♦ 20-25 SERVICING OIL-PRESSURE INDICATORS

Oil-pressure indicators are discussed in ♦20-10. These units require very little service. Defects in either the instrument-panel unit or the engine unit usually require replacement of the defective unit. If the indicator is not functioning normally, a new engine unit may be temporarily substituted for the old one. This will determine whether the fault is in the engine unit or the instrument-panel unit.

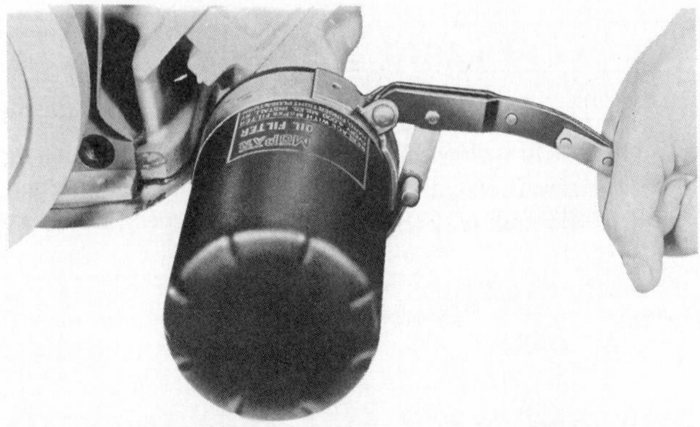

Fig. 20-13 Removing an oil filter with one type of oil-filter wrench. *(Chrysler Corporation)*

♦REVIEW QUESTIONS♦

Select the *one* correct, best, or most probable answer to each question. You can find the answers in the section indicated at the end of each question.

1. In addition to lubricating engine parts and acting as a cooling agent, the lubricating oil must (♦20-2)
   a. improve carburetion, aid fuel pump, and seal
   b. increase clearances, cool engine, and clean
   c. cool engine, improve combustion, and seal bearings
   d. absorb shocks, seal, and clean
2. The two types of oil pumps in automotive engines are (♦20-6)
   a. gear and piston
   b. rotor and piston
   c. gear and rotor
   d. full flow and bypass
3. The most common type of oil-pressure indicator uses a (♦20-10)
   a. gauge
   b. tube
   c. light
   d. meter
4. None of the service classifications for engine oils listed below should be used in automobile engines today *except* (♦20-5)
   a. SB
   b. CB
   c. SA
   d. SF
5. Oil for use in automotive diesel engines must be marked (♦20-5)
   a. high sulfur or low viscosity
   b. heavy duty or synthetic
   c. SF/CC or SF/CD
   d. none of the above
6. The purpose of the relief valve in the lubricating system is to (♦20-7)
   a. ensure minimum pressure
   b. prevent excessive pressure
   c. prevent insufficient lubrication
   d. ensure adequate oil circulation
7. Two types of oil filters used in automotive engines are (♦20-9)
   a. full flow and bypass
   b. open and closed
   c. low pressure and high pressure
   d. full flow and flow through
8. The purpose of crankcase ventilation is to (♦20-12)
   a. remove liquid gasoline and water
   b. remove sludge and sediment
   c. cool the oil
   d. supply oxygen to the crankcase
9. Viscosity can be defined as (♦20-3)
   a. ease of flow
   b. resistance to foaming
   c. resistance to flowing
   d. none of the above
10. The substance added to the oil which helps keep the engine clean is called a (♦20-3)
    a. detergent-dispersant
    b. soap
    c. grease
    d. thickening agent
11. Most of the dilution of the oil in the crankcase takes place during (♦20-4)
    a. high-speed operation
    b. long trips
    c. engine overheating
    d. engine warm-up
12. Oil is lost from the engine in three ways, by passing as a mist through the PCV system, by leaking in liquid form, and by (♦20-15)
    a. evaporating
    b. burning in the combustion chambers
    c. condensing
    d. leaking into the transmission
13. Oil can enter the combustion chambers in two ways—around the valve stems and (♦20-15)
    a. past the float-bowl needle
    b. past the manifold gaskets
    c. through the intake manifold
    d. past the piston rings
14. Common causes of excessive oil consumption include (♦20-15)
    a. heavy oil and tight bearings
    b. high engine speed and worn engine parts
    c. short trips and cold weather
    d. frequent oil changes and weak valve springs
15. Oil-filter elements, according to usual recommendations, should be replaced every (♦20-21)
    a. 2000 miles [3219 km]
    b. oil change
    c. other oil change
    d. two years

CHAPTER 21
Engine Cooling Systems

After studying this chapter, you should be able to:

1. Describe the operation of the two types of engine cooling systems.
2. Explain the operation of the water pump.
3. Define *variable-speed fan*.
4. Discuss the flow of coolant through the two types of automotive radiators.
5. Explain why cars with automatic transmissions are equipped with a transmission oil cooler.
6. Discuss how the heater works.
7. Explain the operation of the thermostat.
8. Name the two valves used in the radiator pressure cap.
9. Define *antifreeze*.

♦ 21-1 PURPOSE OF THE COOLING SYSTEM

The purpose of the cooling system is to keep the engine at its most efficient operating temperature at all speeds and under all operating conditions. During the combustion of the air-fuel mixture in the engine cylinders, temperatures of 4000°F [2200°C] or higher may be reached by the burning gases. Some of this heat is absorbed by the cylinder walls, cylinder head, and pistons. They, in turn, must be provided with some means of cooling so that they will not get too hot.

Cylinder-wall temperature must not go higher than about 400 to 500°F [205 to 260°C]. Temperatures higher than this cause the lubricating-oil film to break down and lose its lubricating properties. However, the engine operates best at temperatures as close to the limits imposed by oil properties as possible. Removing too much heat through the cylinder walls and head lowers the thermal efficiency of the engine. Cooling systems are designed to remove about one-third (30 to 35 percent) of the heat produced in the combustion chambers by the burning of the air-fuel mixture.

The engine is very inefficient while cold. Therefore the cooling system includes devices that prevent normal cooling action during engine warm-up. These devices allow the engine parts to reach their normal operating temperatures more quickly. This shortens the inefficient cold-operating time. When the engine reaches its normal operating temperature, the cooling system begins to function. The cooling system removes excess heat when the engine is hot, and slowly or not at all when the engine is cold or warming up.

Two general types of cooling systems are used on automobile engines. They are *air cooling* and *liquid cooling.* Most automotive engines are liquid-cooled. Most engines for airplanes, snowmobiles, motorcycles, power lawn mowers, and chain saws are air-cooled.

♦ 21-2 AIR-COOLED ENGINES

In air-cooled engines, the cylinders are semi-independent. They are not grouped in a block. There are metal fins on the heads and cylinders to help dissipate the heat from the en-

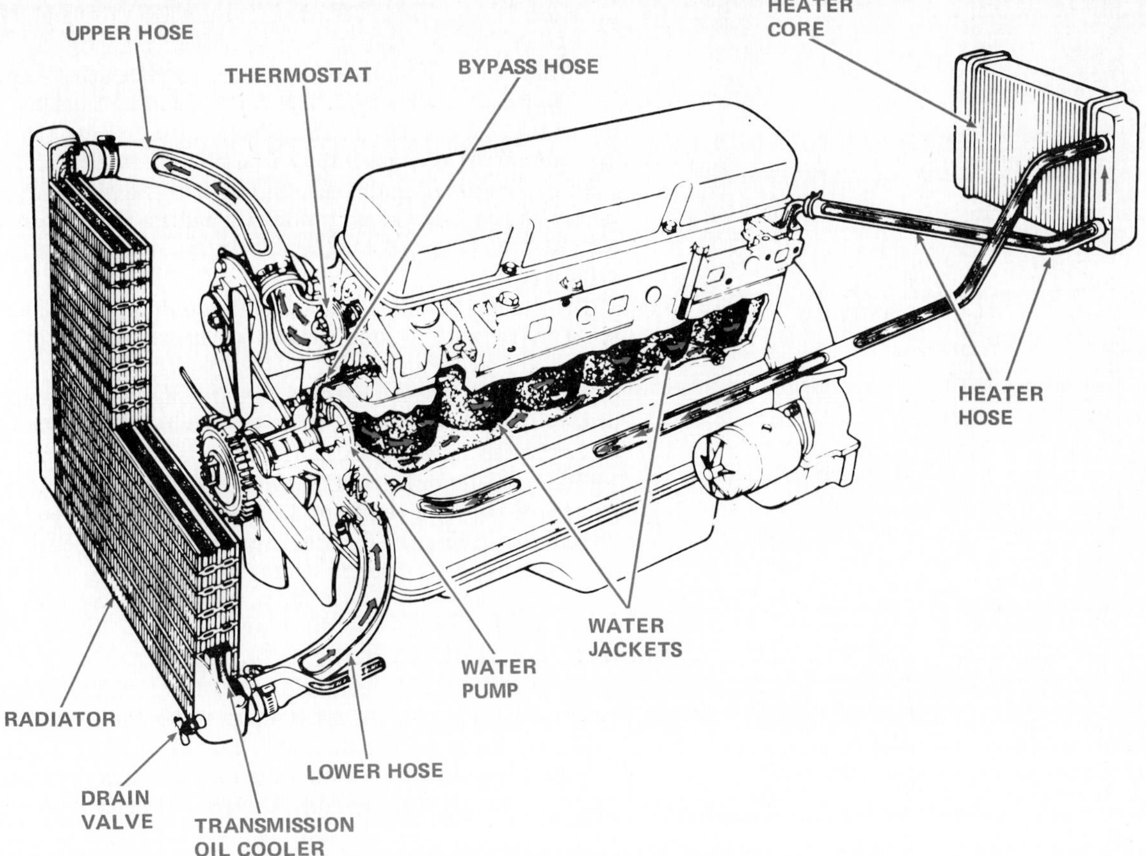

Fig. 21-1 Liquid-cooling system for an automotive engine, including the car heater. The small arrows show the direction of coolant flow. *(Chrysler Corporation)*

gine. Shrouds and fans are used on some air-cooled engines to improve the air circulation around the cylinders and heads.

♦ 21-3 LIQUID-COOLED ENGINES

In operation, the engine cooling system is really a temperature-regulating system. It maintains the engine temperature within certain limits, neither too hot nor too cold. A cold engine is inefficient. Combustion of the air-fuel mixture is incomplete. The engine may run roughly, have excessive fuel consumption and exhaust emissions, and contaminate the crankcase with excessive blowby and liquid fuel dripping into it. The job of the cooling system is to get the engine up to normal operating temperature as quickly as possible and then maintain it at that temperature. The cooling system should not allow overheating or overcooling.

In the liquid-cooling system (Fig. 21-1), a liquid is circulated around the cylinders to absorb heat from the cylinder walls. The liquid is water to which an antifreeze solution (♦21-16) is added to prevent freezing in cold weather. The mixture is called the coolant. The coolant absorbs heat as it passes through the engine. Then the hot coolant flows through a radiator in which the heat in the coolant is passed on to air that is flowing through the radiator. The cooled coolant then flows back through the engine. This circulation of the coolant continually removes heat from the engine. The coolant is kept in circulation by the water pump.

♦ 21-4 WATER JACKETS

Water jackets are designed to keep the cylinder block and cylinder head cool. The water jackets are open spaces between the outside wall of the cylinder and the inside of the cylinder block and head (Fig. 21-2). The coolant can circulate freely around the engine hot spots. These include the valve guides and valve seats, and the upper parts of the cylinder walls where the pistons and rings slide up and down.

When the engine is running at normal temperature, coolant flows into the block and through the water jackets surrounding the cylinders. Then the coolant is forced through the head gasket openings and into the cylinder-head water jackets. In the head, the coolant flows around the combustion chambers and valve seats, picking up additional heat. From the heads, the coolant flows through the upper hose into the radiator. There, the coolant temperature is lowered, and the coolant is drawn again into the engine by the water pump.

♦ 21-5 WATER PUMPS

Water pumps are impeller-type centrifugal pumps. On many engines, they are attached to the front end of the cylinder block between the block and the radiator, as shown in Figs. 21-1 and 21-3. On other engines the water pump is bolted on the side of the block. The pump (Figs. 21-4 and 21-5) can circulate up to 7500 gallons [28,390 L] of coolant per hour between the water jackets and the radiator. The pump consists of a housing, with a coolant inlet and outlet, and an

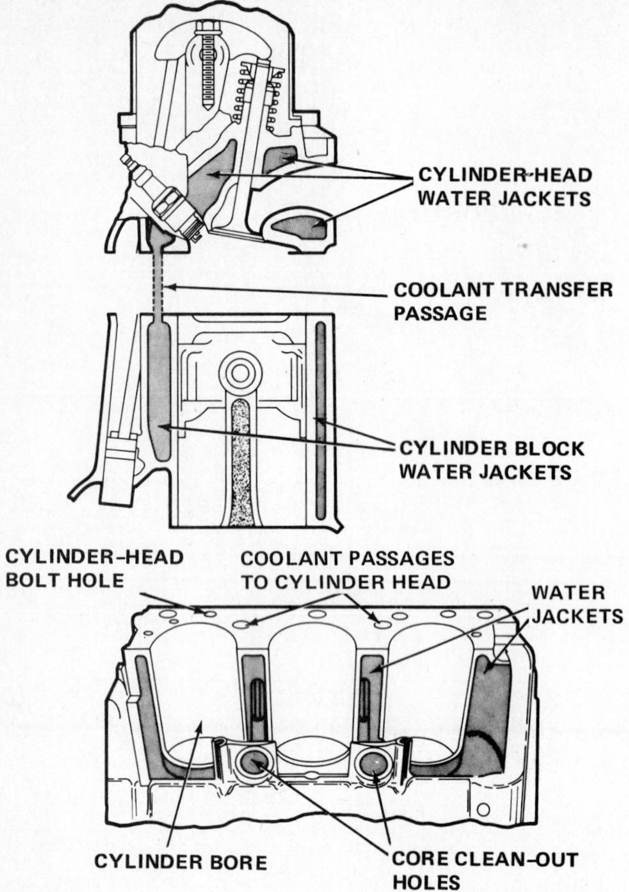

Fig. 21-2 Water jackets in cylinder head and cylinder block.

impeller. The impeller is a flat plate mounted on the pump shaft with a series of flat or curved blades, or vanes.

When the impeller rotates, the coolant between the blades is thrown outward by centrifugal force. Then the coolant is forced through the pump outlet and into the cylinder block. The pump inlet is connected by a hose to the bottom of the radiator. Coolant from the radiator is drawn into the pump to replace the coolant forced through the outlet. The arrows in Fig. 21-1 show how the coolant flows through the cooling system.

The impeller shaft is supported on one or more bearings. A seal prevents coolant from leaking out around the bearings. Most water pumps use sealed bearings which never need lubrication. With a sealed bearing, grease cannot leak out, and dirt and water cannot get in. Older water pumps had bearings that required water-pump lubricant, or soluble oil, mixed with the coolant for lubrication. On longitudinal engines, the pump is driven by a belt from the pulley on the front end of the crankshaft (Fig. 21-6). On transverse engines, the fan is usually driven by an electric motor (♦21-10).

♦ 21-6 ENGINE FAN

The engine fan usually mounts on the water-pump shaft (on longitudinal engines). The fan is driven by the same belt that drives the pump and the alternator (Fig. 21-6). The purpose of the fan is to pull air through the radiator. This improves cooling at slow speeds and idle. At higher car speeds (above about 40 mph [64 km/h]), the air rammed through the radiator by the forward motion of the vehicle provides all the cooling air that is needed.

The fan usually has from four to six blades which in rotating pull air through the radiator. Many cars have a fan shroud

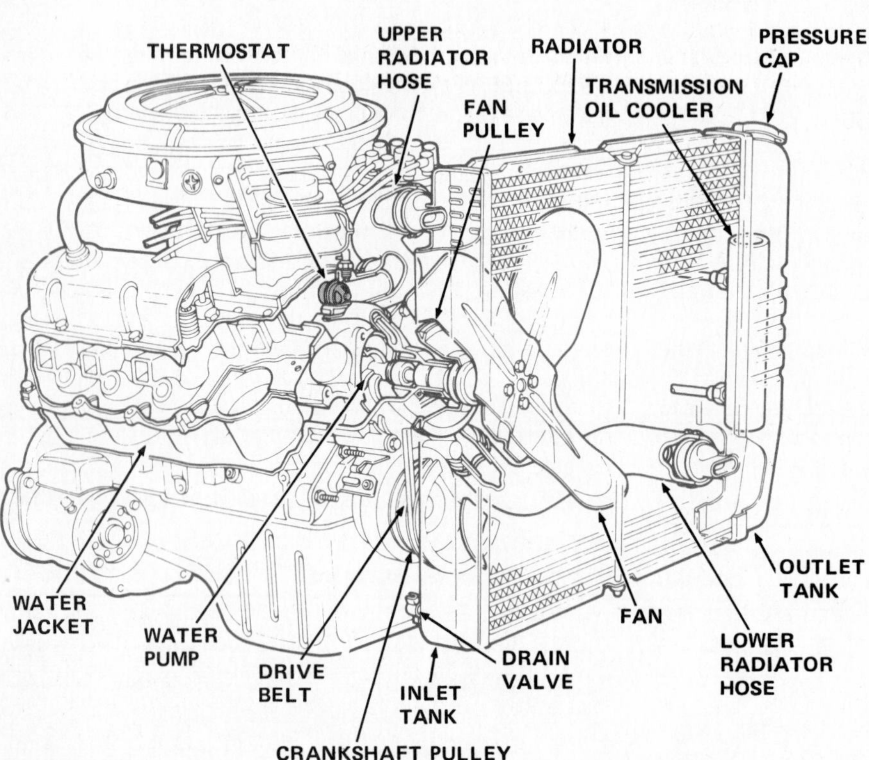

Fig. 21-3 A V-8 engine cut away to show the cooling system. *(Ford Motor Company)*

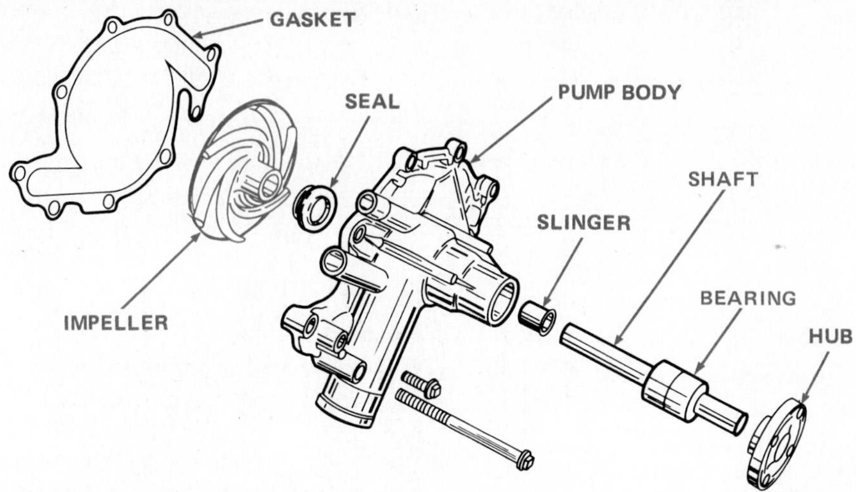

Fig. 21-4 A disassembled water pump. *(Chrysler Corporation)*

that improves fan performance. The shroud increases the efficiency of the fan, since the shroud assures that all air pulled back by the fan must first pass through the radiator.

♦ 21-7 VARIABLE-SPEED FAN

Many engines are equipped with a variable-speed fan. This fan will not exceed a predetermined speed or will rotate only as fast as needed to keep the engine from overheating. The fan control includes a small fluid coupling and a thermostatic device.

One type is shown in Figs. 21-7 and 21-8. It has a small fluid coupling that is partly filled with a special silicone oil. When the temperature of the air coming through the radiator rises, the heat causes the bimetal coil spring on the drive to expand. As the coil expands, it slightly turns the shaft in the center of the drive. This allows more oil to enter the fluid coupling, which now begins to rotate the fan. But when the engine is cold, the fluid coupling slips, and the fan coasts. This reduces noise and saves engine power, thereby reducing fuel consumption.

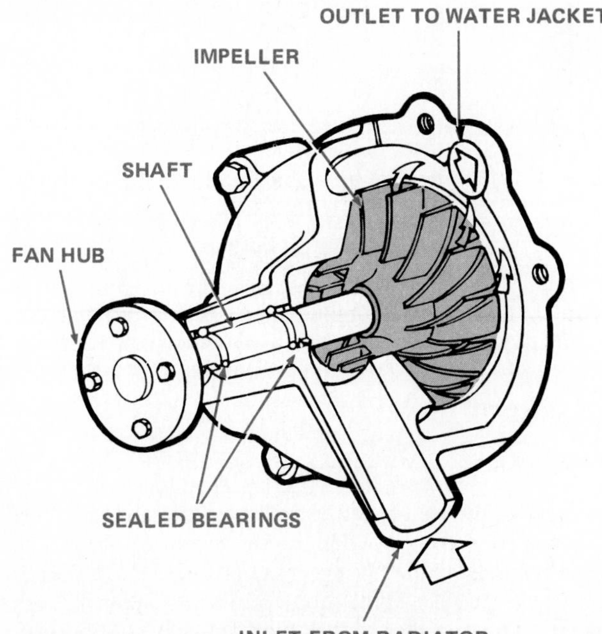

Fig. 21-5 Impeller action in the water pump. *(Chrysler Corporation)*

Another type of temperature-controlled fan uses a flat bimetal strip spring instead of a coil spring. The two outer ends of the strip are attached to the drive face. The center of the strip is attached to the control piston in the center of the drive.

When the temperature in the engine rises, the air flowing through the radiator gets hotter. The hot air causes the metal strip to expand and to bow outward. As the strip bows, it pulls the control piston outward. This allows more oil to flow into the fluid coupling. The fan then speeds up for improved cooling.

As the engine temperature drops, the thermostatic strip straightens, forcing the control piston in. This causes oil to leave the fluid coupling. With the drive uncoupled, the fan slows to its normal coast speed.

The fan bolts to the drive, and then attaches by the drive bolts through the pulley to the hub of the water pump. The pulley, driven by a belt from the crankshaft, turns while the engine is running. Because the flange on the drive shaft is bolted to the pulley, the shaft also turns. But there is no direct mechanical connection through the fluid coupling in the drive. The fluid coupling must fill with oil before the outside housing of the drive rotates. Since the fan bolts to the drive housing, the fan can be driven at normal speed only after the fluid coupling locks up.

♦ 21-8 FLEXIBLE-BLADE FAN

Another way to reduce the power needed to drive the fan, and to reduce fan noise, is to use a fan with flexible plastic blades. Figure 21-9 shows the installation of this type of fan. In operation, the pitch of the blades decreases as fan speed increases. This is a result of a flattening effect on the blades caused by centrifugal force. Each blade therefore takes a smaller bite of air per revolution. Airflow is reduced because of the shallower pitch. This, in turn, saves power and lowers the noise level.

In Fig. 21-9, note the use of the spacer between the water-pump pulley and the fan. Various sizes and shapes of spacers are used to properly position the fan. For example, on radiators without a shroud, one recommendation is to position the fan in the center of the radiator with the blades 2 inches [51 mm] in back of it. The distance between the blades and the radiator allows the engine to shift in its mounts without the blades striking the radiator fins.

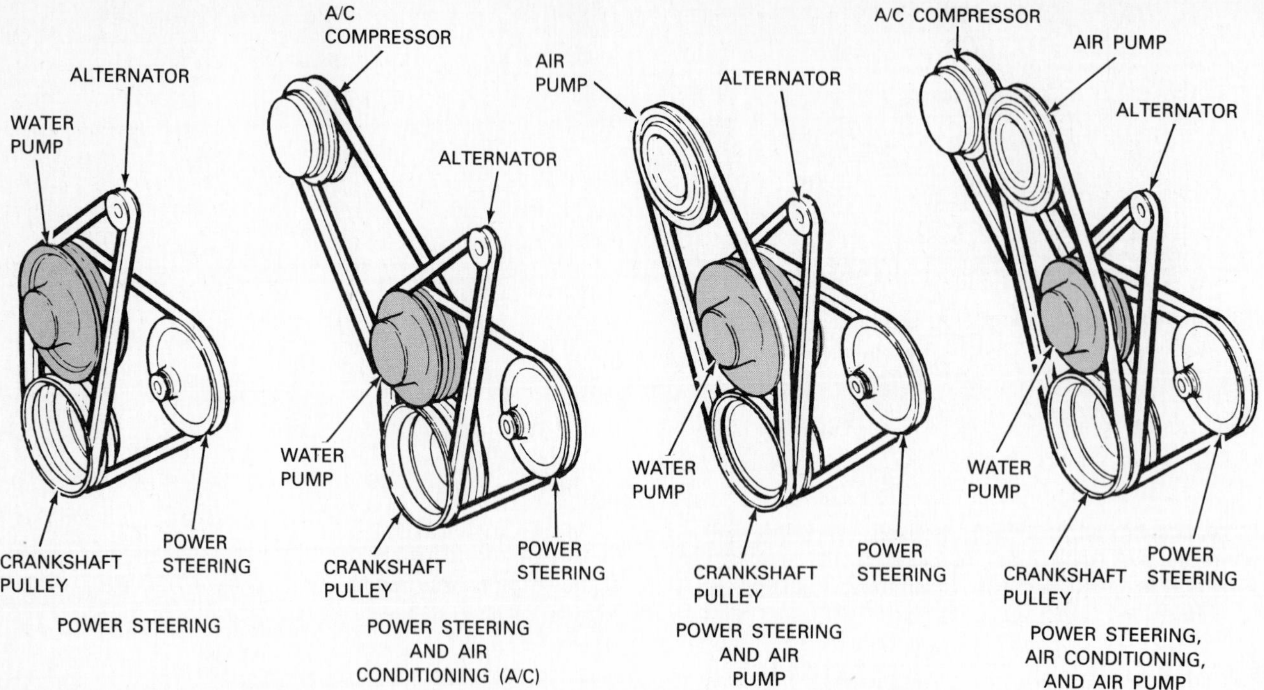

Fig. 21-6 The water pump and fan are driven from the crankshaft pulley by the same belt that drives the alternator. Different belt combinations are used for engines equipped with power steering, air conditioning, and an air pump (an emission control device). *(Pontiac Motor Division of General Motors Corporation)*

Fig. 21-7 A fan-drive clutch, or fluid coupling, installed between the fan and the water-pump shaft. *(Ford Motor Company)*

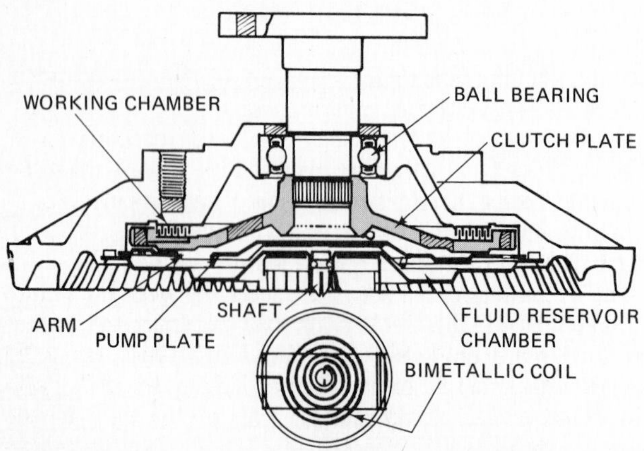

Fig. 21-8 Sectional view of fan fluid coupling. *(Chevrolet Motor Division of General Motors Corporation)*

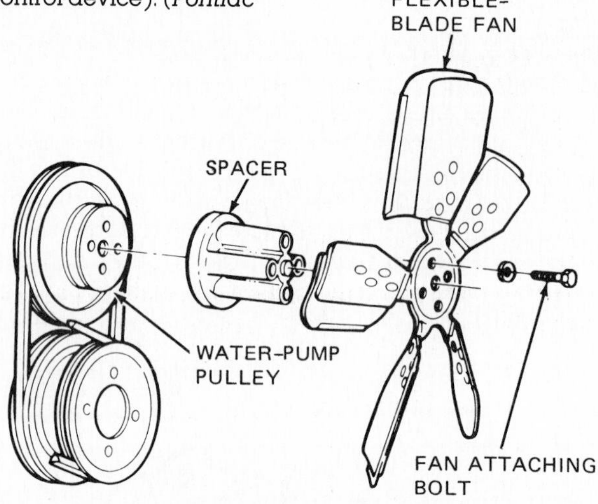

Fig. 21-9 Attachment of a flexible-blade fan to the spacer that mounts on the water-pump pulley. *(Ford Motor Company)*

♦ CAUTION ♦ Fan blades can break and fly off. Whenever the engine is running, never stand directly in line with the rotating fan. Also, avoid allowing your clothing, hands, or tools to get near the fan blades or fan belt.

♦ 21-9 DRIVE BELTS

A fan belt, or drive belt, uses friction, tension, and proper fit in the pulley groove to turn the water pump and fan. Most drive belts are V-shaped. When the belt is properly tightened, it wedges into the pulley groove. Friction between the sides of the belt and the sides of the groove transmits power through the belt from one pulley to the other. Power is *not* transmitted through the bottom of a V-belt.

The V-belt provides a large surface area on its sides in contact with the pulley groove. Because of this, considerable

Fig. 21-10 A single serpentine belt drives all the belt-driven engine accessories. *(Ford Motor Company)*

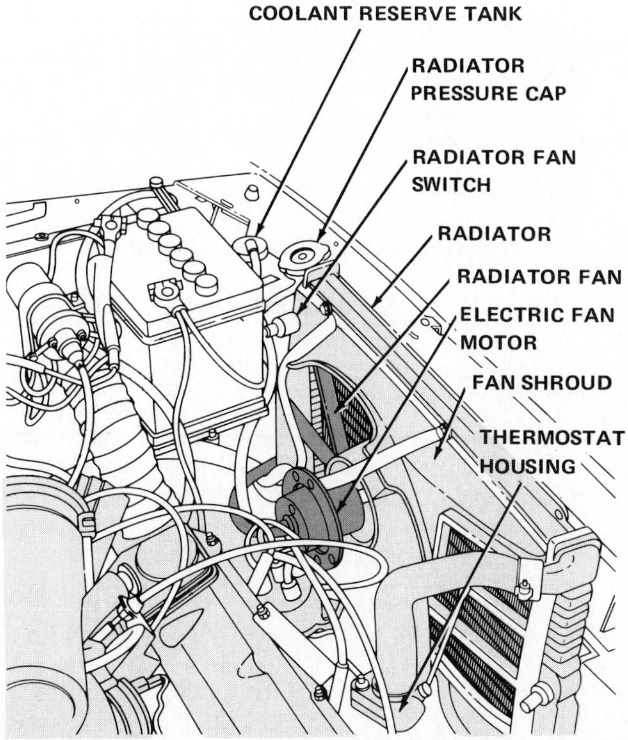

Fig. 21-11 A typical radiator-fan arrangement used with a transverse-mounted engine. The radiator fan is driven by an electric motor. *(Chrysler Corporation)*

power can be transmitted by a V-belt. The wedging action of the belt as it curves into the pulley groove aids in preventing belt slippage. Figure 21-3 shows a V-belt in place on an engine. The belt transmits power from the crankshaft pulley to the water-pump pulley, and to the alternator pulley.

Note Some engines use two V-belts with double pulleys, which are pulleys that have two belt grooves. The added belt provides the power needed to drive the alternator, water pump, and other accessories. These belts are a matched set. If one belt must be replaced, then replace both belts at the same time. Otherwise the new belt will carry most of the load. It will wear rapidly, and may slip. The reason is that the remaining old belt will have stretched slightly. Therefore it will be looser, after the new belt is installed, and will carry less of the load.

Figure 21-10 shows an engine with a *serpentine belt.* On some newer cars, one serpentine belt drives all belt-driven engine accessories. It replaces several V-belts that otherwise would be required. The serpentine belt is wide and has a series of small V-shaped ridges and grooves on the inside surface. The pulleys have matching series of grooves and ridges.

♦ 21-10 ELECTRIC FANS

Engines mounted crosswise at the front (and driving the front wheels) often use another method of driving the engine fan. Figure 21-11 shows a typical arrangement. The radiator is located at the front, as usual. The fan is driven by an electric motor. A thermostatic switch turns the motor on only when it is needed. For example, in one engine, the switch turns the motor on when the coolant reaches 193 to 207°F [89 to 97°C]. It turns the motor off if the coolant temperature drops below these figures. On cars with air conditioning, the thermostatic switch is bypassed and the motor runs all the time the air conditioner is turned on.

The advantage of the electric fan is less power drain on the engine and less fan noise. Also, there is no fan belt to inspect, adjust, or replace. This means less cooling-system maintenance.

Another way to drive the fan when the engine is mounted transversely is to have the drive belt "turn the corner." In this arrangement, the radiator is in its normal position facing the front of the car. Idler pulleys are positioned so that the belt is turned 90 degrees. The crankshaft pulley still drives the belt, but the fan pulley is at a 90 degree angle to the crankshaft pulley.

♦ CAUTION ♦ An electric fan can start running at any time, even with the engine off and key removed from the ignition switch.

♦ 21-11 RADIATOR

In the cooling system, the *radiator* is a heat exchanger that removes heat from coolant passing through it. The radiator holds a large volume of coolant in close contact with a large volume of air so that heat will transfer from the coolant to the air. The radiator core is divided into two separate compartments. Coolant passes through one, and air passes through the other. Several types of radiator cores have been used. The most common type is the tube-and-fin.

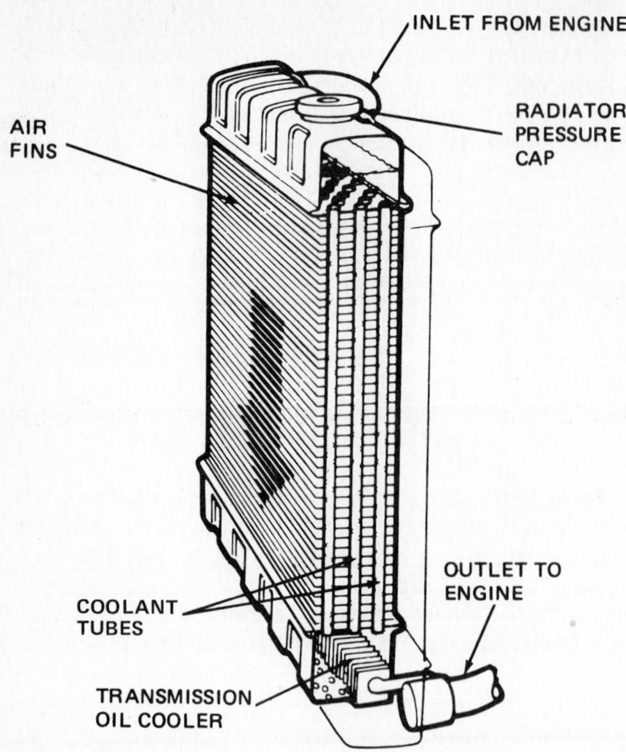

Fig. 21-12 Construction of a down-flow type of tube-and-fin radiator, which includes a transmission oil cooler. *(Chrysler Corporation)*

A tube-and-fin radiator (Fig. 21-12) consists of a series of tubes extending from the top to the bottom or from side to side of the radiator. The tubes run from the inlet tank to the outlet tank. Fins are placed around the outside of the tubes to improve heat transfer. Air passes between the fins. As the air passes by, it absorbs heat from the fins which have, in turn, absorbed heat from the coolant.

In a typical radiator, there are five fins per inch [25.4 mm]. Radiators used in cars that have factory-installed air conditioning have seven fins per inch [25.4 mm]. This provides the additional cooling surface required to handle the additional heat load imposed by air conditioning.

On every radiator there is a small inlet tank at the top or side. Hot coolant from the engine flows into this tank. On the bottom or other side of the radiator, there is another tank, the outlet tank. This tank receives the cool coolant after it has passed through the radiator core. On some vehicles, one of the tanks has a filler cap (the radiator cap) which can be removed to add coolant to the cooling system. On other vehicles, the expansion tank (♦21-12) has the cap which can be removed to add coolant.

Radiators can be classified according to the direction that the coolant flows through them. In some, the coolant flows from top to bottom (down-flow type) as shown in Fig. 21-12. Until about 1970, most automotive radiators were of this type.

In the down-flow radiator, the top tank has a radiator filler neck, sealed by a radiator pressure cap (♦21-15). The cap is removed to add coolant to the cooling system on some cars. On others, the coolant is added to the expansion tank (♦21-12). On cars with automatic transmission, the outlet tank has an oil cooler (♦21-13). Some radiators also have a drain valve in the bottom tank.

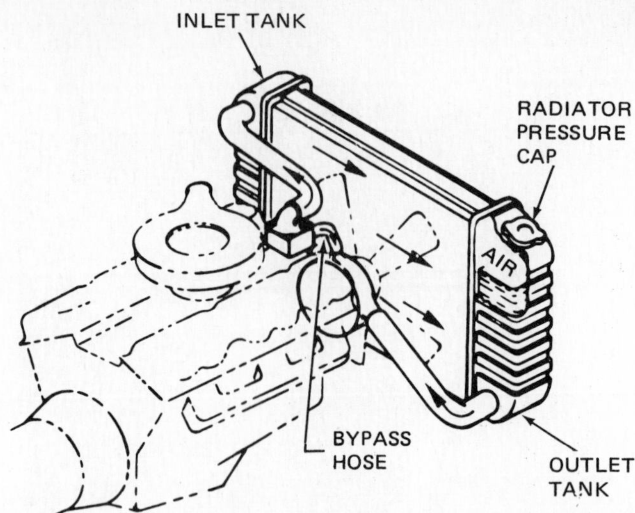

Fig. 21-13 Cooling system using a cross-flow radiator. *(Harrison Radiator Division of General Motors Corporation)*

Most late-model cars have a cross-flow radiator (Figs. 21-11 and 21-13). In this type, the coolant flows horizontally from the inlet tank on one side to the outlet tank on the other side. Basically, the cross-flow radiator is a down-flow radiator turned on its side. This allows the car body to be designed with a lower hood line. The outlet tank contains the transmission oil cooler, where used.

On any radiator, the inlet tank (above or to one side) serves two purposes. It provides a reserve supply of coolant. It also provides a place where the coolant can be separated from any air that might be circulating in the cooling system.

♦ 21-12 EXPANSION TANKS

Many cooling systems have a separate plastic coolant reservoir or expansion tank (Figs. 21-14 and 21-15). The expansion tank, also called the *recovery tank,* is partly filled with coolant and is connected to the overflow tube from the radiator filler neck. The coolant in the engine expands as the engine

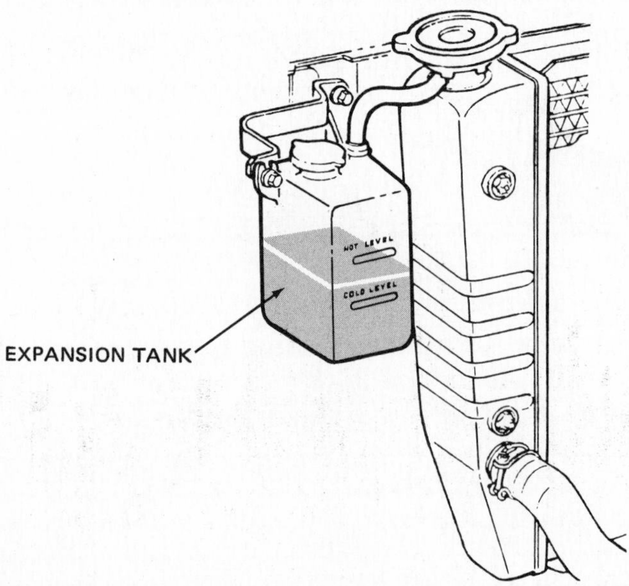

Fig. 21-14 Cooling system using an expansion tank. *(Ford Motor Company)*

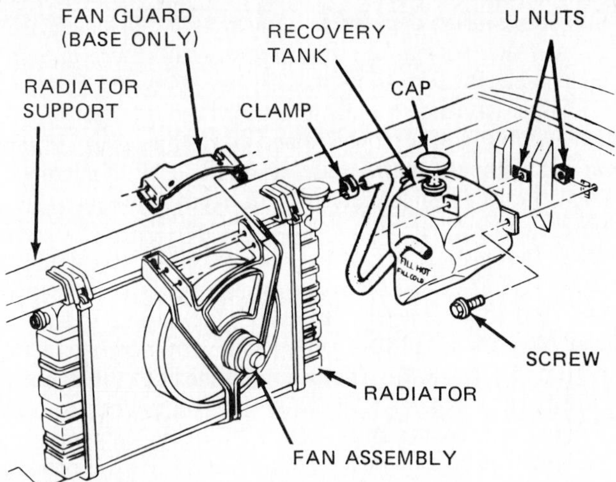

Fig. 21-15 Expansion tank installation for a transverse-mounted engine. *(Chevrolet Motor Division of General Motors Corporation)*

heats up. Instead of dripping out the overflow tube onto the street and being lost from the cooling system completely, the coolant flows into the expansion tank.

When the engine cools, a vacuum is created in the cooling system. The vacuum siphons some of the coolant back into the radiator from the expansion tank. In effect, a cooling system with an expansion tank is a closed system. Coolant can flow back and forth between the radiator and the expansion tank. This occurs as the coolant expands and contracts from heating and cooling. Under normal conditions, no coolant is lost.

An advantage to the use of an expansion tank is that it eliminates almost all air bubbles from the cooling system. Coolant without bubbles absorbs heat much better than coolant with bubbles in it. Although the coolant level in the expansion tank goes up and down, the radiator and cooling system are kept full. This results in maximum cooling efficiency.

♦ 21-13 TRANSMISSION OIL COOLERS

Cars with automatic transmissions are equipped with an oil cooler for the automatic-transmission fluid. The oil cooler is needed because the oil in automatic transmissions can overheat. Overheating reduces transmission performance and can damage the transmission.

The transmission oil cooler is usually a tube located in the outlet (bottom or side) tank of the radiator. Figures 21-1 and 21-3 show the location of the oil cooler in the outlet tank of a cross-flow radiator and a down-flow radiator. The oil-cooler tube, being immersed in the lower-temperature coolant, is cooled by the coolant. This cools the transmission fluid that is passing through. Attached to the outlet tank of the radiator are two metal tubes called the *transmission-cooler lines* (Fig. 21-16). They carry the fluid between the transmission and the oil cooler.

Figure 21-16 shows the use of an auxiliary oil cooler for the automatic-transmission fluid. Cars that are factory-equipped with a trailer-towing package often have an auxiliary oil cooler. It is mounted in front of the radiator and air-conditioner condenser. The auxiliary oil cooler is connected in series with the radiator oil cooler. When both oil coolers are used, the hot oil flows from the transmission through the radiator oil cooler. From there, the oil flows through the aux-

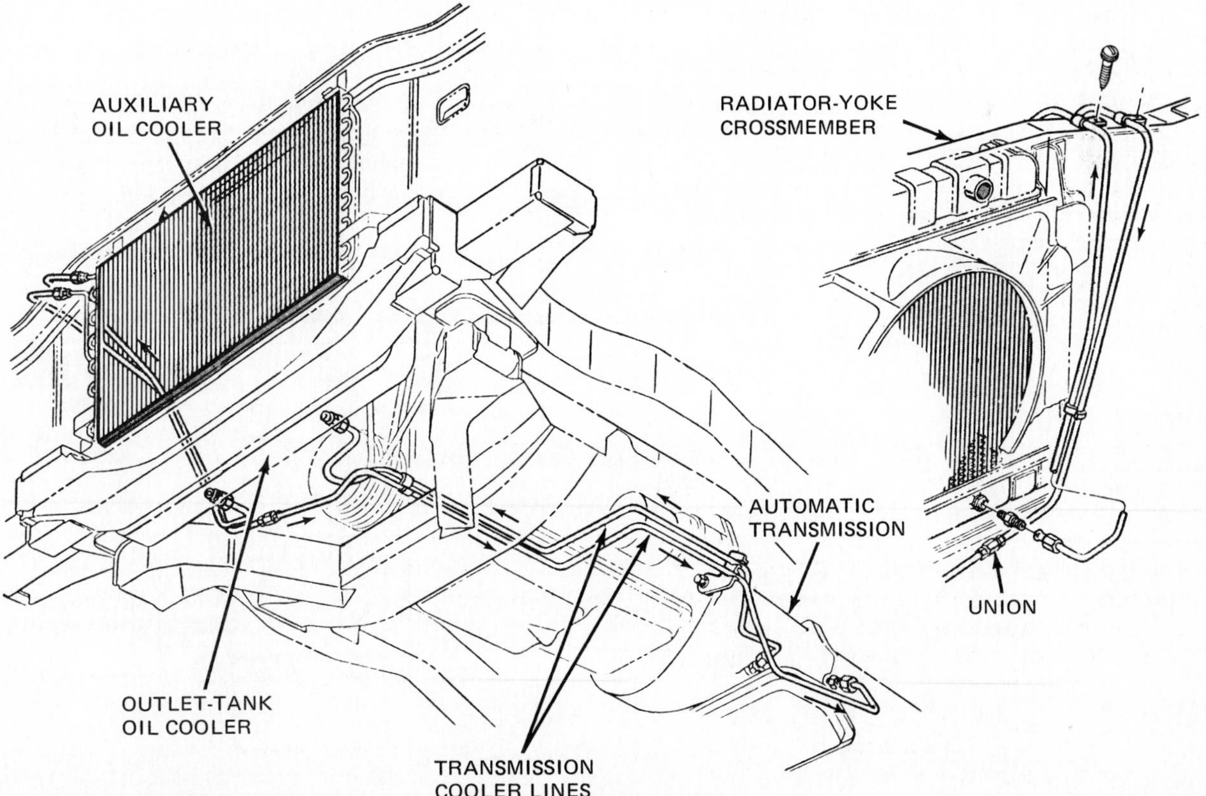

Fig. 21-16 An auxiliary oil cooler for the automatic-transmission fluid, which is installed in series with the oil cooler in the outlet tank of the radiator. *(Chrysler Corporation)*

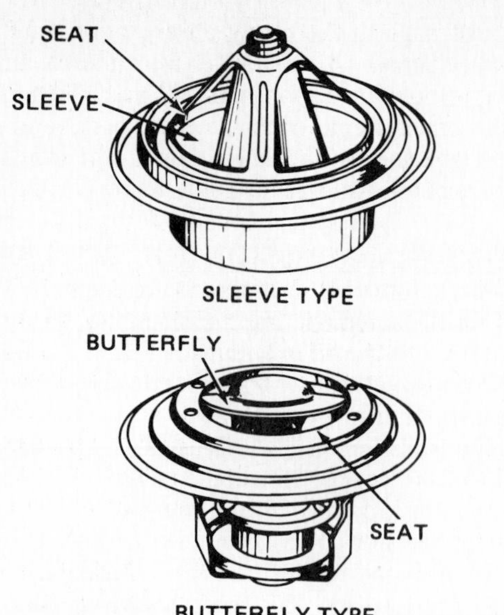

Fig. 21-17 Two types of thermostats for engine cooling systems. *(Chrysler Corporation)*

iliary oil cooler. Then the cool oil returns to the transmission. The small arrows in Fig. 21-16 show the flow of oil through this system.

The auxiliary oil cooler transfers heat from the oil to the cooler air passing through. This is similar to the operation of the radiator, which was discussed earlier. Both types of transmission oil coolers place a greater load on the engine cooling system. The radiator oil cooler transfers heat to the coolant. The auxiliary oil cooler, mounted in front of the radiator, raises the temperature of the air before it reaches the radiator.

♦ 21-14 THERMOSTAT

The thermostat is placed in the coolant passage between the cylinder head and the top of the radiator (Figs. 21-1 and 21-3). Its purpose is to close off this passage when the engine is cold. Now coolant circulation is restricted, causing the engine to reach normal operating temperature more quickly. This reduces the formation of acids, moisture, and sludge in an engine. Also, after warm-up, the thermostat keeps the engine running at a higher temperature than it would, for example, without a thermostat. The higher operating temperature improves engine efficiency and reduces exhaust emissions.

The thermostat consists of a thermostatic device and a valve (Fig. 21-17). Various valve arrangements and thermostatic devices have been used. Most thermostats today are operated by a wax pellet, which expands with increasing temperature to open a valve. The sleeve and the butterfly thermostats shown in Fig. 21-17 both use the wax pellet.

Thermostats are designed to open at specific temperatures. This temperature is known as the *rating* of the thermostat, and may be stamped on it. Two frequently used thermostats have ratings of 185°F [85°C] and 195°F [91°C]. Most thermostats *begin* to open at their rated temperature. They are fully open about 20°F [11°C] higher. For example, a thermostat with a rating of 195°F [91°C] starts to open at that temperature. It is fully open at about 215°F [102°C].

Most engines have a small coolant bypass passage. It permits some coolant to circulate within the cylinder block and head when the engine is cold and the thermostat is closed. This provides equal warming of the cylinders and prevents hot spots. When the engine warms up, the bypass must close or become restricted. Otherwise, the coolant would continue to circulate within the engine itself, and too little would go to the radiator for cooling.

The bypass passage may be an internal passage, or an external bypass hose. In Fig. 21-1, notice the small external bypass hose at the top of the water pump.

One internal bypass system uses a small, spring-loaded valve located in back of the water pump. The valve is forced open by coolant pressure from the pump when the thermostat is closed. As the thermostat opens, the coolant pressure drops within the engine and the bypass valve closes.

Another internal bypass system has a blocking-bypass thermostat (Fig. 21-18). This thermostat operates like those already described, but it also has a secondary, or bypass, valve.

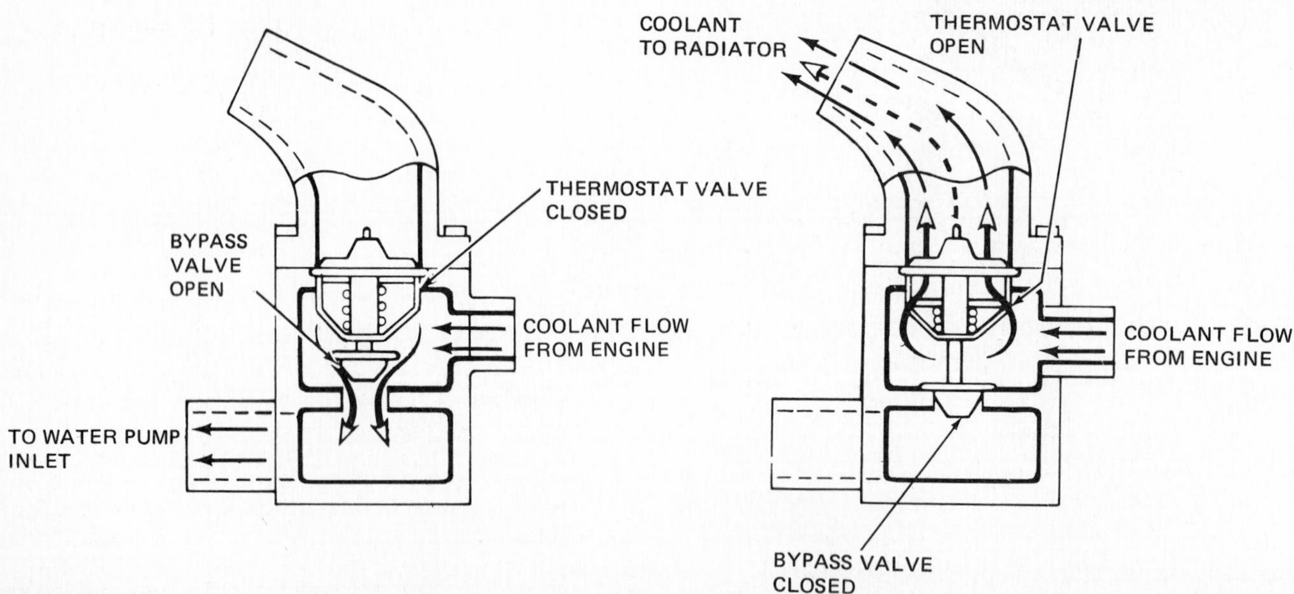

Fig. 21-18 Operation of a blocking-bypass thermostat. *(Chrysler Corporation)*

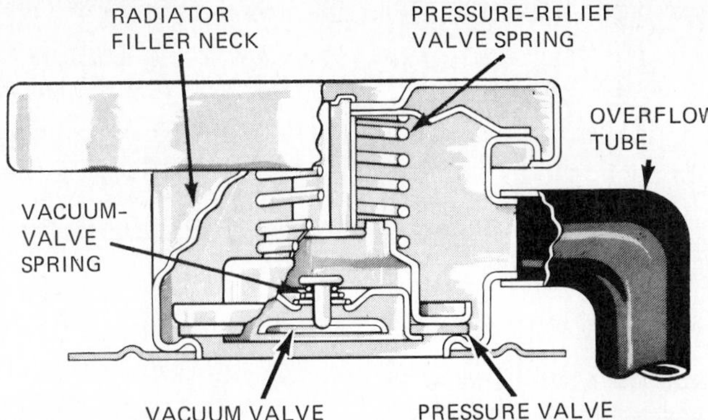

Fig. 21-19 A radiator pressure cap, showing the pressure valve and the vacuum valve. *(Ford Motor Company)*

When the thermostat valve is closed, the circulation to the radiator is shut off. However, the bypass valve is open, permitting coolant to circulate through the bypass. As the thermostat valve opens, permitting coolant to flow to the radiator, the bypass valve closes. This blocks off the engine bypass passage.

♦ 21-15 RADIATOR PRESSURE CAP

The cooling systems on automobile engines today are sealed and pressurized by a radiator pressure cap (Figs. 21-19 and 21-20). There are two advantages to sealing and pressurizing the cooling system. First, the increased pressure raises the boiling point of the coolant. This increases the efficiency of the cooling system. Second, sealing the cooling system reduces coolant losses from evaporation and permits the use of an expansion tank.

At normal atmospheric pressure, water boils at 212°F [100°C]. If the air pressure is increased, the temperature at which water boils is also increased. For example, if the pressure is raised to 15 psi [103 kPa] over atmospheric pressure, the boiling point is raised to about 260°F [127°C]. Every 1-psi [7-kPa] increase in pressure raises the boiling point of water about 3¼°F [1.8°C]. This is the principle upon which the pressurized cooling system works.

As the pressure goes up, the boiling point goes up. Therefore, the coolant can be safely run at a temperature higher than 212°F [100°C] without boiling. The higher the coolant temperature, the greater the difference between it and the outside air temperature. This difference in temperatures is what causes the cooling system to work. The hotter the coolant, the faster the heat moves from the radiator to the cooler passing air. This means that the pressurized, sealed cooling system can take heat away from the engine faster. Therefore, the cooling system works more efficiently when the coolant is under higher pressure.

However, the cooling system can be pressurized too much. If the pressure in the system gets too high, it can damage the radiator and blow off the hoses. To prevent this, the radiator cap has a pressure-relief valve (Figs. 21-19 and 21-20). When the pressure gets too high, it raises the valve so that the excess pressure can escape into the expansion tank (Fig. 21-21).

The radiator pressure cap also has a vacuum vent valve (Figs. 21-19 and 21-20). This valve protects the system from developing a vacuum that could collapse the radiator. When the engine is shut off and cools, the coolant volume is reduced. Cold coolant takes up less space than hot coolant. As the temperature of the coolant drops, a vacuum develops in the cooling system. To prevent excessive vacuum from developing, the vacuum valve opens to allow outside air or coolant from the expansion tank to flow into the cooling system (Fig. 21-21). This relieves the vacuum that could otherwise cause outside air pressure to collapse the radiator.

The radiator pressure cap is removed and installed as follows. When the cap is placed on the filler neck, the locking lugs on the cap fit under the filler-neck flange (Fig. 21-20). As the cap is turned, the cam locking surface of the flange tightens the cap. It also preloads the pressure-relief-valve spring.

To remove the cap, press down and slowly turn the cap back to the safety stop. Never remove the cap when the engine is hot. Boiling coolant and steam can erupt from the filler neck. See the following Caution.

Several types of safety caps are also used on radiators. The safety cap has a button or lever on top of the cap. By pressing the button, or lifting the lever, the pressure is relieved. This eliminates the chance that steam or boiling coolant could scald you when the cap is removed.

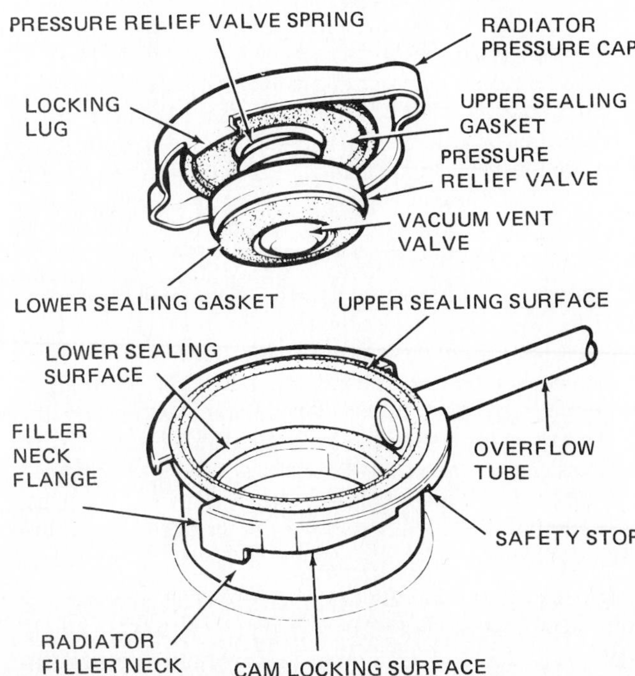

Fig. 21-20 A radiator pressure cap removed from the radiator filler neck. *(Chrysler Corporation)*

♦ CAUTION ♦ Never remove the cap from a hot engine! The coolant can boil over as the pressure is released. The boiling coolant and steam could burn you. In cooling systems with an expansion tank, the radiator pressure cap is more or less permanently installed. Manufacturers' service manuals warn you never to remove the cap except for major service, such as flushing out the system. The cap should not be removed just to check the coolant level or to add coolant. Instead, coolant level is checked visually at the expansion tank. Many tanks are marked to indicate normal hot and cold coolant levels (Fig. 21-21). If coolant is needed, it is poured into the expansion tank.

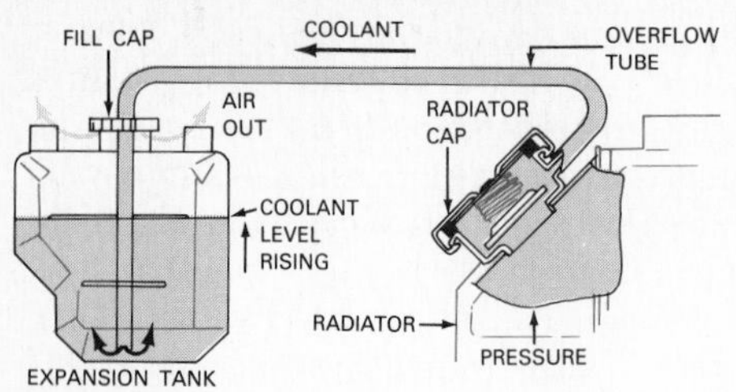

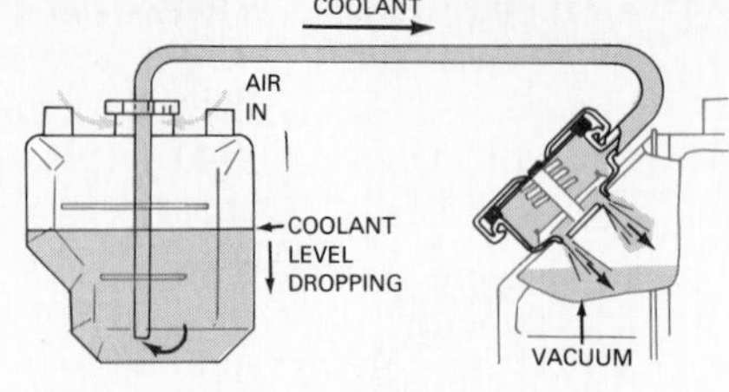

Fig. 21-21 Operation of the radiator cap under pressure conditions (left) and vacuum conditions (right). With an expansion tank, this action assures a full radiator at all times. *(Ford Motor Company)*

♦ 21-16 ANTIFREEZE SOLUTION

Water freezes at 32°F [0°C]. If water freezes in the engine cooling system, it stops coolant circulation. Some parts of the engine will overheat. This could seriously damage the engine. Worse, however, is the fact that water expands when it freezes. Water freezing in the cylinder block or cylinder head could expand enough to crack the block or head. Water freezing in the radiator could split the radiator seams. In either case, there is serious damage. A cracked block or head cannot be repaired satisfactorily. A split radiator is hard to repair.

To prevent freezing of the water in the cooling system, antifreeze is added to form the coolant. The most commonly used antifreeze is ethylene glycol, although an alcohol-base antifreeze has been used in the past. A mixture of half water and half ethylene glycol will not freeze above −34°F [−36.7°C]. This is 34°F below zero and it seldom gets that cold in the United States, except in Alaska. A higher concentration of antifreeze will prevent freezing of the coolant at temperatures as low as −84°F [−64.4°C].

Some antifreeze compounds plug small leaks in the cooling system. These antifreeze compounds contain tiny plastic beads or inorganic fibers which circulate with the coolant. When a leak develops, the beads or fibers jam in the leak and plug it. This is the same action provided by adding "stop-leak" or "sealer" to the cooling system in an emergency. However, if the leak is too large, no chemical can stop it. Also, a chemical cannot stop leaks in hoses, cylinder-head gaskets, or water-pump seals. The only permanent repair for a leak in a cooling-system component is to repair or replace the component. One problem blamed on "sealing" antifreezes and some types of stop-leak is their tendency to plug the heater core.

Corrosion protection is also built into antifreeze solutions. Compounds are added that fight corrosion inside engine water jackets and the radiator. Corrosion shortens the life of metal parts. Also, corrosion forms an insulating layer which reduces the amount of heat transferred from the metal parts to the coolant. In engines with severe corrosion, it is possible for the coolant to maintain normal temperature. But at the same time, the cylinder head and cylinders may be overheating.

Some antifreeze manufacturers add a foam inhibitor to the ethylene glycol. Air does not conduct heat as well as coolant. Any air in the cooling system may cause excess foaming of the coolant as it is whipped up by the water-pump impeller. A foam inhibitor tends to reduce this problem.

Small cans of "rust inhibitor" are widely sold. However, the contents should not be added to new antifreeze. Certain types may contaminate the antifreeze, which already has the proper antirust agent in it. Rust inhibitor usually is added to the cooling system when water is used as the coolant.

Note Automotive cooling systems should never be filled with water only. Indicator lights do not come on until well above the boiling point of water (♦21-17). Therefore, plain water could boil, even though the indicator light does not come on. Severe engine damage could occur before the driver is aware of the problem. Coolant with antifreeze will not boil until a higher temperature is reached.

In addition to having rust and foam inhibitor added, most antifreezes are colored with a dye, such as red, green, or yellow. The dye allows the antifreeze to serve as a leak indicator. The distinctive coloring makes a leak easier to locate.

Antifreeze solutions also serve a purpose during hot-weather operation. They raise the boiling point of the coolant so it is less likely to boil away in hot weather. Also, they continue to fight corrosion.

When the rust and foam inhibitors are used up, the coolant becomes rust-colored. Then the cooling system should be serviced. Car manufacturers recommend that the cooling system be drained, flushed out, and refilled with a fresh mixture of water and antifreeze periodically. One recommendation is that this be done every two years. Another is that it be done every year, preferably in the late fall, just before freezing weather sets in. The procedure is covered later.

When adding antifreeze to an engine that has an aluminum block, cylinder head, or radiator, add only the recommended type of antifreeze. Some commercial antifreeze compounds are not safe for use in engines with aluminum components.

♦ 21-17 TEMPERATURE INDICATORS

The driver should be warned if the temperature of the coolant in the cooling system goes too high. For this reason, a *temperature-indicating light* or *gauge* is installed in the instrument panel of the car. An abnormal heat rise is a warning of abnormal conditions in the engine. The indicator warns the driver to stop the engine before serious damage is done. Temperature gauges are of two general types: the balancing-coil (magnetic) type and the bimetal-thermostat (thermal) type. Instead of a gauge, many cars use one or two temperature-indicator lights.

1. Balancing-Coil Gauge The balancing-coil type of oil-pressure gauge (♦20-10), fuel gauge (♦15-6), and temperature gauge all operate in a similar manner. The instrument-panel units are very similar. Each consists of two coils and an armature to which a pointer is attached. Two coils of wire are placed at an angle of 90 degrees to each other (Fig.

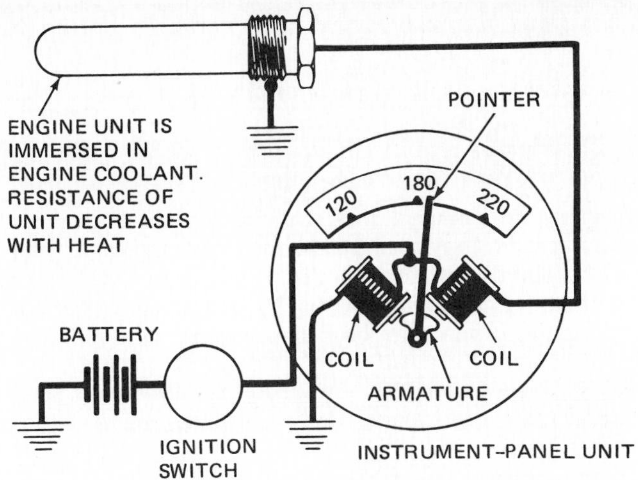

Fig. 21-22 A magnetic, or balancing-coil, type of temperature-gauge system.

21-22). This unit is located in the instrument panel in front of the driver.

An engine unit, or sending unit, that changes resistance with temperature is placed in the engine so that the end of the unit is in the coolant. The resistance of the engine unit drops as coolant temperature goes up. At higher temperatures, it allows more current to flow through the right coil. Then, when the engine is cold, it allows less current. Therefore, the pointer or needle in the gauge moves to the right or left, as the magnetic field around the right coil varies in strength.

With the engine cold, current flows through the ignition switch to the coils in the dash unit. The same amount of current always passes through the left coil, as long as the ignition switch is closed. When the engine unit is cold, it allows only a small current to flow through the right coil of the dash unit. Now, the left coil has more magnetism than the right coil. Therefore, the armature between the two coils is pulled to the left. The pointer is attached to the armature, and moves left with it to indicate that the engine is cold.

As the engine warms up, the engine unit passes more current. More current flows through the right coil of the dash unit. This creates a stronger magnetic field around the right coil. Therefore the pointer swings right to indicate a higher coolant temperature.

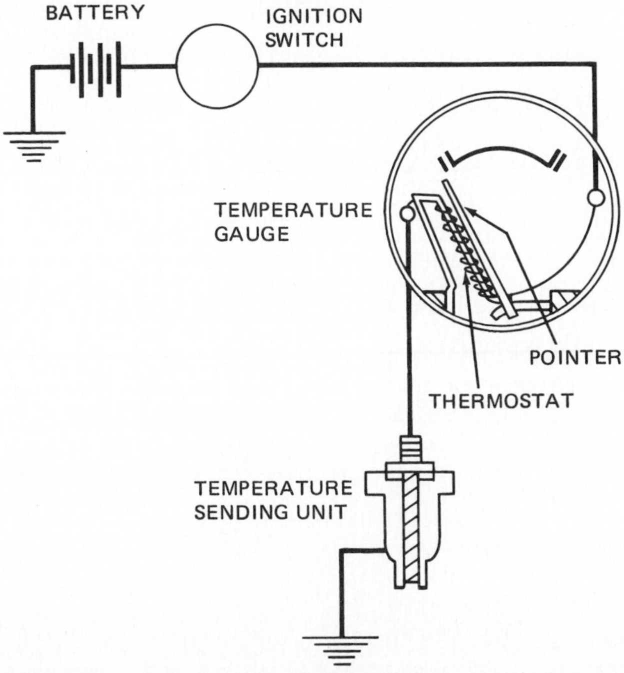

2. Bimetal-Thermostat Gauge The bimetal-thermostat type of temperature gauge is similar to the balancing-coil type except for the use of a bimetal thermostat in the instrument-panel unit (Fig. 21-23). This thermostat is linked to the pointer. As the engine unit warms up and passes more current, the thermostat heats up and bends. This causes the pointer to swing to the right to indicate that the engine temperature is rising.

3. Indicator Light The temperature-indicator-light system has two units. One is the light on the instrument panel and the other is the coolant-temperature switch, or sending unit (Fig. 21-24). The sending unit is mounted on the engine so that the end of the unit is in the coolant. When the temperature goes too high, the sending unit connects the light bulb to the battery. Then the indicator light comes on to signal the driver that the engine is overheating.

The indicator light warns of an overheating condition at about 5 to 10°F [2.8 to 5.6°C] below the coolant boiling point. A "prove-out" circuit is incorporated in the system. When the ignition switch is turned from OFF to RUN, the light should come on, proving that the system is working. If the light does not come on, either the bulb is burned out or the sending unit or connecting wire is defective. The light will go out normally after the engine starts.

BATTERY

IGNITION SWITCH

TEMPERATURE GAUGE

POINTER

THERMOSTAT

TEMPERATURE SENDING UNIT

Fig. 21-23 A thermostatic, or thermal, type of temperature-gauge system. *(Chrysler Corporation)*

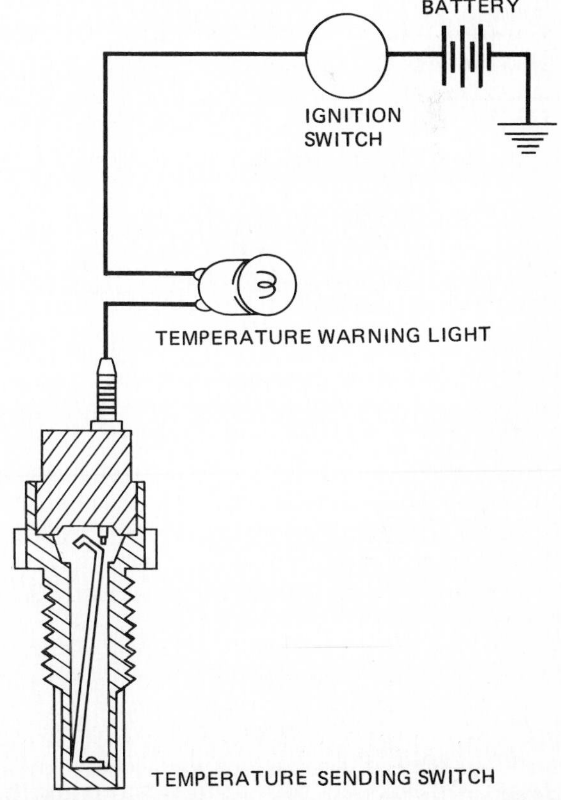

Fig. 21-24 Temperature warning light which indicates engine overheating. *(ATW)*

♦ REVIEW QUESTIONS ♦

Select the *one* correct, best, or most probable answer to each question. You can find the answers in the section indicated at the end of each question.

1. The pump part that rotates to cause coolant circulation between the radiator and engine water jackets is called the (♦21-5)
   a. impeller
   b. fan
   c. body
   d. bypass
2. The purpose of the cooling system is to (♦21-1)
   a. prevent the coolant from boiling
   b. prevent the coolant from freezing
   c. keep the engine running as cold as possible
   d. keep the engine running at its most efficient operating temperature
3. In normal operation, coolant in the down-flow radiator circulates (♦21-11)
   a. from top to bottom
   b. from bottom to top
   c. in a circular path in the radiator
   d. none of the above
4. The part of the cooling-system thermostat that opens and closes the valve is the (♦21-14)
   a. seater
   b. wax pellet
   c. pressure valve
   d. vacuum valve
5. The device in the cooling system that raises the boiling point of the coolant in the system is called the (♦21-15)
   a. pressure cap
   b. vacuum valve
   c. radiator
   d. water jacket
6. A pressure cap contains two valves. They are the (♦21-15)
   a. pressure valve and blowoff valve
   b. atmospheric valve and vacuum valve
   c. pressure valve and vacuum valve
   d. none of the above
7. Two types of antifreeze are (♦21-16)
   a. alcohol base and ethylene glycol
   b. ethylene glycol and permanent
   c. isooctane and ethylene glycol
   d. none of the above
8. Mechanic A says that the thermostat controls maximum coolant temperature. Mechanic B says that an engine that overheats will be repaired by replacing the thermostat with one of a lower temperature. Who is right? (♦21-14)
   a. A only
   b. B only
   c. both A and B
   d. neither A nor B
9. What is the main purpose of the water-pump bypass hose in the engine cooling system? (♦21-14)
   a. to reduce pressure at the water-pump outlet during high engine speeds
   b. to allow coolant flow within the engine while the thermostat is closed
   c. to prevent air pockets in the water-pump housing
   d. to prevent collapse of the lower radiator hose
10. The percent of the heat produced in the combustion chambers that the cooling system removes from the engine is (♦21-1)
    a. 30 to 35 percent
    b. 50 to 60 percent
    c. 10 to 80 percent
    d. 85 to 90 percent

CHAPTER 22
Cooling-System Service

After studying this chapter, and with proper instruction and equipment, you should be able to:

1. Diagnose cooling-system troubles using the pressure tester and the cooling-system analyzer.
2. Test and adjust antifreeze strength.
3. Check the thermostat.
4. Replace and adjust the fan belt.
5. Clean and flush the cooling system.
6. Locate and repair leaks in the cooling system.
7. Replace the water pump.
8. Replace an expansion core plug.

♦ 22-1 WORKING SAFELY ON THE COOLING SYSTEM

There are several safety hazards you must watch for when working on engines and the cooling system:

1. Keep your hand away from the moving fan! When the engine is running, the fan is turning so fast it is a blur. But it can mangle your hand and cut off fingers if your hand should get into the fan.
2. Never stand in a direct line with the fan. A fan blade could break off and fly out from the engine compartment. Anyone standing in line with the fan could be injured or killed. Before starting the engine, examine the fan for cracked or loose blades. If you find any damage, the fan must be replaced.
3. Electrically operated and thermostatically controlled fans may not be shut off when the ignition switch is turned to OFF. Because the coolant-temperature switch turns the fan motor on and off solely on the basis of coolant temperature, the fan may start and run even with the engine stopped. When working near an electric fan, disconnect the lead to the fan motor to avoid injury should the fan start.
4. Keep the fingers away from the moving belt and pulleys! Your fingers could be pinched and cut off if they are caught between the belt and pulley.
5. Never attempt to remove the radiator cap from the cooling system of an engine that is near or above its normal operating temperature. Releasing the pressure may cause instant boiling of the coolant. Boiling coolant and steam spurting from the filler neck can cause scalding and burns. Allow an engine to cool before attempting to remove the cap.
6. Coolant is poisonous! It can cause serious illness or even death if it is swallowed! Always wash your hands thoroughly if you get coolant on them.

♦ 22-2 COOLING-SYSTEM TROUBLE DIAGNOSIS

Three cooling-system complaints are engine overheating, slow warm-up, and cooling-system leaks. Possible causes of these complaints are listed in the Cooling-System Trouble-Diagnosis Chart that follows.

COOLING-SYSTEM TROUBLE-DIAGNOSIS CHART

Condition	Possible Cause	Check or Correction
1. Loss of coolant (♦22-3)	a. Pressure cap and gasket defective	Inspect. Wash gasket and test. Replace only if cap will not hold pressure specified.
	b. Leakage	Pressure test system.
	c. External leakage	Inspect hose, hose connections, radiator, edges of cooling system, gaskets, core plugs, drain plugs, oil-cooler lines, water pump, expansion tank and hoses, heater-system components. Repair or replace as required.
	d. Internal leakage	Check torque of head bolts; retorque if necessary. Disassemble engine as necessary. Check for cracked intake manifold, blown head gasket, warped head or block gasket surfaces, cracked cylinder head or engine block.
2. Engine overheating (♦22-4)	a. Low coolant level	Fill as required. Check for coolant loss.
	b. Loose fan belts	Adjust.
	c. Pressure cap defective	Test. Replace if necessary.
	d. Radiator or air-conditioner condenser obstructed	Remove bugs, leaves, and debris.
	e. Thermostat stuck closed	Test. Replace if necessary.
	f. Fan-drive clutch defective	Test. Replace if necessary.
	g. Ignition faulty	Check timing and advance. Adjust as required.
	h. Temperature gauge or HOT light defective	Check electric circuits. Repair as required.
	i. Inadequate coolant flow	Check water pump and block for blockage.
	j. Exhaust system restricted	Check for restrictions.
3. Engine fails to reach normal operating temperature; slow warm-up (♦22-5)	a. Open or missing thermostat	Test. Replace or install as necessary.
	b. Defective temperature gauge or COLD light	Check electric circuits. Repair as required.

♦ 22-3 CAUSES OF LOSS OF COOLANT

Many leaks can be spotted easily for two reasons. First, the cooling system requires frequent refilling. Second, the point of the leak usually can be found at the top of a telltale stain. Dye is added to most antifreeze to make leak detection easier.

There are two types of coolant leaks, external leaks and internal leaks. External leaks are those where the coolant can drip onto the ground. These can be seen. Typical leak points are from hose and hose connections, heater core, radiator core, and expansion core plugs (freeze plugs) in the block and head.

Internal leaks occur when the coolant leaks from the cooling system into some other part of the engine. These leaks cannot be seen. However, having to add coolant frequently is a clue that an internal leak may exist.

Internal leaks can severely damage the engine. The coolant may contaminate the oil and cause rust. A coolant leak into the combustion chamber while the engine is stopped may fill the combustion chamber. Then, when the engine is cranked, the upward-moving piston could cause the head, piston, or cylinder to crack or the connecting rod to bend.

If coolant leaks from a gasket (cylinder head, water pump), the gasket may require replacement. Attaching bolts should be tightened to the correct torque. If the leak is from the radiator, it should be removed and either repaired or replaced. Oil in the coolant indicates leakage of the transmission oil cooler in the outlet tank of the radiator (Fig. 21-3). If the leak is at a hose connection, the hose connection should be tightened. If a hose is leaking, the hose should be replaced. Pressure testing the cooling system to locate leaks is covered in ♦22-13.

♦ 22-4 CAUSES OF ENGINE OVERHEATING

The driver may notice that the red light stays on or the temperature gauge registers in the overheating zone. Also, the driver may complain that the engine boiled over. Possible causes of engine overheating include:

1. Low coolant level caused by leakage of coolant.
2. Accumulation of rust and scale in the system, which prevents normal circulation of coolant. Antifreeze compounds contain additives which tend to prevent the formation of rust and corrosion.
3. Collapsed hoses which prevent normal coolant circulation. Suction hoses should contain a spring to prevent collapsing.
4. Defective thermostat which does not open normally, blocking circulation of coolant. If the engine overheats without the radiator becoming normally warm, and if the fan belt is properly tightened, then the thermostat is probably at fault. Sometimes on new cars, grains of sand from the sand core for the engine block or head may lodge in the thermostat, preventing it from opening. A thermostat that is installed backwards usually cannot open, and will also cause overheating.
5. Defective water pump which does not circulate enough coolant through the engine. A quick check of water-pump operation can be made by installing a clear plastic hose in place of the upper radiator hose and running the engine. Then you can see how much coolant is circulating, and if any air is in it. A more accurate test of water-pump capacity is made by installing a cooling-system analyzer in place of the upper radiator hose (Fig. 22-1).

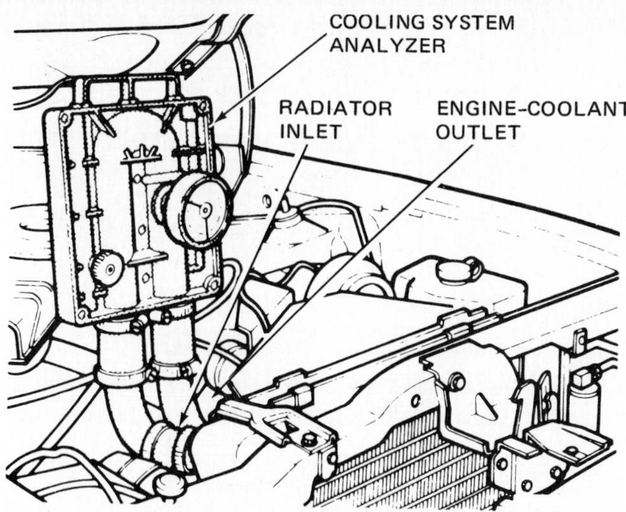

Fig. 22-1 Using a cooling-system analyzer to check the cooling system. *(Ford Motor Company)*

One cause of water-pump bearing failure is an overtight drive belt. A fan belt always should be tightened correctly with a belt-tension gauge (♦22-15). Bearing failure usually makes the water pump noisy. A quick check of the bearing can be made with the fan belt off by grasping the tips of the fan blades and trying to move the fan in and out (Fig. 22-2). Any movement, or a rough and grinding feeling as the fan is slowly turned, indicates a defective bearing. Drops of coolant leaking from the water-pump ventilation hole indicate a leaking water-pump seal. The ventilation hole is below and behind the water-pump pulley (Fig. 22-2).

6. A loose or worn fan belt will not drive the water pump fast enough. The belt should be tightened or replaced. Where a pair of belts is used, both belts should be replaced at the same time, not just the one that appears most worn. When you replace only one belt, all the load is on the new belt. It will wear rapidly. When both belts are replaced with a new matched pair, then each belt will carry half the load.
7. Overheating may be caused by afterboil. This may occur when the coolant starts to boil after the engine has been turned off, for example, after a long hard drive. The engine has so much heat in it that, when the water pump stops circulating coolant, it starts to boil.
8. Boiling can occur if the coolant is frozen. This hinders or stops its circulation. Then the coolant in the engine around the combustion chamber and cylinders becomes so hot that the coolant boils. Freezing of coolant in the radiator, cylinder block, or head may crack the block or head and open up seams in the radiator. Operating an engine in which the coolant is frozen may cause serious engine damage.

Note There are other causes of engine overheating which have nothing to do with conditions in the cooling system. High-altitude operation, insufficient oil, overloading of the engine, hot-climate operation, improper ignition timing, long periods of slow speed or idling, improperly operating emission controls—any of these can cause overheating of the engine.

♦ 22-5 CAUSES OF SLOW WARM-UP

The most likely cause of slow engine warm-up is a thermostat that is stuck open (♦22-9). This allows the coolant to circulate between the engine and the radiator even though the engine is cold. Therefore the engine has to run longer to reach normal operating temperature. As a result, engine wear is greater because the engine operates cold for a longer time.

Another possible cause of slow warm-up is that the thermostat has been removed. Never remove a thermostat and leave it out of the cooling system. This does not improve coolant circulation. It does delay warm-up and increases engine wear and sludge formation.

A quick check for a missing or stuck-open thermostat can be made by squeezing the upper hose immediately after starting a cold engine (Fig. 22-3). Keep your hand away from the fan! No coolant flow through the upper hose should be felt. If you feel movement, the thermostat is missing or open.

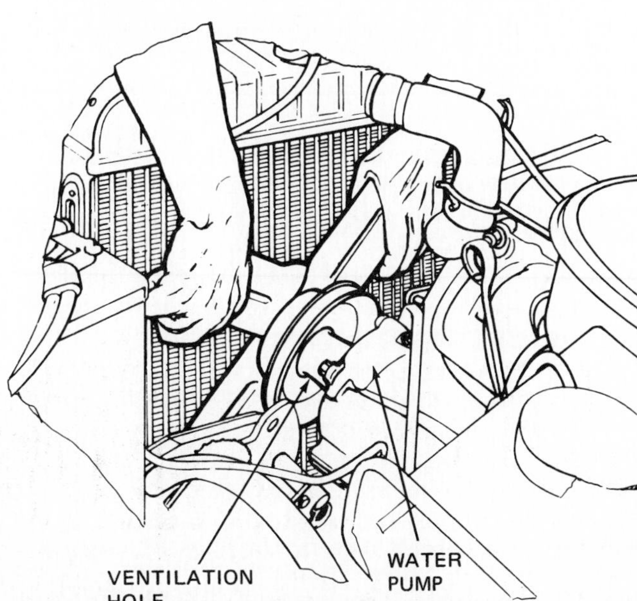

Fig. 22-2 Checking the water-pump bearing. *(Chrysler Corporation)*

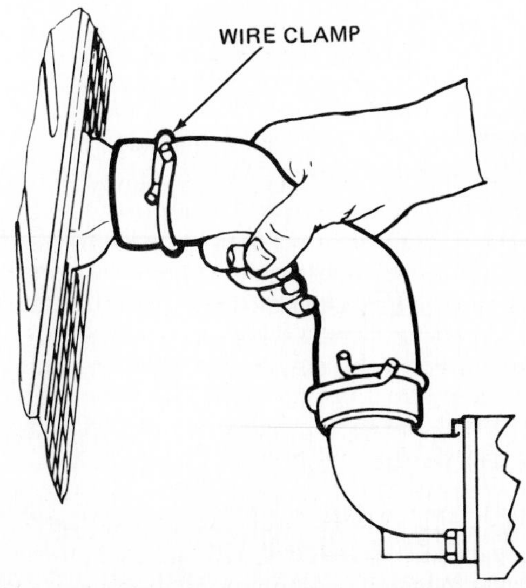

Fig. 22-3 To check the condition of a radiator hose, squeeze it. *(Chrysler Corporation)*

In the winter, the driver's complaint often is that it takes a long time for the car heater to start delivering heat. If you hear this complaint, suspect a defective or missing thermostat. How to test thermostats is covered in ♦22-9.

♦ 22-6 ANALYZING THE COOLING SYSTEM

The trouble-diagnosis chart in ♦22-2 lists possible causes of cooling-system troubles. A cooling-system analyzer (Fig. 22-1) simplifies analysis. Follow the manufacturer's instructions on how to use the analyzer.

♦ 22-7 CHECKING COOLANT LEVEL

A glance at the expansion tank (Figs. 21-14 and 21-15) will show you whether the coolant level is low. On cars without an expansion tank, remove the radiator cap to check coolant level. Never remove the cap when the engine is hot! Boiling coolant and steam can erupt and burn you severely. Wait until the engine cools. See Caution 5 in ♦22-1.

On most vehicles with an expansion tank, coolant is added to the tank. If the engine is cold, fill the tank to the cold-level mark on the tank (Fig. 21-14). If the engine is hot, fill the tank to the hot-level mark. The hot-level mark is above the cold-level mark.

♦ 22-8 TESTING ANTIFREEZE STRENGTH

The amount of antifreeze in the coolant must be great enough to protect against freezing at the lowest temperatures that might occur. The strength of the antifreeze can be checked with the float hydrometer or the ball hydrometer.

1. Float Hydrometer The float-type hydrometer is shown in Fig. 22-4. The higher the float rises in the coolant, the higher the percentage of antifreeze in the coolant. To use the hydrometer, put the rubber tube into the coolant. Then squeeze and release the rubber bulb. Note how high the float rises in the coolant. Check the lower scale which shows the temperature of the coolant and how low the temperature must go before the coolant will freeze.

2. Ball Hydrometer A second tester is the ball-type hydrometer (Fig. 22-5). It has four or five small balls in a small plastic tube. Coolant is drawn into the tube by squeezing and releasing the rubber bulb. The stronger the solution, the more balls that float.

See Caution 6 in ♦22-1 about coolant being poisonous. Wash your hands if you get coolant on them.

♦ 22-9 TESTING THE THERMOSTAT

Different car manufacturers have different testing procedures for checking thermostats. Chevrolet recommends suspending the thermostat in a solution of one-third antifreeze and two-thirds water. The solution should be heated to 25°F [14°C] above the temperature stamped on the thermostat. The thermostat should open. Then submerge the thermostat in the same solution with the temperature at 10°F [5.5°C] below the temperature stamped on the thermostat. The thermostat should close completely. If it does not open and close during the test, it is defective.

♦ 22-10 CHECKING THE HOSE AND HOSE CONNECTIONS

To check the condition of the radiator hose, squeeze the hose (Fig. 22-3). It should not collapse easily when squeezed. The appearance of the hose and hose connections usually indicates their condition. Hose that is soft, hard, rotted, or swollen should be replaced. The hose must be in good condition and connections should be properly tightened to avoid leaks. Air will be drawn into the cooling system if there are leaks in the hose or connections between the radiator and water pump.

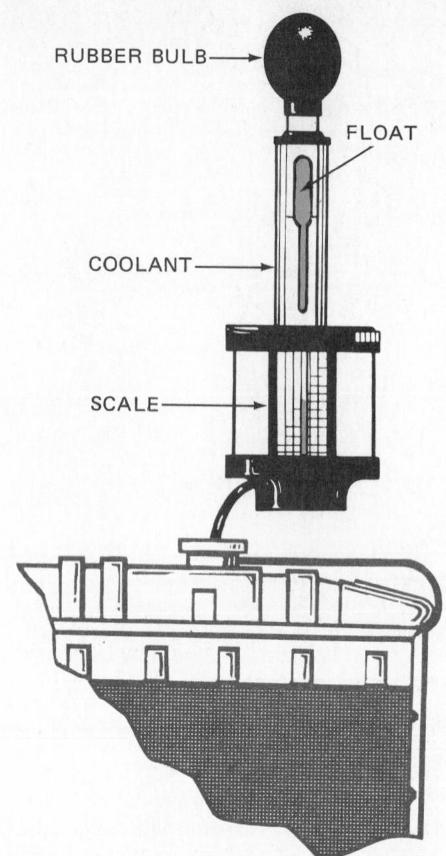

Fig. 22-4 Cooling-system hydrometer being used to check the freezing temperature of the coolant. *(Ford Motor Company)*

♦ 22-11 TESTING THE WATER PUMP

Checking the water-pump bearing and seal is described in ♦22-4. To check the water-pump capacity, substitute a clear plastic pipe for the upper radiator hose. Then, when the engine is running, you can see how much coolant is circulating. A more accurate test of the water pump can be made using a cooling-system analyzer (Fig. 22-1). Always be careful when checking and tightening fan belts (♦22-15). Overtightening the fan belt can cause the water-pump bearing to fail.

♦ 22-12 CHECKING FOR EXHAUST-GAS LEAKAGE INTO COOLING SYSTEM

A defective cylinder-head gasket may allow exhaust gas to leak into the cooling system. This is very damaging. Strong acids can form as the gas unites with the water in the coolant. These acids corrode the radiator and other cooling-system parts. A test for exhaust-gas leakage can be made with a Bloc-Chek tester. It is installed in the radiator filler neck, as shown in Fig. 22-6.

The test is made with the engine running. Squeeze and release the rubber bulb. This draws an air sample from the cooling system up through the test fluid. The test fluid is

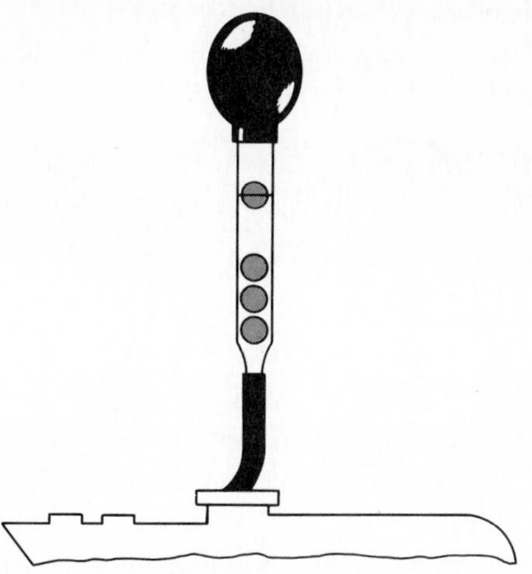

Fig. 22-5 Ball-type hydrometer being used to check antifreeze strength.

ordinarily blue. However, if combustion gas is leaking into the cooling system, the test fluid will change to a yellow color. If a leak is indicated, the exact location can be found by removing one spark-plug wire at a time and retesting. When a leaking cylinder is firing, the liquid will change to yellow. When nonleaking cylinders only are firing, the liquid will remain blue.

Combustion leaks in the valve areas can cause cracked valve seats and cylinder heads. The coolant is forced away from the cracked area during heavy acceleration by the leakage of combustion gases through the leak. This causes excessive heat buildup. When acceleration stops, the diverted coolant rushes back to the overheated area. The sudden cooling of the area can crack the head and valve seat.

♦ 22-13 PRESSURE TESTING THE COOLING SYSTEM

You can use a pressure tester as shown in Fig. 22-7 to check the cooling system for leaks. To use the tester, remove the radiator cap and fill the radiator until the coolant level is about ½ inch [13 mm] below the bottom of the filler neck.

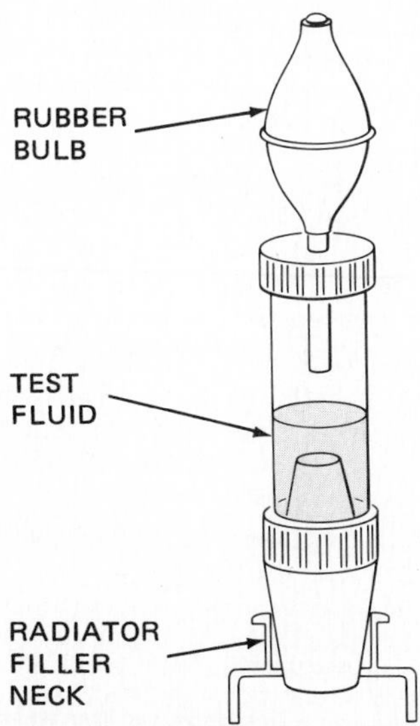

Fig. 22-6 Checking for exhaust-gas leakage into the cooling system with a Bloc-Chek tester.

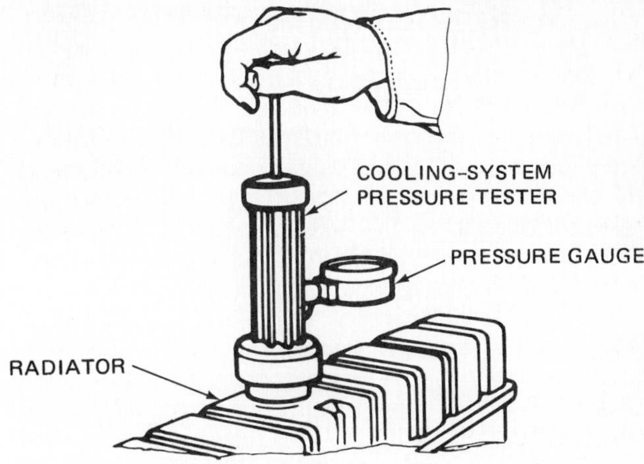

Fig. 22-7 Using a cooling-system pressure tester to check the cooling system for leaks. *(Chrysler Corporation)*

Wipe the neck sealing surface and attach the tester. Then operate the pump to apply a pressure that does not exceed 3 psi [21 kPa] above the manufacturer's specification. If the pressure holds steady, the system is not leaking. If the pressure drops, there are leaks. Look for coolant leaks at hose connections, hose, engine expansion plugs, water-pump and cylinder-head gaskets, water-pump shaft seal, and radiator.

If no external leaks are visible, remove the tester and start the engine. Run the engine until operating temperature is reached. Reattach the tester, apply a pressure of 15 psi [103 kPa], and increase engine speed to about 3000 rpm. If the needle of the pressure gauge fluctuates with engine speed, it indicates an exhaust-gas leak, probably through a cylinder-head gasket. On a V-type engine, you can determine which bank is at fault by grounding the spark plugs in one bank.

If the needle does not fluctuate, sharply accelerate the engine several times and check for abnormal discharge of liquid through the tail pipe. This would indicate a cracked block or head, or a defective gasket.

♦ 22-14 PRESSURE TESTING THE RADIATOR CAP

The cooling-system pressure tester can be used to check the radiator pressure cap. An adapter is attached to the tester pump so that it will fit the cap (Fig. 22-8). Then the pump is operated to apply the rated pressure against the cap. If the cap will not hold its rated pressure, it should be replaced.

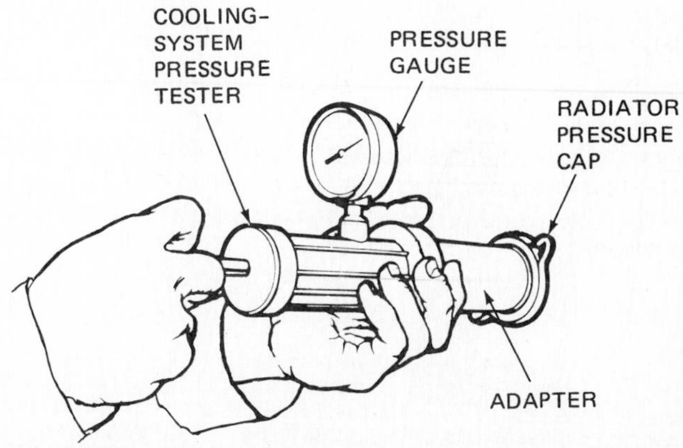

Fig. 22-8 Using the cooling-system pressure tester to check a radiator cap. *(Chrysler Corporation)*

♦ 22-15 TESTING THE FAN BELT

Fan belts, or drive belts, should be checked for wear and tension. Most wear occurs on the underside of the belt. To check a V-belt, be sure the engine is off and will not be cranked. Then twist the belt with your fingers. Check for small cracks, grease, glazing, and tears or splits (Fig. 22-9). Small cracks will enlarge as the belt is flexed. Grease rots rubber and makes the side slick so that belt slips easily. A high-pitched squeal is the typical sound of a loose and slipping belt. Glazed belts result from slippage. Large tears or splits in a belt allow it to be tossed from the pulley. On cars with a set of two fan belts, if one is worn and requires replacement, then both should be replaced.

Use a belt tension gauge (Fig. 22-10) to check and adjust the fan-belt tension. When you do not have a gauge, or if space does not allow use of a gauge, you can make a quick check of belt tension. Press in the middle of a free span, as shown in Fig. 22-11. When the free span is less than 12 inches [305 mm] between pulleys, belt deflection should be ⅛ to ¼ inch [3 to 6 mm]. When the free span is longer than 12 inches [305 mm], belt deflection should be ¼ to ½ inch [6 to 13 mm].

Some cars have a belt tensioner (Fig. 22-12), which provides a visual check of belt tension. Figure 22-13 shows the tensioner installed in a serpentine-belt system. The tensioner can be rotated in the direction of the arrow (in Fig. 22-13) to remove or install a belt. If the pointer is out of its operating range, the belt is incorrect or worn, or the tensioner was improperly installed.

The fan belt should be checked at least every year to make sure it is in good condition. A fan belt that has become worn or frayed, or has separated plies, should be replaced.

A slipping belt can cause engine overheating and a rundown battery. These troubles result because a slipping belt can not drive the water pump and alternator fast enough for normal operation. Sometimes a belt will slip and make noise even after it is adjusted to the proper tension. To help this problem, several types of belt dressing are available which can be applied to the sides of the belt. Belt dressing helps to eliminate noise and increase belt friction.

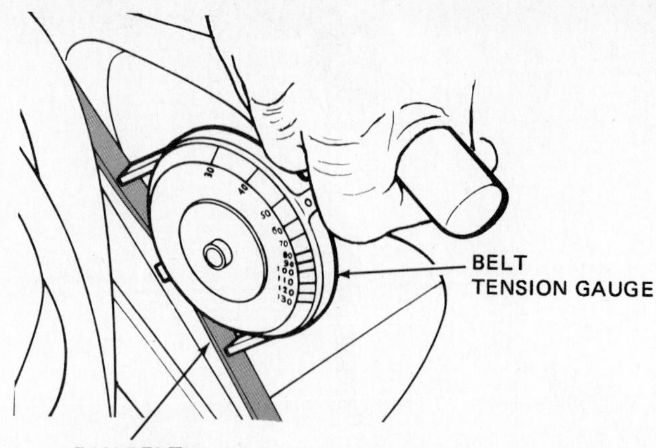

Fig. 22-10 Using a belt tension gauge to check the tension of a drive belt. *(Chrysler Corporation)*

♦ 22-16 CLEANING THE COOLING SYSTEM

The original additives in antifreeze to fight rust and corrosion break down and are ineffective after one to two years. This is because of the continual exposure to the heat in the cooling system. This explains why a permanent antifreeze must be replaced periodically.

When the antifreeze has deteriorated, its color may be used as an indicator of its condition. After the additives in the antifreeze break down, rust begins to form rapidly. Therefore, a rust-colored antifreeze is an indication that cooling-system service is needed.

A quick check of the conditions inside the cooling system can be made. Remove the cap and wipe the inside of the radiator filler neck with you finger. If you find any oil, grease, rust, or scale, the cooling system should be cleaned. Reverse flushing also may be needed.

The cooling system should be cleaned periodically to remove rust, scale, grease, oil, and any acids formed by exhaust-gas leakage into the coolant. Recommendations vary. For example, Chevrolet recommends that the cooling system be drained and flushed with plain water every two years. Then the cooling system should be filled with a solution of anti-

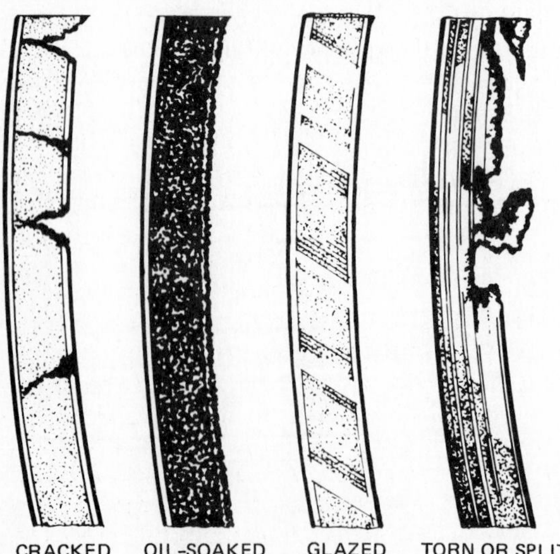

Fig. 22-9 Conditions to look for when inspecting a V-belt. *(Chrysler Corporation)*

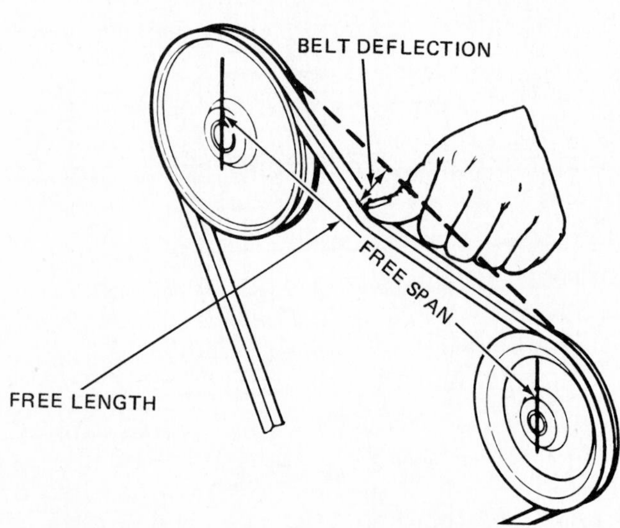

Fig. 22-11 Checking belt deflection. *(Chrysler Corporation)*

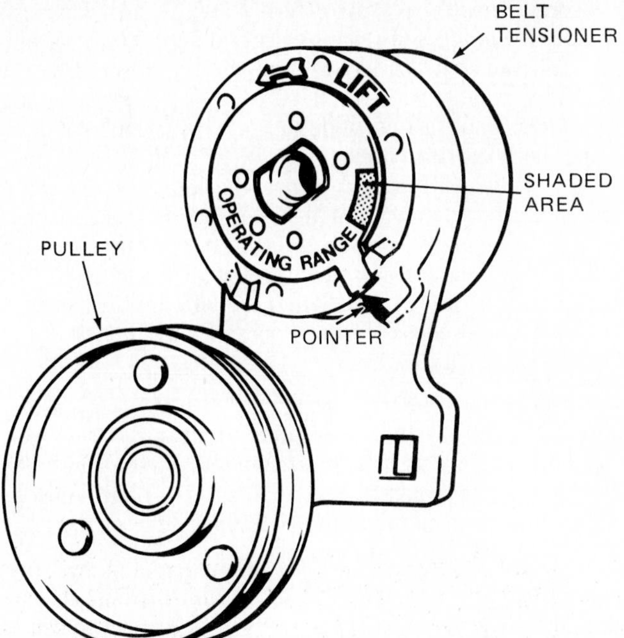

Fig. 22-12 Belt tensioner. Pointer must be pointing toward shaded area with belt installed. *(Buick Motor Division of General Motors Corporation)*

freeze and water. The flushing procedure varies with different vehicles. There is the standard flushing method, the fast-flush method (recommended by Prestone), and the fast-flush-and-fill method, which requires a special machine. The standard method is described below. It is the one that you will normally use in the shop.

First drain the cooling system. Remove the thermostat and reinstall the thermostat housing. If you are using a cooling-system cleaner, pour it into the radiator. Fill the cooling system with water.

Note There are several types of cleaners. Follow the instructions on the cleaner container.

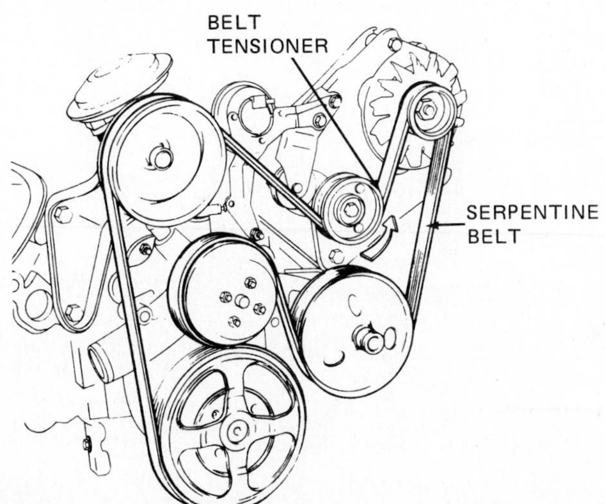

Fig. 22-13 Belt tensioner for a serpentine belt. To remove or install the belt, rotate the tensioner in the direction shown by the arrow. *(Buick Motor Division of General Motors Corporation)*

Run the engine at fast idle for about 20 minutes. Stop the engine if the water begins to boil. Wait for the engine to cool. Then drain out the water-and-cleaner solution.

Use a water hose to put water into the radiator to flush out loosened rust and scale. You can also reverse flush the radiator and engine water jackets with a flushing gun, as shown in Figs. 22-14 and 22-15. To reverse-flush the radiator, connect the flushing gun to the hose from the lower tank (Fig. 22-14). Note the use of a leadaway hose attached to the upper tank. Continue to flush until the water runs clear. Be careful not to use excessive pressure. This could cause leaks to develop.

After draining the system, installing the thermostat, and reconnecting hoses, fill the system with coolant. The coolant should have the proper amount of antifreeze to protect against freezing at the lowest temperature expected. The system will not fill completely to start with because the water, being cold, will cause the thermostat to close. This traps air in back of the thermostat (Fig. 22-16). The engine can be started and run until the thermostat opens. Then the system can be filled.

♦ 22-17 LOCATING AND REPAIRING RADIATOR LEAKS

Radiator leaks are usually easy to find. Telltale marks form below the leak because of the dye in the antifreeze. A cooling-system pressure tester can be used to help locate leaks (♦22-13).

If leaks are found, the radiator should be removed for repair. This is usually a job for the specialty shop that handles radiator repairs. However, if the radiator has several leaks, it may not be worthwhile to repair it.

♦ 22-18 WATER-PUMP SERVICE

Figure 22-2 shows the water pump installed on an engine. The water pump requires no service in normal operation. Older pumps required lubrication through grease fittings or through the use of soluble oil mixed with the coolant. This was known as *water-pump lubricant.*

Today water pumps have sealed bearings which require no lubrication. If the pump develops noise, leaks, or becomes otherwise defective, it is replaced with a new or remanufactured pump. The procedures vary for replacing the water pump. A typical procedure follows.

To remove the water pump, drain the cooling system. Remove the inlet hose and the heater hose from the pump. Remove the fan belt. Then remove the attaching bolts and take off the pump. On some engines, be sure to pull straight out to avoid damaging the impeller and shaft.

In normal operation, the impeller blades may wear away. This may result from abrasive action of sand in the cooling system. Rusting away of the blades may result from use of old antifreeze without active rust and corrosion inhibitors, or from the use of plain water in the cooling system.

Before installing a new water pump, check the impeller size of the new pump against the old pump. Some water pumps look alike and will bolt to more than one engine. However, impellers of different diameters are installed on the pump shaft to change the pump output. Damage to the cooling system or overheating may result from installation of a pump with the wrong impeller.

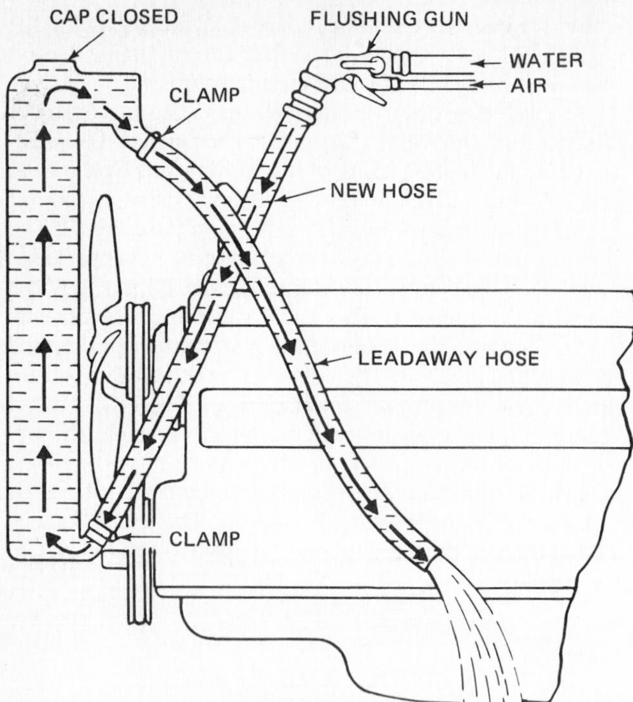

Fig. 22-14 Reverse flushing a radiator.

Install the water pump on the engine, following backward the steps you used to remove it. Then adjust the fan belt to the proper tension (♦22-15). Fill the cooling system before starting the engine.

♦ 22-19 EXPANSION CORE PLUGS

A leaking core plug (Fig. 22-17) must be replaced. To do this, place the pointed end of a pry bar against the center of the plug. Tap the end of the bar with a hammer until the point goes through the plug. Then press the pry bar to one side to pop the plug out.

Another method of core-plug removal is to drill a small hole in the center of the plug. Then pry the plug out. Special tools also are available to remove core plugs. However, removal and installation of certain core plugs on some engines is very difficult to do with the engine in the car, or without removing parts of the engine or accessories.

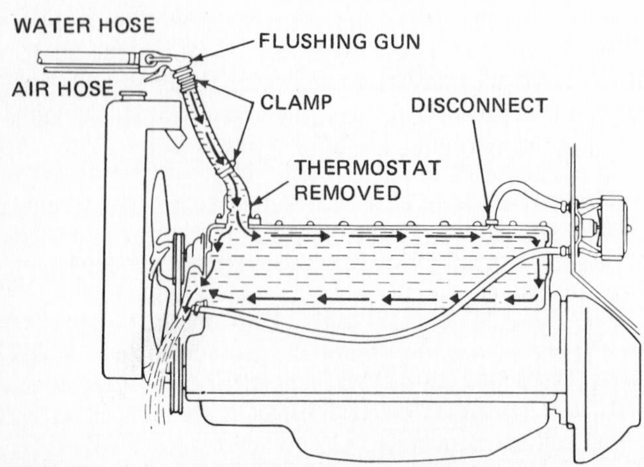

Fig. 22-15 Reverse flushing the water jackets in the engine.

Do not drive the pry bar or drill past the plug. On some engines the plug is only about ⅓ inch [9.5 mm] from a cylinder wall. You could damage the cylinder wall if you drive the pry bar or drill too far in. Do not drive the plug into the water jacket. You would have trouble getting the plug out. Left in, it could block coolant circulation.

After the plug is removed, inspect the bore for rough spots, nicks, or grooves that might allow a new plug to leak. If necessary, bore out the hole to take the next-larger-size plug. Before installing the new plug, coat it with water-resistant sealer. Use the proper installation tool and install the plug. Installation procedures for two types of plugs are shown in Fig. 22-17.

1. Cup Type The cup-type plug, shown at the top of Fig. 22-17, is installed with the flanged edge outward. The proper-size tool must be used. It must not contact the flange. All driving must be against the internal cup. The flange must be driven in until it is below the chamfered edge of the bore.

2. Expansion Type The expansion-type plug, shown at the bottom of Fig. 22-17, is installed with the flanged edge inward. The proper tool must be used. The crowned center part must not be touched when the plug is driven in. Instead, the tool must drive against the outer part of the plug. The plug must be driven in until the center of the crown is below the chamfered edge of the bore.

3. E-Z Seal Expansion Tool This type of expansion plug has a flexible neoprene rubber collar that is held between two washers. To use it, select the proper-size plug and insert it into the bore (Fig. 22-18, left). Then use a wrench to tighten the nut. This pulls the inner washer inward, compressing the rubber collar. The collar expands to fill the bore and it also overlaps on the inside (Fig. 22-18, right). This provides the seal. The rubber is especially compounded to resist damage from heat, oil, antifreeze, and rust inhibitors.

Regardless of type, leaking core plugs are sometimes difficult to replace. They may be located close to the firewall, under the clutch housing or automatic-transmission case, or under a manifold. Replacement may require removing the clutch or automatic transmission. On some cars, you may have to lift or remove the engine from the vehicle.

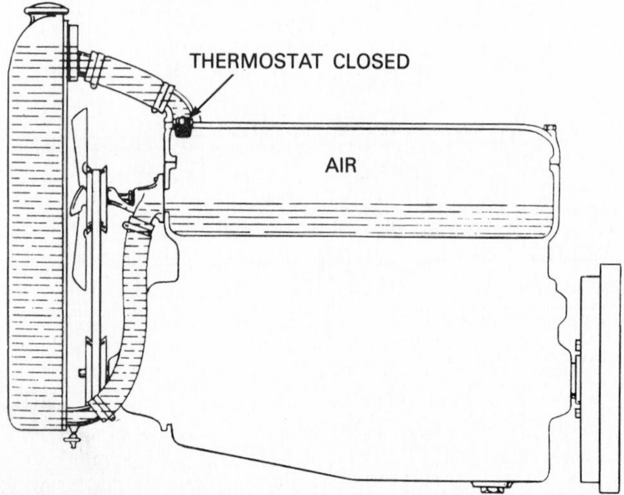

Fig. 22-16 Air is trapped in back of a closed thermostat as the cooling system is filled.

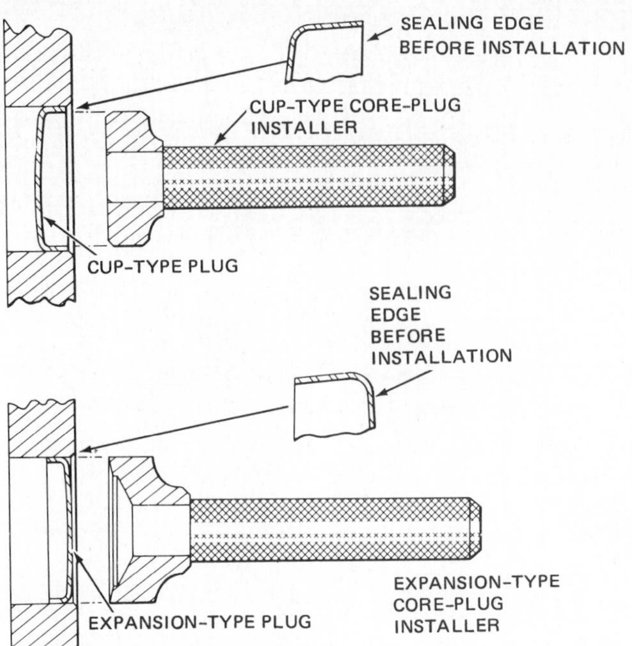

Fig. 22-17 Two types of expansion core plugs and the installation tools required for each. *(Ford Motor Company)*

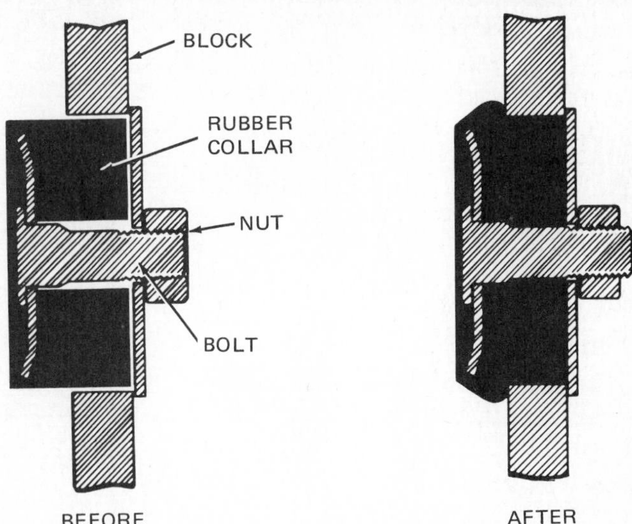

Fig. 22-18 E-Z Seal expansion plug. The plug is inserted into the bore as shown to the left. Then, as the nut is tightened, the rubber collar expands, sealing the opening. *(Hunckler Products, Inc.)*

♦ REVIEW QUESTIONS ♦

Select the *one* correct, best, or most probable answer to each question. You can find the answers in the section indicated at the end of each question.

1. Accumulation of rust and scale in the engine cooling system causes (♦22-4)
   a. slow warm-up
   b. reduced heating capacity
   c. overheating
   d. rough idle
2. If the thermostat is stuck closed, the engine will (♦22-4)
   a. warm up slowly
   b. overheat
   c. fail to start
   d. idle roughly
3. If the thermostat is stuck open, the engine will (♦22-5)
   a. warm up slowly
   b. overheat
   c. fail to start
   d. stall
4. The strength of the antifreeze solution in the cooling system is checked by a special (♦22-8)
   a. micrometer
   b. hydrometer
   c. barometer
   d. thermometer
5. Exhaust-gas leakage into the cooling system is most likely to be caused by a defective (♦22-12)
   a. cylinder-head gasket
   b. manifold gasket
   c. water pump
   d. radiator hose
6. Air will be drawn into the cooling system if there are leaks at any point between the (♦22-10)
   a. water pump and jackets
   b. radiator and water pump
   c. thermostat and radiator
   d. radiator cap and expansion tank
7. If the coolant boils when the engine has been turned off after a long drive, the condition is known as (♦22-4)
   a. overheating
   b. hard running
   c. clogged radiator
   d. afterboil
8. When reverse flushing the radiator, connect the flushing gun to the (♦22-16)
   a. upper tank
   b. pump inlet
   c. intake manifold
   d. lower tank
9. When reverse flushing the engine water jackets, the flushing gun is connected to the (♦22-16)
   a. upper tank
   b. lower tank
   c. thermostat housing
   d. pump inlet
10. Troubles of the engine related directly to the cooling system include (♦22-2)
    a. hard starting and slow warm-up
    b. slow warm-up and overheating
    c. slow cranking and warm-up
    d. none of the above

PART 4

Part 4 of *Automotive Mechanics* discusses electricity and electronics. It explains how the components of the automotive electrical system work and how they are serviced. There are eight chapters in Part 4:

Automotive Electrical and Electronic Equipment

CHAPTER 23
The Automotive Electrical System

After studying this chapter, you should be able to:

1. List the major subsystems in the automotive electrical system and explain the purpose of each.
2. Explain the two ways that electricity is measured.
3. Describe how the ammeter works.
4. Explain the purpose of insulation and why it works.
5. Describe how to make an electromagnet.
6. Describe Ohm's law and why it is important in electric circuits.
7. Explain how the diode and the transistor work.

♦ 23-1 THE AUTOMOTIVE ELECTRICAL SYSTEM

The major components of the automotive electrical system are shown in Fig. 23-1. The major systems include the:

1. Starting system—battery, starting motor, wiring, and switches
2. Charging system—alternator and regulator, with wiring
3. Ignition system
4. Accessory systems—horns, lighting, instrument-panel warning systems

These systems are covered in the following chapters.

♦ 23-2 ATOMS AND ELECTRICITY

Atoms are the "building blocks" of the world (♦8-1). They are extremely tiny. A single drop of water is made up of about 100 billion billion atoms. There are about 100 kinds of atoms. Each kind is an element such as copper, iron, oxygen, and hydrogen. The basic difference among the atoms—the elements—is in the number of *electrons* that circle the center, or nucleus, of the atom (Fig. 23-2). Also, the nucleus is special for each element.

Think of each electron as a very small ball. Each ball carries a charge of negative electricity. When electrons break away from their atoms and are sent moving through a conductor—a wire, for example—there is a flow of electricity. This is called an *electric current*. It takes billions of electrons to make any noticeable amount of electric current. For example, it takes more than 6 billion billion electrons flowing each second to light a small one-ampere light bulb.

♦ 23-3 MEASURING ELECTRICITY

Electricity is measured in two ways. These are by the amount of current (number of electrons) flowing and by the push, or pressure, that causes the current to flow.

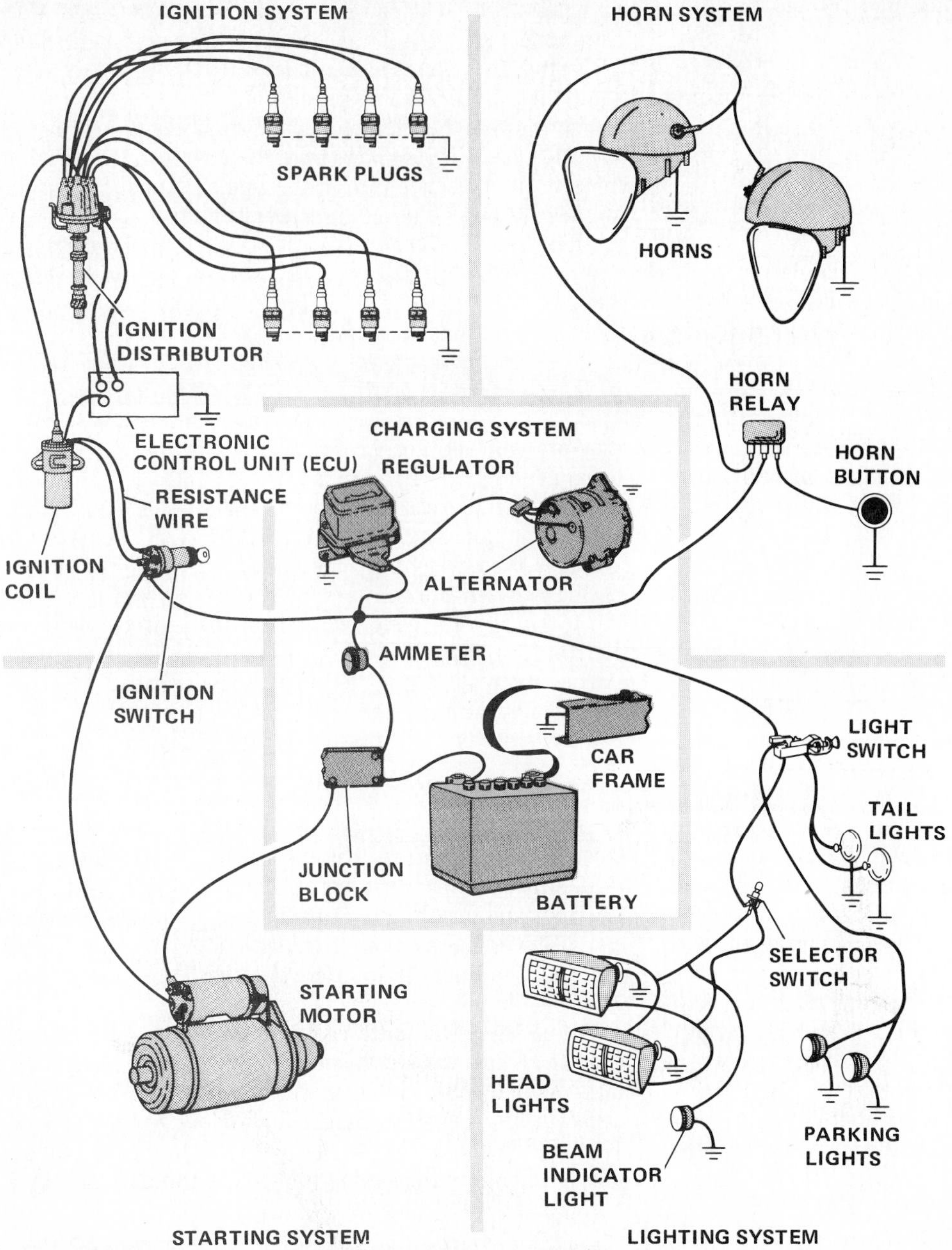

Fig. 23-1 Typical automotive electrical system, showing the major electric units and the connections between them. The symbol ⏚ or ⏚ means ground, or the car body, frame, or engine. Using these metal parts as the return circuit requires only half as much wiring. *(Delco-Remy Division of General Motors Corporation)*

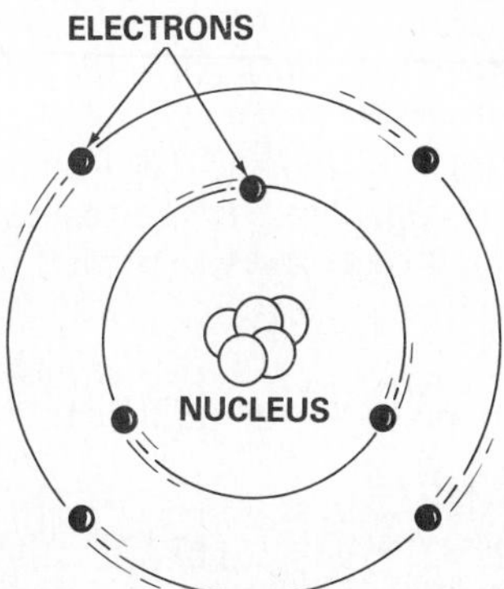

Fig. 23-2 Each atom consists of a core, or nucleus, and electrons circling around the nucleus.

The push, or pressure, is caused by the actions of the electrons. They repel each other. When electrons are concentrated in one place, their negative charges push against each other. If a path is provided for the electrons, they will flow away from the area where they are concentrated.

The pressure to make them move is called *voltage*. If there are many electrons concentrated in one spot, we say that there is a high voltage. With high voltage, many electrons will flow—provided there is a path or conductor through which they can flow. The more electrons that flow, the greater the electric current. Electric current is measured in *amperes*.

These are the two measurements of electricity:

1. Voltage, or the electrical pressure that causes the electricity—electrons—to flow

2. Amperes, or the number of electrons—amount of current—that flows

To measure volts and amperes, two instruments are used, a voltmeter and an ammeter. Both of these are described later. They depend on *magnetism*.

♦ 23-4 MAGNETISM

The voltmeter and ammeter use an effect of electron flow. This effect is that a flow of electrons—an electric current—produces magnetism. Magnetism is a property of magnets. Magnets attract iron and other metals. Magnets make the starting motor, the alternator, the ignition system, and many other electrical devices work.

There are two forms of magnetism—natural and electrical. Both act in the same way. Two important facts about magnets are:

1. Magnets can produce electricity.
2. Electricity can produce magnets. Electrically produced magnets are called *electromagnets*.

♦ 23-5 THE AMMETER

The ammeter uses magnetism to measure the amount of current flowing in a circuit. Figure 23-3 shows the type of ammeter used in many cars.

Here is how the ammeter works: The conductor is connected at one end to the battery. The pointer is mounted on a pivot. There is a small piece of iron, oval shaped, mounted on the same pivot. This oval-shaped piece of iron is called the *armature*. A permanent magnet, almost circular in shape, is placed so that its two ends are close to the armature. The permanent magnet attracts the armature and tends to hold it in a horizontal position. In this position, the pointer or needle points to 0. Nothing is happening. Now suppose the alternator starts sending current to the battery. This current passes through the conductor. The current produces magnetism. This magnetism attracts the armature and causes it to swing clockwise. This moves the pointer to the "charge" side. The more current that flows, the stronger the magnetism and the farther the pointer moves. The meter face is marked off to show the number of amperes flowing.

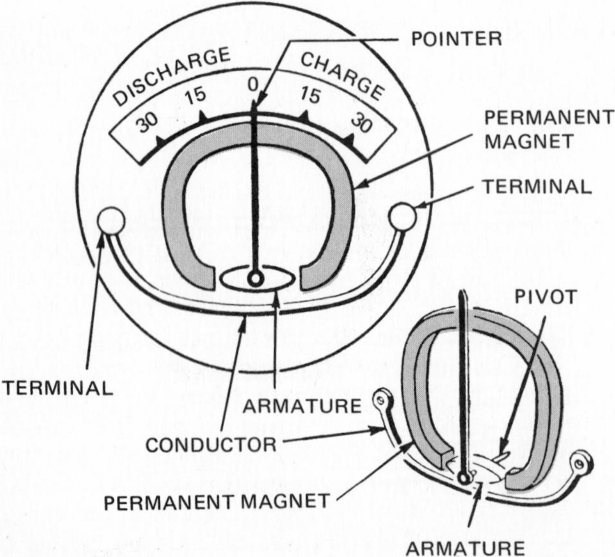

Fig. 23-3 Simplified drawing of a car ammeter, showing its internal construction.

Now suppose the alternator is not working and you turn on the car lights. Current flows from the battery to the lights. It flows in the reverse direction through the conductor in the ammeter. Now the armature is attracted in the opposite direction, and it swings counterclockwise. This moves the pointer to the "discharge" side. The more current being taken out of the battery, the farther the pointer moves.

♦ 23-6 MAKING ELECTRONS MOVE

Electrons try to push away from each other (♦23-3). When they are concentrated in one place, they try to flow away toward another place where there are few electrons. The battery and the alternator are devices that produce this unbalanced condition. They concentrate electrons in one terminal, and take them away from the other. If a conductor is connected between the two terminals, electrons will flow from the terminal with many electrons to the terminal with few electrons.

The "many electrons" terminal is the negative terminal, indicated by a minus sign (−). The "few electrons" terminal is the positive or plus terminal, indicated by a plus sign (+).

♦ 23-7 VOLTAGE

Voltage is the pressure that causes the electric current to flow. Suppose one terminal of an alternator has a great many electrons and the other terminal has a great shortage of electrons. The pressure on the electrons to move from the "many electrons" terminal to the "few electrons" terminal is high. This is a "high" voltage.

The voltage of the batteries in most cars is 12-volts. This is a relatively low voltage. Some ignition systems can produce up to 47,000 volts. This is a relatively high voltage. Devices called *transformers* can change low voltage to high voltage. The ignition coil in the ignition system does this. It changes the 12-volts from the battery to the high voltage in the ignition system.

♦ 23-8 INSULATION

Wires that carry electric current are covered with insulation. The higher the voltage, the stronger the insulation must be. Insulation is a nonconductor. It will not let electrons—electric current—flow through it. In the car, there are wires between the battery, alternator, and other electrical devices. These wires are covered with insulation. The insulation keeps the electric current moving in the proper paths, or circuits. If the insulation goes bad, the electric current will flow where it is not supposed to. It could take a shortcut through the metal of the car frame or body. This is called a *short circuit*, which is discussed later.

♦ 23-9 MAGNETS

Magnets can produce electricity (♦23-4). Electricity can produce magnetism. Magnets are made in many shapes and forms. The simplest is the *bar magnet* (Fig. 23-4). The two ends of the bar magnet are called *poles*. They are the *north*

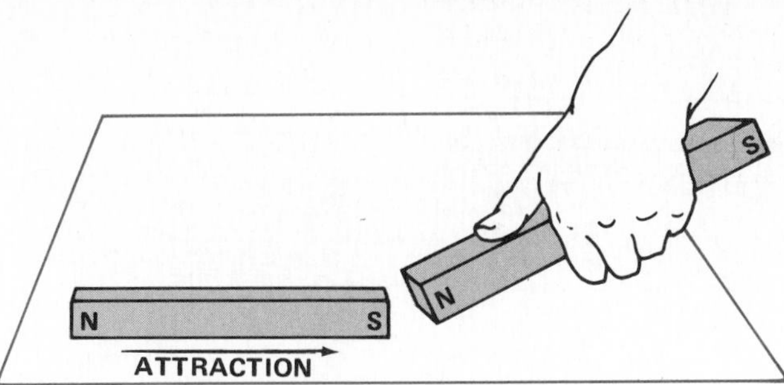

Fig. 23-4 Unlike magnetic poles attract each other.

pole and the *south* pole. Unlike magnetic poles attract each other (Fig. 23-4). Like magnetic poles repel each other (Fig. 23-5). That is:

- North pole attracts south pole
- South pole attracts north pole
- North pole repels north pole
- South pole repels south pole

Like poles repel; unlike poles attract. This is why such electrical devices as alternators and starting motors work. In the starting motor, the rotor (called the armature) is forced to spin because of the actions of unlike magnetic poles in the motor.

♦ 23-10 ELECTROMAGNETS

Electromagnets act just like magnets. An electromagnet can be made by wrapping wire around a tube (Fig. 23-6). Then we connect the ends of the wire (through a switch) to the battery. Now when the switch is closed, there is a strong magnetic field around the coil of wire.

With current flowing through the winding, the winding acts just like a bar magnet. One end of it will either attract or repel a pole of a bar magnet. One end of the winding is a north pole; the other end is a south pole. You can change the poles by reversing the leads to the battery. This shows that when electrons flow through in one direction, it makes one of the poles north. But when the electrons flow through in the reverse direction, the poles reverse. The north pole becomes the south pole, and the south pole becomes the north pole.

An electromagnet, such as that made by winding wire around a tube, is also called a *solenoid*. It is used in several places in the electrical system of the automobile. These are discussed later.

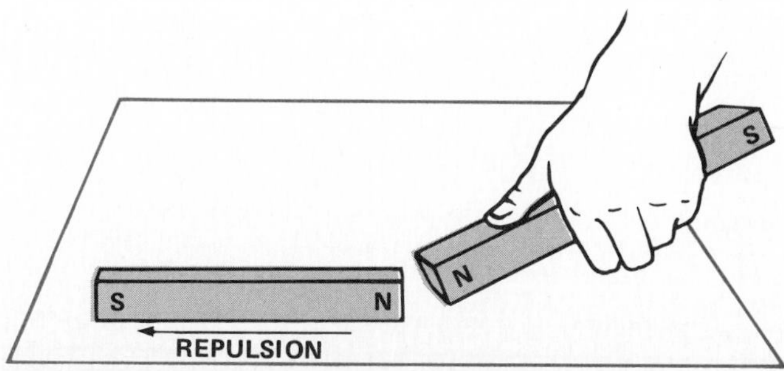

Fig. 23-5 Like magnetic poles repel each other.

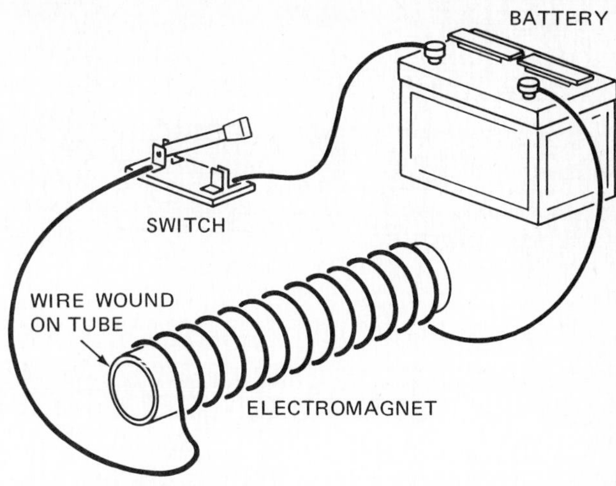

Fig. 23-6 When the switch is closed and current flows from the battery through the winding, it becomes an electromagnet.

♦ 23-11 RESISTANCE

An insulator has a high resistance to the movement of electrons through it. A conductor, such as a copper wire, has a very low resistance. Resistance is a fact of life in all electric circuits. We want resistance in some circuits so that too much current (too many electrons) will not flow. In other circuits, we want as little resistance as possible so that a high current can flow.

Resistance is measure in *ohms*. For example, a 1000-foot [304.8-m] length of number 10 wire (which is about 0.1 inch [2.54 mm] in diameter) has a resistance of 1 ohm. A 2000-foot [605.6-m] length has a resistance of 2 ohms. If the wire is heavier, the resistance drops. For example, a wire 0.2 inch [5.08 mm] in diameter (number 4 wire) has only ¼-ohm resistance per 1000 feet [304.8 m].

The explanation is that the longer the path, or circuit, the farther the electrons have to travel. Therefore, the higher the resistance is to electric current. With the heavier wire, the path is wider. More electrons can flow, so the resistance is lower.

With most substances, copper for example, resistance goes up with temperature. A hot copper wire will carry less current than a cold wire. The reverse is true for a few materials. The substance in the engine unit of the cooling-system temperature indicator (♦21-17), for example, loses resistance as the temperature goes up.

♦ 23-12 OHM'S LAW

There is a definite relation between current (electron flow), voltage (electric pressure), and resistance. As the electric pressure goes up, more electrons flow. Increasing the voltage increases the amperes of current. However, increasing the resistance decreases the amount of current that flows. These relationships can be summed up in a statement known as *Ohm's law*, which is shown below.

Voltage is equal to current times resistance:

$$V = IR$$

where V = voltage
I = current, in amperes
R = resistance, in ohms

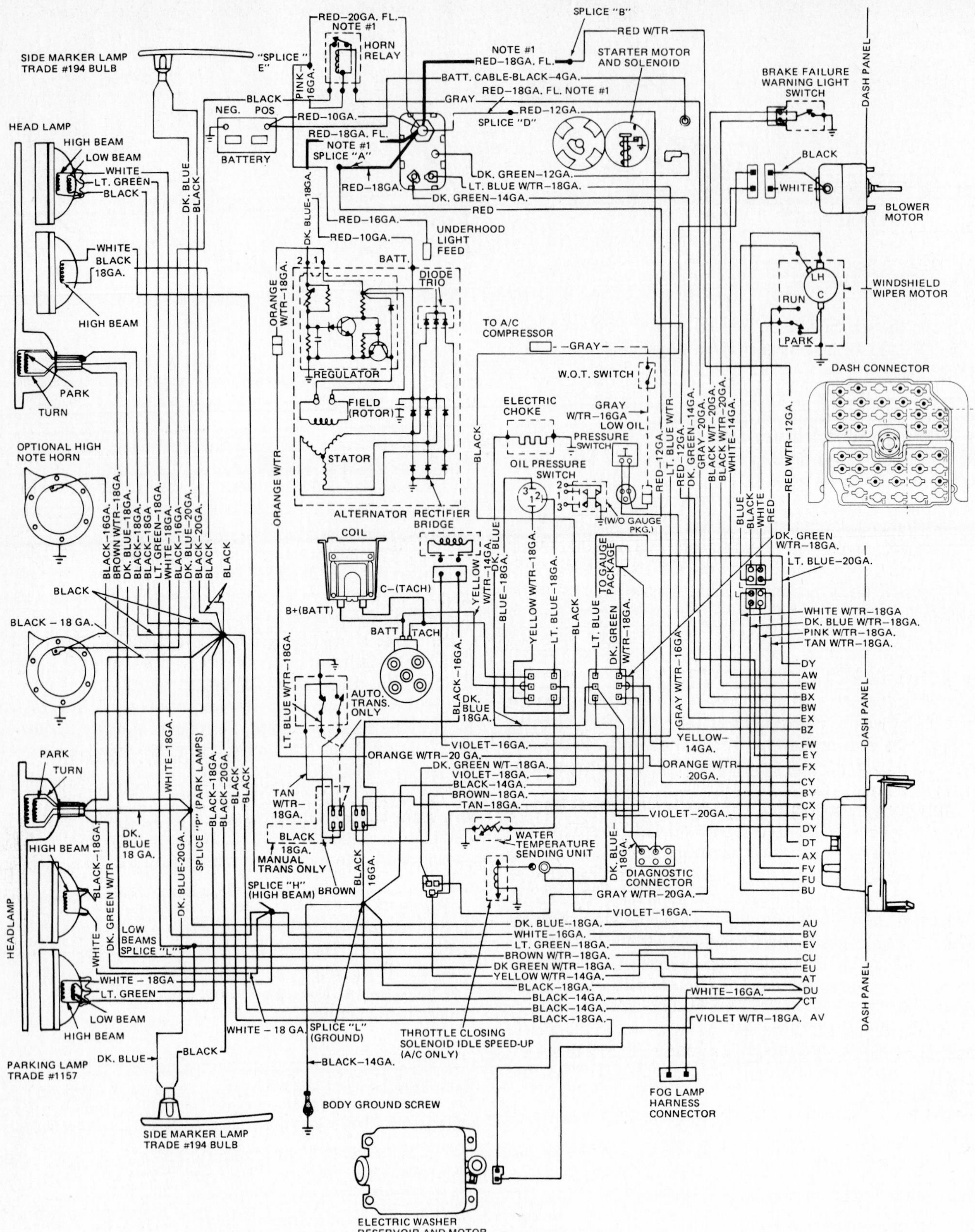

Fig. 23-7 Complete wiring diagram for the front lighting and engine compartment for one vehicle. *(American Motors Corporation)*

Ohm's law can be simply stated: Increasing the resistance in a circuit reduces the current. A major cause of electrical troubles is excessive resistance in circuits, which can be due to poor connections, defective wires, or bad switch contacts.

♦ 23-13 ONE-WIRE SYSTEMS

For electricity to flow, there must be a complete path, or circuit. The electrons must flow from one terminal of the battery or alternator, through the circuit, and back to the other terminal. In the automobile, the engine and car frame are used as the return circuit. Therefore, no separate wires are required for returns from electrical devices to the battery or alternator. The return circuit is called *ground* and is indicated in wiring diagrams by the symbol ⏚ or ⏚. Ground—the engine and car frame and body—is the other half of the circuit. It is the return circuit between the source of electricity (battery or alternator) and the electrical device.

♦ 23-14 ALTERNATING AND DIRECT CURRENT

Most of the electricity generated and used in this world is *alternating current* (ac). The current flows first in one direction and then in the opposite direction. This means it alternates. The current you use in your home is ac. It alternates 60 times per second and is therefore called 60-cycle [60-Hz] ac. (In the metric system of measurement, one cycle per second is called a hertz, abbreviated Hz.)

The automobile cannot use ac. The battery is a *direct-current* (dc) unit. When you discharge it by connecting electrical devices to it, you take current out in one direction only. The current does not alternate, or change directions. Likewise, most other electrical devices in the car operate only on dc.

♦ 23-15 WIRING CIRCUITS

The electrical units in the automobile are connected by wires of different sizes. The size of each wire depends on the amount of current the wire must carry. The heavier the current, the larger the wire must be. The wires are gathered together to form wiring harnesses. Each wire is identified by the color of its insulation. For example, wires are light green, dark green, blue, red, black with a white tracer, and so on. The car manufacturers' shop manuals have illustrations that show the various wires and their colors. If you ever have to trace a particular wire, refer to the shop manual to determine its color. Figure 23-7 shows the wiring diagram for the front lighting and engine compartment for one car. Figure 23-8 shows the wiring harness in the instrument panel for another.

To simplify wiring circuits, symbols are used for the various components in electric circuits. Figure 23-9 shows various symbols and their meanings.

♦ 23-16 PRINTED CIRCUITS

The instrument panel has several indicating devices, switches, and controls (Fig. 23-10). Because the panel is crowded, there can be problems in making connections between the instruments. One solution is the use of printed circuits.

A printed circuit is a flat piece of insulating material on which a series of conducting strips are printed. Figure 23-11 shows part of a printed circuit. When a printed circuit is installed on the instrument panel, the conducting strips carry current between the units. For example, when indicator lamps are installed, the contacts on the lamps rest on the metallic strips to complete the circuit. Likewise, when a switch is installed, its contact terminals connect to the eight strips feeding into the switch.

♦ 23-17 FUSES

Fuses, fusible links, and circuit breakers are installed in circuits to protect the electrical devices in the circuits. Their purpose is to open the circuit in case a short or ground develops and dangerously high currents start to flow. If this should happen, the fuse "blows" or the circuit breaker opens.

A typical "old-style" cartridge fuse is shown partly cut away in Fig. 23-12. It contains a soft metal strip, connected at the ends to the fuse caps. It is connected in series in the circuit. All current in the circuit flows through the fuse. If excessive current flows, the metal strip overheats and melts, or "blows," opening the circuit.This protects the rest of the circuit and connected electrical units from damage. When a fuse blows, the circuit should be checked to see what caused it. Then, after the trouble is fixed, a new fuse should be installed. Figure 23-13 shows the type of fuse block that takes the cartridge fuse.

Instead of using the cartridge fuse, many cars use a blade-type fuse (Fig. 23-14). These were designed for compactness and ease of service.

♦ 23-18 CIRCUIT BREAKERS

Circuit breakers are used in some circuits, such as the headlight circuit. The circuit breaker has a small winding that carries the current in the circuit. When the current is too high, the winding magnetism opens points to open the circuit. The advantage of the circuit breaker is that it keeps resetting itself. Therefore, it gives a warning of trouble but does not completely open the circuit. For example, if excessive current starts to flow in the headlight circuit, the circuit breaker will operate. The lights will flash on and off, warning the driver of trouble. This gives the driver enough light to pull over to the side of the road and stop.

♦ 23-19 FUSIBLE LINKS

A *fusible link* is a short length of insulated wire connected in series in a circuit. It is usually four gauge sizes smaller than the wire it is protecting. In some cars, the fusible links are looped outside the wiring harness. If excessive current flows, the link will burn out, or "blow." This protects the circuit and connected electrical equipment from damage. Wiring harnesses in cars have several fusible links. Some are connected into feed lines supplying electric current to all electrical equipment except the starting motor (Fig. 23-15). If a short or ground should develop in any of these circuits, the fusible link burns out to protect the circuit.

A blown fusible link is usually easy to spot. Bare wire may be sticking out of the insulation. Or the insulation may be bubbled or burned from the heat. Figure 23-16 shows how to repair a blown fusible link.

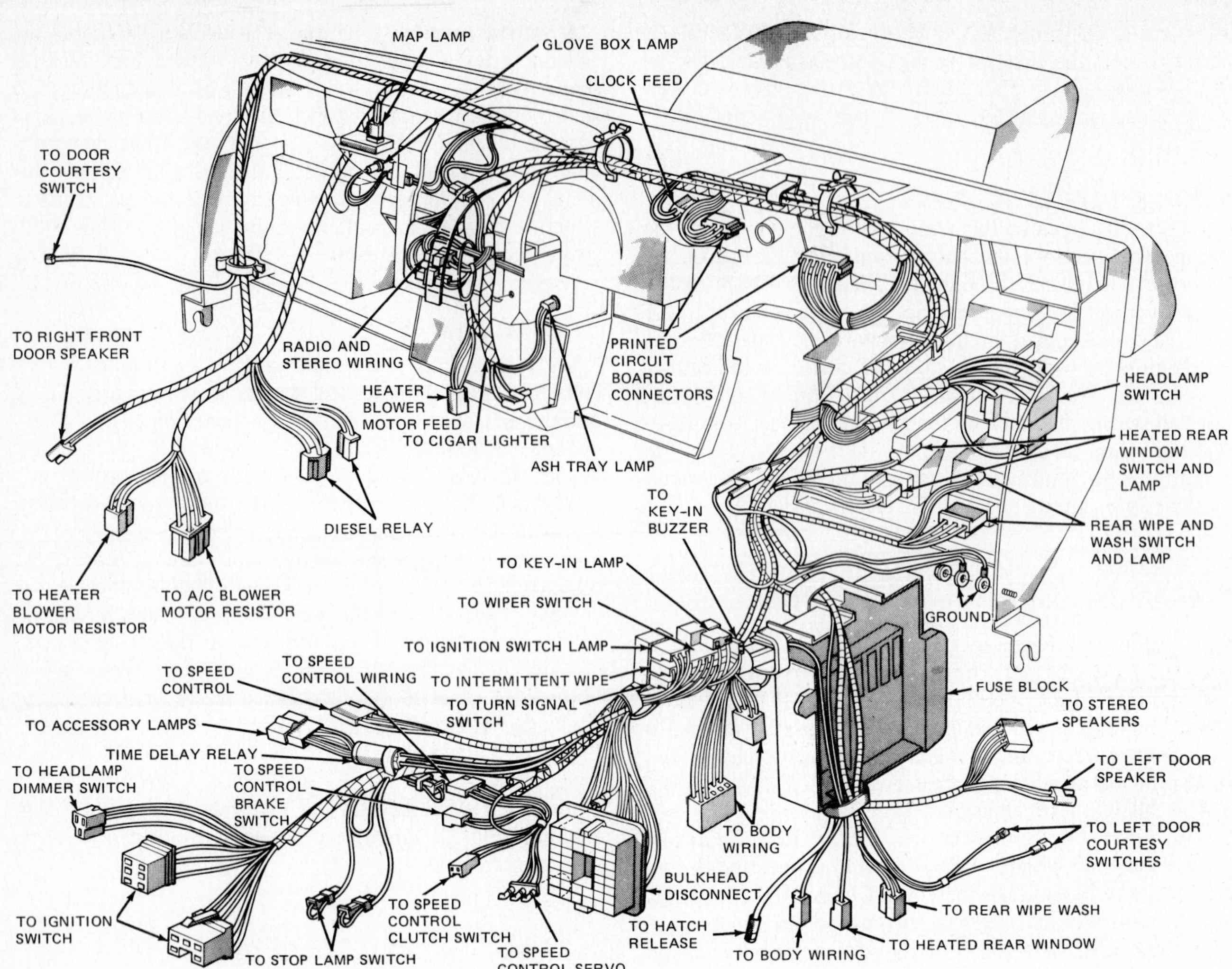

Fig. 23-8 A typical instrument-panel wiring harness. *(Chrysler Corporation)*

♦ 23-20 ELECTRONICS

Electronics is an integral part of modern technology. Radio, television, computers, and electronic games are all electronic devices. The modern automobile uses several electronic devices in the charging system, ignition system, instrument-panel indicating devices, fault detectors, and verbal warning systems. There are two major devices in electronic systems that make them work. These are the *diode* and the *transistor*.

1. Diode The diode is a one-way valve for electrons, or electric current. Automotive electrical systems use only dc. The alternator (Chap. 26) produces ac. See ♦23-14 for a discussion of ac and dc. Diodes convert this ac into dc so the automotive electrical system can use it. A diode allows current to flow one way. But the diode won't allow current to flow in the opposite direction (Fig. 23-17). Chapter 26 on the charging system describes diode action.

2. Transistor The transistor is an electronic device that can serve as an amplifier or as a switch. It is, in effect, a diode with additional material and connections. When the transistor receives a signal—a small current—it becomes a conductor. Also, a very small signal can allow a high current to flow through the transistor. The transistor can be used as an off-on switch, or as an amplifier.

3. Integrated Circuits Modern science has been able to put thousands of transistors and diodes together on a very small piece of material. This material is a *semiconductor*. That means the material is a conductor sometimes and an insulator other times. This combination is called a *chip*. The chip is the "brain" in ECMs and microprocessors. These are used in electronic ignition systems, electronic fuel systems, and electronic engine control systems. These are described in the chapters covering the various systems.

A typical chip is shown in Fig. 23-18. Its actual size is only ¼ inch [6.35 mm] square. Yet it contains thousands of individual transistors and other electronic devices. This is the type of chip used in automotive electronic control systems.

ELECTRICAL SYMBOLS			
SYMBOL	REPRESENTS	SYMBOL	REPRESENTS
ALT	ALTERNATOR	HORN	HORN
A	AMMETER		LAMP OR BULB (Preferred)
	BATTERY-ONE CELL		LAMP OR BULB (Acceptable)
	BATTERY-MULTICELL	MOT	MOTOR-ELECTRIC
12 V + −	(Where required, battery voltage or polarity or both may be indicated as shown in example. The long line is always positive polarity.)	−	NEGATIVE
+ − BAT	BATTERY-VOLTAGE BOX	+	POSITIVE
	BI-METAL STRIP		RELAY
	CABLE-CONNECTED		RESISTOR
	CABLE-NOT CONNECTED		RESISTOR-VARIABLE
	CAPACITOR	IDLE STOP	SOLENOID-IDLE STOP
	CIRCUIT BREAKER	B SOL STARTING MOTOR	STARTING MOTOR
	CONNECTOR-FEMALE CONTACT		
	CONNECTOR-MALE CONTACT		
	CONNECTORS-SEPARABLE-ENGAGED		
	DIODE		
HEI	DISTRIBUTOR		SWITCH-SINGLE THROW
	FUSE		SWITCH-DOUBLE THROW
FUEL	GAUGE-FUEL	TACH	TACHOMETER
TEMP	GAUGE-TEMPERATURE		TERMINATION
	GROUND-CHASSIS FRAME (Preferred)	V	VOLTMETER
	GROUND-CHASSIS FRAME (Acceptable)	OR	WINDING-INDUCTOR

Fig. 23-9 Electrical symbols used in automotive wiring diagrams. *(General Motors Corporation)*

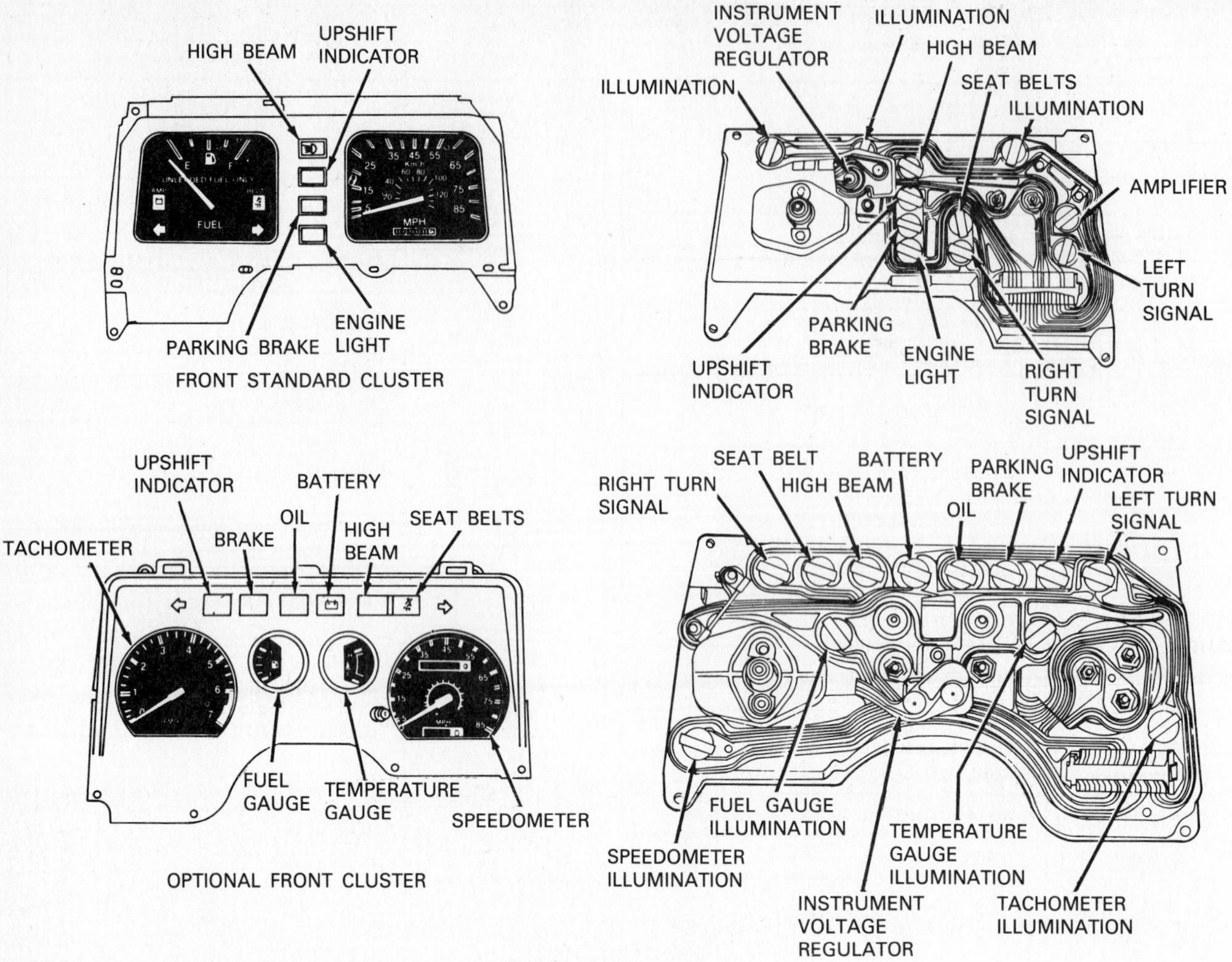

HIGH BEAM
UPSHIFT INDICATOR
PARKING BRAKE
ENGINE LIGHT
FRONT STANDARD CLUSTER
INSTRUMENT VOLTAGE REGULATOR
ILLUMINATION
ILLUMINATION
HIGH BEAM
SEAT BELTS
ILLUMINATION
AMPLIFIER
LEFT TURN SIGNAL
PARKING BRAKE
ENGINE LIGHT
UPSHIFT INDICATOR
RIGHT TURN SIGNAL
UPSHIFT INDICATOR
BATTERY
OIL
BRAKE
HIGH BEAM
SEAT BELTS
TACHOMETER
FUEL GAUGE
TEMPERATURE GAUGE
SPEEDOMETER
OPTIONAL FRONT CLUSTER
SEAT BELT
BATTERY
RIGHT TURN SIGNAL
HIGH BEAM
PARKING BRAKE
UPSHIFT INDICATOR
OIL
LEFT TURN SIGNAL
FUEL GAUGE ILLUMINATION
SPEEDOMETER ILLUMINATION
TEMPERATURE GAUGE ILLUMINATION
INSTRUMENT VOLTAGE REGULATOR
TACHOMETER ILLUMINATION

Fig. 23-10 Instrument panels for different vehicles. *(Ford Motor Company)*

INDICATOR-LIGHT SOCKETS
SWITCH LOCATOR
CONNECTORS

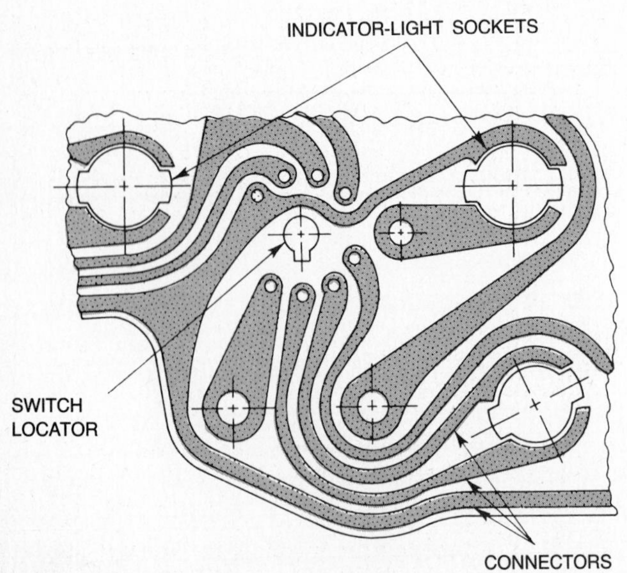

Fig. 23-11 Part of a printed circuit.

FUSE CAP
METAL STRIP

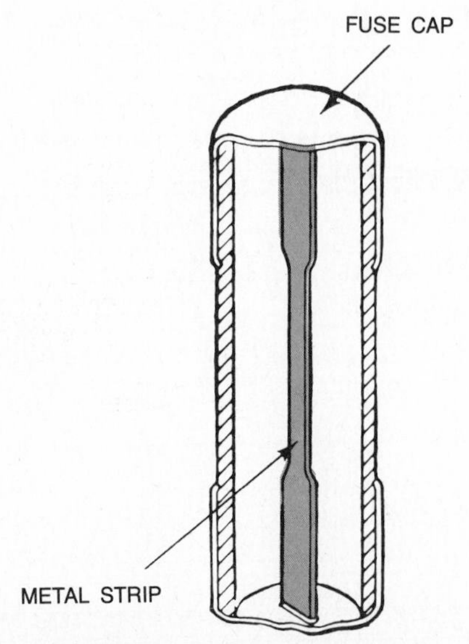

Fig. 23-12 Sectional view of a cartridge fuse.

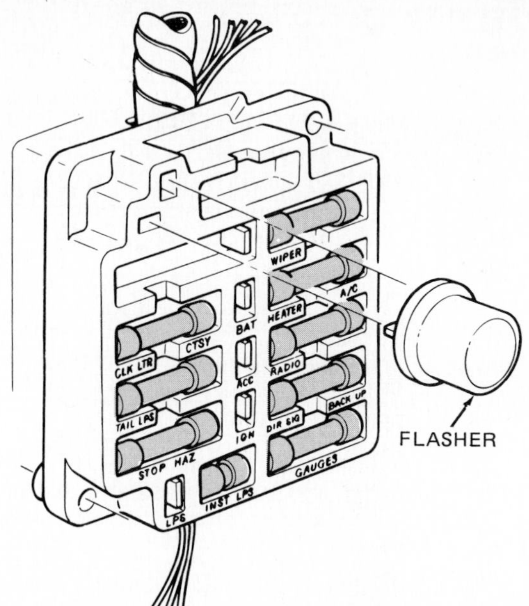

Fig. 23-13 Fuse block with fuses in place. *(Chevrolet Motor Division of General Motors Corporation)*

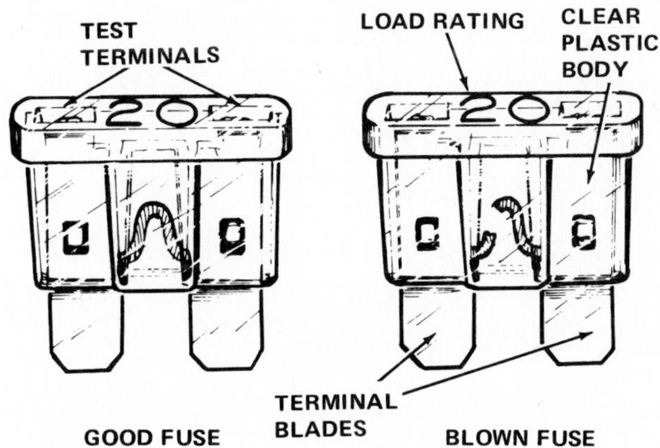

Fig. 23-14 A good and a blown blade-type fuse. Note the terminals to test the fuse. *(General Motors Corporation)*

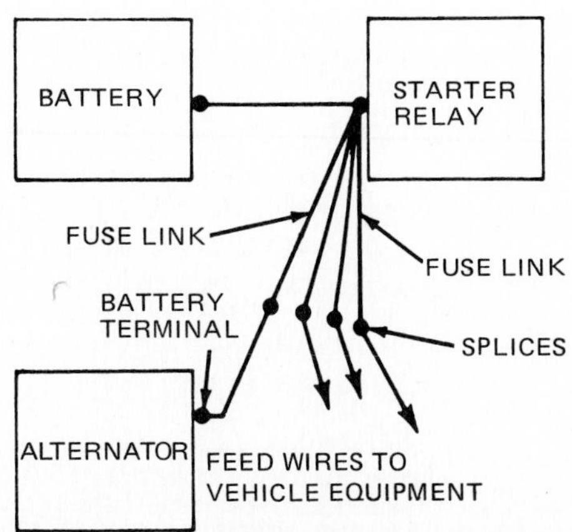

Fig. 23-15 Simplified diagram showing how more than one feed circuit, with fusible links, can be attached to the battery terminal on the starter relay. *(Ford Motor Company)*

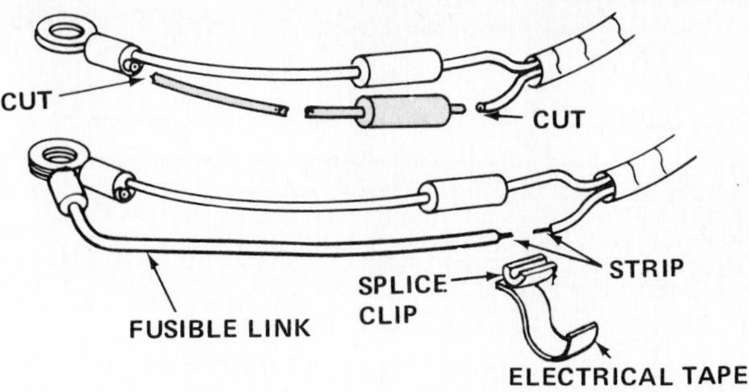

Fig. 23-16 To repair a burned-out fusible link, cut it out as shown in the top view. Strip back the insulation. Splice the wires with a splice clip, and solder in the new fusible link. Tape the splice with a double layer of electrical tape. *(General Motors Corporation)*

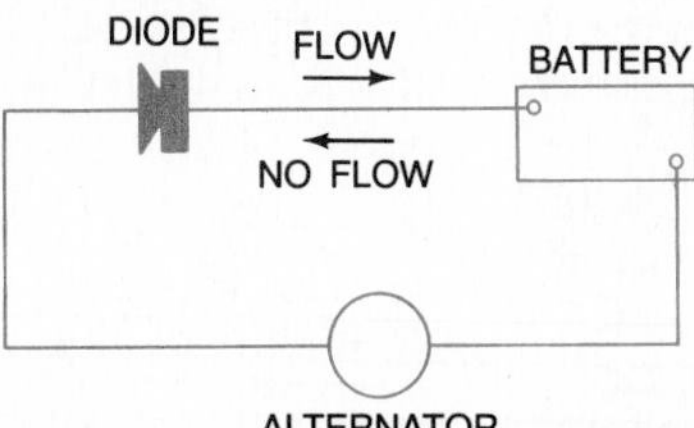

Fig. 23-17 Alternating current from an alternator can be changed to direct current, or rectified, by a diode for charging a battery.

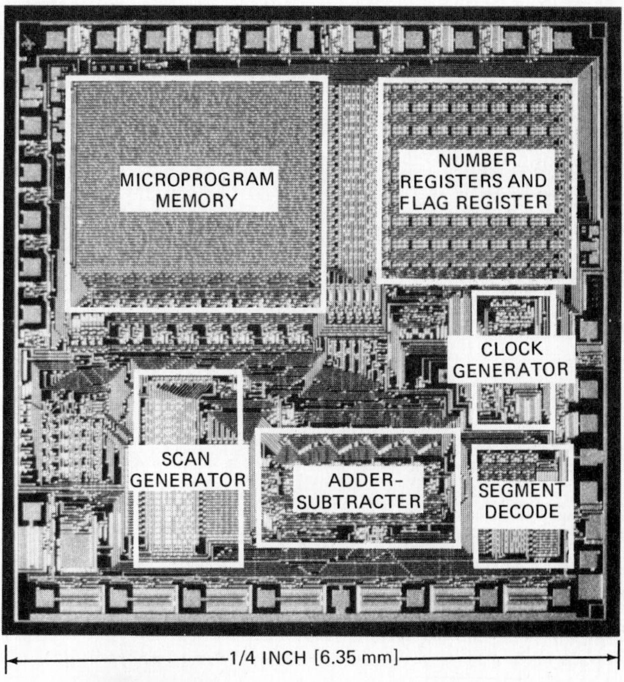

Fig. 23-18 At top, a computer chip, greatly enlarged, with its various major systems labeled. The actual chip is only about ¼ inch (6 mm) in size. At bottom, shown a little larger than actual size, is the package in which the chip is sealed for protection. The pins coming out from the package are the connections to the chip. *(Texas Instruments, Incorporated)*

♦ REVIEW QUESTIONS ♦

Select the *one* correct, best, or most probable answer to each question. You can find the answers in the section indicated at the end of each question.

1. Major systems in the automotive electrical system include the (♦23-1)
   a. horn, lighting, and ignition systems
   b. starting, charging, and ignition systems
   c. accessory, charging, and warning systems
   d. starting, emissions, and ignition systems
2. The pressure to make electrons move is called (♦23-3)
   a. electricity
   b. voltage
   c. amperage
   d. magnetism
3. When the insulation fails on wiring in the automotive electrical system, the result could be (♦23-8)
   a. a short circuit
   b. a long circuit
   c. no current
   d. high voltage
4. If you wrap wire around a tube and send electric current through the wire, the assembly becomes (♦23-10)
   a. an electromagnet
   b. a solenoid
   c. both a and b
   d. neither a nor b
5. Ohm's law is (♦23-12)
   a. $I = VR$
   b. $R = AI$
   c. $V = IR$
   d. $A = RC$
6. The reason that the automotive electrical system is called a one-wire system is that the (♦23-13)
   a. battery has only one terminal
   b. engine and car body and frame furnish the return circuit
   c. alternator can charge in only one direction
   d. system uses direct current
7. Most devices in the automotive electrical system use (♦23-14)
   a. ac
   b. dc
   c. both ac and dc
   d. neither ac nor dc
8. In the modern car, electronic devices are used in the (♦23-20)
   a. sound system
   b. charging system
   c. ignition system
   d. all of the above
9. The device that acts as a one-way check valve for electric current is called a (♦23-20)
   a. triode
   b. diode
   c. monode
   d. transistor
10. The device that can act like a high-speed electronic switch is called a (♦23-20)
    a. triode
    b. diode
    c. monode
    d. transistor

CHAPTER 24
The Automotive Battery: Construction and Service

After studying this chapter, you should be able to:

1. Discuss the construction and operation of a lead-acid storage battery.
2. Describe the chemical actions in the battery during charge and during discharge.
3. Define and discuss battery ratings.
4. Explain why battery terminal voltage varies with temperature, charging rate, discharge rate, and state of charge of the battery.

♦ 24-1 THE BATTERY

The battery (Fig. 24-1) supplies current to operate the starting motor and the ignition system when the engine is being started. It also acts as a voltage stabilizer by supplying current for the lights, radio, and other electrical accessories when the alternator is not handling the load. The battery is an *electrochemical* device. This means it uses chemicals to produce electricity. The amount of electricity it can produce is limited. As the chemicals in the battery are "used up," the battery runs down, or is *discharged.* It can be *recharged* by supplying it with electric current from a battery charger, or from the vehicle alternator. The "used up" chemicals are then returned to their original condition, so the battery becomes recharged.

♦ 24-2 CHEMICALS IN BATTERY

The chemicals in the battery are sponge lead (a solid), lead oxide (a paste), and sulfuric acid (a liquid). These three substances are made to react chemically to produce a flow of current. The lead oxide and sponge lead are held in *plate grids* to form positive and negative plates (Fig. 24-2).

The plate grid is a framework of lead alloy with horizontal and vertical bars. The plate grids are made into plates by applying lead oxide paste. The horizontal and vertical bars hold the paste in the plate.

After the plates are assembled into the battery, the battery is given a "forming" charge. This changes the lead oxide paste in the negative, or minus, plate to sponge lead. It changes the lead oxide paste in the positive, or plus, plate to lead peroxide.

♦ 24-3 BATTERY CONSTRUCTION

In the battery, several similar plates are properly spaced and welded, or *lead-burned,* to a strap. This forms a plate group. Plates of two types are used, one for the positive plate group, the other for the negative plate group. A positive plate group is nested with a negative plate group. Separators are placed between the plates to form an element (Fig. 24-3). The separators hold the plates apart so that they do not touch. At the same time the separators are porous enough to permit liquid to circulate between the plates. Wooden sheets, spun glass matted into sheets, and porous sponge-rubber sheets have been used as separators. Late-model batteries have separators made of acid-resistant polyvinyl chloride or polyethylene-saturated cellulose.

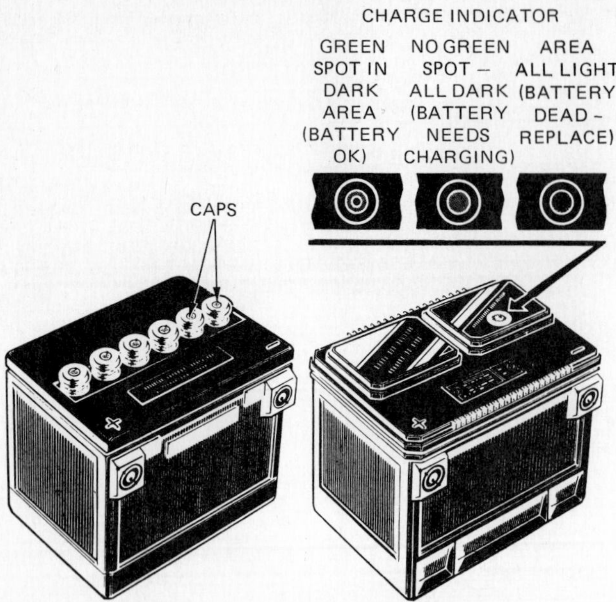

Fig. 24-1 Two types of automotive batteries. The left battery has caps which can be removed to check the battery state of charge and to add water if needed. The right battery with side terminals is a sealed maintenance-free type and requires no water. The charge indicator in the top shows the state of charge of the battery.

The elements are placed in cells in the battery case. Then heavy lead connectors are attached to the cell terminals to connect the cells in series. After the internal connectors are in place, the cover is put on (Figs. 24-1 to 24-3). In many batteries, the cover has openings through which liquid can be added when the filler plugs or vent caps are removed. After the liquid is added and the battery is given an initial charge, it is ready for operation. Maintenance-free batteries have no vent caps.

Some batteries have the two main terminals on the battery

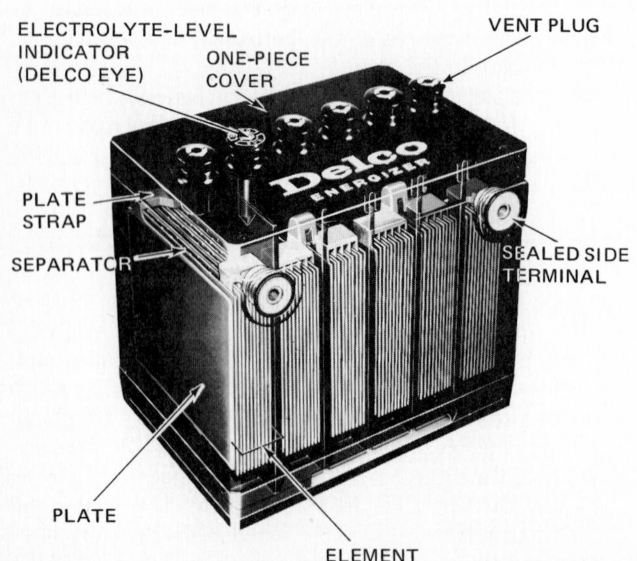

Fig. 24-2 Phantom view of a 12-volt battery with the terminals in the side of the battery case. *(Delco-Remy Division of General Motors Corporation)*

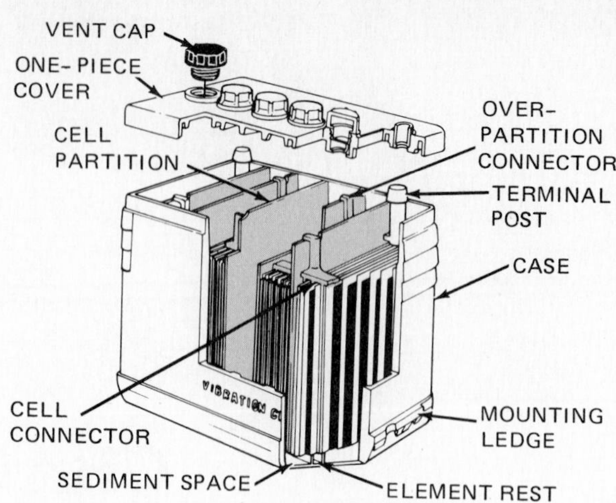

Fig. 24-3 Partly cut away and disassembled 12-volt battery. *(Ford Motor Company)*

cover, as in Fig. 24-3. Other batteries have the terminals in the side of the battery case, as in Fig. 24-2. Figure 24-4 shows how the cables are connected to a side-terminal battery. It also shows the battery-mounting arrangement. On a top-terminal battery, the positive terminal (or "post") is larger than the negative post.

Many batteries are the maintenance-free type. They require no special attention, except for an occasional check of the connections and the built-in charge indicator (Fig. 24-1, right). Other batteries have vent caps (Figs. 24-3 and 24-4) which can be removed. Then you can look down into the battery cells to see whether they need water. Also, a hydrometer can be used to check the battery charge.

♦ 24-4 CHEMICAL ACTIONS IN BATTERY

The liquid in a battery is called the *electrolyte.* It is made up of about 40 percent sulfuric acid and about 60 percent water (in a fully charged battery). When sulfuric acid is placed between the plates, chemical actions take place. These actions remove electrons from one group of plates and collect them at the other. This creates a pressure of 2.1 volts between the two terminals of the battery cell. If the two terminals are not connected by any circuit, no further chemical activity takes place.

However, when the two terminals do become connected by an electric circuit, electrons (current) will flow. They flow

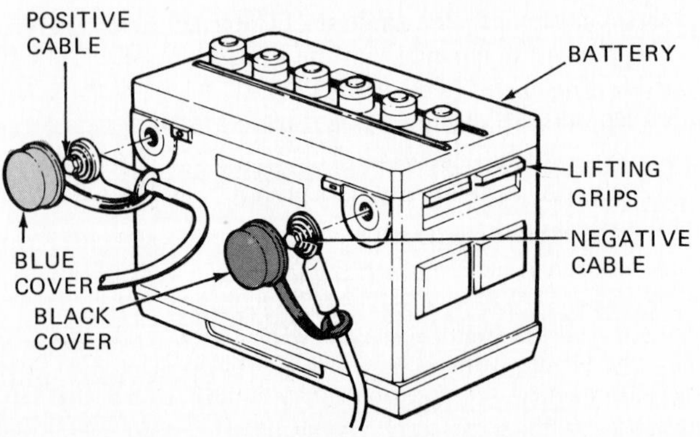

Fig. 24-4 Cable connections for a side-terminal battery. *(Cadillac Motor Car Division of General Motors Corporation)*

from the terminal where chemical activity has collected them. They flow through the circuit to the other terminal, where the chemical activity has removed them. Chemical activities now begin again so the 2-volt pressure is maintained. The current flow continues. The chemical actions "use up" the sponge lead, lead peroxide, and sulfuric acid. After a certain amount of current has been withdrawn, the battery is discharged ("run down" or "dead"). It is not capable of delivering any additional current. When the battery has reached this state, it may be recharged. This is done by supplying it with a flow of current from some external source. The external source forces current back through the battery. This reverses the chemical activities in the battery. The plates are restored to their original composition, and the battery becomes recharged. It is then ready to deliver additional current.

The chemical actions that take place are rather complicated. The sponge lead (negative plate) and lead peroxide (positive plate) change to *lead sulfate* during the discharge process. The sulfate comes from the sulfuric acid. The electrolyte loses acid and gains water as the sulfate goes into the plates. Therefore, discharging the battery changes the two different chemicals in the battery plates to a third chemical, lead sulfate. Recharging the battery changes the lead sulfate back to sponge lead in the negative plates, and to lead peroxide in the positive plates. Meantime, the sulfuric acid reappears in the electrolyte of the battery.

♦ 24-5 CONNECTING CELLS

Automotive batteries are usually 12-volt units. There are six cells in the 12-volt battery. The six cells are connected in series. In series connections, the voltages add. Some special applications use 24-volt batteries; these special-purpose batteries have 12 cells.

Although a battery cell at 80°F [26.7°C] will test on open circuit about 2.1 volts when fully charged, common practice is to call it 2 volts. Therefore, a six-cell battery is said to be a 12-volt battery, rather than a 12.6-volt battery.

♦ 24-6 BATTERY RATINGS

The amount of current that a battery can deliver depends on the total area and volume of active plate material. It also depends on the amount and *strength* of electrolyte. This is the percentage of sulfuric acid in the electrolyte. Factors that influence battery *capacity*—its ability to deliver current—include the number of plates per cell, the size and thickness of the plates, the cell size, and the quantity of electrolyte. The ratings most commonly used in referring to battery capacity are discussed below.

1. Reserve Capacity Reserve capacity is the length of time in minutes that a fully charged battery at 80°F [26.7°C] can deliver 25 amperes. A typical rating would be 125 minutes. This figure tells how long a battery can carry the electrical operating load if the alternator quits.

2. Ampere-Hour Capacity An older method of rating batteries is the 20-hour rate, also called the *ampere-hour capacity.* This is the amount of current that a battery can deliver for 20 hours without the cell voltage dropping below 1.75 volts, with an electrolyte temperature of 80°F [26.7°C]. For example, a battery that can deliver a current of 5 amperes for 20 hours is rated as having a 100 ampere-hour capacity ($20 \times 5 = 100$).

3. Cold-Cranking Rate One of the two cold-cranking rates is the number of amperes that a battery can deliver for 30 seconds when it is at 0°F [−17.8°C] without the cell voltages falling below 1.2 volts. A typical rating for a battery with a reserve capacity of 125 minutes would be 430 amperes. This figure indicates the ability of the battery to crank the engine at low temperatures. The second cold-cranking rate is measured at −20°F [−28.9°C]. In this, the final voltage is allowed to drop to 1 volt per cell. A typical rating for a battery with a reserve capacity of 125 minutes would be 320 amperes.

4. Watts Delco uses an additional rating—watts. This is roughly equivalent to the battery cold-cranking rating.

♦ 24-7 BATTERY EFFICIENCY

The ability of the battery to deliver current varies within wide limits. It depends on temperature and rate of discharge. At low temperature, chemical activities are greatly reduced. The sulfuric acid cannot work as actively on the plates. Therefore the battery is less efficient and cannot supply as much current for as long a time. High rates of discharge will not produce as many ampere-hours as low rates of discharge. At high discharge rates, the chemical activities take place only on the surfaces of the plates. They do not have time to penetrate the plates and to use the materials below the plate surfaces.

The chart below relates battery efficiency to battery temperatures. However, these are only approximations.

Efficiency, Percent	Battery Temperature Degrees F (C)
100	80 [26.7]
65	32 [0]
50	0 [−17.8]
10	−45 [−42.8]

♦ 24-8 VARIATIONS IN TERMINAL VOLTAGE

Because the battery produces voltage by chemical means, the voltage varies according to a number of conditions. These conditions and their effect on battery voltage may be summed up as follows:

1. Terminal voltage, battery being *charged,* increases with:
   a. Increasing charging rate. To increase charging rate (amperes input), the terminal voltage must go up.
   b. Increasing state of charge. As state of charge goes up, voltage must go up to maintain charging rate. For example, a voltage of approximately 2.6 volts per cell is required to force a current through a fully charged battery. This is the reason that voltage regulators are set to operate at 15 volts—slightly below the voltage required to charge a fully charged battery. This setting protects the battery from overcharge.
   c. Decreasing temperature. Lower battery temperatures require a higher voltage to maintain charging rate.
2. The terminal voltage of a battery that is being *discharged* decreases with:
   a. Increasing discharge rate. As the rate of discharge goes

up, chemical activities increase and cannot penetrate plates so effectively. Therefore, voltage is reduced.

b. Decreasing state of charge. With less of the active material and sulfuric acid available, less chemical activity takes place, and voltage drops.

c. Decreasing temperature. With lower temperature, the chemical activities cannot go on as effectively, and the voltage drops.

BATTERY MAINTENANCE AND SERVICE

♦ 24-9 BATTERY MAINTENANCE

Complete battery maintenance includes the following:

1. Visually check the battery.
2. Check electrolyte level in cells on batteries with vent caps.
3. Add water if the level is low (vent-cap batteries).
4. Clean off corrosion around battery terminals and from top of batteries with top terminals (Fig. 24-3).
5. Check battery condition by testing its state of charge (explained later).
6. Recharge battery if it is low.

♦ 24-10 CAUTIONS FOR BATTERY SERVICE

These are the important cautions to observe when working with batteries:

1. The sulfuric acid in the electrolyte is very corrosive. It can destroy most things it touches. It will eat holes in cloth. It can cause serious burns if it gets on your skin. If it gets in your eyes, it can blind you. For this reason, always wear eye protection when working around batteries. If you get battery acid on your skin, flush it off at once with water. Continue to flush for 5 minutes. If you get acid (or any chemical) in your eyes, flush it out at once with water, over and over (Fig. 2-12). Then get to a doctor at once!
2. Gases form in the battery when it is being charged. These gases are highly explosive. Never light a match or a cigarette when working around batteries. You might cause an explosion that could seriously harm you.
3. The battery can supply a very high current. Never wear rings, bracelets, watches, or hanging necklaces when working around batteries. If a metal ring or other ornament should accidentally short a battery, a very high current will flow. This could turn the ring white-hot in an instant, and you could get a serious burn.
4. When jump starting a car, follow the instructions and cautions in ♦25-9. A wrong step can damage the electrical system and cause you to get hurt.
5. When disconnecting a battery, always disconnect the cable from the grounded terminal first. Then, if you should accidentally ground the insulated terminal, or any terminal or wire that is "hot" (connected to the insulated terminal), you will not be making a direct short across the battery.

♦ 24-11 VISUAL INSPECTION OF BATTERY

Look for signs of leakage, a cracked case or top, corrosion buildup on battery terminals and tops, missing vent caps, and loose or missing hold-down clamps. Leakage causes white corrosion on the battery carrier and surrounding metal parts. Leakage is due to a cracked battery case or top. The remedy is to install a new battery.

The top of the battery can be cracked if the wrong wrench is used to disconnect or install cable clamps (top terminals). The case can be cracked if the hold-down clamps are overtightened.

♦ 24-12 CHECKING ELECTROLYTE LEVEL AND ADDING WATER

On vent-cap batteries, the electrolyte level can be checked by removing the caps. Some batteries have a split ring which indicates the electrolyte level (Fig. 24-5). If the level is low, add water. Some batteries have an electrolyte-level indicator (a "Delco Eye"). It gives a visual indication of the electrolyte level (Fig. 24-6). Black means the level is okay. White means the level is low.

Careful Do not add too much water. This can cause electrolyte to leak out and corrode the battery carrier and other metal nearby.

♦ 24-13 CLEANING CORROSION OFF BATTERY

On batteries with the terminals on top, the terminals and cable clamps sometimes corrode. This corrosion can be cleaned off by disconnecting the clamps and cleaning the terminal and clamps with special wire brushes (Fig. 24-7). Battery-top corrosion can be cleaned off by brushing the top with baking-soda solution. After the foaming stops, flush off the battery top with water. Terminals can be coated with an anticorrosion compound to retard corrosion.

♦ 24-14 CHECKING BATTERY CONDITION

On vent-cap batteries, the battery can be checked with a hydrometer to determine its condition. Other methods use testing instruments, as explained later. These other methods are for the maintenance-free batteries but can also be used on the vent-cap type.

♦ 24-15 HYDROMETER TEST

The hydrometer tests the *specific gravity* of the battery electrolyte. There are two types of hydrometer (Fig. 24-8). One uses a series of plastic balls, the other a glass float with a stem on top. To use a ball-type hydrometer, stick the rubber tube in the electrolyte. Then squeeze and release the rubber bulb. This draws electrolyte up into the glass tube. The number of balls that float tells you the battery state of charge. If all balls float, the cell is fully charged. If none float, the cell is discharged.

Careful Do not drip electrolyte on the car or on yourself! Electrolyte will damage the paint on the car and eat holes in your clothes! See Caution 1 in ♦24-10.

The float-type hydrometer (right in Fig. 24-8) has a float with a stem that sticks up above the electrolyte level in the tube. The float stem is marked to indicate the specific gravity

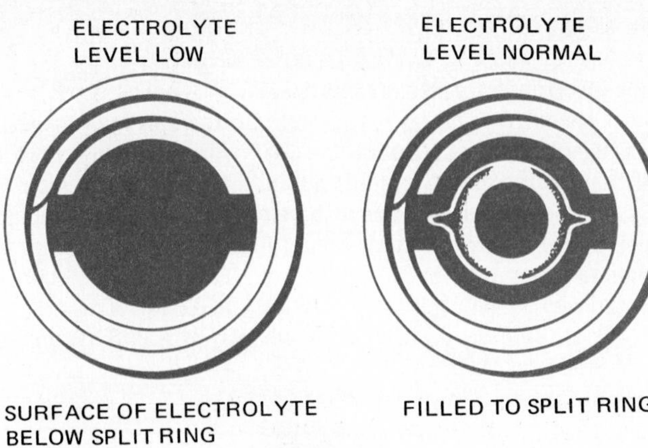

Fig. 24-5 Appearance of the electrolyte and split ring when the electrolyte is too low, and when it is correct. *(Delco-Remy Division of General Motors Corporation)*

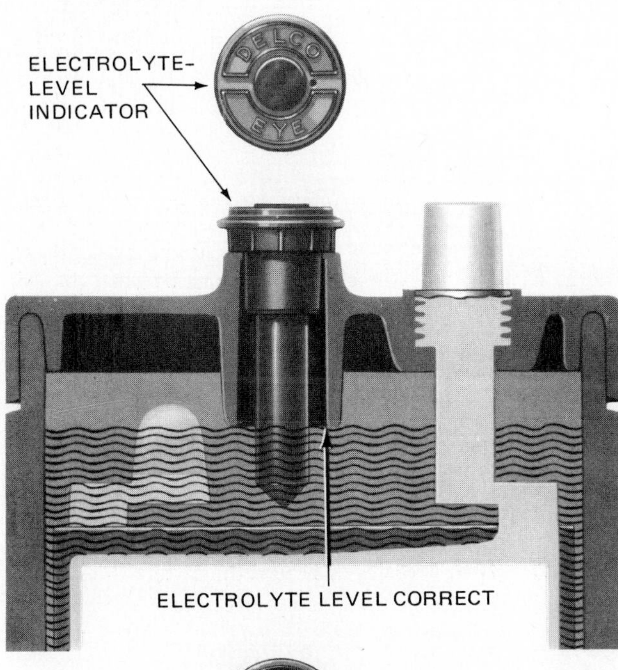

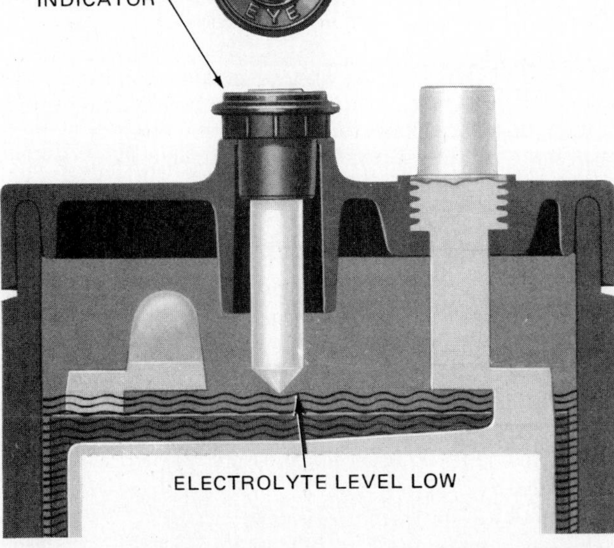

Fig. 24-6 A battery cell with an electrolyte-level indicator, showing (top) the electrolyte at the correct level and (bottom) low electrolyte level. *(Delco-Remy Division of General Motors Corporation)*

of the electrolyte (Fig. 24-9). The height of the stem above the electrolyte tells you the battery state of charge. Here is what the readings mean:

1.265–1.299	Fully charged battery
1.235–1.265	Three-fourths charged
1.205–1.235	One-half charged
1.170–1.205	One-fourth charged
1.140–1.170	Barely operative
1.110–1.140	Completely discharged

If some cells test considerably lower than others, it means there is something wrong with those cells. It could be that a cracked case has allowed electrolyte leakage, or perhaps there is internal damage to the plates or separators. If the variation is only a few specific-gravity points, then there is probably no cause for alarm. But if the low cells measure 50 points lower, then those cells are defective and the battery should be replaced.

Some 12-volt batteries for passenger cars have a slightly lower specific gravity when charged. For example, one type is fully charged with a specific gravity of 1.270. Other batteries, such as those used in hot climates, have a specific gravity of 1.225 when fully charged.

The decimal point is not normally referred to in a discussion of specific gravity. For example, "twelve twenty-five" means 1.225, and "eleven fifty" means 1.150.

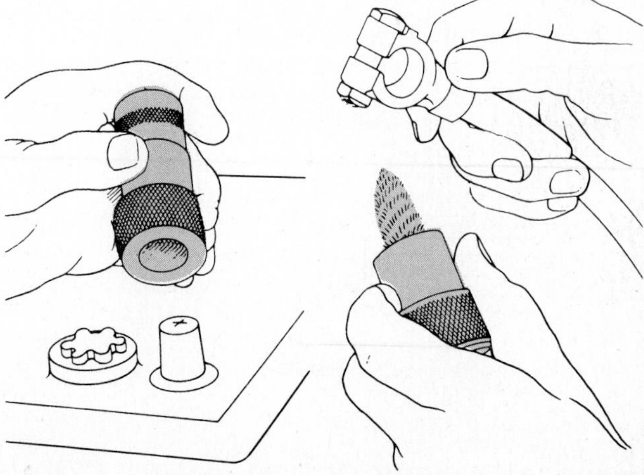

Fig. 24-7 Using a battery-terminal brush to clean the battery terminal posts and cable clamps. *(Buick Motor Division of General Motors Corporation)*

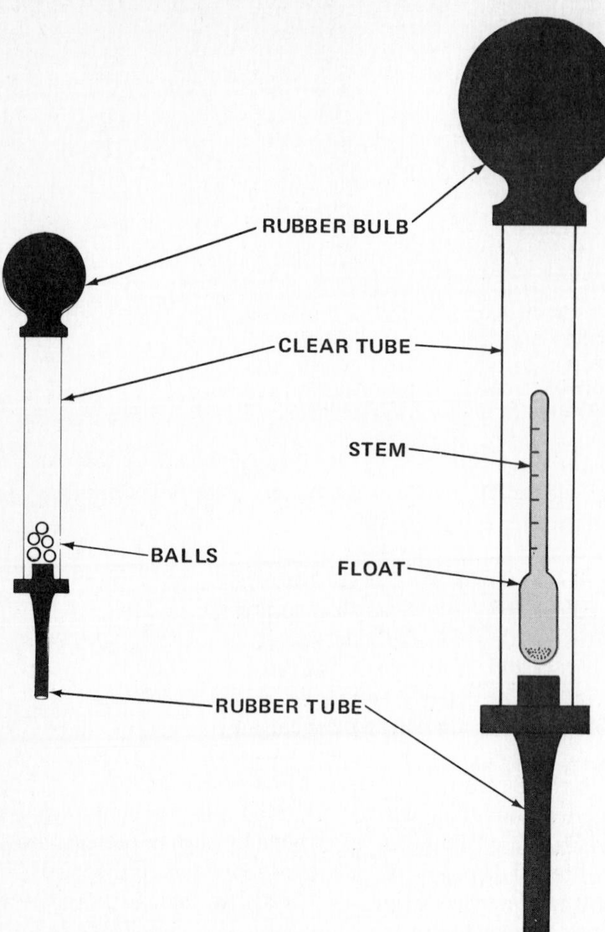

Fig. 24-8 Battery hydrometers to check electrolyte specific gravity.

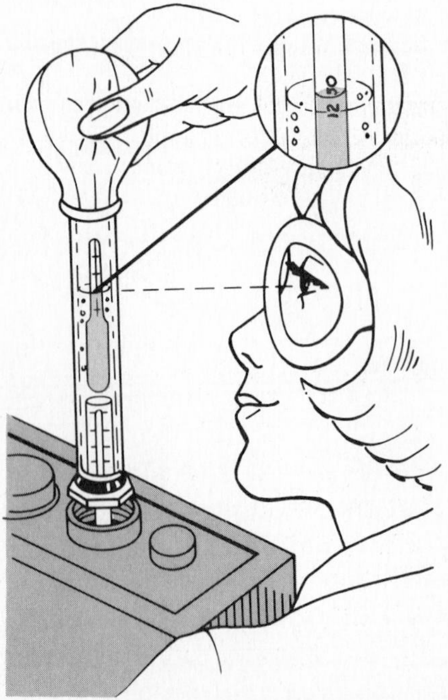

Fig. 24-9 Using a float hydrometer to check a battery cell. Reading must be taken at eye level. The higher the float stem sticks out of the electrolyte, the higher the state of charge of the battery.

♦ 24-16 VARIATIONS OF SPECIFIC GRAVITY WITH TEMPERATURE

The electrolyte thickens (gains specific gravity) as it cools. It loses specific gravity as it gets hot. On the ball hydrometer, this is not important. But on the float hydrometer, correction of the specific-gravity reading should be made if the electrolyte temperature is well above or below the 80°F [27°C] standard. Specific gravity changes about four points for every 10°F change in temperature. A reading of 1.250 at 120°F (48.9°C) would correct to 1.266 (add 0.016, or 4 × 0.004). A reading of 1.230 at 20°F (−6.7°C) would correct to 1.206 (subtract 6 × 0.004, or 0.024)).

♦ 24-17 LOSS OF SPECIFIC GRAVITY FROM AGE

As a battery ages, the electrolyte gradually loses specific gravity. This loss is due to aging and nothing can be done about it. Eventually, the battery wears out and must be replaced.

♦ 24-18 LOSS OF SPECIFIC GRAVITY FROM SELF-DISCHARGE

If a battery stands idle for a long time, it will gradually run down. This is due to the chemical actions between the electrolyte and the battery plates. The higher the temperature, the more rapid this self-discharge.

♦ 24-19 FREEZING TEMPERATURE OF ELECTROLYTE

The higher the specific gravity of the electrolyte, the lower the temperature must be before the electrolyte freezes. The battery must be kept sufficiently charged in cold weather to prevent freezing. Freezing may ruin the battery (Fig. 24-10).

♦ 24-20 CHARGE INDICATOR ON MAINTENANCE-FREE BATTERIES

Many maintenance-free batteries have a charge indicator on their top (Fig. 24-11). The appearance of the charge indicator shows the state of charge of the battery.

♦ CAUTION ♦ If the charge indicator shows yellow, do not attempt to recharge or test the battery. It is close to failure and should be replaced. Never attempt to jump start the car if the charge indicator shows yellow (♦25-9).

♦ 24-21 BATTERY-CAPACITY (HIGH-DISCHARGE) TEST

For this test, a special tester is used. It places a load on the battery so that the battery voltage can be measured while a

Specific Gravity	Freezing Temperature, Degrees F (C)
1.100	18 [−8.2]
1.160	1 [−17.2]
1.200	−17 [−27.3]
1.220	−31 [−35]
1.260	−75 [−59.4]
1.300	−95 [−70.5]

Fig. 24-10 Specific gravities and freezing temperatures for battery electrolyte.

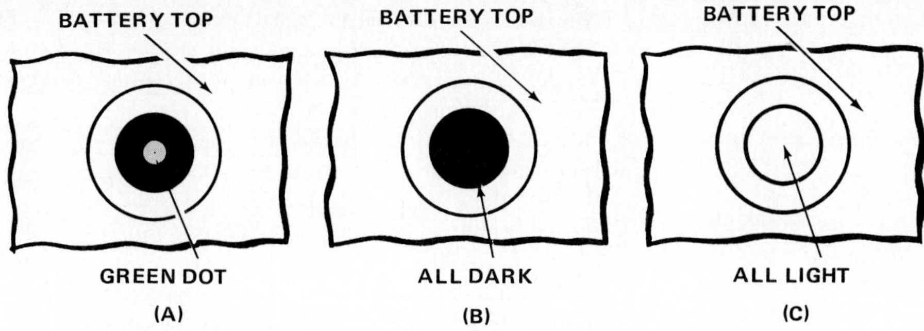

Fig. 24-11 Appearance of the charge indicator in the top of some maintenance-free batteries. (A) If the green dot shows, the battery is charged. (B) If the indicator shows black, the battery is low and should be charged before testing. (C) If the indicator shows light yellow, the battery is dead and should be discarded.

high current is flowing out of the battery. Specifications for this test vary. Follow the procedure in the tester operating instructions. Specifications for voltage and battery capacity are in the manufacturer's service manual.

In general, a load of three times the battery ampere-hour capacity is placed on the battery. You can also use one-half of the cold-cranking rate at 0°F [−17.8°C]. At the end of 15 seconds, a good battery should show a voltage of 9.6 volts or higher. If the voltage is less than 9.6 volts and the battery has removable vent plugs, check the specific gravity of each cell. If there is more than a 50-point difference between cells, replace the battery.

♦ 24-22 BATTERY SERVICE

Battery service can be divided into four parts: visual inspection, testing, charging, and care of batteries in stock.

♦ 24-23 BATTERY TESTING AND ANALYSIS

Battery testing includes a check of the condition of the battery. It should also include analysis of any abnormality found so that corrections can be made. This will prevent a repetition of trouble. Following are various battery troubles and their possible causes. These apply mainly to vent-plug batteries.

1. Overcharging If the battery frequently requires water, it is probably being overcharged. Too much current is being supplied to the battery. This is a damaging condition that overworks the active materials in the battery and shortens battery life. In addition, overcharging causes more rapid loss of water from the battery electrolyte. Unless this water is replaced frequently, the electrolyte level is likely to fall below the tops of the plates. This exposes the plates and the separators to the air and may ruin them. Also, battery overcharge causes the battery plates to buckle and crumble. Therefore, a battery subjected to severe overcharging will soon be ruined. Where overcharging is experienced or suspected, the charging system should be checked. It should be serviced if necessary to prevent overcharging (Chap. 26).

2. Undercharging If the battery is discharged, it should be recharged, as outlined later in this chapter. In addition, an attempt should be made to determine the cause of the trouble. It could be caused by:

a. Charging-system malfunctioning
b. Defective connections in the charging circuit between the alternator and the battery
c. Excessive load demands on the battery
d. A defective battery
e. Permitting the battery to stand idle for long periods so that it self-discharges excessively

In addition, an old battery may have a low specific-gravity reading because it is approaching failure.

3. Sulfation The active materials in the plates are converted into lead sulfate during discharge. This lead sulfate is reconverted into active material during recharge. However, if the battery stands for long periods in a discharged condition, the lead sulfate is converted into hard, crystalline substance. This substance is difficult to reconvert into active materials by normal charging processes. Such a battery should be charged at half the normal rate for 60 to 100 hours. Even though this long charging period may reconvert the sulfate to active material, the battery may still remain in a damaged condition. The crystalline sulfate, as it forms, tends to break the plate grids.

4. Cracked Case A cracked case may result from excessively loose or tight hold-down clamps, from battery freezing, or from flying stones.

5. Bulged Cases Bulged cases result from tight hold-down clamps or from high temperatures.

6. Corroded Terminals and Cable Clamps This condition occurs naturally on batteries. Excessive corrosion should be removed from cable clamps and terminals (♦24-13).

7. Corroded Battery Holder Some spraying of battery electrolyte is natural as the battery is being charged. The battery holder may become corroded from the effects of the electrolyte. Such corrosion may be cleaned off, with the battery removed. Use a wire brush and a solution of baking soda and water.

8. Dirty Battery Top The top of the battery may become covered with dirt and grime mixed with electrolyte sprayed from the battery. This should be cleaned off periodically (♦24-13).

9. Discharge to Metallic Hold-Down If the hold-down clamps are of the uncovered metallic type, a slow discharge may occur from the insulated terminal to the hold-down clamp. This is more apt to occur with a dirty battery top, across which current can leak. The remedy is to keep the battery top clean and dry.

♦ 24-24 REMOVING AND REPLACING BATTERY

To remove a battery from a car, first take off the grounded-battery-terminal cable clamp. This prevents accidental grounding of the insulated terminal when it is disconnected. To remove a nut-and-bolt type of cable, loosen the clamp nut about 3/8 inch [9.5 mm]. Use a box wrench or special cable pliers. Do not use ordinary pliers or an open-end wrench. Either of these might break a cell cover when swung around. If the clamp sticks, use a clamp puller. Do not use a screwdriver or bar to pry on the clamp. This could damage the battery cell or cover. To detach the spring-ring type of clamp, squeeze the ends of the rings apart with Vise-Grip or Channellock pliers.

After the grounded cable is disconnected, disconnect the insulated-terminal cable. Clean both battery terminals and cable clamps (Fig. 24-7). Loosen the battery hold-downs, and take out the battery. When installing a battery, do not reverse the terminal connections. Reconnect the insulated-terminal cable first, and then reconnect the grounded-terminal cable. Apply corrosion inhibitor to clamps and terminals. Install and tighten the hold-downs. Avoid overtightening.

Careful Make sure the cable clamps are tight and make good connections with the terminal posts (top-terminal battery). If the jaws of the clamp touch together, the clamp is not tight on the post. This could mean starting trouble. Correct the condition by disconnecting the clamp from the post. Shave the clamp jaws with a file until there is a gap when the clamp is installed.

♦ 24-25 BATTERY ADDITIVES

Certain chemical compounds, when added to the battery cells, are supposed to restore a battery to a charged condition. Such chemicals should never be added to the battery. Their use may void the battery guarantee and cause battery failure.

♦ 24-26 BATTERY SLOW CHARGING

Two methods of slow charging batteries are in use, the constant-current and the constant-voltage methods. In the constant-current method, the current input to the battery is adjusted to the manufacturer's specifications. The charging is continued until the battery is gassing freely and there is no rise in specific gravity for 2 hours.

In the constant-voltage method, the charging voltage is held at a constant value. The battery, as it approaches a charged condition, increases in resistance to the charging current. At the same time, the current input gradually tapers off. When the battery is fully charged, the current input has been reduced to a few amperes. The battery electrolyte temperature must remain within limits. If the battery electrolyte temperature increases greatly, the resistance of the battery will remain low. Then the battery will be damaged by overcharging, unless it is quickly removed from the charger.

Before charging a battery, check the electrolyte level. If the level is low, the battery can be damaged. If the level is too high, the electrolyte can overflow because of gassing and heat. Here are cautions to observe:

1. The gases released by batteries under charge are very explosive. Be sure the area is well ventilated. Do not smoke or have open flames around charging batteries. This could cause an explosion.
2. Be sure to disconnect the battery ground strap if the battery being charged is in a car. Otherwise, you can damage the electrical equipment in the car.
3. Most manufacturers recommend leaving the cell caps in place. But make sure the vent holes are open. Cover the caps with a cloth during the charging procedure. Some manufacturers recommend removing the caps and covering the openings with a cloth.
4. Do not charge a battery that is frozen. It could explode.
5. Always wear some type of eye protection.
6. If the charge indicator of a maintenance-free battery shows yellow or clear, do not charge it. The electrolyte level is low, and charging it could cause an explosion. The battery should be discarded.
7. Do not turn the charger on until the charger cables are connected to the battery. Turn the charger off before disconnecting the cables.
8. Check the specific gravity and temperature of the electrolyte periodically during charge. If the temperature goes above 125°F [51.7°C], stop the charge.
9. The battery is fully charged when the specific gravity shows no increase for 3 hours. Also, the cells should be gassing freely.
10. After charging, wash and dry the battery top. This removes any electrolyte that might have spewed out during charge.

♦ 24-27 QUICK CHARGERS

The quick charger can be wheeled up to the car and connected to the battery in the car (Fig. 24-12). Here are some special points to watch when using a quick charger. These chargers can supply a fast charge of up to 100 amperes (for some types). Normally, you would set the charging rate for about 40 to 50 amperes and charge the battery for about 30 to 45 minutes. This boosts the battery with up to 38 ampere-hours of charge. A battery in normal condition can stand high charging rates without damage if the electrolyte temperature does not go above 125°F [51.7°C].

Quick charging usually cannot bring the battery up to full charge in a short time. To bring it up to full charge, the battery should be given a slow charge after the quick charge.

Careful A battery with discolored electrolyte (from cycling) or with gravity readings more than 25 points apart should not be quick charged. Likewise, a badly sulfated battery should not be quick charged. Such batteries may be near failure, but they may give additional service if slow charged. However, quick charging might damage them further. During quick charging, check the color of the electrolyte. Stop charging if it becomes discolored as a result of the stirring up of washed-out active plate material. Likewise, cell voltages should be checked every few minutes. Charging should be stopped if cell voltages vary more than 0.2 volt.

Careful When quick charging a battery in a car, disconnect the battery ground strap to protect the electrical system from damage due to high voltage. If the charge indicator in a maintenance-free battery shows yellow, do not try to quick charge it (♦24-20).

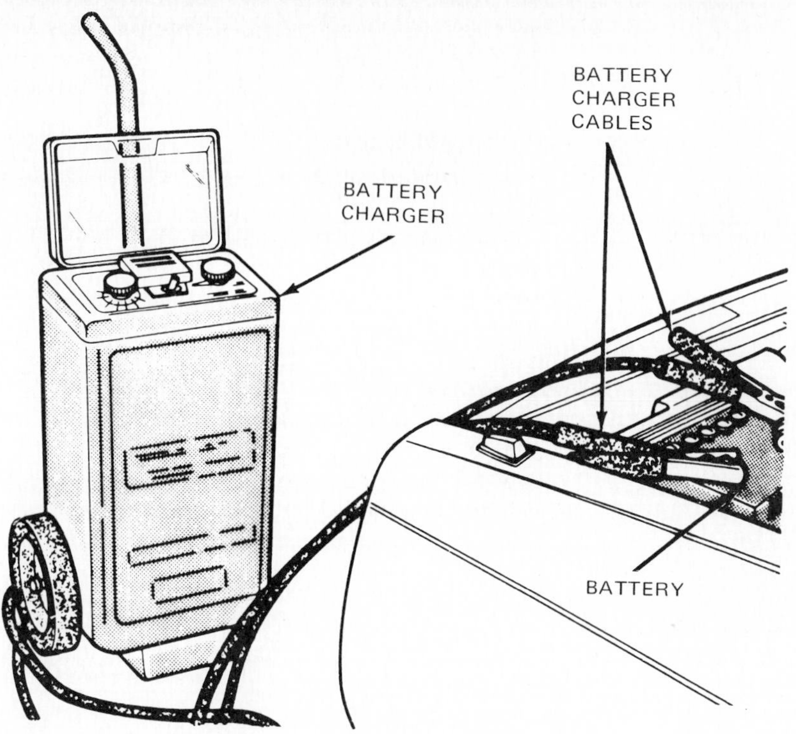

Fig. 24-12 Quick charger connected to a battery in the car. The grounded battery cable should be disconnected before the charger cables are connected. *(Chrysler Corporation)*

A very low battery may not accept a fast charge. The electrolyte in a very low battery does not have very much sulfuric acid in it. Therefore, the conductivity of the electrolyte is too low to allow a high current to flow through the battery. You might think a battery that refuses to take a high charge is worn out. However, it may be possible to restore the battery to a charged condition. First, slow charge it for a few minutes to see whether it starts coming up to charge. If it does, then it can be put on fast charge.

♦ 24-28 CARE OF BATTERIES IN STOCK

Wet batteries (new batteries shipped with electrolyte in them) are perishable. They are subject to self-discharge. If allowed to proceed for too long a time, this can completely ruin them. To prevent this, batteries in stock should be recharged every 30 days. They should not be stacked on top of each other without additional support. The weight of one battery is enough to collapse the plate assemblies.

♦ 24-29 DRY-CHARGED BATTERIES

Dry-charged batteries contain fully charged positive and negative plates but no electrolyte. The batteries are sealed with rubber or plastic seals placed in the vent plugs. Since the batteries contain no moisture, practically no chemical action can take place in them. This means that they will remain in good condition for as long as 36 months.

Dry-charged-battery manufacturers supply ready-mixed electrolyte in special cartons. The carton contains an acid-proof plastic bag which holds the electrolyte. The following steps are performed to activate a dry-charged battery:

1. Remove the vent plugs, and take out the plastic seals.
2. Remove the lid from the electrolyte container. Unfold the top of the plastic bag, and cut a small opening in one corner of the bag.
3. Use a glass or acid-proof funnel and fill each battery cell. *Wear goggles, and observe all cautions already noted regarding sulfuric acid.* Wait a few minutes, and then add more electrolyte if necessary. Some electrolyte will probably be left. Do not attempt to use it all. Do not overfill the battery.
4. Before discarding the container, empty it. Rinse the bag thoroughly with water. Otherwise, someone who handles the carton might be severely burned.

♦ REVIEW QUESTIONS ♦

Select the *one* correct, best, or most probable answer to each question. You can find the answers in the section indicated at the end of each question.

1. When working around batteries, remember that (♦24-10)
   a. battery electrolyte contains a very corrosive acid
   b. explosive gas forms in the battery while it is being charged
   c. the battery can furnish a very high current if it is shorted
   d. all of the above
2. When disconnecting a battery, always (♦24-10)
   a. remove the vent plugs first
   b. disconnect the insulated-terminal cable first
   c. disconnect the grounded-terminal cable first
   d. put the transmission in PARK or REVERSE
3. The battery is an electrochemical device. This means that the battery (♦24-1)
   a. makes chemicals by mechanical means
   b. uses chemical action to provide electricity
   c. has curved instead of flat plates
   d. does not use an electrolyte

4. The purpose of the battery is to (♦24-1)
   a. supply current for cranking the engine
   b. supply current when the charging system can't handle the complete electrical load
   c. both of the above
   d. neither of the above
5. The length of time in minutes that a fully charged battery at 80°F [26.7°C] can deliver 25 amperes is called the (♦24-6)
   a. charging rate
   b. reserve capacity
   c. cold-cranking rate
   d. ampere-hour rate
6. The number of amperes that the battery can deliver for 30 seconds when it is at 0°F [-17.8°C] before the battery voltage falls to 7.2 volts is called the (♦24-6)
   a. charging rate
   b. reserve capacity
   c. cold-cranking rate
   d. ampere-hour rate
7. If you must add water to the battery every few days, the battery probably is (♦24-23)
   a. overloaded
   b. overcharged
   c. sulfated
   d. old
8. A loose battery-cable clamp could cause (♦24-23)
   a. battery overcharge
   b. high battery voltage
   c. overheating
   d. a run-down battery
9. On a top-terminal battery, the negative terminal post is (♦24-3)
   a. smaller than the positive terminal post
   b. larger than the positive terminal post
   c. the same size as the positive terminal post
   d. none of the above
10. You can quick charge the battery at as much as 50 amperes provided (♦24-27)
    a. the electrolyte does not get too hot
    b. you do not charge for more than 5 minutes
    c. you make sure the battery is fully charged
    d. it remains connected to the electrical system

CHAPTER 25
The Starting System: Operation and Service

After studying this chapter, you should be able to:

1. Explain the purpose, construction, and operation of the starting system.
2. Explain why a starting-motor drive mechanism is required and how it works.
3. List the two basic starting-system problems and explain possible causes of each.
4. Describe the procedure for checking the starting system.
5. Describe how to remove, service, and replace starting motors.

♦ 25-1 NEED FOR STARTING SYSTEM

The internal-combustion engine is not capable of self-starting. Some big engines used in trucks, tractors, and off-road and industrial equipment are cranked by compressed air or by a small starting engine. However, automotive engines (both spark-ignition and diesel) are cranked by a small but powerful electric motor. This motor is called the *cranking motor, starter*, or *starting motor*.

The battery sends current to the starting motor when the driver turns the ignition switch to START. This causes a pinion gear in the starting motor to mesh with teeth on the ring gear around the engine flywheel. The starting motor then rotates the engine crankshaft for starting.

The typical starting system includes the battery, the starting motor and drive mechanism, the ignition switch, the starter relay or solenoid, the neutral safety switch, and the wiring which connects these components. Each of the starting-system components is described in following sections.

♦ 25-2 STARTING MOTOR CONSTRUCTION

Figure 25-1 shows the basic principles of the starting motor. Loops of heavy wire are placed between magnetic poles. When current from the battery flows through the loops, a strong magnetic field is produced around the loops. This magnetic field opposes the magnetic field of the stationary magnet. The opposition causes the loops to rotate. Only one loop is shown in Fig. 25-1. In the actual motor, there are many loops assembled in a rotor, called the armature (Fig. 25-2). The stationary magnetic field is produced by field windings in the field-frame assembly (Fig. 25-2). When the motor is connected to the battery, the opposing magnetic fields of the armature and the field windings cause the armature to spin. The starting-motor drive unit then spins the engine crankshaft so that the engine starts.

High magnetic strength is needed for the starting motor to produce the torque required to crank the engine. To obtain the strong magnetic field, a high current must flow through the heavy gauge conductors in the starting motor. Many starting motors are *series wound*. This means that the armature and field windings are connected in series (Fig. 25-3). Some starting motors have four field windings and four brushes (Fig. 25-4). The action is the same. The current flows in through the terminal, passes through the field windings, and then passes through the insulated brush to the commutator. The commutator has a series of segments, insulated

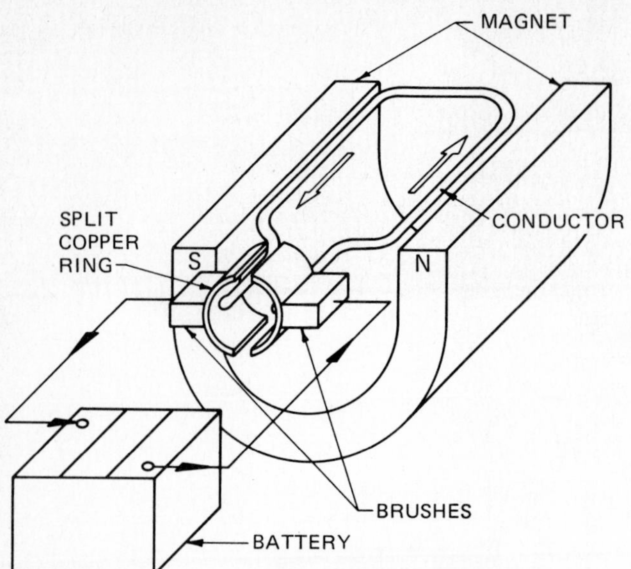

Fig. 25-1 Principle of the electric motor. When current flows through the conductor loop, the current produces a magnetic field which opposes the magnetic field of the magnet. This causes the loop to spin.

from each other and connected to the armature loops. The insulated brush connects to each of these segments as they pass under the brush as the armature rotates. The current then flows through the armature loops and through the grounded brush.

With more field windings and brushes, more current can flow. This creates a more powerful magnetic field for greater cranking torque.

Starting motors are made in different sizes for various engines. Figure 25-5 compares a starting motor for a passenger car using a spark-ignition engine with a starting motor for a diesel engine. The diesel engine requires more cranking torque. Figure 25-6 shows a completely disassembled starting motor. Figure 25-7 is a sectional view of a similar starting motor. All use magnetism to rotate the armature and to mesh the pinion with the flywheel ring gear.

♦ 25-3 STARTING-MOTOR DRIVES

The drive assembly contains a small pinion that, in operation, meshes with teeth on a larger gear on the flywheel or torque-

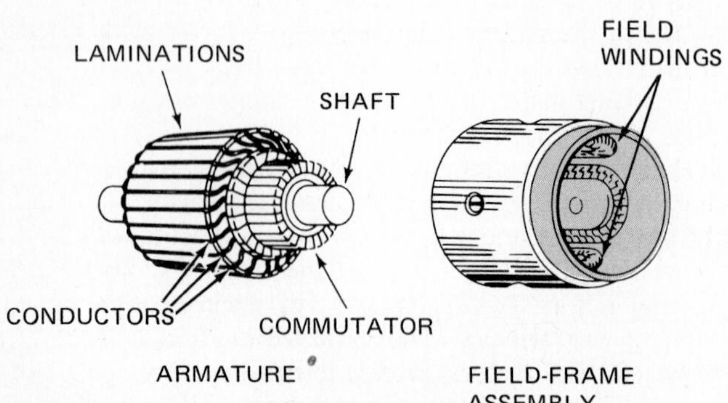

Fig. 25-2 Two major parts of a starting motor, the armature and the field-frame assembly. *(Delco-Remy Division of General Motors Corporation)*

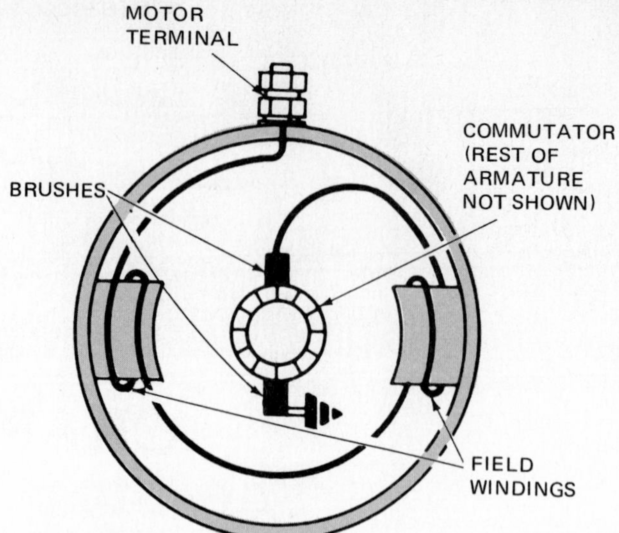

Fig. 25-3 Wiring diagram for a two-pole series-wound starting motor.

converter flex plate. This provides gear reduction, so the armature must rotate about 15 times to cause the flywheel to rotate once. The armature may revolve about 2000 to 3000 rpm when the starting motor is operated. This causes the flywheel to spin at speeds as high as 300 rpm. This is ample for starting the engine.

After the engine starts, it may increase in speed to 3000 rpm or more. If the starting-motor drive pinion remained in mesh with the flywheel, the pinion would be spun at 45,000 rpm because of the 15:1 gear ratio. This means that the armature also would be spun at this very high speed. Centrifugal force would cause the conductors and commutator segments to be thrown out of the armature, ruining it. To prevent such damage, an automatic meshing and de-meshing device called an *overrunning clutch* is used.

♦ 25-4 OVERRUNNING CLUTCH

The overrunning clutch (Fig. 25-7) is operated by a shift lever. The lever pushes the drive pinion into mesh with the

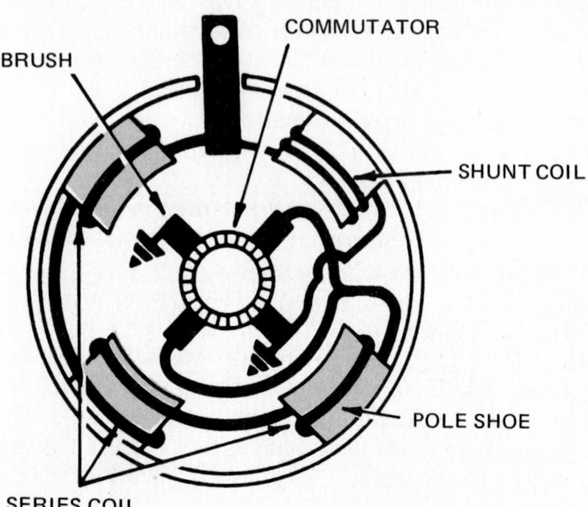

Fig. 25-4 Wiring diagram of a four-pole series-shunt, or compound, starting motor. The shunt coil prevents overspeeding. *(Delco-Remy Division of General Motors Corporation)*

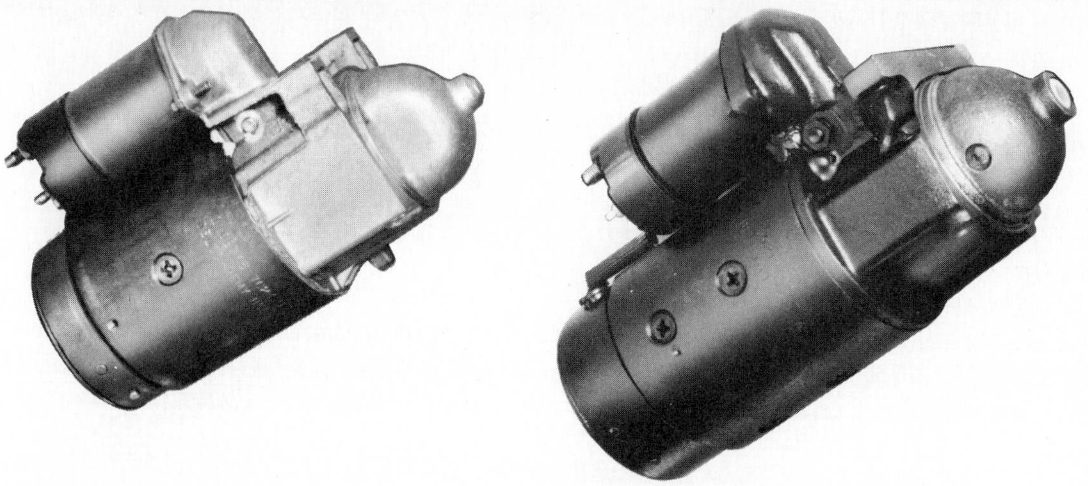

Fig. 25-5 Starting motor for a spark-ignition engine (left) compared with the starting motor for an automotive diesel engine. *(Oldsmobile Division of General Motors Corporation)*

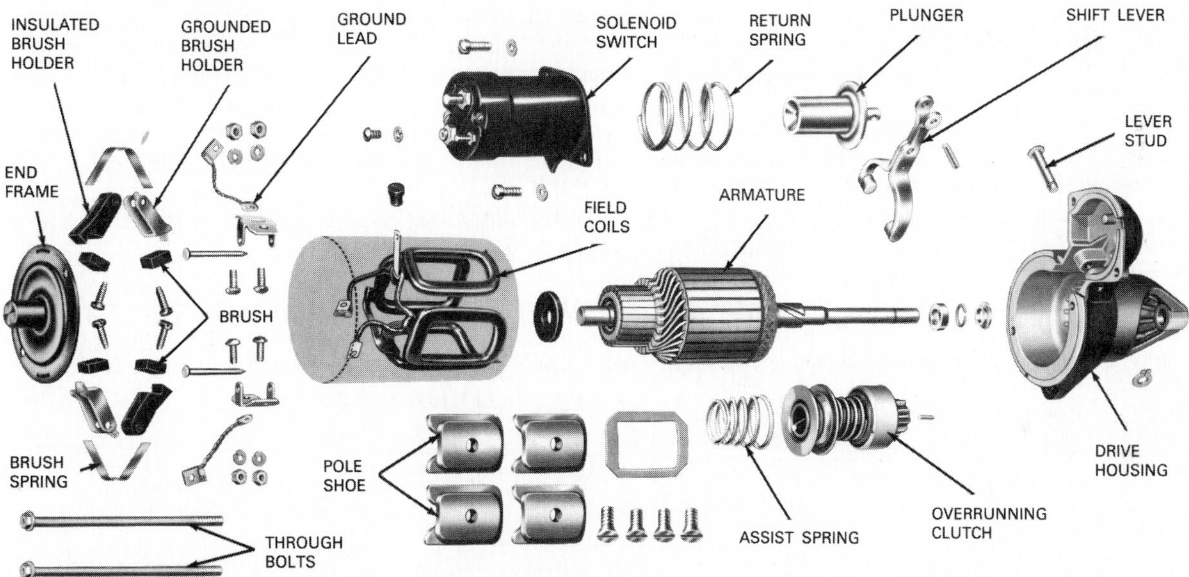

Fig. 25-6 Disassembled starting motor. *(Delco-Remy Division of General Motors Corporation)*

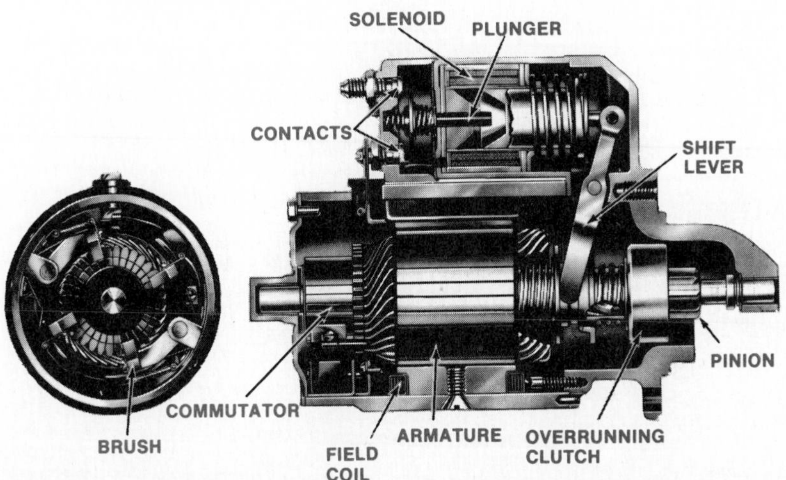

Fig. 25-7 Sectional view of a starting motor. *(Delco-Remy Division of General Motors Corporation)*

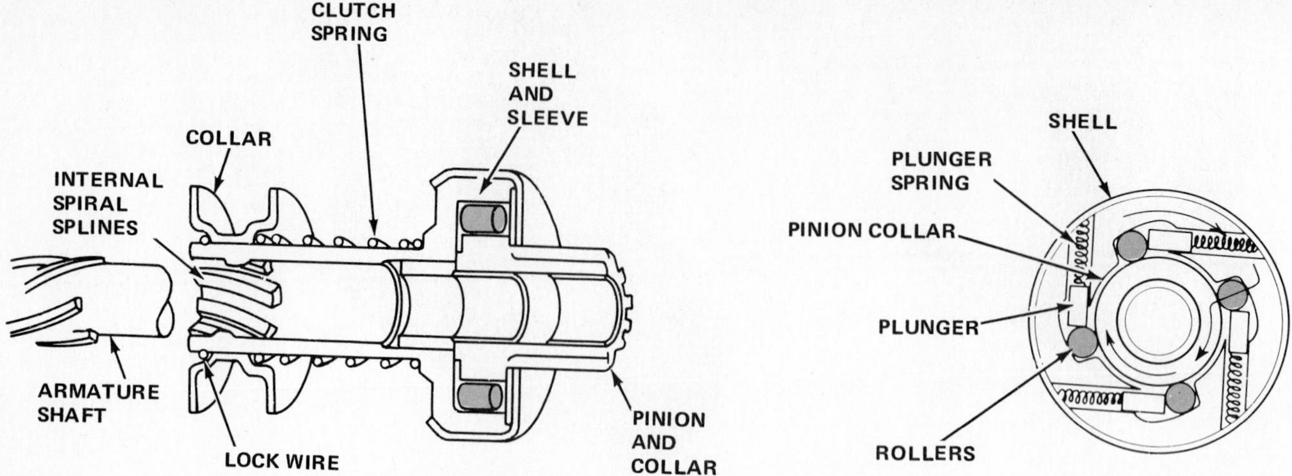

Fig. 25-8 Construction of an overrunning clutch. *(Delco-Remy Division of General Motors Corporation)*

flywheel teeth. As the shift lever completes its travel, it closes the starting-motor switch so that cranking takes place. Straight or spiral splines in the armature shaft and the clutch sleeve cause both to rotate together. A spiral spring is placed between the clutch housing and the shift-lever collar. This spring compresses if the pinion and the flywheel teeth should happen to butt instead of mesh. Then, after the starting-motor switch is closed and the armature starts to rotate, meshing is completed by the spring force.

The clutch (Fig. 25-8) consists of the outer shell and the pinion-and-collar assembly. The outer shell has four hardened-steel rollers fitted into four notches. The notches are not concentric, but are smaller in the end opposite to the plunger springs (Fig. 25-8). When the armature and the shell begin to rotate, the pinion does not. This causes the rollers to rotate into the smaller sections of the notches, where they jam tight. The pinion must now rotate with the armature, cranking the engine. After the engine starts, it spins the pinion faster than the armature is turning. The rollers are rotated into the larger sections of the notches, where they are free. This allows the pinion to spin independently of, or *overrun*, the remainder of the clutch. A spring on the shift lever pulls the pinion back out of mesh when the shift lever is released.

Gear Reduction The starting motor shown in Fig. 25-9 has a gear reduction which increases cranking torque. The shift lever (or fork) is enclosed. When it is actuated by the

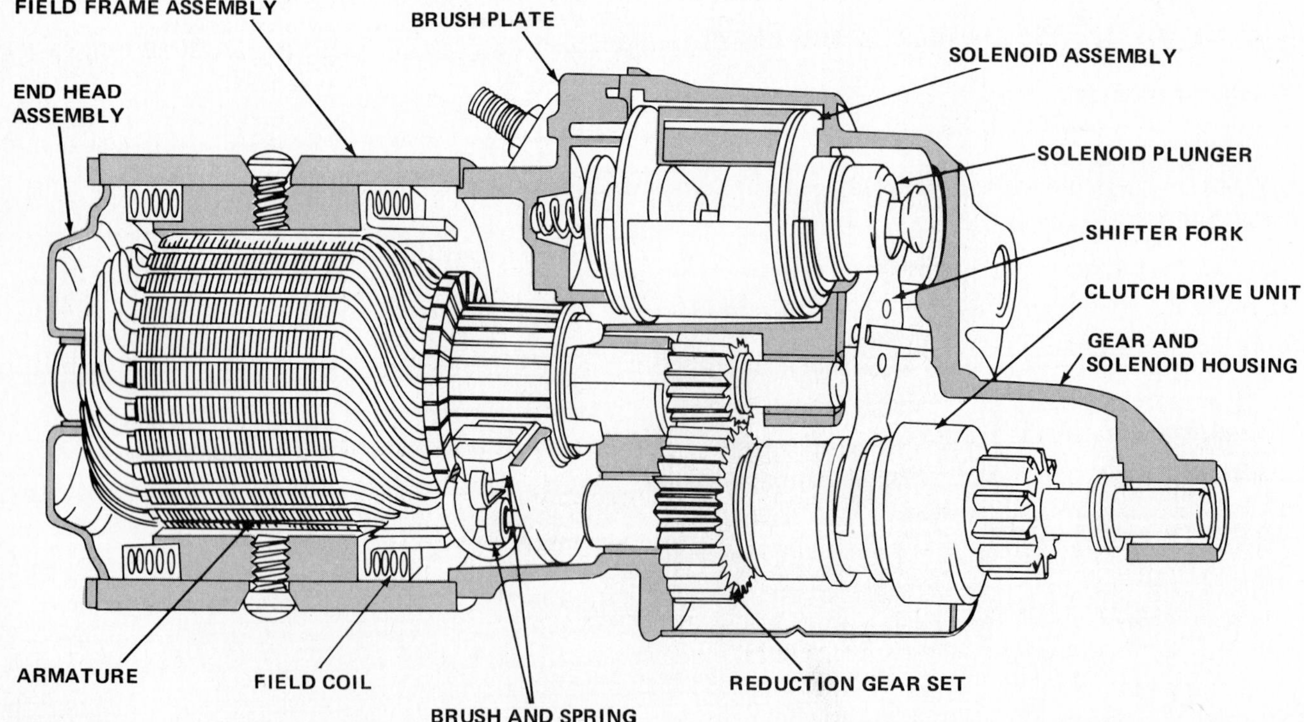

Fig. 25-9 A gear-reduction, overrunning-clutch starting motor. *(Chrysler Corporation)*

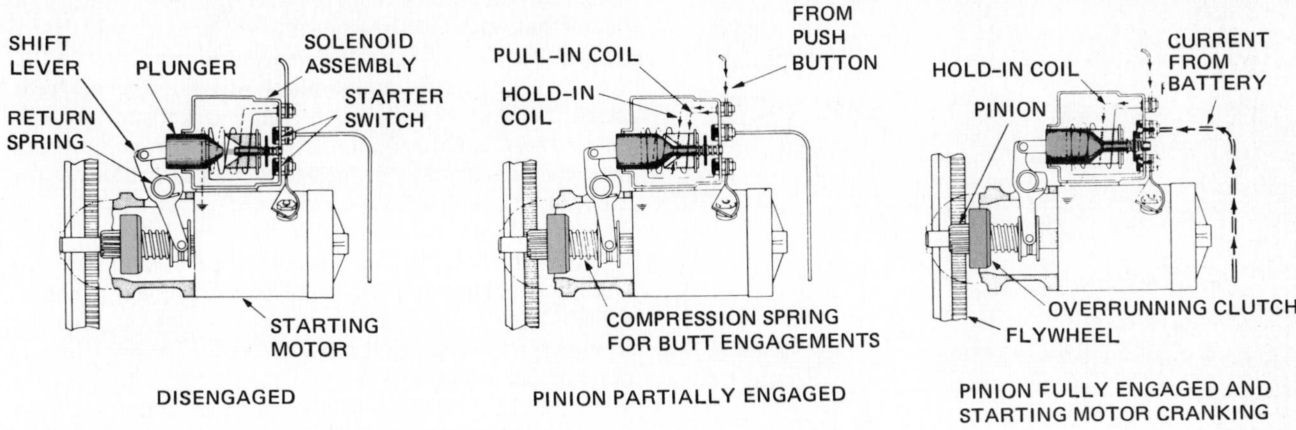

Fig. 25-10 Actions of the solenoid and overrunning clutch as the pinion engages. *(Delco-Remy Division of General Motors Corporation)*

solenoid, it shifts the overrunning-clutch pinion into mesh with the flywheel. The gear ratio between the armature and the flywheel, due to the extra gears in the starting motor, is 45:1. The armature turns 45 times to turn the flywheel, and the engine crankshaft, once. This provides a high cranking torque for starting.

♦ 25-5 STARTING-MOTOR CONTROLS

Starting-motor controls have varied from a simple foot-operated pedal to automatic devices that close the starting-motor circuit when the accelerator pedal is depressed. The system presently used in passenger cars and other vehicles has starting contacts in the ignition switch. When the ignition key is turned against spring force past the ON position to START, the starting contacts close. This connects the starting-motor solenoid or magnetic switch to the battery. After the engine starts and the ignition key is released, spring force returns it to the ON position. This opens the starting contacts.

On starting motors with overrunning clutches, a solenoid is commonly used to produce the clutch-shifting action. The solenoid contains a pair of windings that are connected to the battery when the starting switch is closed. This produces a magnetic field that pulls a plunger in. The plunger movement causes a shift lever to move the overrunning clutch on the armature shaft. This shifts the overrunning-clutch pinion into mesh with the flywheel teeth. At the same time, the plunger movement forces a heavy switch to connect the starting motor directly to the battery. Now, cranking begins. Figure 25-10 shows the sequence of actions. Figure 25-11 is a wiring diagram of a starting-motor system.

The solenoid has two windings, a pull-in winding and a hold-in winding. They work together to pull the plunger in.

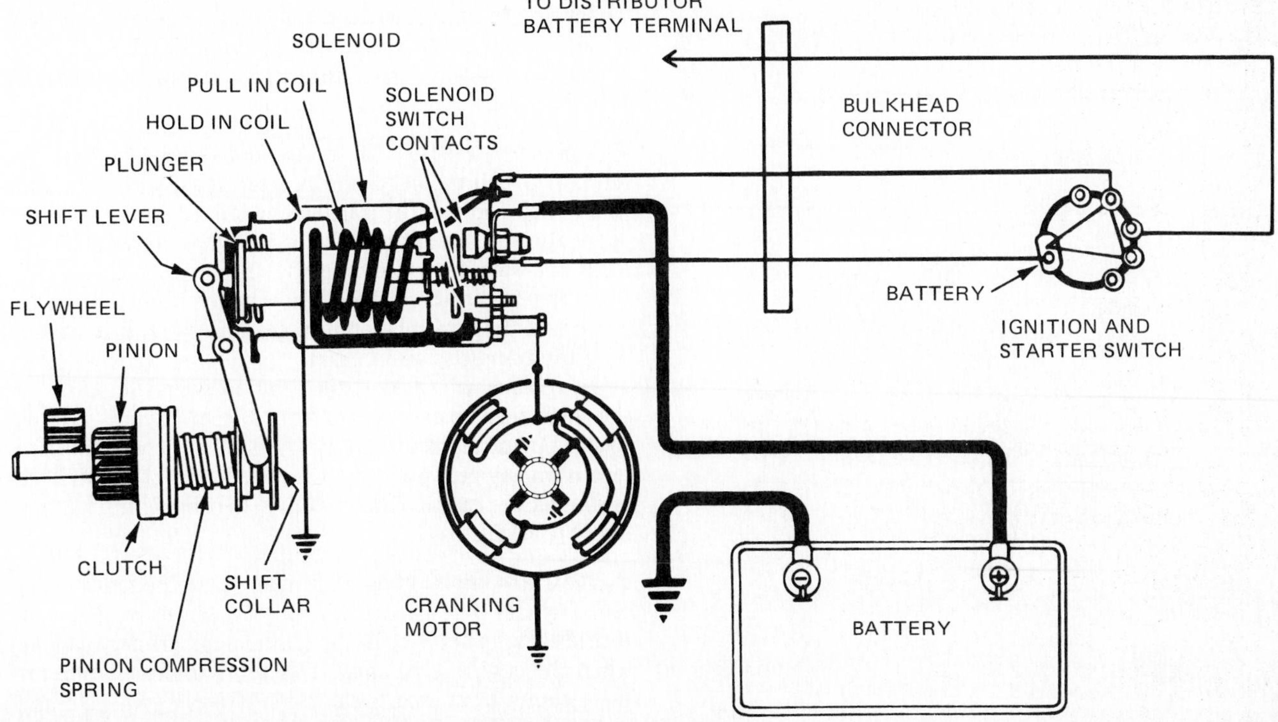

Fig. 25-11 Starting-system circuit. *(Buick Motor Division of General Motors Corporation)*

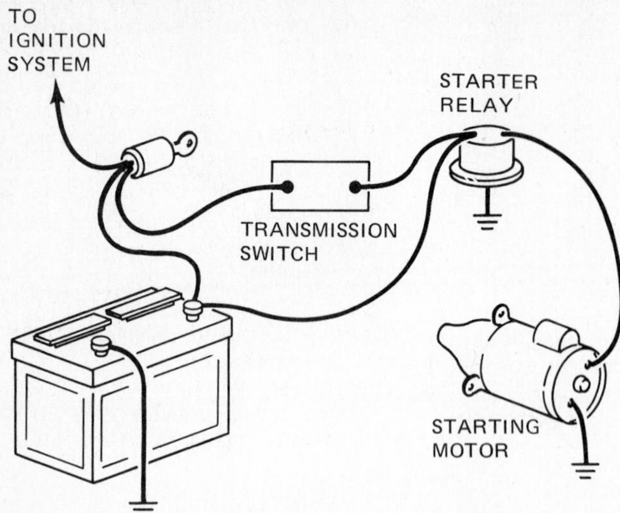

Fig. 25-12 A starting system which uses a separate starter relay (magnetic switch). The starting motor has a sliding pole shoe that moves to shift the pinion into mesh. This action is performed by a solenoid mounted on other starting motors.

This combination of windings provides sufficient magnetic strength to mesh the pinion and close the starting-motor switch. After the pinion is meshed and the switch is closed, less magnetism is required to hold the core in. Consequently, as the switch closes, the pull-in winding is shorted out, since it is connected between the two solenoid terminals. This reduces the load on the battery during cranking.

A second type of starting system is shown in Fig. 25-12. It uses a separate starter relay (a magnetic switch). When the ignition switch is turned to START, the starter relay is connected to the battery. This causes the contact points in the starter relay to close. The action connects the starting motor directly to the battery. Cranking takes place.

The starting system shown in Fig. 25-12 uses the starting motor shown in Fig. 25-13. It has no separate solenoid. Instead, the starting-motor field windings produce the magnetic field which causes both the shifting action and the rotation of the armature. The magnetic field causes a pole shoe to slide in the frame. This action moves the shift lever so that the drive pinion is forced into mesh with the flywheel teeth. When the ignition switch is turned to START and the automatic-transmission neutral safety switch is closed, the magnetic-switch winding is connected to the battery. This causes the magnetic switch (starter relay) to close its contacts. This connects the starting motor directly to the battery.

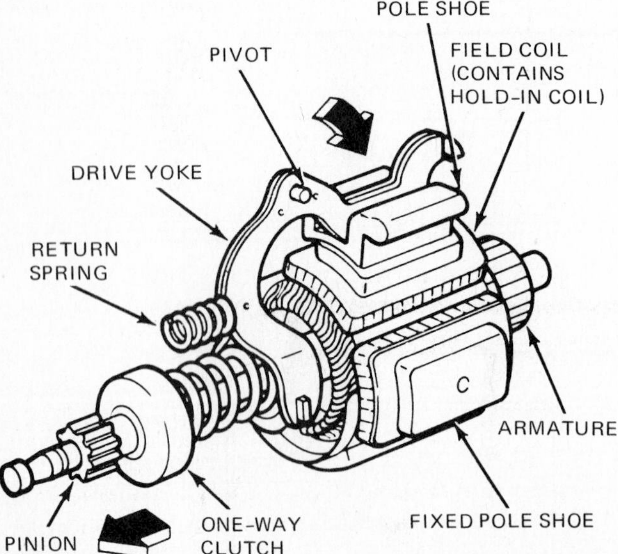

Fig. 25-13 Internal operating mechanism of the starting motor using a sliding pole shoe. *(American Motors Corporation)*

One of the field windings acts as the winding to move the pole shoe. It has two parts, a pull-in winding and a hold-in winding. As the magnetic switch closes its contacts, these two windings are connected directly across the battery (through the contact points). This produces maximum magnetic strength. The other three field windings are connected in series with the armature. The magnetic field of the pole-shoe actuating windings moves the pole shoe and shifts the drive pinion into mesh. At the same time, the armature begins to turn so that the engine is cranked. As the pole shoe and shift lever move, they cause the contact points to open. Now, the pull-in winding is connected in series with the armature. Its magnetism drops. At the same time, however, the hold-in winding retains its full magnetism so that the pole shoe is retained in the cranking position. As soon as the ignition switch is released, the magnetic-switch winding is disconnected from the battery. This allows the contact points in the magnetic switch to open, disconnecting the starting motor from the battery. Cranking stops. The drive pinion is demeshed from the flywheel by the return spring.

1. Automatic Transmissions Years ago, cars with automatic transmissions had a special switch that prevented starting while the car was in gear. The switch (neutral safety switch) is connected between the ignition switch and the solenoid. This switch is open at all transmission-lever positions except PARK or NEUTRAL. Late-model cars with the ignition switch and selector lever on the steering column do not need a safety switch. On these, the ignition switch cannot be turned to CRANK or START unless the transmission selector lever is in PARK or NEUTRAL.

2. Ignition Switch Most automotive vehicles have the ignition switch mounted on the steering column. The ignition switch controls the starting and ignition systems, as described above. In addition, turning the ignition switch to LOCK locks the steering wheel and the transmission shift lever. From the OFF position, the ignition switch cannot be turned to CRANK if the transmission is in gear. This prevents starting with the transmission in gear.

On vehicles which have the ignition switch mounted on the instrument panel, the starting system requires a transmission (neutral safety) switch. This switch is open if the shift lever is in any gear position. It is closed only if the shift lever is in PARK or NEUTRAL. This prevents starting with the transmission in gear.

3. Ignition Resistance In many cars, the ignition system includes a resistance wire in the wiring harness. The resistance is in series with the ignition-coil primary winding when the engine is running. This protects the ignition primary circuit from excessive current. However, during cranking, the ignition switch bypasses the resistance (when the key is turned to START). Now, full battery voltage is imposed on the ignition coil for maximum voltage during cranking. The resistance is also called a *ballast* resistance. On some cars, the resistance is a separately mounted part.

4. Diesel Engines Many diesel engines require a heavy-duty starting motor. Because of the higher compression in diesel engines, it takes more power to crank them. To power these heavy-duty starting motors, many systems use two 12-volt batteries. They work together to provide the heavier electric current needed to crank the diesel engine.

5. Heavy-Duty Controls A heavy-duty system uses a series-parallel switch with two 12-volt batteries. During normal operation with the engine running, the two batteries are connected in parallel, and the system is 12-volt throughout. But for starting, the two batteries are connected in series to supply the starting motor with 24 volts. This higher voltage causes the starting motor to develop a higher cranking torque.

STARTING-SYSTEM SERVICE

♦ 25-6 STARTING-SYSTEM TROUBLES

There are two basic starting-system troubles:

1. The starting motor does not crank the engine.
2. The starting motor cranks slowly, but the engine does not start.

Another possible trouble is that the starting motor cranks the engine normally, but the engine does not start. This condition cannot be blamed on the starting system. Trouble in the fuel or ignition system, or in the engine, is preventing starting.

♦ 25-7 STARTING MOTOR DOES NOT CRANK ENGINE

The most likely cause of this condition is a run-down battery. But there could be other causes. Turn on the headlights, and try cranking. There are five possibilities.

1. No Cranking, No Lights This is probably due to a completely dead battery. It could be caused by a bad connection at the battery or starting motor or an open fusible link (which indicates a short or ground in the system).

2. No Cranking, but Lights Go Out as You Turn the Key to START This usually indicates a bad connection at the battery. It could also be due to a nearly dead (discharged) battery. Try wiggling the battery connections to see if this helps.

3. No Cranking, and Lights Dim Only Slightly as You Try to Start The trouble probably is in the starting motor. The pinion may not be engaging with the flywheel. If the starting-motor armature spins, then the overrunning clutch is slipping.

4. No Cranking, and Lights Dim Heavily as You Try to Start This is most likely due to a run-down battery. It could be low temperature, too. The battery is much less efficient at low temperatures, and the engine oil is much thicker. This combination could prevent cranking, even though the battery is in fairly good condition. Also, the starting motor or engine could be jammed, or locked.

5. No Cranking, and Lights Stay Bright Listen to hear if the magnetic solenoid is pulling the plunger in. You can hear this as a loud click. If nothing happens when you try to start, check the solenoid. Connect one end of a jumper wire to the solenoid battery terminal. Connect the other end to the small terminal on the solenoid that is connected to the ignition switch. If nothing happens, the trouble is in the solenoid. If the solenoid and starting motor work with the jumper wire connected, the trouble is in the ignition switch, the transmission switch, or the wires connecting them.

6. Jump Start You could try to jump start the engine (♦25-9) with a booster battery. Observe the cautions in ♦25-9. This could give you a clue as to the trouble (items 1 to 5 above).

♦ 25-8 ENGINE CRANKS SLOWLY BUT DOES NOT START

This is very likely due to a run-down or defective battery. The battery is unable to spin the starting motor at normal speed. Low temperature could also be a factor here, as described in ♦25-7.

It is also possible that the driver may have run the battery down trying to start. Some condition in the engine, or the fuel or ignition system, is preventing normal starting. The driver continued to try, however, until the battery ran down.

Try to jump start the engine (♦25-9). If the engine still does not start, then the fuel system and ignition system should be checked. But if the engine does start, the battery probably was at fault.

♦ 25-9 JUMP STARTING

To jump start a car, you use the battery from another car. This is called the *booster* battery. You must use certain precautions when jump starting, because a wrong connection can damage the car electrical equipment. Or worse, the battery could explode and you could be injured. Observe the following precautions:

1. Wear eye protection to guard against getting electrolyte in your eyes if a battery should erupt or explode.
2. If the dead battery is the maintenance-free type, check the charge indicator. If it shows light yellow (Fig. 24-11), *do not try to jump start*. The battery could explode!
3. During cold weather, the electrolyte in the battery could freeze—especially if the battery was run down to start with. *Never try to jump start if the battery is frozen*. It could explode!
4. The battery could explode if the booster battery is connected backward (positive to negative, negative to positive). Even if a battery does not explode, both batteries could be ruined.

The jump-start procedure requires two jumper cables (Fig. 25-14).

1. Wear eye protection to guard against getting electrolyte in your eyes.
2. If either battery has vent plugs (Fig. 25-15), remove them. Cover the holes with cloths to prevent splashing of the electrolyte in case of an explosion (Fig. 25-15).
3. Do not allow the two cars to touch each other. This could

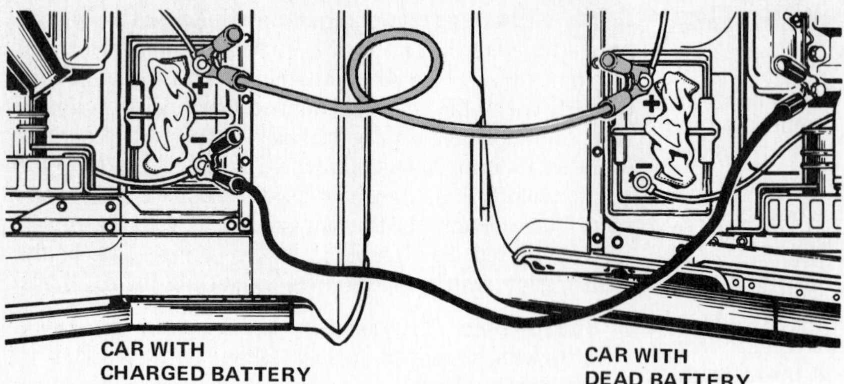

Fig. 25-14 Connections between the booster battery and the dead battery for jump starting a car with a dead battery.

damage electrical equipment if the two electrical systems have different voltage, or different ground polarity.

4. Make sure all electrical equipment except ignition is turned off on the car you are trying to start.
5. Connect the end of one jumper cable to the positive (+) terminal of the booster battery. Connect the other end of this cable to the positive (+) terminal of the dead battery.
6. Connect one end of the second cable to the negative (−) terminal of the booster battery.
7. Connect the other end of the second cable to the engine block of the car you are trying to start. Do not connect it to the negative (−) terminal of the car with the dead battery! A spark from this connection could cause the battery to explode. Do not lean over the battery when making this connection!
8. Start the car with the booster battery. Then start the car with the dead battery. After the dead-battery engine starts, disconnect the booster cables. First disconnect the cable from the engine block. Then disconnect the other end of this negative cable. Finally, disconnect the positive cable.
9. Throw away the cloths used to cover the vent holes. They may have acid on them.

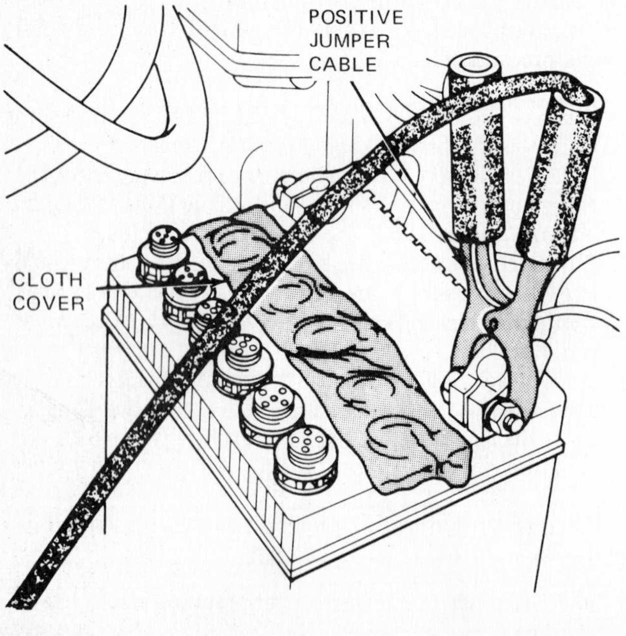

Careful Never operate the starting motor for more than 30 seconds at a time. Pause for a few minutes to allow it to cool off. Then try again. It takes a very high current to crank the engine. You can overheat and damage a starting motor by cranking too long.

Fig. 25-15 Connecting jumper cable and covering the openings in the battery cover with a cloth. The vent plugs are shown removed. Other manufacturers recommend leaving them in after making sure the vents in the plugs are open. *(Chrysler Corporation)*

♦ 25-10 REMOVING AND INSTALLING STARTING MOTOR

If the indications are that the starting motor is defective, it should be removed for repair or for exchange with a new or rebuilt starting motor. First, disconnect the grounded battery-terminal cable (Fig. 25-16). This eliminates the danger of accidentally grounding the insulated side of the battery.

Disconnect the leads and cable from the starting motor. Note carefully how these are connected so that you can reconnect them properly. Remove the bolts holding the starting motor to the flywheel housing. Lift the starting motor out.

If the starting motor has been removed because it is defective, test and repair it, or exchange it for a new or rebuilt

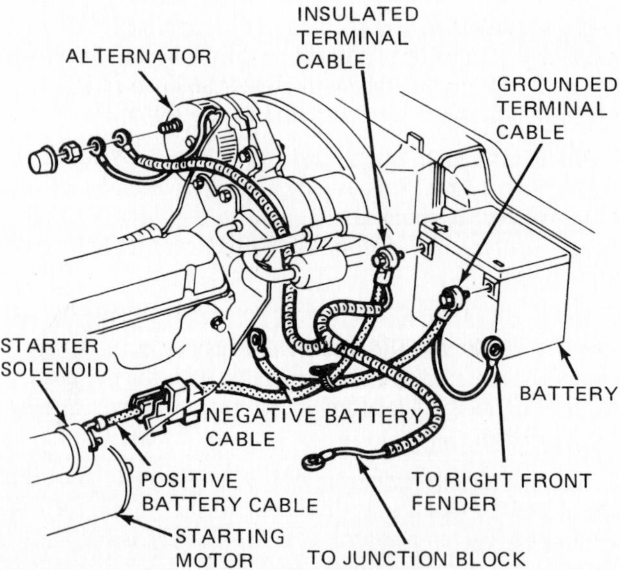

Fig. 25-16 Connections in the starting and charging systems. *(Cadillac Motor Car Division of General Motors Corporation)*

starting motor. When an overrunning clutch is to be reinstalled, wipe it clean with a cloth or shop towel. Do not wash the overrunning clutch in solvent. The overrunning clutch cannot be relubricated.

When installing the starting motor, reconnect the grounded battery cable last. Try the starting motor to make sure it works normally and cranks the engine at normal cranking speed.

♦ REVIEW QUESTIONS ♦

Select the *one* correct, best, or most probable answer to each question. You can find the answers in the section indicated at the end of each question.

1. The starting system includes the (♦25-1)
   a. battery and ignition switch
   b. battery and starting motor
   c. starting motor and wiring
   d. all of the above
2. The purpose of the transmission switch is to (♦25-5)
   a. prevent starting if the transmission is in gear
   b. prevent starting if the transmission is in PARK
   c. allow the starting motor to operate if the engine is not running
   d. allow the transmission to shift into any gear
3. In the starting motor, magnetism (♦25-2)
   a. rotates the armature and demeshes the pinion
   b. rotates the armature and meshes the pinion
   c. prevents high armature speed as the engine starts
   d. sends cranking force in one direction only
4. The overrunning clutch (♦25-4)
   a. transmits cranking force to the engine flywheel
   b. is a one-way clutch
   c. prevents the engine flywheel from driving the starting motor
   d. all of the above
5. If the headlights go out when you try to start, the cause is probably (♦25-7)
   a. a defective light switch
   b. a disconnected battery
   c. a defective connection at the battery
   d. a defective starting motor
6. If the starting motor does not operate and the lights do not come on, the probable cause is that the (♦25-7)
   a. battery is dead or there is a bad connection
   b. lights are burned out
   c. battery is disconnected
   d. battery has been removed from the car
7. If the engine cranks slowly but does not start, a possible cause could be (♦25-8)
   a. a run-down battery
   b. low temperature
   c. the driver may have run the battery down trying to start
   d. all of the above
8. Jump starting a car with a dead battery should not be tried if the dead battery is (♦25-9)
   a. frozen
   b. the maintenance-free type and the indicator shows light yellow
   c. a different voltage than the booster battery
   d. all of the above
9. When removing the battery or starting motor, (♦25-10)
   a. always disconnect the grounded battery cable first
   b. do not disconnect the battery cables
   c. disconnect the insulated battery cable first
   d. remove the vent plugs first
10. The overrunning clutch (♦25-10)
    a. should be oiled
    b. should be repacked with grease
    c. cannot be lubricated
    d. contains no lubricant

CHAPTER 26
The Charging System: Operation and Service

After studying this chapter, you should be able to:

1. Explain how the alternator operates and the function of the diodes in the alternator.
2. Discuss the purpose of the alternator regulator and how it works.
3. List four troubles that might be caused by the charging system and the possible causes of these troubles.
4. Describe how to check an alternator if the battery is run-down.
5. List four charging-system tests.
6. Describe how to replace an alternator.

♦ 26-1 PURPOSE OF CHARGING SYSTEM

The charging system has two jobs:

1. To put back into the battery the current used to start the engine
2. To handle the load of the lights, ignition, radio, and other electrical and electronic equipment while the engine is running

The charging system includes the alternator, regulator, and battery, with connecting wires (Fig. 26-1).

♦ 26-2 FUNCTION OF ALTERNATOR

The alternator (or *ac generator*) converts mechanical energy from the engine into electrical energy. It keeps the battery in a charged condition and handles electrical loads while the engine is running.

♦ 26-3 ALTERNATOR

The alternator (Fig. 26-2) produces alternating current (ac) inside its stator windings. Diodes convert this ac into direct current (dc). Almost all the electrical devices on the car use dc. Alternators have regulators to keep the alternator from producing excessive voltage. These control devices are discussed in following sections.

♦ 26-4 PRODUCING ELECTRICITY

Electrons can be made to move by moving a conductor (wire) through a magnetic field, as shown in Fig. 26-3. As the wire is moved through the magnetic field of the magnet, current will flow in the wire. The direction of current flow can be reversed by reversing the direction the wire moves (from left to right, or from right to left). This is the basis of alternator action.

♦ 26-5 ALTERNATOR PRINCIPLES

In the alternator, the wires, or conductors, are held stationary and a magnetic field is moved through them. Actually, the alternator rotates the magnetic field so that the stationary conductors cut the moving magnetic lines of force.

A simple one-loop alternator is shown in Fig. 26-4. The rotating bar magnet supplies the moving field. At the top, the north pole of the bar magnet passes the upper leg of the loop, and the south pole passes the lower leg of the loop. Current (electron flow) is induced in the loop in the direction shown

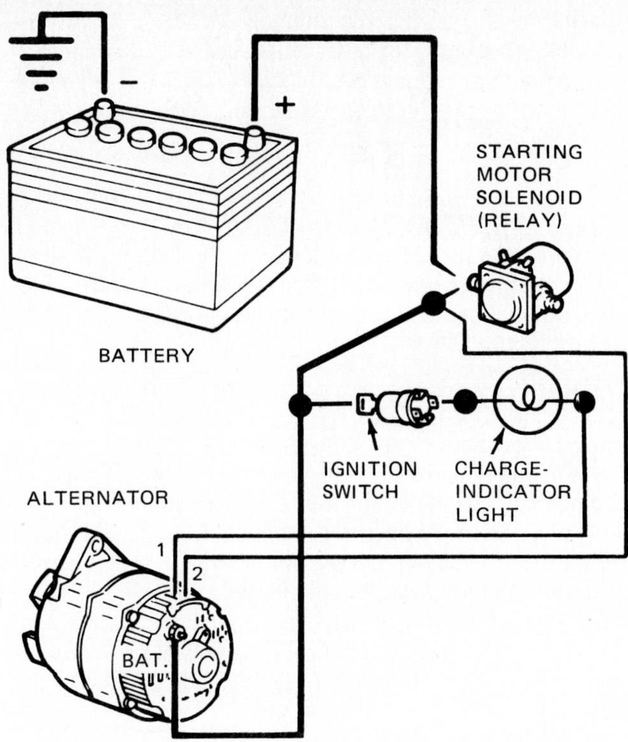

Fig. 26-1 Typical charging system using a charge-indicator light. *(American Motors Corporation)*

by the arrows. At the bottom, the magnet has rotated half a turn. Its south pole is passing the upper leg of the loop, and its north pole is passing the lower leg. Now, magnetic lines of force are being cut by the two legs in the opposite direction. So current (electron flow) is induced in the loop in the opposite direction. Therefore, as the magnet spins and the two poles alternately pass the two legs of the loop, electrons in the loop are pushed first in one direction and then in the other. The electrons alternate in direction. Alternating current flows.

Three things will increase the current (number of electrons) moving in the loop. They are:

1. Increasing the strength of the magnetic field
2. Increasing the speed with which the magnetic field rotates
3. Increasing the number of loops

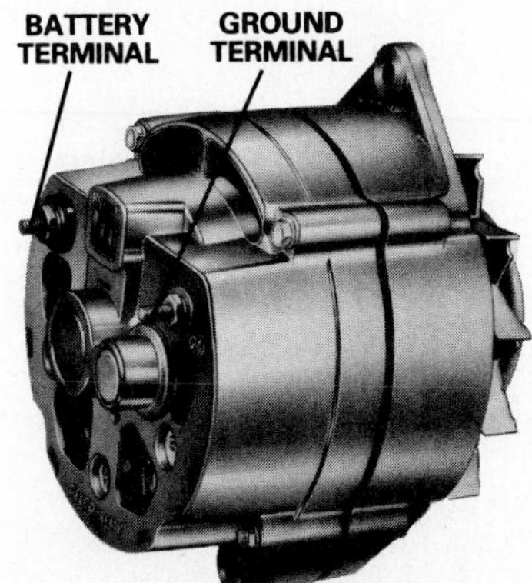

Fig. 26-2 An alternator which has a built-in rectifier and uses a separately mounted regulator. *(Delco-Remy Division of General Motors Corporation)*

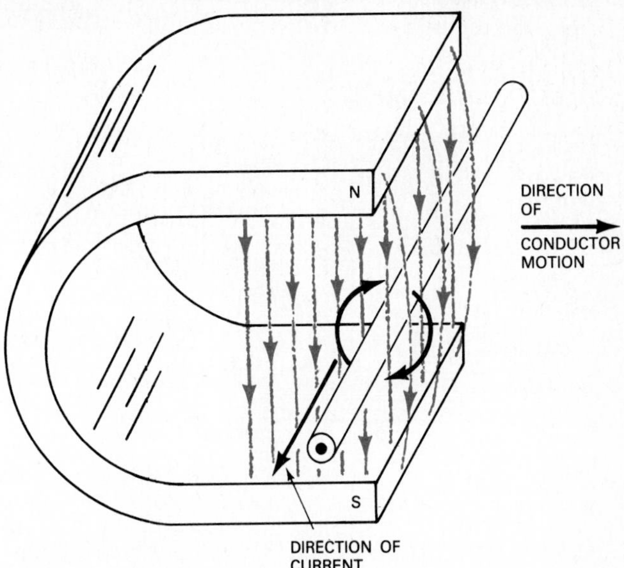

Fig. 26-3 Moving a wire (or conductor) through a magnetic field induces a flow of electric current in the wire.

In the actual alternator, both the strength of the magnetic field and the number of loops are increased. Instead of a bar magnet, two or more pole pieces make up the rotating part of the alternator. The pole pieces are assembled on a shaft over an electromagnetic winding. The electromagnet is made up of many turns of wire. When current flows in the electromagnetic winding, a strong magnetic field is created. The pointed ends of the two pole pieces become, alternately, north and south poles (Fig. 26-5). The winding is connected to the battery through a pair of insulated rings that rotate with the shaft. A pair of stationary brushes ride on the rings. The two ends of the winding are attached to the rings. The brushes make continuous sliding (or slipping) contact with the slip rings (Fig. 26-5).

Figure 26-6 shows the stationary loops of an ac generator

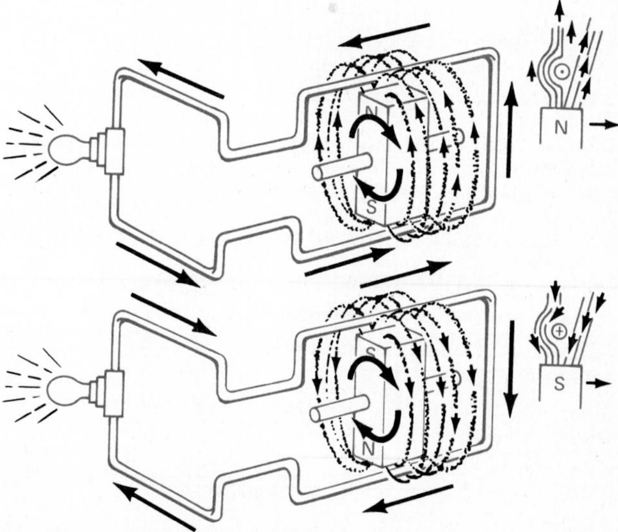

Fig. 26-4 Simplified alternator consisting of a single stationary loop of wire and a rotating bar magnet. The distortion of the moving lines of force around each leg of the wire loop and the direction of current flow are shown at the right.

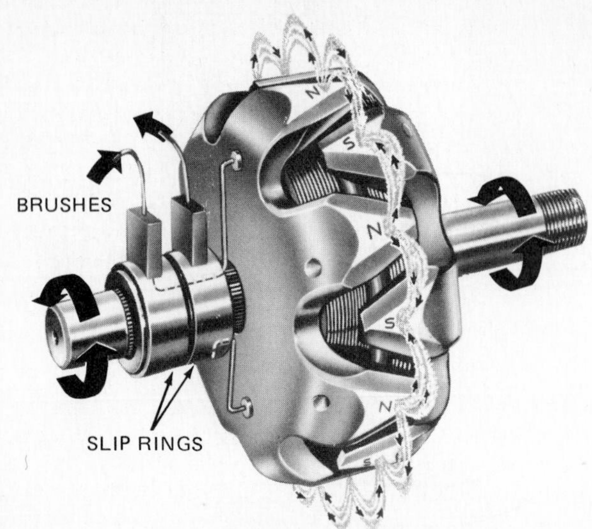

Fig. 26-5 Rotor of an alternator, showing brushes in place on the slip rings. Current flows through the winding, which produces a magnetic field. *(Delco-Remy Division of General Motors Corporation)*

Fig. 26-6 Stator of an alternator. Current is produced in the stator windings as the magnetic field (rotor) turns inside the stator. *(Delco-Remy Division of General Motors Corporation)*

assembled into a frame. The assembly is called a *stator.* The loops are interconnected as explained below so that the current produced in all loops adds together. Since this current is alternating, it must be *rectified,* or converted, into dc.

♦ 26-6 ALTERNATOR OPERATION

The alternator produces ac. However, the battery, ignition system, and most other electrical components on the automobile cannot use ac. Therefore, the ac output must be rectified, or changed, to dc.

1. Rectifying Alternating Current Automotive alternators have built-in diode rectifiers. The *diode* is an electronic device that permits current to flow through it in one direction only. Figure 26-7 illustrates how four diodes can be used to change ac to dc. The four diodes are numbered 1 to 4 in the illustration. To the left, the current from the alternator follows the conductors shown solid. Diodes 1 and 3 permit the current to flow through. But diodes 2 and 4 will not, since the current is flowing in the wrong direction from them. However, when the direction of the current has reversed, as shown to the right in Fig. 26-7, diodes 2 and 4 will pass the current. But diodes 1 and 3 will not.

Note The above explanation and Fig. 26-7 are included here only to show how diodes can change ac to dc. This does not represent the typical automotive alternator.

2. Three-Phase The circuit in Fig. 26-7 is called *single-phase.* There is only a single ac source (the alternator). Such a source would result in a pulsating current. To provide a much smoother flow of current, automotive alternators are built with three stator circuits. These, in effect, give overlapping pulses of ac. When these are rectified, a smooth flow of dc is obtained.

The three stator circuits can be interconnected in either of two ways, with "Y" connections or with "delta" connections (Fig. 26-8). They operate similarly and are serviced similarly. The ac generated in the three legs of the stator passes through the six diodes and is converted into dc.

3. Diode Heat Sinks Diodes are usually mounted in the slip-ring end of the alternator, in a metal bracket called a *heat sink.* The heat sink takes heat from the diodes, which get hot in operation. The heat sink has large surfaces. They transfer the heat into the air surrounding the alternator. Therefore, the diodes do not overheat.

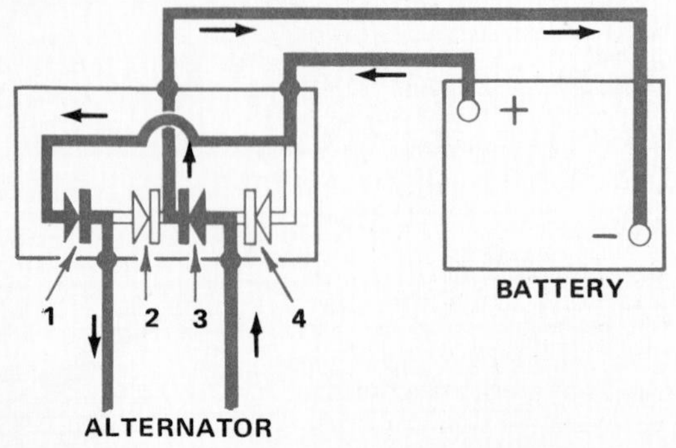

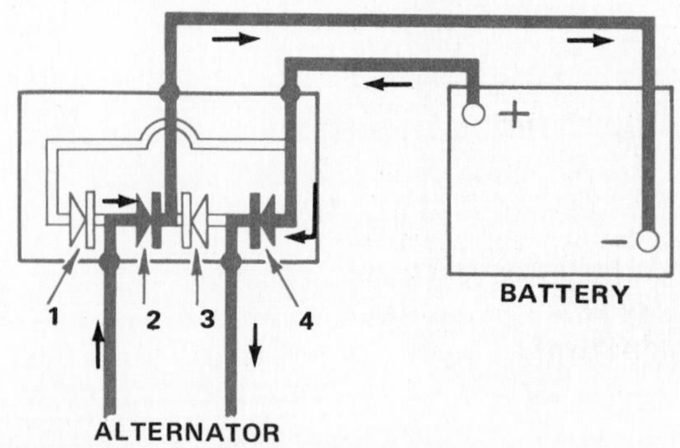

Fig. 26-7 Four diodes connected to an alternator. The diodes rectify the ac, which changes it to dc to charge the battery.

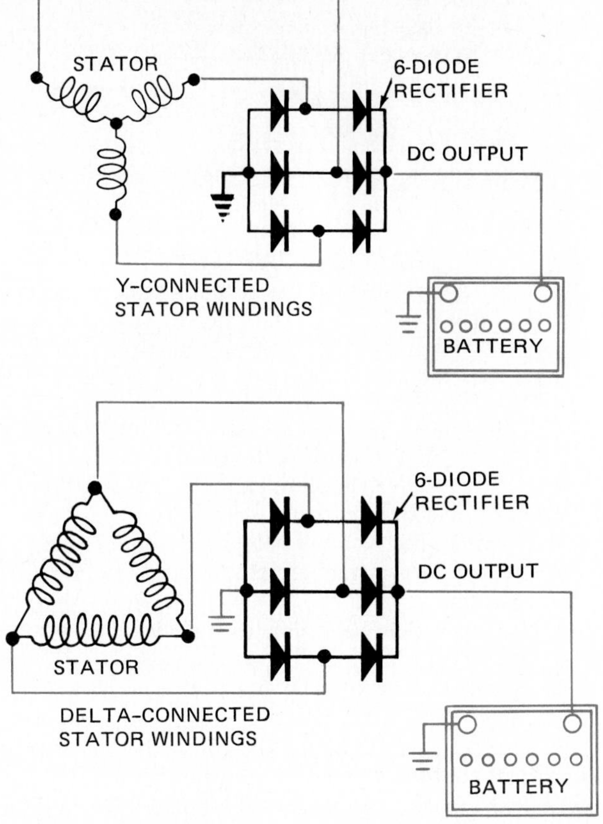

Fig. 26-8 Wiring diagrams showing two ways that the six-diode rectifier can be connected to the stator. With either connection, the ac is rectified to dc.

♦ 26-7 ALTERNATOR REGULATION

A variety of devices has been used to regulate alternators. When alternators were first introduced, many regulators included a field relay, an indicator-light relay, and a voltage regulator. Now alternator-regulator systems have been simplified. For example, some of the latest types have the regulator built into the alternator (Fig. 26-9). The circuit looks like Fig. 26-10. Basically, the regulator limits the alternator field current as necessary to prevent excess alternator voltage. The stator remains permanently connected to the battery through the diodes. They prevent the battery from discharging back through the stator when the alternator is not operating. The field (rotor) is connected to the battery only when the alternator is operating. The connection is made through either the field relay or the ignition switch.

♦ 26-8 REGULATORS FOR ALTERNATORS

Several types of alternator regulators that are mounted outside the alternator have been used. Figure 26-11 shows a typical charging-system wiring circuit which uses an external electronic (solid-state) regulator. The regulator limits alternator voltage by controlling the amount of current flowing in the alternator field (the rotor windings). If the voltage starts to go too high, the regulator reduces the field current. This prevents excessive voltage.

♦ 26-9 CHARGE INDICATORS

There are two types of charge indicator, the ammeter (described in ♦23-5) and an indicator light. The ammeter needle moves off center (or zero) to indicate a flow of current to the battery (charging) or a flow of current from the battery (discharging). The charge-indicator light comes on when first starting to indicate that the alternator is not charging the battery. Then, when the alternator speed increases enough to send a charge to the battery, the indicator light goes off.

The ammeter is connected into the circuit from the alternator to the battery so that it can show what's happening. The charge-indicator light has a different circuit (Fig. 26-10). When the ignition switch is turned on, current can flow through the resistor and the charge-indicator light. However, as soon as the alternator builds up voltage, the voltage on the two sides of the resistor and light becomes the same. The

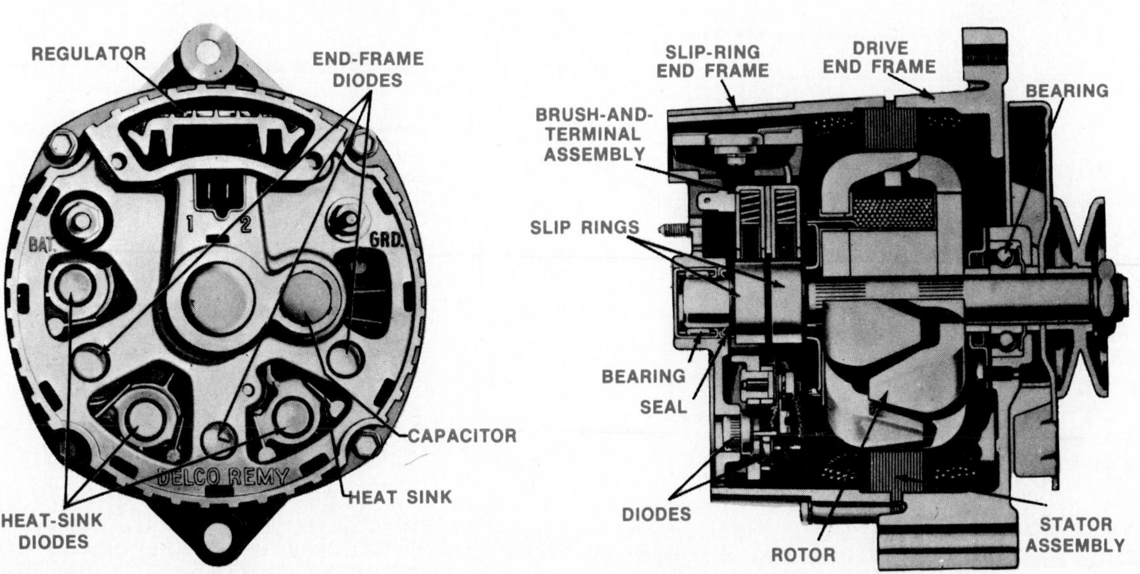

Fig. 26-9 End and sectional views of an alternator with built-in diodes and an integral voltage regulator. This is called a "Delcotron" by the manufacturer. *(Delco-Remy Division of General Motors Corporation)*

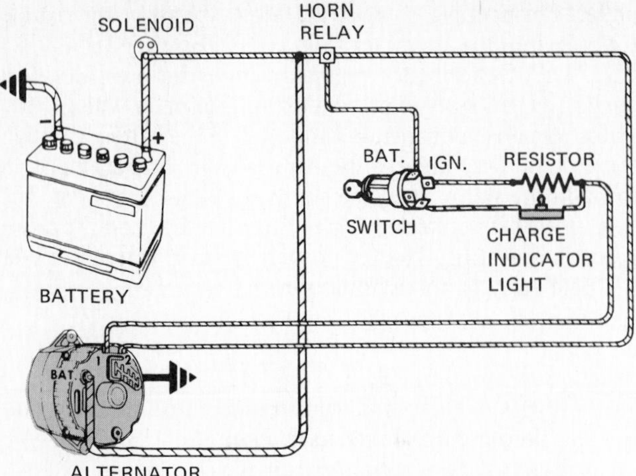

Fig. 26-10 Wiring diagram for a charging system using an alternator with an integral voltage regulator and a charge-indicator light. *(Delco-Remy Division of General Motors Corporation)*

light goes out. This indicates that the alternator is charging the battery. If the alternator fails, its voltage will drop. Then the light will come on to indicate the failure.

CHARGING-SYSTEM SERVICE

♦ 26-10 CHARGING-SYSTEM TROUBLE DIAGNOSIS

There are four troubles that might be caused by the charging system:

1. Run-down battery
2. Overcharged battery
3. Faulty indicator-lamp or charge-indicator action
4. Noisy alternator

Each of these troubles is described in following sections. Their causes and the corrections needed are also discussed.

♦ 26-11 DISCHARGED BATTERY

A discharged battery does not have enough capacity to crank the engine at normal speeds. The trouble could be due to:

1. Start-and-stop driving during which the alternator does not operate long enough to put back into the battery the current taken out in starting the engine
2. Accessories or lights left on when the engine is turned off
3. A loose or defective alternator drive belt
4. An old battery that will not accept a charge normally
5. Defective wiring or connections such as a blown fusible link (♦23-19)
6. Defective alternator, regulator, or connecting wiring

If the run-down battery has not been caused by items 1 and 2 above, and if the battery is not defective, check the drive belt for condition and tension (♦22-15). If the belt is okay and properly adjusted, check the wiring and connections for defects. This includes the battery-cable connections and the engine ground strap. Also, check the wiring harness for blown fusible links (♦23-19).

If all the above check normal, follow the procedure for the

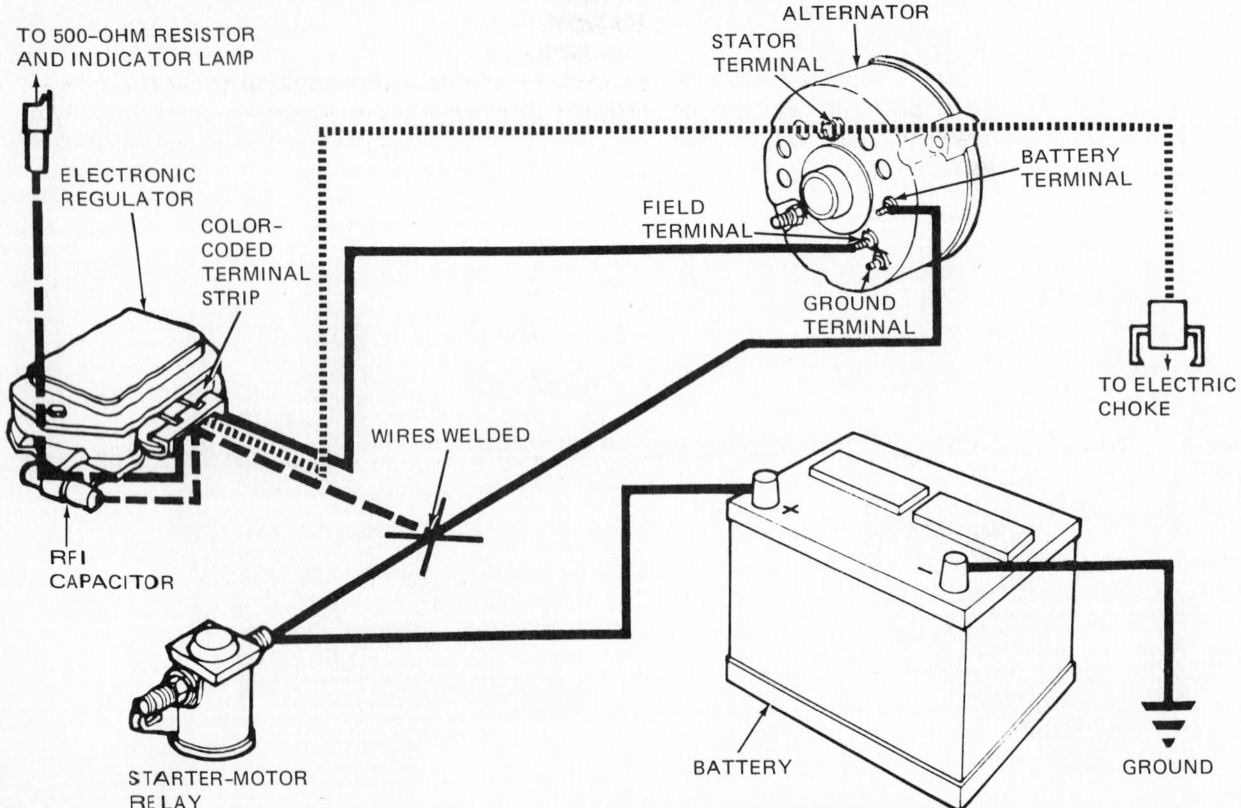

Fig. 26-11 Charging system with separate electronic regulator and charge-indicator light. *(Ford Motor Company)*

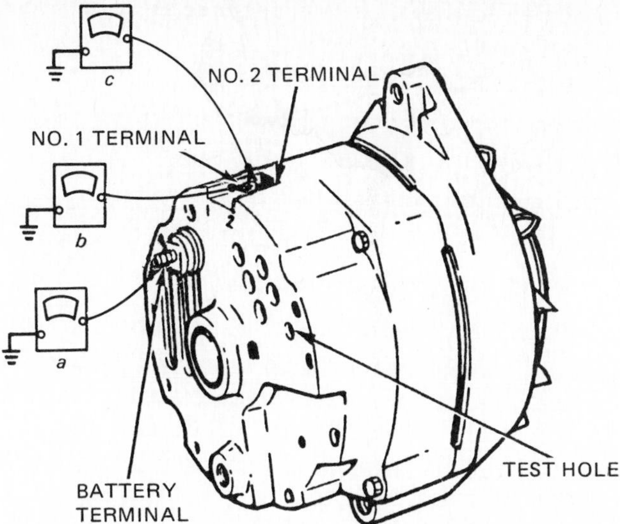

Fig. 26-12 Voltmeter test points on an alternator, and the alternator test hole. *(Delco-Remy Division of General Motors Corporation)*

type of system on the car. Two procedures are covered in following sections. One is for General Motors and other cars which have an alternator with a built-in (internal, or integral) regulator (♦26-12). The other procedure is for Ford and other cars which use an alternator with a separate (external) regulator (♦26-13).

♦ 26-12 CHARGING-SYSTEM TESTS

The following procedure is recommended by General Motors for charging systems which include a Delco-Remy alternator with a built-in regulator. With the ignition switch turned on (engine not running) and all wiring-harness leads connected, use a voltmeter to check, as shown in Fig. 26-12, from:

1. Alternator battery terminal to ground (*a*)
2. Alternator number 1 terminal to ground (*b*)
3. Alternator number 2 terminal to ground (*c*)

The voltmeter should read battery voltage at each of these connections. A zero reading at any of these three checks indicates an open from that connection to the battery. Check the system further to locate and correct the open.

If everything checks, make an alternator output test, as follows.

1. Disconnect the grounded battery cable. Connect an ammeter in the circuit at the alternator battery terminal. Reconnect the grounded battery cable.
2. Turn on the radio, windshield wipers, lights on high beam, and blower motor on high speed. Connect a carbon pile (a variable resistance) across the battery.
3. Operate the engine at about 2000 rpm. Adjust the carbon pile as necessary to get maximum alternator output.
4. If the alternator output is within 10 amperes of the rated output stamped on the alternator frame, the alternator is okay.
5. If the output is low, locate the test hole in the end of the alternator (Fig. 26-12). If it is accessible, insert a screwdriver into the test hole (not deeper than 1 inch). Operate the engine at about 2000 rpm. Adjust the carbon pile to get maximum output. If the output is still low, the alternator is defective. It must be removed for repair or replacement with a new or rebuilt unit.
6. If the test hole is not accessible, remove the alternator so that it can be tested separately and serviced as necessary.

Note There are several types of charging-system testers that can save time in checking out a system having problems. Follow the operating instructions provided by the tester manufacturer.

♦ 26-13 FORD CHARGING-SYSTEM TESTS

This procedure is recommended by Ford for charging systems which have a separate electronic regulator (Fig. 26-11). After checking for the causes of a discharged battery listed in ♦26-11, make a battery-drain check as follows:

1. Battery-Drain Test Disconnect the negative battery cable. Connect a 12-volt test lamp between the cable clamp and the battery terminal (Fig. 26-13). Everything should be turned off and car doors closed so that there should be no current flow from the battery. If the light comes on, there is a drain. Then pull the fuses, one by one, until the light goes out. This locates the circuit that is draining the battery. Check that circuit and make corrections as required.

If there is no battery drain, make a no-load voltage test of the system as follows:

2. No-Load Voltage Test With no electrical load turned on, check the battery voltage. This is the base voltage. Now start the engine and increase its speed to about 1500 rpm. With no load except ignition, the voltage should go up not more than 2 volts above the base voltage. If the voltage goes up more than 2 volts above the base voltage, make the overvoltage check as explained below. If the voltage does not increase, make the undervoltage check as explained below.

Note It may take a few minutes for the voltmeter needle to stop rising. This is the length of time it takes the alternator to bring the battery back up to charge after cranking the engine.

3. Overvoltage Test Connect a jumper wire between the regulator base and the alternator frame. Repeat the no-load test. If the overvoltage condition disappears, there is a bad circuit between the regulator and alternator. Clean the ground connections at the regulator and battery. If the over-

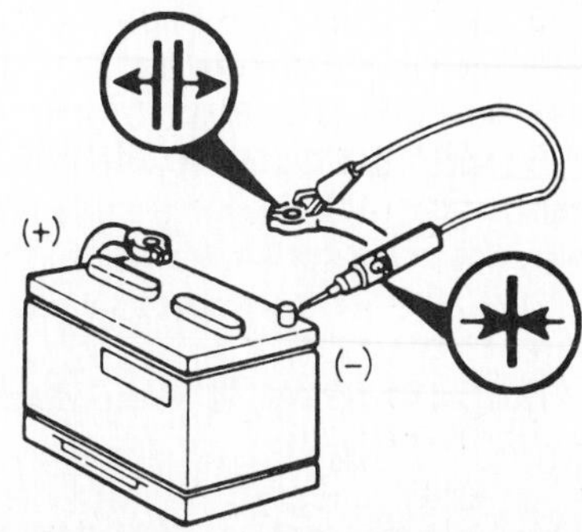

Fig. 26-13 Connecting a test light between the grounded battery post and the negative cable clamp. *(American Motors Corporation)*

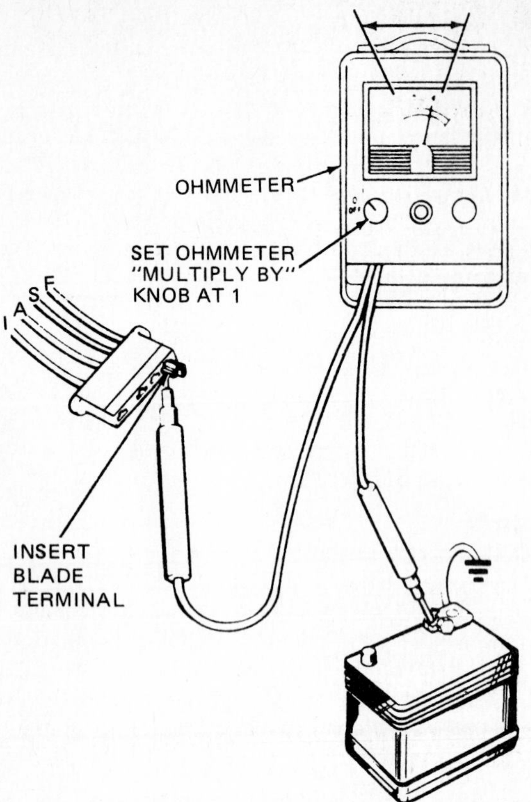

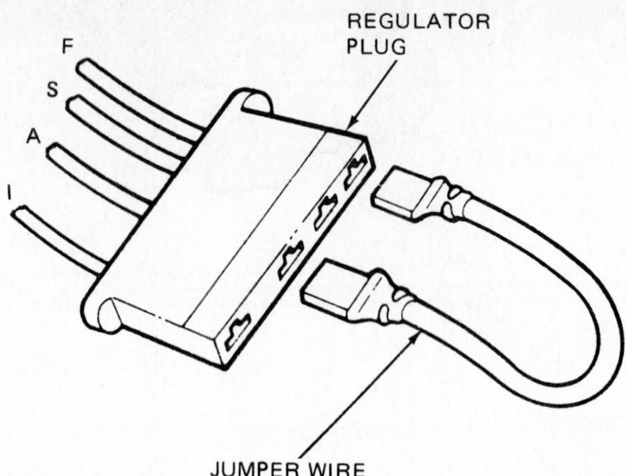

Fig. 26-15 Using a jumper wire to check regulator action by connecting the A terminal to the F terminal at the regulator plug. *(Ford Motor Company)*

Fig. 26-14 Checking the field circuit with an ohmmeter. The meter should indicate between 3 and 250 ohms. A lower reading indicates a shorted or grounded field circuit (including alternator).

voltage condition still exists, disconnect the regulator wiring plug from the regulator and repeat the no-load test. If the overvoltage condition disappears (voltmeter reads base voltage), the regulator is defective. Replace it.

If the overvoltage condition still exists with the regulator wiring plug disconnected, there is a short in the wiring harness between the alternator and regulator. Service as required.

4. Undervoltage Test If the voltage on the no-load test does not increase above the base voltage, the trouble is in the wiring, the regulator, or the alternator. First, check the alternator field circuit with an ohmmeter. The regulator plug is disconnected and the ohmmeter is connected between the F terminal of the plug and ground (Fig. 26-14). The meter should read more than 3 ohms. If it is less, the field circuit is grounded. Check the wiring harness and alternator.

If the ohmmeter reads more than 3 ohms, connect a jumper wire from the A to F terminals of the plug (Fig. 26-15). Repeat the no-load test. If the voltage now goes up above the base voltage, the regulator or wiring is at fault. If the trouble is not in the wiring, the regulator should be replaced.

Careful Do not ground the field circuit during the test. This can damage the regulator.

If the trouble has still not been corrected, remove the jumper wire from the regulator and leave the plug disconnected from the regulator. Disconnect the field terminal on the alternator and pull back the protective cover from the battery terminal. Connect a jumper lead between the field and battery terminals (Fig. 26-16). Repeat the no-load test. If the voltage now rises above the base voltage, the problem is not in the alternator, but in the wiring system.

If the voltage still does not rise, the alternator is at fault. It should be removed for service.

♦ 26-14 OVERCHARGED BATTERY

An overcharged battery will have a relatively short life. On batteries with vent plugs, overcharging can cause electrolyte to spurt or leak from the vent plugs. This lost electrolyte reduces battery capacity. It also spreads across the battery top. There the electrolyte forms a conducting path between the battery terminals so the battery will slowly discharge. This type of slow discharge does not occur with side-terminal batteries.

The escaping electrolyte is very corrosive. It can corrode the battery holder and metal surfaces close by. The condition is not as noticeable with maintenance-free batteries. But if they are overcharged too long, they will get very hot. Also, overcharging means the alternator voltage is going too high. Evidence of this is flaring of the headlights or internal lights. When the engine speed increases from idle, the headlights will increase greatly in brightness. Some brightness increase is normal. But when it is excessive, suspect excessive voltage. The excessive voltage overcharges the battery and shortens the life of the headlights, other lights, and electrical devices.

With alternators having the regulator built in, the cause of high voltage and battery overcharging is in the alternator. It must be removed for service or replacement with a new or rebuilt unit.

Alternators having a separate regulator can be checked by disconnecting the regulator plug. The engine should be running at about 1500 rpm. A voltmeter should be connected as for the no-load voltage test (♦26-13). If the voltage stays high, the trouble is in the alternator (internal shorts). It should be removed for service. If the voltage drops to the base voltage when the plug is disconnected, the trouble is probably in the regulator. It has stopped regulating. Installing a new regulator should correct the problem. If it does not, there is a short in the wiring. Repair the short and reinstall the original regulator. It is probably okay. Recheck the system to make sure it now works normally.

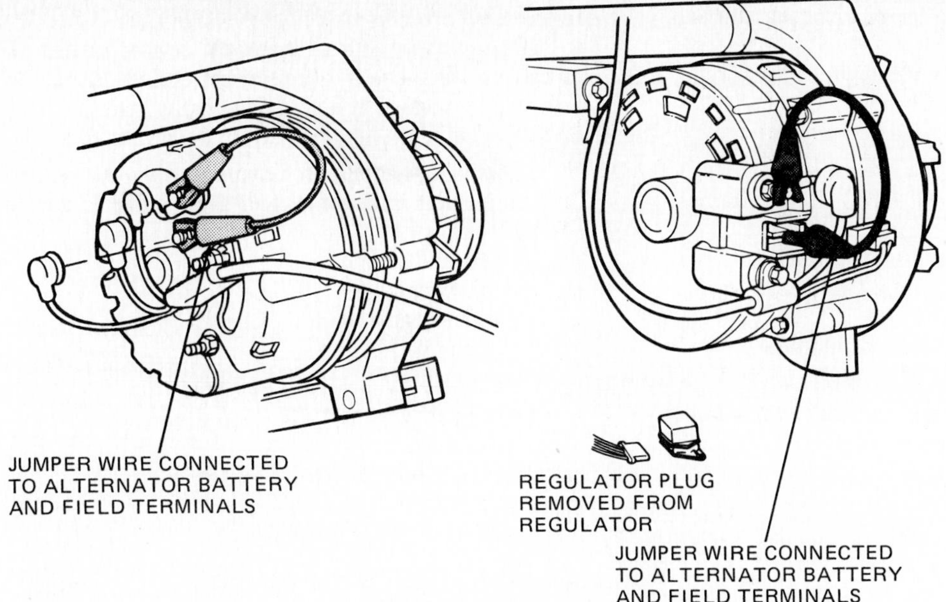

Fig. 26-16 Jumper wire connections at the alternator. *(Ford Motor Company)*

♦ 26-15 ALTERNATOR NOISE

Noise in the alternator is probably due to worn bearings, worn drive belt, loose alternator mounting, damaged rotor or fan, or one or more diodes in the rectifier shorted or open. Shorts or opens in the stator could also cause noise.

♦ 26-16 REMOVING AND INSTALLING ALTERNATOR

To remove an alternator, first disconnect the grounded battery cable. Then disconnect the leads and plug (where used) from the alternator. Note their positions so that you can reconnect them correctly. Loosen the adjusting bolts and move the alternator in so that you can take the drive belt off. Remove the thru bolts that hold the alternator in place and take the alternator off the car.

To install the alternator, attach the alternator to the mounting bracket with the bolts, washers, and nuts. Do not tighten the nuts. Install the drive belt and tighten it to the specified tension. Then tighten the mounting bolts.

Careful Do not pry on the end plates of the alternator. Instead, pry on the center of the unit as necessary to tighten the belt.

Reconnect the leads and plugs. Reconnect the grounded battery cable. Start the engine and check the system for normal operation.

♦ REVIEW QUESTIONS ♦

Select the *one* correct, best, or most probable answer to each question. You can find the answers in the section indicated at the end of each question.

1. The alternator produces electricity in its (♦26-3)
   a. rotor field coil
   b. stator windings
   c. regulator
   d. armature commutator
2. The purpose of the regulator is to (♦26-7)
   a. prevent the alternator voltage from going too high
   b. allow the alternator to produce a high current
   c. keep alternator speed from going too high
   d. keep the alternator voltage high enough to charge the battery
3. The alternator regulator is either inside the alternator or (♦26-8)
   a. mounted on the battery
   b. connected to the lighting system
   c. mounted separately
   d. is not used
4. The cause of a run-down battery could be (♦26-11)
   a. loose alternator drive belt
   b. defective alternator or regulator
   c. start-and-stop driving
   d. all of the above

5. To make the General Motors alternator-voltage checks, the (♦26-12)
   a. ignition should be turned on with engine not running
   b. ignition should be turned off
   c. battery is disconnected
   d. alternator is disconnected
6. To make an output check of the General Motors alternator, (♦26-12)
   a. connect an ammeter into the circuit at the F terminal
   b. connect an ammeter into the circuit at the BAT terminal
   c. turn off all electrical devices except ignition
   d. operate the engine at about 500 rpm
7. The Ford battery-drain check requires that (♦26-13)
   a. all major electrical units be turned on
   b. the engine be running at idle
   c. all electrical devices be turned off
   d. all fuses be pulled
8. In the Ford procedure, the base voltage is the (♦26-13)
   a. battery voltage with all electrical devices turned off
   b. battery voltage with all electrical devices turned on
   c. charging voltage at a specified engine rpm
   d. battery voltage with the battery disconnected
9. If the overvoltage condition disappears when the regulator wiring plug is disconnected, the trouble is due to (♦26-13)
   a. a defective alternator
   b. defective wiring
   c. a defective regulator
   d. an overcharged battery
10. An overcharged battery will (♦26-14)
   a. have a relatively short life
   b. lose electrolyte
   c. result from high alternator voltage
   d. all of the above

CHAPTER 27
Contact-Point Ignition Systems: Operation and Service

After studying this chapter, and with proper instruction and equipment, you should be able to:

1. Explain the operation of the contact-point ignition system.
2. Locate and identify the components of the contact-point ignition system on various vehicles.
3. Explain the construction and operation of centrifugal- and vacuum-advance mechanisms.
4. List three causes of ignition failure and describe the conditions that could produce each.
5. Use the shop testers to check ignition-systems operation and components.
6. Check and adjust ignition timing.
7. Adjust or replace contact points.
8. Remove and install ignition distributors.

♦ 27-1 PURPOSE OF IGNITION SYSTEM

The ignition system (Fig. 27-1) supplies high-voltage surges—as high as 47,000 volts (in some electronic systems)—to the spark plugs in the engine cylinders. These surges produce electric sparks across the spark-plug gaps. The heat from the sparks ignites, or sets fire to, the compressed air-fuel mixture in the combustion chambers. When the engine is idling, the spark appears at the plug gap just as the piston nears TDC on the compression stroke. When the engine is operating at higher speeds, or with part throttle, the spark is advanced. It is moved ahead and occurs earlier in the compression stroke. This gives the compressed mixture more time to burn and deliver its energy to the pistons.

♦ 27-2 TWO KINDS OF IGNITION SYSTEMS

Two kinds of ignition systems are used on cars today. In the older system, movable contact points are used to make and break an electric circuit. This system is still used in some cars today. But most cars now have electronic ignition. In this system, the make-and-break of the electric circuit is done by an electronic device. The electronic system is much faster in action, and more accurate.

This chapter discusses the contact-point system. Chapter 28 describes electronic ignition systems.

♦ 27-3 CONTACT-POINT IGNITION SYSTEM

The contact-point ignition system is shown in Fig. 27-2. It includes the battery, the ignition coil, the ignition distributor, the spark plugs, and the wires and cables that connect them. The distributor has a shaft that is driven by a gear on the camshaft (Fig. 20-6). The upper end of the distributor shaft has a cam with several lobes on it (same number as

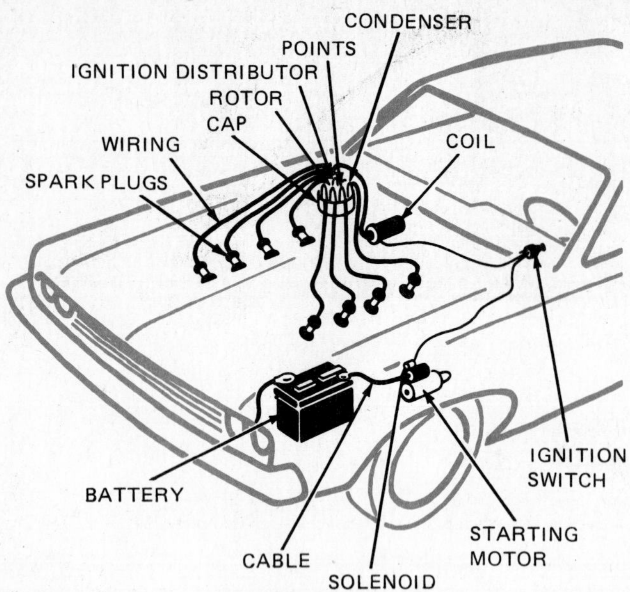

Fig. 27-1 Basic layout of the contact-point ignition system.

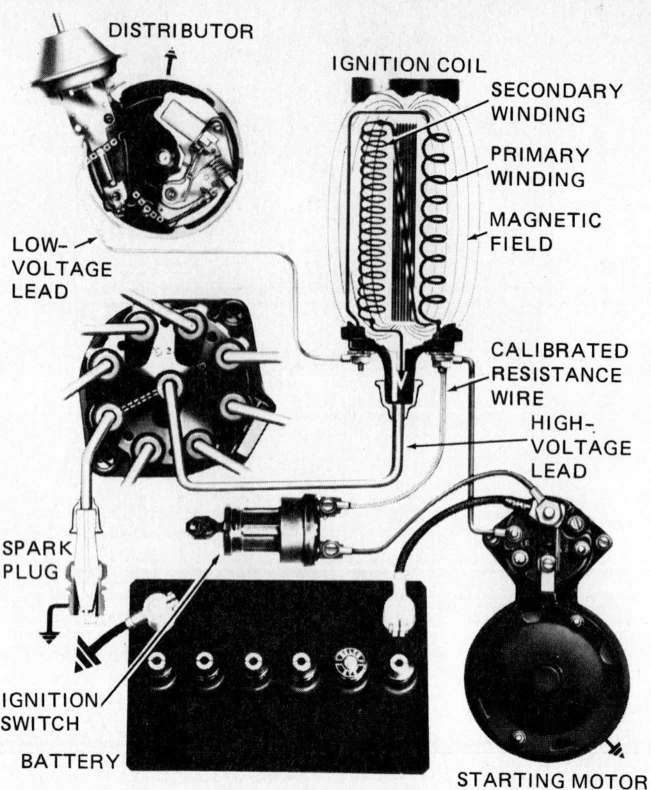

Fig. 27-2 The contact-point ignition system includes the battery, distributor (shown in top view with the cap removed and placed below it), ignition switch, ignition coil, spark plugs (one is shown), and wiring. *(Delco-Remy Division of General Motors Corporation)*

there are cylinders in the engine). As the distributor cam rotates, it causes contact points to open and close.

When the contact points are closed, they connect the primary winding of the ignition coil with the battery. Current flows through the coil, causing a magnetic field to build up around it. Then, when the points are opened, this disconnects the primary winding from the battery. The magnetic field collapses, creating a short pulse of high voltage in the secondary winding of the coil. The high-voltage surge flows through the distributor cap and rotor to the spark plug in the cylinder that is ready to fire. This is the spark plug in the cylinder in which the piston is nearing the end of the compression stroke. Following sections describe each component in the contact-point ignition system.

♦ 27-4 IGNITION DISTRIBUTOR

The contact-point distributor is two separate devices in one. It is a fast-acting mechanical switch and a high-voltage distribution system (Fig. 27-3). There are two separate circuits through the distributor. One is the primary circuit (Fig. 27-4). The other is the secondary circuit (Fig. 27-5).

The primary circuit (Fig. 27-4) includes the battery, the ignition switch, the primary resistance, the primary winding in the ignition coil, the contact points in the distributor, and the primary wiring.

The contact points include a stationary contact point which is connected to ground. This is the return circuit to the battery. The other contact point is mounted on an insulated movable arm. Each contact point consists of a round disk of metal (tungsten or similar material). The distributor cam rotates inside of the movable contact arm. The lobes on the cam move against the rubbing block of the arm. This pushes the arm out so that the contact point is moved away from the stationary point. The points separate. But as soon as the lobe passes out from under the rubbing block, the spring pushes the movable arm in and the points contact again.

There are the same number of lobes on the cam as there are cylinders in the engine. This causes the ignition system to send a high-voltage surge to each cylinder for every revolution of the distributor shaft and cam. The shaft and cam rotate at the same speed as the engine camshaft. This is half the speed of the crankshaft. Therefore, a spark is produced for each cylinder every two crankshaft rotations. It takes two crankshaft rotations to produce one compression and one power stroke (♦9-7 to 9-12).

♦ 27-5 IGNITION COIL

The ignition coil (Fig. 27-6) has two windings. One is the secondary winding, which consists of thousands of turns of a very fine wire. Outside of this winding is the primary winding. It consists of a few hundred turns of a relatively heavy wire. The primary winding is in the primary circuit (Fig. 27-4). The secondary winding is in the secondary circuit (Fig. 27-5).

When current flows through the primary winding, it produces a strong magnetic field. When the current flow is shut off, the magnetic field collapses. This produces the high-voltage surge in the secondary winding.

♦ 27-6 DISTRIBUTOR-CAP AND ROTOR ACTION

The distributor in the contact-point system is two devices in one. First, the contact points act as a switch. It connects and then disconnects the ignition-coil primary winding and the battery. The second device uses the cap and rotor to form a rotary switch. They act as a distributing system. The rotor (Fig. 27-7) has a metal blade. The inner end of the blade has a spring which contacts the center terminal of the cap. The

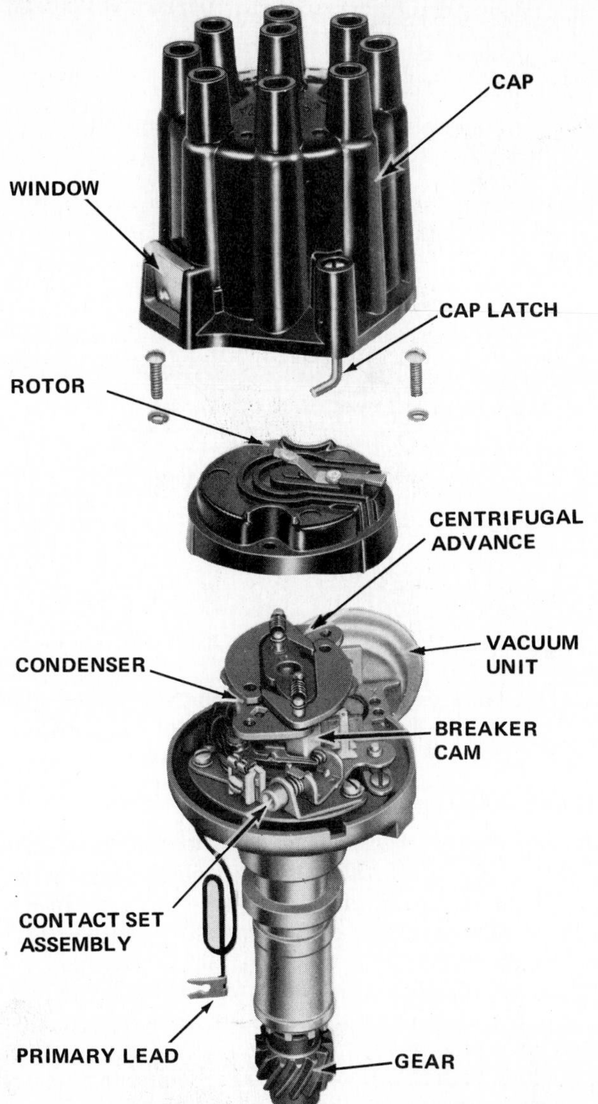

Fig. 27-3 Partly disassembled contact-point distributor. *(Delco-Remy Division of General Motors Corporation)*

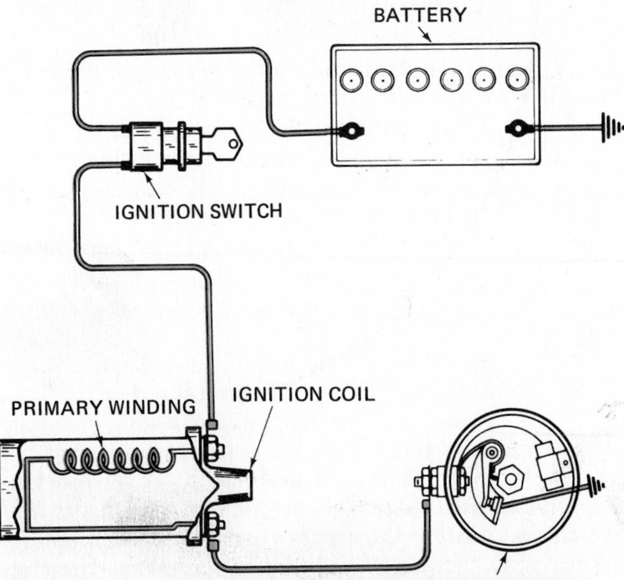

Fig. 27-4 The primary circuit of the contact-point ignition system.

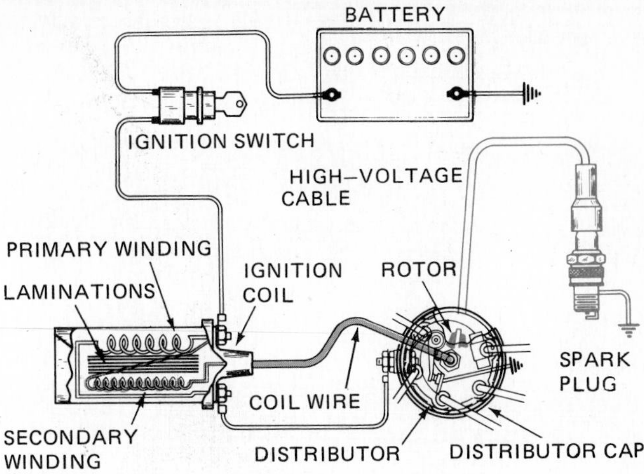

Fig. 27-5 Secondary circuit added to the primary circuit of a contact-point ignition system. Only one spark plug is shown.

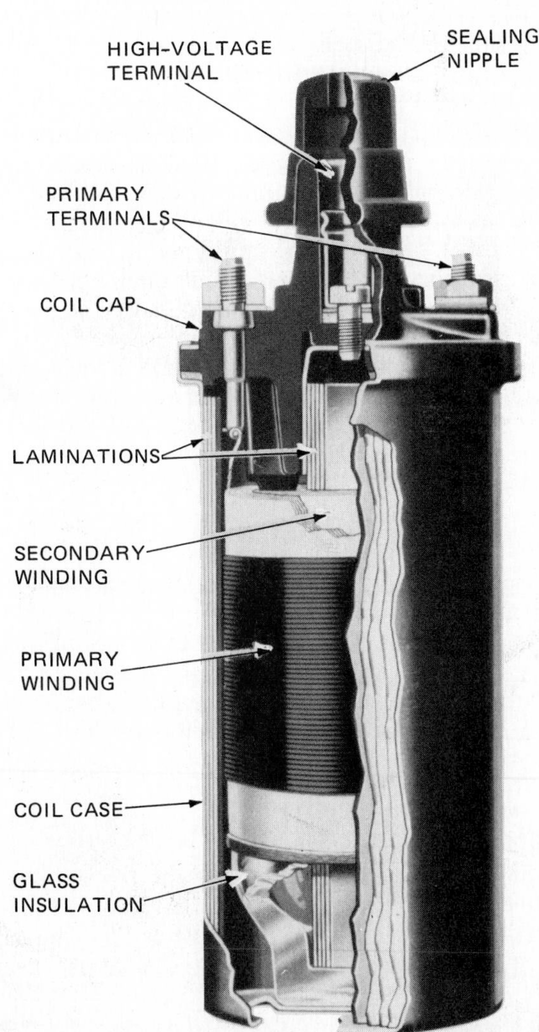

Fig. 27-6 Ignition coil with case partly cut away to show the windings. The primary winding is on the outside of the secondary windings. *(Delco-Remy Division of General Motors Corporation)*

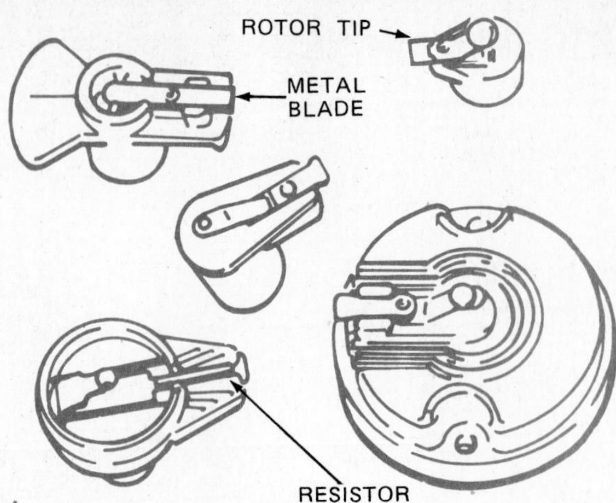

Fig. 27-7 Types of distributor rotors. The one at the lower left has a carbon resistor. The one at the right is attached to the advance mechanism with screws. *(Delco-Remy Division of General Motors Corporation)*

rotor sits on top of the distributor shaft. As the shaft rotates, the blade of the cap moves past the terminals, which are arranged in a circle around the cap. Each of these outside terminals is connected by a spark-plug cable to a spark plug.

When the coil secondary winding produces a high-voltage surge, it passes through a short coil cable to the center terminal of the cap (Fig. 27-5). Then the high-voltage surge goes through the rotor blade and jumps to the outer terminal toward which the rotor blade is pointed. From there, it passes through the outer terminal and through the spark-plug cable to the spark plug (Fig. 27-8). The plug produces a spark which ignites the compressed mixture in the cylinder.

The ignition system is timed so that the spark occurs in each cylinder just as the piston approaches TDC on the compression stroke. The distributor-cap terminals are connected to the spark plugs in the firing order of the engine. In an engine with a firing order of 1–2–4–3, the spark-plug wires would be connected to the cap terminals in that order. The rotor blade passes the cap terminals in that order as the rotor turns.

♦ 27-7 SECONDARY WIRING

The secondary wiring consists of the cables that connect the secondary winding of the coil to the cap center terminal (the *coil cable*), and the outer cap terminals to the spark plugs (the spark-plug cables). These cables have a center conductor of fiberglass impregnated with graphite (a conductor), or a special string coated with carbon (another conductor). They have an insulating jacket of silicone or other insulating material. This jacket must be soft enough for the cables to be bent as required when installed. However, they must be able to contain the high voltage that surges through them.

A high voltage is required because the gap between the electrodes in the spark plugs may be as wide as 0.080 inch [2 mm] in some electronic ignition systems. This wide gap is needed to ignite the lean mixtures on which some engines run.

Careful Silicone insulation is very soft. It is easily damaged if it is bent or if a heavy object falls on it. Any slight puncture or crack can allow high-voltage leakage to ground. Then the spark plug will not fire. A damaged spark-plug cable must be replaced.

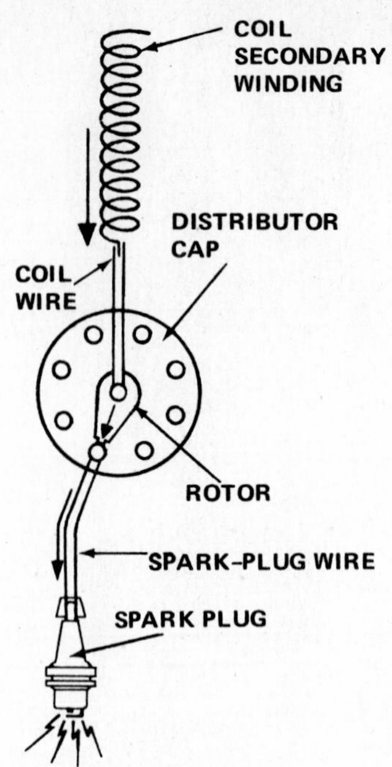

Fig. 27-8 Simplified secondary circuit. The coil secondary winding is connected through the distributor cap, rotor, and wiring to the spark plugs.

♦ 27-8 SPARK PLUGS

The spark plug is a metal shell in which a porcelain insulator is fastened (Fig. 27-9). An electrode extends through the center of the insulator. The metal shell has a short electrode attached to one side. This outer electrode is bent inward to produce the proper gap between it and the center electrode.

Some spark plugs have a built-in resistor which forms part of the center electrode (Fig. 27-9). The resistor reduces television and radio interference (static) from the high-voltage surges in the ignition secondary circuit. This interference is called *radio-frequency interference,* or RFI.

♦ 27-9 SPARK PLUG HEAT RANGE AND REACH

The heat range of the spark plug determines how hot the plug will get. This depends on the shape of the insulator in the spark plug. It also depends on how far the heat must travel from the center electrode to the much cooler cylinder head. Figure 27-10 shows that if the heat path is long, the plug will run hot. If the heat path is short, the plug will run much cooler. However, if the plug runs too cool, sooty carbon will deposit on the insulator around the center electrode. This could soon build up enough to short out the plug. Then the high-voltage surges would leak across the carbon instead of producing a spark across the spark-plug gap. Using a hotter plug will burn this carbon away, or prevent it from forming.

Spark-plug reach is the distance from the seat of the shell to its lower edge (Fig. 27-9). If the reach is too long (Fig. 27-10, left), the spark plug will protrude too far into the combustion chamber. This could interfere with the turbulence of the air-fuel mixture and reduce combustion action. Also, a piston could hit the end and damage the engine. However, the plug must reach into the combustion chamber far enough so that the spark gap will be properly positioned in the combustion chamber.

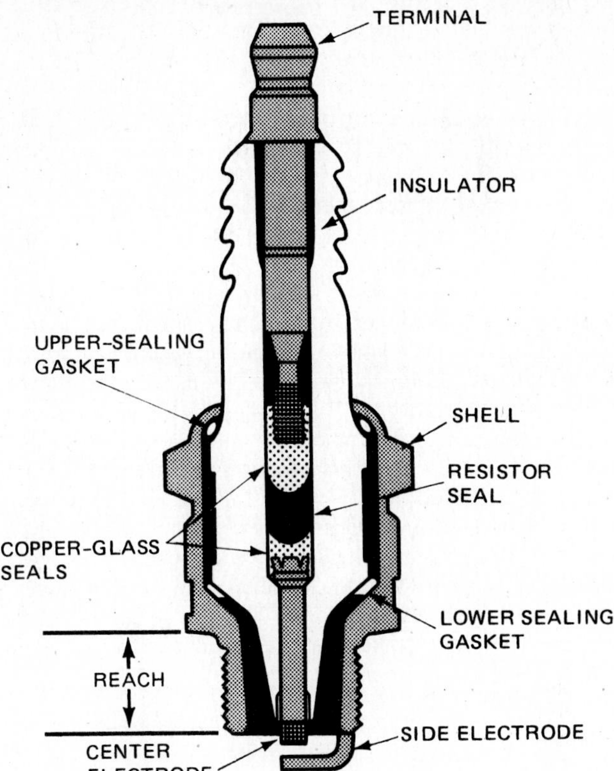

Fig. 27-9 Cutaway resistor-type spark plug. *(AC Spark Plug Division of General Motors Corporation)*

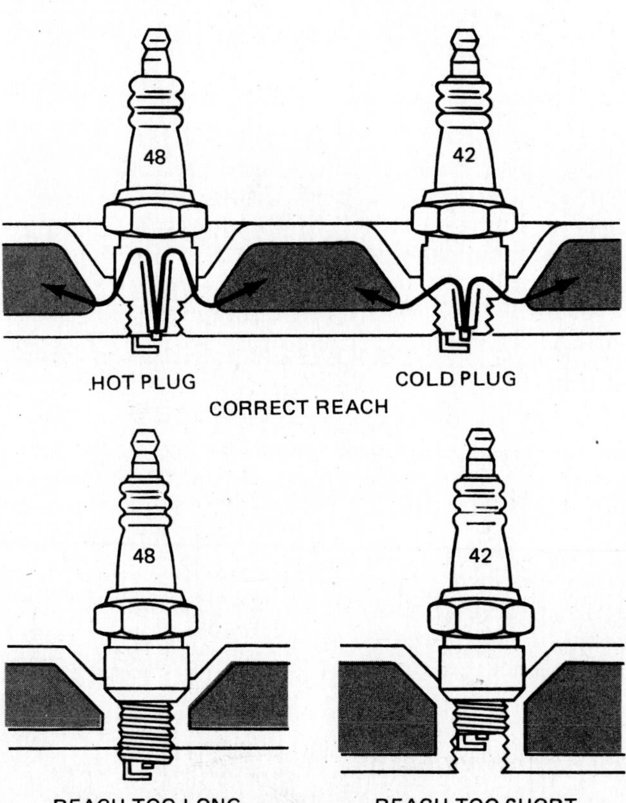

Fig. 27-10 Heat range and reach of spark plugs. Top, the longer the heat path (indicated by arrows), the hotter the plug runs. *(AC Spark Plug Division of General Motors Corporation)*

♦ 27-10 CENTRIFUGAL ADVANCE

As engine speed increases, the spark must occur in the combustion chamber earlier in the cycle. If the engine is idling, the spark must occur just before the piston reaches TDC on the compression stroke. At low speed, the spark has enough time to ignite the compressed air-fuel mixture. The pressure rise due to combustion approaches maximum as the piston passes TDC and starts down on the power stroke.

However, at higher speeds, the spark must occur earlier. The spark must be given enough time to ignite the mixture and initiate combustion. Without this spark advance, the piston would be up over TDC and moving down before the combustion pressures reached a maximum. As a result, the piston would keep ahead of the pressure rise. A weak power stroke would result.

If the spark occurs earlier, the mixture burns and the pressure increases to a maximum just as the piston moves through TDC. The higher the engine speed, the earlier the spark must occur. At higher speeds, everything is happening much faster.

To produce the advance based on engine speed, many distributors have a mechanical centrifugal-advance mechanism. Figure 27-11 shows one type. It has a pair of pivoted advance weights with weight springs. At low speeds, the springs hold the weights in. As speed increases, the centrifugal force on the weights moves them out against the spring tension. This movement causes the cam assembly to move ahead. Figure 27-11 shows how the weights wrap around the oval-shaped base of the cam assembly. With this design, the higher the engine speed, the faster the shaft turns, the farther out the advance weights move, and the farther ahead the cam assembly is moved forward, or advanced. The action of the centrifugal-advance mechanism is shown in Fig. 27-12.

The action of the centrifugal-advance mechanism causes the contact points to open and close earlier. Therefore, the ignition coil produces its high-voltage surges earlier and the sparks appear earlier in the combustion chambers.

Note Some electronic ignition systems do not have mechanical advance mechanisms, like the centrifugal-advance mechanism described above and the vacuum-advance mechanism discussed in the following section. Instead, the spark timing is varied electronically (Chap. 28).

♦ 27-11 VACUUM ADVANCE

When the engine is operating at part throttle, there is a vacuum in the intake manifold. The partly closed throttle valve prevents the maximum amount of air-fuel mixture from entering the intake manifold. So less air-fuel mixture enters the engine cylinders on the intake strokes. With less air-fuel mixture in the cylinders, the mixture is not compressed as much during the compression strokes. There is less mixture to compress.

With lower compression, the mixture does not burn as fast when ignited. Unless there is some additional spark advance when operating at part throttle, full power from the mixture will not be realized. The piston will be up over TDC and moving down before the combustion pressure reaches maximum. As a result, the piston would keep ahead of the pressure rise and the power stroke would be weak.

To produce an advance based on part-throttle operation,

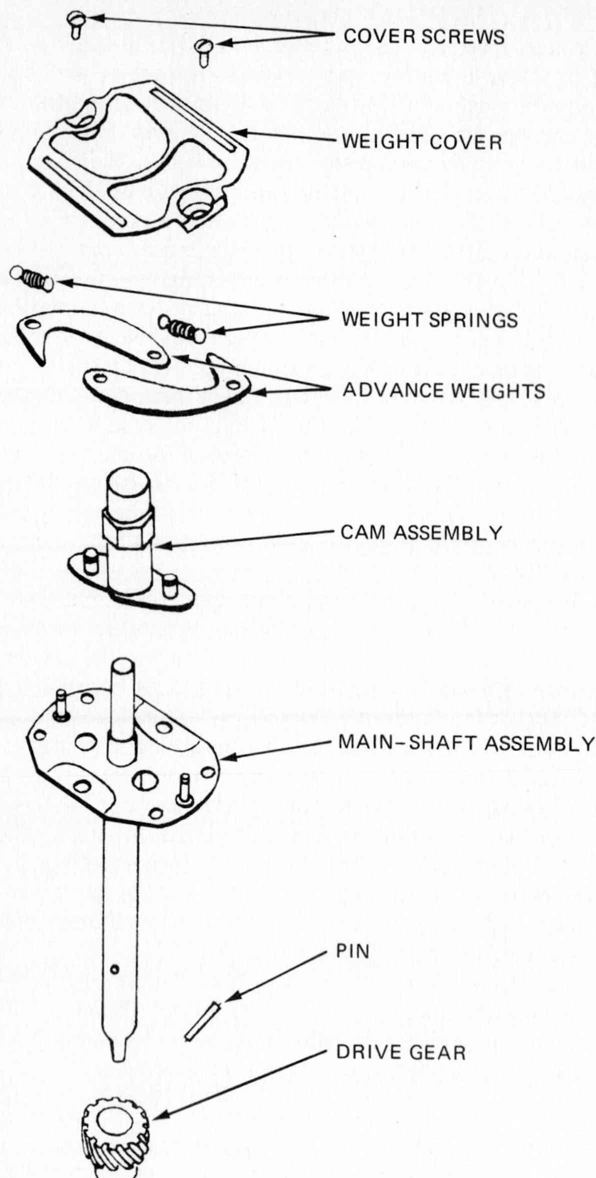

Fig. 27-11 Parts in a centrifugal-advance mechanism.

intake-manifold vacuum is used. Figure 27-13 shows a typical mechanism to produce this action. It has a flexible diaphragm that is spring loaded and linked to a movable breaker plate in the distributor. This is the plate on which the contact points are mounted. A vacuum passage connects the diaphragm chamber to the carburetor, as shown in Fig. 27-14. When the throttle valve is closed, there is no vacuum applied to the vacuum passage. The passage, or port, is above the closed throttle valve, and therefore above intake-manifold vacuum.

When the throttle valve is partly opened (Fig. 27-15), it moves past the opening to the vacuum passage. Now, intake-manifold vacuum is applied, through the vacuum passage, to the diaphragm. The vacuum pulls the diaphragm outward against the spring force. This rotates the breaker plate, as shown in Fig. 27-15. The points are moved ahead so that they open and close earlier. This produces the vacuum advance required to get the most power out of the air-fuel mixture.

The vacuum advance does not produce any advance at full throttle. When the throttle valve is wide open, intake-manifold vacuum is almost zero. Therefore, the vacuum-advance mechanism produces no advance. The spring pushes the diaphragm in so the breaker plate takes the position shown in Fig. 27-14.

♦ 27-12 COMBINED CENTRIFUGAL AND VACUUM ADVANCE

At any engine speed much above idle, there is some spark advance due to the centrifugal-advance action. Above idle and below wide-open throttle, there could be an additional amount of vacuum advance. However, the amount depends on how much (if any) vacuum there is in the intake manifold.

The total advance curve in Fig. 27-16 shows how the centrifugal advance and vacuum advance combine. At 40 mph [64 km/h], the centrifugal-advance mechanism is providing an advance of 15 degrees. With the engine running on part throttle, the vacuum advance supplies an additional 15 degrees. The advances combine to provide the engine with a total spark advance of 30 (15 + 15) degrees. Therefore, the spark is occurring at the spark plug 30 degrees before the piston reaches TDC on the compression stroke.

Figure 27-16 is only typical. The curves would be different for different engines operating under different conditions.

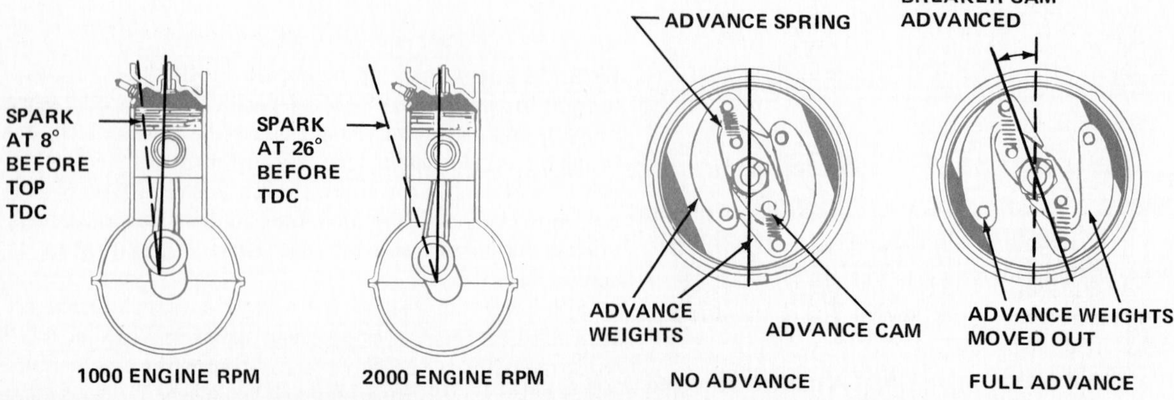

Fig. 27-12 Centrifugal-advance mechanism in no-advance and full-advance positions. In the typical example shown, the ignition is timed at 8 degrees before TDC on idle. There is no centrifugal advance at 1000 engine rpm. There is 26 degrees total advance (18 degrees centrifugal plus 8 degrees due to original timing) at 2000 engine rpm. *(Delco-Remy Division of General Motors Corporation)*

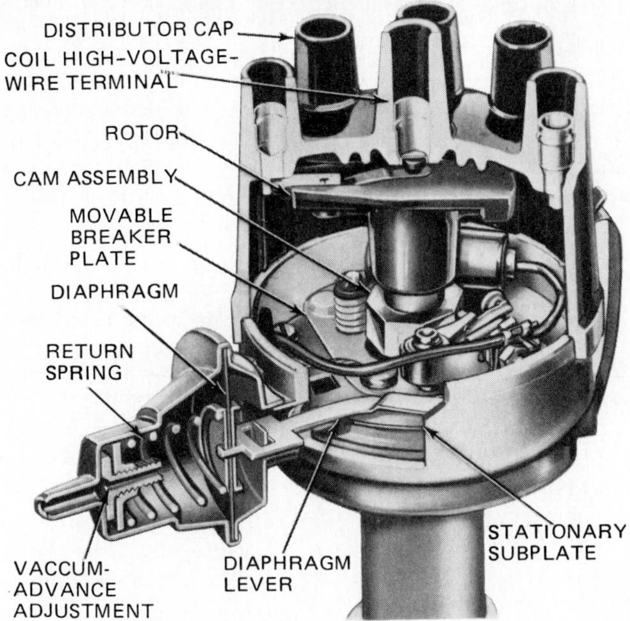

Fig. 27-13 A cutaway vacuum-advance mechanism. The diaphragm is linked to the movable breaker plate. *(Ford Motor Company)*

♦ 27-13 EMISSION CONTROL OF VACUUM ADVANCE

Some engines have an arrangement that prevents vacuum advance in the lower gears. Figure 27-17 shows one such system. When the transmission is in low gear, the transmission switch is open. This means the vacuum-advance solenoid is not connected to the battery. The solenoid plunger is blocking the vacuum line to the distributor vacuum-advance mechanism.

This prevents vacuum advance. *With* vacuum advance, the air-fuel mixture has more time to burn. This increases the time that one of the pollutants from the engine (NO_x) can form. *Without* vacuum advance in the lower gears, the NO_x formation time is shortened.

When the transmission is shifted to high, the transmission switch closes and connects the solenoid to the battery. Now, the plunger is pulled in. This opens the vacuum line so that normal vacuum advance can take place.

General Motors calls this system a transmission-controlled spark (TCS) system. Ford has a similar system called the transmission-regulated spark (TRS) system. Both are described later.

♦ 27-14 IGNITION SWITCH

The ignition switch, mounted on the steering column on many cars, does two jobs. It connects the ignition system to the battery when it is turned to ON. When it is turned past ON to the START position, it connects the starting motor to the battery so that the engine is cranked for starting.

The second job is to lock the steering wheel and the transmission selector lever when the ignition switch is turned to LOCK. Figure 27-18 shows the steering column and the locking mechanism. When the ignition switch is turned from LOCK to OFF, the sector rotates. This pulls the plunger away from the notched disk so that the steering shaft, and wheel, can be turned. When the ignition switch is turned to LOCK, the sector shaft moves the plunger toward the notched disk. If the plunger aligns with a notch, the plunger moves up (as shown in Fig. 27-18) and locks the notched disk and steering wheel. If the plunger does not index with a notch, the plunger is spring-loaded against the disk. Then, when the wheel and disk are turned slightly, the plunger will move up into the notch. This locks the steering wheel.

The ignition switch cannot be turned to LOCK if the transmission selector lever is in any gear position. When the ignition switch is turned to LOCK, the steering wheel is locked. The transmission selector lever is also locked in PARK. This acts as an antitheft deterrent. Also, it prevents starting with the transmission in gear. The ignition switch cannot be turned to START with the shift lever in any gear position.

CONTACT-POINT IGNITION-SYSTEM SERVICE

♦ 27-15 CAUSES OF IGNITION FAILURE

The following sections describe the servicing of contact-point ignition systems. Chapter 28 describes electronic ignition systems. Servicing electronic ignition systems is covered in Chap. 29.

Ignition-system failures in contact-point systems can be grouped into three categories. These are:

1. Loss of energy in the primary circuit. This may be caused by several conditions.
   a. Resistance in the primary circuit due to defective leads, bad connections, burned distributor contact points or switch, or open coil primary winding

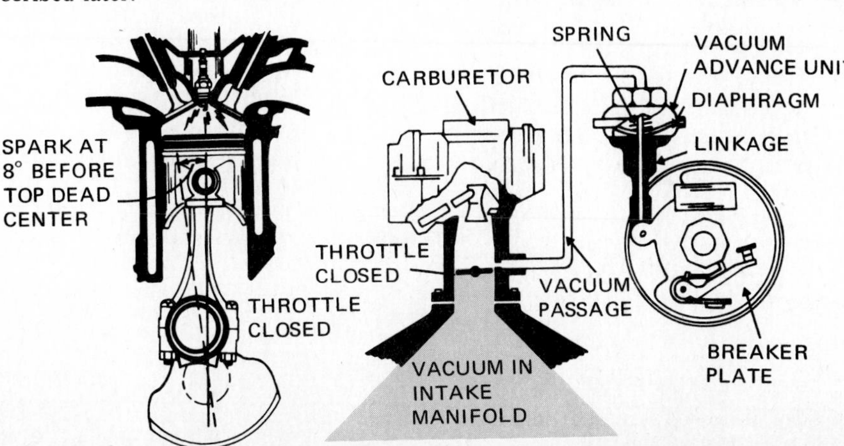

Fig. 27-14 When the throttle valve is closed, there is no vacuum advance.

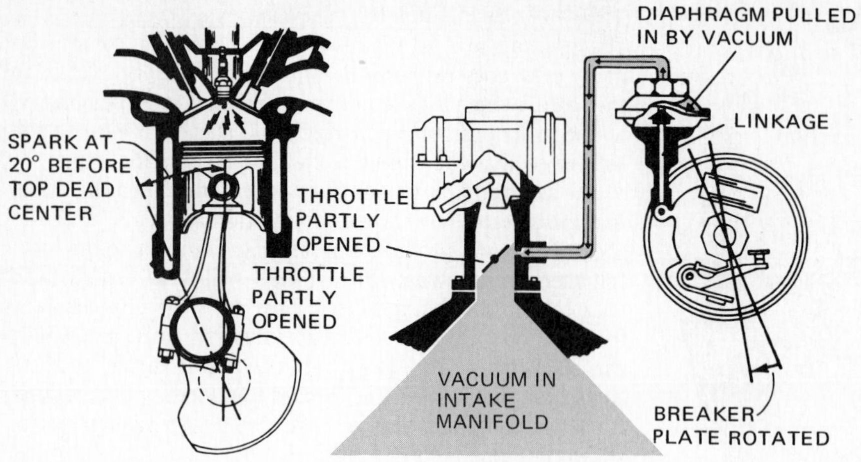

Fig. 27-15 Operation of the vacuum-advance mechanism. When the throttle valve swings past the port, a vacuum is admitted to the vacuum-advance mechanism on the distributor. The breaker plate, with the points, is rotated to advance the spark.

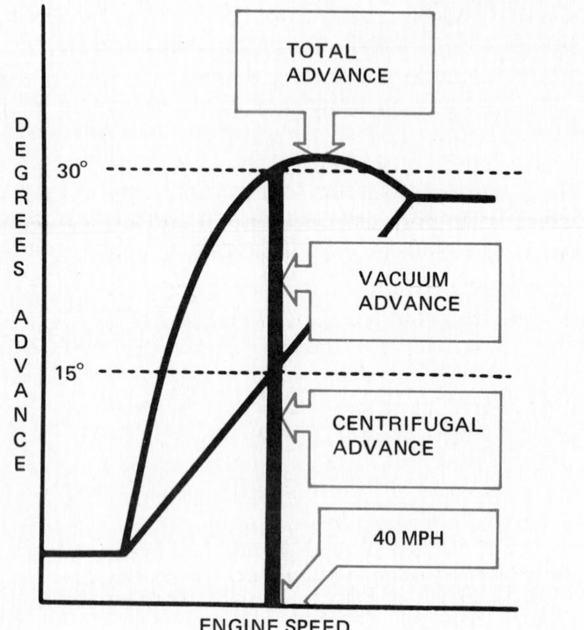

Fig. 27-16 Centrifugal- and vacuum-advance curves in one ignition system.

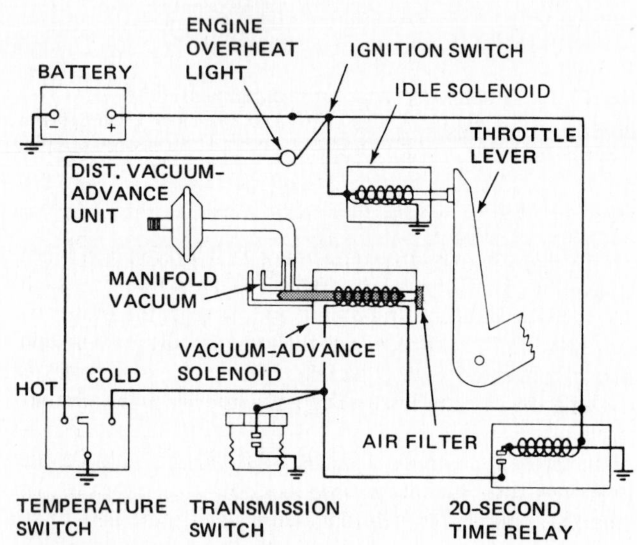

Fig. 27-17 Transmission-controlled spark (TCS) system during low-gear operation. *(Chevrolet Motor Division of General Motors Corporation)*

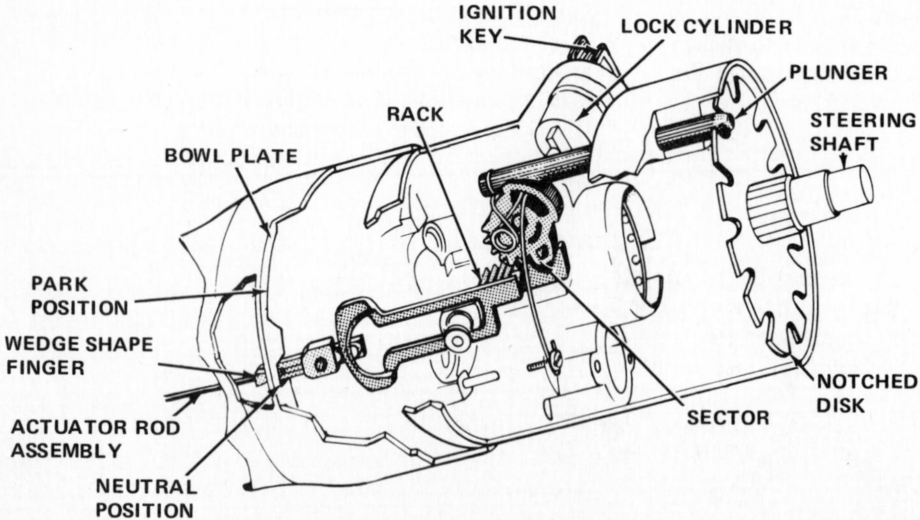

Fig. 27-18 Combination ignition-switch and steering-wheel lock, showing the operating mechanism inside the steering column. *(General Motors Corporation)*

b. Points not properly set
c. Discharged battery or defective alternator
d. Defective condenser (shorted, low insulation resistance, high series resistance)
e. Grounded primary circuit in coil, wiring, or distributor

2. Loss of energy in the secondary circuit.
a. Spark plugs fouled, broken, or improperly gapped
b. Defective high-voltage wiring, which allows high-voltage leaks
c. High-voltage leakage across coil head, distributor cap, or rotor (Figs. 27-19 and 27-20)
d. Defective connections in high-voltage circuits
e. Defective ignition coil

3. Out of time.
a. Timing not set properly
b. Distributor bearing or shaft worn, or shaft bent
c. Vacuum advance defective
d. Centrifugal advance defective
e. Preignition, due to plugs of wrong heat range, fouled plugs, etc. (Fig. 27-21)

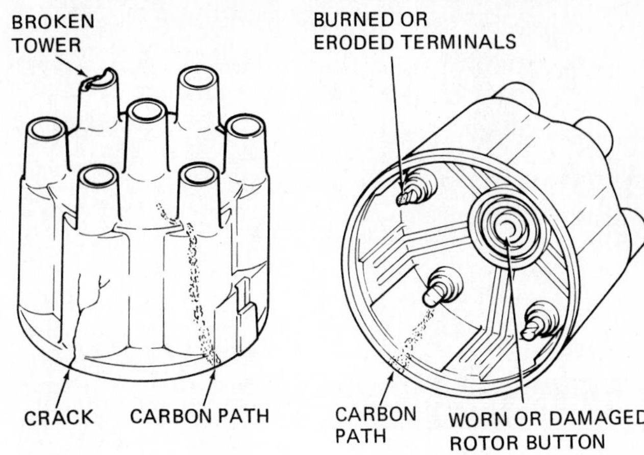

Fig. 27-19 Defects in distributor caps that require them to be discarded. Carbon paths allow voltage leakage to ground, so the spark plugs do not fire. *(American Motors Corporation)*

♦ 27-16 QUICK CHECKS OF IGNITION SYSTEM

Several quick checks can be made to determine whether the ignition system is at fault if the engine does not operate normally. The following applies to the contact-point ignition system. Some of the items listed may apply to electronic ignition systems. However, some electronic systems must be checked by following the procedure in the manufacturer's service manual (Chap. 29).

1. Engine Does Not Run If the engine cranks at normal speed but does not start, the trouble could be in either the ignition or the fuel system. To check the ignition system, disconnect the lead from one spark plug (or from the center distributor-cap terminal). Use insulated pliers to hold it about ¼ inch [6 mm] from the engine block. Crank the engine. If a good spark occurs, the ignition system probably is in running condition (although the timing could be off). If no spark occurs, check the ignition system further.

Connect a test ammeter into the ignition-coil primary circuit. Watch the ammeter while cranking the engine. If there is a small, steady reading that fluctuates slightly, the primary circuit is probably all right. The trouble is probably a defective coil secondary winding or secondary leads, a defective condenser, or high-voltage leakage across the cap, rotor, or coil head.

If the ammeter shows a fairly high and steady reading:

a. The contact points are out of adjustment.
b. The condenser is shorted.
c. The coil primary circuit is grounded.

If there is no ammeter reading, the primary circuit is open. This could be due to out-of-adjustment contact points, a loose connection, defective wiring or switch, or an open coil primary winding.

2. Engine Misses Missing is caused by such defects in the ignition system as:

a. Worn or out-of-adjustment contact points
b. Defective condenser
c. Centrifugal- or vacuum-advance malfunctioning
d. Defective secondary wiring
e. Defective ignition coil
f. Poor connections
g. High-voltage leakage across ignition-coil head, rotor, or cap (Figs. 27-19 and 27-20).
h. Defective spark plugs (Fig. 27-21). If no change in engine speed occurs when a spark plug is shorted out, then that cylinder is not firing.

The wrong ignition coil for the engine, or reversed connections to the ignition coil, may cause missing. Installing a battery backwards can also cause missing. This reverses the polarity of the coil.

Careful If a battery is installed backwards, the charging system and other electronic devices on the car may be damaged.

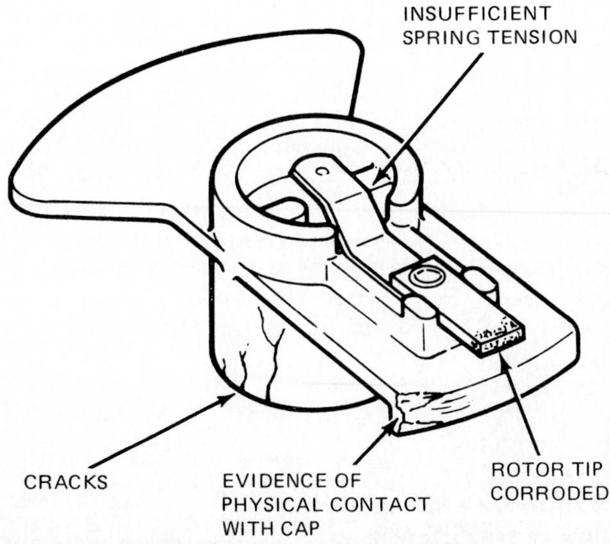

Fig. 27-20 Rotor defects that require the rotor to be discarded. *(American Motors Corporation)*

NORMAL

Brown to grayish tan color and slight electrode wear. Correct heat range for engine and operating conditions.

RECOMMENDATION: Service and reinstall. Replace if over recommended mileage.

WORN

Center electrode worn away too much to be filed flat.

RECOMMENDATION: Replace with new spark plugs of proper heat range.

PREIGNITION

Improper heat range, incorrect ignition timing, lean air-fuel mixture, or hot spots in combustion chamber has caused melted electrodes. Center electrode generally melts first and ground electrode follows. Normally, insulators are white, but may be dirty due to misfiring or flying debris in combustion chamber.

RECOMMENDATION: Check for correct plug heat range, overadvanced ignition timing, lean air-fuel mixture, clogged cooling system, leaking intake manifold, and lack of lubrication.

DETONATION

Insulator has cracked and broken away as a result of the shock waves created by detonation.

RECOMMENDATION: Cause of detonation must be found and corrected. Check for use of low-octane fuel, improper air-fuel mixture, incorrect ignition timing, overheating, and increased octane requirement in the engine.

MECHANICAL DAMAGE

Something such as a foreign object in cylinder, or the piston, has struck the ground electrode. It has been forced into the center electrode, which has bent, and broken off the insulator.

RECOMMENDATION: Check that plug has the correct reach for the engine, that the gap was set properly, and that no foreign object remains in cylinder.

GAP BRIDGED

Deposits in the fuel have formed a bridge between the electrodes, eliminating the air gap and grounding the plug.

RECOMMENDATION: Sometimes plug can be serviced and reinstalled. Check for excess additives in fuel.

CARBON DEPOSITS

Dry soot, frequently caused by use of spark plug with incorrect heat range.

RECOMMENDATION: Carbon deposits indicate rich mixture or weak ignition. Check for clogged air cleaner, high float level, sticky choke or worn contact points. Hotter plugs will provide additional fouling protection.

OIL DEPOSITS

Oily coating.

RECOMMENDATION: Caused by poor oil control. Oil is leaking past worn valve guides or piston rings into the combustion chamber. Hotter spark plug may temporarily relieve problem, but correct the cause with necessary repairs.

SPLASHED DEPOSITS

Spotted deposits. Occurs shortly after long-delayed tuneup. After a long period of misfiring, deposits may be loosened when normal combustion temperatures are restored by tuneup. During a high-speed run, these materials shed off the piston and head and are thrown against the hot insulator.

RECOMMENDATION: Clean and service the plugs properly and reinstall.

ASH DEPOSITS

Poor oil control, use of improper oil, or use of improper additives in fuel or oil has caused an accumulation of ash which completely covers the electrodes.

RECOMMENDATION: Eliminate source of ash. Install new spark plugs.

OVERHEATED

Blistered, white insulator, eroded electrodes and absence of deposits.

RECOMMENDATION: Check for correct plug heat range, overadvanced ignition timing, low coolant level or restricted flow, lean air-fuel mixture, leaking intake manifold, sticking valves, and if car is driven at high speeds most of the time.

HIGH-TEMPERATURE GLAZING

Insulator has yellowish, varnish-like color. Indicates combustion chamber temperatures have risen suddenly during hard, fast acceleration. Normal deposits do not get a chance to blow off. Instead, they melt to form a conductive coating.

RECOMMENDATION: If condition recurs, use plug type one step colder.

Fig. 27-21 Appearance of spark plugs related to causes. *(Champion Spark Plug Company)*

With reversed coil polarity, the electrons have to jump from the relatively cool, outer spark-plug electrode to the hotter center electrode. This requires a higher secondary voltage. It increases the possibility of the engine missing, especially at high speed. Normally, the coil and battery are connected so that electrons jump from the hot center electrode to the outer electrode. With the emitting electrode hot, the electrons can jump the spark-plug gap more easily. Voltage requirements are considerably lower. Coil terminals are usually marked to prevent incorrect connections.

Reversed polarity can be checked by looking at the coil primary connections. The negative terminal should be connected to the wire to the distributor. If the polarity is reversed, the oscilloscope pattern will be upside-down (♦27-18).

3. Overheating This condition may be caused by improper ignition timing. Retarded, or late, ignition timing sends more heat through the cylinder walls into the coolant. This occurs because with later timing the air-fuel mixture is burning later in the power stroke.

♦ 27-17 CONTACT-POINT IGNITION-SYSTEM TESTERS

Various testers can be used to check condensers and ignition coils. The results of testing ignition coils with different defects on an oscilloscope are shown in Fig. 27-22.

Distributor testers are used to check distributors that have been removed from the engine. The tester drives the distributor at various speeds. A vacuum pump in the tester can be used to check the operation of the vacuum-advance unit. In the shop, you will be instructed on how to use the distributor tester.

The contact-point opening decreases gradually due to wear on the contact-arm rubbing block. The opening should be checked periodically and adjusted if necessary. There are two ways to do this, with a thickness gauge or with a dwell meter. Dwell is the number of degrees of cam rotation that the points are closed (Fig. 27-23). As the point opening is increased, the dwell angle decreases. Contact points are adjusted by loosening a hold down screw and moving the stationary point. The adjustment varies with different distributor designs.

♦ 27-18 OSCILLOSCOPE TESTERS

The oscilloscope (also called the "scope") is a high-speed voltmeter that uses a televisionlike picture tube to show ignition voltages. Figure 27-24 shows an engine performance tester which includes an oscilloscope. It draws a picture of the ignition voltages on the face of the tube. Their "pattern" shows the voltage changes in either the primary or the secondary circuit of the ignition system.

The basic pattern for one spark plug is shown in Fig. 27-25. This is the pattern that the scope would show if it were drawing the voltage changes in the secondary circuit required to fire one spark plug. To start with (in Fig. 27-25), the contact points separate and the high-voltage surge reaches the spark plug. The voltage goes up (A to B), reaching several thousand volts. The height of the firing line shows the voltage required to start the spark jumping across the spark plug gap. Once the spark has started, less voltage is needed to sustain it. Therefore, the voltage drops (from B to C). The spark line from C to D indicates how long the spark is occurring. This is a very short time, measured in milliseconds (thousandths of a second). But the spark lasts for about 20 degrees of crankshaft rotation. This is long enough to ignite the compressed air-fuel mixture in the engine cylinder.

After most of the electromagnetic energy stored in the ignition coil (the energy in the magnetic field) has been used up, the spark dies out (D to E in Fig. 27-25). The remaining energy causes some diminishing oscillations (ripples) of voltage in the coil-condenser section. Then the contact points close (at E). From E to F, current flows through the coil primary winding. The magnetic field builds up again, loading the coil with electromagnetic energy. This is the dwell section (E to F). At F, the points open, and the whole cycle occurs again (from A to F).

♦ 27-19 OSCILLOSCOPE PATTERNS

The oscilloscope is connected to check the voltage variations in the ignition system as each spark plug fires. These individual patterns can be shown on the scope face in any of three ways, parade or display, stacked or raster, and superimposed (Figs. 27-26, 27-27, and 27-28). Figure 27-29 shows how the scope leads are attached or clamped on to make a test. Figure 27-30 shows various abnormal traces and their causes.

♦ 27-20 IGNITION TIMING

The ignition system must be timed so that the sparks jump across the spark-plug gaps at exactly the right time. Adjusting the distributor on the engine so that the spark occurs at this correct time is called *setting the ignition timing*. The ignition timing is normally set at idle, or at a speed specified by the engine manufacturer. On engines with idle-speed solenoids, the solenoid must be energized. Also, the vacuum hose from the intake manifold should be disconnected so that the vacuum advance does not work. Plug the hose with a golf tee or similar plug. Adjustment is made by turning the distributor in its mounting.

Note The contact-point opening (dwell ♦27-17) should be adjusted before the ignition is timed.

Turning the distributor in the direction opposite to distributor shaft rotation advances the spark. Turning the distributor in the direction of shaft rotation retards the spark.

To time the ignition, use a timing light. Connect the timing-light lead to the number 1 spark plug by clamping a pickup around the number 1 spark-plug cable. Many timing lights use a clamp similar to the ones shown in Fig. 27-29. Each time number 1 plug fires, the timing light flashes. This happens so quickly that when the light is pointed at the crankshaft pulley, the pulley appears to stand still (Fig. 27-31). Note the markings on the timing indicator and the timing mark on the pulley.

To make the adjustment, loosen the clamp screw that holds the distributor in its mounting. Turn the distributor one way or the other until the timing mark aligns with the proper mark on the timing indicator. Then tighten the clamp screw.

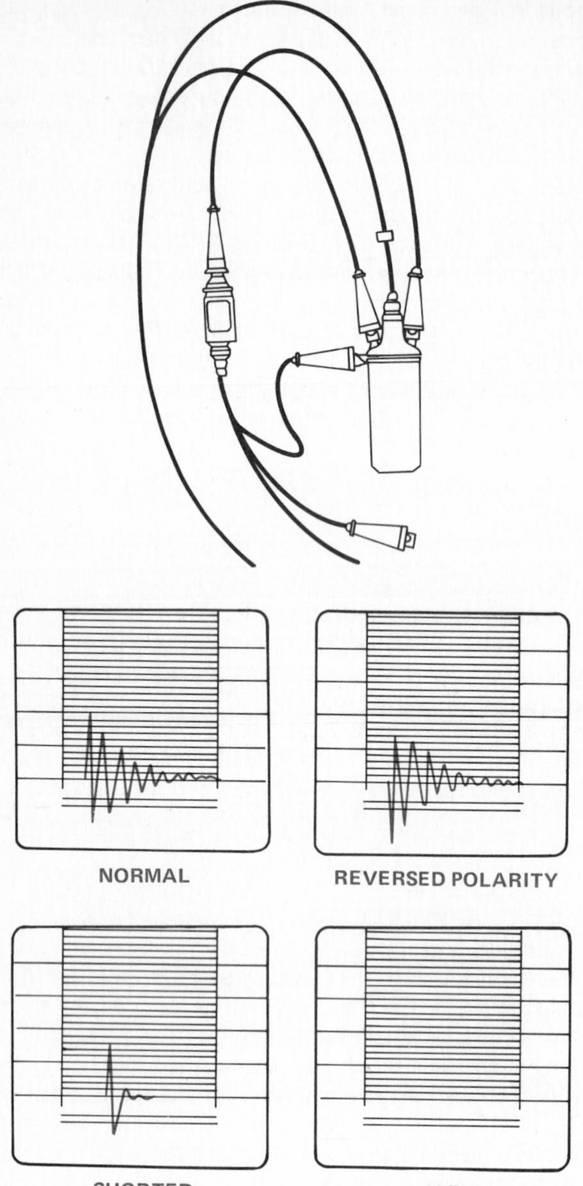

Fig. 27-22 Top, leads from scope to test coil. Bottom, scope patterns for various coil conditions. *(Sun Electric Corporation)*

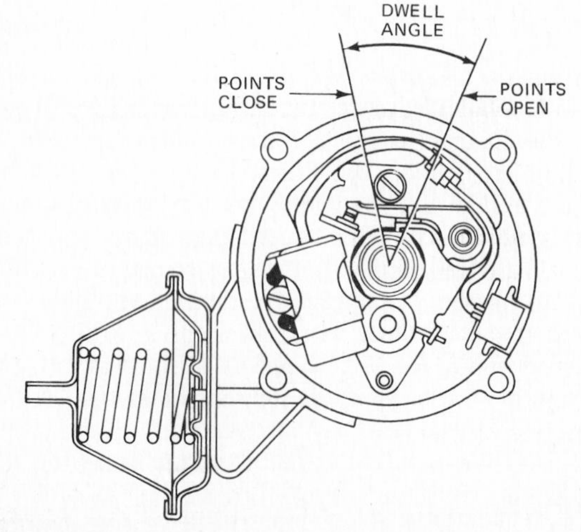

Fig. 27-23 Dwell angle.

♦ **CAUTION** ♦ Keep your hands and the timing-light leads away from the engine fan and drive belts. You could be seriously injured if your hand got in the way of the fan or a belt. Never stand in line with the engine fan. There have been serious injuries and deaths caused by fan blades breaking off and hitting someone standing nearby.

♦ 27-21 DIAGNOSTIC TIMING COMPUTER

Figure 27-32 shows a computerized timing light that not only does the timing job, but also gives a readout of the engine rpm and the dwell. It has a series of buttons plus a small viewing screen. This allows you to check engine rpm and dwell while you are timing the ignition.

♦ 27-22 SPARK-PLUG SERVICE

Spark plugs will foul or wear rapidly if they are not of the correct heat range (♦27-9). Figure 27-21 shows various spark-plug conditions and their causes. There are spark-plug cleaners that sandblast the electrodes and porcelain tip. Then, after filing the electrodes flat, the gap can be reset if the plug is in otherwise good condition. However, with the cost of labor so high, many mechanics recommend replacing spark plugs at periodic intervals, instead of cleaning and regapping them. Always select and install the correct spark plugs for the engine. First, inspect and gap the plugs to the manufacturer's specifications.

♦ 27-23 IGNITION WIRING

The insulation is soft and can be damaged by bending it sharply. Never puncture cables or boots with test probes. This can ruin the cable by allowing high-voltage leakage. Then the spark plug will not fire.

Figure 27-33 shows the wrong and right ways of disconnecting a cable from a spark-plug. If you pull on the cable, the conductor inside the cable may break. This ruins the cable. Figure 27-34 shows the use of insulated pliers to remove a cable from a spark plug.

Figure 27-35 shows how to reconnect cables to the distributor cap and ignition-coil tower. Grasp the boot and clip end of the cable. Push the cable clip into the cap or coil tower. Pinch the larger diameter of the boot to release trapped air. Then push the cable and boot until the cable clip is all the way in.

If you are replacing a set of cables or a distributor cap, replace one cable at a time (Fig. 27-36). This avoids getting them mixed. If all cables have been removed, first determine which direction the rotor turns and the firing order. From these, you can determine how the cables are to be connected.

♦ 27-24 CONTACT-POINT SERVICE

Burned or worn contact points should be replaced. They are supplied in sets; one stationary point, one movable point on the contact arm. Some sets are supplied assembled, with an attached condenser. Figure 27-37 shows a contact set being installed. Adjustment is made by turning an eccentric or by moving the stationary contact-point support.

♦ 27-25 DISTRIBUTOR SERVICE

The contact-point distributor requires periodic checking of the contact points, and their replacement and adjustment as necessary. Also, the centrifugal and vacuum advances should be checked to make sure they are working properly. These

Fig. 27-24 An engine performance tester that includes an oscilloscope. *(Sun Electric Corporation)*

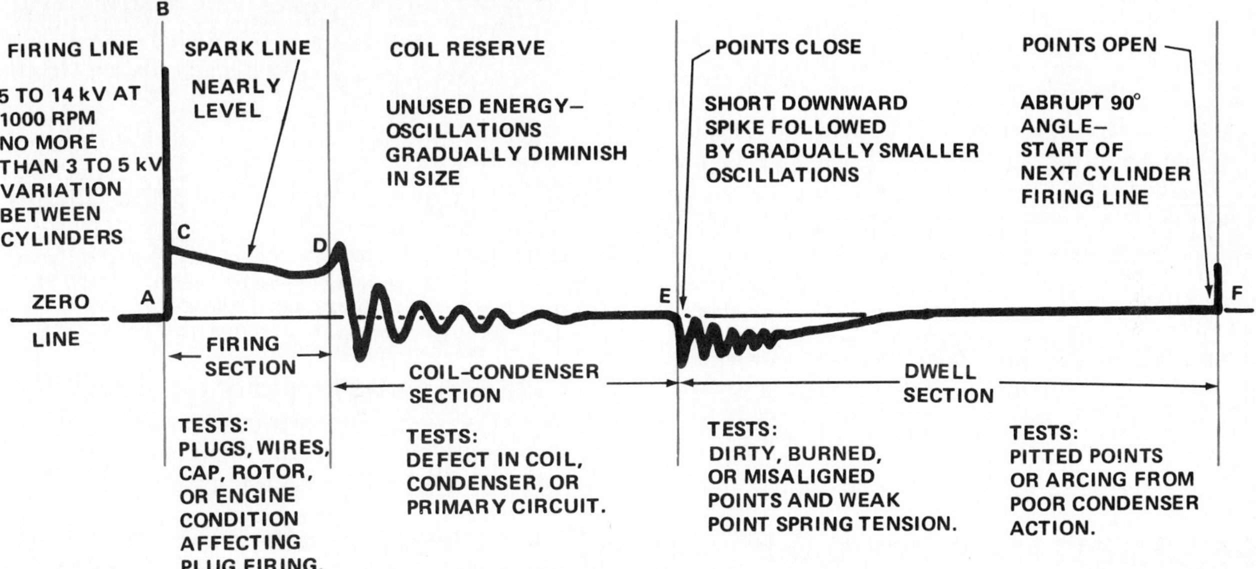

Fig. 27-25 A waveform, or trace, showing one complete spark-plug firing cycle. The "dwell section" is the time during which the points are closed. In electronic ignition systems, dwell is the time that the ECU allows current to flow through the primary winding of the coil. *(Sun Electric Corporation)*

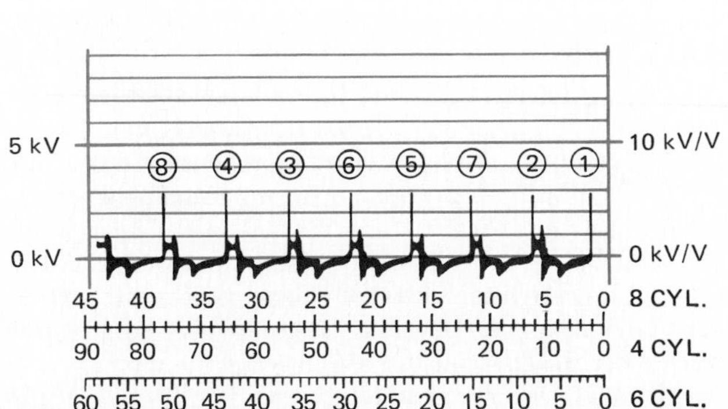

Fig. 27-26 A parade, or display, pattern of the ignition secondary voltages in an eight-cylinder engine. The numbers above the firing spikes show the order in which the cylinders are firing (engine firing order). *(Sun Electric Corporation)*

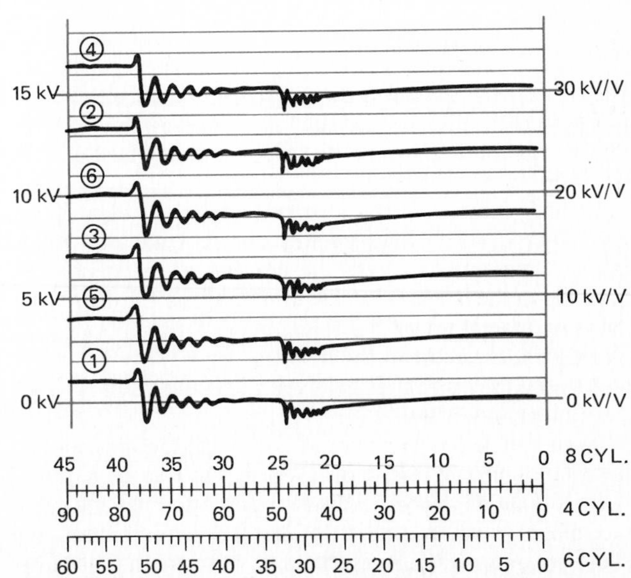

Fig. 27-27 A stacked, or raster, pattern of the ignition secondary voltages in a six-cylinder engine. The pattern is read from the bottom up, in the firing order. *(Sun Electric Corporation)*

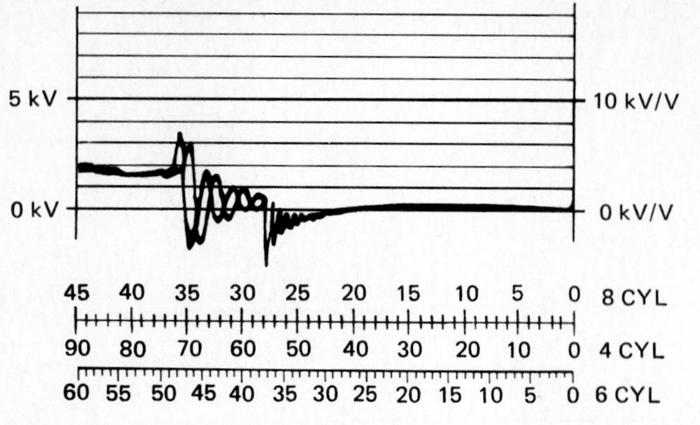

Fig. 27-28 Superimposed pattern of the ignition secondary voltages in a six-cylinder engine. *(Sun Electric Corporation)*

checks require removal of the distributor for testing in a distributor tester. The tester drives the distributor at various speeds and vacuums to check the operation of the centrifugal and vacuum advances. If trouble is found, the distributor may require complete servicing or replacement.

Some distributors have provisions for adjusting the centrifugal and vacuum advances.

Note The *Workbook for Automotive Mechanics* has step-by-step service procedures for servicing distributors, replacing and adjusting contact points, and checking the ignition system with an oscilloscope.

♦ 27-26 DISTRIBUTOR REMOVAL AND INSTALLATION

Distributor removal and installation is a simple job if the engine is left undisturbed while the distributor is out. However, if the engine is cranked so that crankshaft and camshaft are turned with the distributor out, then the installation requires more steps.

1. Distributor Removal Remove the air cleaner, and disconnect the vacuum hose or hoses from the distributor. Disconnect the primary lead running from the ignition coil to the distributor. Remove the distributor cap, and push the cap and wire assembly aside.

Scratch a mark on the distributor housing. Scratch another mark, which lines up with the first, on the engine block. These marks locate the position of the distributor housing in the block. Scratch a third mark on the distributor housing exactly under the rotor tip. This mark locates the position of the rotor in the housing.

Remove the distributor hold-down bolt and clamp. Lift the distributor out of the block.

If the engine is not cranked while the distributor is out, the distributor can be easily installed in the correct position. Simply align the marks on the distributor housing and cylinder block. As you push the distributor down into place, the shaft and rotor will turn as the spiral gears mesh. Therefore, start with the rotor turned back from the installed position so that when you install the distributor, the rotor will turn into the proper position. Then the tip will line up with the mark on the distributor housing.

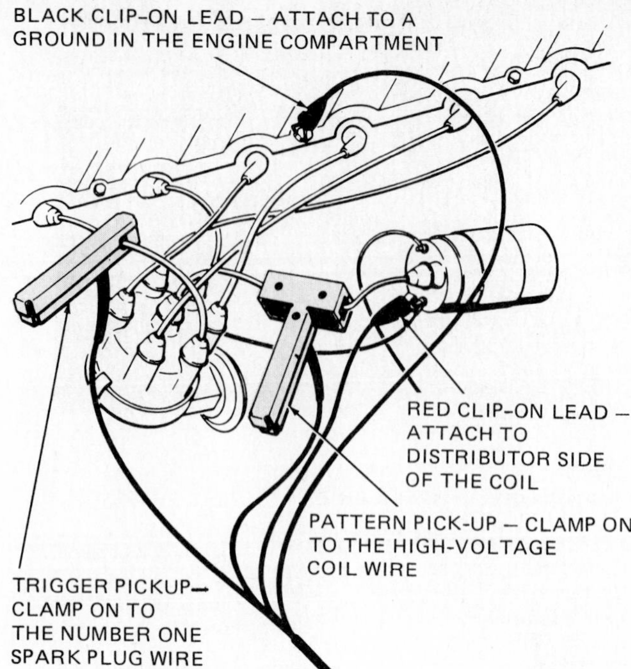

Fig. 27-29 Test leads are clipped to terminals, and pickup sensors are clamped on high-voltage cables, to test an ignition system with an oscilloscope. *(Autoscan, Inc.)*

2. DISTRIBUTOR INSTALLATION If the engine has been cranked with the distributor out, timing has been lost. The engine must be retimed. This is necessary to establish the proper relationship between the distributor rotor and the number 1 piston.

Remove number 1 spark plug from the cylinder head. Place a shop towel over your finger, and cover the spark-plug hole. Crank the engine until you feel compression pressure on your finger.

Bump the engine with the starting motor until the timing marks on the crankshaft pulley and timing cover are aligned. This means that number 1 piston is in firing position.

Now, the distributor can be installed in the cylinder block. Make sure to align the marks you made on the distributor housing and cylinder block. Check to make sure that the distributor gasket or rubber O-ring is in place when you install the distributor.

Four different distributor drives are shown in Fig. 27-38. You may have to turn the rotor slightly to engage the drive.

Make sure the distributor housing is fully seated against the cylinder block. If it is not, the oil-pump shaft is not engaging. Hold the distributor down firmly, and bump the engine a few times until the distributor housing drops into place. Then bump the engine again to realign the timing marks.

Install, but do not tighten, the distributor clamp and bolt. Rotate the distributor until the contact points just start to open to fire number 1 cylinder. Hold the distributor cap in place above the distributor. Make sure that the rotor tip lines up with number 1 terminal on the cap. Install the cap with wires. Connect the primary wire from the ignition coil to the distributor.

Start the engine. Set the ignition timing (♦27-20). Connect the vacuum hose or hoses to the distributor. Install the air cleaner.

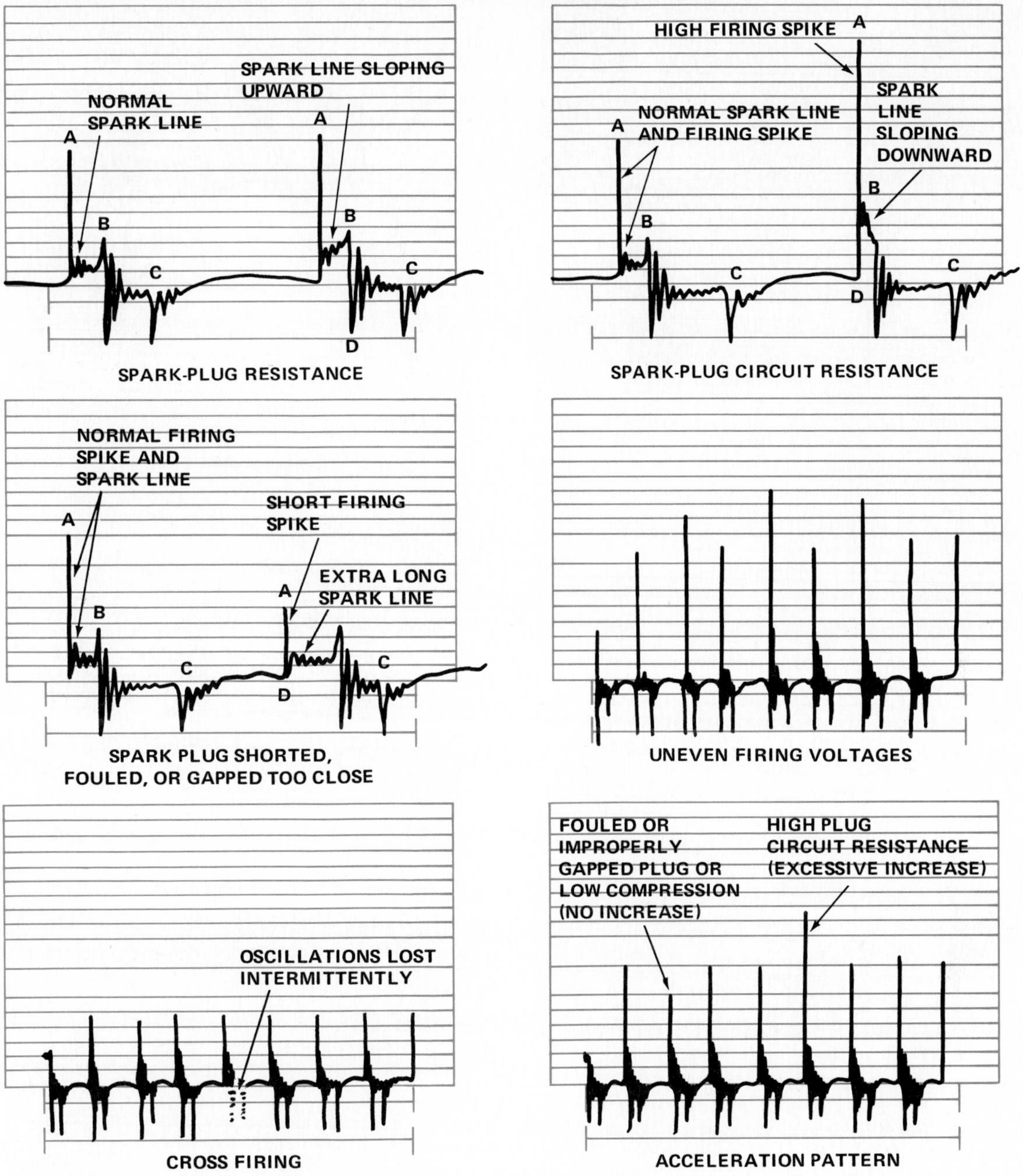

SPARK LINE SLOPING UPWARD
NORMAL SPARK LINE
A B C A B C D

SPARK-PLUG RESISTANCE

HIGH FIRING SPIKE A
NORMAL SPARK LINE AND FIRING SPIKE
SPARK LINE SLOPING DOWNWARD
A B C B D C

SPARK-PLUG CIRCUIT RESISTANCE

NORMAL FIRING SPIKE AND SPARK LINE
SHORT FIRING SPIKE
EXTRA LONG SPARK LINE
A B C A D C

SPARK PLUG SHORTED, FOULED, OR GAPPED TOO CLOSE

UNEVEN FIRING VOLTAGES

OSCILLATIONS LOST INTERMITTENTLY

CROSS FIRING

FOULED OR IMPROPERLY GAPPED PLUG OR LOW COMPRESSION (NO INCREASE)
HIGH PLUG CIRCUIT RESISTANCE (EXCESSIVE INCREASE)

ACCELERATION PATTERN

Fig. 27-30 Abnormal scope traces and their causes. *(Ford Motor Company)*

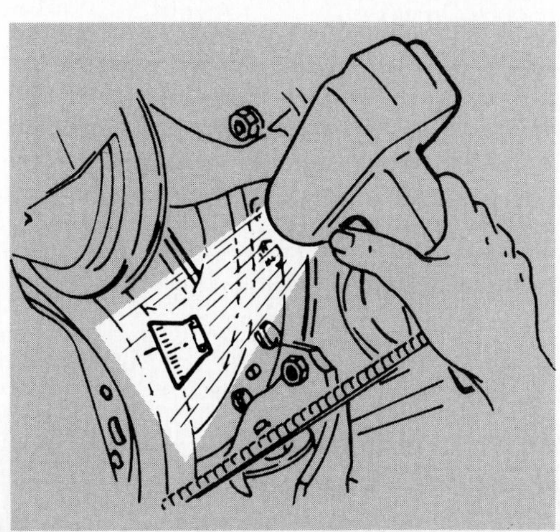

Fig. 27-31 The timing light flashes every time number 1 spark plug fires.

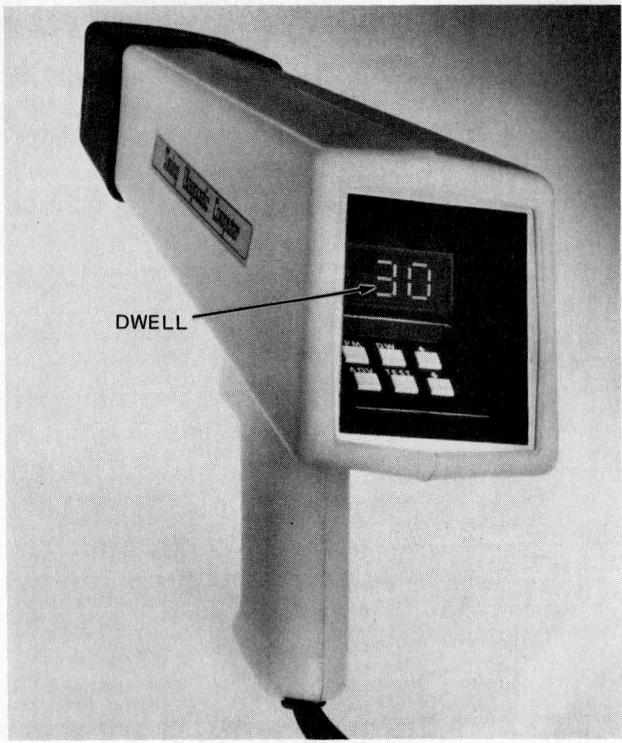

Fig. 27-32 Diagnostic timing computer. This electronic timing light checks timing and provides readouts on engine rpm, dwell, and spark advance. *(Chrysler Corporation)*

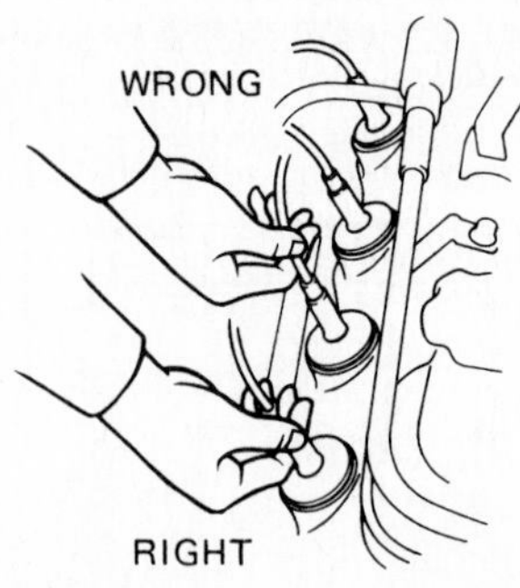

Fig. 27-33 Wrong and right ways to remove cables from spark plugs. Pull on the rubber boot, not on the cable. *(Toyota Motor Sales, Inc.)*

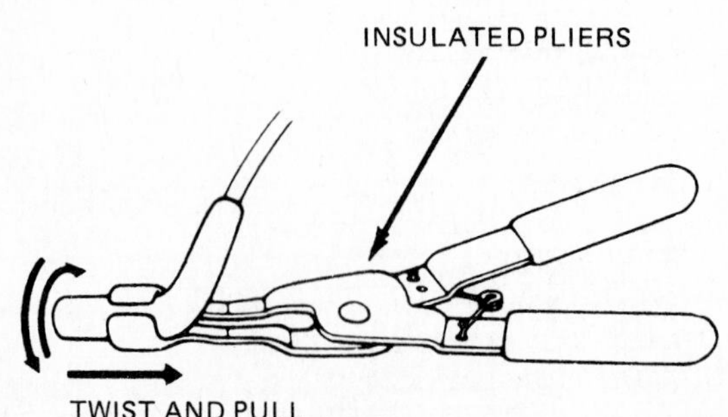

Fig. 27-34 Using cable pliers to remove a cable from a spark plug. *(Ford Motor Company)*

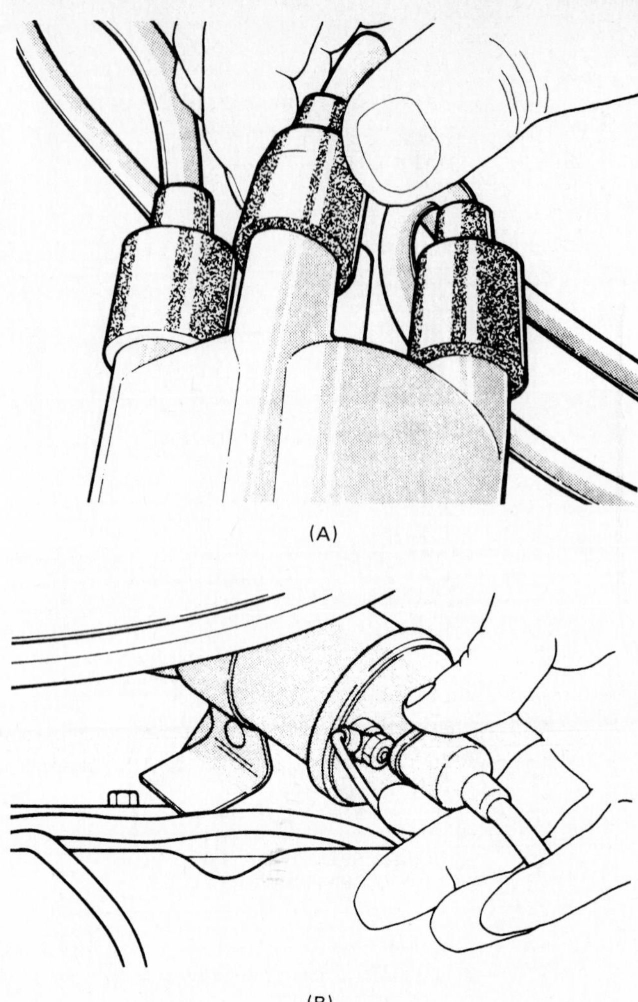

Fig. 27-35 Installing a cable and boot on (A) a distributor cap and (B) a coil tower. *(Chrysler Corporation)*

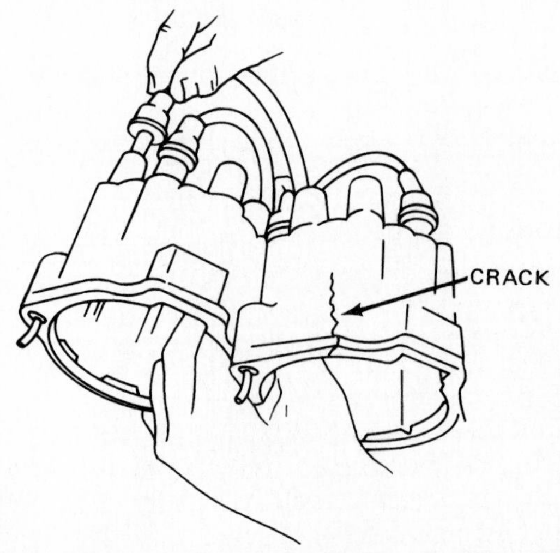

Fig. 27-36 Changing the cables from an old cap to a new cap.

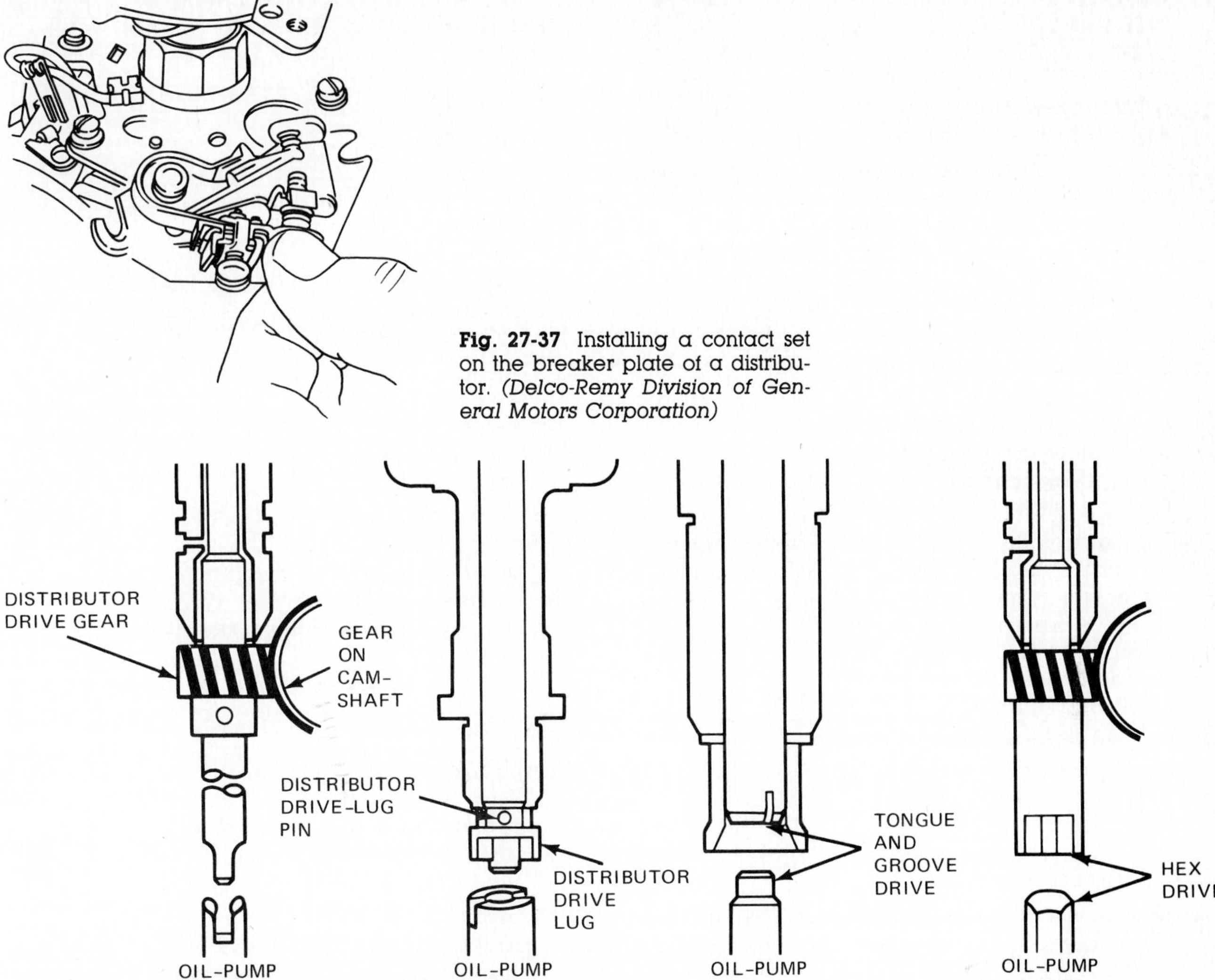

Fig. 27-37 Installing a contact set on the breaker plate of a distributor. *(Delco-Remy Division of General Motors Corporation)*

Fig. 27-38 Distributor drives and installation methods.

♦ REVIEW QUESTIONS ♦

Select the *one* correct, best, or most probable answer to each question. You can find the answers in the section indicated at the end of each question.

1. The ignition coil has (♦27-5)
   a. one winding
   b. two windings
   c. three windings
   d. four windings
2. The primary winding of the ignition coil is electrically connected to the battery through the (♦27-5)
   a. spark-plug cables
   b. distributor cap and rotor
   c. distributor gearing
   d. distributor contact points
3. The secondary winding of the ignition coil is connected to the spark plugs through the (♦27-7)
   a. distributor cap and rotor
   b. spark-plug cables
   c. secondary wiring
   d. all of the above
4. The distributor shaft is driven by (♦27-4)
   a. gears on the shaft and the crankshaft
   b. gears on the shaft and the camshaft
   c. chain or belt from the camshaft
   d. water pump
5. The high-voltage surge is produced in the ignition-coil secondary winding when the (♦27-4)
   a. contact points close
   b. contact points open
   c. centrifugal-advance mechanism works
   d. vacuum-advance mechanism works
6. The contact-point distributor is two separate devices in one, a fast-acting switch, and (♦27-4)
   a. a condenser
   b. a high-voltage distribution system
   c. an electronic signaling system
   d. an oil-pump drive

7. The rotary switch in the distributor is formed by the (♦27-6)
   a. cap and rotor
   b. centrifugal and vacuum mechanisms
   c. cam and camshaft
   d. contact points and rotor
8. Spark-plug heat range is determine by how far the heat must travel (♦27-9)
   a. from the outer terminal to the center terminal
   b. from the outer electrode to the center electrode
   c. from the electrode to the cylinder head
   d. and how deeply the plug enters the combustion chamber
9. The device in the distributor that moves the breaker cam ahead as engine speed increases is called the (♦27-10)
   a. vacuum-advance mechanism
   b. centrifugal-advance mechanism
   c. full-advance mechanism
   d. vacuum-brake mechanism
10. The device on the distributor that shifts the position of the breaker plate to produce a change in the timing of the spark is operated by (♦27-11)
    a. engine intake-manifold vacuum
    b. engine speed
    c. centrifugal advance
    d. throttle opening
11. Failure to start with normal cranking is usually due to trouble in the (♦27-16)
    a. ignition or starting system
    b. ignition or engine
    c. ignition or fuel system
    d. battery or fuel system
12. If you cannot get a spark during the spark test, the trouble probably is in the (♦27-16)
    a. engine
    b. fuel system
    c. ignition system
    d. spark plugs
13. If no spark occurs during the spark test and the ammeter shows a small fluctuating reading, the trouble probably is in the (♦27-16)
    a. secondary circuit
    b. primary circuit
    c. ignition switch
    d. starting system
14. If no spark occurs and the ammeter shows a fairly high and steady reading, the trouble probably is in the (♦27-16)
    a. secondary circuit
    b. primary circuit
    c. spark plugs
    d. wiring
15. If no change in engine speed occurs when a spark plug is shorted out, then that (♦27-16)
    a. spark plug is okay
    b. cylinder is delivering power
    c. means everything is okay
    d. cylinder is not firing
16. In normal operation, the contact-point opening decreases due to (♦27-17)
    a. cam wear
    b. rubbing-block wear
    c. shaft bearing wear
    d. centrifugal-advance wear
17. Dwell is the (♦27-17)
    a. length of time it takes the points to close
    b. length of time the points are open
    c. number of degrees of cam rotation that the points are closed
    d. number of degrees of cam rotation that the points are open
18. The scope pattern for spark-plug voltage shows that the highest voltage occurs (♦27-18)
    a. when the plug fires
    b. to sustain the spark across the gap
    c. when the points close
    d. none of the above
19. To adjust the ignition timing, (♦27-20)
    a. turn the cam on the camshaft
    b. turn the distributor in its mounting
    c. install different centrifugal-advance springs
    d. readjust the contact points
20. Contact-point opening is adjusted by (♦27-17)
    a. shifting the stationary point
    b. shifting the movable point
    c. changing the spring tension
    d. turning the distributor in its mounting

CHAPTER 28
Electronic Ignition Systems

After studying this chapter, you should be able to:

1. Explain the basic difference between contact-point and electronic ignition systems.
2. Describe the sensor coil and what it does.
3. Discuss the Hall-effect distributor and how it works.
4. Explain how the Duraspark II and IV systems work and the difference between them.
5. Describe the operation of the General Motors high-energy ignition system.
6. Explain the purpose and operation of electronic spark-timing systems.

♦ 28-1 ELECTRONIC IGNITION SYSTEMS

The basic difference between the contact-point and the electronic ignition systems is in the primary circuit (Fig. 28-1). The primary circuit in the contact-point system is opened and closed by contact points. In the electronic system, the primary circuit is opened and closed by the electronic control unit (ECU).

The secondary circuits are practically the same for the two systems. The difference is that the distributor, ignition coil, and wiring are altered to handle the higher voltage that the electronic ignition system produces.

One advantage of this higher voltage—up to 47,000 volts—is that spark plugs with wider gaps can be used. This results in a longer spark which can ignite leaner air-fuel mixtures. As a result, engines can run on leaner mixtures for better fuel economy and lower emissions.

Another difference is that some electronic ignition systems have no mechanical advance mechanisms—centrifugal or vacuum. Instead, the spark timing is adjusted electronically.

♦ 28-2 DISTRIBUTOR PRIMARY SIGNAL

The ECU responds to signals from a *sensor coil* (also called a *magnetic pickup*) or from a *magnetic sensor*. The sensor coil is affected by an armature (also called a *reluctor*). The magnetic sensor is affected by a rotor with curved shutter blades.

♦ 28-3 SENSOR COIL (MAGNETIC PICKUP)

Figures 28-2 and 28-3 show the distributor that uses a sensor coil. It is also called a magnetic-pickup assembly. Each time a tip on the armature passes through the magnetic field of the pickup assembly, a voltage is produced in the sensor coil. The coil is connected to the ECU as shown in Fig. 28-1. This voltage is an instant signal to the ECU to cut off the primary current flow. When the current stops flowing, the ignition-coil secondary winding produces a high-voltage surge. This is carried through the distributor cap and rotor to the spark plug in the cylinder that is ready to fire (♦27-6). The secondary circuits for both the contact-point ignition system and the electronic ignition system work in exactly the same way.

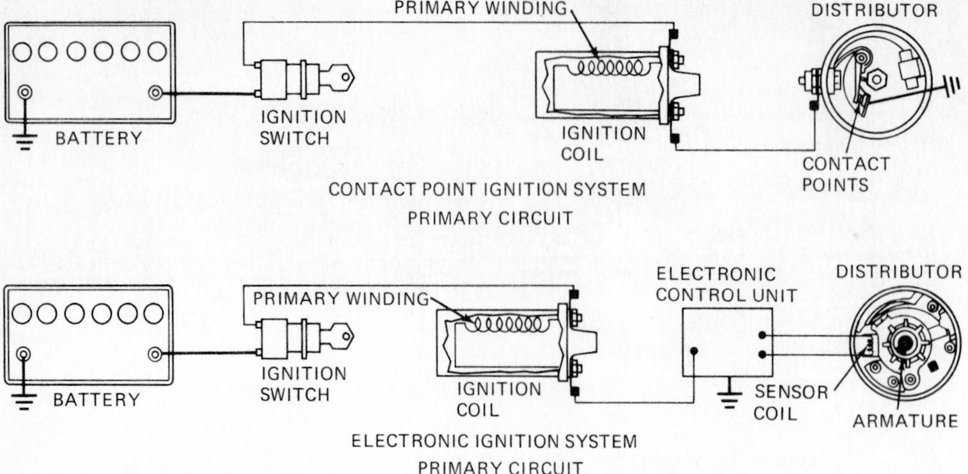

Fig. 28-1 Comparison of the primary circuits of a contact-point ignition system and an electronic ignition system.

The armature has the same number of tips as there are cylinders in the engine. The armature in Figs. 28-2 and 28-3 has eight tips, so it is for an eight-cylinder engine. This is the same as the cam in the contact-point distributor. It has the same number of lobes as there are cylinders in the engine.

♦ 28-4 HALL-EFFECT SENSOR

Figure 28-4 illustrates the *Hall effect*. (It was discovered by a scientist named Hall.) Some electronic distributors have a magnetic sensor using the Hall effect (Figs. 28-5 and 28-6). When a steel shutter moves between the two poles of a magnet, it cuts off the magnetism between the two poles. The Hall-effect distributor has a rotor with curved plates, called *shutters* (Fig. 28-6). These shutters are curved so they can pass through the air gap between the two poles of the magnetic sensor when the rotor turns. There are the same number of shutters as there are cylinders in the engine.

Each time a shutter moves through the air gap between the two poles of the magnetic sensor, it cuts off the magnetism between the poles. This provides a signal to the ECU. When a shutter is not in the way, the magnetic sensor is producing a voltage. This voltage is signaling the ECU to allow current to flow through the ignition-coil primary winding. But when a shutter moves to cut off the magnetic field, the signal voltage drops to zero. Then the ECU cuts off the flow of current to the ignition-coil primary winding. The magnetic field collapses, causing the coil secondary winding to produce a high-voltage surge.

In addition to carrying shutters between the magnetic poles, the rotor also sends the high-voltage surges to the proper spark plugs. There is a counterweight on the rotor (Fig. 28-5), opposite the segment. The segment carries the spring and metal blade. They make the rotating connection between the center terminal and the outer terminals of the cap. The counterweight balances the weight of the segment. This allows the rotor to turn without vibration. Any out-of-balance condition might affect the clearance between the shutters and the poles of the magnetic sensor.

♦ 28-5 SPARK ADVANCE

Many electronic distributors have mechanical advance mechanisms—centrifugal and vacuum—as in the contact-point distributors (♦27-10 and 27-11). However, many of the latest electronic ignition systems use electronic spark advance, as explained later in the chapter. The following sections describe various electronic systems used by different manufacturers that have mechanical advance mechanisms.

♦ 28-6 CHRYSLER ELECTRONIC IGNITION SYSTEMS

Chrysler has been using electronic ignition systems on all cars made in the United States since 1973. The type of distributor used with some variations in Chrysler and some Ford cars is shown in Figs. 28-2 and 28-3. The operation of this distributor is described in ♦28-3.

♦ 28-7 FORD ELECTRONIC IGNITION SYSTEM

Ford has used four variations of the basic electronic ignition system—Duraspark I, Duraspark II, Duraspark III, and Duraspark IV. Duraspark I and II are very similar and are discussed in this section. Duraspark III and IV have electronic control of the spark advance. These are described later in the chapter.

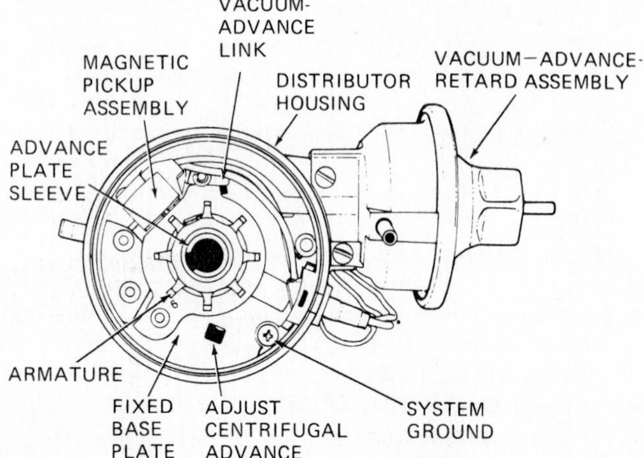

Fig. 28-2 Ignition distributor for an electronic ignition system. This type of distributor is used in Ford and Chrysler cars. Ford calls the rotating part an armature. Chrysler calls it a reluctor. *(Ford Motor Company)*

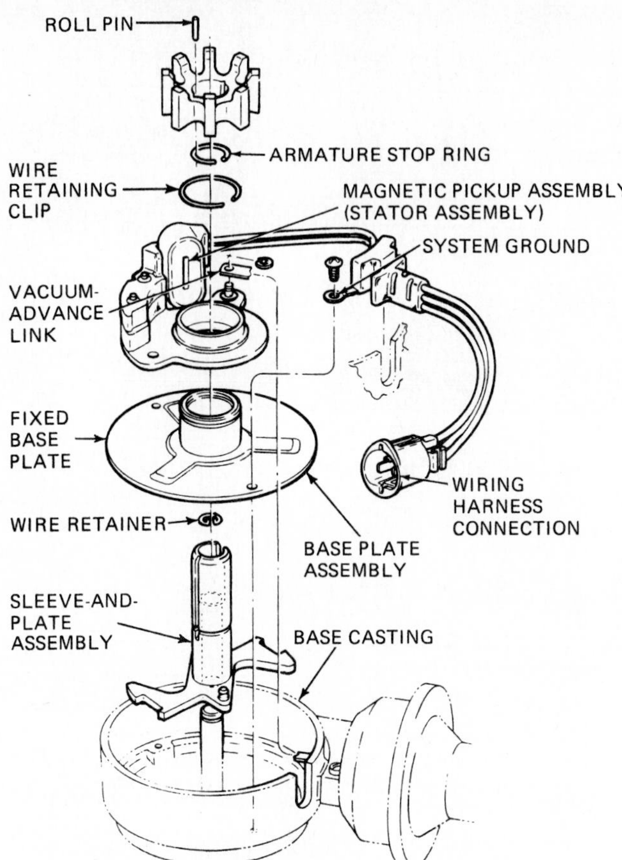

Fig. 28-3 Disassembled electronic distributor. *(Ford Motor Company)*

Figure 28-7 is the wiring circuit of the Duraspark II system. Ford calls the ECU an *ignition module*. The parts of the distributor are shown in Figs. 28-2 and 28-3. The system works as previously described in ♦28-3. As the armature rotates, a voltage is generated in the sensor coil as each tip passes it. The voltage signals the ignition module to stop the flow of current to the ignition-coil primary winding. Then the magnetic field collapses, producing a high-voltage surge in the secondary winding of the coil. The high-voltage surge flows through the distributor cap, rotor, and wiring to the spark plug in the cylinder that is ready to fire.

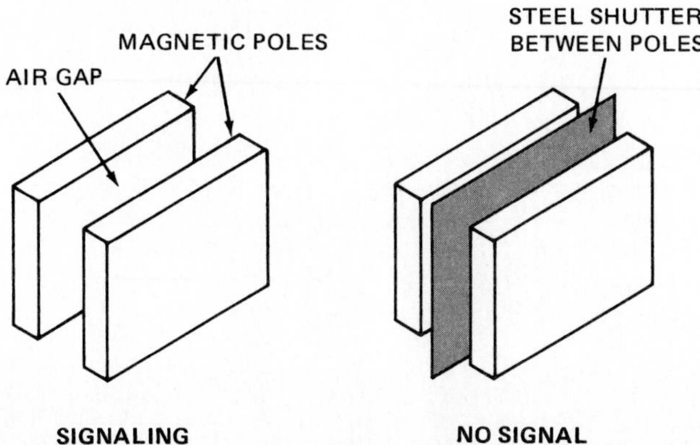

Fig. 28-4 When there is no shutter between the magnetic poles, they send a signal to the computer. When the shutter moves between the poles, the signal is cut off.

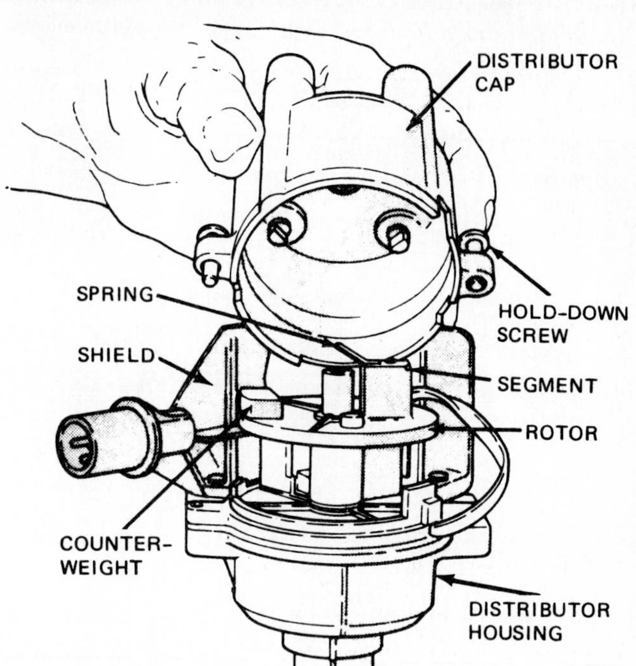

Fig. 28-5 Hall-effect distributor with cap removed. *(Chrysler Corporation)*

The distributor (Fig. 28-8) has centrifugal- and vacuum-advance mechanisms. The cap is larger than in contact-point distributors. This provides room for the sensor coil and other electronic parts in the distributor. Also, the terminals are the post type. These two changes enable the secondary circuit to handle the high voltages (up to 47,000 volts) that the system can produce. The bigger cap allows the terminals to be spaced further apart. This reduces the possibility of high-voltage leakage between terminals. The post-type terminals are not subject to moisture gathering in the terminals, as are the push-in type (Fig. 27-3).

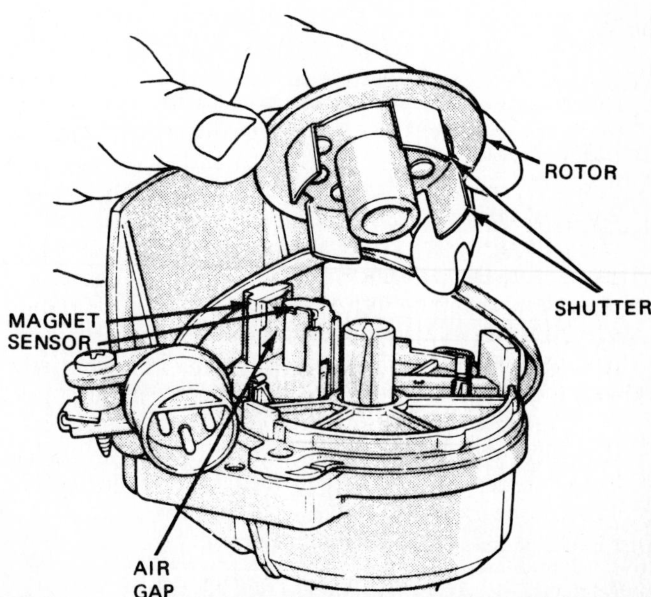

Fig. 28-6 Removing the rotor from a Hall-effect distributor. *(Chrysler Corporation)*

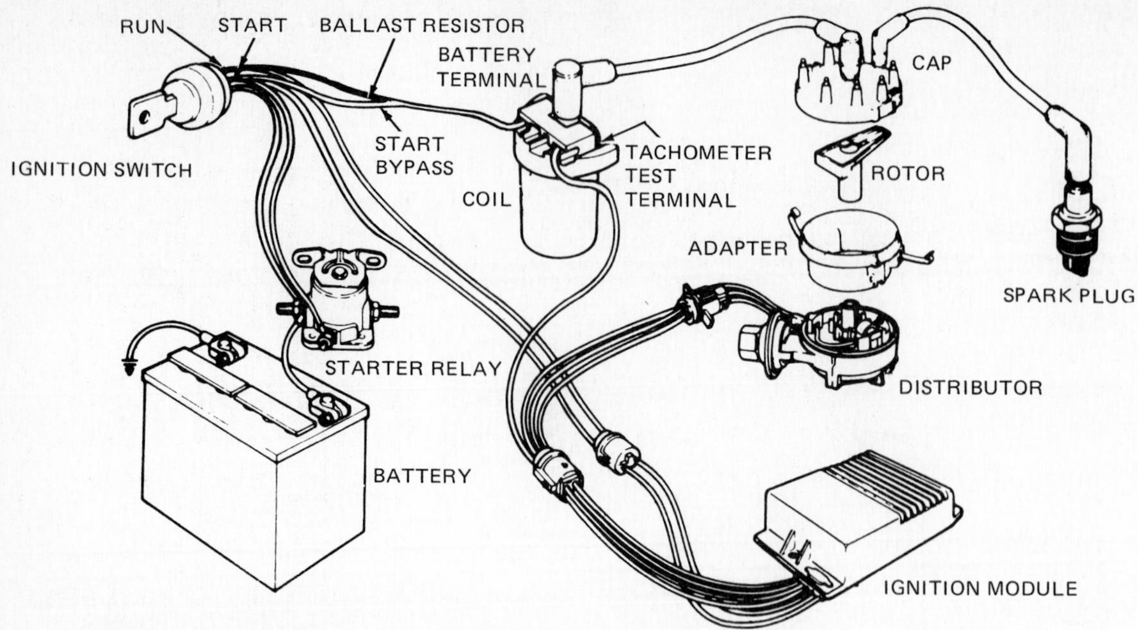

Fig. 28-7 Duraspark II ignition system. *(Ford Motor Company)*

♦ 28-8 GENERAL MOTORS HIGH-ENERGY IGNITION (HEI) SYSTEM

General Motors made this system standard equipment on its cars in 1975. GM adopted the name because the system produces high voltage (high energy). In one version of this system, the ignition coil is assembled in the top of the distributor cap (Fig. 28-9). This makes the wiring system very simple (Fig. 28-10). The connections between the ignition coil and distributor are inside the system. Also, the ECU, which General Motors calls the *module*, or *electronic module*, is inside the distributor, as shown in Fig. 28-9. The distributor uses an armature with tips, and a pickup coil. The operation is similar to the other electronic distributors that are described in ♦28-3.

♦ 28-9 ELECTRONIC SPARK CONTROL (ESC)

This system is also called a *detonation-control system*. It was developed to prevent excessive detonation in engines using turbochargers (♦15-14).

When an engine is running with the turbocharger providing boost, detonation may occur. This happens if the turbocharger forces too much air-fuel mixture into the engine cylinders for the amount of spark advance. To prevent detonation, many turbocharged engines have a detonation sensor, usually mounted on the intake manifold. When detonation starts, the detonation sensor signals the electronic controller. It then signals the distributor so that the spark timing is retarded. This reduces detonation. Figure 28-11 shows the ESC system.

♦ 28-10 ELECTRONIC SPARK-ADVANCE CONTROLS

All the electronic ignition systems described so far are similar in operation. A pickup coil senses the rotation of a trigger wheel or armature. The pickup coil sends voltage pulses to the ECU, or electronic module. These pulses cause the ECU to close and open the circuit between the battery and the ignition-coil primary winding. This action is like that of the contact points in the contact-point ignition system.

Some electronic ignition systems do not have mechanical centrifugal- and vacuum-advance mechanisms. Instead, various sensors feed information to the ECU or system computer. This information includes engine speed and piston position, engine temperature, and manifold vacuum. The ECU then produces the spark timing that the engine needs. Four systems—Ford, General Motors, Renault, and distributorless—are described in following sections.

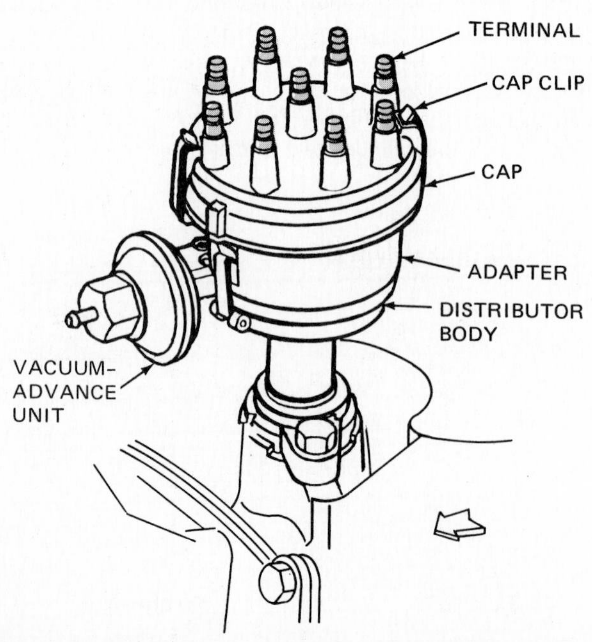

Fig. 28-8 Ford electronic distributor for the Duraspark II system. *(Ford Motor Company)*

♦ 28-11 FORD DURASPARK III AND IV IGNITION SYSTEMS

These electronic ignition systems are used with the Ford electronic engine control (EEC) system (♦28-12). It controls the ignition system and various other systems related to engine operation. These include the *exhaust-gas recirculation* (EGR) system and the *air-injection* (AIR) system. They are described in later chapters on emission controls.

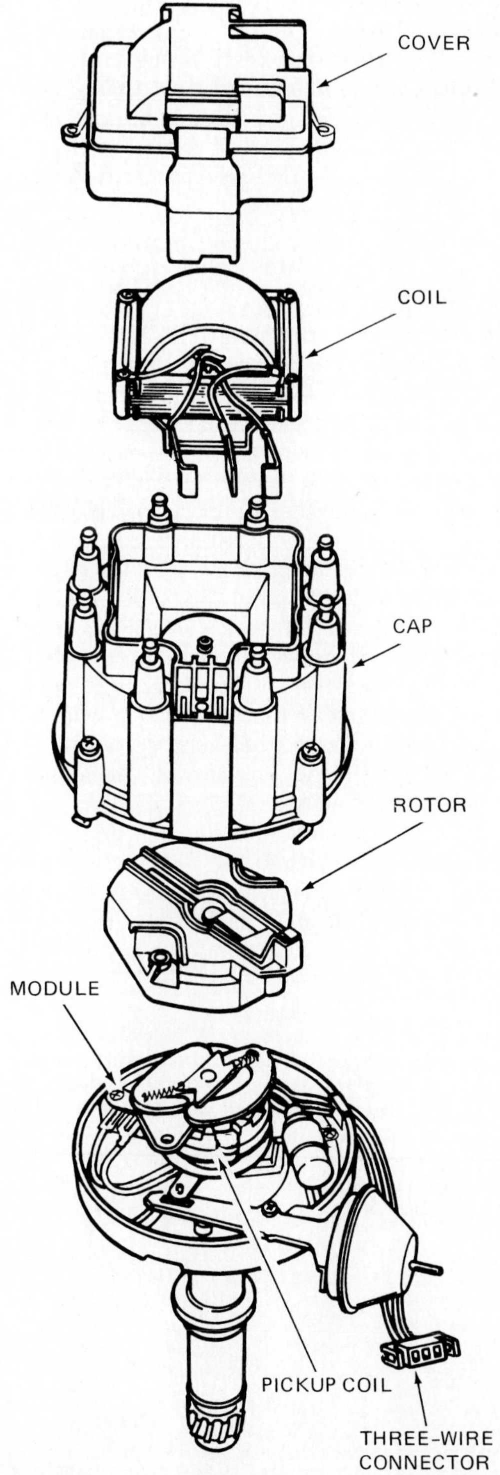

Fig. 28-9 Partly disassembled HEI distributor which shows how the ignition coil mounts in the cap. *(Delco-Remy Division of General Motors Corporation)*

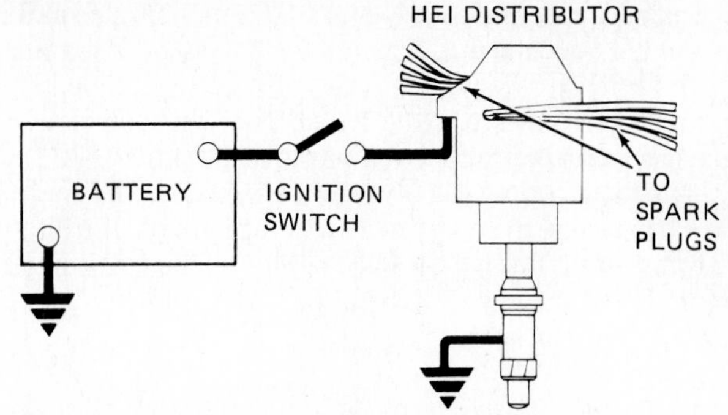

Fig. 28-10 Basic wiring diagram for the HEI system. *(Delco-Remy Division of General Motors Corporation)*

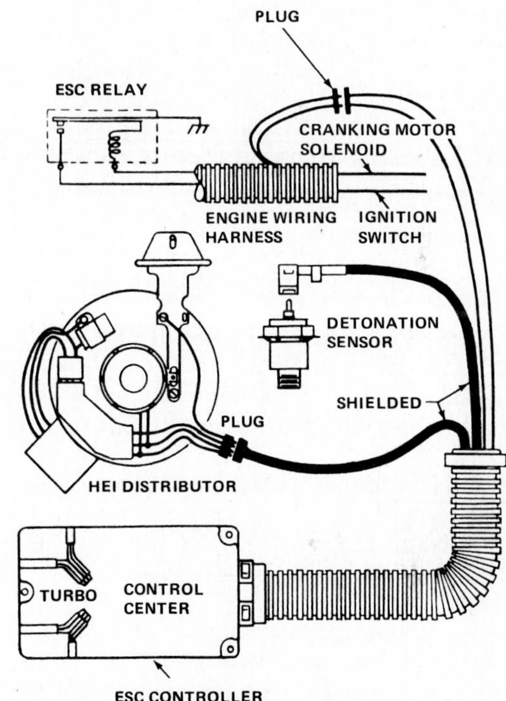

Fig. 28-11 Components of the ESC system. *(Buick Motor Division of General Motors Corporation)*

Figure 28-12 shows the Duraspark III ignition system. The Duraspark III and IV systems control spark advance in response to several engine sensors. These sensors report the following information to the computer (called the *microprocessor* in Fig. 28-12):

1. Temperature of air entering the air cleaner.
2. Throttle movement and position.
3. Engine-coolant temperature.
4. Air pressure, which changes as altitude changes.
5. Intake-manifold vacuum.
6. Crankshaft and therefore piston position. This signal originates at a disk or pulse ring with four teeth 90 degrees apart on the rear of the crankshaft (Fig. 28-13). The signal is carried by the crankshaft-position sensor to the computer.

7. Engine speed. The signals from the crankshaft pulse ring also tell the computer how fast the crankshaft is turning.

The computer correlates the signals and then provides the best spark advance for the operating conditions.

Since the centrifugal and vacuum advances are taken care of electronically, the only function of the ignition distributor is to distribute the high-voltage surges to the spark plugs. Figure 28-14 shows the distributor. It is an empty shell except for the shaft and rotor.

♦ 28-12 FORD ELECTRONIC ENGINE CONTROL (EEC)

The Ford EEC system (Fig. 28-15) includes the Duraspark III electronic ignition system, and also controls the EGR system and the AIR system. These systems are discussed in a following chapter on emission controls. The EGR system sends some exhaust gas back through the engine under some conditions to reduce the formation of the exhaust pollutant nitrogen oxides (NO_x).

The AIR system sends fresh air into the exhaust manifold under some operating conditions. The oxygen in the fresh air helps to burn up any unburned or partly burned fuel in the exhaust gases. Under other operating conditions, the EEC system cuts off this fresh air.

♦ 28-13 GENERAL MOTORS ELECTRONIC SPARK TIMING (EST)

This system is similar to the ESC system previously described. The General Motors system uses four sensors:

1. Pickup coil in the distributor, which senses engine speed and piston positions
2. Engine-coolant sensor
3. Manifold-vacuum sensor
4. Atmospheric-pressure sensor

The computer puts these signals together. Then it advances or retards the spark the proper amount to suit operating conditions.

♦ 28-14 RENAULT INTEGRAL ELECTRONIC IGNITION SYSTEM

This system (Fig. 28-16) has a different appearance from the other electronic ignition systems described earlier. However, it has the same basic components and operates in the same general manner. It receives signals from various sensors. Then it puts these signals together to produce the proper spark advance for the operating conditions.

The TDC sensor is located at the flywheel or torque converter, where it senses both TDC and BDC. A separate ring of special teeth is located in back of the ring gear for the starting-motor drive pinion (Fig. 28-17). However, two of the special teeth have been machined off, 180 degrees apart. These are 90 degrees from TDC and BDC. As the engine runs, the TDC sensor senses these gaps. Therefore, it gives the ignition-control module a continuous report on engine speed and piston positions.

♦ 28-15 DISTRIBUTORLESS IGNITION SYSTEMS

Cars with electronic spark timing (♦28-13) use the distributor only for distributing the high-voltage sparks from the coil. Spark timing is controlled electronically, and not by centrifugal- and vacuum-advance mechanisms on the distributor. Therefore, the ignition system on some of these cars has been redesigned to eliminate the distributor. Camshaft and crankshaft sensors provide the electronic control module with information on engine speed and piston position.

Figure 28-18 shows an ignition system for a four-cylinder engine that has no distributor. The system fires two spark plugs at the same time. One spark plug is in a cylinder during its exhaust stroke. Therefore, the spark has no effect. This is called the *waste-spark method*. The other spark occurs in the cylinder near the end of the compression stroke. Normal combustion follows. There are two primary windings and one secondary winding in the ignition coil. The action is controlled by a rotating sensor on the crankshaft. The trigger uses two triggering points, one for each primary winding. Each triggering point signals the ECU to stop the flow of current through one of the primary windings. When this happens, the magnetic field of the primary winding collapses and a high voltage is produced in the secondary winding.

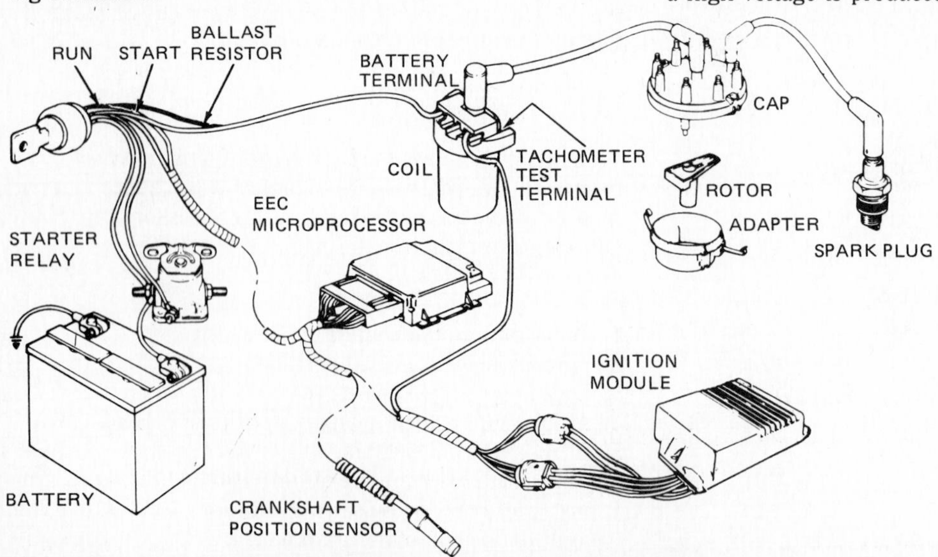

Fig. 28-12 Duraspark III ignition system. *(Ford Motor Company)*

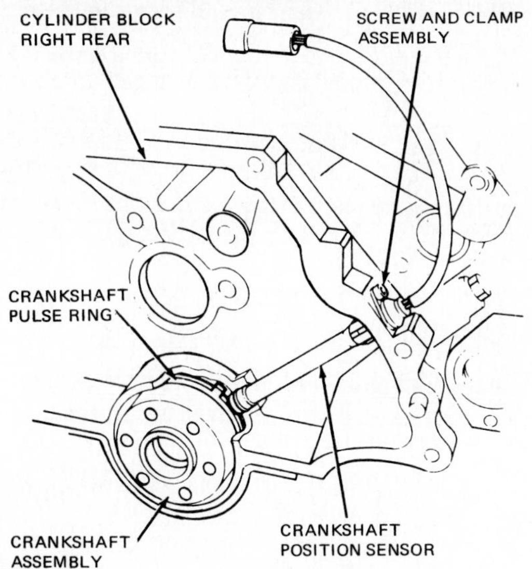

Fig. 28-13 Location of the crankshaft pulse ring. *(Ford Motor Company)*

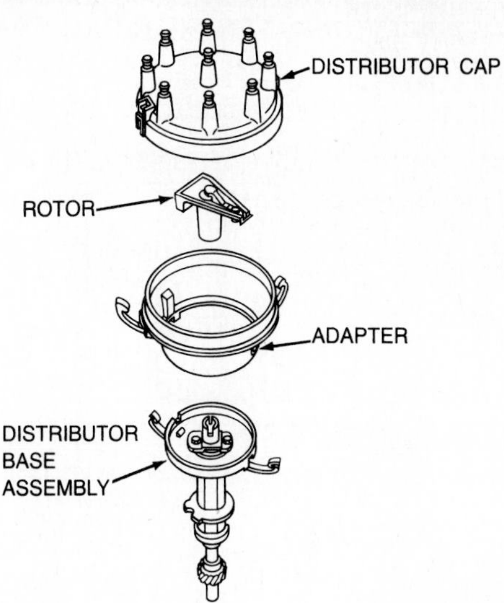

Fig. 28-14 Distributor for the Duraspark III ignition system. Its only function is to distribute the high-voltage surges to the spark plugs in the proper firing order. *(Ford Motor Company)*

The polarity of this high voltage, and the high-voltage diodes, determines which two plugs fire. For example, suppose that the upper end of the secondary winding (in Fig. 28-18) is negative. Then the electrons in the high-voltage pulse can flow to ground through plug 2. From ground the electrons flow through plug 3 and back to the other end of the secondary winding. Current cannot flow through plug 1 because the polarity of the high-voltage diode for plug 1 prevents it.

A similar distributorless ignition system, called *computer-controlled coil ignition* (C^3I), is used by Buick on turbocharged V-6 engines (Fig. 28-19). The system includes an ignition module and coilpack, crankshaft sensor, and camshaft sensor. There are three coils in the coilpack for the six-cylinder engine. As the crankshaft rotates, a Hall switch detects the signal from the crankshaft sensor. This signal is sent to the ignition module, which then triggers each coil at the proper time.

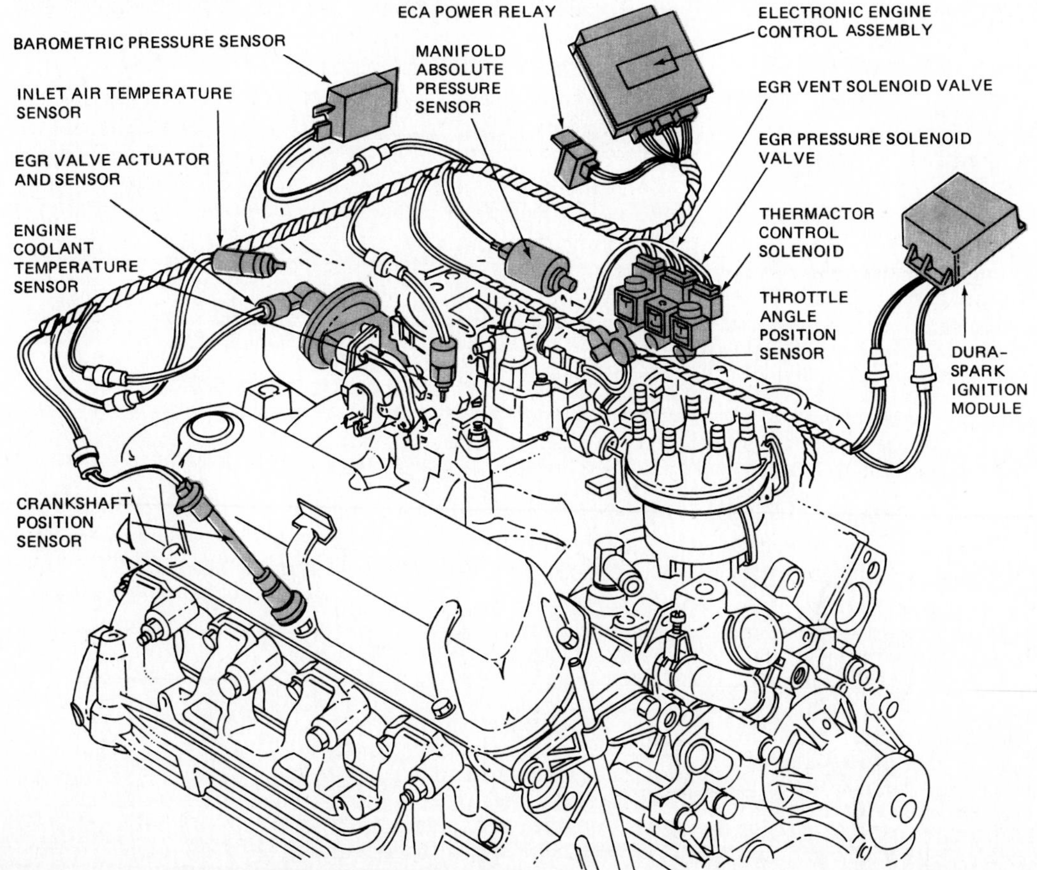

Fig. 28-15 Ford EEC system. *(Ford Motor Company)*

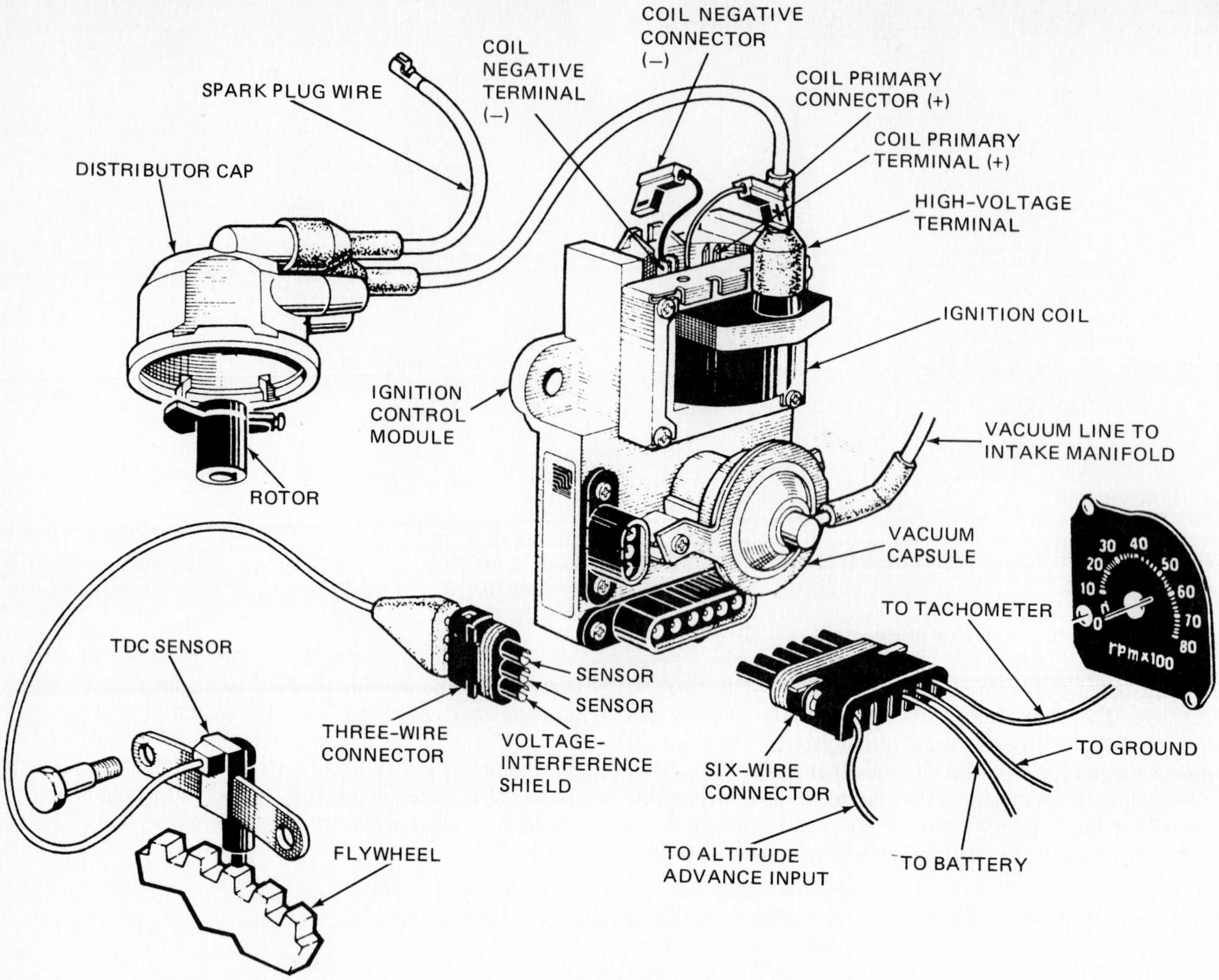

Fig. 28-16 Layout of electronic ignition system which includes automatic spark advance. *(American Motors Corporation)*

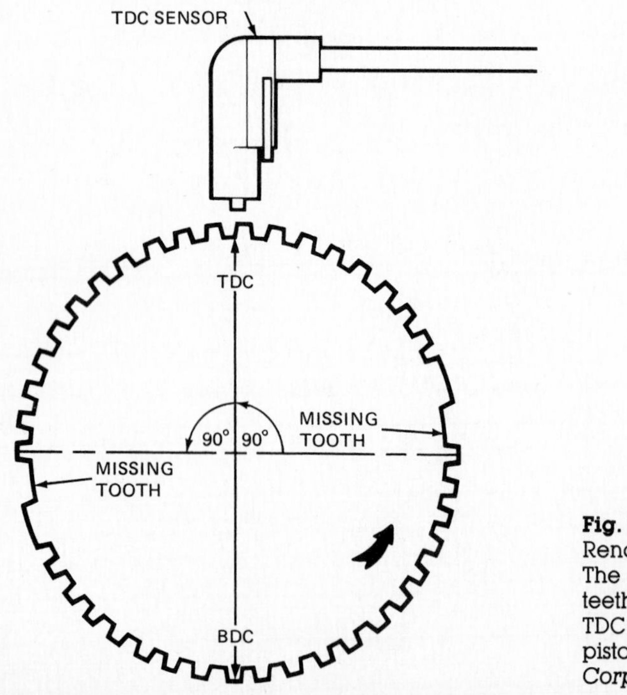

Fig. 28-17 Sensor-teeth ring for the Renault electronic ignition system. The gaps formed by the missing teeth provide information to the TDC sensor on engine speed and piston position. *(American Motors Corporation)*

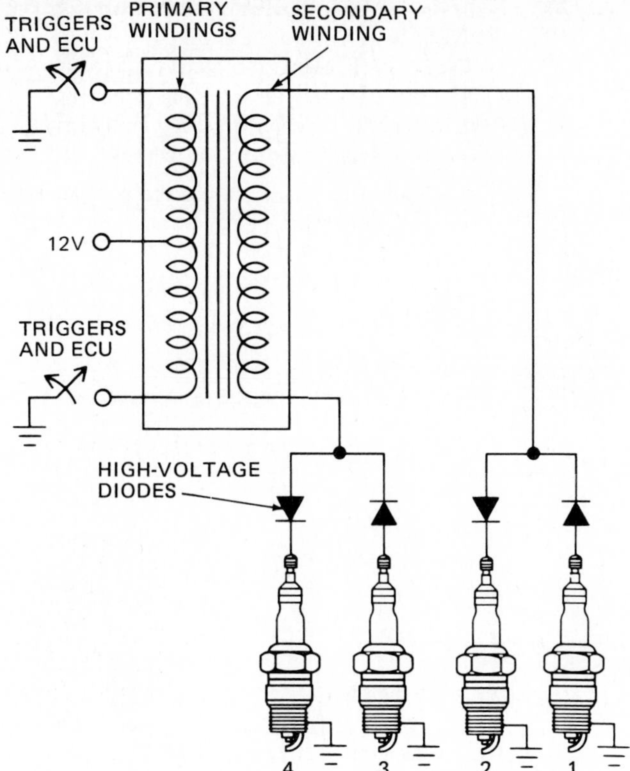

Fig. 28-18 Wiring diagram of a distributorless ignition system for a four-cylinder engine with a firing order of 1–3–4–2. *(Ford Motor Company)*

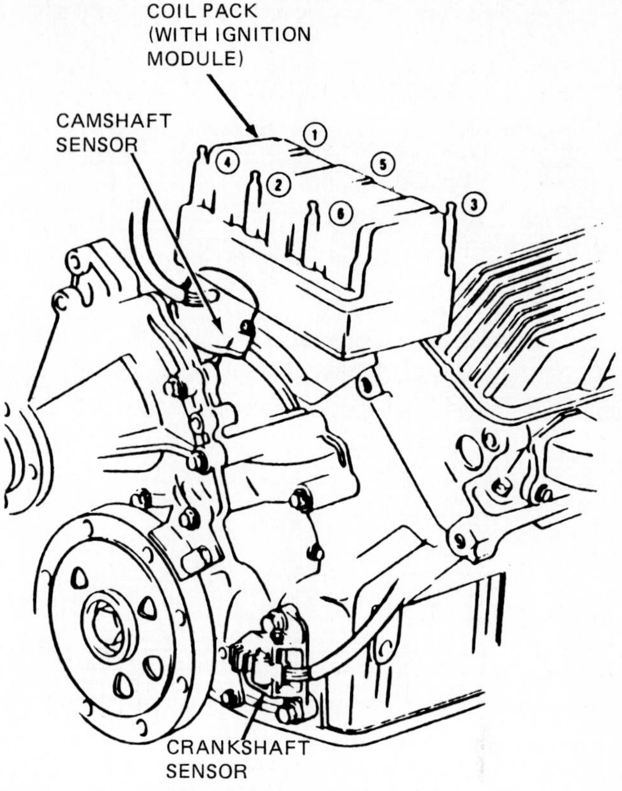

Fig. 28-19 Components in the *computer-controlled coil ignition* (C^3I) system. *(Buick Motor Division of General Motors Corporation)*

♦ REVIEW QUESTIONS ♦

Select the *one* correct, best, or most probable answer to each question. You can find the answers in the section indicated at the end of each question.

1. In the electronic ignition system, the circuit between the battery and ignition-coil primary winding is closed and opened by (♦28-1)
   a. contact points
   b. a field relay
   c. a switch
   d. an ECU
2. The secondary circuit of the electronic ignition system has been altered to (♦28-1)
   a. handle the quicker response of the system
   b. handle higher voltages
   c. fit the overhead-cam engines
   d. work faster
3. In some electronic ignition systems, spark advance is produced by (♦28-1)
   a. sensors in the fuel system
   b. a mechanical centrifugal-vacuum-advance unit
   c. an electronic device
   d. the higher voltages of the system
4. The tips on the rotating armature or reluctor (♦28-3)
   a. open and close the contact points
   b. cause the magnetic field to create a voltage signal in the sensor coil
   c. repeatedly connect to the ECU as the rotor turns
   d. connect the ignition-coil secondary winding to the spark plugs
5. The Hall-effect distributor has shutters that move between the (♦28-4)
   a. contact points
   b. sensor coils
   c. two poles of the magnetic sensor
   d. terminals of the distributor cap
6. The distributor for the Duraspark I and II systems has (♦28-7)
   a. electronic spark advance
   b. control of the catalytic converter
   c. mechanical centrifugal- and vacuum-advance mechanisms
   d. all of the above

7. The distributor caps in electronic ignition systems are larger than on contact-point distributors to (♦28-7)
   a. handle the higher voltages
   b. reduce possibility of high-voltage leakage between terminals
   c. provide room for the sensor coils and other electronic parts in the distributor
   d. all of the above
8. The distributor for the General Motors HEI system (♦28-8)
   a. does not require an ignition coil
   b. includes the ignition coil
   c. uses a pair of ignition coils
   d. none of the above
9. The electronic spark control used on some turbocharged engines (♦28-9)
   a. retards the spark if detonation begins
   b. takes the place of mechanical advance mechanisms
   c. advances the spark to suit operating conditions
   d. reduces spark voltage if detonation begins
10. In ignition systems with electronic spark advance, sensors report to the spark-control computer on (♦28-10)
   a. engine speed and piston position
   b. engine temperature
   c. manifold vacuum
   d. all of the above

CHAPTER 29
Electronic Ignition-System Service

After studying this chapter, and with proper instruction and equipment, you should be able to:

1. List nine abnormal operating conditions that can be caused by trouble in the electronic ignition system and explain the possible causes and corrections of each.
2. Demonstrate how to safely make a spark test.
3. Explain how oscilloscope patterns for electronic ignition systems differ from those for contact-point systems.
4. Demonstrate how to obtain a readout from the instrument-panel display or CHECK ENGINE light.

♦ 29-1 ELECTRONIC IGNITION-SYSTEM SERVICE

In many ways, the servicing of electronic ignition systems and of contact-point ignition systems are similar. However, there are special procedures for electronic ignition systems. These are covered later in this chapter. Following sections list possible causes of electronic ignition-system failure.

♦ 29-2 POSSIBLE CAUSES OF ELECTRONIC IGNITION-SYSTEM FAILURE

Ignition-system failures can be grouped into three categories. These are:

1. Loss of energy in the primary circuit. This could result from:
   a. Resistance in the primary circuit due to defective leads, bad connections or ignition switch, or open ignition-coil primary winding.
   b. Discharged battery or defective alternator
   c. Grounded primary circuit in ignition coil, wiring, or distributor
   d. Defective ECU or sensor-coil circuit to ECU
2. Loss of energy in the secondary circuit, due to:
   a. Spark plugs fouled, defective, or improperly gapped
   b. Defective high-voltage wiring which allows high-voltage leakage
   c. High-voltage leakage across ignition-coil head, distributor cap, or rotor
   d. Defective connections in high-voltage circuits
3. Out-of-time ignition, due to:
   a. Timing not set properly
   b. Centrifugal or vacuum advance defective
   c. Preignition from spark plugs of wrong heat range,

fouled plugs, or carbon in combustion chambers

d. Electronic control unit (ECU) defective

♦ 29-3 ELECTRONIC IGNITION-SYSTEM TROUBLE-DIAGNOSIS CHART

The chart that follows covers, in general, various possible troubles that might be caused by conditions in the electronic ignition system. However, many of these conditions could result from troubles in other components and systems of the engine.

♦ 29-4 IGNITION-SYSTEM QUICK CHECKS

Several checks have been used in the past to help locate the cause of various troubles. For example, one test is the spark test. Remove a cable from a spark plug, insert a plug extender into the cable end, and hold the extender about ⅜ inch [10 mm] from the engine block with insulated pliers. Crank the engine and check for sparking (Fig. 29-1). If no sparking occurs, there is trouble in the ignition system.

If there is a good spark, the failure to start is probably in the fuel system. This test is not recommended for use on all electronic ignition systems.

ELECTRONIC IGNITION-SYSTEM TROUBLE-DIAGNOSIS CHART*

Condition	Possible Cause	Check or Correction
1. Engine cranks normally but fails to start (♦29-5)	a. No voltage to ignition system	Check battery, ignition switch, wiring
	b. ECU ground lead open, loose, or corroded	Repair as needed
	c. Primary-wiring connectors not fully engaged	Clean, firmly seat connectors
	d. Ignition coil open or shorted	Test coil, replace if defective
	e. Damaged armature (trigger wheel, reluctor) or sensor	Replace damaged part
	f. ECU faulty	Replace
	g. Defective distributor cap or rotor	Replace defective part
	h. Fuel system faulty	
	i. Engine faulty	
2. Engine backfires but fails to start (♦29-6)	a. Incorrect timing	Check and adjust timing
	b. Moisture in distributor cap	Dry cap
	c. Cap faulty—voltage leakage across carbon paths	Replace cap (Fig. 27-19)
	d. High-voltage cables not connected in firing order	Reconnect cables correctly
3. Engine runs but misses—does not run smoothly (♦29-7)	a. Spark plugs fouled or faulty	Clean and regap, or replace
	b. Distributor cap or rotor faulty	Replace
	c. High-voltage cables defective	Replace
	d. Defective (weak) coil	Replace
	e. Bad connections	Clean, tighten
	f. High-voltage leakage	Check distributor cap, rotor, cables (Figs. 27-19 and 27-20)
	g. Advance mechanisms defective	Check advances, repair or replace distributor
	h. Defective fuel system	
	i. Defects in engine such as loss of compression or faulty valve action	
4. Engine runs but backfires	a. Ignition timing off	Retime
	b. Ignition cross-firing	Check high-voltage cables, distributor cap, and rotor for leakage paths (Figs. 27-19 and 27-20)
	c. Faulty antibackfire valve	Replace valve
	d. Spark plugs of wrong heat range	Install correct plugs
	e. Defective air-injection system	Check system
	f. Engine overheating	See item 5
	g. Fuel system not supplying proper air-fuel ratio	
	h. Engine defects such as hot valves, carbon, etc.	
5. Engine overheats (♦29-9)	a. Late ignition timing	Retime
	b. Lack of coolant or other trouble in cooling system	
	c. Late valve timing or other engine conditions	

*See ♦ 29-4 to 29-13 for detailed explanations of trouble causes and checks or corrections to be made.

ELECTRONIC IGNITION-SYSTEM TROUBLE-DIAGNOSIS CHART (*Continued*)

Condition	Possible Cause	Check or Correction
6. Engine lacks power (♦29-10)	a. Ignition timing off	Retime
	b. Troubles listed in item 3	
	c. Exhaust system restricted	Clear
	d. Heavy engine oil	Use correct viscosity oil
	e. Wrong fuel	Use correct fuel
	f. Excessive rolling resistance	Check tires, brakes, wheel bearings, alignment
	g. Engine overheats	See Item 5
7. Engine detonates, or pings (♦29-11)	a. Improper timing	Time ignition
	b. Wrong fuel	Use correct fuel
	c. Spark plugs of wrong heat range	Install correct plugs
	d. Advance mechanism faulty	Rebuild or replace distributor
	e. Carbon buildup in cylinders	Service engine
8. Spark plugs defective (♦29-12)	a. Cracked insulator	Careless installation, install new plug
	b. Plug sooty	Install hotter plug, correct condition in engine causing oil burning or high fuel consumption
	c. Plug white or gray, with blistered insulator	Install cooler plug
	d. See also analyzing defective spark plugs in ♦27-22	
9. Engine runs on, or diesels (♦29-13)	a. Idle-stop solenoid out of adjustment or defective	Readjust, replace as necessary
	b. Hot spots in combustion chambers	Service engine
	c. Engine overheating	See item 5
	d. Advanced timing	Retime ignition

Today, most shops have an oscilloscope (♦27-18 and 27-19). Using the oscilloscope is a quick way to analyze ignition systems and help pinpoint trouble causes.

Often, the first step in trouble diagnosis is to recharge or replace the battery. The driver may have run it down attempting to start the engine.

Following sections discuss the various troubles and possible causes listed in the trouble-diagnosis chart.

♦ 29-5 ENGINE CRANKS NORMALLY BUT FAILS TO START

If there is no spark during the spark test, or if the oscilloscope fails to show a secondary-voltage pattern, there are several possible causes. See item 1 in the trouble diagnosis chart (♦29-3).

If you get a good spark, the ignition primary and secondary circuits probably are okay. The failure to start could be due to fouled spark plugs or out-of-time ignition. However, failure to start with a good spark is more likely due to trouble in the fuel system. The fuel system is not delivering the correct amount or ratio of air-fuel mixture. Other conditions could prevent starting, such as malfunctioning valves, loss of engine compression, and other engine troubles.

Many manufacturers recommend making the spark test only by using a spark tester (Fig. 29-2). It is inserted between the end of the cable clip and the spark-plug terminal. This prevents excessively high voltage from arcing across the rotor or cap and damaging them.

♦ CAUTION ♦ The electronic ignition systems can deliver a very high voltage, so proceed with care when making the spark test. Use rubber gloves or insulated pliers. The 47,000 volts some of these systems can produce can give you a very painful and dangerous shock!

♦ 29-6 ENGINE BACKFIRES BUT FAILS TO START

This can be caused by ignition or valve timing that is considerably off, by a faulty or wet distributor cap or rotor that allows high-voltage leakage from one terminal to another, or by the high-voltage cables being incorrectly connected.

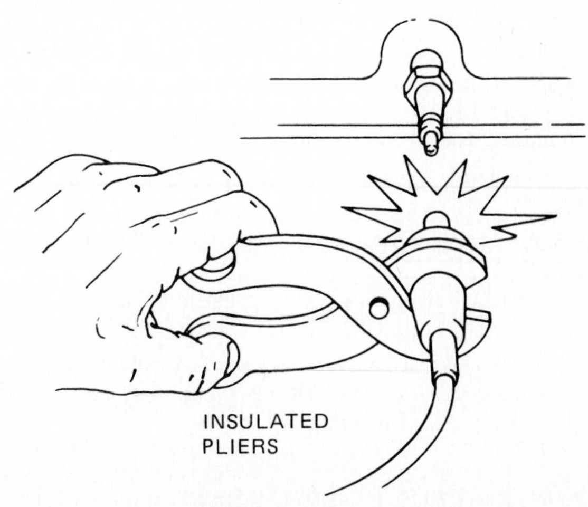

Fig. 29-1 Checking for spark from spark-plug-cable clip to engine block. Do not touch any other ignition-system wire. *(American Motors Corporation)*

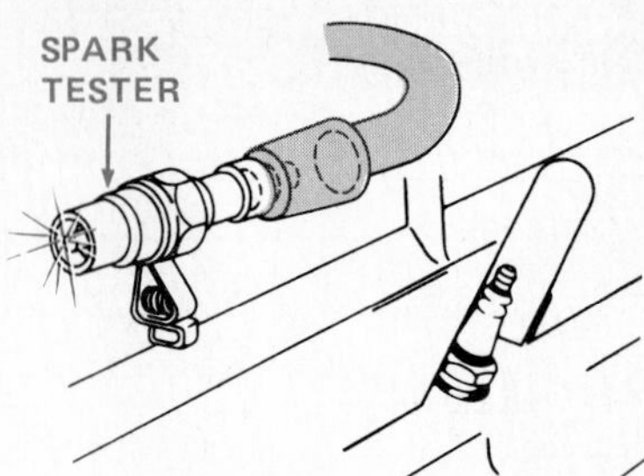

Fig. 29-2 To make a spark test using a spark tester, attach the spark-plug cable to the tester, and clamp the tester to a good ground. Then crank the engine. *(ATW)*

♦ 29-7 ENGINE RUNS BUT MISSES

An engine that misses runs unevenly and does not develop full power. It is sometimes difficult to tell by listening whether one cylinder is missing, or whether the miss is intermittent and jumping around from one cylinder to another. If one cylinder is missing, the cause could be a defective spark plug or high-voltage cable, a bad connection, or high-voltage leakage across the distributor cap or through the cable insulation. In the engine, the miss could be due to a stuck or burned valve or loss of compression resulting from broken piston rings.

If the miss jumps around, the cause could be defects in the electronic part of the system—the ECU and the sensor coil in the distributor. This requires checking out as explained in later sections on specific systems.

Another cause of a jump-around miss is the advance mechanisms not working properly so the advance is erratic. Also, a defective fuel system that is delivering too-rich or too-lean a mixture can cause missing. An excessively lean mixture will not fire. An excessively rich mixture can wet or foul the plugs, causing them to misfire.

♦ 29-8 ENGINE RUNS BUT BACKFIRES

Backfiring is a "pop" or "bang" in the exhaust manifold or intake manifold. It can be caused by several conditions in the ignition system. If the ignition timing is considerably off, or if ignition cross-firing occurs, ignition may result before the intake valve closes. This produces a backfire. There will be a pop back through the air cleaner. Cross-firing is spark jumpover from one terminal to another, or from one high-voltage cable to another. Cracked or damaged cable insulation can allow spark jumpover.

If a spark plug runs too hot, it may glow enough to ignite the air-fuel mixture before the intake valve closes. This produces a backfire, or pop back through the air cleaner. This action is called *preignition*. It can also be caused by excessively hot valves or carbon deposits in the combustion chamber.

Incorrect air-fuel ratios can also cause backfiring. A lean mixture tends to cause backfiring through the air intake. A rich mixture and burned exhaust valves can cause backfire in the exhaust system. A defective air-injection system can also cause backfiring in the exhaust system.

♦ 29-9 ENGINE OVERHEATS

Most engine overheating is caused by loss of coolant through leaks in the cooling system. Other causes include a loose or broken fan belt, a defective water pump, clogged water jackets in the engine, a defective radiator hose, and a defective thermostat or fan clutch.

Late ignition or valve timing, lack of engine oil, overloading the engine, or high-speed, high-altitude, or hot-climate operation can cause engine overheating. Freezing of the coolant can cause lack of coolant circulation, resulting in local hot spots and boiling. Also, if a faulty TCS system prevents vacuum advance in any gear, or if the distributor vacuum advance is defective, overheating may result.

♦ 29-10 ENGINE LACKS POWER

Many conditions can cause the engine to lose power. The wrong ignition timing, or any of the conditions discussed in ♦29-7 which cause the engine to miss, will reduce engine power. Also, a restricted exhaust system can create excessive back pressure which will prevent normal exhaust flow from the engine. The cylinders will retain pressure and will not take in a full air-fuel charge during the intake strokes. Heavy engine oil, the wrong fuel, or excessive rolling resistance can also give the impression of low engine power.

♦ 29-11 ENGINE DETONATES, OR PINGS

Detonation, or pinging (also called *spark knock*), is often blamed on the ignition system. But there are many other possible causes. In the ignition system, detonation may be caused by excessively advanced ignition timing, faulty advance mechanisms (which can cause excessive advances), and spark plugs of the wrong heat range. Fuel with an octane rating too low for the engine can cause pinging, or detonation. Carbon buildup in the engine combustion chambers can result in detonation in two ways. First, the carbon may glow or become so hot that it can cause preignition. This can result in ping. Second, the carbon buildup increases the compression ratio. This can also cause detonation.

♦ 29-12 SPARK PLUGS DEFECTIVE

Possible defects in spark plugs and their causes are described in ♦27-22. Basically, plugs that run too cold will foul. Plugs that run too hot will wear rapidly and burn away. This means the plug gap increases rapidly due to the eroding effect of the spark combined with the excessive temperature of the electrodes.

♦ 29-13 ENGINE DIESELS, OR RUNS ON

Engines with emission controls require a fairly high hot-idle speed for best operation. This makes run-on, or dieseling, possible. Hot spots in the combustion chambers, along with enough air-fuel mixture getting past a slightly opened throttle valve, can keep the engine running. The hot spots act as the spark plugs, igniting the mixture in the combustion chambers. Hot spots could be from hot spark plugs or exhaust valves, or from carbon deposits in the combustion chambers. Dieseling can damage an engine.

To prevent dieseling, many engines have an idle-stop solenoid. It closes the throttle valve completely when the ignition switch is turned off. If an engine runs on, or diesels, first check the idle-stop solenoid. Make sure it is releasing when the ignition switch is turned off to allow the throttle valve to close completely. Make sure the engine idle speed is not set too high. Engine run-on could also be caused by advanced ignition timing.

♦ 29-14 SERVICING ELECTRONIC IGNITION SYSTEMS

Although all electronic ignition systems operate in a similar manner, they require somewhat different checking and servicing procedures. The General Motors HEI system, with its coil mounted in the distributor, requires its own special procedure. Likewise, the Ford Duraspark systems and the Chrysler electronic system require their own individual testing and servicing procedures. For detailed, step-by-step procedures on how to test and service these systems, refer to the manufacturers' shop manuals. Some suggestions on the use of the oscilloscope and special points to watch when servicing electronic ignition systems follow.

♦ 29-15 OSCILLOSCOPE PATTERNS

The oscilloscope patterns for the various electronic ignition systems are not always the same. However, all have the common characteristic of increasing the dwell with speed. The term *dwell* is a carryover from the contact-point system. In that system, dwell is the number of degrees that the contact points are closed. This is the number of degrees of distributor-shaft rotation that the primary winding of the ignition coil remains connected to the battery. In the electronic ignition system, it means the same thing—the number of degrees of distributor-shaft rotation that the ignition-coil primary winding remains connected to the battery. Dwell in the electronic ignition system increases with engine (and distributor-shaft) speed. Figure 29-3 illustrates this for the General Motors HEI system. Figure 29-4 shows how, in this system, the dwell can vary considerably from cylinder to cylinder. This is considered normal in the HEI system.

In some HEI ignition systems, the rotor air gap (gap between the rotor tip and cap inserts) has been increased to 0.120 inch [3 mm]. This was done to reduce radio noise. When testing these systems with the oscilloscope, you will see an increase in the voltage pattern when you check for cap-rotor wear.

Note Always follow the oscilloscope operating instructions when checking electronic ignition systems. The instruction booklets illustrate the various patterns for the different ignition systems.

♦ 29-16 SERVICE TIPS FOR ELECTRONIC IGNITION SYSTEMS

Many of the service procedures required for servicing electronic ignition systems are similar to those used in servicing contact-point ignition systems. However, there are some special points to watch when working on electronic ignition systems. The similarities and differences are described below.

1. Spark Plugs How to analyze or "read" spark-plug condition is described in ♦27-22. Lead fouling should not be found if the engine has been running on unleaded gasoline.

2. Timing The timing procedure for electronic ignition systems is similar to that for timing contact-point ignition systems (♦27-20 and 27-21). The vehicle emission control information label in the engine compartment lists, step by step, the approved procedure for adjusting ignition timing. Follow these steps.

3. High-Voltage Cables These cables should be handled with care. While the insulation is electrically strong to contain the 47,000 volts the electronic systems can produce, it is relatively soft. These cables should never be punctured or bent sharply so the insulation is damaged. Any pinhole or other damage will allow the high voltage to leak and jump to

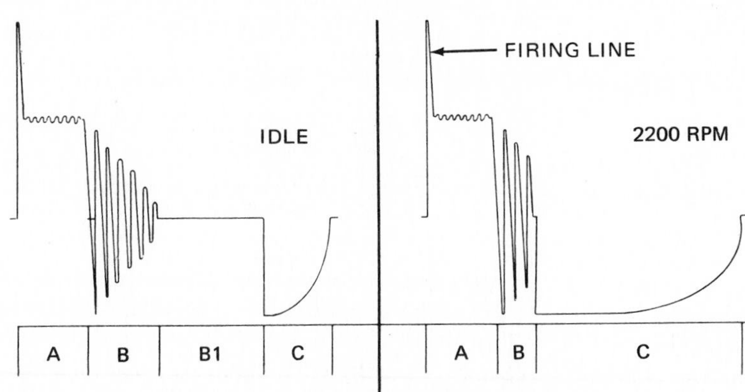

SCOPE INSTRUCTIONS:

1. SCOPE PATTERN PICKUP CANNOT BE CONNECTED BECAUSE CENTER COIL TERMINAL IS INSIDE DISTRIBUTOR.
2. CONNECT TRIGGER PICKUP TO NO. 1 SPARK-PLUG CABLE.
3. CONNECT PRIMARY PICKUP TO TACHOMETER TERMINAL OF DISTRIBUTOR. THIS WILL DISPLAY PRIMARY PATTERN IN PARADE ONLY. (NOTE: A SPECIAL ADAPTER PLACED ON TOP OF THE COIL-CAP ASSEMBLY MAY BE USED WITH SOME SCOPES TO VIEW THE SECONDARY PATTERN. THE OUTPUT VOLTAGE WILL READ LOW WITH THE ADAPTER.)

SCOPE PATTERN:

A. SPARK ZONE - SPARK-PLUG ARCING
B. COIL – CONDENSER ZONE
B1. FIRING ZONE – NO PLUG ARC
C. DWELL ZONE – MODULE ON, CURRENT THROUGH COIL PRIMARY WINDING

Fig. 29-3 Scope primary pattern for one cylinder of an HEI system, showing how the dwell increases with engine speed. *(Oldsmobile Division of General Motors Corporation)*

the nearest ground so the plug will not fire. The cables should be disconnected only if they are suspected of being faulty or if other tests of the system must be made. Figure 27-34 shows the use of cable pliers to disconnect a cable, with boot, from a spark plug. Figure 27-35 shows how to install a cable and boot on a distributor cap and the coil tower.

4. Silicone Grease Silicone grease is used to coat the rotor segment and cap electrodes on many distributors. A similar grease is used to coat the insides of the spark-plug boots. It is also used in harness connectors. The purpose of the grease in the distributor is to reduce radio interference from the high-voltage arc as it jumps the rotor gap from the rotor segment to the cap electrodes. The grease greatly reduces this radio-frequency interference (static). The grease also serves as added insulation against high-voltage leakage, and protects the connectors against corrosion.

5. Visual Inspection Before proceeding with actual tests of the system, always make a visual inspection. Inspect the wiring and cables, distributor cap, coil, and retainers that hold the cables in place. Check all connections for tightness. If there are no obvious defects, then proceed with the tests.

The Chrysler, Ford, and General Motors electronic ignition-system test procedures are different because the various systems are different. But all these systems work in the same general way. Refer to the manufacturer's shop manual for instructions on the specific system you are working on. Figure 29-5 shows an electronic ignition-system tester used on some Chrysler models. Figure 29-6 shows special adapters and procedures for testing different electronic ignition systems.

6. Oscilloscope Some oscilloscope patterns for electronic ignition systems are different from those for contact-point systems. Also, the patterns for the various electronic ignition systems may differ from one another. Follow the oscilloscope operating instructions. They illustrate the patterns, normal and abnormal, for the different systems.

7. Cable Boot When it is necessary to replace only a boot on a cable, cut off the old boot. Apply silicone lubricant to that part of the old cable that will be under the new boot. Use the special tool as shown in Fig. 29-7 to install the new boot. Push the tool through the new boot and into the cable clip. Slide the boot onto the cable and remove the tool.

8. Cables to Distributor Cap on Hall-Effect Distributors These cables are clipped into the cap terminals (Fig. 29-8). Hall-effect distributors are described in ♦28-4. To remove a cable, take off the cap. Then use needle-nose pliers to pinch the clip together so that the cable can be removed. The cable is installed by pinching the clip together and then pushing it into the cap terminal. As the clip clears the bottom of the terminal, it will snap open, thereby holding the cable securely in place. Push the boot down over the cap terminal. Install the cap.

♦ 29-17 CHECKING ELECTRONIC SPARK-TIMING SYSTEMS

These systems require special testers for a complete diagnosis. Figure 29-9 shows a typical tester. The face of the tester has two lists. One lists the steps to be taken to make the complete test. The other lists the numbers that may appear on the fault-code display on the tester face. This list is shown in Fig. 29-10. Note that the tester (Fig. 29-9) has a magnetic timing probe. This probe is inserted into the probe receptacle (Fig. 29-11). There the probe senses the rotation of the pulley and therefore engine speed and piston position.

Many cars have a CHECK ENGINE light on the instrument panel of the car (Fig. 18-22). If something goes wrong, the light comes on. This informs the driver that the system should be checked by a mechanic promptly. If a major electronic component fails, the system has a backup, or "limp-in," mode which enables the car to be driven to the mechanic. Then the mechanic can obtain a readout of the trouble-code numbers on a display panel (Fig. 18-21) or by the flashing of the CHECK ENGINE light (Fig. 18-22). The number shown tells the mechanic what has failed, or where the fault lies.

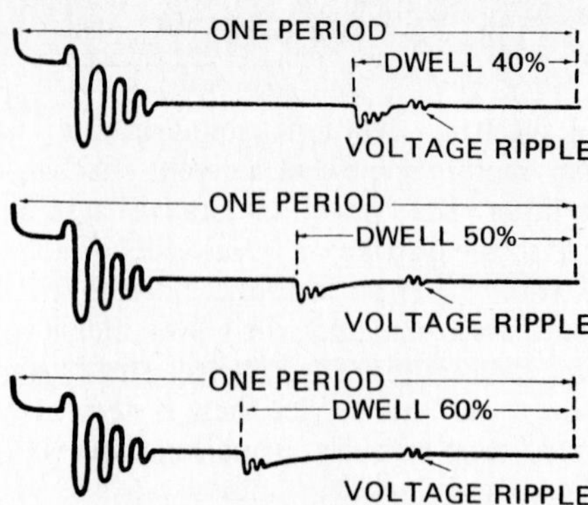

Fig. 29-4 Typical scope secondary patterns for different cylinders in an engine with an HEI system. Dwell varies considerably from cylinder to cylinder. The voltage ripple shown may or may not be seen. Either is normal. *(Oldsmobile Division of General Motors Corporation)*

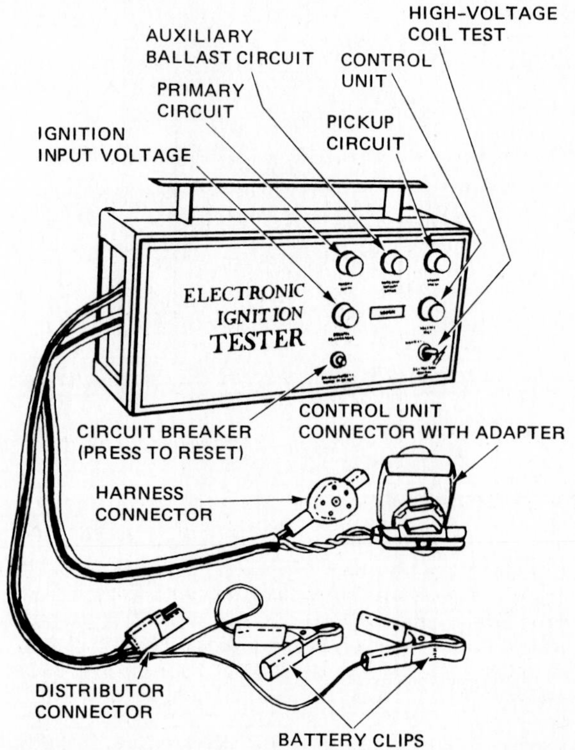

Fig. 29-5 Electronic ignition-system tester. *(Chrysler Corporation)*

NOTE: Before you begin testing, make sure the ignition key is in the "OFF" position and you have selected the correct adapter for the system being tested.

CHRYSLER ELECTRONIC IGNITION
(Except LEAN BURN)

1. Disconnect harness at control unit and install tester adapter between wire harness and control unit.
2. Connect Black clip (ground) to engine block.
3. Connect Red clip to (+) battery terminal.
4. Connect adapter to tester.
5. Test system (always press number switches in order).

FORD SOLID STATE IGNITION
and DURA SPARK SYSTEMS
(Including AMC with Ford Ignition System)

1. Disconnect harness at control unit and install tester adapter between wire harness and control unit.
2. Connect adapter to tester
3. Test system (always press number switches in order).

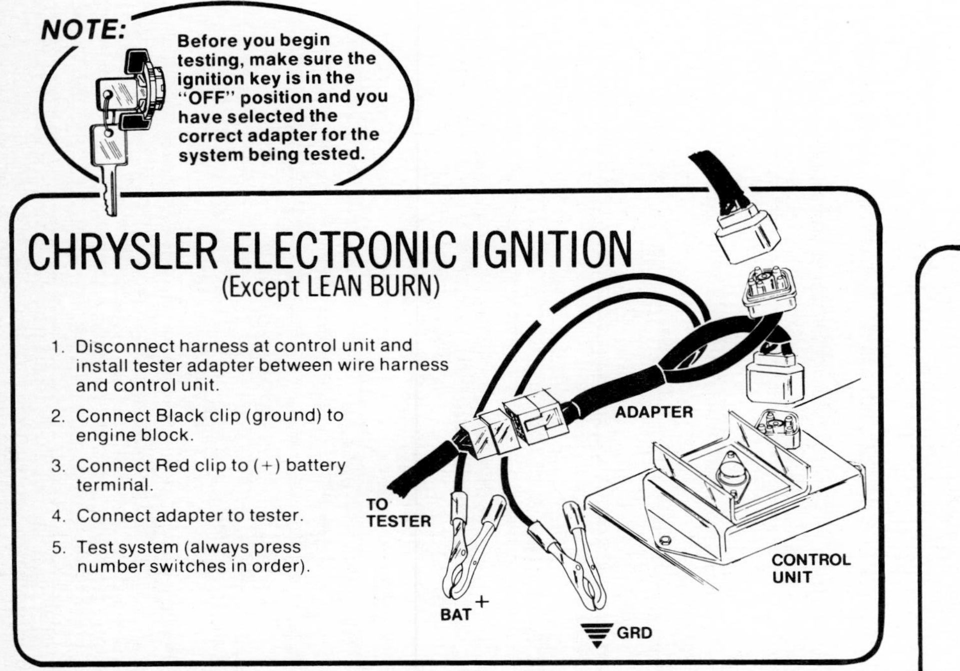

GENERAL MOTORS HIGH ENERGY IGNITION
(Except MISAR)

MODELS WITH COIL IN CAP:

1. Remove Red battery feed wire and 3 wire connector from distributor cap.
2. Remove cap and rotor from distributor. (leave spark plug wires in cap and all leads attached to module).
3. Connect Green adapter clip to module terminal "G".
4. Connect White adapter clip to module terminal "W". (position clips so distributor shaft turns freely).
5. Install adapter 3 wire connectors between distributor and cap.
6. Connect adapter Red jumper wire to cap terminal "BATT".
7. Connect Red battery feed wire to adapter.
8. Connect adapter to tester.
9. Test system (always press number switches in order).

MODELS WITH EXTERNAL COIL:

1. Remove connector from coil and install adapter between coil and harness.
2. Connect Black clip (ground) to engine block.
3. Remove distributor cap and rotor. (leave spark plug wires in cap and all leads attached to module).
4. Position cap to clear advance weights.
5. Connect adapter Green clip to module terminal "G".
6. Connect adapter White clip to module terminal "W". (position clips so distributor shaft turns freely).
7. Connect adapter to tester.
8. Test system (always press number switches in order).

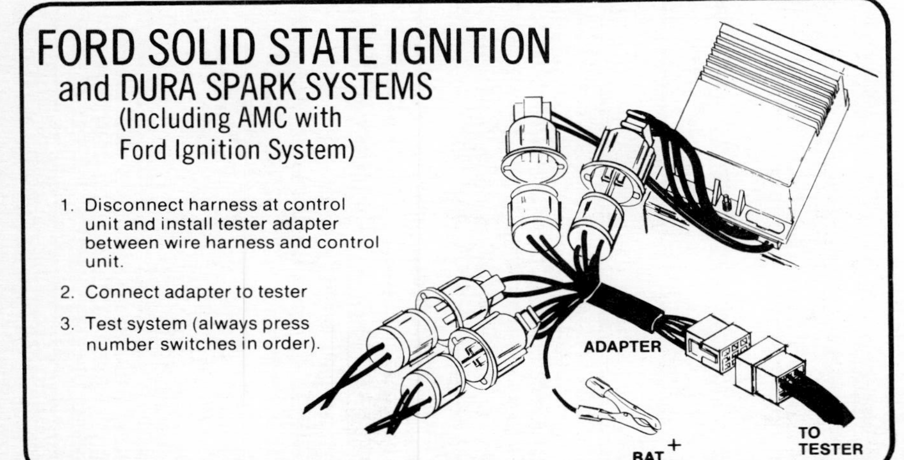

Fig. 29-6 Special adapters and procedures for testing different electronic ignition systems. *(Guaranteed Parts Company)*

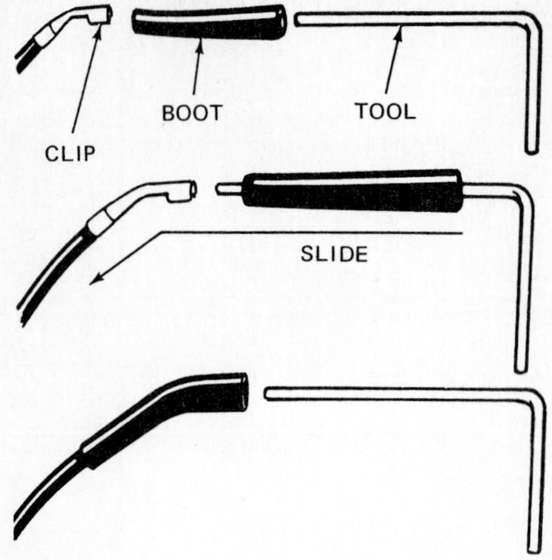

Fig. 29-7 Using a special tool to install a boot on a cable. *(Ford Motor Company)*

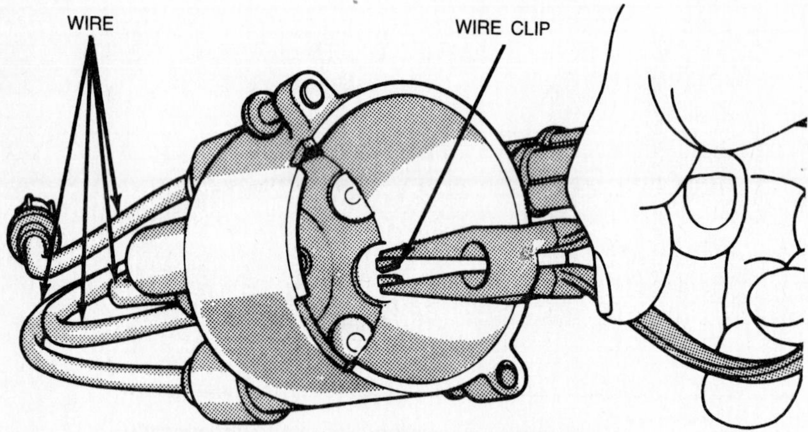

Fig. 29-8 Removing spark-plug cables from a Hall-effect-distributor cap. *(Chrysler Corporation)*

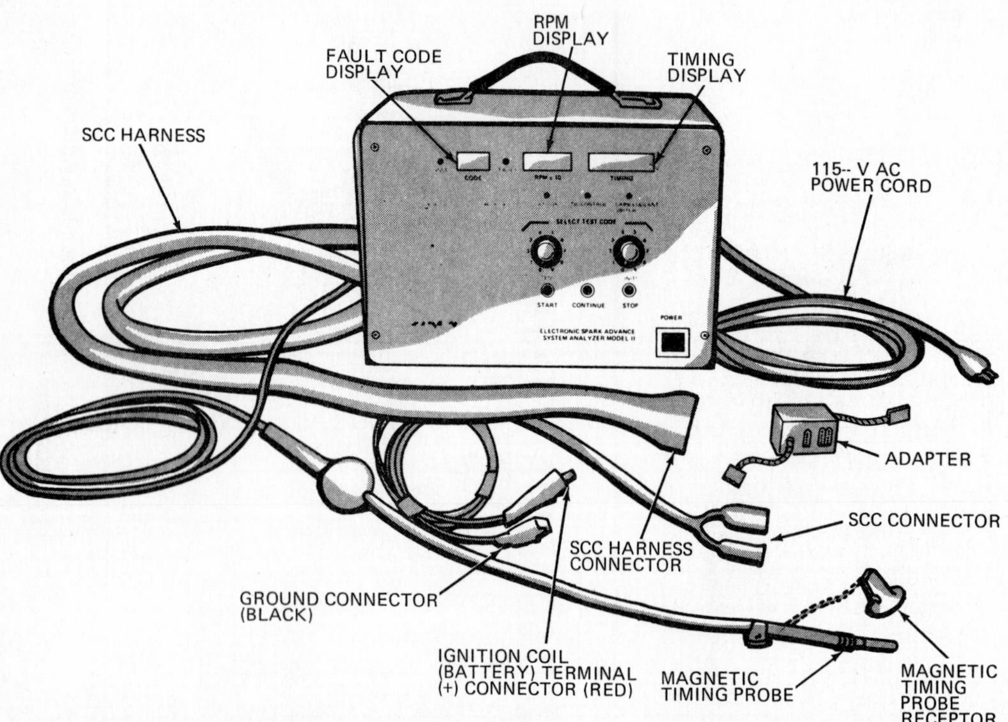

Fig. 29-9 Tester for electronic spark-timing ignition systems. *(Chrysler Corporation)*

Fault Code	Malfunction	Remarks
20	Invalid SCC code has been selected	Test code is stamped on side of SCC.
21–25	Analyzer fault	Rerun the test; if fault code continues, contact authorized service center.
28	Analyzer fault or operator error	Code 28 appears if the RED clip is not making proper contact at coil BAT (+) terminal.
29	Spare code	
30	Distributor bad	The relationship between the start and run pickups is incorrect. Replace pickup coils assembly.
31	Ignition voltage	Check all harness connections, battery, ignition switch, and wiring.
32	5-ohm ballast resistor out of tolerance	Check for continuity from ballast resistor to SCC connector. If okay replace the ballast resistor.
33	0.5-ohm ballast resistor out of tolerance	Check for continuity from ballast resistor to coil (+). If okay replace the ballast resistor.
34	SCC fault	Power or ground connections, SCC open, or SCC bad.
35	Coil resistance out of tolerance	Check connections to coil and wiring. If okay replace the coil.
36	Harness ground bad	Engine harness connector not grounded properly.
37	Start pickup bad	Check the pickup gap, wiring, and connections. If okay replace the pickup.
38	Start pickup phase bad	Wiring is reversed to start pickup. Check the wiring.
39	Run pickup bad	Check the pickup gap, wiring, and connections. If okay replace the pickup.
40	Run pickup phase bad	Wiring to run pickup is reversed; correct the wiring.
41	Ignition switch bad	Ignition switch is bad in the START position. Check the ignition bypass circuit wiring. If okay replace the switch.
42	Throttle transducer bad	Check wiring to transducer; if okay replace the transducer.
43, 44	Spare codes	
45–85	SCC fault	Record this code and attach to failed SCC.

Fig. 29-10 Fault-code listing and trouble-diagnosis guide. *(Chrysler Corporation)*

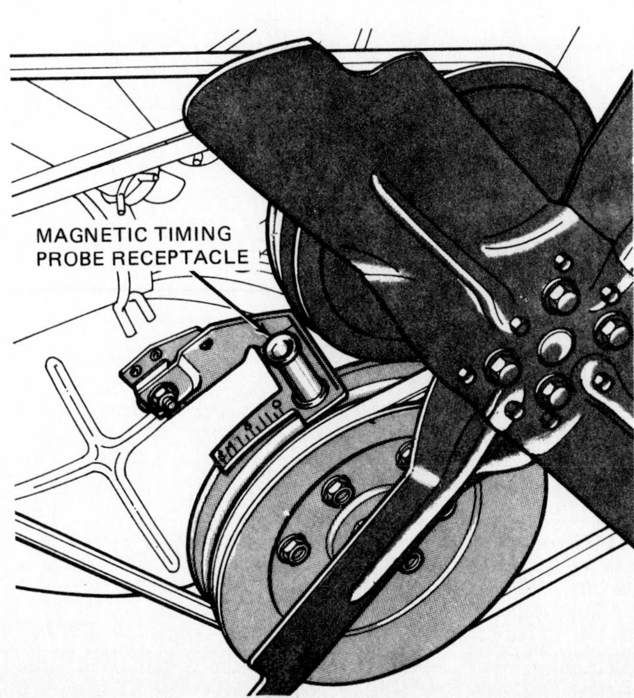

Fig. 29-11 Magnetic-timing-probe receptacle for an in-line engine. *(Chrysler Corporation)*

♦ REVIEW QUESTIONS ♦

Select the *one* correct, best, or most probable answer to each question. You can find the answers in the section indicated at the end of each question.

1. Failure of the engine to start although it cranks normally could be due to (♦29-5)
   a. no voltage to the ignition system
   b. ignition coil open
   c. ECU faulty
   d. any of these
2. Engine overheating can be due to (♦29-9)
   a. low battery
   b. early ignition timing
   c. late ignition timing
   d. high voltage setting
3. One cause of engine run-on, or dieseling, is (♦29-13)
   a. low battery
   b. idle-stop solenoid defective
   c. defective ECU
   d. defective EGR
4. You must be careful when making the spark test on electronic ignition systems because these systems can produce (♦29-5)
   a. 470 volts
   b. 4700 volts
   c. 47,000 volts
   d. 470,000 volts
5. Backfire can take place in (♦29-8)
   a. both the intake and exhaust manifolds
   b. only in the intake manifold
   c. only in the exhaust manifold
   d. only through the carburetor
6. In a comparison of the oscilloscope patterns for contact-point and electronic ignition systems, the dwell (♦29-15)
   a. is the same for both
   b. increases with speed in the electronic system
   c. decreases with speed in the electronic system
   d. none of the above
7. In the electronic ignition system, dwell means (♦29-15)
   a. the length of time the points are closed
   b. the length of time the points are open
   c. the number of degrees of distributor-shaft rotation during which current flows through the ignition-coil primary winding
   d. the number of degrees of distributor-shaft rotation between plug firings
8. The purpose of silicone grease is to (♦29-16)
   a. reduce radio interference
   b. insulate against high-voltage leakage
   c. protect against corrosion
   d. all of the above
9. To remove the cables from the Hall-effect distributor cap, (♦29-16)
   a. remove the cap and pinch the clips together with needle-nose pliers
   b. pull them out gently with a twisting motion
   c. use the special cable-removal pliers
   d. cannot be done because they are permanently fastened to the cap
10. The magnetic probe is inserted into the receptacle to sense (♦29-17)
    a. engine speed
    b. piston position
    c. engine rpm
    d. all of the above

CHAPTER 30
Other Electrical and Electronic Devices

After studying this chapter, and with proper instruction and equipment, you should be able to:

1. Replace a headlamp.
2. Aim headlights.
3. Describe the purpose and operation of the dimmer switch.
4. List the exterior lights on the car and describe the purpose and operation of each.
5. List the interior lights in the car and describe the purpose and operation of each.
6. Explain the operation of the horn relay.
7. Describe the operation of the speed-control system.

♦ 30-1 HEADLIGHTS

A typical lighting system for a car is shown in Fig. 23-7. The complete system includes headlights, parking lights, turn signals, side marker lights, stoplights, backup lights, taillights, and interior lights. The interior lights include instrument-panel lights, various warning and indicator lights, and courtesy lights that turn on when a car door is opened.

Figure 30-1 shows a headlight. It has a reflector and a tungsten filament at the back, and a lens at the front. When the filament is connected to the battery through the light switch, current flows through the filament. It glows white hot. The light is concentrated by the reflector into a forward beam and is focused by the lens.

Headlights are made in two types and four sizes, two round and two rectangular (Fig. 30-2). The round sizes are 5¾ inches [146 mm] in diameter and 7 inches [178 mm] in diameter. The rectangular sizes are 4 by 6½ inches [100 by 165 mm] and 5.6 by 9.7 inches [142 by 200 mm]. Both sizes are identified by the number 1 or 2 molded into the glass at the top of the lens. Type 1 has only one filament. Type 2 has two filaments, one for the high beam and the other for the low beam. The high beam is for driving on the highway when there is no car approaching from the other direction. The low beam is for city driving and for passing a car coming in the opposite direction. The use of the low beam in passing prevents the oncoming driver from being temporarily blinded by the high beam.

The rectangular headlights have become increasingly popular in recent years. They fit the tapered front-end style of the modern car.

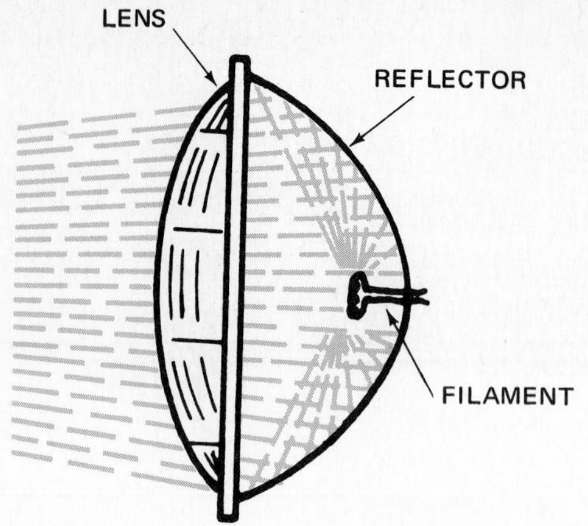

Fig. 30-1 Parts of a sealed-beam headlamp.

Some cars have only one pair of headlights. These are type 2. Other cars have two pairs of headlights—one pair of type 1 and one pair of type 2.

The driver uses the dimmer switch to select the filaments that will glow. For example, on a car having only one pair of headlights (type 2), the driver operates the dimmer switch to select either the high or the low beam. On a car having two sets of headlights (one set of type 1 and one set of type 2), the arrangement is different. When the driver operates the dimmer switch for low-beam driving, one of the filaments in the type-2 lights comes on. When the driver changes the dimmer switch to high beam, the other filament of the type-2 lights comes on. At the same time, the single filament of type-1 lights comes on.

Some foreign cars use headlights which have a separate small light bulb enclosed in a reflector, with a separate lens in front (Fig. 30-3). Certain advantages are claimed for this arrangement.

Many cars are now equipped with halogen or quartz headlamps. These lamps provide considerably more light. The headlamps are interchangeable with the standard type. However, cars with these headlamps should not have the standard headlamps installed if replacement is necessary. Standard headlamps on a car originally equipped with the halogen or quartz headlamp will not have the same appearance and light. Some customers could be dissatisfied.

♦ 30-2 HEADLIGHT COVER

Some cars have vacuum-operated headlight covers that move upward to expose the headlights when they are turned on. The system lowers the covers when the headlights are turned off. Either an electric motor or a vacuum motor is used to operate the covers. Figure 30-4 shows a system using a separate vacuum motor, or actuator, at each headlight, linked to a cover.

When the headlights are off, vacuum overcomes the tension of a spring to hold the covers in the lowered position. Pulling the light switch all the way out to turn on the headlights operates a distribution valve. The valve is mounted on the back end of the light switch. Now the distribution valve directs vacuum to the other side of the diaphragms in the vacuum motors. This causes the vacuum motors to lift the headlight covers.

When the headlights are turned off, the distribution valve allows atmospheric pressure to enter the vacuum side of the vacuum motors. Now, springs on the headlight covers cause them to drop to cover the headlights.

A vacuum reservoir holds sufficient vacuum for several cover operations. It is used if the headlights are turned on and off when the engine is not running. Also, the headlight covers can be operated manually in case the vacuum system fails. This is done by turning on the headlights and then lifting the covers by hand.

Springs are built into the vacuum motors to automatically open the covers if the system looses vacuum. The spring force may also open the covers if the engine is not started for about 18 hours or longer.

Note In some cars, the entire headlamp module rotates to swing the headlamps up into position when the light switch is turned on.

♦ 30-3 HEADLIGHT AIMING

The headlights must be aimed correctly. If they are aimed too high or to the left, they might blind an oncoming driver and cause a serious accident. Incorrect aiming can also reduce the driver's ability to see the road properly. This could also lead to an accident.

Headlights have three aiming buttons on the front lens. These are for mounting a mechanical aimer. Aiming adjustments are made by turning screws in a spring-loaded holder. There is one spring-loaded screw at the top for up-and-down adjustment, and one at the side for left-to-right adjustment (Fig. 30-2).

There are several methods of checking the aiming of headlights. These include the use of an aiming screen, mechanical aimer, and an optical aimer. The simplest method uses a screen set 25 feet [7.6 m] in front of the vehicle and a perfectly level floor (Fig. 30-5). With the car aligned perpendicular to the screen, the low beam and high beam can be checked

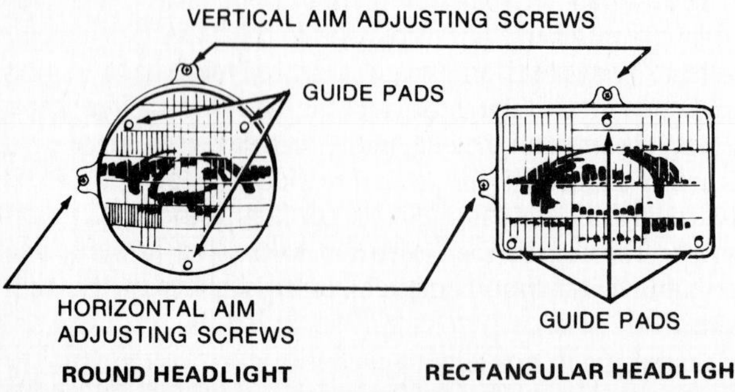

Fig. 30-2 Shapes of round and rectangular headlamps.

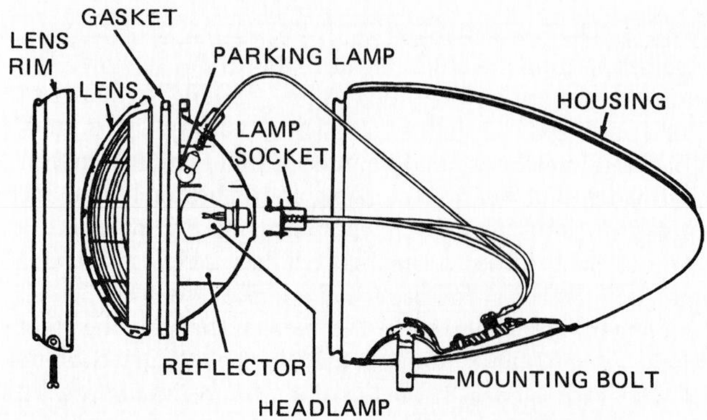

Fig. 30-3 Type of headlight using a separate headlamp and lens.

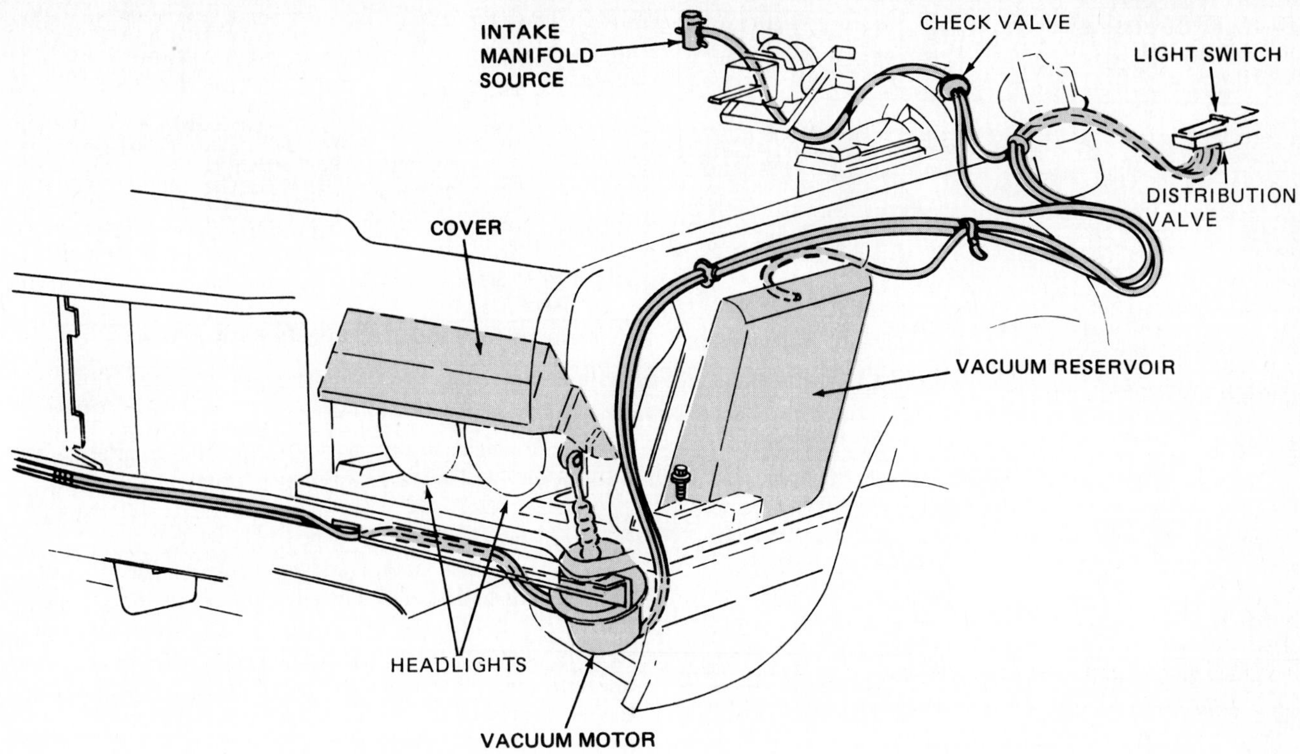

Fig. 30-4 A headlight-cover control system. *(Ford Motor Company)*

(Fig. 30-6). Note that the centerlines of the lights and the centerline of the car are indicated. Also shown are the hot-spot (or high-intensity-area) reference lines. The low beams are adjusted so that the hot spot is below and to the right of the reference lines.

Some manufacturers, in their aiming instructions, recommend a full fuel tank and an empty car. Others recommend the car loaded as it is normally driven with the driver behind the steering wheel. Tires must be inflated to the specified pressure. Before checking the aim, after the car has been positioned, bounce each corner of the car a couple of times to equalize the suspension system.

A variety of headlight-aiming instruments are available. Learn to use as many different types as possible.

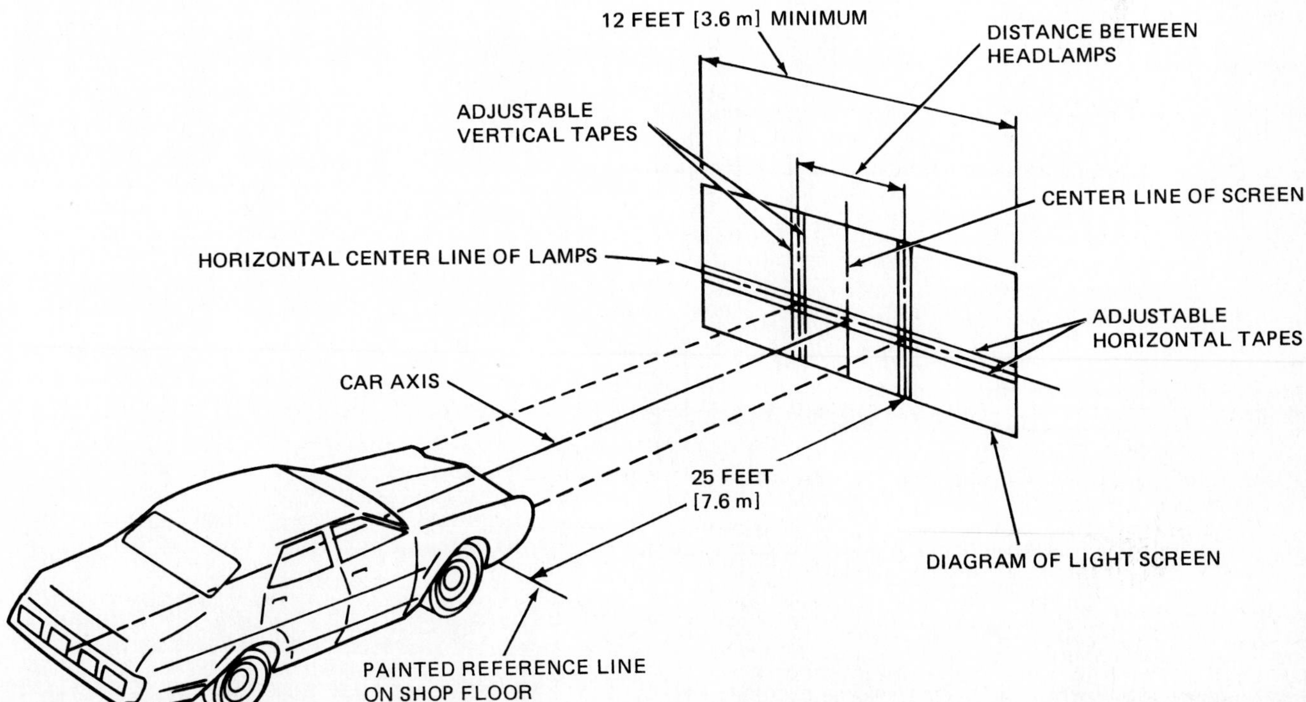

Fig. 30-5 Floor-type headlight aiming screen.

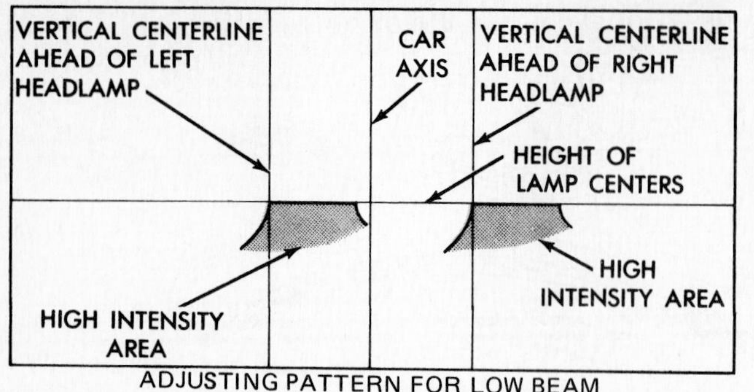

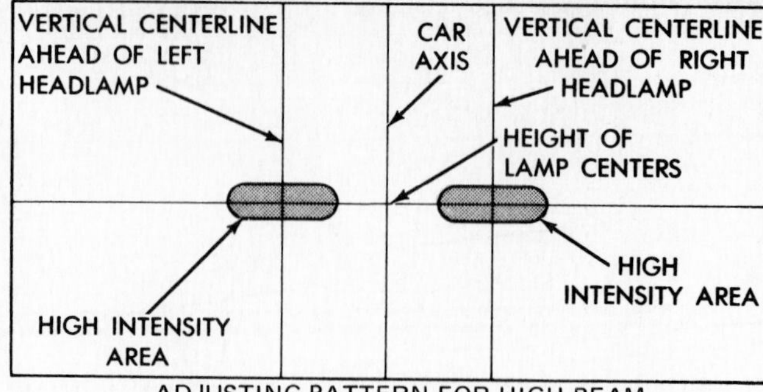

Fig. 30-6 Headlight patterns for low beam (left) and high beam (right). *(Chrysler Corporation)*

♦ 30-4 AUTOMATIC HEADLAMP DIMMER

This is a system which electronically selects the proper headlamp beam for country driving. It holds the lights on upper, or high, beam until a car approaches from the other direction. Then, the headlamps of the approaching car trigger the system so that it shifts the headlamps to the lower beam. When the other car has passed, the system electronically shifts the headlamps back to the upper beam.

Figure 30-7 is the wiring diagram of the system. The sensor amplifier combines a light-sensing optical device and a transistorized amplifier. It is located at the front of the car where it can sense the headlights of oncoming cars. The driver has a sensitivity control (center in Fig. 30-7) which allows adjustment of the system to the surrounding light. The dimmer switch is an override switch which allows the driver to manually control the system. This might be used, for example, if an oncoming car does not dim and the driver needs more light to see by. Operating the dimmer switch returns the headlamps to the upper beam.

♦ 30-5 AUTOMATIC ON-OFF AND TIME-DELAY HEADLAMP CONTROL

This is a device that electronically controls the on-off operation of the headlamps and taillights. It has various names such as Twilight Sentinel and Safeguard Sentinel. It is mounted under the instrument-panel grille, facing upward so that it is exposed to direct outside light through the windshield (Fig. 30-8). The internal resistance of the photocell varies according to the amount of light striking it. As the amount of light is reduced, the internal resistance of the photocell increases until it finally causes the amplifier to turn the lights on. It turns the lights off if the amount of external light increases enough.

The amplifier contains a transistorized amplifier unit, a

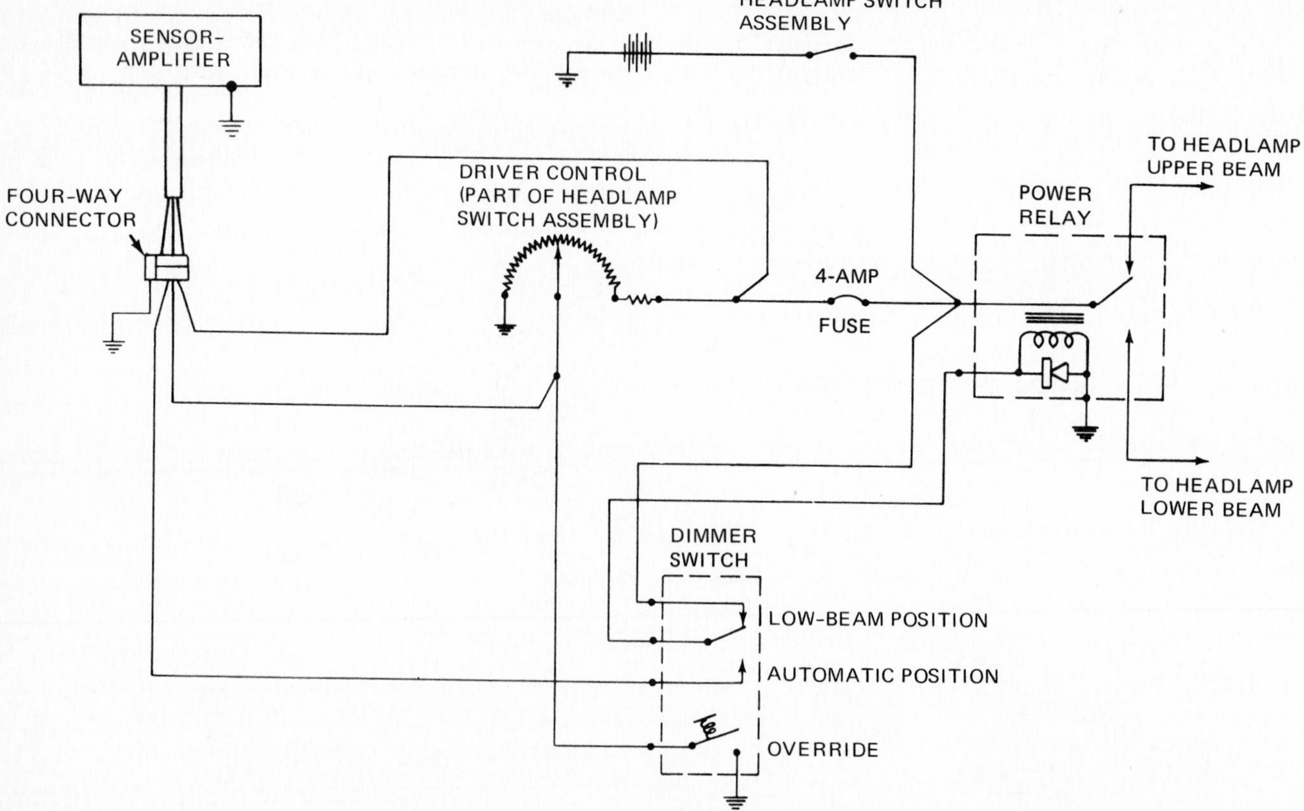

Fig. 30-7 Wiring diagram for an electronic headlamp-dimmer system. *(Ford Motor Company)*

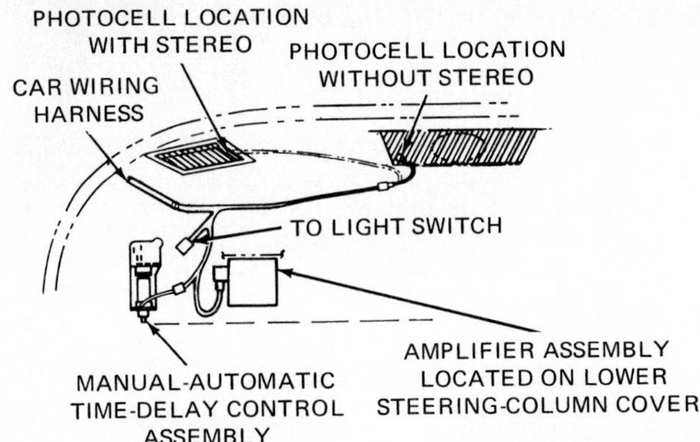

Fig. 30-8 Locations of the photocell in an electronic headlamp-dimmer system. *(Cadillac Motor Car Division of General Motors Corporation)*

sensitive relay, a power relay, and a transistorized time-delay unit. The time-delay unit delays the turning on or off of the lights. For example, it delays the turning on of the lights anywhere from 10 to 60 seconds. This keeps the lights from coming on in the daytime when the car is passing under an underpass or trees. The time delay for turning off the lights is adjustable. There is a control lever on the light switch which can be swung in one direction or the other to change the delay period from a few seconds to 4½ minutes. This permits the driver to drive into the garage, leave the lights on, get out of the car, lock the garage, enter the house, and lock the house door, all in the light from the car headlamps before they are turned off automatically.

♦ 30-6 HEADLAMPS-ON WARNING BUZZER

This system is usually combined with the open-door warning-buzzer system. When the headlamps are on and the driver's door is opened, a warning buzzer or chime sounds. This warns the driver to turn off the headlamps before leaving the car.

♦ 30-7 HEADLAMP REPLACEMENT

Headlamps are installed in a variety of ways. The assembly includes an adjustment ring in back of the headlamp and a retaining ring in front of the headlamp. To replace a headlamp, remove the headlamp door, retainer ring, and headlamp bulb. Be sure to replace the old bulb with a new one of the proper type. After installation, aim the headlamps (♦30-3).

♦ CAUTION ♦ The halogen headlamp bulb contains gas under pressure. The bulb may shatter if the glass envelope is scratched or if the bulb is dropped. Handle the bulb carefully. Grasp the bulb only by its plastic base. Avoid touching the glass envelope. Keep the bulb out of the reach of children.

♦ 30-8 HEADLIGHT SWITCH

The headlight switch in most cars controls the headlamps, parking lamps, marker lamps, taillamps, license-plate lamp, courtesy lamps, instrument-panel lamps, and ashtray lamps.

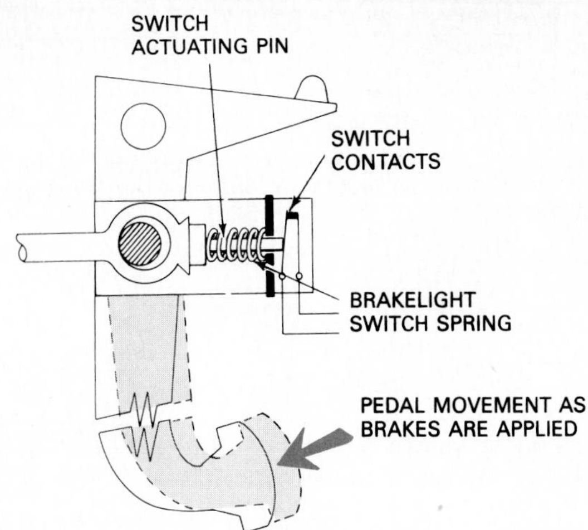

Fig. 30-9 Mechanical brakelight switch. When brakes are applied, as shown, the brake-pedal movement allows the switch contacts to close. *(Ford Motor Company)*

♦ 30-9 BRAKELIGHT SWITCH

A mechanical brakelight switch is used with the dual-brake system on todays cars. Figure 30-9 shows one type. When the pedal is pushed for braking, it carries the switch contacts with it. This brings the switch contacts together so that the stoplights come on.

♦ 30-10 DIRECTIONAL SIGNALS

Directional signals permit the driver to signal an intention to turn right or left. They are operated by a switch on the steering column. When the switch lever is moved (right for a right turn, left for a left turn), circuits are completed between the battery and the appropriate lights. There is a flasher in the circuit which intermittently closes and opens contacts. This causes the turn-indicator lights to flash on and off. This intermittent flashing makes the lights more noticeable.

The flasher contains a thermostatic blade and a heater. The heater carries the circuit current and heats up. This heating causes the blade to bend, opening the contacts. Now, current stops flowing, and the blade cools, straightens, and closes the circuit. The cycle is repeated as long as the turn-signal switch is closed.

♦ 30-11 BACKUP LIGHTS

The backup lights come on when the driver shifts into reverse. The switch contacts are closed when the shift lever is moved to R (reverse). This connects the backup lights at the rear of the vehicle with the battery.

♦ 30-12 EMERGENCY OR HAZARD FLASHER

The emergency-flasher, or hazard-warning system, is designed to signal following cars that a car has stopped or stalled or has pulled up to the side of the road. When the driver operates the flasher switch, it causes all four directional signal lights to flash on and off. The system includes a flasher similar to the one used for directional signals. The system is operated by a switch usually located on the side of the steering column.

♦ 30-13 COURTESY LIGHTS

Courtesy lights come on when the car doors are opened so that passengers or driver can see to get in or out of the car. The courtesy lights are operated by switches in the door posts or pillars. When a door is opened, the switch closes to complete the circuit by providing a ground for the domelight or side lights.

♦ 30-14 FIBER-OPTIC MONITOR SYSTEMS

In many cars, the instrument panel requires lights at many places—to illuminate the speedometer, the indicating gauges, and the various controls. Because of the small spaces available, it becomes a problem to locate light bulbs where they are needed. To eliminate this problem, some cars use *fiber-optic conductors.* These conductors are made up of a very large number of very fine and flexible threads, or fibers, of glass which are bound together into a bundle, or cord. Each fiber has the property of being able to conduct light, even around bends or corners. As light starts down the fiber, it is reflected off the outer surfaces of the fiber. If the fiber is curved, the light keeps bouncing off the outer surfaces with little loss. By the time the light comes out the other end of the fiber, it is almost as strong as when it entered.

Now, to utilize this effect, fiber bundles (each with many fibers) are run from a central light source to the various outlets on the instrument panel where light is needed. Therefore, only one light bulb is needed to provide light at many places. Installation and servicing problems are made easier to solve by the use of the fiber bundles. Only one light bulb needs to be replaced if a burnout occurs, and the fiber bundles can be bent almost any way without damaging them.

Fiber optics are also used in a *lights-out* warning system. This allows the driver to check the operation of the exterior lights without leaving the car. Fiber-optic conductors are connected at one end to the headlamps, at the other to a readout mounted in a housing on the top of the fenders (Fig. 30-10). If a headlight is on, light from it passes through the fiber-optic conductor to the readout. The high-beam-headlamp readout is blue (on one car), which matches the high-beam indicator in the speedometer face. The low-beam readout is clear, not colored.

The taillamp monitor system has fiber-optic conductors running from the taillights to a readout in a housing in the headliner or roof of the car, just above the rear window. When the taillights are on, the spots of light in the readout tell the driver that they are on.

♦ 30-15 HORNS AND HORN RELAYS

The automotive horn is of the vibrating type. It has a field coil, a set of contact points, and a metal diaphragm. When the horn button is pressed, this closes the circuit and current starts flowing through the field coil. This produces a magnetic field that pulls the diaphragm down. The diaphragm movement produces a click. As the diaphragm moves down, the contacts separate so no current can flow. The magnetic field in the field coil collapses, and the diaphragm is released. It moves up with another click. This action is repeated, so rapidly that the separate clicks blend to form the sound you hear.

The horn relay (Fig. 30-11) has a single winding that is connected through the horn button to the battery. Current flowing through the winding produces a magnetic field that pulls the armature down and closes the contact points. This connects the horns directly to the battery so the horns sound (Fig. 30-11).

In today's cars, the horn relay has a second job. If the driver leaves the ignition key in the ignition switch and then opens the car door, the horn relay buzzes. The circuit for this arrangement is shown in Fig. 30-11. When the ignition key is left in the ignition switch, the warning switch remains closed. The warning switch is located in the ignition switch and is connected to the door switch. The door switch is closed as the door is opened. This completes the circuit to the horn-relay winding. The circuit runs through a special set of contact points above the armature. When the circuit to the winding is completed, the winding magnetism pulls the armature

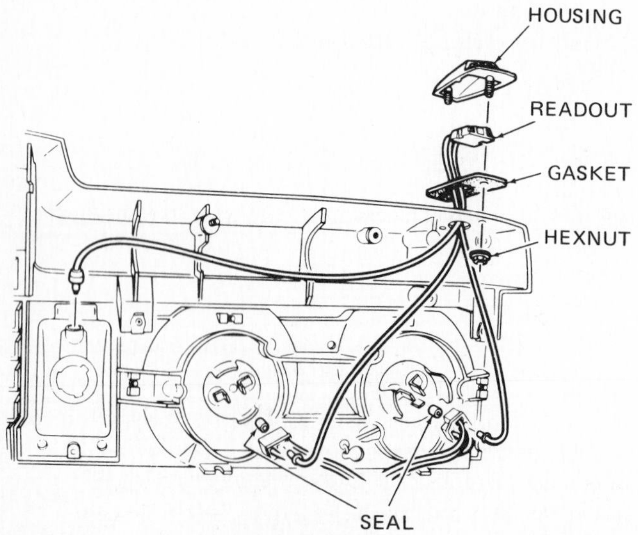

Fig. 30-10 Fiber-optic monitoring system for headlamps. *(Cadillac Motor Car Division of General Motors Corporation)*

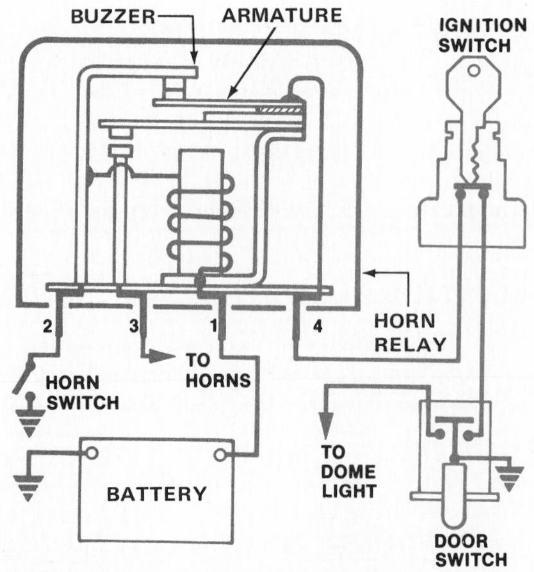

Fig. 30-11 Horn relay which includes a key-in-switch warning buzzer.

down. This opens the upper points to open the winding circuit. The magnetic field collapses and the armature moves back up. The points close, and the action is repeated. This action produces a buzzing sound that warns the driver to remove the ignition key.

In some cars, the buzzer will also sound if the driver leaves the car with the headlights on. This is a reminder to turn off the headlights so that the battery will not run down while the driver is absent from the car.

Note In many late-model cars, chimes sound instead of a buzzer. Also, some cars respond with a vocal warning (♦30-16, item 3).

♦ 30-16 INDICATING DEVICES

Cars have an ammeter or charge indicator, a fuel gauge, an oil-pressure gauge or indicator, and an engine-temperature indicator. These have all been discussed in earlier chapters. Some cars have monitor panels that warn of problems in the brakes, lights, and other components. Other cars have voice alert systems that warn, vocally, of certain troubles.

1. Monitor Panels Figure 30-12 shows one monitor system which has a series of lights. To make a test of the various components, the test bar is pressed. If all lights go on, everything is okay. If a light does not go on, something is wrong with the system that the light monitors.

2. Trip Computer Figure 30-13 shows the test buttons in a trip computer. By touching various buttons, the driver can call up the following information:

- ♦ Miles per gallon of fuel
- ♦ Average speed
- ♦ Time of day
- ♦ Elapsed time since starting

Fig. 30-13 Cadillac trip computer. At top, the driver is touching the MPG button. The display indicates that the car is getting 16 miles per gallon of fuel. At bottom are the 12 buttons of the system. *(Cadillac Motor Car Division of General Motors Corporation)*

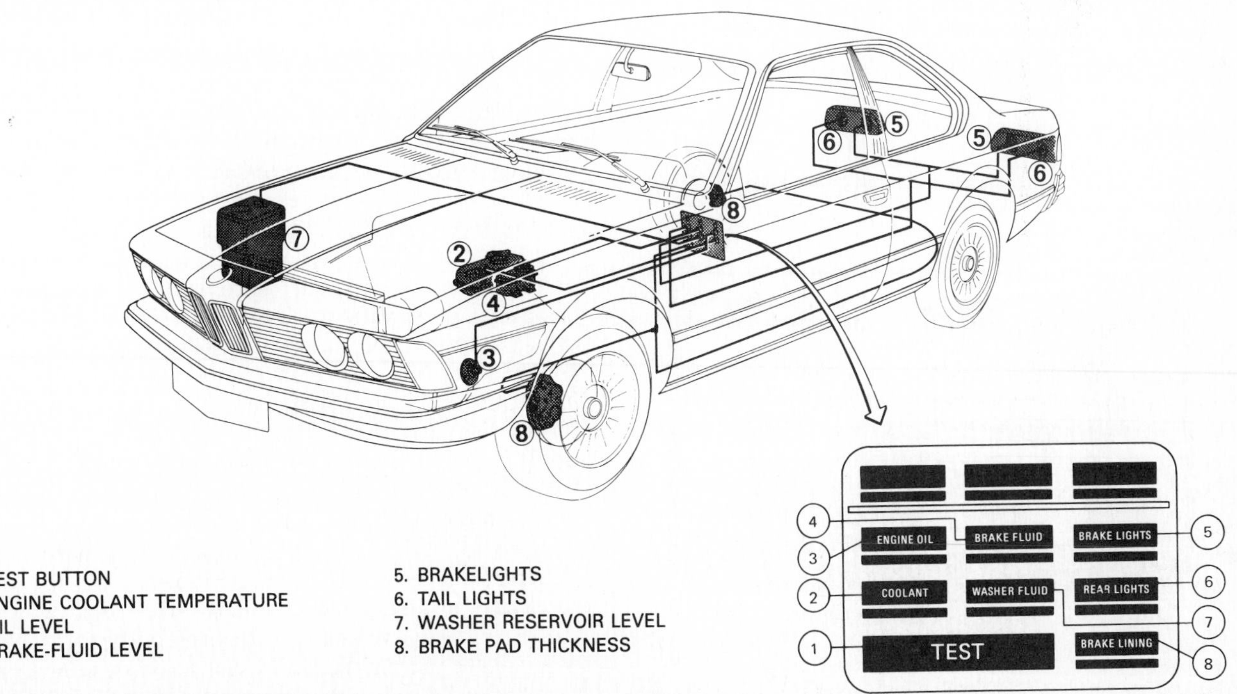

Fig. 30-12 Location of the monitor on the instrument panel and the components it monitors. *(BMW of North America, Inc.)*

♦ Driving range in miles on remaining fuel supply
♦ Miles to predetermined destination
♦ Estimated arrival time
♦ Engine rpm
♦ Engine temperature
♦ System voltage

3. Vocal Alert System Some cars have a system that gives a vocal warning if the driver leaves the car with the key in the ignition, or if the door is left ajar. The system is equipped with several vocal warning messages.

A variation on this is a voice command system which responds to the driver's voice. The driver can tell the system to open the driver's window, and to turn to any of five stations on the radio.

♦ 30-17 SPEEDOMETER AND ODOMETER

The mechanical speedometer and odometer are not electrical devices. However, they are mounted on the car instrument panel, along with the other instruments. The speedometer tells the driver how fast the car is going. The odometer tells the driver the distance the car has traveled. Figure 30-14 is a cutaway view of the assembly.

There is a small magnet mounted on a shaft inside the speedometer. This magnet is driven by a flexible cable from the transmission. The faster the car goes, the faster the magnet spins. This action produces a rotating magnetic field that drags on the metal ring surrounding the magnet. The faster the spinning, the more drag on the ring. The spinning causes the ring to swing around against the tension of a spring. This, in turn, moves a pointer attached to the ring, which indicates car speed.

The odometer is operated by a pair of gears from the same rotating flexible cable that drives the speedometer. The motion is carried through the gears to the mileage rings on the odometer indicator. These rings turn to show how many miles the car has been driven.

The cable is usually driven from a pair of gears in the rear extension housing of the transmission. One of these gears is on the main shaft of the transmission. The other is on the end of the flexible cable.

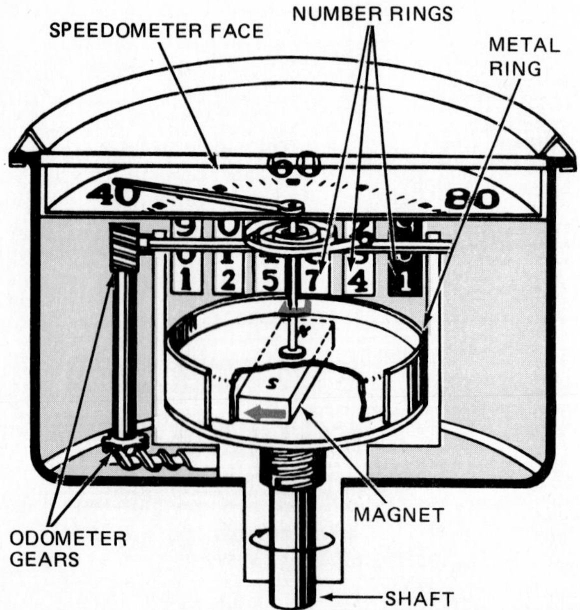

Fig. 30-14 A speedometer-odometer assembly.

♦ 30-18 WINDSHIELD WIPER

Windshield wipers are driven by an electric motor. The motor, through gearing, causes the wiper blades to move back and forth on the windshield. Most cars have a windshield washer as part of the windshield-wiper system. When the driver presses a button, a liquid squirts on the windshield so that the blades can clean more effectively.

Many cars have a controlled-action wiper. One name for this system is the *multiplex pulse-wiper system.* It includes a delay control which can be adjusted from zero delay to 12 seconds' delay between blade movements. The wiper blades will move across and back, pause, and then repeat the wiper action.

♦ 30-19 ANTILOCK BRAKING SYSTEM

The antilock brake system reduces the tendency of the tires to skid by preventing wheel lock. This system, which has electronic components, is covered later in ♦51-29.

♦ 30-20 AUTOMATIC LEVEL CONTROL

The automatic-level-control system takes care of changes in the load at the rear of the car. The system and its electronic components are described in ♦48-23.

♦ 30-21 SPEED- (CRUISE-) CONTROL SYSTEM

This system allows the driver to pick a speed (anything above about 30 mph), set a control, and take his or her foot off the accelerator pedal. The system will then hold the car to the speed set. If the car starts to slow down when going up a hill, the throttle valve automatically opens. If the car starts to speed up coasting down a hill, the throttle valve automatically closes.

The system includes two switches—SET and OFF-ON-RESUME—a throttle actuator, a speed sensor, an amplifier, wires, check-valve assembly, vacuum reserve tank, and vacuum hoses. The switches are mounted either in the spokes of the steering wheel, or on the turn-signal lever (Fig. 30-15).

The speed sensor is located in back of the speedometer. An electronic device "reads" the car speed by noting how fast the speed indicator in the speedometer is rotating. This information is fed to the electronic controller. The controller then compares this with the speed the driver has set in the memory. If car speed is too low, the controller signals the vacuum-control valve. The valve then admits more air to the power unit. This causes the diaphragm in the power unit to move. The motion carries to the throttle through linkage, and the throttle opens to increase engine power so car speed goes up. When the preset car speed is reached, the power unit eases off to prevent any further increase in car speed.

When the OFF-ON switch is ON, the system memory is ready to accept a speed value set by the driver. When the driver accelerates to the desired speed and presses the SET switch, the system takes over and holds that speed. The speed may be decreased by moving the OFF-ON switch to OFF, or by tapping the brake pedal. When the driver moves the OFF-ON switch to RESUME, the system automatically accelerates the car to the preset speed.

Fig. 30-15 The cruise-control system is operated by two switches, a SET button in the left end of the turn-signal lever and an OFF-ON-RESUME switch mounted on the lever. *(Pontiac Motor Division of General Motors Corporation)*

♦ 30-22 SEAT ADJUSTER

Many cars have a motor-powered seat adjuster. Figure 30-16 shows one type. It is a six-way adjuster, which moves the seat forward or backward or up or down and tilts the seat forward or backward. The mechanism includes a drive motor, drive cables, jack screws, and a transmission. Depending on which lever the driver operates, the movement of the motor puts into action one or another of the drive cables. The mechanism causes the seat to move in the direction selected by the driver.

♦ 30-23 WINDOW REGULATORS

Many cars have power window regulators. They are operated by an electric motor that causes the window to be raised or lowered.

♦ 30-24 POWER DOOR LOCKS

Power door locks use a motor to operate the locking mechanism (Fig. 30-17). The system is actuated by a control switch at each front door. However, each door lock can be operated manually by lifting or pushing down the door-lock knob, or by moving the door-lock lever forward or backward.

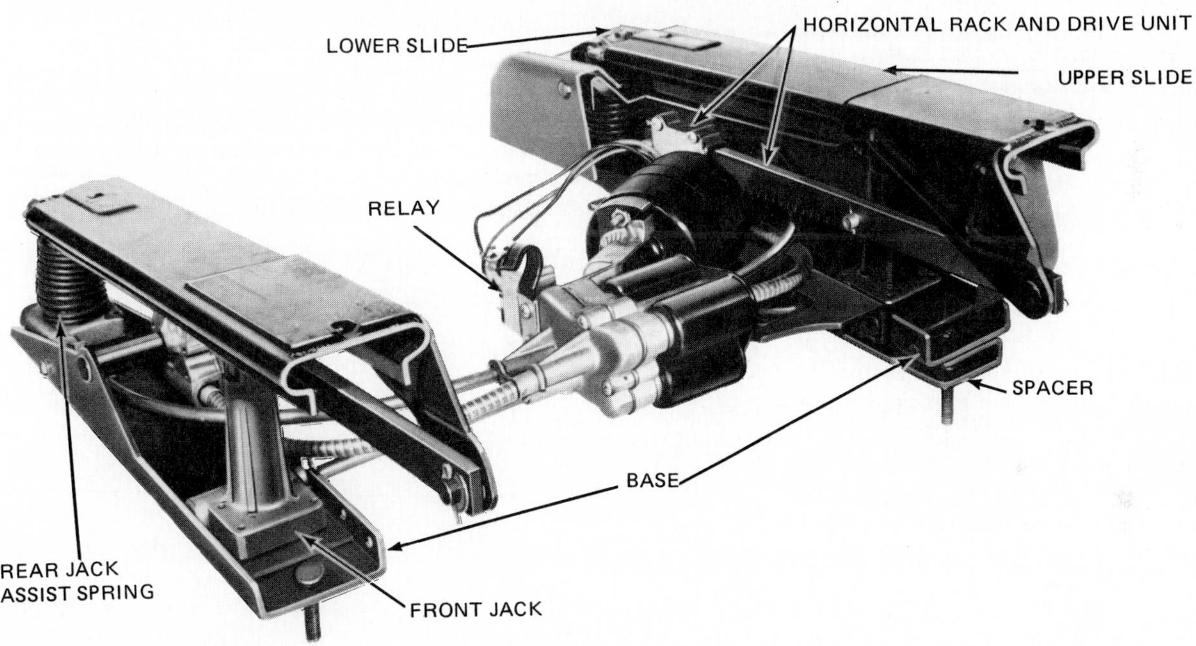

Fig. 30-16 Six-way power seat adjuster. *(Chrysler Corporation)*

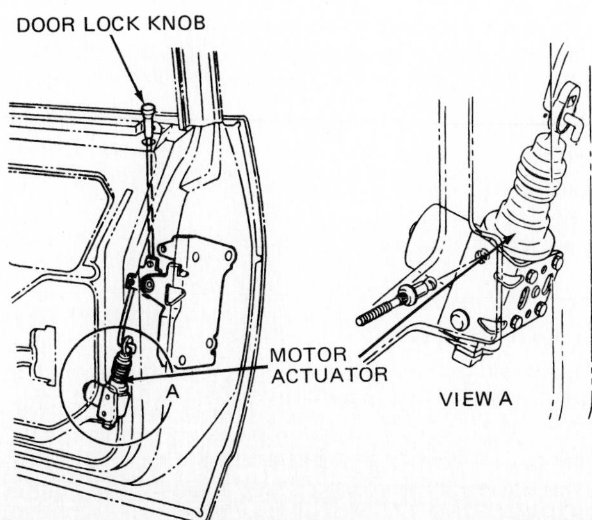

Fig. 30-17 Power door-lock actuator. *(Chevrolet Motor Division of General Motors Corporation)*

♦ REVIEW QUESTIONS ♦

Select the *one* correct, best, or most probable answer to each question. You can find the answers in the section indicated at the end of each question.

1. After replacing a headlamp, the driver complains that the new headlamp is dim. Mechanic A says a halogen headlamp has been installed in place of a standard headlamp. Mechanic B says a standard headlamp has been installed in place of a halogen headlamp. Who is right? (♦30-1)
   a. mechanic A
   b. mechanic B
   c. both A and B
   d. neither A nor B
2. The headlamp-beam dimmer has (♦30-1)
   a. one position
   b. two positions
   c. three positions
   d. four positions
3. The four sizes that headlamps are made in include (♦30-1)
   a. three square and one round
   b. three round and one rectangular
   c. two round and two rectangular
   d. all rectangular
4. A car with two sets of headlamps has (♦30-1)
   a. one pair of type 1 and one pair of type 2
   b. one round pair and one rectangular pair
   c. four headlamps each with one filament
   d. four headlamps each with two filaments
5. Headlight aiming is done by (♦30-3)
   a. moving the light bulb in back of the lens
   b. turning spring-loaded adjustment screws
   c. rotating the headlamps in their sockets
   d. bending the adjustment brackets
6. The automatic on-off and time-delay headlamp control (♦30-5)
   a. turns the headlamps off as the driver gets out of the car
   b. times the flashing of the lights when the hazard system is energized
   c. turns the headlamps off after a preset time delay following the turning off of the engine
   d. turns the headlamps off 13 minutes after the driver leaves the car
7. The horn has (♦30-15)
   a. vibrating contact points
   b. a vibrating diaphragm
   c. an electromagnet
   d. all of the above
8. The purpose of the automatic level control is to (♦30-20)
   a. increase the car height as it is loaded
   b. maintain the correct level of the rear as it is loaded or unloaded
   c. pump up the shock absorbers if a rough pavement is encountered
   d. instantly correct the car level as a load is applied
9. Information that the trip computer can supply includes (♦30-16)
   a. miles per gallon of fuel
   b. miles to predetermined destination
   c. estimated arrival time
   d. all of the above
10. Mechanic A says the odometer reports the total miles or kilometers the car has traveled. Mechanic B says the speedometer reports the miles or kilometers per hour the car is traveling. Who is right? (♦30-17)
    a. mechanic A
    b. mechanic B
    c. both A and B
    d. neither A nor B

PART 5

Part 5 of this book discusses the devices installed on engines and automobiles, as well as design changes in the automobile, to reduce automotive pollution.

Tampering with automotive emission controls is unlawful. If an employer or employee is caught tampering, the employer can be fined $2500.00 for each vehicle that has been tampered with. *Tampering* is defined as removing, disconnecting, damaging, or in any way rendering ineffective any emission control device installed on a motor vehicle or motor-vehicle engine.

There are two chapters in Part 5:

Automotive Emission Controls

CHAPTER 31
Automotive Emission Control Systems

After studying this chapter, you should be able to:

1. Identify the components of the PCV and fuel-vapor control systems on several vehicles.
2. Describe the construction and operation of the PCV valve.
3. Explain the operation of the evaporative emission control system.
4. Describe three methods of cleaning the exhaust gas.
5. Explain the purpose and operation of the EGR system.
6. Describe the construction and actions in two-way and three-way catalytic converters.
7. Explain the operation of electronic engine control systems that include an oxygen sensor and feedback system.

♦ 31-1 ATMOSPHERIC POLLUTION AND THE AUTOMOBILE

There are four possible sources of atmospheric pollution from the automobile (Fig. 31-1). Without emission controls, the carburetor and fuel tank can emit fuel vapors, the crankcase can emit blowby gases and fuel vapor, and the tail pipe can give out engine exhaust gas with pollutants in it. All of the pollutants put various harmful substances into the air. Some contribute to the formation of smog.

To reduce these pollutants, automotive vehicles are equipped with several emission controls:

1. Positive crankcase ventilation (PCV)
2. Evaporative emission control systems
3. Heated-air systems
4. Exhaust-gas recirculation (EGR)
5. Air-injection systems
6. Catalytic converters

In addition, several changes have been made in the engine, fuel system, and ignition system which have helped to reduce the pollutants in the exhaust gas. All of these are described in this chapter. The following chapter discusses emission-control-system service.

♦ 31-2 POSITIVE CRANKCASE VENTILATION SYSTEM

The engine crankcase must be ventilated (♦20-12). Outside air must flow through the crankcase to remove blowby. This blowby gets past the piston rings during the compression and power strokes. Unless it is cleared from the crankcase, it will cause trouble. It can form sludge and acids. The sludge can

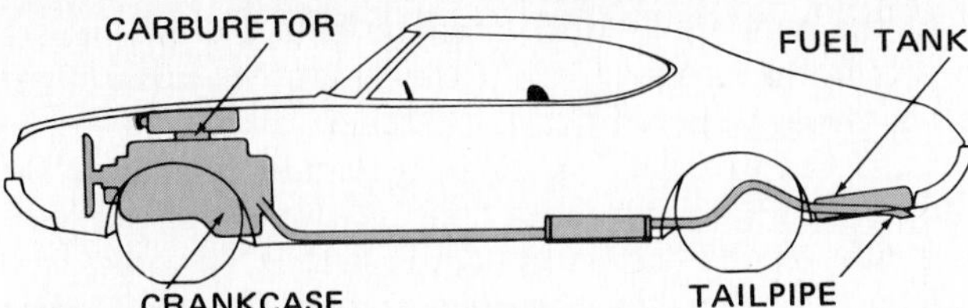

Fig. 31-1 Four possible sources of atmospheric pollution from the automobile.

clog oil lines and starve the lubricating system. This could ruin the engine. Acids corrode metal parts, and this can also ruin the engine.

The removal process requires that the engine must first heat up enough to vaporize the liquid gasoline and water that has collected in the crankcase. Then the circulating air can remove them, along with the blowby gases.

In older engines, the crankcase was ventilated by an opening at the front of the engine and a vent tube in the back. The forward motion of the car and the rotation of the crankshaft moved air through the crankcase, as shown in Fig. 31-2. The air passing through removed the water, fuel vapors, and blowby. However, discharging these gases into the atmosphere caused air pollution.

To prevent this pollution, engines now have a positive crankcase ventilating (PCV) system. Figure 31-3 shows a typical PCV system for a V-type engine. Filtered air from the carburetor air cleaner is drawn through the crankcase. In the crankcase, the air picks up the water and fuel vapors, and blowby. The air then flows back up to the intake manifold and enters the engine. There, unburned fuel is burned.

Too much air flowing through the intake manifold during idling could upset the air-fuel ratio. This could cause poor engine idling and even stalling. To prevent this, a flow-control valve is used. The valve is called a PCV valve. The PCV valve allows only a small amount of air to flow through during idle. But as engine speed increases, reduced intake-manifold vacuum allows the valve to open more. This allows more air to flow through. Figure 31-4 shows the operation of the valve.

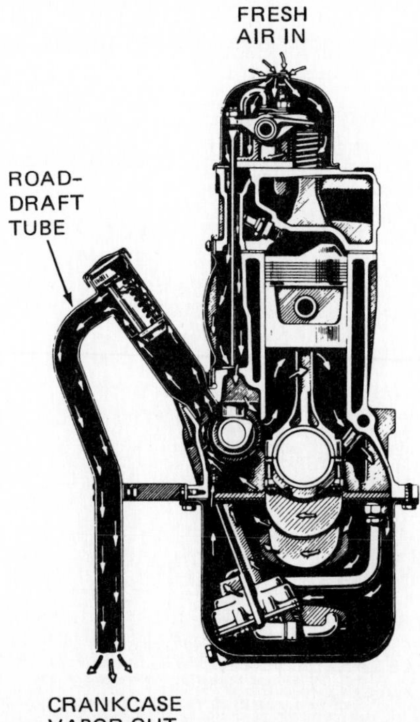

Fig. 31-2 Open crankcase ventilating system.

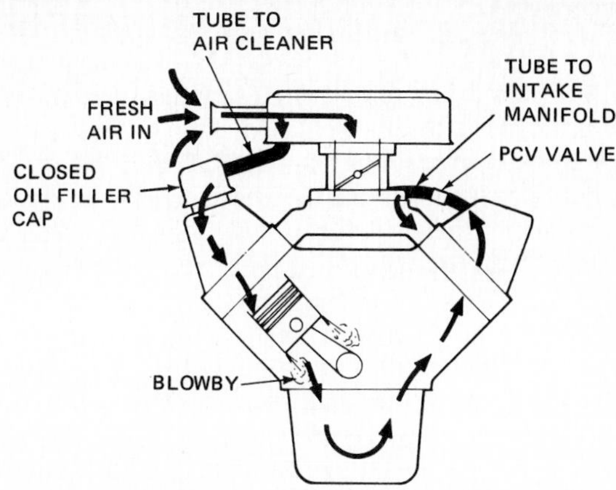

Fig. 31-3 A typical closed PCV system on a V-type engine. *(Ford Motor Company)*

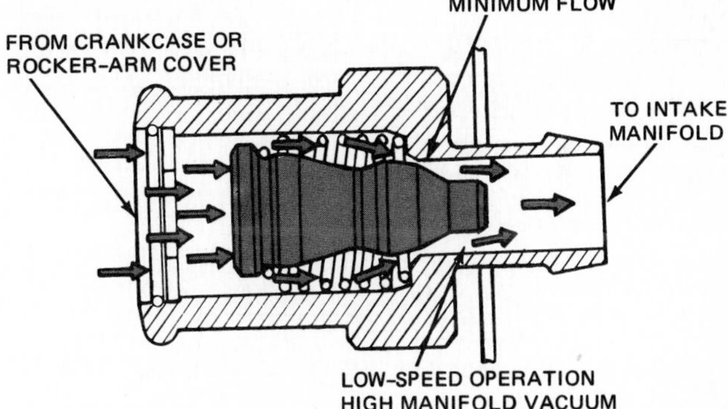

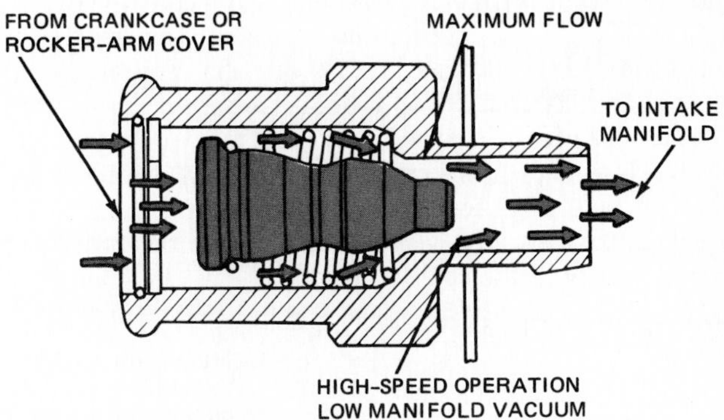

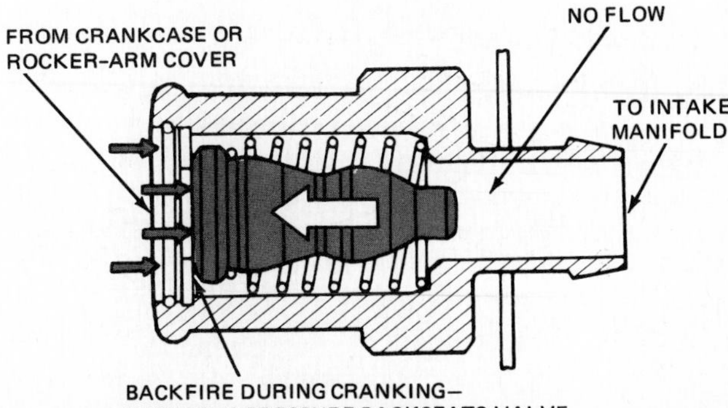

Fig. 31-4 Three positions of the PCV valve. Top, valve position during low engine speed; center, valve position during high engine speed; bottom, valve position during engine backfire. *(Ford Motor Company)*

♦ 31-3 NEED FOR FUEL-VAPOR EMISSION CONTROLS

Both the fuel tank and the carburetor can lose gasoline vapor to the atmosphere, causing pollution, if the car does not have a vapor emission control system. The fuel tank "breathes" as temperature changes. As the tank heats up, the air inside it expands. Part of the air is forced out through the tank vent tube, or through the vent in the tank cap. This air is loaded with gasoline vapor. Then, when the tank cools, the air inside contracts. More air enters the tank from outside. This breathing of the tank causes a loss of gasoline. The higher the tank temperature goes (for example, when the car is parked in the sun), the more gasoline vapor is lost.

The carburetor also can lose gasoline by evaporation. The carburetor float bowl is full whenever the engine is running. When the engine stops, engine heat evaporates some or all of the gasoline stored in the float bowl. Without a vapor-recovery system, this gasoline vapor would pass into the atmosphere.

A fuel-vapor emission control system captures these gasoline vapors and prevents them from escaping into the air. It thereby tends to reduce atmospheric pollution. All cars built since 1970 are equipped with such systems. They are called by various names: evaporation control system (ECS), evaporation emission control (EEC), vehicle vapor recovery (VVR), and vapor saver system (VSS). All work in the same general way.

♦ 31-4 FUEL-VAPOR EMISSION CONTROL SYSTEMS

Figures 31-5 and 31-6 show typical vapor emission control systems. The canister is filled with activated charcoal. Just after the engine is shut off, heat continues to enter the carburetor. This vaporizes gasoline in the carburetor float bowl. The vapor passes through the line into the canister, where the vapor is adsorbed by the charcoal. "Adsorbed" means that the gasoline vapor is trapped by sticking to the outside of the charcoal particles. Some carburetor float bowls have a canister vent hole (Fig. 16-6). It is connected by a tube to the charcoal canister. The vent and tube carry the float-bowl vapor directly to the canister.

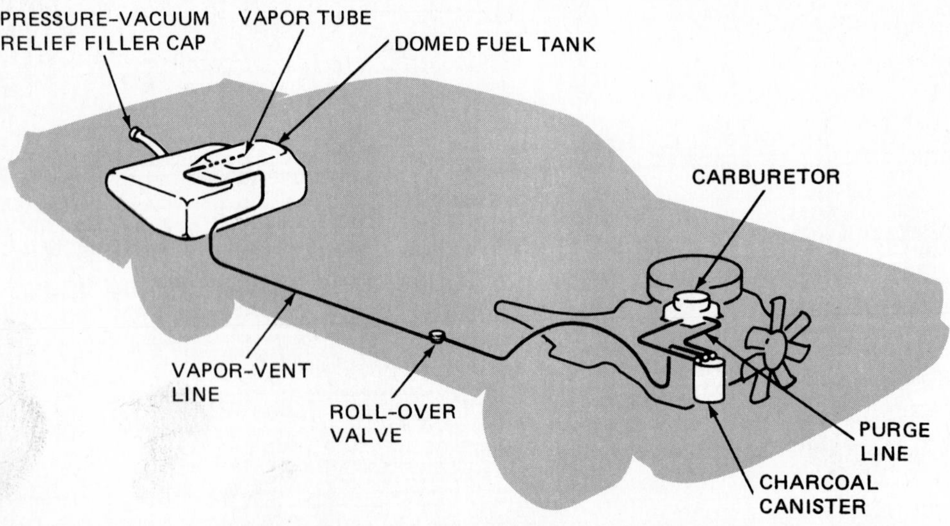

Fig. 31-5 A fuel-vapor (evaporative) emission control system on a car. *(Chrysler Corporation)*

At the same time, vapor-laden air from the fuel tank is carried by the vapor-vent line to the canister. As the air passes down through the canister, the gasoline vapor is trapped by the charcoal particles. The air exits from an opening in the canister, leaving the hydrocarbon (HC) vapor behind. There is a filter at the bottom of some canisters. It comes into action during the *purge* phase of operation. This occurs when the engine is started. Now, intake-manifold vacuum draws fresh air up through the canister. This fresh air removes, or purges, the gasoline vapor from the canister. The air carries the HC through the purge line to the carburetor.

♦ 31-5 FUEL-VAPOR RETURN LINE

Figure 31-6 shows that a fuel-return line parallels the fuel-supply line. This return line connects the pressure side of the fuel pump to the fuel tank. Any excess gasoline being pumped by the fuel pump is returned to the fuel tank. This action removes any vapor that might develop in the fuel pump. It also maintains a flow of fuel through the fuel pump. This keeps the fuel pump relatively cool and helps prevent vapor lock. In some systems, there is a check valve in the fuel-vapor return line (Fig. 15-18). Its purpose is to prevent fuel from feeding back to the carburetor from the fuel tank through the return line. There is more on vapor-return lines in ♦15-9.

♦ 31-6 CHARCOAL CANISTER

Figure 31-7 shows a charcoal canister in sectional view. This canister is used on some V-6 General Motors engines. The downward-pointing arrows to the left show the flow of fuel vapor from the carburetor float bowl and from the fuel tank during the time that the engine is not running. When the engine is running, the action is as shown on the right side in Fig. 31-7. Air is pulled up through the charcoal. This purges the gasoline vapor from the charcoal. Ford calls the canister a *carbon* canister. Charcoal is a special form of carbon.

Note Some early charcoal canisters did not have the connection for a vapor line from the carburetor float bowl.

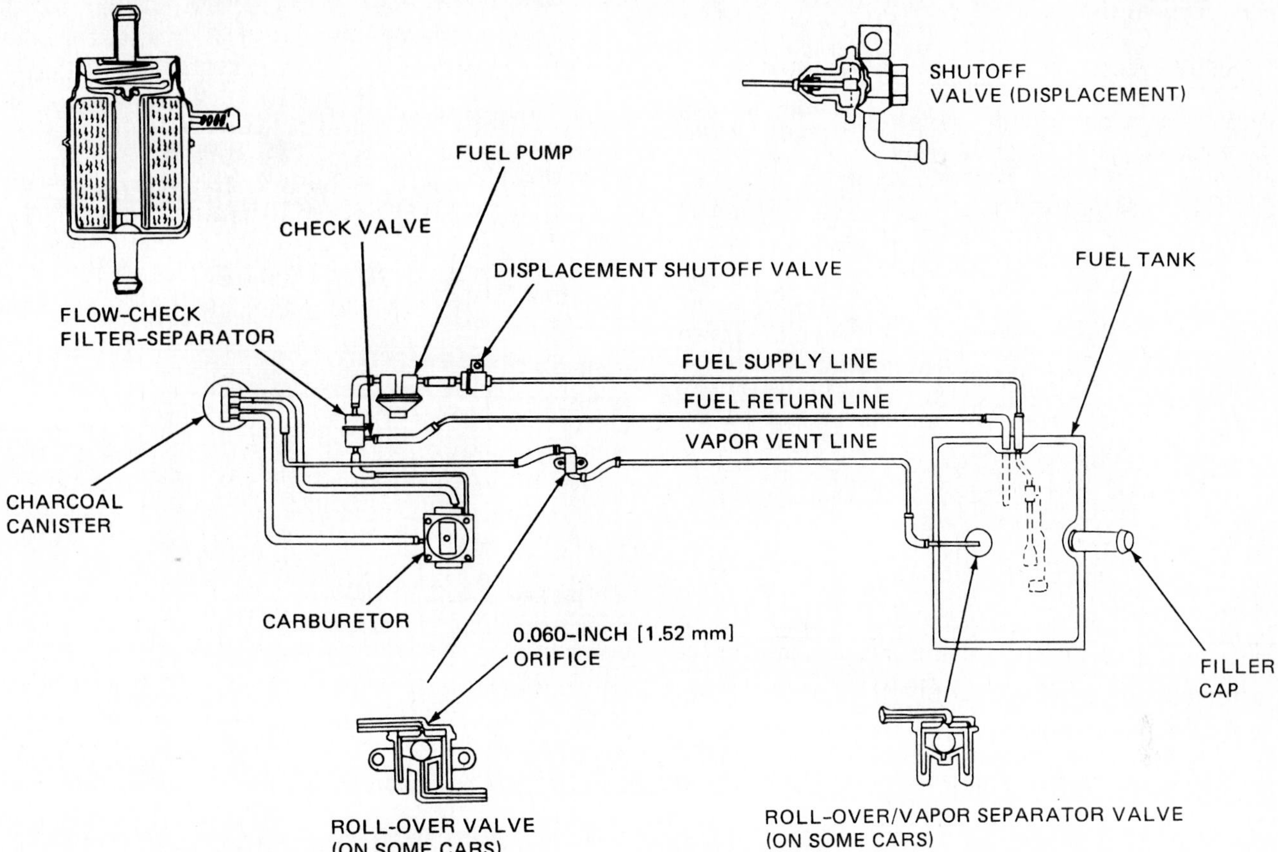

Fig. 31-6 Fuel-vapor emission control system that has rollover valves to shut off fuel flow if the car rolls over. *(Chrysler Corporation)*

They had just two connections at the top. One was for the tube from the fuel tank. The other was for the purge line to the carburetor.

Figure 31-8 shows a typical canister hose routing for a V-8 engine. Hose routings vary from car to car.

Figure 31-9 is a canister for a four-cylinder engine. A purge valve has been added (to upper left). This valve limits the flow of vapor and air to the carburetor during idle. But it allows

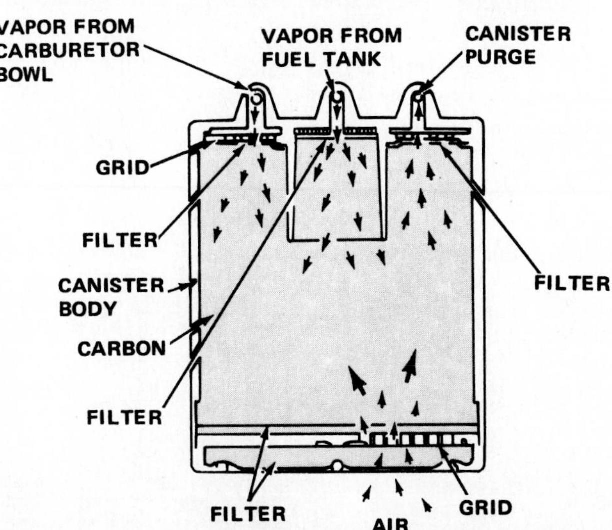

Fig. 31-7 Charcoal canister for a V-6 engine. *(Buick Motor Division of General Motors Corporation)*

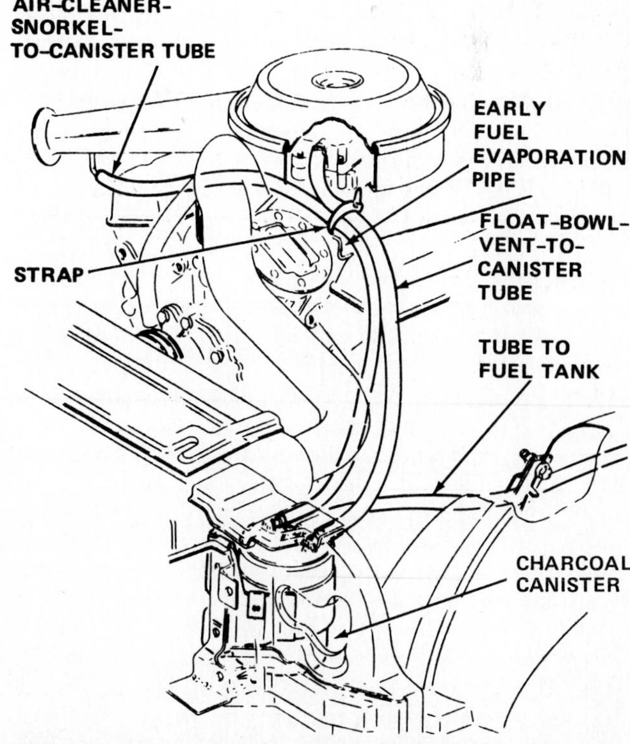

Fig. 31-8 Hose routing for a fuel-vapor emission control system. *(Buick Motor Division of General Motors Corporation)*

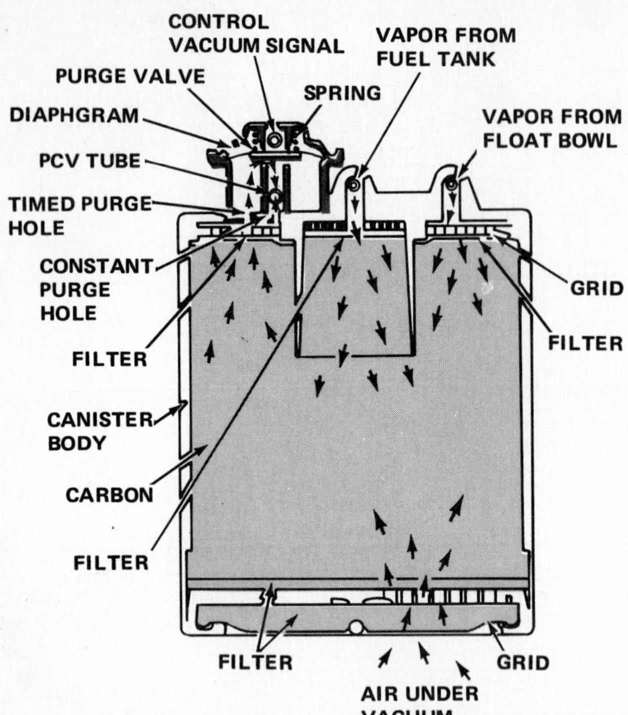

Fig. 31-9 Charcoal canister for a four-cylinder engine. *(Pontiac Motor Division of General Motors Corporation)*

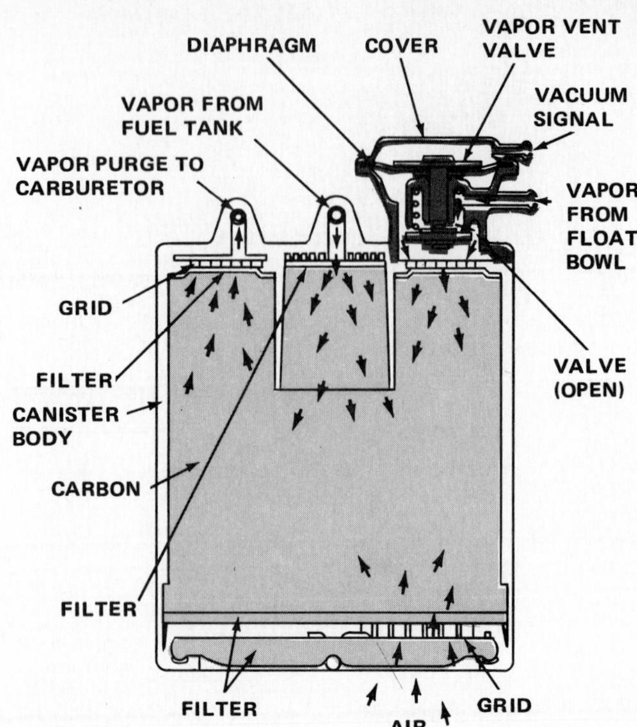

Fig. 31-10 Charcoal canister with added vapor-vent valve. *(Pontiac Motor Division of General Motors Corporation)*

full air and vapor flow during part- to full-throttle operation. In the four-cylinder engine, full flow of the fuel-laden vapor could upset engine idle. However, at higher engine speeds, full flow can be tolerated. The valve is operated by a vacuum signal from the carburetor.

During idle, a small amount of purging takes place through a small *constant purge hole*. When the throttle valve is opened, it passes a drilled hole in the carburetor that allows intake-manifold vacuum to lift the purge valve off its seat. Now, additional air and vapor can flow through the purge hole. This hole is connected by a tube to the PCV system. The air and vapor flow through a tube that is connected to the PCV hose on the engine.

Figure 31-10 shows in sectional view a charcoal canister similar to the one in Fig. 31-7. However, a vapor-vent valve has been added. This valve is operated by a vacuum signal from the carburetor. When the engine is off, there is no manifold vacuum and no vacuum signal. Therefore, the spring pushes the valve down so that it is open. Now, any vapor forming in the carburetor float bowl can flow through a tube and into the canister. When the engine starts, vacuum from the intake manifold pulls the vapor-vent valve closed. This prevents any air leak from the canister back into the float bowl. Such a leak could upset the carburetor calibration. Allowing air leaks to the float bowl could unbalance the carburetor and result in an excessively rich air-fuel mixture. This vapor-vent valve has a more positive action than the pressure-relief valve used in some carburetors.

Another type of charcoal canister has both the purge valve, shown in Fig. 31-9, and the vapor-vent valve, shown in Fig. 31-10. This canister is used on cars sold in areas demanding the most rigid fuel-vapor control.

The canisters described so far are round, as shown in Fig. 31-8. Some cars have a rectangular canister (Fig. 31-11). Some cars with big engines and large fuel tanks use two canisters (Fig. 31-12). One canister handles the fuel-tank vapors, the other handles the carburetor float-bowl vapors. Float-bowl venting on many Ford models is handled by a purge control valve, a solenoid vent valve, or a thermal vent valve.

The purge control valve (Fig. 31-13) is similar to the unit previously described (Fig. 31-9). But it also handles the vapor from the fuel tank. When the engine is running above idle, the vacuum opens the purge valve to allow a free flow of air through the canister. When the engine is idling, the valve closes to reduce the airflow.

The float-bowl solenoid vent valve (Fig. 31-14) is located in the line from the bowl to the canister. When the engine is turned off, the valve is open to allow vapor to flow from the bowl to the canister. When the engine is turned on, the sole-

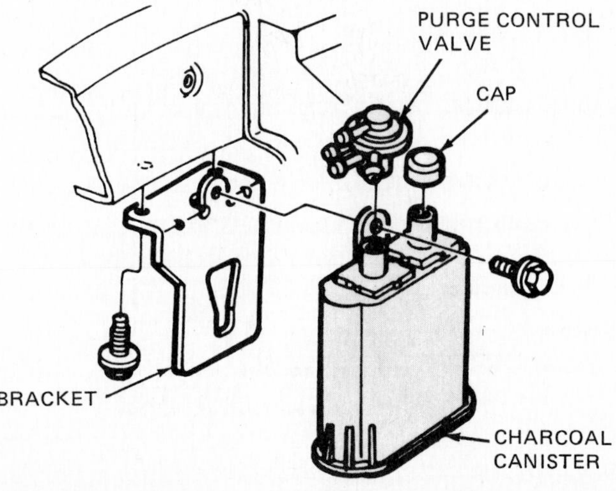

Fig. 31-11 Charcoal canister used by Ford. *(Ford Motor Company)*

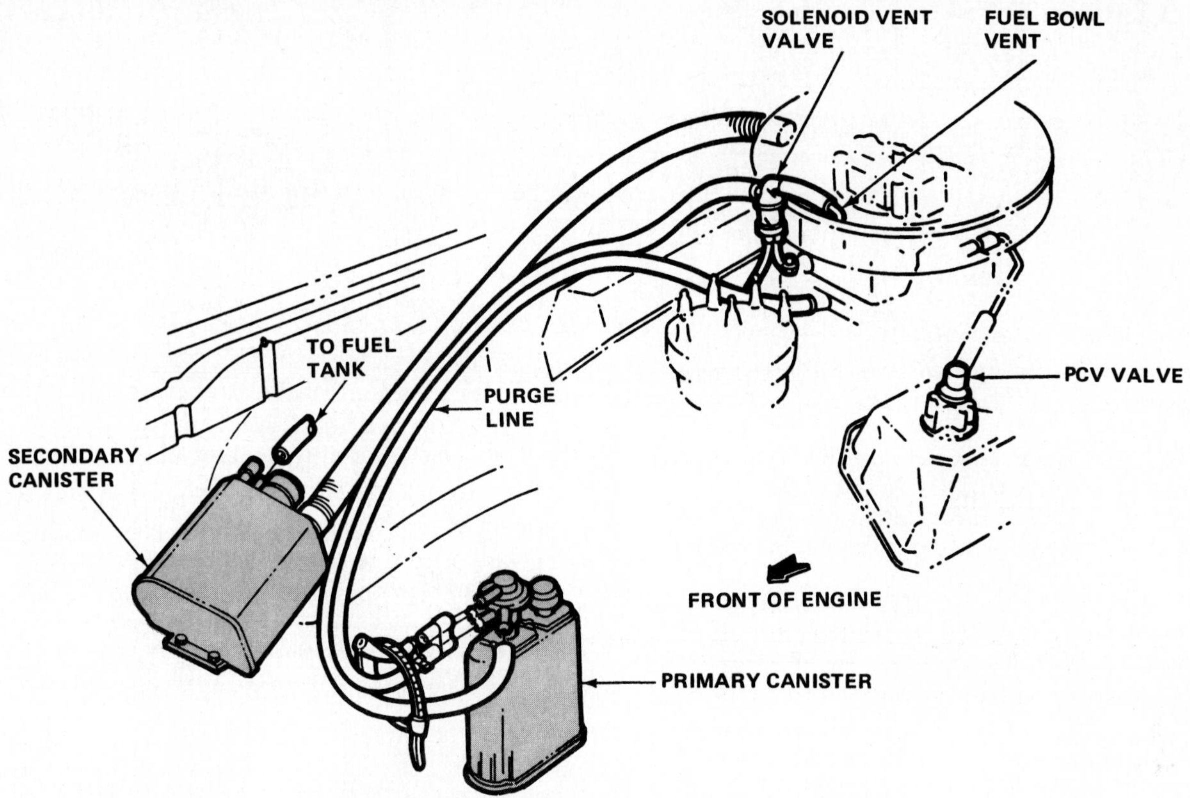

Fig. 31-12 Dual canister fuel-vapor emission control system. *(Ford Motor Company)*

noid is connected to the battery through the ignition switch. This causes the solenoid to close the valve. It does the same job as the vapor-vent valve, shown in Fig. 31-10. However, the solenoid vent valve does the job electrically. The valve in Fig. 31-10 does the job by vacuum.

The float-bowl thermal vent valve is connected in the line from the float bowl and the canister (Fig. 31-15). Its purpose is to prevent fuel-tank vapors from being vented through the carburetor float bowl. This tends to happen when the engine is cold and the fuel tank is hot. For example, a car sitting out all night will be cold. But when the sun strikes it, the fuel tank will warm up much faster than the engine. This could send some fuel vapor from the tank through the float bowl. However, the bimetal plate in the thermal vent valve distorts when cold and seals off the line. When the engine warms up, the bimetal plate opens the valve so that normal float-bowl venting takes place.

♦ 31-7 SEPARATING VAPOR FROM FUEL

A variety of devices have been used to prevent liquid gasoline from leaking back through the vapor-vent line from the tank to the canister. An early system used a standpipe assembly

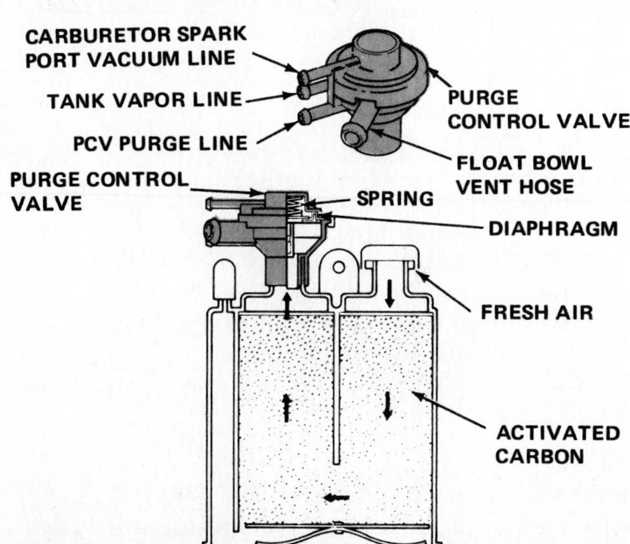

Fig. 31-13 Charcoal canister with purge control valve. *(Ford Motor Company)*

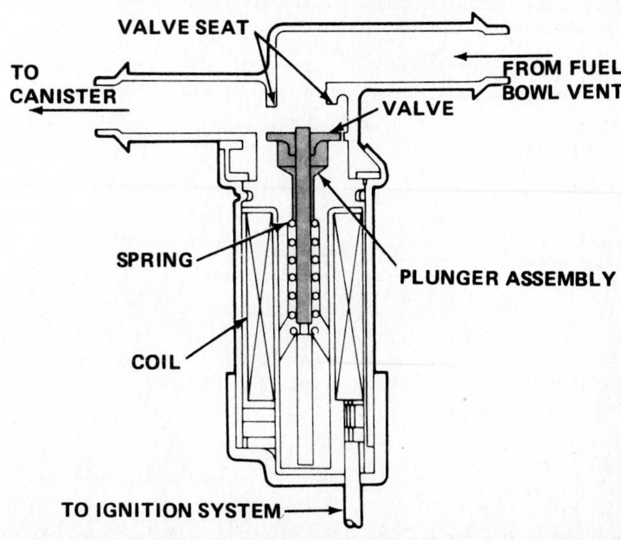

Fig. 31-14 Float-bowl solenoid vent valve. *(Ford Motor Company)*

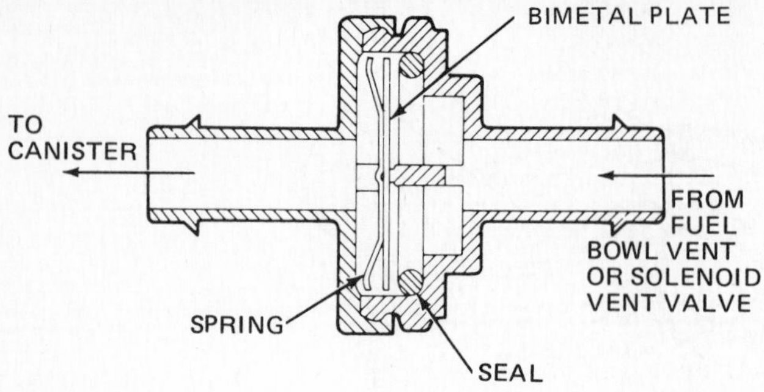

Fig. 31-15 Float-bowl thermal vent valve. *(Ford Motor Company)*

(Fig. 31-16). It contains a series of pipes that are connected to three vents in the fuel tank (Fig. 31-17). One of the vents is always above the fuel level. This vent can feed vapor through the standpipe assembly to the vent pipe connected to the canister.

Another type of vapor-fuel separator is shown in Fig. 31-18. In this system, a liquid check valve is used. The liquid check valve will pass air or vapor, but not liquid. This prevents liquid gasoline from getting to the canister.

A different type of vapor separator is shown in Fig. 31-19. It is mounted on top of the tank. It is filled with filter material that will pass vapor but not liquid.

Figure 31-20 shows another type of vapor-fuel separator. As long as vapor flows through, the float stays down. But if liquid enters, the float goes up and closes the orifice to the tube leading to the canister.

Many fuel tanks that mount horizontally have a dome, (Fig. 31-5), or the filler pipe on the tank is slightly below the

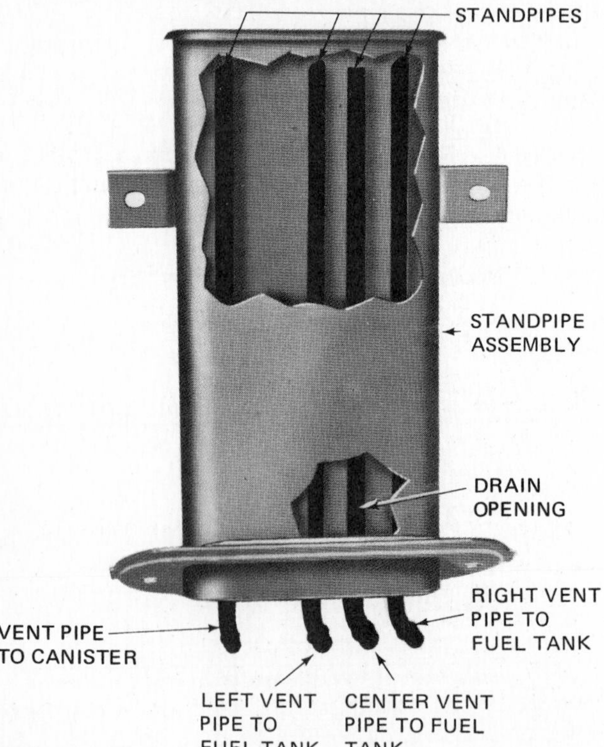

Fig. 31-16 Standpipe assembly used in some evaporative control systems. *(Oldsmobile Division of General Motors Corporation)*

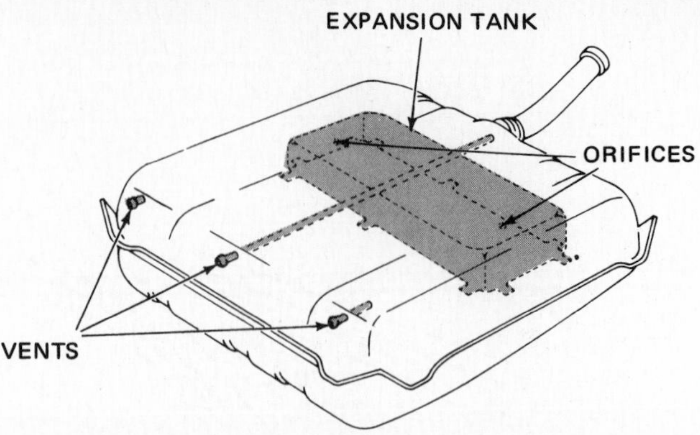

Fig. 31-17 Fuel tank for a fuel-vapor emission control system. *(Pontiac Motor Division of General Motors Corporation)*

top of the tank. In either case, the tank cannot be completely filled. If it were possible to completely fill the tank, then expansion of the gasoline as it warmed up would send it spilling out through the fuel-tank cap or through the canister. Some fuel tanks have an internal expansion tank (Figs. 31-17 and 31-18). It gives the gasoline some space in which to expand as it warms up.

Many vapor-vent lines have a rollover check valve (Figs. 31-5 and 31-6). Its purpose is to block the line if the car rolls over and is upside down. This prevents gasoline leakage back through the line to the carburetor or canister.

♦ 31-8 SEALED FUEL TANK

The fuel tank must be sealed to prevent the escape of gasoline vapor into the atmosphere as the tank "breathes" (Figs. 31-5 and 31-6). Figure 31-21 shows the special fuel-tank filler cap used. It has a two-way relief valve, operating on both pressure and vacuum. When gasoline is withdrawn from the tank, a slight vacuum develops. The vacuum valve opens to admit air. If the pressure builds up excessively, the pressure valve opens to relieve the pressure. Normally, excessive pressure would not develop. Any pressure is relieved through the fuel-vapor emission control system. Some filler caps have a rollover check valve which closes if the car is in an accident that rolls it over. When the car and cap are upside down, the check valve closes to prevent leakage of gasoline from the tank.

♦ 31-9 CARBURETOR INSULATION

Some carburetors use an insulator to reduce heat flow from the engine to the float bowl. The insulator is placed between the carburetor and the intake manifold. Also, some carburetors have an insulator between the throttle body and the float bowl. Either position forms a heat barrier to the float bowl. This reduces fuel evaporation from the float bowl after the engine is turned off. Another arrangement uses an aluminum heat-dissipating plate which sticks out to prevent heat from reaching the float bowl.

♦ 31-10 VAPOR STORAGE IN CRANKCASE

Some cars have used the crankcase to store gasoline vapors from the fuel tank and carburetor. When the engine is stopped, gasoline vapors from the vapor separator at the fuel tank flow to the crankcase air cleaner. From there, they flow

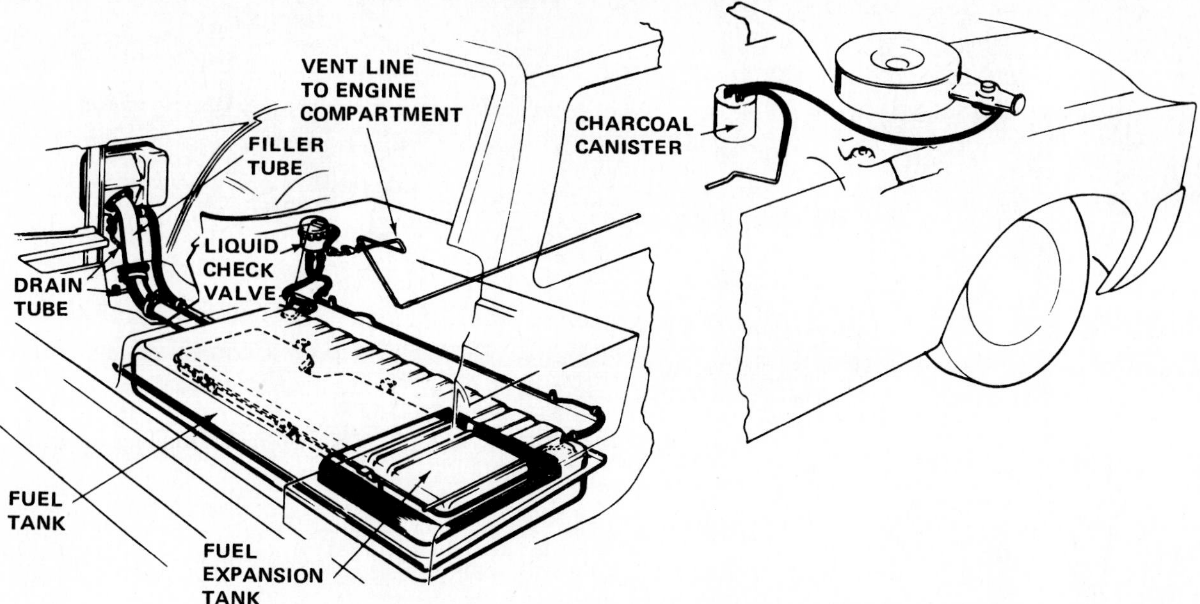

Fig. 31-18 Fuel-vapor emission control system using a liquid check valve. *(American Motors Corporation)*

down into the crankcase. At the same time, fuel vapors from the carburetor float bowl flow down into the crankcase. The vapors are two to four times as heavy as air. Therefore, they sink to the bottom of the crankcase. Then, when the engine is started, the PCV system clears the crankcase of the vapors. The vapors are carried up into the intake manifold and then into the engine, where they are burned.

♦ 31-11 EXHAUST EMISSIONS

Exhaust gases from the engine contain unburned fuel (HC), partly burned fuel (carbon monoxide—CO), and nitrogen oxides (NO_x). These compounds pollute the air. Because air pollution is a threat to health, laws limit the amount of each pollutant that an engine can emit.

To reduce these exhaust emissions, cars are equipped with air-injection systems, exhaust-gas recirculation, and catalytic converters. The air-injection system sends air into the exhaust manifold. The oxygen in this air helps burn the HC and CO. The catalytic converter converts most of these pollutants that do get out of the exhaust manifold. The exhaust-gas recirculation system circulates part of the exhaust gas back through the engine. This lowers combustion temperatures and reduces the amount of NO_x that is formed. Many catalytic converters have extra materials that help convert NO_x into harmless nitrogen and oxygen. These three systems are covered later in the chapter.

♦ 31-12 CLEANING THE EXHAUST GAS

There are three ways of cleaning the exhaust gas:

1. Controlling the air-fuel mixture
2. Controlling the combustion process
3. Treating the exhaust gas

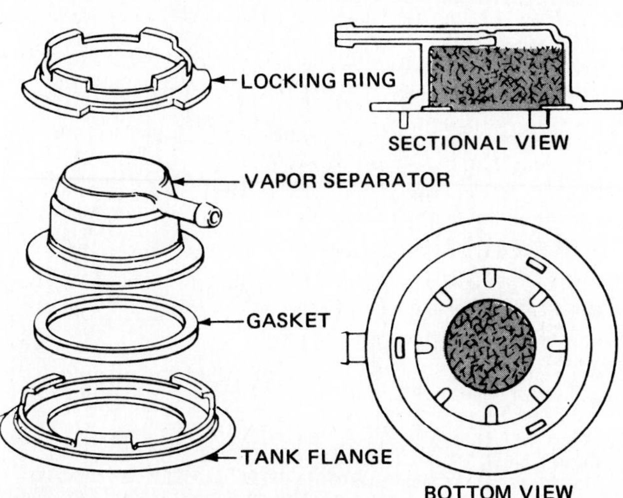

Fig. 31-19 Vapor separator using filter material. When assembling, rotate the locking ring fully against the tank-flange stops. *(Ford Motor Company)*

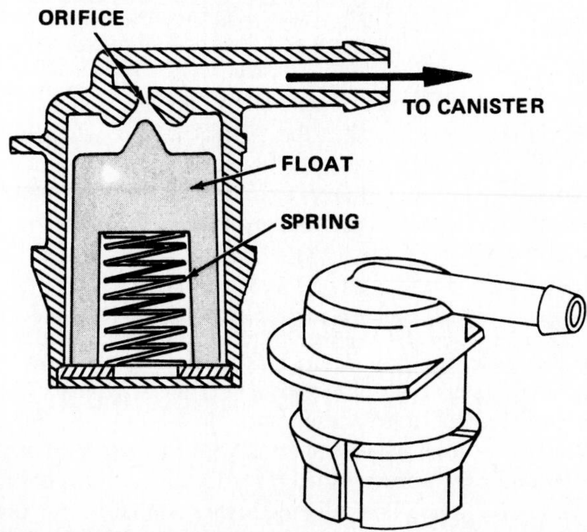

Fig. 31-20 Vapor separator using a float with an internal spring. *(Ford Motor Company)*

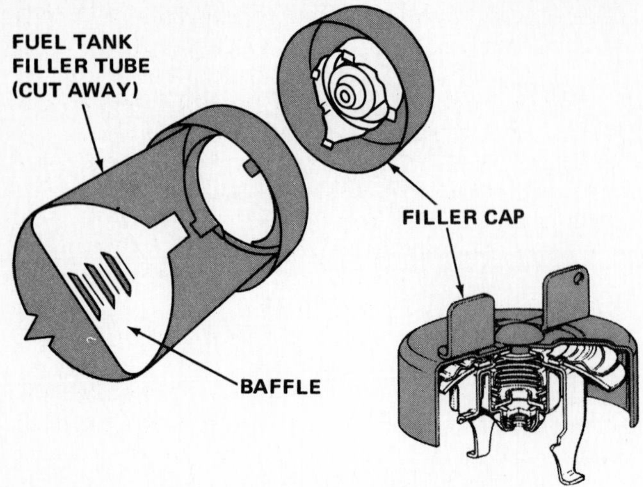

Fig. 31-21 Fuel-tank filler cap and filler tube cut away to show baffle. *(Chrysler Corporation)*

♦ 31-13 CONTROLLING THE AIR-FUEL MIXTURE

Gasoline has been changed to make it burn cleaner. One of the changes is the elimination of lead. Originally, tetraethyl lead was added to gasoline to control detonation and permit higher compression ratios (Chap. 14). However, gasoline without lead is required for cars having catalytic converters. The use of unleaded gasoline has required that the compression ratios be lowered. Basically, controlling the air-fuel mixture means:

1. Modifying the carburetor to deliver a leaner air-fuel mixture
2. Providing faster warm-up and quicker choke opening

On late-model cars, feedback carburetors and fuel-injection systems precisely control the air-fuel ratio of the mixture being delivered to the cylinders. The operation of feedback carburetors is discussed in ♦16-22 to 16-24. The operation of gasoline fuel-injection systems is described in Chap. 18. The feedback system includes an oxygen sensor that monitors the exhaust gas flowing through the exhaust manifold. If the exhaust gas is low in oxygen, the air-fuel mixture is too rich. If the exhaust gas is high in oxygen, the air-fuel mixture is too lean.

By continuous measurement of the oxygen content, the system determines whether the air-fuel ratio needs adjustment. If it does, the system automatically adjusts the amount of fuel entering the intake air. The action is produced by electronic controls described in ♦16-22 to 16-24 and in Chap. 18.

♦ 31-14 LEANER IDLING AIR-FUEL MIXTURE

The idle setting of carburetors on modern engines is on the lean side. Wider spark-plug gaps and higher secondary voltages are required to fire these leaner air-fuel mixtures. Some years ago, to discourage tampering with the leaner settings, *idle-limiter caps* were installed (Fig. 31-22). The caps permit some limited amount of idle-mixture adjustment. But they prevent adjustments outside the range allowed by law. In more recent years, the idle-mixture screws are factory set and sealed with a hardened-steel plug to eliminate tampering.

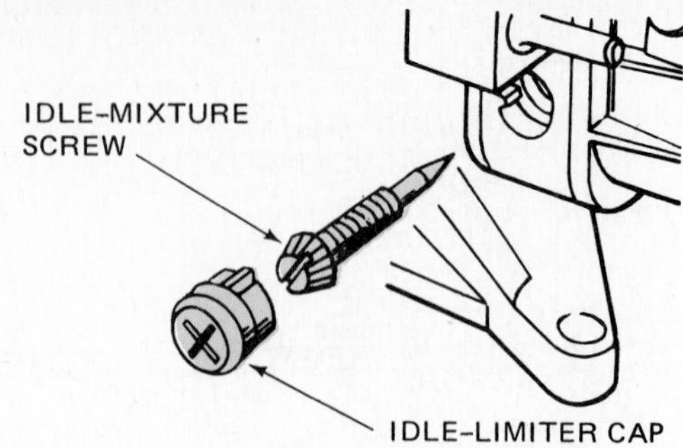

Fig. 31-22 Idle-limiter cap on a carburetor idle-mixture screw.

These cannot be adjusted without disassembling the carburetor.

♦ 31-15 FASTER WARM-UP

If the air-fuel mixture coming from the carburetor is cold, only part of the fuel will vaporize. This means that an extra-rich mixture is needed. Otherwise, the engine will not get enough gasoline vapor for it to run. The situation changes as soon as the engine begins to run. Then, special devices quickly come into operation to add heat to the ingoing air-fuel mixture.

One device is the manifold heat-control valve, described in ♦16-28. This device sends hot exhaust gases circulating around a portion of the intake manifold to add heat to the air-fuel mixture. However, this is so slow that it causes excessive exhaust emissions during warm-up. Most carbureted engines now use the heated-air system described in ♦31-16. This system uses heat from the exhaust manifold to heat the air going into the air cleaner and carburetor when the engine is cold. Cold-engine performance is improved.

♦ 31-16 HEATED-AIR SYSTEM

This system has a thermostatic air cleaner (Fig. 31-23). The purpose of the system is to send heated air to the carburetor

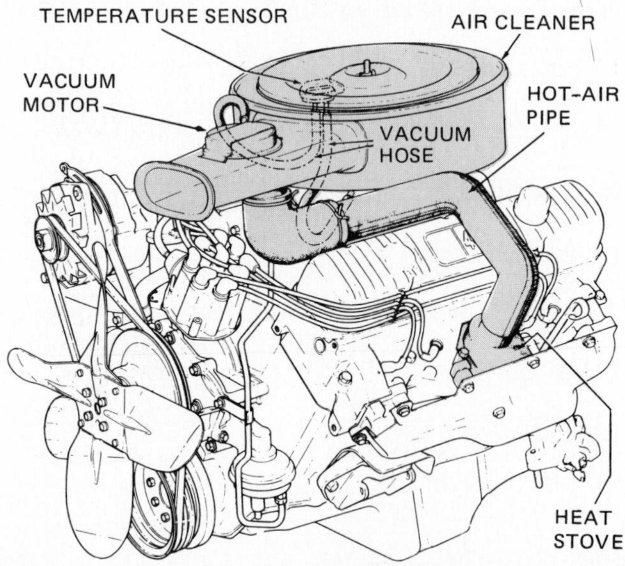

Fig. 31-23 Heated-air system on a V-8 Engine. *(Buick Motor Division of General Motors Corporation)*

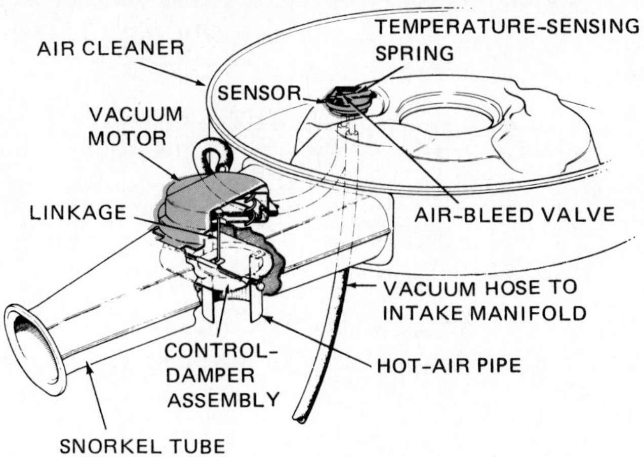

Fig. 31-24 Thermostatically controlled air cleaner. *(Chevrolet Motor Division of General Motors Corporation)*

during cold-engine operation. This improves engine performance after a cold start and during engine warm-up. An advantage of this quick heating of the ingoing air to the carburetor is that leaner air-fuel mixtures can be used. This results in less HC coming out the tail pipe.

The air cleaner shown in Fig. 31-23 has a thermostatic spring which reacts to the temperature of the air entering the carburetor through the air cleaner. This spring controls an air-bleed valve (Fig. 31-24). When the entering air is cold, the sensing spring holds the bleed valve closed. Now, intake-manifold vacuum is applied to the vacuum motor. Its diaphragm is pushed upward by atmospheric pressure, and the diaphragm spring is compressed. In this position, linkage from the diaphragm raises the control-damper assembly. The snorkel tube is blocked off. All air now has to enter from the hot-air pipe (view B in Fig. 31-25). This pipe is connected to the heat stove on the exhaust manifold. Therefore, as soon as the engine starts and the exhaust manifold begins to warm up, hot air is delivered to the carburetor and engine. This improves cold and warm-up operation.

As the engine begins to warm up, the under-hood temperature increases. If the under-hood temperature goes above 128°F [53.3°C] (in the application shown), the conditions are as shown in view C in Fig. 31-25. The temperature-sensing spring has bent enough to open the air-bleed valve. This reduces the vacuum above the diaphragm so that the diaphragm spring pushes the control damper all the way down. Now, all air entering the carburetor comes from under the hood. None flows through the hot-air pipe.

If the temperature under the hood stays between 85 and 128°F [29.4 and 53.3°C], conditions are as shown in view D in Fig. 31-25. The temperature-sensing spring holds the air-bleed valve partly open. Therefore, some vacuum gets to the vacuum chamber above the diaphragm. This vacuum holds the control damper partly open. In this position, some cool air enters from under the hood. Some air flows through the hot-air pipe, from the heat stove around the exhaust manifold.

A similar thermostatically controlled air cleaner is shown in Fig. 31-26. It has a thermostatic bulb that acts directly on the valve plate. When the engine is cold, the thermostatic bulb positions the valve plate as shown in Fig. 31-26, left. Now, all ingoing air must come from the hot-air duct, which is connected to a shroud around the exhaust manifold. As the

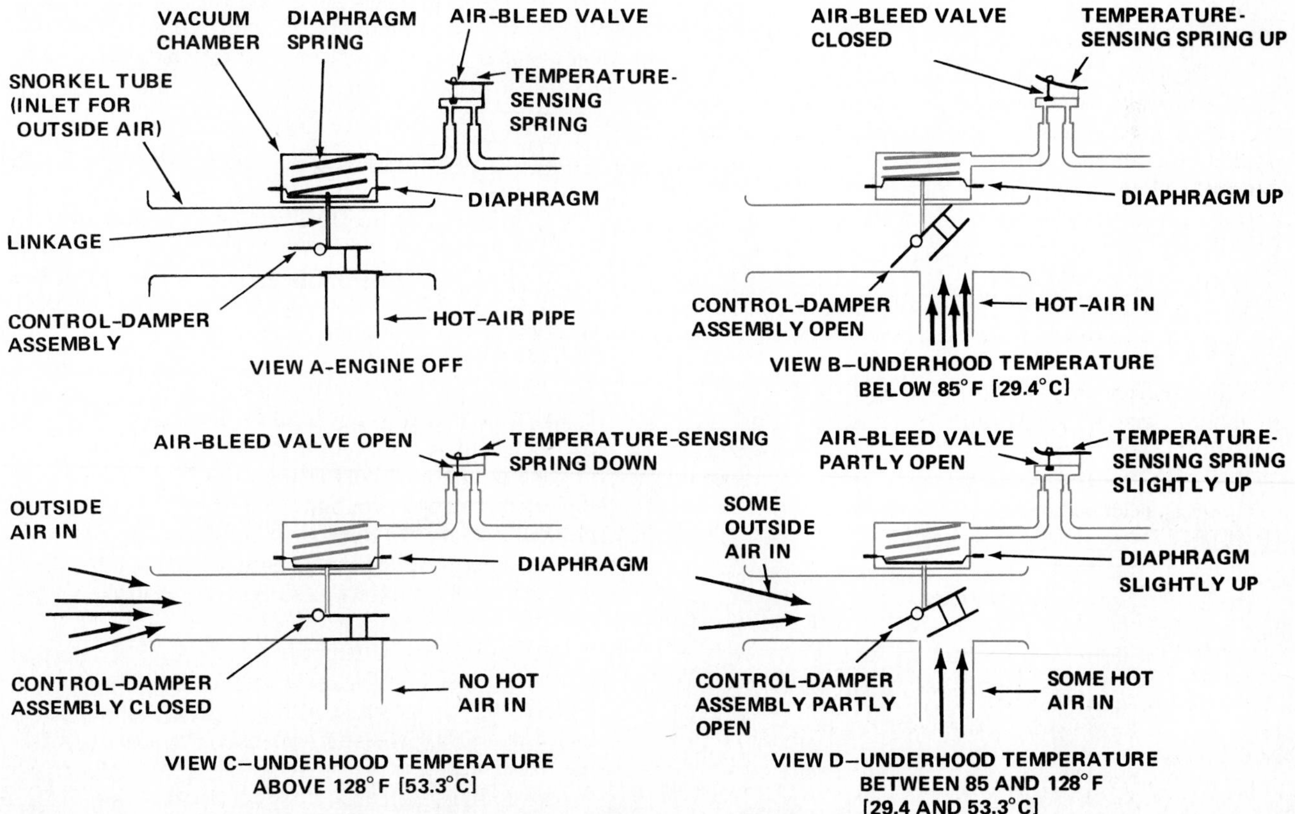

Fig. 31-25 Operating modes for the thermostatically controlled air cleaner. *(Chevrolet Motor Division of General Motors Corporation)*

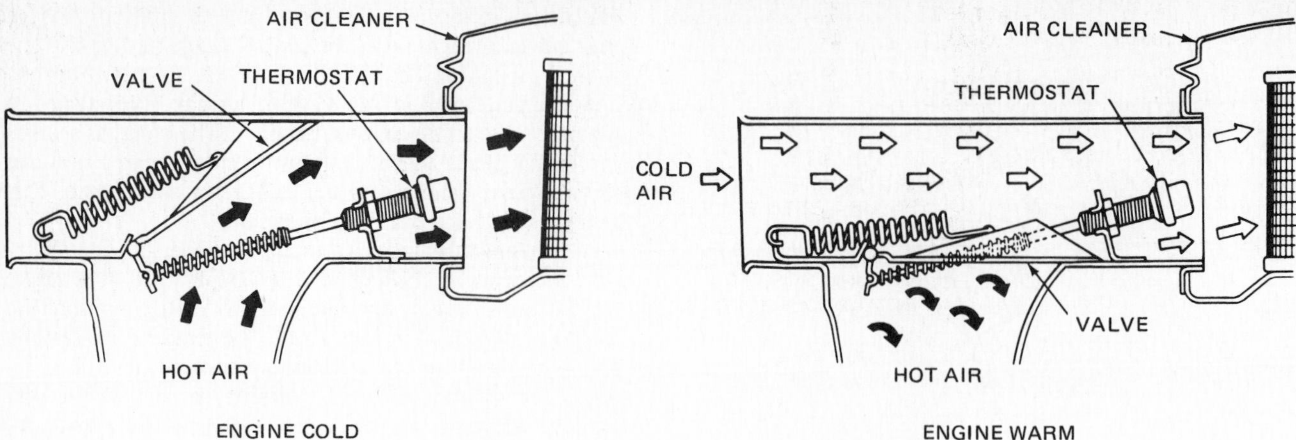

Fig. 31-26 Operation of thermostatic air cleaner with a thermostatic bulb. Left, when the engine is cold and heated air is being taken from the heat stove around the exhaust manifold. Right, when the engine is warm and cooler air is being taken in from under the hood. *(Ford Motor Company)*

engine warms up, the hotter air from the shroud causes the thermostatic bulb to move the valve plate. Therefore, some air begins to enter from the engine compartment. With further increases of temperature, the valve plate moves farther so that more compartment air enters. When the engine compartment becomes hot, then most or all ingoing air comes from the engine compartment (Fig. 31-26, right).

One type of air cleaner includes a vacuum override motor. This motor operates on intake-manifold vacuum. During cold-engine acceleration, when additional air is needed, the motor overrides the thermostatic control. This opens the system to both engine-compartment and heated air so that adequate air is delivered to the carburetor.

Some air cleaners also have a vacuum motor that opens and closes an auxiliary hole in the air-cleaner housing. This type of air cleaner is shown in Fig. 31-27. The vacuum motor operates if a partial vacuum develops in the air cleaner. During cold-weather acceleration, not enough air may get through the heat stove to satisfy engine requirements. When this happens, the partial vacuum in the air cleaner operates

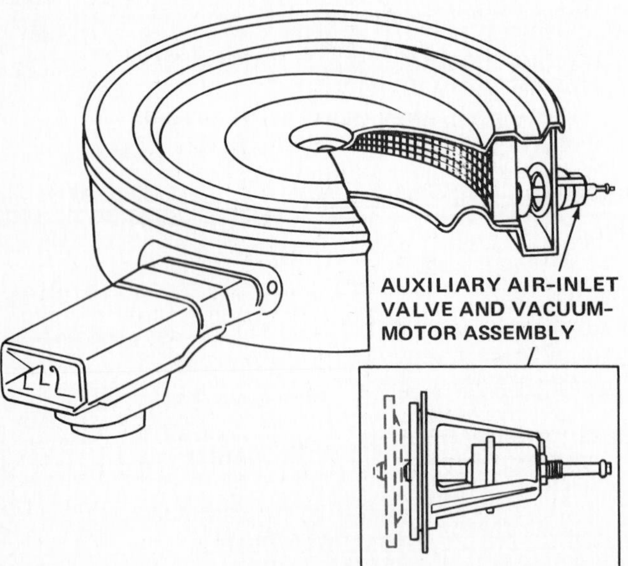

Fig. 31-27 Air cleaner with auxiliary air-inlet valve and vacuum motor. *(Ford Motor Company)*

the vacuum motor. It then opens the auxiliary air-inlet passage. Now, extra air can enter the air cleaner.

♦ 31-17 FASTER-ACTING CHOKE

In older cars, the automatic chokes operated only from the engine heat (♦16-27). But in newer cars, *electric-assist chokes* are used. (These are described in ♦16-27). The choke thermostat is subjected to heat from an electric heating element in the choke and heat from the exhaust manifold. This reduces the amount of time during which the engine operates in a choked condition. With the choke valve closed, the engine is fed a very rich air-fuel mixture. Therefore, the exhaust gas is loaded with unburned HC or CO. The electric-assist choke reduces the length of time during which these pollutants are fed into the atmosphere.

♦ 31-18 CONTROLLING THE COMBUSTION PROCESS

In the combustion process a mixture of air and gasoline vapor is compressed in the combustion chamber. A spark ignites it. The mixture burns and produces the high pressure that pushes the piston down. However, combustion is a complicated process. Here are some of the factors involved:

1. The layers of air-fuel mixture next to the relatively cool cylinder head and piston head do not burn. The metal surfaces chill these layers below the combustion point. Therefore, unburned fuel is swept out of the cylinder on the exhaust stroke. This adds polluting HC to the atmosphere. There are two methods of combating this problem. One is to use stratified charge or fuel injection. The other is to reduce the surface area surrounding the combustion chamber.
2. Increasing the engine temperature improves combustion of the fuel. One way of achieving higher temperatures is to increase the temperature setting of the cooling-system thermostat. This increases combustion temperature. But the higher temperature produces more NO_x, which produces another problem.
3. Vacuum advance gives the air-fuel mixture a longer time to burn when the engine operates at part throttle. But it also gives more time for NO_x to form under certain oper-

ating conditions. Therefore a means must be provided to prevent vacuum advance during these special operating conditions.

4. Carbon buildup in the combustion chambers increases the HC in the exhaust gas. The carbon has pores that fill up during the compression and combustion strokes. Then, during the exhaust stroke, when pressure is released, the HC in the carbon pores escapes and exits with the exhaust gas.

♦ 31-19 REDUCING COMBUSTION-CHAMBER SURFACE AREA

Actually, reducing combustion-chamber surface area means reducing the *surface-to-volume (S/V) ratio*. The *S/V* ratio is the ratio between the surface area and the volume of the combustion chamber. A sphere has the lowest possible *S/V* ratio. The wedge combustion chamber (Fig. 11-8) has a higher *S/V* ratio than the sphere. The hemispheric combustion chamber has a small surface area. It has less surface to chill the air-fuel mixture. Therefore, the hemispheric combustion chamber should produce less unburned HC in the exhaust gas. However, the hemispheric design does not produce good turbulence to aid air-fuel mixing and complete combustion at low speeds. As a result, most manufacturers have modified the wedge chamber into an "open" design.

♦ 31-20 STRATIFIED CHARGE

The stratified-charge engine has a means of concentrating a rich mixture near the spark plug in the combustion chamber (Fig. 31-28). During combustion, the burning rich mixture spreads outward and moves into areas where the mixture is lean and harder to ignite. With stratified charging, a much leaner air-fuel mixture, on the average, can be used. Therefore, the amount of pollutants (CO, HC, and NO_x) is reduced.

There are several ways of achieving stratified charging. These include the proper placement and shape of the intake port, using a turbulence-generating pot, having a third valve, or using a precombustion chamber. These and similar arrangements are described in ♦11-10 and 11-11.

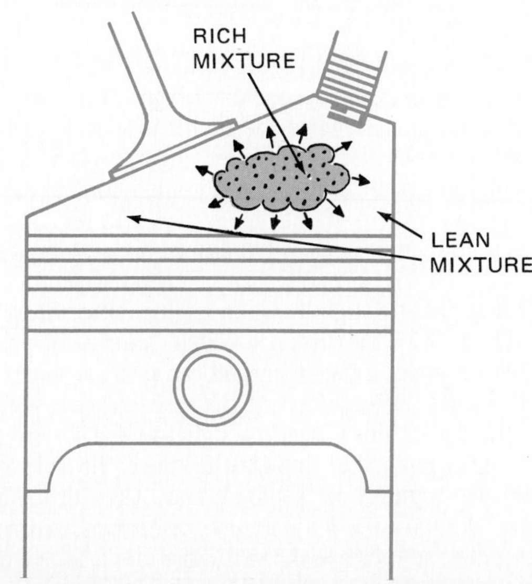

Fig. 31-28 Principle of stratified charging.

♦ 31-21 FUEL INJECTION

Fuel-injection systems for gasoline engines are described in Chap. 18. Fuel injection can improve combustion and reduce HC and CO in the exhaust gas. The fuel-injection system more accurately meters the fuel. In the port-injection system, it supplies the same amount of fuel to each cylinder. When a carburetor is used, some cylinders can get a richer mixture than others.

♦ 31-22 INCREASING COMBUSTION TEMPERATURE

Increasing the combustion temperature reduces CO and unburned HC in the exhaust. But it increases the formation of NO_x. One method of reducing NO_x is to reduce the compression ratio. This reduces peak combustion temperatures and the amount of NO_x that is formed.

Other NO_x-reduction methods are used. One of these reduces NO_x during normal running of the engine. This is the exhaust-gas recirculation (EGR) system. A second method reduces NO_x during acceleration in lower gears and during part-throttle, high-vacuum conditions. A third method uses a three-way catalytic converter that converts the NO_x into nitrogen (N_2) and oxygen (O). All of these are described in following sections.

♦ 31-23 EXHAUST-GAS RECIRCULATION (EGR)

If a small part of the exhaust gas is sent back through the engine, it reduces the combustion temperature and lowers the formation of NO_x. The amount sent through the engine should vary according to operating conditions (Fig. 31-29). A special passage connects the exhaust manifold with the intake manifold. This passage is opened or closed by an EGR valve. The upper part of this valve is sealed. It is connected by a vacuum line to a vacuum port in the carburetor. When there is no vacuum at the port, there is no vacuum in the EGR valve. The spring holds the valve closed. No exhaust gas recir-

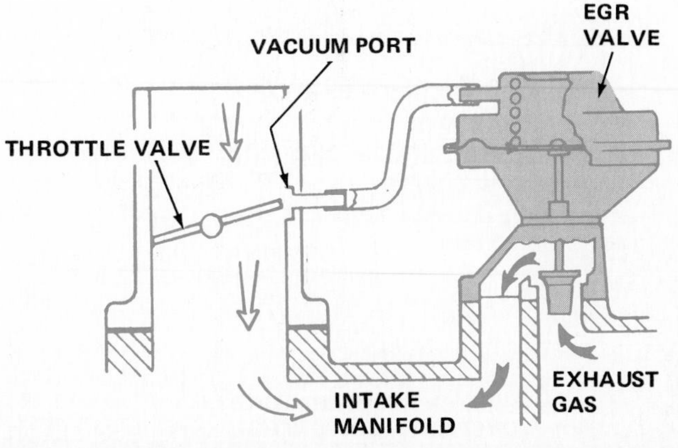

Fig. 31-29 An exhaust-gas recirculation (EGR) system. *(Chevrolet Motor Division of General Motors Corporation)*

culates. This is the situation during engine idling, when NO_x formation is near a minimum.

However, when the throttle is opened, it passes the signal port. This allows the intake-manifold vacuum to operate the EGR valve. The vacuum raises the diaphragm in the valve. This lifts the valve off the seat. Now exhaust gas can pass into the intake manifold. There, it mixes with the air-fuel mixture and enters the engine cylinders. The exhaust gas lowers the combustion temperature and thereby reduces the formation of NO_x. At wide-open throttle, there is little vacuum in the intake manifold. This produces a denser mixture which burns cooler. Therefore, at wide-open throttle, there is less need for EGR. With low vacuum, the EGR valve is nearly closed.

Figure 31-29 shows the EGR valve in the fully open position. A thermal vacuum switch on many cars prevents exhaust-gas recirculation until engine temperature reaches about 100°F [37.8°C]. The thermal vacuum switch is also called a *coolant temperature override* (CTO) switch. The CTO switch is mounted in a cooling-system water jacket, so the switch senses coolant temperature. If this temperature is below 100°F [37.8°C], the switch remains closed. This prevents the vacuum from reaching the EGR valve, so exhaust gas does not recirculate. This improves cold-engine performance immediately after starting. After the engine warms up to where it can tolerate EGR, the CTO valve opens. Now vacuum can get to the EGR valve so that exhaust gas can recirculate.

Figures 31-30 and 31-31 show the EGR system for a six-cylinder and an eight-cylinder engine. There are other mounting arrangements for the EGR valve and CTO switch. Note that the EGR valve is a combination EGR valve and back-pressure sensor. The valve functions as previously described (Fig. 31-29). The back-pressure sensor has been added to control the flow of exhaust gas into the intake manifold in accordance with the amount of exhaust-gas back pressure in the exhaust manifold.

When the engine load is light and back pressure is low, the back-pressure sensor keeps the EGR valve closed. There is no EGR. But when power demands are made on the engine, the exhaust back pressure goes up. Now the back-pressure sensor allows the EGR valve to open so that recirculation can proceed.

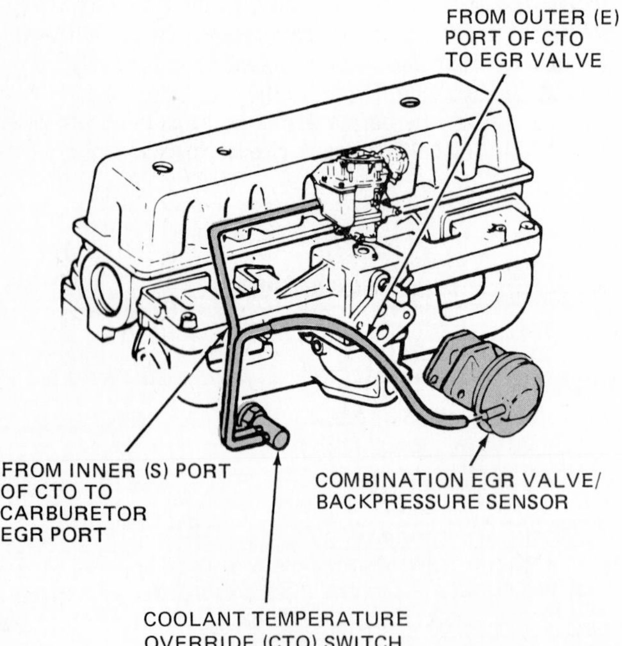

Fig. 31-30 EGR system on a six-cylinder engine. *(American Motors Corporation)*

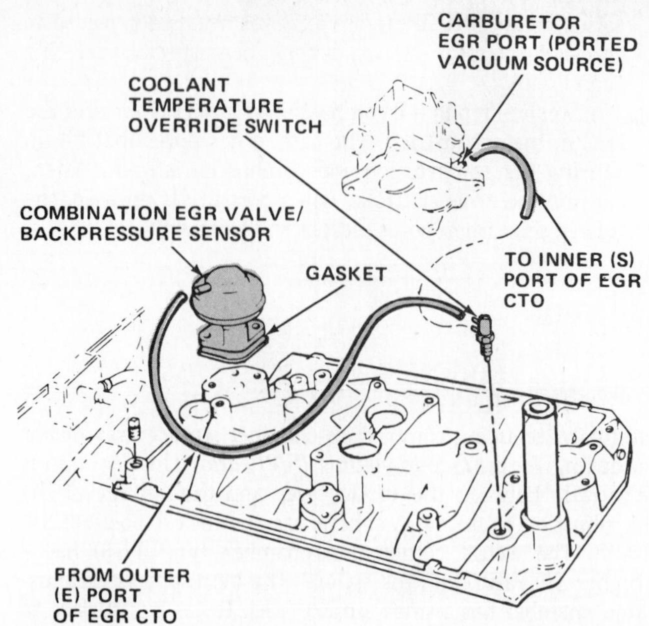

Fig. 31-31 EGR system on a V-8 engine. *(American Motors Corporation)*

Figure 31-32 shows how it works. The left illustration shows the condition when the engine load is light. The back pressure from the exhaust is low. It enters through the small hole in the stem of the EGR valve. It works on the space between the diaphragm and the plate under it. But it is not great enough to have any influence. Air flows as shown by the arrows in Fig. 31-32 (left) to satisfy the vacuum. Figure 31-32 (right) shows what happens when the engine load goes up. The back pressure is higher—high enough to lift the diaphragm above the plate. As it moves up, it closes the control valve. Now, no air can flow up to the vacuum chamber. The vacuum pulls up the actuating diaphragm and plate. This lifts the EGR valve off its seat, so exhaust gas can flow to the intake manifold.

There are several variations of this basic system. For example, some EGR valves have a second diaphragm. Its purpose is to produce increased EGR when the engine is heavily loaded, as during hard acceleration. Also, some engines use an additional modulator system to provide additional control based on car speed. One system of this type is shown in Fig. 31-33. The modulator system is enclosed in dashed lines. It includes a solenoid valve that is normally open, allowing intake-manifold vacuum to pass through it. When engine temperature is high enough to open the thermal switch, and the throttle valve is partly opened, intake manifold can operate the EGR valve. EGR results. However, when car speed reaches a certain level, the speed sensor sends a signal to the electronic amplifier. This causes the amplifier to close the solenoid valve. Now the vacuum line is closed, and EGR stops.

Figure 31-34 shows a still different arrangement. It uses a vacuum amplifier to increase the vacuum enough to operate the EGR valve. At wide-open throttle, recirculation is eliminated by a dump diaphragm inside the amplifier. The ampli-

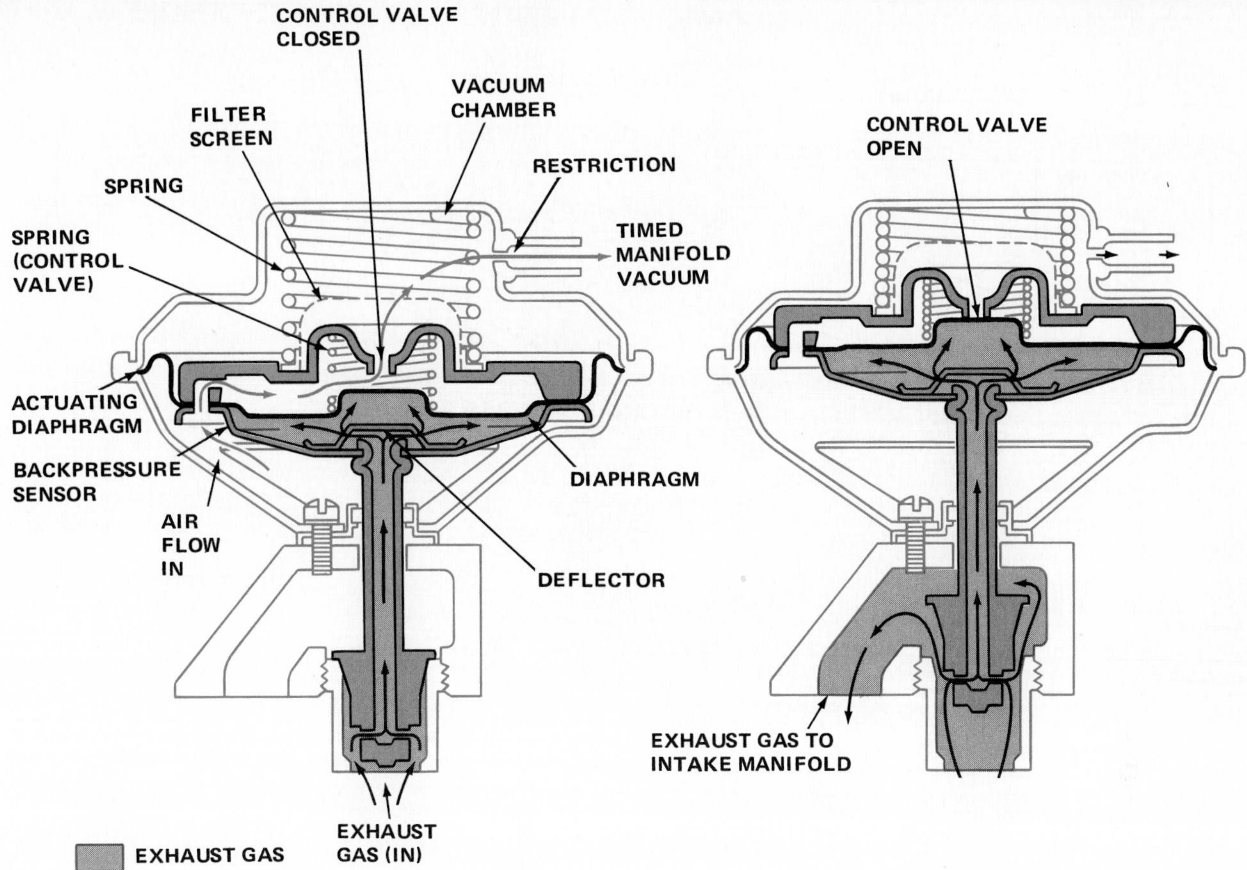

Fig. 31-32 Combination EGR and back-pressure sensor valve. *(Chevrolet Motor Division of General Motors Corporation)*

fier continually compares the vacuum signal going to the EGR valve and the vacuum in the intake manifold. When the comparison shows that the throttle valve is wide open (vacuum practically the same), the amplifier stops amplification of the vacuum signal. Now, it is not great enough to hold the EGR valve open so it closes, halting recirculation.

The EGR delay timer and EGR delay solenoid (upper left in Fig. 31-34) delay EGR action for 35 seconds after starting a cold engine. This assures reliable starting and initial operation of the engine. After the engine has begun to warm up, EGR can begin without upsetting the operation of the engine. If EGR started immediately after the engine began to run, the engine could stumble and even stall. The EGR delay system prevents this.

The reasons for the variety of EGR systems lies in the differing characteristics of the engines and exhaust systems with which they are used. Each EGR system must be tailored to fit the engine it will work with. Several variations of the basic system have been described above. But there are others.

♦ 31-24 VALVE OVERLAP

One of the features of the complete emission control system used in many engines is additional valve overlap. Additional valve overlap does the same thing as the EGR system, but in a different way. Increased valve overlap leaves more of the exhaust gas in the cylinders. The intake valve opens while there is still some exhaust gas in the cylinder. This exhaust gas mixes with the air-fuel mixture entering the cylinder. The result is that the peak combustion temperatures are reduced,

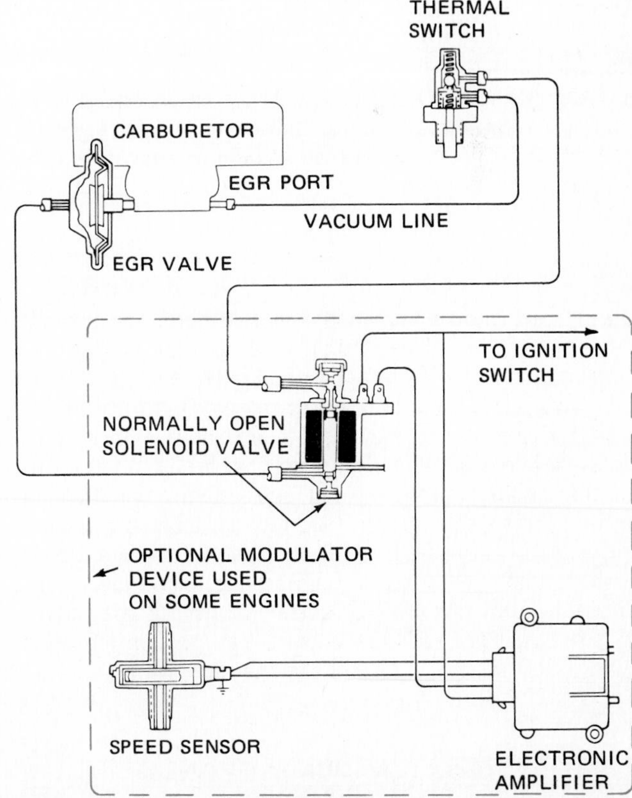

Fig. 31-33 EGR system, showing modulator device used on some engines (shown inside dashed line). *(Ford Motor Company)*

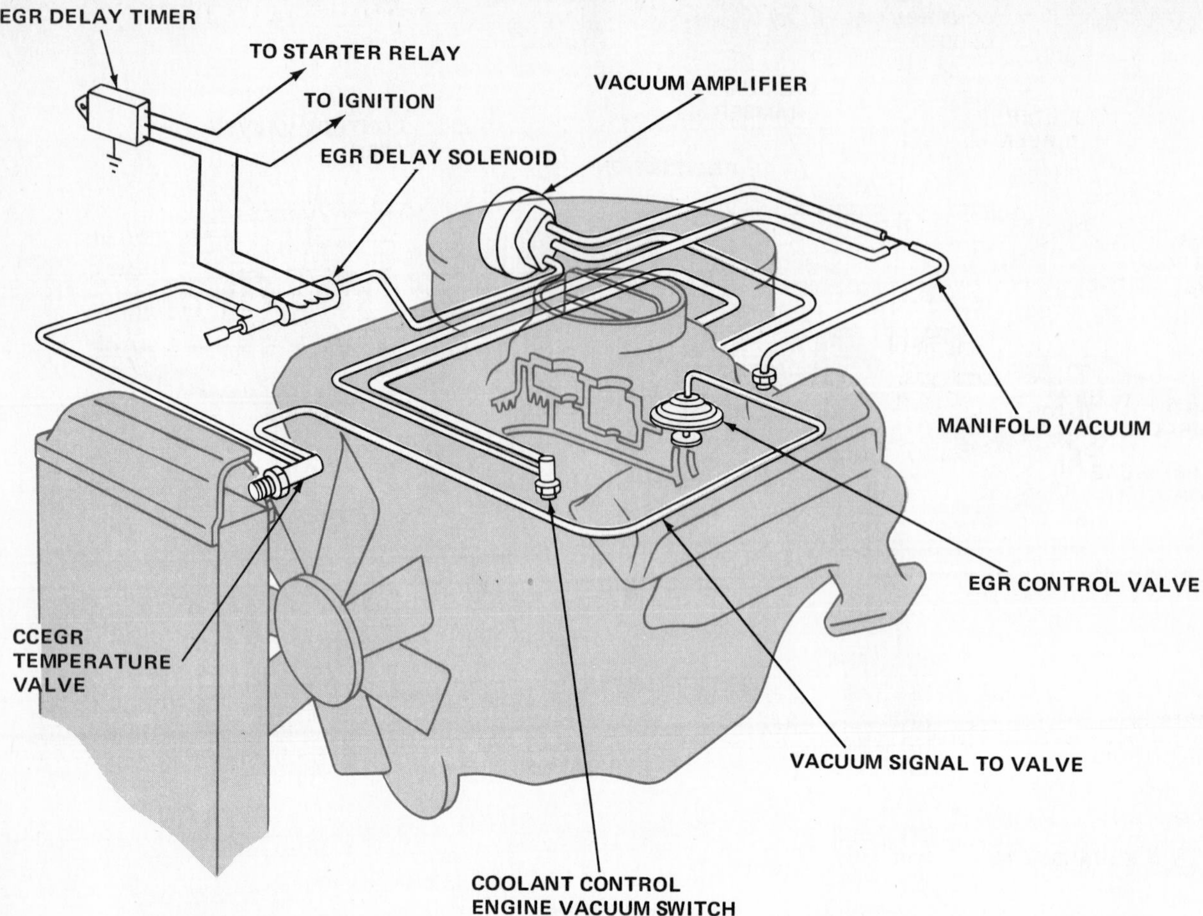

Fig. 31-34 EGR system with vacuum amplifier and EGR delay timer and solenoid. *(Chrysler Corporation)*

and there is less NO_x formation. However, increased valve overlap can cause rough idling. In these engines, the normal idle speed may be set higher than on other engines. This is to provide a smooth idle.

♦ 31-25 CONTROL OF VACUUM ADVANCE

During part-throttle operation, the distributor vacuum advance operates as intake-manifold vacuum changes. This provides more time for the leaner air-fuel mixture to burn. However, this added time also allows more NO_x to develop. A variety of controls have been used to prevent vacuum advance under certain conditions. Some cars have been equipped with a transmission-controlled spark (TCS) system, as described in ♦27-13. This system prevents vacuum advance when the car is operated in reverse, neutral, or low-forward gears. Under these special conditions, vacuum advance could greatly increase the formation of NO_x. General Motors calls its system a TCS system. Ford calls its system a transmission-regulated spark (TRS) system. Both work in the same way.

♦ 31-26 OTHER VACUUM-ADVANCE CONTROL SYSTEMS

There are other vacuum-advance controls. Most are specially designed for the engines and vehicles with which they are used. Some cars produced by Chrysler use an *orifice spark-advance control* (OSAC). It includes a very small hole, or orifice. This delays any change in the application of vacuum to the distributor by about 17 seconds, between idle and part throttle. Therefore, there is a delay in vacuum advance until acceleration is well under way. This is a critical time, during which vacuum advance could produce high NO_x.

Ford has a similar system, called the *spark-delay-valve system.* This system delays vacuum advance during some vehicle-acceleration conditions. The spark-delay valve is connected in series with the vacuum supply from the vacuum-advance port in the carburetor and the distributor vacuum advance. During mild acceleration, the vacuum signal to the distributor can increase only gradually. This is because the spark-delay valve only allows the vacuum to pass through slowly. During deceleration or heavy acceleration, the change in vacuum is great enough to open a check valve. This valve allows the vacuum to bypass the spark-delay valve. This produces vacuum advance during these critical times, for better engine performance. If engine temperatures are low, the temperature switch actuates the solenoid valve. The actuated valve then passes vacuum directly to the distributor vacuum advance (through the check valve). This provides vacuum advance when the engine is cold.

♦ 31-27 ELECTRONIC SPARK ADVANCE

Some late-model cars have electronic ignition systems with electronic spark advance (♦28-10 to 28-13). Electronic control of spark advance provides more accurate spark timing for every operating condition. This, combined with electronic ignition systems, permits the engines to burn leaner mixtures. Fuel efficiency is increased (more miles per gallon), and the amount of pollutants in the exhaust gas is reduced.

♦ 31-28 ELECTRONIC CONTROL OF AIR-FUEL RATIO

Many engines have electronic control of the air-fuel mixture being delivered to the engine cylinders. These systems are described in ♦16-22 to 16-24 and in Chap. 18. They monitor the oxygen content of the exhaust gas. The amount of oxygen in the exhaust gas indicates whether the correct air-fuel mixture is being delivered to the cylinders. If the mixture needs adjustment, the system automatically changes the amount of fuel entering the intake air.

♦ 31-29 ELECTRONIC ENGINE CONTROL SYSTEMS

Late-model cars have an EEC system that controls the air-fuel ratio (♦31-28) and the spark timing (♦31-27). In addition, the system controls the flow of air from the air-injection pump. The air may be sent to the exhaust manifold or to the catalytic converter, as operating conditions require (Figs. 16-21 to 16-24). These systems use a dual-bed or three-way catalytic converter (♦31-34). EEC systems are described in ♦16-22 to 16-24 and in Chap. 28. Air-injection systems are described in ♦31-31.

♦ 31-30 TREATING THE EXHAUST GAS

Exhaust emissions may be reduced by controlling the air-fuel mixture (♦31-13), controlling the combustion process (♦31-18), and treating the exhaust gas. To meet today's emission standards, most engines must use all three methods. These various combinations are required to limit the amounts of HC, CO, and NO_x in the exhaust gas.

Treatment of the exhaust gas means that some "cleaning" action must occur after the exhaust gases leave the engine cylinders and before they exit the tail pipe and enter the atmosphere. Two methods are widely used. These are the air-injection system (♦31-31) and the catalytic converter (♦31-33).

♦ 31-31 AIR-INJECTION SYSTEM

One method used to reduce CO and unburned HC in the exhaust gas is to blow fresh air into the exhaust manifold. This is called the *air-injection system* (Fig. 31-35). It provides the additional oxygen needed to burn the hydrocarbons remaining in the exhaust gases as they flow through the exhaust manifold. The result is that less HC and CO are emitted into the atmosphere.

The air is supplied by a vane-type pump which is belt-driven from the engine crankshaft (Fig. 31-35). Most air pumps have a centrifugal air-filter fan mounted on the pump-rotor shaft. As the fan spins, centrifugal force throws out the dust particles, which are heavier than air. Then only clean air is drawn into the pump.

A diverter valve, or other backfire-suppressor valve, is used to prevent backfire in the exhaust system (Fig. 31-36). When

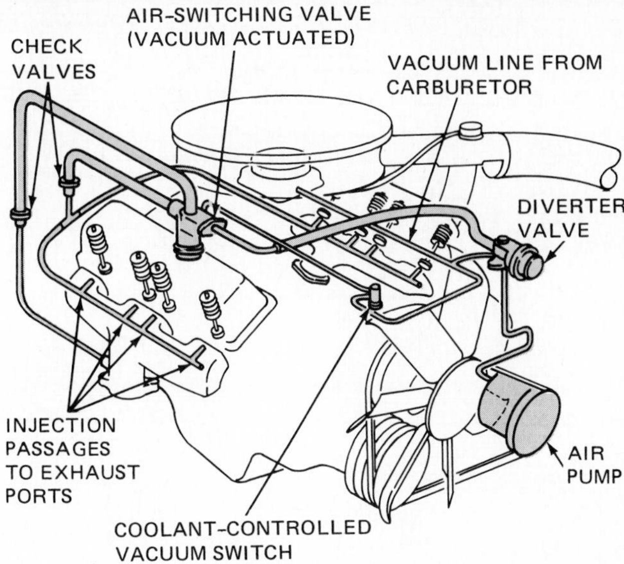

Fig. 31-35 Air-injection system using a belt-driven air pump to deliver air to the exhaust ports. *(Chrysler Corporation)*

the throttle closes suddenly, the throttle valve prevents air from entering the carburetor. However, the inertia of the fuel causes it to continue flowing for about a second and temporarily creates an air-fuel mixture too rich to burn in the combustion chamber.

When the mixture enters the exhaust manifold and combines with the injected air, the next-firing cylinder will ignite the air-fuel charge in the exhaust manifold. A loud "bang" in the exhaust system, sometimes called a backfire, results. By diverting the fresh air, or oxygen, from the exhaust manifold during deceleration, backfiring is prevented.

Some air-injection systems have an *air-switching valve* (Fig. 31-35). This valve switches the point of injection to either the exhaust manifold or to the catalytic converter. Where the air is injected must be changed as the engine warms up to prevent increasing NO_x emissions.

A check valve is located in the air lines between the pump and the nozzles (Fig. 31-35). The check valve prevents any hot exhaust gases from backing up into the hoses and pump. This could happen if the pump drive belt fails, if abnormally high pressure develops in the exhaust system, or if an air hose leaks.

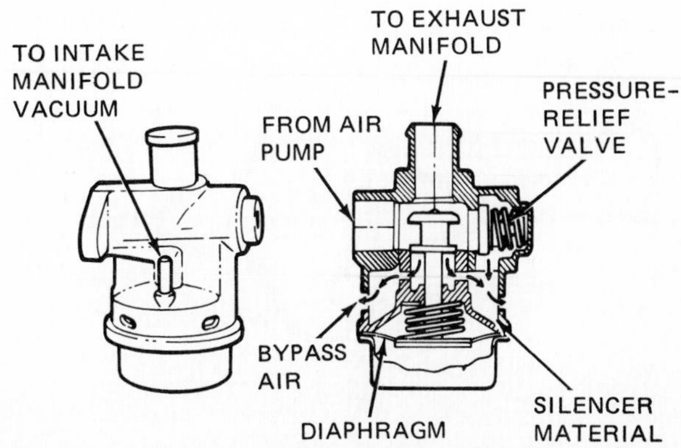

Fig. 31-36 Diverter valve, used to prevent backfire on engines equipped with air injection. *(Chrysler Corporation)*

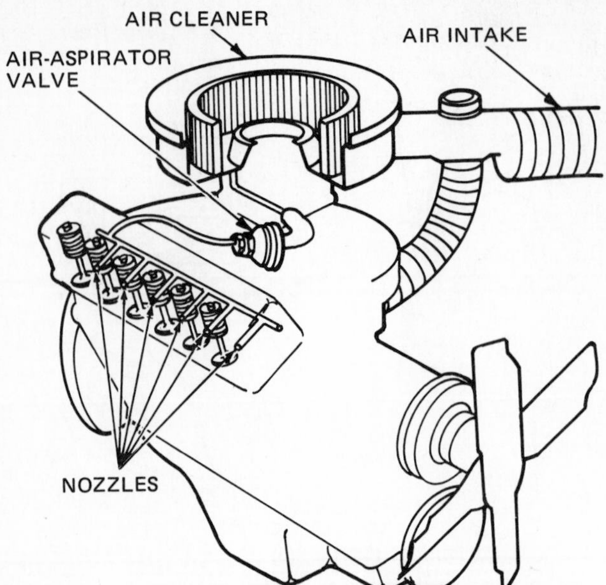

Fig. 31-37 Air-aspirator system, which is operated by pulses in the exhaust system. *(Chrysler Corporation)*

Air injection also is used with certain types of catalytic converters (♦31-33). Many EEC systems also control the air-injection system (♦16-22 to 16-24).

♦ 31-32 AIR-ASPIRATOR SYSTEM

Some vehicles equipped with a catalytic converter (♦31-33) use an air-aspirator system (Fig. 31-37). It furnishes additional air to reduce HC and CO emissions. At idle there is usually little oxygen in the exhaust gas that can react with HC and CO in the catalytic converter. The air-aspirator system delivers the needed air.

The air-aspirator valve is located in the tube between the air cleaner and the exhaust manifold (Fig. 31-37) or catalytic converter. Basically the aspirator valve is a one-way check valve operated by the negative-pressure (vacuum) pulses in the exhaust system. These occur in the exhaust ports and manifold each time an exhaust valve opens at idle.

Figure 31-38 shows the operation of the aspirator valve. At idle and slightly off idle, the negative-pressure pulses apply a slight vacuum at the ends of the nozzles. This vacuum causes the valve to open, admitting air from the clean-air side of the air cleaner. The valve snaps shut as soon as a pressure rise occurs in the exhaust system. This prevents any backflow of exhaust gas into the air cleaner.

♦ 31-33 CATALYTIC CONVERTERS

A second method of treating exhaust gas uses catalytic converters (Fig. 31-39). These convert harmful pollutants into harmless gases. A catalyst is a material that causes a chemical change without entering into the chemical reaction. In effect, the catalyst encourages two chemicals to react with each other. For example, in the HC/CO catalytic converter, the catalyst encourages the HC to unite with oxygen to produce H_2O (water). It encourages the CO to change to CO_2 (carbon dioxide). The catalyst in the NO_x converter splits the nitrogen from the oxygen. The NO_x therefore becomes harmless nitrogen and oxygen.

Figure 31-40 shows a two-way type of catalytic converter that converts only HC and CO. Figure 31-40 also shows the flow of the exhaust gas through the converter. The converter is filled with BB-shot-size metal pellets. They are coated with a thin layer of platinum or similar catalytic metal. The pellets form a matrix through which the exhaust gas must pass. As the exhaust gas flows through, the catalyst produces the chemical reaction. Another type of catalytic converter has a catalyst-coated honeycomb through which the gas must pass.

Cars with catalytic converters must use nonleaded gasoline. If the gasoline contains lead, the lead will coat the catalyst and the converter will stop working. If this happens to the pellet-type converter, there is a way to remove the old pellets and put in a charge of fresh pellets. But on the honeycomb type, the complete catalytic converter must be replaced.

The converter gets hot. Therefore, the floor pan above it must be insulated. This prevents heat from warming the passenger compartment (Fig. 31-41).

♦ 31-34 DUAL-BED AND THREE-WAY CATALYTIC CONVERTERS

There are three general categories of catalytic converters. These are oxidizing, reducing, and three way. The oxidizing converter (♦31-33) handles HC and CO, using platinum and palladium as the catalyst.

To control NO_x, rhodium is used as a reducing catalyst. It changes NO_x to harmless N_2. Instead of having two separate catalytic converters in the exhaust system, one for HC and CO and the other for NO_x, most manufacturers use either a *dual-bed catalytic converter* or a *three-way catalytic converter*.

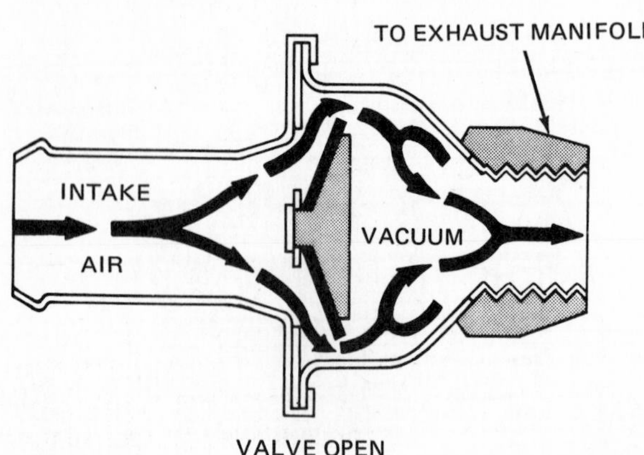

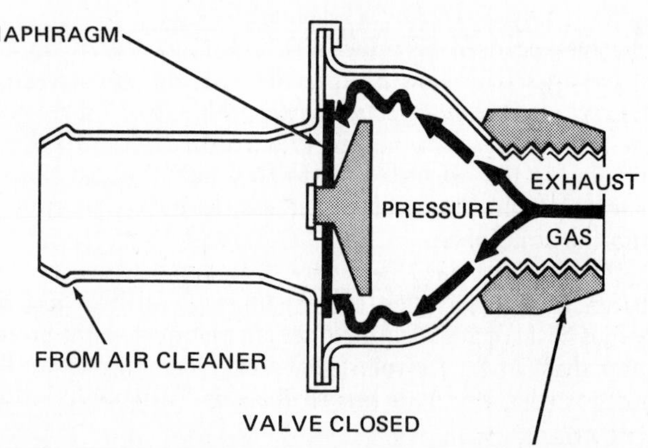

Fig. 31-38 Open and closed positions of the aspirator valve. *(Chrysler Corporation)*

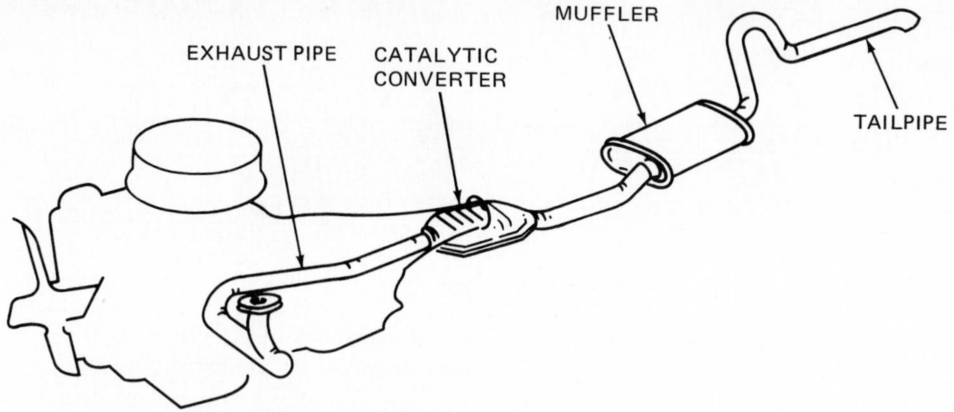

Fig. 31-39 The catalytic converter removes pollutants from exhaust gas flowing through the exhaust system. *(ATW)*

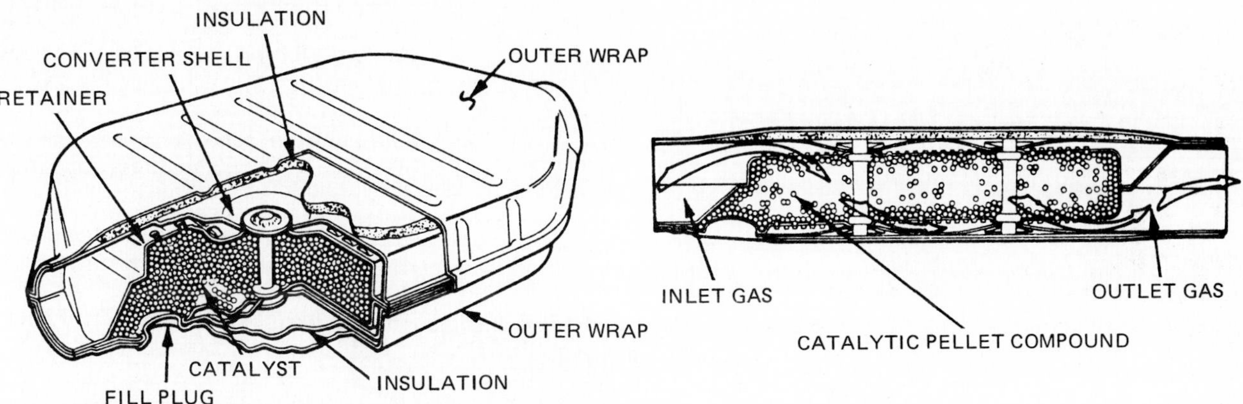

Fig. 31-40 Construction of a two-way pellet- (or bead-) type catalytic converter. Right, the flow of exhaust gas through the converter is shown by arrows. *(Buick Motor Division of General Motors Corporation)*

The dual-bed converter (Fig. 31-42) is like two bead-type converters in one housing with an air chamber between them. The exhaust gas first passes through the upper bed, reducing the NO_x and oxidizing some of the HC and CO. Then the exhaust gas flows through the air chamber to the lower bed, where the air pump is adding sufficient air for final oxidizing of the HC and CO.

A three-way catalyst is a mixture of platinum and rhodium (sometimes mixed with palladium). It acts on all three of the regulated pollutants (HC, CO, and NO_x), but only when the air-fuel-mixture ratio is precisely controlled (Fig. 31-43).

If the engine is operated with the ideal or *stoichiometric* air-fuel ratio of 14.7:1, the three-way catalyst is very effective. It strips oxygen away from the NO_x to form harmless water (H_2O), carbon dioxide (CO_2), and nitrogen (N_2). However, the air-fuel mixture must be precisely controlled if this action is to occur. For this reason, a closed-loop fuel-metering system (either feedback carburetor or fuel injection) must be used.

There are two types of three-way catalytic converters. They have a mesh or honeycomb (Fig. 31-44) coated with catalyst. The front section (in the direction of gas flow) handles NO_x and partly handles HC and CO. The partly treated exhaust gas then flows through the air chamber into the rear section of the converter. There the gas mixes with the air being pumped in by the air pump (♦31-31). This is called *secondary air*. It puts more oxygen in the exhaust gas so that the two-way catalyst can take care of the HC and CO.

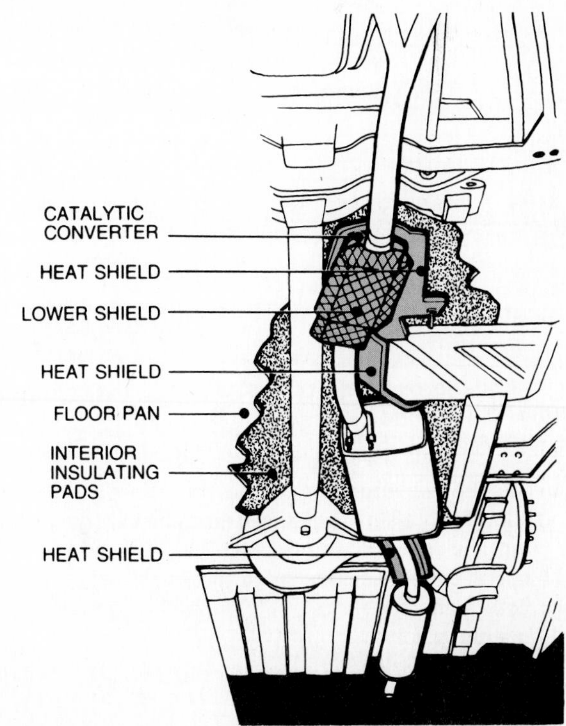

Fig. 31-41 Heat shields and insulating pads surrounding the catalytic converter. *(Chrysler Corporation)*

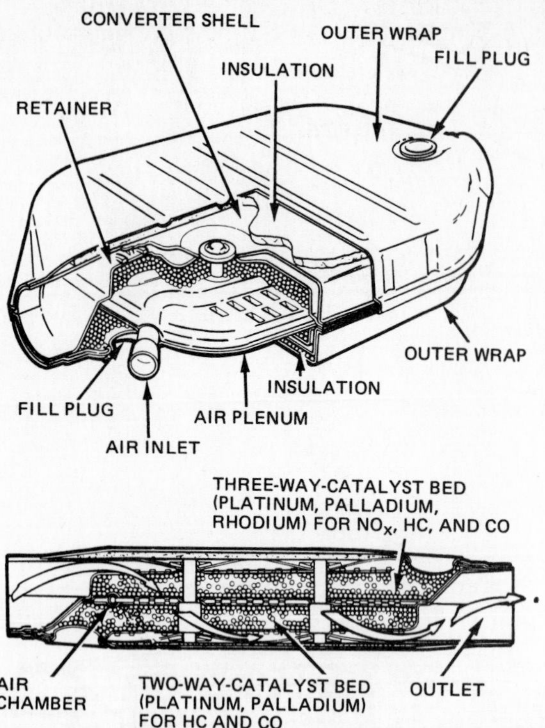

Fig. 31-42 A dual-bed catalytic converter. *(Pontiac Motor Division of General Motors Corporation)*

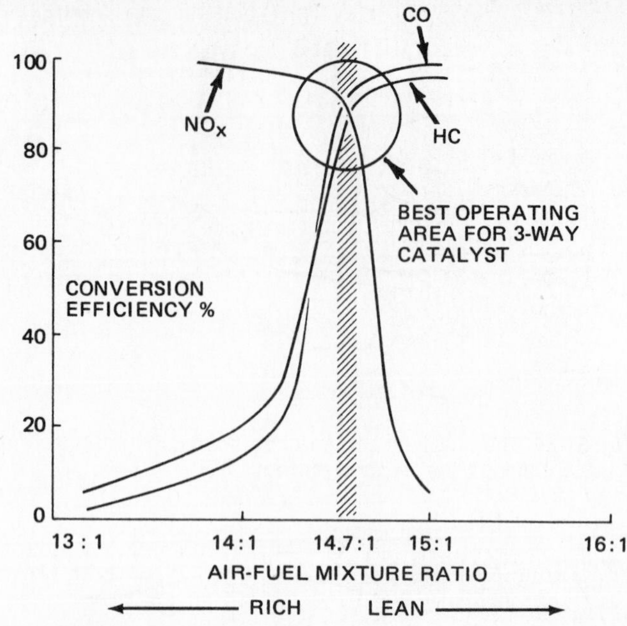

Fig. 31-43 The air-fuel-mixture ratio "window," within which the air-fuel ratio must remain if the three-way catalyst is to work. *(General Motors Corporation)*

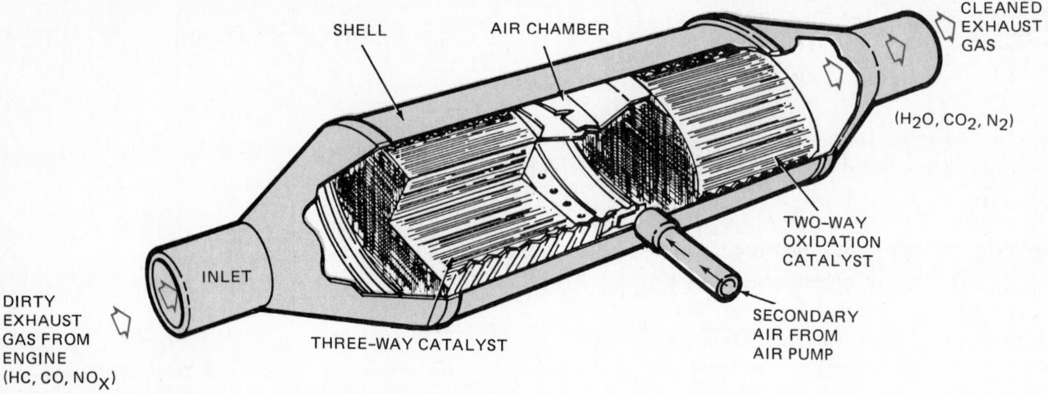

Fig. 31-44 A three-way catalytic converter using a *monolith,* or honeycomb, coated with catalyst. *(Ford Motor Company)*

♦ REVIEW QUESTIONS ♦

Select the *one* correct, best, or most probable answer to each question. You can find the answers in the section indicated at the end of each question.

1. In operation on a running engine, the PCV valve (♦31-2)
   a. allows only a small volume of air to flow through during idle
   b. opens more as engine speed increases
   c. both a and b
   d. neither a nor b
2. An uncontrolled fuel tank breathes as (♦31-3)
   a. car speed changes
   b. temperature changes
   c. engine speed increases
   d. all of the above
3. The fuel-vapor emission control system stores gasoline vapor in (♦31-4)
   a. a can
   b. the fuel tank
   c. a charcoal canister
   d. none of the above
4. The automobile equipped with an evaporative emission control system has a (♦31-8)
   a. sealed fuel gauge
   b. pressure-vacuum fuel-tank cap
   c. combination vapor and gasoline fuel pump
   d. all of the above

5. The purpose of the sealed fuel tank is to (♦31-8)
   a. allow the tank to expand to hold more fuel
   b. control vapor loss
   c. prevent overfilling of the tank
   d. none of the above
6. The device that prevents excessive airflow through the intake manifold during engine idle is called the (♦31-2)
   a. manifold valve
   b. crankcase valve
   c. PCV valve
   d. vacuum valve
7. The burned gases and air-fuel mixture that escape past the piston rings are called (♦31-2)
   a. lost gas
   b. blow up
   c. blowby
   d. smoke
8. The two fuel-system components that vent gasoline vapor to the charcoal canister are the (♦31-4)
   a. carburetor and fuel tank
   b. crankcase and manifold
   c. air cleaner and carburetor
   d. fuel pump and carburetor
9. The purpose of the fuel-vapor return line is to return to the fuel tank any vapor that (♦31-5)
   a. escapes the charcoal canister
   b. develops in the fuel pump
   c. leaks from the carburetor
   d. escapes from the fuel-tank cap
10. Some charcoal canisters have a purge valve which (♦31-6)
    a. purges the canister after the engine is stopped
    b. limits the flow of vapor and air to the carburetor during idle
    c. limits the flow of vapor and air to the carburetor during wide-open-throttle operation
    d. prevents fresh air from entering
11. The pollutant formed by high combustion temperatures is (♦31-22)
    a. HC
    b. CO
    c. NO_x
    d. H_2O
12. Three ways of cleaning the exhaust gas are to control the air-fuel mixture, control the combustion, and (♦31-12)
    a. control the carburetor
    b. treat the exhaust gas
    c. add tetraethyl lead
    d. none of the above
13. Two ways of treating the exhaust gas to reduce the amount of pollutants in it are by supplying additional air to the exhaust gases as they leave the cylinders and running the gases through (♦31-30)
    a. an air injector
    b. a catalyst
    c. charcoal canisters
    d. none of the above
14. All of these statements are true about the operation of the EGR emission control system *except* (♦31-23)
    a. the EGR valve allows exhaust gases to enter the intake manifold
    b. the EGR valve is open at engine idle
    c. the EGR system reduces NO_x emissions
    d. the EGR system reduces combustion-chamber temperatures
15. To control the air-fuel mixture and reduce atmospheric pollution, carburetors have been modified to deliver a leaner air-fuel mixture during (♦31-14)
    a. idling
    b. acceleration
    c. high-speed operation
    d. none of the above
16. Reducing the compression ratio of an engine reduces the combustion temperature, and this (♦31-22)
    a. reduces the amount of NO_x formed
    b. increases the amount of NO_x formed
    c. reduces the amount of HC formed
    d. increases the amount of H_2O formed
17. The transmission-controlled spark system prevents normal spark advance when the car is operating (♦31-25)
    a. in lower gears
    b. in higher gears
    c. under extreme load
    d. under light load
18. Reducing the combustion-chamber surface area (♦31-19)
    a. reduces the amount of unburned HC in the exhaust gas
    b. increases the amount of unburned HC in the exhaust gas
    c. reduces the amount of NO_x in the exhaust gas
    d. none of the above
19. One method of reducing NO_x in the exhaust gas is to (♦31-24)
    a. increase valve overlap
    b. reduce valve overlap
    c. prevent valve overlap
    d. all of the above
20. If the exhaust gas is low in oxygen, the air-fuel mixture is (♦31-13)
    a. lean
    b. rich
    c. about right
    d. none of the above

CHAPTER 32
Servicing Emission Control Systems

After studying this chapter, and with proper instruction and equipment, you should be able to:

1. Explain why emission control service is required.
2. Diagnose and service PCV systems.
3. Diagnose and service fuel-vapor emission control systems.
4. Diagnose and service air-injection systems.
5. Diagnose and service heated-air systems.
6. Diagnose and service EGR systems.
7. Diagnose and service catalytic converters.

♦ 32-1 REQUIREMENT FOR SERVICING EMISSION CONTROLS

Control of emissions from automobiles is required by the Clean Air Act of 1970. In addition, other laws impose penalties for modifying or tampering with any emission control system. The federal government requires that each car sold in the United States meets the standards of the Clean Air Act. To enforce compliance, some states and communities require periodic inspection of all motor vehicles. These inspections may include a visual and operational inspection of all emission control systems and a check of exhaust-gas emissions.

To ensure that the car continues to run without polluting excessively, the car manufacturer issues an *emissions-performance warranty* to the new-car purchaser. The warranty states that the car will pass any emissions tests approved by the Environmental Protection Agency (EPA) during the first five years or 50,000 miles [80,467 km] of vehicle service. However, the car must be operated and maintained according to the instructions in the owner's manual.

During the first 24 months or 24,000 miles [38,624 km], any defective component necessary to assure the continued performance of an emission control device or system for the remainder of the vehicle's useful life will be adjusted, repaired, or replaced by the manufacturer. But if more than 24 months or 24,000 miles have passed, the manufacturer is responsible for making repairs and adjustments only on certain components. These must have been installed in or on the car for the sole purpose of reducing vehicle emissions and were not in general use prior to the 1968 models.

The car owner is responsible for keeping the car properly maintained for continuing emissions control. This means that all cars should be serviced regularly according to the manufacturer's maintenance schedules. Each required operation must be done following the manufacturer's procedure and specifications.

The law also requires that whenever parts are replaced during these operations, any new parts installed must be equivalent in performance and durability to the original-equipment parts.

To assist the car owner and service technician, the manufacturer provides detailed information on engine identification, maintenance schedules, and service procedures in the owner's manual and the service manual. However, the vehicle emission control information label in the engine compartment contains the vacuum-hose diagram, specifications, and ignition-timing and idle-adjustment procedures. If any specifications or procedures on the decal differ from those in the service manual, follow the engine-compartment labels.

Careful It is against the law to tamper with emission controls on vehicles or vehicle engines. The penalty is a fine of $2,500 for every vehicle tampered with. When servicing engines and emission controls, do *only* those jobs specified in the manufacturer's shop manuals. If you remove, disconnect, damage, or render ineffective any emission control, you are breaking the law.

♦ 32-2 POSITIVE CRANKCASE VENTILATION TROUBLES

Several engine troubles can result from defective conditions in the PCV system. These troubles can also result from faults in other systems as explained in Part 6, Engine Trouble Diagnosis and Tuneup. This section describes only troubles arising from the PCV system.

Rough idle and frequent stalling could result from a plugged or stuck PCV valve or a clogged PCV air filter. In either case, the remedy is to replace the valve or filter.

Vapor flow into the air cleaner and oil in the air cleaner can result from *backflow*. Instead of filtered air flowing into the crankcase, vapors from the crankcase are flowing into the air cleaner. The cause is a plugged PCV valve or plugged or leaking condition somewhere in the PCV system. It could also be caused by worn piston rings or cylinder walls. The worn parts allow more blowby than the PCV system can handle.

Sludge or oil dilution in the crankcase can result from a plugged PCV valve or line that prevents normal circulation.

♦ 32-3 PCV-SYSTEM SERVICE

The PCV valve should be checked periodically and replaced at specified intervals. There are special testers that can be used to check the operation of the PCV system. A simple way to check the system and the valve is to remove the valve or valve connection, with the engine running. Place your hand over the opening. You should feel a slight vacuum pull against your hand. If there is no vacuum, or if you can feel a positive pressure, then something is wrong. Check the PCV valve, hoses, and connections.

Another method is the *rpm-drop test*. With the engine running at idle, connect a tachometer and note the engine idle speed. Then remove the PCV valve from the engine grommet, with hose attached. Block the opening of the valve, and note the change in engine speed. A decrease of less than 50 rpm indicates a plugged PCV valve or hose. A quick check of the PCV system can be made with the engine idling at normal operating temperature. Remove the PCV valve from the engine. The valve should hiss. You should be able to feel a strong vacuum when your finger is placed over the valve inlet. Reinstall the PCV valve, and remove the crankcase inlet-air cleaner. Hold a piece of stiff paper over the opening of the rocker-arm cover. After a few seconds, the paper should be drawn against the opening. Then stop the engine. Remove the PCV valve from the rocker-arm cover, and shake it. It should rattle, indicating that the valve is free. If the system does not meet these tests, install a new PCV valve and try again. If the system still does not pass the tests, the hose may be clogged. It should be cleaned out or replaced. It may be necessary to remove the carburetor and clean the vacuum passage with a ¼-inch drill. Also, clean the inlet vent on the crankcase inlet-air cleaner that is connected by the hose to the engine air cleaner.

♦ 32-4 FUEL-VAPOR EMISSION CONTROL TROUBLES

Here are troubles that might be caused by conditions in the fuel-vapor emission control or fuel system.

Fuel odor or loss of fuel could be caused by several of the conditions listed below.

1. Overfilled fuel tank
2. Leaks in fuel, vapor, or vent line
3. Wrong or faulty fuel-tank cap
4. Faulty liquid-vapor separator
5. Excessively high fuel volatility
6. Vapor-line restrictor missing
7. Canister drain cap or hose missing

A collapsed fuel tank can result if the wrong fuel-tank cap is installed or if the vacuum valve in the cap sticks. Then no air can enter to replace fuel being withdrawn by the fuel pump. the result could be a vacuum in the fuel tank great enough to allow atmospheric pressure to crush the tank.

Excessive pressure in the fuel tank could result from a combination of high temperatures and a plugged vent line, liquid-vapor separator, or canister. Pressure can be released by turning the tank filler cap just enough to allow the pressure to slowly escape.

Many different engine-idling problems can result from faulty or improper connections of a hose in the control system. A plugged canister, vapor-line restrictor missing, or high-volatility fuel can also cause poor idle.

♦ 32-5 FUEL-VAPOR EMISSION CONTROL SERVICE

These systems require little service. About the only troubles that occur are restriction of the fuel flow (so the engine is starved and stalls) and a collapsed fuel tank (♦32-4). Some types of charcoal canisters require periodic replacement of the air filter in the bottom of the canister (Fig. 32-1).

♦ 32-6 AIR-INJECTION-SYSTEMS TROUBLES

Troubles related to the air-injection system (Fig. 32-2) include noise, no air supply, backfire, and high HC and CO levels in the exhaust.

Noise from the belt or air pump could result from a loose belt, loose air-pump mounting bolts, worn pump bearings or other internal trouble, or air leaks from the system. The pump is not repairable. It must be replaced if damaged. The air pump should be adjusted so that the belt has the correct tension. Do not pry on the pump housing because this can damage the pump. If you have to use a pry bar, pry as close to the pulley end as possible. If they are available, use a belt tensioner and a belt-tension gauge (Fig. 32-3).

Air leaks should be stopped by tightening hose connections

Fig. 32-1 Replacing the air filter in the charcoal canister.

and replacing any hose that is defective.

If no air is getting to the air manifold, the exhaust gas will probably be high in HC and CO. Causes of no or inadequate air include a loose belt, frozen pump, leaks in the hoses or connections, and failure of the diverter or check valve. A defective pump or valve must be replaced. They are not repairable.

Backfire is usually caused by a defective diverter valve. It fails to block off the air supply under conditions of high intake-manifold vacuum. The same thing happens if the vacuum hose becomes disconnected or blocked.

♦ 32-7 AIR-INJECTION-SYSTEM SERVICE

No routine service is required on the air-injection system except to replace the air filter every 12,000 miles [19,312 km]

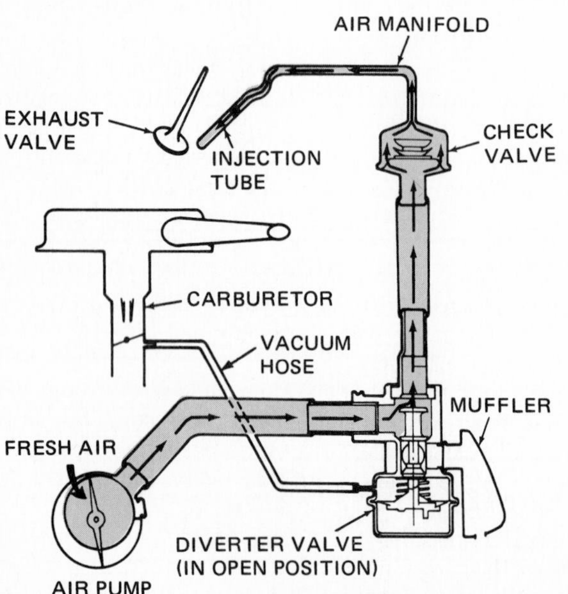

Fig. 32-2 Components of the air-injection system. *(Chevrolet Motor Division of General Motors Corporation)*

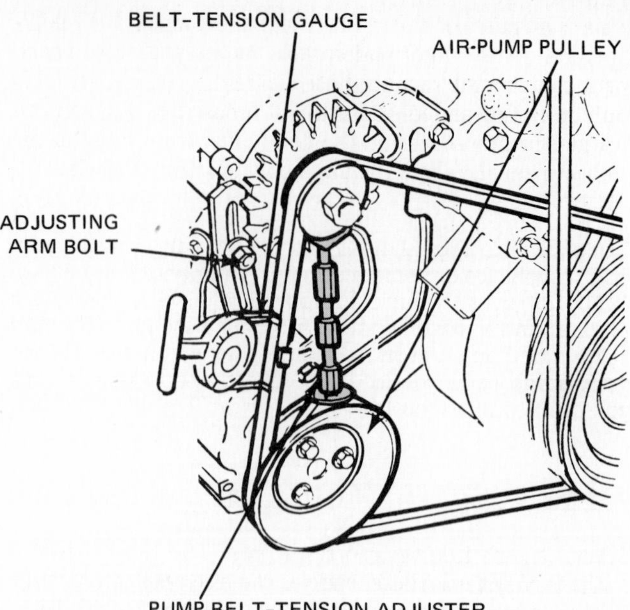

Fig. 32-3 A belt-tension adjuster being used to hold the air pump in place while the adjustment bolts are tightened. *(Ford Motor Company)*

on those systems using a separate air filter. Most air pumps use a centrifugal filter which requires no separate service.

The air-pump drive belt should be checked periodically for wear and tension. Replace any belts that are worn, cracked, or brittle. Do not overtighten the air-pump drive belt. Overtightening can cause premature pump-bearing failure.

♦ 32-8 HEATED-AIR-SYSTEM TROUBLES

The heated-air system adds heat to the air entering the carburetor almost as soon as the engine is started. This means the mixture is warm and the gasoline vaporizes better. If the heated-air system is not working to add heat, the control damper has closed off the hot-air pipe. The hot-air pipe may be disconnected or have holes in it. As a result, the mixture is not getting warmed. The cold engine will hesitate or stumble, or even stall, because the mixture is too cold and lean to fire consistently. If the damper does not shut off the hot-air pipe when the engine is hot, the mixture will be overheated. Then not enough mixture will get into the cylinders for full power.

Check the operation of the damper by starting the engine cold. See whether the damper in the snorkel is closed or open. It should be wide open. Then, when the engine starts, it should close. This opens the hot-air pipe. As the engine warms up, the damper should move to the open position. An accurate check can be made with a temperature gauge or thermometer. If the damper does not perform properly, the trouble could be in the thermostat or vacuum motor, or in the linkage. You may need a small mirror to observe the damper action (♦31-16).

The heated-air system requires no routine service.

♦ 32-9 EXHAUST-GAS-RECIRCULATION TROUBLES

Trouble in the EGR system may cause poor engine performance. For example, rough engine idle and stalling could be caused by a leaky EGR valve or valve gasket that allows exhaust gas or air to enter the intake manifold during idling. A defective thermal vacuum switch could cause vacuum to operate the EGR valve when it should not.

Poor part-throttle performance, poor fuel economy, and rough running on light acceleration could also be caused by a defective thermal vacuum switch. In addition, a sticking or binding EGR valve, or deposits in the EGR passages, could cause these conditions. If deposits have clogged the EGR passages, remove the manifold to clean out the passages.

If the engine stalls on deceleration, it could be due to a restricted vacuum line that is preventing the EGR valve from closing promptly. Detonation at part throttle could be caused by insufficient EGR. This could be due to clogged or damaged hoses, EGR valve, or a defective thermal vacuum switch.

♦ 32-10 TESTING EGR SYSTEMS

There are three types of EGR valves. These are the ported-vacuum valve, the positive-back-pressure valve, and the negative-back-pressure valve. Many ported-vacuum EGR valves have the stem visible under the diaphragm (Fig. 32-4). Check this type of valve with the engine warmed up and idling. With the transmission in neutral, snap the throttle open to bring the engine rpm up to about 2,000. The valve stem should move up, indicating the valve has opened. If it does not, connect a hand vacuum pump to the vacuum tube of the valve. With the engine warmed up and idling, apply about 8 inches Hg [200 mm Hg] of vacuum to the valve. The valve should open. If it does not, it is either defective or dirty. You can tell when the valve opens because the engine will idle roughly and may even stall.

To test the thermal vacuum switch, connect a vacuum gauge and a hand vacuum pump as shown in Fig. 32-5. With the engine cold, no vacuum should pass through the switch. When the engine warms up, vacuum should pass through.

Before testing a back-pressure EGR valve, you must determine whether it is a positive-back-pressure valve or a negative-back-pressure valve. There is a difference in the design of the diaphragm for each type. Illustrations and identifying marks are in the manufacturer's service manual. The negative-back-pressure valve can be checked following the same procedure described above for the ported-vacuum valve. To check the positive-back-pressure valve, the valve must be removed from the engine. With a hand-vacuum pump, apply a vacuum to the vacuum port. If the valve opens, it is defective and should be replaced.

♦ 32-11 EGR-SYSTEM SERVICE

A typical service interval for EGR systems is every 12 months or 12,000 miles [19,312 km] if the car is running on leaded gasoline. If the car is running on unleaded gasoline, check the system half as often (24 months or 24,000 miles [38,624 km]). Some cars have an EGR-maintenance reminder light that comes on at 15,000 miles [24,140 km] to remind the driver to have the system checked. However, many cars do not require any regular check. Instead, if trouble develops, the system should be checked and the trouble corrected.

♦ 32-12 TRANSMISSION-CONTROLLED SPARK (TCS) SYSTEM TROUBLES

This system allows vacuum advance only when the transmission is in high gear. Engine stall at idle, car creeping excessively in idle, and engine dieseling can all be due to a defective or improperly adjusted idle-stop solenoid.

Poor high-gear performance, stumble or stall on cold starts, poor fuel economy, and backfiring during deceleration could be due to an inoperative vacuum-advance solenoid, a defective temperature or transmission switch, or failure of the time relay to energize.

If vacuum advance is obtained in all gears and there are high levels of HC and NO_x in the exhaust, consider the four possible causes discussed above. There may be a defective

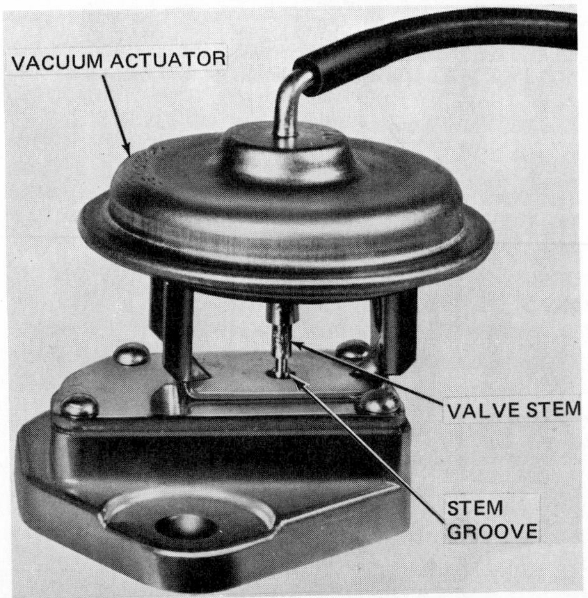

Fig. 32-4 EGR valve with exposed valve stem. *(Chrysler Corporation)*

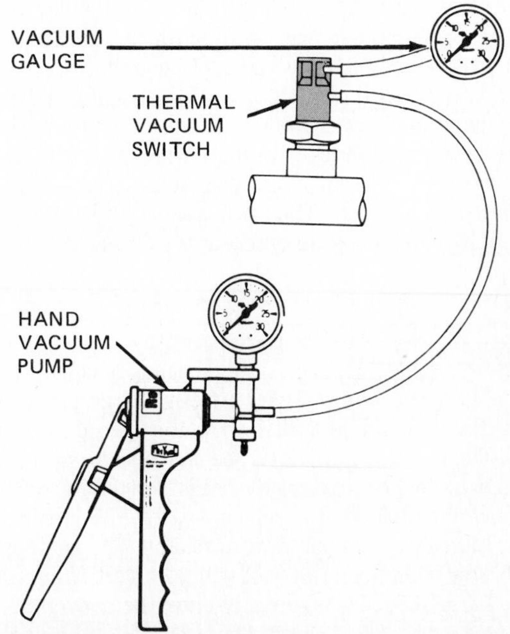

Fig. 32-5 Testing the thermal vacuum switch in the EGR system. *(Ford Motor Company)*

transmission switch, temperature switch, time relay, or vacuum-advance solenoid.

♦ 32-13 OTHER VACUUM-ADVANCE CONTROLS

The Ford TRS and the Chrysler OSAC systems work in a similar manner to the TCS system (♦31-26). Similar troubles can occur in all these systems. Failure of the system to work causes the conditions described in ♦32-12.

♦ 32-14 SERVICE INTERVALS FOR VACUUM-ADVANCE CONTROLS

No special servicing checks are usually required by the manufacturers. However, when a tuneup is performed, the system operation should be checked and the idle-stop solenoid adjusted. Inspection includes checking hoses for cracks, brittleness, or poor connections.

♦ 32-15 SERVICING ELECTRONIC SPARK ADVANCE CONTROLS

These systems develop very little trouble because everything is done electronically. Theoretically, transistors and diodes do not wear out. The manufacturers have developed special testers to check the components of EEC systems. Procedures and equipment are covered in the service manuals of the manufacturers.

♦ 32-16 CATALYTIC-CONVERTER TROUBLES

Catalytic-converter troubles are indicated by noise, BB-size particles coming out the tail pipe, a rotten-egg smell, high CO and HC levels in the exhaust gas, and possible loss of power due to a restricted converter.

Noise could be due to loose exhaust-pipe joints, a damaged converter, or a loose or missing catalyst replacement plug in a bead-type converter.

BB-size particles coming out the tail pipe means the converter has been overheated so the catalyst support has warped. This allows the beads to be blown out by the exhaust gas. The condition can happen only on the bead-type converter. The remedy is converter replacement.

A rotten-egg smell comes from hydrogen sulfide (H_2S) that the catalytic converter is producing. The S, or sulfur, is in the gasoline. Some gasolines have more than others. Advise the driver to try a different brand of gasoline. Also, check the carburetor adjustments. The smell is more noticeable when a momentarily rich mixture enters a hot converter.

♦ 32-17 CATALYTIC-CONVERTER SERVICE

Damaged or overheated converters must be replaced. They are not repairable. However, on the bead-type converter, the old beads can be removed and a fresh charge of beads installed. Figures 32-6 and 32-7 show the special devices required. The vacuum pump is turned on while the vibrator and the can are attached. It keeps the beads from falling out when the converter filler plug is removed. After the vibrator and can are attached, the vacuum is turned off. The air supply to the vibrator is turned on. Beads will now start falling in the can. It takes about 10 minutes to clear the converter.

To install new beads, dump the old beads and fill the can with new beads. Attach the can to the vibrator. Turn on the air and vacuum lines. After the beads stop flowing, disconnect the air hose and remove the vibrator. The converter should be filled flush with the full-plug hole. The vacuum pump is keeping them from falling out. Install the plug and remove the vacuum pump. Use antiseize compound on the plug threads.

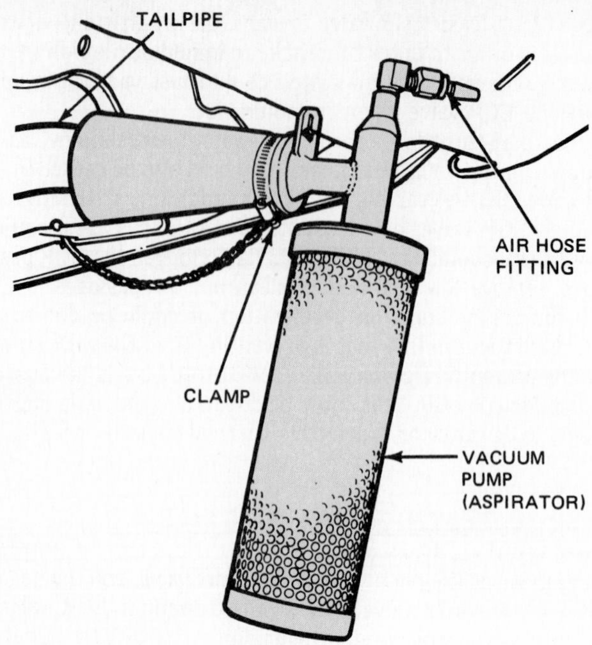

Fig. 32-6 Vacuum pump, or aspirator, mounted on the tail pipe of a car to change the beads in a pellet-type catalytic converter. *(American Motors Corporation)*

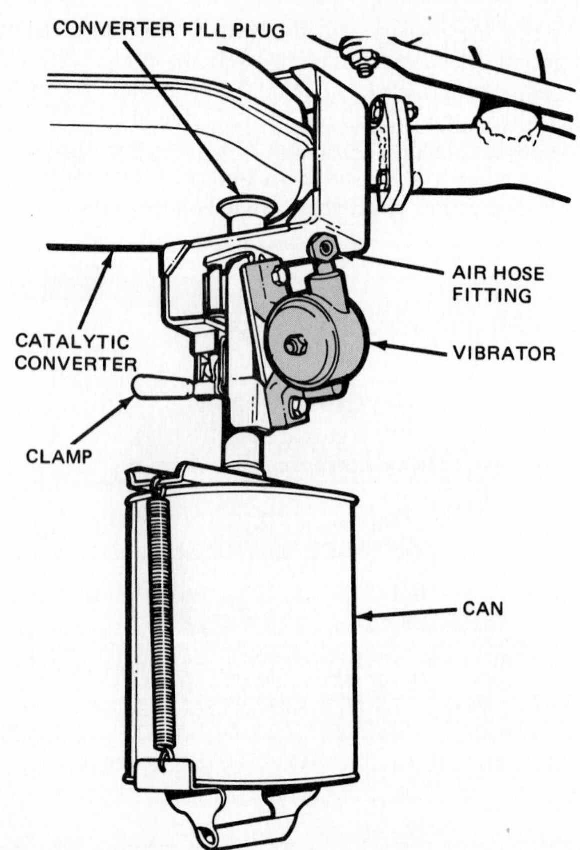

Fig. 32-7 Vibrator mounted on a catalytic converter. *(American Motors Corporation)*

♦ REVIEW QUESTIONS ♦

Select the *one* correct, best, or most probable answer to each question. You can find the answers in the section indicated at the end of each question.

1. The fine for tampering with any emission control is (♦32-1)
   a. $25
   b. $250
   c. $2,500
   d. $25,000
2. A plugged PCV valve can cause (♦32-2)
   a. rough idle
   b. stalling
   c. oil in the air cleaner
   d. all of the above
3. If a PCV valve is defective, it should be (♦32-3)
   a. disassembled for repair
   b. replaced
   c. washed in clean solvent
   d. blown out with compressed air
4. A strong fuel odor could be caused by (♦32-4)
   a. overfilled fuel tank
   b. leaks in fuel, vapor, or vent line
   c. faulty fuel-tank cap
   d. any of the above
5. The only service that some fuel-vapor emission control systems require is (♦32-5)
   a. replacement of the canister filter
   b. replacement of the fuel-tank cap
   c. adjustment of the canister valves
   d. removal of the vapor-line restrictor
6. In the EGR system, rough engine idle and stalling could be due to (♦32-9)
   a. a leaky EGR valve or valve gasket
   b. a restricted vacuum line
   c. a defective thermal vacuum switch
   d. any of the above
7. To adjust the air-pump belt, use (♦32-6)
   a. a pry bar to pry on the center of the pump housing
   b. a belt tensioner and belt-tension gauge
   c. a hose clamp
   d. none of the above
8. In the air-injection system, backfire is usually caused by (♦32-6)
   a. a defective air pump
   b. misadjustment of the diverter valve
   c. a defective diverter valve
   d. air leaks
9. Catalytic-converter trouble is indicated by (♦32-16)
   a. BB-size particles coming out the tail pipe
   b. a rotten-egg smell
   c. high CO and HC levels in the exhaust gas
   d. any of the above
10. On the bead-type catalytic converter, the beads can be replaced (♦32-17)
    a. with a vibrator and vacuum pump
    b. with a pressure pump and guide
    c. by removing the converter and turning it upside down
    d. with a torch and cutters

PART 6

Part 6 of this book explains how to diagnose troubles in the engine. It describes engine-testing instruments and how to use them, and how to perform a complete tuneup on a car. Some of the testing instruments were discussed in earlier chapters. They are reviewed here so that this part of the book covers the complete procedure for engine trouble diagnosis and tuneup. There are two chapters in Part 6:

Engine Trouble Diagnosis and Tuneup

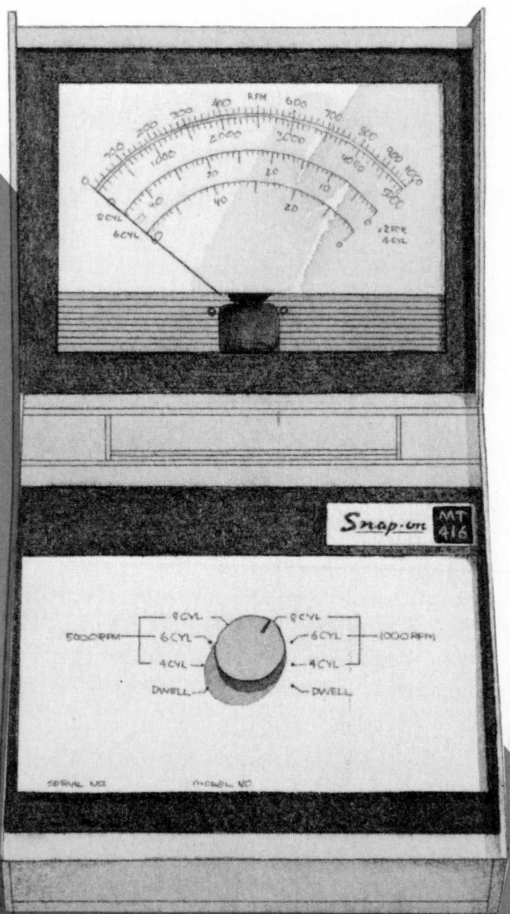

CHAPTER 33
Engine-Testing Instruments and Tuneup

After studying this chapter, and with proper instruction and equipment, you should be able to:

1. Use the tachometer.
2. Make a cylinder compression test.
3. Make a cylinder leakage test.
4. Use the vacuum gauge.
5. Measure engine exhaust emissions.
6. Check ignition timing.
7. Connect the oscilloscope and interpret its patterns.
8. Use the dynamometer to measure engine performance.
9. Define *tuneup*.
10. List the major steps in an engine tuneup.
11. Perform an engine tuneup.

♦ 33-1 ENGINE-TESTING PROCEDURES

Engine-testing procedures are of two types. One type is used when there is an obvious specific trouble that seems related to the engine. For example, if there is a miss in the engine or a complaint of excessive fuel or oil consumption, then there are definite trouble-diagnosis checks that can be made to pinpoint the cause of trouble.

The second type of engine-testing procedure is a general approach. Every engine system is tested as the procedure is carried out, and any worn condition, subnormal operation, or other defect will be detected. This general-approach procedure is often referred to as *engine tuneup*. The correction of troubles found during the testing procedure "tunes up" the engine and improves engine performance.

Both types of engine-testing procedures have their place in automotive-service business. When you encounter a specific trouble, you want to follow a specific procedure to find the cause and correct it. At other times, it is proper procedure to make a complete check of the engine and its components. For example, some recommendations are that the engine and its components be checked periodically (for example, every 10,000 miles [16,093 km] or at least once a year). Such an engine analysis will detect worn parts or improper adjustments that soon might cause trouble. Correction can then be made before serious trouble develops. In this way, the general-procedure diagnosis eliminates trouble before it happens. This is called *preventive maintenance*. You prevent trouble by maintaining the engine in good operating condition.

♦ 33-2 ENGINE-TESTING INSTRUMENTS

The engine-testing instruments covered in this chapter are:

1. Tachometer, which measures engine speed in revolutions per minute (rpm).
2. Cylinder compression tester, which measures the ability of the cylinders to hold compression pressure.

3. Cylinder leakage tester, which finds places where there is leakage of compression pressure.
4. Engine vacuum gauge, which measures intake-manifold vacuum.
5. Exhaust-gas analyzer, which measures the amount of pollutants in the exhaust gas.
6. Ignition timing light, which is used to set the ignition timing and check the spark advance.
7. Oscilloscope, which shows the overall operating condition of the ignition-system circuits.
8. Chassis dynamometer, which checks the engine and vehicle components under actual operating conditions.
9. Engine analyzer, which combines several testing instruments and makes several tests at once, or in sequence. Some computerized analyzers provide a printed record of the test results.

There are also instruments to test the battery, starting system, charging system, and cooling system. There are other instruments to test ignition coils, condensers, spark plugs, distributor contact-point dwell (on contact-point systems), and distributor advance mechanisms. Many of these tests are combined in the engine analyzer so that separate testing instruments are not required.

♦ 33-3 TACHOMETER

The tachometer is usually connected to the ignition system and operates electrically (Fig. 33-1). The tachometer measures engine speed in rpm. It measures the number of times the primary circuit is interrupted (on spark-ignition engines). The tachometer is an essential instrument because the idle speed must be adjusted to a specific rpm. Also, many tests must be made at specific engine speeds. The tachometer has a selector knob or buttons that can be set to the number of cylinders in the engine being tested.

Diesel engines, having no electric ignition system, require a different type of tachometer. Some are triggered by a magnet on the crankshaft which is sensed by a magnetic probe (Fig. 33-2).

Some cars have tachometers mounted on the instrument panel (Fig. 33-3). They indicate how fast the engine is turning. The driver can therefore keep the rpm within the range at which the engine develops maximum torque. This lets the driver get the best performance from the engine. Many of these tachometers have a red line at the top rpm on the dial. The red line marks off the danger point for engine speed.

Some tachometers are mechanical instead of electrical. They are driven off a gear on the ignition distributor shaft. They operate somewhat like the speedometer.

♦ 33-4 CYLINDER COMPRESSION TESTER

The cylinder compression tester measures the ability of the cylinders to hold compression. Pressure, operating on a diaphragm in the tester, causes the needle on the face of the tester to move around to indicate the pressure being applied. Figure 33-4 shows a compression tester used to measure the pressure in an engine cylinder.

To use the tester, first remove all the spark plugs. A recommended way to do this is to disconnect the wires and loosen the plugs one turn. Next, reconnect the wires and start the engine. Then, run the engine for a few seconds at 1000 rpm. The combustion gases will blow out of the plug well any dirt

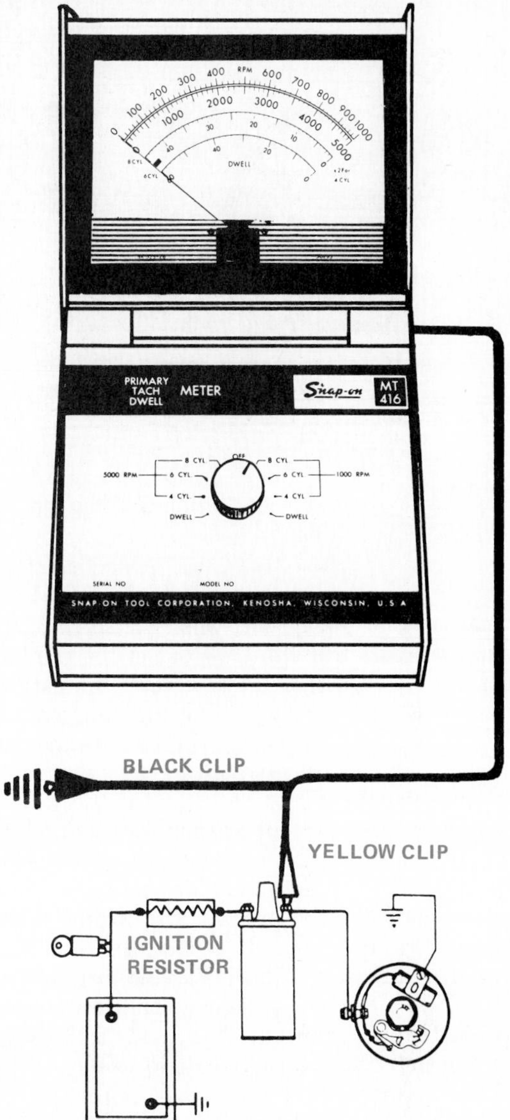

Fig. 33-1 Tachometer connected to a contact-point ignition system. *(Snap-on Tools Corporation)*

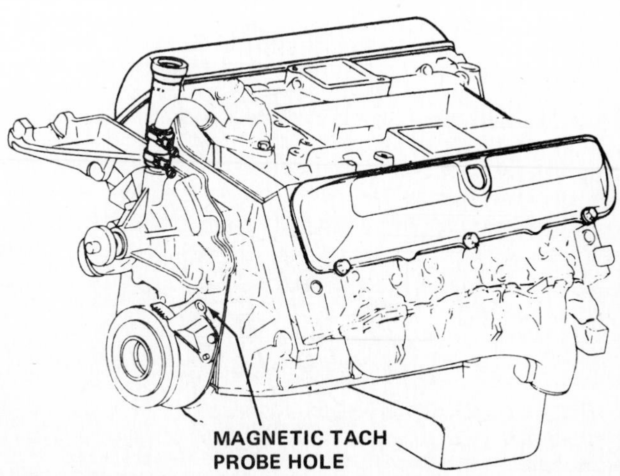

Fig. 33-2 Probe hole into which the magnetic tachometer probe is inserted to check diesel-engine rpm. *(Oldsmobile Division of General Motors Corporation)*

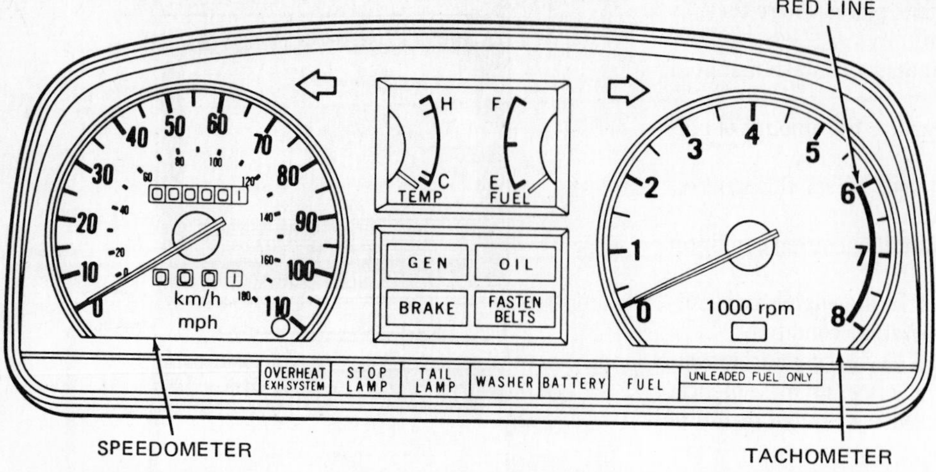

Fig. 33-3 Engine tachometer mounted in the car instrument panel. *(Mazda Motors of America, Inc.)*

that could fall into the cylinder when the spark plugs are removed. The gases will also blow out of the combustion chamber any loosened carbon that is caked around the exposed threaded end of the plug. This procedure prevents carbon and dirt particles from lodging under a valve and holding it open during the compression test.

After removing the plugs, block the throttle valve wide open. This is to make sure the maximum amount of air will get into the cylinders.

Next, screw the compression-tester adapter into the spark-plug hole of number 1 cylinder (Fig. 33-4). To protect the coil from high voltage, disconnect the primary lead from the negative terminal of the coil. (This is the lead that goes to the distributor.) On electronic ignition systems, disconnect the positive lead to the control unit. Then, hold the throttle wide open and operate the starting motor to crank the engine through four compression strokes (eight crankshaft revolutions). The needle will move around to indicate the maximum compression pressure in the cylinder.

Write down this figure. Test the other cylinders the same way. The lowest compression reading should be more than 75 percent of the highest.

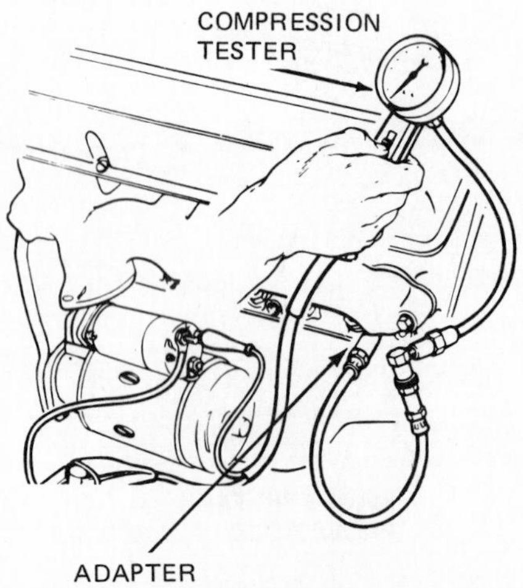

Fig. 33-4 Using a cylinder compression tester. *(Sun Electric Corporation)*

♦ 33-5 DIESEL-ENGINE COMPRESSION TEST

The compression test is different for diesel engines. The procedure for one engine is as follows. First, remove the air cleaner and install a manifold cover to keep dirt out of the engine. Disconnect the wire from the fuel-shutoff-solenoid terminal of the injection pump (Fig. 19-11). This prevents delivery of fuel during the test. Disconnect the glow-plug wires and remove the glow plugs. Screw the compression-tester fitting into the glow-plug hole of the cylinder to be checked. Then crank the engine for at least 12 crankshaft revolutions (six "puffs").

Check all cylinders the same way. The lowest compression reading should be not less than 70 percent of the highest. No cylinder should read less than 275 psi [1892 kPa].

In addition to the pressure reached, note the following. If everything is normal, the compression builds up quickly and evenly. If there is leakage past the piston rings, the compression is low on the first strokes but will tend to build up toward normal with later strokes. However, it does not reach normal and the pressure is rapidly lost after cranking stops.

♦ 33-6 RESULTS OF THE COMPRESSION TEST

The engine manufacturer's specifications tell you what the compression pressure of the cylinders should be. If the results of the compression test show that compression is low, there is leakage past the piston rings, valves, or cylinder-head gasket. To correct the trouble, you must remove the cylinder head and inspect the engine parts.

Before you do this, however, you can make one more test to pinpoint the trouble. Squirt a small quantity of engine oil through the spark-plug hole into the cylinder. Then retest the compression. If the pressure increases to a more normal figure, the low compression is due to leakage past the piston rings. Adding the oil helps seal the rings temporarily so that they can hold the compression pressure better. The trouble is caused by worn piston rings, a worn cylinder wall, or a worn piston. The trouble could also be caused by rings that are broken or stuck in the piston-ring grooves.

If adding oil does not increase the compression pressure,

the leakage is probably past the valves. This could be caused by:

1. Broken valve springs
2. Incorrect valve adjustment
3. Sticking valves
4. Worn or burned valves
5. Worn or burned valve seats
6. Worn camshaft lobes
7. Dished or worn valve lifters

It may also be that the cylinder-head gasket is "blown." This means the gasket has burned away so that compression pressure is leaking between the cylinder head and the cylinder block. Low compression between two adjacent cylinders is probably caused by the head gasket blowing between the cylinders.

Whatever the cause—rings, pistons, cylinder walls, valves, or gasket—the cylinder head must be removed so that the trouble can be fixed. The exception would be if the trouble is due to incorrect valve adjustment. Adjusting the valves does not require head removal. Engine service is discussed in later chapters.

♦ 33-7 CYLINDER LEAKAGE TESTER

The cylinder leakage tester checks the compression but in a different way. It applies air pressure to the cylinder with the piston at TDC on the compression stroke. In this position, both valves are closed. Very little air should escape from the combustion chamber. Figure 33-5 shows a cylinder leakage tester. Figure 33-6 shows how the tester pinpoints places where leakage can occur.

To use the cylinder leakage tester, first remove all the spark plugs. Then remove the air cleaner, crankcase filler cap or dipstick, and radiator cap. Set the throttle wide open, and fill the radiator to the proper level.

Connect the adapter, with the whistle, to the spark-plug hole of number 1 cylinder. Crank the engine until the whistle sounds. This means the piston is moving up on the compression stroke. Continue to rotate the engine until the TDC timing marks on the engine align. When the marks align, the piston is at TDC. Disconnect the whistle from the adapter hose and connect the tester, as shown in Figs. 33-5 and 33-6.

Next, apply air pressure from the shop supply. Note the gauge reading, which shows the percentage of air leakage from the cylinder. Specifications vary, but a reading above 20 percent means there is excessive leakage. If the air leakage is excessive, check further by listening at the air intake, tail pipe, and oil filler hole. If air is blowing out of an adjoining-cylinder spark-plug hole, the head gasket is blown between the cylinders.

Figure 33-6 shows what it means if you can hear air escaping at any of the three listening places. Also, if air bubbles up through the radiator, then the trouble is a blown cylinder-head gasket, a cracked cylinder head, or a cracked cylinder block. These conditions allow leakage from the cylinder to the cooling system.

Check the other cylinders in the same manner. A test light connected across the contact points can be used to position the next cylinder in the firing order at TDC. When you use the tester, follow the manufacturer's operating instructions.

♦ 33-8 ENGINE VACUUM GAUGE

The engine vacuum gauge is a tester for locating troubles in an engine that does not run as well as it should. This gauge measures intake-manifold vacuum. The intake-manifold vacuum changes with the load on the engine, the position of the throttle valve, and different engine defects. The way the intake-manifold vacuum varies from normal indicates what is wrong inside the engine.

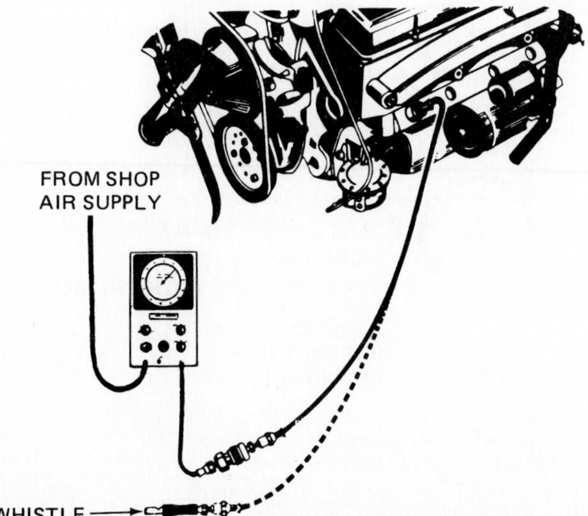
FROM SHOP
AIR SUPPLY
WHISTLE

Fig. 33-5 Cylinder leakage tester. The whistle is used to locate TDC in number 1 cylinder. *(Sun Electric Corporation)*

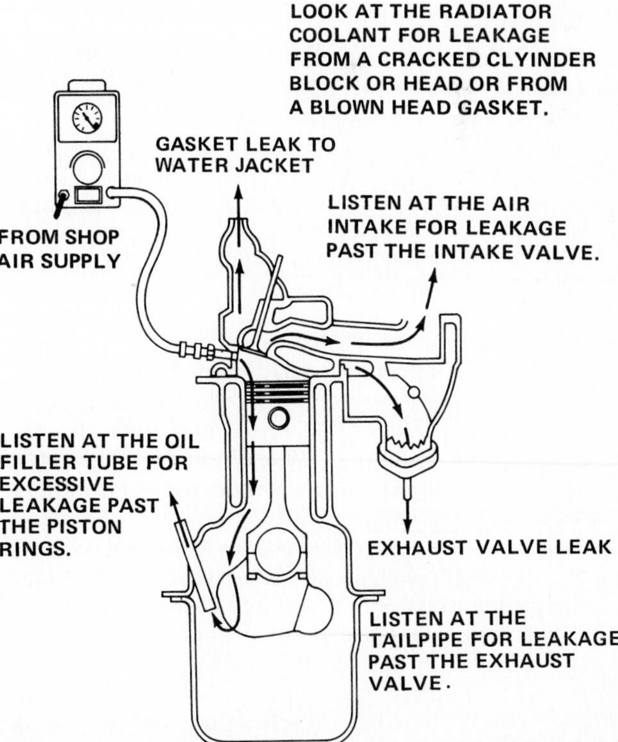
LOOK AT THE RADIATOR
COOLANT FOR LEAKAGE
FROM A CRACKED CLYINDER
BLOCK OR HEAD OR FROM
A BLOWN HEAD GASKET.
GASKET LEAK TO
WATER JACKET
LISTEN AT THE AIR
INTAKE FOR LEAKAGE
PAST THE INTAKE VALVE.
FROM SHOP
AIR SUPPLY
LISTEN AT THE OIL
FILLER TUBE FOR
EXCESSIVE
LEAKAGE PAST
THE PISTON
RINGS.
EXHAUST VALVE LEAK
LISTEN AT THE
TAILPIPE FOR LEAKAGE
PAST THE EXHAUST
VALVE.

Fig. 33-6 The cylinder leakage tester applies air pressure to the cylinder through the spark-plug hole with the piston at TDC and both valves closed. Points where air is leaking can then be pinpointed. *(Sun Electric Corporation)*

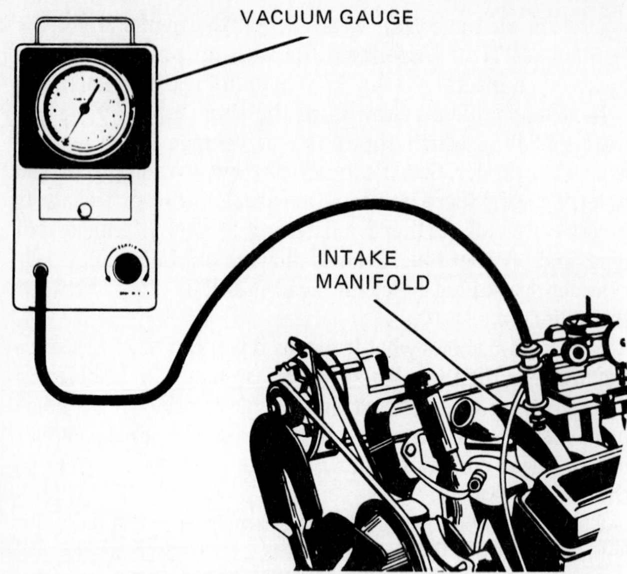

Fig. 33-7 Vacuum gauge connected to the intake manifold for a manifold vacuum test. *(Sun Electric Corporation)*

Figure 33-7 shows the vacuum gauge connected to the intake manifold. With the gauge connected, start the engine. The test must be made with the engine at operating temperature. Operate the engine at idle and at other speeds, as explained below. The meanings of the various readings are given in Fig. 33-8.

1. A steady and fairly high reading on idle indicates normal performance. Specifications vary with different engines, but a reading somewhere between 17 and 22 inches [432 and 559 mm] of mercury (Hg) indicates the engine is okay. The reading will be lower at higher altitudes because of lower atmospheric pressure. For every 1000 feet [305 m] above sea level, the reading will be reduced about 1 inch [25.4 mm] of mercury.

Note "Inches or millimeters of mercury" refers to the way the vacuum is measured. There is no mercury in the gauge. However, the reading compares with the changes that a vacuum would produce on a column of mercury in a barometer.

2. A steady and low reading on idle indicates late ignition or valve timing, or possibly leakage around the pistons. Leakage around pistons—excessive blowby—could be due to worn or stuck piston rings, worn cylinder walls, or worn pistons. Each of these conditions reduces engine power. With reduced power, the engine does not "pull" as much vacuum.
3. A very low reading on idle indicates a leaky intake manifold or throttle-body gasket, or possible leakage around the throttle shaft. Air leakage into the manifold reduces the vacuum and engine power.

Note Some engines, with high-lift cams and more valve overlap, are likely to have a lower and more uneven intake-manifold vacuum. Also, certain emission control systems lower the intake-manifold vacuum.

4. Back-and-forth movement of the needle that increases with engine speed indicates weak valve springs.

	READING	DIAGNOSIS
1	Average and steady at 17–21.	Everything is normal.
2	Extremely low reading—needle holds steady.	Air leak at the intake manifold or throttle body; incorrect timing.
3	Needle fluctuates between high and low reading.	Blown head gasket between two side-by-side cylinders. (Check with compression test.)
4	Needle fluctuates very slowly, ranging 4 or 5 points.	Idle mixture needs adjustment, spark-plug gap too narrow, sticking valves
5	Needle fluctuates rapidly at idle—steadies as RPM is increased.	Worn valve guides.
6	Needle drops to low reading, returns to normal, drops back, etc., at a regular interval.	Burned or leaking valve.
7	Needle drops to zero as engine RPM is increased.	Restricted exhaust system.
8	Needle holds steady at 12 to 16—drops to 0 and back to about 21 as you open and release the throttle.	Leaking piston rings. (Check with compression test.)

Fig. 33-8 Vacuum-gauge readings and their meaning. *(Champion Spark Plug Company)*

5. Gradual falling back of the needle toward zero with the engine idling indicates a clogged exhaust line.
6. Regular dropping back of the needle indicates that a valve is sticking open or a spark plug is not firing.
7. Irregular dropping back of the needle indicates that valves are sticking only part of the time.
8. Floating motion or slow back-and-forth movement of the needle indicates that the air-fuel mixture is too rich.

A test can be made for loss of compression due to leakage around the pistons. This condition would be the result of stuck or worn piston rings, worn cylinder walls, or worn pistons. Race the engine and then quickly release the throttle. The needle should swing around to 23 to 25 inches [584 to 635 mm] as the throttle valve closes. This indicates good compression. If the needle fails to swing around this far, there is loss of compression. Further checks should be made.

Note This test does not apply to engines equipped with a deceleration valve as part of the emission control system.

Another test using the vacuum gauge is the *cranking vacuum test*. It is made with the ignition disabled and the throttle valve closed. An even cranking vacuum at normal cranking speed indicates that the engine is mechanically sound. Any unevenness in the movement of the vacuum-gauge needle indicates an air leak into one or more cylinders.

♦ 33-9 EXHAUST-GAS ANALYZER

At one time, the major use of the early type of exhaust-gas analyzer (sometimes called a *combustion-efficiency meter*) was to adjust the carburetor. It is still used for that purpose. However, the new infrared type of exhaust-gas analyzer is also used to check out the emission controls on a car. Emission controls are described in Chaps. 31 and 32. The main purpose of the emission controls is to reduce the amounts of CO, HC, and NO_x in the exhaust gas.

Figure 33-9 shows the infrared type of exhaust-gas analyzer. To use it, you insert a probe into the tail pipe of the car (Fig. 33-9). The probe draws out some of the exhaust gas and carries it through the analyzer. Two meters, a digital display, or a printout indicate how much HC and CO are contained in the exhaust gas. The HC meter reports in parts per million. The CO meter reports in a percentage. Federal and state laws set the maximum legal limits on the amount of HC and CO permitted in the exhaust gas.

In addition to HC and CO, some exhaust-gas analyzers also measure the amount of carbon dioxide (CO_2) and oxygen (O_2) in the exhaust gas. A tester that measures three gases (for example, HC, CO, and CO_2) is called a *three-gas analyzer*. A tester that measures four gases is a *four-gas analyzer*.

A different kind of tester is required for NO_x, but it works in the same general way. It draws exhaust gas from the tail pipe and runs the gas through the analyzer. The finding is reported in grams per mile. Generally, NO_x testers are available only in testing laboratories. They are not widely used in automotive-service shops.

♦ 33-10 IGNITION TIMING LIGHT

The sparks must occur at the spark plugs in the cylinders at exactly the right time. They must occur a specific number of degrees before TDC on the compression stroke. Adjusting the distributor to make the sparks occur at the right time is called *ignition timing*. The procedure is covered in ♦27-20. A decal in the engine compartment lists, step-by-step, the procedure for timing the ignition. These instructions must be followed if the ignition timing is to be set correctly.

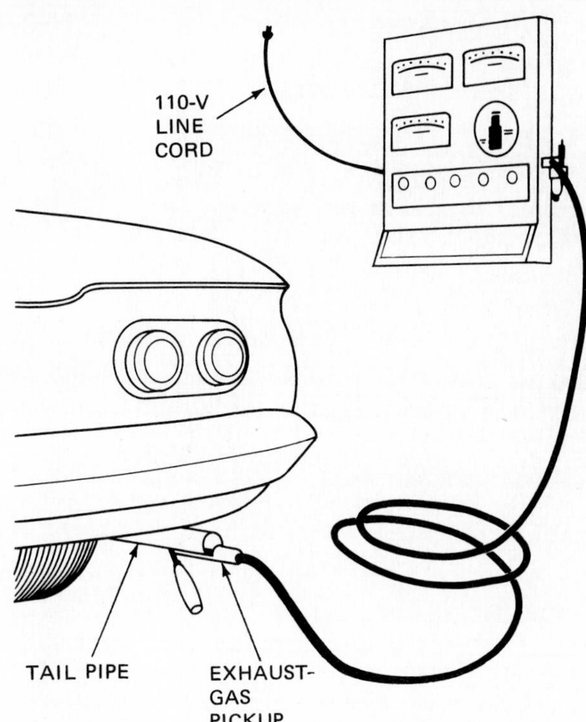

Fig. 33-9 Exhaust-gas analyzer connected for test.

♦ 33-11 OSCILLOSCOPE

The oscilloscope is a high-speed voltmeter that draws pictures of ignition voltages on the face of a televisionlike picture tube. These pictures, or scope patterns, show any troubles in the ignition system. Sections 27-18 and 27-19 explain how to use the oscilloscope on contact-point ignition systems. Section 29-15 explains how to use the oscilloscope to check electronic ignition systems.

♦ 33-12 DYNAMOMETER

The chassis dynamometer can test the engine-power output under various operating conditions. It can duplicate any kind of road test at any load or speed desired. The part of the dynamometer that you can see consists of two heavy rollers mounted at about floor level (Fig. 33-10). The car is driven onto these rollers, as shown in Fig. 33-10, so that the drive wheels can drive the rollers. Next, the engine is started, and the transmission is put into gear. The car is then operated as though it were out on an actual road test.

Under the floor is a power absorber that can place various loads on the rollers. This allows the technician to test the engine under various operating conditions. For example, the technician can find out how the engine would do during acceleration, cruising, idling, and deceleration. The test instruments, such as the scope, dwell-tach meter, and vacuum gauge, are hooked to the engine. These instruments then show how the engine is performing during various operating conditions.

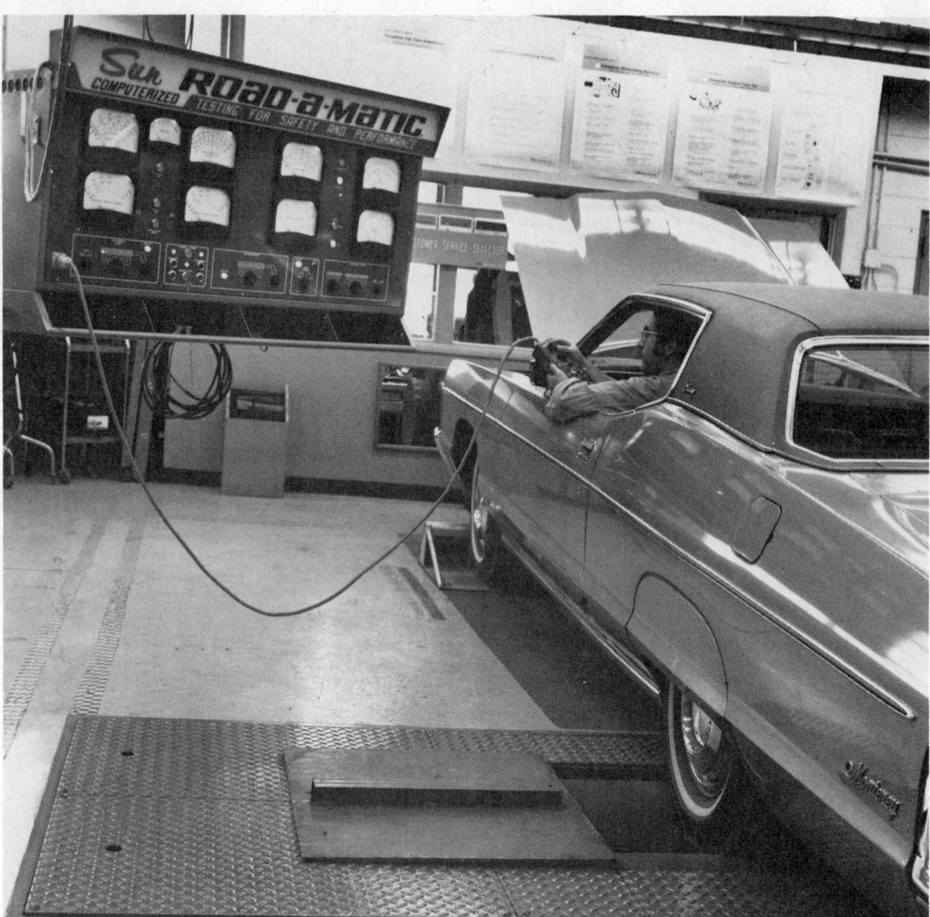

Fig. 33-10 Automobile in place on a chassis dynamometer. The drive wheels rotate the dynamometer rollers. At the same time, instruments on the console measure car speed, brake horsepower, and manifold vacuum. When front-drive cars are being tested, the front wheels are placed on the rollers. *(Sun Electric Corporation)*

The dynamometer can also be used to check the transmission and the differential. For example, the shift points and other operating conditions of an automatic transmission can be checked. Special diagnostic dynamometers are available in some schools and shops. These units have many instruments attached to them and have motored rollers that permit testing of wheel alignment, suspension, brakes, and steering.

♦ 33-13 ENGINE ANALYZERS AND COMPUTERIZED TESTERS

Figure 27-24 shows an engine analyzer. It includes an oscilloscope and other instruments for making tests of the engine parts and systems. Once you have learned how to use this equipment, you can make a complete engine analysis in a very short time.

There are also testers that run many of the tests almost automatically. They produce a printed record of the tests and the test results.

Some cars have a socket, or diagnostic connector, wired to the various components in the automobile. For a diagnostic check of the components, special computerized test equipment is required. It includes a plug that can be inserted into the car socket to complete connections between the car components and the test equipment. A series of connections are made all at once this way. Making a series of separate connections is not necessary.

Figure 33-11 shows one type of diagnostic computer. This computer runs the tests. Then it furnishes a printed record that tells the mechanic and customer what work is needed to bring the car up to specifications. Some models of this computer can read and report on exhaust emissions (HC and CO). Programmed tape cassettes provide the computer with the vehicle specifications against which the actual conditions of the vehicle are compared. This pinpoints any possible trouble area so that it can be serviced.

♦ 33-14 ON-BOARD DIAGNOSTIC DEVICES

The car has always carried some indicating devices that warn the driver if some action or correction is necessary. For example, all cars have fuel gauges, speedometers, and battery-charge, engine-temperature, and oil-pressure indicators. Car manufacturers have added to these basic indicators. For example, some cars have sensors connected to various engine components which signal with a light if something is wrong. Figure 33-3 shows an instrument panel with a series of signal lights that will come on if some component needs attention—low fluid in the windshield washer, for example.

Many cars now have internal diagnostic computers that flash a light if there is a problem with the engine. For example, the *check engine* light might come on. This is a warning that the driver should get the car to a service facility without

Fig. 33-11 The computerized tester tests the engine and electrical system, and then compares the readings against the manufacturer's specifications. *(Hamilton Test Systems)*

delay. The mechanic can then activate the diagnostic system. The flashing code numbers tell the mechanic what is wrong. In effect, the car diagnoses itself (♦29-17). This is called *self-diagnosis.*

♦ 33-15 COOLING-SYSTEM TESTERS

There are four cooling-system testers:

1. Antifreeze strength tester
2. Cooling-system pressure tester
3. Belt-tension tester
4. Exhaust-gas leakage into the cooling-system tester

All of these are described in detail in the chapter on cooling-system service (Chap. 22).

♦ 33-16 OTHER TESTERS

Several other testers are described earlier in the book: PCV testers (♦32-3), fuel-system testers (Chaps. 17 and 19), and electrical-system testers (Chaps. 24 to 27, and 29). Refer to these chapters for information on these testers and how to use them.

TUNEUP

♦ 33-17 DEFINITION OF TUNEUP

Engine tuneup means different things to different people. To some, it means a quick check of the engine that covers only frequent trouble sources. To others, it means use of test instruments to do a careful, complete analysis of all engine systems. In addition, it means adjusting everything to specifications and repairing or replacing all worn parts.

Note In this chapter, many steps are given as part of the general engine tuneup procedure. Details of how to perform most of these steps are covered in other chapters in the book. For further information on each step, refer to the chapter covering that part or system, or turn to the entry in the index at the back of the book.

♦ 33-18 TUNEUP PROCEDURE

An engine tuneup follows a procedure. Many mechanics use a printed form supplied by automotive or test-equipment manufacturers (Fig. 33-12). By following the form and checking off the items listed, one by one, the mechanic is sure of not overlooking any part of the procedure. However, not all tuneup forms are the same. Different companies have different ideas about what should be done and the order in which it should be done. In addition, the tuneup procedure depends on the equipment available. If the shop has an oscilloscope or a dynamometer, it is used as part of the tuneup procedure. If these test instruments are not available, then the tuneup is performed differently.

♦ 33-19 TUNEUP AND CAR CARE

The tuneup procedure restores driveability, power, performance, and economy that have been lost through wear, corrosion, and deterioration of engine parts. These changes take place gradually in many parts during normal car operation. Because of federal and state laws limiting automotive emissions, the tuneup procedure should include checks of all emission controls. Here is the procedure.

1. If the engine is cold, operate it for at least 20 minutes at 1500 rpm or until it reaches operating temperature. Note any operational problems during this warm-up time.
2. Connect the oscilloscope and exhaust-gas analyzer and perform a diagnosis. Check for any abnormal condition and, if possible, the cylinder(s) in which it appears.
3. Remove all spark plugs. Fully open the throttle and choke valves. Disconnect the distributor lead from the coil primary terminal to prevent excessive secondary voltage.
4. Check the compression of each cylinder. Record the readings. If one or more cylinders read low, squirt about

CUSTOMER ________ PHONE ________ DATE ________

MAKE-YEAR-MODEL ________ MILEAGE ________

SIGNATURE ________

PLEASE CHECK DESIRED SERVICES

Complete Infra-Red Engine Tune-Up

(CONSISTS OF ALL ITEMS LISTED BELOW)

1 ☐ USE INFRARED EXHAUST GAS ANALYSIS & VISUAL INSPECTION TO TEST & ADJUST CARBURETOR & CHOKE
(Needed for smooth idle, economy, good performance & low exhaust emissions)

2 ☐ TEST CYLINDER BALANCE & COMPRESSION FACTORS
(Needed for smooth idle, economy & good performance)

3 ☐ CLEAN OR REPLACE SPARK PLUGS AS NEEDED

4 ☐ REMOVE DISTRIBUTOR FROM ENGINE FOR SERVICE—TEST MECHANICAL & ELECTRICAL CONDITION—CALIBRATE ADVANCE MECHANISM
(Has direct bearing on gasoline mileage, exhaust emissions & performance—also permits proper installation of ignition points)

5 ☐ REPLACE IGNITION POINTS IF NEEDED & SET INITIAL TIMING

6 ☐ SERVICE MANIFOLD HEAT CONTROL VALVE
(Affects warm-up & running performance)

7 ☐ REPLACE FUEL FILTER ELEMENT IF NEEDED & TEST FUEL PUMP OPERATION
(To guard against unpredictable road failures)

8 ☐ SERVICE AIR POLLUTION CONTROL MECHANISM—P.C.V.
(Affects fuel air ratio and idle performance)

9 ☐ CLEAN, TEST & SERVICE SPARK PLUG WIRES, IGNITION COIL, DISTRIBUTOR CAP & ROTOR
(Helps prevent starting & missing problems due to dampness and/or resistance)

10 ☐ TEST GENERATOR OR ALTERNATOR & REGULATOR—REPLACE DRIVE BELTS IF NEEDED
(To insure battery will be kept fully charged for starting)

11 ☐ SERVICE & TEST BATTERY—CLEAN BATTERY TOP & CABLES—RECHARGE IF NEEDED—TEST STARTER
(Helps prevent unpredicted starting failures)

12 ☐ SERVICE AIR FILTER—REPLACE IF NEEDED
(Affects general performance of engine)

13 ☐ USE CYLINDER & CRANKCASE CHEMICAL TUNE-UP ADDITIVES
(Helps prevent sticking valves, rings & lifters)

14 ☐ ROAD TEST
(For over-all automobile performance & general condition)

15 ☐ TELEPHONE IF ADDITIONAL SERVICES ARE FOUND NECESSARY

Sun SERVICE CONTROL SYSTEM

Fig. 33-12 Printed tuneup form. *(Sun Electric Corporation)*

1 tablespoon [15 cc] of engine oil through the spark-plug hole. Recheck the compression. Record the new readings. For further diagnosis of engine mechanical problems, perform a cylinder leakage test.

Note If engine mechanical problems are found, tell the owner the engine cannot be tuned without overhaul or repair.

5. Clean, inspect, file, gap, and test the spark plugs. Discard worn or defective spark plugs (Fig. 27-21). Many mechanics install all new plugs instead of servicing the old plugs. Gap all plugs, old and new. Then install the plugs.
6. Inspect and clean the battery case, terminals, cables, and hold-down brackets. Test the battery. Add water, if necessary. If severe corrosion is present, clean the battery and cables with brushes and a solution of baking soda and water.
7. Test the cranking voltage. If the battery is in good condition but cranking speed is low, the cause can be the battery or the starting system. Test the starting system.
8. If the battery is low or the customer complains that the battery keeps running down, check the charging system (alternator and regulator). If the battery is old, it may have worn out. A new battery is required.
9. Check the drive belts. Replace any that are in poor condition. If you have to replace one belt of a two-belt drive, replace both belts. Tighten the belts to the correct tension, using a belt-tension gauge.
10. Inspect the distributor rotor, cap, and primary and high voltage (spark-plug) wires.
11. Clean or replace and adjust distributor contact points by setting the point gap. Lubricate the distributor breaker cam. On distributors with round cam lubricators, turn the cam lubricator 180 degrees or replace it if required.
12. Check the distributor cap and rotor (Figs. 27-19 and 27-20). Check the centrifugal and vacuum advances. Set the dwell and then adjust ignition timing. Make sure the idle speed is not too high. This could produce centrifugal advance during timing adjustment.

13. Use the oscilloscope to recheck the ignition system. Any abnormal conditions that appeared in step 2 should now have been eliminated.
14. Check the manifold heat-control valve. Lubricate with heat-valve lubricant. Free or replace the valve if necessary.
15. Check the fuel-pump operation with a fuel-pump tester. Replace the fuel filter. Check the fuel-tank cap, fuel lines, and connections for leakage and damage.
16. Clean or replace the air-cleaner filter. If the engine is equipped with a thermostatic air cleaner, check the operation of the vacuum motor (♦32-8).
17. Check the operation of the choke and the fast-idle cam. Check the throttle valve for full opening, and the throttle linkage for free movement. Clean all external carburetor linkage.
18. Inspect all engine vacuum fittings, hoses, and connections. Replace any brittle or cracked hose.
19. Clean the engine oil-filler cap if a filter-type oil-filler cap is used.
20. Check the cooling system. Inspect all hoses and connections, the radiator, water pump, and fan clutch if used. Check the strength of the antifreeze and record the reading. Pressure-check the system and radiator cap. Squeeze the hoses to check them (Fig. 22-3). Replace any defective hose (collapsed, soft, cracks, etc.).
21. Check and replace the PCV valve if necessary. Clean or replace the PCV-system air filter if required. Inspect the PCV hoses and connections. Replace any cracked or brittle hose. Test the system for vacuum.
22. If the engine is equipped with an air pump, replace the pump inlet air filter if used. Inspect the system hoses and connections. Replace any brittle or cracked hose.
23. Replace the charcoal-canister air filter, if used, in the evaporative emission control system.
24. Check the transmission-controlled spark system if the vehicle is so equipped.
25. Inspect and clean the EGR valve. Inspect and clean the EGR discharge port. Test system operation by applying vacuum to the valve with engine at idle. Engine should run rough, then smooth out with no vacuum applied.
26. If the engine is equipped with a turbocharger, check the operation of the wastegate (♦15-14). Follow the procedure in the manufacturer's service manual.
27. Tighten the intake-manifold and exhaust-manifold bolts to the specified torque in the proper sequence.
28. Adjust the engine valves if necessary.
29. Adjust the engine idle speed. Use an exhaust-gas analyzer, and propane if required, and adjust the idle-mixture screw (on older cars; on newer cars, the screw is sealed and cannot be adjusted without disassembling the carburetor). Check the amounts of HC and CO in the exhaust gas. (Checking the HC and CO both before and after a tuneup shows how much the tuneup has reduced these pollutants.)
30. Road-test the car on a dynamometer or on the road. Check for driveability, power, and idling. Any abnormal condition should be noted on the repair order before you return the car to the customer.
31. Check the lubrication or maintenance sticker to determine whether an oil and oil-filter change is due. Also note the schedule for chassis lubrication. Recommend an oil change and a chassis lubrication if they are due. Car manufacturers recommend changing the oil filter every other time the oil is changed.
32. Whenever the car is raised, check the exhaust system for leaks which could admit carbon monoxide into the passenger compartment. Also check for loose bolts, rust spots, and other under-the-car damage.

Note Items 33 to 38 are not actually part of the engine tuneup. They are included because they are important steps in a vehicle maintenance program. Each is quickly checked during the road test.

33. Check the brakes for even and adequate braking.
34. Check the steering system for ease and smoothness of operation. Check for excessive play in the system. Record any abnormal conditions.
35. Check the tires for inflation and for abnormal wear. Abnormal wear can mean suspension trouble, and wheel alignment should be recommended.
36. Check the suspension system for looseness, excessive play, and wear.
37. Check the front wheels and ball joints for excessive wear or loose bearings. Adjust the bearings if necessary.
38. Check the headlights and horns to make sure they work. Check all other lights. Replace any burned-out light bulbs. Check headlight aim if possible.

Note The preceding list for tuneup and maintenance covers most conditions on the vehicle that could cause trouble. The complete procedure will uncover most problems that might affect driveability and performance. If all necessary corrections are made, vehicle performance and reliability will be improved.

♦ REVIEW QUESTIONS ♦

Select the *one* correct, best, or most probable answer to each question. You can find the answers in the section indicated at the end of each question.

1. When connected to the engine, the tachometer measures (♦33-2)
   a. engine vacuum
   b. engine speed
   c. engine compression
   d. engine power
2. The first step in using the compression tester is to (♦33-4)
   a. disconnect the battery
   b. adjust engine idle
   c. remove the spark plugs
   d. shift transmission into low gear

3. While making a compression test, the purpose of squirting a small amount of engine oil through the spark-plug hole is to (♦33-6)
   a. lubricate the cylinders and pistons
   b. see whether the compression pressure increases
   c. measure manifold vacuum
   d. make it easy to install the spark plug
4. If squirting engine oil through the spark-plug hole does not increase the compression pressure (♦33-6)
   a. there is leakage past the valves
   b. a valve spring is broken
   c. the valves are sticking or burned
   d. any of the above
5. The cylinder leakage tester applies air pressure to the cylinder with the piston (♦33-7)
   a. at TDC and both valves closed
   b. at BDC and both valves open
   c. at TDC and both valves open
   d. starting the compression stroke
6. If the vacuum-gauge needle swings around to 23 to 25 inches [584 to 635 mm] of mercury as the throttle valve is quickly closed after the engine has been raced, this indicates (♦33-8)
   a. stuck valves
   b. low compression
   c. satisfactory compression
   d. leaky valves
7. A compression test has been made on an in-line six-cylinder engine. Cylinders 3 and 4 have readings of 10 psi. The other cylinders all have readings between 130 and 135 psi. Mechanic A says this is probably due to a blown head gasket. Mechanic B says this could be caused by wrong valve timing. Who is right? (♦33-6)
   a. A only
   b. B only
   c. either A or B
   d. neither A nor B
8. A steady but low vacuum reading with the engine idling indicates that the engine (♦33-8)
   a. is losing power
   b. has a stuck valve
   c. exhaust pipe is plugged
   d. none of the above
9. The device which can give a very close approximation of a road test in the shop is called (♦33-12)
   a. a dynamometer
   b. a tachometer
   c. an engine tester
   d. a computer
10. To check diesel-engine compression, the compression-tester fitting is screwed into the (♦33-5)
    a. spark-plug hole
    b. exhaust-manifold hole
    c. glow-plug hole
    d. intake-manifold hole
11. You determine when the engine oil was last changed by (♦33-19)
    a. checking the color of the oil
    b. checking the oil level
    c. looking at the lubrication sticker
    d. none of the above
12. A tuneup includes (♦33-19)
    a. checking the air cleaner, spark plugs, and ignition system
    b. checking the thermostatic air cleaner, PCV valve, and idle speed
    c. both a and b
    d. neither a nor b
13. If you replace one belt of a two-belt set (♦33-19)
    a. tighten belts based on tension of new belt
    b. replace both belts
    c. check pulleys to see what is wrong
    d. none of the above
14. When the car is up on the lift, check the exhaust system for leaks that could cause (♦33-19)
    a. carbon monoxide to leak into the passenger compartment
    b. excess noise
    c. exhaust smoke
    d. all of the above
15. A quick way to check the radiator hose is to (♦33-19)
    a. squeeze the hose
    b. look under the car for coolant
    c. remove and inspect the hose
    d. all of the above
16. During the tuneup procedure, you (♦33-19)
    a. clean and reinsulate the spark plugs
    b. clean, gap, and install old spark plugs
    c. install new spark plugs
    d. either b or c
17. If the vehicle is equipped with an evaporative emission control system, you may have to replace the (♦33-19)
    a. charcoal canister
    b. canister air filter
    c. purge valve
    d. charcoal in the canister
18. To see how much the tuneup has reduced pollutants in the exhaust gases, many mechanics check the HC and CO readings (♦33-19)
    a. after the tuneup is done
    b. before the tuneup is done
    c. both a and b
    d. neither a nor b
19. While cranking the engine with a voltmeter connected across the battery terminals, the reading on the voltmeter is below specs. Mechanic A says a low reading can be caused by a bad battery. Mechanic B says a low reading can be caused by a bad starter. Who is right? (♦33-19)
    a. A only
    b. B only
    c. both A and B
    d. neither A nor B
20. Possible causes of a low or discharged battery include (♦33-19)
    a. a defective alternator
    b. a defective regulator
    c. a defective battery
    d. all of the above

CHAPTER 34
Spark-Ignition-Engine Trouble Diagnosis

After studying this chapter, and with proper instruction and equipment, you should be able to:

1. Demonstrate how to use spark-ignition-engine trouble-diagnosis charts.
2. List the basic engine troubles.
3. Describe the checks and corrections for at least six troubles in spark-ignition engines.
4. Demonstrate how to make quick checks of the fuel and ignition systems.
5. Identify the source of various engine noises.
6. Describe the causes of blue, black, and white exhaust smoke.
7. Diagnose the causes of troubles in several spark-ignition engines.

♦ 34-1 HOW TO STUDY THIS CHAPTER

There are different ways to study this chapter. You can go through it page by page, just as you have studied the previous chapters. Another way is to take one complaint at a time (as listed in the Engine Trouble-Diagnosis Chart), read through the possible causes and checks or corrections, and then study the section later in the chapter that discusses the complaint. For example, you could take complaint 1, "Engine will not crank." After reading the causes and checks or corrections listed in the second and third columns in the chart, you would turn to ♦34-4 (referred to in the first column) and study it.

♦ 34-2 NEED FOR LOGICAL PROCEDURE

After a trouble has been located in an engine, the mechanic usually makes the corrections necessary to eliminate the conditions causing the trouble. Chapters 35 to 38 discuss the various engine services and explain the corrections to be made to eliminate different causes of engine trouble.

This chapter is devoted to *trouble diagnosis*. This is the detective work that a mechanic must do when presented with a case of engine trouble. See the chart in ♦34-3 for possible causes and corrections of engine trouble.

Note This chapter discusses spark-ignition-engine troubles. Chapter 19, Diesel Fuel-Injection Systems, describes how to diagnose and correct troubles in the diesel-engine fuel-injection system (♦19-18 to 19-31). Aside from the special problems discussed in Chap. 19 for diesel fuel-injection systems, diesel-engine troubles are similar to troubles in the spark-ignition engine. Both types of engine could have similar problems with valves, pistons, rings, bearings, and other engine parts.

♦ 34-3 ENGINE TROUBLE-DIAGNOSIS CHART

Most engine complaints can be grouped under a few basic headings. These include engine will not crank; engine cranks but will not start; engine runs but misses; engine lacks power, acceleration, or high-speed performance; engine overheats; engine uses excessive oil or gasoline; or engine is noisy. The Engine Trouble-Diagnosis Chart that follows lists the various engine troubles, together with their possible causes, checks to be made, and corrections needed. Some causes of trouble are in the engine itself. Later chapters in the book cover engine-service operations that correct these troubles.

Note The trouble-diagnosis chart has several references to the Diesel Fuel-Injection-System Trouble-Diagnosis Chart (in Chap. 19 on page 208). This is because certain additional or special troubles occur only in diesel engines.

ENGINE TROUBLE-DIAGNOSIS CHART*

Complaint	Possible Cause	Check or Correction
1. Engine will not crank (♦34-4)	a. Run-down battery	Recharge or replace battery; start engine with jumper battery and cables
	b. Starting circuit open	Find and eliminate the open; check for dirty or loose cables
	c. Starting-motor drive jammed	Remove starting motor and free drive
	d. Starting motor jammed	Remove starting motor for disassembly and repair
	e. Engine jammed	Check engine to find trouble
	f. Transmission not in neutral or neutral switch out of adjustment	Check and adjust neutral switch if necessary
	g. Seat belt not fastened or interlock faulty	Check interlock
	h. See also causes listed under item 3; driver may have run battery down trying to start	
2. Engine cranks slowly but will not start (♦34-5)	a. Partly discharged battery	Recharge or replace battery; start engine with jumper battery and cables
	b. Defective starting motor	Repair or replace
	c. Bad connections in starting circuit	Check for undersize, loose, or dirty cables; replace, or clean and tighten
	d. See also causes listed under item 3; driver may have run battery down trying to start	
3. Engine cranks at normal speed but will not start (♦34-6)	a. Defective ignition system	Try spark test; check timing, ignition system
	b. Defective fuel pump or overchoking	Prime engine; check accelerator-pump discharge, fuel pump, fuel line, choke, carburetor
	c. Air leaks in intake manifold or carburetor	Tighten mounting; replace gaskets as needed
	d. Defect in engine	Check compression or leakage (♦33-4 to 33-7), valve action, timing
	e. Ignition coil or resistor burned out	Replace
	f. Plugged fuel filter	Clean or replace
	g. Plugged or collapsed exhaust system	Replace collapsed parts
	h. See also item 1 in Diesel Fuel-Injection-System Trouble-Diagnosis Chart†	
4. Engine runs but misses: one cylinder (♦34-7)	a. Defective spark plug	Clean or replace
	b. Defective distributor cap or spark-plug cable	Replace
	c. Valve stuck open	Free valve; service valve guide
	d. Broken valve spring	Replace
	e. Burned valve	Replace
	f. Bent pushrod	Replace
	g. Flat cam lobe	Replace camshaft
	h. Defective piston or rings	Replace; service cylinder wall as necessary
	i. Defective head gasket	Replace
	j. Intake-manifold leak	Replace gasket; tighten manifold bolts
	k. See also item 5 in Fuel-Injection-System Trouble-Diagnosis Chart†	

*See ♦34-4 to 34-18 for detailed explanation of the trouble causes and corrections listed.
†You can find the Diesel Fuel-Injection-System Trouble-Diagnosis Chart on page 208.

ENGINE TROUBLE-DIAGNOSIS CHART *(Continued)*		
Complaint	**Possible Cause**	**Check or Correction**
5. Engine runs but misses: different cylinders (♦34-7)	a. Defective distributor advance, coil, condenser	Check distributor, coil, condenser
	b. Defective fuel system	Check fuel pump, flex line, carburetor
	c. Cross-firing plug wires	Replace or relocate
	d. Loss of compression	Check compression or leakage (♦33-4 to 33-7)
	e. Defective valve action	Check compression, leakage, vacuum (♦33-4 to 33-8)
	f. Worn pistons and rings	Check compression, leakage, vacuum (♦33-4 to 33-8)
	g. Overheated engine	Check cooling system
	h. Manifold heat-control valve stuck	Free valve
	i. Restricted exhaust	Check exhaust, tail pipe, muffler; eliminate restriction
6. Engine lacks power, acceleration, or high-speed performance: hot or cold (♦34-8)	a. Defective ignition	Check timing, distributor, wiring, condenser, coil, and plugs
	b. Defective fuel system; secondary throttle valves not opening	Check carburetor, choke, filter, air cleaner, and fuel pump
	c. Throttle valve not opening fully	Adjust linkage
	d. Restricted exhaust	Check tail pipe and muffler; eliminate restriction
	e. Loss of compression	Check compression or leakage (♦33-4 to 33-7)
	f. Excessive carbon in engine	Service engine
	g. Defective valve action	Check with compression, leakage, vacuum testers (♦33-4 to 33-8)
	h. Excessive rolling resistance from low tires, dragging brakes, wheel misalignment, etc.	Correct the defect causing rolling resistance
	i. Heavy oil	Use correct oil
	j. Wrong or bad fuel	Use correct octane fuel
	k. Transmission not downshifting, or defective torque converter	Check transmission
	l. See also item 6 in Diesel Fuel-Injection-System Trouble-Diagnosis Chart†	
7. Engine lacks power, acceleration, or high-speed performance: hot only (♦34-8)	a. Engine overheats	Check cooling system
	b. Choke stuck partly open	Repair or replace
	c. Sticking manifold heat-control valve	Free valve
	d. Vapor lock	Use different fuel or shield fuel line
8. Engine lacks power, acceleration, or high-speed performance: cold only (♦34-8)	a. Automatic choke stuck open	Repair or replace
	b. Manifold heat-control valve stuck open	Free valve
	c. Cooling-system thermostat stuck open	Repair or replace
	d. Engine valves stuck open	Free valves; service valve stems and guides as needed
9. Engine overheats (♦34-9)	a. Lack of coolant	Add coolant; look for leak
	b. Ignition timing late	Adjust timing
	c. Loose or broken fan belt	Tighten or replace
	d. Thermostat stuck closed	Replace
	e. Clogged water jackets or radiator core	Flush and clean
	f. Defective radiator hose	Replace
	g. Defective water pump	Repair or replace
	h. Insufficient oil	Add oil
	i. High-altitude, hot-climate operation	Drive more slowly; keep radiator filled
	j. Defective fan clutch	Replace
	k. Valve timing late; slack timing chain has allowed chain to jump a tooth	Retime, adjust or replace
	l. No vacuum advance in any gear	TCS system or distributor defective
10. Engine idles roughly (♦34-10)	a. Incorrect idle adjustment	Readjust idle mixture and speed
	b. PCV or EGR valve stuck open	Replace
	c. See also other causes listed under items 6 to 8	

ENGINE TROUBLE-DIAGNOSIS CHART (*Continued*)

Complaint	Possible Cause	Check or Correction
11. Engine stalls cold or as it warms up (♦34-11)	a. Choke valve stuck closed or will not close	Open choke valve; free or repair automatic choke
	b. Fuel not getting to or through carburetor	Check fuel pump, lines, filter, float, and idle systems
	c. Manifold heat-control valve stuck	Free valve
	d. Throttle solenoid improperly set	Adjust
	e. Idling speed set too low	Increase idling speed to specified rpm
	f. PCV or EGR valve stuck open	Replace
	g. Damper in thermostatic air cleaner in cold-air mode stuck closed	Free; repair or replace control motor
	h. See also item 2 in Diesel Fuel-Injection-System Trouble-Diagnosis Chart †	
12. Engine stalls after idling or slow-speed driving (♦34-11)	a. Defective fuel pump	Repair or replace fuel pump
	b. Overheating	See item 9
	c. High carburetor float level	Adjust
	d. Incorrect idling adjustment	Adjust
	e. Malfunctioning PCV or EGR valve	Replace
	f. Throttle solenoid improperly set	Adjust
13. Engine stalls after high-speed driving (♦34-11)	a. Vapor lock	Use different fuel or shield fuel line
	b. Carburetor venting or idle-compensator valve defective	Check and repair
	c. Engine overheats	See item 9
	d. PCV or EGR valve stuck open	Replace
	e. Improperly set idle solenoid	Adjust
14. Engine backfires (♦34-12)	a. Ignition timing off	Adjust timing
	b. Spark plugs of wrong heat range	Install correct plugs
	c. Excessively rich or lean mixture	Repair or readjust fuel pump or carburetor
	d. Engine overheats	See item 9
	e. Carbon in engine	Clean
	f. Valves hot or stuck	Adjust, free, clean; replace if bad
	g. Cracked distributor cap	Replace
	h. Inoperative antibackfire valve	Replace
	i. Cross-firing plug wires	Replace
15. Engine run-on, or dieseling (♦34-13)	a. Incorrect idle-solenoid adjustment	Adjust; fix solenoid
	b. Engine overheats	See item 9
	c. Hot spots in cylinders	Check plugs, pistons, cylinders for carbon; check valves for defects and faulty seating
	d. Timing advanced	Adjust
	e. In diesel engine, could be due to injection-pump solenoid not turning off fuel valve	
16. Too much HC and CO in exhaust gas (♦34-14)	a. Ignition miss	Check plugs, wiring, cap, coil, etc.
	b. Incorrect ignition timing	Time ignition
	c. Carburetor troubles	Check choke, float level, idle-mixture screw, etc., as listed in item 20
	d. Faulty air injection	Check pump, hoses, manifold
	e. Defective TCS system	Check system
	f. Defective catalytic converters	Replace converters or catalyst
17. Smoky exhaust		
1. Blue smoke	Excessive oil consumption	See item 18
2. Black smoke	Excessively rich mixture	See item 20
3. White smoke	Steam in exhaust	Replace gasket; tighten cylinder-head bolts to eliminate coolant leakage into combustion chambers
4. See also item 8 in Diesel Fuel-Injection-System Trouble-Diagnosis Chart†		
18. Excessive oil consumption (♦34-15)	a. External leaks	Correct seals; replace gaskets
	b. Burning oil in combustion chamber	Check valve-stem clearance, piston rings, cylinder walls, rod bearings

†You can find the Diesel Fuel-Injection-System Trouble-Diagnosis Chart on page 208.

ENGINE TROUBLE-DIAGNOSIS CHART (*Continued*)

Complaint	Possible Cause	Check or Correction
	c. High-speed driving	Drive more slowly
19. Low oil pressure (♦34-16)	a. Worn engine bearings	Replace
	b. Engine overheating	See item 9
	c. Oil dilution or foaming	Replace oil
	d. Lubricating-system defects	Check oil lines, oil pump, relief valve
20. Excessive fuel consumption (♦34-17)	a. Jackrabbit starts	Drive more reasonably
	b. High-speed driving	Drive more slowly
	c. Short-run operation	Drive longer distances
	d. Excessive fuel-pump pressure or pump leakage	Reduce pressure; repair pump
	e. Choke partly closed after warm-up	Open; repair automatic choke
	f. Clogged air cleaner	Clean or replace filter element
	g. High carburetor float level	Adjust
	h. Stuck or dirty float needle valve	Free and clean
	i. Worn carburetor jets	Replace
	j. Stuck metering rod or power piston	Free
	k. Idle too rich or too fast	Adjust
	l. Stuck accelerator-pump check valve	Free
	m. Carburetor leaks	Replace gaskets; tighten screws; etc.
	n. Cylinder not firing	Check coil, condenser, timing, plugs, contact points, wiring
	o. Automatic transmission slipping or not upshifting	Check transmission
	p. Loss of engine compression (worn engine)	Check compression or leakage (♦33-4 to 33-7)
	q. Defective valve action (worn camshaft, belt or chain slack, or jumped tooth)	Check with compression, leakage, or vacuum tester (♦33-4 to 33-8)
	r. Excessive rolling resistance from low tires, dragging brakes, wheel misalignment, etc.	Correct the defects causing the rolling resistance
	s. Clutch slippage	Adjust or repair
21. Engine noises (♦34-18)		
1. Regular clicking	Valve and lifter	Readjust valve clearance or replace noisy hydraulic lifters
2. Ping on load or acceleration	Detonation due to low-octane fuel, carbon, advanced ignition timing, or causes listed under item 14	Use higher-octane fuel; remove carbon; adjust ignition timing
3. Light knock or pound with engine floating	Worn connecting-rod bearings or crankpin; misaligned rod; lack of oil	Replace bearings; service crankpins; replace rod; add oil
4. Light, metallic double knock, usually most audible during idle	Worn or loose pin or lack of oil	Service pin and bushing; add oil
5. Chattering or rattling during acceleration	Worn rings, cylinder walls, low ring tension, or broken rings	Service walls; replace rings
6. Hollow, muffled bell-like sound (engine cold)	Piston slap due to worn pistons or walls, collapsed piston skirts, excessive clearance, misaligned connecting rods, or lack of oil	Replace or resize pistons; service walls; replace rods; add oil
7. Dull, heavy, metallic knock under load or acceleration, especially when cold	Regular noise: worn main bearings; irregular noise: worn thrust bearing knock on clutch engagement or on hard acceleration	Replace or service bearings and crankshaft
8. Miscellaneous noises (rattles, etc.)	Loosely mounted accessories: alternator, horn, oil pan, front bumper, water pump, etc.	Tighten mounting
9. See also items 4, 7, and 8 in Diesel Fuel-Injection-System Trouble-Diagnosis Chart†		

♦ 34-4 ENGINE WILL NOT CRANK

If the engine will not crank when starting is attempted, make sure the shift lever is in neutral (N) or park (P). Or, if the car has a manual transmission, make sure that the clutch pedal is depressed. Check the battery and cables. How to locate the cause of this trouble is described in ♦25-7.

♦ 34-5 ENGINE CRANKS SLOWLY BUT WILL NOT START

The causes of this condition could be a rundown battery, a defective starting motor, or mechanical trouble in the engine. This condition is described further in ♦25-8.

♦ 34-6 ENGINE CRANKS AT NORMAL CRANKING SPEED BUT WILL NOT START

The battery and starting motor are in normal condition. The cause of trouble is probably in the ignition or fuel system. The difficulty could be due to overchoking. Failure to start with a hot engine may be caused by a defective choke that fails to open properly as the engine warms up. This condition would cause flooding of the engine (delivery of too much gasoline). Open the throttle valve wide and crank the engine. If the engine does not start, make a spark test (♦27-16). If a good spark occurs, the ignition system is probably operating normally. However, the ignition or valve timing could be off.

If the ignition system seems to be operating normally, check the fuel system. On a carbureted engine, prime the engine by operating the accelerator pump several times. Now crank the engine.

♦ CAUTION ♦ Gasoline is highly explosive. Install the air cleaner before cranking.

If the engine starts and runs for a few seconds, the fuel system is probably faulty. It is not delivering fuel to the engine. Make sure there is gasoline in the fuel tank. Temporarily disconnect the fuel line to the carburetor. Also disconnect the primary lead from the negative terminal of the ignition coil so that the engine will not start. Hold a container under the fuel line to catch fuel, and crank the engine to see whether fuel is delivered (Fig. 34-1). If fuel is not delivered, there is no fuel in the fuel tank, the fuel pump is defective, or the fuel line is clogged. If fuel is delivered, the fuel filter is probably at fault, the automatic choke is not working correctly, or possibly there are air leaks into the intake manifold or carburetor.

If the fuel and ignition systems check okay, test the mechanical condition of the engine with the cylinder compression tester and the cylinder leakage tester (♦33-4 to 33-7). Also, a restricted exhaust system can build up back pressure (Fig. 34-2). The inner layer of a laminated exhaust pipe could collapse (Fig. 34-3). Any restriction could prevent normal exhaust and intake so that the engine will not start, or will not run much above idle.

♦ 34-7 ENGINE RUNS BUT MISSES

A missing engine is a rough engine. If one or more cylinders fail to fire, the engine is thrown out of balance. The result is roughness and loss of power. It is sometimes difficult to locate a miss. The miss might occur at some speeds and not others. Also, the miss may skip around. To check out one or more misfiring cylinders, use an oscilloscope and a dynamometer. Sections 27-18 and 27-19 explain how to use the oscilloscope on contact-point ignition systems. Section 29-15 explains how to use the oscilloscope to check electronic ignition systems. The dynamometer is discussed in ♦33-12. If these testing instruments are not available, then the test can be made as follows.

Use insulated pliers to disconnect each spark-plug cable in turn to locate the missing cylinder. (Disconnecting the cable

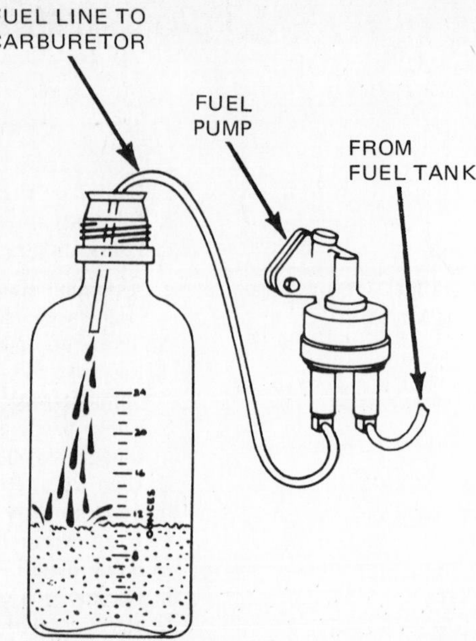

Fig. 34-1 To check whether the fuel system is delivering fuel, disconnect the fuel line from the carburetor, place the fuel line in a safe container, and crank the engine. Fuel should discharge from the line. *(Ford Motor Company)*

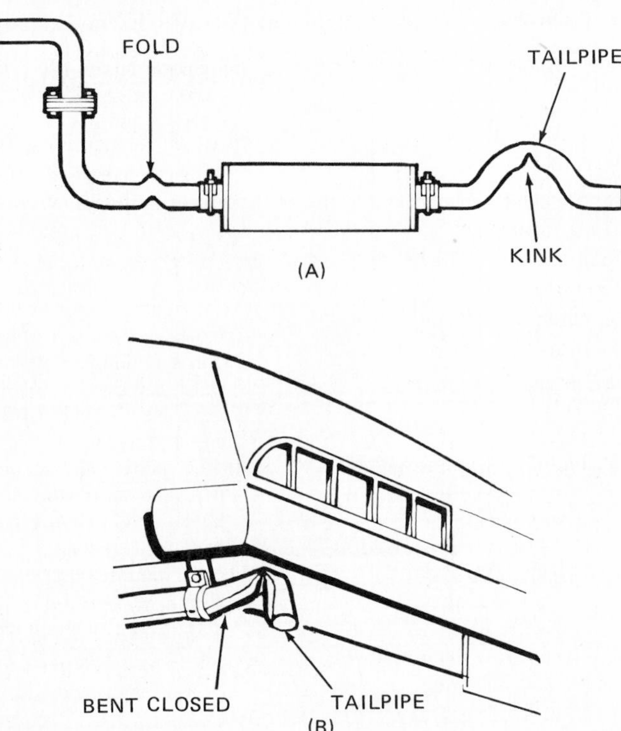

Fig. 34-2 A clogged or restricted exhaust system may cause the engine, hot or cold, to lack power. *(Ford Motor Company)*

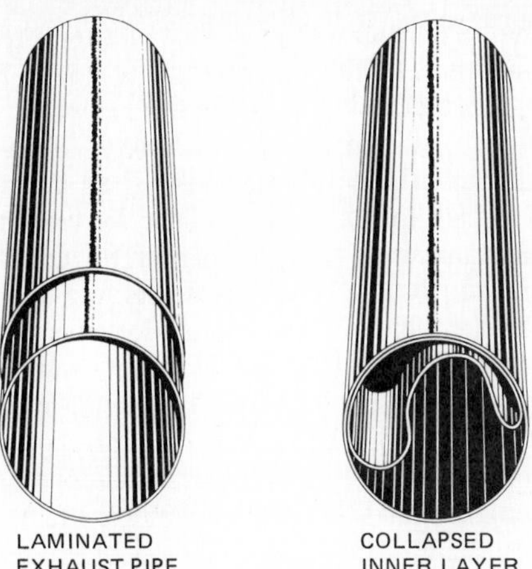

Fig. 34-3 Collapse of the inner layer of a laminated exhaust pipe will prevent normal intake and exhaust. *(ATW)*

prevents the spark from occurring at the spark plug.) If disconnecting the cable changes the engine speed, then the cylinder was delivering power before you disconnected the cable. But if there is no change in engine speed, then that cylinder was missing before you disconnected the cable.

Note This test may not be recommended for use on engines equipped with certain types of electronic ignition systems. Consult the manufacturer's service manual for ignition-testing procedures. Most manufacturers recommend the use of a spark tester.

Check a missing cylinder further by holding the spark-plug lead clip close to the engine block while the engine is running (Fig. 29-1). If no spark occurs, there is probably a high-voltage leak due to a bad lead or a cracked or burned distributor cap. If a good spark occurs, install a new spark plug in the cylinder (or swap plugs between two cylinders). Then reconnect the lead, and determine whether the cylinder still misses. If it does, the cause of the trouble is probably defective engine parts, such as valves or rings.

If the miss is hard to locate, one approach is to perform an engine tuneup (Chap. 33). A tuneup will locate and possibly eliminate various causes of missing. These could include defects in the ignition system or the fuel system, loss of engine compression, sticky or damaged engine valves, overheated engine, sticky manifold heat-control valve, and a restricted exhaust system.

A cylinder balance test using a tachometer and a vacuum gauge can be used to check that a cylinder is delivering power. With most engine analyzers that include an oscilloscope, you can make a cylinder balance test that will quickly pinpoint the missing cylinder. When the analyzer is connected to the running engine, you turn a knob or push a button and the cylinders are shorted out, one by one, in the firing order. The scope then shows which cylinder is shorted out. If shorting out a cylinder causes the manifold vacuum or the engine speed to drop by 50 rpm or more (Fig. 34-4), the cylinder was delivering power. But if little or no change in rpm or vacuum takes place, then the cylinder was not delivering power.

Note In the diesel engine, a miss usually is caused by failure of the fuel system to deliver an adequate amount of fuel to the cylinder. This could be due to a clogged nozzle, air in the line, or injection-pump problems.

♦ 34-8 ENGINE LACKS POWER

As a first step, determine whether the engine lacks power when both hot and cold, only when hot, or only when cold. Then find out if the problem developed suddenly or if the power gradually fell off over a period of many months or many miles of operation. A chassis dynamometer (♦33-12) and an oscilloscope (♦27-18, 27-19, and 29-15) can be used to help locate the cause of the trouble.

1. Engine Lacks Power Hot or Cold The fuel system may not be enriching the mixture as the throttle is opened. This condition could be due to a faulty accelerator pump or defective main-metering or power system in the carburetor. Also, the fuel system could be supplying an excessively lean or rich mixture. This condition could be due to a defective fuel pump, clogged lines, clogged filter, worn carburetor jets or lines, air leaks at the carburetor or manifold joints, and a sticking PCV valve. Carburetors and fuel-system action can be checked with an exhaust-gas analyzer (♦33-9).

Another condition could cause lack of power when the engine is hot or cold. This condition is an improper linkage adjustment that prevents full-throttle opening. Also, the ignition system may be causing trouble, owing to conditions such as incorrect timing, a "weak" coil, reversed coil polarity, and wrong spark-plug heat range. The wrong fuel or oil for the engine could reduce performance. Numerous other engine conditions could cause loss of power. These include carbon deposits, lack of compression (faulty valves, rings, worn cylinder walls, worn pistons), and defective bearings. A clogged exhaust system (bent or collapsed exhaust or tail pipe, or a clogged muffler or catalytic converter) could create back pressure that would cause poor engine performance (Fig. 34-2). Also, any excessive rolling resistance would absorb engine power and hold down engine acceleration and speed. This would include dragging brakes, underinflated tires, misaligned wheels, and excessive friction in the transmission or power train. Also, the automatic transmission may not be downshifting, or the torque converter may be defective.

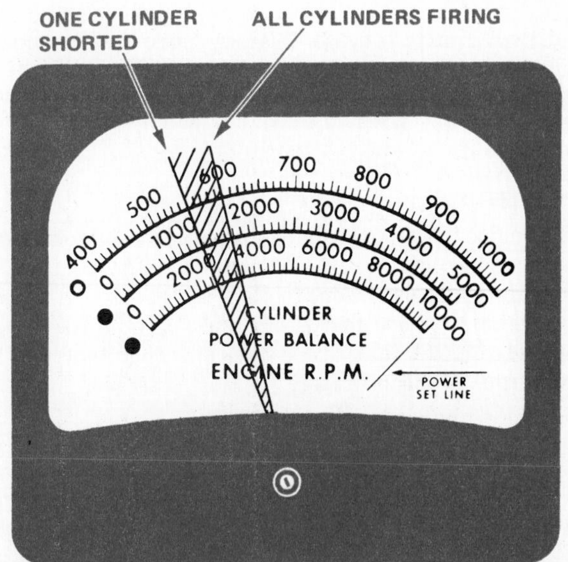

Fig. 34-4 To check for a misfiring engine, use the engine analyzer to short out each spark plug in turn. If the cylinder is firing, shorting out the cylinder will cause the engine speed to drop at least 50 rpm. *(Sun Electric Corporation)*

2. Engine Lacks Power Only when Hot The engine may be overheating (♦34-9). Also, the automatic choke may not be opening normally as the engine warms up. If this condition is indicated, operation of the automatic choke can be checked with a choke tester. Also, the manifold heat-control valve may be stuck. Or there may be a vapor lock in the fuel pump or line.

3. Engine Lacks Power Only when Cold (or Reaches Operating Temperature Too Slowly) The automatic choke may be leaning out the mixture too soon (before the engine warms up). The manifold heat-control valve may not be closed (so not enough heat reaches the intake manifold). Or the cooling system thermostat may be stuck open. Then coolant circulates between the engine and radiator even with the engine cold, delaying warm-up. Occasionally, engine valves may stick open when the engine is cold. But as the engine warms up, the valves become free and work normally.

♦ 34-9 ENGINE OVERHEATS

Most engine overheating is caused by loss of coolant due to leaks in the cooling system. Other causes include a loose or broken fan belt, a defective water pump, clogged water jackets or radiator tubes, blocked radiator airflow, a defective radiator hose, and a defective thermostat or fan clutch. Also, late ignition or valve timing, lack of engine oil, overloading the engine, or high-speed, high-altitude, or hot-climate operation can cause engine overheating. Freezing of the coolant could cause lack of coolant circulation so that local hot spots and boiling develop. If a faulty TCS system prevents vacuum advance in any gear, or if the distributor vacuum advance is defective, overheating may result. Idling and slow-moving traffic can also cause the engine to overheat when combined with other problems.

Note When the car is in slow-moving traffic and the engine begins to overheat, open the car windows and turn off the air conditioner. Turn on the heater to maximum. This lightens the load on the engine and also takes some of the heat away from the engine. This may make the passengers temporarily uncomfortable. However, taking some heat from the engine may make the difference between radiator boil-over and possible stalling, or simply a hot engine.

♦ 34-10 ENGINE IDLES ROUGHLY

If the engine idles roughly but runs normally above idle, the idle speed and idle mixture may be incorrectly adjusted. A rough idle could also be due to other causes, such as a loose vacuum hose (or one that is disconnected from the intake manifold) or a PCV valve that is stuck open. See items 6 to 10 in the trouble-diagnosis chart.

♦ 34-11 ENGINE STALLS

If the engine starts and then stalls, note whether the stalling takes place before or as the engine warms up, after idling or slow-speed driving, or after high-speed or full-load driving. If the PCV valve becomes clogged or sticks, it will cause poor idling and stalling. Stalling also may be caused by improper operation or adjustment of the idle solenoid.

1. Engine Stalls before It Warms Up This condition could be due to an improperly set fast or slow idle or to improper adjustment of the idle-mixture screw in the carburetor. Also, it could be due to a low carburetor float setting or to insufficient fuel entering the carburetor. This condition could result from a defective thermostatic air cleaner, dirt or water in the fuel lines or filter, a defective fuel pump, or a plugged fuel-tank vent. Also, the carburetor could be icing. Certain ignition troubles could cause stalling after starting. But if ignition troubles are bad enough to cause stalling, they usually would prevent starting. However, burned contact points might permit starting but fail to keep the engine running. Another cause of stalling is an open primary resistance. When the engine is cranked, the resistance is bypassed. Then, when the engine starts and cranking stops, the resistance is inserted into the ignition primary circuit. If the resistance is open, the engine will stall.

2. Engine Stalls as It Warms Up This condition could result if the choke valve sticks closed. The mixture becomes too rich for a hot engine, and the engine stalls. If the manifold heat-control valve sticks closed, the air-fuel mixture might become overheated and too lean, causing the engine to stall. If the hot-idle speed is too low, the engine may stall as it warms up because the idling speed drops too low. Also, stalling may be caused by overheating of the engine, which could cause vapor lock. In addition, if the damper in the thermostatic air cleaner sticks closed, the air-fuel mixture can become overheated and too lean. This will cause the engine to stall.

3. Engine Stalls after Idling or Slow-Speed Driving This condition could occur if the fuel pump has a cracked diaphragm, weak spring, or defective valve. The pump fails to deliver enough fuel for idling or slow-speed operation (although it could deliver enough for high-speed operation), and the engine stalls. If the carburetor float level is set too high or the idle adjustment is too rich, the engine may "load up" and stall. A lean idle adjustment may also cause stalling. The engine may overheat during sustained idling or slow-speed driving. With this condition, the air movement through the radiator may not be sufficient to keep the engine cool. Overheating, in turn, could cause vapor lock and engine stalling. (See ♦34-9 for causes of overheating.)

4. Engine Stalls after High-Speed or Full-Load Driving This condition could occur if enough heat accumulates to cause vapor lock. The remedy is to shield the fuel line and fuel pump or use a less volatile fuel. Failure of the venting or idle-compensator valve in the carburetor may also cause stalling after high-speed or full-load operation. Excessive overheating of the engine also may cause stalling (♦34-9)

♦ 34-12 ENGINE BACKFIRES

Backfiring could be caused by a faulty antibackfire valve in the air-injection system (♦31-31) or by faulty accelerator-pump action. It could also be caused by late ignition timing or ignition cross-firing (caused by the spark jumping across the distributor cap or through the cable insulation). In addition, backfiring could be due to spark plugs of the wrong heat range (which overheat and cause preignition), excessively rich mixture (caused by fuel-pump or carburetor troubles), overheating of the engine (♦34-9), carbon in the engine, hot valves, or intake valves that stick or seat poorly. Carbon in the engine, if excessive, may retain enough heat to cause the air-fuel mixture to preignite as it enters the cylinder so that backfiring occurs. Carbon also increases the compression ratio and the tendency for detonation and preignition. Hot plugs may cause preignition; cooler plugs should be installed. If intake valves hang open, combustion may be carried back into the carburetor. Valves that have been ground excessively so that they have sharp edges, valves that seat poorly, or valves that are coated with carbon deposits overheat and often produce backfiring.

♦ 34-13 ENGINE RUN-ON, OR DIESELING

Spark-ignition engines with emission controls require a fairly high, hot-idle speed for best operation. This makes run-on, or dieseling, possible. If there are hot spots in the combustion chambers, the engine could continue to run if the throttle valve is not completely closed. The hot spots take the place of the spark plugs. If the throttle valve is slightly open, enough air-fuel mixture could get past it to keep the engine running. Ignition in the combustion chambers is caused by the hot spots.

Many engines have an idle-stop solenoid to close the throttle valve completely when the ignition switch is turned off (Fig. 16-38). If an engine runs on, or diesels, check the idle-stop solenoid to make sure it is releasing when the ignition is turned off. It could require adjustment to permit the throttle to close completely. Be sure the engine idle speed is not set too high. Engine run-on could also be caused by advanced ignition timing. Correction of engine overheating is covered in ♦34-9. Correction of hot spots may require spark-plug service or removing the cylinder head for cleaning plus valve service.

♦ 34-14 TOO MUCH HC AND CO IN EXHAUST GAS

If the exhaust-gas analyzer (♦33-9) shows that there is too much HC and CO in the exhaust gas, correction should be made. Excessive CO usually is caused by a rich air-fuel mixture. At idle, it can usually be brought within specifications by proper adjustment of the carburetor. Excessive HC usually is caused by a misfiring spark plug. Correction of the excessive HC usually can be made by proper servicing and adjusting of the ignition system, and by replacement of worn parts in the engine. Excessive HC may also result from an excessively lean air-fuel mixture caused by intake manifold or vacuum leaks. This, in turn, could cause lean misfire which dumps the unburned air-fuel mixture into the exhaust manifold. Some states require exhaust-gas testing of all cars during state inspection. Cars that emit too much HC and CO must be repaired before they can be licensed. This removes cars that pollute excessively from the highways. Possible causes of excessive HC and CO are:

1. Missing due to ignition problems such as faulty plugs, high-voltage wiring, distributor cap, ignition coil, condenser, or contact points. (It can also be caused by excessive carbon deposits in the combustion chamber or stuck or burned valves.)
2. Incorrect ignition timing.
3. Carburetor troubles such as the choke partly closed, worn jets, high float level, and other conditions listed in ♦34-17.
4. Faulty air-injection system which does not inject enough air into the exhaust manifold to completely burn the HC and CO. (This could be caused by a faulty air pump or a leaking hose or air manifold.)
5. Defective TCS system which permits vacuum advance in all gear positions instead of high and reverse only.
6. Defective catalytic converters which must be replaced or serviced to restore the catalytic action.
7. Excessive carbon deposits in the combustion chamber, or stuck or burned valves. Carbon deposits often act like sponges, absorbing air-fuel mixture during compression and then releasing it during exhaust.

♦ 34-15 EXCESSIVE OIL CONSUMPTION

Oil may be lost from the engine in three ways: by leakage in liquid form, by burning in the combustion chambers, and by passing out of the crankcase through the PCV system as mist or vapor (♦20-15).

♦ 34-16 LOW OIL PRESSURE

Low oil pressure is often a warning of a worn oil pump or worn engine bearings. The bearings can pass so much oil that the oil pump cannot maintain oil pressure. Further, the end bearings will probably be oil-starved and may fail. Other causes of low oil pressure are a weak relief-valve spring, worn oil pump, broken or cracked oil line, and clogged oil line. Oil dilution, foaming, sludge, insufficient oil, or oil made too thin by engine overheating will cause low oil pressure.

♦ 34-17 EXCESSIVE FUEL CONSUMPTION

This condition can be caused by almost anything in the car, from the driver to underinflated tires or a defective choke. A fuel-mileage tester can be used to accurately check fuel consumption. The compression or leakage tester and the vacuum gauge (♦33-4 to 33-8) will help determine whether the trouble is in the engine, fuel system, ignition system, or elsewhere. Also, the exhaust-gas analyzer, dynamometer, and fuel-mileage tester are useful in analyzing the problem.

If the trouble seems to be in the fuel system, consider the following:

1. A driver who pumps the accelerator when idling and insists on being the first to get away when the traffic light changes will use excessive amounts of fuel.
2. Operation with the choke partly closed after warm-up will use excessive amounts of fuel.
3. Short-run operation means that the engine will operate on warm-up most of the time. This causes high fuel consumption.

These three conditions are due to the type of operation. Changing operating conditions is the only cure. However, if excessive fuel consumption is not due to any of these conditions, then the fuel pump should be checked for excessive pressure. High fuel-pump pressure will cause a high carburetor float-bowl level and a rich mixture. A fuel-pump tester is used to check fuel-pump pressure.

4. If excessive fuel consumption is not due to operating conditions or high fuel-pump pressure, the trouble is probably in the carburetor. It could be any of the following conditions:

a. The automatic choke may not be opening rapidly enough during warm-up or may not open fully. The automatic choke should be checked by removing the air cleaner and observing choke operation during warm-up.
b. A clogged air cleaner that does not admit sufficient air can act somewhat like a partly closed choke valve. The air-cleaner element should be cleaned or replaced.
c. If the float level in the carburetor float bowl is high, it will cause flooding and delivery of excessive fuel. The needle valve may be stuck open or may not be seating fully. The float level should be checked and adjusted. Also, the float should be weighed to determine if it is too heavy from having absorbed fuel.
d. If the idle mixture is set too rich or the idle speed too high, excessive fuel consumption will result. These should be checked and adjusted as necessary. On late-model cars, the idle-mixture screw is set at the factory and cannot be readjusted without disassembling the carburetor.
e. Where the accelerator-pump system has a check valve, failure of the check valve to close properly may allow fuel to feed through into the carburetor. The carburetor should be disassembled for repair.
f. The metering rod stuck in the full-power position or the power piston stuck open permits the power system to function, supplying an excessively rich mixture. The carburetor should be disassembled for repair.
g. Worn jets permit the discharge of too much fuel. The jets should be replaced during carburetor rebuilding.

5. Faulty ignition can also cause excessive fuel consumption. The ignition system could cause engine miss and failure of the engine to burn all the fuel. This trouble would also be associated with loss of power, acceleration, or high-speed performance (♦34-8). Conditions in the ignition system that might contribute to the trouble include a defective coil or condenser, incorrect timing, faulty advance-mechanism action, dirty or worn plugs or contact points, and defective wiring.
6. A worn engine can produce excessive fuel consumption. Possible causes are loss of engine compression from worn or stuck rings, worn or stuck valves, or a loose or burned cylinder-head gasket. Power is lost under these conditions, and more fuel must be burned to achieve the same speed. (Refer to ♦33-4 to 33-7 for compression- and leakage-checking procedures.)
7. Excessive fuel consumption can also result from conditions that require additional power for the engine to move the car. Such factors as low tires, dragging brakes, defective automatic transmission, and misalignment of wheels increase the rolling resistance of the car. The engine must use more fuel to overcome this excessive rolling resistance.

♦ 34-18 ENGINE NOISES

Some engine noises may have little significance. Other noises may indicate serious engine trouble that requires prompt attention to prevent major damage to the engine.

A listening rod, or *stethoscope*, is helpful in locating the source of a noise (Fig. 34-5). When one end of the rod is placed at your ear and the other end at some part of the engine, noises from that part of the engine are carried along the rod to your ear. A long screwdriver also can be used.

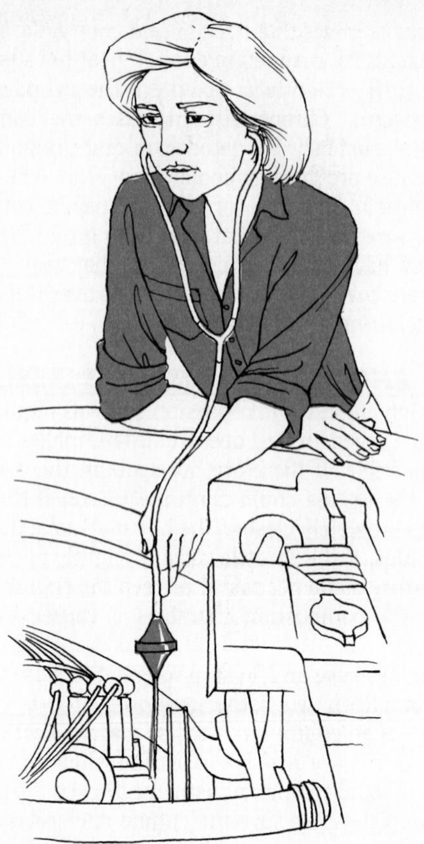

Fig. 34-5 A mechanic's stethoscope being used to locate engine noise.

Move the listening rod or stethoscope around the engine until the noise is loudest. By determining the approximate source of the noise, you can locate a broken and noisy ring in a particular cylinder, a main-bearing knock, and other conditions.

♦ CAUTION ♦ Keep away from the moving fan belt and fan when using the stethoscope.

Following are various engine noises, along with tests that may be necessary to confirm a diagnosis.

1. Valve and Tappet Noise This is a regular clicking sound occurring at half engine speed. It sometimes disappears at higher engine speeds. The cause is usually excessive valve clearance or a defective hydraulic valve lifter. A thickness gauge inserted between the valve stem and lifter or rocker arm will reduce the clearance. If the noise also is reduced, then the cause is excessive clearance. The clearance should be readjusted. If inserting the thickness gauge does not reduce noise, the noise is the result of such conditions in the valve mechanism as weak springs, worn lifter faces, lifters loose in the block, and rough cams.

2. Detonation Spark knock, or detonation, is a pinging or chattering sound that is most noticeable during part-throttle acceleration or when the car is climbing a hill. Some spark knock is considered normal under such operating conditions. But when detonation becomes excessive, it may be due to such conditions as fuel with too low an octane rating for the engine, carbon deposits in the engine which increase the compression ratio, and advanced ignition timing.

3. Connecting-Rod Noise Connecting-rod noise usually has a light knocking or pounding sound. It is most noticeable when the engine is "floating" (not accelerating or decelerating). The sound becomes more noticeable as the accelerator is eased off with the car running at medium speed. To locate the connecting-rod noise, short out the spark plugs one at a time. It is difficult to short out some of the plugs on some engines. The easiest way to make the check is to use an engine analyzer which can short out cylinders with the turn of a knob or push of a button. The noise will be considerably reduced when the cylinder that is responsible is not delivering power. A worn bearing or crankpin, a misaligned connecting rod, inadequate oil, or excessive bearing clearances can cause connecting-rod noise.

4. Piston-Pin Noise Piston-pin noise is similar to valve and tappet noise. However, piston-pin noise has a unique metallic double knock. It is usually most audible during idle with the spark advanced. On some engines, the noise becomes most audible at car speeds of about 30 mph [48 km/h]. A check can be made by running the engine at idle with the spark advanced. Then short out the spark plugs as explained in item 3, above. Piston-pin noise will be reduced when a plug in a noisy cylinder is shorted out. Causes of this noise are a worn or loose piston pin, a worn bushing, and lack of oil.

5. Piston-Ring Noise Piston-ring noise is also similar to valve and tappet noise since it has a clicking, chattering, or rattling sound. However, this noise is most evident on acceleration. Low ring tension, broken rings, worn rings, or worn cylinder walls produce this noise. Since the noise can sometimes be confused with other engine noises, a test can be made as follows: Remove the spark plugs and add 1 tablespoon [15 cc] of engine oil to each cylinder. Crank the engine for several revolutions to work the oil down past the rings. Then install the plugs and start the engine. If the noise has been reduced, the rings probably are at fault.

6. Piston Slap Piston slap is a muffled, hollow, bell-like sound. It is caused by the piston rocking back and forth in the cylinder. If it occurs only when the engine is cold, it is not serious. When it occurs under all operating conditions, further investigation is in order. Piston slap is caused by inadequate oil, worn cylinder walls, worn pistons, collapsed piston skirts, excessive piston clearances, or misaligned connecting rods.

7. Crankshaft Knock Crankshaft knock is a heavy, dull metallic knock. It is most noticeable when the engine is under a heavy load or accelerating, particularly when cold. When the noise is regular and more of a rumble, it probably results from worn main bearings. Worn rod bearings produce a more distinct knock. When the noise is irregular and sharp, it is probably due to a worn thrust bearing. The latter condition, when unusually bad, will cause the noise to be produced each time the clutch is released and engaged and also when accelerating.

8. Miscellaneous Noises Other noises result from loosely mounted accessory parts, such as the alternator, starting motor, horn, water pump, manifolds, flywheel, crankshaft pulley, and oil pan. In addition, other automotive components, such as the clutch, transmission, and differential, may develop various noises.

♦ REVIEW QUESTIONS ♦

Select the *one* correct, best, or most probabie answer to each question. You can find the answers in the section indicated at the end of each question.

1. An engine is using too much oil. Mechanic A says that gasket leaks could be the cause. Mechanic B says that burning in the combustion chambers could be the cause. Who could be right? (♦34-15)
   a. A only
   b. B only
   c. either A or B
   d. neither A nor B
2. An engine may crank slowly because of (♦34-5)
   a. a defective water pump
   b. vapor lock
   c. a run-down battery
   d. excessive fuel-pump pressure
3. Failure of an engine to start even though it cranks at normal speed could be due to a (♦34-6)
   a. run-down battery
   b. defective starting motor
   c. sticking engine valve
   d. defective ignition component
4. Missing in one cylinder may result from (♦34-7)
   a. a clogged exhaust
   b. an overheated engine
   c. vapor lock
   d. a defective spark plug
5. Irregular missing in different cylinders may result from (♦34-7)
   a. a defective starting motor
   b. a defective carburetor
   c. an open cranking circuit
   d. an overcharged battery
6. Loss of engine power as the engine warms up is most likely caused by (♦34-8)
   a. fuel starvation
   b. excessive rolling resistance
   c. the throttle valve not closing fully
   d. heavy oil
7. An engine will lose power (hot or cold) if it has (♦34-8)
   a. incorrect idle speed
   b. an automatic choke valve that is stuck open
   c. worn rings and cylinder walls
   d. none of the above

8. An engine may stall as it warms up if the (♦34-11)
   a. ignition timing is off
   b. choke valve sticks closed
   c. battery is run down
   d. throttle valve does not open fully
9. An engine will overheat if the (♦34-9)
   a. automatic choke sticks
   b. fan belt breaks
   c. fuel pump is defective
   d. battery is run down
10. The most probable cause of an engine stalling after idling or slow-speed driving is (♦34-11)
   a. loss of compression
   b. a defective fuel pump
   c. sticking engine valves
   d. all of the above
11. Stalling of an engine after high-speed driving is probably caused by (♦34-11)
   a. vapor lock
   b. incorrect ignition timing
   c. worn carburetor jets
   d. all of the above
12. Engine backfiring may result from (♦34-12)
   a. spark plugs of wrong heat range
   b. vapor lock
   c. a run-down battery
   d. worn piston rings
13. A smoky blue exhaust may be due to (♦34-3, item 17)
   a. an excessively rich mixture
   b. burning of oil in the combustion chamber
   c. a stuck choke valve
   d. incorrect valve adjustment
14. A smoky black exhaust may be due to (♦34-3, item 17)
   a. worn piston rings
   b. excessively rich mixture
   c. spark plugs of wrong heat range
   d. none of the above
15. A light knock or pound with the engine floating can result from worn (♦34-18)
   a. main bearings
   b. connecting-rod bearings
   c. rings
   d. none of the above
16. A light double knock during idle can result from (♦34-18)
   a. piston slap
   b. spark knock
   c. incorrect ignition timing
   d. loose or worn piston pin
17. A rattling or chattering sound during acceleration may be due to (♦34-18)
   a. worn or collapsed pistons
   b. worn main bearings
   c. broken piston rings
   d. sticking engine valves
18. A hollow, muffled, bell-like sound, with the engine cold, is probably due to (♦34-18)
   a. worn or collapsed pistons
   b. worn main bearings
   c. loose oil pan
   d. sticking engine valves
19. A dull, heavy knock under load or acceleration is probably due to worn (♦34-18)
   a. piston rings
   b. main bearings
   c. piston pins
   d. pistons
20. A V-6 engine uses too much oil and idles roughly. The rings and valves are good and properly installed. Which of these may be true? (♦34-10)
   I. The intake-manifold gasket leaks.
   II. The engine has an oil leak.

   a. I only
   b. II only
   c. either I or II
   d. neither I nor II

PART 7

Part 7 of this book describes service operations on automotive engines. Included in Part 7 are servicing details on engine valves and valve trains, pistons and related parts, crankshafts, and cylinder heads and cylinder blocks. The servicing procedures cover both spark-ignition and diesel engines. There are four chapters in Part 7:

Chapter 35 Engine Service: Valves and Valve Trains
Chapter 36 Engine Service: Connecting Rods, Rod Bearings, Pistons, and Rings
Chapter 37 Engine Service: Crankshafts and Cylinder Blocks
Chapter 38 Diesel-Engine Service

Automotive Engine Service

CHAPTER 35
Engine Service: Valves and Valve Trains

After studying this chapter, and with proper instruction and equipment, you should be able to:

1. Describe the causes of various valve troubles.
2. Adjust valve clearance on various engines.
3. List the steps in performing a complete valve job.
4. Replace a rocker-arm stud in a cylinder head.
5. Reface valves.
6. Refinish valve seats.
7. Knurl valve guides.
8. Perform a complete valve job on various engines.
9. Replace the camshaft and camshaft bearings.

♦ 35-1 CLEANLINESS

The major enemy of good engine-service work is dirt. A speck of dirt left on a bearing or cylinder wall can ruin a perfect service job. Be sure you do not leave abrasive or dirt in the engine or on engine parts after a service job.

Before any major service job, the engine should be cleaned. Electrical and air-conditioning units should be removed or protected if the engine is to be steam-cleaned. Otherwise, they could be damaged.

♦ 35-2 VALVE TROUBLES

Figure 35-1 shows a cylinder head from a V-8 engine with the valve train for one cylinder disassembled. Valves must be properly timed. They must seat tightly and operate without lag. On engines with mechanical valve lifters, the clearance between the rocker arm and valve stem must be correct. Clearance between the valve stems and guides must be within specifications. Hydraulic valve lifters must perform properly to take up clearance in the valve train when the valve is closed. Failure to meet any of these requirements means valve and engine trouble.

As an example, suppose there is too much clearance between valve stems and guides. Then, on every intake stroke, oil can be pulled past the valve stems and guides (Fig. 35-2). The oil enters the combustion chamber, where it burns. This, in turn, leads to excessive oil consumption, engine deposits, preignition, stuck piston rings, fouled spark plugs, and probably burned or otherwise damaged valves and valve seats. Therefore, what seems like a minor fault can lead to serious engine trouble.

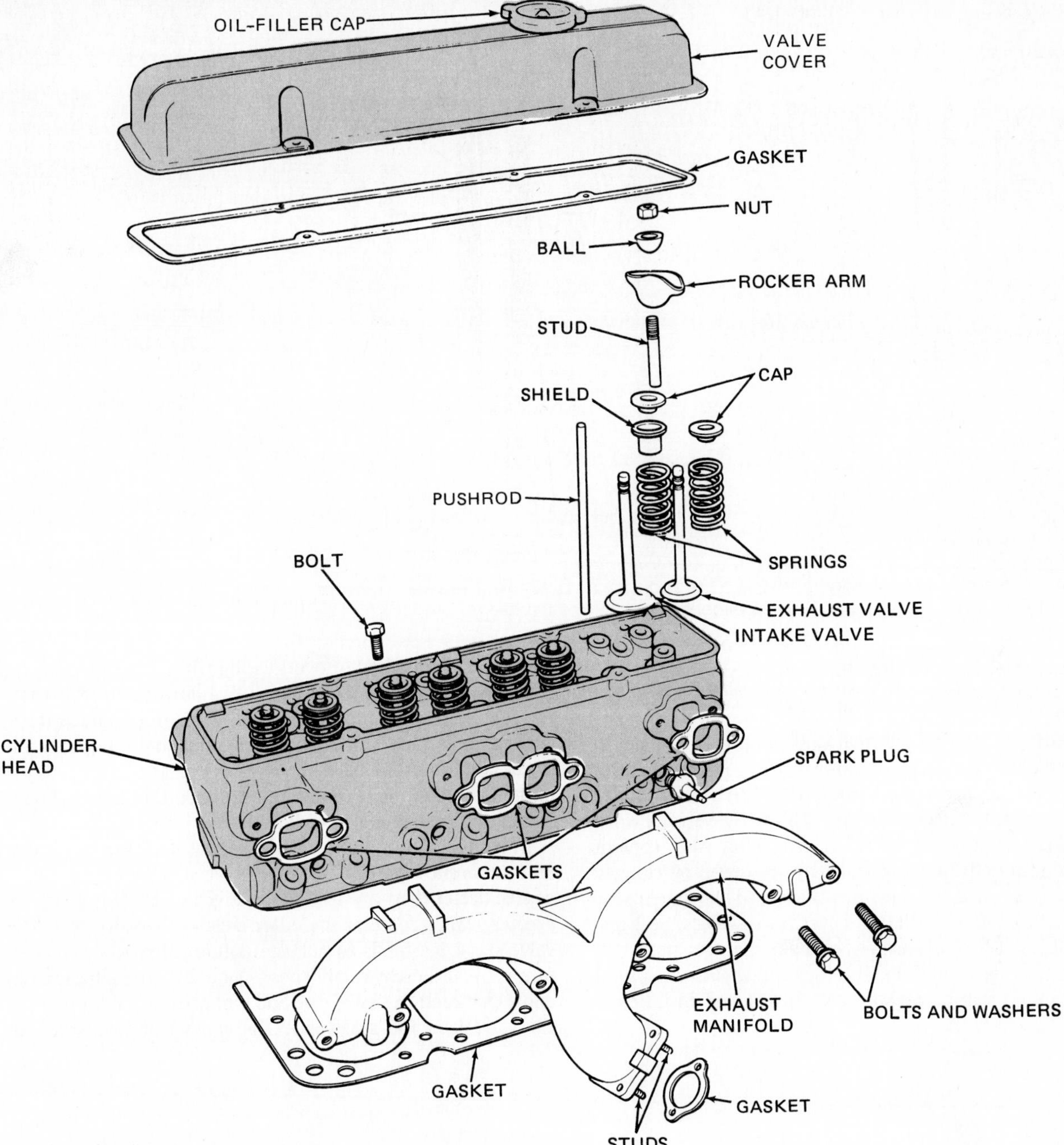

Fig. 35-1 Cylinder head from a V-8 engine with related parts. *(Chevrolet Motor Division of General Motors Corporation)*

Valve troubles include sticking, burning, breakage, wear, deposits, and valve-seat recession. These are discussed in following sections.

♦ 35-3 VALVE STICKING

Gum or carbon deposits on the valve stem (♦35-9) cause valve sticking (Fig. 35-3A). Excessive valve-stem clearance (worn valve guides) causes carbon deposits on the valves and stems. Another cause of valve sticking is warped stems. These could result from overheating, an eccentric seat (which throws side force on the valve), or a cocked spring or retainer (which tends to bend the stem). Insufficient oil also causes valve sticking. Sometimes valves stick when cold, but work free as the engine warms up.

♦ 35-4 CLEANING VALVE STEMS WITH SPECIAL CLEANER

When valves and piston rings are badly clogged with deposits, an engine overhaul is usually required. However, there are certain compounds that can be put into the fuel or oil to help free rings and valves. One type of cleaner is in a spray can. The chemical is sprayed into the running engine through the air intake after the air cleaner has been removed. When parts are not too badly worn, and the major trouble seems to be from deposits, use of these compounds may postpone an engine overhaul.

Read the instructions on the can to make sure the cleaner will not harm the catalytic converter on the car. Do not use a compound unless you are sure it will not damage the converter.

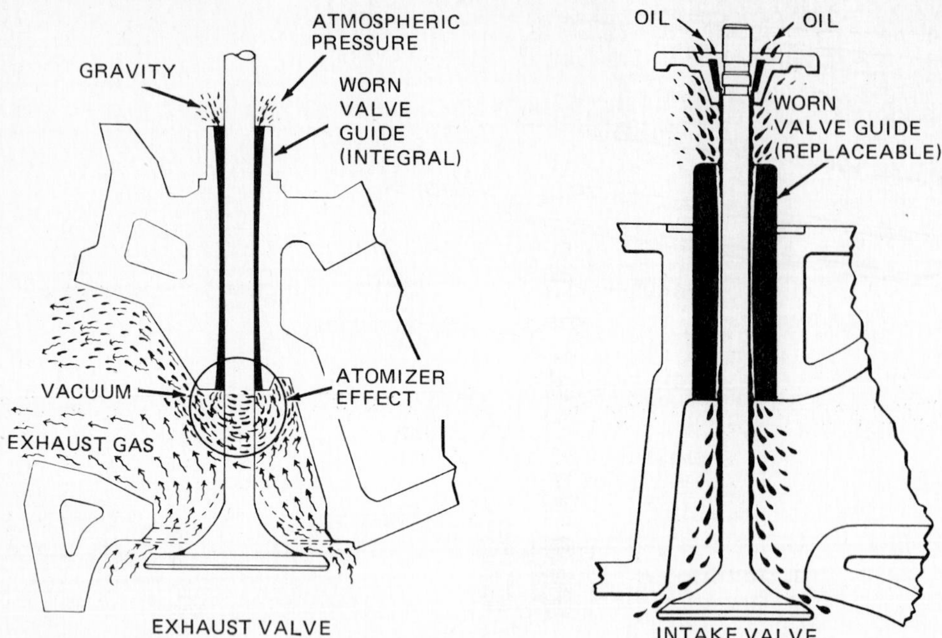

Fig. 35-2 Oil is forced by gravity, and the difference in air pressure, between a worn valve guide and the valve stem. *(Dana Corporation)*

♦ 35-5 VALVE BURNING

This is usually an exhaust-valve problem. However, intake valves may also burn. Any condition that overheats the air-fuel mixture, or prevents normal valve seating, may lead to valve burning (Figs. 35-3B and 35-3C). The poor seating prevents normal valve cooling through the valve seat. It also allows hot gases to blow by, further heating the valve. A worn valve guide also prevents normal valve cooling. Also, coolant circulation around the valve seat may be slowed by clogged passages. Then local hot spots may develop. This could cause seat distortion, poor seating, and overheated valves. Seat distortion can also result from improper cylinder-head-bolt tightening. Other conditions that could prevent normal seating include a weak or cocked valve spring and insufficient valve-tappet clearance.

Engine overloading or overheating will cause hot valves. A lean air-fuel mixture may cause valve burning. In this case, the fuel system should be serviced. Preignition and detonation produce high combustion pressures and temperatures. These conditions are hard on valves as well as on other engine parts. Correction is by cleaning out carbon, retiming the ignition, and using higher-octane fuel.

In many engines, valve-seat leakage is prevented by use of an *interference angle*. The valve is faced at an angle ¼ to 1 degree flatter than the seat angle (Fig. 12-11). This produces greater sealing force at the outer edge of the valve seat. The valve-seat edge tends to cut through any deposits that have formed, and thereby establishes a good seal. The difference in angles between the valve face and seat gradually disappears as the valve face and seat wear. The contact between the two

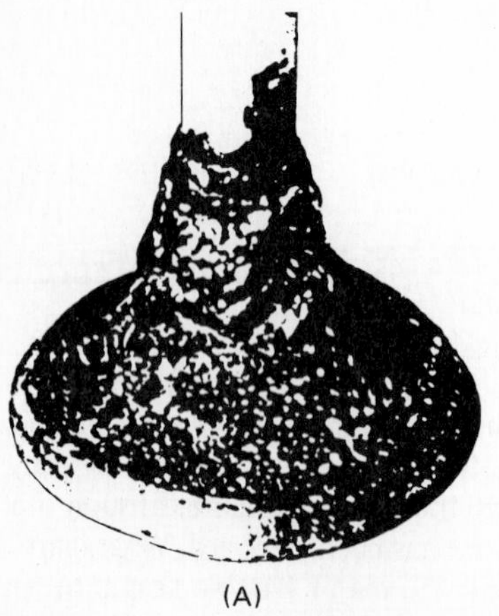

(A)

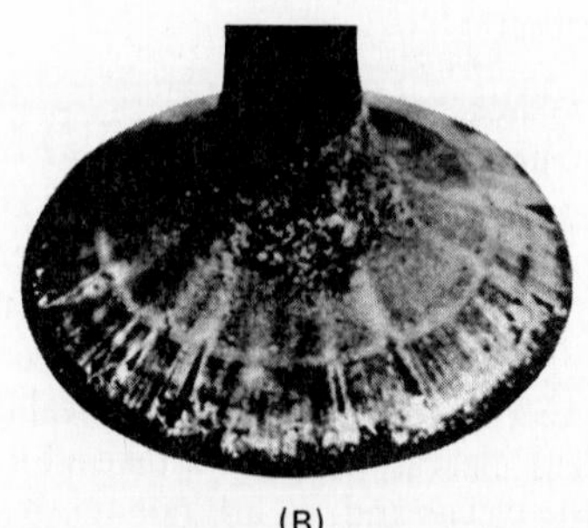

(B)

(C)

Fig. 35-3 (A) Gummed intake valve, with deposits under valve head. (B) Valve burning due to failure of the valve to seat fully. Note that the valve is uniformly burned all the way around its face. (C) Valve burning due to guttering. This is caused by accumulation of deposits on the valve face and seat. Parts of the deposits break off to form paths through which exhaust gas can pass when the valve is closed. This soon burns channels in the valve face. *(Clayton Manufacturing Company; TRW, Inc.)*

changes from line contact (interference angle) to area contact.

Figure 12-11 shows one manufacturer's recommendations for the valve and seat interference angles. This manufacturer does not recommend an interference angle on stellite-faced exhaust valves and induction-hardened exhaust-valve seats. These surfaces are so hard that seating would not be improved by using an interference angle.

♦ 35-6 VALVE BREAKAGE

Any condition that causes the valve to overheat (♦35-5) may cause it to break. Heavy pounding (as from excessive tappet clearance or from detonation) may cause valves to break. Excessive tappet clearance permits heavy impact seating. An off-center seat or cocked valve spring or retainer will force the valve to the side every time it seats. This may cause the valve to fatigue and break. A scratch on the stem may serve as a starting point for a crack and a break in the stem.

♦ 35-7 VALVE-FACE WEAR

Excessive tappet clearance or dirt on the valve face or seat can cause valve-face wear. Excessive tappet clearance causes heavy impact seating. This wears the valve and may cause valve breakage (♦35-6). Dirt may cause valve-face wear if the engine operates in dusty conditions, or if the air cleaner is not functioning properly. The dust enters the engine with the intake air. Some dust settles on the valve seat. Dust also causes bearing, cylinder-wall, and piston and ring wear.

♦ 35-8 VALVE-SEAT RECESSION

Valve-seat recession is the gradual wearing of the valve seat so that it recedes away from the combustion chamber. This decreases valve-tappet clearance. As the valve seat recedes, the clearance in the valve train is reduced. In an engine with mechanical valve lifters, the result can be a complete loss of clearance. The valve can no longer close completely. Then valve and seat burning result.

Lead has been removed from the gasoline used in converter-equipped cars. Lead additives in the gasoline form a lubricant between the valve face and valve seat. This prevents iron particles that flake off the valve seat from sticking to the valve face. However, without this lead coating, particles of iron tend to stick on the valve face. Gradually, as these particles embed in the valve face, they build up into tiny bumps, or warts. This turns the valve face into a cutting surface. Then, the valve seat is gradually cut away and seat recession results. The warts also cause leakage, and loss of compression and power.

To prevent valve-seat recession in engines run on lead-free gasoline, the valve faces are given a very thin coating of aluminum, nickel, or other metal. This coating is less than 0.002 inch [0.05 mm] in thickness. It keeps particles from sticking to the valve face and thereby causing valve-seat recession. The coating gives the valve face a dull, almost rough, appearance. However, *coated valves should not be refaced or lapped.* This would remove the coating and deny the valve seat the protection of the coating. The result could be very short valve and seat life.

♦ 35-9 VALVE DEPOSITS

A fuel with excessive amounts of gum may deposit some of the gum on the intake valve. This happens as the air-fuel mixture passes the valve on the way to the engine cylinder. Carbon deposits may form because of an excessively rich mixture or because of oil passing through a worn valve guide. Improper combustion may be due to a rich mixture, defective ignition system, loss of compression in the engine, or a cold engine. Whatever its cause, it will result in carbon deposits on the exhaust valves. Dirty or improper oil will cause deposits to form on the valves.

♦ 35-10 VALVE SERVICE

Valve service includes adjusting valve-tappet clearances (also called *adjusting valves*); cleaning carbon from all parts; refinishing valves and valve seats; installing new valve-seat inserts; cleaning, knurling, reaming, or replacing valve guides; replacing the camshaft and camshaft bearings; and replacing the camshaft gear, or camshaft sprocket and chain or belt. A complete valve job, or *valve grind,* includes:

1. Refinishing the valve faces and valve seats
2. Checking the runout of the valve head and valve seat
3. Checking the clearance between the valve stem and valve guide
4. Checking valve-spring tension
5. Checking valve-spring installed height
6. Adjusting the valves (if applicable)
7. Adjusting ignition timing
8. Adjusting idle speed

On a typical six-cylinder in-line overhead-valve engine, the complete valve job requires about 5 hours. Replacing the camshaft takes about 4.5 hours or more. The actual time depends on how much power equipment the car has attached to the front of the engine. Replacing the camshaft bearings on the same engine would take from 6 to 10 hours. These times vary greatly from engine to engine, and from car to car.

♦ 35-11 VALVE-LIFTER CLEARANCE

The procedure for checking and adjusting valve-tappet (or valve-lifter) clearance depends on the type and model of engine. Some engines with hydraulic valve lifters normally require no clearance adjustment. Others require checking and adjustment whenever valve-service work has been performed. The following procedures for pushrod engines and for overhead camshaft engines are typical.

♦ 35-12 PUSHROD ENGINE WITH MECHANICAL VALVE LIFTERS

First, remove the valve cover. Measure the clearance between the valve stem and rocker arm, as shown in Fig. 35-4. Most specifications call for making the check with the engine hot and not running. The clearance is measured with the valve lifter on the base circle of the cam. Turn the crankshaft by bumping the engine with the starting motor until the base circle of the cam is under the valve lifter. On many engines, exhaust-valve clearance is greater than intake-valve clearance. This is because exhaust valves run hotter and expand more.

The shaft-mounted type of rocker arm (Fig. 35-4) often has

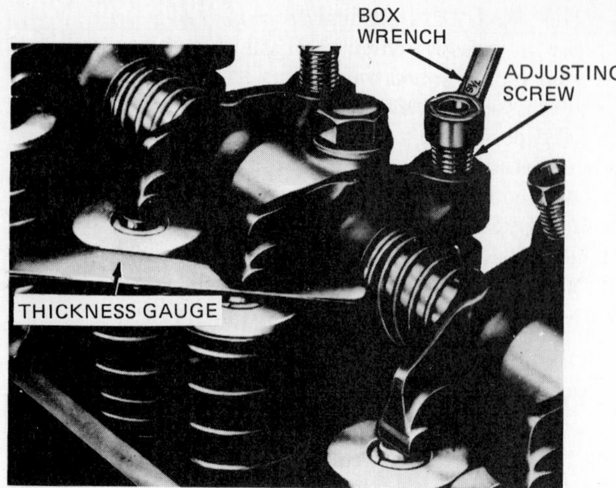

Fig. 35-4 Adjusting valve clearance on an overhead-valve engine which has shaft-mounted rocker arms. *(Ford Motor Company)*

an adjustment screw. This screw is usually self-locking and does not require a lock nut. Use a box wrench to turn the adjustment screw. Adjust the clearance to specifications.

On ball-stud-mounted rocker arms (Fig. 35-5), turn the self-locking rocker-arm-stud nut to make the adjustment. Turning the nut down reduces clearance.

Too much valve clearance can cause poor engine performance. Too little valve clearance can cause valve burning.

♦ 35-13 PUSHROD ENGINE WITH HYDRAULIC VALVE LIFTERS

On some engines with hydraulic valve lifters, no adjustment is provided in the valve train. In normal service, no adjustment is necessary. The hydraulic valve lifter takes care of any

Fig. 35-5 Adjusting valve clearance on an overhead-valve engine with stud-mounted rocker arms. Backing the nut out increases clearance. *(Chevrolet Motor Division of General Motors Corporation)*

Fig. 35-6 Checking valve-train clearance on an engine with hydraulic valve lifters and shaft-mounted rocker arms. The hydraulic valve lifter is bled down from force applied with the special tool. *(Ford Motor Company)*

small changes in valve-train length. However, adjustment may be needed if valves and valve seats are refinished. Unusual and severe wear of the pushrod ends, rocker arm, or valve stem may also require adjustment. Then some correction may be required to reestablish the correct valve-train length. Typical checking and correcting procedures follow.

♦ 35-14 FORD PUSHROD ENGINES WITH HYDRAULIC VALVE LIFTERS

Ford engines use two types of rocker arms, the shaft mounted (Fig. 35-6) and the ball-stud mounted (Fig. 35-7). On both types the clearance in the valve train is checked with the valve

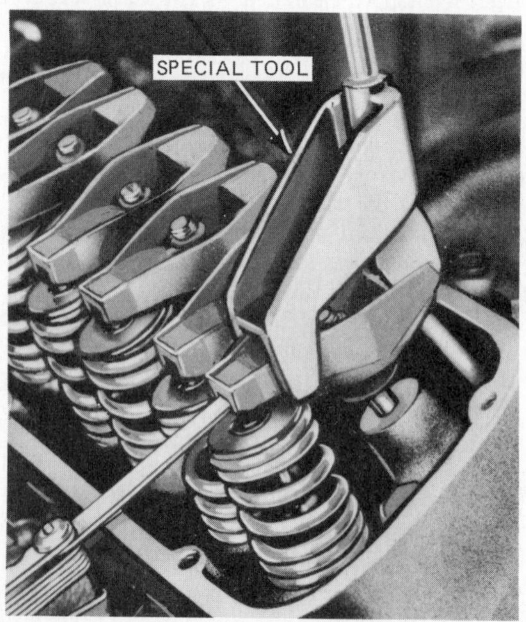

Fig. 35-7 Checking valve-train clearance on an engine with hydraulic valve lifters and stud-mounted rocker arms. The hydraulic valve lifter is bled down from force applied with the special tool. *(Ford Motor Company)*

lifter bled down so that the valve-lifter plunger is bottomed. First, the crankshaft must be turned until the lifter is on the base circle or low part of the cam (rather than on the lobe). This is done by setting the piston in number 1 cylinder at TDC at the end of the compression stroke. Then check both valves in number 1 cylinder. The crankshaft can then be rotated to put other lifters on the base circles of their cams so that they can be checked.

To make the check, use a special tool to apply force against the rocker arm (Figs. 35-6 and 35-7). This gradually forces oil out of the valve lifter until the plunger bottoms. Then, the thickness gauge is used to check the clearance between the valve stem and rocker arm. If the clearance is too small, a shorter pushrod should be installed. If the clearance is excessive, install a longer pushrod. The clearance might be too small if valves and seats have been ground. The clearance might be excessive due to wear of the valve-train parts. This includes wear of the pushrod ends, valve stem, and rocker arm.

♦ 35-15 PLYMOUTH PUSHROD ENGINES WITH HYDRAULIC VALVE LIFTERS

The procedure for setting Plymouth valves is typical of Chrysler-manufactured engines. It is necessary only when valves and valve seats have been ground. When this happens, the increased height of the valve stem above the cylinder head should be checked. With the valve seated, place the special gauge over the valve stem. If the height is excessive, the end of the valve stem must be ground off to reduce the height to within limits. The hydraulic valve-lifter plunger will now be near its center position. It would be near the bottom if the valve stem was excessively high.

♦ 35-16 CHEVROLET PUSHROD ENGINES WITH HYDRAULIC VALVE LIFTERS

The procedure for Chevrolet engines is typical of General Motors engines using the ball-pivot type of rocker arm (Fig. 35-8). With the valve lifter on the base circle of the cam, back off the adjustment nut until the pushrod is loose. Then slowly turn the adjustment nut down. At the same time, rotate the pushrod with your fingers until the pushrod is tight and you cannot easily rotate it. Then turn the adjustment nut down one additional full turn. This positions the valve-lifter plunger in the center of its travel.

♦ 35-17 OVERHEAD-CAMSHAFT ENGINE

OHC engines have several arrangements for carrying the cam action to the valve stems. In some engines, cam action is carried directly to the valve stem through a cap, called a *bucket tappet* or *cam follower* (Fig. 12-1). This cap fits over the valve stem and spring. In others, the cam action is carried through a rocker arm.

1. Type with Bucket Tappet. On this type (Fig. 12-1), there is a shim between the end of the valve stem and the bucket tappet. Shims of various thicknesses are available to change the valve clearance. The clearance is measured between the tappet and the cam with the valve closed. If the clearance is excessive, the camshaft must be removed so that the bucket tappet can be taken off. Then a new shim of the thickness needed to provide the correct clearance is installed.

2. Type with Rocker Arm and Mechanical Lifters. One engine of this type has rocker arms which float between a stationary stud on one side and the valve stem on the other. The center of the rocker arm rests on the cam. Figure 35-9 shows the use of a thickness gauge to check the clearance between the base circle of the cam and the rocker arm. Adjustment is made by loosening the lock nut. Use an open-end wrench to turn the adjustment screw in or out. Turning the screw in increases the clearance. Tighten the lock nut securely after the adjustment. Then recheck the clearance.

3. Type with Rocker Arm and Hydraulic Lifters. In one type of valve-train arrangement, the rocker arm floats between the end of the valve stem and an automatic valve-lash adjuster. The automatic valve-lash adjuster is a type of

Fig. 35-8 Adjusting the valve rocker-arm-stud nut to properly position the plunger of the hydraulic valve lifter. *(Chevrolet Motor Division of General Motors Corporation)*

Fig. 35-9 Checking valve clearance on an OHC engine using rocker arms. *(Ford Motor Company)*

hydraulic valve lifter. It automatically takes up any clearance between the cam and the rocker arm. Normally, no valve adjustment is necessary on this engine.

♦ 35-18 STEPS IN THE COMPLETE VALVE JOB

A complete valve job requires the steps listed below. Valve and valve-seat servicing are described in detail in the sections that follow.

1. Drain cooling system. Disconnect upper radiator hose from engine.
2. Remove air cleaner. Disconnect throttle linkage, fuel line, and air and vacuum hoses from the carburetor or fuel-injection system.
3. Remove or move aside lines and hoses as necessary to get at the cylinder head.
4. Disconnect spark-plug cables and temperature-sending-unit wire.
5. Remove PCV hoses. On air-injection systems, disconnect the air hose at the check valve. Then remove the air-supply-tube assembly.
6. On many in-line engines, it is not necessary to remove manifolds. But on V-type engines, the carburetor or throttle body and intake manifold must be removed.
7. Remove valve cover.
8. On engines with stud-mounted rocker arms, the rocker arms and pushrods can now be removed. If they are left on, the nuts should be loosened so that the rocker arms can be moved aside and the pushrods removed. Pushrods should be placed in a rack, in order. Then they can be reinstalled in their proper positions. On an overhead-valve engine, the condition of the pushrods and valve lifters should be checked.
9. On engines with shaft-mounted rocker arms, remove the shaft assembly (♦35-21). Then remove the pushrods, in order.
10. Remove head bolts. Take the head off the engine.
11. Remove the valves and springs from the head. (Keep them in order so that they can be put back in their proper positions.)

Careful If a valve-stem end has mushroomed, the valve cannot be pulled out by hand. The mushroom must be removed, as explained in ♦35-23. Otherwise, if you try to force the valve, the guide may break.

12. Check valve guides for wear. Clean, replace, or knurl and ream for same-size valve stem if necessary. Or ream for a larger-diameter valve stem.
13. Check valves and valve seats. Clean valve heads and stems on a wire wheel. Refinish valve seats, and reface valves as necessary. Check valve seating. Refinish valve-stem ends if necessary.

    If you are installing new valves of the coated type, do not reface them. Refacing or lapping coated valves removes the protective coating. This will shorten valve and seat life.
14. Check rocker arms for wear. Service or replace as necessary.
15. Replace valves and springs in head.
16. Install head, pushrods, rocker arms, rocker-arm cover, and other parts removed during head removal.
17. Check and adjust valve clearance as necessary.

♦ 35-19 REMOVING, CLEANING, AND INSTALLING CYLINDER HEADS

On some cars, the manifolds must be removed before the cylinder heads can be taken off. On other cars, the manifolds may remain in place.

1. Removing the Cylinder Head Follow the general instructions in ♦35-18 to remove the cylinder head. Slightly loosen all cylinder-head bolts first, to ease the tension on the head. Then remove the bolts. If the head sticks, carefully pry it loose. Do not insert the pry bar too far between the head and block. This could mar the mating surfaces and lead to leaks. Lift the head off, and place it in a head-holding fixture.

Careful Never remove a cylinder head from a hot engine. Wait until the engine cools. If the head is removed hot, it can warp so that it cannot be used again.

2. Cleaning the Cylinder Head After the valves and other parts are removed from the head (as explained later), it should be cleaned and inspected. Clean carbon from the combustion chambers and valve ports. Use a wire brush driven by an air or electric motor (Fig. 35-10). Keep the wire brush away from valve seats, because it could scratch the seating surfaces. Scratched seats can cause poor valve seating. Blow out all dust with an air hose.

Some mechanics recommend temporarily reinstalling the valves before cleaning the combustion chambers. This protects the valve seats and at the same time cleans the valve heads.

♦ CAUTION ♦ Always wear eye protection while using a wire brush, compressed air, or similar equipment. You must protect your eyes from flying particles.

Clean gasket surfaces with a flat scraper or spray gasket remover. Be careful not to scratch the gasket surface. All traces of gasket material and sealer should be removed.

Remove dirt and grease from the cylinder head. Then clean the water jackets and passages by soaking the head in a boil tank. Flush the water jackets as recommended by the manufacturer of the cleaning compound.

3. Inspecting the Cylinder Head As you remove the head, examine the gasket and mating surface for traces of

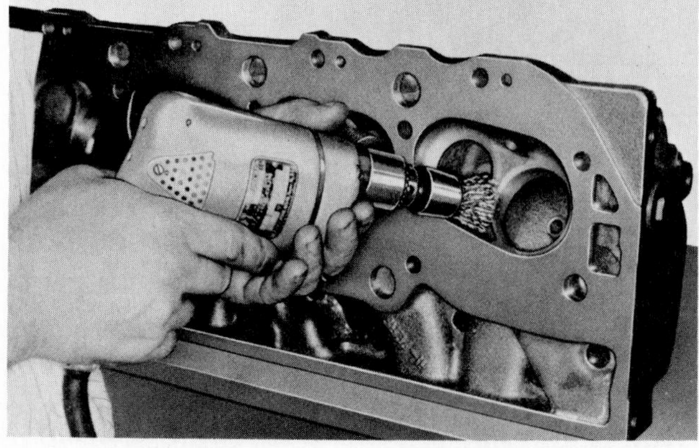

Fig. 35-10 Cleaning combustion chamber and valve ports with a wire brush. *(Chevrolet Motor Division of General Motors Corporation)*

leakage or cracks. If cracks are suspected, have the head checked with Magnaflux or similar equipment. A blown gasket or coolant leakage could result from a warped head or improper gasket installation. In the head, cracks usually occur between valve seats. If they are not too bad, they can be repaired by *cold welding.* This is usually a job for the automotive machinist. The job is done by drilling a small hole in the crack and threading it. Then screw in a threaded rod and cut it off. Next, drill a second hole overlapping the first one and thread it. Screw in the threaded rod and cut it off. Repeat until the crack is completely treated. Make sure to get to both ends of the crack. This relieves the stress that caused the crack. Sometimes it is best to install a seat insert when the crack runs into the valve seat. This is done by making an undercut in the head and pressing in the insert (♦35-28).

Clean and inspect valve guides. Note the condition of valve seats and rocker-arm studs (on heads using them). Servicing of valve guides, rocker-arm studs, and seats is described later.

Warpage is detected by laying a machinist's straightedge against the gasket surface of the head (Fig. 35-11). Check crossways and longways. One specification calls for 0.005 inch [0.127 mm] maximum out of straight. More than this requires either a new head or resurfacing the head gasket surface.

Check the gasket surface of the head for nicks or rough spots. These can be removed with a fine-cut mill file.

If one head from a V-type engine requires machining, then the other head should be machined the same amount. Otherwise, uniform compression and manifold alignment will be lost. Removing metal from the head gasket surface lowers the head with respect to the intake manifold. Therefore, a compensating amount may have to be machined from the manifold to restore alignment.

4. Installing the Cylinder Head Reassemble the cylinder head (as explained later) so that valve springs, rocker arms, and other parts are in place. Then install the head, using a new gasket, as follows.

Before installing the head, check the cylinder block. Make sure the gasket surface is flat and in good condition. All traces of gasket material must be removed from the block. Boltholes in the block (where present) should be cleaned out. Head bolts may bottom in dirty boltholes, preventing head-to-block sealing. Cylinder-block studs (where present) should be in good condition. Cylinder-head bolts should be cleaned with a wire brush or wheel. Cylinder-block studs (where present) can be cleaned with a thread chaser if the threads are damaged.

Use care when handling the gasket. If it is of the lacquered type, do not chip the lacquer. If the block has studs, put the gasket into place, correct side up. Some are marked TOP so that you know which side goes up. Use gasket sealer only if specified by the manufacturer. For example, one manufacturer recommends using sealer on steel gaskets but not on composition steel-asbestos gaskets.

If the block does not have studs, use two pilot pins set into two boltholes to assure gasket alignment. Then lower the head into position. Substitute bolts for pilot pins (if used). Run on the nuts or bolts finger tight.

Make sure that all boltholes in the block have been cleaned out. If they are not clean, and the bolts bottom on foreign material, then the head will not be tight.

Use a torque wrench to tighten the nuts or bolts. They must be tightened in the proper sequence and to the proper torque. If they are not, head or block distortion, gasket leakage, or bolt failure may occur. Refer to the sequence chart for the engine being serviced, and note the specified torque. Figure 35-12 shows the sequence for one head of a V-8 engine. Each bolt should be tightened in two or more steps. The complete sequence should be made at least twice, with each bolt or nut being drawn down little by little. After engine assembly is completed, the engine should be run until it is warm. Then the torques should be checked. Some engines using aluminum heads must be warmed up and allowed to cool. Then the bolt or nut torques must be checked again.

Some torque specifications call for clean, dry threads. Others call for lightly lubricated threads. Antiseize compound should be used on bolts in aluminum blocks.

Careful If the rocker arms are in place, tighten the bolts slowly. This gives the hydraulic valve lifters time to bleed down to their operating length. If the bolts are tightened too rapidly, excessive pressure will be put on the lifters. They could be damaged, and the pushrods could be bent. On most engines, the head bolts cannot be tightened if the rocker arms are in place.

On overhead-valve cylinder heads with the rocker arms and

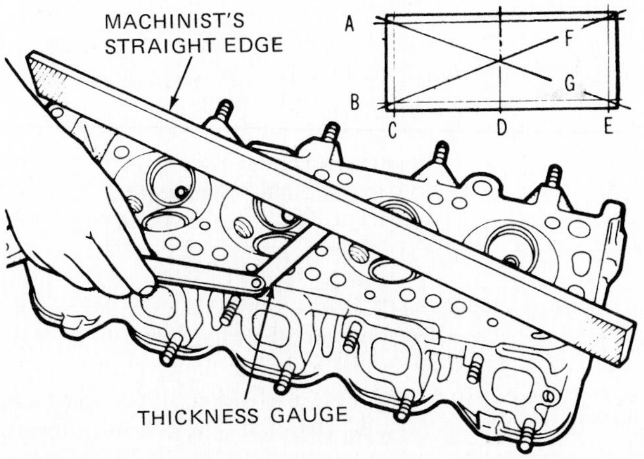

Fig. 35-11 Checking cylinder head for warpage with a machinist's straightedge and a thickness gauge. *(Chrysler Corporation)*

Fig. 35-12 Sequence in which cylinder-head bolts should be tightened on a V-8 engine. *(Chrysler Corporation)*

shaft in place, make sure the pushrods are in position. The lower ends of the pushrods should be in the valve-lifter sockets.

♦ 35-20 ROCKER-ARM-STUD SERVICE

If a rocker-arm stud is loose in the cylinder head, has damaged threads, or has begun to pull out, the stud should be replaced. The old stud is removed with a special puller. The stud hole is then reamed to a larger size to take the oversize stud. Some studs are threaded into tapped holes.

♦ 35-21 SERVICING ROCKER-ARM ASSEMBLIES

When removing rocker arms and pushrods, keep them together to identify their location in the head. These parts are "wear mated" and should not be interchanged.

If the rocker arms are mounted on separate studs (Fig. 35-8), removing the adjusting nut permits removal of the rocker arm. The studs can be replaced if they are loose or if the stud threads are damaged (♦35-20).

On some shaft-mounted rocker arms, the shaft with rocker arms and brackets is removed as an assembly. On others, remove the shaft-lock plug and slide the shaft out of the head. Then the rocker arms can be removed from the head.

After rocker arms are removed, they should be inspected for wear or damage. Rocker arms with bushings can be rebushed if the old bushings are worn. On some rocker arms, worn valve ends can be refinished on the valve-refacing machine. Excessively worn rocker arms should be discarded.

When reinstalling rocker arms and shafts on the cylinder head, make sure that the oilholes (in shafts so equipped) are on the underside. Otherwise, they will not feed oil to the rocker arms. Be sure all springs and rocker arms are in their original positions when the shafts are reattached to the head.

♦ 35-22 PUSHROD SERVICE

Pushrods should be inspected for wear at the ends. Roll the rods on a flat surface to check for straightness. Replace defective rods. Rods on some engines have one tip hardened and marked with a stripe of color. The pushrod should be installed so that the hardened end is toward the rocker arm. Always make sure that the lower end of the pushrod is seated in the valve-lifter socket.

Special short-length pushrods are available for some engines. These may be used in some engines after valves and valve seats have been refinished.

♦ 35-23 VALVE REMOVAL

After the head is off, and rocker-arm mechanisms are removed from the head, the valves are taken out. Valves and valve parts must not be interchanged. Each valve, with its own spring, retainer, and lock, should be put back in the valve port from which it was removed. For this reason use of a special valve rack is recommended (Fig. 35-13).

Before removing a valve from the cylinder head, examine the valve stem. Look for burrs at the retainer-lock grooves and for mushrooming on the end. Burrs and mushrooming must be removed with a file or a small grinding stone in a drill motor. If they are not removed, the valve guide could be damaged or broken when the valve is forced out. This would mean extra work in refinishing or replacing the valve guide.

Fig. 35-13 Valve rack for holding valves and associated parts. *(Chevrolet Motor Division of General Motors Corporation)*

After everything else is off the head, the valves can be removed. This requires a valve-spring compressor. Figure 35-14 shows how a spring compressor is used. First, be sure that the locks are not locked in the retainer. Then install the valve-spring compressor. As the handle is pressed, the spring is compressed. This allows removal of the retainer lock. Then the spring can be released so that the retainer and spring can be removed. Valve-stem seals or shields (Figs. 12-15 and 12-16) should always be replaced. Many manufacturers recommend installation of new seals or shields whenever valves or valve springs are removed.

A single valve spring, stem seal, or shield can be replaced without removing the head on many engines. An example is the engine with rocker-arm studs. A special spring compressor is installed in place of the rocker arm to compress the spring. To hold up the valve while the spring is being compressed, compressed air from the shop supply is introduced into the cylinder through the spark-plug hole (Fig. 35-15). A special air-hose adapter, which can be screwed into the spark-plug hole, is required. The air pressure holds the valve closed while the spring is compressed. If the air pressure does not hold the valve closed, then the valve is stuck or damaged. The head must be taken off for an inspection.

Careful The air pressure may push the piston to BDC. If it does, the valve could drop into the cylinder if the air pressure is released. To prevent this, wrap a rubber band or tape around the valve stem.

On some engines with shaft-mounted rocker arms, it is possible to bleed down the hydraulic valve lifter. To do this, force is applied with a special tool (Fig. 35-6). Then you can remove the pushrod, and move the rocker arm to one side. With the rocker arm wired out of the way, a valve-spring compressor can be used to compress the spring. Now you can remove the retainer, spring, and seal. Air pressure must be applied to the cylinder (as explained above) to hold the valve on its seat when the spring is compressed.

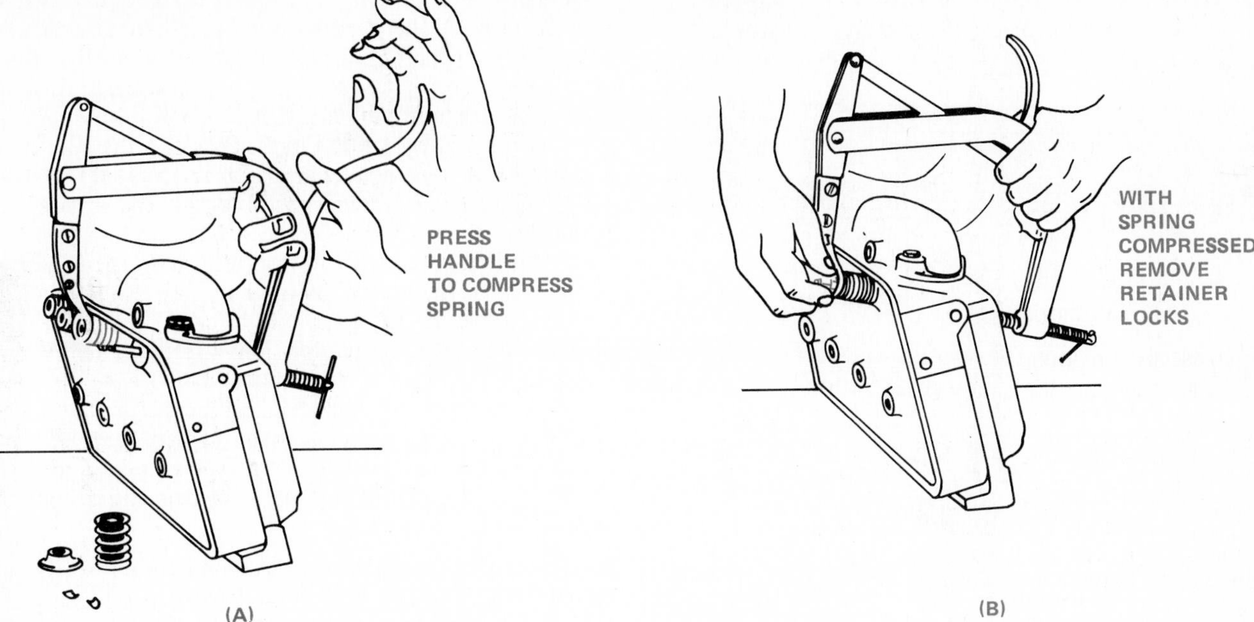

Fig. 35-14 Using a valve-spring compressor: (A) compressing the spring; (B) removing the retainer locks.

♦ 35-24 VALVE INSPECTION

As you take the valves out of the head, inspect each valve. Decide whether or not it can be serviced and used again (♦35-2 to 35-9). If the valve looks good enough to use again, put it into its proper place in the valve rack (Fig. 35-13). If the valve looks too bad to be cleaned up for further service, discard it. Put a new valve in the appropriate place in the valve rack.

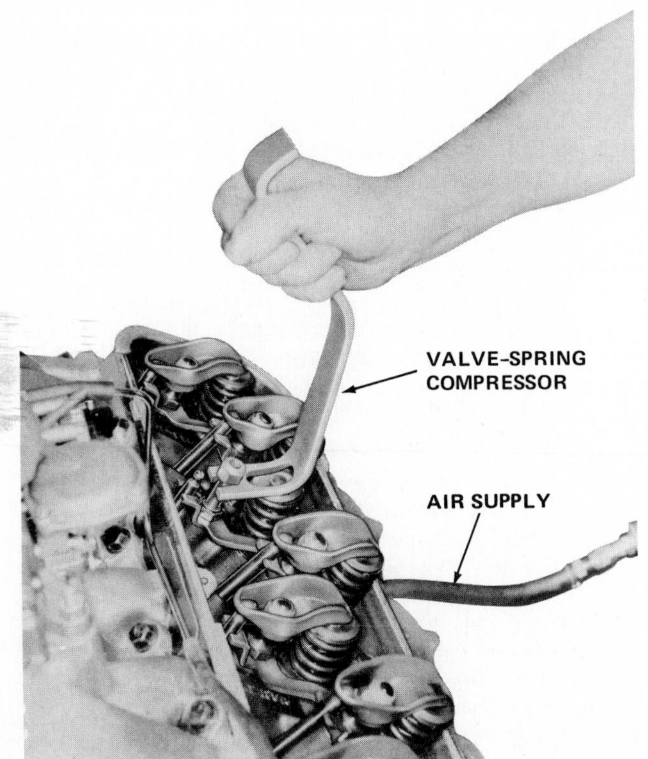

Fig. 35-15 Compressing a valve spring while the valve is held closed with air pressure. *(Chevrolet Motor Division of General Motors Corporation)*

♦ 35-25 SERVICING VALVES

Once all the valves are out of the head, remove them one by one from the valve rack. Clean each valve. Remove the carbon with a wire wheel. (Wear goggles to protect your eyes from flying particles of metal and dirt!) Polish the valve stems, if necessary, with a fine grade of emery cloth. Do not take off more than the dirty coating on the surface. Do not take metal off the stems.

Careful Do not scratch the valve-seating surface or valve stem with the wire brush or emery cloth.

As you clean the valves, reexamine them to make sure all are usable. Small pits or burns in the valve face can be removed by refinishing the valve. Larger pits or burns require that the valve be discarded. Figure 35-16 shows specific parts of the valve to be examined. For reuse after grinding, the valve margin must be at least 1/32 inch [0.79 mm]. If the margin is any thinner, the valve will run too hot. Some engine manufacturers recommend the use of a runout gauge to check for a bent valve stem. Eccentricity can also be checked in the valve refacer. If the runout or eccentricity is excessive, discard the valve.

After cleaning the valves, install them temporarily in their valve guides to check for guide wear. This procedure, and valve-guide service, is covered in ♦35-27.

1. Refacing or Grinding Valves The next step, if the valves are good enough to reuse, is to reface, or grind, them. This requires a valve-refacing machine (Fig. 35-17). The valve-refacer has a grinding wheel, a coolant-delivery system, and a chuck which holds the valve for grinding. Set the chuck to grind the valve face at the specified angle. This angle must just match the valve-seat angle, or make an interference angle of 1/4 to 1 degree (Fig. 12-11). Then put the valve into the chuck and tighten the chuck. The valve should be placed in the chuck so that the part of the stem that runs in the valve guide is gripped by the chuck.

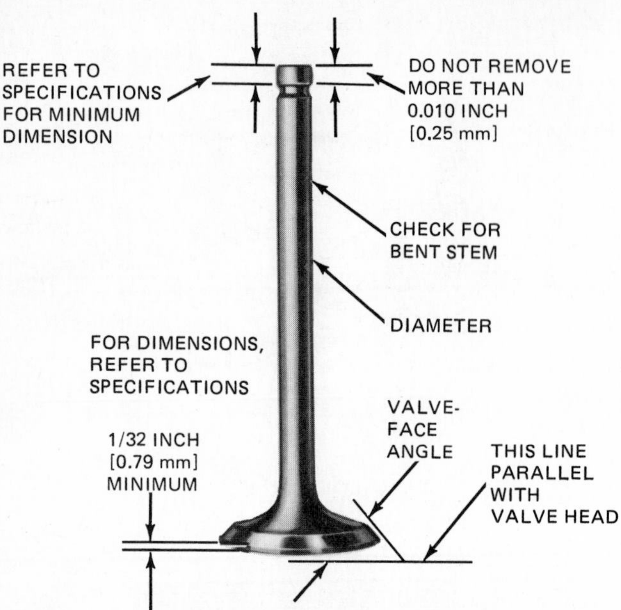

Fig. 35-16 Valve parts to be checked. On the valve shown, the stem is hardened on the end. Therefore, not more than 0.010 inch (0.25 mm) should be removed. *(Ford Motor Company)*

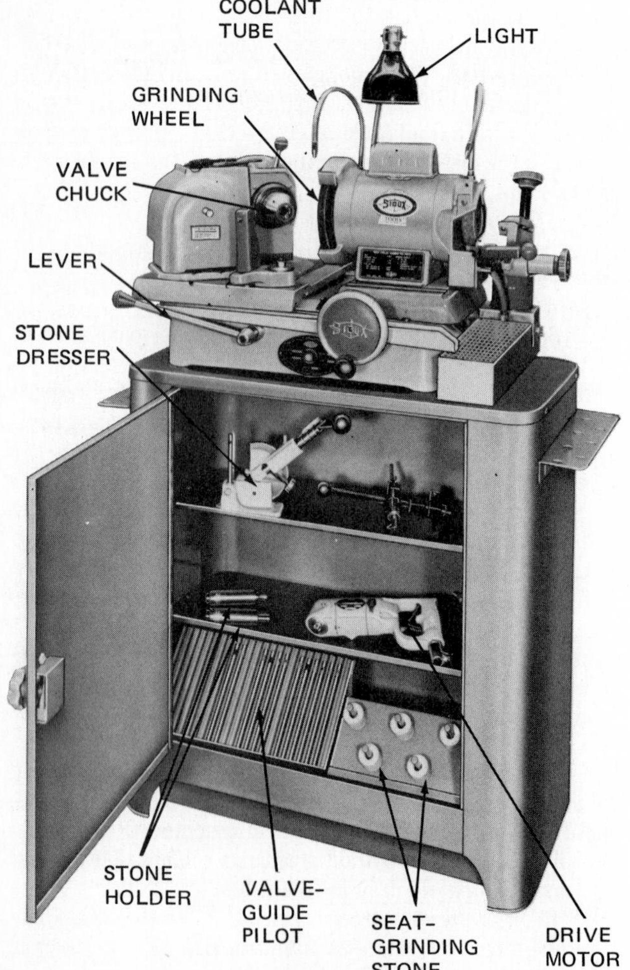

Fig. 35-17 A valve-refacing machine. The valve-seat grinding set is shown in the cabinet under the valve refacer. *(Sioux Tools, Inc.)*

To start the operation, align the coolant feed so that it sprays coolant on the rotating valve face. Then start the machine. Move the lever to carry the valve face across the grinding wheel. The first cut should be a light one. If this cut removes metal from only one-half or one-third of the face, the valve may not be centered in the chuck. Or the valve stem is bent, and the valve should be discarded. Cuts after the first should remove only enough metal to true the surface and remove pits. Do not take heavy cuts. If so much metal must be removed that the margin is lost, discard the valve. Loss of the margin causes the valve to run hot. Then it will soon fail.

If new valves are required, reface them lightly, provided they are not of the coated type. *Never reface or lap coated valves!*

Follow the operating instructions of the valve-refacer manufacturer. Dress the grinding wheel as necessary with the diamond-tipped dressing tool. As the tool is moved across the rotating face of the grinding wheel, the diamond cleans and aligns the grinding face.

2. Refacing Valve-Stem Tips The tip of a valve stem should be ground lightly. Use the special attachment furnished with the valve-refacing machine. One recommendation is to grind off as much from the stem end as you ground off the valve face. That way, you make up for the amount the valve sinks into the seat.

The ends of some valve stems are hardened. These should have no more than 0.010 inch [0.25 mm] ground off (Fig. 35-16). Excessive grinding exposes soft metal so that the stem wears rapidly.

♦ 35-26 VALVE INSTALLATION

As the valves are refaced and cleaned, they should be ready for installation in the cylinder head. However, the valve guides and valve seats must be serviced (♦35-27 and 35-28). Also, the other components of the valve train—pushrods, rocker arms, and valve lifters—must be checked and serviced as necessary.

New shields or seals should be installed if the old ones are worn or if the manufacturer recommends them. To avoid damage to the seals, special plastic caps can be placed over the ends of the valve stems. The seals will then slip on without being damaged by the sharp edges of the stem and lock grooves.

If valves and seats have been refinished, the effective tension of the valve spring will not be great enough. To restore normal spring tension, a valve-spring shim must be installed.

Using a spring compressor (Fig. 35-14), install the springs, spring retainers, and locks. With the valve closed, measure the spring installed height (Fig. 35-18). Note that the end of the steel scale has been cut away. If the spring height is excessive, a spring shim is required. The shim is installed between the spring and cylinder head.

Do not install a shim that reduces the spring height below the specified minimum. This results in excessive spring force and rapid wear of valve-train parts.

Install the valve springs with the proper side against the cylinder head. The close-spaced coils go next to the head (on springs with differential spacing of coils). Also, a damper spring (where used) must be placed inside the valve spring in an exact relationship with the spring coils. One typical example is that the coil end of the damper spring should be 135 degrees counterclockwise from the coil end of the valve spring.

Fig. 35-18 Measuring valve-spring installed height. *(Chevrolet Motor Division of General Motors Corporation)*

♦ 35-27 VALVE-GUIDE SERVICE

The valve guide must be clean and in good condition for proper valve seating. It must be serviced before the valve seats are refinished. As a first step, the valve guide should be cleaned with a wire brush or adjustable-blade cleaner. Then, the guide should be checked for wear. If it requires service, the type of service depends on whether the guide is integral or replaceable. If replaceable, the old guide should be pressed out. Then a new guide is installed and reamed to size. If integral, the guide can be serviced in either of two ways. One, ream it to a larger size, and install a valve with an oversize stem. Two, knurl and ream the guide. These services are described below.

1. Testing the Guide for Wear Valve guides wear oval shaped and bell mouthed. One method of testing the guide uses a dial indicator (Fig. 35-19). With the valve in place, the dial indicator is attached to the cylinder head. The button just touches the edge of the valve head. Use a sleeve, as shown in Fig. 35-20, to hold the valve off its seat. Then rotate the valve and move it sideways to determine the amount of guide wear. On some engines, the recommendation is to check valve movement from the stem end with the valve seated.

Another checking method is to insert a tapered pilot into the guide until the pilot is tight. Then, pencil-mark the pilot at the top of the guide, and remove it. Measure the pilot diameter ½ inch [12.7 mm] below the pencil mark. This gives the guide diameter, which can then be compared with the valve-stem diameter.

Another method that is more accurate than the two described above uses a dial-indicating valve guide gauge (Fig. 35-21). The gauge is preset to register zero. There are two buttons near the end of the probe. When the probe is inserted in the valve guide, the buttons move in or out of the probe to show variations in guide diameter. Any movement of the buttons moves the needle to show the amount of wear, taper, and out-of-round in the guide.

2. Removing Valve Guide (Replaceable Type) On overhead-valve engines, the valve guide can be pressed out of the head with a shop press.

3. Installing Valve Guide (Replaceable Type) Guides can be installed with a shop press. Guides must be installed to the proper depth in the block or head. Then, they must be reamed to size.

4. Reaming the Guide The guide is reamed to take a valve with an oversize stem. After reaming, the sharp edge at the top of the guide should be removed with a scraper.

5. Knurling the Valve Guide In the guide-knurling operation, a knurling tool is run down the guide. Then the guide is reamed. At the end of the procedure, the original diameter has been restored to the guide. Now a standard-size valve can be installed.

6. Checking Concentricity with Seat After the valve guide is serviced and checked for size, check its concentricity with the valve seat. The seat is always ground whenever the guide is serviced (♦35-28).

♦ 35-28 VALVE-SEAT SERVICE

For effective valve seating and sealing, the valve face must be concentric with the valve stem. Also, the valve guide must be concentric with the valve face. In addition, the valve-face angle must match the valve-seat angle (or have an interfer-

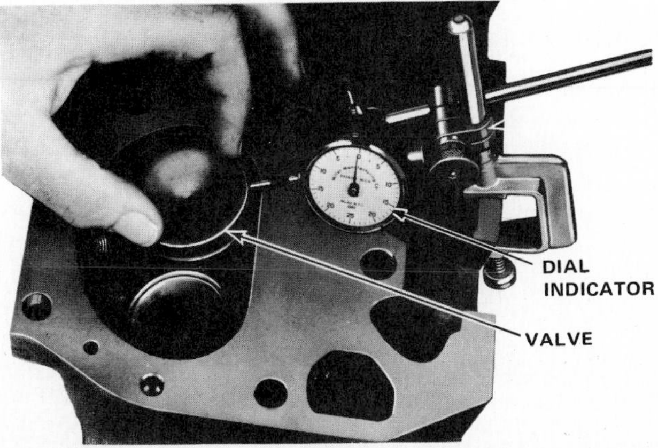

Fig. 35-19 Dial indicator used to measure valve-guide wear. *(Chrysler Corporation)*

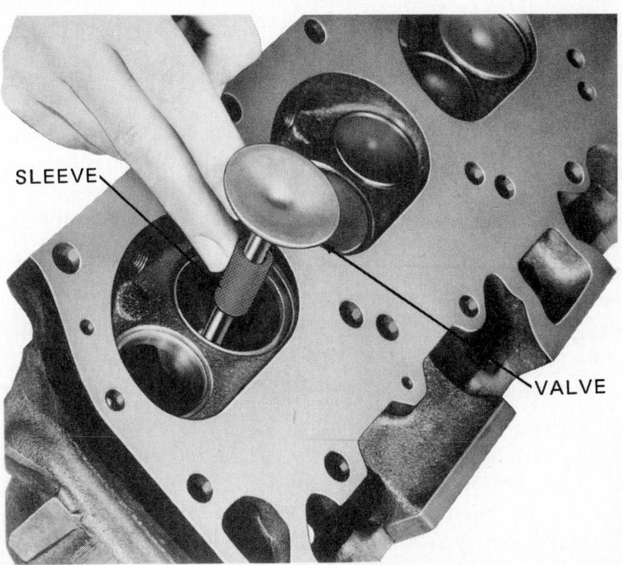

Fig. 35-20 Using a sleeve on the valve stem to hold the valve off its seat while measuring valve-guide wear. *(Chrysler Corporation)*

Fig. 35-21 Using a dial-indicating valve-guide gauge to check for wear. Movement of the probe in and out of the guide will cause the needle to move if the guide is irregularly worn. *(Sunnen Products Company)*

ence angle). Therefore, as a first step in valve-seat service, the valve guides must be cleaned and serviced (♦35-27).

Valve seats are of two types, the *integral* type and the *insert* type (♦12-5). Replacing seat inserts and grinding seats are described below.

1. Replacing Valve-Seat Inserts A valve-seat insert may be badly worn. Or it may have been refinished previously so that there is insufficient metal. With either condition, the seat must be replaced. The old seat must be removed with a special puller. If a puller is not available, the insert is punch-marked on two opposite sides. An electric drill is then used to drill holes almost through the insert. Then, a chisel and hammer can be used to break the insert into halves so that it can be removed. Care must be used so that the counterbore is not damaged. If the new insert fits too loosely, the counterbore must be rebored oversize. Then an oversize insert is installed. It may be necessary to chill the seat and heat the head before the seat is driven into place. After installation, the valve seat should be ground.

2. Refinishing Valve Seats The valve-seat grinder rotates a grinding stone of the proper shape on the valve seat (Fig. 35-22). The stone is kept concentric with the valve seat by a pilot installed in the valve guide (Fig. 35-23). This means that the valve guide must be cleaned and serviced (♦35-27) before the seat is ground. In the seat grinder shown in Fig. 35-22, the stone is automatically lifted about once a revolution. This permits the stone to clear itself of grit and dust by centrifugal force.

After the seat is ground, it may be too wide. Narrow the seat by using upper and lower grinding stones to grind away the upper and lower edges of the seat. If the seat is too high, it can be lowered by grinding with a 15 degree stone. A steel scale can be used to measure seat width. When a new valve is installed, it may sit too high above the seat. Then grind the seat slightly with a stone of the same seat angle. This will lower the valve further into the seat, and raise the seat contact on the valve face.

Many shops are now using a hand-operated carbide cutter.

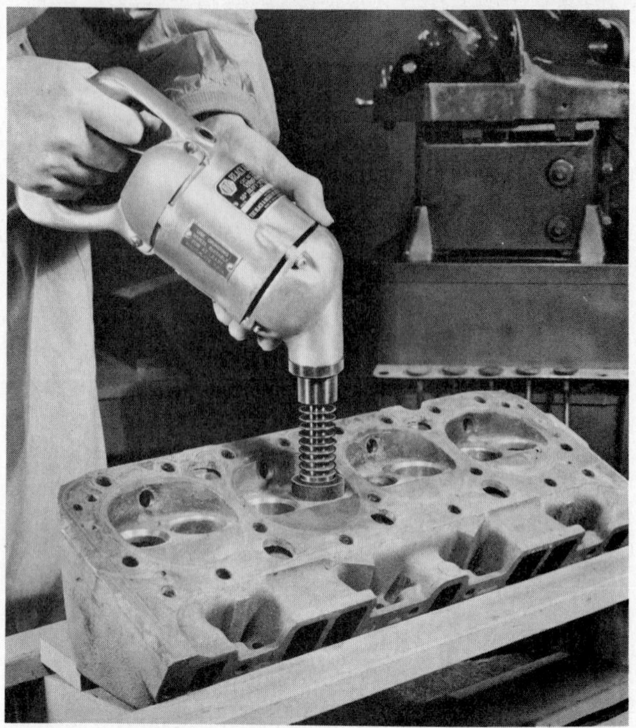

Fig. 35-22 Using the valve-seat grinder. The stone rotates at high speed. About once every revolution, it lifts off the valve seat to throw off loose grit and grindings. *(The Black and Decker Manufacturing Company)*

This device takes the place of the motor-driven stones in refinishing the valve seats.

Careful Be sure to follow the operating instructions for the valve-seat cutter or grinder you are using. The grinding stone must be dressed frequently with the diamond-tipped dressing tool.

3. Checking Valve Seats for Concentricity After the valve guides are serviced and valve seats ground, the con-

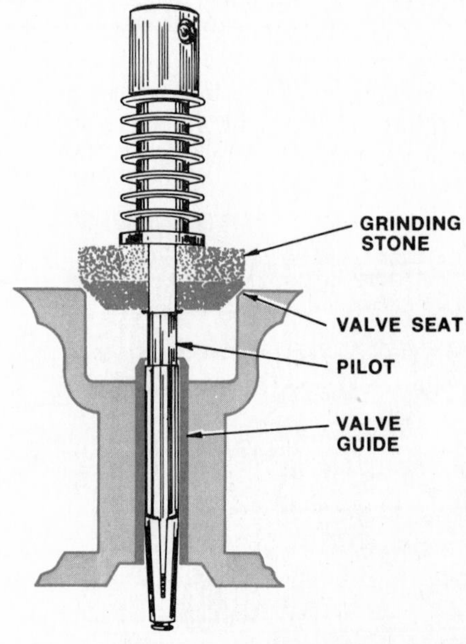

Fig. 35-23 Pilot on which grinding stone rotates. The pilot keeps the stone concentric with the valve seat. *(The Black and Decker Manufacturing Company)*

centricity of the two can be checked with a valve-seat dial gauge. The gauge is mounted in the valve guide and is rotated so that the indicator finger sweeps around the valve seat. Any eccentricity (or runout) of the seat shows on the gauge dial.

4. Testing Valve Seating Contact between the valve face and seat may be tested as follows: Mark lines with a soft pencil about 1/4 inch [6.35 mm] apart around the entire valve face. Then put the valve in place. While applying light force, rotate the valve half a turn to the left and then half a turn to the right. If this removes the pencil marks, the seating is good.

The seating can also be checked with prussian blue or other material that will coat the valve face. Coat the valve face lightly. Put the valve on its seat, and turn it while applying light force. If the coating appears all the way around the valve seat, the valve seat and guide are concentric with each other. Now, check the concentricity of valve face with valve stem. Remove the coating from valve and seat. Lightly coat the seat and then lightly rotate the valve on the seat. If the coating transfers all the way around the valve face, the valve face and stem are concentric. This check is similar to the check that uses a runout gauge.

♦ 35-29 VALVE-SPRING INSPECTION

Valve springs should be checked for proper tension, squareness, and installed height. A special valve-spring tester checks tension. To check for squareness, stand the spring closed-coil end down, on a flat surface. Hold a steel square next to the spring, as shown in Fig. 35-24. Rotate the spring slowly to see whether the top coil moves away from the square more than 1/16 inch [1.6 mm]. If the spring is excessively out of square, or has lost tension, discard it. One manufacturer's recommendation is to replace all valve springs during the complete valve job.

♦ 35-30 CAMSHAFT SERVICE

Camshaft removal varies from engine to engine. It is less complex in an OHC engine. The general procedure in a camshaft-in-block engine begins with removal of the radiator. Then remove the pulley from the crankshaft. Remove the gear or timing-chain cover. Detach the camshaft thrust plate (where present). Take off the camshaft sprocket and chain (where used). The distributor or oil pump (whichever has the driven gear) must be removed so that the gear will not interfere with camshaft removal.

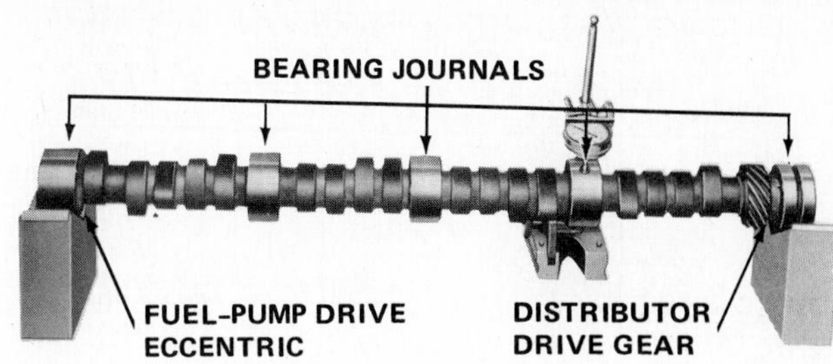

Fig. 35-25 Checking alignment of a camshaft with a dial indicator. *(Chevrolet Motor Division of General Motors Corporation)*

The pushrods must be removed so that the valve lifters can be raised up out of the way. Now, the camshaft can be pulled forward and out. Be careful to prevent the journals and cams from scratching the camshaft bearings. Support the rear of the camshaft as it is pulled out so that the bearings are not damaged. On some engines, you must remove the crankshaft to get to the camshaft and support it.

1. Checking the Camshaft Check for camshaft alignment by rotating it in V blocks and using a dial indicator (Fig. 35-25). Journal diameters should be checked with a micrometer, and the bearings with a telescope gauge. The two dimensions can be compared to determine whether bearings are worn and require replacement.

2. Checking for Cam Wear Figure 35-26 shows normal and abnormal cam wear. Normal cam wear is close to the center of the cam because the cam, in most engines, is slightly tapered. Also, the lifter foot is slightly spherical, or crowned, in shape. Therefore, the normal contact pattern is as shown in Fig. 35-27. If wear shows across the full width of the cam, a new camshaft is required. The lifter should also be checked (♦35-31). The general rule is: If new lifters are required, install a new camshaft. If a new camshaft is required, install new lifters.

Cam-lobe lift can be checked with the camshaft in or out of the engine. Figure 35-28 shows the check with the camshaft in the engine. The setup in Fig. 35-25 can also be used to measure the lobe lift.

3. Replacing Camshaft Bearings (Camshaft in Block) A special camshaft-bearing removal-and-installa-

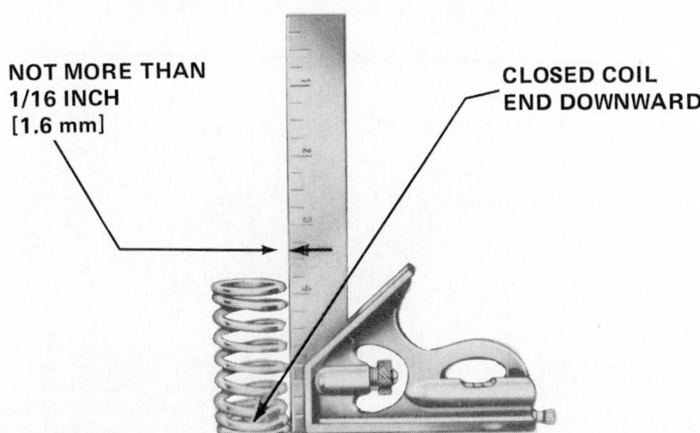

Fig. 35-24 Checking valve-spring squareness. *(Ford Motor Company)*

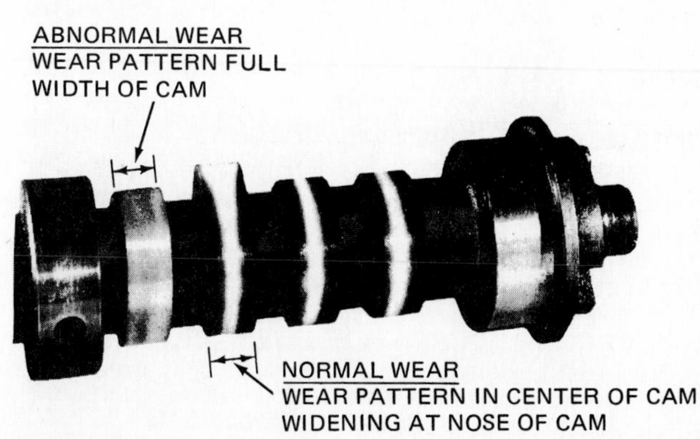

Fig. 35-26 Normal and abnormal cam wear. *(Oldsmobile Division of General Motors Corporation)*

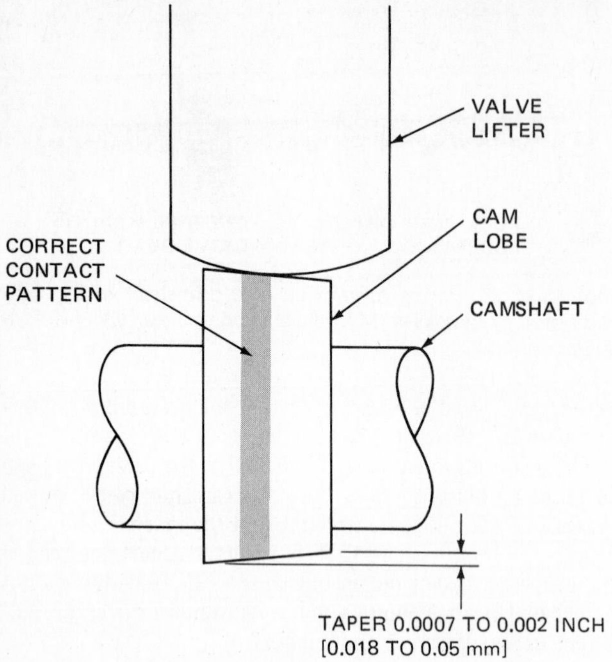

tion tool set is required to do this job. For some engines, the removal tool is threaded, and a nut is turned to remove the bearings. For others, a hammer is used to drive against the bar and force the bearings out. Oilholes in the new bearings should align with the oilholes in the block. Also, new bearings should be staked in place if the old bearings were staked. OHC bearings are removed by first removing the valve cover. Then the camshaft is removed. Now the bearings can be replaced.

4. Timing the Valves Crankshaft and camshaft gears or sprockets are marked for proper positions and correct valve timing (Figs. 12-3 and 12-4). To get to these markings, you must partly disassemble the front of the engine. Some engines have another marking system for checking valve timing. This marking is on the flywheel or vibration damper, near the ignition-timing markings. When this marking is visible or registers with a pointer, a designated valve should be just opening. Or it should have opened a specified amount. Valve action is observed by removing the valve cover.

5. Timing Gear and Chain Gear runout can be checked by mounting a dial indicator on the block. The indicating finger should rest on the side of the gear (Fig. 35-29). Runout will then be indicated as the gear is rotated. Gear backlash is measured by inserting a narrow thickness gauge between the meshing teeth. Excessive runout or backlash requires gear replacement.

Excessive slack in the timing chain indicates a worn chain, and possibly worn sprockets. Timing-chain wear can be checked without disassembling the engine. Rotate the crankshaft in its normal direction. Stop when the timing mark on the vibration damper is aligned with the TDC mark on the timing scale. Now remove the distributor cap, and rotate the crankshaft in the opposite direction. Stop the instant the distributor rotor begins to move. Look at the timing scale to see

VALVE LIFTER
CAM LOBE
CORRECT CONTACT PATTERN
CAMSHAFT
TAPER 0.0007 TO 0.002 INCH [0.018 TO 0.05 mm]

Fig. 35-27 Offset between the cam lobe and lifter face (which is crowned) gives a wide, centered contact area. (The taper of the cam and the crown of the lifter face are shown exaggerated.) *(Service Parts Division of Dana Corporation)*

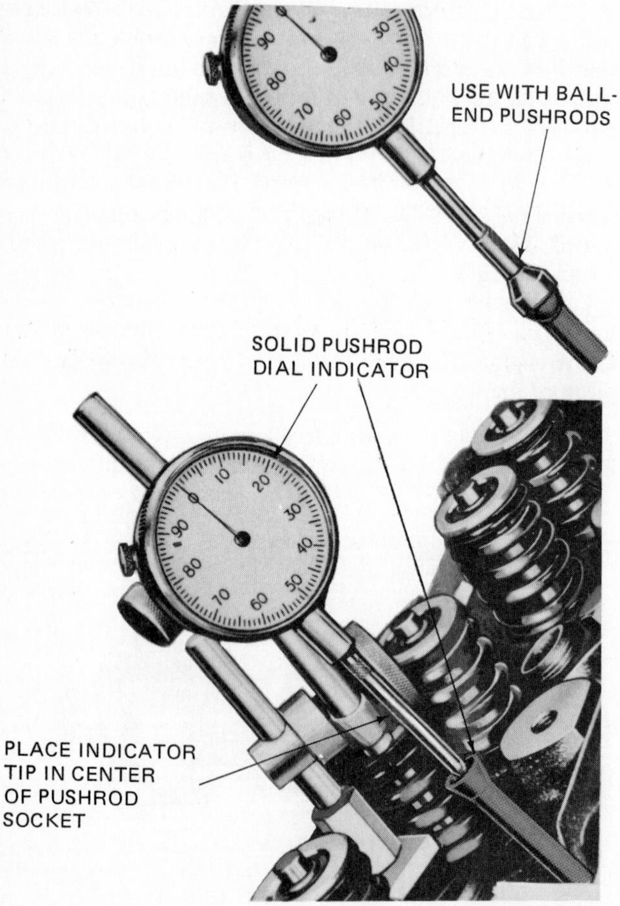

USE WITH BALL-END PUSHRODS
SOLID PUSHROD DIAL INDICATOR
PLACE INDICATOR TIP IN CENTER OF PUSHROD SOCKET

Fig. 35-28 Checking cam-lobe lift with the camshaft in the engine. *(Ford Motor Company)*

Fig. 35-29 Checking timing-gear runout, or eccentricity, with a dial indicator. *(Buick Motor Division of General Motors Corporation)*

which mark aligns with the timing mark on the vibration damper. This is the chain slack in degrees of crankshaft rotation. If the slack exceeds 10 degrees, the chain is excessively worn and should be replaced.

♦ 35-31 VALVE LIFTERS

The solid and the hydraulic valve lifters require different servicing procedures.

1. **Solid Lifter** Solid lifters are removed from the camshaft side on some engines. This requires camshaft removal as a first step. In most engines, lifters are removed from the valve or pushrod side. Lifters should be kept in order so that they can be reinstalled in the bores from which they were removed. Oversize valve lifters may be installed on some engines if the lifter bores have worn. Before this is done, the lifter bores must be reamed oversize.

2. **Hydraulic Valve Lifter** On some engines, a "leak-down" test is used to determine the condition of the hydraulic valve lifters. One way to make this test is to insert a thickness gauge between the rocker arm and valve stem. Then, note the time it takes the valve lifter to leak enough oil to seat the valve. As the valve seats, the thickness gauge becomes loose. This indicates the end of the test. If the leak-down time is too short, the valve lifter is defective.

A more accurate leak-down test is made with the lifter out of the engine and installed in a leak-down tester. With this tester, the time required for a standard weight on the end of a lever to force the lifter plunger to bottom is measured. If the time is too short, the lifter is defective.

To remove a hydraulic valve lifter from some engines, you must remove the intake manifold, valve cover, and rocker-arm assembly. Then the pushrod is taken out. On some engines with shaft-mounted rocker arms, the rocker arm can be moved by compressing the spring. This allows the pushrod to be removed. The rocker-arm assembly does not have to be taken off these engines.

A valve-lifter remover is shown in Fig. 35-30.

3. **Servicing Hydraulic Valve Lifters** If a hydraulic valve lifter is defective, it is probably cheaper to replace it than to disassemble and service it. To service the lifter, disassemble it and clean all parts in solvent. If any part is defective, the complete lifter should be replaced. On reassembly, fill the lifter with clean, light engine oil.

Work on only one lifter at a time so that you do not mix lifter parts. Each lifter should be reinstalled in the bore from which it was removed.

Careful Be very careful to keep everything clean when servicing and handling hydraulic valve lifters. It takes only one tiny particle of dirt to cause a lifter to malfunction.

4. **Checking Lifter Foot** As shown in Fig. 35-27, the foot of the lifter should be slightly spherical, or crowned. If it is worn or pitted, it can sometimes be reground and then reused. However, most lifters that are worn or pitted must be replaced. To check the lifter, place a straightedge across the base, or foot, of the lifter. Then hold the lifter up to the light. The proper crown will show light through at each side of the lifter.

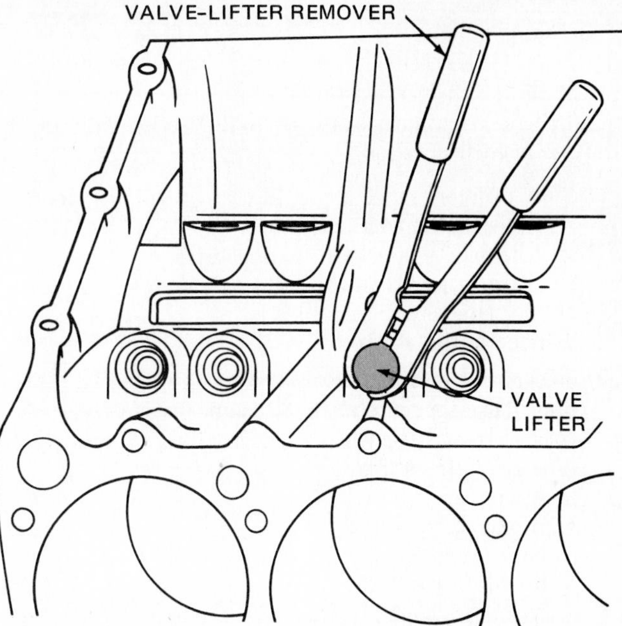

Fig. 35-30 Removing a valve lifter. *(Oldsmobile Division of General Motors Corporation)*

♦ 35-32 REMOVING AND INSTALLING INTAKE MANIFOLDS

Remove the throttle body or carburetor. Handle it with care to avoid damaging it or spilling gasoline from the float bowl. Disconnect vacuum lines, emission control hoses, and any other pipe or wire connected to the manifold. Remove the nuts or bolts, and take the manifold off.

When installing the intake manifold, be sure all old gasket material has been removed from the manifold and cylinder head. Use new gaskets. Tighten nuts or bolts to the proper torque and in the proper sequence (Fig. 35-31).

Instead of using preformed gaskets, some intake manifolds are installed with room-temperature-vulcanizing (RTV) silicone rubber gasket compound (♦5-20). It is squeezed out of the tube onto the gasket surfaces (Fig. 5-26). Follow the instructions on the tube. When the intake manifold is installed, the bead of silicone rubber spreads to form the gasket.

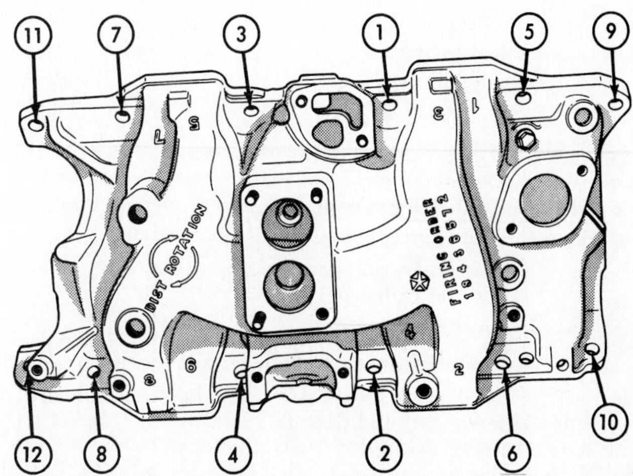

Fig. 35-31 Sequence for tightening the intake-manifold bolts on a V-8 engine. *(Chrysler Corporation)*

♦ REVIEW QUESTIONS ♦

Select the *one* correct, best, or most probable answer to each question. You can find the answers in the section indicated at the end of each question.

1. With an interference angle, the valve would be faced at an angle that is (♦35-25)
   a. 1 degree greater than the seat angle
   b. 1 degree less than the seat angle
   c. the same as the seat angle
   d. none of the above
2. Mechanic A says that a valve-spring shim is used to improve valve-spring rotation. Mechanic B says that a valve-spring shim is used to correct installed spring height. Who is right? (♦35-26)
   a. A only
   b. B only
   c. both A and B
   d. neither A nor B
3. On pushrod engines, valve-tappet clearance is measured between the valve stem and the (♦35-12)
   a. rocker arm
   b. valve retainer
   c. adjustment screw in lifter
   d. adjustment screw in rocker arm
4. When a new valve with a 45 degree face is placed in a 45 degree valve seat, the valve head is too high above the seat. To raise the seat contact on the valve face, the seat should be ground with a (♦35-28)
   a. 60 degree stone
   b. 45 degree stone
   c. 30 degree stone
   d. 15 degree stone
5. To adjust valve clearance in the pushrod engine, turn the adjustment screw in the (♦35-12)
   a. rocker arm
   b. pushrod
   c. valve lifter
   d. valve stem
6. When doing a valve job, check the valve springs for (♦35-29)
   a. tension
   b. squareness
   c. installed height
   d. all of the above
7. On Ford pushrod engines with hydraulic valve lifters, insufficient clearance after bleed-down requires correction by (♦35-14)
   a. turning the adjustment screw
   b. installing shorter pushrods
   c. grinding the valves
   d. grinding the camshaft
8. On Chevrolet pushrod engines with hydraulic valve lifters, lifter adjustment is made by backing off the adjustment nut until the pushrod is loose, tightening it until looseness is gone, and then (♦35-16)
   a. backing off the nut one turn
   b. installing longer pushrods
   c. turning the nut down one turn
   d. none of the above
9. An engine that requires no valve adjustment uses (♦35-11)
   a. free valves
   b. hydraulic valve lifters
   c. L-head valves
   d. I-head valves
10. When inspecting the width of the margin on a set of valves from an engine, the mechanic finds that all but two of the valves have a 1/32-inch margin. One valve has a 1/64-inch margin. The second valve has a margin of less than 1/64-inch. The mechanic should (♦35-25)
    a. replace both valves with less than 1/32-inch margin
    b. replace the one valve with less than 1/64-inch margin
    c. replace all of the valves because the margins are too small
    d. reinstall all of the valves
11. If the cylinder-block boltholes are not cleaned out, the head bolts may bottom in the boltholes and prevent (♦35-19)
    a. head-to-block sealing
    b. normal valve action
    c. adequate oil-ring action
    d. loss of compression
12. Mechanic A says that if the valve seat is too high, it should be lowered by using a 15 degree stone. Mechanic B says that if the valve seat is too high, it should be lowered by using a 75 degree stone. Who is right? (♦35-28)
    a. A only
    b. B only
    c. both A and B
    d. neither A nor B
13. Exhaust-valve clearance usually is greater than intake-valve clearance. Mechanic A says that this is because exhaust valves usually run hotter than intake valves and therefore expand more. Mechanic B says that this is due to valve overlap. Who is right? (♦35-12)
    a. A only
    b. B only
    c. both A and B
    d. neither A nor B
14. If a valve is ground down so much that its margin is lost, the valve will (♦35-25)
    a. run too hot
    b. run too high
    c. run too cool
    d. run too loose
15. When you measure the assembled valve-spring height (installed height), the valve must be (♦35-26)
    a. open
    b. closed
    c. in any position
    d. none of the above

CHAPTER 36
Engine Service: Connecting Rods, Rod Bearings, Pistons, and Rings

After studying this chapter, and with proper instruction and equipment, you should be able to:

1. Remove the cylinder ring ridge.
2. Replace piston rings.
3. Replace connecting-rod bearings.
4. Install piston pins.
5. Diagnose the cause of bearing failure.
6. Diagnose the cause of piston failure.
7. Measure bearing clearance with Plastigage.
8. Check for bent or twisted connecting rods.

♦ 36-1 PREPARING TO REMOVE RODS

Connecting rods and pistons are removed from the engine as assemblies. Removing, servicing, and installing connecting rods requires about 5 to 8 hours, depending on the type of engine. About 3 additional hours are required to install new piston rings. Additional time is needed for such services as piston-pin-bushing replacement. On most engines, the piston-and-rod assemblies are removed from the top of the engine. Therefore, the first step is to remove the cylinder head (♦35-19). Cylinders should be examined for wear. If wear has taken place, there will be a ridge at the top of the cylinder. This ridge, called the *ring ridge,* marks the upper limit of piston-ring travel. If this ridge is not removed, the top ring could jam under it as the piston is moved upward. This could break the rings or the lands on the piston (Fig. 36-1). Therefore, any ridge must be removed.

A quick way to check for a ring ridge is to see whether your fingernail catches under it. If your fingernail catches on the ring ridge, so will the piston rings. A more accurate way to check is to use an inside micrometer. Measure the diameter on the ring ridge and then immediately below the ring ridge. If the difference is more than 0.004 inch [0.1 mm], the ridge must be removed.

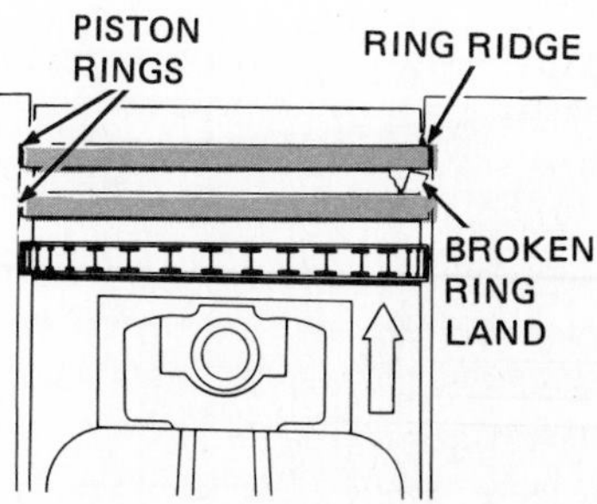

Fig. 36-1 When the ring ridge is not removed, forcing the piston out of the cylinder will often break the piston-ring lands. *(ATW)*

♦ 36-2 REMOVING RING RIDGE

To remove the ring ridge, use a ring-ridge remover, as shown in Fig. 36-2. With the piston near BDC, stuff a cloth into the cylinder, and install the ridge remover.

There are several different kinds of ridge removers. Follow the instructions for the ridge remover you use.

Adjust the cutter blades to take off just enough metal to remove the ridge. Cover the other cylinders to keep cuttings from getting into them. Rotate the tool to cut the ridge away.

Careful Turn the ridge remover by hand, not with an impact wrench! Do not remove too much metal. Do not undercut the top of the cylinder deeper than the material next to the ring ridge. Do not run the cutting tool above the cylinder. This would taper the edge.

Remove the tool, take the cloth out, and wipe the cylinder clean. Repeat the procedure for the other cylinders.

♦ 36-3 REMOVING OIL PAN

The oil pan must be removed so that the connecting rods can be detached from the crankshaft. First, drain the engine oil (Fig. 20-12). On some cars, the steering idler arm or other steering linkage must be removed. Note how the linkage is attached, and the number and location of any shims (if used). On some cars, the oil pan is easier to remove if the engine mounting bolts are removed and the engine is raised slightly. Other parts may require removal before the oil pan can be taken off. These include the exhaust pipe, dipstick tube, brake-return spring, and starting motor. Then, the nuts or bolts holding the oil pan to the engine cylinder block can be removed. Steady the pan as the last two nuts or bolts are removed so that it does not drop. If the pan does not clear the crankshaft, turn the crankshaft slightly.

If the oil pan does not break loose when the last bolt is out, tap the sides of the oil pan with a rubber mallet. If the pan still sticks, carefully force the claw or flat edge of a pry bar or scraper between the edge of the pan and cylinder block. Try to get the flat edge on the pan side of the gasket to avoid scratching the block. You can tap the pry bar with a hammer to help free the oil pan.

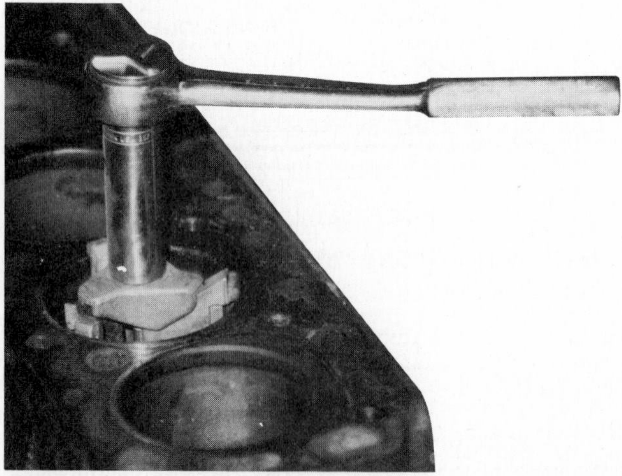

Fig. 36-2 Ring-ridge remover in place in the top of a cylinder. As the tool is turned in the cylinder, cutting blades remove the ring ridge. *(ATW)*

Many engines have metal reinforcements under the corner bolts of the oil pan. They help seal around the rear main bearing and the bottom of the timing cover. Don't lose these as you remove the bolts.

Clean the oil pan, oil screen, and oil pump before installing the oil pan. Make sure that the gasket material is scraped off the pan and block gasket surfaces. Check the flatness of the oil-pan gasket surfaces. Make sure that the boltholes have not been dished out by overtightening of the bolts. The gasket surfaces can be straightened by laying the oil pan on a flat surface and tapping the gasket-surface flanges with a hammer.

Apply new gasket cement, if specified. Lay the gasket, or gaskets, in place on the oil pan. Be sure that the boltholes in the gasket and pan line up. Install the pan, and tighten the bolts or nuts to the proper torque.

Oil pans can be installed with formed-in-place gasket material (♦5-20).

♦ 36-4 REMOVING PISTON-AND-ROD ASSEMBLIES

Now you are ready to remove the piston-and-rod assemblies.

Careful Handle pistons and rods with care. They can be easily damaged. Never clamp a rod tightly in a vise. This can bend the rod and ruin it. Never clamp a piston in a vise. This can nick or break the piston. Do not allow the pistons to hit against each other or against other metal surfaces. Distortion of the piston or nicks in the soft aluminum piston material may result from careless handling. These can ruin the piston.

With the head and pan off, rotate the crankshaft so that the piston of number 1 cylinder is near the bottom. Examine the rod and rod cap for identifying marks. If none can be seen, use a small watercolor brush and a little white metal paint, or a marking pen, to mark a "1" on the rod cap and the rod. Marks are needed to make sure that the parts go back into the cylinders from which they were removed. Each piston should also be numbered.

Careful Do not mark the rod caps and rods with metal numbering dies or a center punch and hammer. This can distort or break the rods and caps.

Remove the rod nuts and caps. Slide the rod-and-piston assembly up into the cylinder, away from the crankshaft. Use guide sleeves on the rod bolts, as shown in Fig. 36-3. They prevent the bolt threads from scratching the crankshaft journals. Also, the long handle permits easy removal and replacement of the piston-and-rod assembly. Short pieces of rubber hose, split and slipped over the rod bolts, will protect the crankshaft journals.

Turn the crankshaft as you go from rod to rod, so you can reach the rod nuts. Many mechanics recommend that you remove each cap and piston-and-rod-assembly in sequence. As you do so, lay the cap and rod out on a cloth spread on the bench. Or put them in a wooden piston box. Make sure each rod, cap, and piston is marked with the number of the cylinder from which it was removed.

♦ 36-5 SEPARATING RODS AND PISTONS

There are five basic piston-pin arrangements, as shown in Fig. 11-30. The rods and pistons are separated by removing

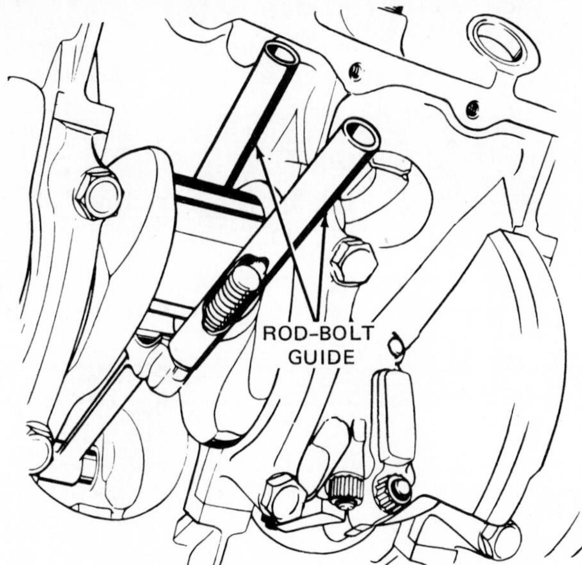

Fig. 36-3 Using short pieces of hose or guide sleeves to protect the connecting-rod journals. *(Oldsmobile Division of General Motors Corporation)*

the piston pins. If the pin is free-floating, the pin is removed by removing the retainer rings and sliding the pin out. If the pin is locked to the connecting rod or piston with a lock bolt, loosen the lock bolt and slide the pin out.

If the pin is a press fit in the connecting rod, the pin must be pressed out. This requires attaching a special tool to the piston and pin, as shown in Fig. 36-4. Then the shop press is used to force the pin out.

The rods and pistons are usually not separated unless new pistons are to be installed on the old rods. Check the rods, rings, and pistons as explained in the following sections.

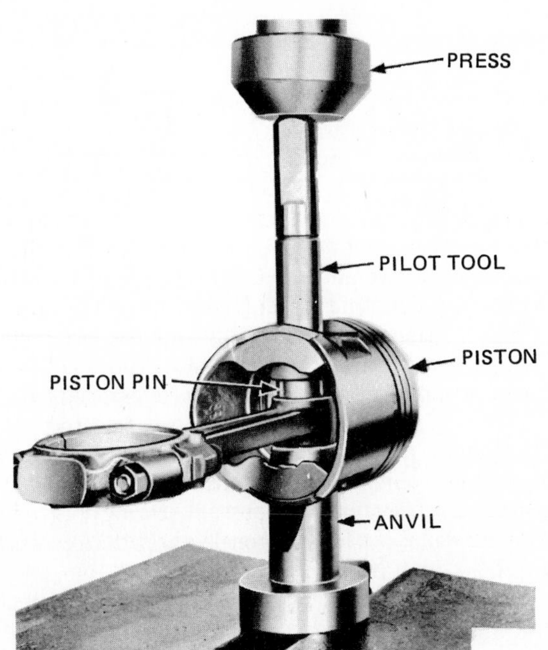

Fig. 36-4 Piston-pin removal-and-installation tool being used with a press to remove the pin from a piston. *(Chrysler Corporation)*

Careful Avoid nicking or scratching the pistons, piston rings, or rods (♦36-4).

♦ 36-6 ATTACHING RODS AND PISTONS

After rods and pistons have been cleaned, serviced, and checked, place them on a clean bench in their engine order. Make sure parts match as in the original assembly. The pistons should then be attached to the rods with the piston pins, as follows. Make sure the piston is in the correct position on the rod as the two are attached. On many engines, the piston notches face to the front of the engine. Also, the rod oilhole faces toward the outside of the block.

To attach the piston and connecting rod where a lock bolt is used, put the pin through the piston and rod. Tighten the lock bolt to hold the parts together. On the free-floating type, install the pin and retainer rings.

On the type with a press fit of the pin in the rod, a special tool (Fig. 36-5) is required to press the pin in. Force is applied to the tool in a shop press to push the pin into place. Some shops heat the rods in a heater before installing a press-fit pin. Plymouth recommends a fit test after the pin has been installed. This is done by placing the assembly in a vise. Then a torque wrench is used to apply 15 lb-ft [20 N-m] of torque to the nut on the end of the tool. If this amount of torque causes the connecting rod to move down on the piston pin, the press fit is too loose. The connecting rod must be discarded. If the rod does not move, the fit is satisfactory.

♦ 36-7 INSTALLING PISTON-AND-ROD ASSEMBLIES

After rods are reattached to their pistons, the piston rings are installed (♦36-24). Then the assemblies go back into the engine. Rings should be positioned so that the ring gaps are uniformly spaced around the piston, or as specified by the

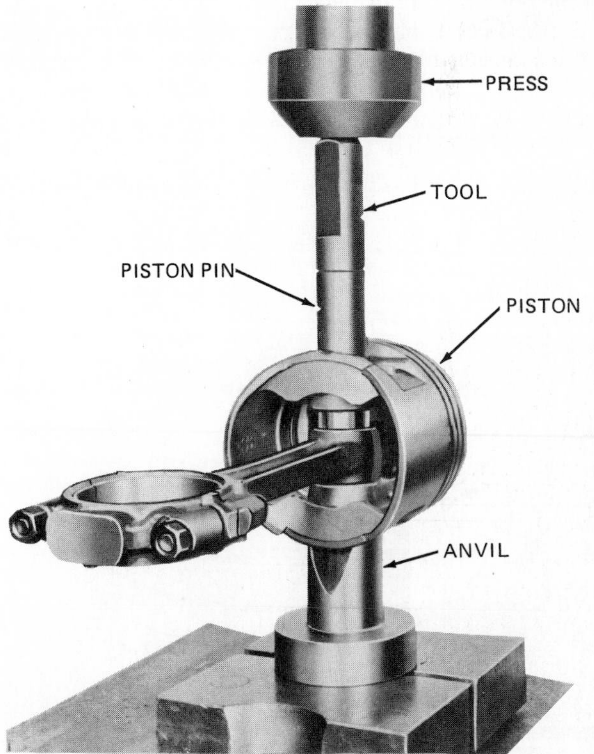

Fig. 36-5 Installing a piston pin that is a press fit in the connecting rod. *(Chrysler Corporation)*

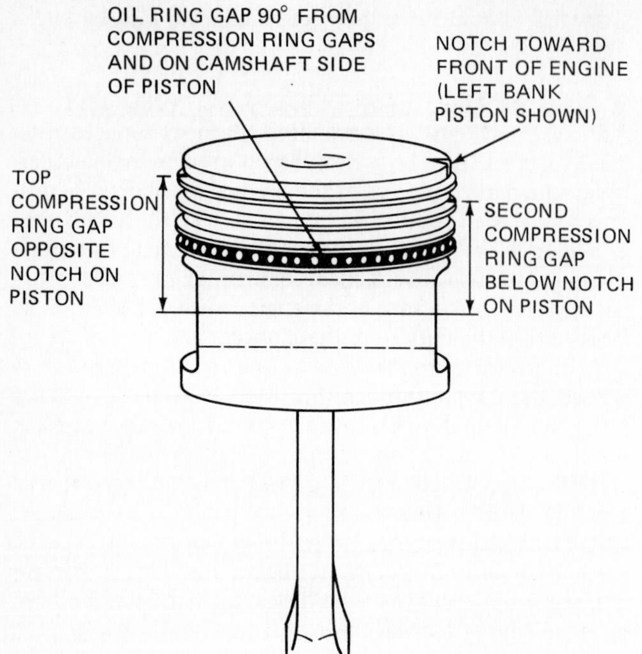

Fig. 36-6 Proper positioning of piston-ring gaps for one engine. *(Chevrolet Motor Division of General Motors Corporation)*

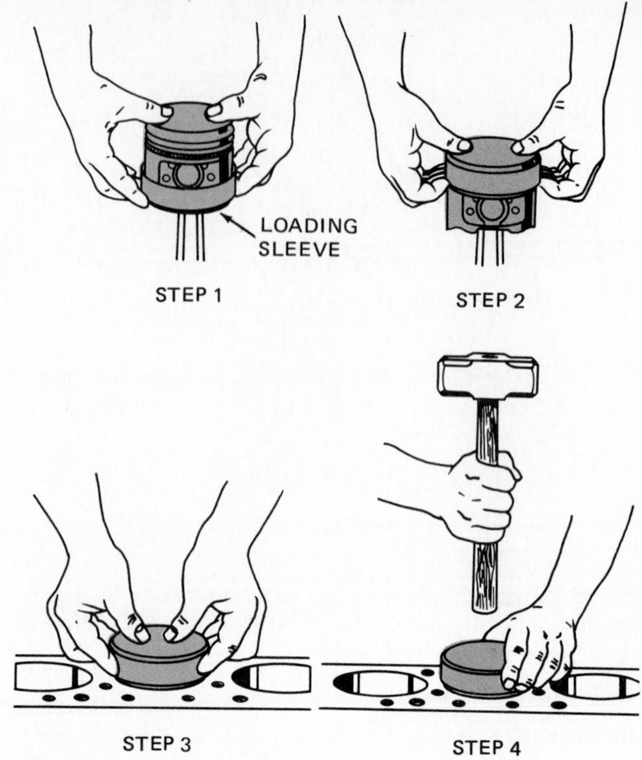

Fig. 36-8 Using a loading sleeve to install a piston-and-ring assembly in the cylinder. *(Chrysler Corporation)*

manufacturer. For example, Fig. 36-6 shows a Chevrolet recommendation.

Dip the piston assembly above the piston pin in SAE30 oil. Drain excess oil. Use a loading sleeve or piston-ring compressor to compress the rings into the piston-ring grooves (Figs. 36-7 and 36-8). Install guide sleeves on the rod bolts (Fig. 36-3), or cover the rod bolts with rubber hose. Then push the piston down into the cylinder. Tapping the head of the piston with the wooden handle of a hammer helps get the piston started. Make sure the assembly is installed with the piston facing in the right direction. Many pistons have a notch or other mark that should face toward the front of the engine (Fig. 36-7). On V-type engines, the connecting rods must be installed with the oil spit holes in the rods facing the opposite cylinders.

Attach the rod cap with the nuts turned down lightly. Then tap the cap on its crown lightly, to help center it. Tighten the nuts to specifications with a torque wrench.

Bearing clearances must be checked (♦36-13).

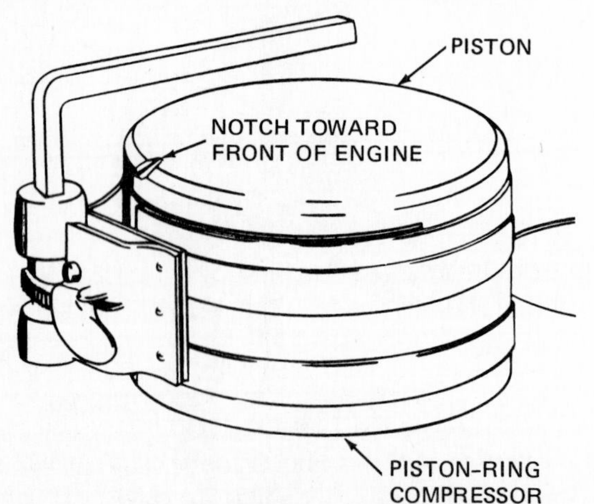

Fig. 36-7 Piston-ring compressor installed on a piston, compressing the rings into the ring grooves of the piston. *(Oldsmobile Division of General Motors Corporation)*

♦ 36-8 CHECKING ROD SIDE CLEARANCE

Make sure that the rods are centered on the crankshaft crankpins. If a rod is offset to one side, the rod-and-piston assembly has probably been installed incorrectly. It has been turned 180 degrees from its correct position. Also, offset could mean a bent rod (♦36-9). Clearance between connecting rods on V-type engines should also be checked (Fig. 36-9). Incorrect side clearance means a bent rod.

♦ 36-9 CHECKING CONNECTING RODS

After rods are detached from the pistons, the rod and rod caps should be cleaned and inspected. Blow out the oilholes in the rods with compressed air.

Inspect the rod big-end bearings (♦36-12). If the rod has a bushing in the small end, check its fit with the piston pin. If it is not correct, service is required (♦36-10).

Check rod alignment. Figure 36-10 is an exaggerated view showing the effects of a misaligned connecting rod. Heavy loading at points A and B on the bearing would cause bearing failure at these points. The heavy-pressure spots C and D on the piston cause heavy wear and possibly scoring of the piston and cylinder wall. Look for a diagonal wear pattern across the faces of the pistons. If any are found, the piston, pin, and rod should all be discarded. A misaligned (bent or twisted) connecting rod may cause piston slap.

Examine the rod bearings for side wear. If a rod is out of line, the upper bearing half will wear on one side and the lower half will wear on the other side (Fig. 36-10).

To accurately check rod alignment out of the engine, you

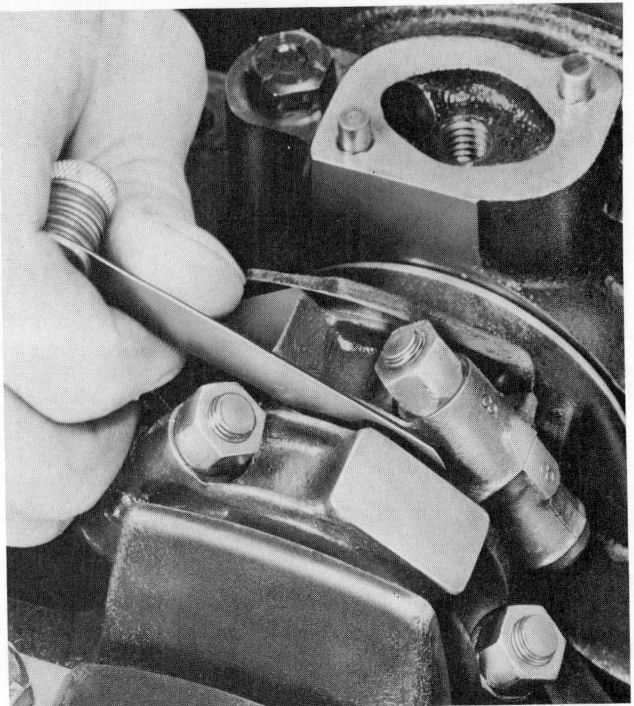

Fig. 36-9 Checking connecting-rod side clearance on a V-type engine. *(Ford Motor Company)*

need a special alignment fixture. This fixture has an arbor and a faceplate. With the rod on the arbor and the piston pin in the rod, a V block is placed on the pin and the rod is moved back and forth across the face plate. This shows any misalignment.

If the rod is out of line, check the crankpin for taper (♦36-14). A tapered crankpin can cause the rod to bend. Bent rods should be replaced. They cannot be satisfactorily straightened.

With the cap on and torqued to specifications, check the

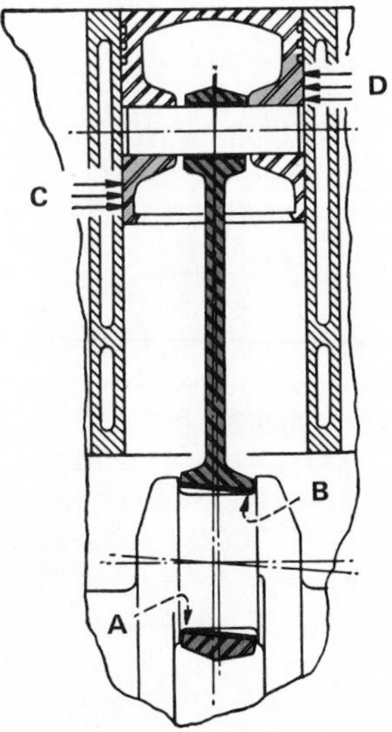

Fig. 36-10 Heavy pressure areas due to a bent connecting rod. The bent condition is exaggerated. Areas of heavy pressure (A, B, C, and D) wear rapidly so that early bearing failure results.

big-end bore for out-of-round. If the bore is out of round, the rod must be reconditioned or replaced.

♦ 36-10 PISTON-PIN BUSHINGS IN RODS

When the rod has a piston-pin bushing (Fig. 11-30), check the fit of the pin. If the fit is correct, the pin will not drop (of its own weight) through the bushing when held vertical. A light push will be required to force the pin through. If the fit is too loose, the bushing should be replaced. Sometimes a worn bushing is reamed or honed for an oversize pin.

Aluminum pistons usually have no bushings. They are supplied with prefitted piston pins as a matched set. If the pin is worn, or is too loose a fit in the piston, a new pin-piston set is required.

On some rods, the bushing cannot be replaced. If the bushing is so worn that it cannot be reamed or honed for an oversize pin, the complete rod must be replaced. On other rods, worn bushings can be replaced. The new bushings can be reamed or honed to fit the present pins (if they are in good condition) or new standard-sized pins.

♦ 36-11 CONNECTING-ROD BEARINGS

Connecting-rod big-end bearings are *precision-insert* bearings. The insert-type bearing is usually not adjustable. However, it can be replaced, if the rod, crankpin, and other engine components are in good condition. When a rod bearing fails, an analysis should be made to determine the cause. Then the cause can be eliminated so that the failure will not be repeated (♦36-12).

♦ 36-12 ANALYSIS OF BEARING FAILURES

Various types of bearing failure are described below.

1. Bearing Failure Due to Lack of Oil (Fig. 36-11A) When insufficient oil flows to a bearing, actual metal-to-metal contact results. The bearing overheats, and the bearing metal melts or is wiped out of the bearing shell. Welds may form between the rotating journal and bearing shell. There is a chance that the engine will "throw a rod." This means the rod will "freeze" to the crankpin and break, and parts of the rod will punch a hole through the engine block. Oil starvation of a bearing could result from clogged oil lines, a defective oil pump or pressure regulator, or insufficient oil in the crankcase. Also, bearings with excessive clearance may pass all the oil from the pump, so other bearings are starved and will fail.

2. Fatigue Failure of Bearings (Fig. 36-11B) Repeated application of loads on a bearing fatigue the bearing metal. It starts to crack and flake out. Craters, or pockets, form in the bearing. As more and more of the metal is lost, the remainder carries a greater load and fatigues at a faster rate. Then complete bearing failure occurs.

Fatigue failure seldom occurs under average operating conditions. However, certain conditions will cause this type of failure. For example, if a journal is worn out of round, the bearing will be overstressed with every crankshaft revolution. Also, if the engine is idled or operated at low speed much of the time, the center part of the upper rod-bearing half will carry most of the load and will "fatigue out." If the engine is "lugged" by operating at maximum torque with wide-open throttle, then most or all of the upper bearing half will fatigue out. High-speed operation tends to cause fatigue failure of the lower bearing half.

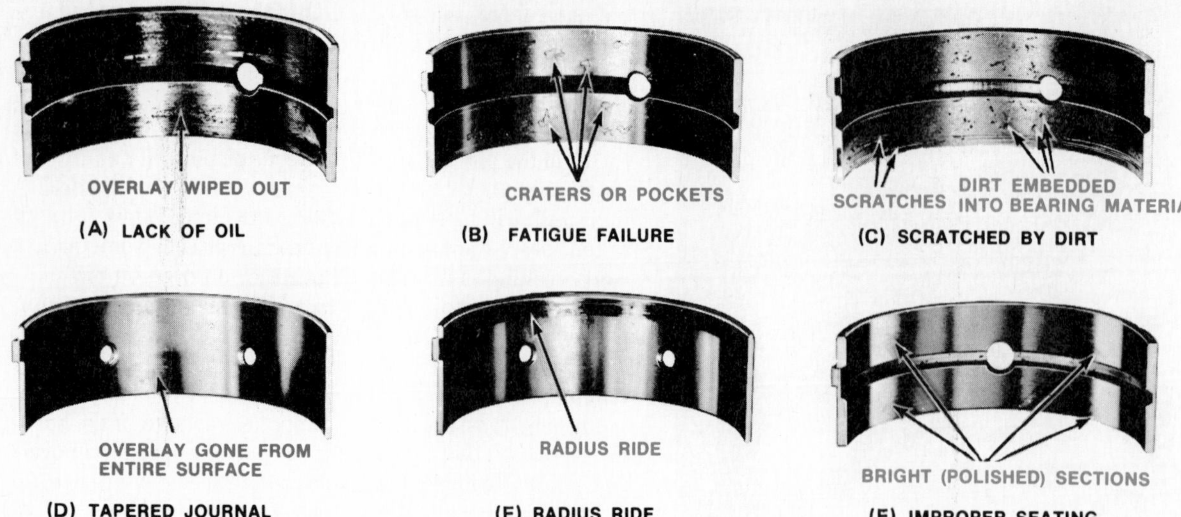

Fig. 36-11 Types of engine-bearing failure. The appearance of a bearing usually indicates the cause of failure. *(Ford Motor Company)*

3. Bearing Scratched by Dirt in the Oil (Fig. 36-11C) Embeddability (♦11-21) enables a bearing to protect itself by allowing particles to embed in the bearing. Then they will not gouge out bearing material or scratch the rotating journal. Figure 36-12A shows what happens when a particle embeds. The metal is pushed up around the particle, reducing oil clearance in the area. Usually the metal can flow outward enough to restore adequate oil clearance. However, if the dirt particles are too large, they do not embed completely. They are carried with the rotating journal, gouging out scratches in the bearing. Also, if the oil is very dirty, the bearing becomes overloaded with particles. In either case, bearing failure soon occurs.

4. Bearing Failure due to Tapered Journal (Fig. 36-11D) If the journal is tapered, one side of the bearing carries most or all of the load. This side will overheat and lose its bearing metal. With a tapered journal, both bearing halves fail on the same side. With a bent rod, failure will be on opposite sides (A and B in Fig. 36-10).

5. Bearing Failure from Radius Ride (Fig. 36-11E) If the journal-to-crank-cheek radius is not cut away sufficiently when the crankshaft is machined, the edge of the bearing rides on the radius. This causes cramming of the bearing, possibly poor seating, rapid fatigue, and early failure. Radius ride is most likely to occur on a reground crankshaft.

6. Bearing Failure from Improper Seating (Fig. 36-11F) Improper seating of the bearing shell in the bore causes local high spots where oil clearances are too small. Figure 36-12B shows what happens when particles of dirt are trapped between the bearing shell and the bearing bore. This reduces oil clearance (as at X). Also, an air space exists which prevents proper cooling of the bearing (A). The combination can lead to premature bearing failure.

7. Bearing Failure from Ridging Crankpin ridging, or "camming," may cause failure of a partial-oil-groove type of replacement bearing installed without removal of the ridge. The ridge forms on the crankpin because of uneven wear between the part of the crankpin in contact with the partial oil groove and the part that runs on the solid bearing. The original bearing wears around this ridge. However, when a new bearing is installed, the center zone may be overloaded (at the ridge) and may soon fail. A ridge so slight that it can hardly be detected (except with a micrometer) may be enough to cause failure. Failures of this sort have been reported in engines having ridges of less than 0.001 inch [0.025 mm].

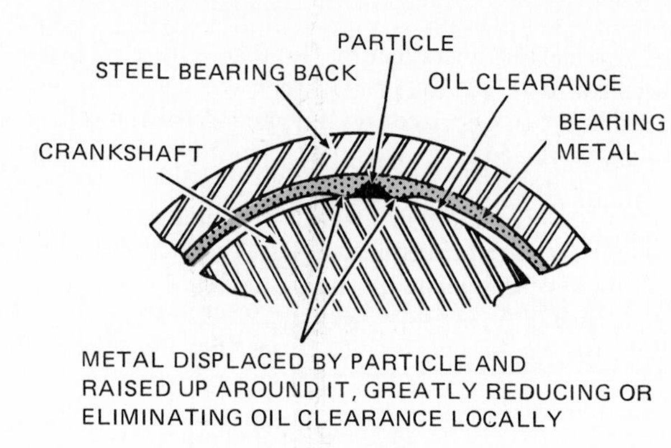

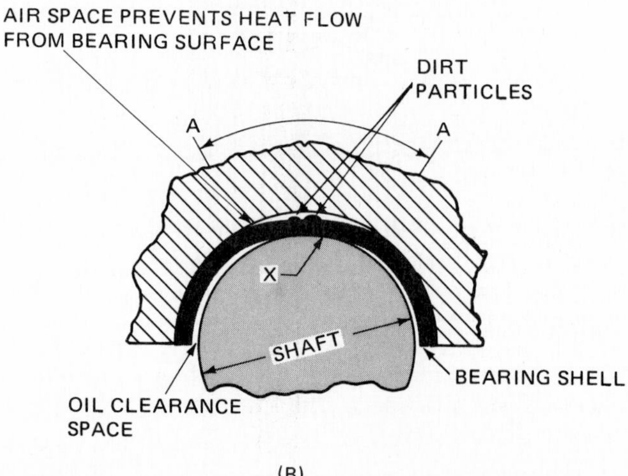

Fig. 36-12 (A) Effect of a particle embedded in the bearing metal. (B) Effect of dirt particles left under bearing shell during installation. *(Federal-Mogul Corporation)*

♦ 36-13 CHECKING CONNECTING-ROD-BEARING CLEARANCE

Careful Before installing new bearings, the crankpins should be checked for taper and out-of-round (♦36-14).

The clearance of the crankshaft bearings can be checked with Plastigage, shim stock, or with micrometer and telescope gauge.

1. Plastigage Plastigage is a plastic thread that flattens when pressure is applied to it. A strip of the material is put into the bearing cap. The cap is installed, and the rod nuts are tightened to the specified torque. Then, the cap is removed, and the amount of flattening is measured. If the Plastigage is flattened only a little, then oil clearances are large. If it is flattened considerably, oil clearances are small. Actual clearance is measured with the scale printed on the Plastigage package (Fig. 36-13).

The bearing cap and crankpin should be wiped clean of oil before the Plastigage is used. The crankshaft should be turned so that the crankpin is about 30 degrees back of BDC. Do not move the crankshaft while the cap nuts are tight. This would further flatten the Plastigage and throw off the clearance measurement.

2. Shim Stock The shim-stock method is seldom used today because the Plastigage method is faster and easier. When using shim stock, lay a strip of 0.001-inch [0.025-mm] stock in the cap, install the cap, and tighten the nuts to see whether the rod tightens on the crankpin. Repeat the procedure, adding strips until the rod tightens. The amount of clearance is the thickness of the shim stock required to barely lock the rod.

3. Micrometer and Telescope Gauge Check the crankpin diameter with a micrometer. Check the bearing diameter (cap in place) with a telescope gauge and micrometer (or an inside micrometer). Compare the two diameters to determine the difference, or bearing clearance. At the same time, the crankpin can be checked for taper or out-of-round. Measure the diameter at several places along the crankpin to check for taper. Then measure around the crankpin to check for out-of-round. Typical allowable taper or out-of-round is 0.001 inch [0.025 mm].

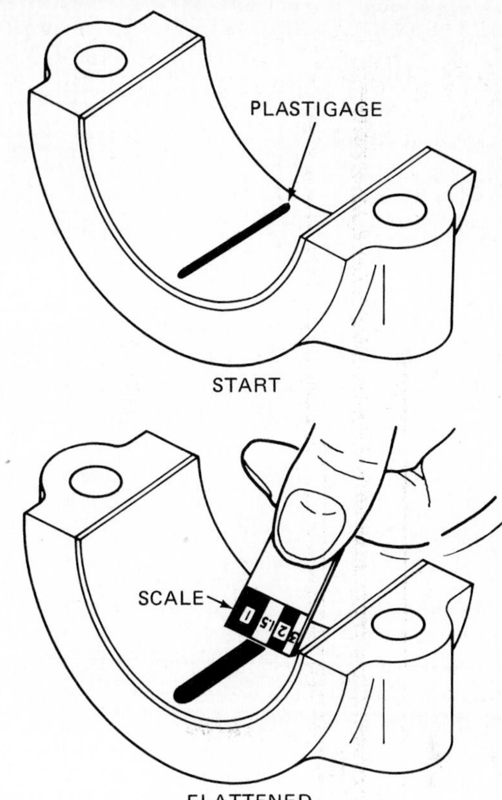

Fig. 36-13 Checking bearing clearance with Plastigage. Top, Plastigage in place before tightening the cap; bottom, measuring the amount of flattening (or bearing clearance) with the scale on the Plastigage package. *(Chrysler Corporation)*

♦ 36-14 INSTALLING CONNECTING-ROD BEARINGS

New connecting-rod bearings are required if the old bearings are defective (♦36-12) or have worn so much that clearances are excessive. They are also required if the crankpins have worn out of round or have tapered so much that they must be reground. Then new undersize bearings are required. Engine rebuilders usually replace the bearings in an engine when it is rebuilt whether or not the old bearings are in bad condition. Slightly undersize bearings may be installed when the clearance is excessive but the crankpins are not tapered or out-of-round.

1. Inspecting Crankpins Crankpins should always be inspected with a micrometer for taper and out-of-round. Measurements should be taken at several places along the crankpin to check for taper. Diameter should be checked all the way around for out-of-round. If crankpins are out-of-round or tapered more than 0.001 inch [0.025 mm], the crankshaft must be replaced or the crankpins reground. Bearings running against taper or out-of-round of more than 0.001 inch [0.025 mm] will fail prematurely.

2. Installing New Bearings When new bearings are to be installed, make sure your hands, the workbench, tools, and all engine parts are clean. Keep the new bearings wrapped up until you are ready to install them. Then handle them carefully. Wipe each with a fresh piece of cleaning tissue just before installing it. Be sure that the bores in the cap and rod are clean and not excessively out of round. (Some manufacturers recommend a check of bore symmetry with the bearing shells removed. The cap should be attached with nuts drawn up to specified torque. Then a telescope gauge and micrometer or a dial-indicating bore gauge can be used to check the bore. Maximum allowable bore out-of-round is 0.001 inch [0.025 mm].) Then put the bearing shells in place. Do *not* oil the back of the bearing during installation. If the bearing has locating tangs, make sure that the tangs enter the notches in the rod and cap. Important points about bearing spread and crush are discussed below. Check clearance after installation (♦36-13).

Do not attempt to correct the clearance by filing the rod cap. This destroys the original relationship between cap and rod and will lead to early bearing failure.

3. Bearing Spread Bearing shells are usually manufactured with "spread." The shell diameter is slightly greater

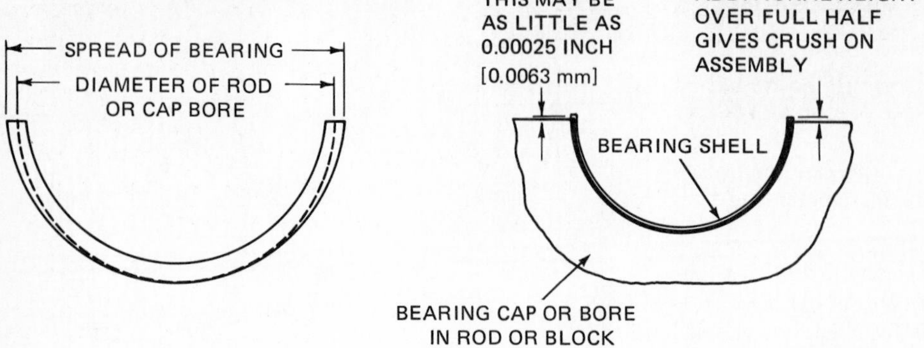

Fig. 36-14 Left, bearing spread; right, bearing crush.

than the diameter of the rod cap or rod bore into which the shell will fit (Fig. 36-14). When the shell is installed into the cap or rod, the bearing half snaps into place and holds its seat during later assembling operations.

4. Bearing Crush To make sure that the bearing shell will seat into its bore in the rod cap or rod when the cap is installed, the bearings have "crush" (Fig. 36-14). They are manufactured to have a slight additional height over a full half. This additional height must be crushed down when the cap is installed. Crushing down the additional height forces the shells into the bores in the cap and rod. This ensures firm seating and snug contact with the bores.

Never file off the edges of the bearing shells in an attempt to remove crush. When you select the proper bearings for an engine (as recommended by the engine manufacturer), they will have the correct crush. Precision-insert bearings should not be tampered with to make them "fit better." This usually leads only to rapid bearing failure.

PISTONS AND RINGS

♦ 36-15 PISTON SERVICE

After the piston-and-rod assemblies are removed from the engine, the pistons and rods should be separated (♦36-5). Then the rings can be removed from the pistons. The rings can also be removed from the pistons before the pistons and rods are separated. A piston-ring expander can be used for ring removal. The tool has two small tangs that catch under the ends of the ring (Fig. 36-15). When force is applied to the tool handles, the ring is opened slightly so it can be lifted out of the ring groove and off the piston. Install new rings during an engine overhaul. Once the ring break-in coating and tool marks are worn off, the ring will not reseat itself if it is reinstalled.

♦ 36-16 PISTON CLEANING

Remove carbon and varnish carefully from piston surfaces. Do not use a caustic cleaning solution or wire brush! These could damage the piston-skirt finish. You may decide to reinstall the pistons in the engine. Therefore, do not damage them. Use the cleaning method provided in your shop to clean the pistons. Clean ring grooves with a clean-out tool. You can also use the end of a broken piston ring filed to a sharp edge. Oil-ring slots, or holes, must be clean so that oil can drain back through them. Use a drill of the proper size. Do not remove metal when cleaning the slots or holes.

♦ 36-17 PISTON INSPECTION

Examine the pistons carefully for wear, scuffs, scored skirts, worn ring grooves, and cracks. Look for cracks at the ring lands, skirts, pin bosses, and heads. Any defects require replacement of the piston, with these exceptions: Worn ring grooves can sometimes be repaired by cutting the grooves larger and using ring-groove spacers (♦36-18). Piston-skirt wear or collapse (reduction in skirt diameter) can sometimes be corrected by knurling the piston skirt (♦36-19).

Check the fit of the piston pins in the pistons or piston bushings. One way of doing this is to use a small-hole gauge to check the piston-bushing bores, and a micrometer to measure the pin diameter. On the type of piston without a bushing in which the pin oscillates (Fig. 11-30), the piston and pin are supplied in matched sets. If the fit is too loose, or there are other pin or piston defects, the pin and piston are replaced as a matched set. The piston-pin clearance should be no greater than 0.001 inch [0.025 mm].

Measure the size of each piston with a micrometer (Fig.

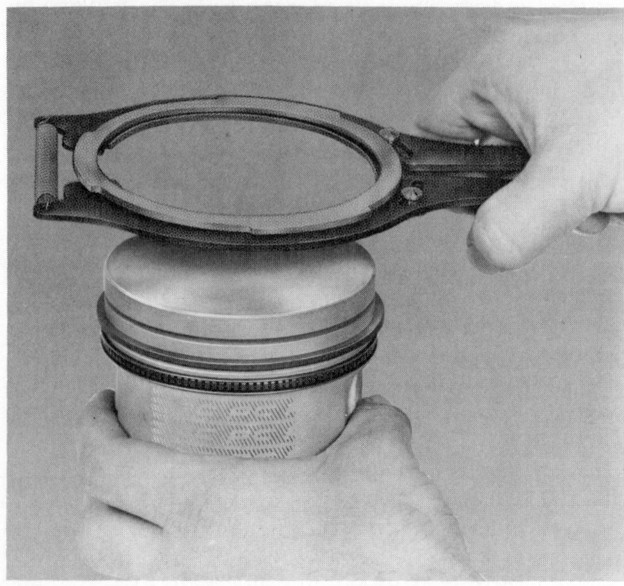

Fig. 36-15 Using a piston-ring expander to remove or install a compression ring on a piston. *(Service Parts Division of Dana Corporation)*

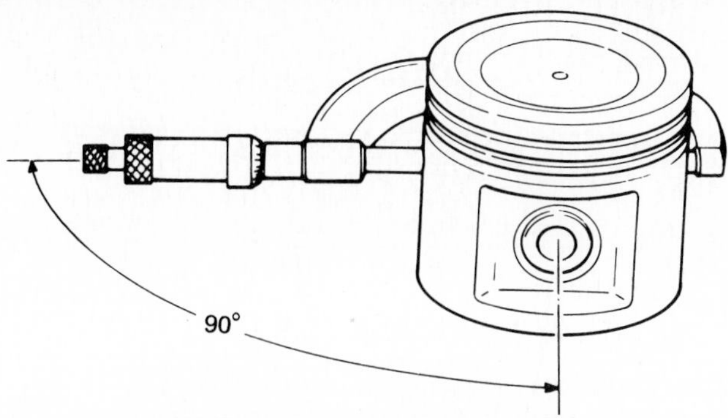

Fig. 36-16 Using a micrometer to measure piston diameter. *(Chevrolet Motor Division of General Motors Corporation)*

36-16). The measurement should be made on the piston skirt 90 degrees from the piston pin, and ¼-inch [6-mm] below the bottom of the oil-ring groove. This is called the *sizing point.* (Some manufacturers specify taking the measurement at a slightly different place on the piston skirt.) Compare the piston measurement made at the sizing point with the cylinder diameter. This measurement may be made with a telescope gauge and micrometer, a cylinder-bore gauge, or an inside micrometer. If the cylinder wall is excessively worn or tapered, it will require refinishing (Chap. 37). When the cylinder is refinished, then a new oversize piston must be installed.

On some engines, the manufacturer recommends fitting the pistons by using a long thickness gauge (Fig. 36-17). Place the thickness gauge and the piston (upside down) in the cylinder. Use a 0.0025-inch [0.06-mm] thickness gauge for used pistons, and a 0.002 inch [0.05-mm] thickness gauge for new pistons. With the bottom of the piston skirt about 1 inch [25 mm] below the top of the block, the piston should hang (not fall free) on the thickness gauge. If the piston falls through the bore, the piston is too small for the cylinder.

♦ 36-18 RING-GROOVE REPAIR

If a piston is in good condition except for excessive ring-groove wear, the groove can be repaired in some pistons. The top ring groove is the groove that wears the most. During engine overhaul, many pistons will be found with excessively worn top ring grooves. The ring groove may be checked with a *wear gauge.* If the ring groove is worn 0.006 inch [0.15 mm] or more, it can be machined to a larger width with a special hand-operated lathe. This squares up the top and bottom sides of the ring groove. Then the new piston ring is installed with a spacer, as shown in Fig. 36-18.

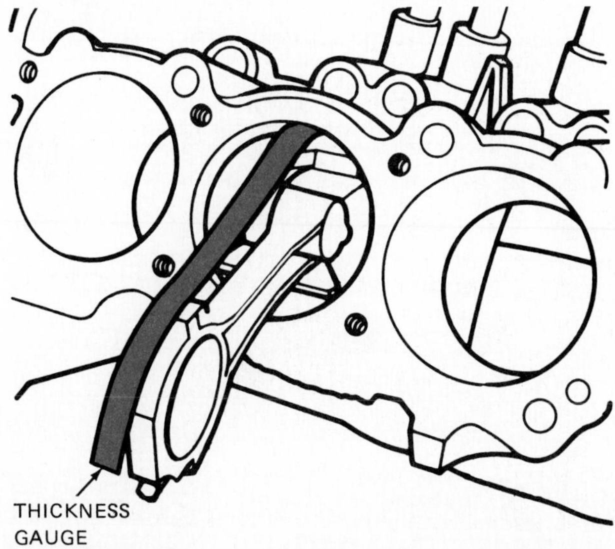

Fig. 36-17 Using a long thickness gauge to check piston clearance. *(Chevrolet Motor Division of General Motors Corporation)*

♦ 36-19 PISTON RESIZING

Resizing of collapsed or worn pistons is not recommended by automotive manufacturers. Excessive resizing can weaken the piston. One piston-reconditioning procedure for older pistons is *knurling.* The piston skirt is run between a supporting wheel and a knurling tool. This displaces metal which expands the diameter of the piston skirt. The indentations form little pockets that can hold lubricating oil to reduce scuffing.

♦ 36-20 SELECTING NEW PISTONS

New pistons are ready for assembly and installation when they are removed from the box. They are available in various sizes for each engine, usually known as "standard oversizes." When oversize pistons are used, the cylinders are rebored and then finished to fit the pistons. Engine manufacturers supply oversize pistons of the same weight as the original pistons. This eliminates any balance problem with the engine when different-size pistons are installed. Some engine repairs may require the reboring of only one or two cylinders. When this is done, maximum cylinder-size difference should not exceed 0.010 inch [0.25 mm].

Aluminum pistons are usually supplied with new piston pins already fitted and packaged in the same box. This ensures the proper clearance between the pin and the pin bore in the piston.

New pistons have a special finish to prevent scuffing during initial start-up. They must not be buffed with a wire wheel. This would remove the finish and increase the chances of scuffing during break-in.

♦ 36-21 FITTING PISTON PINS IN PISTONS

On pistons with piston-pin bushings, worn bushings may be replaced. The new bushings are honed to size to fit the piston

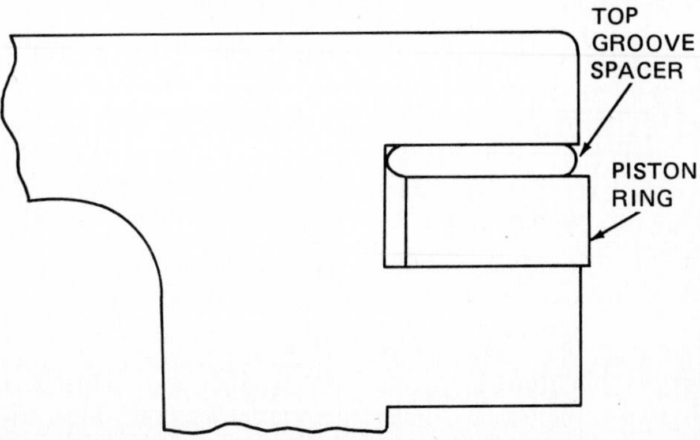

Fig. 36-18 Top-ring-groove spacer in place above the ring. *(Federal-Mogul Corporation)*

pins (Fig. 11-30). Piston-pin clearance should be no greater than 0.001 inch [0.025 mm].

Aluminum pistons are supplied with prefitted piston pins as a matched set. If a pin is worn or has too loose a fit in the piston, a new piston-pin set is required. However, some automotive machine shops hone the piston-pin holes and install oversize piston pins, provided the piston is otherwise in good condition. This is not recommended by most engine manufacturers.

♦ 36-22 ROD-AND-PISTON ALIGNMENT

After the rod and piston have been reassembled, but before the rings are installed on the piston, alignment should be checked with a rod aligner. A V block is held against the piston. If the V block does not line up with the faceplate as the piston is swung across it, the connecting rod is bent or twisted.

Some connecting rods can be straightened. Others should not be straightened. Check with the manufacturer's service manual before reusing a straightened connecting rod.

♦ 36-23 PISTON-RING SERVICE

When an engine is disassembled for service, the old piston rings should not be reinstalled. Rings that have been used, even for a very short time, usually will not reseat properly.

Selection of new piston rings depends on the condition of the cylinder walls and how they are to be reconditioned. The measurement of cylinder walls for wear and taper is described in Chap. 37. If the cylinder walls are only slightly tapered or out of round, then new standard rings can be used. Consult the manufacturer's specifications for maximum allowable taper and wear. The size of the new rings selected for an engine may depend on the cylinder reconditioning procedure that has been used.

Automotive manufacturers generally recommend refinishing the cylinder walls (Chap. 37) before piston-ring installation to "break the glaze." Cylinder walls take on a hard, smooth glaze after the engine has been running. In many automotive shops, this glaze is removed by running a glaze breaker up and down the cylinder a few times before installing new rings. However, some piston-ring manufacturers state that, for some rings, honing does not have to be done, provided the cylinder walls are not wavy or scuffed. The glaze is a good antiscuff surface and will not retard the seating of certain types of new rings. However, the cylinder walls must be reasonably round and in good condition. If you question the condition of the cylinder walls, then deglaze them.

When a cylinder is honed, the proper honing job leaves a crosshatch pattern on the cylinder walls (Fig. 36-19). The hone marks should intersect at about a 60 degree angle. This leaves the surface needed for oil retention and quick seating of new rings.

♦ 36-24 FITTING PISTON RINGS

Piston rings must be fitted to the cylinder and to the ring grooves in the piston. Rings come in packaged sets in graduated sizes to fit various sizes of cylinders. Most packages include instructions that describe how to install the rings. Follow these instructions carefully.

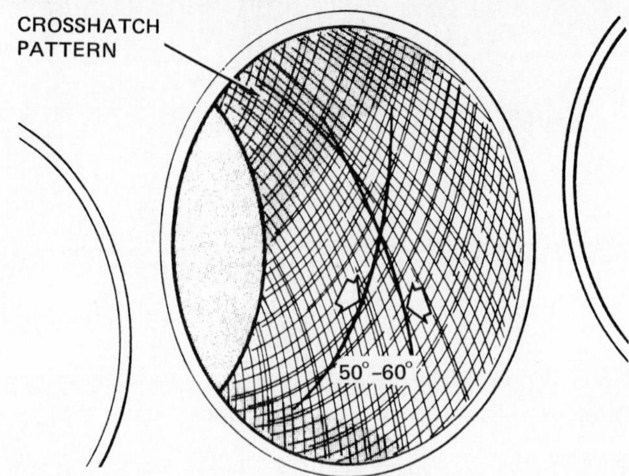

Fig. 36-19 Crosshatch pattern left on the cylinder wall after honing. *(Chrysler Corporation)*

Careful Never throw the instructions away until you have finished the ring-installation job.

As a first step, the ring should be pushed down into the cylinder with a piston, and the ring gap measured. The ring gap is the space between the ends of the ring. It is measured with a thickness gauge. Figure 36-20 shows the gap being measured with the ring pushed down to the lower limit of ring travel. If the cylinder is worn, that is where the ring gap will be smallest. If the ring gap is too small, check the package the rings came in. You may have the wrong ring set for the engine, you may have incorrectly measured cylinder diameter, or the wrong rings may have been packaged in the box. Typical piston-ring end gap in an automotive engine is from 0.010 to 0.020 inch [0.25 to 0.51 mm].

If the cylinder is tapered, the diameter at the lower limit of ring travel (in the cylinder) will be smaller than the diameter at the top. Therefore, the ring must be fitted to the diameter at the lower limit of ring travel. If it is fitted to the upper part

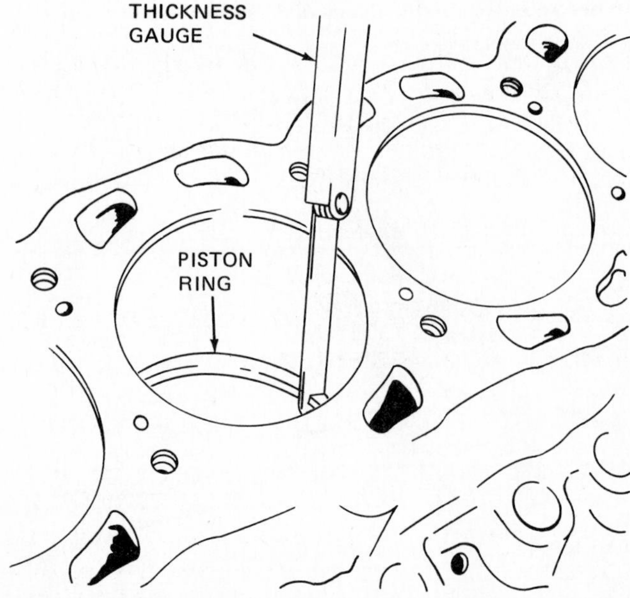

Fig. 36-20 Ring gap is measured with the ring at the lower limit of its travel in the cylinder. *(Oldsmobile Division of General Motors Corporation)*

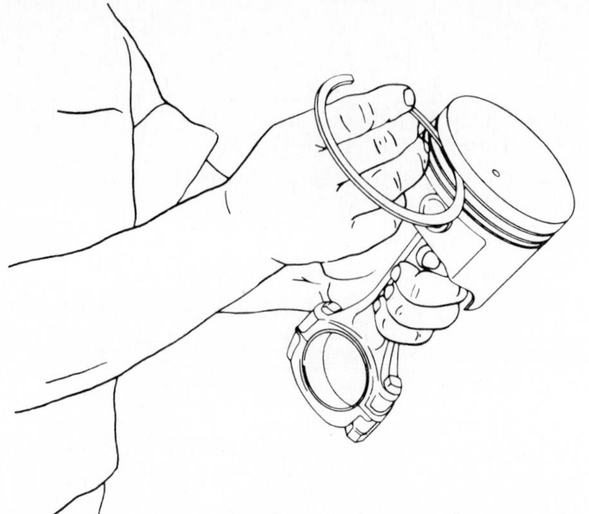

Fig. 36-21 Checking the fit of the piston ring in the ring groove. *(Chevrolet Motor Division of General Motors Corporation)*

of the cylinder, the ring gap will not be great enough at the lower limit of ring travel. As a result, the ring ends will touch together. The ring will be broken, and the cylinder wall will be scored. Always measure the ring gap with the ring pushed down to its minimum diameter at the lower limit of ring travel. The clearance should be wide enough so that the ring ends do not butt together at normal engine temperatures.

If the ring gap is correct, insert the outside surface of the ring into the proper ring groove in the piston (Fig. 36-21). Then roll the ring around in the groove to make sure the ring has a free fit around the entire piston. An excessively tight fit probably means that the ring groove is dirty. Another possibility is that the ring groove has been nicked or burred with the blade of the ring-groove cleaner. Some companies recommend using the end of a broken ring which has been filed to a sharp edge to clean the ring grooves. This is preferred by some technicians because the piece of ring will not cause nicks or burrs.

Install the rings in the ring grooves, using a piston-ring expander (Fig. 36-15). Then measure the piston-ring side clearance. The clearance should be at least 0.001 inch [0.025 mm] and not more than 0.004 inch [0.10 mm] for most engines. Check the shop manual for the specifications on the engine you are servicing.

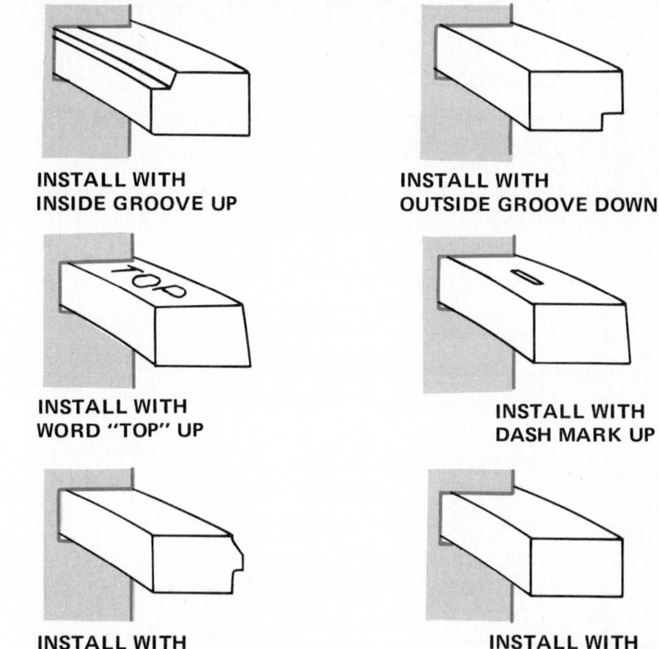

Fig. 36-23 Types of compression rings, and the proper way to install them. *(Federal-Mogul Corporation)*

♦ 36-25 CAUTIONS ON INSTALLING PISTON RINGS

The three-part oil ring is installed one part at a time (Fig. 36-22). Various types of compression rings and their proper installed positions are shown in Fig. 36-23. Never spiral the compression rings into the grooves. (Spiraling the rails of an oil ring is shown in Fig. 36-22.) This could bend or break the compression ring and cause loss of compression and blowby. Instead, always install compression rings with a piston-ring expander (Fig. 36-15). Never overexpand compression rings while installing them. They may break. Piston rings installed upside down may cause excessive oil consumption.

The piston rings must be fitted to the ring grooves in the piston and also to the cylinder. If you are fitting rings to a tapered cylinder, check the ring end gap at the lower limit of ring travel. Installing the piston-and-ring assemblies is described in ♦36-7.

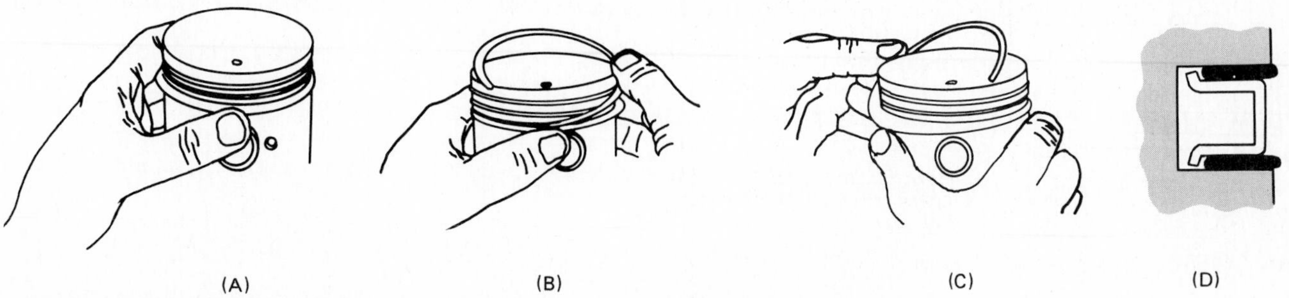

Fig. 36-22 Installation of a three-piece oil-control ring. (A) Place expander spacer in oil-ring groove with ends of spacer above the solid part of the ring-groove bottom. (B) Hold ends of spacer together, and install steel rail above the spacer. (C) Install other steel rail on lower side of the spacer. Make sure the ends of the spacer are not overlapping. (D) Sectional view of the three parts fitted into the groove. *(Perfect Circle Division of Dana Corporation)*

♦ REVIEW QUESTIONS ♦

Select the *one* correct, best, or most probable answer to each question. You can find the answers in the section indicated at the end of each question.

1. Which, if any, of these could happen when piston rings are installed upside down? (♦36-25)
   a. increased oil consumption
   b. broken piston-ring lands
   c. increased cylinder-wall wear
   d. none of the above
2. One reason for removing the cylinder ring ridge before taking out the piston is to avoid (♦36-1)
   a. damaging the piston pin
   b. breaking the piston-ring lands
   c. scratching the cylinder wall
   d. breaking the connecting rod
3. If the piston-pin bushings are so worn that they must be reamed or honed, then the mechanic must install (♦36-21)
   a. the old piston pins
   b. new undersize piston pins
   c. new oversize piston pins
   d. cast-iron piston pins
4. The material which flattens varying amounts to indicate the amount of bearing clearance is called (♦36-13)
   a. shim stock
   b. feeler stock
   c. Plastigage
   d. steel wire
5. If both bearing halves fail on the same side, the failure is probably due to (♦36-12)
   a. a bent rod
   b. a tapered journal
   c. lack of oil
   d. heavy loads
6. Where bearing clearance is excessive but the crankpin is not tapered or out of round, the correct clearance may be obtained by installing (♦36-14)
   a. new undersize bearing
   b. new oversize bearing
   c. a new crankshaft
   d. shims under the bearing
7. The amount by which the bearing-shell diameter is greater than the diameter of the bore into which it is placed is called the bearing (♦36-14)
   a. crush
   b. spread
   c. diameter
   d. bore
8. The additional height over a full half that the bearing shell has is called bearing (♦36-14)
   a. crush
   b. spread
   c. diameter
   d. bore
9. Aluminum pistons are supplied with (♦36-10)
   a. prefitted bushings
   b. prefitted piston pins
   c. connecting rods as an assembly
   d. none of the above
10. New pistons (♦36-20)
    a. should be ground to fit the cylinders
    b. should never be buffed or finished to a smaller size
    c. have special finishes that must not be removed
    d. both b and c
11. After the ring groove has been machined to a larger width, (♦36-18)
    a. a spacer must be installed with the ring
    b. a reinforcing ring should be used
    c. a wider ring should be used
    d. none of the above
12. Piston-pin clearance should be no greater than (♦36-21)
    a. 0.100 inch [2.54 mm]
    b. 0.010 inch [0.25 mm]
    c. 0.001 inch [0.025 mm]
    d. none of the above
13. When new oversize pistons are installed, (♦36-20)
    a. the cylinder walls are finished to fit the pistons
    b. the pistons are finished to fit the cylinders
    c. both the pistons and cylinder walls are finished
    d. none of the above
14. When an engine is rebuilt, (♦36-23)
    a. the old piston rings should not be reinstalled
    b. the old piston rings should be cleaned and reinstalled
    c. both a and b
    d. neither a nor b
15. Which of these statements is (are) true when checking the end-gap clearance of a new piston ring? (♦36-24)
    I. The ends should just butt together to prevent blowby.
    II. Clearance should be enough so that the ends do not butt together at normal operating temperatures.
    a. I only
    b. II only
    c. either I or II
    d. neither I nor II

CHAPTER 37
Engine Service: Crankshafts and Cylinder Blocks

After studying this chapter, and with proper instruction and equipment, you should be able to:

1. Remove and install an engine.
2. Replace an engine mount.
3. Replace the crankshaft main bearings.
4. "Mike" a cylinder block.
5. Recondition a cylinder block.
6. Bore a cylinder.
7. Install cylinder sleeves.
8. Repair certain types of cracks or porosity in cylinder blocks.

♦ 37-1 THE SHORT BLOCK

This chapter discusses the servicing of the cylinder block, crankshaft, and main bearings. If a major servicing job is required, such as boring the cylinders and turning the crankshaft, it is often cheaper to buy a short block. The labor costs of major service may be greater than the cost of the short block. Figure 37-1 shows a short block. It includes the block with all internal parts—pistons, piston pins, piston rings, connecting rods, bearings, and crankshaft. If relatively minor service is required, such as replacing main bearings and oil seals, then it is probably cheaper to perform these services rather than buy a short block.

REMOVING AND INSTALLING ENGINES AND MOUNTS

♦ 37-2 REMOVING AN ENGINE

Many engine-service jobs can be performed with the engine in the vehicle. Other jobs, such as boring cylinders or main-bearing bores, require removal of the engine from the vehicle. Specific removal procedures vary. Always check the manufacturer's shop manual for the car you are servicing before starting the job. The following is a typical procedure.

1. Drain the cooling system including the engine block. Mark the hood hinge position with a scribe and remove the hood.
2. Disconnect the battery cables, negative cable first, and remove the battery.
3. Remove the air cleaner.
4. Disconnect the radiator hoses from the engine outlet and

Fig. 37-1 A short block. *(Ford Motor Company)*

water pump. Disconnect the heater hoses from the engine.
5. Disconnect the automatic transmission oil-cooler lines from the radiator.
6. Remove the fan shroud (if present) and the radiator.

Note Before proceeding, mark all wires and hoses as to their specific locations on the engine. Labels for this may be made from masking tape. Be exact so that all wires and hoses can be reinstalled correctly. You may also want to label certain bracket locations.

7. Disconnect the oil-pressure and coolant-sending-unit wires.
8. Disconnect the fuel-pump hoses. Plug the hoses to prevent leaks.
9. Disconnect the throttle linkage and the engine ground strap.
10. Such units as the alternator, air-conditioner compressor, and power-steering pump need not be removed from the engine compartment. Instead, they can be detached from the engine and moved to one side out of the way with all wires and hoses attached.

♦ CAUTION ♦ Do not disconnect the air-conditioner pressure hoses unless necessary. These hoses hold refrigerant under pressure. Disconnecting them would allow refrigerant to escape. This can be dangerous. Also, it would then be necessary to recharge the system with refrigerant.

11. Disconnect the hoses from the water pump, intake manifold, and choke housing (if present).
12. Disconnect the primary wire from the ignition coil. Remove the wiring harness from the top of the engine.
13. Raise the vehicle. Drain the engine oil.
14. Disconnect the starting motor cable from the starting motor. (The starting motor may also have to be removed.)
15. Disconnect the exhaust pipe(s) from the exhaust manifold(s). Wire up the exhaust system to support it.
16. Disconnect the engine mounts from the frame brackets.
17. Disconnect the clutch or automatic transmission. Make sure the torque converter in the automatic transmission is secure in its housing. Push the torque converter back into the transmission to make sure it is free from the engine.

Note It is sometimes easier to remove the transmission with the engine.

18. Lower the vehicle. Support the transmission with a jack. Use a pad to prevent damage to the automatic-transmission oil pan.
19. Attach the engine-lift cable to the engine. Raise the engine enough to remove the mount bolts (Fig. 37-2). Make a final check to make sure all wiring, hoses, and other parts are free and clear of the engine.
20. Raise the engine enough to clear the mounts. Then alternately raise the transmission jack and engine until the engine separates from the transmission.
21. Carefully pull the engine forward from the transmission. Then lift the engine out of the engine compartment. Tie the transmission to the vehicle fire wall. Reinstall the hood (for safe storage).

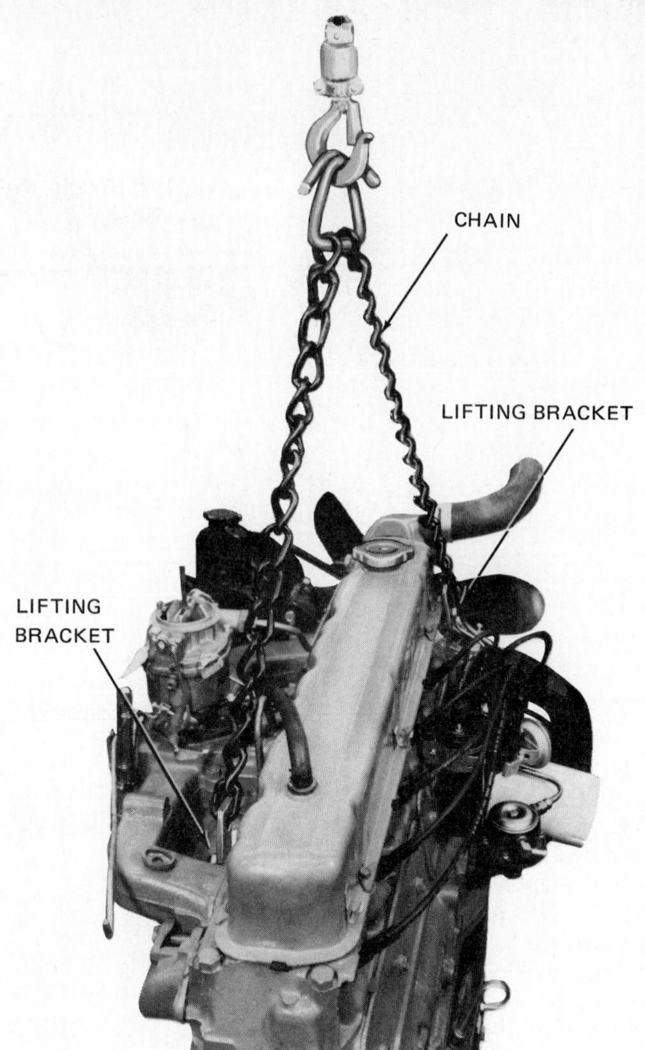

Fig. 37-2 Hoisting an engine from a car. *(Chevrolet Motor Division of General Motors Corporation)*

Note Engine installation is essentially the reverse of the preceding procedure.

♦ 37-3 REPLACING ENGINE MOUNTS

Typical engine mounts are shown in Figs. 11-26 to 11-28. Broken or deteriorated engine mounts put extra strain on other mounts and the drive line. They should be replaced with new mounts. Always check the manufacturer's shop manual before attempting to replace an engine mount. Typical procedures follow.

1. Replacing a Front Engine Mount Support the engine with a jack and wood block under the engine oil pan, or with an engine-support tool (Fig. 37-3). Then raise the engine enough to take the weight off the mount. Remove the nut and through bolt. Remove the mount-and-frame bracket assembly from the cross member. Install the new mount on the cross member. Use new self-locking bolts and nuts. Do not tighten the bolts until you have lowered the engine. Then tighten them to the specified torque.

2. Replacing a Rear Engine Mount Support the engine with a jack and wood block under the oil pan, or (on a lift) place a transmission jack under the transmission. As you lower the transmission, align and start the cross-member-to-mount bolts. Tighten all bolts to the proper torque.

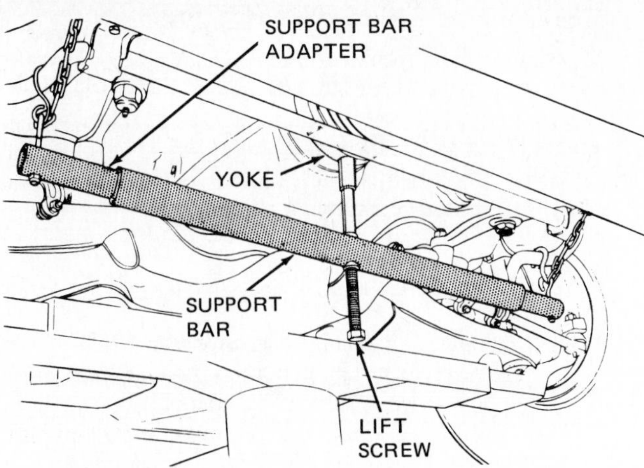

Fig. 37-3 Engine-supporting tool used to raise or support the engine. *(Oldsmobile Division of General Motors Corporation)*

CRANKSHAFT AND MAIN-BEARING SERVICE

♦ 37-4 CRANKSHAFT AND BEARING SERVICE

Sometimes main bearings can be replaced without removing the crankshaft. However, bearing replacement will not fix plugged oil passages, worn crankshaft journals, a damaged crankshaft, or a block in which a bearing has spun. This occurs when the bearing and crankshaft journal become so hot from lack of oil that they weld momentarily. Then the bearing spins with the crankshaft and gouges the bearing bore in the cylinder block. Bearing spin will damage the cylinder block. Block replacement may be required.

If all bearings have worn evenly, then all that is normally required is to check the journals and install new bearings. However, all bearings do not usually wear the same amount. Some bearings wear more than others. This is acceptable provided none of the bearings wear beyond manufacturer's specifications. The lower bearing half wears the most. It carries the weight of the crankshaft and combustion pressures through the rods and cranks. Uneven bearing wear can result from oil-pump wear. As the oil pump wears, oil pressure and volume drop. The main and rod bearings farthest from the oil pump get less oil. Therefore they wear the most. Also, a clogged oil passage will starve bearings. When this happens, the bearings fail and may spin in the block or rod bores.

If main bearings have worn very unevenly, remove the crankshaft from the engine block. Then check the block and crankshaft for damage and clogged oil lines.

♦ 37-5 REMOVING MAIN-BEARING CAPS

Sometimes it is difficult to remove all of the main-bearing caps with the engine in the car. In some cars, the front cross member is so close to the engine that the engine must be lifted off its mounts to get enough clearance to remove the cap. Also, the rear-main-bearing cap may be difficult to remove because of interference with other parts.

When you are planning to check only the journals on the crankshaft and the bearings, remove the main-bearing caps one at a time to make the checks. If the crankshaft is to come out of the engine, the connecting rods can be detached and all main-bearing caps removed, after the engine is out of the car.

Main-bearing caps should be marked so that they can be installed in the same location from which they were removed. (Marking rod caps is described in ♦36-4.) To take off the main-bearing cap, remove the nuts or bolts. Bend back the lock-washer tangs, if used. (Use new lock washers on reassembly.) Disconnect the oil pump or any oil lines that are in the way. Now the cap should pull straight off.

If a cap sticks, work it loose carefully to avoid nicking or cracking it. In some engines, a screwdriver or pry bar can be used to work the cap loose. Sometimes tapping the cap lightly on one side and then the other with a brass or plastic hammer will loosen it.

Note Heavy hammering or prying can nick or crack the cap, bend the dowel pins, or damage the dowel holes. Then the bearing may not fit when the cap is replaced, and early bearing failure will occur. Bearing caps are made of cast iron and are brittle. A hard blow can crack or break a cap. A damaged cap will have to be discarded, and a new cap used. A cracked cap can break if it is reinstalled on the engine.

When a bearing cap is damaged or lost, a new cap is required. It will be necessary to take the engine out of the car, disassemble the engine, and install the caps, without bearings, back in the block. Then the block will have to be *line bored* to reestablish bearing-bore alignment.

♦ 37-6 CHECKING CRANKSHAFT JOURNALS

Both the crankpins and the main-bearing journals on the crankshaft should be inspected whenever the main-bearing caps and rod caps are removed. Inspecting crankpins is discussed in ♦36-13.

Usually crankshaft bearing clearance and journal taper are measured with Plastigage, when the crankshaft is in the engine. (The use of Plastigage is discussed in ♦36-13.) However, a crankpin can be checked with an outside micrometer. Main-bearing journals can be measured with the crankshaft in the block by using a special micrometer, or a crankshaft gauge (Fig. 37-4). Measurements should be taken at several places along the journal to check for taper. Then, the crankshaft should be rotated by one-quarter or one-eighth turns to check for out-of-round. (Damage caused by a tapered, ridged, or out-of-round journal is described in ♦36-13). If the journals are tapered or out of round by more than 0.001 inch [0.025 mm], they should be reground. Any out-of-round or taper shortens bearing life.

To check only the main-bearing journals, remove the oil pan (♦36-3) and the main-bearing caps one at a time (♦37-5). It is not necessary to detach the connecting rods from the crankshaft. However, the crankshaft may be turned more easily if the spark plugs are removed.

♦ 37-7 INSPECTING MAIN BEARINGS

Crankshaft main bearings should be replaced if they are worn, burned, scored, pitted, rough, flaked, cracked, or otherwise damaged. (Bearing failures are described in ♦36-12.) Always check the crankshaft journals (♦37-4 and 37-6) before installing new bearings. If the journals are not in good condition, the new bearings may soon fail. Also bearings

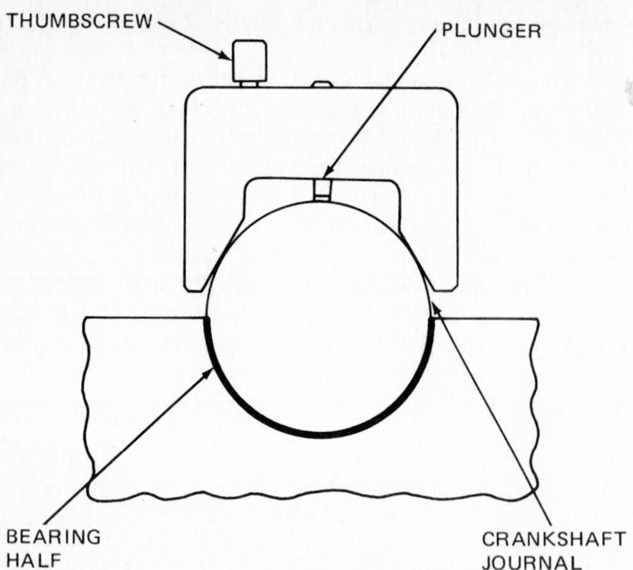

Fig. 37-4 Using a crankshaft gauge to check a main-bearing journal with the crankshaft in the engine. *(Federal-Mogul Corporation)*

may have worn unevenly, and some bearings may be worn more than others. If so, or if a bearing is damaged, check for a bent crankshaft or clogged oil passages. Remove the crankshaft so that the oil passages in the crankshaft and block can be inspected.

If one main bearing requires replacement, then all main bearings usually should be replaced even though the others appear to be in good condition. If only one main bearing is replaced, crankshaft alignment might be lost. This could overload other bearings, causing them to fail rapidly.

♦ 37-8 MEASURING MAIN-BEARING CLEARANCE

Main-bearing clearance should always be checked after new bearings are installed. The clearance should also be checked whenever the condition of the bearings is inspected. (Crankshaft-journal condition should also be inspected at the same time.) Bearing clearance can be checked with shim stock or Plastigage.

1. With Shim Stock The shim-stock method is seldom used today. The Plastigage method is faster and easier. To use shim stock, lay a strip of shim stock on the journal and install the cap. Tighten the cap nuts. Repeat, adding shim stock until the crankshaft will not turn. When the crankshaft locks, count the strips and subtract one. That is the clearance.

Note Do not attempt to rotate the crankshaft with shims in the bearing. This could damage the bearing. Instead, try to turn the crankshaft about 1 inch (25.4 mm) one way or the other.

2. With Plastigage Wipe the journal and bearing clean of oil. Put a strip of Plastigage lengthwise in the center of the journal (Fig. 37-5). Install and tighten the cap. Then remove the cap and measure the amount the Plastigage has flattened (Fig. 37-6). Do not turn the crankshaft with the Plastigage in place. (See ♦36-13 for more information on Plastigage.)

The crankshaft must be supported so that its weight will not cause it to sag. A sagging crankshaft could result in an incorrect measurement. One way to support it is to position a small jack under the crankshaft. Let the jack bear against the counterweight next to the bearing being checked. Another method is to put shims in the bearing caps of the two adjacent main bearings. Then tighten the cap bolts. This lifts and supports the crankshaft. If the engine is out of the car and inverted, this is not necessary.

♦ 37-9 MEASURING CRANKSHAFT END PLAY

Crankshaft end play will become excessive if the thrust bearing in the engine is worn. This condition produces noticeably sharp, irregular knock. In a car with a manual transmission, if the wear is excessive, the knock will occur every time the clutch is disengaged and engaged. This action causes sudden endwise movements of the crankshaft. Check end play by forcing the crankshaft endwise as far as it will go. Then measure the clearance at the thrust bearing with a thickness gauge or with a dial indicator (Fig. 37-7). Consult the engine manufacturer's service manual for allowable end play.

♦ 37-10 INSTALLING MAIN BEARINGS

Before the main bearings are installed, the crankshaft journals should be checked (♦37-5 and 37-6). After the bearings are installed, the bearing clearance should be checked (♦37-8). In some engines, the main bearings can be replaced without removing the crankshaft. However, with uneven bearing wear, the crankshaft should be removed for further inspection (♦37-4 to 37-6).

To install a main bearing without removing the crankshaft, use a bearing roll-out tool (Fig. 37-8). The tool is inserted into the oilhole in the crankshaft journal. Then the crankshaft is rotated. The tool forces the bearing shell to rotate with the crankshaft so that the bearing is turned out of the bore. The crankshaft must be rotated in the proper direction so that the lock, or tang, in the bearing is raised up out of the notch in the cylinder block.

Fig. 37-5 Measuring main-bearing clearance with Plastigage. *(Chevrolet Motor Division of General Motors Corporation)*

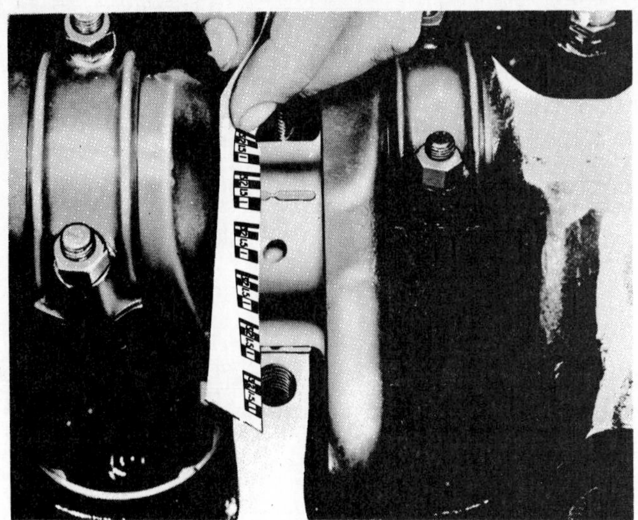

Fig. 37-6 Checking flattening of Plastigage to determine main-bearing clearance. *(Chevrolet Motor Division of General Motors Corporation)*

To install a new bearing half, coat the journal surface of the bearing with engine oil. Leave the outside of the bearing shell dry. Make sure that the bore, or bearing seat, in the block is clean. Do not file the edges of the shell. This would remove its crush. Use the tool to slide the bearing shell into place as shown in Fig. 37-8. Make sure that the tang on the bearing shell seats in the notch in the block. Then place a new bearing shell in the cap, and install the cap. Tighten the cap bolts or nuts to the specified torque. Tap the crown of the cap lightly with a brass or plastic hammer while tightening it. This helps to align the bearing properly. After all bearings are in place, check the bearing clearance.

If the crankshaft is removed, it is easier to install the main bearings. Then you can wipe the bearing bores in the block and make sure they are clean and in good condition. The bearing halves can be easily slid into position.

Some bearings have oil grooves in only one bearing half. Others have grooves in both halves. Some do not use grooves. Be sure to check the service manual for the engine you are servicing to determine what kind of bearing half goes where.

Some crankshaft journals have no oilhole. For example, the rear main journals of many in-line engines do not. To remove and install the upper bearing half on these journals, first start the bearing half with a wood tongue depressor or similar installing tool. Then use a pair of pliers with taped jaws to hold the bearing half against the oil slinger. Rotate the crankshaft (Fig. 37-9). This movement will pull the old bearing out. The new bearing is put into position in the same manner. The last fraction of an inch can be pushed into place by holding only the oil slinger with the pliers while rotating the crankshaft. Or the bearing may be tapped down with a soft tool and hammer. Be careful that you do not damage the bearing.

While removing and installing the upper bearing shell of a rear main bearing, hold the oil seal in position in the cylinder block. Otherwise, the seal may move out of position (♦37-11).

Bearing clearance should be checked after all bearing caps have been installed (♦37-8). If excessive clearances are found with the new bearings, the journals are worn. This means the crankshaft must be removed for replacement or regrinding (♦37-12 to 37-14). Then, undersize bearings must be installed. (Bearings are available in several undersizes.) The journals should be ground down enough to remove imperfections. Then the journals are ground an additional amount to fit the next undersize bearings.

Precision-insert bearings usually are installed without shims. Never use shims on these bearings unless the engine manufacturer specifies them. Similarly, bearing caps must not be filed in an attempt to improve bearing fit. Shimming and filing the cap usually lead to premature bearing failure. At the same time, the main-bearing journal on the crankshaft may be damaged.

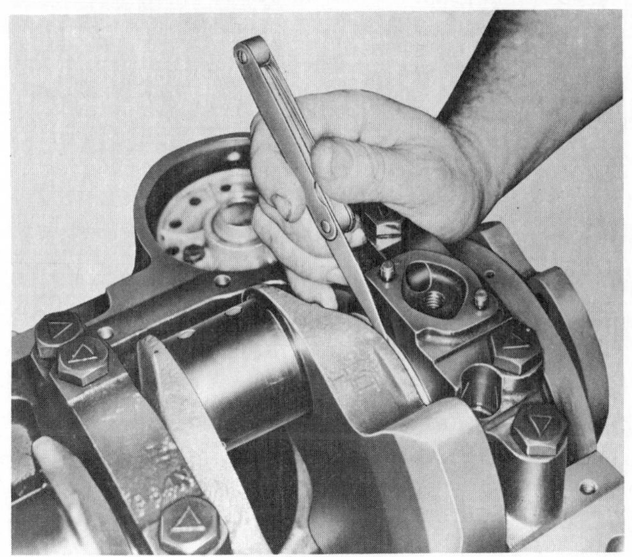

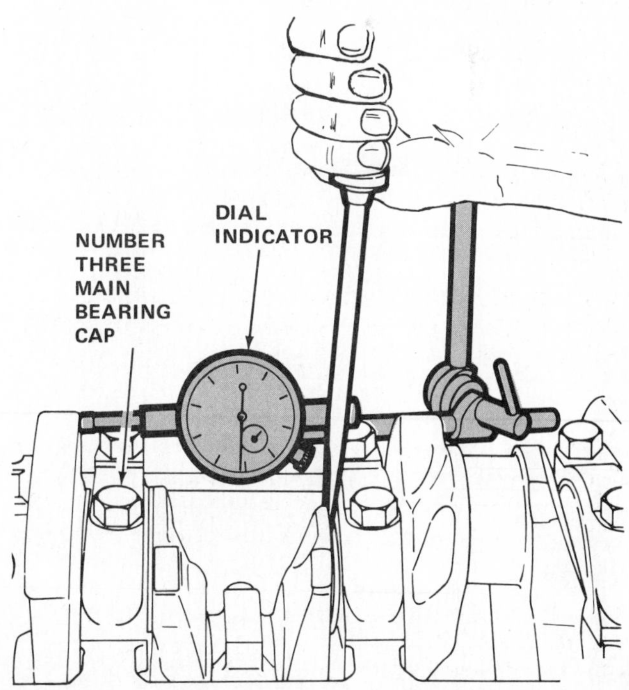

Fig. 37-7 Crankshaft end play being checked at the thrust bearing with a thickness gauge (top), and with a dial indicator (bottom). *(Chevrolet Motor Division of General Motors Corporation)*

Fig. 37-8 Using a bearing roll-out tool to install a new upper main bearing. Arrow shows the direction in which crankshaft is turned to roll the bearing into place. *(Federal-Mogul Corporation)*

♦ 37-11 REPLACING REAR-MAIN-BEARING OIL SEAL

An oil seal is required at the rear main bearing to prevent oil leakage through it. When main-bearing service is being performed, or whenever oil leaks from the rear main bearing, the oil seal must be replaced.

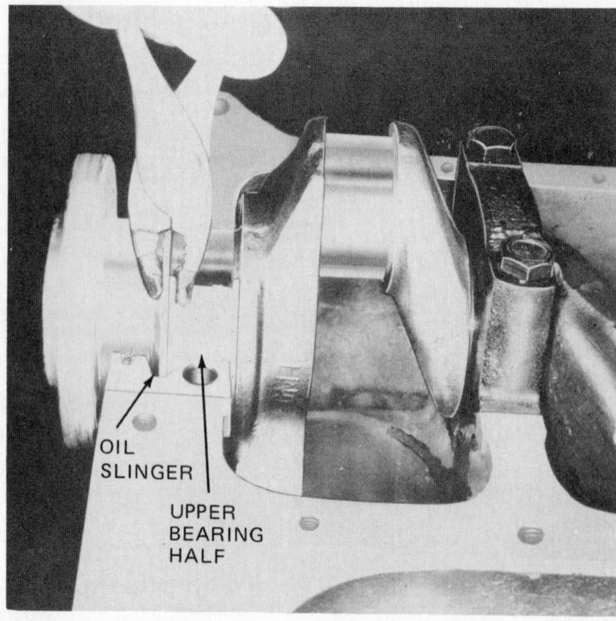

Fig. 37-9 Replacing a rear-main-bearing half with pliers. In this engine, the crankshaft rear-main-bearing journal has no oilhole. *(Chevrolet Motor Division of General Motors Corporation)*

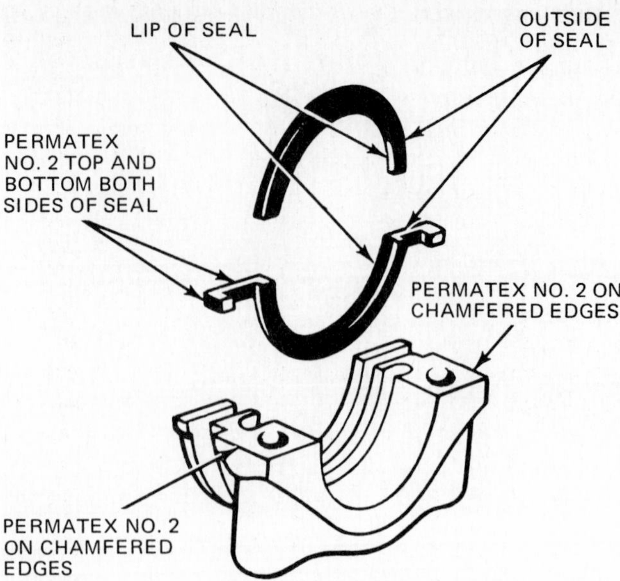

Fig. 37-10 Crankshaft rear-main-bearing oil seal. The upper part fits in a groove in the block. *(American Motors Corporation)*

Replacement of the main-bearing oil seal varies with different designs. Many engines use a two-piece oil seal as shown in Fig. 37-10. The lower half of the oil seal is installed in the bearing cap. The upper half may be removed without removing the crankshaft. First, a punch and hammer are used to start the oil seal out. Then pliers are used to pull the oil-seal half completely out.

A new upper half of the oil seal can be worked into place. The new oil seal should be coated with oil so that it will slide into position.

To install the upper half of a rope-type oil seal in the block, usually you must remove the crankshaft. Then a bearing-seal installer is used to compress the new seal into the cylinder-block groove (Fig. 37-11). The seal is then trimmed flush with the block. The oil-seal half in the cap can be replaced by removing the cap, installing the oil seal, and trimming it flush.

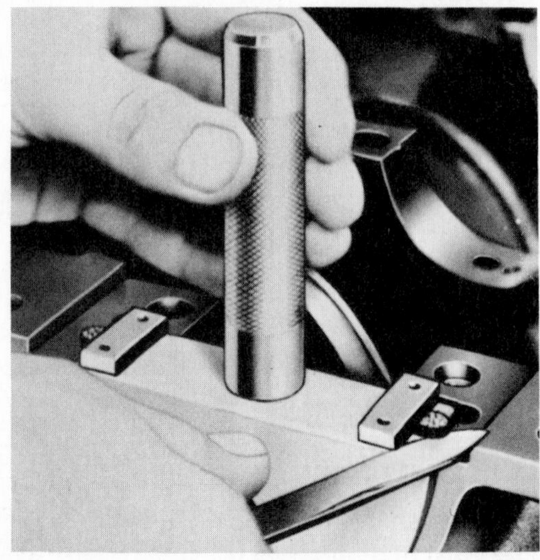

Fig. 37-11 Using a bearing-seal installer to compress the upper half of a rear-main-bearing oil seal in the cylinder block. *(Ford Motor Company)*

Some engines have a one-piece oil seal which fits around the crankshaft flange (item 16 in Fig. 11-2). To remove this type of seal, first remove the flywheel from the engine. Then the old seal can be pried out with a screwdriver. Be careful not to nick the seal surface on the crankshaft flange. To install a new seal, position the seal on the crankshaft. A seal driver is placed over the seal. Then a plastic hammer is used to drive the seal into place.

♦ 37-12 REMOVING THE CRANKSHAFT

Such parts as the oil pan, timing-gear or timing-chain cover, crankshaft timing gear or sprocket, interfering oil lines, and oil pump must be removed before the crankshaft can be taken out. Also, on some engines, the flywheel must be detached from the crankshaft. With other parts off, the bearing caps are removed to release the crankshaft.

♦ CAUTION ♦ The crankshaft is heavy! Support it adequately as you remove the bearing caps.

For a complete engine overhaul, the cylinder head and piston-and-rod assemblies must be removed. However, if only the crankshaft is coming out, the piston-and-rod assemblies need not be removed. Instead, they can be detached from the crankshaft and pushed up out of the way. Be careful not to push them up too far. If you do, the top piston ring may move up beyond the cylinder block. Then the ring will catch on the top of the block. You will be unable to pull the piston-and-rod assembly back down. The cylinder head will have to be removed. Then a ring compressor must be used to get the ring back into the cylinder again.

♦ 37-13 SERVICING THE CRANKSHAFT

During engine rebuilding, and whenever the crankshaft is removed from the block, check the crankshaft for end play, alignment or "binding," and bearing clearance. This includes checking for main-bearing journal and crankpin wear. If the crankshaft is out of line, a new or reground crankshaft should be used. If journal or crankpin wear, taper, or out-of-round exceed allowable limits, or if they are rough, scratched, pitted, or otherwise damaged, they must be ground. Then new undersize bearings must be installed.

♦ 37-14 GRINDING CRANKSHAFTS

A special lathe, or *crankshaft grinder*, is required to grind the main-bearing journals and crankpins on a crankshaft. This machine is found in some automotive machine shops and in some engine-rebuilding shops. The grinding wheel must refinish the bearing surfaces with great accuracy and to extreme smoothness.

CYLINDER BLOCKS

♦ 37-15 CYLINDER WEAR

The piston and ring movement, the high temperatures and pressures of combustion, the washing action of gasoline entering the cylinder—all these tend to cause cylinder-wall wear. At the start of the power stroke, pressures are the greatest. The compression rings are forced with the greatest pressure against the cylinder wall. At the same time, the temperatures are highest while the oil film is least effective in protecting the cylinder walls. Therefore, the most wear ("taper") takes place at the top of the cylinder. As the piston moves down on the power stroke, the combustion pressure and temperature decrease. Less wear takes place. The result is that the cylinder wears irregularly, as shown in Fig. 37-12.

The cylinder also tends to wear slightly oval shaped. This is due to the side thrust of the piston as it moves down in the cylinder on the power stroke. The side thrust results from the swing, from vertical, of the connecting rod. However, cylinders that are oval when cold will probably be less oval when hot. Another factor is the washing action of the gasoline. At times the air-fuel mixture is not perfectly blended. Small droplets of gasoline, still unvaporized, enter the cylinder. They strike the cylinder wall (at a point opposite the intake valve) and wash away the oil film. Therefore, this area wears more rapidly.

♦ 37-16 CLEANING AND INSPECTING CYLINDER BLOCK

As a first step, make a visual inspection of the block. Major damage, resulting from a main bearing spinning or a broken connecting rod going through the block, means the block must be discarded. If the engine has overheated and there are cracks in the cylinder walls, discard the block.

During an engine overhaul, the cylinder block should be thoroughly cleaned. Remove the expansion plugs, oil-gallery plugs, and any switches or other removable parts. Then the block should be cleaned in a hot tank or other suitable method.

Inspect the cylinder bores for cracks, grooves, scratches, or discoloration. Inspect the crankshaft bearing bores for signs of a spun bearing. Inspect for cracks across the top of the block between cylinders, between boltholes, on the outside of the block, and in the main-bearing bore webs. Some limited types of block damage are repairable. (See ♦37-23 on block repair.) If everything looks okay, measure the cylinder bores with a micrometer. If the bores are not excessively worn, and

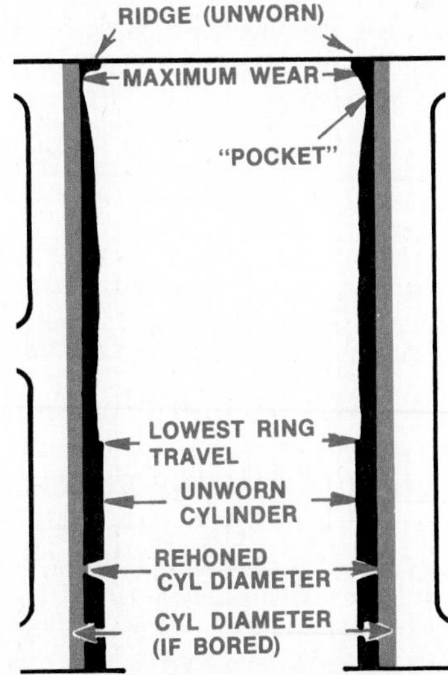

Fig. 37-12 Taper of an engine cylinder (shown exaggerated). Maximum wear is at the top, just under the ring ridge. Honing the cylinder usually requires removal of less metal than boring. Metal to be removed by honing is shown solid. Metal to be removed by boring is shown both solid and shaded. *(Sunnen Products Company)*

can be honed or rebored within specified limits, then clean and service the block. However, cylinder-block service is often performed by an automotive machine shop.

♦ 37-17 CHECKING BEARING BORES

Very uneven bearing wear with some bearings worn much more than others may mean out-of-round bearing bores or a warped block. To check for out-of-round bores, you must remove the crankshaft and main bearings. The bearing bores must be cleaned and the caps installed with the bolts torqued to specifications. Then an inside micrometer is used to check for out-of-round. If the bores are out of round or out of line, they must be line bored to restore roundness and bore alignment.

♦ 37-18 CYLINDER SERVICE

Up to certain limits, cylinders may wear tapered or out of round and not require refinishing. Some replacement piston-ring sets can help control compression leakage and oil pumping in tapered and out-of-round cylinders. But when wear becomes excessive, loss of compression, high oil consumption, poor performance, and heavy carbon accumulations in the cylinders will result. The cylinders must be refinished by honing or boring to restore proper engine operation. When the cylinders are honed or bored to a larger size, new pistons must be installed at the same time.

♦ 37-19 INSPECTING CYLINDER WALLS

Wipe walls and examine them for scores and spotty wear (which shows up as dark, unpolished spots). Hold a light at the opposite end of the cylinder so that you can see the walls better. Scores or spots will require honing or boring of the cylinder.

Next, measure the cylinders for wear, taper, and out-of-round (Fig. 37-13). This can be done with an inside micrometer, a telescope gauge and an outside micrometer, or a cylinder-bore gauge. The cylinder-bore gauge is moved up and down in the cylinder and rotated at various positions to detect wear.

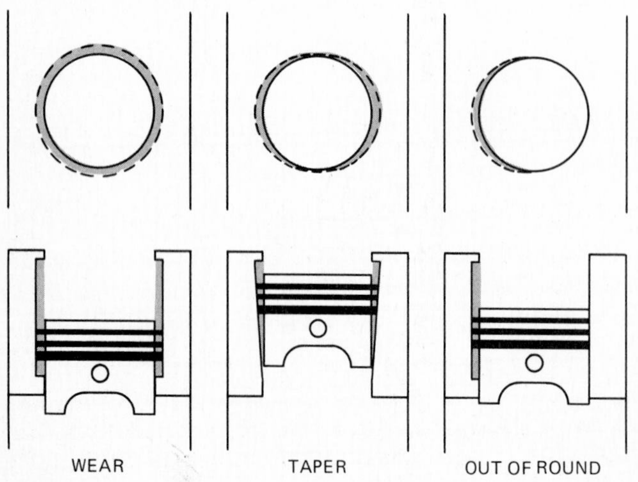

Fig. 37-13 A cylinder must be checked for wear, taper, and out-of-round.

A quick way to estimate cylinder taper is to push a compression ring down to the lower limit of ring travel. Measure the ring gap with a thickness gauge (Fig. 36-20). Then pull the ring up to the upper limit of ring travel. Remeasure the ring gap. The top gap minus the bottom gap divided by three gives you the approximate cylinder taper.

♦ 37-20 REFINISHING CYLINDERS

As a first step, the block should be cleaned. The decision on whether the cylinders are to be honed or bored depends on the amount of cylinder wear. Figure 37-12 shows the amount of metal removed by the hone and by boring. The hone (Fig. 37-14) uses a set of abrasive stones which are turned in the

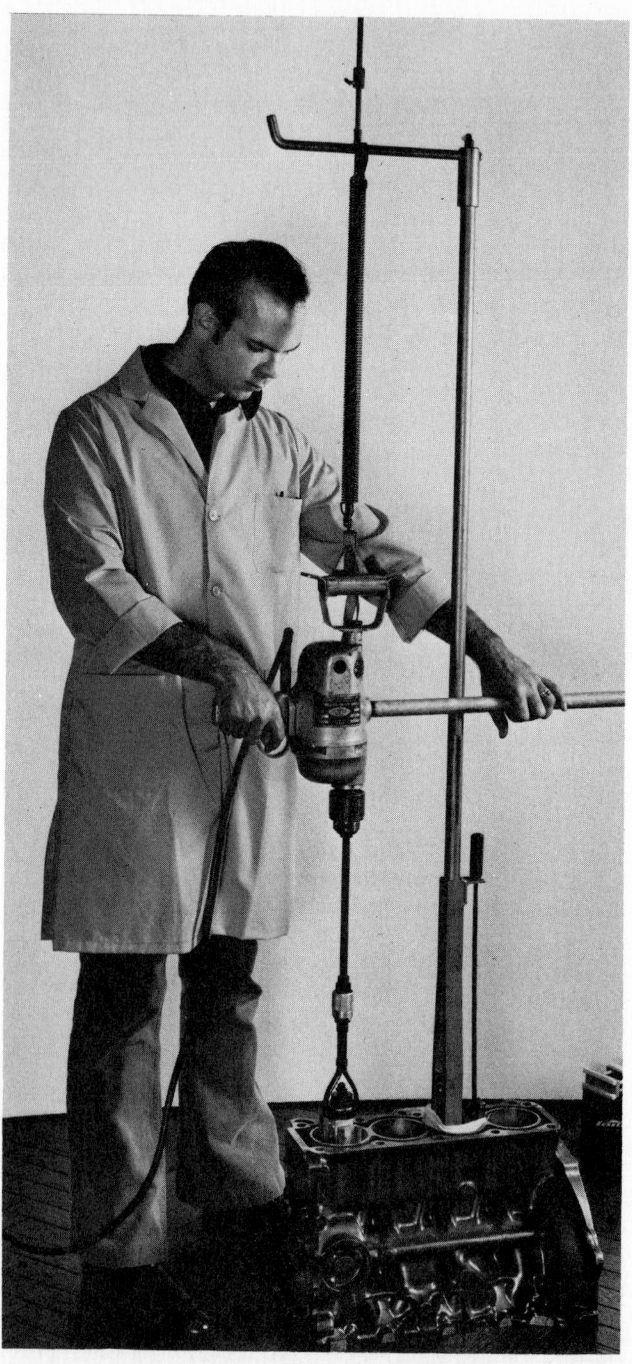

Fig. 37-14 Honing a cylinder. *(Sunnen Products Company)*

cylinder. Honing should leave a crosshatch pattern in the cylinder (Fig. 36-19). The boring bar (Fig. 37-15) uses a revolving cutting tool. Where cylinder wear is not excessive, only honing is necessary. But if wear is excessive, then the cylinder must be rebored and oversize pistons installed. Before cylinders are bored, the main-bearing cap should be installed and the bolts torqued to specifications.

♦ 37-21 CLEANING CYLINDERS

Cylinders must be cleaned thoroughly after honing or boring. Even slight traces of grit or dust on the cylinder walls may cause rapid ring and wall wear and early engine failure. As a first step, some engine manufacturers recommend wiping down the cylinder walls with very fine crocus cloth. This loosens embedded grit and also knocks off "fuzz" left by the honing stones or cutting tool. Then use a stiff brush and hot soapy water to wash the walls. All abrasive material must be cleaned off the cylinder walls.

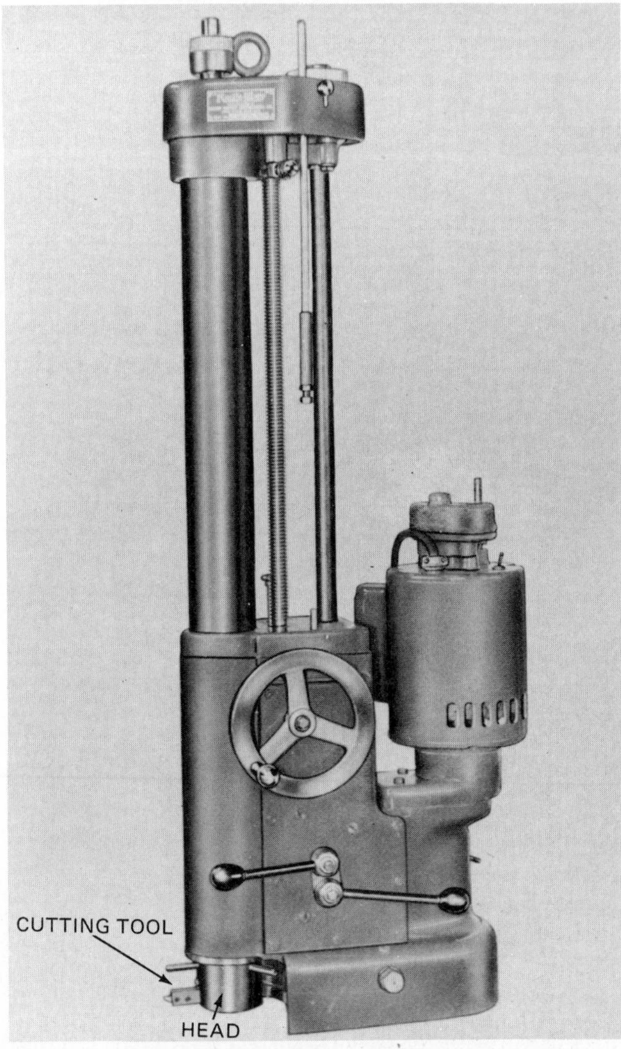

Fig. 37-15 Cylinder boring bar. The cutting tool is carried in a rotating head that feeds down into the cylinder as the head rotates. This causes the cutting tool in the rotating head to remove metal from the cylinder wall. *(Kwik-Way Manufacturing Company)*

Gasoline, kerosene, and solvent will not remove all of the grit from cylinder walls. Their use to clean cylinder walls after honing or boring is not recommended by some manufacturers.

After washing the walls, wipe them several times with a cloth dampened with light engine oil. Wipe off the oil each time with a clean, dry cloth. At the end of cleaning and oiling, a clean cloth should come away from the walls showing no trace of dirt.

Clean out all coolant and oil passages in the block, and stud and boltholes, after the walls are cleaned (see ♦37-16).

♦ 37-22 INSTALLING CYLINDER SLEEVES

There are two types of cylinder sleeves: wet and dry. The wet sleeve is sealed to the block at the top and bottom. It is in direct contact with the coolant. The dry sleeve is pressed into the cylinder. This type is in contact with the cylinder wall from top to bottom.

Cracked blocks, scored cylinders, and excessively worn cylinders can often be repaired by installing cylinder sleeves (Fig. 37-16). As a first step, the cylinders are bored oversize to take the sleeves. Then the sleeves are pressed into place.

Figure 37-16 shows a technician using a pneumatic hammer to drive sleeves into place. The cylinders have been bored, and the sleeves have been positioned on the cylinder block. The pneumatic hammer uses compressed air, which operates the driving head. It hammers the sleeves down into place in the cylinder block. The sleeves are finished to the proper size to take a standard piston and set of rings.

♦ 37-23 REPAIRING CYLINDER-BLOCK CRACKS OR POROSITY

Sometimes a cylinder block is in good condition except for some cracks or sand holes (left in the block during casting). Areas which are not subjected to temperatures of more than 500°F [260°C] or pressure (from coolant, oil, or cylinders) can be repaired with a metallic plastic or epoxy. Permissible repair areas for one manufacturer's engines are shown in Fig. 37-17.

Fig. 37-16 Installing dry sleeves in a cylinder block. *(Automotive Rebuilders, Inc.)*

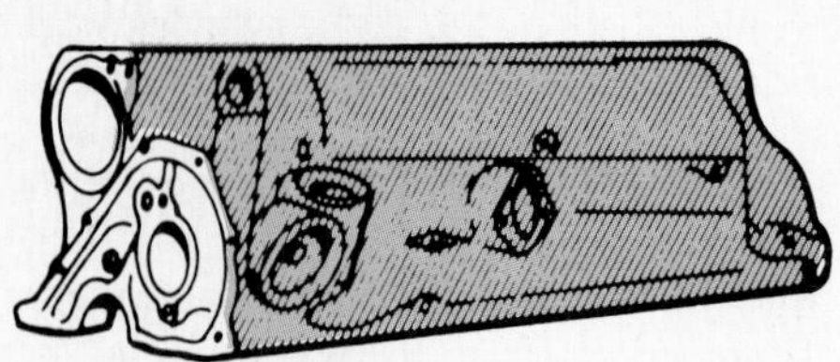

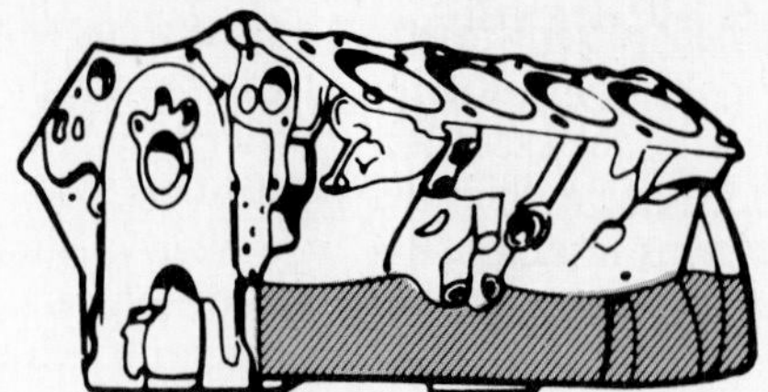

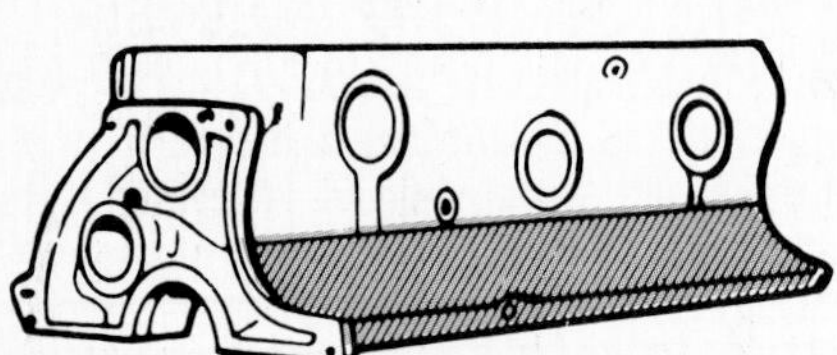

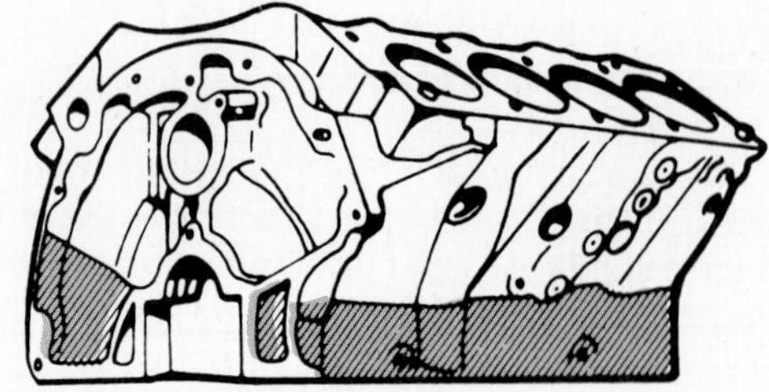

Fig. 37-17 Shaded areas of cylinder blocks that can be repaired with epoxy. *(Ford Motor Company)*

♦ REVIEW QUESTIONS ♦

Select the *one* correct, best, or most probable answer to each question. You can find the answers in the section indicated at the end of each question.

1. Mechanic A says that main-bearing oil clearance can be checked with Plastigage. Mechanic B says that main-bearing oil clearance can be checked with a thickness gauge. Who is right? (♦37-8)
   a. A only
   b. B only
   c. both A and B
   d. neither A nor B
2. Crankshaft journals should be reground if they are tapered or out-of-round by more than (♦37-6)
   a. 0.0003 inch [0.0076 mm]
   b. 0.001 inch [0.025 mm]
   c. 0.030 inch [0.76 mm]
   d. 0.300 inch [7.60 mm]
3. If a main bearing requires replacement, you should replace (♦37-7)
   a. adjacent main bearings
   b. all main bearings
   c. camshaft bearings
   d. rod bearings
4. Excessive thrust-bearing wear will cause excessive (♦37-9)
   a. crankshaft end play
   b. crankshaft deflection
   c. crankshaft torsional vibration
   d. none of the above
5. Before installing new main bearings, you should always (♦37-10)
   a. check crankshaft journals
   b. check connecting-rod journals
   c. grind crankshaft journals
   d. replace the crankshaft
6. After installing new main bearings, you should always (♦37-10)
   a. check crankshaft journals
   b. check bearing clearance
   c. check bore eccentricity
   d. shim-adjust the bearings
7. Main-bearing clearance can be checked with either (♦37-8)
   a. shims or dial indicator
   b. feeler stock or micrometer
   c. micrometer or Plastigage
   d. shim stock or Plastigage
8. During engine rebuilding, the crankshaft should be checked for (♦37-13)
   a. end play
   b. binding
   c. bearing clearance
   d. all of the above
9. If all bearings have worn evenly, all that is normally required is to (♦37-4)
   a. grind journals and replace bearings
   b. replace the bearing caps
   c. check the journals and install new bearings
   d. none of the above

10. Taper in the cylinder is greatest at the (♦37-15)
    a. top of the cylinder
    b. center of the cylinder
    c. bottom of the cylinder
    d. none of the above
11. A wear spot on the cylinder wall opposite the intake valve may be caused by (♦37-15)
    a. high-speed operation
    b. heavy compression-ring pressure
    c. washing by gasoline droplets
    d. none of the above
12. Cylinders that are oval when cold will probably be (♦37-15)
    a. more oval when hot
    b. less oval when hot
    c. about the same when hot
    d. none of the above
13. One major cause of cylinder-wall wear is the force of the (♦37-15)
    a. piston on the walls
    b. compression rings on the walls
    c. combustion gases on the walls
    d. none of the above
14. Two methods of resizing cylinder bores are (♦37-18)
    a. grinding and honing
    b. reaming and boring
    c. roughing and finishing
    d. honing and boring
15. When considerable metal must be removed to eliminate the taper in a cylinder bore, the cylinder should be (♦37-20)
    a. bored
    b. honed
    c. ground
    d. reamed
16. When you are boring cylinders, the main-bearing caps should be (♦37-20)
    a. off
    b. lightly attached
    c. attached with normal bolt torque
    d. attached with heavy bolt torque
17. After cylinders are bored or honed, they should be cleaned with (♦37-21)
    a. soapy water and light oil
    b. light oil and gasoline
    c. gasoline and kerosene
    d. none of the above
18. Cracked blocks and scored and badly worn cylinders can sometimes be repaired by (♦37-22)
    a. reboring cylinders
    b. rehoning cylinders
    c. installing cylinder sleeves
    d. none of the above
19. Crankshaft main-bearing journals should be checked for (♦37-6)
    a. high spots and stretch
    b. taper, ridges, and out-of-round
    c. connecting-rod bearing fit
    d. all of the above
20. An engine has a heavy knock at idle. When the clutch is engaged, the knock stops. Mechanic A says that the crankshaft thrust bearing is worn. Mechanic B says that the rod bearings are loose. Who is right? (♦37-9)
    a. A only
    b. B only
    c. both A and B
    d. neither A nor B

CHAPTER 38
Diesel-Engine Service

After studying this chapter, and with the proper instruction and equipment, you should be able to:

1. Time the injection pump on various engines.
2. Remove and install the injection pump.
3. Service the valve train on various diesel engines.
4. Replace a precombustion chamber.
5. Replace a fuel-injection nozzle.
6. Replace a fuel line.
7. Adjust the idle speed on various diesel engines.

♦ 38-1 DIESEL-ENGINE TROUBLES

Except for the fuel system and the spark-ignition system, both the diesel engine and the spark-ignition engine can have many of the same troubles. Much of the engine trouble-diagnosis chart in Chap. 34 applies to both diesel and spark-ignition engines. Also, Chap. 19 contains a trouble-diagnosis chart on diesel-engine fuel-injection systems. Used together, these two charts give you the causes and checks or corrections to be made when diesel-engine trouble occurs. Following sections cover special servicing procedures required on diesel engines.

The causes of such troubles as engine will not crank and engine cranks slowly but does not start are the same for both engine types. The diesel engine should crank at a speed of at least 100 rpm cold and 240 hot.

♦ 38-2 DIESEL-ENGINE SERVICE

The engine described in the following diesel-engine servicing procedures is the General Motors V-8 automotive diesel engine. The manufacturer's service manual will provide you with the detailed procedures and specifications.

♦ 38-3 TIMING THE INJECTION PUMP

The marks on the top of the injection-pump adapter and the flange of the injection pump must align (Fig. 38-1). This cor-

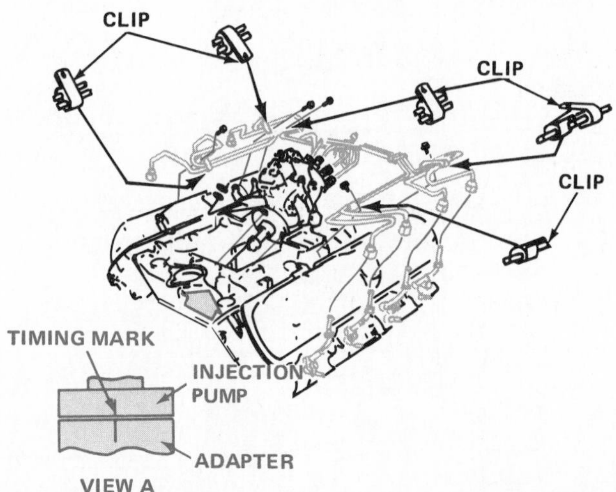

Fig. 38-1 Timing marks and injection-pump lines. *(Oldsmobile Division of General Motors Corporation)*

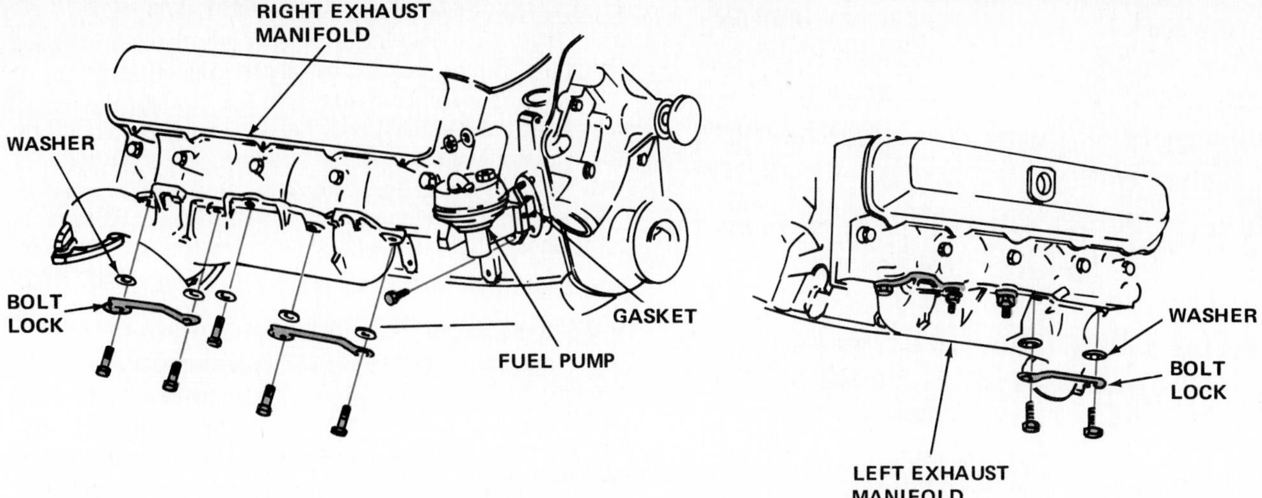

Fig. 38-2 Exhaust-manifold attaching bolts and bolt locks. After bolts are torqued to specifications, bend lock tabs around bolt heads. *(Oldsmobile Division of General Motors Corporation)*

rects the timing of the fuel delivery to the nozzles. The adjustment is made with the engine not running. To make the adjustment, loosen the three pump retaining nuts and align the marks. Torque nuts to specifications. Use an open-end wrench on the boss at the front of the pump to help rotate the pump into alignment.

♦ 38-4 EXHAUST-MANIFOLD REMOVAL AND INSTALLATION

Figure 38-2 shows how the exhaust manifolds are attached. Bolt locks are used to prevent the bolts from loosening. The diesel engine is subjected to greater vibrational stress. Therefore, the bolts must be secured with locks.

♦ 38-5 LINKAGE ADJUSTMENTS

Injection-pump timing must be correct before linkage is adjusted. The following may require adjustment:

1. Throttle rod (Fig. 38-3).
2. Transmission throttle-valve (TV) or detent cable (Fig. 38-4).
3. Transmission vacuum valve.
4. Slow-idle speed. This requires a magnetic tachometer inserted in the tach hole (Fig. 38-5). The slow-idle adjustment screw is shown in Fig. 38-6.

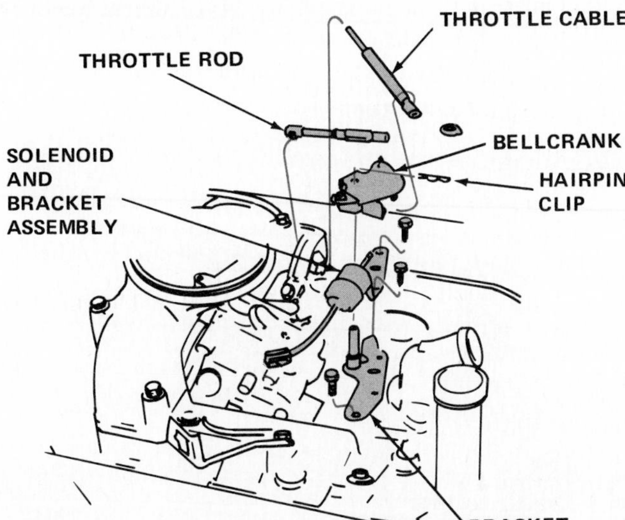

Fig. 38-3 Throttle linkage. *(Oldsmobile Division of General Motors Corporation)*

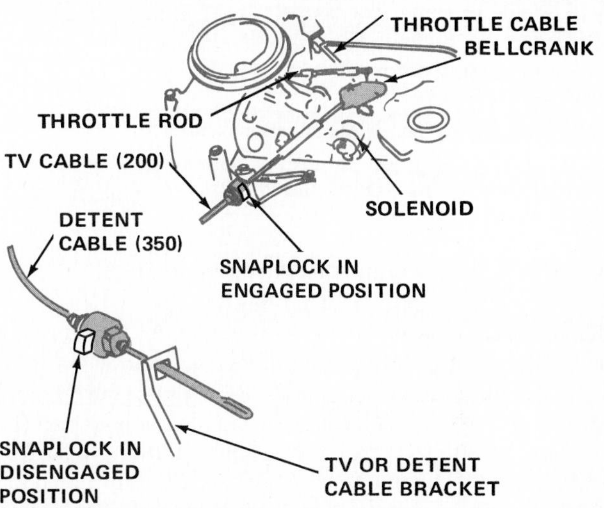

Fig. 38-4 Transmission throttle-valve (TV) cable adjustment. *(Oldsmobile Division of General Motors Corporation)*

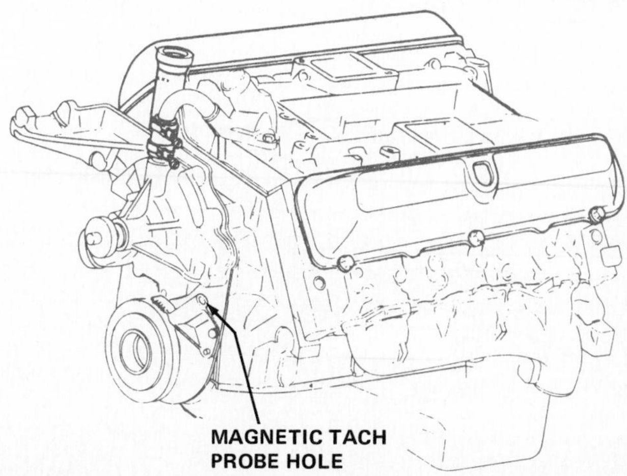

Fig. 38-5 Probe hole into which the magnetic tachometer probe is inserted to check engine rpm. *(Oldsmobile Division of General Motors Corporation)*

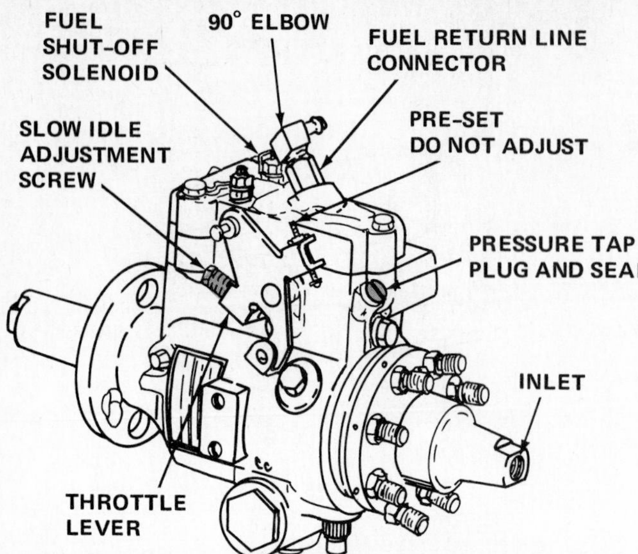

Fig. 38-6 Injection pump, showing locations of the slow-idle adjustment screw and the pressure tap plug. *(Oldsmobile Division of General Motors Corporation)*

5. Fast-idle solenoid. The fast-idle speed is controlled by the plunger in the fast-idle solenoid. It should be checked with the air conditioner on and the compressor wires disconnected.
6. Cruise-control servo relay rod.

Details are in the manufacturer's shop manual.

♦ 38-6 CHECKING INJECTION-PUMP-HOUSING FUEL PRESSURE

Remove the air cleaner and crossover (Fig. 38-7). Cover the manifold openings with screened covers. Remove the pressure tap plug from the pump (Fig. 38-6). Attach low-pressure gauge by screwing adapter into the tap-plug hole. Install the magnetic-pickup tachometer (Fig. 38-5). Check pressure at 1000 rpm (transmission in PARK). It should be 8 to 12 psi [55 to 83 kPa] with not more than 2 psi [14 kPa] fluctuation. If incorrect, remove the pump for repair.

Reinstall air crossover after removing the screened covers. Install the air cleaner.

♦ 38-7 INJECTION-PUMP FUEL LINES

When these lines (Fig. 38-1) are removed, the lines, nozzles, and pump fittings must be capped to keep dirt out.

To remove the lines, remove the air cleaner, filters, and pipes from valve covers, and the air crossover. Take off the line clamps and disconnect the fuel lines. Cap the open lines, nozzles, and pipe fittings (♦19-28).

On reassembly, do not bend or twist the lines. Torque the pump ends of the lines and the nozzle ends to specifications.

♦ 38-8 REMOVING INJECTION PUMP

After removing the fuel lines (♦38-7), disconnect the throttle rod and return springs. Take off the bellcrank, throttle, and throttle-valve (TV) cables. Remove fuel filter. Disconnect the fuel line at the fuel pump and the fuel-return line from the injection pump.

With everything disconnected from the pump, remove the nuts holding the pump and lift the pump off the engine. Cap all open lines and nozzles.

On reinstallation, remove the protective caps. Line up the offset tang on the pump drive shaft with the pump-driven gear (♦19-31) and install the pump. Adjust timing (♦38-3). Reconnect lines and linkages and adjust linkages as necessary.

♦ 38-9 VALVE-COVER INSTALLATION

With everything out of the way (injection pump, lines, air crossover) remove the attaching screws and lift the valve cover off (Fig. 38-8). After cleaning the valve compartment and cover, run a bead of sealer around the edge of the valve cover. Install the cover and tighten the screws. Reinstall all parts that were removed.

♦ 38-10 INTAKE-MANIFOLD REMOVAL AND INSTALLATION

Figure 38-9 shows the intake manifold with related parts, raised above the cylinder block. To remove the manifold, the radiator must be drained and everything above the manifold (injection pump, air crossover, vacuum pump, and all interfering lines) must be removed. The air-conditioner lines at

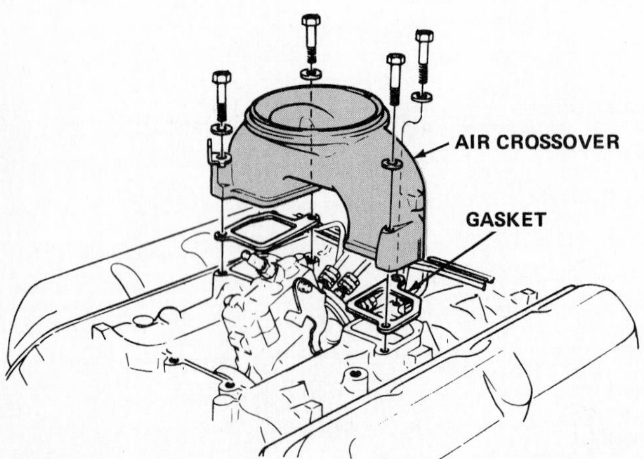

Fig. 38-7 Removing the air crossover from intake manifold. *(Oldsmobile Division of General Motors Corporation)*

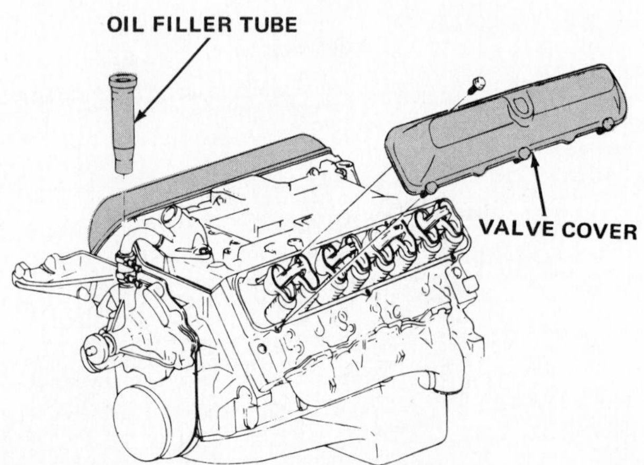

Fig. 38-8 Valve covers with oil-filler tube removed. *(Oldsmobile Division of General Motors Corporation)*

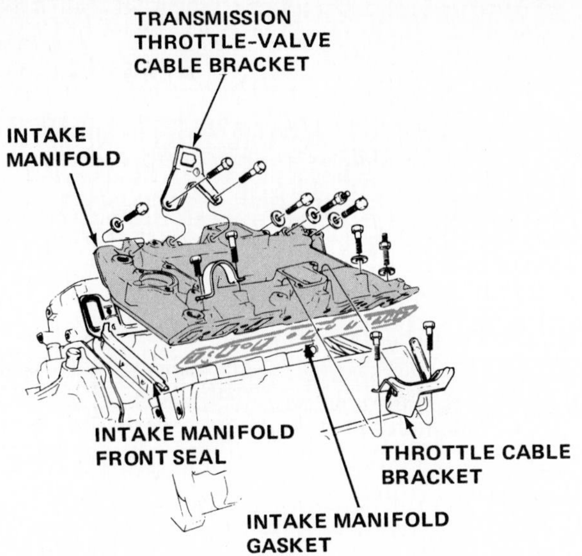

Fig. 38-9 Intake manifold and gasket. *(Oldsmobile Division of General Motors Corporation)*

the compressor are flexible. They can be moved to one side without disconnecting them from the compressor.

Do not bend the injection-pump lines when removing or installing them.

Figure 38-10 shows the bolt-tightening sequence for installing the intake manifold.

♦ 38-11 VALVE, VALVE-TRAIN, AND CYLINDER-HEAD SERVICE

Figure 38-11 shows a cylinder head removed and the rocker-arm attachment method. Figure 38-12 shows the valves, springs, and rocker arms for one cylinder. Note that valve rotators are used.

To remove a cylinder head, first remove the intake manifold. To do this, you must also remove everything above the manifold, as explained in ♦38-10.

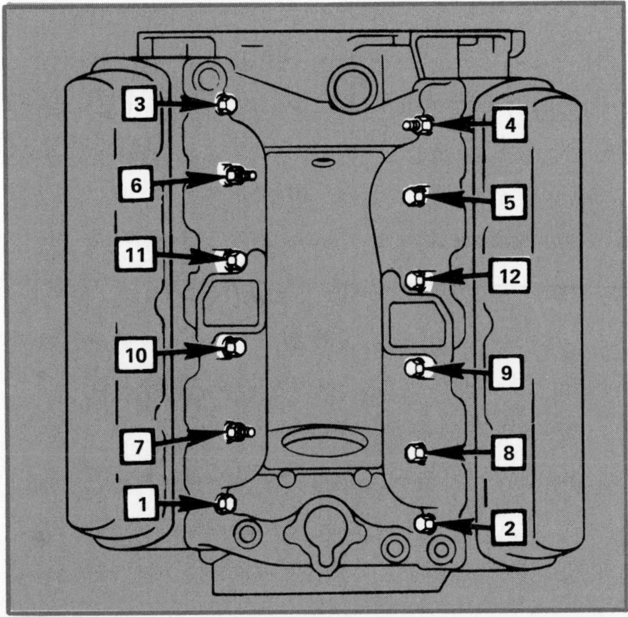

Fig. 38-10 Sequence for tightening intake-manifold bolts. *(Oldsmobile Division of General Motors Corporation)*

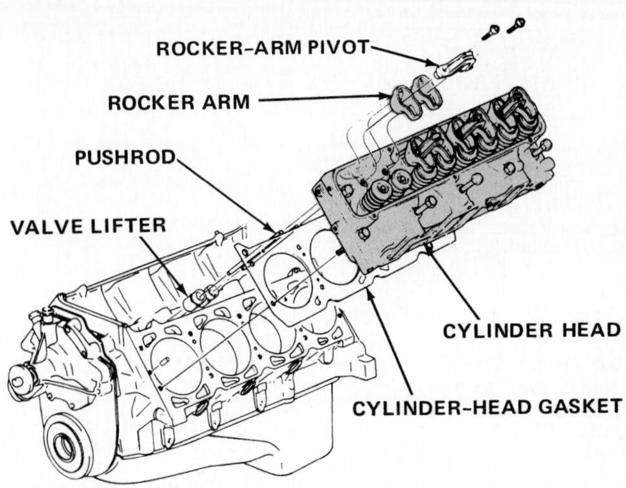

Fig. 38-11 Cylinder head removed from block with two rocker arms detached from head. *(Oldsmobile Division of General Motors Corporation)*

When installing the cylinder head, make sure the gasket surfaces are in good condition and clean. Do not damage the finish while handling the gasket. Dip all head bolts in oil and torque to specifications.

Valves, valve seats, springs, pushrods, and valve lifters are all checked and serviced as for the same parts in spark-ignition engines (Chap. 35).

♦ 38-12 PRECOMBUSTION-CHAMBER REPLACEMENT

The precombustion chambers (or *pre-chambers*) can be removed if necessary (Fig. 38-13). First remove the glow plug

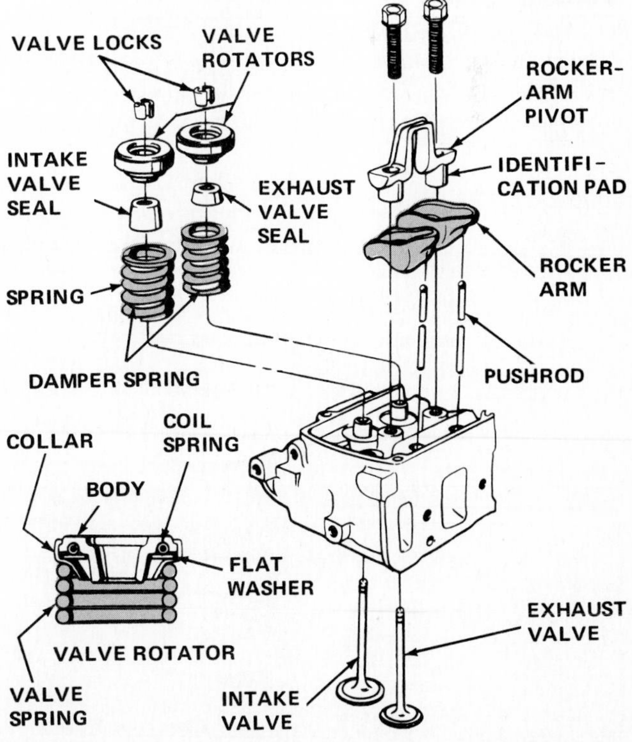

Fig. 38-12 Valve springs and rocker arms detached from head. *(Oldsmobile Division of General Motors Corporation)*

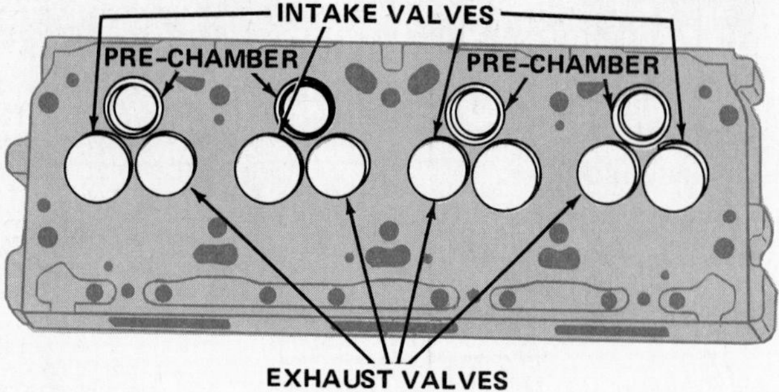

Fig. 38-13 Valve and pre-chamber locations. *(Oldsmobile Division of General Motors Corporation)*

and nozzle. Then tap out the pre-chamber with a small drift punch. The pre-chamber can be installed in only one position (Fig. 38-14). Tap it into place with a small soft-head mallet. Then install the glow plug and nozzle.

♦ 38-13 MEASURING VALVE-STEM HEIGHT

Removing, servicing, and installing valves is the same as for spark-ignition engines (Chap. 35). Also, valve seats are serviced in the same way. The valve face and seat are finished to provide an interference angle (Fig. 12-11). Figures 38-15 and 38-16 show a gauge being used to check the installed height of the valve stem and the valve rotator. If the height is excessive, grind the valve tip. If grinding the tip reduces the distance between the tip and rotator excessively, install a new valve.

♦ 38-14 ROD BEARINGS AND SIDE CLEARANCE

Connecting rods are checked and replaced as for spark-ignition engines (Chap. 36). It is especially important for the crankpins to be in good condition because of the high bearing

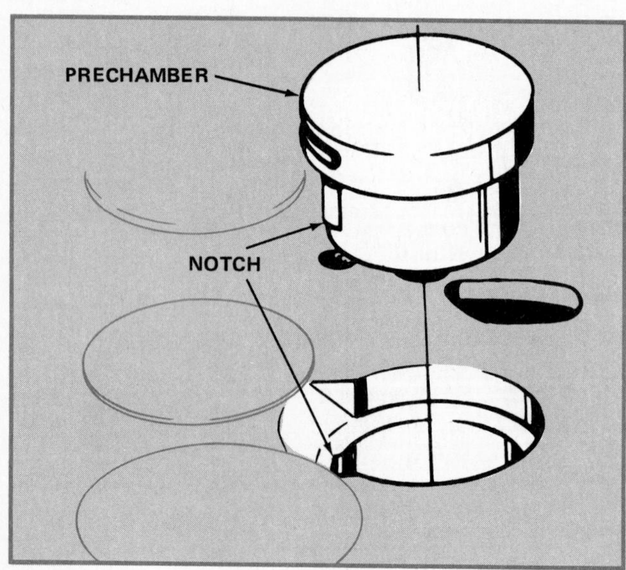

Fig. 38-14 Installation of pre-chamber in cylinder head. *(Oldsmobile Division of General Motors Corporation)*

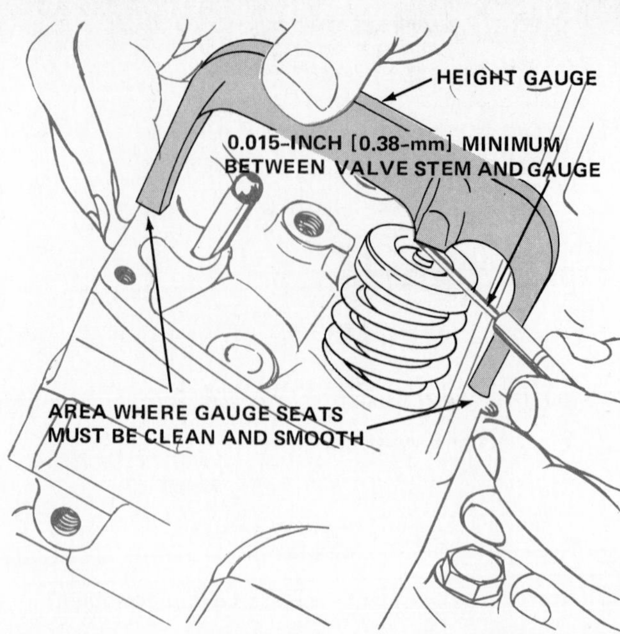

Fig. 38-15 Measuring valve-stem height with a special height gauge. *(Oldsmobile Division of General Motors Corporation)*

pressure. This is due to the high compression ratio of the engine and the high combustion pressures developed in the cylinders.

Figure 38-17 shows the procedures for checking rod side clearance. If it is not correct, the probable cause is a bent rod. Bent rods must be discarded. They cannot be straightened satisfactorily.

When the engine is being disassembled for service, always check the wear pattern across the face of the piston. A diagonal wear pattern on a used piston indicates a bent connecting rod.

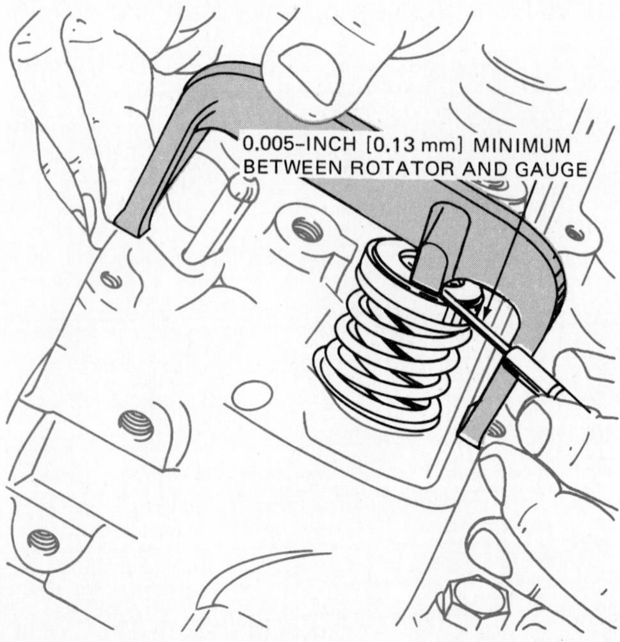

Fig. 38-16 Measuring distance valve stem rises above rotator. After grinding the valve stem, if the stem tip is less than 0.005 inch (0.13 mm) above rotator, install a new valve. *(Oldsmobile Division of General Motors Corporation)*

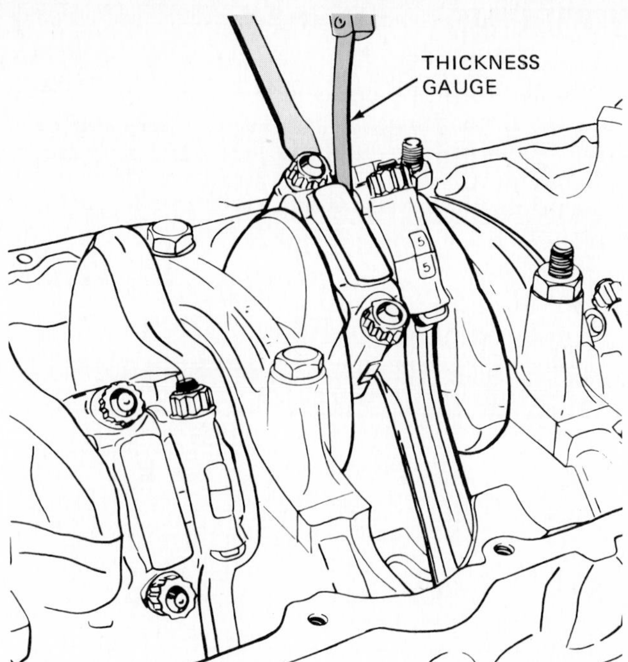

Fig. 38-17 Checking side clearance between connecting rods. Spread the rods apart with a screwdriver, and measure the clearance with a thickness gauge. The clearance should be 0.006 to 0.020 inch (0.15 to 0.5 mm). *(Oldsmobile Division of General Motors Corporation)*

♦ 38-15 DIESEL PISTON AND RING SERVICE

The basic piston and ring servicing and replacement procedures used in spark-ignition engines also apply to diesel engines. When installing the pistons, correctly locate the larger valve depressions (Fig. 38-18). The large valve depression on cylinders 1, 2, 3, and 4 goes to the front. On cylinders 5, 6, 7, and 8, it goes to the back.

♦ 38-16 SERVICING CAMSHAFT AND INJECTION-PUMP GEARS

The injection pump is driven by a pair of gears from the camshaft (Fig. 19-10). This compares with the way the ignition distributor is driven. To remove the camshaft from the diesel engine, you take off the same parts as with the spark-ignition engine. Figure 38-19 shows the engine front cover and other parts removed to show the timing chain and camshaft and crankshaft sprockets.

During reassembly, line up the timing marks on the two sprockets. Also, line up "0" marks on the pump drive and driven gears. Then time the injection pump.

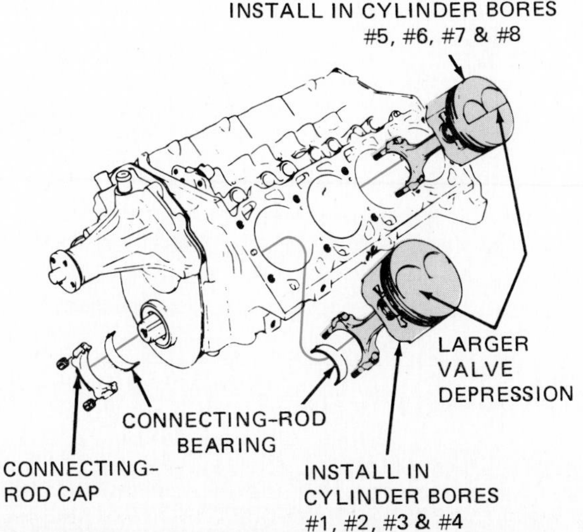

Fig. 38-18 Proper positioning of pistons in cylinder block. *(Oldsmobile Division of General Motors Corporation)*

♦ 38-17 CRANKSHAFT AND MAIN-BEARING SERVICE

The service and replacement procedure for the crankshaft and main bearings is the same as for spark-ignition engines (Chap. 37). Before the bearing-cap bolts are tightened, the flanges of the thrust bearings are aligned (Fig. 38-20). Use a block of wood to bump the shaft in each direction. This centers the thrust bearing. Then tighten all bearing-cap bolts.

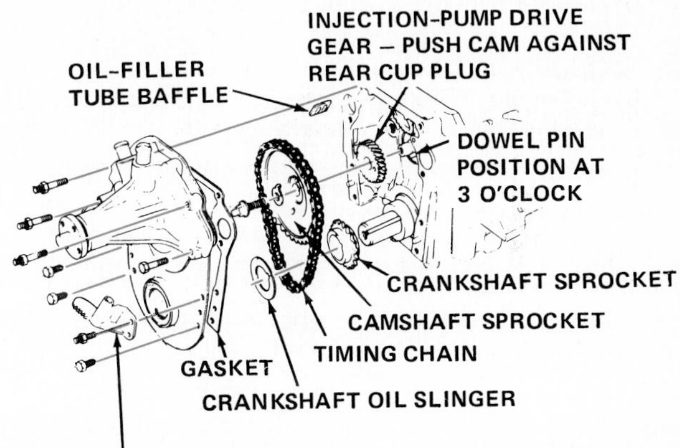

Fig. 38-19 Relationship of front cover, sprockets and timing chain, and injection-pump drive gear on camshaft. *(Oldsmobile Division of General Motors Corporation)*

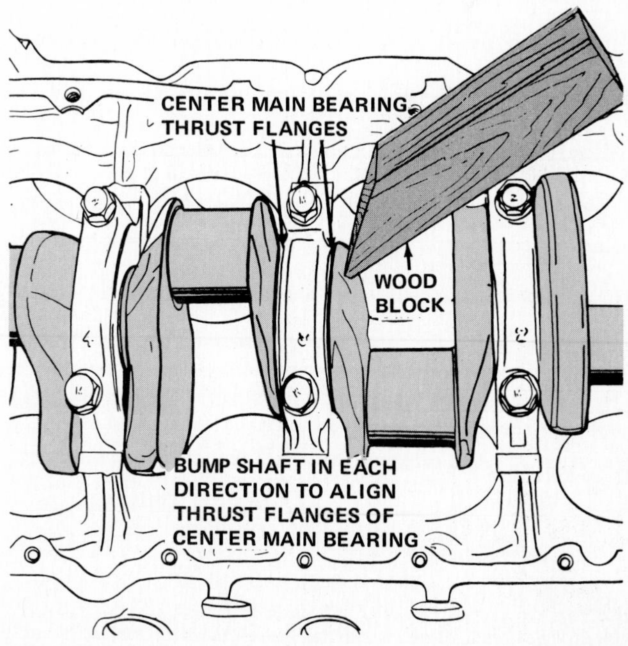

Fig. 38-20 Aligning thrust bearings by bumping the crankshaft with a wood block. *(Oldsmobile Division of General Motors Corporation)*

♦ REVIEW QUESTIONS ♦

Select the *one* correct, best, or most probable answer to each question. You can find the answers in the section indicated at the end of each question.

1. Normal cranking speed for starting a diesel engine is at least (♦38-1)
   a. 240 rpm cold, 480 rpm hot
   b. 100 rpm cold, 240 rpm hot
   c. 100 rpm cold, 110 rpm hot
   d. none of the above
2. The timing of the injection pump is made with the (♦38-3)
   a. engine running
   b. engine not running
   c. pump on a test bench
   d. none of the above
3. Before linkage adjustments can be made, be sure that (♦38-5)
   a. the glow plugs work
   b. the engine cranks at normal speed
   c. the injection-pump timing is correct
   d. none of the above
4. The fast-idle solenoid should be checked with (♦38-5)
   a. the air-conditioner compressor wires disconnected
   b. the transmission throttle-valve or detent cable disconnected
   c. the throttle rod disconnected
   d. all of the above
5. Fuel pressure in the injection-pump housing should be (♦38-6)
   a. not more than 2 psi [14 kPa]
   b. fluctuating more than 2 psi [14 kPa]
   c. checked with the engine off
   d. 8 to 12 psi [55 to 83 kPa]

PART 8

This part describes the automotive power train, which carries power from the engine to the car wheels. It consists of the clutch (in some cars), transmission, drive line, and differential. In addition to front- and rear-wheel drive, vehicles with four-wheel drive have also become common. All of these drive arrangements are described in the following chapters. There are nine chapters in Part 8:

Automotive Power Trains

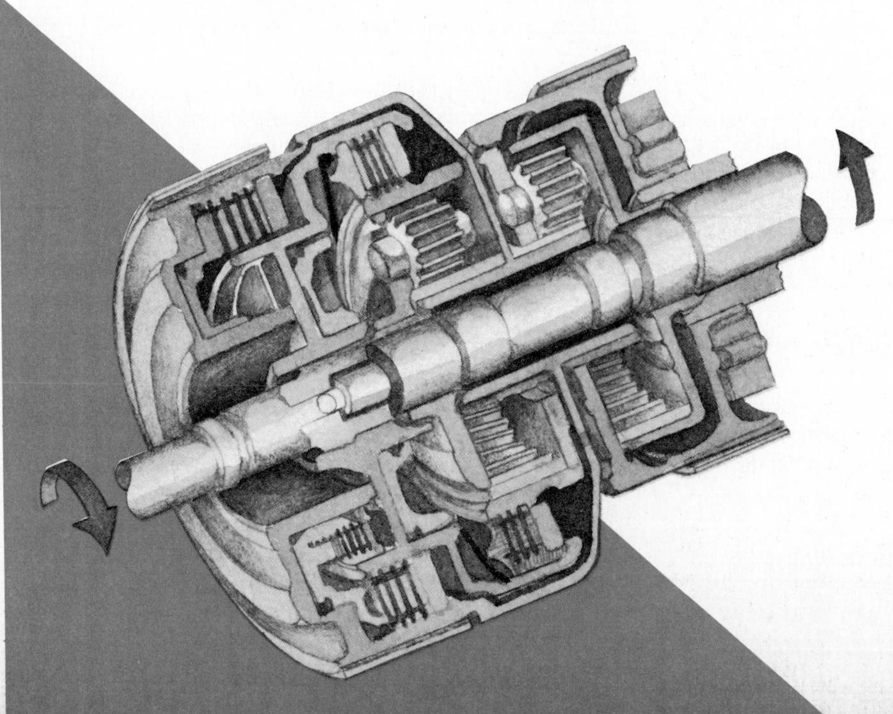

CHAPTER 39
Automotive Clutches: Operation and Service

After studying this chapter, and with proper instruction and equipment, you should be able to:

1. Explain the purpose of a clutch on cars with manual transmissions.
2. Describe the construction and operation of a typical clutch.
3. Explain the difference between a coil-spring and a diaphragm-spring clutch.
4. Describe the operation of a hydraulic clutch.
5. List nine clutch troubles and explain possible causes and checks and corrections for each.
6. Service and adjust clutch linkage.
7. Remove and replace a clutch.

♦ 39-1 TWO DRIVE-TRAIN ARRANGEMENTS

There are two basic drive-train arrangements in common use today. These are front-engine rear-wheel drive and front-engine front-wheel drive (Fig. 39-1). Figure 39-2 shows the locations of the major power-train components for the front-engine rear-wheel-drive car. Figure 39-3 shows the power train for the front-engine front-wheel-drive arrangement. In both systems, the clutch is located between the engine and transmission.

The clutches in both arrangements are basically the same. The purpose, construction, and operation of the various clutches used in automobiles with manual transmissions are described in the following sections. *Manual transmissions* are gearboxes that are shifted by hand, or "manually."

♦ 39-2 PURPOSE OF CLUTCH

The clutch is used on cars with transmissions that are shifted by hand. It allows the driver to couple the engine to, or uncouple the engine from, the transmission. Figure 39-4 shows the clutch arrangement for a front-engine rear-wheel-drive car. The clutch is linked to the clutch pedal in the passenger compartment. When the driver presses down on the clutch pedal, the linkage causes the clutch to disengage. This uncouples the engine from the transmission. When the driver releases the clutch pedal, springs in the clutch cause it to engage again. Now power can flow from the engine, through the clutch, to the transmission and power train.

♦ 39-3 OPERATION OF CLUTCH

Basically, the clutch consists of three parts. These are the engine flywheel, a friction disk, and a pressure plate (Fig. 39-5). When the engine is running, the flywheel is rotating. The pressure plate is attached to the flywheel so the pressure plate also rotates. The friction disk is located between the two. When the clutch is released (left in Fig. 39-5), the driver has pushed down on the clutch pedal. This action forces the

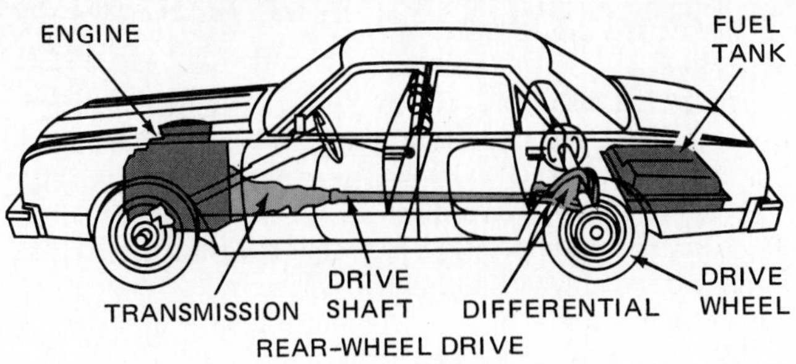

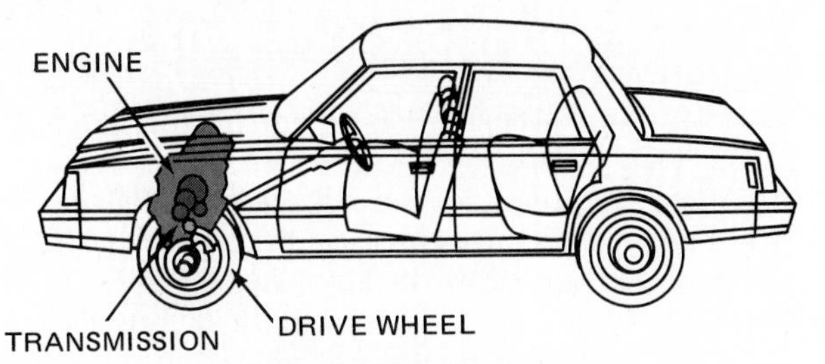

Fig. 39-1 Comparison of the power-train layout in a car with rear-wheel drive (top) and a car with front-wheel drive (bottom). *(Chrysler Corporation)*

pressure plate to move away from the friction disk. There are now air gaps between the flywheel and the friction disk, and between the friction disk and the pressure plate. No power can be transmitted through the clutch.

When the driver releases the clutch pedal, power can flow through the clutch. Springs in the clutch force the pressure plate against the friction disk. This action clamps the friction disk tightly between the flywheel and the pressure plate (right, Fig. 39-5). Now, the pressure plate and friction disk rotate with the flywheel. The friction disk is assembled on a splined shaft that carries the rotary motion to the transmission. This shaft is called the *clutch shaft*, or *transmission input shaft*.

♦ 39-4 CLUTCH CONSTRUCTION

Two basic types of clutch are the coil-spring clutch and the diaphragm-spring clutch. The difference between them is in

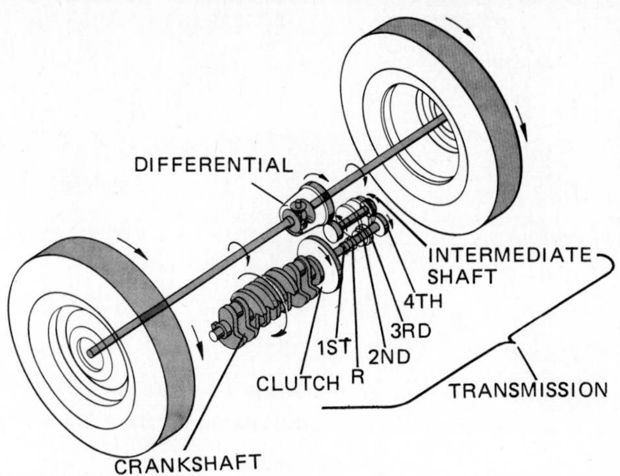

Fig. 39-3 Power flow from the engine crankshaft to the wheels of a front-wheel-drive car. The transmission and differential are combined into an assembly called a *transaxle. (Chrysler Corporation)*

the type of spring used. The clutch shown in Fig. 39-4 uses nine coil springs as pressure springs (only one pressure spring is shown). The clutch shown in Fig. 39-6 uses a diaphragm spring. These two basic types of clutches are described in following sections.

♦ 39-5 COIL-SPRING CLUTCH

The coil-spring clutch has a series of coil springs set in a circle. Figures 39-4 and 39-7 show the coil-spring clutch in sectional and cutaway views. Figure 39-8 shows a disassembled coil-spring clutch. The flywheel and transmission shaft are not shown. The friction disk (or *driven plate*) is about 12 inches [305 mm] in diameter, and it is mounted on the transmission input shaft. The disk has splines in its hub that match the splines on the input shaft (Figs. 39-4 and 39-7). These splines consist of two sets of teeth. The internal teeth in the hub of the friction disk match the external teeth on the shaft. When the friction disk is driven, it turns the transmission input shaft. The end of the transmission input shaft rides

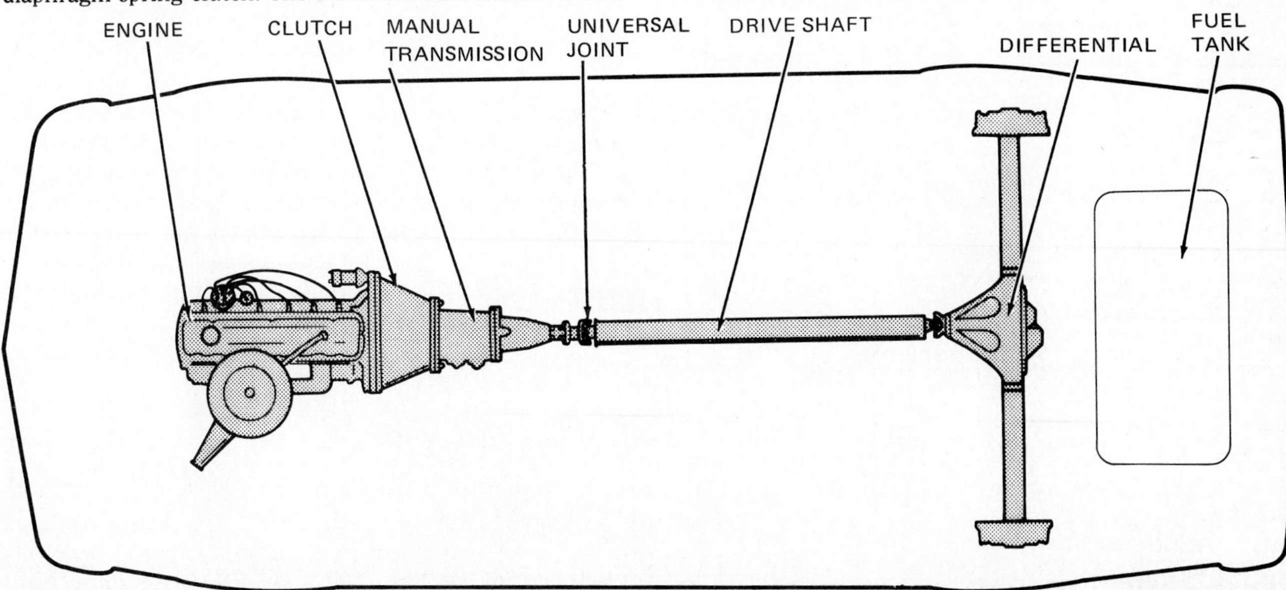

Fig. 39-2 Power train for a rear-wheel-drive car with a manual transmission.

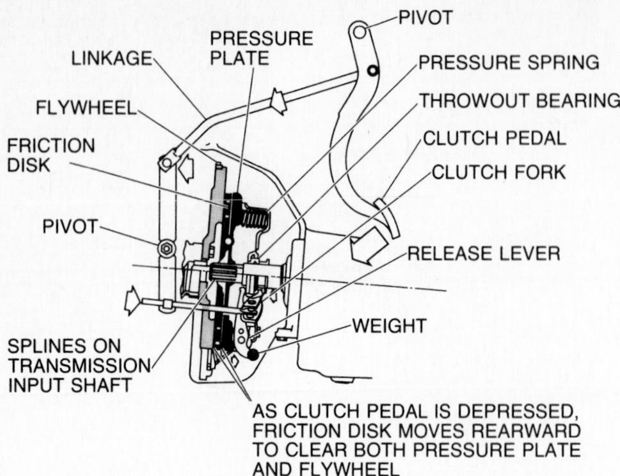

Fig. 39-4 Sectional view of a clutch, showing the linkage to the clutch pedal. *(Buick Motor Division of General Motors Corporation)*

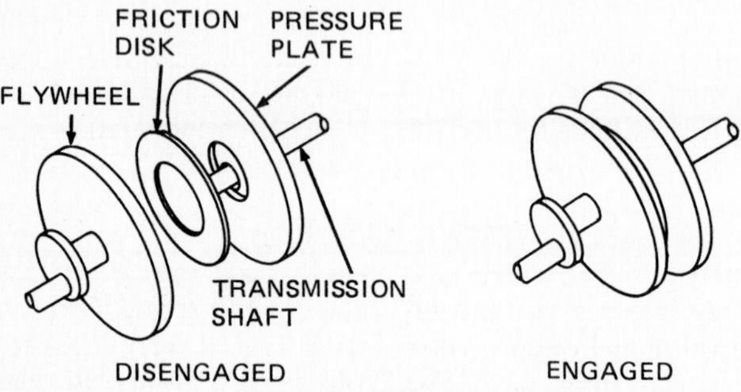

Fig. 39-5 Basic clutch action. Left, clutch disengaged. Pressure plate and friction disk are separated and have moved away from the flywheel. Right, clutch engaged. Pressure plate is clamping the friction disk to the flywheel so all rotate together.

in a pilot bearing or bushing in the end of the crankshaft (Fig. 39-9).

The clutch also has a pressure-plate-and-cover assembly, which includes a series of coil springs. The cover is bolted to the engine flywheel (Fig. 39-6). The springs provide the force to hold the friction disk against the flywheel. Then, when the flywheel turns, the pressure plate and the friction disk also turn. However, when the clutch is disengaged, the spring force is relieved so that the friction disk and the flywheel can rotate separately.

♦ 39-6 OPERATION OF COIL-SPRING CLUTCH

Figure 39-9 shows a typical coil-spring clutch. The major parts are shown disassembled at the right. Nine springs are used, although only three are shown in Fig. 39-9. The springs are held between the clutch cover and the pressure plate. In the engaged position shown in Fig. 39-9, the springs are clamping the friction disk (driven plate) tightly between the flywheel and the pressure plate (Fig. 39-5). This forces the friction disk to rotate with the flywheel.

When the driver operates the clutch pedal to disengage the clutch, the linkage from the pedal forces the release bearing (or *throwout bearing*) inward (to the left in Fig. 39-10). As the release bearing moves to the left, it pushes against the inner ends of three release levers (or *release fingers*). The release levers are pivoted on pins and eyebolts, as shown in Figs. 39-8 and 39-9. When the inner ends of the release levers are pushed in by the release bearing, the outer ends are moved out.

This motion is carried by struts to the pressure plate (Figs. 39-8 to 39-10). Therefore, the pressure plate is moved to the right (in Fig. 39-10), and the springs are compressed. With the spring force off the friction disk, a small airspace appears between the disk and the flywheel, and between the other side of the disk and the pressure plate (Figs. 39-5 and 39-10). Now the clutch is disengaged. The flywheel can rotate without transmitting power through the friction disk.

To engage the clutch, release the clutch pedal. This re-

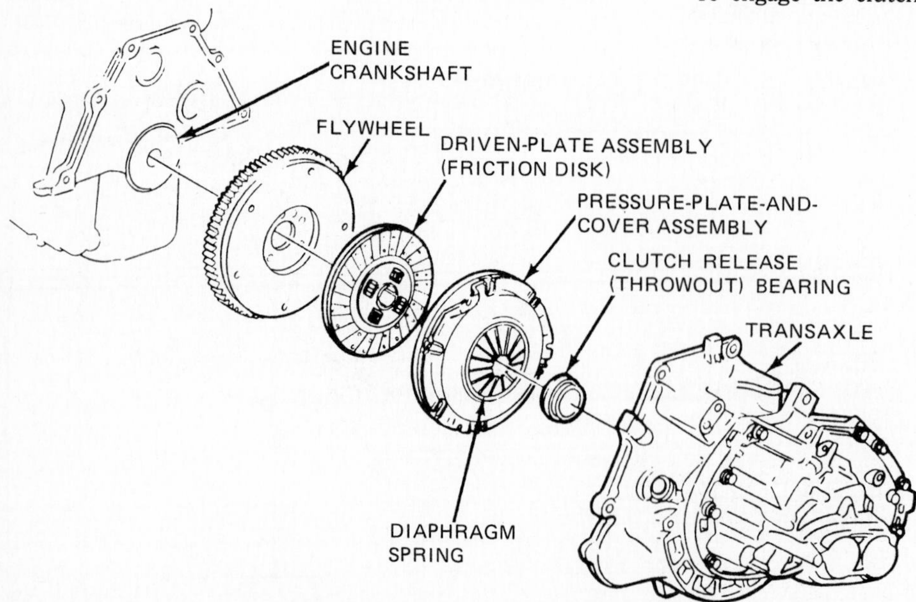

Fig. 39-6 Disassembled diaphragm-spring clutch used in a front-wheel-drive car with a transaxle. *(Chevrolet Motor Division of General Motors Corporation)*

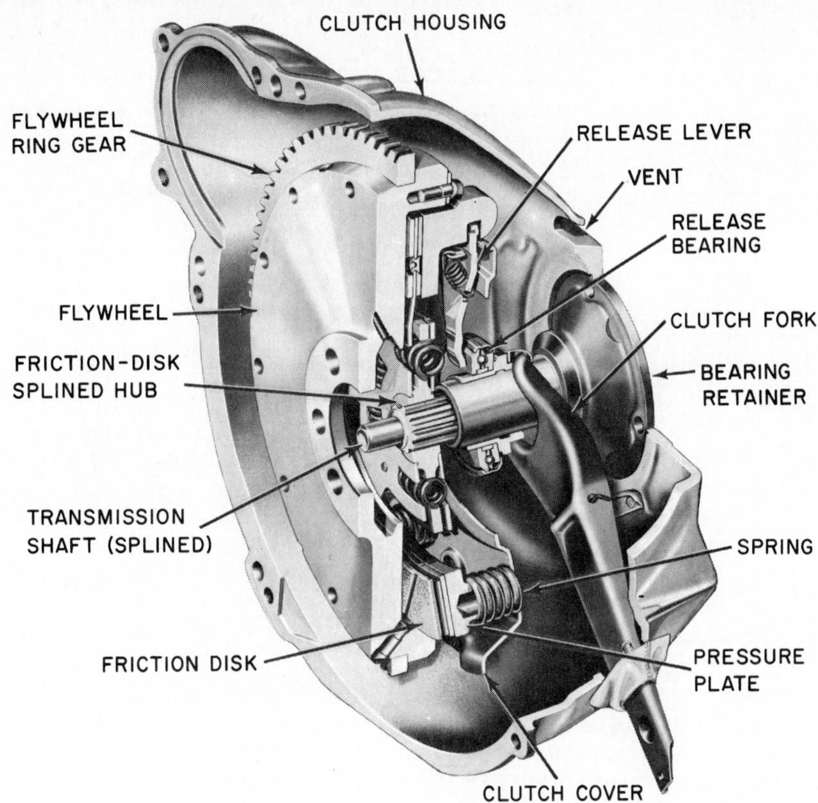

Fig. 39-7 Partially cutaway coil-spring clutch. *(Ford Motor Company)*

moves the linkage force from the release bearing. The springs push the pressure plate to the left (in Fig. 39-9). The friction disk is again clamped tightly between the flywheel and the pressure plate. The friction disk must again rotate with the flywheel.

In this system, there must be some free play in the clutch linkage. The clutch pedal must move about 1 inch [25 mm] before all the free play is taken up. Only then does the release bearing come into contact with the release levers. Without this free play, the release bearing would be riding on the ends of the release levers. This would cause rapid wear of the bearing and release levers.

♦ 39-7 FRICTION DISK

The friction disk, or driven plate, is shown partly cut away in Fig. 39-11. It consists of a hub and a plate, with facings attached to the plate. The friction disk has cushion springs and dampening springs. The cushion springs are slightly waved,

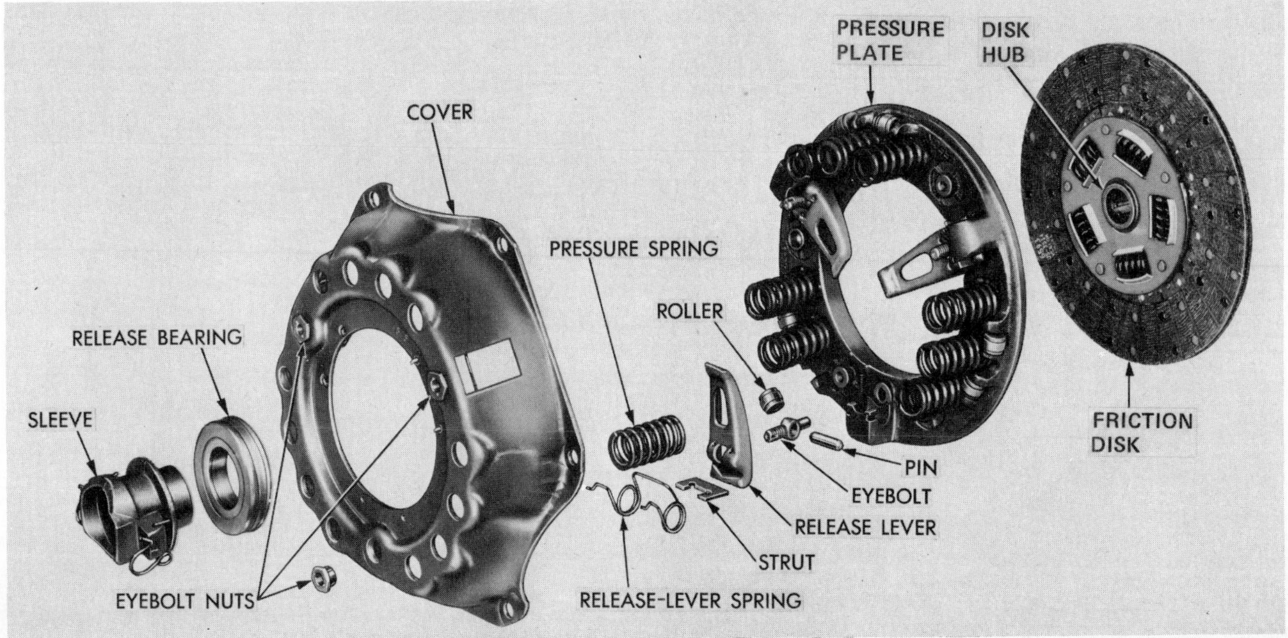

Fig. 39-8 Disassembled coil-spring clutch. *(Chrysler Corporation)*

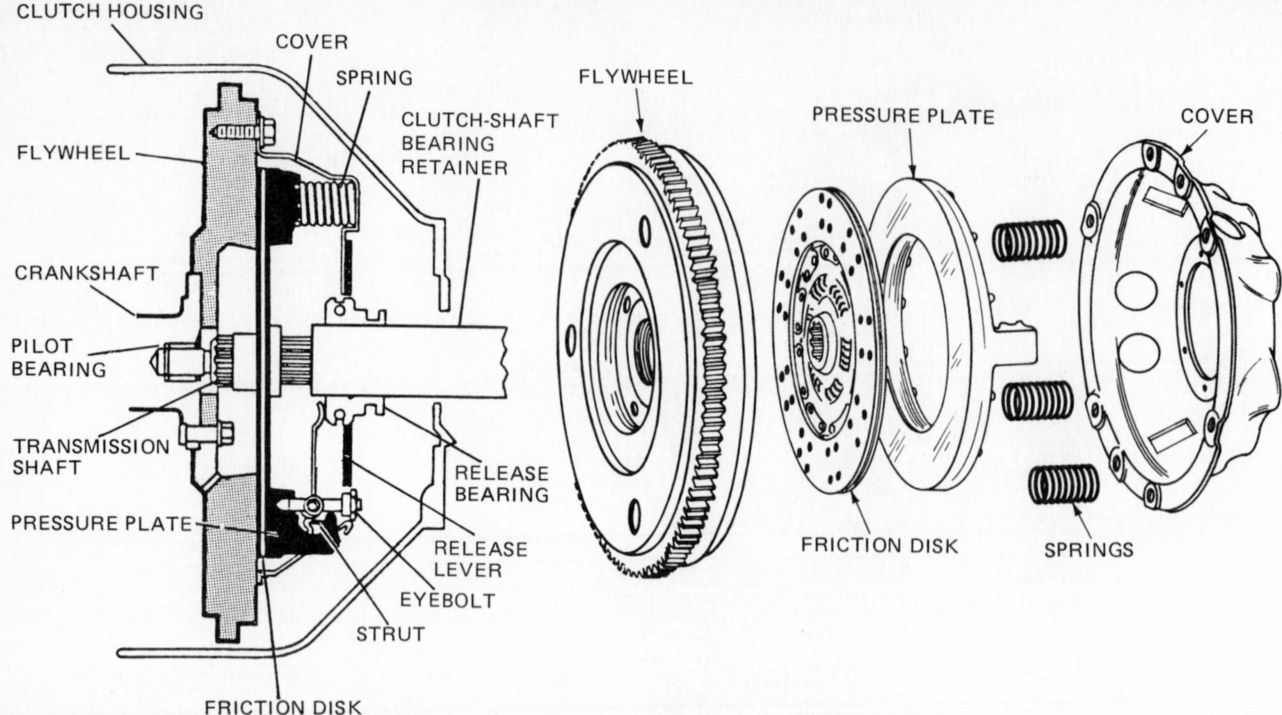

Fig. 39-9 Coil-spring clutch in the engaged position (at left). Major clutch parts are shown to the right.

or curled. The cushion springs are attached to the plate, and the friction facings are attached to the springs. When the clutch is engaged, the cushion springs compress slightly to take up the shock of engagement. The dampening springs are heavy coil springs set in a circle around the hub. The hub is driven through these springs. They help to smooth out the torsional vibration (the power pulses from the engine) so that the power flow to the transmission is smooth.

There are grooves in both sides of the friction-disk facings. These grooves prevent the facings from sticking to the flywheel face and pressure plate when the clutch is disengaged. The grooves break any vacuum that might form and cause the facings to stick to the flywheel or pressure plate.

The facings on many friction disks are made of cotton and asbestos fibers woven or molded together and impregnated with resins or other binding agents. In many friction disks, copper wires are woven or pressed into the facings to give then added strength. However, asbestos is being replaced with other materials in many clutches. Some friction disks have ceramic-metallic facings (Fig. 39-12).

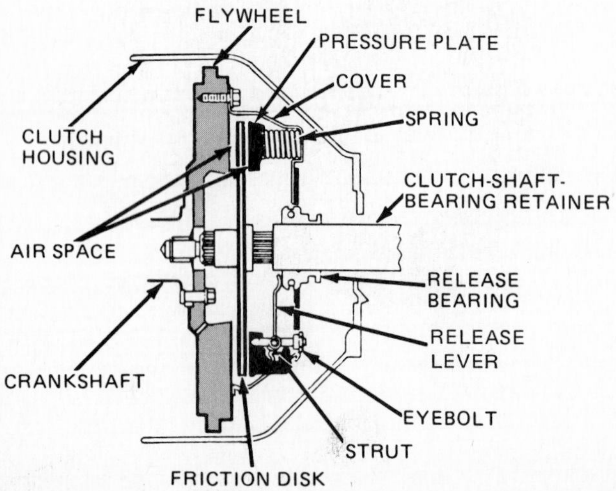

Fig. 39-10 Coil-spring clutch in the disengaged position.

♦ CAUTION ♦ Asbestos is used in the facings of many friction disks because it can withstand the high pressures and temperatures inside the clutch. However, authorities claim that breathing asbestos dust can cause lung cancer. For this reason, be careful when working around clutches. Do not blow the dust out of the clutch housing with compressed air. This dust may contain powdered asbestos. The compressed air could send the dust up into the air around

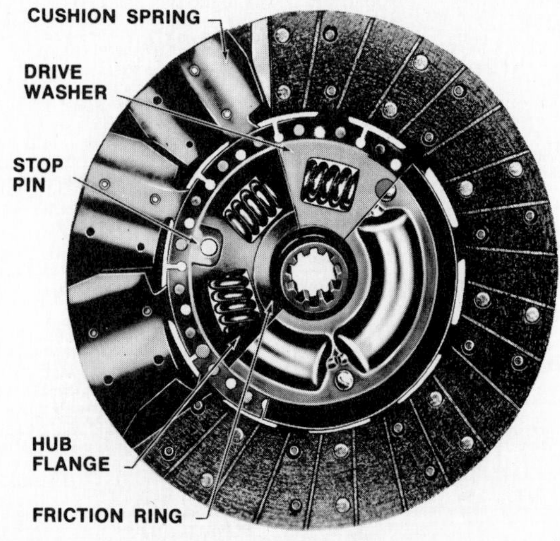

Fig. 39-11 Friction disk, or driven plate. Facings and drive washer have been partly cut away to show springs. *(Buick Motor Division of General Motors Corporation)*

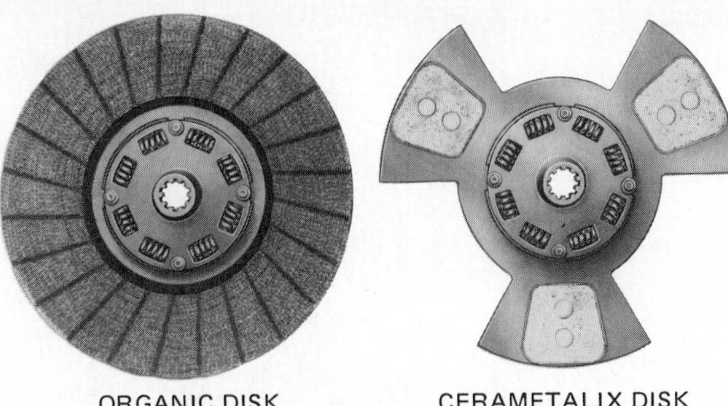

Fig. 39-12 Organic or asbestos friction disk compared with a ceramic-metallic (Cerametalix) friction disk. *(Chevrolet Motor Division of General Motors Corporation)*

you and you could inhale it. Instead, use a vacuum suitable for asbestos, or cleaning should be done wet. Use damp cloths to wipe out the clutch housing. Then, after working on a clutch, always wash your hands thoroughly to remove any trace of asbestos dust.

♦ 39-8 SEMICENTRIFUGAL COIL-SPRING CLUTCH

The coil springs must be strong enough to prevent disk slippage. However, the stronger the coil springs, the harder the driver must push on the clutch pedal. One solution to this problem is to use semicentrifugal clutches. These clutches use relatively light coil springs. They exert enough force for low- to medium- speed operation. Then, as engine speed increases, centrifugal devices in the clutch increase the force on the friction disk. Therefore, there will be no slippage when power demands through the clutch are high.

One type of centrifugal clutch has weights on the ends of the release levers. As speed increases, centrifugal force causes the weights to add to the force of the springs. A second system uses rollers (Fig. 39-8). As speed increases, the rollers move out between the clutch cover and pressure plate. This adds to the force on the friction disk.

♦ 39-9 DIAPHRAGM-SPRING CLUTCH

A diaphragm-spring clutch (Fig. 39-6) is widely used on cars with small- to medium-size engines. It has a diaphragm spring (Fig. 39-13) that supplies the force to hold the friction disk against the flywheel. The diaphragm spring also acts as the release lever to take up the spring force when the clutch is disengaged. Figure 39-14 shows the diaphragm-spring clutch in sectional view.

The diaphragm spring is a Belleville spring that has a solid ring on the outer diameter. It has a series of tapering fingers pointing inward toward the center of the clutch (Fig. 39-13). The action of the clutch diaphragm is like the action that takes place when the bottom of an oil can is depressed. It "dishes" inward. When the throwout bearing moves in against the ends of the fingers, the entire diaphragm is forced against a pivot ring, causing the diaphragm to dish inward. This moves the pressure plate away from the friction disk.

Figures 39-15 and 39-16 illustrate the two positions of the diaphragm spring and clutch parts. In the engaged position (Fig. 39-15), the diaphragm spring is slightly dished, with the tapering fingers pointing slightly away from the flywheel. This position places spring force against the pressure plate around the entire circumference of the diaphragm spring. The diaphragm spring is shaped to exert this initial force.

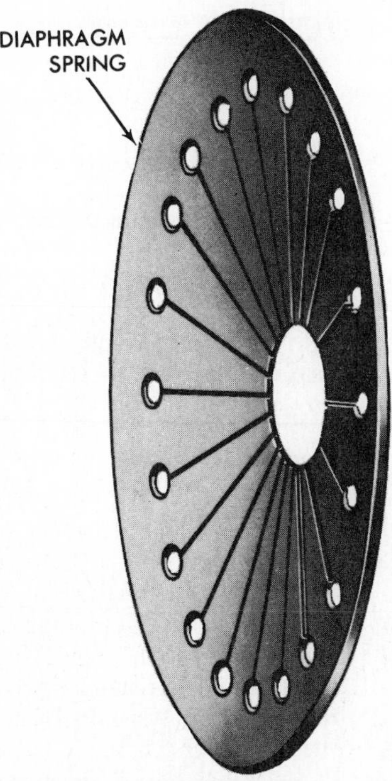

Fig. 39-13 A diaphragm spring used in the diaphragm-spring clutch.

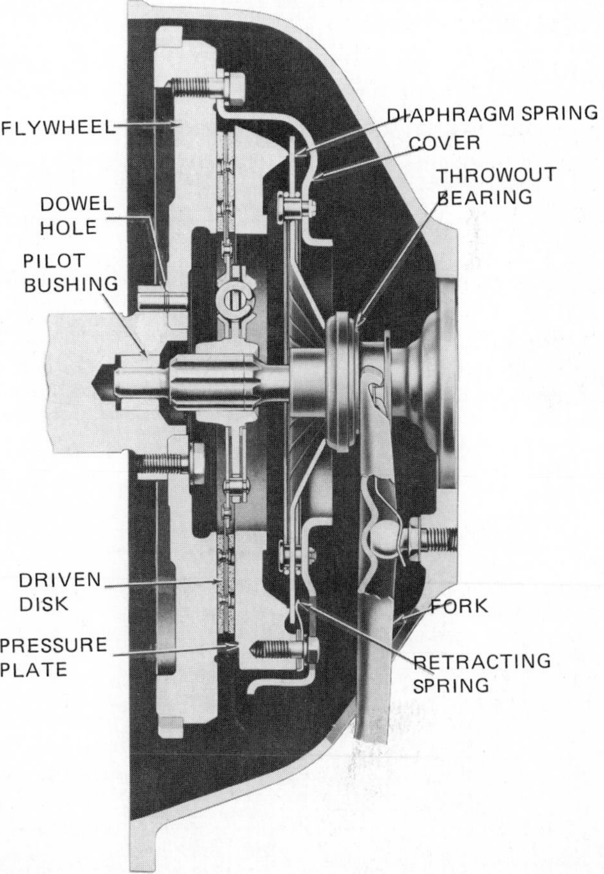

Fig. 39-14 Diaphragm-spring clutch with bent tapering fingers. *(Chevrolet Motor Division of General Motors Corporation)*

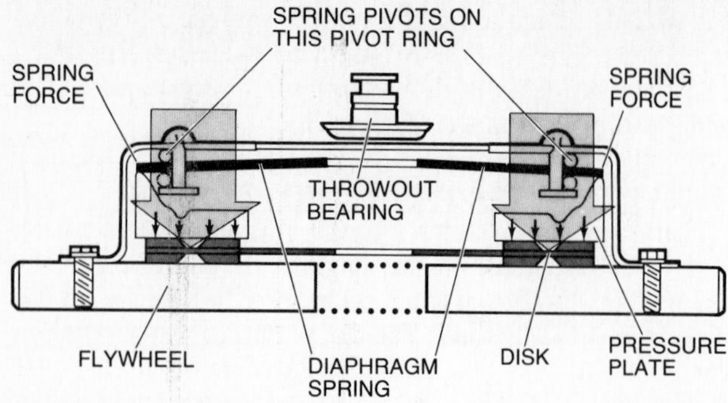

Fig. 39-15 Diaphragm-spring clutch in the engaged position. *(Chevrolet Motor Division of General Motors Corporation)*

When the throwout bearing is moved inward against the spring fingers (as the clutch pedal is depressed), the spring is forced to pivot about the inner pivot ring, dishing in the opposite direction. The outer circumference of the spring now lifts the pressure plate away through a series of retracting springs placed about the outer circumference of the pressure plate (Fig. 39-16).

The clutch shown in Figs. 39-15 and 39-16 has a differently shaped spring than the clutch shown in Fig. 39-14. On the diaphragm spring shown in Fig. 39-14, the inner ends of the tapered fingers are bent outward toward the throwout bearing. In Figs. 39-15 and 39-16, the fingers are flat. However, both clutches function in a similar manner. Spring force varies according to the size and thickness of the diaphragm spring.

♦ 39-10 DOUBLE-DISK CLUTCHES

Sometimes a clutch with greater holding power is required. This may be because more power must go through the clutch but size is limited. Therefore, a clutch with a larger diameter cannot be installed. Then a clutch with two friction disks may be used.

Figure 39-17 shows a coil-spring clutch which has two friction disks separated by an intermediate, or center, drive plate. Figure 39-18 is a sectional view of a diaphragm-spring clutch using two friction disks and an intermediate pressure plate. The added pressure plate and friction disk give the double-disk clutch greater holding power. This makes it suitable for use with higher-output engines. Double-disk clutches are widely used in medium and heavy trucks.

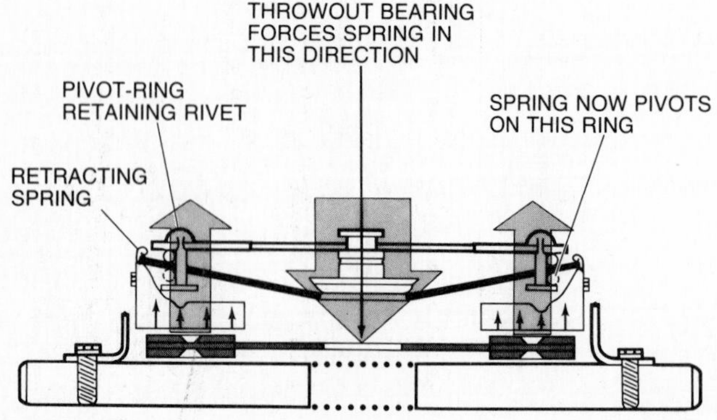

Fig. 39-16 Diaphragm-spring clutch in the disengaged position. *(Chevrolet Motor Division of General Motors Corporation)*

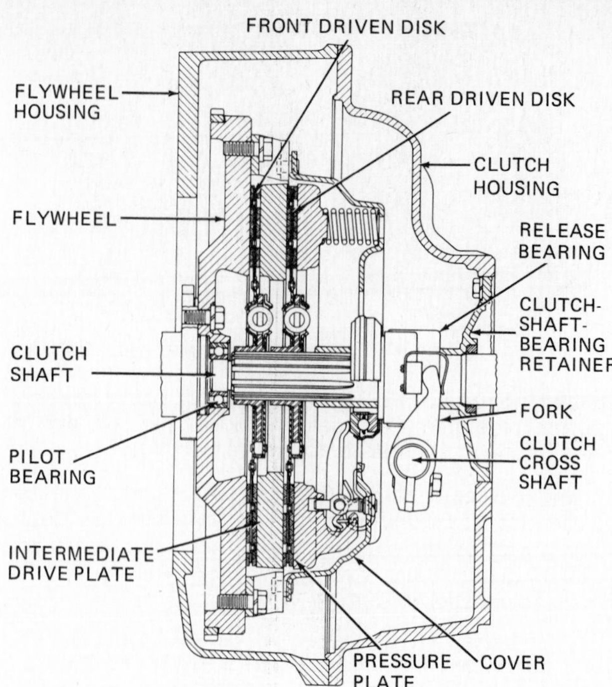

Fig. 39-17 A coil-spring clutch with two friction disks and an intermediate drive plate. *(Chevrolet Motor Division of General Motors Corporation)*

Use of the second friction disk adds clutch-plate area, thereby providing greater torque-carrying capacity. When the clutch is engaged, each friction disk transmits half of the flywheel torque to the clutch shaft. These clutches are operated and work in the same way as single-disk coil-spring clutches (♦39-6 and 39-9).

♦ 39-11 CLUTCH FOR TRANSAXLE

The *transaxle* is a combined transmission and differential, with a clutch on manually shifted transmissions. In front-wheel-drive cars, the transaxle is attached to the engine. Figure 39-3 shows in simplified form the power flow from the engine crankshaft, through the clutch, to the transmission, differential, and wheel. Transaxles are covered in detail in a later chapter.

The clutch in most small cars with a transaxle is a single-disk diaphragm-spring clutch (♦39-9). Instead of being contained within a clutch housing, the clutch has been adapted to fit into the transaxle assembly.

♦ 39-12 CLUTCH LINKAGE

Control of the clutch is maintained by the driver through the foot pedal and suitable linkage. Three types of clutch linkages are used in cars. These are the rod type, the cable type, and the hydraulic type. Although the operation of each type of clutch linkage varies, the purpose is the same. That is to convert a light force applied to the clutch pedal (which travels a relatively long distance) into a greatly increased force that moves the pressure plate a very short distance.

The linkage starts at the clutch pedal and ends at the clutch release lever, or clutch "fork." Various views of how the levers are shaped and located are shown in Figs. 39-4, 39-7, 39-17, and 39-18.

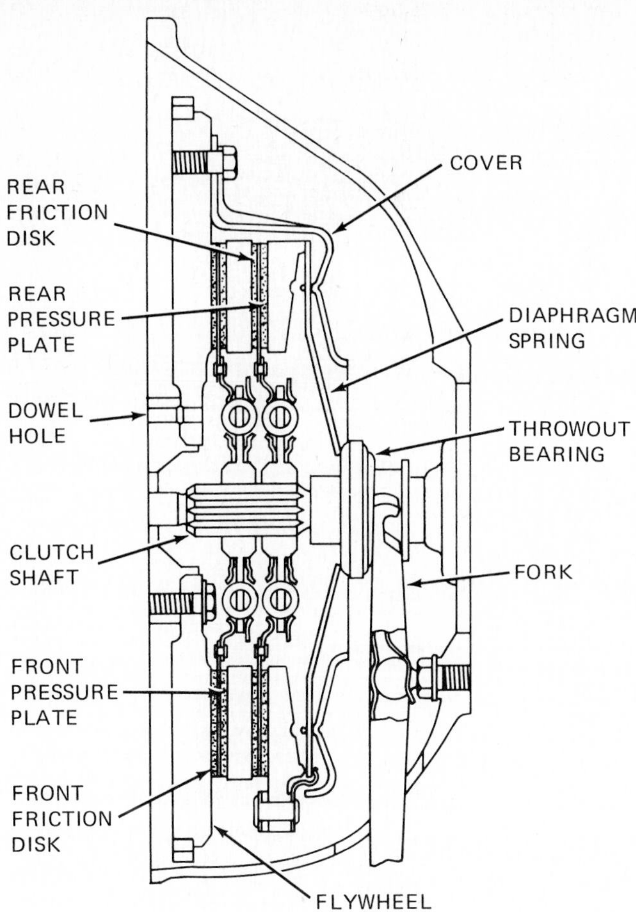

Fig. 39-18 Sectional view of a diaphragm-spring clutch using two pressure plates. *(Chevrolet Motor Division of General Motors Corporation)*

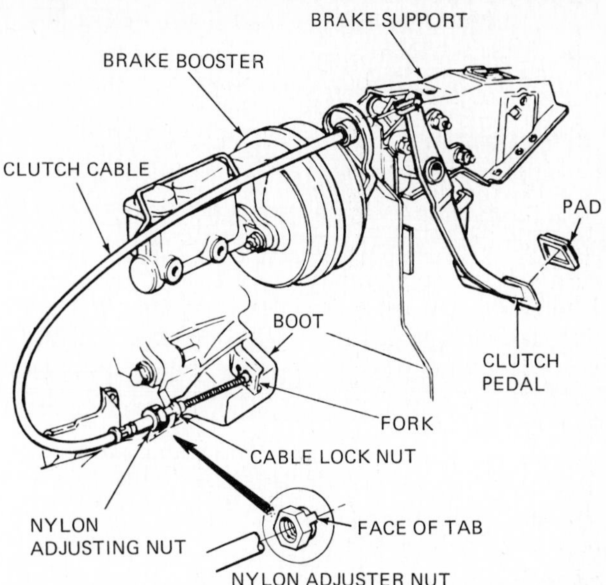

Fig. 39-19 A clutch fork operated by a cable from the clutch pedal. The cable routing should follow a smooth arc and maintain adequate clearance to all other components. *(Ford Motor Company)*

Figure 39-4 shows a rod-type clutch linkage. The system includes an overcenter spring. Its purpose is to reduce the force the driver must apply to the foot pedal to operate it.

A typical cable-operated clutch linkage is shown in Fig. 39-19. On many cars, it is easier for the manufacturer to install a cable system than to try to develop a workable rod arrangement.

♦ 39-13 HYDRAULIC CLUTCH LINKAGE

A hydraulically operated clutch linkage (Fig. 39-20) is used when the clutch is located so that it would be difficult to run rods or cable from the foot pedal to the clutch. This type of linkage is also used on high-output engines, which require heavy pressure-plate springs. When a clutch is designed to transmit high torque, strong springs are used to provide sufficient force on the friction disk. With insufficient force, the pressure plate and flywheel would slip on the friction disk, quickly ruining it.

However, heavy spring force increases the force that must be applied to the clutch fork. This, in turn, increases the force that the driver must apply to the clutch pedal. To reduce the force required to operate the clutch pedal, a hydraulic system is used.

Figure 39-20 shows a hydraulically operated clutch. The clutch pedal does not work the release lever directly through rods or cable. Instead, when the driver pushes down on the clutch pedal, a pushrod is forced into a master cylinder. This forces hydraulic fluid out of the master cylinder, through a tube, and into a servo, or "slave," cylinder. This is similar to the action in a hydraulic brake system when the brakes are applied (Chap. 51).

As the fluid is forced into the servo cylinder, the fluid forces a piston and pushrod out. This movement causes the clutch fork to move, pushing the throwout bearing against the release levers on the pressure plate.

The hydraulic system can be designed to multiply the driver's efforts so that a light force applied to the foot pedal produces a much greater force on the clutch fork. A small piston in the master cylinder travels a relatively long distance with only a low input force. This moves a larger piston in the servo cylinder a short distance, transmitting a greater force. There

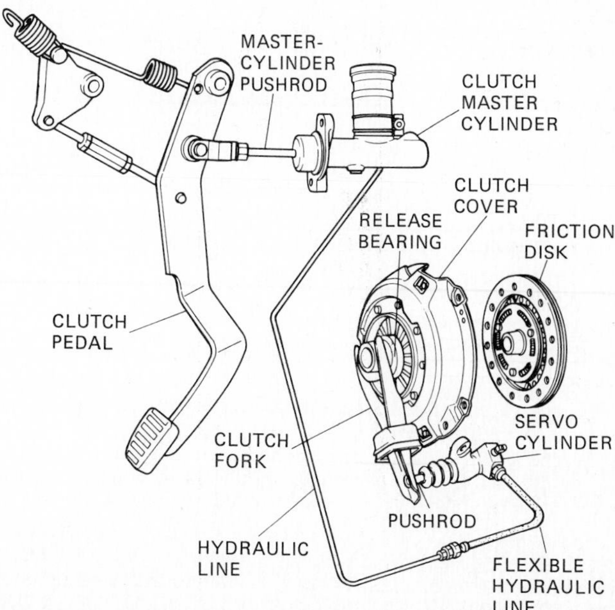

Fig. 39-20 A hydraulically operated clutch linkage. *(Nissan Motor Compnay, Ltd.)*

is the additional advantage that no mechanical linkage between the two is required. Only hydraulic lines are required. These lines may be preformed to any angle, or they may be sections of flexible tubing. Many trucks have hydraulically operated clutches.

♦ 39-14 CLUTCH SAFETY SWITCH

Late-model cars using clutches have a clutch safety switch that prevents starting if the clutch is engaged. The clutch pedal must be depressed at the same time that the ignition switch is turned to START. The movement of the clutch pedal closes the safety switch so that the circuit to the starting motor can be completed.

The purpose of the clutch safety switch is to prevent starting with the transmission in gear and the clutch engaged. If this happened, the car might move before the driver is ready. This could lead to an accident.

♦ 39-15 SELF-ADJUSTING CLUTCH

A self-adjusting clutch (Fig. 39-21) is used on many 1981 and later model cars. This device eliminates the need for routine clutch adjustment. The adjusting device is a spring-loaded quadrant gear (Fig. 39-21) that is attached to the clutch pedal through a shaft. A pawl located at the top of the quadrant gear engages the quadrant-gear teeth. As the clutch pedal is depressed, the quadrant gear rotates, pulling the pawl and cable through its travel. This disengages the clutch.

In the released, or disengaged, position, the spring-loaded quadrant gear takes up excess free play in the clutch mechanism. This compensates for movement of the release lever, clutch cable, and quadrant gear as the clutch disk wears. When sufficient wear occurs, the pawl engages a new tooth on the quadrant gear. This automatically keeps the clutch pedal in the proper position for proper clutch operation.

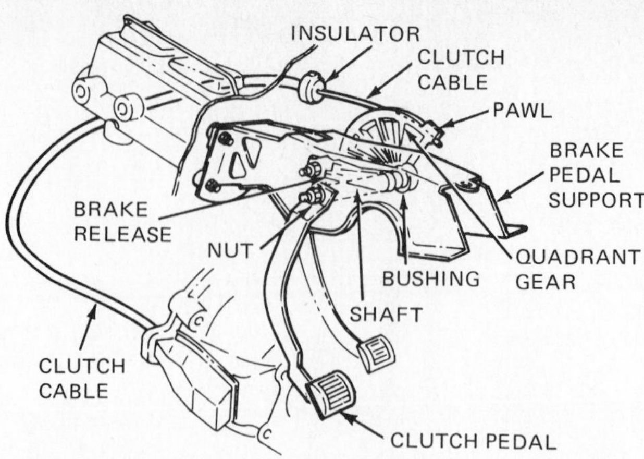

Fig. 39-21 Clutch linkage which includes a self-adjusting device. *(Ford Motor Company)*

CLUTCH SERVICE

♦ 39-16 CLUTCH TROUBLE-DIAGNOSIS CHART

Several types of clutch troubles may occur. Usually, the trouble falls into one of the following categories: slipping, chattering or grabbing when engaging, spinning or dragging when disengaged, clutch noises, clutch-pedal pulsations, and rapid wear of the friction-disk facing. The following Clutch Trouble-Diagnosis chart lists possible causes of each of these troubles. The chart also gives the number of the section in this book that explains how to locate and eliminate the trouble.

CLUTCH TROUBLE-DIAGNOSIS CHART*

Complaint	Possible Cause	Check or Correction
1. Clutch slips while engaged (♦39-17)	a. Incorrect pedal-linkage adjustment	Readjust
	b. Broken or weak pressure springs	Replace
	c. Binding in clutch-release linkage	Free, adjust, and lubricate
	d. Broken engine mount	Replace
	e. Worn friction-disk facings	Replace facings or disk
	f. Grease or oil on disk facings	Replace facings or disk
	g. Incorrectly adjusted release levers	Readjust
	h. Warped clutch disk	Replace
2. Clutch chatters or grabs when engaged (♦39-18)	a. Binding in clutch-release linkage	Free, adjust, and lubricate
	b. Broken engine mount	Replace
	c. Oil or grease on disk facings or glazed or loose facings	Replace facings or disk
	d. Binding of friction-disk hub on clutch shaft	Clean and lubricate splines; replace defective parts
	e. Broken disk facings, springs, or pressure plate	Replace broken parts
	f. Warped clutch disk	Replace
3. Clutch spins or drags when disengaged (♦39-19)	a. Incorrect pedal-linkage adjustment	Readjust
	b. Warped friction disk or pressure plate	Replace defective part
	c. Loose friction-disk facing	Replace defective part
	d. Improper release-lever adjustment	Readjust
	e. Friction-disk hub binding on clutch shaft	Clean and lubricate splines; replace defective parts
	f. Broken engine mount	Replace

*See ♦39-17 to 39-24 for detailed explanations of the trouble causes and corrections listed.

CLUTCH TROUBLE-DIAGNOSIS CHART *(Continued)*

Complaint	Possible Cause	Check or Correction
4. Clutch noises with clutch engaged (♦39-20)	a. Friction-disk hub loose on clutch shaft	Replace worn parts
	b. Friction-disk dampener springs broken or weak	Replace disk
	c. Misalignment of engine and transmission	Realign
5. Clutch noises with clutch disengaged (♦39-20)	a. Clutch throwout bearing worn, binding, or out of lubricant	Lubricate or replace
	b. Release levers not properly adjusted	Readjust or replace assembly
	c. Pilot bearing in crankshaft worn or out of lubricant	Lubricate or replace
	d. Retracting spring (diaphragm-spring clutch) worn	Replace pressure-plate assembly
6. Clutch-pedal pulsations (♦39-21)	a. Engine and transmission not aligned	Realign
	b. Flywheel not seated on crankshaft flange or flywheel bent (also causes engine vibration)	Seat properly, straighten, replace flywheel
	c. Clutch housing distorted	Realign or replace
	d. Release levers not evenly adjusted	Readjust or replace assembly
	e. Warped pressure plate or friction disk	Realign or replace
	f. Pressure-plate assembly misaligned	Realign
	g. Broken diaphragm	Replace
7. Friction-disk-facing wear (♦39-22)	a. Driver "rides" clutch	Keep foot off clutch except when necessary
	b. Excessive and incorrect use of clutch	Reduce use
	c. Cracks in flywheel or pressure-plate face	Replace
	d. Weak or broken pressure springs	Replace
	e. Warped pressure plate or friction disk	Replace defective part
	f. Improper pedal-linkage adjustment	Readjust
	g. Clutch-release linkage binding	Free, readjust, and lubricate
8. Clutch pedal stiff (♦39-23)	a. Clutch linkage lacks lubricant	Lubricate
	b. Clutch-pedal shaft binds in floor mat	Free
	c. Misaligned linkage parts	Realign
	d. Overcenter spring out of adjustment	Readjust
	e. Bent clutch pedal	Replace
9. Hydraulic-clutch troubles (♦39-24)	a. Hydraulic clutches can have any of the troubles listed elsewhere in this chart	Inspect the hydraulic system; check for leakage
	b. Gear clashing and difficulty in shifting into or out of gear	Inspect the hydraulic system; check for leakage

♦ 39-17 CLUTCH SLIPS WHILE ENGAGED

Clutch slippage is very noticeable during acceleration. It is extremely hard on the clutch facings and mating surfaces of the flywheel and pressure plate. The slipping clutch generates excessive heat. As a result, the clutch facings wear rapidly and may char and burn. When the flywheel face and pressure plate wear, they may groove, crack, and score. The heat in the pressure plate can cause the springs to lose their tension, which makes the situation worse.

Several conditions can cause clutch slippage. The pedal linkage may be incorrectly adjusted (Figs. 39-19, 39-22, and 39-23). This may reduce free play so much that the throwout bearing presses against the release fingers, even with the pedal released. Then the pressure plate cannot apply full spring force to lock the friction disk to the flywheel. The result is that the clutch slips while engaged. Readjusting the linkage may correct the problem.

Binding linkage or a broken return spring may prevent full return of the linkage to the engaged position. Replace the spring if it is broken. Lubricate the linkage. Much of the clutch linkage is pivoted in nylon or neoprene bushings. These should be lubricated with silicone spray, SAE 10 oil, or multipurpose grease, depending on the manufacturer's recommendation.

Note If the linkage is not at fault, the slippage could be caused by a broken engine mount. This could allow the engine to shift enough to prevent good clutch engagement. The correction is to replace the mount.

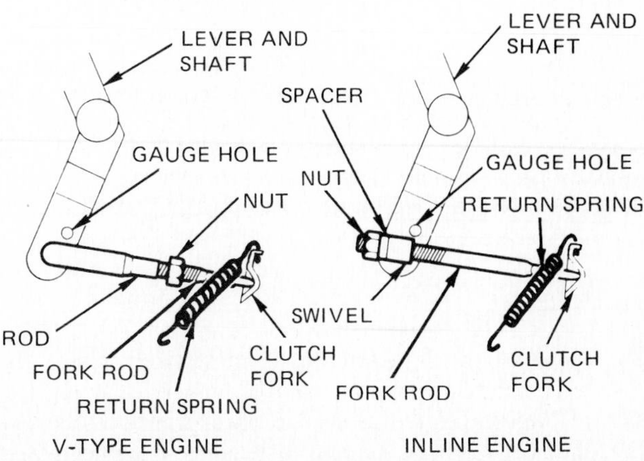

Fig. 39-22 Adjustment of clutch-pedal free travel on various cars built by Chevrolet. *(Chevrolet Motor Division of General Motors Corporation)*

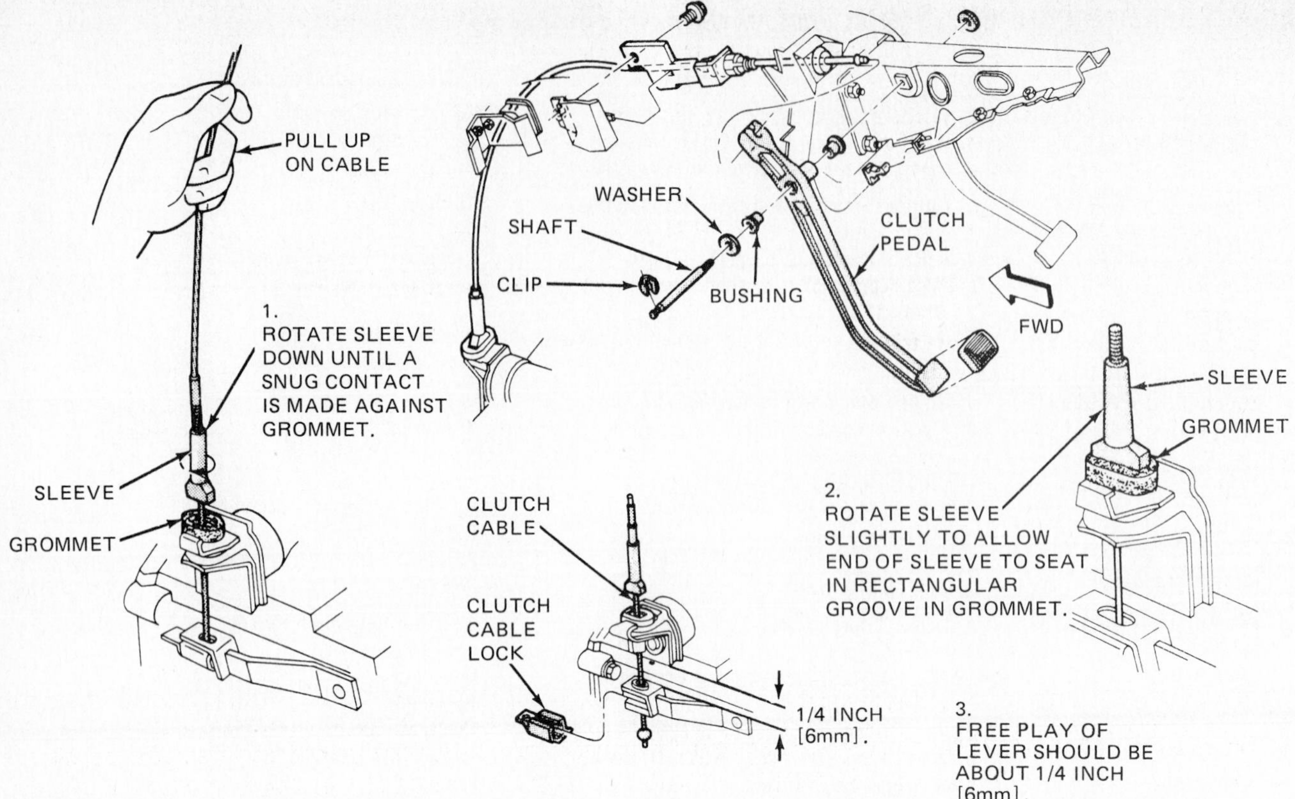

Fig. 39-23 Adjusting clutch-pedal free play on a front-wheel-drive car with a transversely mounted engine and transaxle. *(Chrysler Corporation)*

If none of the above is causing the slipping, then the clutch should be removed for service. Conditions in the clutch that could cause slipping include worn friction-disk facings, broken or weak pressure-plate or diaphragm springs, grease or oil on the disk facings, incorrectly adjusted release levers, or a warped clutch disk.

One clue to a slipping clutch is metal and facing material in the bottom of the clutch housing. This condition can be detected by removing the inspection cover from under the clutch and flywheel.

The recommendation of most manufacturers is to replace the disk and pressure-plate assembly if there is internal wear or damage or weak springs. Pressure-plate assemblies can be rebuilt, but this usually is a job for a clutch rebuilder.

Careful If the clutch disk and pressure-plate assembly are replaced, the flywheel should be inspected carefully for damage such as wear, cracks, grooves, and checks. Any of these conditions, if well advanced, will require replacement of the flywheel. Putting a new disk facing against a damaged flywheel will lead to rapid facing wear.

♦ 39-18 CLUTCH CHATTERS OR GRABS WHEN ENGAGED

The cause of clutch chattering is most likely inside the clutch. The clutch should be removed for service or replacement. However, before this is done, check the clutch linkage to make sure it is not binding. If it binds, it could release suddenly to throw the clutch into quick engagement, with a resulting heavy jerk.

A broken engine mount can also cause chattering because the engine is free to move excessively. This can cause the clutch to grab or chatter when engaged. The remedy is to replace the mount.

Inside the clutch, the trouble could be due to oil or grease on the disk facings or to glazed or loose facings. If this is the case, the disk should be replaced. The trouble could also be due to binding of the friction-disk hub on the splines of the clutch shaft. This condition requires cleaning and lubrication of the splines in the hub and on the shaft.

Note Clutch chatter after removal and reinstallation of an engine may be caused by a misaligned clutch housing. Some clutch housings have small shims that can be lost during engine or clutch-housing removal. These shims must be reinstalled in the original positions to ensure housing alignment. It is also possible for dirt to get between the clutch housing and cylinder block, or either could be nicked or burred. Any of these conditions can throw off the housing alignment.

Other clutch problems, such as glazed or loose facings or oil or grease on the facings, require replacement of the friction disk and pressure plate.

♦ 39-19 CLUTCH SPINS OR DRAGS WHEN DISENGAGED

The clutch friction disk spins briefly after disengagement when the transmission is in neutral. This normal spinning should not be confused with a dragging clutch. When the clutch drags, the friction disk is not releasing fully from the flywheel or pressure plate as the clutch pedal is depressed.

Therefore, the friction disk continues to rotate with or rub against the flywheel or pressure plate. The common complaint of drivers is that they have trouble shifting into gear without clashing. This is because the dragging disk keeps the transmission gears rotating.

The first thing to check with this condition is the pedal-linkage adjustment. If there is excessive pedal lash, or free travel, even full movement of the pedal will not release the clutch fully. If linkage adjustment does not correct the problem, the trouble is in the clutch.

Internal clutch troubles could be due to a warped friction disk or pressure plate or to a loose friction-disk facing. One cause of loose friction-disk facings is abuse of the clutch. This abuse includes "popping" the clutch for a quick getaway (letting the clutch out suddenly with the engine turning at high rpm), slipping the clutch for drag-strip starts, and modifying the engine for increased power output.

The pressure-plate release levers may be incorrectly adjusted so they do not fully disengage the clutch. Also, the friction-disk hub may be binding on the clutch shaft. This condition may be corrected by cleaning and lubricating the splines.

Note A broken engine mount can also cause clutch spinning or dragging. The engine is free to move excessively, which can cause the clutch to spin or drag when disengaged. The remedy is to replace the mount.

♦ 39-20 CLUTCH NOISE

Clutch noises are usually most noticeable when the engine is idling. To determine the cause, note whether the noise is heard when the clutch is engaged, when it is disengaged, or during pedal movement to engage or disengage the clutch.

Noises heard while the pedal is in motion are probably due to dry or dirty linkage pivot points. Clean and lubricate the linkage (♦39-17).

Noises that are heard when the transmission is in neutral but disappear when the pedal is depressed are transmission noises. (These noises could also be due to a dry or worn pilot bearing in the crankshaft.) They are usually rough-bearing sounds. The cause is worn transmission bearings, sometimes caused by clutch-popping and shifting gears too fast. These conditions throw an extra load on the transmission bearings and on the gears.

Noises heard while the clutch is engaged could be due to a friction-disk hub that is loose on the clutch shaft. This condition requires replacement of the disk or clutch shaft, or perhaps both if both are excessively worn. Friction-disk dampener springs that are broken or weak will cause noise. This condition requires replacement of the disk. Misalignment of the engine and transmission will cause a backward-and-forward movement of the friction disk on the clutch shaft. The alignment must be corrected.

The throwout bearing is the most frequently replaced part of the clutch. Noises heard while the clutch is disengaged could be due to a clutch throwout bearing that is worn or binding, or has lost its lubricant. Such a bearing squeals when the clutch pedal is depressed and the bearing begins to spin. The bearing should be lubricated (Fig. 39-24) or replaced. If the release levers are not properly adjusted, they will rub against the friction-disk hub when the clutch pedal is depressed. The release levers should be readjusted or the assembly should be replaced.

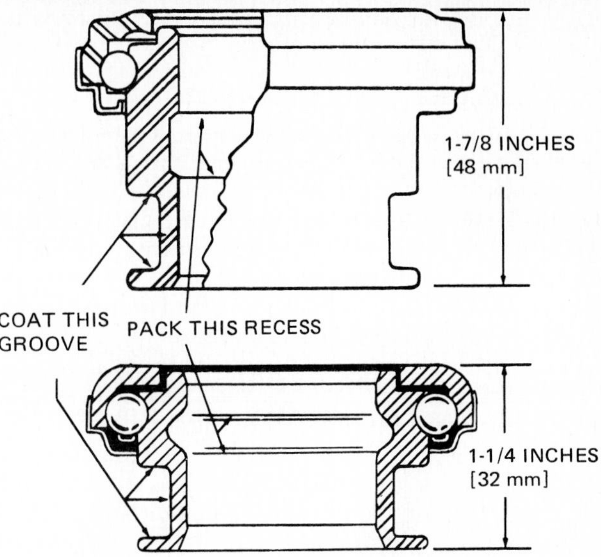

Fig. 39-24 Release-bearing lubrication points. *(Chevrolet Motor Division of General Motors Corporation)*

If the pilot bearing in the crankshaft is worn or lacks lubricant, it will produce a high-pitched whine when the transmission is in gear, the clutch is disengaged, and the car is stationary. Under these conditions, the clutch shaft (which is piloted in the bearing in the crankshaft) is stationary, but the crankshaft and bearing are turning. The bearing should be lubricated or replaced.

In the diaphragm-spring clutch, worn or weak retracting springs will cause a rattling noise when the clutch is disengaged and the engine is idling. Eliminate the noise by replacing the pressure-plate assembly.

♦ 39-21 CLUTCH PEDAL PULSATES

Clutch-pedal pulsations are noticeable when a slight force is applied to the clutch pedal with the engine running. The pulsations can be felt by the foot as a series of slight pedal movements. As pedal force is increased, the pulsations cease. This condition often indicates trouble that must be corrected before serious damage to the clutch results.

One possible cause is misalignment of the engine and transmission. If the two are not in line, the friction disk or other clutch parts will move back and forth with every revolution. The result will be rapid wear of clutch parts. Correction is to detach the transmission, remove the clutch, and then check the housing alignment with the engine and crankshaft. At the same time, the flywheel can be checked for wobble or runout. A flywheel that is not seated on the crankshaft flange will also produce clutch-pedal pulsations. The flywheel should be removed and remounted to make sure that it seats evenly.

If the clutch housing is distorted or shifted so that alignment between the engine and transmission has been lost, it is sometimes possible to restore alignment. This is done by installing shims between the housing and engine block and between the housing and transmission case. Otherwise, a new clutch housing is required.

Note Clutch-pedal pulsations caused by such conditions as a bent flywheel, a flywheel that is not seated on the crankshaft flange, or housing misalignment usually result only from faulty reassembly after a service job. They do not normally develop during operation of the vehicle.

Another cause of clutch-pedal pulsations is uneven release-lever adjustment (so that release levers do not meet the throwout bearing and pressure plate together). Release levers of the adjustable type should be readjusted. Still another cause is a warped friction disk or pressure plate. A warped friction disk must be replaced. If the pressure plate is out of line because of a distorted clutch cover, the cover sometimes can be straightened to restore alignment. However, usually the pressure-plate assembly is replaced.

In the diaphragm-spring clutch, a broken diaphragm will cause clutch-pedal pulsations. The clue here is that the pulsations develop suddenly, just as the diaphragm breaks.

♦ 39-22 FRICTION-DISK FACINGS WEAR RAPIDLY

Rapid wear of the friction-disk facings is caused by slippage between the facings and the flywheel or pressure plate. If the driver has the habit of "riding" the clutch by resting the left foot on the pedal, part of the pressure-plate spring force will be taken up so that slippage may take place. Likewise, frequent use of the clutch, incorrect clutching and declutching, overloading the clutch, and slow clutch engagement and disengagement increase clutch-facing wear. Speed shifting ("snap" shifting), increasing engine output, and drag-strip starts shorten clutch life. Also, the installation of wide, oversize tires increases the load on the clutch. Some manufacturers will not warranty the clutch if oversize tires are installed.

Rapid facing wear after installation of a new friction disk can be caused by heat checks and cracks in the flywheel and pressure-plate faces. The sharp edges act like tiny knives. They shave off a little of the face during each engagement and disengagement. Because of this, the pressure-plate assembly should also be replaced whenever the friction disk is replaced. In addition, the flywheel face should be inspected, and if it is damaged the flywheel should be replaced.

Several conditions in the clutch itself can cause rapid friction-disk-facing wear. For example, weak or broken pressure springs will cause slippage and facing wear. In this case, the springs or the pressure-plate assembly must be replaced. If the pressure plate or friction disk is warped or out of line, it must be replaced or realigned. In addition, an improper pedal-linkage adjustment or binding of the linkage may prevent full spring force from being applied to the friction disk. With less than full spring force, slippage and wear may take place. The linkage must be readjusted and lubricated.

♦ 39-23 CLUTCH PEDAL STIFF

A clutch pedal that is stiff or hard to depress is likely to result from lack of lubricant in the clutch linkage, from binding of the clutch-pedal shaft in the floor mat, or from misaligned linkage parts that are binding. In addition, the overcenter spring (on cars so equipped) may be out of adjustment or broken. Also, if the clutch pedal has been bent so that it rubs on the floorboard, it may not operate easily. The remedy for each of these troubles is to lubricate or readjust the clutch parts as necessary, or to replace the clutch pedal.

♦ 39-24 HYDRAULIC-CLUTCH TROUBLES

The hydraulic clutch can have any of the troubles described previously plus several in the hydraulic system. These special troubles include gear clashing and difficulty in shifting into or out of gear. The cause is usually loss of fluid from the hydraulic system. Fluid loss prevents the system from completely declutching for gear shifting. The hydraulic system should be checked and serviced in the same way as the hydraulic system in hydraulic brakes. Leaks may be in the master cylinder, the servo cylinder, or in the line or connections between the two.

♦ 39-25 CLUTCH SERVICE

The major clutch services include clutch-linkage adjustment, clutch removal and replacement, and clutch disassembly, inspection, adjustment, and reassembly. If a clutch defect develops, you must do more than just replace a worn part. You must determine what caused the part to wear and fix the trouble so that the new part will not wear rapidly.

One of the most common causes of rapid disk-facing wear and clutch failure is improper pedal lash, or free travel. If pedal free travel is not sufficient, the clutch will not engage completely. If will slip and wear rapidly. In addition, the throwout bearing will be operating continuously and will soon wear out.

♦ CAUTION ♦ Asbestos is used in the facings of many friction disks because it can withstand the high pressures and temperatures inside the clutch. However, authorities claim that breathing asbestos dust can cause lung cancer. For this reason, be careful when working around clutches. Do not blow the dust out of the clutch housing with compressed air. This dust may contain powdered asbestos. The compressed air could send the dust up into the air around you and you could inhale it. Instead, use a vacuum suitable for asbestos, or cleaning should be done wet. Use damp cloths to wipe out the clutch housing. Then, after working on a clutch, always wash your hands thoroughly to remove any trace of asbestos dust.

♦ 39-26 CLUTCH-LINKAGE ADJUSTMENT

Clutch-linkage adjustment may be required at intervals to compensate for wear of the facing or lining on the friction disk. In addition, certain points in the linkage or pedal support may require lubrication. The adjustment of the linkage changes the amount of clutch-pedal free travel, or free play. The free travel of the pedal is the distance that the pedal moves before the throwout bearing makes contact with the release levers in the pressure plate. After this happens, there is a noticeable increase in the force required for further pedal movement. From this position on, pedal movement causes release-lever movement and compression of the springs in the pressure plate. In normal operation, free travel is lost as the friction disk wears. Free travel seldom increases.

A test of pedal free travel should be made with your finger and not with your foot. Your finger can detect the increase in force more accurately than your foot.

There are two checks to make on the clutch and linkage:

1. Make sure the clutch fully disengages.
2. Make sure there is adequate free travel.

♦ 39-27 CHECKING FOR CLUTCH DISENGAGEMENT

With the engine idling and the brakes firmly applied, hold the clutch pedal about ½ inch [13 mm] from the floor mat. Move the shift lever between first and reverse several times. If this can be done smoothly, the clutch is disengaging fully. If the shifts are not smooth, the clutch is not disengaging. Adjust the linkage. If the adjustment does not cure the problem, check for the following:

1. Clutch-pedal bushing may be sticking or worn.
2. Fork or yoke may not be properly installed on its pivot or ball stud. Lack of lubrication can cause the fork or yoke to be pulled off.
3. Linkage may be bent or damaged.
4. Loose or broken engine mounts may allow the engine to shift enough to cause the clutch linkage to bind. Make sure there is clearance between the linkage and any mounting brackets.
5. Clearance between the end of the release bearing and release levers in the pressure plate may be insufficient due to worn parts.

♦ 39-28 REMOVING, SERVICING, AND INSTALLING CLUTCHES

When you are required to adjust clutch linkage, or service or replace the clutch, refer to the manufacturer's service manual. Clutch linkages and service procedures differ among the various car models and model years.

If the trouble is located in the clutch itself, replacement of the complete clutch assembly is recommended by most manufacturers. At one time, manufacturers' service manuals carried instructions on the disassembly and repair of clutches. But now complete replacement is usually performed. However, some shops specialize in rebuilding clutches. They replace any parts in the pressure plate that are worn or damaged. This may include replacing weak springs and refinishing the friction-disk surface.

♦ REVIEW QUESTIONS ♦

Select the *one* correct, best, or most probable answer to each question. You can find the answers in the section indicated at the end of each question.

1. The friction disk is splined to a shaft which extends into the (♦39-3)
   a. transmission
   b. drive shaft
   c. differential
   d. engine
2. The friction disk is positioned between the flywheel and the (♦39-3)
   a. engine
   b. crankshaft
   c. pressure plate
   d. differential
3. When the clutch is engaged, spring force clamps the friction disk between the pressure plate and the (♦39-3)
   a. flywheel
   b. differential
   c. reaction plate
   d. clutch pedal
4. When the clutch pedal is depressed, the throwout bearing moves in and causes the pressure plate to release its force on the (♦39-6)
   a. throwin bearing
   b. pressure springs
   c. friction disk
   d. flywheel
5. The clutch cover is bolted to the (♦39-5)
   a. friction disk
   b. flywheel
   c. car frame
   d. engine block
6. To make engagement as smooth as possible, the friction disk has a series of waved (♦39-7)
   a. cushion pads
   b. cushion bolts
   c. cushion springs
   d. disks
7. The release levers in the coil-spring clutch pivot on (♦39-6)
   a. springs
   b. levers
   c. threaded bolts
   d. pins and eyebolts
8. In the friction disk, torsional vibration is absorbed by the use of a series of heavy (♦39-7)
   a. cushion bolts
   b. coil springs
   c. waved pads
   d. friction pads
9. In the diaphragm-spring clutch, inward movement of the throwout bearing causes the diaphragm spring to (♦39-9)
   a. dish inward
   b. expand
   c. contract
   d. flatten
10. In the semicentrifugal clutch, the force of the pressure plate against the friction disk increases with vehicle speed because of weights located on the (♦39-8)
    a. pressure plate
    b. flywheel
    c. release levers
    d. clutch shaft

11. Clutch slippage while the clutch is engaged is particularly noticeable (♦39-17)
    a. during idle
    b. at low speed
    c. when starting the engine
    d. during acceleration
12. Clutch chattering or grabbing is noticeable (♦39-18)
    a. during idle
    b. at low speed
    c. when engaging the clutch
    d. during acceleration
13. Clutch dragging is noticeable (♦39-19)
    a. when the clutch is disengaged
    b. at road speed
    c. during acceleration
    d. at high speed
14. Clutch noises are usually most noticeable when the engine is (♦39-20)
    a. accelerating
    b. decelerating
    c. idling
    d. being started
15. Clutch-pedal pulsation is noticeable when the engine is running and (♦39-21)
    a. accelerating
    b. a slight force is applied to the pedal
    c. decelerating
    d. the car is moving at steady speed
16. Slippage between the friction-disk facings and the flywheel or pressure plate will cause (♦39-22)
    a. clutch-pedal pulsation
    b. rapid facing wear
    c. excessive acceleration
    d. rapid pressure-plate wear
17. Clutch-pedal free travel, or pedal lash, is the distance the pedal moves before the release bearing makes contact with the (♦39-26)
    a. release levers
    b. flywheel
    c. floorboard
    d. stop
18. Heat checks or cracks on the flywheel and pressure-plate faces will cause (♦39-22)
    a. excessive clutch slippage
    b. rapid flywheel and pressure-plate wear
    c. rapid wear of friction-disk facings
    d. excessive pedal pulsation
19. The reason for the caution about the asbestos in the clutch facings is that asbestos dust can (♦39-25)
    a. make you sneeze
    b. cause eye irritation
    c. cause lung cancer
    d. damage the clutch
20. Clutch slippage can be caused by all of the following *except* (♦39-17)
    a. incorrect linkage adjustment
    b. loose friction-disk facings
    c. grease on the facings
    d. broken or weak pressure springs

CHAPTER 40
Manual Transmissions and Transaxles

After studying this chapter, you should be able to:

1. Discuss the purpose and operation of typical manual transmissions.
2. Explain the difference between three-speed, four-speed, and five-speed transmissions.
3. Describe how shifting is accomplished.
4. Explain the purpose of overdrive and how it is achieved.
5. Explain the purpose and operation of manual transaxles.

♦ 40-1 LOCATION OF TRANSMISSION OR TRANSAXLE

The manual transmission in front-engine rear-wheel-drive vehicles is located between the clutch and the drive shaft (Fig. 39-2). In a car with a front engine and front-wheel drive, the transmission is located between the clutch and the differential in the transaxle (Fig. 39-3). Figure 39-1 compares the two power-train layouts.

♦ 40-2 PURPOSE OF TRANSMISSION

A *manual transmission* is an assembly of gears and shafts that transmit power from the engine to the drive axle. The transmission allows the engine crankshaft to turn fast while the wheels turn slowly. The transmission can then change the ratio of crankshaft speed to car speed as car speed increases. Therefore, with a three-speed transmission, the engine crankshaft may turn about four, eight, or twelve times for each wheel revolution. In addition, the transmission includes a reverse gear so that the car can be backed. Each of these gear ratios is selected manually by the driver.

♦ 40-3 GEAR RATIOS

Many different gears are used in automobiles. All are basically similar. They all have teeth that mesh to transmit force and motion from one gear to another. The simplest gear is the spur gear, which is like a wheel with teeth. The gears used in transmissions are *helical* gears (Figs. 40-1 and 40-2). In these gears, the teeth are set at an angle to the gear centerline. The teeth have a wiping action which improves their contact and lubrication.

When one meshing gear rotates, the teeth of that gear cause the teeth of the other gear to move so that the other gear also rotates. The relative speed of the two meshing gears (the *gear ratio*) is determined by the number of teeth of the two gears. For example, when two gears have the same number of teeth (Fig. 40-1), they both rotate at the same speed. But if one gear has 12 teeth and the other gear has 24 teeth (Fig. 40-2), the smaller gear will rotate twice for every revolution of the larger gear. When the smaller gear is driving the larger gear, this is a two-to-one gear ratio (written 2:1). If the 12-tooth gear was driving with a 36-tooth gear, the gear ratio would be 3:1.

Fig. 40-1 Two meshing helical gears with the same number of teeth.

♦ 40-4 TORQUE

The *torque*, as well as the gear ratio, changes with the relative number of teeth in the meshing gears. Torque is twisting or turning force. It is measured in pound-feet (lb-ft) or newton-meters (N-m).

Any shaft or gear that is being turned has torque applied to it. The engine pistons and connecting rods push on the cranks on the crankshaft. This applies torque to the crankshaft and causes it to turn. The crankshaft applies torque to the gears in the transmission so that the gears turn. This turning force, or torque, is carried through the power train to the driving wheels so that they turn.

♦ 40-5 TORQUE IN GEARS

Torque on shafts and gears is measured as a straight-line force at a distance from the center of the shaft or gear. For example, suppose we want to measure the torque in the gears shown in Fig. 40-2. If we could hook a spring scale to the gear teeth and get a measurement of the pull on the scale, we could determine the torque. However, a spring scale is not actually used because the gear teeth are moving.

In a circle, the distance from the center to the outside edge is the *radius*. On a gear, the radius is the distance from the center to the point on the tooth where the force is applied. Now, suppose that a tooth on the driving gear is pushing against a tooth on the driven gear with a force of 25 pounds [111 N]. When the force is applied at a distance of 1 foot [0.31 m], which is the radius of the driving gear, a torque of 25 lb-ft [33.9 N-m] is applied to the driven gear.

The 25 pounds [111 N] force from the teeth of the smaller (driving) gear is applied to the teeth of the larger (driven) gear. But the force is applied at a distance of 2 feet [0.61 m] from the center. Therefore the torque on the shaft at the center of the driven gear is 50 lb-ft (25 × 2) [67.8 N-m (33.9 × 2)]. The same force is acting at twice the distance from the shaft center.

♦ 40-6 TORQUE AND GEAR RATIO

In Fig. 40-2, when the smaller gear is driving the larger gear, the gear ratio is 2:1. However, the *torque ratio* is 1:2. The larger gear turns at half the speed of the smaller gear. As a result, the larger gear will have twice the torque of the smaller gear.

In gear systems, *speed reduction means torque increase*. For example, when a typical three-speed transmission is in first gear, there is a speed reduction (or gear reduction) of 12:1 from the engine to the drive wheels. The crankshaft turns 12 times to turn the wheels once. Ignoring losses resulting from friction, this means that the torque increases 12 times. If the engine produces a torque of 100 lb-ft [135.6 N-m], then a torque of 1200 lb-ft [1627 N-m] is applied to the drive wheels.

Figure 40-3 shows how this torque forces the car forward. In the example shown, the engine is delivering a torque of 100 lb-ft [135.6 N-m]. The gear reduction from the engine to the drive wheels is 12:1 with a torque increase of 1:12. The wheel radius is assumed to be 1 foot [0.31 m], for ease of figuring.

With the torque acting on the ground at a distance of 1 foot [0.31 m] (the radius of the wheel), the force of the tire pushing against the ground is 1200 pounds [5338 N]. Consequently, the push on the wheel axle and therefore on the car is 1200 pounds [5338 N].

Actually, the torque is split between the two drive wheels. Each tire pushes against the ground with a force of 600 pounds [2669 N]. Both tires together push with a force of 1200 pounds [5338 N], giving the car a forward thrust of 1200 pounds [5338 N].

♦ 40-7 TYPES OF GEARS

Many types of gears are used in the automobile (Fig. 40-4). The most basic type is the *spur gear*. On the spur gear, the

24 TEETH

12 TEETH

Fig. 40-2 Two meshing helical gears with different numbers of teeth. The smaller gear is called the *pinion* gear.

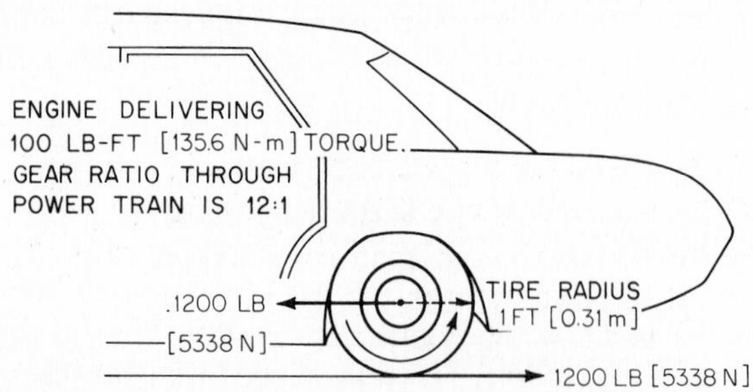

Fig. 40-3 How torque at the drive wheels pushes the car forward. The tire is turned with a torque of 1200 lb-ft (1627 N-m). Since the tire radius is 1 foot (0.31 m), the push of the tire against the ground is 1200 lb-ft (1627 N-m). As a result, the car is pushed forward with a force of 1200 pounds (5338 N).

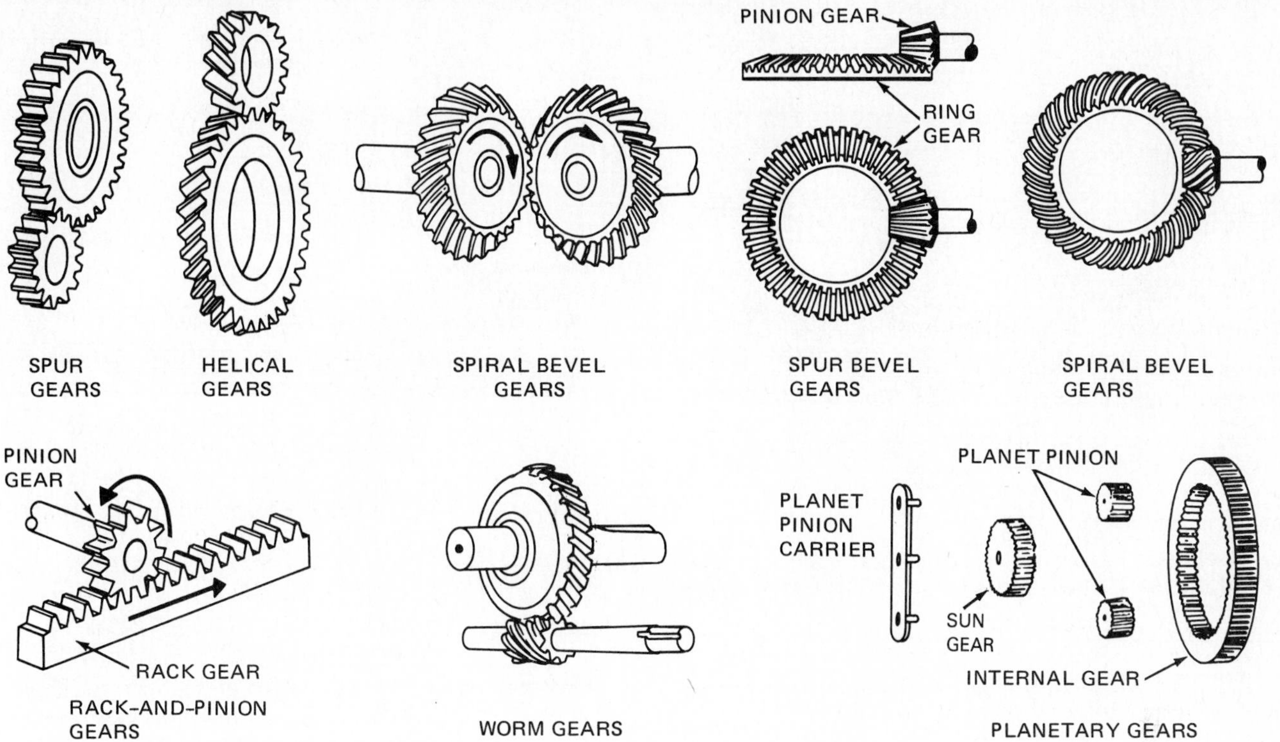

Fig. 40-4 Various types of gears. *(ATW)*

teeth are parallel to and align with the center of the gear. Other types of gears differ from the spur gear mainly in the shape and alignment of the gear teeth.

For example, *helical gears* are like spur gears except that the teeth are formed at an angle to the gear centerline. *Bevel gears* are shaped like imaginary cones with the tops cut off. The teeth point inward toward the apex, or peak, of the cone. Bevel gears are used to transmit motion through angles. Some gears, called *internal gears*, have their teeth pointing inward. These gears form part of the planetary gearset which is used in automatic transmissions.

♦ 40-8 TYPES OF MANUAL TRANSMISSIONS

Today, more cars have four-speed transmissions than three-speed transmissions. In recent years, about 17 percent of cars built in the United States had four-speed manual transmissions. About ½ percent had five-speed transmissions. A few had three-speed transmissions. The other 82 percent had automatic transmissions (covered later in the book).

♦ 40-9 THREE-SPEED MANUAL TRANSMISSION

The three-speed manual transmission is the simplest automotive transmission in construction and operation. However, there are many similarities with four-speed and five-speed transmissions and transaxles.

The gears in the three-speed transmission are shown in Fig. 40-5. There are four gears on the countergear assembly, all rigidly attached to the shaft. Four independent gears mesh with these four gears, the clutch gear, the second-speed gear, the first-speed gear, and the reverse gear. There is also an idler gear. It is placed between the small gear on the end of the countergear assembly and the reverse gear.

In gear shifting, the gears themselves are not actually moved. Connecting devices, called *synchronizers*, lock one gear or another to the transmission main shaft. Figure 40-6 shows the complete gear train removed from the transmission case. Shift forks fit into each of the two synchronizers. All elements are in the neutral position in Fig. 40-6. No power is flowing through the transmission. The heavy arrow running through the clutch gear to the countergear shows the power flow. But since none of the gears on the transmission main shaft are locked to it, these gears rotate freely on the shaft. The main shaft does not turn. The actions when

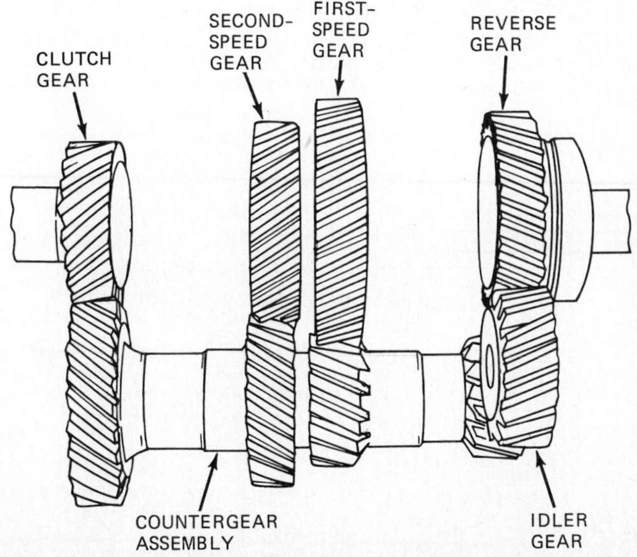

Fig. 40-5 The gears in a three-speed transmission.

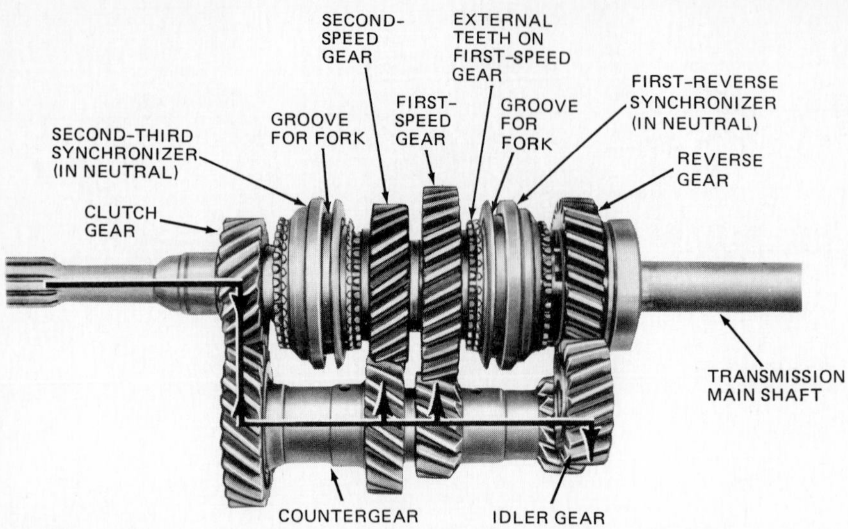

Fig. 40-6 Gear train and shafts in the three-speed transmission. *(Chevrolet Motor Division of General Motors Corporation)*

gearshifts are made to first, second, third, and reverse are described below. How linkages from the gearshift lever on the floor or steering column produce shifts is discussed later.

1. First Speed To shift into first gear, depress the clutch pedal to disengage the clutch. Then move the gearshift lever into the first position. This action causes linkage to the transmission to select the first-reverse synchronizer sleeve and move it to the left (in Fig. 40-6). Figure 40-7 shows the result. The hub of the synchronizer is splined to the main shaft. The synchronizer sleeve can slide back and forth on the splined hub. When the sleeve is moved to engage the row of small external teeth on the first-speed gear, the synchronizer is locked to the gear. Now the gear is locked to the main shaft, through the synchronizer. When the gear turns, it makes the main shaft turn. (Synchronizers are discussed further in ♦40-11.)

In first speed, there is high torque multiplication and speed reduction through the transmission. The clutch gear (in Fig. 40-7) is smaller than the gear on the countergear assembly that it drives. This provides some gear reduction. There is further gear reduction as the small first gear on the countergear drives the big first-speed gear. In a typical three-speed transmission, the total gear reduction is about 3:1 (2.99:1 in one model). This means that the crankshaft turns three times to turn the transmission main shaft, and the drive shaft, once. There is further gear reduction in the final drive of the drive axle.

2. Second Speed To shift into second (Fig. 40-8), disengage the clutch and move the shift lever from first to second. This causes linkage to the transmission to move the first-reverse synchronizer sleeve back to its central, or neutral, position. Then the second-third synchronizer is selected and the

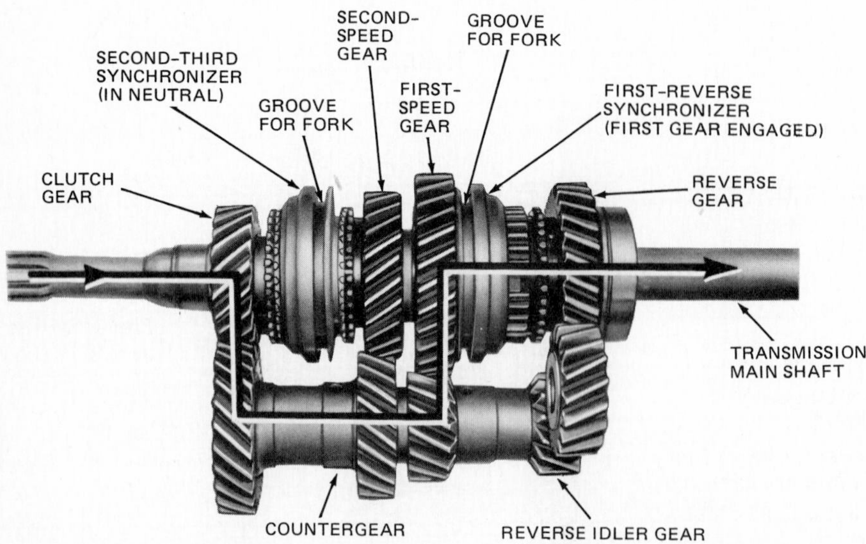

Fig. 40-7 Power flow through the gear train of a three-speed transmission in first gear. *(Chevrolet Motor Division of General Motors Corporation)*

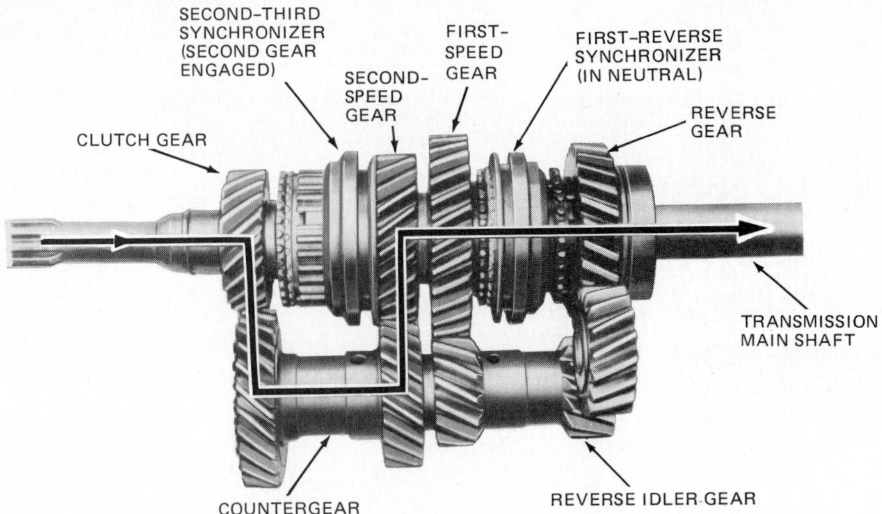

Fig. 40-8 Power flow through a three-speed transmission in second gear. *(Chevrolet Motor Division of General Motors Corporation)*

sleeve moved to the right, as shown in Fig. 40-8. Internal teeth on the second-third synchronizer sleeve mesh with external teeth on the second-speed gear. This locks the second-speed gear, through the synchronizer, to the main shaft. Now the power flow is shown by the arrow in Fig. 40-8. The power enters through the clutch gear to drive the countergear assembly. The medium-size gear on the countergear drives the second-speed gear. This drives the second-third synchronizer and the main shaft.

There is less gear reduction in second because of the relative sizes of the second-speed gear and the gear on the countergear that drives it. Gear reduction in second is a little less than 2:1 in most three-speed transmissions (1.83:1 in one model).

3. Third Speed To shift into third (Fig. 40-9), disengage the clutch and move the shift lever from second to third. This moves the second-third synchronizer sleeve to the left. It disengages from the second-speed gear, passes through its neutral position, and then engages the teeth on the clutch gear. Now, power flows directly from the clutch gear, through the second-third synchronizer, to the transmission main shaft. There is no gear reduction. This is called *direct drive*. The gear ratio is 1:1.

4. Reverse When a shift is made into reverse (Fig. 40-10), the second-third synchronizer sleeve is moved into the centered, or neutral, position. The first-reverse synchronizer sleeve is moved to the right where it engages with the reverse gear. This gear will turn in the reverse direction when the clutch is engaged because of the reverse idler gear. This idler gear carries the power flow from the small gear on the countergear to the reverse gear. Now, when the clutch is engaged, the power flow is as shown in Fig. 40-10. However, the transmission main shaft turns in the reverse direction so that the car will move backward, instead of forward. The gear ratio is

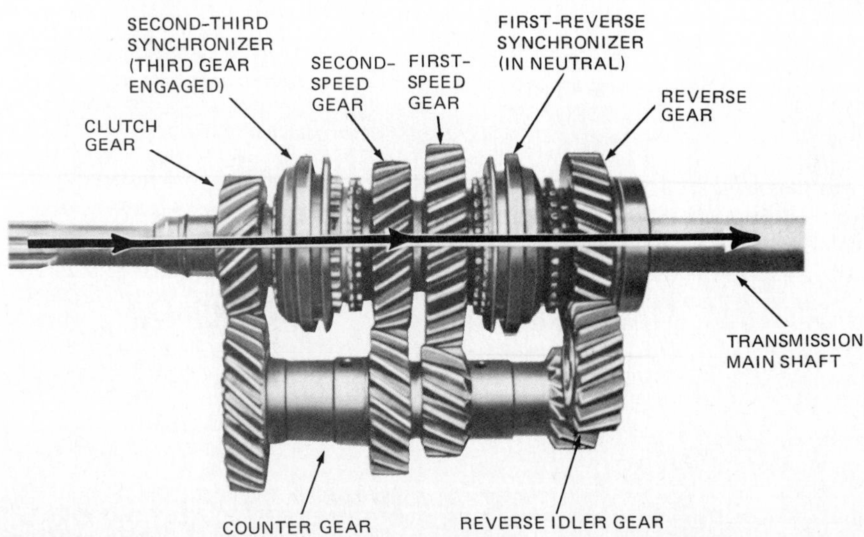

Fig. 40-9 Power flow through a three-speed transmission in third gear. *(Chevrolet Motor Division of General Motors Corporation)*

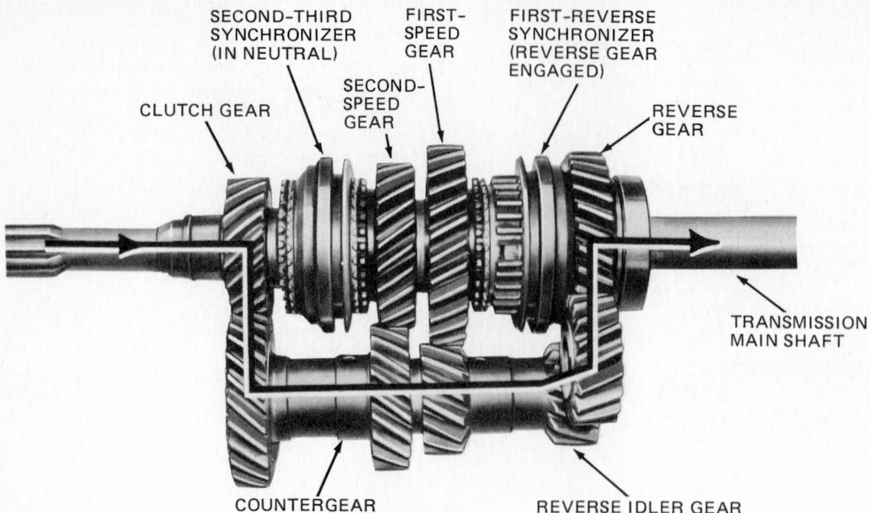

Fig. 40-10 Power flow through a three-speed transmission in reverse. *(Chevrolet Motor Division of General Motors Corporation)*

about 3:1 in most three-speed transmissions (3.00:1 in one model).

♦ 40-10 INTERLOCK DEVICES

To prevent both synchronizers from moving at the same time, the linkage and transmission has an interlock device. The mechanisms that move the synchronizers are U-shaped shift forks set in the side of the transmission (Fig. 40-11). One of the shift forks fits into a groove in the first-reverse synchronizer sleeve. The other shift fork fits into a groove in the second-third synchronizer sleeve. Figure 40-6 shows the grooves in the synchronizer sleeves into which the forks fit.

The interlocking device in Fig. 40-11 uses a double-ended detent cam. When one of the shift forks is moved, the detent cam is raised at that end. This causes the other end of the detent cam to move down, locking the other shift fork.

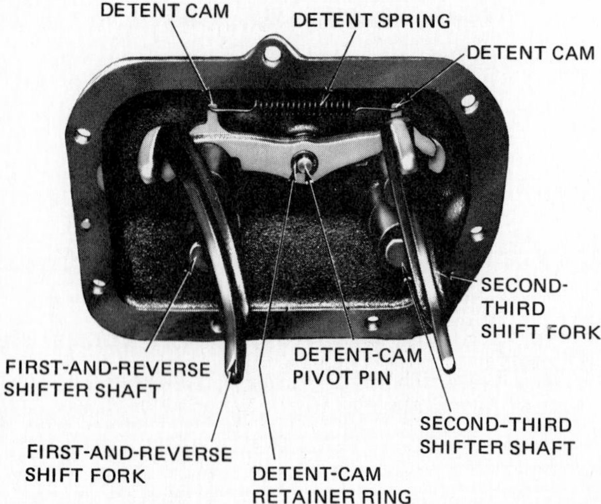

Fig. 40-11 Transmission side cover viewed from inside the transmission. The shift forks are mounted on the ends of levers attached to shafts. The shafts can rotate in the side cover, moving the shift forks to left or right. A detent cam and spring prevent more than one of the shift forks from moving at the same time. *(Chevrolet Motor Division of General Motors Corporation)*

Other types of interlock devices are also used. One type consists of a spring-loaded ball or plunger. It is moved into a notch or hole in the inoperative shift-fork shaft by movement of the shift lever.

♦ 40-11 SYNCHRONIZERS

To prevent gear clash during shifting while the car is in motion, synchronizing devices are used in automotive transmissions. These devices ensure that gears that are about to mesh will be rotating at the same speed, so they will engage smoothly.

One type of synchronizer uses synchronizing cones on the gears and on the synchronizing hub (Fig. 40-12). In the neutral position, the sliding sleeve is held in place by spring-loaded balls resting in detents in the sleeve. When a shift starts, the hub and sleeve, as an assembly, are moved toward

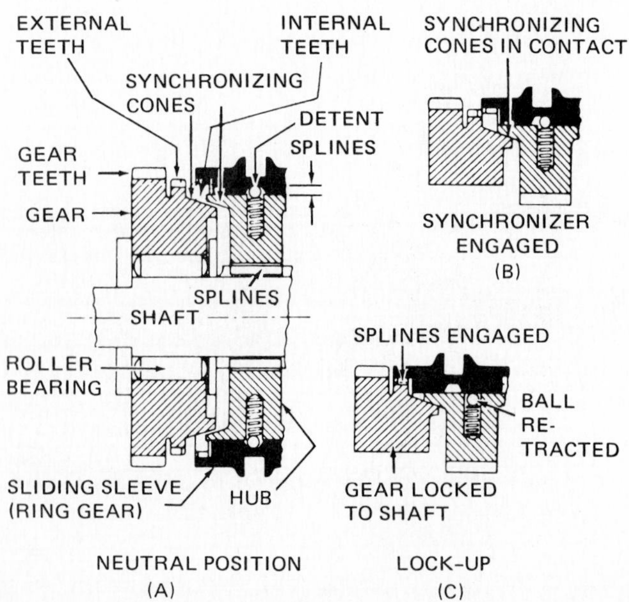

Fig. 40-12 Operation of a synchronizing device that uses cones.

the selected gear (to the left in Fig. 40-12). The first contact is between the synchronizing cones on the gear and hub. As the two cones are forced together (upper right in Fig. 40-12), they are brought into synchronization. Both rotate at the same speed.

Further movement of the shift fork forces the sliding sleeve on toward the selected gear. The internal teeth on the sliding sleeve match the external teeth on the gear. With both the gear and the main shaft rotating at the same speed, the sleeve slides over the teeth on the gear without clashing. Now the gear is locked to the main shaft through the sliding sleeve (lower right in Fig. 40-12) and the shift is completed. Notice that the sliding sleeve moves off center from the hub and over the balls for engagement. This pushes the balls down against their springs.

Note The synchronizer sliding sleeve is called a *synchronizer sleeve* and a *clutch sleeve*.

Another type of synchronizer is shown partly disassembled in Fig. 40-13. Instead of retracting balls, as in Fig. 40-12, this synchronizer has three keys and a pair of ring-shaped synchronizing springs. The keys are assembled in slots in the hub, which is splined to the main shaft. Assembled outside the hub is the synchronizing sleeve. The hub has external splines that fit the internal splines of the sleeve. The three keys have raised sections that fit in the annular groove of the sleeve.

Note The sleeve shown in Fig. 40-13 has external teeth, but the synchronizing sleeves in the transmission shown in Fig. 40-6 do not. However, the action is the same.

Synchronizing is a four-stage action: First, the sleeve is moved toward the first-speed gear (when shifting to first). The sleeve slides on the hub splines and carries the three keys with it. Second, the keys move up against the synchronizer ring and push the ring toward the first-speed gear. The ring presses against the cone of the first-speed gear. Third, further sleeve movement causes the keys to be pressed out of the annular groove in the sleeve. The sleeve continues to move toward the first-speed gear. The friction between the synchronizing ring and the first-speed gear brings the two into synchronous rotation. Fourth, movement of the sleeve allows the internal teeth of the sleeve to engage the external teeth of the first-speed gear. Meshing is completed. Similar actions take place in the shifts to second and third.

♦ 40-12 FOUR-SPEED TRANSMISSION

The four-speed transmission has four forward speeds, neutral, and reverse. In some four-speed transmissions, fourth speed, or "gear," is an *overdrive ratio*. This means that the main shaft is overdriven. It turns more than one complete revolution for every complete revolution of the crankshaft and clutch gear. The advantage of overdrive is discussed in ♦40-13.

The gears and shafts for one type of four-speed transmission which does not have overdrive are shown in Fig. 40-14. The gears are shown in neutral in Fig. 40-14. In this transmission, fourth gear is direct drive, or 1:1.

1. First Gear In first gear, the first-second synchronizer sleeve has been moved to the right. Its internal splines engage the external splines of the first-speed gear.

2. Second Gear In second gear, the first-second synchronizer sleeve has been moved to the left. Its internal splines engage the external splines of the second-speed gear.

3. Third Gear In third gear, the third-fourth synchronizer sleeve has been moved to the right. Its internal splines

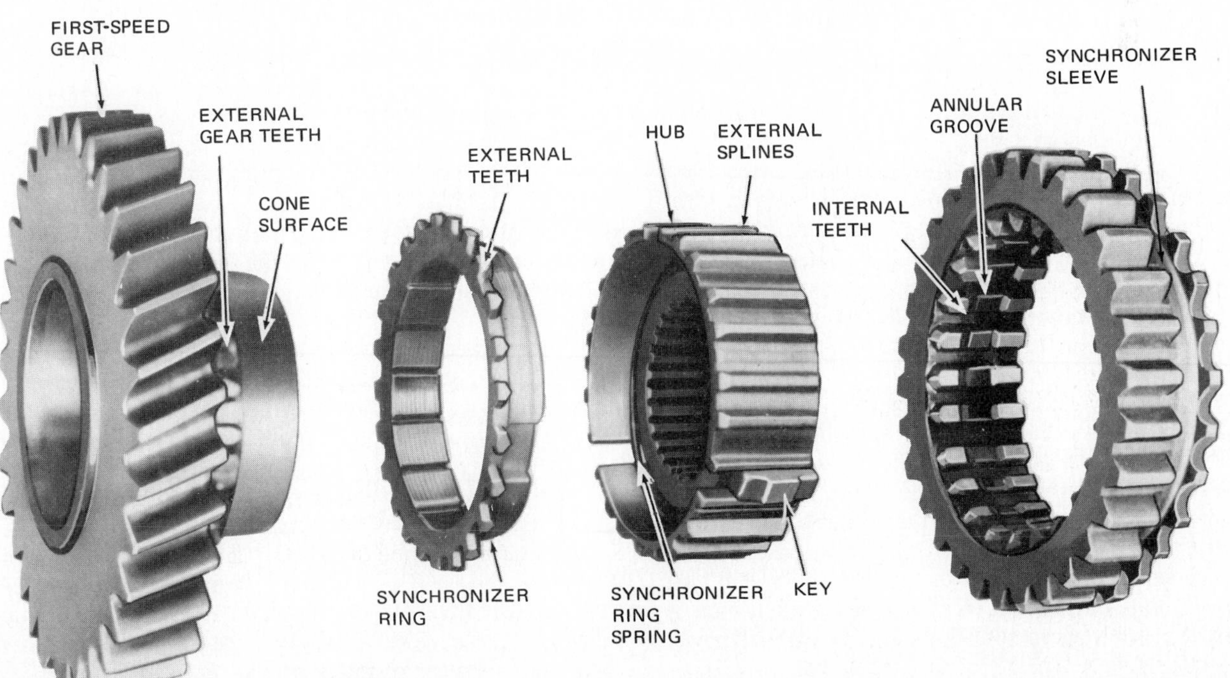

Fig. 40-13 A disassembled synchronizer. *(Chevrolet Motor Division of General Motors Corporation)*

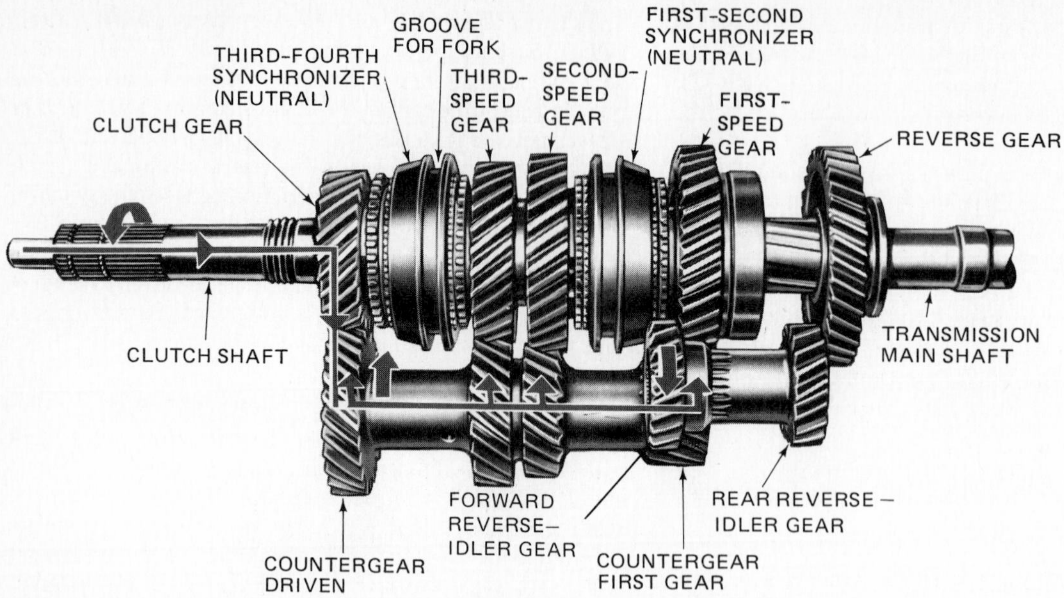

Fig. 40-14 Gear train and shafts of a four-speed transmission, shown in neutral. *(Chevrolet Motor Division of General Motors Corporation)*

engage the external splines of the third-speed gear.

4. Fourth Gear In fourth gear, the third-fourth synchronizer sleeve has been moved to the left. Its internal splines engage the external splines of the clutch gear.

5. Reverse In reverse, both synchronizer sleeves are in neutral. The reverse gear, which is a sliding gear splined to the transmission main shaft, is moved to the left (in Fig. 40-14) so that it engages the rear reverse idler gear. Now, the rear reverse idler gear causes the main shaft to turn in the reverse direction, and the car moves backward.

♦ 40-13 OVERDRIVE

In the transmissions discussed so far, the high-gear position produces a 1:1 ratio between the clutch gear and the transmission output shaft. There is neither gear reduction nor gear increase through the transmission. This is direct drive.

At intermediate and high car speeds, it is sometimes desirable to have the transmission output shaft turn faster than the clutch gear and engine crankshaft. Therefore, some transmissions are designed with gears that provide an overdrive ratio. A transmission is in overdrive when the transmission output shaft is turning faster than the transmission input shaft, or clutch gear.

Years ago, many cars were equipped with a separate overdrive unit which was attached to the rear of the transmission. Today, the overdrive is built into the transmission. Many four-speed and five-speed transmissions have overdrive. The top gear position causes the transmission main shaft to overdrive (turn faster) than the clutch gear.

The advantage of overdrive is that it allows a lower engine speed to maintain the car at highway speed. Once the car is moving at a steady speed, it does not require as much power to keep it moving. Therefore, the engine can turn more slowly, produce less power, and still maintain car speed. This saves fuel and reduces wear on the engine and accessories. In overdrive, a typical overdrive transmission can maintain a car speed of 55 mph [89 km/h] while allowing the engine to turn at the equivalent of only 44 mph [71 km/h].

♦ 40-14 FOUR-SPEED TRANSMISSION WITH OVERDRIVE

From the outside, a four-speed transmission with overdrive may look like any other four-speed transmission. Even when you look inside at the gears, you might not be able to see the difference immediately. Figure 40-15 shows a four-speed transmission with overdrive, partly cut away so that the gears can be seen. Notice the different sizes of the clutch gear, the countergear-assembly driven gear, and the overdrive gear.

In the overdrive transmission with four forward speeds, the gear ratios in first, second, and third can be compared with the gear ratios of a standard three-speed transmission. First is low, second is intermediate, and third is high, or direct drive with a 1:1 gear ratio. Fourth is overdrive. Figure 40-16 shows the shift pattern for a four-speed transmission with overdrive. Fourth, or overdrive (OD), is at the lower right corner.

When the transmission is shifted into overdrive, the clutch synchronizing sleeve locks the overdrive gear to the main shaft. The power flow is then as shown in Fig. 40-17. Note that the clutch gear is smaller than the countergear-assembly driven gear. This means there is gear reduction because the countergear assembly turns more slowly than the clutch gear.

However, the overdrive gear on the countergear assembly is larger than the overdrive gear on the mainshaft. As the countergear rotates and drives the mainshaft, this provides a gear-ratio increase. The mainshaft overdrive gear turns faster than the overdrive gear on the countergear assembly. The result is that the mainshaft overdrive gear and the mainshaft turn faster than the clutch gear. This is *overdrive.*

As an example, let us assume that the four gears in a four-speed transmission with overdrive have the following number of teeth:

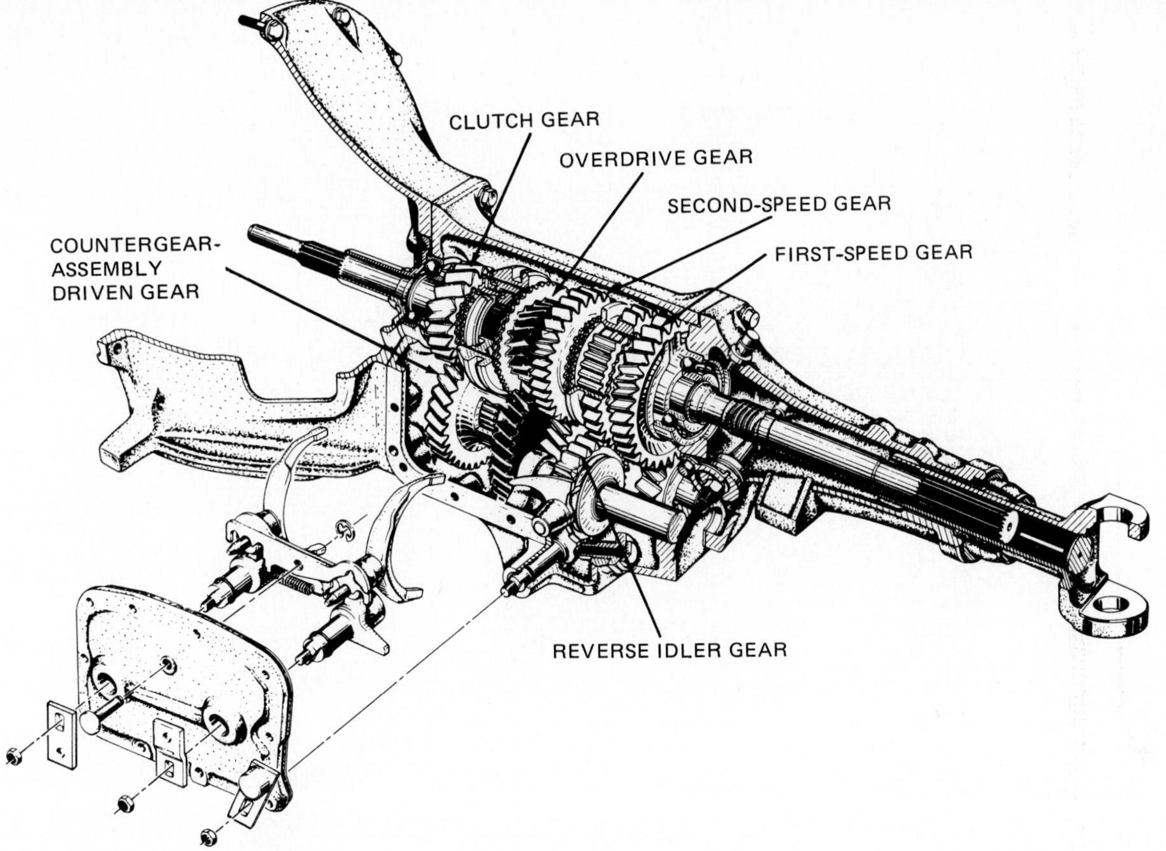

Fig. 40-15 A four-speed transmission which provides overdrive in fourth gear. *(Chrysler Corporation)*

Clutch gear	16
Countergear driven gear	24
Overdrive driving gear (on countergear)	28
Overdrive gear (on main shaft)	14

Let's find out how much faster the main shaft will turn than the clutch gear. The gear ratio between the clutch gear and the countergear driven gear is 24:16, or 3:2. This means that the clutch gear turns three times to turn the countergear assembly two times.

Meantime, the gear ratio between the gear on the countergear assembly that is driving the overdrive gear, and the overdrive gear, is 14:28, or 1:2. Therefore, when the countergear turns one time, it turns the overdrive gear on the main shaft two times.

Here is how this works out in terms of revolutions per minute. Suppose the engine speed is 3000 rpm. The clutch gear turns at the same speed of 3000 rpm. The gear reduction between the clutch gear and countergear driven gear is 3:2. Therefore the countergear assembly turns at 2000 rpm.

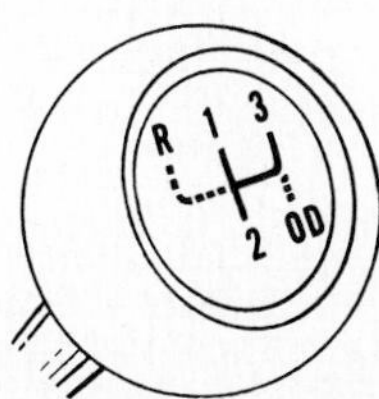

Fig. 40-16 Shift pattern for a four-speed transmission with overdrive.

The gear ratio between the driving gear on the countergear assembly and the overdrive gear is 1:2. This means that when the countergear assembly turns at 2000 rpm, it turns the overdrive gear and the main shaft at 4000 rpm.

When the transmission is in overdrive, the overall gear ratio is 3:4 (3000 to 4000). In practical terms, this means the engine speed can drop 25 percent and the car can still maintain highway speed. The only thing that is lost in overdrive is acceleration. With the higher gear ratio through the transmission, there is less torque being delivered to the wheels. However, a downshift can be made into third gear if more torque is desired for passing.

Note The example above is only typical. Gear sizes vary from one transmission to another so that different gear ratios are achieved in overdrive. Figure 40-18 shows the gear ratios of one model of four-speed transmission with overdrive.

♦ 40-15 FIVE-SPEED TRANSMISSION WITH OVERDRIVE

Several five-speed transmissions with overdrive have been designed. One widely used type is built as either a four-speed or a five-speed transmission. Figure 40-19 shows the gear ratios in the five-speed transmission with overdrive. Figure 40-20 shows the gears for both the four-speed and the five-speed transmission. The basic gear arrangement is the same for both. The extra gearing required in the five-speed transmission is shown in boxes at the top and bottom of Fig. 40-20.

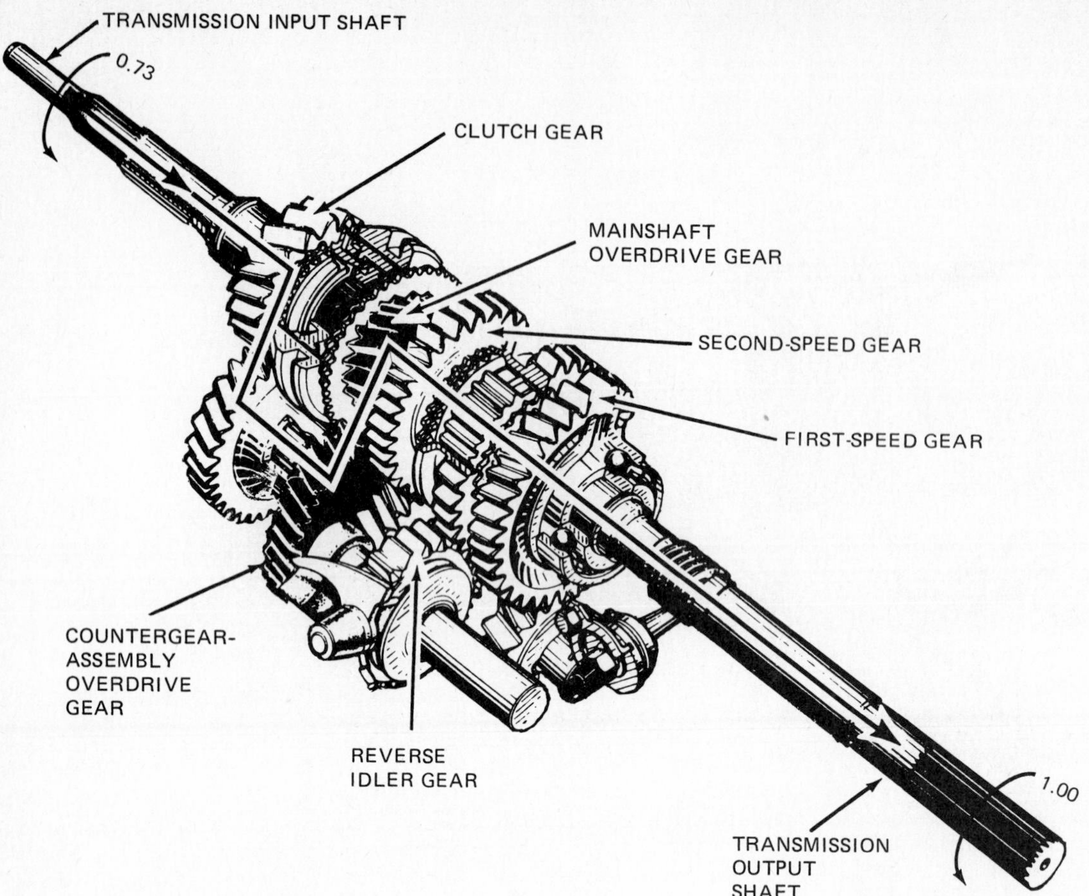

Fig. 40-17 Power flow in overdrive. *(Chrysler Corporation)*

♦ 40-16 TRANSAXLE

The manual transaxle includes the transmission, final-drive gearing, and differential. It is mounted on the engine (Fig. 40-21). The engine and transaxle assembly is mounted transversely at the front of most cars today (Figs. 10-3 and 10-10). Figure 39-3 is a simplified view of the power flow from the engine to the front wheels.

Figure 40-22 is a sectional view of a transaxle. Identify the various gears, the two synchronizers, and the other parts in the transmission. The synchronizers work the same way as in other manual transmissions.

Figure 40-23 shows the power flow in the four forward speeds and reverse. Compare the power flow in each gear position with the illustration in Fig. 40-22. Note how the movement of the synchronizers in one direction or the other locks the different gears to either the main shaft or the output shaft (in first, second, third, and fourth). In reverse, the reverse idler gear causes the output shaft to rotate in the reverse direction.

Gear Position	Gear Ratio	
	Input Shaft	Output Shaft
First	3.090	1.000
Second	1.670	1.000
Third	1.000	1.000 (Direct drive)
Fourth	0.730	1.000 (Overdrive)

Fig. 40-18 Gear ratios of the four-speed overdrive transmission. *(Chrysler Corporation)*

♦ 40-17 TRANSAXLE WITH DUAL-SPEED RANGE

A dual-range transaxle provides the usual four forward speeds plus reverse as in other transaxles. In addition, some models of this transmission have an extra set of gears that can supply overdrive in each gear position. This provides two speed ranges which the manufacturer calls the *economy range* and the *power range*. With four forward speeds in each range, this transaxle provides a total of eight forward speeds.

The gear ratios for both the single-range transaxle and the dual-range transaxle are shown in Fig. 40-24. In fourth gear, the power range of the two-speed transaxle provides a gear ratio of 1.105:1. The crankshaft turns 1.105 times to turn the output shaft once. When shifted into the economy range, fourth gear has a gear ratio of 0.855:1, which is overdrive. For

Gear Position	Gear Ratio	
	Input Shaft	Output Shaft
First	3.587	1.000
Second	2.022	1.000
Third	1.384	1.000
Fourth	1.000	1.000 (Direct drive)
Fifth	0.861	1.000 (Overdrive)
Reverse	3.384	1.000

Fig. 40-19 Gear ratios of the five-speed transmission, showing the number of times the output shaft turns for each revolution of the input shaft. *(Toyota Motor Sales Co., Ltd.)*

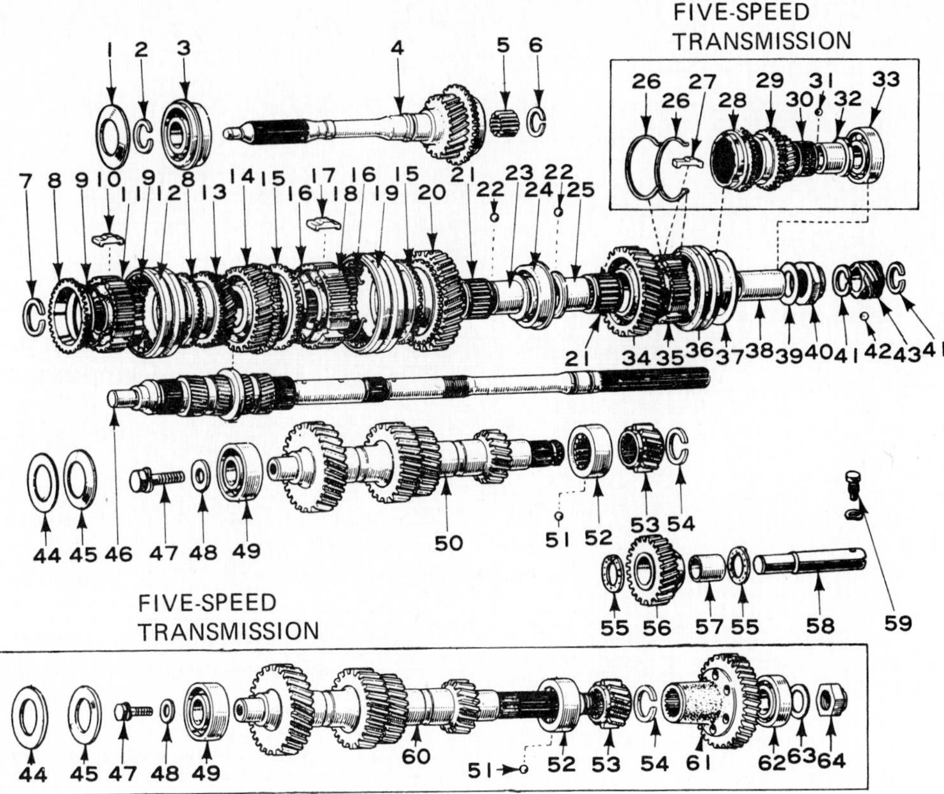

1. GEAR-THRUST-CONE SPRING
2. SHAFT SNAP RING
3. RADIAL BALL BEARING
4. INPUT SHAFT
5. ROLLER
6. HOLE SNAP RING
7. SHAFT SNAP RING
8. NO. 1 SYNCHRONIZER RING
9. NO. 1 SYNCHROMESH-SHIFTING-KEY SPRING
10. NO. 2 SYNCHROMESH-SHIFTING KEY
11. NO. 2 TRANSMISSION-CLUTCH HUB
12. NO. 2 TRANSMISSION-HUB SLEEVE
13. THIRD-GEAR SUBASSEMBLY
14. SECOND-GEAR SUBASSEMBLY
15. NO. 2 SYNCHRONIZER RING
16. NO. 1 SYNCHROMESH-SHIFTING-KEY SPRING
17. NO. 1 SYNCHROMESH SHIFTING KEY
18. NO. 1 TRANSMISSION-CLUTCH HUB
19. NO. 1 TRANSMISSION-HUB SLEEVE
20. FIRST-GEAR SUBASSEMBLY
21. NEEDLE ROLLER BEARING
22. BALL
23. FIRST-GEAR BUSHING
24. RADIAL BALL BEARING
25. REVERSE-GEAR BUSHING
26. NO. 1 SYNCHROMESH-SHIFTING-KEY SPRING
27. NO. 3 SYNCHROMESH SHIFTING KEY
28. NO. 1 SYNCHRONIZER RING
29. FIFTH-GEAR SUBASSEMBLY
30. NEEDLE ROLLER BEARING
31. BALL
32. FIFTH-GEAR BUSHING
33. RADIAL BALL BEARING
34. REVERSE GEAR
35. NO. 3 TRANSMISSION-CLUTCH HUB
36. NO. 3 TRANSMISSION-CLUTCH HUB
37. SPACER
38. SPACER
39. SHIM
40. NUT
41. SHAFT SNAP RING
42. BALL
43. SPEEDOMETER DRIVE GEAR
44. SHIM
45. GEAR-THRUST-CONE SPRING
46. OUTPUT SHAFT
47. BOLT WITH WASHER
48. PLATE WASHER
49. RADIAL BALL BEARING
50. COUNTERGEAR
51. BALL
52. CYLINDRICAL ROLLER BEARING
53. COUNTERSHAFT REVERSE GEAR
54. SHAFT SNAP RING
55. REVERSE-IDLER-GEAR THRUST WASHER
56. REVERSE IDLER GEAR
57. BIMETAL-FORMED BUSHING
58. REVERSE-IDLER-GEAR SHAFT
59. SHAFT RETAINING BOLT
60. COUNTERGEAR
61. FIFTH-GEAR COUNTER-SHAFT
62. RADIAL BALL BEARING
63. SHIM
64. NUT

Fig. 40-20 Gears and shafts in the four- and five-speed transmissions. The boxes enclose the additional parts in the five-speed transmission. *(Toyota Motor Sales Co., Ltd.)*

each complete revolution of the output shaft, the crankshaft turns less than one complete revolution (only 0.855 of one revolution).

Figures 40-25 and 40-26 are sectional views of the two transaxles. Notice that there is an intermediate-gear assembly (item 5 in Fig. 40-25 and item 8 in Fig. 40-26). This is the same as the countergear assembly in standard transmissions. The intermediate-gear assembly is located between the input

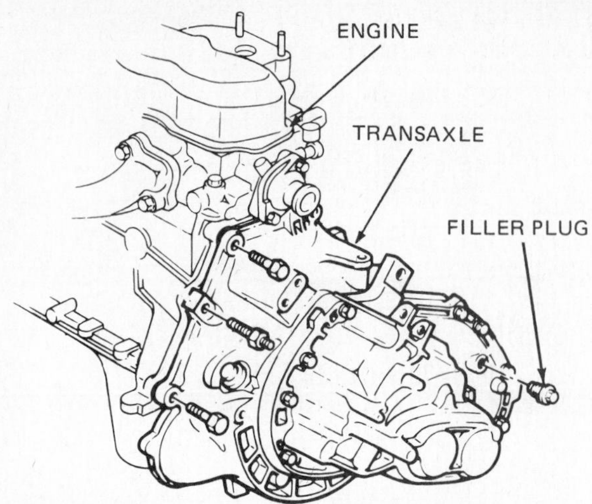

Fig. 40-21 Exterior view of the transaxle. Bolts are shown removed to detach the transaxle from the engine. *(Chevrolet Motor Division of General Motors Corporation)*

shaft and the output shaft. By changing the design of the input shaft and the intermediate shaft, it is possible to use the same case for either the single-range or the dual-range gearset.

Basically, the dual-range transaxle has an additional gear and a synchronizer on the input shaft. This can be seen by comparing the dual-range input shaft shown in Fig. 40-26 with the input shaft for the single-range transaxle shown in Fig. 40-25.

Shifting between the two ranges requires a separate lever. Figure 40-27 shows the shift patterns for both the range-selector lever and the gearshift lever. By moving the range-selector lever, the range-selector synchronizer (Fig. 40-27) can be shifted back and forth along the input shaft. This locks either the input low gear or the input high gear to the shaft. The range-selector lever is placed in either low range for power or high range for economy.

♦ 40-18 DIFFERENTIAL AND DRIVE SHAFTS FOR TRANSAXLES

The drive shafts for front-engine cars with a transaxle are covered in a later chapter. These drive shafts have universal joints that permit them to carry rotary motion through angles. Slip joints allow the effective length of the drive shaft to change. These joint actions allow the front wheels to move up and down as they meet irregularities in the road and to be swung from side to side for steering.

The differential allows the two front wheels to rotate at different speeds as the car travels around a curve or makes a turn. The outer wheel always has to travel farther and therefore must turn faster (Fig. 40-28). The differential in the transaxle allows this to happen.

When both wheels are turning at the same speed, the differential acts like a solid coupling. The differential case turns both drive shafts and wheels at the same speed. However, when the car makes a turn, the differential pinions rotate on the pinion shaft. This causes the side gear that is splined to the outer-wheel drive shaft to turn faster. A later chapter describes differential construction and operation.

♦ 40-19 STEERING-COLUMN AND FLOOR SHIFT LEVERS

Years ago, the gearshift lever was always located on the floor of the driver's compartment. The lower end of the lever was attached to the shifting devices in the transmission case. Later, the gearshift lever was moved up to the steering column, where it was more readily accessible to the driver. This position also provided more leg room for the center passenger in the front seat.

When the transmission shift lever is mounted on the steering column, the car is said to have a *column shift*. When a transmission has a floor shift lever (instead of the column-shift lever), it is called a *floor shift*. A four-speed transmission with a floor shift is often called "four on the floor." Today, many cars with manual transmissions have the gearshift lever back on the floor.

Regardless of the position of the shift lever, movement of the shift lever operates linkages to the transmission. Figure 40-29 shows the shift patterns for a steering-column shift lever and a floor shift lever for a three-speed manual transmission. The linkage from the shift lever (or *selector lever*) ends at the transmission shift forks (♦40-10). It is the movement of these shift forks that moves the synchronizers and produces the different gear speeds.

Two separate movements of the gearshift lever are required. The first movement of the gearshift lever selects the synchronizer-sleeve shift fork (for example, first-and-reverse, or second-and-third). Once this selection is made, the second movement moves the synchronizer to produce the desired gear (♦40-9).

♦ 40-20 STEERING-COLUMN SHIFT LEVER

Figures 40-30 and 40-31 show one type of steering-column shift and its linkages to the transmission. To shift into first or reverse, the driver depresses the clutch pedal to momentarily disconnect the engine from the transmission. Then, the driver lifts the selector lever and moves it up for reverse or down for first gear. When the lever is lifted, it pivots on its mounting pin, which forces a tube or rod downward in the steering column. This downward movement pushes downward on a crossover blade at the bottom of the steering column. A slot in the blade engages a pin on the first-and-reverse shift lever (Fig. 40-31).

When the shift lever is moved into first, the first-and-reverse lever is rotated. This movement is carried by the linkage to the transmission (Fig. 40-32). At the transmission, the movement causes the first-and-reverse shift lever to move. This lever is on a shaft that extends through the transmission side cover (Fig. 40-11). There is a lever on the inside end of this shaft, and a shifter fork is mounted on this lever.

When the shaft is rotated by movement of the first-and-reverse shift lever, it causes the shifter fork to move backward or forward inside the transmission. This motion causes the first-and-reverse synchronizer to engage a gear (♦40-9).

To shift into second or third, the driver moves the shift lever into the positions shown in Fig. 40-29. This movement causes the slot in the blade to engage the second-and-third shift lever at the bottom of the steering column. The lever is moved to actuate the second-and-third linkage to the transmission.

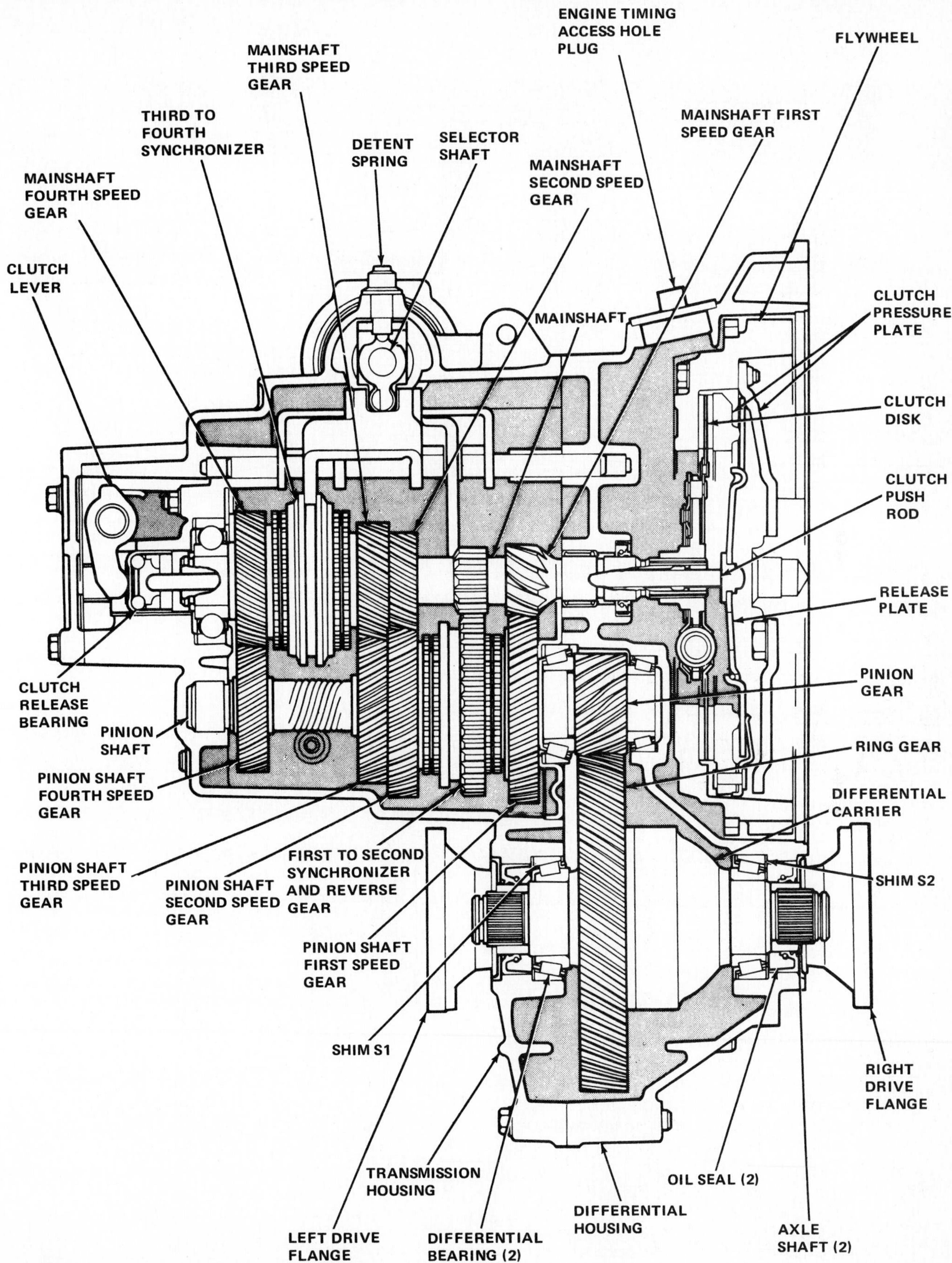

Fig. 40-22 Four-speed transaxle for a front-wheel-drive car. *(Chrysler Corporation)*

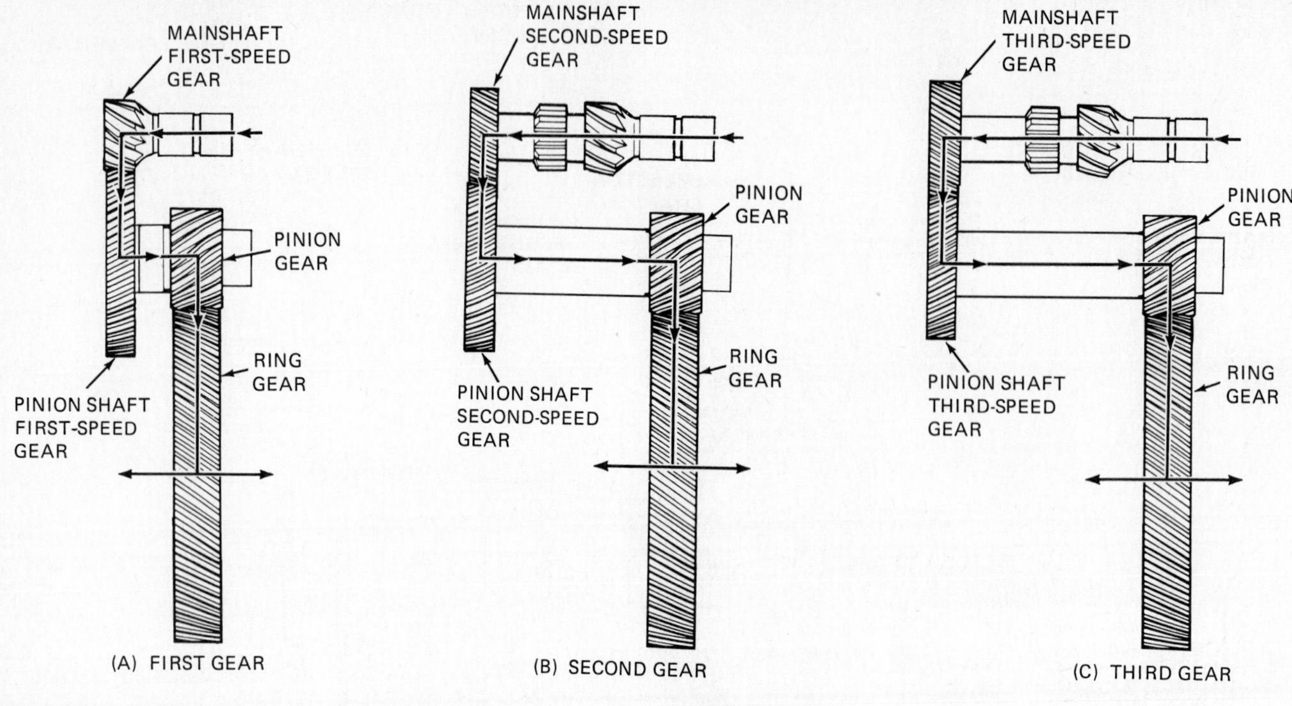

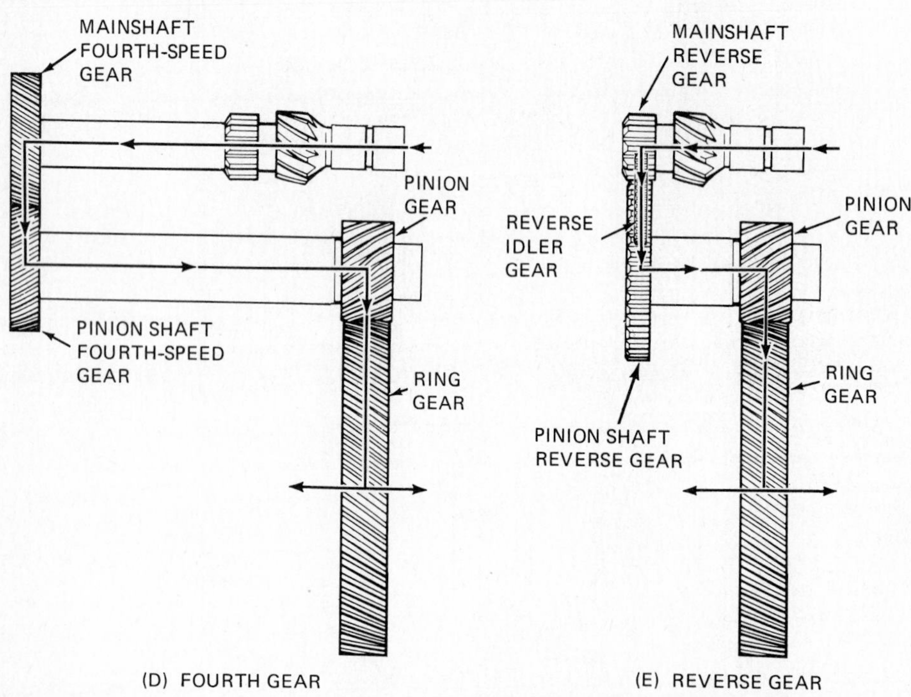

Fig. 40-23 Power flow through the four-speed transaxle in each gear position.

Gear Ratios	Standard Single-Range Transaxle	Dual-Range Transaxle Power Range	Economy Range
1st	4.226	4.226	3.272
2nd	2.365	2.365	1.831
3rd	1.467	1.467	1.136
4th	1.105	1.105	0.855
Reverse	4.109	4.109	3.181

Fig. 40-24 Gear ratios of a single-range transaxle compared with gear ratios in each range of a dual-range transaxle. *(Chrysler Corporation)*

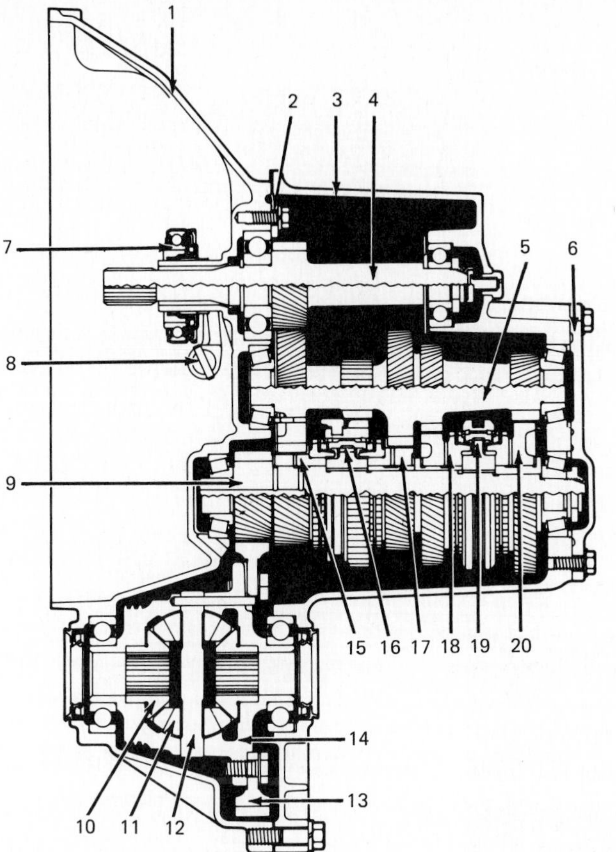

1. CLUTCH HOUSING
2. BEARING RETAINER
3. TRANSAXLE CASE
4. INPUT SHAFT
5. INTERMEDIATE-GEAR ASSEMBLY
6. REAR COVER
7. CLUTCH RELEASE BEARING
8. CLUTCH RELEASE FORK
9. OUTPUT SHAFT
10. DIFFERENTIAL SIDE GEAR
11. DIFFERENTIAL PINION
12. PINION SHAFT
13. DIFFERENTIAL DRIVE GEAR
14. DIFFERENTIAL CASE
15. FOURTH-SPEED GEAR
16. THIRD- AND FOURTH-SPEED SYNCHRONIZER
17. THIRD-SPEED GEAR
18. SECOND-SPEED GEAR
19. FIRST- AND SECOND-SPEED SYNCHRONIZER
20. FIRST-SPEED GEAR

Fig. 40-25 Sectional view of a single-range four-speed transaxle. *(Chrysler Corporation)*

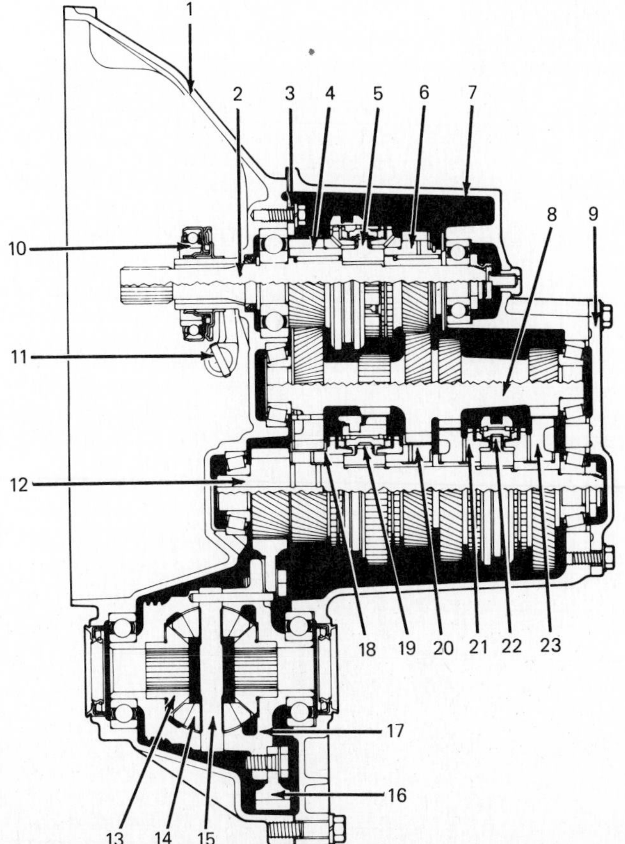

1. CLUTCH HOUSING
2. INPUT SHAFT
3. BEARING RETAINER
4. INPUT LOW GEAR
5. SYNCHRONIZER ASSEMBLY
6. INPUT HIGH GEAR
7. TRANSAXLE CASE
8. INTERMEDIATE-GEAR ASSEMBLY
9. REAR COVER
10. CLUTCH RELEASE BEARING
11. CLUTCH RELEASE FORK
12. OUTPUT SHAFT
13. DIFFERENTIAL SIDE GEAR
14. DIFFERENTIAL PINION
15. PINION SHAFT
16. DIFFERENTIAL DRIVE GEAR
17. DIFFERENTIAL CASE
18. FOURTH-SPEED GEAR
19. THIRD- AND FOURTH-SPEED SYNCHRONIZER
20. THIRD-SPEED GEAR
21. SECOND-SPEED GEAR
22. FIRST- AND SECOND-SPEED SYNCHRONIZER
23. FIRST-SPEED GEAR

Fig. 40-26 Sectional view of a dual-range transaxle. Note the additional gearing (in upper part). *(Chrysler Corporation)*

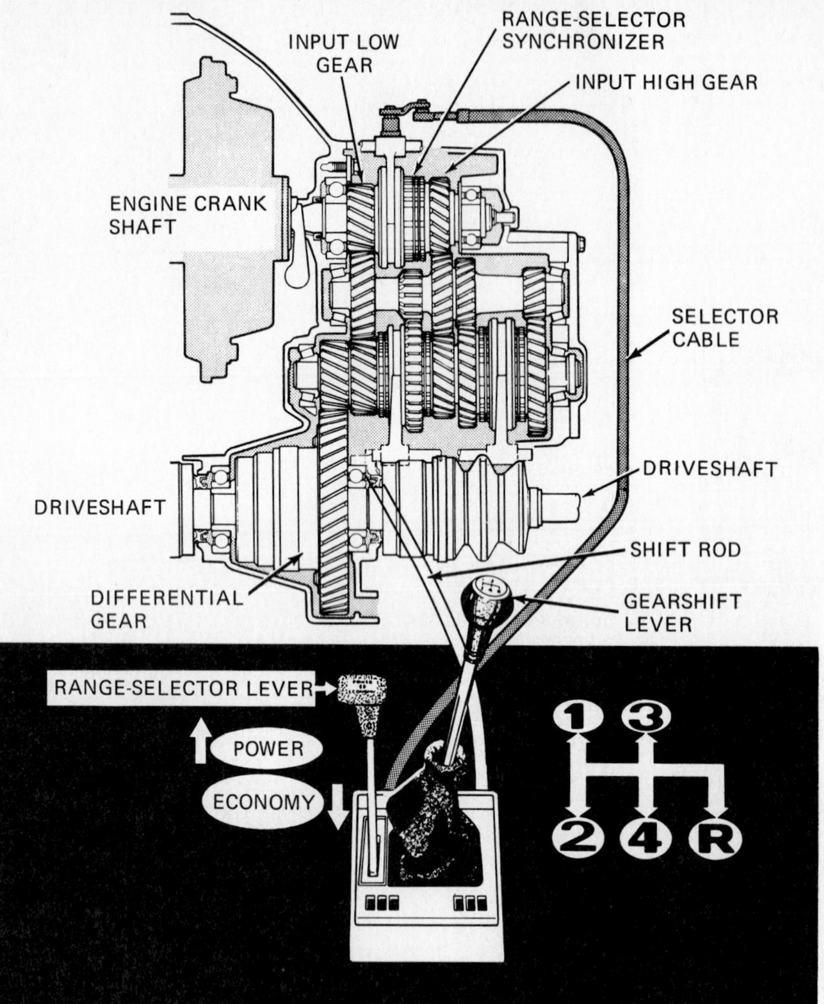

Fig. 40-27 Control levers and linkage for the dual-range transaxle. *(Chrysler Corporation)*

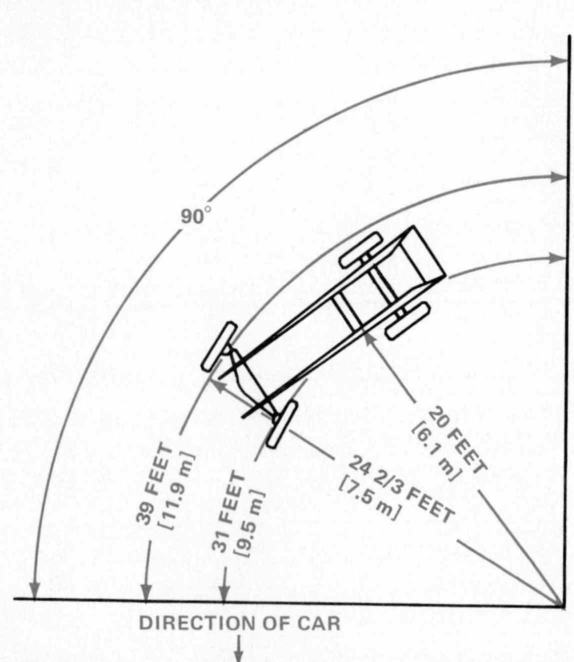

Fig. 40-28 During a turn, the outer wheel travels farther than the inner wheel.

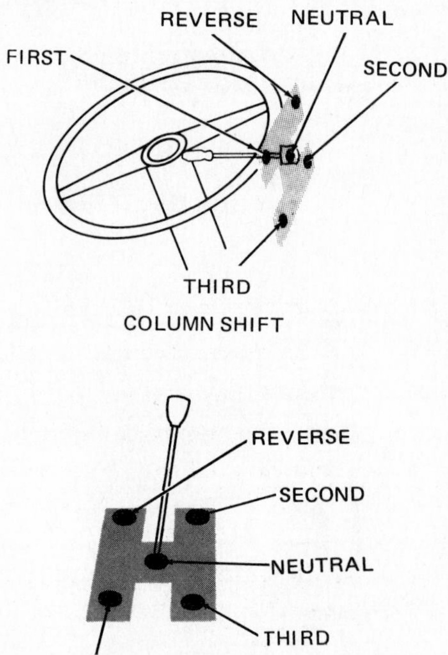

Fig. 40-29 Gearshift patterns for steering-column and floor shift levers.

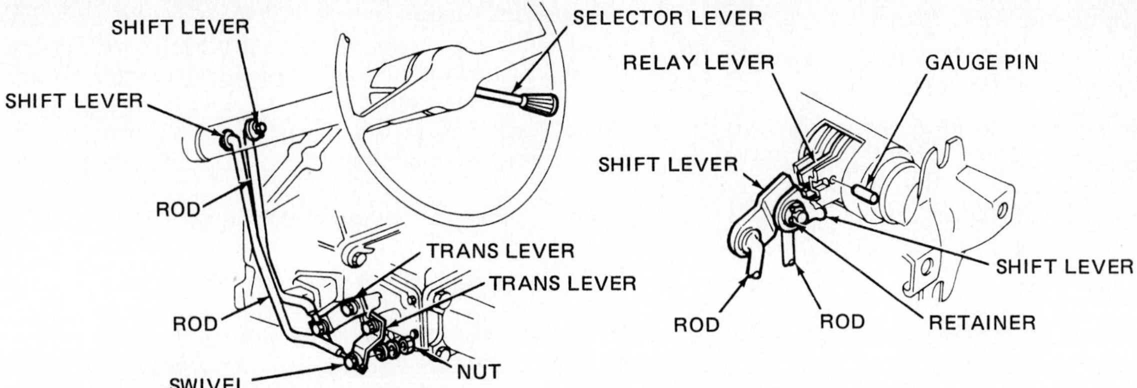

Fig. 40-30 Linkage between steering-column shift selector lever and transmission. *(Chevrolet Motor Division of General Motors Corporation)*

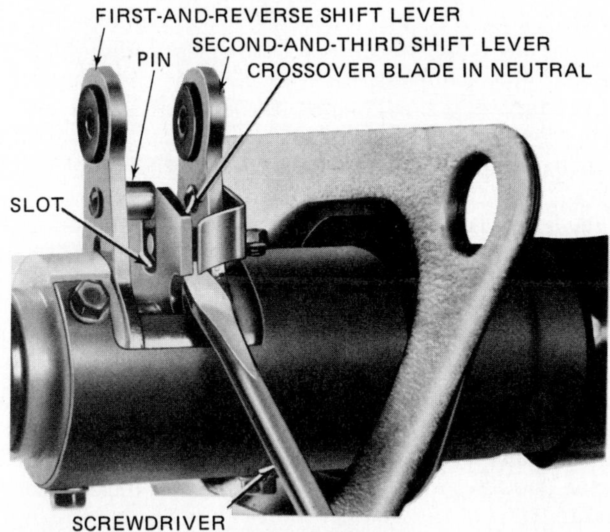

Fig. 40-31 Shift levers and crossover blade at the bottom of the steering column. The screwdriver holds the crossover blade in neutral for an adjustment check. *(Chrysler Corporation)*

♦ 40-21 FLOOR SHIFT LEVER

Many cars have a floor-mounted shift lever. A transmission using a floor shift lever is shown in Fig. 40-33. Figure 40-34 shows the linkage on an automobile that has a *console*. In the automobile, a console is a small cabinet or raised decorative centerpiece on the floor of the front compartment between the two front bucket seats. It houses the shift lever and sometimes includes a glove compartment, ashtray, gauges, and various controls, such as electric window switches and heater and air-conditioner controls.

In the shift-lever support there are levers attached to each rod. These levers have slots that are selected by a tongue on the lower end of the shift lever as it is moved into the various gear positions. The first movement of the shift lever makes this selection. Then the second movement of the shift lever causes the selected lever and rod to move. This causes the transmission lever to move and thereby shift the selected gear into the selected gear position.

Note Although it is common practice to describe movement of the shift lever as "shifting gears," the gears actually do not move in most transmissions. Instead, synchronizers are moved to engage gears that are always in constant mesh (♦40-9).

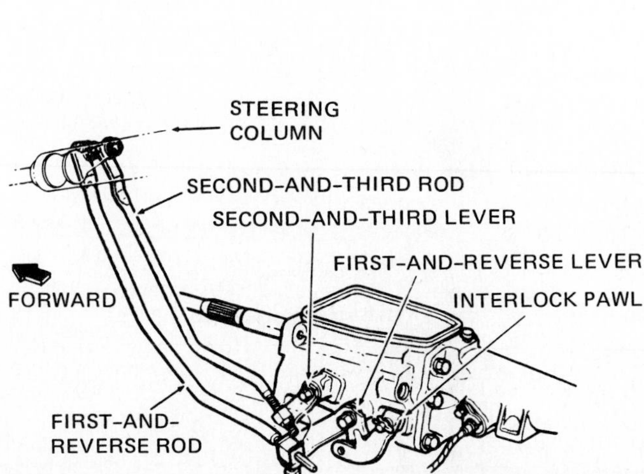

Fig. 40-32 Gearshift linkage between the shift levers at the bottom of the steering column and the transmission levers on the side of the transmission. *(Chrysler Corporation)*

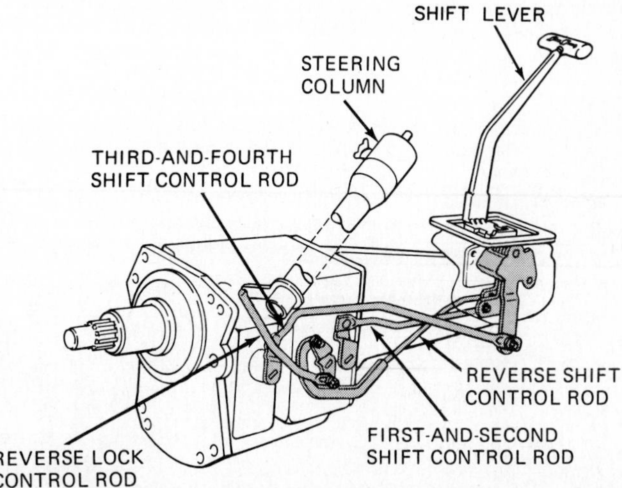

Fig. 40-33 A four-speed floor-shift transmission, showing the linkage from the shift lever to the transmission. *(Ford Motor Company)*

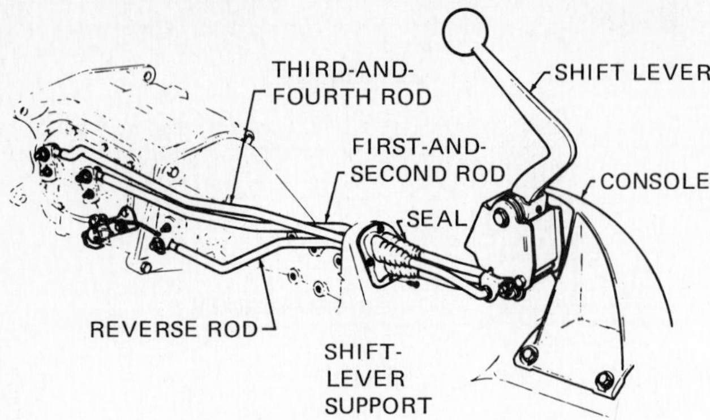

Fig. 40-34 Linkage between the four-speed transmission and a console-mounted shift lever. *(Ford Motor Company)*

Some transmissions do not use linkage rods between the gearshift lever and the transmission. Instead, they use a single-rail or shifter shaft (Fig. 40-35). The lower end of the gearshift lever moves into the bracket on the end of the shifter shaft when a gear is selected. Then further movement of the gearshift lever causes the shifter shaft to move. This then moves the fork that has been selected. The fork and the synchronizer sleeve it surrounds move in the proper direction to produce the selected gear position. Operation of the synchronizer is covered in ♦40-9 and 40-11.

♦ 40-22 GEAR LUBRICANTS

Manual transmissions, transaxles, and transfer cases are all various types of gearboxes. They are very similar in three ways. All have:

1. Gears that transmit power
2. Splined shafts that rotate while other parts are sliding on them
3. Bearings that support the shafts and transfer the load to the case or housing

In the gearbox, the moving metal parts must not touch each other. They must be continuously separated by a thin film of lubricant to prevent excessive wear and premature failure.

As gear teeth mesh, there is a sliding or wiping action between the contact faces. This action produces friction and heat. Without lubrication, the gears would wear quickly and fail. However, lubrication provides a fluid film between the contact faces. This prevents metal-to-metal contact. Therefore, all gearboxes have some type of lubricant or *gear oil* in them.

A gear oil has five jobs to do. These are:

1. To lubricate all moving parts and prevent wear
2. To reduce friction and power loss
3. To protect against rust and corrosion
4. To keep the interior clean
5. To cool the gearbox

In addition, the oil must have adequate load-carrying capacity to prevent puncturing of the oil film. Chemical additives are mixed with gear oil to improve its load-carrying capacity. An oil that has an additive in it to increase the load-carrying capacity is called an *extreme-pressure* (EP) lubricant. Other additives are also added to the oil to improve the viscosity (thickness), to prevent channeling (solidifying), to improve stability and oxidation resistance, to prevent foaming, to prevent rust and corrosion, and to prevent damage to the seals.

The typical gear oil is a straight mineral oil (refined from crude oil) with the required additives in it. Today, some gear oils are made from synthetic oil. Regardless of type, gear oil for use in most cars and light trucks has a classification of SAE 75W, 75W-80, 80W-90, 85W-90, 90, or 140.

Gear oil is *not* recommended for use in all gearboxes by all manufacturers. Gears which are lightly loaded, such as the planet-pinion gears in a planetary gearset, do not require gear oil. Therefore, some transfer cases are filled with SAE 10W-30 engine oil. Other transfer cases use automatic-transmission fluid (ATF).

ATF is also used as the factory fill in some manual transmissions built by Chrysler. If excessive gear rattle is heard at idle or during acceleration in direct drive or in overdrive, the ATF may be drained out and the transmission filled with a multipurpose gear oil, such as SAE 85W-90. Some manual transaxles are also filled with ATF.

To prevent the lubricant from leaking out, the gearbox has

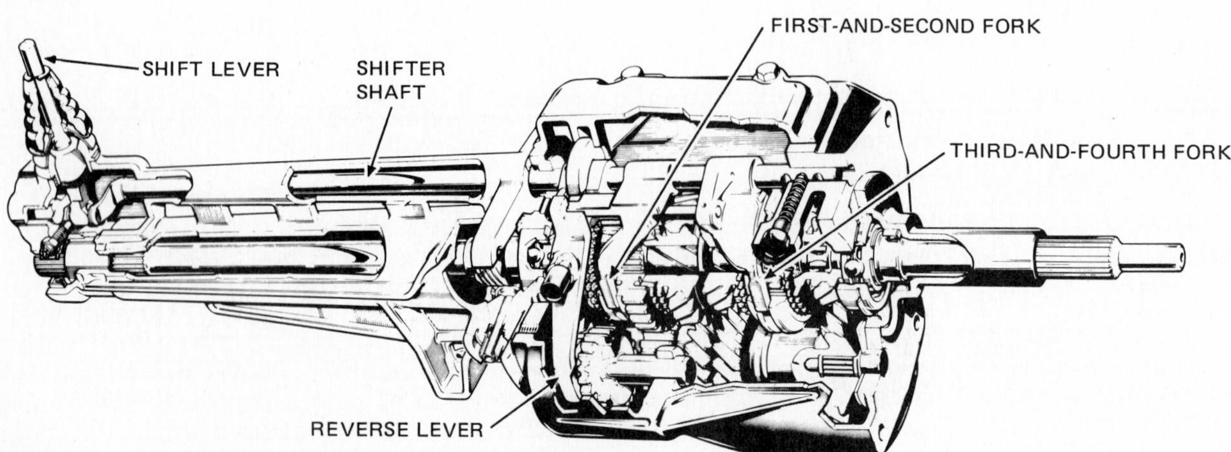

Fig. 40-35 A single-rail four-speed transmission, showing how the shifter shaft is connected to the shift forks. *(Ford Motor Company)*

an oiltight case. Seals are used around each cover and shaft. In addition, seals are provided around the input shafts and the output shafts. The clutch shaft on many transmissions does not have a separate seal. Instead, an oil slinger is used to throw back any oil that reaches it. Other designs have a passage in the clutch-shaft-bearing retainer that returns to the case any oil passing through the bearing.

♦ 40-23 TRANSMISSION-CONTROLLED-SPARK SWITCH

One of the emission control systems on many cars today is the TCS system. This system prevents ignition vacuum advance in any gear but high. The switch that controls the system is threaded into a boss on the side cover of the transmission (Fig. 40-36).

The TCS switch is open in all gears but high, thereby preventing vacuum advance. In high, the switch is closed, allowing vacuum advance. Vacuum advance in other gear positions increases the engine exhaust emissions coming from the tail pipe.

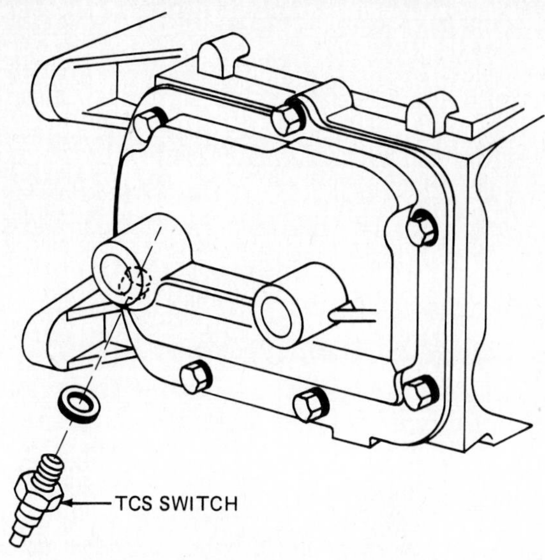

Fig. 40-36 Location of the TCS switch on a manual transmission side cover. *(Pontiac Motor Division of General Motors Corporation)*

♦ 40-24 BACKUP LIGHTS

When the gearshift lever is moved to reverse, the linkage closes a switch that connects the backup lights to the battery. With this arrangement the lights come on automatically. This is a warning that the car is about to back up. The backup lights also allow the driver to see behind the car at night.

♦ 40-25 SPEEDOMETER DRIVE

On most cars, the mechanical speedometer is driven by a pair of gears in the transmission-extension housing. One of these gears is mounted on the transmission main shaft. The other gear is mounted on the end of the flexible shaft connecting the speedometer to the transmission gear.

♦ REVIEW QUESTIONS ♦

Select the *one* correct, best, or most probable answer to each question. You can find the answers in the section indicated at the end of each question.

1. Synchronizing devices are normally used when shifting into (♦40-11)
   a. first
   b. second
   c. third
   d. all of the above
2. The gearshift lever requires two separate motions to shift gears, and the first movement (♦40-19)
   a. moves the synchronizer
   b. selects the synchronizer
   c. meshes the gears
   d. operates the clutch
3. On the main shaft of the four-speed transmission described in ♦40-12, there are
   a. three gears
   b. four gears
   c. five gears
   d. six gears
4. The three-speed transmission has (♦40-9)
   a. one shift fork
   b. two shift forks
   c. three shift forks
   d. four shift forks
5. The gearshift lever on the steering column is normally connected to the transmission by (♦40-20)
   a. a single link
   b. two or three linkages
   c. four linkages
   d. a shifter rod

CHAPTER 41
Manual-Transmission Trouble Diagnosis and Service

After studying this chapter, and with proper instruction and equipment, you should be able to:

1. Explain why the car should be road tested with the customer.
2. Discuss three types of transmission noise and the causes of each.
3. List several transmission troubles, the possible causes of each, and how to correct it.
4. Adjust manual transmission and transaxle linkage.
5. Describe the conditions that can result from adding the wrong lubricant to the transmission.
6. List the places where oil can leak from a transmission.
7. Service a manual transmission.

♦ 41-1 MANUAL-TRANSMISSION DIAGNOSIS PROCEDURE

The type of trouble a transmission has is often a clue to the cause of that trouble. Before attempting any repairs, try to determine the cause. Driver complaints, their possible causes, and the checks or corrections to be made are listed and discussed in later sections. Internal transmission troubles are fixed by disassembling the transmission. Or the old transmission can be replaced with a new, rebuilt, or used transmission.

To accurately diagnose a complaint about a manual transmission, a procedure must be followed. Road test the car with the owner to verify that the complaint exists (Fig. 41-1). Road testing with the owner gives you the opportunity to identify the condition that the owner wants corrected. There are two general types of manual-transmission troubles. They are (1) noise and (2) improper operation.

♦ CAUTION ♦ Do not drive a car unless you have a valid driver's license. Do not make this test unless you have your instructor's permission. Then fasten your safety belt and conduct the test where designated by your instructor.

During the road test, determine any related symptoms that may be occurring. Get all the facts and service history possible from the owner. Then, as you drive, determine when, where, and how the symptoms occur.

Immediately begin to analyze the symptoms. Now you are performing the diagnosis step by answering the question "What is wrong?" As soon as you know, tell the owner what to

Fig. 41-1 When possible, road test the car with the customer.

expect. Then, with the owner's permission, perform the required adjustments or repairs. When the job is completed, road test the car again. This time make sure that the trouble you found on the first road test no longer exists.

When proper operation has been restored, you know that the trouble has been corrected. This second road test serves as a quality-control check. As far as possible, it assures both you and the car owner that the job has been done right the first time. This helps prevent shop "come-backs" that may result from incorrect diagnosis, installation of defective parts, or faulty workmanship.

♦ 41-2 MANUAL-TRANSMISSION TROUBLE DIAGNOSIS

Most internal transmission problems can be accurately diagnosed before disassembling the transmission. For example,

Noise	Cause
Periodic clunk	Broken teeth
Growl or whine	Defective bearing or worn teeth
Gear clash	Defective synchronizer

Fig. 41-2 Three types of manual-transmission noise and their causes.

there are three general types of noise from a manual transmission (Fig. 41-2). The noise provides you with information about what is taking place inside the transmission. A periodic clunking noise indicates broken teeth. A growl or whine indicates a defective bearing or worn contact faces on the gear teeth. Gear clash during shifting or when shifting is attempted indicates a defective synchronizer.

Note Certain clutch problems produce symptoms similar to the symptoms of transmission problems. Follow the trouble-diagnosis procedures for the transmission you are servicing to determine the actual cause of the problem before attempting any repair. It may be that what you thought was transmission trouble is actually a trouble located in some other part of the car.

♦ 41-3 MANUAL-TRANSMISSION TROUBLE-DIAGNOSIS CHART

The chart that follows lists the various manual-transmission troubles together with their possible causes, and checks and corrections to be made. Most transmission troubles can be listed under a few headings, such as "hard shifting," "slips out of gear," and "noises."

MANUAL-TRANSMISSION TROUBLE-DIAGNOSIS CHART

Complaint	Possible Cause	Check or Correction
1. Hard shifting into gear	a. Gearshift linkage out of adjustment	Adjust
	b. Gearshift linkage needs lubrication	Lubricate
	c. Clutch not disengaging	Adjust
	d. Excessive clutch-pedal free play	Adjust
	e. Shifter fork bent	Replace or straighten
	f. Sliding gears or synchronizer tight on shaft splines	Replace defective parts
	g. Gear teeth battered	Replace defective gears
	h. Synchronizing unit damaged or springs improperly installed (after overhaul)	Replace unit or defective parts; install springs properly
	i. Shifter tube binding in steering column	Correct tube alignment
	j. End of transmission input shaft binding in crankshaft pilot bushing	Lubricate; replace bushing
2. Transmission sticks in gear	a. Gearshift linkage out of adjustment or disconnected	Adjust; reconnect
	b. Gearshift linkage needs lubrication	Lubricate
	c. Clutch not disengaging	Adjust
	d. Detent balls (lockouts) stuck	Free; lubricate
	e. Synchronizing unit stuck	Free; replace damaged parts
	f. Incorrect or insufficient lubricant in transmission	Replace with correct lubricant and correct amount
	g. Internal shifter components damaged	Remove transmission to inspect and service shifter parts
3. Transmission slips out of gear	a. Gearshift linkage out of adjustment	Adjust
	b. On floor shift, shift boot stiff or shift-lever binding	Replace boot; adjust console to relieve binding

MANUAL-TRANSMISSION TROUBLE-DIAGNOSIS CHART *(Continued)*

Complaint	Possible Cause	Check or Correction
	c. Weak lockout springs	Replace
	d. Bearings or gears worn	Replace
	e. End play of shaft or gears excessive	Replace worn or loose parts
	f. Synchronizer worn or defective	Repair; replace
	g. Transmission loose on clutch housing or misaligned	Tighten mounting bolts; correct alignment
	h. Clutch housing misaligned	Correct alignment
	i. Pilot bushing in crankshaft loose or broken	Replace
	j. Input-shaft retainer loose or broken	Replace
	k. Broken engine mount	Replace
4. No power through transmission	a. Clutch slipping	Adjust
	b. Gear teeth stripped	Replace gears
	c. Shifter fork or other linkage part broken	Replace
	d. Gear or shaft broken	Replace
	e. Drive key or spline sheared off	Replace
5. Transmission noisy in neutral	a. Gears worn or tooth broken or chipped	Replace gears
	b. Bearings worn or dry	Replace; lubricate
	c. Input-shaft bearing defective	Replace
	d. Pilot bushing worn or loose in crankshaft	Replace
	e. Transmission misaligned with engine	Realign
	f. Countershaft worn or bent, or thrust plate or washers damaged	Replace worn or damaged parts
6. Transmission noisy in gear	a. Clutch friction disk defective	Replace
	b. Incorrect or insufficient lubricant	Replace with proper amount of correct lubricant
	c. Rear main bearing worn or dry	Replace or lubricate
	d. Gears loose on main shaft	Replace worn parts
	e. Synchronizers worn or damaged	Replace worn or damaged parts
	f. Speedometer gears worn	Replace
	g. Any condition noted in item 5	See item 5
7. Gears clash during shifting	a. Synchronizer defective	Repair or replace
	b. Clutch not disengaging; pedal free play incorrect	Adjust
	c. Hydraulic system (hydraulic clutch) defective	Check cylinder; add fluid, etc.
	d. Idle speed excessive	Readjust
	e. Pilot bushing binding	Replace
	f. Gearshift linkage out of adjustment	Adjust
	g. Lubricant incorrect	Replace with correct lubricant
8. Transmission noisy in reverse	a. Reverse idler gear or bushing worn or damaged	Replace
	b. Reverse gear on main shaft worn or damaged	Replace
	c. Countergear worn or damaged	Replace
	d. Shift mechanism damaged	Repair, replace defective parts, readjust
9. Oil leaks	a. Foaming due to incorrect lubricant	Replace with correct lubricant
	b. Oil level too high	Use proper amount, no more
	c. Gaskets broken or missing	Replace
	d. Oil seals damaged or missing	Replace
	e. Oil slingers damaged, improperly installed, or missing	Replace correctly
	f. Drain plug loose	Tighten
	g. Transmission retainer bolts loose	Tighten
	h. Transmission or extension case cracked	Replace
	i. Speedometer-gear retainer loose	Tighten
	j. Side cover loose	Tighten
	k. Extension-housing seal worn or drive-line yoke worn	Replace

♦ 41-4 MANUAL-TRANSMISSION REMOVAL AND INSTALLATION

A variety of arrangements are used to mount manual transmissions in vehicles. Basically, they are very similar. Several bolts attach the transmission to the clutch housing (Fig. 41-3). The rear end of the transmission output shaft or transmission main shaft is attached to the drive shaft (Fig. 39-2). A cross member supports the transmission. Shift rods connect the selector lever to the shift levers on the side of the transmission (Figs. 40-30 and 40-33).

Following is a manual-transmission removal procedure that applies, in general, to all vehicles. However, because of variations from car to car, refer to the manufacturer's shop manual that covers the specific vehicle you are working on. Manuals have step-by-step procedures, together with specifications for checking parts and tightening attaching bolts and nuts.

Note You may have to remove the catalytic converter and its support bracket to provide space for transmission removal.

1. On floor-shift cars (Fig. 40-32), remove the shift lever.
2. Raise the vehicle on a lift.
3. Mark the rear-axle flange and drive shaft so that the drive shaft can be reinstalled in the same position when reconnected.
4. Drain the lubricant from the transmission.
5. Disconnect the speedometer cable from the transmission. Some manufacturers specify that you remove the speedometer-cable adapter and gear from the transmission housing at this time.
6. Disconnect the wiring to the backup-light and TCS switches, if present.
7. Disconnect the drive shaft. Some manufacturers recommend removing the drive shaft completely.
8. Support the engine with a jack or engine support. Remove the bolts attaching the transmission support to the cross member. Then remove the bolts attaching the cross member to the body or frame and remove the cross member.
9. Remove the upper bolts attaching the transmission to the clutch housing and install guide pins in the holes. The purpose of these guide pins is to prevent damage to the clutch friction disk. If the transmission were removed hanging down at an angle, the weight on the disk hub could damage it.
10. Remove the other transmission-attaching bolts. Then slide the transmission rearward until the clutch shaft clears the clutch.

♦ CAUTION ♦ The transmission is heavy. Always use a transmission jack if available (Fig. 41-4). Place the jack under the transmission to support it. If a jack is not available, get another person to help you. Move the transmission to the rear until it is free, then lower it and move it out from under the car.

With the transmission out, inspect the clutch and flywheel condition and tightness. Check the clutch-shaft pilot bushing in the end of the engine crankshaft. If the pilot bushing is worn, replace it as explained earlier.

11. In general, installation is the reverse of removal. Just before installation, shift the transmission into each gear, and turn the input shaft to check that the transmission works as it should. Be sure the matching faces of the transmission and the flywheel housing are clean. Place a small amount of lubricant on the splines of the input shaft. Prealign the splines on the input shaft and the friction-disk hub by turning the input shaft so the splines line up. Install guide pins, and lift the transmission. Slide the transmission forward into position. Turn the shaft, if necessary, to secure alignment of the shaft and the friction-disk hub splines. Put the bolts in place, and tighten them. Replace the guide pins with bolts, and tighten them.

Careful If the transmission does not fit snugly against the flywheel housing, or you cannot move the transmission easily into place, do not force it. The splines on the shaft and hub may not be aligned. Or perhaps roughness or dirt, or a loose retainer ring in the transmission, may be blocking the transmission. If the bolts are tightened under such circumstances, there will not be proper alignment, and the transmission case may be broken.

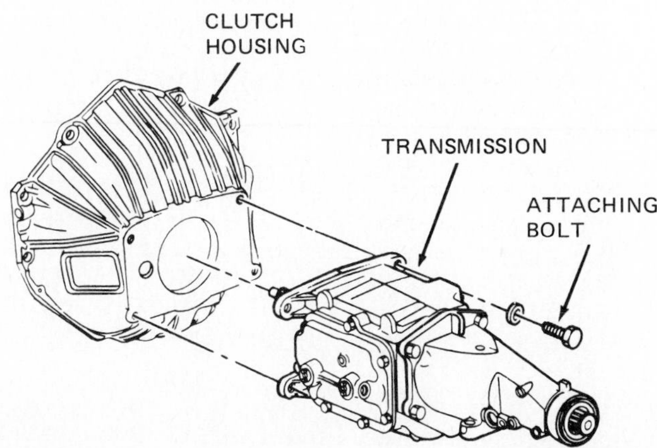

Fig. 41-3 Typical manual-transmission mounting on the clutch housing. *(Chevrolet Motor Division of General Motors Corporation)*

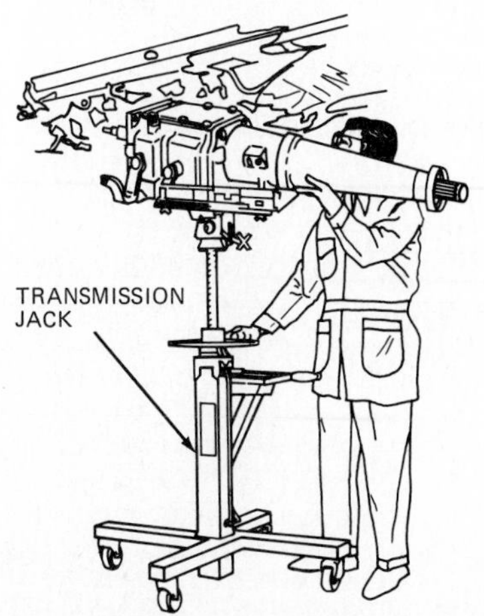

Fig. 41-4 Using a transmission jack to support the transmission. *(ATW)*

As a final step in the procedure, fill the transmission with the proper type and amount of lubricant.

♦ 41-5 MANUAL-TRANSMISSION OVERHAUL

Manual-transmission construction varies considerably from car to car. Therefore, removal and servicing procedures also vary. Before attempting to disassemble a manual transmission, carefully study both the transmission and the transmission section in the manufacturer's service manual. If possible, locate the illustrations and exploded views for the transmission you are working on. Overhaul procedures differ for different transmissions. Follow the procedures for disassembly, service, and reassembly that are in the manufacturer's service manual.

♦ 41-6 SHIFT-LINKAGE ADJUSTMENTS

The linkage between the selector lever and the shift levers on the transmission must be properly adjusted. This permits proper selection of gears and completion of the shifts. Typically, the adjustment is made with the transmission levers positioned in neutral. Then position the selector lever in neutral. The rods that were disconnected may require minor adjustment, but they will usually slip in and clip in. However, the rods may not fit if you disconnected the linkage at the wrong points, bent the rods, or unnecessarily turned the threaded clevis pins.

If the linkage has been tampered with, or the rods do not fit into the shift levers on the transmission, then a linkage adjustment must be made. Follow the procedures in the manufacturer's service manual.

♦ REVIEW QUESTIONS ♦

Select the *one* correct, best, or most probable answer to each question. You can find the answers in the section indicated at the end of each question.

1. Hard shifting into gear may be caused by (♦41-3)
   a. gearshift linkage out of adjustment
   b. clutch not disengaging
   c. excessive clutch-pedal freeplay
   d. all of the above
2. The transmission may stick in gear because (♦41-3)
   a. the gearshift linkage is out of adjustment
   b. the clutch is not disengaging
   c. a synchronizer is stuck
   d. all of the above
3. Noise from the transmission when it is in neutral could be caused by (♦41-3)
   a. failure of the clutch to engage
   b. worn or dry bearings
   c. main-shaft gears having chipped or broken teeth
   d. all of the above
4. The transmission may slip out of gear because (♦41-3)
   a. the gearshift linkage is out-of-adjustment
   b. the clutch housing is misaligned
   c. the lockout springs are weak
   d. all of the above
5. Noise from the transmission when it is in gear could be caused by (♦41-3)
   a. a slipping clutch
   b. excessive lubricant
   c. worn or damaged gears
   d. all of the above
6. Gear clash while shifting could be caused by (♦41-3)
   a. gears loose on the main shaft
   b. clutch not disengaging
   c. gearshift linkage disconnected
   d. all of the above
7. Transmission oil leaks could be caused by (♦41-3)
   a. foaming due to incorrect lubricant
   b. excessive lubricant
   c. damaged or missing oil seals or slingers
   d. all of the above
8. Noise from the transmission in reverse could be caused by (♦41-3)
   a. worn or damaged reverse idler gear
   b. defective synchronizer
   c. clutch not disengaging
   d. all of the above
9. Hard shifting into gear could be caused by any of the following *except* (♦41-3)
   a. clutch not disengaging
   b. synchronizer damaged
   c. bearings worn
   d. shifter fork bent
10. Gear clash during shifts can be caused by any of the following *except* (♦41-3)
    a. synchronizer defective
    b. idle speed too low
    c. clutch not disengaging
    d. incorrect shift-linkage adjustment

CHAPTER 42
Manual-Transaxle Trouble Diagnosis and Service

After studying this chapter, and with proper instruction and equipment, you should be able to:

1. List and describe several kinds of transaxle noise, and explain the possible causes and corrections of each.
2. Explain what might cause the transaxle to slip out of gear and how to correct the trouble.
3. Discuss what might cause difficulty shifting into gear and how to correct the trouble.
4. Explain what might cause the transaxle to stick in gear and how to correct the trouble.
5. Discuss what might cause the gears to clash in shifting and how to correct the trouble.
6. Explain where the transaxle might leak lubricant and how to stop the leaks.
7. Adjust the shift linkage.
8. Service a manual transaxle.

♦ 42-1 MANUAL-TRANSAXLE TROUBLE-DIAGNOSIS PROCEDURE

The type of trouble a transaxle has is often a clue to the cause of that trouble. Before attempting any repairs, try to determine the cause. Driver complaints, their possible causes, and the checks or corrections to be made are listed and discussed in later sections. Internal transaxle troubles are fixed by disassembling the transaxle. Or the old transaxle can be replaced with a new, rebuilt, or used transaxle.

To accurately diagnose a complaint about a manual transaxle, a procedure must be followed. This procedure is the same as the manual-transmission diagnosis procedure described in ♦41-1. Turn back to that section to review the diagnosis procedure.

♦ CAUTION ♦ Do not drive a car unless you have a valid driver's license. Do not make this test unless you have your instructor's permission. Then fasten your safety belt and conduct the test in the area designated by your instructor.

♦ 42-2 TRANSAXLE TROUBLE-DIAGNOSIS CHART

Before attempting any repair of the clutch, transaxle, or their linkages, for any reason except an obvious failure, try to identify the problem and its possible cause. Many clutch and transaxle problems show up as shifting difficulties. These include excessive effort needed to shift, gear clashing and grinding, and the inability to shift into some gears.

In addition, there may be noise problems. These vary with vehicle size, type and size of engine, and amount of body insulation used. The fact that the entire drive train is located at the front of the car almost under the feet of the driver makes any drive-train noise more audible to the driver. But noises that you might believe are coming from the drive train could be coming from the tires, road surfaces, wheel bearings, engine, or exhaust system. For this reason, a thorough and careful check should be made to locate the cause of the noise before removing and disassembling the transaxle.

The chart that follows lists various transaxle troubles and their possible causes and corrections. For example, if a knock at low speeds is caused by drive-axle joints, the remedy is to service the joints. If vibration is being caused by rough wheel bearings, the correction is to replace the bearings.

Note The chart applies especially to the General Motors transaxle used in their smaller, front-drive cars. However, the chart can be used as a guide when diagnosing trouble in any transaxle.

TRANSAXLE TROUBLE-DIAGNOSIS CHART*

Complaint	Possible Cause
1. Noise is the same in drive or coast (♦42-3)	a. Road noise b. Tire noise c. Front-wheel bearing noise d. Incorrect drive-axle angle (standing height)
2. Noise changes on different types of road (♦42-3)	a. Road noise b. Tire noise
3. Noise tone lowers as car speed is lowered (♦42-3)	Tire noise
4. Noise is produced with engine running, whether vehicle is stopped or moving (♦42-3)	a. Engine noise b. Transaxle noise c. Exhaust noise
5. Knock at low speeds (♦42-3)	a. Worn drive-axle joints b. Worn side-gear-hub counterbore
6. Noise loudest during turns (♦42-3)	Differential-gear noise
7. Clunk on acceleration or deceleration (♦42-3)	a. Loose engine or transaxle mounts b. Worn differential pinion shaft in case, or side-gear-hub counterbore in case worn oversize c. Worn or damaged drive-axle inboard joints
8. Clicking noise in turns (♦42-3)	Worn or damaged outboard joint
9. Vibration (♦42-3)	a. Rough wheel bearing b. Damaged drive-axle shaft c. Out-of-round tires d. Tire unbalance e. Worn joint in drive-axle shaft f. Incorrect drive-axle angle
10. Noisy in neutral with engine running (♦42-3)	Damaged input-gear bearings
11. Noisy in first only (♦42-3)	a. Damaged or worn first-speed constant-mesh gears b. Damaged or worn 1-2 synchronizer
12. Noisy in second only (♦42-3)	a. Damaged or worn second-speed constant-mesh gears b. Damaged or worn 1-2 synchronizer
13. Noisy in third only (♦42-3)	a. Damaged or worn third-speed constant-mesh gears b. Damaged or worn 3-4 synchronizer
14. Noisy in high gear only (♦42-3)	a. Damaged or worn 3-4 synchronizer b. Damaged fourth-speed gear or output gear
15. Noisy in reverse only (♦42-3)	a. Worn or damaged reverse idler gear or idler bushing b. Worn or damaged 1-2 synchronizer sleeve
16. Noisy in all gears (♦42-3)	a. Insufficient lubricant b. Damaged or worn bearings c. Worn or damaged input gear (shaft) and/or output gear (shaft)
17. Transaxle slips out of gear (♦42-5)	a. Worn or improperly adjusted linkage b. Transmission loose on engine housing c. Shift linkage does not work freely; binds d. Bent or damaged cables e. Input-gear-bearing retainer broken or loose f. Dirt between clutch cover and engine housing g. Stiff shift-lever seal

*See ♦42-3 to 42-9 for explanations of the trouble causes and their corrections.

TRANSAXLE TROUBLE-DIAGNOSIS CHART *(Continued)*

Complaint	Possible Cause
18. Hard shifting into gear (♦42-6)	a. Gearshift linkage out of adjustment or needs lubricant b. Clutch not disengaging c. Clutch linkage needs adjustment d. Internal trouble in transaxle
19. Transaxle sticks in gear (♦42-7)	a. Gearshift linkage out of adjustment, disconnected, or needs lubricant b. Clutch not disengaging c. Internal trouble in transaxle
20. Gears clash in shifting (♦42-8)	a. Incorrect gearshift-linkage adjustment b. Clutch not disengaging c. Clutch linkage needs adjustment d. Internal trouble in transaxle
21. Lubricant leaks out (♦42-9)	a. Axle-shaft seals faulty b. Excessive amount of lubricant in transmission c. Loose or broken input-gear (shaft)-bearing retainer d. Input-gear-bearing-retainer O ring and/or lip seal damaged e. Lack of sealant between case and clutch cover or loose clutch cover f. Shift-lever seal leaks

♦ 42-3 NOISES

Transaxle gears may produce some noise. If the noise is annoying, try the following steps to determine whether or not it is excessive, and what is causing it.

1. Drive the car on a smooth, level asphalt road. This will reduce tire and road noise to a minimum.
2. Drive the vehicle long enough to warm up all lubricant.
3. Note the speed at which the noise occurs and in which gear range.
4. Stop the vehicle and see if the noise is still present with transaxle in neutral. Then listen with the transaxle in gear with clutch pedal depressed.
5. Determine during which of the following driving conditions the noise is most noticeable:
   a. Driving—light acceleration or heavy pull.
   b. Float—constant vehicle speed with light throttle on a level road.
   c. Coast—partly or fully closed throttle with transaxle in gear.
   d. All of the above.
6. After road testing the vehicle, consider the following:
   a. If the noise is the same in drive or coast, it could be due to excessive drive-axle angle. The front suspension may be binding, or the springs may be weak. This could cause the drive-axle universal joints to be driving through an excessive angle.
   b. A knock at low speed could be caused by worn drive-axle universal joints or by worn counterbores in the side-gear hubs.
   c. A clunk on acceleration or deceleration could be caused by loose engine or transaxle mounts or by items 7b and 7c in the trouble-diagnosis chart (♦42-2).
   d. Refer to the manual-transaxle trouble-diagnosis chart for other possible causes of noise.
   e. Bearing noises are described in the following section (♦42-4).

Note Chapters 46 and 47 cover drive lines and universal joints, and differentials and drive axles.

♦ 42-4 BEARING NOISE

Defective bearings usually produce a rough growl or grating noise rather than the whine that is typical of gear noise. If bearing noise is suspected (see items 1, 9, 10, 15, and 16 in the trouble-diagnosis chart), it will be necessary to remove the transaxle and disassemble it so that the bearings can be inspected.

Clean the bearing assembly in solvent and allow it to dry. If the bearing has become magnetized, normal cleaning methods will not remove any metal particles attached to the bearing. The bearing must be demagnetized first. Bearings fail by lapping, spalling, or locking.

Lapping is caused by fine particles of abrasive material, such as scale, sand, or emery. These particles circulate with the oil and cause wear of the roller and race surfaces. Bearings which are worn lose but appear smooth without spalling or pitting have been running with dirty oil.

Spalling is caused by overloading or faulty assembly. Bearings that fail by spalling have either flaked or pitted rollers or races. Spalling can be caused by faulty assembly, such as cocking of bearings, misalignment, or excessively tight adjustments.

Locking is caused by large particles of dirt or other material wedging between rollers and race, usually causing one of the races to turn. If a race spins, it will wear the housing in which it is assembled. Then the housing must be replaced.

Careful Preloading a tapered-roller bearing higher than specified can cause the bearing to lock up.

1. Bearing Noise Side bearings are preloaded. Therefore noise will not disappear or diminish when the differential is run with the wheels off the ground. Noise in this area can easily be confused with wheel-bearing noise.

2. Wheel-Bearing Noise A rough bearing produces a vibration or growl which continues when the vehicle is coasting with the transmission in neutral. Wheel bearings are not preloaded, and so noise should diminish if the differential is run with the wheels off the ground. A brinnelled bearing causes a knock or click about every two wheel revolutions.

The bearing race is brinnelled when it has an indentation caused by a roller or ball. This could occur if the vehicle is transported over a rough highway on a vehicle carrier. The repeated pounding of the race by the roller or ball because of the rough road causes the race to be indented.

To check for brinnelling, spin the wheel by hand while listening at the hub for brinnelling or rough-bearing noise. Wheel bearings are not always serviceable as a separate item. In some cars, they must be replaced as an integral part of the hub and spindle.

♦ 42-5 TRANSAXLE SLIPS OUT OF GEAR

This could be caused by a worn or out-of-adjustment linkage, a transaxle that is loose on the engine housing, a binding shift lever, bent or damaged shift cables, a stiff shift-lever seal, or trouble inside the transaxle (see item 17 in the trouble-diagnosis chart).

♦ 42-6 HARD SHIFTING INTO GEAR

This can be caused by the linkage being out of adjustment or needing lubrication, by the clutch not disengaging, by the clutch linkage being out of adjustment, or by trouble inside the transaxle.

♦ 42-7 TRANSAXLE STICKS IN GEAR

This could be caused by the shift linkage being out of adjustment or disconnected or needing lubrication. It could also be caused by the clutch not disengaging or by trouble inside the transaxle.

♦ 42-8 GEARS CLASH IN SHIFTING

When gears clash during a shift, it could be that the shift linkage is not properly adjusted. Also, it could be that the clutch is not disengaging or the clutch linkage is out-of-adjustment. If not due to any of these conditions, the trouble is probably inside the transaxle.

♦ 42-9 LUBRICANT LEAKS

If the transaxle is overfilled with lubricant, some will leak out. If this is not the cause, then one or more of the seals are faulty. (See item 21 in the manual-transaxle trouble-diagnosis chart.)

♦ 42-10 BROKEN TRANSAXLE MOUNTS

Raise the car on a lift. Push up and pull down on the transaxle case while watching the mounts. If the rubber separates from the metal plate of the mount or if the case moves up but not down (mount is bottomed out), replace the mount. If there is movement between a metal plate of the mount and its attaching point, tighten the screws or nuts that attach the mount to the case or cross member.

♦ 42-11 LOW LUBRICANT LEVEL

Figure 42-1 shows the location of the filler plug for the transaxle. To check the level of the lubricant in the transaxle, remove the plug. The lubricant level should be within ½ inch [13 mm] of the lower edge of the filler opening. If lubricant is needed, add Dexron®-II, or the specified lubricant, to bring the lubricant up to the proper level. Install the plug and torque it to specifications.

♦ 42-12 TRANSAXLE SERVICE

Several different transaxle designs are used on cars with front-wheel drive. Various suspension and transaxle installation methods are also used. The manufacturer's service manual for the car you are servicing contains the procedures on trouble diagnosis (♦42-2). It also explains how to remove, service, and reinstall the transaxle you are working on. Follow the procedures in the service manual.

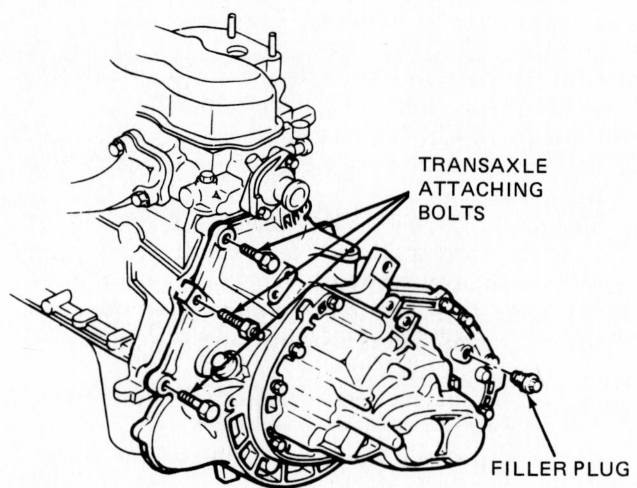

Fig. 42-1 Filler-plug position and position of the transaxle attaching bolts. *(Pontiac Motor Division of General Motors Corporation)*

♦ REVIEW QUESTIONS ♦

Select the *one* correct, best, or most probable answer to each question. You can find the answers in the section indicated at the end of each question.

1. If the noise is the same in drive or coast, it could be (♦42-2)
   a. road noise
   b. tire noise
   c. front-wheel-bearing noise
   d. any of the above
2. If the noise is a knock at low speed, the cause is probably (♦42-3)
   a. worn drive-axle joints
   b. a worn side-gear-hub counterbore
   c. defective tires
   d. a and b

3. If the transaxle is noisy in one gear position, the cause is probably (♦42-2)
   a. worn constant-mesh gears or synchronizer
   b. worn crankshaft main bearings
   c. a defective clutch
   d. a damaged clutch gear
4. If the transaxle slips out of gear, the cause could be (♦42-2)
   a. improperly adjusted linkage
   b. transmission loose on engine
   c. dirt between the clutch cover and engine
   d. any of the above
5. Hard shifting into gear can be caused by failure of the clutch to disengage, linkage out of adjustment, or (♦42-6)
   a. worn pilot bearing in engine crankshaft
   b. internal transaxle troubles
   c. worn wheel bearings
   d. loose extension housing
6. Gear clash when shifting could be caused by an incorrect linkage adjustment, internal transaxle problems, or (♦42-8)
   a. a damaged drive axle
   b. internal differential problems
   c. the clutch not disengaging
   d. worn constant-mesh gears
7. A clunk on acceleration or deceleration could be caused by (♦42-3)
   a. loose engine or transaxle mounts
   b. worn parts in the differential
   c. worn joints in drive axles
   d. any of above
8. Defective bearings produce a (♦42-4)
   a. whine
   b. thump
   c. growl or grating noise
   d. constant noise regardless of car speed
9. Types of manual-transaxle troubles are (♦42-1)
   a. noise
   b. improper operation
   c. both a and b
   d. neither a nor b
10. Transaxle noise problems may vary with (♦42-2)
    a. vehicle size
    b. type and size of engine
    c. amount of body insulation
    d. all of the above

CHAPTER 43
Four-Wheel Drive and Transfer Cases: Operation and Service

After you have studied this chapter, and with proper instruction and equipment, you should be able to:

1. Discuss the advantages of four-wheel drive.
2. Explain the two driving modes of a four-wheel-drive vehicle.
3. Describe the operation of a part-time and a full-time transfer case and explain how each operates.
4. Describe the operation of a planetary gearset.
5. Describe and list the four basic complaints about transfer cases, and explain the possible causes of each.
6. Remove and install a transfer case.
7. Adjust transfer-case linkages.

♦ 43-1 PURPOSE OF TRANSFER CASE

The transfer case is an auxiliary or second gearbox (attached to the back of the main transmission) through which power flows to the drive line. However, the transfer case can send engine power to the rear wheels of the vehicle, or to all four wheels (Figs. 43-1 and 43-2). The system has a selective arrangement which permits the driver to shift from rear-wheel drive only to four-wheel drive, and back, according to driving requirements. Following sections describe the basic types of transfer cases and how they work.

♦ 43-2 FOUR-WHEEL DRIVE WITH TRANSFER CASE

Many utility vehicles, some trucks, and a few cars have four-wheel drive (Figs. 43-1 and 43-2). Engine power can flow to all four wheels. With all four wheels driving, the vehicle can travel over rugged terrain and up steep grades. It can go through rough or muddy ground where two-wheel-drive cars would stall or get stuck. A *transfer case* is required on vehicles with four-wheel drive.

The transfer case is an auxiliary transmission mounted in back of the main transmission. By shifting gears in the transfer case, engine power is divided and transferred to both the front and rear differentials. Transfer cases in automotive vehicles are classed as *full-time* or *part-time*, depending on whether or not the front axle is engaged automatically as soon as the rear wheels begin to spin. With part-time four-wheel drive, the transfer-case shift lever must be moved to engage or disengage the front differential.

Figures 43-1 and 43-2 show how the design of the transfer case allows the front drive shaft to be placed to one side of the engine crankcase. By not running the front drive shaft under the crankcase, vehicle ground clearance is increased. A *skid plate* is usually mounted under the transfer case to protect it from hitting rocks and from getting snagged in rough terrain.

The typical transfer case may be operated in either of two modes. In one, both the front and the rear wheels are driven.

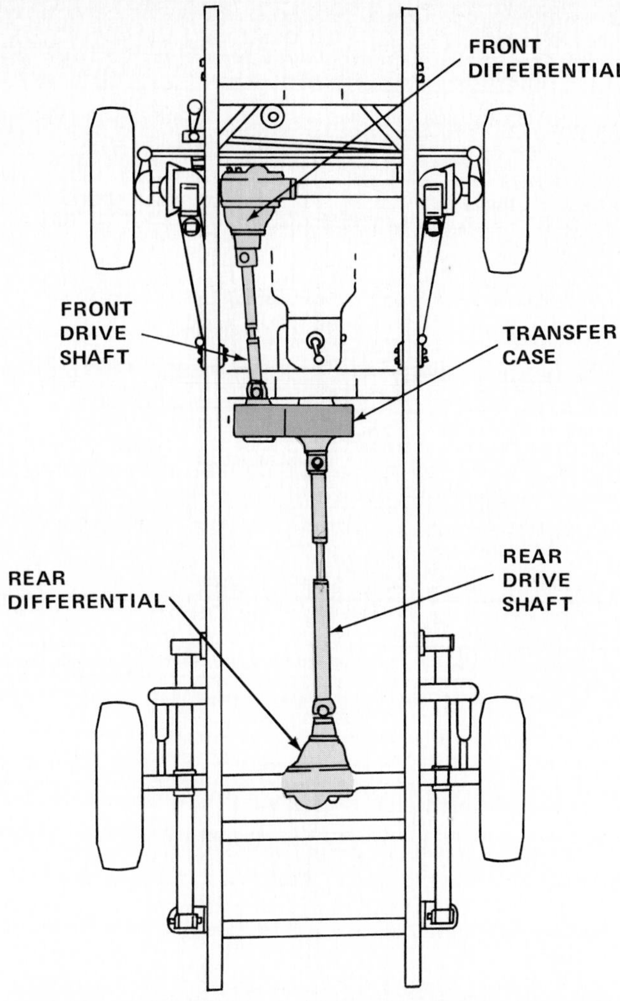

Fig. 43-1 Suspension and drive-train parts for a typical four-wheel-drive vehicle. The transfer case allows the driver to select rear-wheel drive or four-wheel drive. *(Ford Motor Company)*

In the other, only the rear wheels are driven. In most vehicles, a transfer case also provides the driver with a selection of either of two drive speeds, or *ranges*, high or low. The change from the high-speed range to the low-speed range is made when the driver moves the transfer-case shift lever. As the shift lever is moved, it moves a gear on the main drive shaft in the transfer case from engagement with the high-speed drive gear to engagement with the low-speed drive gear. High speed in the transfer case provides direct drive, or a gear ratio of 1:1. Low speed usually produces a gear ratio of about 2:1.

The front axle is engaged by shifting a sliding gear or a "dog" clutch into engagement with a driven gear on the front-wheel drive shaft inside the transfer case. The sliding gears and clutches are positively driven by splines on the shafts.

Figure 43-3 shows in simplified view the power flow through one type of transfer case. Notice that there are two parts that are moved by the transfer-case shift lever to provide the various gear combinations. A sliding gear on the main shaft locks either the low-speed gear or the high-speed gear to the main shaft. In many transfer cases, this shift cannot be make unless the transmission is in neutral (or in park in an automatic transmission). Otherwise, the transfer-case main shaft and the sliding gear, which is splined to it, will be turning. Gear clash will result.

To engage and disengage the front axle, another sliding gear or a clutch is used to lock the front-axle drive gear to the front-axle drive shaft. In most transfer cases in automotive vehicles today, the front axle can be engaged or disengaged while the vehicle is moving. All the driver must do is release the accelerator pedal to remove the torque load through the gears. Then the shift lever can be moved as desired. However, both the front and rear wheels must be turning at the same speed. If the rear wheels have lost traction and are spinning, or if the brakes are applied and either the front or rear wheels are locked and sliding, gear clashing will occur when engagement of the front axle is attempted.

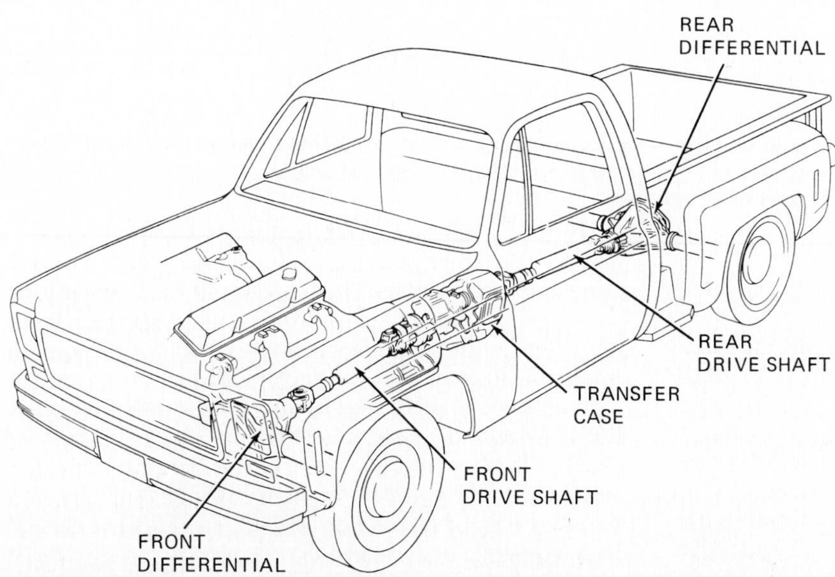

Fig. 43-2 Pickup truck with four-wheel drive. *(Chevrolet Motor Division of General Motors Corporation)*

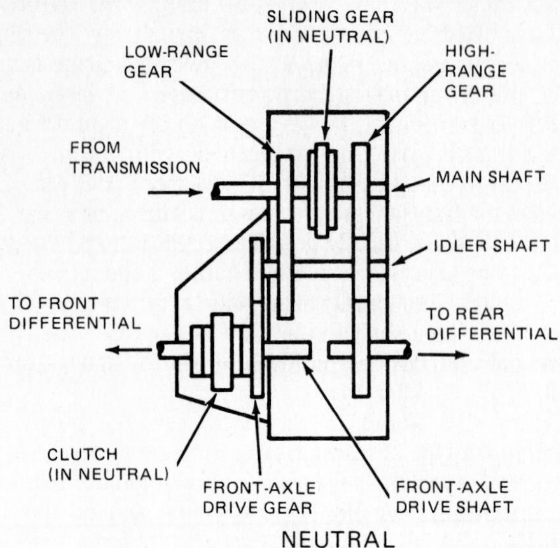

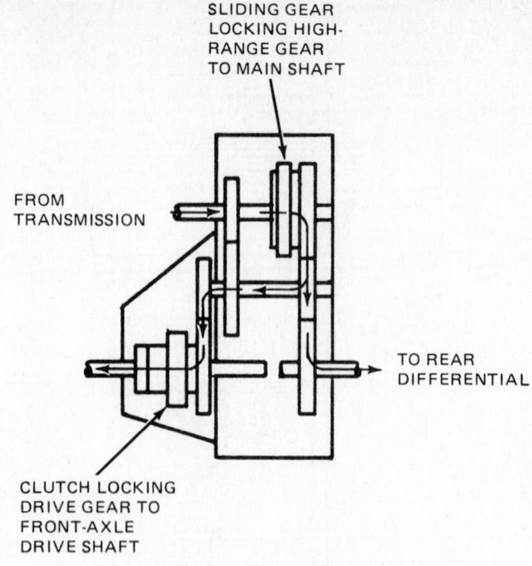

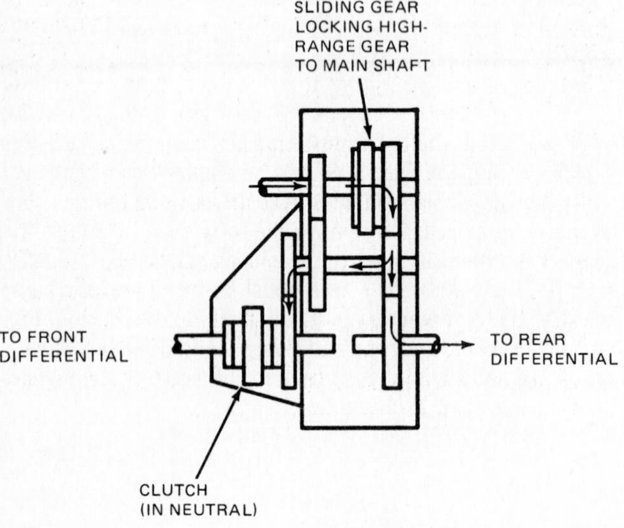

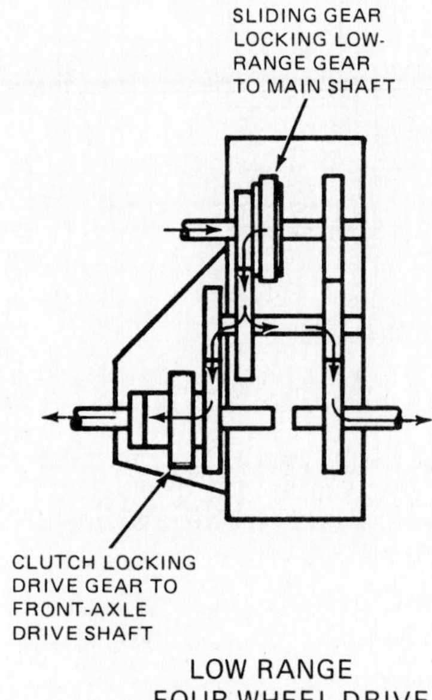

Fig. 43-3 Basic operation of a transfer case. *(ATW)*

Various types of transfer cases are installed in automotive vehicles. Some have all gears, as described above. Others are full-time units which have a chain that drives the front drive shaft instead of gears. This reduces the weight of the transfer case, improving fuel economy. In addition, some full-time transfer cases, such as those used in the American Motors Eagle models, have no low range.

While there is a slight fuel-mileage penalty when full-time four-wheel drive is used, it has certain advantages. For example, when a wheel spins on a vehicle equipped with part-time four-wheel drive, power continues to flow to the spinning wheel. As a result, so little power may reach the other axle that under certain conditions the vehicle may not move. With full-time four-wheel drive, the transfer case transfers the power from the axle with the spinning wheel to the other axle. This improves traction and keeps the vehicle moving.

The operation of various types of transfer cases is covered in following sections. Sections 43-7 to 43-13 explain the trouble diagnosis and service of the various types of transfer cases used in four-wheel-drive vehicles.

♦ 43-3 TYPES OF TRANSFER CASES

Transfer cases have been used with a variety of manual and automatic transmissions. They provide full-time or part-time four-wheel drive, a low range which doubles the number of gear ratios in the transmission, and a power-takeoff point to operate auxiliary equipment, such as a winch or pump.

The transfer case is used in all vehicles with four-wheel drive. Figure 43-1 shows its location in a vehicle. Figure 43-2 shows its location in a pickup truck. The main purpose of the transfer case is to provide a means of sending engine power to both the front and rear wheels (♦43-2). There are two general types, part-time or full-time.

Note *Four-wheel-drive vehicles are designated as 4WD by some manufacturers. FWD means front-wheel drive.*

In the full-time transfer case, power is available to all four wheels at any time. The transfer case has a gearshift which provides for either direct drive through the transfer case (high range) or gear reduction (low range). Gear reduction means torque increase at the wheels. The transfer case is shifted into low range by the driver when additional torque is needed, such as for climbing steep hills. Some cars with four-wheel drive (for example, the American Motors Eagle) have a transfer case with only a high range.

The part-time transfer case can be shifted into gear reduction, just as in the full-time unit. In addition, the part-time unit also has a gearshift that sends power to only the rear wheels, or to both the front and rear wheels (♦43-2).

The gear positions in a typical transfer case for the various transfer-case shift-lever positions are shown in Figs. 43-4 to 43-7. The transfer-case shift lever is located on the floor of the passenger compartment. Figure 43-8 shows two different shift patterns, which vary according to the design of the transfer case.

♦ 43-4 FULL-TIME TRANSFER CASE WITH DIFFERENTIAL

If a vehicle has full-time four-wheel drive, a controllable differential is built into the transfer case (Fig. 43-9). The purpose of this interaxle differential is to compensate for any difference in front-wheel and rear-wheel travel while the transfer case is in high range and four-wheel drive. The differential allows the front and rear axles to operate at their own speeds, without forcing wheel slippage in normal dry-road driving. Differential construction and operation are covered in later chapters.

When differential action is not wanted, such as for maximum engine braking or maximum wheel torque, the driver moves the transfer-case shift lever either to low range or to the LOCK position. This locks together the output shafts in the transfer case for the front and rear axles. Now differential

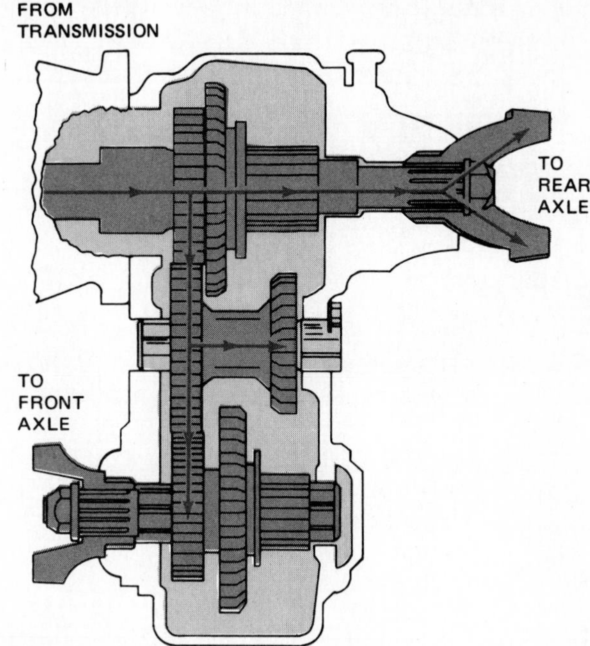

Fig. 43-5 Gearing in the transfer case in two-wheel drive, high range. Only the rear wheels are being driven. *(American Motors Corporation)*

action cannot occur. Equal torque is delivered to both the front and rear axles.

During turns, the front wheels travel a greater distance than the rear wheels. This is because the front wheels move through a wider arc than the rear wheels (Fig. 40-28). With full-time four-wheel drive, it is the differential in the transfer case (Fig. 43-9) that allows the front wheels to travel farther or turn faster than the rear wheels without slipping. As a result, power-train and tire wear are reduced while the advantages of full-time four-wheel drive are retained.

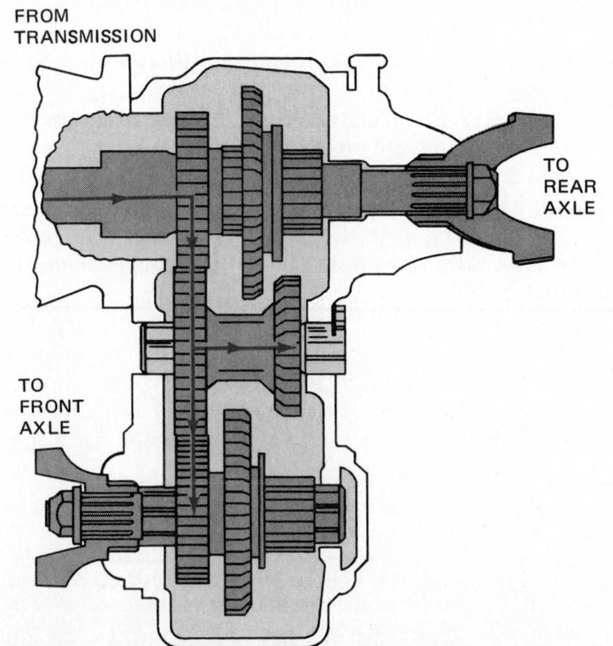

Fig. 43-4 Gearing in the transfer case with the gears in neutral. *(American Motors Corporation)*

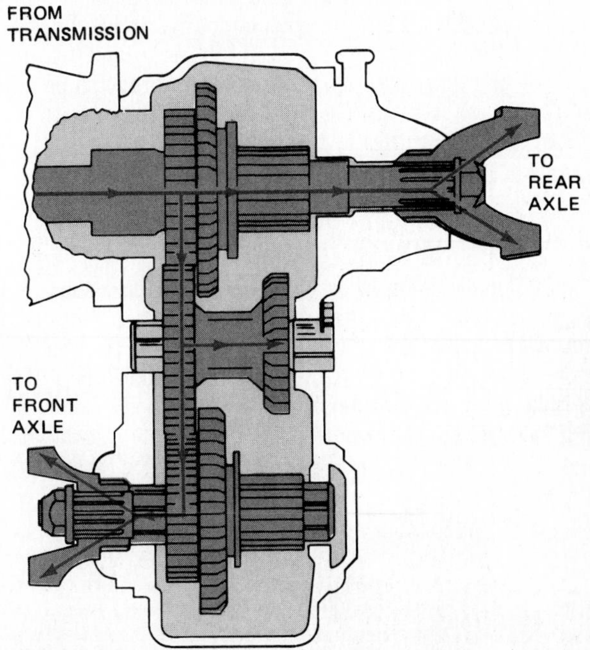

Fig. 43-6 Gearing in the transfer case in four-wheel drive, high range. All four wheels are being driven. *(American Motors Corporation)*

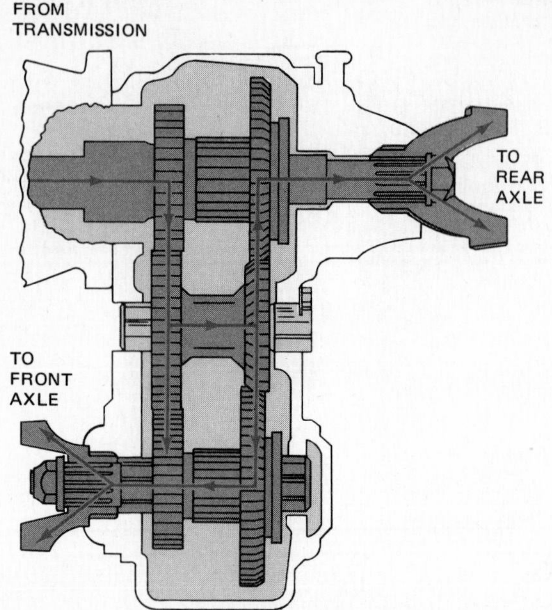

Fig. 43-7 Gearing with the transfer case in four-wheel drive, low range. *(American Motors Corporation)*

This differs from the operation of a part-time transfer case. It delivers equal power to each axle while in four-wheel drive. During turns made on a dry surface, one of the axles must slip. Therefore, the driver must shift the part-time transfer case out of four-wheel drive on returning to a dry surface.

There is a disadvantage to a transfer case with a simple differential in it. If the wheels on either axle lose traction and begin to spin, the transfer-case differential continues to deliver maximum torque to the axle with the minimum traction. As a result, insufficient torque to move the vehicle may be provided to the wheels that still have traction.

To overcome this problem, most full-time transfer cases have some type of limited-slip differential in them. The slip-limiting device may be a viscous coupling, brake cones, or clutch plates. Regardless of type, its job is to divert torque from the spinning axle. This causes more torque to be supplied to the wheels with the most traction. The operation of limited-slip differentials is covered in Chap. 47.

♦ 43-5 PART-TIME TRANSFER CASE WITH PLANETARY GEARS

In 1980, a new design of part-time transfer case for automotive vehicles was introduced by the New Process Gear Division of Chrysler Corporation (Fig. 43-10). This transfer case has an aluminum case, chain drive, and a planetary gearset for reduced weight and increased mechanical efficiency. The internal parts of the transfer case do not rotate in two-wheel drive. This leaves most of the lubricant undisturbed and reduces the power loss caused by dragging parts through it.

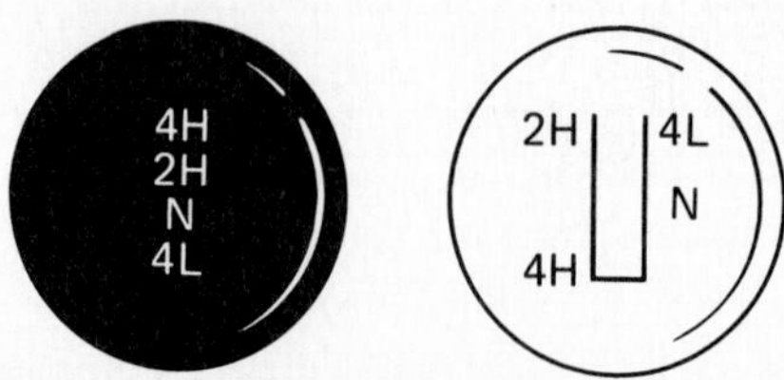

Fig. 43-8 Shift patterns for transfer cases. *(American Motors Corporation)*

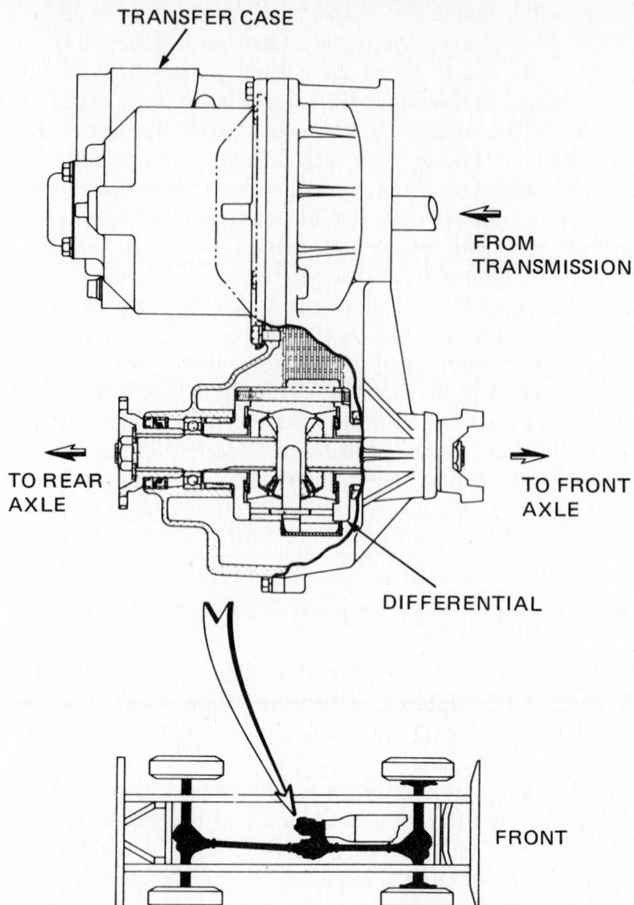

Fig. 43-9 Transfer case for a full-time, four-wheel drive which includes a controllable differential. *(American Motors Corporation)*

An internal oil pump turns with the rear output shaft to maintain lubrication to critical bearings and bushings whenever the rear drive shaft is turning. This provides improved lubrication during normal operation. In addition, the oil pump allows towing of the vehicle with the transfer case in neutral at any safe speed up to 55 mph [89 km/h] and for any distance. It is not necessary to disconnect the drive shafts. However, the front-wheel locking hubs should be in UNLOCK or AUTO (depending upon the hub type) to prevent unnecessary rotation of the drive-train components. Locking hubs are discussed in ♦47-8.

Figure 43-11 shows the power flow through the transfer case for each drive condition. In neutral, rotation of the input shaft spins only the planetary gears (♦43-6) and the ring gear around them. With both the planetary gears and the ring gear spinning freely, no power is transmitted through the planetary gearset.

♦ 43-6 PLANETARY GEARSET OPERATION

In its simplest form, a planetary gearset is made up of three gears (Fig. 43-12). In the center is the sun gear. All gears in the set revolve around the sun gear. Meshing with the sun gear are two or more planet-pinion gears (only one is shown in Fig. 43-12). In any set of two or more gears, the smallest gear is often called the *pinion gear*. The planet pinions are

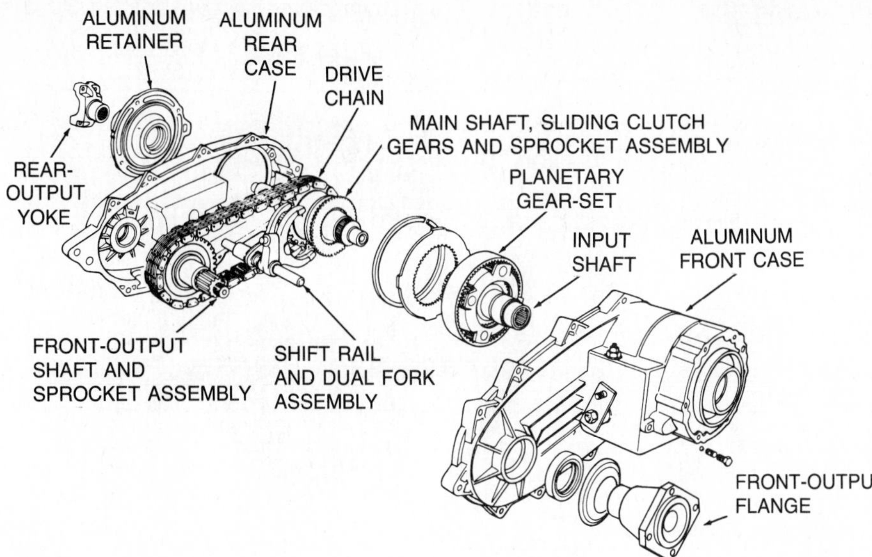

Fig. 43-10 A partly disassembled part-time, four-wheel-drive transfer case which uses a planetary gear-set to achieve gear reduction. *(Chrysler Corporation)*

fastened together by the planet-pinion carrier, or planet carrier (Fig. 43-13). This holds the gears in place while allowing them to rotate around their pins or shafts. There are usually three or more pinion gears fastened to the carrier.

On the outside of the planet pinions is the internal gear, or ring gear (Figs. 43-13 and 43-14). It has teeth around its inside circumference that mesh with the teeth on the planet pinions.

Figure 43-14 shows how the planet pinions are held in place in the internal gear. Each planet pinion is mounted on a pin, and the pins are set in the planet-pinion carrier. Notice in Figs. 43-13 and 43-14 that the carrier is mounted on a separate shaft from the sun-gear shaft. The planetary gear used in the transfer case discussed in ♦43-5 has four planet pinions instead of the three shown in Figs. 43-13 and 43-14. Additional planet pinions provide the strength needed to handle the higher torque flow through the transfer case.

A planetary gearset can provide any of five conditions. These are:

1. A speed increase with a torque decrease (overdrive)
2. A speed decrease with a torque increase (reduction)
3. Direct drive (lockup)
4. Neutral
5. Reverse

In the transfer case, the two drive conditions used are speed reduction and direct drive. Speed reduction provides the low range. Direct drive provides the high range. To get speed reduction, the internal gear, or ring gear, is locked in a stationary position. For direct drive, the internal gear and the planet-pinion carrier are locked together.

Gear reduction is provided when the sun gear on the input shaft is the driving gear (Fig. 43-15). In the transfer case, this shaft is coupled directly to the transmission. When the internal gear is held stationary and the sun gear is driving, the planet pinions rotate on their pins. As the planet pinions rotate, they must "walk around" the internal gear since they are meshed with it. This action causes the pinion carrier to rotate in the same direction as the sun gear.

However, the planet-pinion carrier turns more slowly than the sun gear. This is because of the gear reduction between the planet pinions and the sun gear. As the pinions move around the inside of the internal gear, the shaft attached to the planet carrier is driven in the same direction as the sun

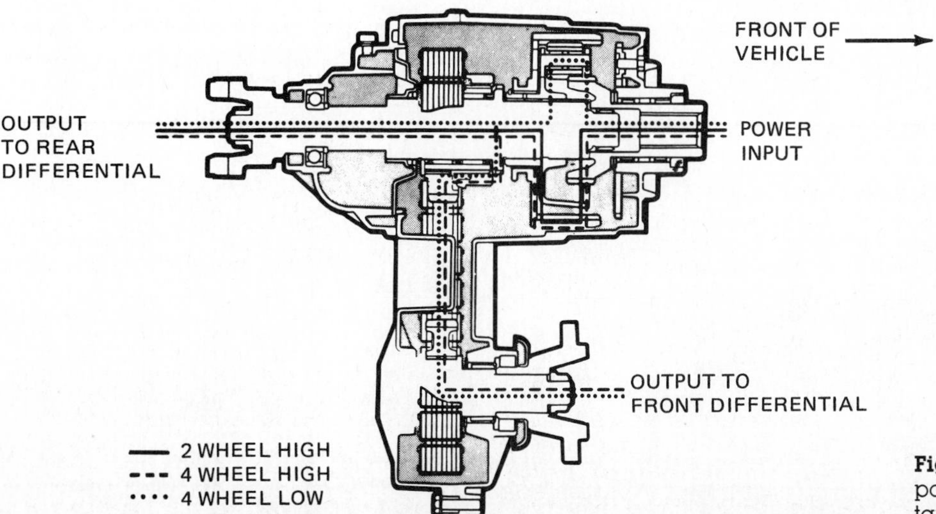

Fig. 43-11 Power flow through the part-time transfer case with planetary gears. *(Chrysler Corporation)*

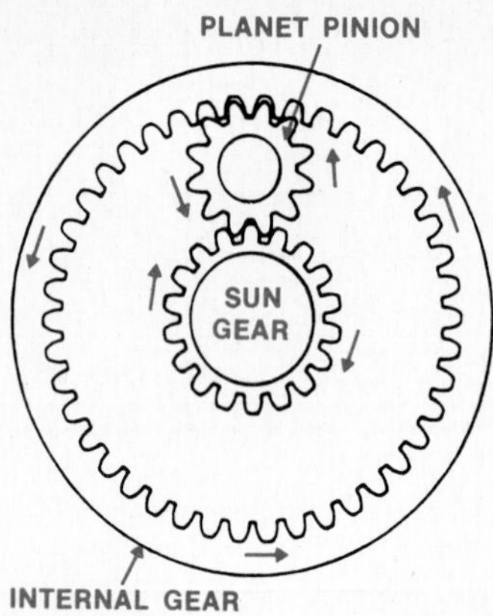

Fig. 43-12 The three gears that make up the basic planetary gearset.

gear, but at a lower speed. This action provides the transfer case with low range. The output shaft turns more slowly than the input shaft, but with increased torque.

Note Figure 43-15 shows the internal gear held stationary by the clamping action of a brake band wrapped around it. In the transfer case using a planetary gearset (♦43-5), when the selector lever is moved to low range, the planetary-gear assembly slides forward on its shaft. In this position, a locking plate bolted to the case engages the teeth of the internal gear to hold it stationary.

To get direct drive, the internal gear and the planet-pinion carrier are locked together. Now, the whole planetary-gear assembly turns as a solid unit because the planet pinions cannot rotate on their pins. This provides direct drive. The output shaft turns at the same speed as the input shaft.

TRANSFER-CASE SERVICE

♦ 43-7 PREPARING FOR TRANSFER-CASE REPAIR

Various transfer-case troubles and the possible causes of each are described in following sections. Then how to remove a transfer case and reinstall it is described. Also described is the procedure for adjusting the transfer-case linkage.

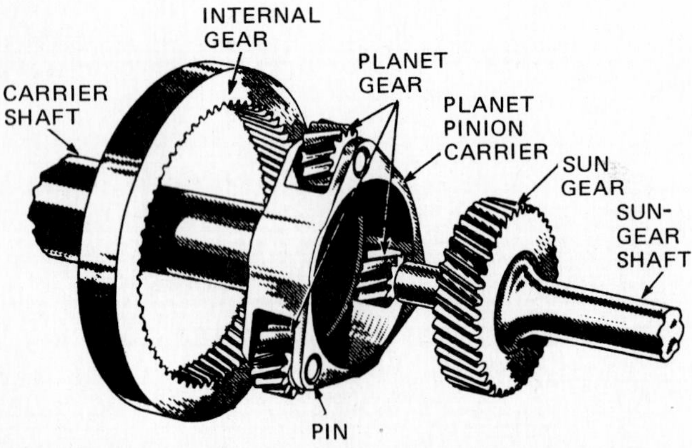

Fig. 43-13 A planetary gearset, showing how the planet-pinion gears rotate on pins that are part of the carrier.

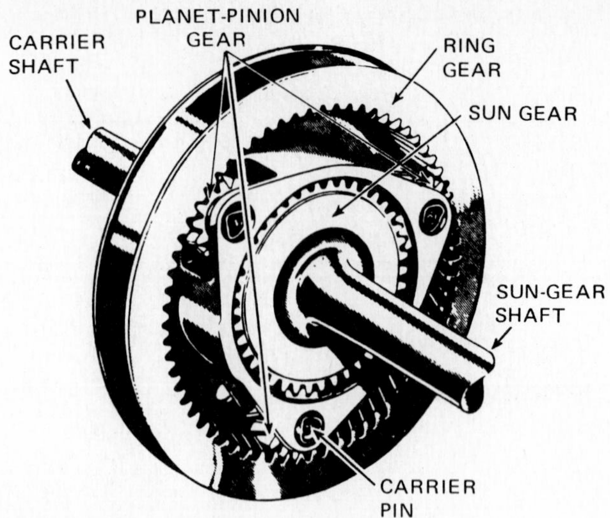

Fig. 43-14 An assembled planetary gearset.

♦ 43-8 TRANSFER-CASE TROUBLE DIAGNOSIS

The chart that follows lists transfer-case complaints, possible causes, and checks or corrections. Typical troubles with transfer cases include excessive noise, shift lever hard to move, gears slip out of engagement, and leaking lubricant.

TRANSFER-CASE TROUBLE-DIAGNOSIS CHART

Complaint	Possible Cause	Check or Correction
1. Excessive noise	a. Lubricant level low	Fill as required
	b. Worn or damaged bearings	Replace
	c. Worn or damaged chain	Replace
	d. Misalignment of drive shafts or universal joints	Align
	e. Yoke bolts loose	Torque to specifications
	f. Loose adapter bolts	Torque to specifications
2. Shift lever difficult to move	a. Dirt or contamination on linkage	Clean and lubricate
	b. Binding inside transfer case	Repair as required
3. Gears disengage from position	a. Linkage misadjusted or loose	Readjust or tighten
	b. Gears worn or damaged	Replace
	c. Shift rod bent	Replace
	d. Missing detent ball or spring	Replace
4. Lubricant leaking	a. Excessive lubricant in case	Adjust level
	b. Leaking seals or gaskets	Replace
	c. Loose bolts	Tighten
	d. Scored yoke in seal-contact area	Refinish or replace

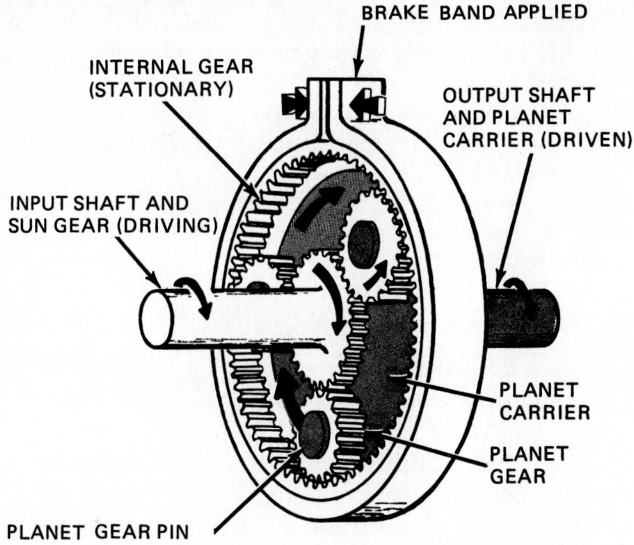

Fig. 43-15 With the internal gear held stationary by the band, the planet carrier and output shaft turn more slowly than the sun gear. This provides the transfer case with low range. *(Ford Motor Company)*

♦ 43-9 TRANSFER-CASE REMOVAL

The purpose and operation of the transfer case is covered in ♦43-1 to 43-6. To remove the transfer case:

1. Raise the vehicle on a lift. Drain the lubricant from the transfer case (♦43-11). Transfer-case attachment points are shown in Fig. 43-16.
2. Disconnect the speedometer cable and remove the skid plate and cross-member supports (Fig. 43-17) as necessary. On cars with automatic transmissions, remove the strut rod (Fig. 43-18).
3. Disconnect both the rear drive shaft and the front drive shaft from the transfer case. Tie them up out of the way.
4. Disconnect the shift-lever rod from the shift-rail link. On some vehicles, the shift levers are disconnected at the transfer case.
5. Place a transmission jack under the transfer case to support it. Remove the bolts attaching the transfer case to the transmission adapter. Move the transfer case to the rear until the input shaft clears the adapter. Lower the transfer case from the vehicle and take it to the work bench.

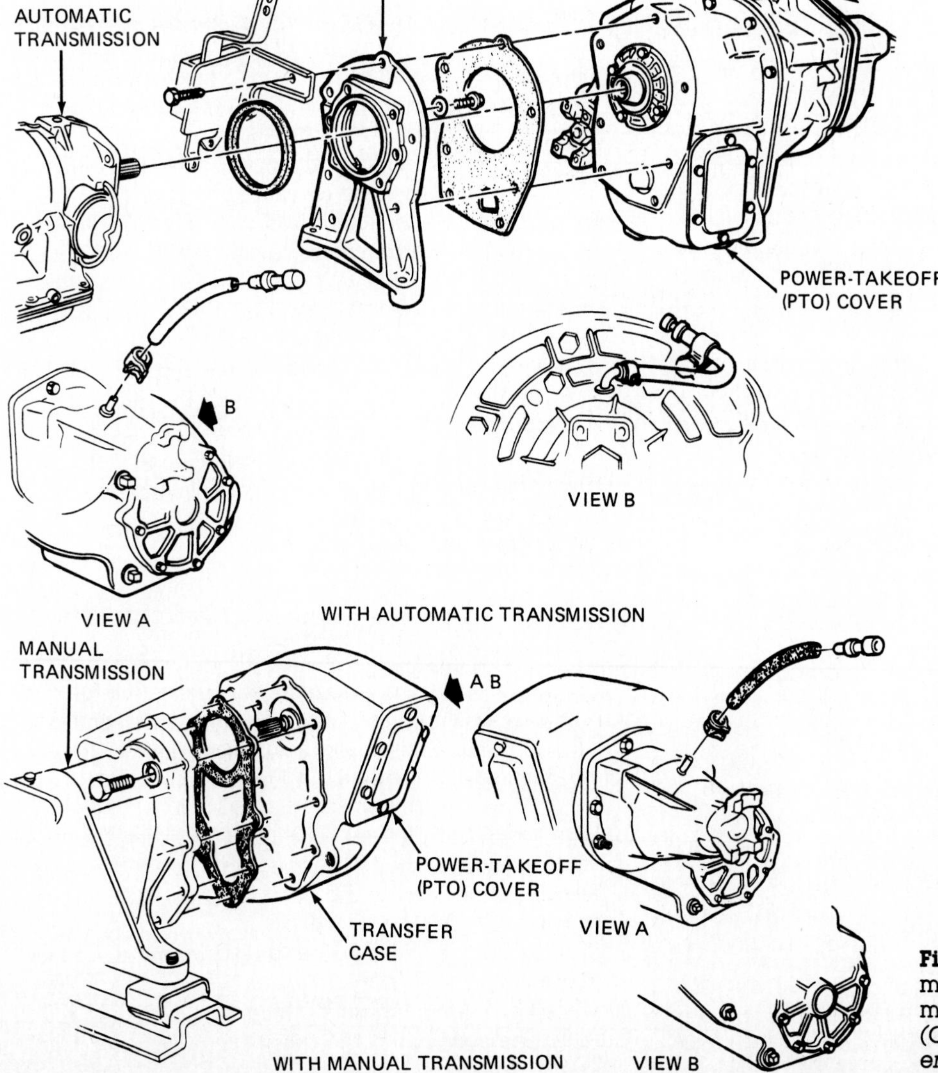

Fig. 43-16 Transfer-case attachments for both manual- and automatic-transmission installations. *(Chevrolet Motor Division of General Motors Corporation)*

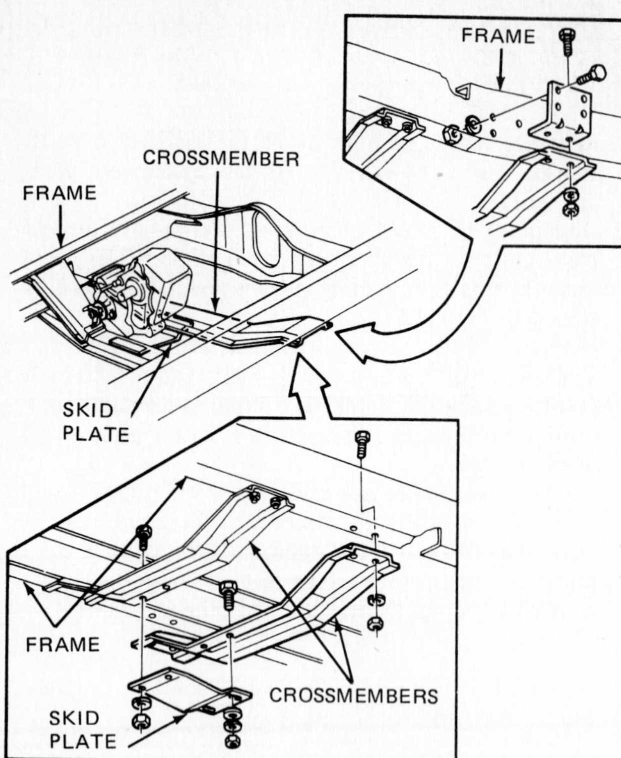

Fig. 43-17 A skid plate helps protect the transfer case from damage. *(Chevrolet Motor Division of General Motors Corporation)*

♦ 43-10 INSTALLING THE TRANSFER CASE

Support the transfer case with a transmission jack. Raise the transfer case into position and slide it forward so that the input shaft enters properly. Move the transfer case up against the transmission adapter. Install the bolts attaching the case to the adapter and torque to specifications. Remove the jack and proceed as follows:

1. Install the connecting rod to the shift-rail link or connect the shift levers to the transfer case, according to the design. On the GM Model 203 full-time transfer case, be sure the nylon spacer is in place before installing the levers.

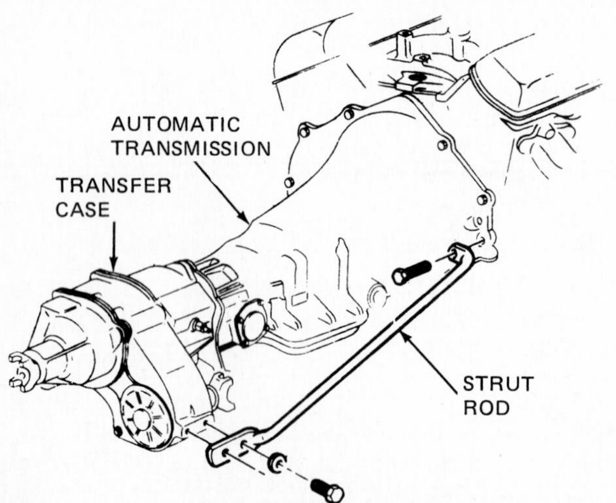

Fig. 43-18 A strut rod is sometimes used to help support the transfer case. *(Chevrolet Motor Division of General Motors Corporation)*

2. Connect the front and rear drive shafts.
3. Install the cross-member support and skid plate, if removed.
4. Connect the speedometer cable.
5. Fill the transfer case to the proper level with the specified lubricant (♦43-11).
6. Check and adjust the shift linkage (♦43-12).
7. Lower the vehicle. Road test it to make sure the transfer case is operating properly.

♦ 43-11 TRANSFER-CASE LUBRICATION

The transfer-case lubricant should be changed at the intervals recommended by the manufacturer. The frequency of change depends on the type of operation. For example, one manufacturer recommends changing the oil every 24,000 miles [38,624 km] for normal off-on road work. For heavy-duty work, such as snowplowing or pulling a trailer, change the oil every 12,000 miles [19,312 km]. For severe use, change the oil every 1000 miles [1609 km]. If the vehicle is used in very severe work, where the transfer case is submerged, the oil should be changed every day.

Always use the type of oil or lubricant recommended by the manufacturer. Figure 43-19 shows the items to be checked. To change the oil in the transfer case:

1. Operate the vehicle on a rough road to agitate the lubricant so that it reaches normal operating temperature.
2. Raise the vehicle on a lift. Remove the lubricant filler plug. Have a container in place to catch the lubricant.
3. Remove the lowest bolt from the front-output-shaft rear bearing (Fig. 43-19). Allow the lubricant to drain.
4. Remove the bolts holding the power-takeoff (PTO) cover in place and remove the cover (Fig. 43-19).
5. Remove the speedometer-driven gear (Fig. 43-19).
6. Use a suction gun to remove as much of the lubricant as possible.
7. Install the speedometer gear, the PTO cover, and the lower bolt removed in item 3, above.
8. Add about 7 pints [3.3 liters] of the recommended oil through the filler-plug opening. Check the fluid level and add more oil if necessary to raise the level to about ½ inch [13 mm] below the filler-plug opening. Install the plug. Then wipe the surfaces of the case and the skid plate to remove excess oil.
9. Lower the vehicle to the floor.

♦ 43-12 TRANSFER-CASE LINKAGE ADJUSTMENT

Figure 43-20 shows the linkage for the Model 203 full-time transfer case used in many General Motors trucks and other vehicles. To make the linkage adjustment, follow the steps below. Each part is identified in Fig. 43-20.

1. With rods C and H removed (Fig. 43-20), align the gauge holes in levers A and B with the gauge hole in the shifter assembly and insert gauge pin J. This positions levers A and B in neutral.
2. Position arms F and G in the straight down "six-o'clock" position.
3. With swivel E and locknuts D loosely assembled in rod C, rotate the swivel until the ends of rod C will enter both lever B and arm A at the same time.

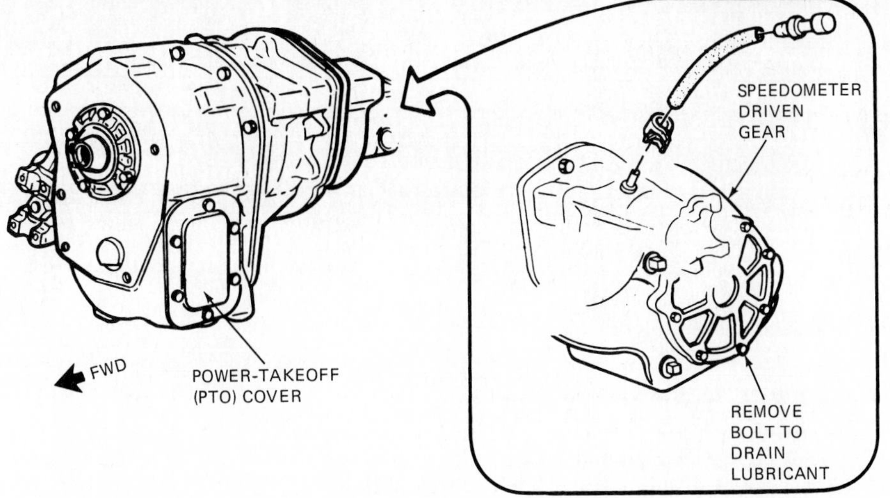

Fig. 43-19 Places to drain lubricant from a full-time transfer case. *(Chevrolet Motor Division of General Motors Corporation)*

4. Lock rod C in place with retainer K.
5. Tighten locknuts D against swivel E to specified torque. Be careful not to change the position of arm F.
6. Repeat steps 3, 4, and 5 for rod H when installing the rod to lever A and arm G.
7. Remove gauge pin J.

For adjustments of other transfer-case linkages, refer to the manufacturer's service manual.

♦ 43-13 TRANSFER-CASE SERVICE

There are a variety of transfer-case designs and constructions. Therefore, disassembly, checks of internal parts, reassembly, and testing differ. When you have to service a transfer case, refer to the manufacturer's service manual covering that model of transfer case.

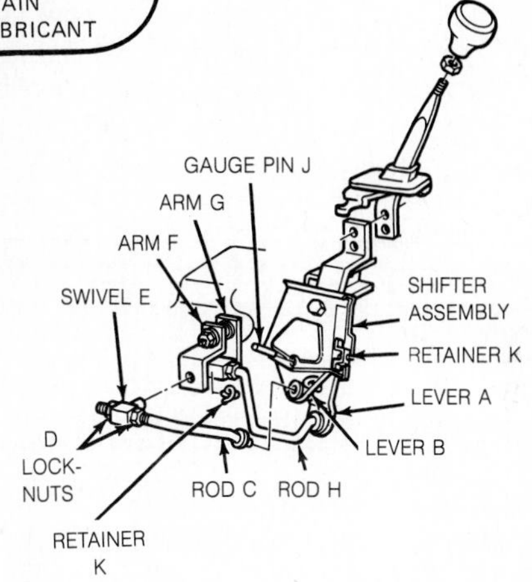

Fig. 43-20 Linkage adjustments on Model 203 full-time transfer case. *(Chevrolet Motor Division of General Motors Corporation)*

REVIEW QUESTIONS

Select the *one* correct, best, or most probable answer to each question. You can find the answers in the section indicated at the end of each question.

1. Excessive noise from the transfer case could be caused by (♦43-8)
   a. low lubricant level
   b. misaligned drive shafts
   c. worn bearings
   d. all of the above
2. If the shift lever is difficult to move, it could be caused by (♦43-8)
   a. excessive lubricant in the case
   b. binding inside transfer case
   c. worn gears
   d. a defective synchronizer
3. If the shift lever will not stay in position, the cause could be (♦43-8)
   a. linkage loose or out of adjustment
   b. gears worn or damaged
   c. missing detent ball or spring
   d. all of the above
4. The purpose of the skid plate is to (♦43-2)
   a. prevent the vehicle from skidding
   b. protect the transfer case from rocks or other snags it might encounter in rough areas
   c. help support the transfer case
   d. strengthen the frame
5. Lubricant leaks may be caused by (♦43-8)
   a. excessive lubricant in the case
   b. leaking seals or gaskets
   c. loose bolts
   d. all of the above

CHAPTER 44
Automatic Transmissions and Transaxles

After studying this chapter, you should be able to:

1. Describe the purpose, construction, and operation of a torque converter.
2. Explain how the lock-up torque converter works.
3. Discuss planetary-gear construction and operation.
4. Explain how bands and clutches control the planetary gears.
5. Describe the hydraulic control system and explain how it controls the clutches and bands.

♦ 44-1 AUTOMATIC TRANSMISSIONS

This chapter describes the construction and operation of automatic transmissions used in front-wheel drive and rear-wheel drive vehicles. Although automatic transmissions vary in detail from one model to another, all operate in the same basic way. All have a torque converter and planetary-gear system with clutches and bands controlled by the hydraulic system (Fig. 44-1). Almost all modern automatic transmissions have three or four forward speeds and reverse. Some four-forward-speed automatic transmissions provide overdrive in fourth.

Note Most automatic transmissions for front-drive vehicles are called *automatic transaxles* (Fig. 44-2).

♦ 44-2 FUNCTION OF AUTOMATIC TRANSMISSION

Automatic transmissions shift gears without assistance from the driver. They start the car moving in first. Then they shift into second, third, and fourth (if used) as car speed increases and engine load decreases. The shifts are produced by hydraulic pressure acting through the automatic-transmission fluid (ATF), or oil, in the transmission.

There are three basic parts to the automatic transmission, the torque converter, the gear system, and the hydraulic control system (Fig. 44-1). The torque converter transmits the engine power to the gear system. Hydraulic pressure acts on the gear system to produce the shifts. In most automatic transmissions, the shifts are produced in response to car speed and throttle opening, as explained later.

♦ 44-3 TORQUE CONVERTER

The torque converter is a special sort of fluid coupling. In the torque converter, the vanes are curved as shown in Fig. 44-3. Curving the vanes reduces "bounce-back" of the fluid. With flat vanes, the fluid, as it hit the vanes of the driven member, would tend to bounce back into the driving member. This would remove some of the driving torque, and power would be lost.

But with curved vanes, the fluid is unable to bounce back.

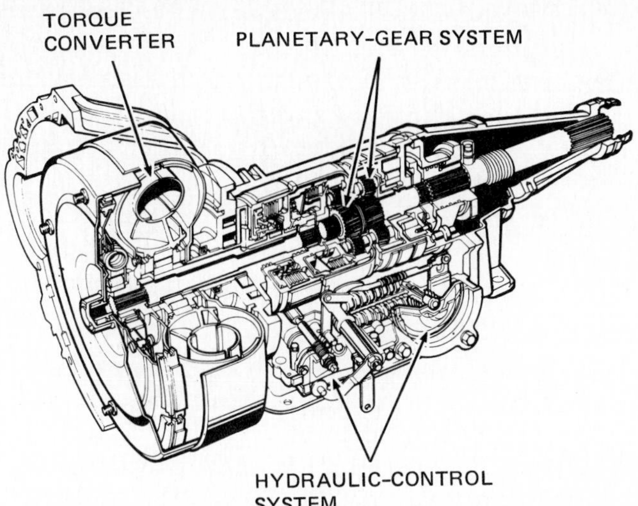

Fig. 44-1 The automatic transmission has three basic parts—the torque converter, the planetary-gear system, and the hydraulic control system. *(Ford Motor Company)*

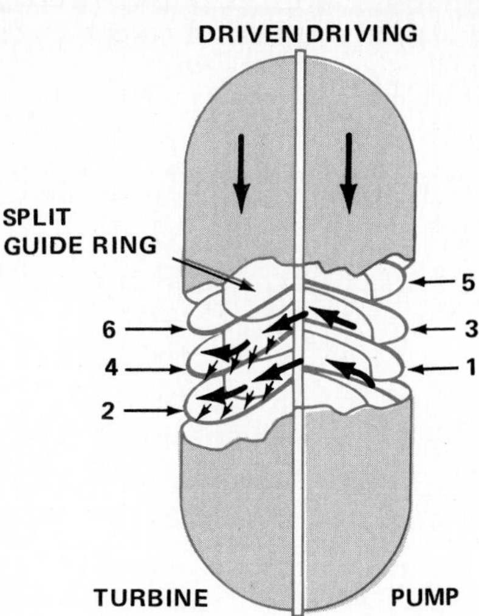

Fig. 44-3 Simplified cutaway view of two members of a torque converter. The heavy arrows show how oil circulates between the driving-member and the driven-member vanes. In operation, the oil is forced by vane 1 downward toward vane 2. Therefore, the oil pushes downward against vane 2, as shown by the small arrows. Oil then passes around behind the split guide ring and into the driving member again, or between vanes 1 and 3. Then, it is thrown against vane 4 and continues this circulatory or vortex flow, passing continuously from one member to the other.

As shown in Fig. 44-3, the split guide ring is a smaller dough-nut-shaped ring. It tends to keep the fluid in the outer part of the driving and driven members, where it can do the most good.

Note From this point on, we will use the terms *pump* for the driving member and *turbine* for the driven member.

♦ 44-4 THE STATOR

The coupling shown in Fig. 44-3 would not be very efficient. As the fluid leaves the inner part of the turbine, the fluid would be thrown back into the pump in the wrong direction, opposing the rotation of the pump (Fig. 44-4). The arrows in Fig. 44-4 show how the fluid hits the pump vanes in the direction opposite to the direction in which the vanes are moving. This greatly reduces the efficiency of the fluid coupling. The pump has to overcome this opposing force to get the oil moving in the right direction again.

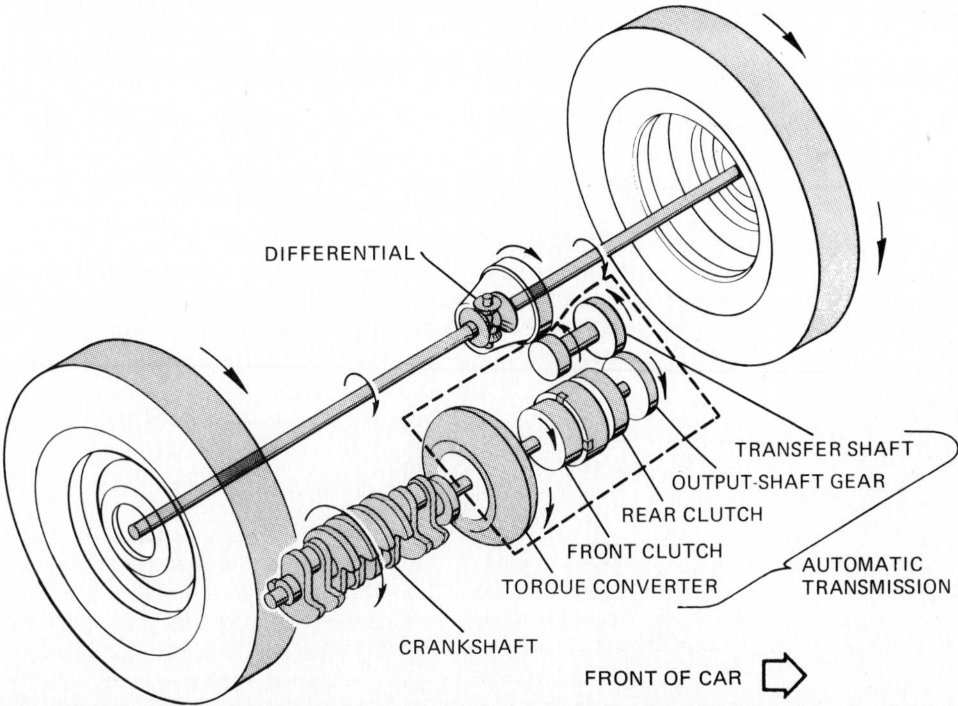

Fig. 44-2 Power flow from the engine crankshaft to the drive wheels for a car with front-wheel drive. *(Chrysler Corporation)*

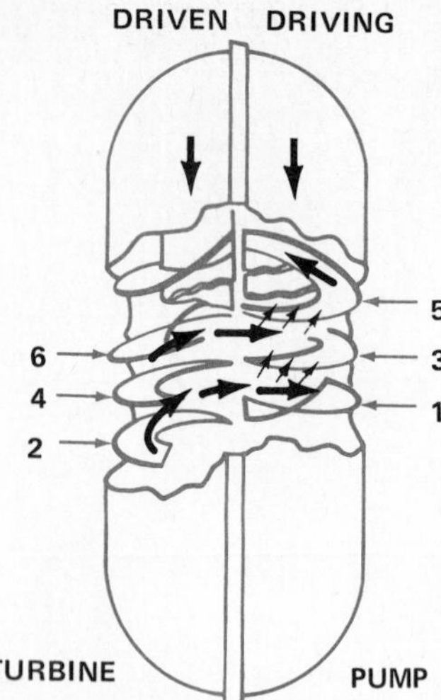

Fig. 44-4 This shows the actions if the vanes in Fig. 44-3 were continuous. Actually, the inner ends of the vanes are not as shown here, but as in illustrations that follow. Here, the split guide ring and outer ends of the vanes have been cut away. If the vanes were as shown here, the oil leaving the trailing edges of the turbine would be thrown upward against the forward faces of the pump vanes. Therefore, this oil would oppose the driving force. This effect, shown by the small arrows, would waste energy and torque.

To eliminate this, a third member, called a *stator,* is installed between the inner ends of the pump and the turbine vanes (Figs. 44-5 to 44-7). The stator has curved vanes. They change the direction of the fluid coming out of the turbine to one that helps the rotation.

Figure 44-7 shows torque-converter action in detail. The stator is mounted on a one-way clutch which locks up when the oil hitting it tries to make it turn backward. However, the clutch unlocks to allow the stator to spin in a forward direction when the torque converter approaches the *coupling point.* Coupling occurs when the turbine is turning nearly as fast as the pump.

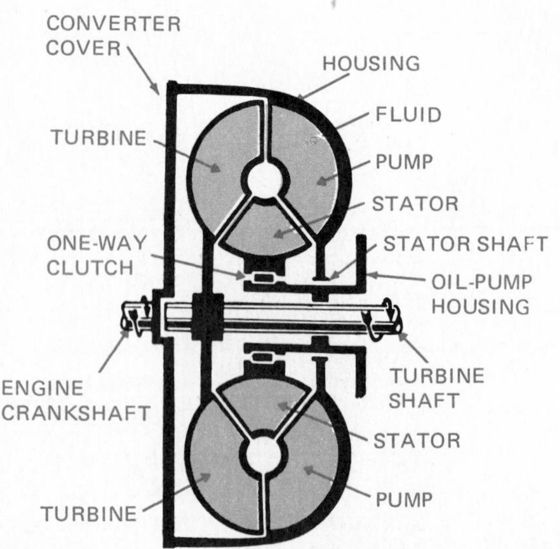

Fig. 44-5 Locations of the pump, stator, and turbine in the torque converter. *(Chevrolet Motor Division of General Motors Corporation)*

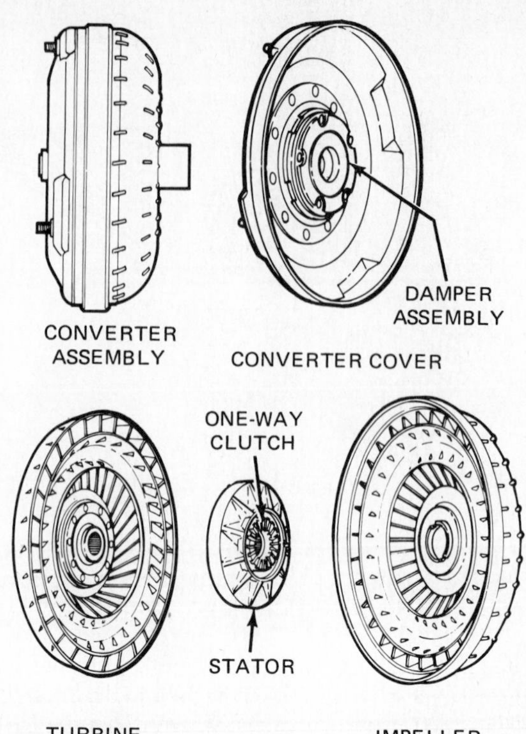

Fig. 44-6 Assembled and disassembled torque converter. *(Ford Motor Company)*

Torque multiplication occurs in the torque converter because of the action of the oil and the stator vanes. As the oil leaves the turbine, the oil hits the stator vanes. There, the oil is redirected into the pump in a helping direction. The pump then throws the oil back into the turbine. This is a continuous action. The repeated pushes of the fluid on the turbine vanes increase the torque on the turbine. In many automotive torque converters, the torque is more than doubled. For each 1 pound-foot [1.35 N-m] of torque entering the pump, the turbine delivers more than 2 pound-feet [2.7 N-m] of torque to the transmission output shaft. This is *torque multiplication.*

♦ 44-5 STATOR ACTION

The stator causes the torque converter to multiply torque *when the pump is turning faster than the turbine.* This speed difference and increase in torque have the same effect as a low gear in the manual transmission. They allow the engine to turn fast while the car wheels are turning slowly. This is the condition needed for acceleration.

However, as the car reaches cruising speed, the turbine begins to "catch up" with the pump. When this happens, the fluid leaving the trailing edges of the turbine vanes is moving at about the same speed as the pump. Therefore, the fluid could pass directly into the pump in a helping direction, without being given an assist by the stator. In fact, under these conditions, the stator vanes are in the way. The fluid begins to hit the back sides of the stator vanes. To allow the stator vanes to move out of the way, the stator is mounted on a *freewheeling* mechanism called a *one-way clutch.*

The one-way clutch allows the stator to revolve freely, or "freewheel," in only one direction. The clutch locks the stator when it tries to turn in the opposite direction. Figure 44-5

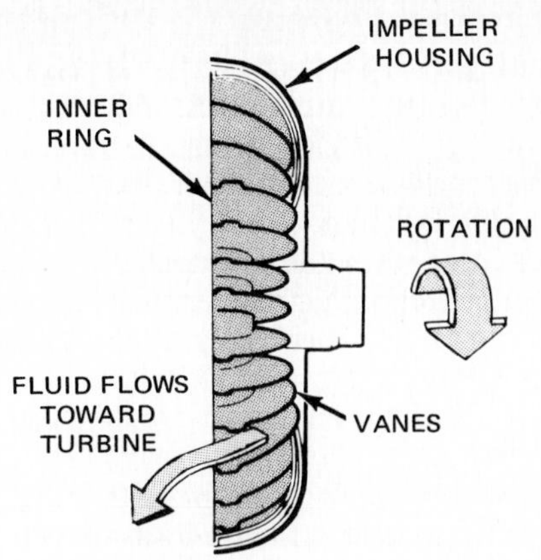

(A) IMPELLER OPERATION

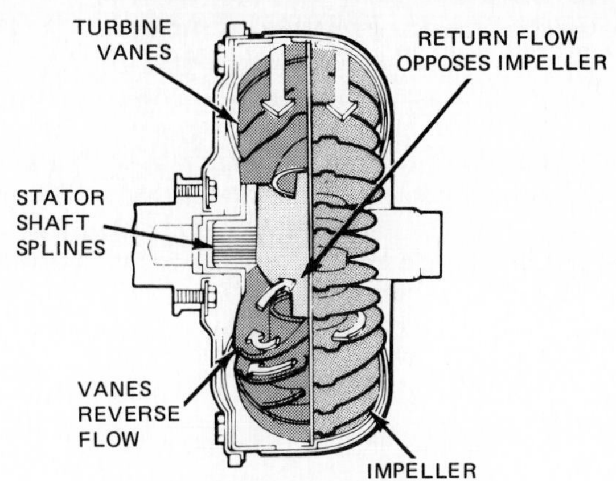

(B) DRIVING THE TURBINE

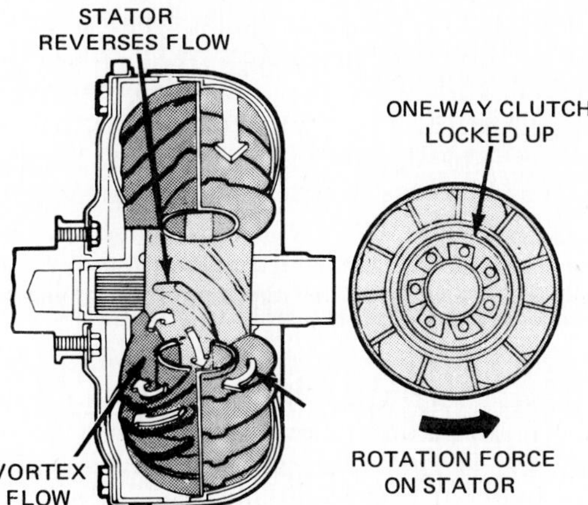

(C) TORQUE MULTIPLICATION

(D) COUPLING PHASE

Fig. 44-7 Torque-converter action. (A) The impeller, or pump, sends a flow of oil toward the turbine. (B) The turbine vanes receive the flow and this spins the turbine. The vanes reverse the direction of oil flow and send it back toward the impeller. (C) The stator reverses the flow of the oil into a helping direction and this multiplies the torque. (D) When the turbine speed nears the impeller speed, the oil strikes the backs of the stator vanes, causing the stator to spin forward. This prevents the stator vanes from interfering with the oil flow. *(Ford Motor Company)*

shows the location of the one-way clutch in the torque converter. The clutch includes a hub, an outer ring that is part of the stator, and a series of rollers. The rollers are located in notches in the outer ring. The outer ring is called the overrunning-clutch *cam* in Fig. 44-8. The notches are smaller at one end than at the other. The rollers have springs behind them. When there is a push on the front of the stator vanes from the fluid leaving the turbine, the stator attempts to roll backward. This causes the rollers to roll into the smaller ends of the notches. There, they jam and lock the stator to the hub. Now the stator cannot turn backward. Instead, the stator vanes change the direction of the fluid into a helping direction.

However, as the turbine speed approaches the pump speed, the direction of the fluid no longer has to be changed as it leaves the turbine. The fluid now begins to hit the other side of the stator vanes. The stator begins to revolve in a forward direction. The rollers roll out of the smaller ends of the notches and into the larger ends. There, they cannot jam, and the stator is able to freewheel. The vanes simply move forward to get out of the way of the fluid.

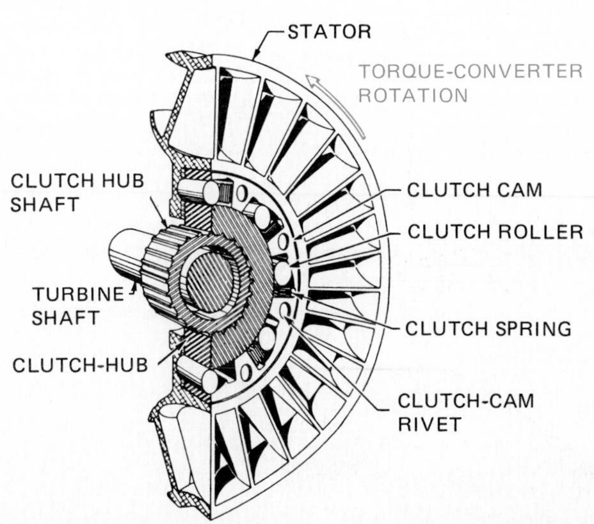

Fig. 44-8 A roller type of one-way clutch used to support the stator in a torque converter. *(Chrysler Corporation)*

Some one-way clutches use *sprags* instead of rollers. Sprags are shaped like flattened rollers. A series of sprags is placed between inner and outer races (Fig. 44-9). The sprags are held in place by two cages and small springs. During overrunning, when stator action is not needed, the outer race is unlocked (Fig. 44-9A). It spins freely around the stator shaft (Fig. 44-5). When stator action is needed, the oil is directed into the stator vanes. This attempts to spin the stator backward (Fig. 44-9B). When this happens, the sprags jam between the outer and inner races to lock the stator to the stator shaft.

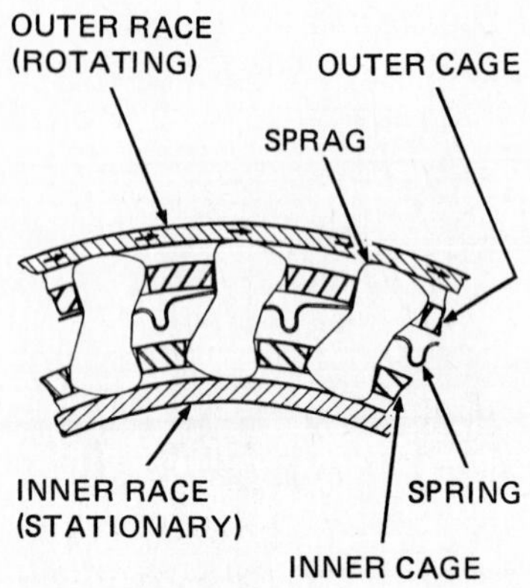

(A) FREE-WHEELING

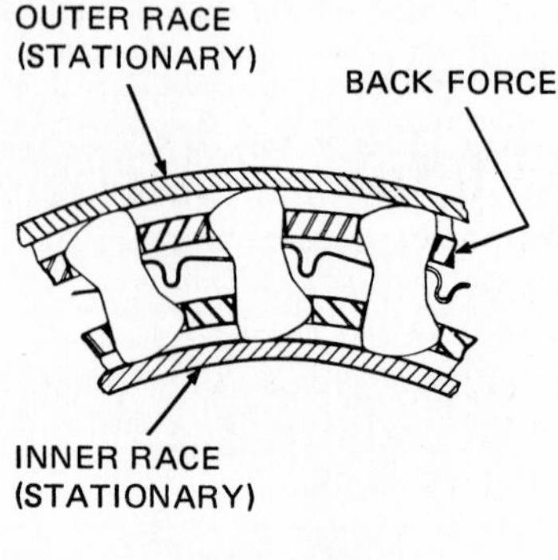

(B) LOCKED

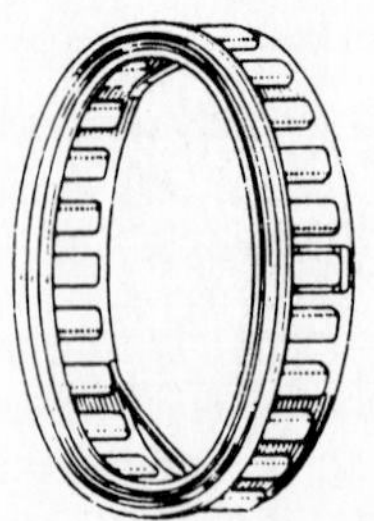

(C) ASSEMBLED

Fig. 44-9 Operation of a sprag one-way clutch.

♦ 44-6 LOCKING THE TORQUE CONVERTER

When the car is cruising at highway speed, the impeller is turning slightly faster than the turbine. This difference in speed represents a power loss. Therefore, many new cars have a torque converter that includes a *torque-converter clutch.* When engaged, the clutch locks the impeller to the turbine. The torque converter locks when the car reaches cruising speed and is neither accelerating nor decelerating. Locking the torque converter improves fuel economy. It also lowers the temperature of the ATF.

Figure 44-10 shows a lockup torque converter. It has a clutch and a clutch apply piston. The isolator springs on the clutch help dampen the shock of engagement as the torque converter locks up. These isolator springs also dampen out the power pulses from the engine when the transmission is in direct drive and the converter is locked. They do the same job as the torsional springs in the standard clutch disk.

Figure 44-11 shows the operation of one type of torque-converter clutch. In Fig. 44-11B, the torque converter is unlocked. The piston is disengaged. Figure 44-11C shows the locked position. The torque-converter cover has a ring of friction material bonded to it. To lock the torque converter, oil pressure is applied in back of the piston. This forces it to the left (in Fig. 44-11C). Now the piston and the output shaft must turn together. Since the torque input is to the cover, the torque converter turns as a unit. The arrows in Fig. 44-11B and C show the power flow through the torque converter with the clutch unlocked and locked.

♦ 44-7 PLANETARY GEARS

An automatic transmission or transaxle has two or more planetary gearsets (♦43-6). The simple planetary gearset shown in Fig. 44-12 can provide any one of five conditions. It can

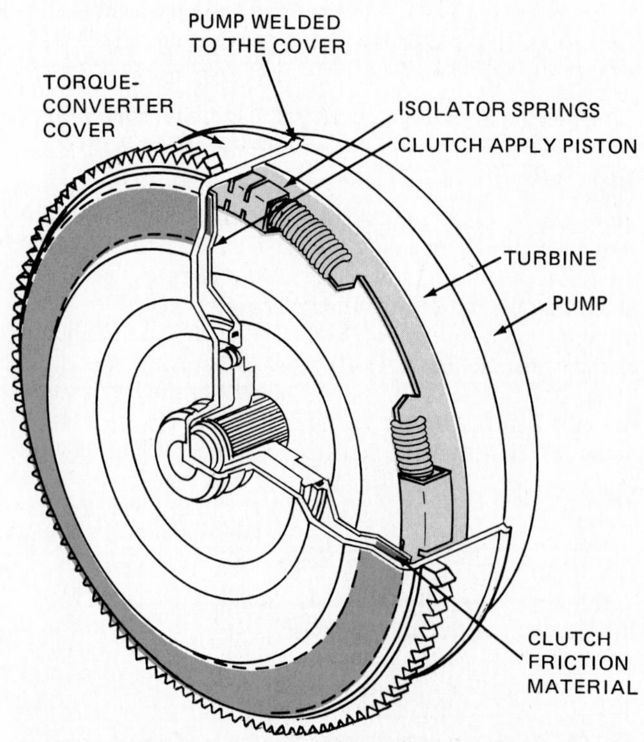

Fig. 44-10 A torque converter with locking clutch. *(Chrysler Corporation)*

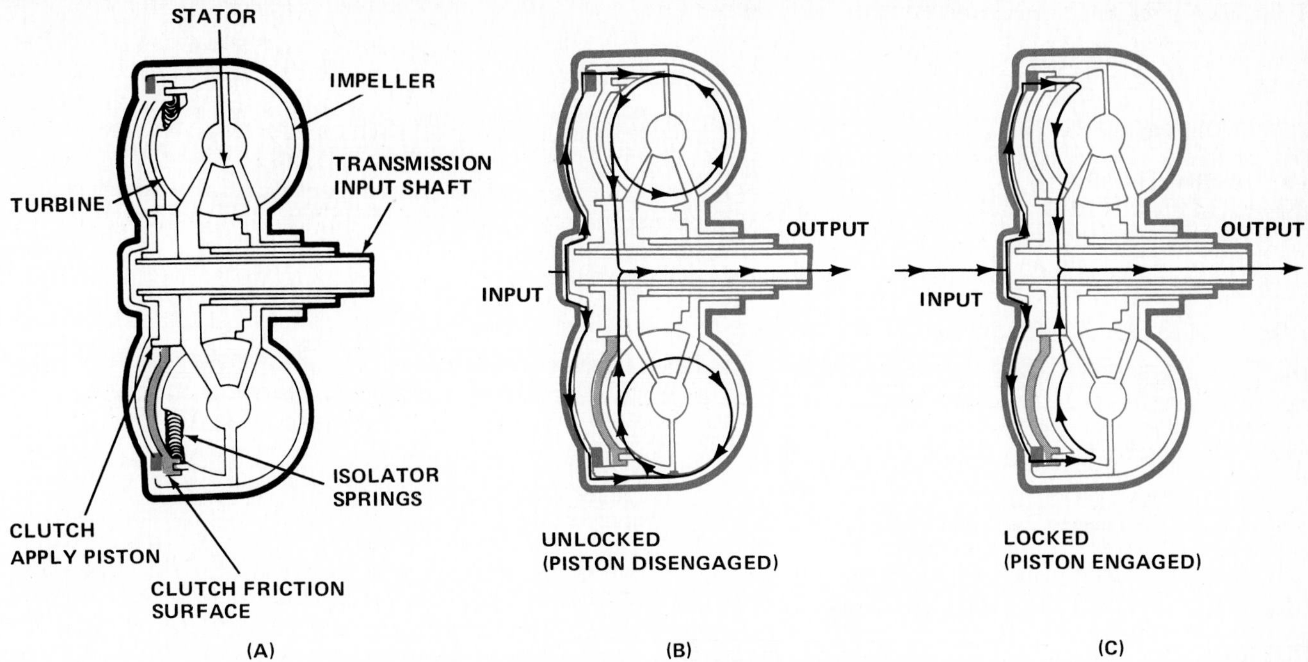

Fig. 44-11 Operation of a locking torque converter. *(Chrysler Corporation)*

1. increase speed and decrease torque (overdrive)
2. decrease speed and increase torque (reduction)
3. reverse the direction of rotation (reverse)
4. act as a solid shaft (direct drive)
5. disconnect the driving shaft from the driven shaft (neutral)

The planetary gearset includes an internal gear (or *ring gear*), a sun gear, and two or more planet pinions on a carrier and a shaft. Although planetary gears look different from other gear combinations, both types follow the same basic principles of operation.

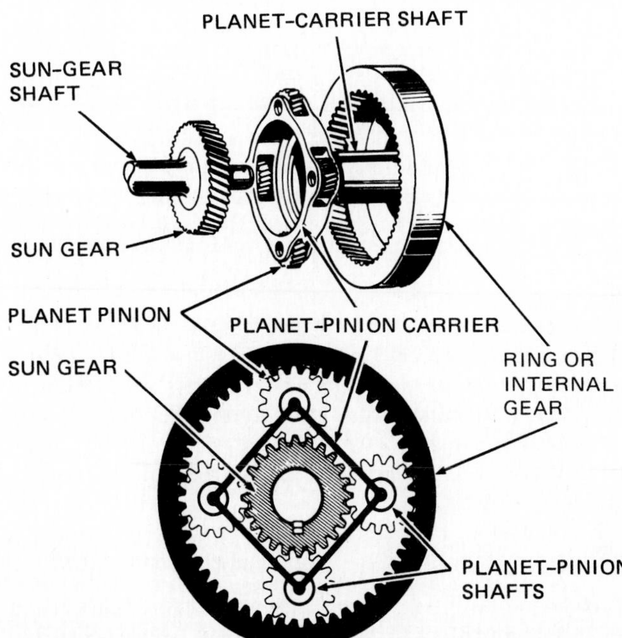

Fig. 44-12 A single planetary gearset.

♦ 44-8 GEAR COMBINATIONS

When two gears are in mesh, as shown in Fig. 40-1, they turn in opposite directions. But if another gear is put into the gear train, as shown in Fig. 44-13, the two outside gears turn in the same direction. The middle gear is called an *idler* gear. It doesn't work—it is idle.

To get a combination of two gears to rotate in the same direction, you can use one internal gear. The internal gear or ring gear, has teeth on the inside. The spur gear and the internal gear both rotate in the same direction (Fig. 44-14).

If another spur gear is added in the center and meshed with the small spur gear, the combination is a simple planetary-gear system (Fig. 44-15). The center gear is called the *sun gear* because the other gears revolve around it. This is similar to the way the planets in our solar system revolve around the sun. The spur gear between the sun gear and the internal gear is called the *planet pinion.* This is because it revolves around the sun gear, just as planets revolve around the sun.

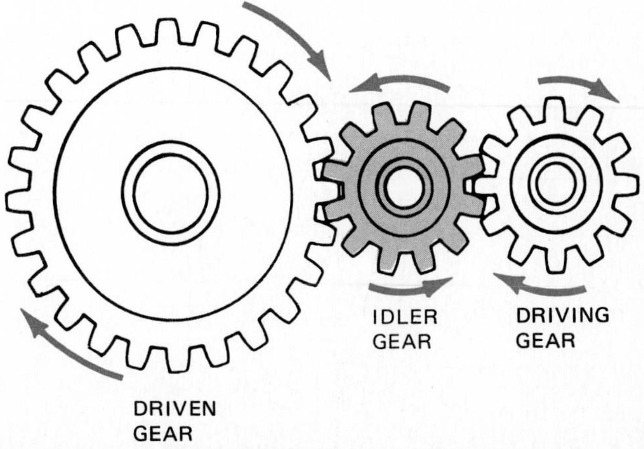

Fig. 44-13 The idler gear causes the driven gear to turn in the same direction as the driving gear.

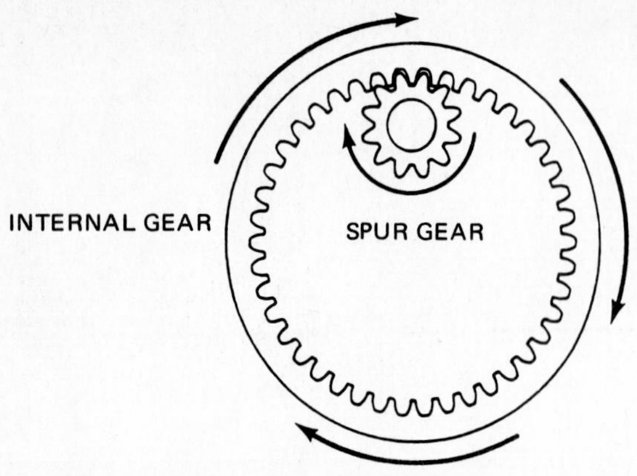

Fig. 44-14 If one internal gear is used with one external gear, both gears turn in the same direction.

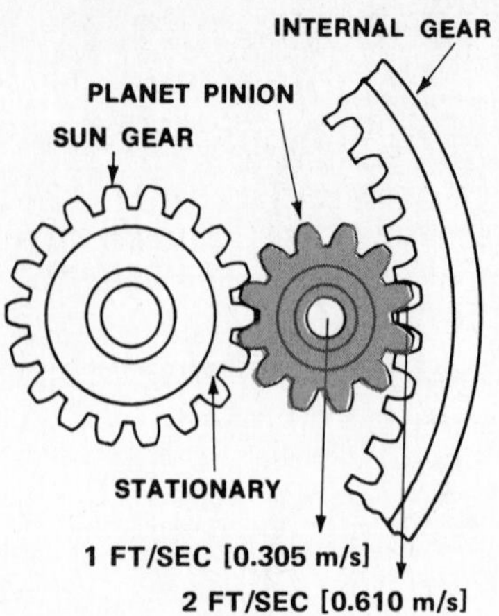

Fig. 44-16 If the sun gear is held stationary and the planet-pinion carrier is turned, the ring gear will turn faster than the carrier. The planet pinion pivots around the stationary teeth. If the center of the pinion shaft is moving at 1 foot per second (0.3 m/s), the tooth opposite the stationary tooth must move at 2 feet per second (0.6 m/s) because it is twice as far away from the stationary tooth as the center of the shaft.

♦ 44-9 PLANETARY-GEARSET OPERATION

The planetary gearsets used in automatic transmissions usually have three or more planet pinions. However, the explanation that follows uses the basic two-planet-pinion gearset shown in Fig. 44-15.

The two planet pinions rotate on shafts that are a part of a planet-pinion carrier (Fig. 44-15). There are three members in the planetary gearset: internal gear, sun gear, and planet-pinion carrier assembly. The planetary gearset can increase speed and reduce torque, reduce speed and increase torque, reverse the direction of rotation, act as a solid shaft, and disconnect the driving and driven shafts.

If one member is held stationary and another is turned, there is either a speed increase, a speed reduction, or reverse. If two members are locked together, the gearset acts like a solid shaft. If no members are locked, no power is transmitted through the gearset.

1. Speed Increase 1 Suppose the sun gear is held stationary, and the planet-pinion carrier is turned. Then the ring gear increases in speed. This occurs because as the carrier revolves, it carries the planet pinions around with it. This movement makes the planet pinions rotate on their shafts. As the pinions rotate, they cause the internal gear to rotate also (Fig. 44-16). Note the conditions. The sun gear is stationary. The planet-pinion carrier is moving, carrying the pinions around with it. The planet pinions walk around the sun gear, which means they rotate on their shafts. The inside pinion tooth, meshed with the sun gear, is stationary because the sun gear is stationary. That means the outside pinion tooth, meshed with the internal gear, is moving twice as fast as the shaft on which the planet pinion is turning. If the planet-pinion shaft is moving at 1 foot per second [0.305 m/s], the outer tooth is moving at 2 feet per second [0.610 m/s]. Therefore, speed increases.

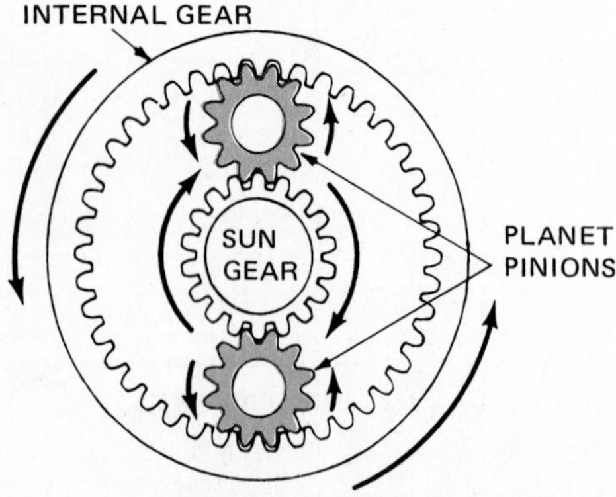

Fig. 44-15 A complete single planetary gearset using two planet-pinion gears. Two or more pinion gears are used to balance the forces so that the gears will rotate smoothly.

This condition is used in some automatic transmissions to provide an "overdrive" fourth gear.

2. Speed Increase 2 Another combination is to hold the internal gear stationary and turn the planet-pinion carrier. Now the sun gear is forced to rotate faster then the planet-pinion carrier. The result is a speed increase.

This condition usually is not used in automatic transmissions.

3. Speed Reduction 1 If the internal gear is turned while the sun gear is held stationary, the planet-pinion carrier turns more slowly than the internal gear. (This is just the opposite of the conditions described in speed increase 1.) With the internal gear turning the planet-pinion carrier, the planetary gearset acts as a speed-reducing system. This is the way second gear is achieved in many automatic transmissions.

4. Speed Reduction 2 If the internal gear is held stationary and the sun gear is turned, the planet-pinion carrier is driven at reduced speed. The planet pinions must rotate on their shafts. They must also walk around the internal gear, since they are in mesh with it. As the pinions rotate, the planet carrier rotates, but slower than the speed at which the sun gear is turning.

This condition provides the greatest increase in torque. It is used to obtain first gear in automatic transmissions.

5. Reverse 1 To get reverse, the planet-pinion carrier can be held stationary and the internal gear turned. Now the planet pinions act as idlers. They drive the sun gear in the reverse direction, and faster than the internal gear.

This condition is not used to obtain reverse in automatic transmissions.

6. Reverse 2 A second way to get reverse is to hold the planet-pinion carrier stationary and turn the sun gear. The internal gear turns in the reverse direction, but slower than the sun gear. This is the condition used to provide reverse gear in an automatic transmission.

7. Direct Drive If any two members are locked together, then the entire planetary-gear system acts as a solid shaft. Locking two members together locks the planetary gearset. There is no change of speed or direction through the system, and the gear ratio is 1:1. This condition is used for third gear in automatic transmissions.

8. Neutral When all of the members in a planetary gearset are free, no power can be transmitted through it. This condition provides the transmission with a neutral position for starting the engine without load.

The conditions discussed above are listed in the chart in Fig. 44-17. The conditions used in most three-speed automatic transmissions are listed in columns 3, 4, and 6.

♦ 44-10 AUTOMATIC-TRANSMISSION PLANETARY GEARSETS

The preceding description explains the operation of a single planetary gearset as shown in Fig. 44-12. However, to get all the conditions required from the automatic transmission, more complex systems are required. There are two major types, and both are often referred to by the names of their designers. The *simple,* or *Simpson,* planetary-gear train has two separate sets of planetary gears (Figs. 44-1 and 44-18). They revolve around one common sun gear. The other design is a *compound,* or *Ravigneaux,* planetary gear train (Fig. 44-19). It has one ring gear, two sets of planet gears (long and short), two sun gears (forward and reverse), and a planet carrier.

♦ 44-11 HYDRAULIC SHIFT CONTROLS

Previous sections in this chapter describe how power flows through the torque converter into the transmission, and how planetary gears operate. There are two controls, a band and a clutch. The band surrounds a metal drum. The drum may be attached to the sun gear, as shown in Fig. 44-20, or it may be the outer surface of the planetary ring (internal) gear. The clutch consists of a series of clutch plates. Half the plates are splined to an outer ring, called the *clutch drum,* and the other half are splined to the clutch hub. The clutch hub is

Conditions	1	2	3	4	5	6
Ring gear	D	H	T	H	T	D
Carrier	T	T	D	D	H	H
Sun gear	H	D	H	T	D	T
Speed	I	I	L	L	IR	LR

D—Driven (output)
H—Held (stationary)
I—Increase speed
L—Decrease speed
T—Turned or driving (input)

Fig. 44-17 Various conditions that are possible in the single planetary gearset if one member is held and another is turned.

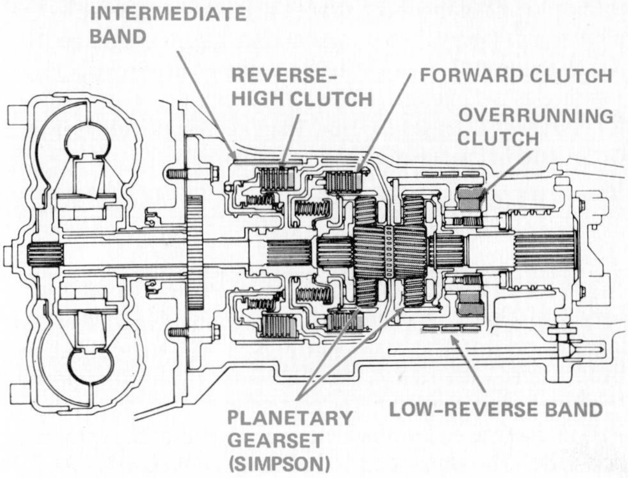

Fig. 44-18 An automatic transmission using a Simpson planetary gearset. *(Ford Motor Company)*

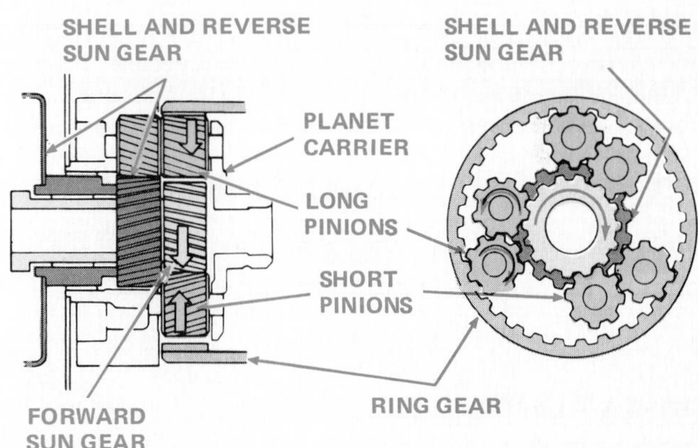

Fig. 44-19 A compound, or Ravigneaux, planetary gearset. *(Ford Motor Company)*

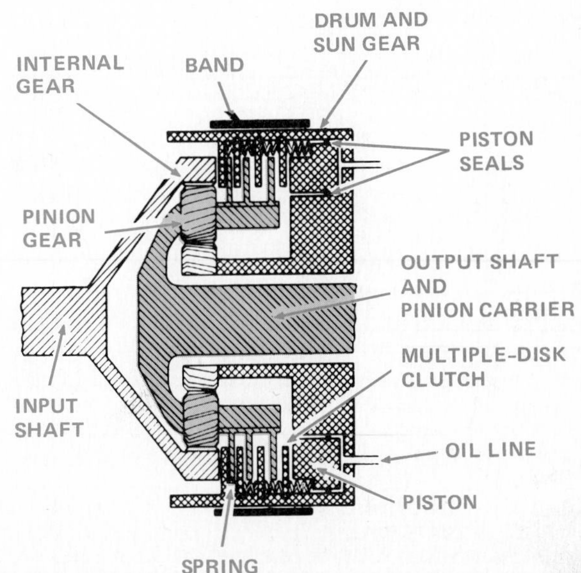

Fig. 44-20 The two hydraulically operated control devices for the planetary gearset. One device consists of a drum and band. The other device is the multiple-disk clutch.

splined to one of the members of the planetary gearset. When oil pressure forces the two sets of clutch plates together, the clutch is engaged. That means the planetary gearset is locked, rotating as a single unit. When the oil pressure is released, the clutch is disengaged. Now the two sets of clutch plates can rotate independently.

♦ 44-12 BAND

Figure 44-20 is a sectional view of a planetary gearset with a band and clutch. A band is shown in Fig. 44-21. It wraps around the drum as shown in Figs. 44-20 and 44-22. The band is lined with friction material. When the band is applied, or tightened on the drum, the drum and the sun gear are held stationary. One end of the band is anchored to the transmission case. The other end is linked to a *servo* (Fig. 44-22) which has a piston that can be applied hydraulically.

When oil pressure is directed to the *apply side* of the piston, it moves to the left (in Fig. 44-22). This applies the band so that the drum and sun gear are held stationary. Now the planetary gearset is a speed reducer. The internal gear is turning because it is attached to the input shaft. This action forces the pinion gears to rotate. They walk around the stationary sun gear, carrying the pinion carrier around with them. The carrier-and-output shaft rotates at a slower speed than the internal gear.

To release the band, the oil pressure is relieved. This allows a spring to push the piston back. In most transmissions, the spring is assisted by redirecting the oil pressure from the apply side to the *release side* of the piston (Fig. 44-22).

♦ 44-13 CLUTCH

If the clutch is engaged instead of the band, oil pressure is directed into the assembly through the oil line (Fig. 44-20). This forces the annular piston to the left (in Fig. 44-20). The action pushes the clutch plates together so that the clutch is engaged. This locks the planet-pinion carrier and the sun

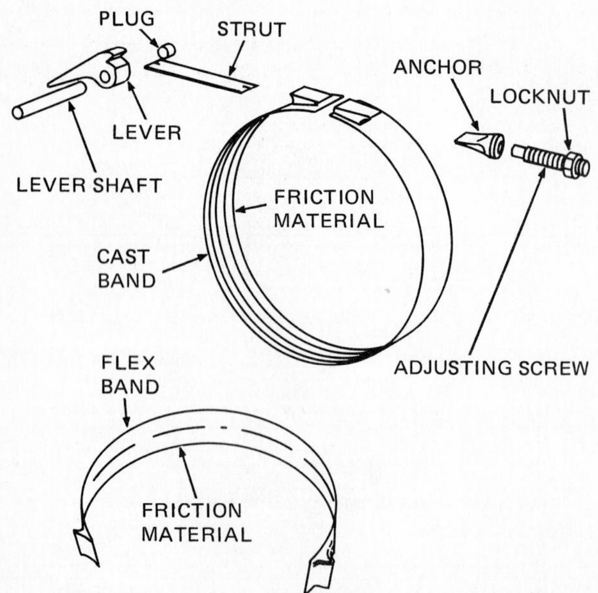

Fig. 44-21 Band with related linkages for an automatic transmission. *(Chrysler Corporation)*

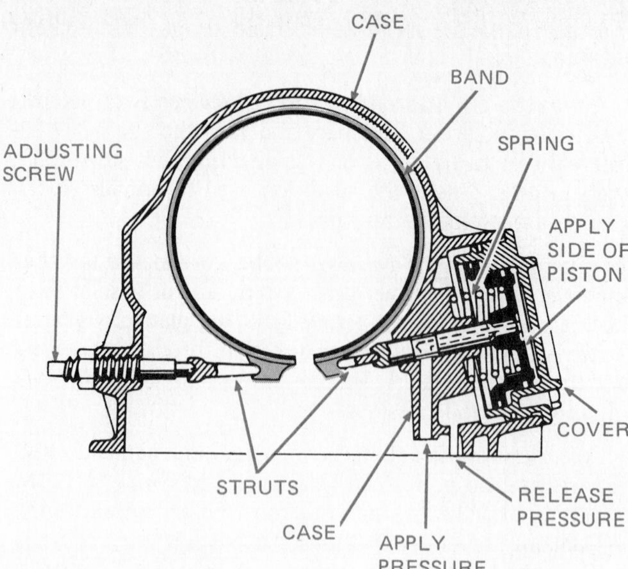

Fig. 44-22 Band and servo that operates it. *(Ford Motor Company)*

gear together. The planetary gearset is now in direct drive. Figure 44-23 shows a disassembled clutch. The clutch plates are alternately splined to the drum and to the clutch hub. Compression springs prevent the clutch from engaging too quickly. When the oil pressure is released, the springs push the piston away from the clutch plates. This allows the two sets of plates to rotate independently of each other. The clutch is now disengaged.

The arrangement shown in Fig. 44-20 and described above is only one of several arrangements used in automatic transmissions. In other transmissions, when the band is applied, it holds the internal gear or the pinion carrier stationary. Other transmissions lock different members together when the clutch is engaged. However, the principle is the same in all automatic transmissions. In the planetary gearset (in a three-speed automatic transmission), there is speed reduction through the planetary gearset when the band is applied. There is direct drive when the clutch is engaged.

♦ 44-14 HYDRAULIC CIRCUITS

Figure 44-24 is a simplified diagram of a hydraulic control system for a single planetary gearset in an automatic transmission. Its job is to apply and release a band and to engage and disengage a clutch. By these actions, the hydraulic system controls the shift from speed reduction to direct drive. The timing of the shift depends on car speed and throttle opening, or engine load. These two factors produce two varying oil pressures that work against the two ends of the shift valve.

The shift valve (Fig. 44-24) is a *spool valve* inside a bore, or hole, in the valve body. A spool valve is a metal rod with cutaway sections. It controls oil flow in the automatic transmission. The smaller diameter is called a *valve groove*. The larger diameters are called *valve lands*.

Pressure at one end of the shift valve comes from the governor (♦44-15). This is called *governor pressure*. Pressure at the other end of the shift valve changes as vacuum in the intake manifold changes. This is called *throttle pressure*.

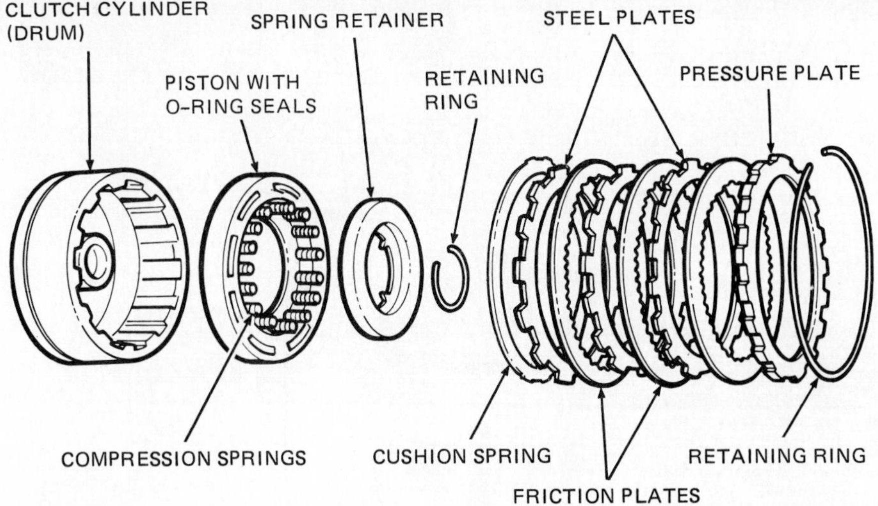

Fig. 44-23 A disassembled multiple-disk clutch. *(Ford Motor Company)*

♦ 44-15 GOVERNOR

The governor controls, or *governs*, gear shifting in relation to car speed. The governor is driven by the output shaft in the transmission (Fig. 44-25). As output-shaft speed and car speed go up, governor pressure increases. This pressure works against one end of the shift valve (Fig. 44-24).

Governor pressure is actually modified line pressure. An oil pump in the front of the transmission (Fig. 44-25) produces the line pressure. As car speed increases, the governor spins faster. This allows more pressure to pass through. The result is a modified pressure that changes with car speed which pushes against the right end of the shift valve (Fig. 44-24).

♦ 44-16 THROTTLE PRESSURE

Working on the other end of the shift valve is a pressure that changes as intake-manifold vacuum changes. Line pressure enters the modulator valve (upper right in Fig. 44-24). The modulator valve has a spool valve attached to a spring-loaded diaphragm. Vacuum increases in the intake manifold when the throttle valve is partly closed, or when engine load decreases. This vacuum pulls the diaphragm and modulator valve to the right (Fig. 44-24).

As the modulator valve moves to the right, it closes off the line-pressure passage from the oil pump. This reduces oil pressure on the left end of the shift valve. Now governor pressure pushes the shift valve to the left. Line pressure passes through the shift valve to the servo and clutch for the planetary gearset. This pressure releases the band and engages the clutch. The planetary gearset shifts to direct drive.

♦ 44-17 SHIFT ACTION

When the planetary gearset shifts from reduction to direct drive, the action is called an *upshift*. When the planetary gearset shifts from direct drive to gear reduction, it is called a *downshift*.

The reason for varying the pressure at the left end of the shift valve (Fig. 44-24) is to change the upshift to meet driving conditions. When the car is accelerating, high torque is needed. The gearset should stay in reduction. But when cruising speed is reached, less torque is needed. The driver eases up on the throttle. This increases intake-manifold vacuum so that an upshift results.

If the driver again wants fast acceleration, the throttle is pushed to the floor. This reduces intake-manifold vacuum,

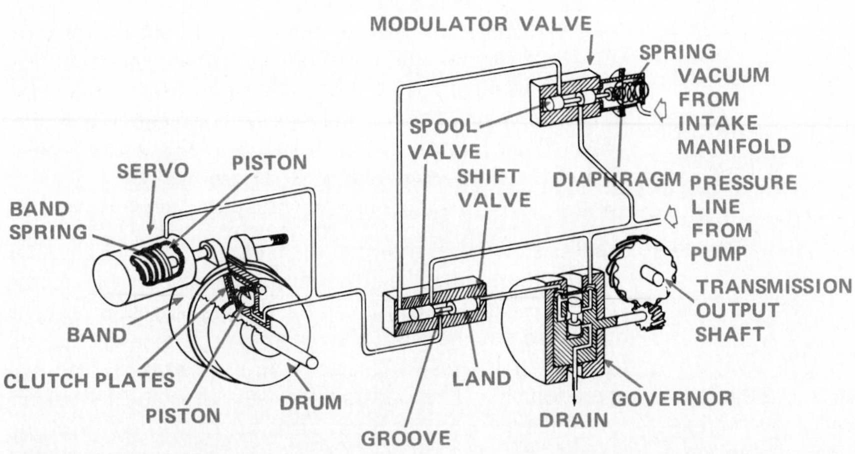

Fig. 44-24 Schematic diagram showing one type of hydraulic control system that has been used to operate the band and clutch.

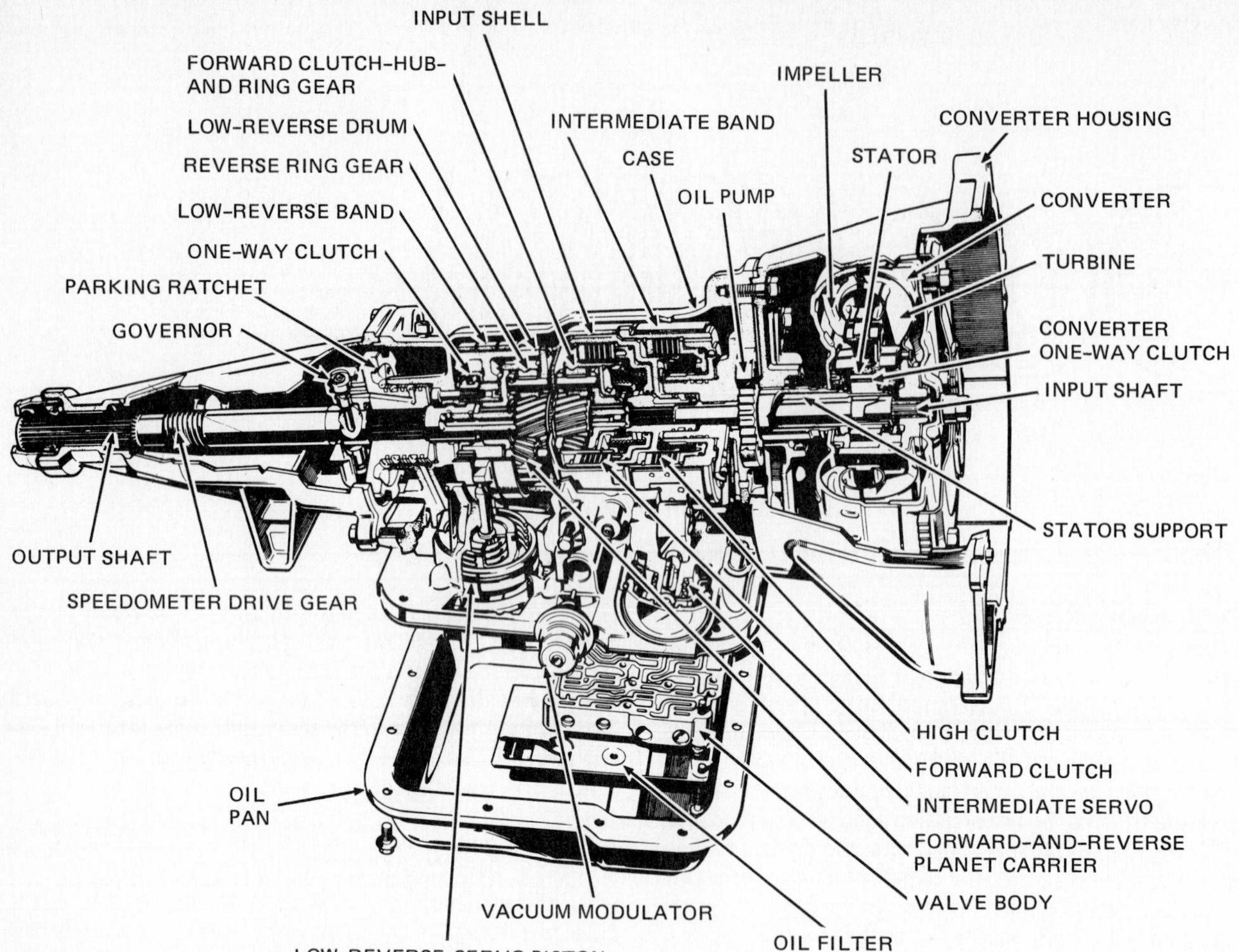

Fig. 44-25 A complete automatic transmission with the major parts named. *(Ford Motor Company)*

causing the planetary gearset to downshift into reduction. This increases torque to the drive wheels.

♦ 44-18 HYDRAULIC VALVES

The preceding sections provide a simplified explanation of how upshifts and downshifts are controlled. The actual valves are more complicated. They have springs to properly position them when oil pressure is not acting on them. Other valves help ease the shifts, regulate pressures, and time the downshifts. There is also the *manual valve*. This is the valve that moves as the driver moves the selector lever to the desired driving range.

In the automatic transmission, the hydraulic valves are contained in a main-control assembly called a *valve body* (Fig. 44-25). It directs fluid under pressure to the torque converter, servos, clutches, and governor to control transmission operation. The valve body contains several multiple-land spool valves.

♦ 44-19 AUTOMATIC-TRANSMISSION FLUID

The automatic-transmission fluid (ATF) used in automatic transmissions is a special oil. It has several additives such as viscosity-index improvers, oxidation and corrosion inhibitors, extreme-pressure and antifoam agents, detergents, dispersants, friction modifiers, pour-point depressants, and fluidity modifiers. The oil usually is dyed red. If leakage occurs, you can easily tell whether it is engine oil or transmission fluid.

There are several types of ATF. This is because various models of automatic transmissions have different requirements for the lubricating oil. Most automotive automatic transmissions require either type F or Dexron-II ATF.

♦ 44-20 FORD C6 AUTOMATIC TRANSMISSION

Most transmissions provide three forward speeds, one reverse, and neutral and park. All have torque converters. The Ford C6 automatic transmission has been used in many car models for years. The transmission is shown cut away in Fig. 44-26. The hydraulic system is shown in Fig. 44-27. Operation of the C6 transmission is described below.

1. The Manual Shift Valve The action starts at the manual shift valve, shown at the bottom in Fig. 44-26. After starting the engine, the driver moves the selector lever on the steering column or console to the driving range desired. The driver can put the selector lever in R, or reverse, to back up the car. Or the driver can put it in D, or drive, for normal operation. In D the transmission will automatically upshift from first to second to third. If the driver does not want the transmission to shift up to third, the driver can move the selector lever to 2 or 1. In 2 the transmission will remain in second. In 1 the transmission will remain in first. The driver

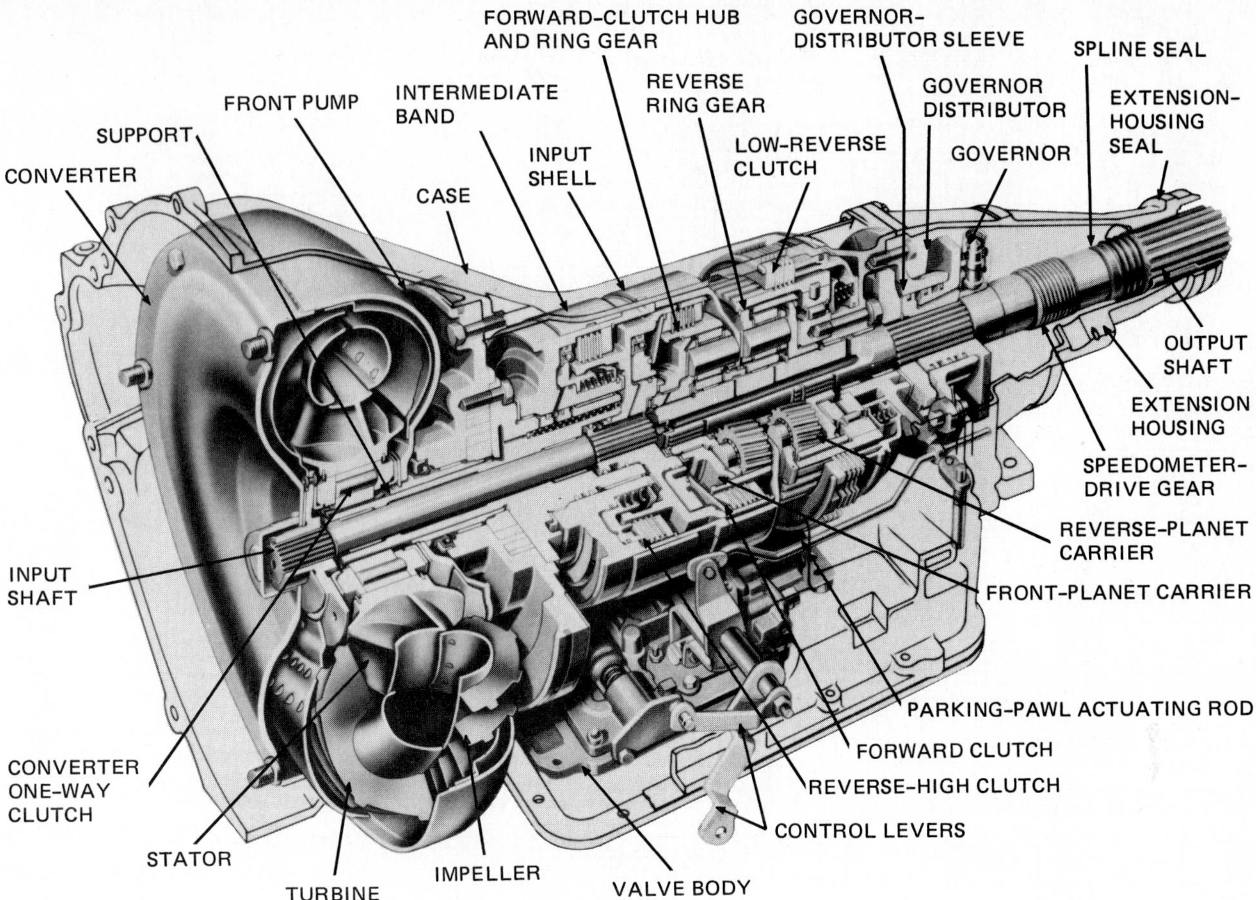

Fig. 44-26 A Ford C6 automatic transmission. *(Ford Motor Company)*

may need 1 or 2 for slowing down while descending a long hill.

2. Upshifting Let's assume the driver selects D. Starting out, the transmission is in first. Then, as car speed increases, the shifts are made from low to second and from second to third. Shift points are determined by car speed and manifold vacuum.

3. First Gear In first gear, the forward clutch is engaged. This locks the front planetary ring gear to the input shaft, so they turn together. As the ring gear rotates, it drives the planet pinions. They, in turn, drive the sun gear. This produces a gear reduction through the front planetary gearset. As the sun gear turns, it drives the rear planet pinions. They drive the ring gear of the rear planetary gearset. The ring gear is splined to the output shaft, so it turns. There also is gear reduction in the rear planetary gearset. With gear reduction in both sets, the transmission is in low, or first gear.

4. Second Gear When the upshift to second gear takes place, the hydraulic system applies the intermediate band. It holds the sun gear and the reverse-and-third clutch drum stationary. Now there is gear reduction in the front gearset only. The transmission is in second.

5. Third Gear, or Direct Drive When the hydraulic system produces the shift into third, both the forward clutch and the reverse-and-third clutch are engaged. This locks the planetary gearsets so that there is direct drive through both. The transmission is in third gear.

6. Reverse In reverse, the reverse-and-third clutch and the low-and-reverse clutch are both engaged. This condition causes gear reduction through both planetary gearsets. Also, the direction of rotation is reversed in the rear set.

♦ 44-21 OTHER AUTOMATIC TRANSMISSIONS

A variety of automatic transmissions is used in automobiles. However, they all work in about the same way. All have planetary gearsets that are controlled by bands and clutches.

AUTOMATIC TRANSAXLES

♦ 44-22 PURPOSE OF AUTOMATIC TRANSAXLES

The transaxle is used with engines mounted transversely at the front of the car (Figs. 10-3 and 10-10). The transaxle is also used in rear-engine cars with rear-wheel drive. Figure 44-2 shows the major components of an automatic transaxle with the engine crankshaft and drive train. The major difference between an automatic transaxle and a manual transaxle (Chap. 40) is that the automatic transaxle uses an automatic transmission instead of a manual transmission. However, the automatic transmission works the same as described earlier in this chapter.

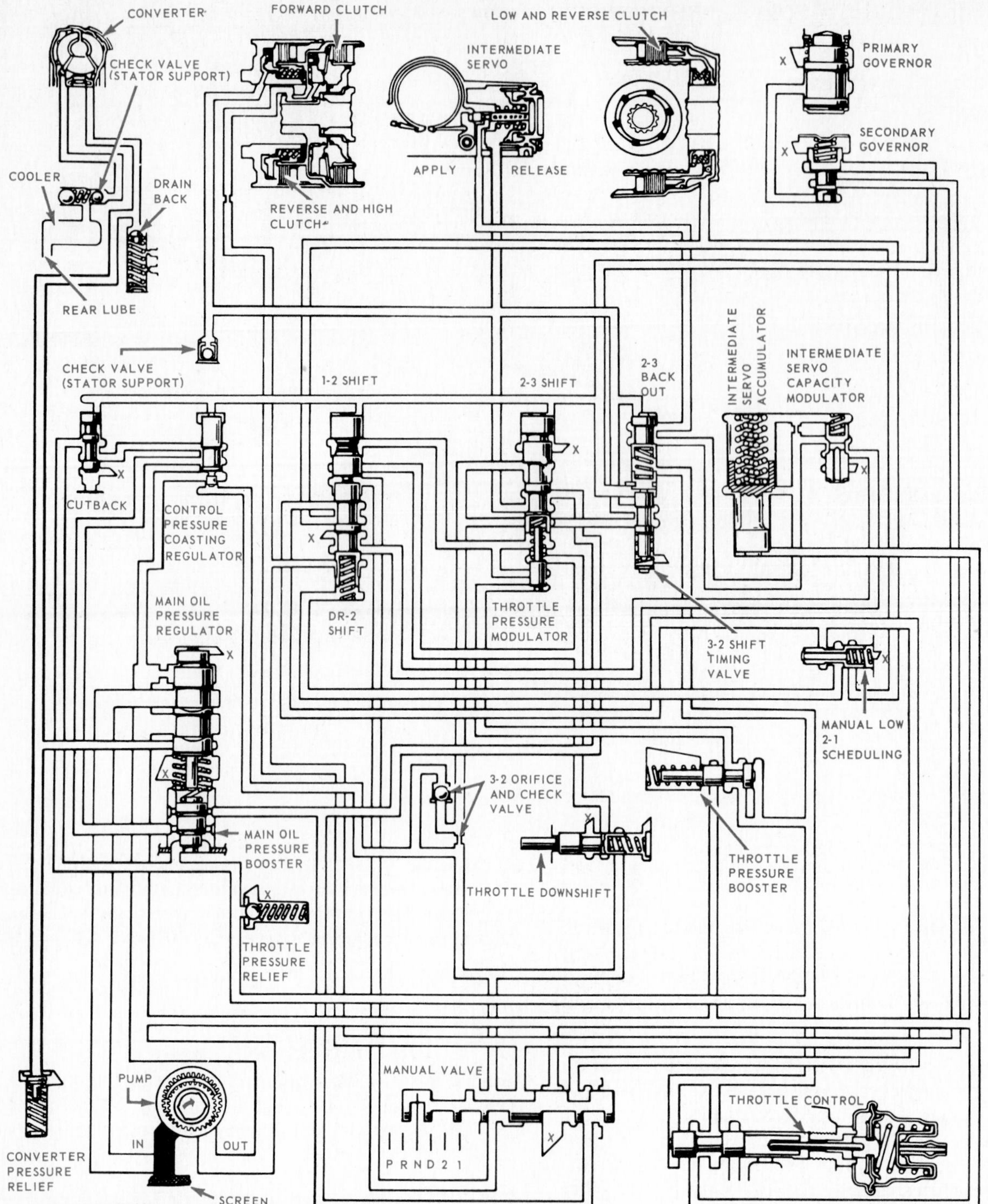

Fig. 44-27 Complete hydraulic system of the Ford C6 automatic transmission. *(Ford Motor Company)*

♦ 44-23 CHRYSLER AUTOMATIC TRANSAXLE

Figure 44-28 shows the automatic transaxle and drive train for a front-wheel-drive car built by Chrysler. The transmission section of the transaxle uses the Chrysler TorqueFlite automatic transmission. The complete Chrysler TorqueFlite transaxle is shown in sectional view in Fig. 44-29. Construction and operation of the differential are described in a later chapter.

The Chrysler TorqueFlite transaxle includes a differential and two couplings for the drive shafts to the two front wheels. It also includes the TorqueFlite automatic transmission, adapted to fit into the transaxle case.

The automatic transmission in the transaxle has two multiple-disk clutches, two bands with servos, an overrunning clutch, and a Simpson type of planetary gearset. These units, and their operation, are described earlier in this chapter. A gear on the end of the output shaft is meshed with a gear on the end of the transfer shaft. The transfer shaft carries the governor. The other end of the transfer shaft, next to the torque converter, has a pinion gear that meshes with the ring gear of the differential. The differential takes care of any difference in rotational speed of the two front wheels when the car makes a turn.

♦ 44-24 TORQUEFLITE TRANSAXLE OPERATION

The driver moves the selector lever to the desired operating range. This moves the manual valve, through linkage, to the proper operating position. Then car speed and throttle-valve position take over to produce the shift pattern that has been selected. The governor and throttle position control the hydraulic system so that upshifts and downshifts are made to suit the operating condition.

♦ 44-25 ELECTRONIC CONTROL OF AUTOMATIC TRANSMISSION

Figure 44-30 shows the layout of a Bosch electronic control system for a three-speed automatic transmission. It is being used on some BMW cars. It has sensors for engine and road speeds, engine load, position of the selector lever, status of the driving program, and the kickdown switch. The driving program is stored in the memory of the microcomputer. Using the input signals from the sensors, the microcomputer determines the best gear and then shifts the transmission into it. Solenoid valves operate to control the hydraulic system in the transmission. As the solenoid valves operate, they apply or release the bands and engage or disengage the clutches to produce the shift.

♦ 44-26 STARTING CONTROLS

When a vehicle is equipped with an automatic transmission, the engine cannot be started unless the selector lever is in N or P. The reason for this is to prevent accidents. If the engine could be started with the selector lever in a driving range, the car would suddenly start to move. This could result in an accident.

There are two different starting-control systems. One is for the steering column–mounted selector lever and quadrant. The other may be used with either floor shift or column shift.

Figure 27-18 shows an ignition switch which is also a steering-wheel lock and a starting switch. As the ignition key and switch are turned to the start position, the steering wheel is unlocked. Then the starting motor operates to crank the engine. When the engine starts, the driver releases the ignition key and the ignition switch moves back to the ON, or RUN, position. The ignition key and switch cannot be turned to CRANK or START if the transmission is in any driving range.

The floor-mounted selector lever has a different arrangement. This includes a neutral safety switch which is open in all positions except P and N. If the selector lever is in any other position, the safety switch is open so the starting motor cannot operate. But if the selector lever is moved to the N or P position, the safety switch points are closed. Then, turning the ignition switch to start will allow the starting motor to operate.

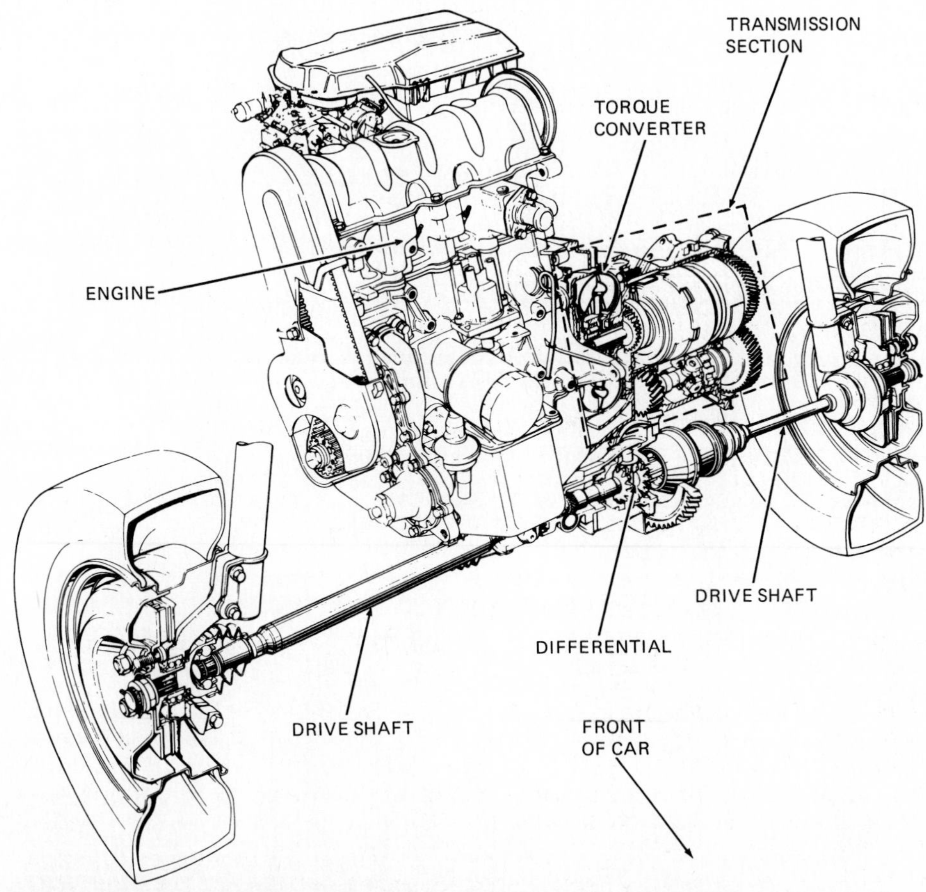

Fig. 44-28 Automatic transaxle and drive train for a front-wheel-drive car. *(Chrysler Corporation)*

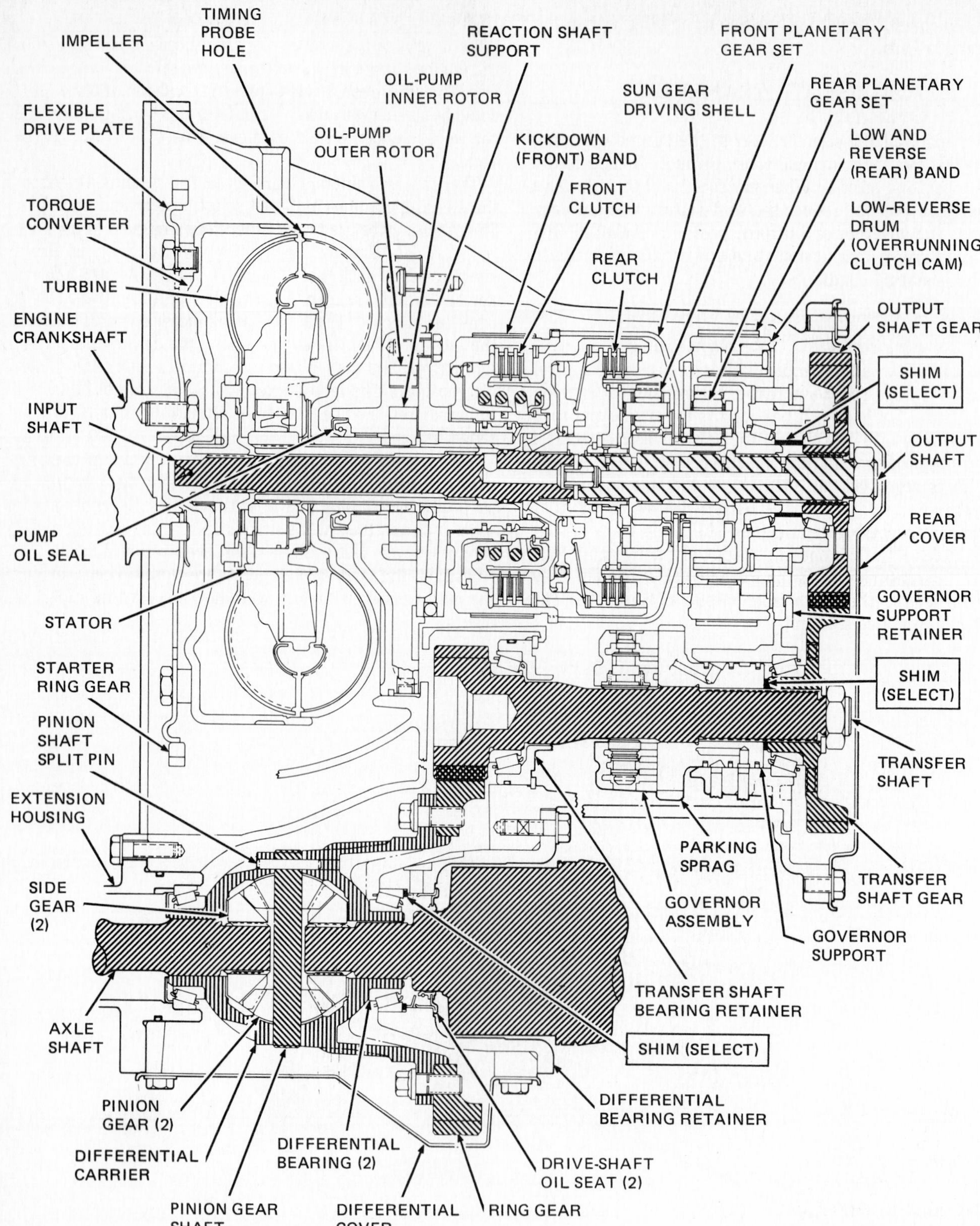

Fig. 44-29 Sectional view of the Chrysler TorqueFlite automatic transaxle. *(Chrysler Corporation)*

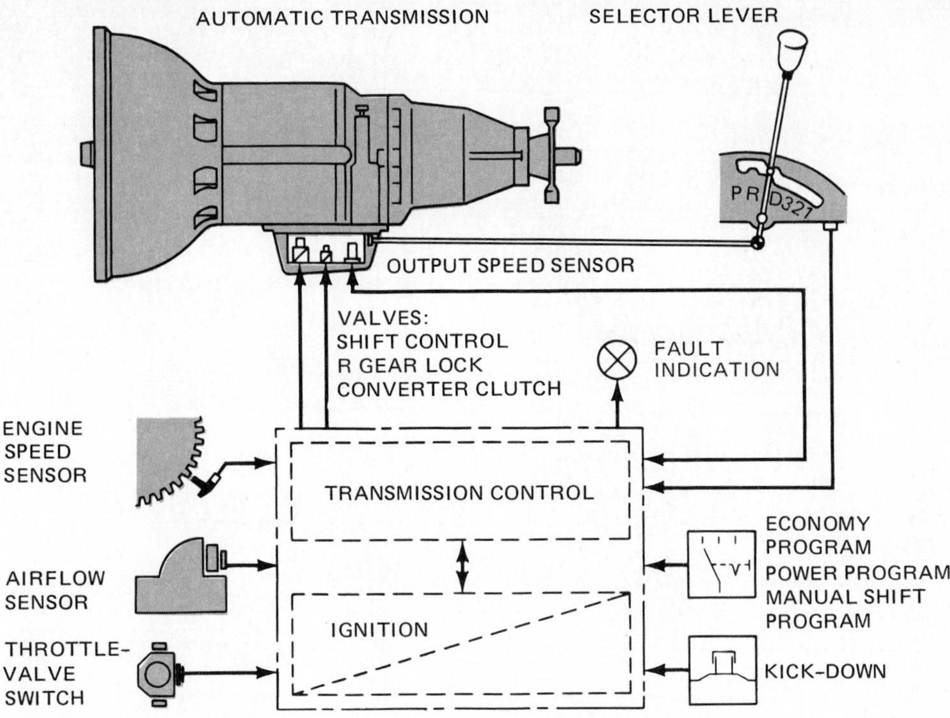

Fig. 44-30 An electronic control system for a three-speed automatic transmission. *(Robert Bosch Corporation)*

♦ REVIEW QUESTIONS ♦

Select the *one* correct, best, or most probable answer to each question. You can find the answers in the section indicated at the end of each question.

1. In the automatic transmission, a band is applied by (♦44-12)
   a. an accumulator
   b. a clutch
   c. a servo
   d. a governor
2. The element in an automatic transmission that is used most often to obtain direct drive is (♦44-13)
   a. an accumulator
   b. a clutch
   c. a servo
   d. a governor
3. The governor is driven by (♦44-15)
   a. the stator clutch
   b. the transmission input shaft
   c. the engine crankshaft
   d. the transmission output shaft
4. The servo is released by (♦44-12)
   a. spring force
   b. oil pressure
   c. both a and b
   d. neither a nor b
5. A torque converter will be locked up to provide mechanical drive when (♦44-6)
   a. the impeller is locked to the cover
   b. the impeller is locked to the turbine
   c. the impeller is locked to the stator
   d. none of the above
6. The clutch disks are forced together by movement of the (♦44-13)
   a. clutch springs
   b. clutch piston
   c. servo piston
   d. clutch fork
7. In the torque converter, the stator is placed between the pump and the (♦44-4)
   a. clutch
   b. rotor
   c. turbine
   d. flywheel
8. Two types of one-way clutches are (♦44-5)
   a. ball and roller
   b. sprag and roller
   c. needle bearing and friction bearing
   d. taper bearing and antifriction bearing
9. Two controlling devices in the automatic transmission operated by hydraulic pressure are the bands and (♦44-11)
   a. pistons
   b. clutches
   c. gears
   d. planetary gearsets
10. The two pressures which act on opposite ends of the shift valve to produce gear shifting are (♦44-14)
    a. hydraulic pressure and governor pressure
    b. car speed and governor pressure
    c. intake-manifold pressure and throttle pressure
    d. governor pressure and throttle pressure

CHAPTER 45
Automatic-Transmission and Transaxle Service

After studying this chapter, and with proper instruction and equipment, you should be able to:

1. Check the fluid in an automatic transmission and in an automatic transaxle.
2. Check the transmission or transaxle for fluid leaks.
3. Diagnose troubles in various models of automatic transmission and transaxle.
4. Perform a pressure test and interpret the results.
5. Make a stall test and interpret the results.
6. Perform linkage and band adjustments.
7. Remove and install an automatic transmission and transaxle.
8. Rebuild an automatic transmission and transaxle.

♦ 45-1 SERVICING AUTOMATIC TRANSMISSIONS AND TRANSAXLES

Automatic-transmission and transaxle service can be divided into four parts. These are normal maintenance, trouble diagnosis, on-the-car repairs, and transmission overhaul.

Before you undertake any of these procedures, refer to the manufacturer's shop manual for the model of automatic transmission or transaxle you are about to work on. Then follow the procedures outlined. Because there are many models and designs, there are variations in the service procedures. Always follow the procedure recommended for the unit you are working on.

♦ 45-2 NORMAL MAINTENANCE

Normal maintenance includes:

1. Changing fluid and filter
2. Checking fluid level
3. Adding fluid if necessary
4. Checking throttle and shift linkages
5. Adjusting the neutral starting switch
6. Possibly band adjustment

The level of the automatic-transmission fluid (also called *ATF* or *oil*) should be checked every time the engine oil is changed. In addition, many car manufacturers recommend changing the transmission fluid and filter at periodic intervals. The length of the intervals depends on how the car is used.

For example, Chevrolet recommends changing the fluid and filter every 100,000 miles [160,000 km] for normal service. For severe service, as in taxis, in vehicles hauling trailers, in police vehicles, in stop-and-go city driving, or in delivery service, Chevrolet recommends changing the fluid and filter every 15,000 miles [24,000 km].

Linkages and bands may require relatively frequent adjustment if the vehicle is in severe service. General adjustment procedures are described in ♦45-10.

♦ 45-3 CHECKING THE FLUID

Check the fluid level in the transmission and also the condition of the fluid. This should be done at every engine-oil change.

1. Checking Fluid Level Clean dirt from around the dipstick cap. Pull the dipstick, wipe it, reinsert it, and pull it out again. Note the level of the fluid on the dipstick. The level will vary under normal operating conditions as much as ¾ inch [19 mm] from cold to hot. For example, as the temperature of the fluid goes up from 60°F [16°C] to 180°F [82°C], the level of the fluid will rise as much as ¾ inch [19 mm]. This is the reason that many dipsticks are marked to indicate proper levels at different temperatures (Fig. 45-1).

Note On the General Motors 125 automatic transaxle, the fluid level goes *down* as temperature increases. The COLD mark on the dipstick is *above* the FULL mark.

Do not add too much fluid. Normally, not more than 1 pint [0.5 L] should be required. Too much fluid will cause foaming. Foaming fluid cannot operate clutches and bands effectively. They will slip and probably burn. This could result in an expensive transmission overhaul.

You can get a general idea of the fluid temperature by cautiously touching the transmission end of the dipstick to find out if it feels cool, warm, or hot. If the fluid feels cool, the fluid level on the dipstick should be on the low side of the dipstick. If the fluid feels warm, or hot (too hot to hold), the fluid level on the dipstick should be on the high side. Specific instructions for the various automatic transmissions are given in later chapters.

2. Condition of the Fluid The transmission fluid is normally red in color. If it is brown or black and has a burned odor, the transmission may be in trouble. Bands and clutch plates may have overheated and burned. Then, particles of friction material from the bands and clutch plates have probably circulated through the transmission and oil cooler. These particles can build up and cause valves in the valve body to hang. Shifts will be noisy or will not take place. Servos and clutches can malfunction. The transmission shifts may be jerky or may not take place. Or the transmission may slip.

Use a piece of absorbent white paper such as facial tissue to wipe the dipstick (Fig. 45-2). Examine the stain for evidence of solids (specks on the paper) or for evidence of antifreeze leakage (gum or varnish on the dipstick).

If the fluid is dark in color, or if you find specks on the paper or gum or varnish on the dipstick, remove the transmission-oil pan and look for further evidence of trouble. If you find contamination in the oil pan, it is added proof of transmission trouble. The transmission must be removed from the car for overhaul. Overhaul of specific transmission models is covered in the manufacturer's service manuals.

When the fluid is contaminated, the oil cooler and lines must be flushed out. Also, the torque converter must be flushed out or replaced. Most manufacturers state that a torque converter cannot be flushed out to remove all debris. It must be replaced with a new or remanufactured torque converter.

3. Engine Coolant in Transmission Fluid If engine coolant has leaked into the transmission fluid, the trans-

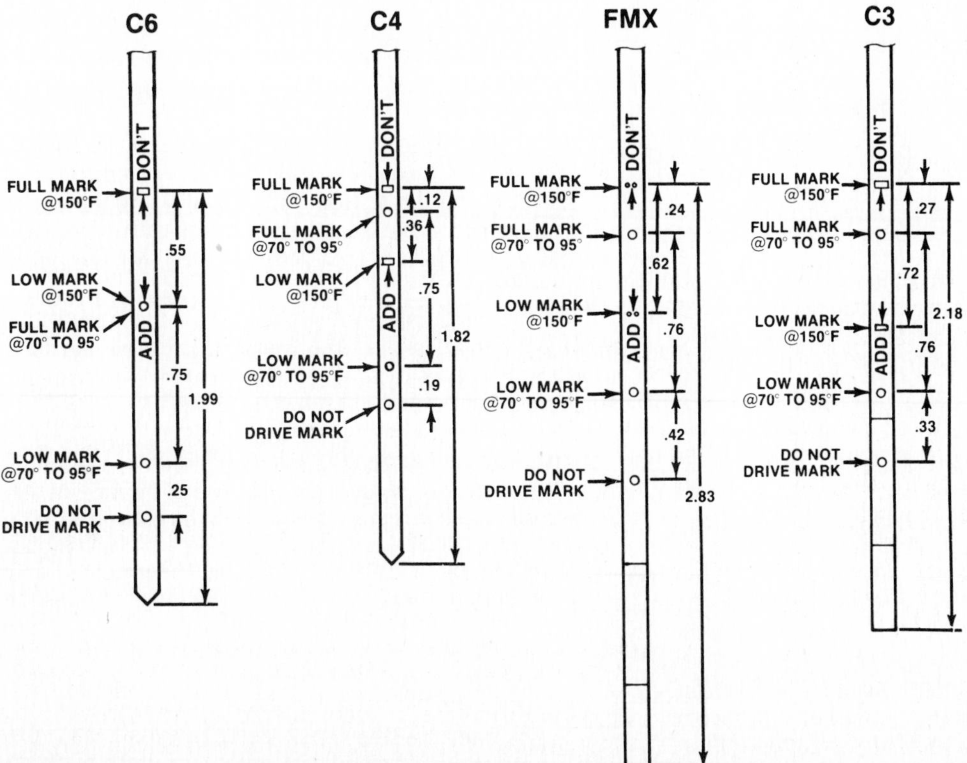

Fig. 45-1 Dipsticks used in various automatic transmissions.

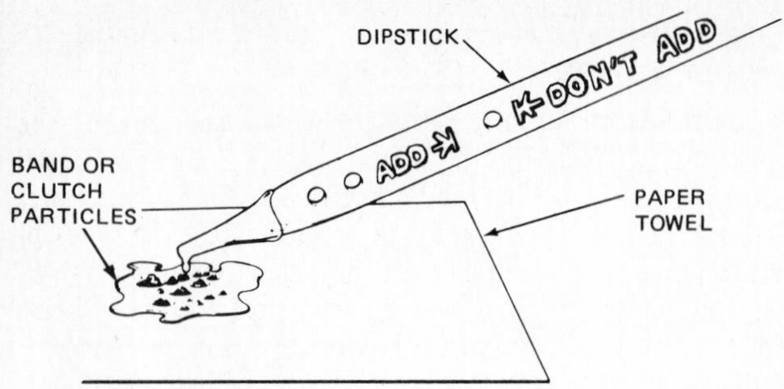

Fig. 45-2 Checking the condition of the fluid in an automatic transmission or transaxle. *(Ford Motor Company)*

mission must be removed for complete overhaul. This includes cleaning and replacement of seals, composition-faced clutch plates, nylon washers, and speedometer and governor gears. All these parts can be affected by coolant. The converter should be flushed out. The cooler in the engine radiator must be repaired or replaced. Then the cooler lines must be flushed out.

4. Checking for Transmission Leaks If the transmission fluid is low, suspect leakage. Some fluid might be lost through the vent if it has foamed. The fluid is red and this helps in finding any place where a leak occurs. Here is one procedure.

1. Clean the suspected area with solvent to remove all traces of fluid.
2. Remove the converter shield, if present, to expose as much of the converter as possible.
3. Spray the cleaned area with a spray can of white foot powder. This will show up the red fluid at the leak points.
4. Start and run the engine at high idle.
5. If the leak does not show up immediately, have the owner bring the car back the next day or later. Then recheck the suspected area.
6. Repair any leak points found by replacing gaskets or seals, tightening attached bolts, or replacing porous castings.

♦ 45-4 TROUBLE DIAGNOSIS

Each make and model of automatic transmission and transaxle has its own specific trouble-diagnosis guides. These are included in the manufacturer's service manual. The guides list the diagnostic procedure step by step. Locate the trouble-diagnosis section in the service manual covering the unit you are about to service.

Before making operating tests on an automatic transmission or transaxle, be sure the engine is in good condition and operating normally. A sluggish or missing engine will not allow the transmission or transaxle to perform normally.

Identify the make and model of transmission or transaxle in the vehicle you are about to check. Then study the specific reference that applies to that unit. Follow the procedure carefully. Any deviation from the procedure could damage the unit.

You may be required by the service manual to make stall checks or operating checks. Observe the cautions outlined in the references. Some manufacturers do not recommend stall tests. Others forbid them. Be very quick if you do make a stall test.

The stall test checks the torque converter and the holding ability of the transmission or transaxle clutches. It is performed by applying the car brakes, blocking the wheels, and measuring engine speed with the transmission in drive and the throttle wide open. The check must be made within 5 seconds. To take longer risks damage to the transmission or transaxle.

♦ 45-5 CAUSES OF FAILURE

The usual causes of automatic-transmission trouble are abuse and neglect. Most often these are overloading the transmission and not checking transmission fluid and adding or changing it if necessary.

Extremes of heat and cold are also hard on transmissions. Long periods of idling in stop-and-go traffic can overheat the transmission fluid if the transmission is left in a driving range. This continues to whip the fluid in the torque converter and adds to the heat. The remedy is to shift to neutral or park if the car is to idle for more than a few seconds.

The car, and transmission, may be overloaded when the car pulls a trailer, or when it is carrying extra weight. Revving the engine with the brakes applied and the transmission in a driving range also overloads the transmission.

For mountain driving, either with a heavy load or when pulling a trailer, use the 2 (second) or 1 (low) selector-lever position when on upgrades requiring a heavy throttle for ½ mile [0.8 km] or more. This reduces transmission and converter heating.

Working the transmission hard and overloading it can overheat the transmission fluid. The heat can eventually cause the fluid to deteriorate. Gum and varnish may form in the transmission. These could cause poor valve action and slippage of the bands and clutches. This can lead to further trouble.

Neglect is the other abuse that can damage an automatic transmission. If the transmission is being worked hard (pulling a trailer, or in police, taxicab, or door-to-door delivery service), the fluid and filter should be changed frequently. The bands should also be adjusted frequently.

Another problem may be caused by operation in consistently low temperatures. For example, Ford recommends that the fluid be changed every 7500 miles [12,000 km] if the vehicle is operated for more than 60 days in temperatures consistently below 10°F [−12°C] and with short trips of less than 10 miles [16 km].

Even in normal, conservative operation, some manufacturers recommend changing the transmission fluid and filter periodically. For example, Chevrolet recommends changing these every 100,000 miles [160,000 km].

Careful When towing a rear-wheel-drive car with automatic transmission, lift the rear of the car or remove the drive shaft. The car can be towed with the rear wheels on the ground and the drive shaft connected for short distances at low speed. But the transmission is lubricated only when the engine is running. To guard against transmission damage, lift the rear end or remove the drive shaft. For front-wheel-drive cars, lift the front end.

♦ 45-6 TROUBLE-DIAGNOSIS CHARTS

The purpose of trouble-diagnosis charts is to help you pinpoint the cause of trouble quickly. Trouble-diagnosis charts take different forms. Each manufacturer distributes a chart

Trouble	Possible Cause
No drive in drive range.	Low oil pressure (low oil, defective pump, screen or line plugged), incorrect manual-valve-linkage adjustment, defective clutch or band, or defective overrunning clutch.
Does not shift automatically.	Throttle-valve cable misadjusted, vacuum modulator defective, governor not working properly, 1-2 or 2-3 shift valves sticking, band or clutch defective, low oil pressure (low oil, defective pump, screen or line plugged).
Transmission slips.	Low oil pressure, defective band or clutch, misadjusted linkage, vacuum modulator defective.

Fig. 45-3 A simplified trouble-diagnosis chart for automatic transmissions and transaxles.

Fig. 45-4 An automatic-transmission tester. It combines in one unit a tachometer, vacuum gauge, and oil-pressure gauge. *(Ford Motor Company)*

for each type of transmission it makes. The chart in Fig. 45-3 lists three typical transmission troubles and several possible causes of each. This is a very general chart. Do not use it to diagnose troubles on specific automatic transmissions.

There are other possible troubles, such as no braking in low 1 or low 2, no drive in reverse, rough shifts, and failure to hold in park. Each trouble can be caused by certain specific conditions. When you are trouble-diagnosing a transmission, your job is to first verify that the trouble exists. Road test the car if necessary. Then consider the various possible conditions that could cause the trouble. If the transmission has been recently overhauled, perhaps some part was left out or not assembled properly. If the transmission is older and has had service, suspect worn bands, clutches, and bearings.

Some troubles might be fixed with a minor repair or adjustment. Other troubles mean removing the transmission for a complete overhaul.

♦ 45-7 TROUBLE-DIAGNOSIS PROCEDURE

Most automatic-transmission troubles require some testing to verify the trouble and also to check oil pressures under different operating conditions. The tests can be made on the road, on a chassis dynamometer, or in the shop with the driving wheels either on or off the floor (Fig. 45-4). All manufacturers do not recommend the same procedures. Before any test is made, the fluid level and condition should be checked and the control linkages checked and adjusted if necessary.

Note Be sure the engine is in good condition before making any tests of the transmission. If the engine lacks power, the transmission probably will not work properly.

The oil pressure in the hydraulic system is usually checked during the test procedure. This requires a pressure gauge with a long enough hose that it can be connected to the transmission and read from the driver's seat. Test procedures for some transmissions also require a tachometer to measure engine rpm. This is because some tests are made at specific engine speeds. An engine that lacks power can make a transmission perform improperly even though nothing is wrong with it.

The purpose of the diagnostic test is to operate the car at various speeds and check each gear position for slipping or incorrect shifting. For example, weak clutch springs cause the clutch to engage quickly and disengage slowly. Note the oil pressures and whether the shifts are harsh or spongy. Note also the speeds and throttle positions at which the shifts take place. Then compare the readings with the specifications in the manufacturer's service manual. Figure 45-5 is a typical pressure test-diagnosis chart.

1. Flare-up Slipping or flare-up (engine speedup or runaway on a shift) usually means clutch, band, or overrunning-clutch trouble. Flare-up is a sudden surge or increase in engine speed before or during a shift. You can usually determine which band or clutch is slipping by pinpointing the gear position in which the slipping occurs. Figure 45-6 is a typical band-and-clutch application chart. It shows the combinations of bands and clutches that are applied, or ON, in each gear position. To correct a slipping clutch, the transmission must be overhauled. Some bands can be adjusted. However, when an adjustment fails to stop a band from slipping, the band servo should be checked.

2. Torque Converter Poor acceleration could be caused by burned engine valves or by a clogged exhaust system. It could also be caused by a slipping stator clutch in the torque converter. To check this out, accelerate the engine in N. If it reaches high rpm normally, the trouble could be that the stator clutch is slipping. Check for poor performance in R and in D.

If the one-way clutch is locked up, the result will be limited high vehicle speed and high engine rpm. The engine and torque converter may overheat. The overheating may actually cause the converter to turn blue. If this happens, the torque converter must be replaced. Also, a locked-up or slipping one-way clutch requires a new torque converter. A one-way clutch cannot be repaired.

♦ 45-8 STALL TEST

The stall test checks the holding ability of the torque-converter stator clutch and the transmission clutches. The test is made by applying the car brakes, shifting to different selector-lever positions, and opening the throttle wide to check

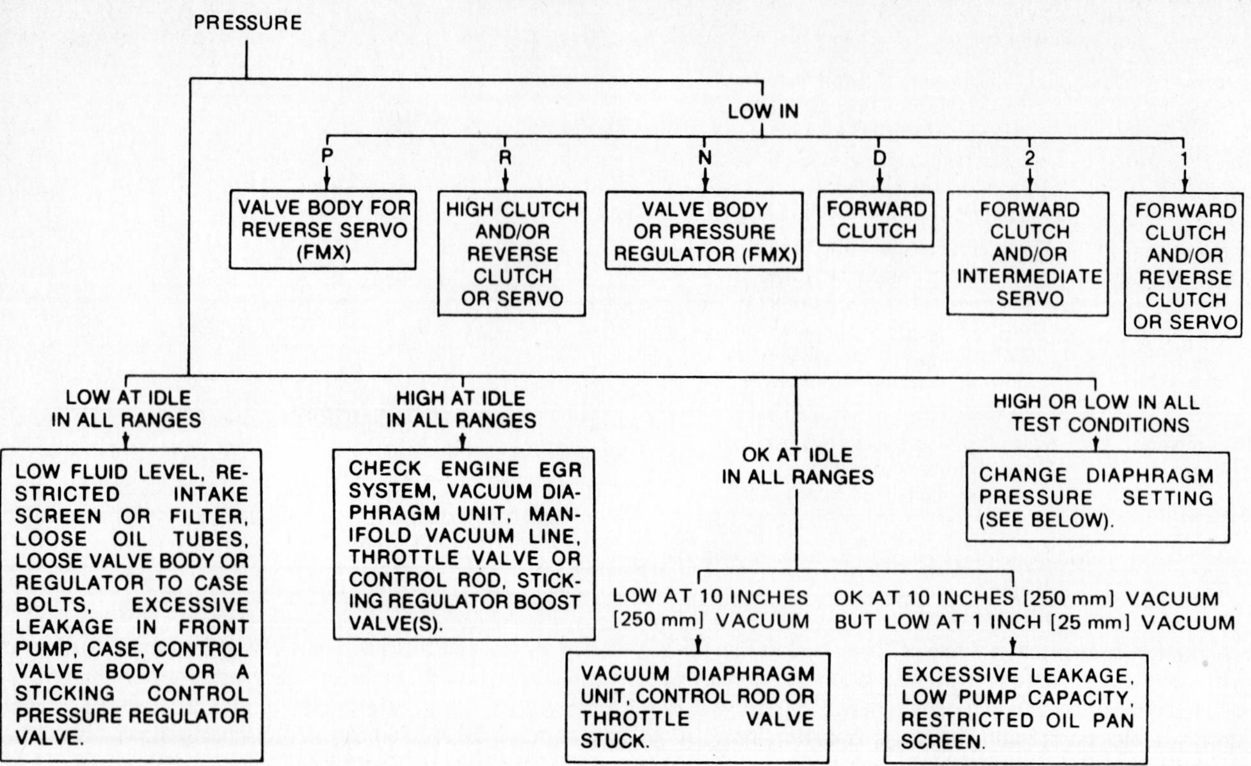

Fig. 45-5 A pressure-test diagnosis chart. *(Ford Motor Company)*

Range	Gear	Intermediate Clutch	Direct Clutch	Forward Clutch	Low-and-Reverse Clutch	Intermediate Overrun Roller Clutch	Low-and-Reverse Roller Clutch	Intermediate Overrun Band
Park, neutral		Off	Off	Off	Off	Ineffective	Ineffective	Off
Drive	First	Off	Off	On	Off	Locked	Locked	Off
	Second	On	Off	On	Off	Locked	Overrunning	Off
	Third	On	On	On	Off	Overrunning	Overrunning	Off
Second	First	Off	Off	On	Off	Locked	Locked	Off
	Second	On	Off	On	Off	Locked	Overrunning	On
Low	First	Off	Off	On	On	Locked	Locked	Off
Reverse		Off	On	Off	On	Ineffective	Ineffective	Off

Fig. 45-6 Clutch-engagement and band-application chart. *(Oldsmobile Division of General Motors Corporation)*

	Selector Position	Stall Speeds High (Slip)	Stall Speeds Low
Stall Test Results	D only	Transmission one-way clutch	
	D, 2, & 1	Forward clutch	
	D, 2, 1, & R	Control pressure test	Converter stator one-way clutch or engine performance
	R only	High and/or reverse clutch or reverse band	

Fig. 45-7 A stall-speed diagnosis chart. *(Ford Motor Company)*

the maximum engine speed or rpm (or the oil pressure, on some transmissions).

A tachometer is required to accurately measure engine speed. The throttle should not be held wide open for more than a few seconds. Also, the front wheels should be blocked in addition to applying the car brakes. Be sure no one is standing in front or in back of the car! It could lurch forward or backward as engine speed reaches maximum, in spite of the brakes and blocks. (Not all manufacturers recommend the stall test.)

Figure 45-7 is a chart showing how to diagnose band, clutch, and torque-converter troubles based on stall-test results.

♦ 45-9 ON-THE-CAR REPAIRS

Each make and model of automatic transmission and transaxle has certain repairs that can be made with the transmission in the car. The manufacturer's service manual for the unit you are about to service describes the operations that can be performed without removing the transmission or transaxle from the car. On most cars, these include linkage and band adjustments (♦45-10). However, if the repair is not one that can be done on the car, the unit must be removed.

♦ 45-10 LINKAGE AND BAND ADJUSTMENTS

There are basically two linkage adjustments. These are the shift-linkage adjustment and the throttle- or kickdown-linkage adjustment.

1. Shift-Linkage Adjustment When the selector lever is moved into any position, the manual valve in the transmission valve body must be exactly centered in its selected position. If the manual valve is not properly positioned, delays in shifting and slipping while operating can occur. Slipping of clutch plates or bands can soon damage the transmission so seriously that an overhaul will be required. Adjustment procedures vary with different car and transmission models. The procedure is in the manufacturer's service manual.

2. Throttle- or Kickdown-Linkage Adjustment The throttle or kickdown linkage causes the transmission to downshift if the throttle is opened wide (within a certain speed range). Basically, the adjustment is correct if the throttle linkage causes the kickdown linkage to produce the downshift when the throttle is opened to the full-throttle-stop position.

3. Band Adjustments Some automatic transmissions require periodic band adjustments if the transmission is used in severe service. Also, bands are always adjusted when the transmission is overhauled. Bands are adjusted by tightening and then loosening the band-adjusting screw (Fig. 44-22) a specified number of turns. However, all bands are not adjustable. If a band is loose and worn through the lining, the band must be replaced.

♦ 45-11 AUTOMATIC-TRANSMISSION OR TRANSAXLE OVERHAUL

Rebuilding procedures vary considerably from model to model. These procedures also vary according to whether you are replacing one defective component or performing a complete overhaul. In the complete overhaul, you will discard all old gaskets, oil seals, O rings, metal sealing rings, clutch friction and steel disks, filters, and modulators. You will also inspect and replace, if needed, such parts as bands, bushings, pumps, gears, governor, linkage, and converter. The complete procedures are in the manufacturer's service manual.

♦ 45-12 REMOVING AND INSTALLING AUTOMATIC TRANSMISSIONS AND TRANSAXLES

The removal and installation procedure varies with different cars. In general, here are the items you may have to disconnect or remove. First, open the hood and disconnect the battery ground cable. This eliminates the possibility of causing an electrical short-circuit. Other parts that may have to be removed or disconnected include the following:

- ♦ Starting motor
- ♦ Backup-light wire
- ♦ Neutral-park starting switch
- ♦ Vacuum line to modulator
- ♦ Oil lines to oil cooler
- ♦ Shift linkages
- ♦ Drive shaft
- ♦ Torque-converter flexplate bolts
- ♦ Housing-to-engine bolts

There may be additional items to be disconnected or removed. Specific removal procedures are described in the manufacturer's service manual.

♦ REVIEW QUESTIONS ♦

Select the *one* correct, best, or most probable answer to each question. You can find the answers in the section indicated at the end of each question.

1. Mechanic A says you cannot check the fluid level if the transmission is hot. Mechanic B says you can check fluid level either hot or cold. Who is right? (♦45-3)
   a. A only
   b. B only
   c. both A and B
   d. neither A nor B

2. Mechanic A says that if the transmission fluid is brown or black, the clutch plates probably have overheated and burned. Mechanic B says the bands have overheated and burned. Who is right? (♦45-3)
   a. A only
   b. B only
   c. both A and B
   d. neither A nor B

3. Which statement is correct? (♦45-10)
   a. some bands are not adjustable
   b. bands need adjustment in severe service
   c. if a band is loose and worn through the lining, replace the band
   d. all of the above
4. Mechanic A says the purpose of the throttle or kickdown linkage is to cause the throttle to kick back when the transmission upshifts. Mechanic B says the linkage causes the transmission to down-shift if the throttle is wide open. Who is right? (♦45-10)
   a. A only
   b. B only
   c. both A and B
   d. neither A nor B
5. Severe service that requires periodic changing of the fluid and filter includes (♦45-2)
   a. taxi service
   b. trailer pulling
   c. stop-and-go delivery service
   d. all of these
6. The purpose of the road test is to check for slipping, flare-up, or improper shifting with the selector lever in (♦45-7)
   a. direct drive
   b. second
   c. reverse
   d. all of these
7. Mechanic A says that if the fluid is black and has particles of dirt in it, most manufacturers recommend that the torque converter be discarded. Mechanic B says you can always flush out the converter and use it again. Who is right? (♦45-3)
   a. A only
   b. B only
   c. both A and B
   d. neither A nor B
8. The shift linkage is connected between the (♦45-10)
   a. selector lever and throttle
   b. throttle lever and kickdown lever
   c. selector lever and manual valve
   d. manual valve and throttle valve
9. Weak clutch springs cause the clutch to (♦45-7)
   a. engage quickly
   b. disengage slowly
   c. both a and b
   d. neither a nor b
10. Bands are adjusted by (♦45-10)
    a. tightening the adjusting screw a specified number of turns
    b. installing an oversize drum
    c. tightening and then loosening the adjusting screw a specified number of turns
    d. installing an oversize band

CHAPTER 46
Drive Lines and Universal Joints

After studying this chapter, and with the proper instruction and equipment, you should be able to:

1. Explain the basic differences between the drive lines for rear-wheel-drive and front-wheel-drive vehicles.
2. Discuss the purpose and operation of drive lines.
3. Describe the construction of different types of universal joints and explain how each works.
4. Explain the construction and purpose of slip joints.
5. Replace a universal joint.

♦ 46-1 FRONT-WHEEL DRIVE AND REAR-WHEEL DRIVE

Rear-wheel-drive vehicles have a long drive line, or shaft, extending from the transmission at the front to the differential at the rear. Front-wheel-drive vehicles usually have the engine-transaxle assembly mounted sideways (transversely). They have short drive shafts extending from the transaxle to the front wheels.

Figures 39-1 to 39-3 compare the two arrangements. Figures 10-3 and 10-10 show the mounting and drive arrangements for transversely mounted engines in front-wheel-drive vehicles. Following sections describe the construction of each type of drive line and the power flow through it.

♦ 46-2 DRIVE LINE—REAR-WHEEL DRIVE

The drive line (Fig. 39-2) is also called the *propeller shaft* and the *drive shaft*. In the front-engine, rear-wheel-drive vehicle, it connects the transmission main or output shaft to the differential. The differential connects to the two rear wheels. Therefore, the rotary motion of the transmission output shaft is carried to the differential and, from there, to the wheels. The drive-line design must take two facts into account. First, the engine and transmission are more or less rigidly connected to the car body or frame. Second, the rear-axle housing (with wheels and differential) is attached to the car body or frame through springs. This means two things:

1. The drive line must change in length as the wheels move up and down.
2. The angle of drive must change as the wheels move up and down.

Figure 46-1 shows how the length of the drive line and the angle of drive change as the wheels move up and down. At the top, the wheels and differential are shown at their up position. The angle of drive is small, and the drive line is at its maximum length. In the lower part of Fig. 46-1, the differential and wheels have moved to their lowest position. This is the position they take when the wheels drop into a depression in the road. In this position, the drive angle is increased. Also, the drive-line length is reduced. This occurs because as the

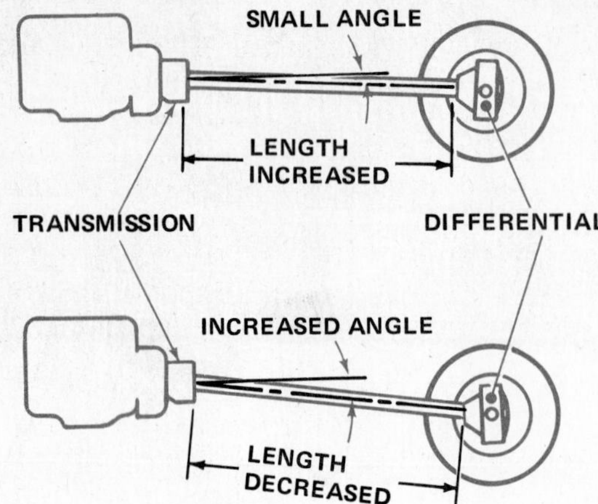

Fig. 46-1 The rear axle housing, with differential and wheels, moves up and down. As it does, the angle between the transmission output shaft and the drive line changes. The length of the drive shaft also changes. The drive shaft shortens as the angle increases because the rear axle and differential move in a shorter arc than the drive shaft. The center point of the axle-housing arc is the rear spring or control-arm attachments to the frame.

rear wheels and differential swing down, they also move forward. The axle housing is attached to the car body or frame by springs or control arms. This makes them move in a shorter arc than the drive line.

To allow for these two variations, two different kinds of joints are necessary. The drive line must include two or more *universal* joints and a *slip* joint (Fig. 46-2). The universal joints take care of the change in drive angle. The slip joint takes care of the change in length. These joints are described in following sections.

The drive shaft is a hollow shaft, or tube, on most cars with a front-mounted engine and rear-wheel drive. Some drive lines have two-piece drive shafts, with a third universal joint between each section (Fig. 46-3). The center universal joint in Fig. 46-3 is a *constant-velocity universal joint.* It includes a center bearing and support assembly which supports the center of the drive line. Many larger cars have a single-piece drive shaft with constant-velocity universal joints at each end.

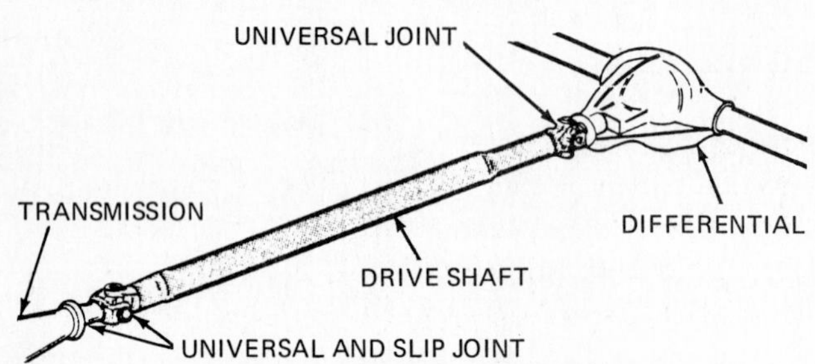

Fig. 46-2 Relationship of the drive shaft to the transmission and differential. This is a one-piece drive shaft supported at the front and rear by universal joints.

♦ 46-3 UNIVERSAL JOINTS

The universal joint allows driving power to be carried through two shafts that are at an angle to each other. A simple universal joint is shown in Fig. 46-4. It is a double-hinged joint consisting of two Y-shaped yokes and a cross-shaped member called the *spider.* One of the yokes is on the driving shaft, and the other is on the driven shaft. The four arms of the spider, called *trunnions,* are assembled into bearings in the ends of the two shaft yokes. The driving shaft and yoke cause the spider to rotate. The other two trunnions of the spider cause the driven shaft to rotate. When the two shafts are at an angle to each other, the bearings in the yokes permit the yokes to swing around on the trunnions with each revolution. A variety of universal joints have been used on automobiles. The types include the spider-and-two-yoke, ball-and-trunnion, and several types of constant-velocity universal joints.

The spider-and-two-yoke design is essentially the same as the simple universal joint discussed above, except that needle

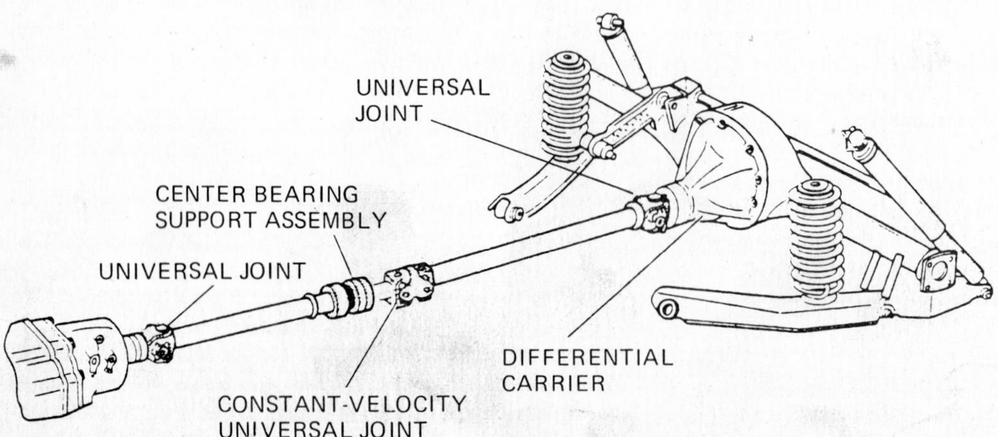

Fig. 46-3 A two-piece drive shaft that uses three universal joints. The front section is supported at its rear by a center support bearing. A constant-velocity universal joint is located behind the bearing. *(Buick Motor Division of General Motors Corporation)*

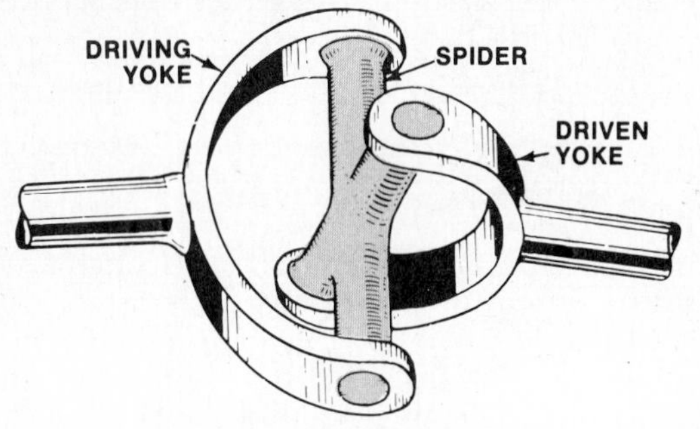

Fig. 46-4 A simple universal joint.

bearings are used (Fig. 46-5). There are four needle bearings, one for each trunnion of the spider. The bearings are held in place by snap rings that drop into undercuts in the yoke-bearing holes.

The universal joints shown in Fig. 46-5 are not constant-velocity joints. If the two shafts are at an angle, the driven shaft will be given a variable speed, or velocity. With each revolution, the shaft will speed up and slow down slightly two times. The greater the angle, the greater the velocity variation. This can cause wear on the bearings and gears in the differential because the load on them varies, or pulsates. To eliminate this, constant-velocity joints are used. In these, the driven shaft turns at the same speed as the drive shaft, with no variation in the velocity, or speed.

Constant-velocity universal joints are shown on drive lines in Figs. 46-3 and 46-6. The drive line in Fig. 46-6 has three universal joints, one at each end and one in the center connecting the two drive shafts. Figure 46-7 is a disassembled constant-velocity universal joint.

The double-Carden type of constant-velocity universal joint shown in Fig. 46-7 includes two separate universal joints.

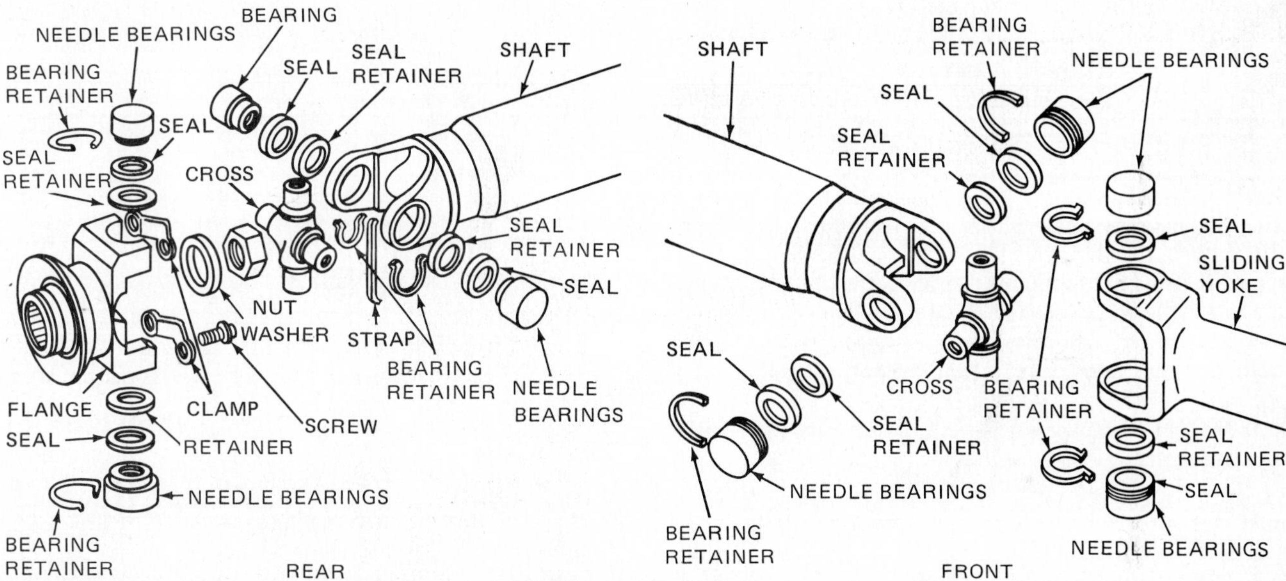

Fig. 46-5 Disassembled universal joints at the front and rear of a drive shaft. *(Chrysler Corporation)*

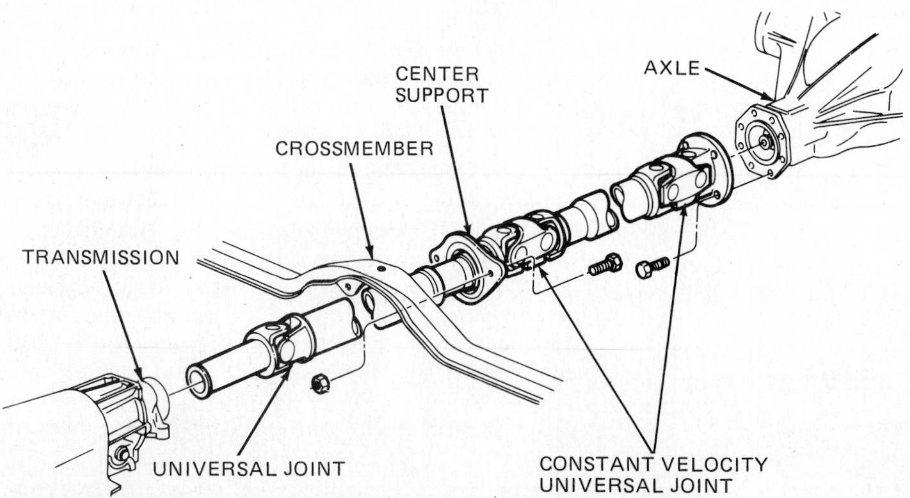

Fig. 46-6 A two-piece drive shaft with three universal joints. Two are constant-velocity units. *(Cadillac Motor Car Division of General Motors Corporation)*

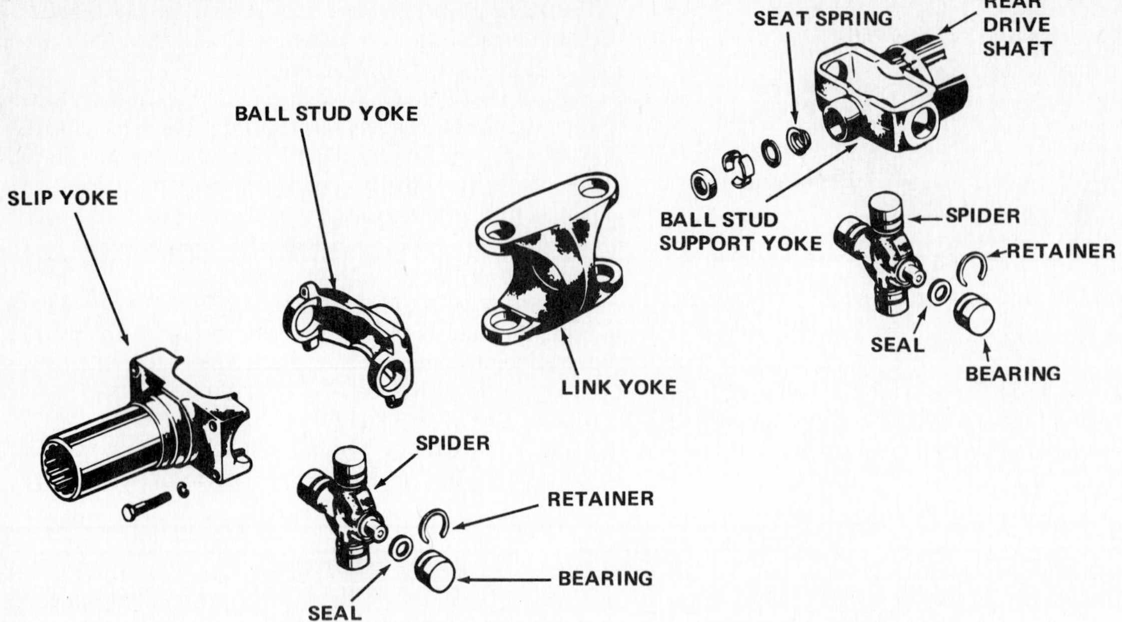

Fig. 46-7 Disassembled constant-velocity universal joint. *(Cadillac Motor Car Division of General Motors Corporation)*

They are linked by a ball and socket. The ball and socket split the angle of the two drive shafts between the two universal joints. Because the two joints operate at the same angle (half the total), the variations that could result from a single joint are canceled out. The acceleration resulting at any instant from the action of one joint is nullified by the deceleration of the other joint.

Other types of universal joints used in front-wheel-drive vehicles are described in ♦46-6.

♦ 46-4 SLIP JOINT

A slip joint is shown in Fig. 46-8. The slip joint has outside splines on one shaft and matching internal splines on a mating hollow shaft. The splines cause the two shafts to rotate together but permit the two to move endwise in relation to each other. This allows changes in the length of the drive shaft as the rear axle moves toward or away from the car body or frame.

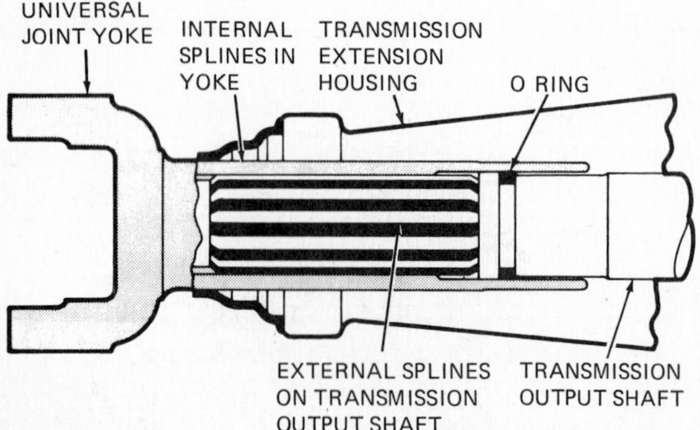

Fig. 46-8 Slip joint used to compensate for changes in length of the drive line.

♦ 46-5 CENTER SUPPORT

Many vehicles have drive lines which are supported at the center, as shown in Fig. 46-6. The center support prevents angular movement, or "whipping" of the drive line. The center support has a bearing in which the drive line can rotate. However, this arrangement requires an additional universal joint back of the center support.

♦ 46-6 DRIVE LINES—FRONT-WHEEL DRIVE

Front-wheel drive has become very popular in recent years. With front-wheel drive, the long drive shaft to the rear axles is eliminated. This also eliminates the need for the long tunnel in the floor pan of the car required for the drive shaft. Figure 46-9 shows the engine and power-train layout used in many cars today. Figure 46-10 shows the front suspension and drive shafts for a front-wheel-drive car.

Figure 46-11 shows the drive shafts in complete detail. The left shaft has two constant-velocity (CV) joints. The right shaft has three universal joints. The two outer ones are CV joints.

Driving the front wheels makes the front drive and suspension systems more complicated. The front wheels must swing from side to side so that the car can be steered. Regardless of how far the front wheels are turned from straight ahead, power must continue to flow smoothly to them. Also, they must be supported so that they can drive. The arrangement requires two CV universal joints in each front-wheel drive shaft (Fig. 46-11). These joints are protected by rubber boots (called *bellows* in Fig. 46-11).

The inner ends of the two drive shafts are connected to the differential in the transaxle. Differentials are covered in Chap. 47.

Figure 46-12 shows one type of CV universal joint used in many front-wheel drive cars. The outer races have long grooves that the balls can roll in to accommodate changes in length of the shafts. The drive is through the inner race and

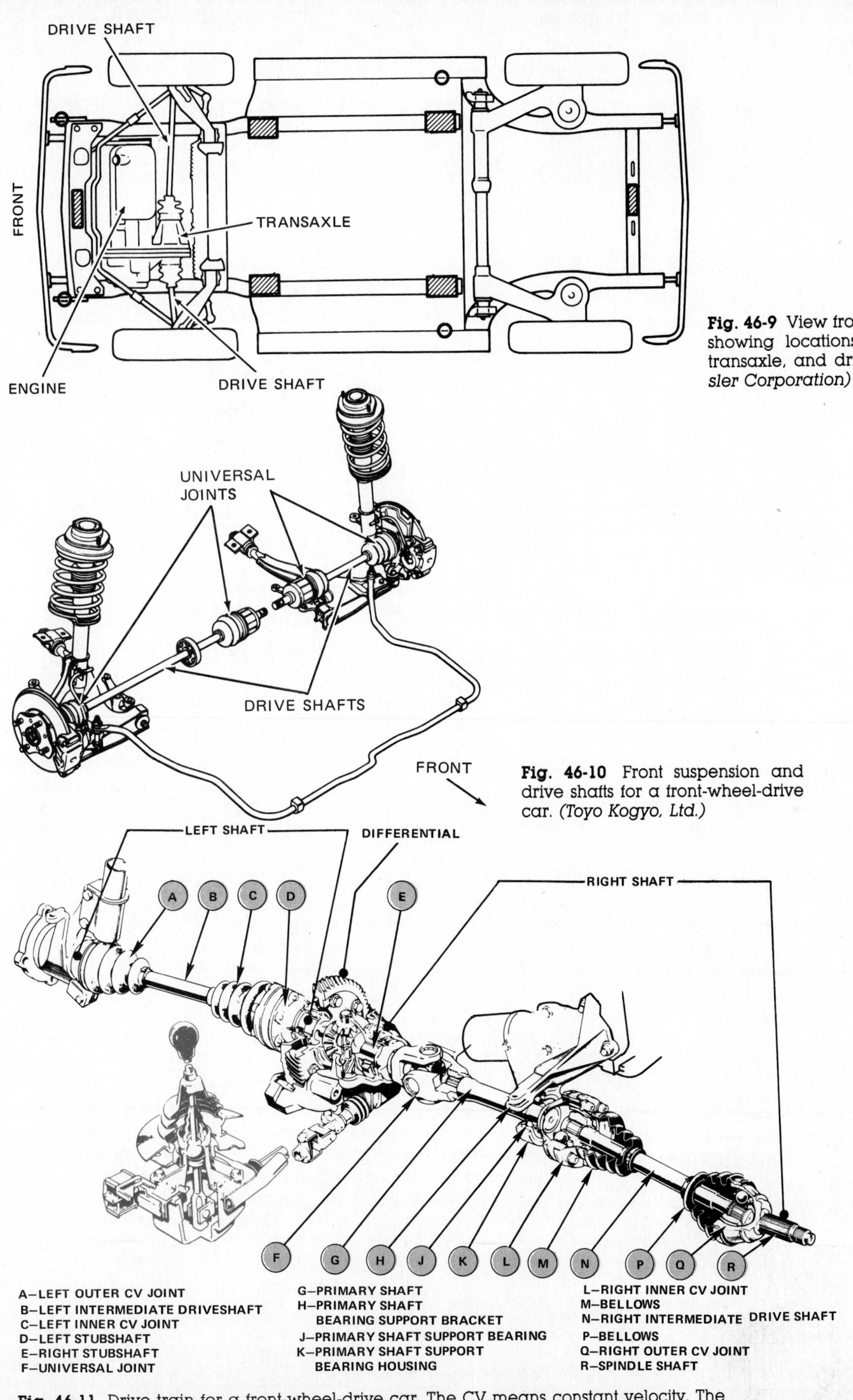

Fig. 46-9 View from under the car, showing locations of the engine, transaxle, and drive shafts. *(Chrysler Corporation)*

Fig. 46-10 Front suspension and drive shafts for a front-wheel-drive car. *(Toyo Kogyo, Ltd.)*

Fig. 46-11 Drive train for a front-wheel-drive car. The CV means constant velocity. The drive train includes the transmission and differential. *(Ford Motor Company)*

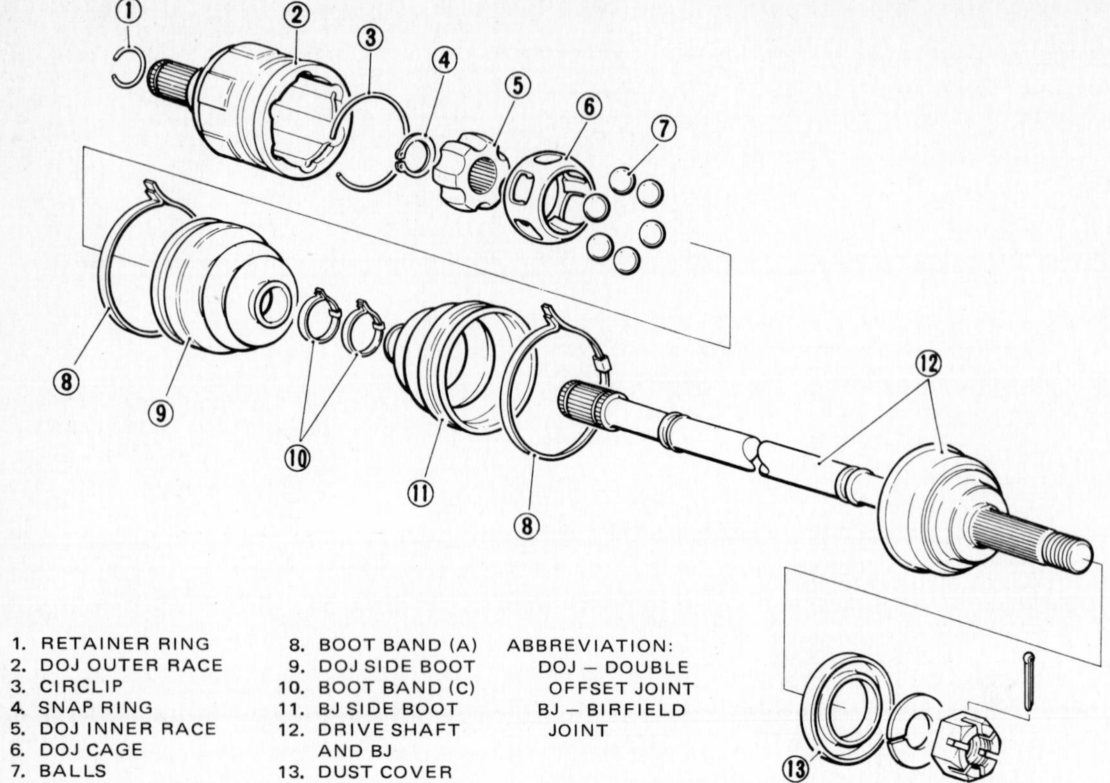

Fig. 46-12 Disassembled drive shaft for a front-wheel-drive car. *(Chrysler Corporation)*

the balls. The cage holds the balls in place in the inner race. As the drive angle changes, the balls assume new positions on the inner race. Power flows from the transaxle to the outer race, then through the balls to the inner race. The inner race is splined to the drive shaft. Typically, two different types of universal joints are used on each drive shaft. The inboard universal joint usually also acts as a slip joint.

♦ 46-7 REAR DRIVE WITH REAR-MOUNTED ENGINE

Some cars have the engine mounted at the rear. They use short stub shafts and universal joints to carry the engine power to the two rear wheels. The rear wheels are independently suspended. Some Volkswagen models use this type of rear-wheel drive (Fig. 46-13).

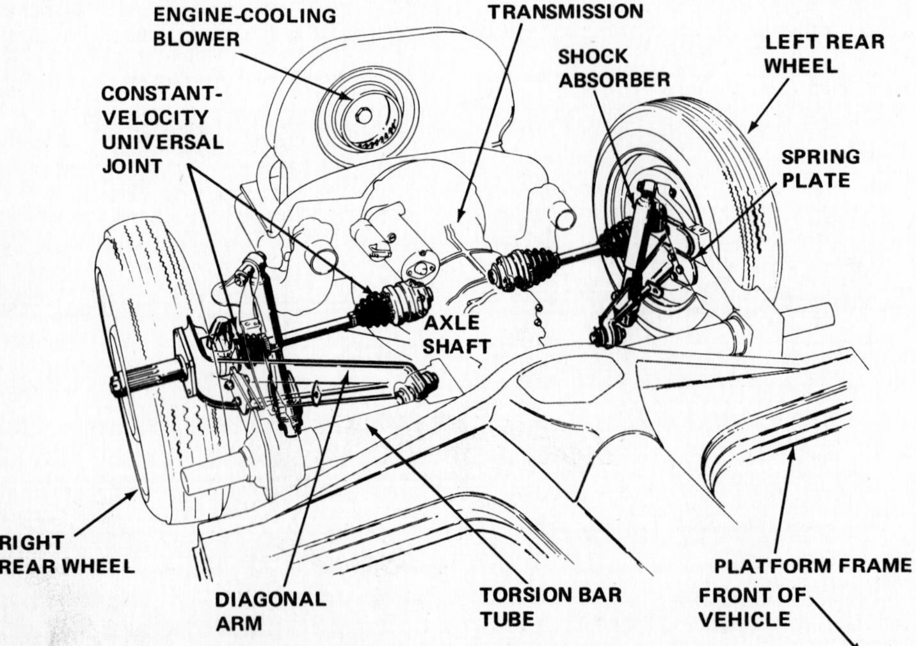

Fig. 46-13 Rear-suspension and drive-line components used in some Volkswagen models that have rear-mounted engines and rear-wheel drive. *(Volkswagen of American, Inc.)*

♦ 46-8 FOUR-WHEEL DRIVE

Some vehicles, especially those used off-road, can drive all four wheels (Figs. 43-1 and 43-2). Each drive shaft to the front and rear axles has universal joints and slip joints.

SERVICE

♦ 46-9 SERVICING THE UNIVERSAL JOINT AND DRIVE SHAFT

Universal joints and drive shafts do not require service in normal use. Some universal joints are prelubricated for life during original assembly. When wear or noise occurs, the universal joints should be replaced. However, some manufacturers recommend lubricating the universal joints every time a chassis lubrication is performed.

The drive shaft and universal joints are carefully balanced during original assembly. Always mark the position and alignment of the parts before disassembly. Then after reassembly and installation, they should still be in balance. A drive shaft that is out of balance can often be balanced by installation of two worm-type hose clamps.

When you have a drive shaft or universal joint to service, follow the procedures in the manufacturer's service manual.

♦ REVIEW QUESTIONS ♦

Select the *one* correct, best, or most probable answer to each question. You can find the answers in the section indicated at the end of each question.

1. The drive shaft in the front-engine, rear-wheel-drive car has at least (♦46-2)
   a. one universal joint
   b. two universal joints
   c. three universal joints
   d. two slip joints
2. Most drive shafts have two types of joints (♦46-2)
   a. U and universal
   b. transmission and differential
   c. slip and spline
   d. universal and slip
3. The basic purpose of the universal joint is to allow the (♦46-2)
   a. drive shaft length to change
   b. drive shaft to be supported at the middle
   c. drive angle to change
   d. drive shaft to be removed and installed easily
4. The two most common types of drive arrangements are the front-engine, rear-wheel drive and the (♦46-1):
   a. front-engine, front-wheel drive
   b. rear-engine, rear-wheel drive
   c. front-engine, four-wheel drive
   d. rear-engine, four-wheel drive
5. The typical front-engine, front-wheel-drive system has (♦46-6):
   a. two universal joints
   b. three universal joints
   c. four universal joints
   d. five universal joints

CHAPTER 47
Differentials and Drive Axles: Construction and Service

After studying this chapter, and with the proper instruction and equipment, you should be able to:

1. Explain the construction and operation of a differential.
2. Identify the parts of a disassembled differential.
3. Describe the operation of a limited-slip differential.
4. Discuss the construction and action of the front axle in a four-wheel-drive vehicle.
5. Name three types of hubs used on the front wheels of four-wheel-drive vehicles, and describe the operation of each.
6. Determine the cause of noise from a drive axle.
7. Disassemble, assemble, and adjust a differential.

♦ 47-1 DIFFERENTIAL APPLICATIONS

Differentials are used in the rear drive axle of front-engine, rear-wheel-drive vehicles. Differentials are also used in the transaxles on front-engine, front-wheel-drive vehicles. Also, four-wheel-drive vehicles have differentials at both the front and the rear wheels. In addition, some four-wheel-drive vehicles have a third differential in the transfer case (♦43-4 and Fig. 43-1).

This chapter discusses the differential used in the front-engine, rear-wheel-drive vehicle. This differential is located at the rear of the vehicle, between the rear wheels (Fig. 39-2). The drive line that carries engine power from the transmission to the differential is described in Chap 46.

Note Manufacturers' service manuals usually refer to the differential and drive axle as the *rear axle* or *rear-axle assembly*. By this, they mean the rear-axle housing, the wheel axles, and the differential assembly. In this book, the term *differential* is used to designate the differential assembly itself.

♦ 47-2 FUNCTION OF DIFFERENTIAL

A differential is required to compensate for the difference in distance that the drive wheels travel when the car rounds a curve. If a right-angle turn were made with the inner rear wheel turning on a 20-foot [6.1-m] radius, the inner rear wheel would travel about 31 feet [9.5 m] while the outer rear wheel would travel about 39 feet [12 m] (Fig. 47-1). The differential permits power flow to both drive wheels while allowing the wheels to turn different distances when the car is rounding a curve.

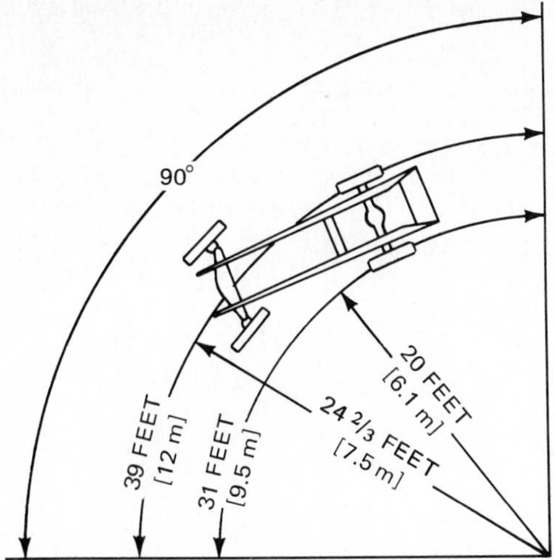

Fig. 47-1 Difference in wheel travel as a rear-wheel-drive car makes a 90 degree turn with the inner wheel turning on a 20-foot (6.1-m) radius.

Suppose both rear wheels are attached to the ends of a solid shaft. When the shaft is turned by gearing from the drive shaft, each tire would skid an average of 4 feet [1.22 m] while making the turn described above. The tires will wear quickly and the car will be difficult to control during turns. The rear wheels will slide and start a skid. However, the differential avoids these problems. It allows the outer rear wheel to turn faster than the inner rear wheel during a turn.

Front-wheel-drive cars have a differential between the front wheels. It usually is part of the transaxle (Figs. 10-3 and 10-10). A differential is also used in the transfer case of many vehicles that have full-time four-wheel drive (Chap. 43).

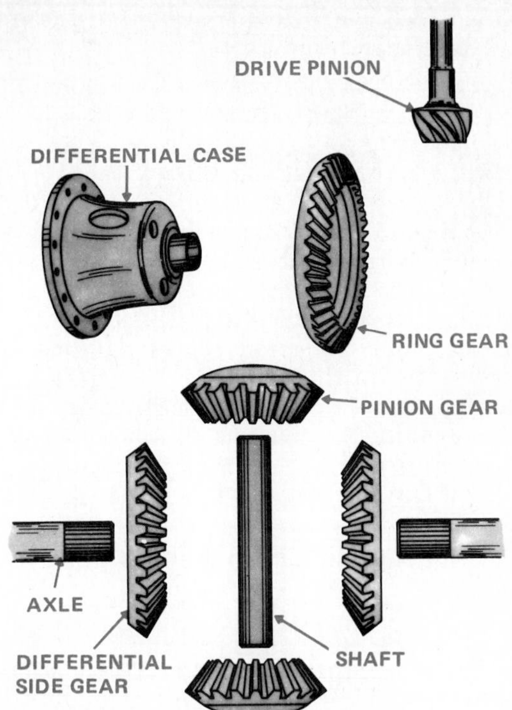

Fig. 47-3 Basic parts of a differential.

♦ 47-3 CONSTRUCTION OF DIFFERENTIAL

Figure 47-2 is a cutaway differential. Figure 47-3 shows the basic parts of the type of differential used in rear-wheel-drive cars. On the inner ends of each axle is a small bevel gear called a *differential side gear*. When any two bevel gears are put together so that their teeth mesh, the driving and driven shafts can be at a 90° angle (Fig. 47-3).

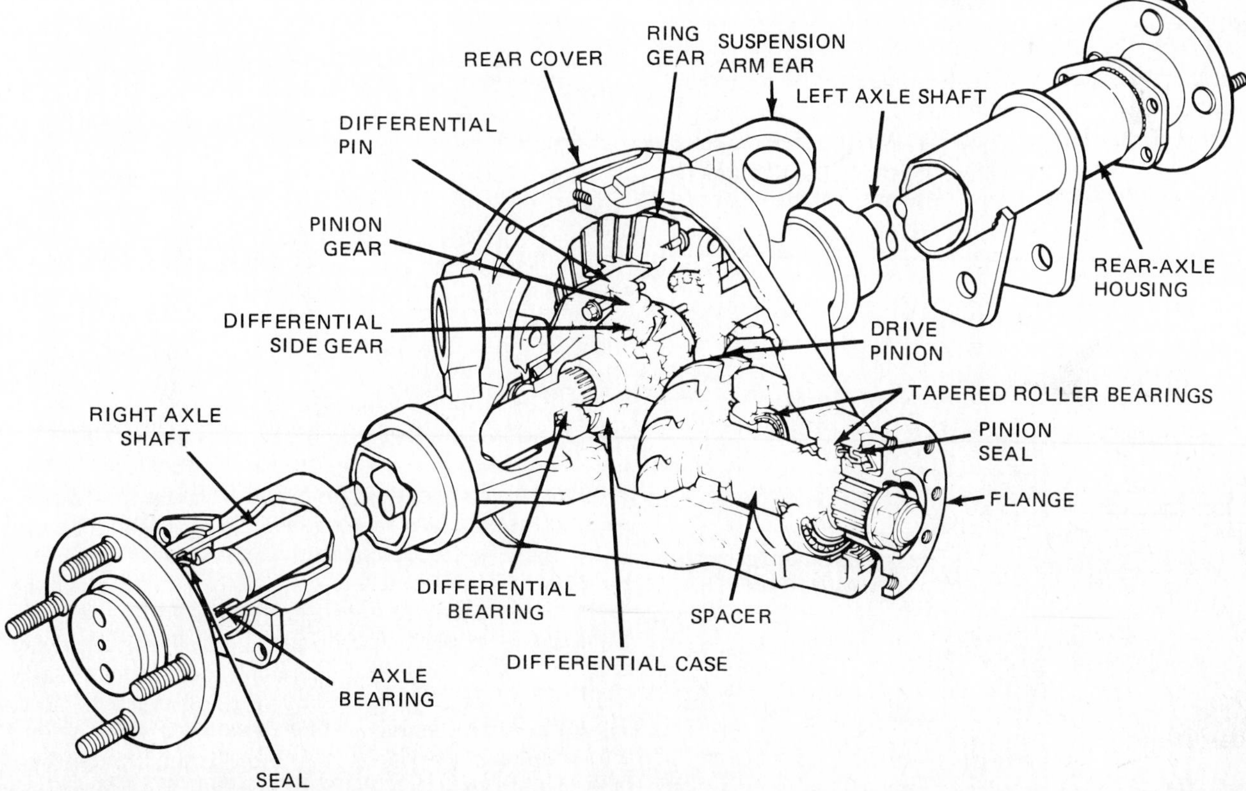

Fig. 47-2 A cutaway differential and rear axle. *(Ford Motor Company)*

Figure 47-3 shows the essential parts of a differential. The parts are separated so that they can be seen clearly. Refer to Fig. 47-3 as assembling the differential is described.

First, we add the differential case to the two wheel axles and differential side gears. The differential case has bearings that permit it to rotate independently of the two axles. Next, we add the two pinion gears and their supporting shaft, called the *pinion shaft*. The shaft fits into the differential case. The two pinion gears are held in place by the pinion shaft. They mesh with the two differential side gears, which are attached to the inner ends of the axle shafts.

Next, we add the ring gear. It is bolted to a flange on the differential case. The differential case is rotated by the ring gear which is attached to the case. Finally, we add the drive pinion (Fig. 47-4). The drive pinion is assembled into the differential housing (Fig. 47-2) or *carrier*. A universal joint connects the drive shaft with the drive pinion, which meshes with the ring gear. Therefore, when the drive shaft turns, the drive pinion rotates. This rotates the ring gear.

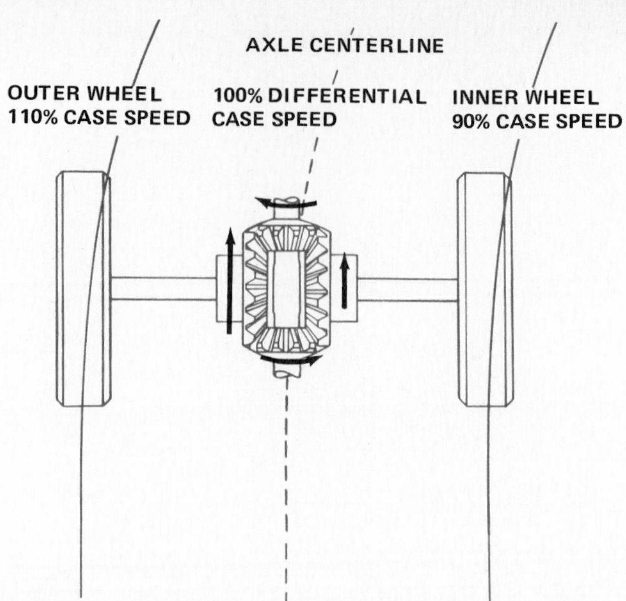

Fig. 47-5 Differential action on turns. *(Chevrolet Motor Division of General Motors Corporation)*

♦ 47-4 OPERATION OF DIFFERENTIAL

When the car is on a straight road, the ring gear, differential case, differential pinion gears, and two differential side gears all turn as a unit. The two differential pinion gears do not rotate on the pinion shaft. This is because they exert equal force on the two differential side gears. As a result, the side gears turn at the same speed as the ring gear, which causes both drive wheels to turn at the same speed also.

However, when the car begins to round a curve, the differential pinion gears rotate on the pinion shaft. This permits the outer wheel to turn faster than the inner wheel.

Suppose one wheel turns slower than the other as the car rounds a curve. As the differential case rotates, the pinion gears must rotate on their shaft. This occurs because the pinion gears must walk around the slower-turning differential side gear. Therefore, the pinion gears carry additional rotary motion to the faster-turning outer wheel on the turn. The action in a typical turn is shown in Fig. 47-5. The differential-case speed is considered to be 100 percent. The rotating action of the pinion gears carries 90 percent of this speed to the slower-rotating inner wheel. It sends 110 percent of the speed to the faster-rotating outer wheel.

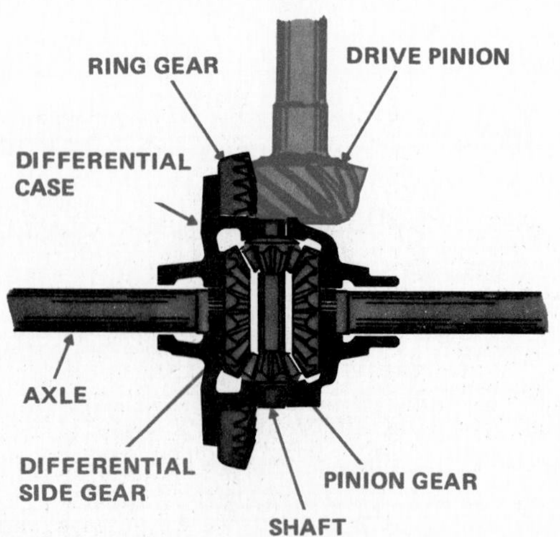

Fig. 47-4 The basic differential, assembled.

This is how the differential allows one drive wheel to turn faster than the other. Whenever the car goes around a turn, the outer drive wheel travels a greater distance than the inner drive wheel. The two pinion gears rotate on their shaft and send more rotary motion to the outer wheel.

When the car moves down a straight road, the pinion gears do not rotate on their shaft. They apply equal torque to the differential side gears. Therefore, both drive wheels rotate at the same speed.

♦ 47-5 DIFFERENTIAL GEARING

Since the ring gear has many more teeth than the drive pinion, a considerable gear reduction is produced in the differential. The gear ratios vary on different cars, depending on car and engine design. Ratios from 2:1 upward to about 4:1 are used on passenger cars. This means that the ring gear has from two to four times as many teeth as the drive pinion. Therefore the drive pinion must rotate from two to four times (according to gear ratio) to cause the ring gear to rotate once. For heavy-duty trucks, ratios of about 9:1 may be used. These high ratios require the use of *double-reduction* gearing (Fig. 47-6).

The gear ratio in the differential is usually referred to as the *axle ratio*. However, it would be more accurate to call it the *differential ratio*.

Figure 47-7 shows gear-tooth nomenclature. The mating teeth to the left illustrate clearance and backlash. The tooth to the right has the various tooth parts named. *Clearance* is the distance between the top of the tooth of one gear and the valley between adjacent teeth of the mating gear. *Backlash* is the distance between adjacent meshing teeth in the driving and driven gears. It is the distance one gear can rotate backward, or backlash, before it will cause the other gear to move. The *toe* is the smaller section of the gear tooth which is nearest the center of the ring gear. The *heel* is the larger section of the gear tooth, which is farthest from the center of the ring gear. the toe end of the tooth is smaller than the heel end.

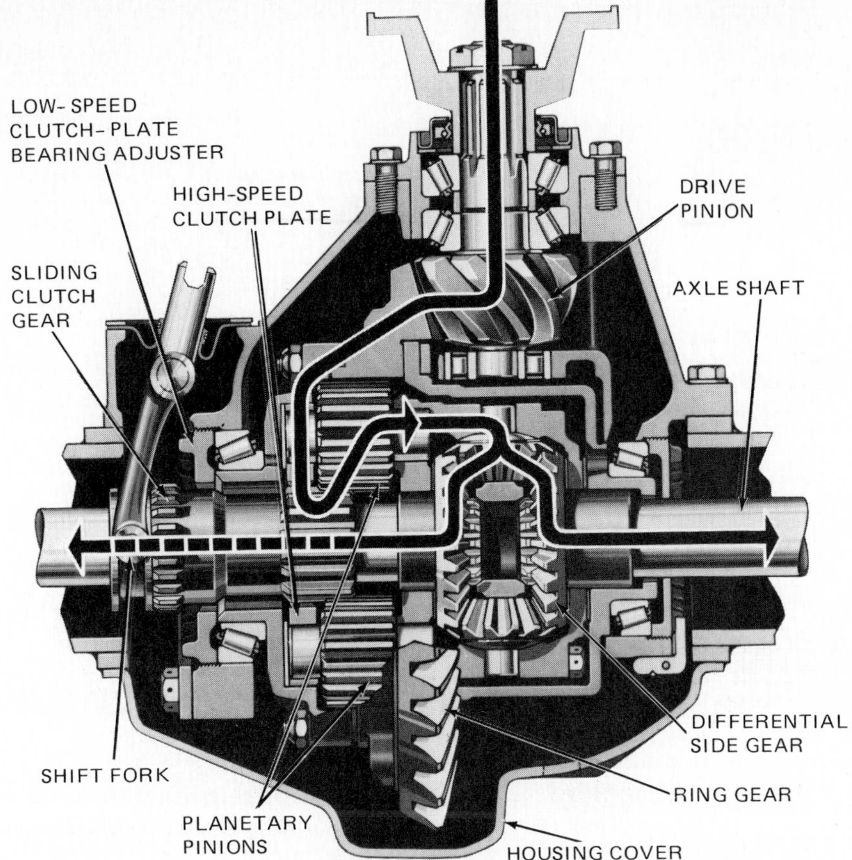

Fig. 47-6 A double-reduction differential. The extra gearing and the shifter allow the differential to offer two gear reductions, the standard, and the double-reduction with the extra gears in the gear train. *(Axle Division, Eaton Corporation)*

♦ 47-6 LIMITED-SLIP DIFFERENTIALS

The standard differential delivers the same amount of torque to each rear wheel when both wheels have equal traction. When one wheel has less traction than the other—for example, when one wheel slips on ice—the other wheel cannot deliver torque. All the turning effort goes to the slipping wheel. To provide good traction even though one wheel is slipping, a *limited-slip differential* is used in many cars. It is very similar to the standard unit but has some means of preventing wheel spin and loss of traction. The standard differential delivers maximum torque to the wheel with minimum traction. The limited-slip differential delivers maximum torque to the wheel with maximum traction.

One type of limited-slip differential is shown in Fig. 47-8. It has two sets of clutch plates. The ends of the pinion-gear shafts lie loosely in notches in the two halves of the differential case. Figure 47-9 is a sectional view of the limited-slip differential. During normal straight-road driving, the power flow is as shown in Fig. 47-10.

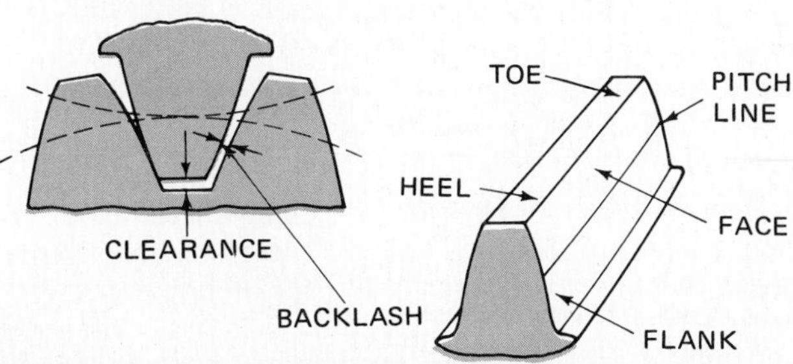

Fig. 47-7 Gear-tooth nomenclature. *(Chrysler Corporation)*

Note In Figs. 47-8 to 47-10, the ring gear is called the *axle drive gear*. The pinion gears are called the *differential pinions*. Different manufacturers use different names for the same parts.

The rotating differential case carries the pinion-gear shafts around with it. Because there is considerable side thrust, the pinion shafts tend to slide up the sides of the notches in the two halves of the differential case. As the pinion shafts slide up, they are forced outward. This force is carried to the two sets of clutch plates. The clutch plates lock the axle shafts to the differential case. Therefore both wheels turn.

Suppose one wheel encounters a patch of ice or snow and loses traction, or tends to slip. Then the force is released on

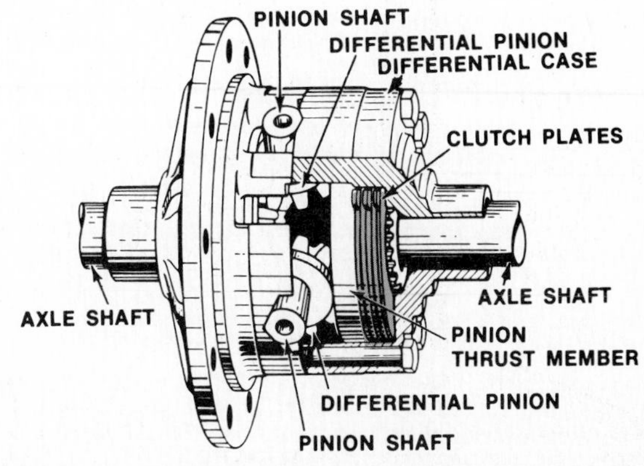

Fig. 47-8 A cutaway limited-slip differential. *(Chrysler Corporation)*

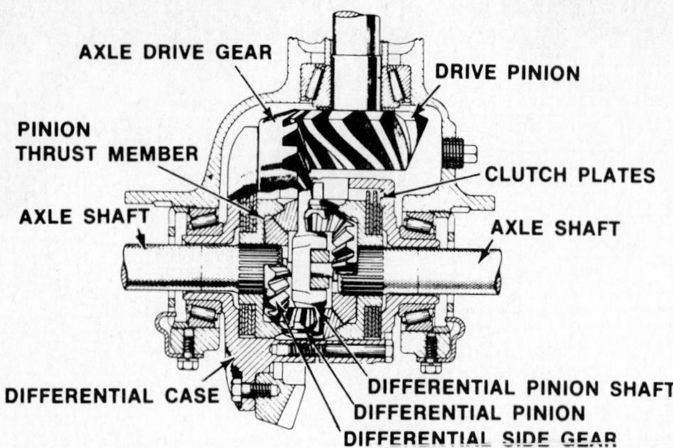

Fig. 47-9 Sectional view of a limited-slip differential. *(Chrysler Corporation)*

the clutch plates feeding power to that wheel. The torque goes to the other wheel, and the wheel on the ice does not slip.

During normal driving, if the car rounds a curve, force is released on the clutch for the inner wheel. Just enough force is released to permit some slipping. Figure 47-11 shows the action. This release of force permits the outer wheel to turn faster than the inner wheel.

There are two types of late-model limited-slip differentials. Some have spring-loaded clutch plates. Others have spring-loaded clutch cones. Figure 47-12 shows a disassembled clutch-cone differential. The action is the same as in the limited-slip differential described above. The basic difference is that the plates or cones are preloaded by springs. This provides a quicker locking action.

♦ 47-7 FOUR-WHEEL-DRIVE DRIVE AXLES

The most widely used application of four-wheel drive is in multipurpose utility vehicles, such as the American Motors Jeep, Chevrolet Blazer, and Ford Bronco. These vehicles, along with most other four-wheel-drive cars and trucks, normally drive the rear axle. But when four-wheel drive is en-

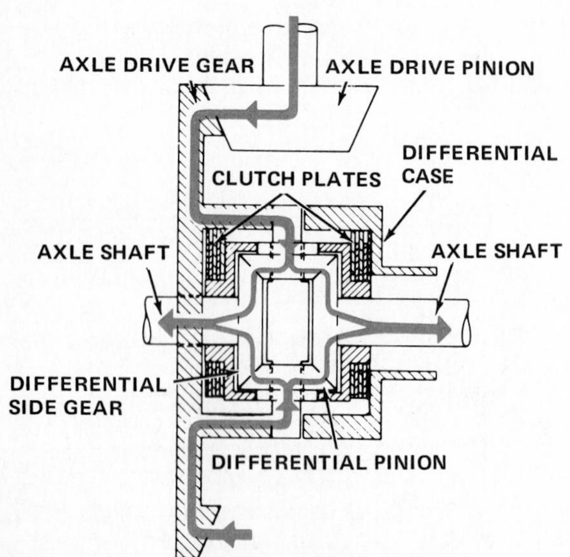

Fig. 47-10 Power flow through a limited-slip differential on a straightaway. *(Chrysler Corporation)*

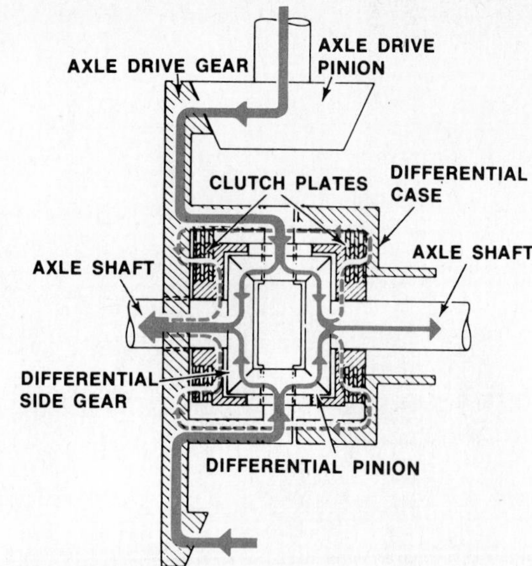

Fig. 47-11 Power flow through a limited-slip differential when rounding a turn. Heavy arrows show greater torque to the left axle. *(Chrysler Corporation)*

gaged through the transfer case, power is also delivered to the front axle.

Chapter 43 covers various types of transfer cases used in these vehicles. Differential action is covered earlier in this chapter. The design and operation of various types of universal joints are discussed in Chap. 46.

The front axle on most four-wheel-drive vehicles is very similar to the rear axle on vehicles with rear-wheel drive. As in the rear axle, the differential and front-axle shafts are carried inside a rigid tube, or axle housing (Fig. 47-13). However, the front axle is the steering axle. Therefore universal joints are located in the right and left axle shafts so that the outer ends can swing with the steering knuckle.

♦ 47-8 LOCKING HUBS

Three different types of hubs are used on front-drive axles of four-wheel-drive vehicles. These are the manual locking hub, the automatic locking hub with manual override, and the hub used with full-time four-wheel drive. Locking hubs are also called *freewheeling hubs*. Their purpose is to disengage the front axle and allow it to freewheel while the vehicle is operating in two-wheel drive. As a result, fuel economy and tire life are increased, and engine and transfer case wear is decreased.

A locking hub is a type of clutch that disengages the outer ends of the front axle shafts from the front wheel hubs. Then the axle shafts cannot turn, or *back-drive*, the front differential. Therefore the front differential does not turn the front drive shaft (from the transfer case to the front differential). The engine power saved by disengaging the front hubs has been reported to increase fuel economy by as much as 2 miles per gallon.

Many vehicles with part-time four-wheel drive are equipped with manual locking hubs. These hubs are locked or unlocked by the driver, who must turn a knob or lever at each wheel (Fig. 47-14). Another type of hub is the automatic locking hub. It engages when the transfer-case shift lever is moved to four-wheel drive. When the lever is in two-wheel drive, the front axle and drive shaft are not engaged and do

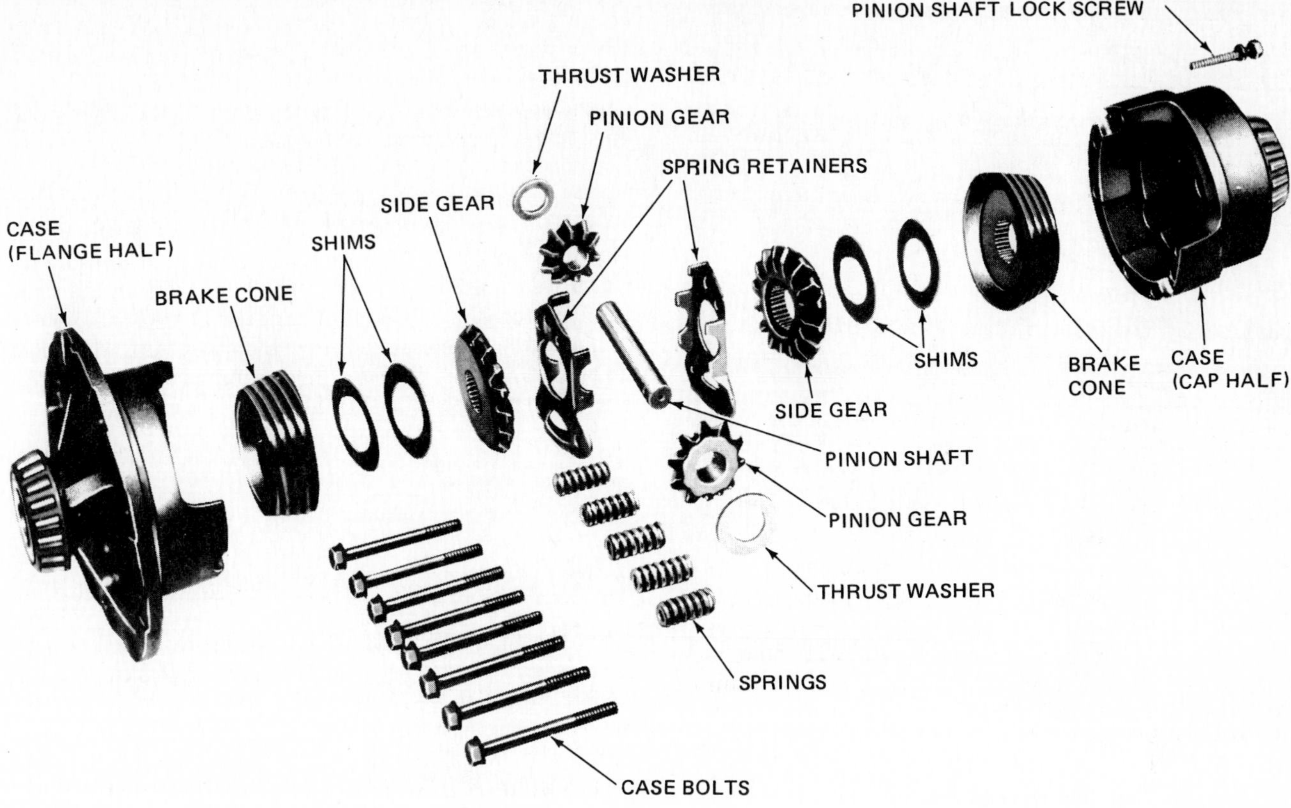

Fig. 47-12 Disassembled limited-slip differential using clutch cones. *(Chevrolet Motor Division of General Motors Corporation)*

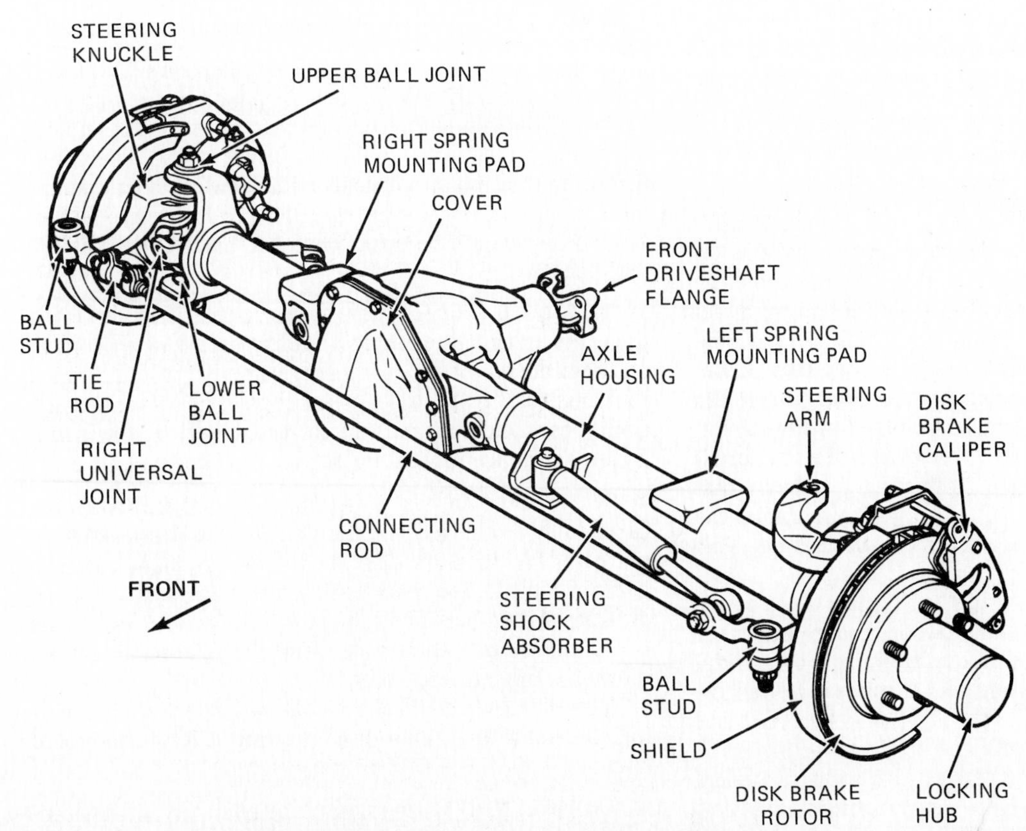

Fig. 47-13 Typical front-drive axle used on a four-wheel-drive vehicle. *(Chevrolet Motor Division of General Motors Corporation)*

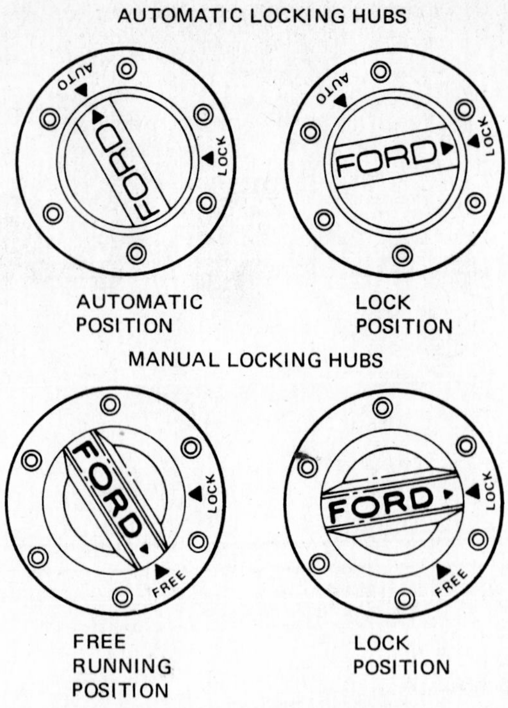

AUTOMATIC LOCKING HUBS

AUTOMATIC POSITION

LOCK POSITION

MANUAL LOCKING HUBS

FREE RUNNING POSITION

LOCK POSITION

Fig. 47-14 Positions of control knob on manual and automatic locking hubs. *(Ford Motor Company)*

not rotate. Automatic locking hubs can be manually locked by turning the control knob at each wheel (Fig 47-14).

Locking hubs are not used with some types of full-time front-wheel drive. The hubs are always engaged with the axle shafts. The interaxle differential in the transfer case prevents damage and undue wear of power-train parts. Other front hubs use overrunning clutches. These lock automatically when four-wheel drive is engaged and the front axle shafts rotate.

♦ 47-9 REAR-END TORQUE

The rotation of the drive shaft sends torque through the differential to the rear wheels. The wheels rotate and the car moves because torque is applied to them. To move the car forward, the torque rotates the wheels in the forward direction. At the same time, the torque also tries to rotate the differential housing in the opposite direction. This occurs because the ring gear is connected through other gears to the rear-wheel axles. The torque applied through the drive pinion forces the ring gear and wheels to rotate. It is the side thrust of the drive-pinion teeth against the ring-gear teeth that makes the ring gear rotate. This side thrust also causes the drive-pinion shaft to push against the shaft bearing. This pushes the differential housing up.

Here's another way to look at it. The ring gear gets a downward push from the drive pinion when in forward drive. Therefore, the drive pinion pushes up against the differential housing. The differential housing tries to rotate in a direction opposite to wheel rotation. The effect is called *rear-end torque*. The differential and axle housing must be braced to resist this. If the rear suspension uses leaf springs, the leaf springs provide the bracing. If the rear suspension uses coil springs, a set of control arms provides the bracing. This is covered in Chap. 48.

However, despite the bracing, the differential and axle housing and rear of the car react to rear-end torque. You can see this any time a driver accelerates the car away from a traffic light or stop sign. The rear end of the car will drop down momentarily as the torque arrives at the differential. This action is called *rear-end squat*.

♦ 47-10 FRONT-DRIVE DIFFERENTIALS

The differentials for front-drive cars are similar to those for rear-drive cars. Both are constructed and work in the same way. Figures 10-3, 10-10, 44-28, and 44-29 are various views of transaxles. Note especially how the outer front wheel in a turn travels farther than the inner front wheel (Fig. 40-28). The differential permits this difference in wheel travel.

SERVICE

♦ 47-11 DIFFERENTIAL TROUBLE DIAGNOSIS

The first sign of differential trouble is usually noise. The kind of noise you hear can help you determine what is causing the trouble. However, be sure that the noise is actually coming from the differential. Sometimes you may be fooled by universal-joint, wheel-bearing, or tire noise. Determine if the noise is a hum, a growl, or a knock. Note whether the noise is produced when the car is operating on a straight road or only on turns. Note whether the noise is louder when the engine is driving the car or when the car is coasting. It is difficult to diagnose differential noise by running the car in the shop with the wheels raised. Power must be flowing through the differential to the wheels, or the car must be coasting.

1. Humming A humming noise is often caused by incorrect internal adjustment of the drive pinion or the ring gear. Incorrect adjustment prevents normal tooth contact and can cause rapid tooth wear and early failure of the differential. The humming noise will take on a growling sound as the wear progresses. Follow the procedure in the service manual for the car you are servicing when you make differential adjustments.

2. Noise on Acceleration Noise that is louder when the car is accelerating probably means there is heavy contact on the heel ends of the gear teeth. Noise that is louder when the car is coasting probably means there is heavy toe contact. Both these conditions must be corrected. Refer to the manufacturer's service manual for servicing procedures.

3. Noise on Curves If the noise is heard only when the car is going around a curve, the trouble is inside the differential case. Pinion gears tight on the pinion shaft, damaged gears or pinions, too much backlash between gears, or worn differential-case bearings can cause this trouble. When the car rounds a curve, these parts inside the differential case are moving relative to each other.

Noise on curves could also be due to a defective axle bearing. During a turn, the outside bearing takes an increased load.

4. Limited-Slip Differential The limited-slip differential requires a special type of lubricant. The wrong lubricant can cause the clutch surfaces to grab. This will produce chattering noise during a turn. The remedy is to drain the old lubricant and put in the specified lubricant designed for limited-slip differentials.

Wheel spin can occur under some conditions, even if the differential is in good condition. For example, if one wheel is on dry pavement and the other is on smooth ice, sudden acceleration can produce wheel spin. The remedy is to open the throttle *slowly* to allow the differential to lock.

Careful On cars with limited-slip differentials use the same type of tire on both rear wheels. Both tires should have the same pattern, the same air pressure, and the same amount of wear. If one tire is larger than the other, the differential will be working all the time. This will shorten the life of the differential.

♦ 47-12 SERVICING THE DIFFERENTIAL

Repair and overhaul procedures on rear axles and differentials vary among car models. Make sure you have the manufacturer's service manual that covers the car you are working on when you repair a differential.

♦ REVIEW QUESTIONS ♦

Select the *one* correct, best, or most probable answer to each question. You can find the answers in the section indicated at the end of each question.

1. In the differential, the ring gear is bolted to the (♦47-3)
   a. differential housing
   b. differential case
   c. drive pinion
   d. axle shaft
2. The drive pinion is assembled into the (♦47-3)
   a. carrier or axle housing
   b. differential case
   c. axle case
   d. drive shaft
3. The distance between adjacent meshing teeth of mating gears is called (♦47-5)
   a. clearance
   b. pitch line
   c. backlash
   d. flank
4. In a differential with a gear ratio of 4:1, the drive pinion would revolve four times to cause the ring gear to rotate (♦47-5)
   a. one time
   b. two times
   c. four times
   d. 16 times
5. In the basic differential illustrated in Fig. 47-4, there are (♦47-3)
   a. nine gears
   b. eight gears
   c. seven gears
   d. six gears
6. Universal joints are used at the outer ends of the front-axle shafts of a four-wheel-drive vehicle to allow (♦47-7)
   a. spring action
   b. steering action
   c. braking action
   d. none of the above
7. The use of locking hubs on a four-wheel-drive vehicle improves (♦47-8)
   a. ride quality
   b. traction
   c. fuel economy
   d. all of the above
8. Noise on acceleration probably means (♦47-11)
   a. heavy contact on the heel ends of the gear teeth
   b. differential case bearings are bad
   c. worn drive pinion
   d. worn ring gear
9. Noise when going around a curve indicates the trouble is (♦47-11)
   a. due to heavy contact on the heel ends of the bevel gears
   b. due to heavy contact on the toe ends of the bevel gears
   c. in the differential case
   d. due to slippage of the clutch surfaces
10. A humming noise can be caused by (♦47-11)
   a. incorrect adjustment of the drive pinion
   b. incorrect adjustment of the ring gear
   c. worn gear teeth
   d. all of the above

PART 9

This part of the book describes the automotive chassis, which includes the car frame, springs and shock absorbers, steering system, brakes, tires, and wheels. There are six chapters in Part 9:

Automotive Chassis

CHAPTER 48
Springs and Suspension Systems

After studying this chapter, you should be able to:

1. Describe the basic types of springs used in automobiles.
2. Explain why low unsprung weight is desirable.
3. Describe the construction and operation of several types of front-suspension systems.
4. Describe the construction and operation of several types of rear-suspension systems.
5. Explain the purpose, construction, and operation of shock absorbers.
6. Describe the operation of automatic-level-control systems.

♦ 48-1 FUNCTION OF SPRINGS

The car body or frame supports the weight of the engine, the power train, and the passengers. The body or frame is supported by the springs. There is a spring at each wheel. The weight of the car frame, body, and attached parts applies an initial compression to the springs. The springs compress further as the car wheels hit bumps or expand as the wheels drop into holes in the road. The springs cannot do the complete job of absorbing road shocks. The tires absorb some of the irregularities in the road. The springs in the car seats also help to absorb shock. However, little shock from road bumps and holes is felt by the passenger.

♦ 48-2 TYPES OF SPRINGS

There are three basic types of automotive springs: leaf, coil, and torsion bar (Figs. 48-1, 48-2, and 48-3). In addition, air suspension is used in some cars, trucks, and buses. Air suspension was offered for passenger cars some years ago. However, few people wanted it then so the option was dropped.

♦ 48-3 CHARACTERISTICS OF SPRINGS

The ideal spring for automotive suspension would absorb road shock rapidly and then return to its normal position slowly. However, this is difficult to attain. An extremely flexible, or soft, spring allows too much movement. A stiff, or hard, spring gives too rough a ride. However, satisfactory riding qualities are attained by using a fairly soft spring with a shock absorber (♦48-21).

The softness or hardness of a spring is referred to as its *rate*. The rate of a spring is the weight required to deflect it 1

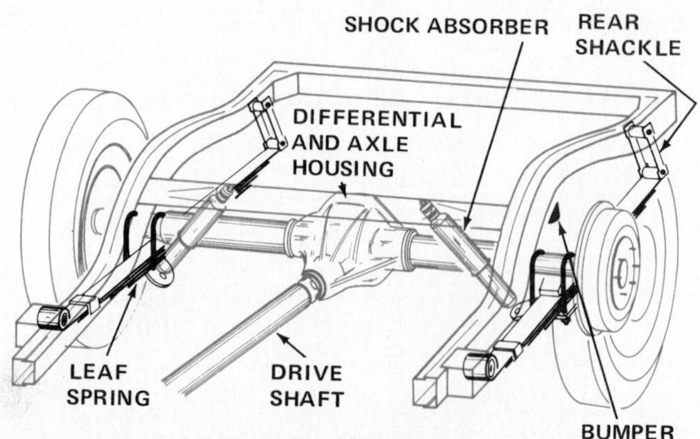

Fig. 48-1 Rear suspension using leaf springs.

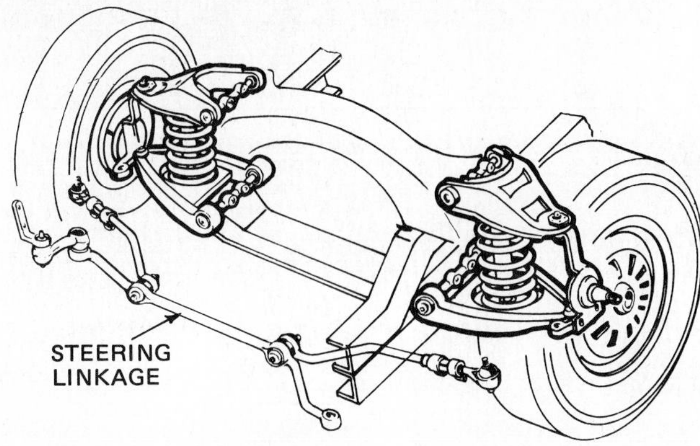

Fig. 48-2 Front-suspension system using coil springs. *(Moog Automotive, Inc.)*

inch [25.4 mm]. The rate of most automotive springs is almost constant through their operating range, or deflection, in the car. This is stated by Hooke's law, as applied to coil springs: The spring will compress in direct proportion to the weight applied. Therefore, if 600 pounds [272 kg] will compress the spring 3 inches [76 mm], then 1,200 pounds [544 kg] will compress the spring twice as far, or 6 inches [152 mm].

♦ 48-4 SPRING ARRANGEMENTS

Springs are used in various ways to achieve satisfactory ride quality and control. For example, torsion bars are used in two ways. Coil springs are used in four ways. The greatest variety of spring arrangements is found in front-suspension systems.

FRONT SUSPENSION

♦ 48-5 FRONT-SUSPENSION SYSTEMS

The front suspension system allows the front wheels to move up and down and absorb road shocks. But the system must also allow the front wheels to swing from side to side so the car can be steered. Coil, leaf, and torsion-bar springs are used in seven basic arrangements:

1. Coil spring between lower control arm and a seat in the car frame.
2. Coil spring between the upper control arm and a seat in the car body.
3. Coil spring between an I-beam axle and a seat in the frame.
4. Coil spring between a seat on a strut rod which is attached to the lower control arm and a seat in the car body. There is no upper control arm. This is called *MacPherson-strut suspension*.
5. Torsion bar connected longitudinally between the lower control arm and the car frame.
6. Torsion bar connected transversely between the lower control arm and the car body.
7. Leaf springs between an I-beam axle and a seat in the frame.

The construction and operation of each of these arrangements are described in the following sections.

♦ 48-6 COIL SPRING ON LOWER CONTROL ARM

This arrangement is shown in Figs. 48-2 and 48-4. The coil springs are between the lower control arm and the car frame. The control arms are pivoted on the car frame so that they can swing up and down. On their outer ends, the control arms are connected by ball joints to steering knuckles which support the front wheels (Fig. 48-5). The wheel is attached to the spindle on the steering knuckle.

Figure 48-6 shows front-suspension ball joints. The ball joints support the weight of the car. They allow the wheels to move up and down, and to swing from side to side.

Figures 48-7 and 48-8 show how the suspension system allows the front wheels to move up and down. Figure 48-7 shows what happens when a front wheel hits a bump in the road. The wheel moves up, as shown by the dashed lines. As the wheel moves up, the two control arms pivot upward. This action compresses the spring between the lower control arm and the car frame.

When a front wheel meets a hole in the road (Fig. 48-8), the wheel moves down. The control arms pivot downward. This allows the spring to expand. In both expansion and compression (Figs. 48-7 and 48-8), notice how far the spring can allow the wheel to travel, but how little movement is carried to the frame. The passengers have a relatively smooth ride.

In the coil-spring front-suspension system shown in Figs. 48-4 to 48-8, the shock absorbers are centered in the coil springs. Shock absorbers are covered later in the chapter.

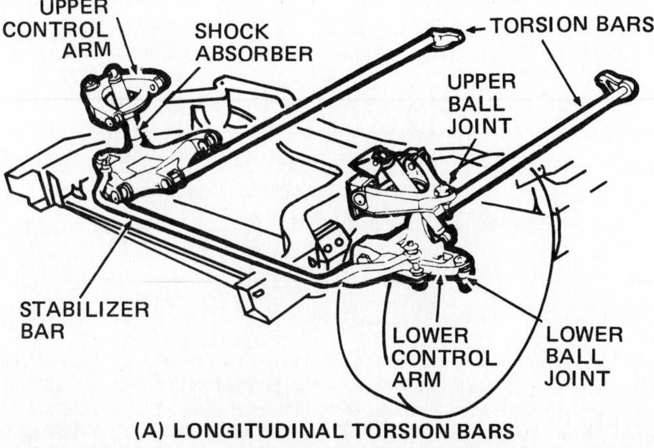

(A) LONGITUDINAL TORSION BARS

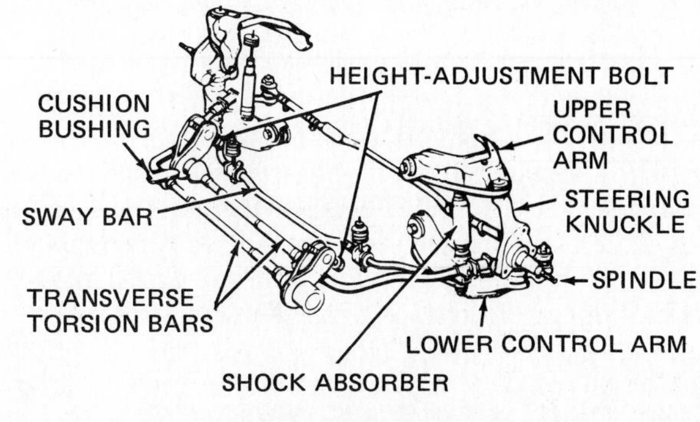

(B) TRANSVERSE TORSION BARS

Fig. 48-3 Front-suspension systems using torsion bars. *(Moog Automotive, Inc.; Chrysler Corporation)*

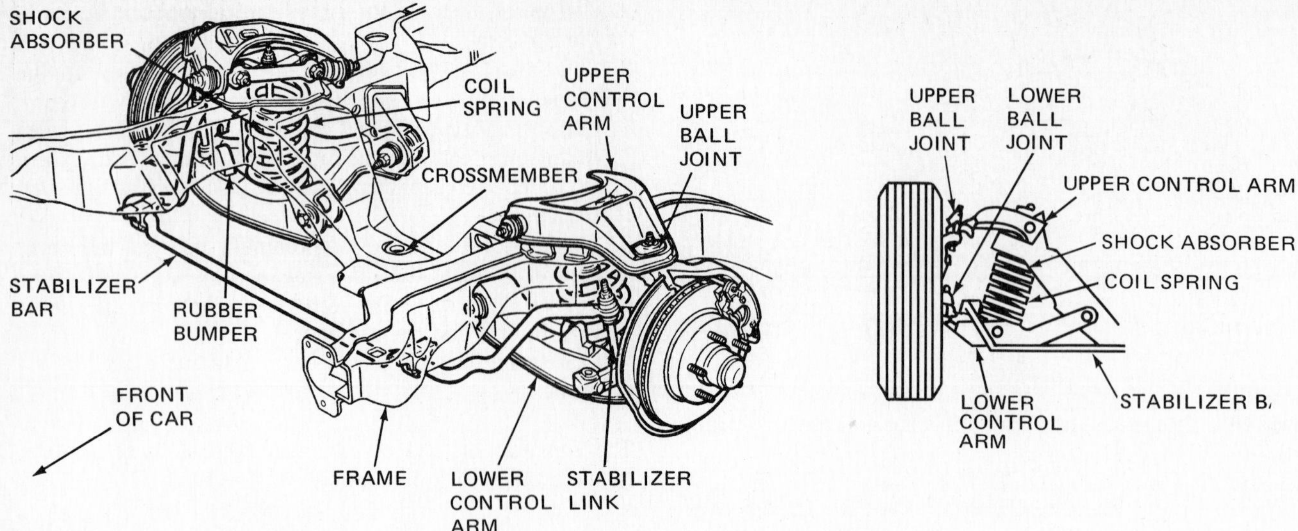

Fig. 48-4 Coil-spring front-wheel suspension. The wheels mount on bearings on the tapered spindles of the steering knuckles. The support of one wheel is shown to the right. *(Buick Motor Division of General Motors Corporation)*

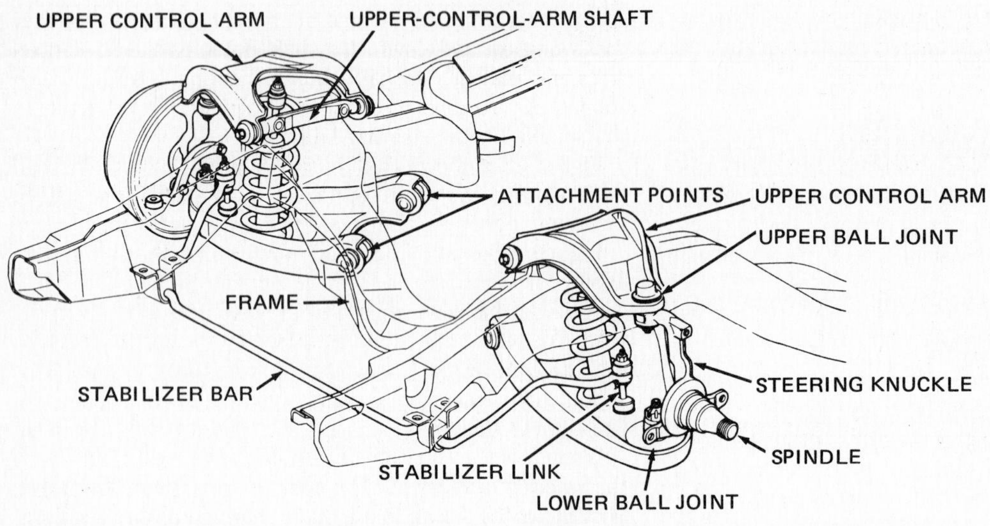

Fig. 48-5 Front-suspension system with the coil springs between the car frame and the lower control arms. The lower control arms pivot on two attachment points to the frame. *(Chevrolet Motor Division of General Motors Corporation)*

In Fig. 48-5, the two control arms are the "A" type. They have the shape of an A, with two points of attachment to the frame. This gives them, and the wheels, forward-and-back stability. The wheels can move up and down and can swing from side to side, but they cannot move forward or backward.

Some cars have lower control arms with only one point of attachment to the frame (Fig. 48-9). With this system, an extra part is required. It is called a *strut* or a *brake-reaction rod*. The strut is connected between the outer end of the lower control arm and the car frame. The strut prevents the outer end of the lower control arm from swinging forward or backward during braking or when the wheel hits holes or bumps in the road.

♦ 48-7 COIL SPRING ON UPPER-CONTROL ARM

This arrangement is shown in Fig. 48-9. The coil spring is between the upper control arm and a spring tower which is

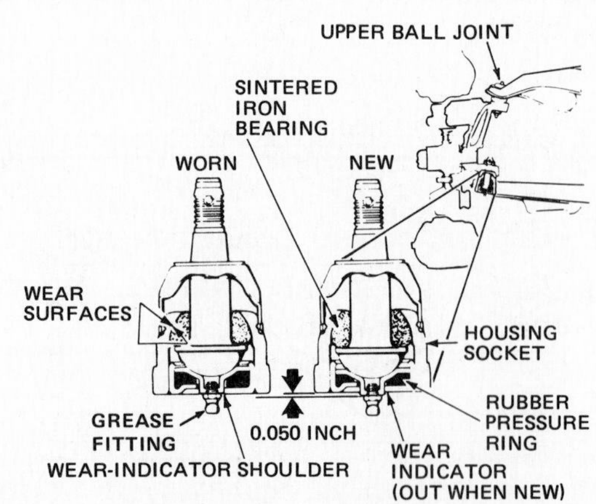

Fig. 48-6 Ball joint, showing how it attaches the steering knuckle to the lower control arm. When ball-joint wear causes the wear-indicator shoulder to recede within the socket housing, replacement is required.

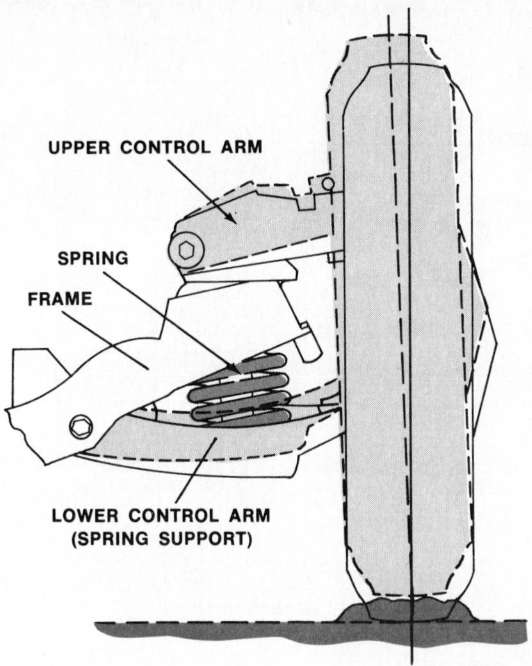

Fig. 48-7 Front-wheel suspension at one wheel, showing the actions as the wheel meets a bump in the road. The upward movement of the wheel, shown dashed, raises the lower control arm, causing the spring to compress.

part of the car body. The action is the same as for the coil-spring front-suspension system which has the spring sitting on the lower control arm (♦48-6). When the lower control arm has one point of attachment to the car body or frame, a strut rod is used. It prevents the outer end of the lower control arm from swinging forward or backward. The strut rod also provides a means of adjusting caster (♦50-20) in this type of suspension system.

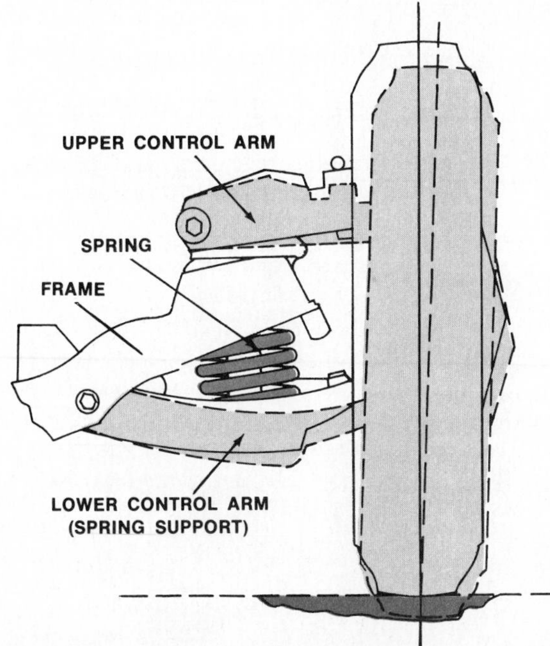

Fig. 48-8 Front suspension at one wheel, showing the actions as the wheel drops into a hole in the road. The downward movement of the wheel, shown dashed, lowers the lower control arm, permitting the spring to expand.

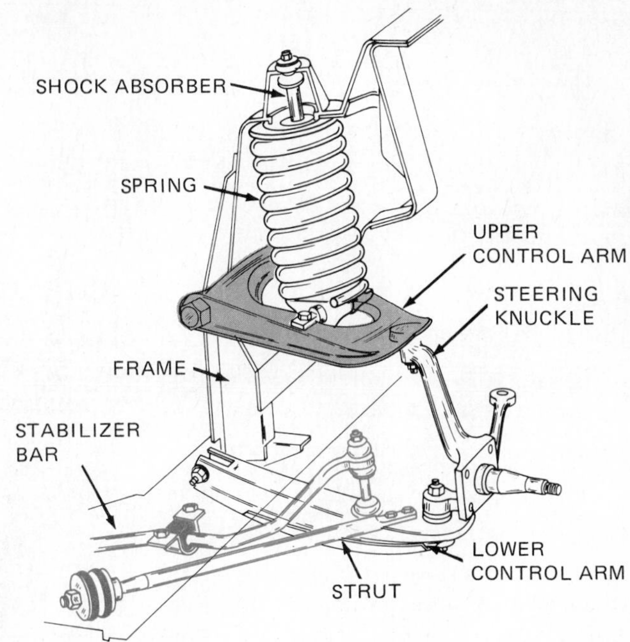

Fig. 48-9 Coil-spring front-suspension system with the spring mounted on the upper control arm.

♦ 48-8 COIL SPRING BETWEEN I BEAM AND FRAME

This system (Fig. 48-10) uses a pair of I beams. One end of each beam is attached through a pivot to the vehicle frame. The other end carries the spindle on which the wheel is mounted. There is a coil spring above each I beam, positioned between the I beam and a seat on the frame. As the wheels meet bumps and holes in the road, the I beams pivot up or down, compressing or expanding the springs. This system is used on several models of Ford trucks.

♦ 48-9 MACPHERSON-STRUT FRONT SUSPENSION

In this suspension system (Fig. 48-11), no upper control arm is used. The upper end of the assembly rests in a spring tower

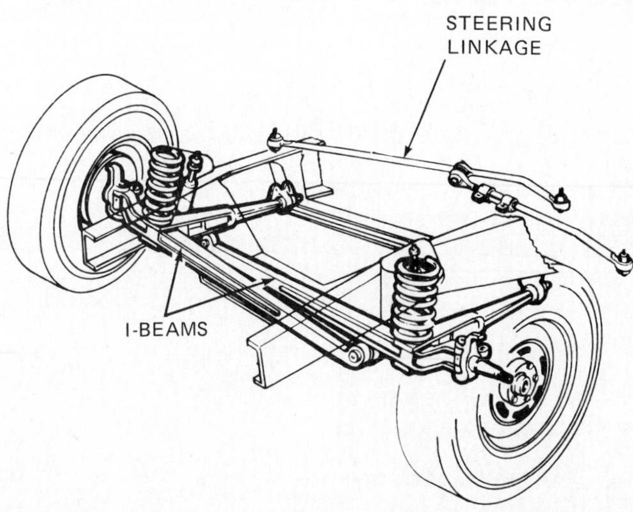

Fig. 48-10 Twin I-beam front suspension. *(Moog Automotive, Inc.)*

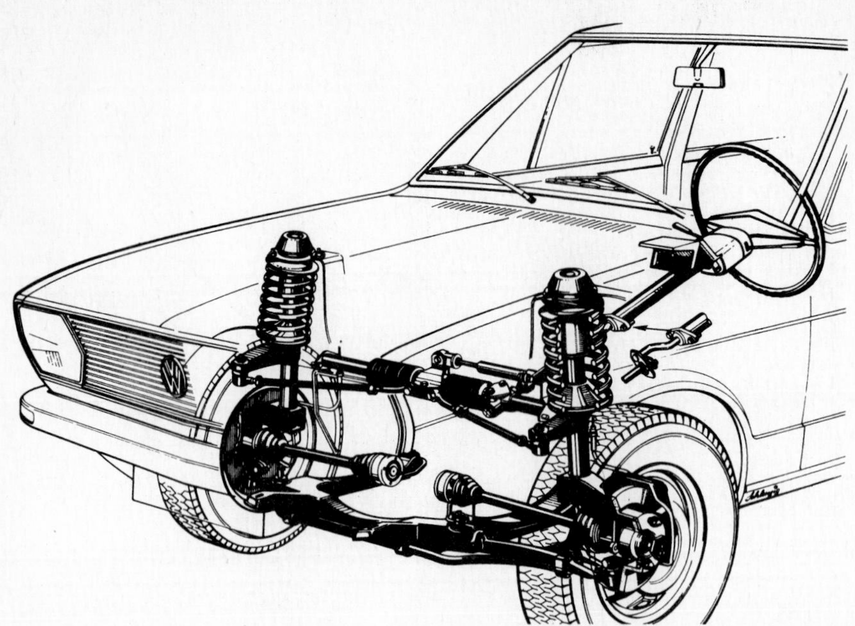

Fig. 48-11 MacPherson-strut front-suspension system. *(Volkswagen of America, Inc.)*

in the front-end of the body. The strut assembly includes a coil spring and a shock absorber (Fig. 48-12). Figure 48-13 shows the system isolated from the car. The lower control arms are pivoted at two points to the car frame. The outer ends of the control arms are attached to the bottom of the strut. The strut carries the wheel spindle. As the wheel meets holes or bumps in the road, the spring expands or compresses to allow the wheel to move down or up. Figure 48-13 also shows a MacPherson-strut rear-suspension system.

♦ 48-10 TORSION-BAR FRONT SUSPENSION

In one torsion-bar front-suspension system, two long steel bars serve as the springs (Fig. 48-14). The rear ends of the bars are locked to a cross member of the frame. The front ends of the bars are attached to the lower control arms. In operation, the lower control arms pivot up and down, twisting the torsion bars. The effect is very similar to the actions of coil and leaf springs. The car weight places an initial twist on the torsion bars, just as it places an initial compression on the coil springs of cars with coil-spring suspension. Figure 48-15 shows the car-leveling device at the rear end of one

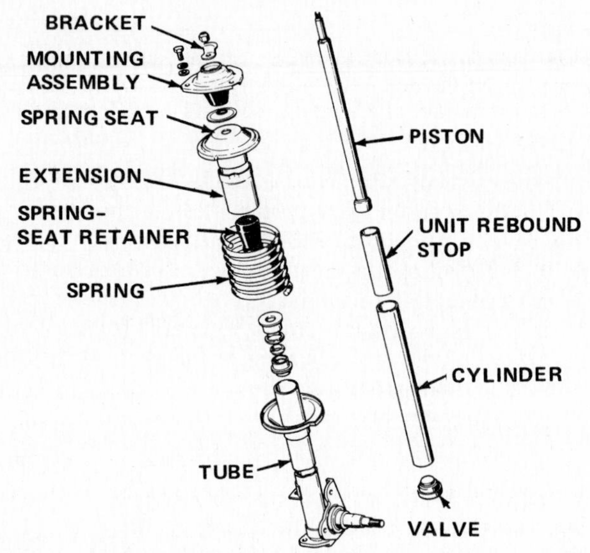

Fig. 48-12 A disassembled MacPherson strut for a front-suspension system. *(Volkswagen of America, Inc.)*

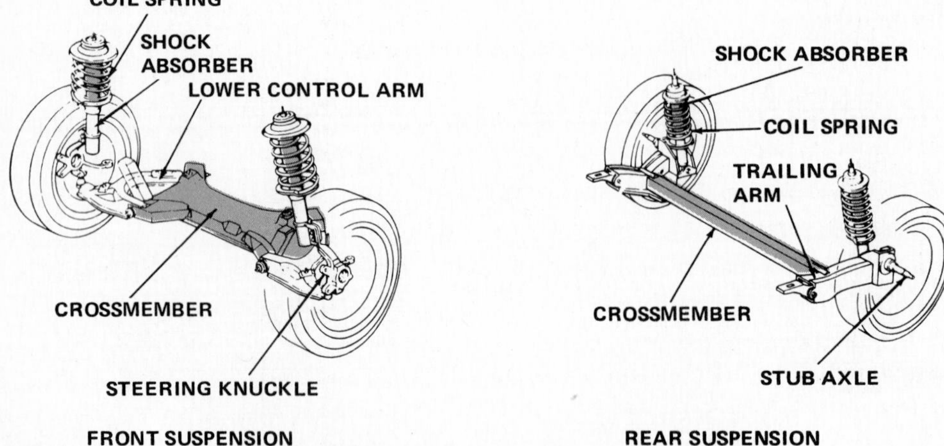

Fig. 48-13 Left, front suspension system using MacPherson struts. Right, modified MacPherson-strut rear-suspension system. *(Chrysler Corporation)*

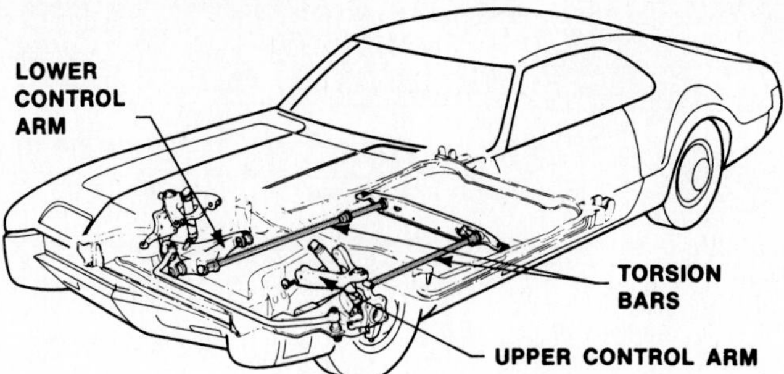

Fig. 48-14 Front suspension of a front-drive car using longitudinal torsion bars. The bars are locked at the rear to the frame. They are attached at the front to the inner ends of the lower control arms. *(Oldsmobile Division of General Motors Corporation)*

torsion bar. Turning the height-adjustment bolt causes the hub-and-anchor assembly to turn. This rotates the rear end of the torsion bar so that the front end of the car is raised or lowered.

Torsion bars that are mounted from front to rear, or the long way of the car (Fig. 48-14), are called *longitudinal torsion bars*. Some torsion bars on smaller cars are mounted sideways, or transversely (Fig. 48-16). Torsion bars work the same way in either position. They twist more or less as the wheels meet bumps and holes in the road.

♦ 48-11 LEAF-SPRING FRONT SUSPENSION

Leaf springs are used at the rear for some passenger cars. Trucks often have leaf springs at the front. A leaf spring is made up of several long plates or leaves. Figure 48-1 shows leaf springs at the rear of a car. Figure 48-17 shows how a leaf spring is attached, front and back, to the car frame.

Because the leaf spring is made up of a series of thin leaves, one on top of another, it does not break when bent. The plates, or leaves, are held together at the center by a center bolt which passes through holes in the leaves. Clips are placed at intervals along the spring, as shown in Fig. 48-17. They keep the leaves in alignment and prevent separation during rebound. Many leaf springs have special inserts between the leaves to permit the leaves to slip over one another more easily when the spring bends.

Some vehicles use single-leaf springs at the rear (Fig. 48-18). They are tapered from the center to the ends so that they work in the same way as multileaf springs.

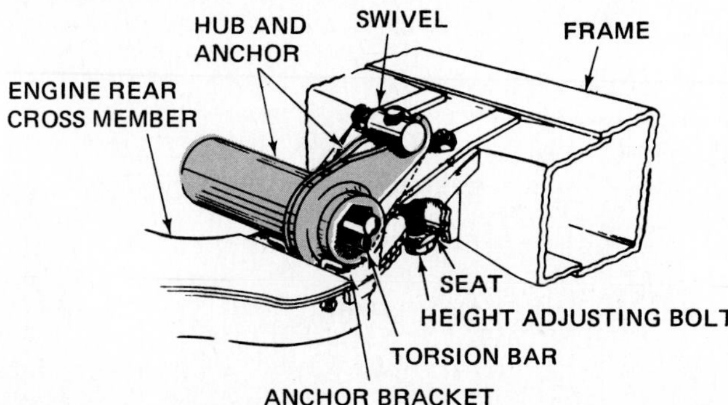

Fig. 48-15 Method of attaching the rear of a longitudinal torsion bar to the frame. The hub and anchor, swivel, and adjusting bolt are for adjusting the height of the car. *(Chrysler Corporation)*

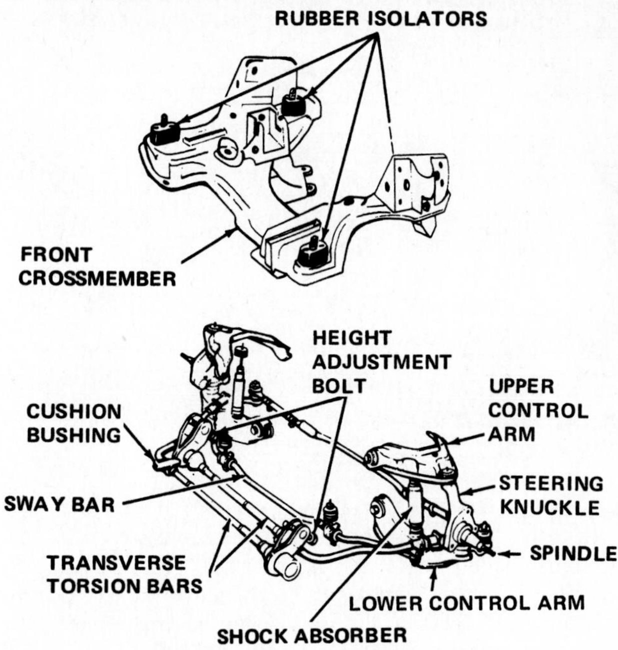

Fig. 48-16 Front-suspension system using transverse torsion bars. *(Chrysler Corporation)*

Figure 48-19 shows the leaf-spring arrangement for a truck using an I-beam front axle and two leaf springs. Methods of attachment of leaf springs are described in ♦48-14.

♦ 48-12 STABILIZER BAR

Figure 48-4 shows a *stabilizer bar* installed across the front of the frame. This bar is a long steel rod, fastened at each end to the two lower control arms. The stabilizer bar is sometimes called a *sway bar*.

When the car goes around a curve, centrifugal force tends to keep the car moving in a straight line. Therefore, the car "leans out" on the turn. This *lean out* is also called *body roll*. With lean out, or body roll, additional weight is thrown on the outer spring. This puts additional compression on the outer spring, and the control arm pivots upward. As the control arm pivots upward, it carries its end of the stabilizer bar up with it. At the inner wheel on the turn, there is less weight on the spring. Weight has shifted to the outer spring because of centrifugal force. Therefore, the inner spring tends to expand. The expansion of the inner spring tends to pivot the lower control arm downward. As this happens, the lower control arm carries its end of the stabilizer bar downward.

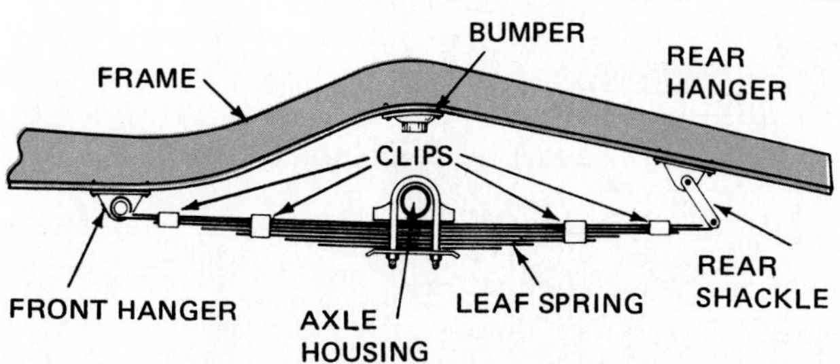

Fig. 48-17 Rear suspension using leaf springs, showing arrangement at one wheel. The leaf spring is anchored at three places: the front hanger, the rear shackle, and the axle housing.

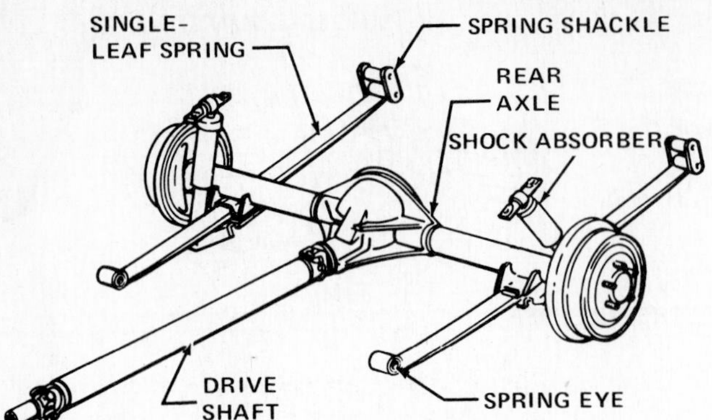

Fig. 48-18 Rear suspension using tapered-plate, or single-leaf, springs. *(Chrysler Corporation)*

The outer end of the stabilizer bar is carried upward by the outer control arm. The inner end is carried downward. This combined action twists the stabilizer bar. The resistance of the bar to twisting combats the tendency of the car to lean out on turns. There is less suspension movement and body roll than there would be without the stabilizer bar.

REAR SUSPENSION

♦ 48-13 REAR-SUSPENSION SYSTEMS

In addition to the leaf-spring rear suspension, shown in Fig. 48-1, many vehicles use coil-spring rear suspension. Rear coil springs are arranged in a variety of ways. With coil springs, special links or bars must be used to keep the wheels tracking. The wheels must be allowed to move up and down but must be prevented from moving sideways or back and forth.

Various rear-suspension systems are described in following sections.

♦ 48-14 LEAF-SPRING REAR-SUSPENSION SYSTEMS

Figures 48-1, 48-17, and 48-19 show leaf springs and how they are located. The spring leaves are of graduated length.

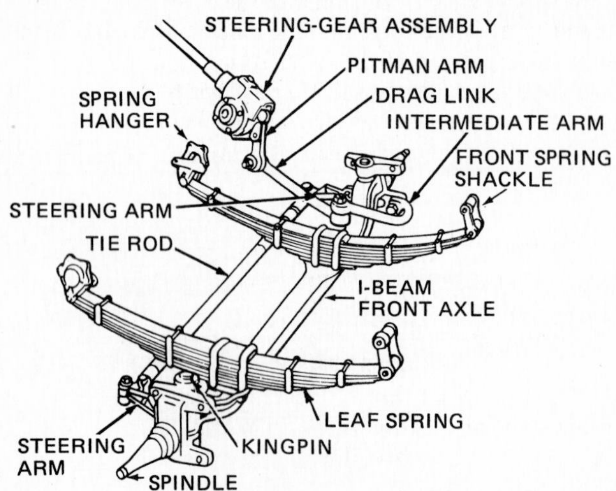

Fig. 48-19 Truck front-suspension system using a solid I-beam front axle and two leaf springs. *(Ford Motor Company)*

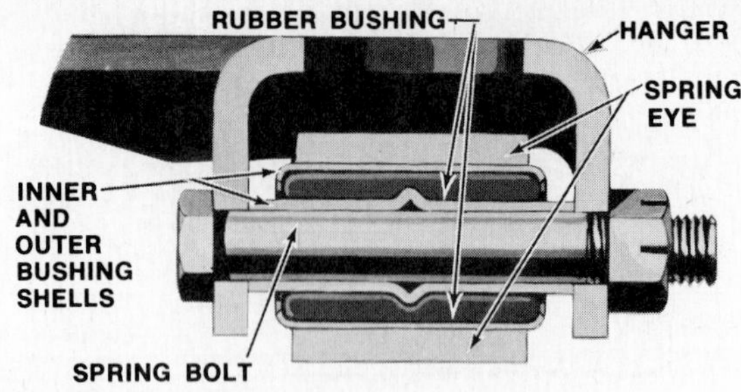

Fig. 48-20 Details of the bushing in a spring eye through which the leaf spring is attached to the hanger on the car.

The front end of the longest leaf is bent into a circle to form a spring eye. The spring eye is attached to the spring hanger by a bolt. Rubber bushings insulate the bolt from the spring hanger (Fig. 48-20). The rubber bushings serve two purposes: They absorb vibration and therefore prevent it from reaching the car frame and body. The bushings also allow the spring eye to twist back and forth as the leaf spring bends.

The rear end of the spring is also bent to form a spring eye. This spring eye is attached to the car frame through a spring shackle. The shackle allows for changes in the length of the leaf spring as it bends. As the spring is pushed upward or downward by bumps or holes in the road, the distance between the two spring eyes changes. The shackle forms a swinging support that permits this change in length. Figure 48-21 shows a disassembled spring shackle. The shackle is always located at the rear of the spring, as shown in Fig. 48-1. The shackle also includes rubber bushings. They absorb vibration and prevent it from reaching the car frame and body.

The center of the spring is hung from the rear-axle housing by a pair of U bolts. The rear of the car is, in effect, hung from the axle housing by two pairs of U bolts. Rubber bumpers are mounted on the car frame above the axles. The purpose of these bumpers is to absorb the shock that would result if the axle housing actually moved up far enough to hit the frame. The axle housing would move up this far only if the wheels hit a large bump, or if the rear were carrying a heavy load.

Two shock absorbers, one for each spring, are shown in Fig. 48-1. Shock absorbers are described later in the chapter.

♦ 48-15 COIL-SPRING REAR-SUSPENSION SYSTEMS

In the typical coil-spring rear-suspension system, there is a coil spring at each wheel. The wheels must be allowed to

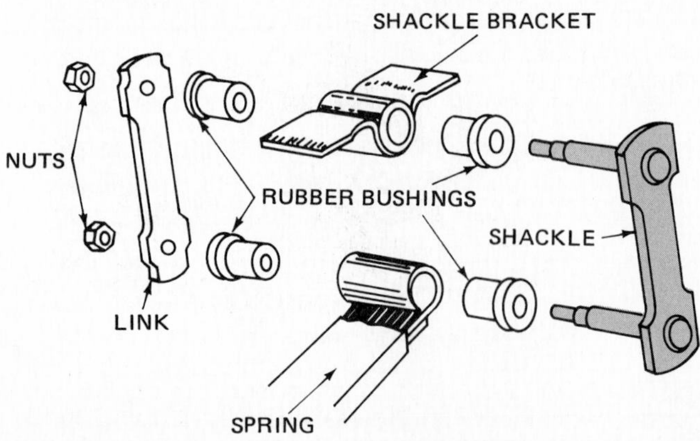

Fig. 48-21 Disassembled shackle for a leaf spring.

move up and down, compressing and expanding the springs. But the wheels must not move sideways, or forward and back. To prevent these movements, control arms are required. Figure 48-22 shows a coil-spring rear-suspension system that has four control arms. Two prevent forward or backward movement of the rear-axle housing and the wheels. The other two prevent sideward movement of the rear-axle housing and the wheels.

The coil springs are located between spring seats in the car body or frame and spring seats on the control arms, or on pads on the axle housing. A shock absorber is located at each wheel (Fig. 48-22).

♦ 48-16 MACPHERSON-STRUT REAR SUSPENSION

MacPherson struts are used in many rear-suspension systems (Figs. 48-13 and 48-23). The MacPherson strut combines the spring and shock absorber into a single unit. The unit is constructed and works in the same way as the MacPherson struts used in front-suspension systems (♦48-9).

♦ 48-17 REAR-END TORQUE

Whenever the rear wheels are being driven by the power train, they rotate, as shown in Fig. 48-24. At the same time, the rear-axle housing tries to rotate in the opposite direction (Fig. 48-24). The twisting motion applied to the axle housing is called *rear-end torque*. It is absorbed by the rear springs and control arms. On the leaf-spring rear suspension, the leaf spring absorbs the rear-end torque. On the coil-spring rear suspension, the control arms absorb the rear-end torque.

One effect of rear-end torque is rear-end "squat" on acceleration (Fig. 48-25). When a car is accelerated from a standing start, the drive pinion tries to climb the teeth of the ring gear. Therefore, the drive pinion and the differential carrier move upward. The result is that the rear springs are pulled downward, or compressed, so the rear end of the car moves down, or squats. On braking, the rear of the car moves up. This is because of the inertia of the car.

Many cars with front-wheel drive exhibit similar tendencies. During acceleration, torque reaction and inertia cause the front end to lift. On braking, inertia causes the rear of the car to move up.

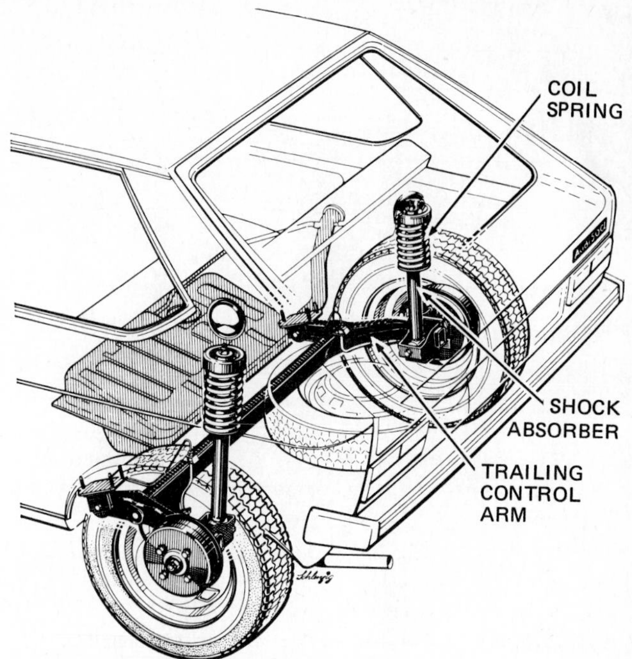

Fig. 48-23 MacPherson-strut rear-suspension system. *(Volkswagen of America, Inc.)*

AIR AND HYDROPNEUMATIC SUSPENSION

♦ 48-18 AIR SUSPENSION

Air suspension was offered as an option years ago by some car manufacturers. However, it was not widely accepted for use in passenger cars. In recent years, some heavy-duty trucks and buses have used air suspension. Now, with *electronic air suspension*, air springs are making a comeback.

In air-suspension systems, the four steel springs are replaced by four rubber cylinders, or *air springs* (Fig. 48-26). Each rubber cylinder is filled with compressed air, which supports the car weight. When a wheel encounters a bump in the road, the air is further compressed and absorbs the shock.

The electronic air-suspension system is shown in Fig. 48-26. It includes an electric air compressor, a microcomputer

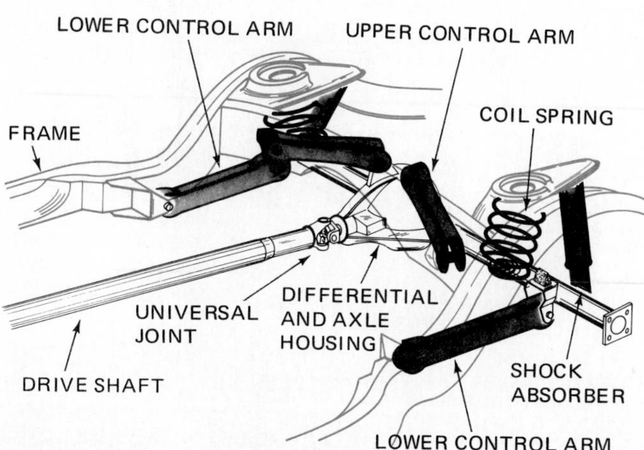

Fig. 48-22 Coil-spring rear-suspension system using four control arms.

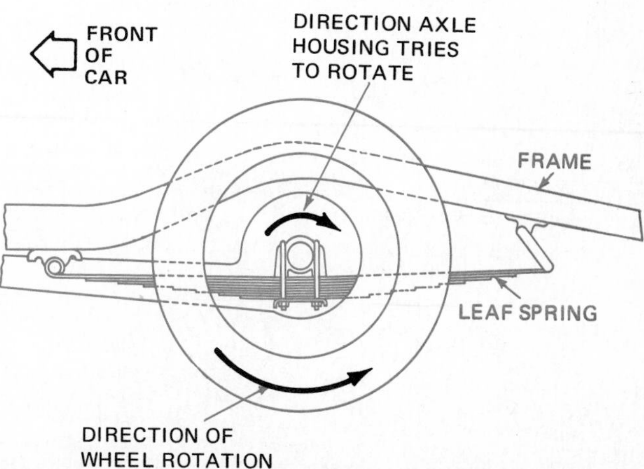

Fig. 48-24 Axle housing tries to rotate in a direction opposite to wheel rotation.

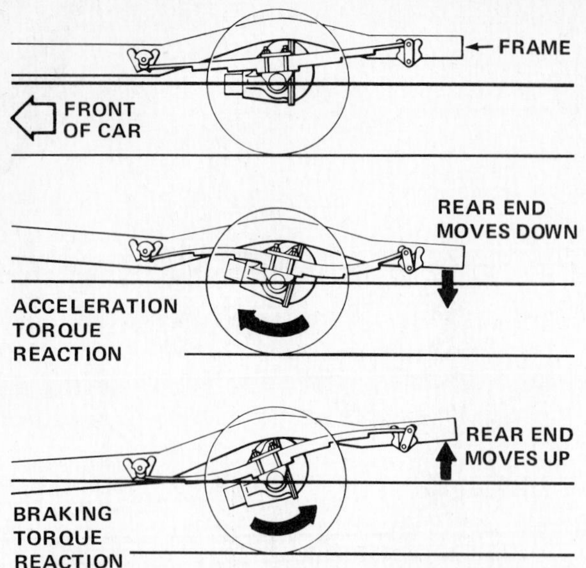

Fig. 48-25 Actions of the spring and rear end when the car is accelerated or braked. *(Ford Motor Company)*

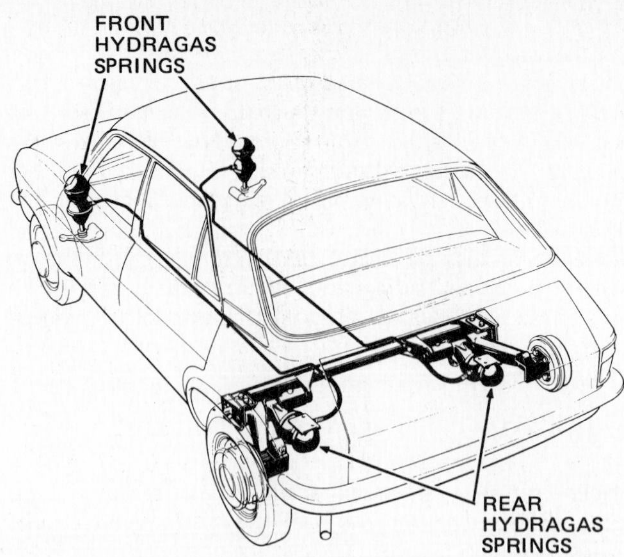

Fig. 48-27 A hydropneumatic suspension system, using Hydragas springs at each wheel. *(British Leyland)*

control module (MCM), four air springs with built-in solenoid valves, three height sensors (two front and one rear), and the air-distribution system of lines and fittings.

The height sensors monitor the riding height, or *vehicle trim height*. They signal the control module of any change. If the height is too high, the control module opens the solenoid valves in the springs with too much air. This allows some of the air to escape, lowering the car. If the height is too low, the control module turns on the air compressor. Then the control module opens the solenoid valves in the springs needing more air. Now more air flows into the spring until proper trim height is restored.

The control and operation of the system are very similar to that of the automatic-level-control system (♦48-23). However, the air-suspension system provides springing for all four wheels, instead of only the two rear wheels as with automatic level control.

♦ 48-19 HYDROPNEUMATIC SUSPENSION

Various suspension systems using air, gas, fluid, or a combination of these have been developed. One of the best known is the Hydragas suspension system. It uses gas-filled spring units (called Hydragas springs), one at each wheel (Fig. 48-27). Each unit has a sealed chamber containing a quantity of nitrogen gas at high pressure. Below this chamber is a displacement chamber filled with water-based fluid (Fig. 48-28). When the wheel meets a bump, the fluid is pushed upward, compressing the gas. This action provides the springing effect.

In addition, the two units on each side of the car are interconnected front to back (Fig. 48-27). Therefore, when the left front wheel meets a bump, for example, part of the fluid from the left front unit is forced through a pipe to the left rear unit. This action raises the left rear wheel also. Therefore the shock is distributed between the left front and the left rear wheels. This improves the ride.

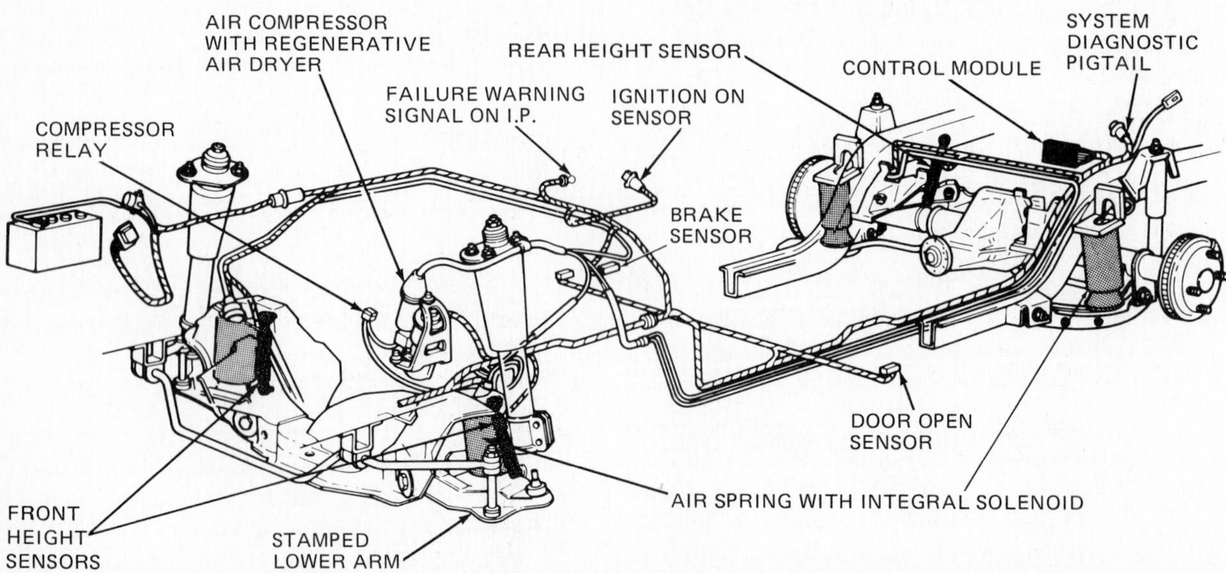

Fig. 48-26 Air suspension system, using an air spring at each wheel. *(Ford Motor Company)*

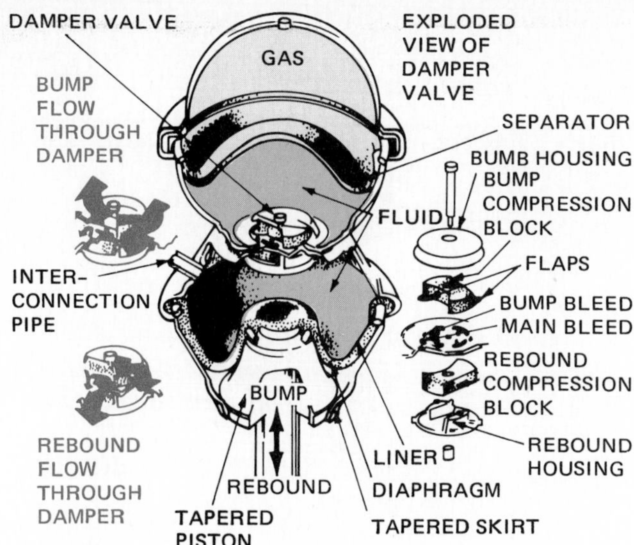

Fig. 48-28 Partly cutaway Hydragas spring. *(British Leyland)*

♦ 48-20 SPRUNG AND UNSPRUNG WEIGHT

The term *sprung weight* refers to the part of the car that is supported on springs. The term *unsprung weight* refers to the part that is not. The frame and the parts attached to the frame are sprung. Their weight is supported on the car springs. However, the wheels and wheel axles (and axle housing and differential) are not supported on springs. They represent unsprung weight.

Unsprung weight should be kept as low as possible. The reason is that unsprung weight increases the roughness of the ride. For example, consider a single wheel. If it is light, it can move up and down without causing much reaction of the car body and frame. But if the weight of the wheel is increased, then its movement becomes more noticeable to the car occupants. Suppose the unsprung weight at the wheel is equal to the sprung weight above the wheel. Then the sprung weight tends to move as much as the unsprung weight. The unsprung weight, which must move up and down over road irregularities, tends to cause a like motion of the sprung weight. This is the reason for keeping the unsprung weight as low as possible.

♦ 48-21 SHOCK ABSORBERS

Shock absorbers are necessary because springs do not "settle down" fast enough. After a spring has been compressed and released, it continues to shorten and lengthen, or *oscillate,* for a time.

This is what happens if the spring at the wheel is not controlled. When the wheel hits a bump, the spring compresses. Then the spring expands after the wheel passes the bump. The expansion of the spring causes the car body and frame to be thrown upward. But, having overexpanded, the spring shortens again. This action causes the wheel to move up and momentarily leave the road at the same time that the car body and frame drops down. The action is repeated until the oscillations gradually die out.

Such spring action on a car would produce a very bumpy and uncomfortable ride. It could also be dangerous, because a bouncing wheel makes the car difficult to control. Therefore, a dampening device is needed to control the spring oscillations. This device is the shock absorber.

♦ 48-22 OPERATION OF SHOCK ABSORBERS

The shock absorber absorbs the shock of the wheel meeting a bump or hole. As soon as the wheel passes the hole or bump, the shock absorber returns the wheel to contact with the road. This prevents the wheel from bouncing.

The shock absorber is the direct-acting tubular or telescope type (Fig. 48-29). In operation, the shock absorbers lengthen and shorten as the wheels meet irregularities in the road. As they do this, a piston inside the shock absorber moves in a cylinder filled with fluid. Therefore the fluid is put under high pressure and forced to flow through small openings. The fluid can only pass through the openings slowly. This slows the piston motion and restrains spring action.

Figure 48-29 shows a shock absorber in cutaway view. Figure 48-30 shows two sectional views of a shock absorber during compression and rebound. With either action, the piston is moving. The fluid in the shock absorber is being forced through small openings which restrains spring movement. Actually, there are small valves in the shock absorber that open when the internal pressure becomes excessive. When the valves are open, a slightly faster spring movement occurs. However, restraint is still imposed on the spring.

A quick check of shock-absorber action can be made by bouncing each corner of the car. With the car on a level surface, push down on one corner of the car and then release it. The car should come back up to its original height and stay

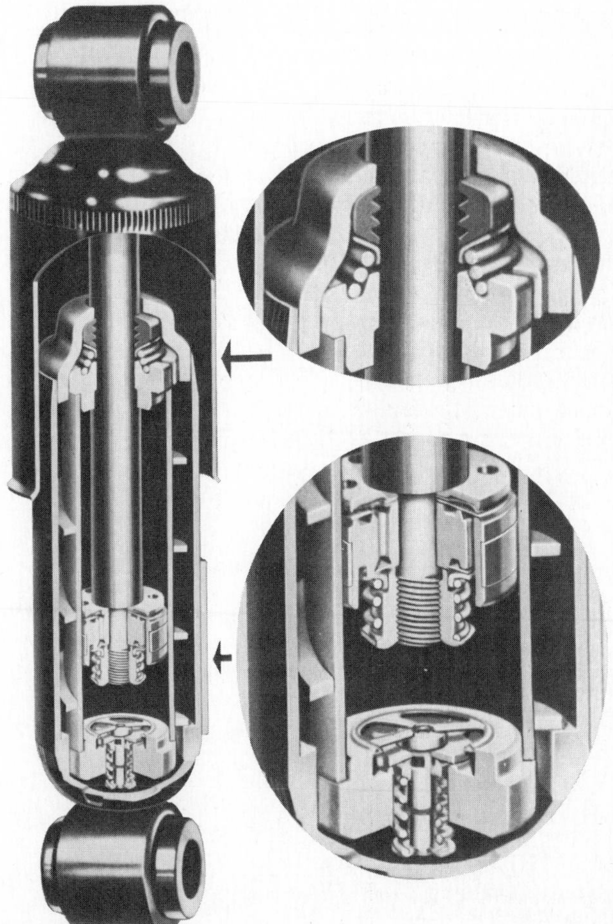

Fig. 48-29 A cutaway shock absorber. *(Monroe Auto Equipment Company)*

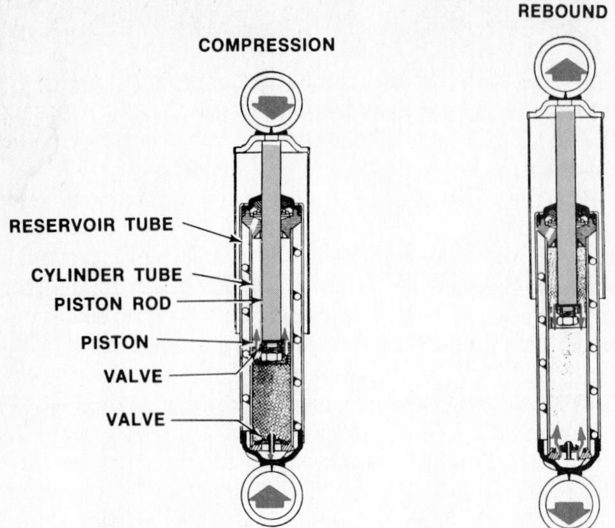

Fig. 48-30 Operation of the shock absorber during compression *(left)* and extension *(right)*. Fluid movement is shown by the arrows. *(Chrysler Corporation)*

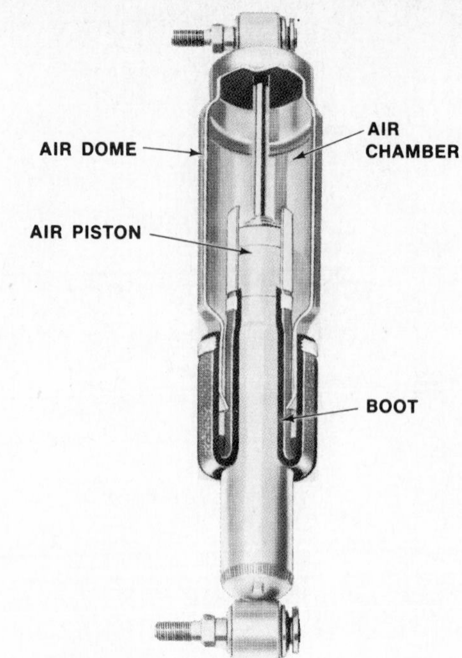

Fig. 48-32 A cutaway air shock absorber used in the automatic-level-control system.

there. If the car continues to bounce up and down more than two times after you release it, the shock absorber is probably defective and should be replaced.

♦ 48-23 AUTOMATIC LEVEL CONTROL

The automatic-level-control system takes care of changes in the load at the rear of the car. In a car without automatic level control, adding weight at the rear will make the rear end of the car squat. This changes the handling characteristics of the car. It also causes the headlights to point upward. The automatic level control prevents this by automatically raising the rear end of the car to level when a load is added. The system also automatically lowers the rear end to level when the load is removed.

The automatic-level-control system includes a compressor, an air-reserve tank, a height-control valve, and two special shock absorbers with built-in air chambers (Figs. 48-31 and 48-32). On some cars, the compressor is operated by engine intake-manifold vacuum. The vacuum operates a pump that builds up air pressure in the reserve tank. When a load is added to the rear of the car, additional air passes through the height-control valve to the two rear shock absorbers. This raises the air domes (Fig. 48-32). The movement brings the rear up to normal level.

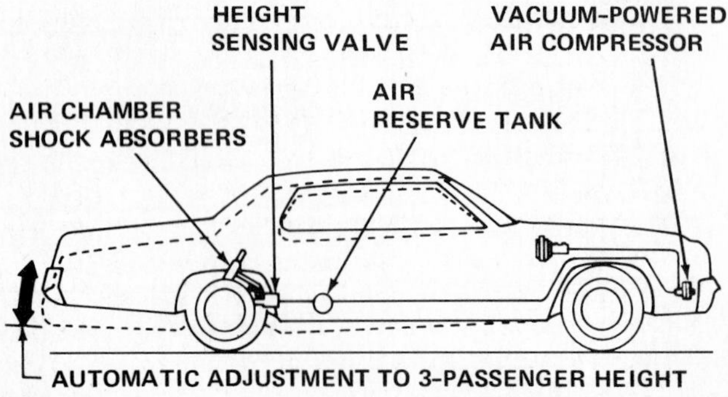

Fig. 48-31 An automatic-level-control system. The dotted lines show the lower height of the car before the automatic level control restores the rear end to the correct height.

On other cars, the compressor is driven by an electric motor installed in the left rear of the car (Fig. 48-33). The shock absorbers and the routing of the air lines to them are also shown in Fig. 48-33.

The height-control valve (Fig. 48-34) has a linkage to the rear-suspension system. When the linkage is operated by the addition of a load, it opens the intake valve. This admits air to the shock absorber. Note that the height-control arm is moved upward by the addition of a load. When the load is removed, the rear of the car moves up, and the height-control valve operates. The exhaust valve opens to allow air to exit from the shock absorber. The rear of the car then settles down to normal height.

The height-control valve has a time-delay mechanism that allows the valve to operate only after several seconds. This mechanism prevents fast valve action, which could operate the system after each bump or hole in the road. Therefore, the automatic-level-control system works only when loads are added or removed from the rear of the car.

Figure 48-35 is the wiring circuit of an electronic automatic-level-control system. It has a height sensor that operates electronically rather than mechanically. The electronic height sensor has a shutter connected to the control arm. The control arm is connected to a suspension arm so that the control arm moves as the height of the car rear end changes. The shutter can interrupt a beam of light inside the height sensor if the height is not correct. If it is either too low or too high, the sensor signals either the compressor relay or the solenoid exhaust valve. If the height is low, it triggers the compressor relay. This causes the compressor relay to close its points so the compressor is connected to the battery. It runs and supplies air to the rear shock absorbers. This raises the rear of the car to level.

If the height is too high, as it would be after a load is removed from the rear seat or trunk, the sensor triggers the

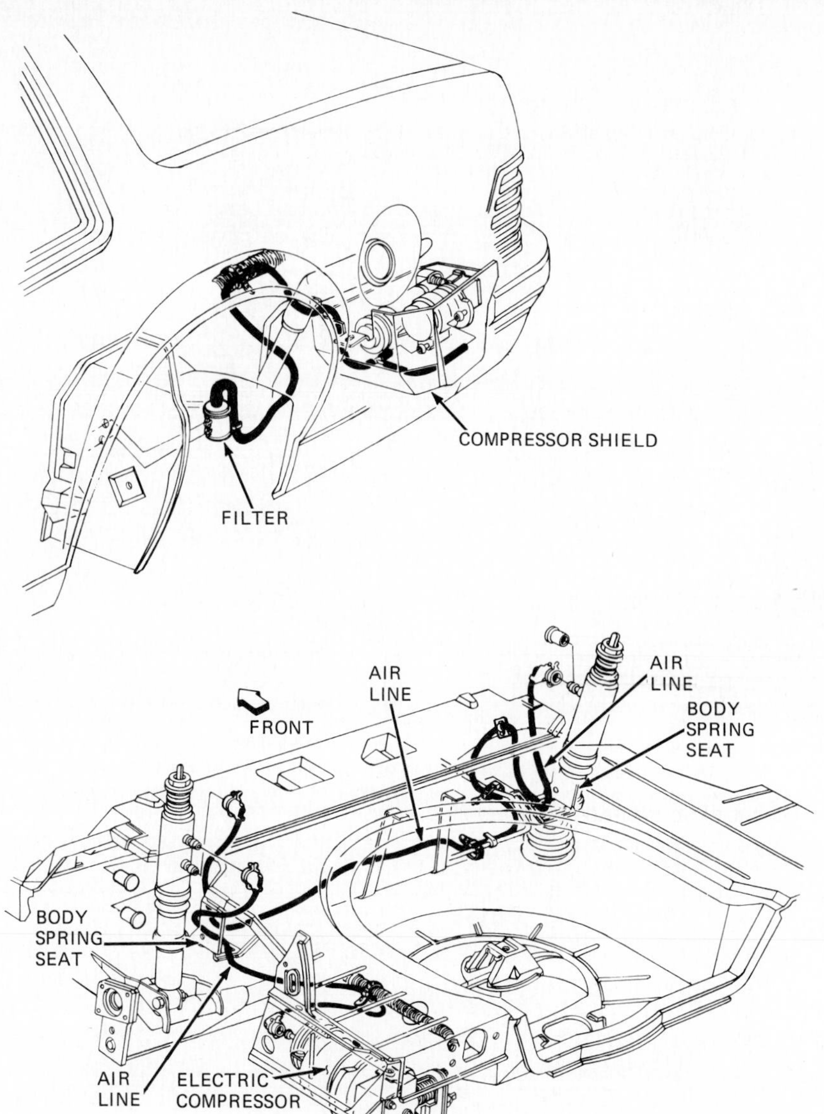

Fig. 48-33 Locations of components and air-line routing between the compressor and shock absorbers. *(Buick Motor Division of General Motors Corporation)*

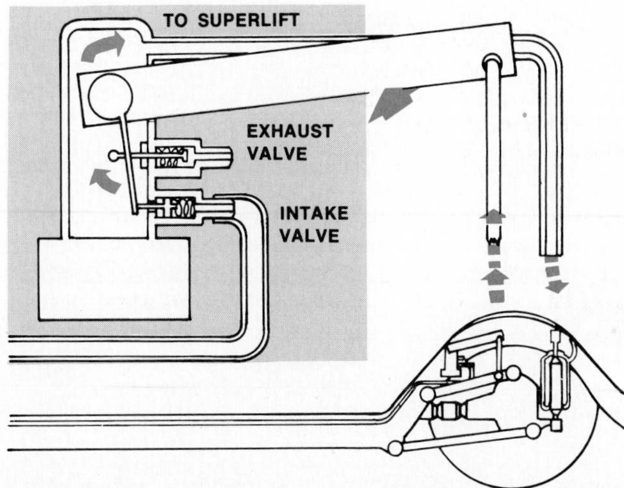

Fig. 48-34 A mechanical height-control valve showing the actions that take place when a load is added to the car. The components are not in proportion. *(Buick Motor Division of General Motors Corporation)*

solenoid exhaust valve. The solenoid then opens the valve to allow some of the air in the shock absorbers to escape. This lowers the car rear end to level.

The air dryer has a supply of chemical which absorbs any moisture in the air being pumped in by the compressor. This assures a supply of dry air for the shock absorbers. When air is released from the shock absorbers by the solenoid-exhaust-valve action, this air passes back through the air dryer where it absorbs the moisture from the chemical as it exits. The air dryer also has a valve that maintains some air pressure in the shock absorbers to improve ride characteristics.

Figure 48-36 shows an electronic automatic-level control system that is similar to the one shown in Fig. 48-35. The major difference is that the electronic control is installed inside one of the rear shock absorbers. A photo-optic sensor (an electric eye) is built into the shock absorber. The sensor signals the electronic control module when any change in height has occurred. Then the electronic module either sends air to the shock absorbers or releases air from them to adjust the height to the correct level. This system also has a time delay, similar to those previously discussed.

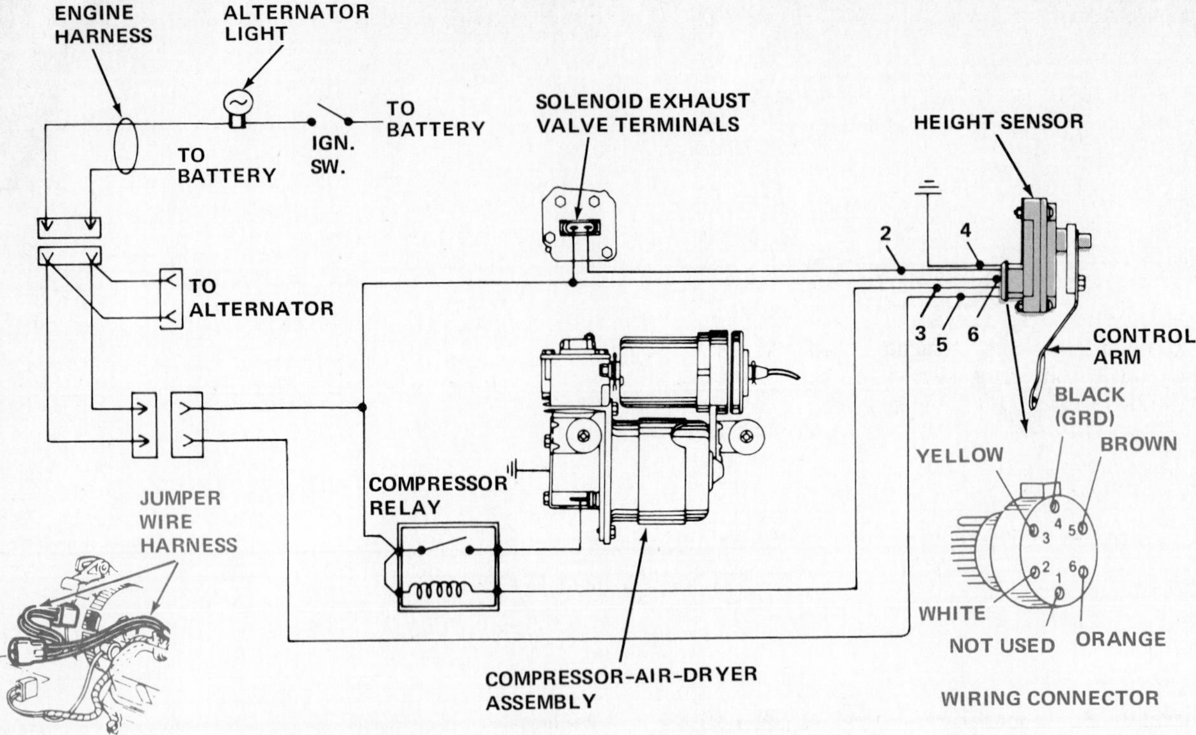

Fig. 48-35 Schematic wiring diagram of an electronic automatic-level-control system using an electronic height sensor. *(General Motors Corporation)*

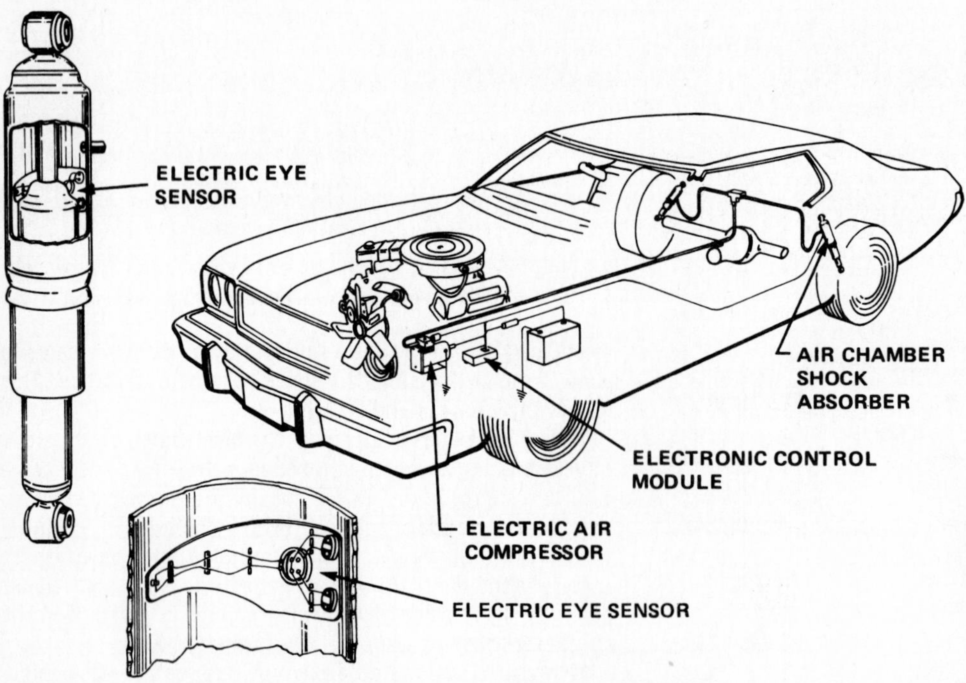

Fig. 48-36 An electronic automatic-level-control system. It uses an electric eye in one of the shock absorbers to switch the air compressor on and off. *(Monroe Auto Equipment Company)*

The sensor circuitry provides an 8- to 14-second delay before it triggers either the compressor or the exhaust-valve solenoid. This keeps normal ride motions from operating the system. In addition, the operating time is limited to a maximum of 3½ minutes. This time limit prevents continued operation in case of some trouble in the system (leakage, malfunction of solenoid). Turning the ignition off and on resets the system.

♦ REVIEW QUESTIONS ♦

Select the *one* correct, best, or most probable answer to each question. You can find the answers in the section indicated at the end of each question.

1. A front stabilizer bar is used to (♦48-12)
   a. increase vehicle load-carrying capacity
   b. provide a softer ride
   c. control suspension movement and body roll
   d. all of the above
2. Which of the following statements is *true* about a MacPherson-strut front suspension? (♦48-9)
   a. Upper and lower control arms are used.
   b. Two ball joints are used with each strut.
   c. Only an upper control arm is used.
   d. The shock absorber is built into the strut.
3. In the coil-spring rear-suspension system, the axle housing is kept in place by (♦48-15)
   a. U bolts
   b. the stabilizer bar
   c. control arms
   d. none of the above
4. The bolt on the end of a torsion bar is used for (♦48-10)
   a. locating the control-arm end of the torsion bar
   b. holding the back end of the torsion bar to the chassis
   c. adjusting the ride height of the vehicle
   d. caster adjustment
5. Three types of springs used in automotive suspension systems are (♦48-2)
   a. coil, leaf, and torsion bar
   b. coil, torsion bar, and air
   c. leaf, air, and gas
   d. all of the above
6. The rubber bushing in the eye of a leaf spring (♦48-14)
   a. absorbs vibration
   b. should be oiled regularly
   c. can be left out when the spring is replaced
   d. all of the above
7. A strut rod, or brake-reaction rod, is used with (♦48-6)
   a. each MacPherson strut
   b. leaf springs
   c. lower control arms using a single frame-attachment point
   d. upper control arms have two frame-attachment points
8. The steering knuckle is attached to the upper and lower control arms by (♦48-6)
   a. the kingpin
   b. upper and lower ball joints
   c. the stabilizer bar
   d. the spindle
9. The front-suspension system that uses leaf springs and an I-beam front axle usually is found on (♦48-11)
   a. race cars
   b. passenger cars
   c. trucks
   d. none of the above
10. Rear-end squat occurs (♦48-17)
    a. in front-engine, rear-drive cars
    b. when the car accelerates rapidly
    c. as a result of rear-end torque
    d. all of the above
11. A standard shock absorber will do all of the following *except* (♦48-21)
    a. dampen the action of the spring
    b. support the weight of the car
    c. help hold the tire to the road during driving
    d. help in controlling and steadying the car
12. In a coil-spring suspension system, as the wheel passes over a bump, the shock absorber is (♦48-22)
    a. expanded
    b. extended
    c. compressed
    d. none of the above
13. When the shock absorber is compressed or telescoped, fluid passes through the piston orifices and (♦48-22)
    a. out of the reservoir
    b. into the upper part of the cylinder
    c. into the piston rod
    d. out of the piston rod
14. Automatic level control takes care of changes in the (♦48-23)
    a. load in the rear of the car
    b. speed of the car
    c. air pressure in the tires
    d. load in the front of the car
15. To prevent the automatic-level-control system from reacting too quickly, the system includes (♦48-23)
    a. a height-control valve
    b. an air compressor
    c. an air dryer
    d. a time-delay mechanism

CHAPTER 49
Automotive Steering Systems

After studying this chapter, you should be able to:

1. Explain the purpose, construction, and operation of manual steering systems.
2. Discuss front-end geometry and the various angles involved.
3. Describe the construction and operation of a rack-and-pinion manual-steering gear.
4. Define *steering ratio.*
5. Explain the purpose, construction, and operation of power-steering systems.
6. Describe the difference between the integral and linkage types of power-steering systems.
7. Discuss the construction and operation of tilt and telescoping steering wheels.
8. Explain the purpose and construction of three types of collapsible steering columns.

♦ 49-1 PURPOSE OF STEERING SYSTEM

The steering system allows the driver to guide the car along the road and turn left or right as desired. The system includes the steering wheel, which the driver controls, the steering gear, which changes the rotary motion of the wheel into straightline motion, and the steering linkages. Most systems were manual until a few years ago. Then power steering became popular. It is now installed on about 90 percent of cars manufactured in the United States today.

♦ 49-2 TYPES OF STEERING SYSTEMS

Figure 49-1 shows a simplified pitman-arm type of steering system. The rack-and-pinion type is shown in Fig. 49-2. Chapter 48 describes how the wheels are supported on steering knuckles. The steering knuckles are attached to the steering arms by ball joints. The ball joints at each wheel permit the steering knuckle to swing from side to side. This movement turns the front wheels left or right so that the car can be steered.

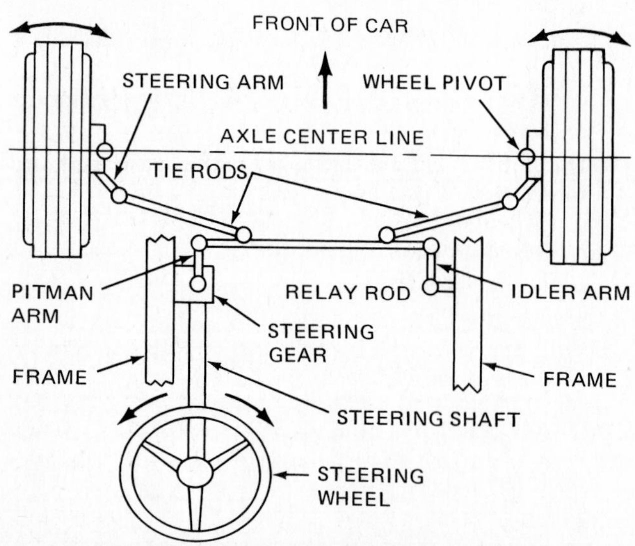

Fig. 49-1 A simplified pitman-arm steering system.

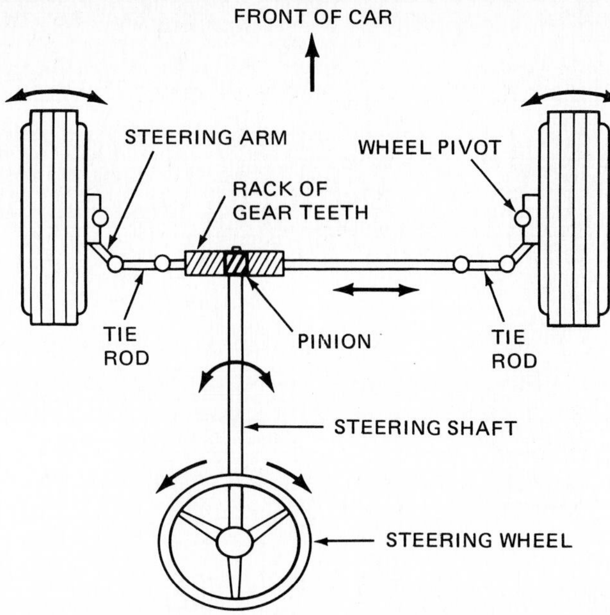

Fig. 49-2 A simplified rack-and-pinion steering system.

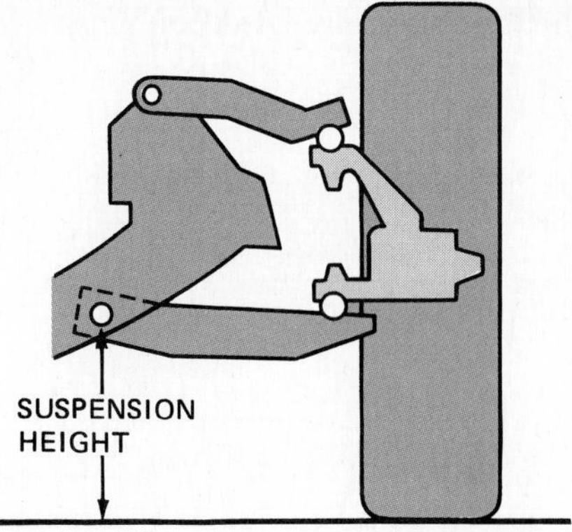

Fig. 49-3 Measurement of suspension height. *(Chrysler Corporation)*

The various angles in the front-suspension and steering systems are described in the following sections. Later sections describe steering gears and steering linkage.

Note Cars with independent rear-wheel suspension and cars with front-wheel drive may require rear-wheel alignment. This is described further in ♦50-23.

♦ 49-3 FRONT-END GEOMETRY

Front-end geometry (also called *wheel-alignment angularity*) is the relationship of the angles among the front wheels, the front-wheel attaching parts, and the ground. The various factors that enter into front-end geometry are

1. Front-suspension height
2. Camber
3. Steering-axis inclination
4. Caster
5. Toe
6. Turning radius

♦ 49-4 SUSPENSION HEIGHT

This is the distance measured from some specific point on the body, frame, or suspension to the ground (Fig. 49-3). If suspension height is not correct, it can affect the angles in the suspension system (the *wheel alignment*). Incorrect height could result from sagging coil or leaf springs, or incorrect torsion-bar adjustment.

♦ 49-5 CAMBER

Camber is the tilting in or out of the front wheels from the vertical when viewed from the front of the vehicle (Fig. 49-4). If the top of the wheel tilts out, it has positive camber. If the top of the wheel tilts in, it has negative camber. The amount of tilt, measured in degrees from the vertical, is called the *camber angle.*

On many cars, the wheels are given a slight outward tilt when camber is correctly set. Then, when the car is loaded and rolling along the road, the wheels should run straight up and down in a true vertical position. This "zero camber" position puts the full width of the tire tread on the road surface. The load and wear are evenly distributed across the tread.

However, an average "running" camber of zero does not occur at all times during driving. This is because camber changes as the body and wheels move up and down. Figure 49-5 shows how the camber goes negative when the tire hits a bump. When the tire drops into a hole in the road, the camber changes from zero to slightly positive (Fig. 49-6). In wheel-alignment and suspension work, the upward movement of the wheel is called *jounce.* The downward movement of the wheel is called *rebound.*

Any amount of camber, either positive (+) or negative (−), tends to cause uneven and more rapid tire wear. Tilting the wheel puts more load on one side of the tire tread than on the other side. This is why camber is called a *tire-wear angle.* Excessive positive camber causes the outside of the tire tread to wear. Excessive negative camber causes the inside of the tire tread to wear. The camber angle specified by the manufacturer is the setting that will normally provide maximum tire life.

If the vehicle is rolling on a perfectly level road, the ideal camber would be the same for both front wheels. This ideal camber would be just sufficient to bring the front wheels to the vertical position when the vehicle is normally loaded and moving. However, roads are seldom perfectly level.

Most roads are crowned, slightly higher at the center than on the two sides. When a car is moving along one side of a crowned road, the car tends to lean out slightly and drift to the right. This could cause the outside of the tread on the right front tire to wear excessively because it carries more than its share of car weight. To overcome this, some vehicles have camber settings that give the left front wheel ¼ degree more positive camber than the right. Since the car tends to pull toward the wheel with the most positive camber, the effects of the average crowned road are overcome. The car travels forward in a straight line.

Another reason for the increased positive camber on the left front wheel is that most cars are operated with only the driver in them. When the driver gets in the car, the driver's weight tends to slightly lower the left side of the car. This, in turn, reduces the positive camber of the left front wheel.

Camber changes have little effect on directional stability when the car is traveling straight ahead. However, in cornering, centrifugal force causes the body to lean toward the out-

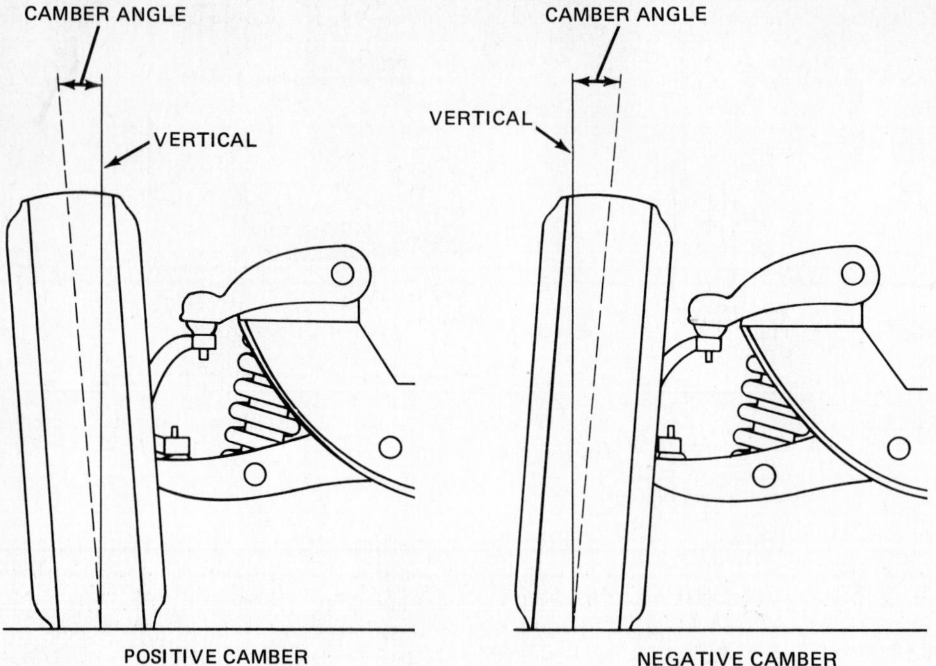

Fig. 49-4 Camber is the amount in degrees that the top of the wheel tilts outward (left) or inward (right) from vertical. *(Hunter Engineering Company)*

side of the turn (Fig. 49-7). Tire traction resists the resulting tendency to skid. As a result, the side forces against the bottom of the tires cause their tops to tilt toward the inside of the turn. In relation to the car frame, the outer wheel tilts in at the top, producing negative camber. The inner wheel tilts out at the top, producing positive camber.

Improper camber on both wheels can cause hard steering, unstable steering, and wander. Excessive, unequal camber between front wheels contributes to low-speed shimmy.

Incorrect chassis height, caused by sagging front or rear springs, can also affect camber. When a rear spring is weak, it can cause a camber change at the diagonally opposite front wheel. In some cars, this camber change may be as much as ¾ degree for each inch [25 mm] of rear-spring sag. This is the reason that front-suspension height is checked and corrected before the front wheels are aligned.

Camber is an adjustable angle (except on some cars with Macpherson-strut front suspension).

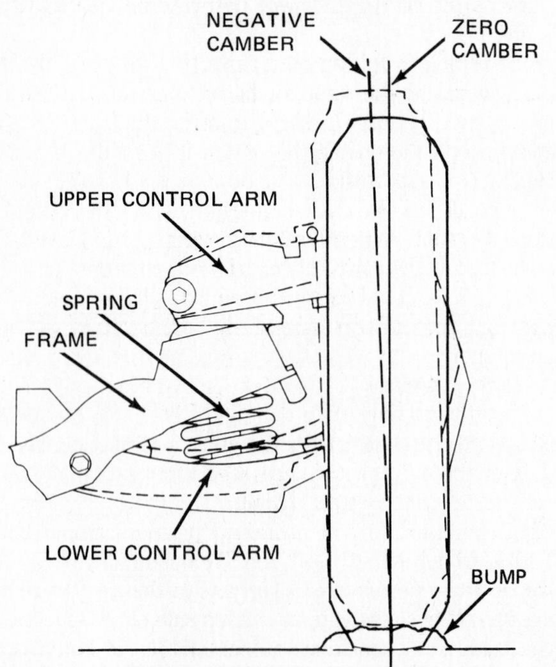

Fig. 49-5 Front-wheel camber goes negative when the tire hits a bump on the road.

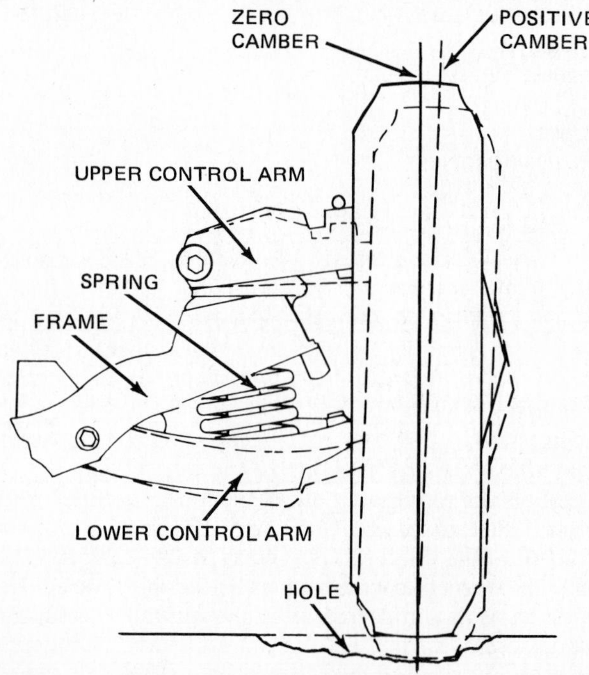

Fig. 49-6 Camber goes slightly positive when the tire drops into a hole in the road.

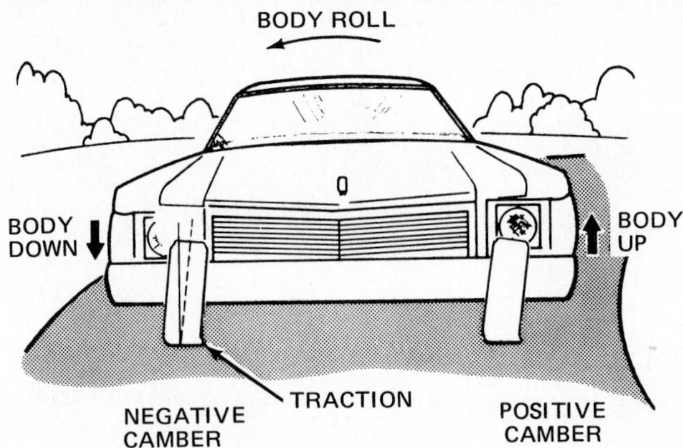

Fig. 49-7 Changes in camber during a left turn. *(Ford Motor Company)*

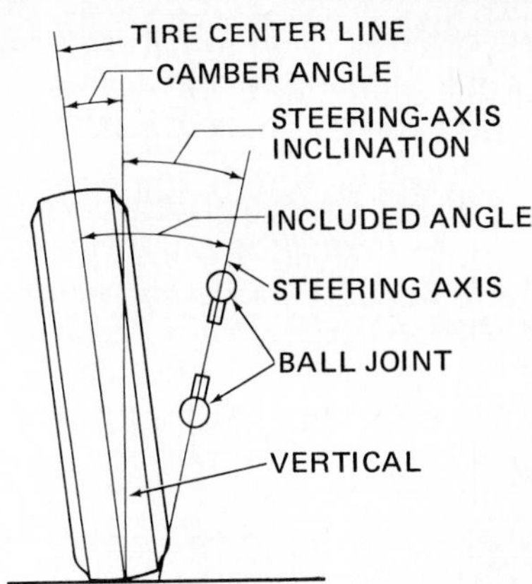

Fig. 49-9 Camber angle and steering-axis inclination (SAI). Positive camber is shown.

♦ 49-6 STEERING-AXIS INCLINATION

In older cars, all steering systems had a kingpin that attached the steering knuckle to a support (Fig. 49-8). Then ball joints (♦48-6) were adopted, allowing the steering knuckle and steering-knuckle support to be combined into a single part. With ball joints, no kingpin is used. Instead, the steering knuckle is supported through ball joints by upper and lower control arms.

A line drawn through the centers of the ball joints is called the *steering axis* (Fig. 49-9). This is the line around which the steering knuckle pivots as the wheel swings to the right or left. Therefore, *steering axis* is defined as *the center line around which the front wheel swings for steering.*

Steering-axis inclination (SAI) is also called *ball-joint inclination* [*or kingpin inclination* (KPI) on vehicles that have kingpins]. It is the angle, measured in degrees, between the vertical and a line drawn through the centers of the ball joints, when viewed from the front of the vehicle (Fig. 49-10). Another definition is the inward tilt of the ball-joint center line from the vertical (Fig. 49-9).

Inward tilt of the ball-joint center line is desirable for several reasons. First, it helps provide steering stability by returning the wheels to a straight-ahead position after the car has turned. This is called *returnability*. Second, SAI reduces steering effort, particularly when the car is stationary. Third, it tends to keep the front wheels rolling straight ahead. This is because the inward tilt of the steering axis causes the front end to raise slightly as the front wheels are swung away from the straight-ahead position.

When the front wheel is in the straight-ahead position, the wheel spindle is at its highest point. As the spindle pivots forward or backward, it begins to drop. This is because the spindle pivots around the steering axis, which is tilted inward. However, the tire is already in contact with the ground and cannot move down. So, as the wheel is swung away from the straight-ahead position, the steering knuckle, ball joints, and car body and frame are lifted upward. The lift is slight—only about 1 inch [25 mm] or less. But it is enough for the weight of the car to help return the wheels to straight ahead after the turn is completed.

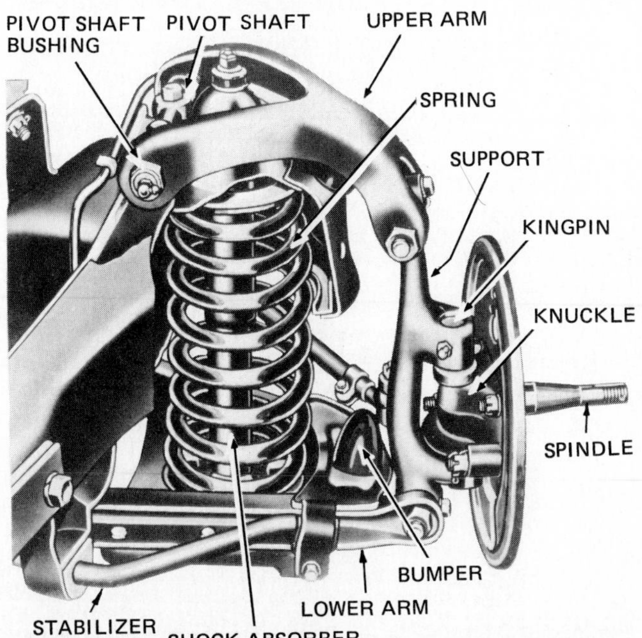

Fig. 49-8 Coil-spring front suspension using a kingpin. *(Ford Motor Company)*

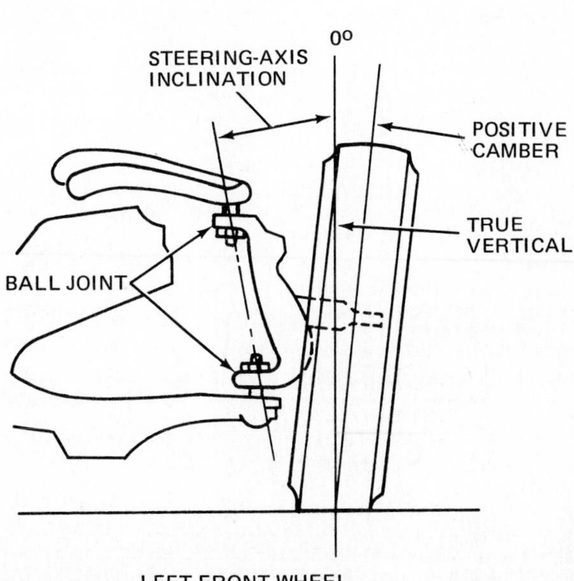

Fig. 49-10 Steering-axis inclination (SAI). *(Chevrolet Motor Division of General Motors Corporation)*

This same action tends to make rolling wheels resist any small force trying to move them away from the straight-ahead position. The resistance is not enough to cause hard steering, only good steering stability.

SAI is not adjustable. The amount of the inclination, or tilt, is designed into the steering knuckle. Generally, if camber can be adjusted to specifications, the SAI is correct. However, a change in camber will cause a similar change in SAI. This is discussed further in the following section on included angle (♦49-7). When the SAI is not within specifications, the spindle, ball joints, or other parts are bent or worn and should be replaced.

♦ 49-7 INCLUDED ANGLE

In front-end geometry, the *included angle* is the camber angle plus the SAI angle (Figs. 49-10 and 49-11). This can be written as

$$\text{Included angle} = \text{camber} + \text{SAI}$$

The included angle determines the point of intersection of the tire center line with the steering axis or ball-joint center line (Fig. 49-11). This point determines whether the rolling wheel tends to toe in or toe out.

Toe-in (♦49-9) is the amount that the front wheels point inward (Fig. 49-12). It is measured in inches, millimeters, or degrees. Toe-out is just the opposite. A wheel that has toe-out points outward as it rolls. The tire on a wheel that rolls with toe-in or toe-out will wear rapidly. The tire has to travel in the direction that the car is moving. But if the tire has toe-in or toe-out, it is dragged sideways as it rolls forward. The more toe-in or toe-out the tire has, the more it is dragged sideways and the faster it wears. Ideally, the wheel rolls straight ahead, with neither toe-in nor toe-out.

Figure 49-11 shows two opposing forces acting on the wheel. One is the forward push through the ball joints. The other is the road resistance to the tire. If these two forces are exactly in line, the wheel will have no tendency toward toe-in or toe-out. However, the two forces are in line with each other only when the point of intersection of the tire center line with the steering axis is at the road surface.

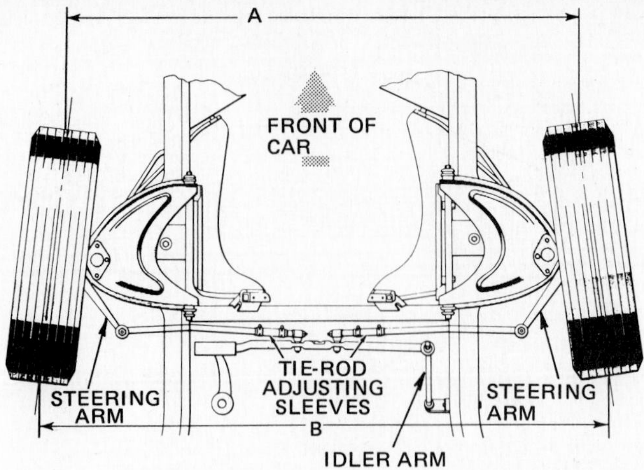

Fig. 49-12 Toe-in. The wheels are viewed from above, with the front of the car at the top of the illustration. Distance *A* is less than distance *B*. *(Bear Manufacturing Company)*

When the point of intersection is below the road surface (Fig. 49-11A), the wheel tends to toe out. Since the forward push through the center line of the ball joints is inside the tire center line at the road surface, the wheel attempts to swing outward. When the point of intersection is above road level (Fig. 49-11B), the wheel attempts to swing inward, or toe in.

♦ 49-8 CASTER

Caster is the angle (measured in degrees) formed by the forward or rearward tilt of the steering axis from vertical, when viewed from the side of the wheel (Fig. 49-13). The angle is positive (+) when the steering axis tilts backward (Fig. 49-14, left). With positive caster, the upper ball joint is behind the lower ball joint. Caster is negative (−) when the steering axis tilts forward (Fig. 49-14, right). Then the upper ball joint is

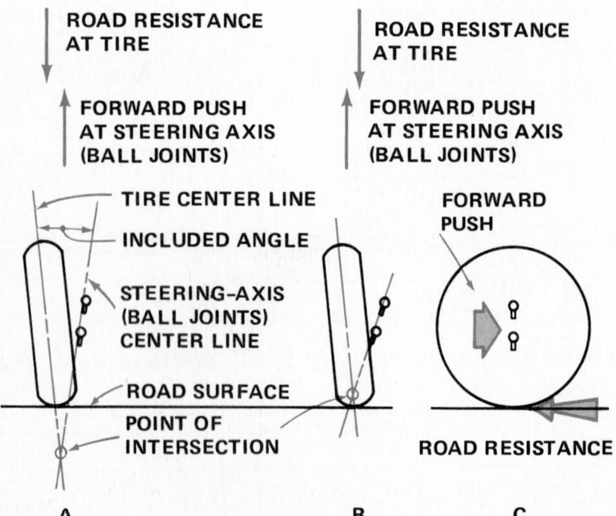

Fig. 49-11 Effect when the point of intersection is (A) below the road surface and (B) above the road surface. The left front wheel as viewed from the driver's seat is shown in (A) and (B). (C) is a side view of the wheel to show the two forces acting on the tires and ball joints.

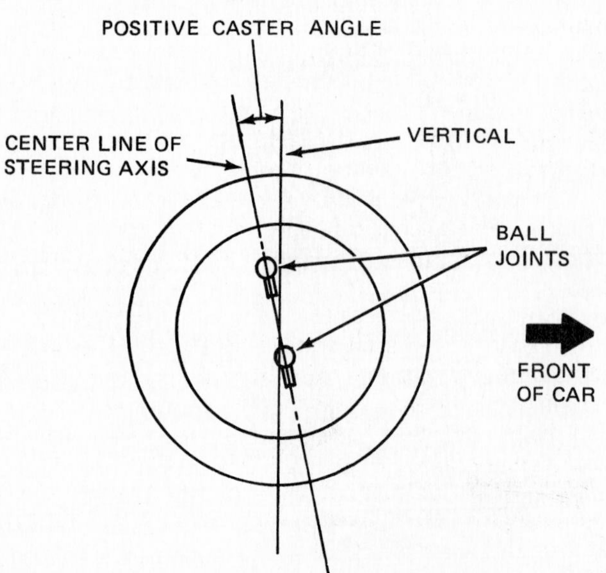

Fig. 49-13 Caster of the left-front wheel as viewed from the driver's seat. The view is from the inside so that the backward tilt of the steering axis from the vertical can be seen. This backward tilt is called *positive caster*.

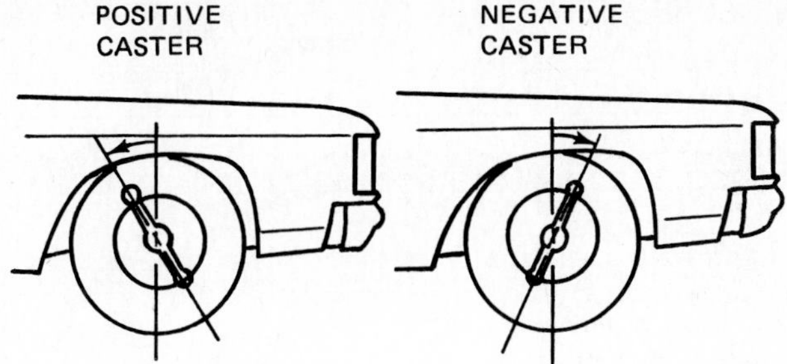

Fig. 49-14 Left, positive caster angle. Right, negative caster angle. *(Ford Motor Company)*

ahead of the lower ball joint. When the upper ball joint is directly above the lower one, there is zero caster.

Although the caster angle is adjustable (except on some cars with MacPherson struts), caster has little effect on tire wear. There are three reasons why caster is used. (Note that tire wear is *not* one of them.) These are:

1. To maintain directional stability and control
2. To increase steering returnability
3. To reduce steering effort

Positive caster aids directional stability. The center line of the ball joints passes through the road surface ahead of the center line of the wheel (Fig. 49-13). Therefore, the push on the ball joints is ahead of the road resistance to the tire (Fig. 49-11). The tire is trailing behind. A car wheel that is pulled has greater directional and steering stability than a wheel that is pushed. In addition, positive caster tends to keep the wheels pointed straight ahead. It helps overcome any tendency for the car to wander or steer away from the straight-ahead position.

Negative caster does not aid directional stability. SAI contributes more than caster to directional stability of the tire. This is why cars with manual steering, which often have a slight negative caster, still retain good directional stability.

Caster also affects the tendency of the steering wheel to return to its center position after making a turn. This is called *returnability*. Positive caster increases steering-wheel returnability. Therefore, cars with power steering often have slightly more positive caster than manual-steering cars. The additional positive caster helps overcome the tendency of the power steering to hold the front wheels in a turn. Although the additional positive caster requires greater turning effort, the driver does not notice it because of the power steering.

Positive and negative caster have different effects on the actions of the wheels and car body. When both front wheels have positive caster, the body leans toward the outside of the turn (Fig. 49-7). But if the front wheels have negative caster, the car tends to lean into the turn (Fig. 49-15).

When the left front wheel has positive caster, during a left turn the left front wheel tries to move down against the road surface. This causes the ball joints to be lifted, which lifts the left side of the car body. At the same time, positive caster causes the right side of the car to be lowered slightly during a left turn.

When the right side of the car is lowered and the left side is lifted as a left turn is made, the car leans out on the turn (Fig. 49-7). This is just the opposite of what is desirable, since it adds to the effect of centrifugal force on the turn. By using negative caster, which tilts the center line of the ball joints

Fig. 49-15 Negative caster counteracts the roll-out effects of centrifugal force during a high-speed turn. *(McQuay-Norris, Inc.)*

forward as in Fig. 49-14 (right), the car can be made to lean in on a turn. For example, with negative caster, the left side of the car drops during a left turn and the right side lifts. This decreases the roll-out effect of centrifugal force.

Caster has another effect. Positive caster tends to make the front wheels toe in. With positive caster, the car is lowered as the front of the wheel pivots inward. Therefore, the weight of the car is always exerting force to make the wheels toe in. With negative caster, the wheels tend to toe out, although the car steers easier. Then the force to return the front wheels to straight ahead after a turn is provided by SAI.

Positive caster increases the force required to steer because it tries to keep the wheels running straight ahead. SAI also tried to keep the wheels straight ahead (♦49-6). Therefore, to make a turn, the steering effort must overcome the effects of both positive caster and SAI.

Excessive positive caster may cause too much steering effort, snap-back of the steering wheel after a turn, wander, low-speed shimmy, and an increased amount of road shock. Improper caster may occur as a result of spring sag. For example, front-spring sag decreases positive caster (Fig. 49-16). For this reason, front-suspension height is checked and corrected before the front end is aligned.

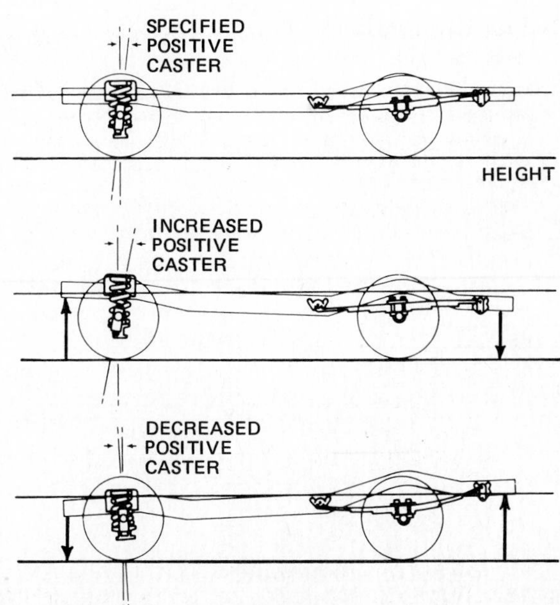

Fig. 49-16 Effects of sagging springs on caster. *(Ford Motor Company)*

♦ 49-9 TOE

Toe is the amount in inches, millimeters, or degrees by which the front wheels point inward or outward. Its purpose is to ensure parallel rolling of the front wheels, to stabilize steering, and to prevent side-slipping and excessive wear of the tires. With toe-in, the front wheels attempt to roll inward instead of straight ahead. This is because the tires are slightly closer together at the front than at the rear (Fig. 49-12). A typical toe-in setting is about ⅛ inch [3.2 mm].

Front-wheel toe is set with the car standing still. The toe offsets the play in the steering linkage, which is eliminated when the car is moving forward (Fig. 49-17). This change in toe is due to the rolling resistance of the tires against the road. The amount of toe should be sufficient to prevent any toe-out or toe-in while the car is moving forward. This is called *running toe* and may be measured by some types of dynamic wheel analyzers. Ideally, the toe setting will bring the actual running toe to zero when the car is moving.

As a car with toe-in begins to move forward, the backward push of the ground against the tires (shown in Fig. 49-17) tries to force the wheels to toe out. Since the tie rods are behind the wheels, this force compresses the steering linkage and takes up any play. As a result, the tires become parallel and roll straight ahead.

Toe greatly affects tire wear. Therefore toe can be adjusted. Excessive toe-in or toe-out causes the tire tread to scrub the road. This happens when the car is moving straight, with the tire turned slightly in or out. Whatever the toe setting, the wheels must not run with toe-out. This causes wander.

♦ 49-10 TURNING RADIUS

Turning radius is sometimes called *toe-out during turns* and *turning angle.* It is the difference between the two angles formed by the two front wheels and the car frame during turns. Then the inner wheel is following the radius of a smaller circle than the outer wheel (Fig. 49-18). Therefore the inner wheel must toe out more to prevent tire sideslip and excessive wear.

When the front wheels are steered to make the turn shown in Fig. 49-18, the inner wheel turns at an angle of 23 degrees, while the outer wheel turns only 20 degrees. This permits the inner wheel to follow a shorter radius than the outer wheel. The two front wheels turn on concentric circles, which have a common center (Fig. 49-18).

Toe-out during turns is achieved by the proper relationship between the steering arms, tie rods, and steering gear (Fig. 49-19). For this reason, the steering arms are angled inward (Fig. 49-12). If the steering arms were parallel to the wheels, then the wheels would remain parallel during a turn. This would scuff the tires. The relationship of the parts ensures that the inner wheel on a curve always has more toe-out than the outer wheel.

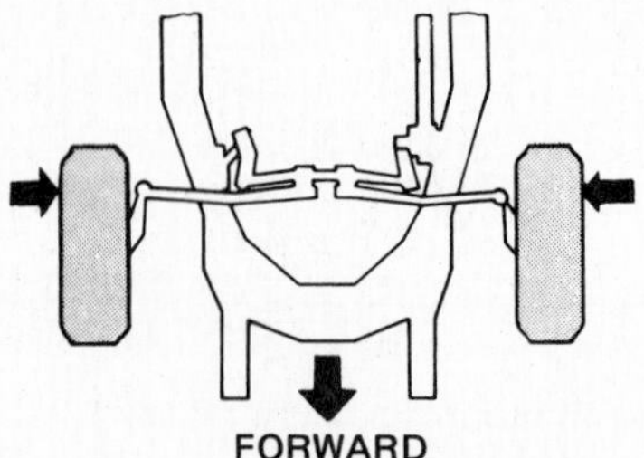

Fig. 49-17 As the car moves forward, the tires try to toe out, which compresses the steering linkage. *(Ford Motor Company)*

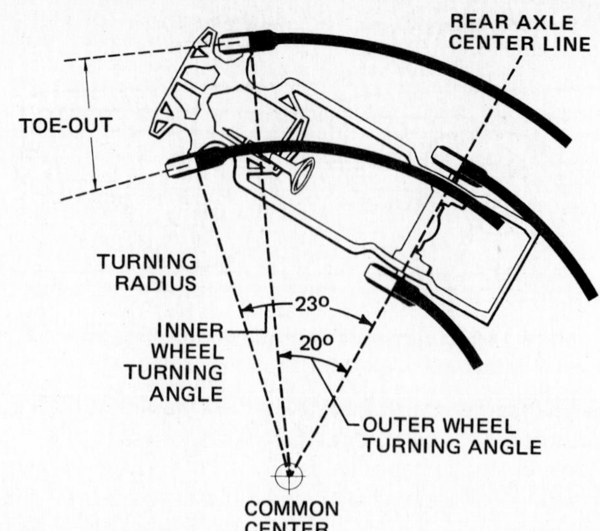

Fig. 49-18 Turning radius, or toe-out on turns. *(Hunter Engineering Company)*

As the dotted line in Fig. 49-19 shows, when the tie rod is moved to the left during a right turn, it pushes at almost a right angle against the left steering arm. At the same time, the right end of the tie rod moves to the left and swings forward. This turns the right wheel an additional amount. When a left turn is made, the left wheel is turned more than the right wheel.

Figure 49-19 shows a parallelogram type of steering linkage. Other types of steering linkage provide a similar toe-out of the inner wheel during turns. Turning radius cannot be adjusted. If it is not correct, one or both of the steering arms are bent and should be replaced.

♦ 49-11 STEERING LINKAGE

Many types of steering linkages have been used to connect the front-wheel steering knuckles to the steering gear. In the system using the pitman-arm steering gear (Fig. 49-1), the pitman arm swings from one side to the other as the steering wheel is turned. (In some cars, the pitman arm swings forward or backward.) This movement is carried to the wheel spindles by the steering linkage.

In systems using rack-and-pinion steering (Fig. 49-2), rotation of the steering wheel and pinion causes the rack to move to left or right. This motion is carried to the wheel spindles by the steering linkage.

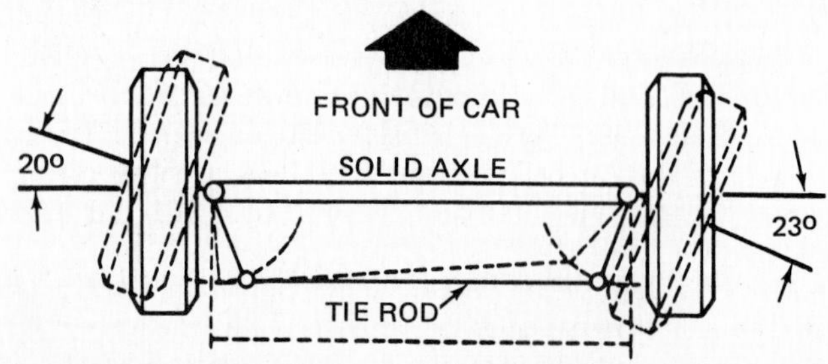

Fig. 49-19 How toe-out on turns is obtained. *(Chevrolet Motor Division of General Motors Corporation)*

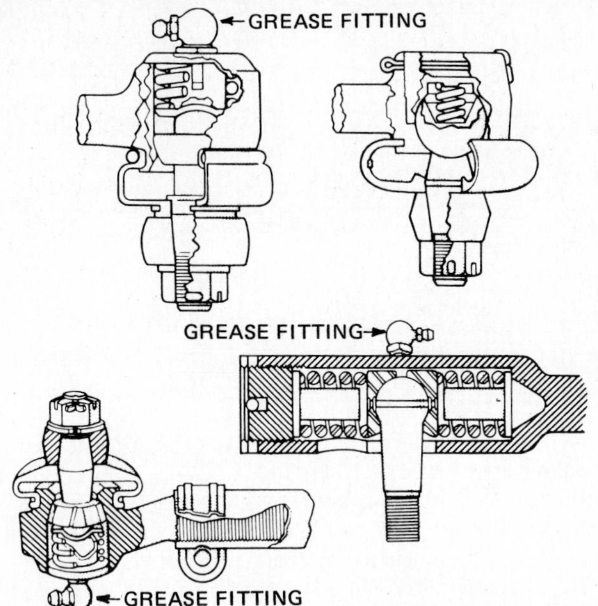

Fig. 49-20 Ball sockets and tie-rod ends. *(Ford Motor Company)*

The linkage must be designed so that the driver can easily and accurately control the position of the wheels. However, the position of the wheels, and any up-and-down motion they have, should not affect the driver's control of the car.

♦ 49-12 BALL SOCKETS AND TIE-ROD ENDS

The various parts of the steering linkage are connected by ball sockets or tie-rod ends of several kinds (Fig. 49-20). Some have a grease fitting for lubrication. Others are prelubricated for life at the time of manufacture.

On many cars, the idler arm is connected through rubber bushings. These bushings twist as the idler arm swings to one side or the other. They then supply some force to help return the wheels to center after a turn is completed. Ball sockets are smaller, but similar in construction to the front-suspension ball joints (♦48-6).

MANUAL STEERING

♦ 49-13 STEERING GEARS

The steering gear converts the rotary motion of the steering wheel into straight-line motion of the linkage. There are two basic types of steering gears, the pitman-arm type and the rack-and-pinion type. Either type can be used in a manual-steering system or a power-steering system.

The pitman-arm type has a gear box at the lower end of the steering shaft (Fig. 49-21). The rack-and-pinion type has a small gear (a *pinion*) at the lower end of the steering shaft (Fig. 49-22). The action is the same in either system. When the steering wheel and shaft are turned by the driver, the rotary motion is changed into straight-line motion. This causes the front wheels to pivot or swing from one side to the other to steer the car.

♦ 49-14 PITMAN-ARM STEERING GEARS

Steering linkages using pitman arms are shown in Figs. 49-1 and 49-21. The steering gear at the lower end of the steering shaft consists essentially of two parts. These are a worm on

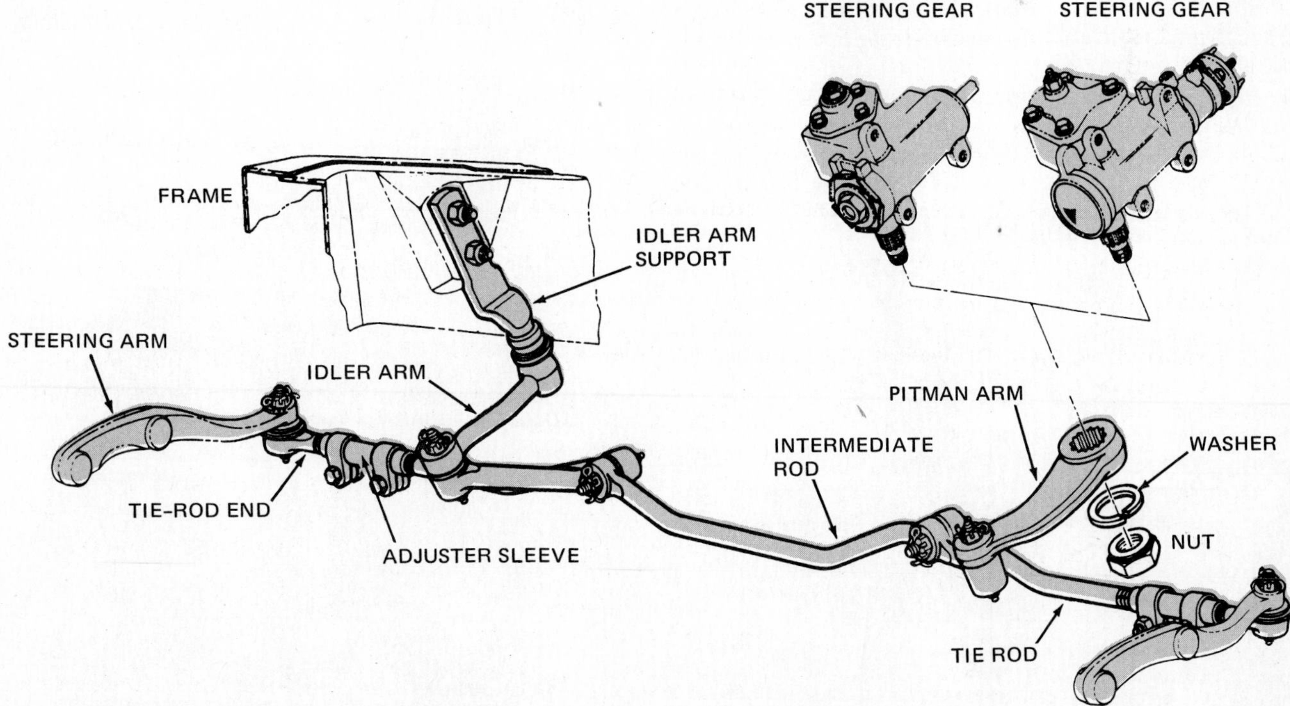

Fig. 49-21 Steering linkage for pitman-arm steering gears. The linkage is the same for both manual and power steering gears. *(Pontiac Motor Division of General Motors Corporation)*

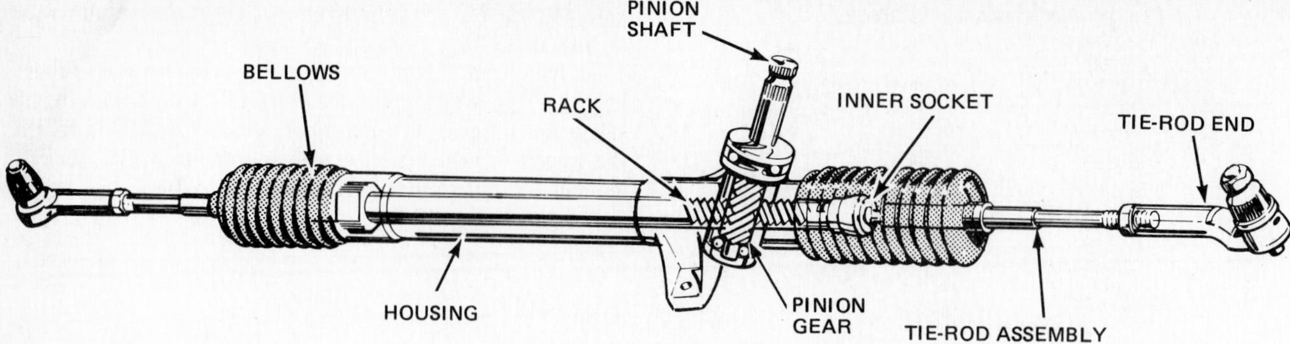

Fig. 49-22 Phantom view of a manual rack-and-pinion steering gear. *(Chrysler Corporation)*

the end of the steering shaft, and a pitman-arm shaft (or cross-shaft) on which there is a gear sector, a toothed roller, or a stud.

The gear sector, toothed roller, or stud meshes with the worm (Fig. 49-23). In Fig. 49-23, the steering gear uses a toothed roller. The roller and worm teeth mesh. When the worm is rotated (by rotation of the steering wheel), the roller teeth must follow along. This action causes the pitman-arm shaft to rotate. The other end of the pitman-arm shaft carries the pitman arm. Rotation of the pitman-arm shaft causes the arm to swing in one direction or the other. This motion is then carried through the linkage to the steering knuckles at the wheels.

Note The pitman-arm shaft is also called the *cross-shaft, pitman shaft, roller shaft, steering-arm shaft,* and *sector shaft.*

The recirculating-ball steering gear is shown in Fig. 49-24. In these units, the worm gear on the end of the steering shaft has a special "nut" running on it. The nut rides on rows of small recirculating balls. The recirculating balls move freely through grooves in the worm and inside the nut. As the steering shaft is rotated, the balls force the nut to move up and down the worm gear. A short rack of gear teeth on one side of the nut mesh with the sector gear. Therefore, as the nut moves up and down the worm, the sector gear turns in one direction or the other for steering.

The recirculating balls are the only contacts between the worm and the nut. This greatly reduces friction and the turning effort or force applied by the driver for steering.

The balls are called recirculating balls because they continuously recirculate from one end of the ball nut to the other end through a pair of ball-return guides. For example, suppose the driver makes a right turn. Then the worm gear is rotated in a clockwise direction when viewed from the driver's seat. This causes the ball nut to move upward. The balls roll between the worm and the ball nut. As the balls reach the upper end of the nut, they enter the return guide and then roll back to the lower end. There they reenter the groove between the worm and the ball nut.

♦ 49-15 RACK-AND-PINION STEERING GEARS

The rack-and-pinion steering gear has become increasingly popular for today's smaller cars. It is simpler, more direct acting, and may be straight mechanical or power-assisted in operation. Figure 49-25 shows a complete rack-and-pinion

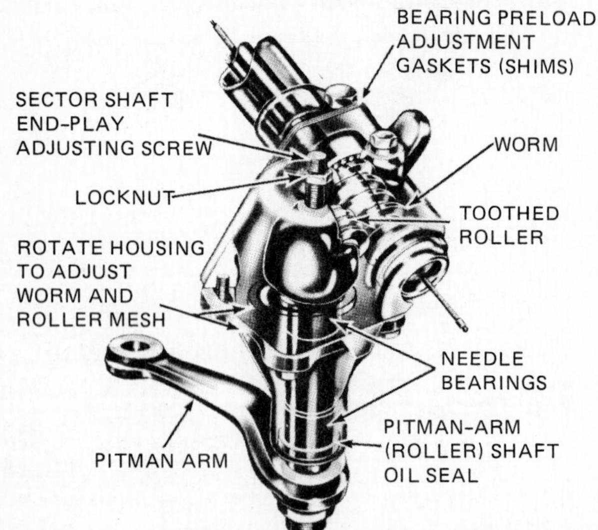

Fig. 49-23 A pitman-arm steering gear using a toothed roller attached to the pitman-arm shaft. The teeth on the worm mesh with the teeth on the roller. *(Ford Motor Company)*

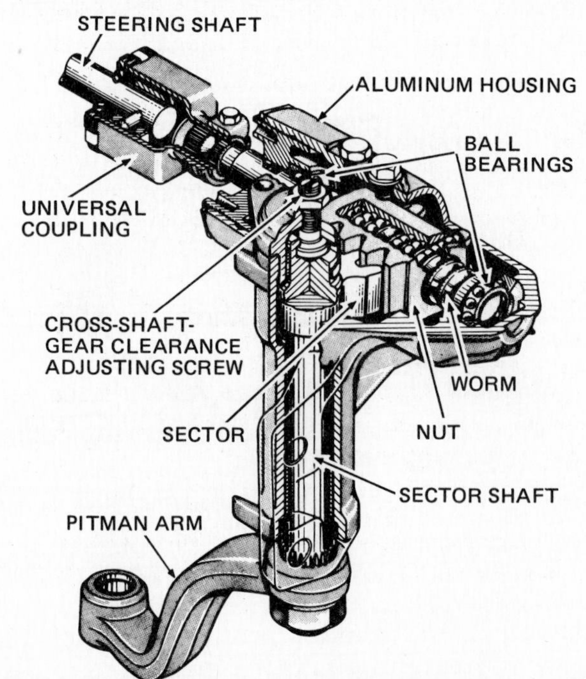

Fig. 49-24 A cutaway recirculating-ball steering gear. *(Chrysler Corporation)*

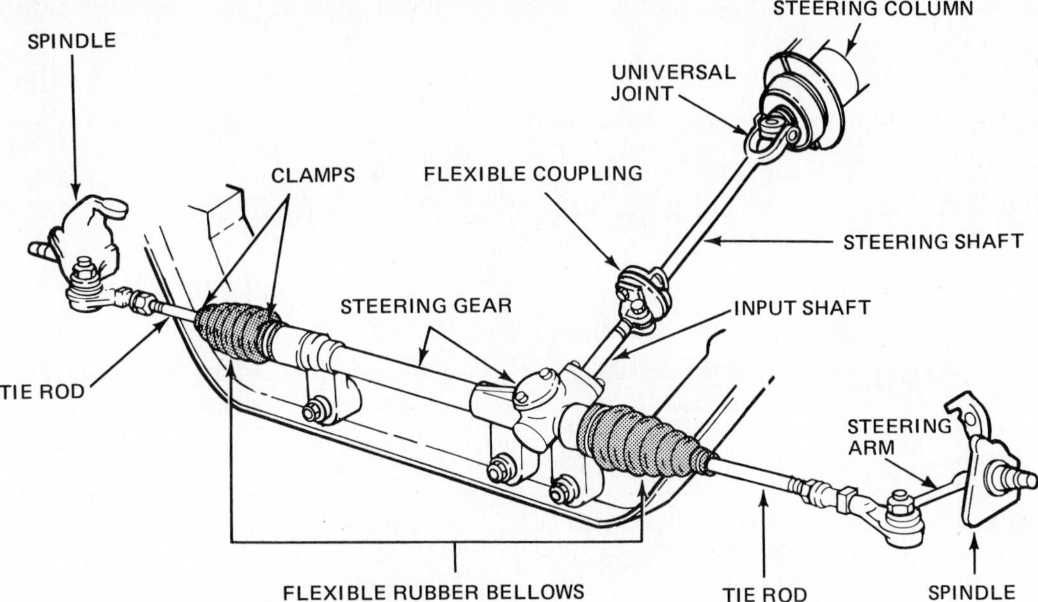

Fig. 49-25 Rack-and-pinion steering gear showing linkages to the wheel spindles. *(Ford Motor Company)*

steering system, set apart from the rest of the car. As the steering wheel and shaft are turned, the rack moves from one side to the other. This pushes or pulls on the tie rods, forcing the wheel spindles to pivot on their ball joints. This turns the wheels to one side or the other so that the car is steered.

Figure 49-25 shows the steering gear and tie rods. The universal joint at the upper end of the steering shaft and the flexible coupling at the lower end help eliminate road shock. They also prevent road shocks and noise from passing up through the steering column to the driving compartment.

Figure 49-26 shows a completely disassembled manual rack-and-pinion steering gear. The pinion gear is integral with the input shaft. The inner ends of the tie rods have balls which fit into the ball sockets on the two ends of the rack. The ball-and-socket arrangement allows the spindle ends of the tie rods to move up and down with the spindles and wheels.

Rack-and-pinion steering is basically a very simple type of steering gear. It can be used in small cars where the steering forces are light. However, the gear ratio is limited by the diameter of the steering wheel and the pinion gear. On larger, heavier vehicles, this can be a disadvantage. Therefore, other types of steering gears, such as the recirculating ball (♦49-14), are usually found on larger cars and trucks. The greater mechanical advantage possible with the other types reduces the effort required to steer a larger vehicle.

In a small car, rack-and-pinion steering is quick and easy. It provides the maximum amount of road feel as the tires meet irregularities in the road. There is no damping out of road shocks and vibration. Other types of steering systems usually provide some damping action. They can also be adjusted to eliminate almost all play in the steering wheel. Mechanical advantage and the meaning of "quick" steering are discussed in ♦49-16.

♦ 49-16 STEERING RATIO

One of the jobs of the steering gear is to provide mechanical advantage. In a machine or mechanical device, this is the ratio of the output force to the input force applied to it. This means that a relatively small applied force can produce a much greater force at the other end of the device.

In the steering system, the driver applies a relatively small force to the steering wheel. This results in a much larger steering force at the front wheels. For example, in one steering system, applying a force of 10 pounds [44.5 N] to the steering wheel can produce up to 270 pounds [1201 N] at the wheels. This increase in steering force is produced by the steering ratio.

The steering ratio is the number of degrees that the steering wheel must be turned to pivot the front wheels 1 degree. For example, in Fig. 49-27 the steering wheel must be turned 17.5 degrees to pivot the front wheels 1 degree. Therefore, the steering ratio is 17.5:1. In cars built in the United States, manual-steering ratios typically vary between 15:1 and 33:1. The higher the steering ratio (33:1, for example), the easier it is to steer the car, all other things being equal. However, the higher the steering ratio, the more the steering wheel has to be turned to achieve steering. With a 30:1 steering ratio, the steering wheel must turn 30 degrees to pivot the front wheels 1 degree.

Actual steering ratios vary greatly, depending on the type of vehicle and the type of operation. Many cars with manual steering use steering ratios as high as 28:1 to minimize steering effort. Some lightweight sports cars have steering ratios as low as 10:1. High steering ratios are often called *slow* steering because the steering wheel has to be turned many degrees to produce a small steering effect. Low steering ratios, called *fast* or *quick* steering, require much less steering-wheel movement to produce the desired steering effect.

Steering ratio is determined by two factors: steering-linkage ratio and the gear ratio in the steering gear. The steering-linkage ratio is determined by the relative length of the pitman arm and the steering arm. The steering arm is bolted to the wheel spindle at one end and connected to the steering linkage at the other end.

When the effective lengths of the pitman arm and the steering arm are equal, the linkage has a ratio of 1:1. If the

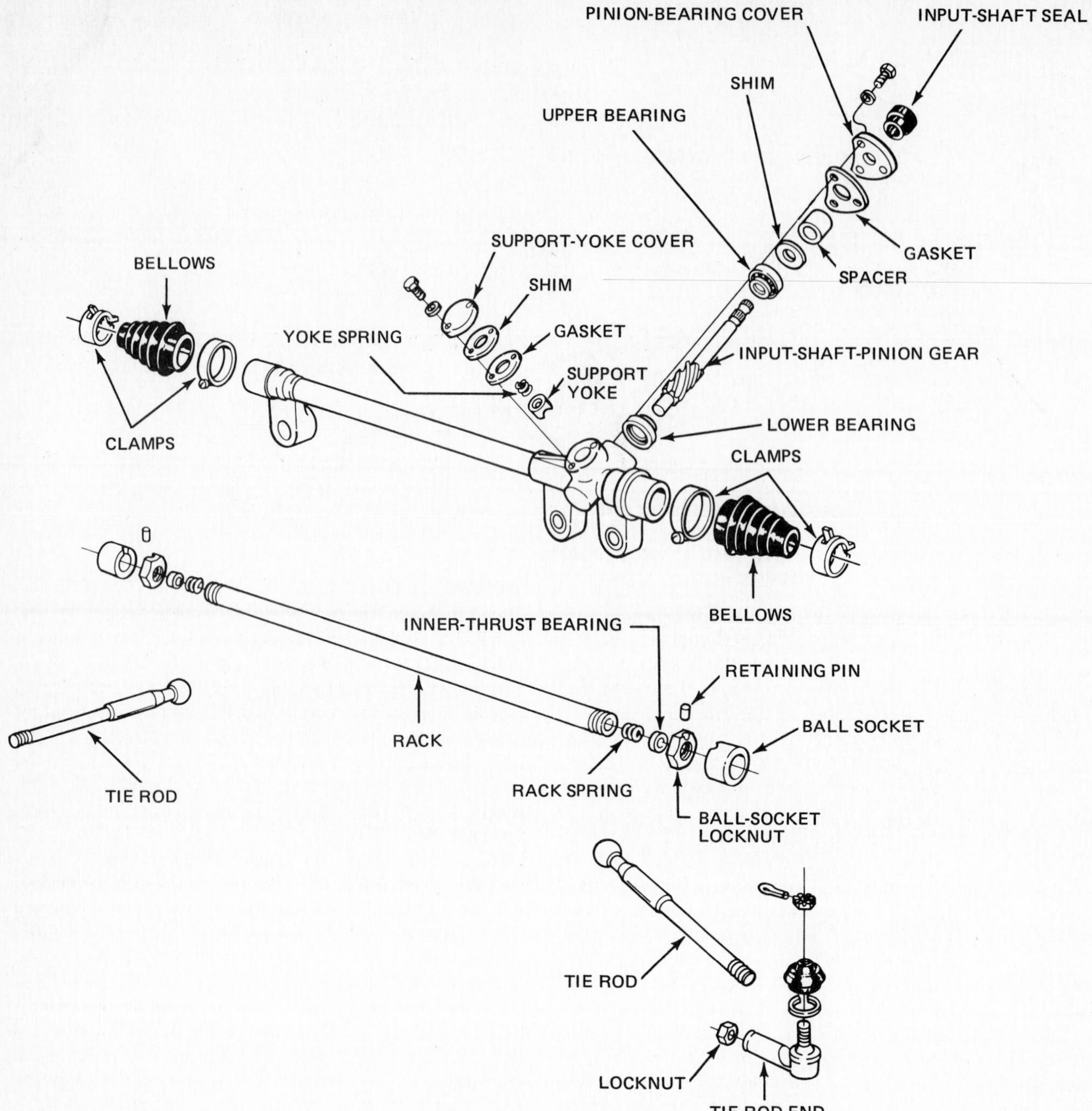

Fig. 49-26 Disassembled rack-and-pinion steering gear. *(Ford Motor Company)*

pitman arm is shorter than the steering arm, the ratio is less than 1:1. In Fig. 49-27, for example, the pitman arm is about twice as long as the steering arm. This means that for every degree the pitman arm swings, the wheels will pivot about 2 degrees. Therefore, the steering-linkage ratio is about 1:2.

Most of the steering ratio is developed in the steering gear. The ratio is due to the angle or pitch of the teeth on the worm gear and to the angle or pitch of the teeth on the sector gear. Steering ratio is also determined to some extent by the effective length and shape of the teeth on the sector gear.

In a rack-and-pinion steering system (♦49-15), the steering ratio is determined largely by the diameter of the pinion gear. The smaller the pinion, the higher the steering ratio. However, there is a limit to how small the pinion can be made.

POWER STEERING

♦ 49-17 POWER-STEERING FUNDAMENTALS

When a car has manual steering, the driver supplies all the steering force. Then, through the mechanical advantage of the steering gear and the linkage, the front wheels are pointed to the right or left as desired by the driver.

However, there are some disadvantages to manual steering. This became evident as cars and their engines became bigger and heavier, and as more cars were equipped with wider, low-pressure tires. To steer these cars, the steering ratio had to be increased. This meant that more turns of the steering wheel were required to move it from lock to lock, and the steering was slower.

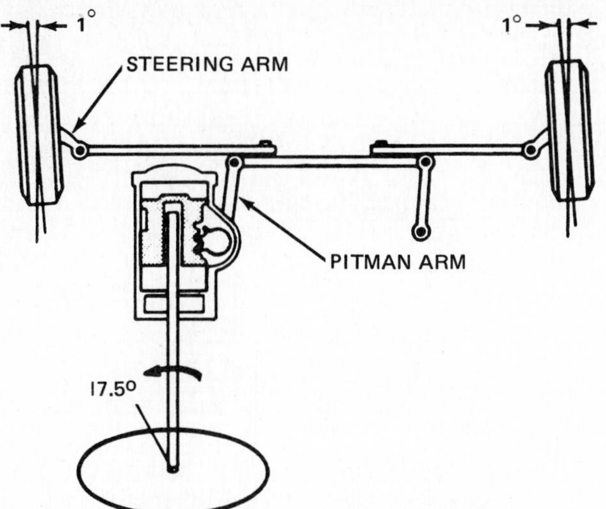

Fig. 49-27 Steering ratio. In the typical arrangement shown, turning the steering wheel 17.5 degrees is required to pivot the front wheels 1 degree.

Larger tires, with a heavier weight on them, made parking the car more difficult. In fact, parking became a task that often required a great deal of physical strength on the part of the driver. To overcome this problem, power-assisted steering was introduced (Fig. 49-28). It supplements the steering force supplied by the driver with an assisting force from some other source.

Automobiles usually do not have full power steering. They have power-assisted steering, which is called *power steering*. When there is no mechanical connection between the steering wheel and the front wheels, the car cannot be steered if the engine stalls or the power-steering system fails. With power-assisted steering, the car can be steered manually if the power assist is not available.

Power-steering systems have used compressed air, electrical devices, and hydraulic pressure. Some trucks still have air-operated power steering. However, today hydraulic (oil) pressure is used in all car and most truck power-steering systems.

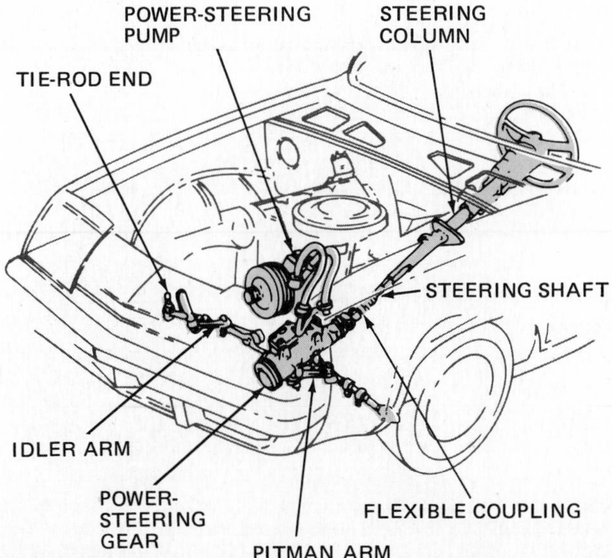

Fig. 49-28 Power-steering installation on a car with an in-line engine.

♦ 49-18 POWER-STEERING OPERATION

The power-steering system used in automobiles is basically a modified manual-steering system (Fig. 49-28). The steering column, steering gear, and steering linkage are little changed from the manual-steering systems described earlier. The main difference is that a power booster has been added to assist the driver.

In the basic system of power-assisted steering, the booster is set into operation when the steering shaft is turned. Then, after the steering effort exceeds a certain force, the booster takes over and provides most of the force required for steering. For example, some cars have power assistance when the force applied at the steering wheel is more than about 3 pounds [13 N]. The force varies with the make and model of the car, but for most cars it ranges from 1.5 to 7 pounds [7 to 31 N].

In the power-steering system, a continuously operating pump provides hydraulic pressure when needed (Fig. 49-29). As the steering wheel is turned, valves are operated to admit this hydraulic pressure to a cylinder that contains the power piston. Then the pressure causes the piston to move, and it provides most of the steering force.

Two general types of power-steering systems are found in automobiles (Fig. 49-30). In the integral type, the power piston is built into the steering gear. In the linkage type, the power piston is attached between the frame of the vehicle and the steering linkage. The integral type is the more widely used. Usually it must be installed at the factory as the vehicle is built. The linkage type is widely used on trucks. It also can be installed after manufacture on some vehicles.

♦ 49-19 POWER-STEERING PUMP

The power-steering pump runs continuously to supply steering-system oil at high pressure to the steering gear. Figure 49-29 shows a typical arrangement. The pump is driven by a belt from the engine crankshaft.

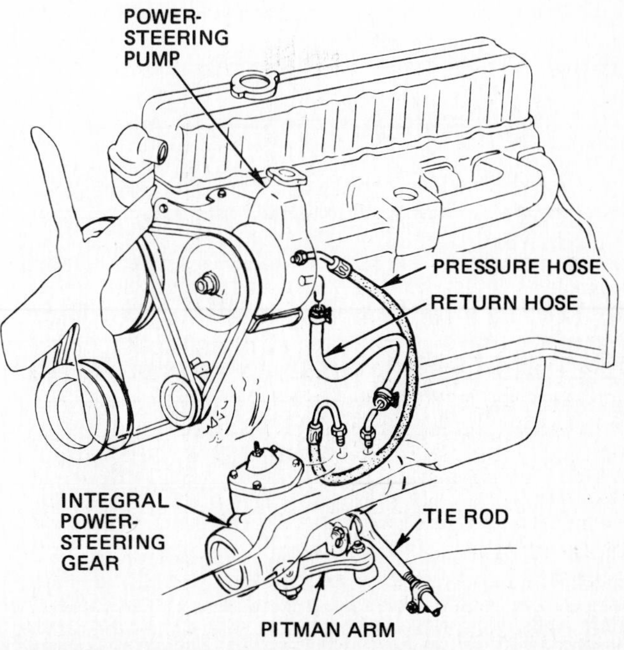

Fig. 49-29 A typical power-steering installation. *(Chevrolet Motor Division of General Motors Corporation)*

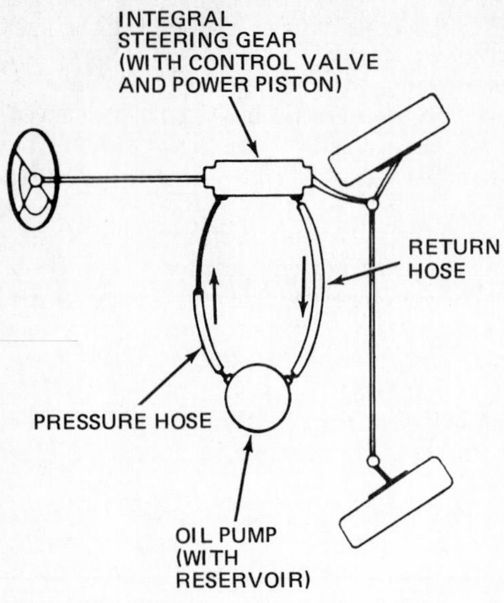

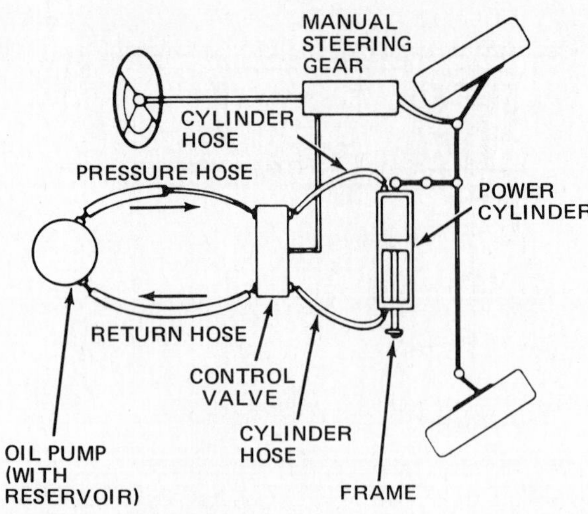

Fig. 49-30 Two types of power steering used with pitman-arm steering gears. *(Lempco Industries, Inc.)*

Figure 49-31 shows the internal working parts of one type of power-steering pump. The rotor rotates, and the vanes move in and out of slots in the rotor. As the vanes move out, the spaces between the vanes increase. Oil is drawn into these spaces. Then, as the rotor turns further, the vanes are pushed back into the rotor. This decreases the spaces between the vanes. The fluid is forced out under high pressure. It goes through hoses to the steering gear. The pump has a pressure-relief valve which opens if the pressure goes too high.

There are two other types of power-steering pumps. They are the slipper type and the roller type. All work on the same principle. Spaces increase and then decrease so oil is put under pressure and pumped to the steering gear.

♦ 49-20 INTEGRAL POWER-STEERING GEAR

An integral power-steering gear is shown in Figs. 49-32 and 49-33. This unit includes a spool valve, a rotary valve, and a torsion bar. The torsion bar is attached to the end of the worm shaft. When the steering wheel is turned, the torsion bar twists. This moves the spool valve slightly, opening passages which move fluid to the side of the piston where hydraulic assist is required.

When there is no power-assist action, the spool valve is in the neutral or straight-ahead position. Fluid flows from the

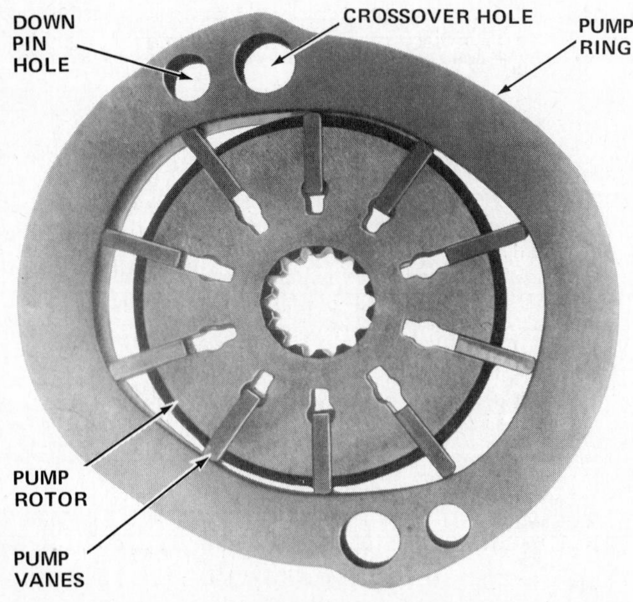

Fig. 49-31 Internal working parts of a vane-type power-steering pump. The vanes follow the oval shape of the housing and move in and out of the slots in the rotor. *(Chevrolet Motor Division of General Motors Corporation)*

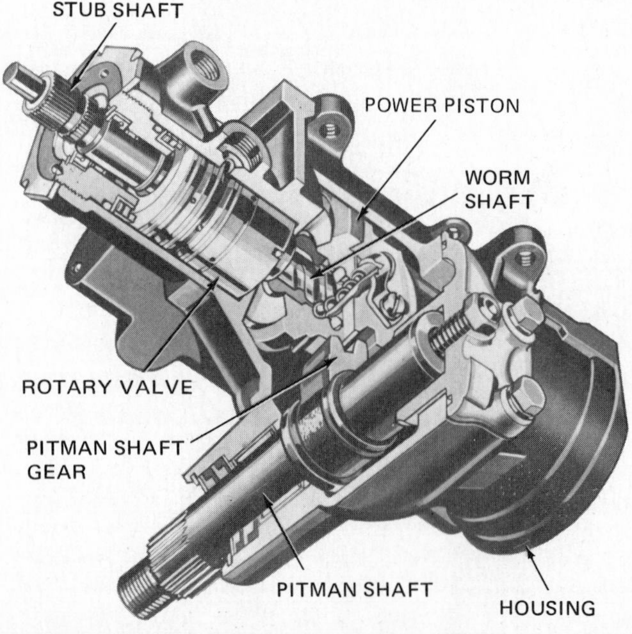

Fig. 49-32 Cutaway integral power-steering gear using a rotary valve. *(General Motors Corporation)*

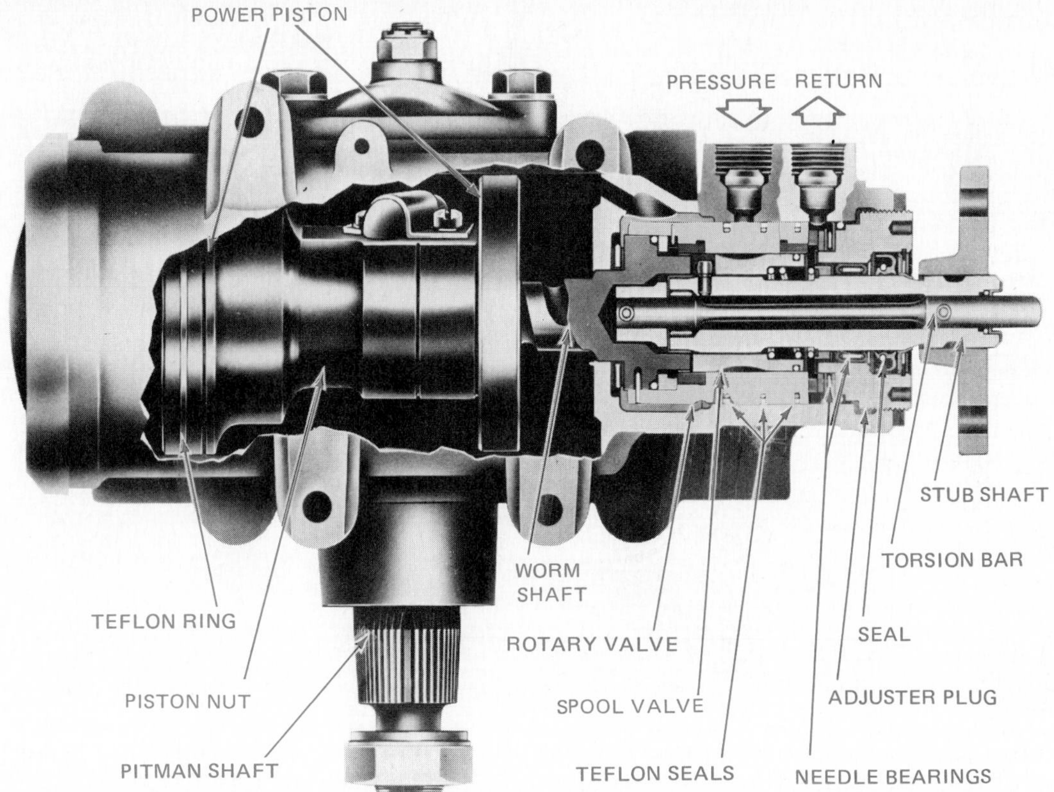

Fig. 49-33 Rotary-valve power-steering gear, partly cut away. *(General Motors Corporation)*

pump, through the open spool valve. The fluid then returns to the pump. Oil pressure of about 30 psi (207 kPa) fills the spool valve, housing, and power-piston cylinder (Fig. 49-33) with oil. However, with the same pressure on both sides of the piston, the oil can not force it to move.

When a turn is made, the worm shaft resists turning because of the normal resistance of the front wheels to turning. Therefore, the torsion bar begins to twist. This moves the spool valve, allowing high-pressure oil from the pump to enter the power cylinder and push against one side of the power piston. The fluid on the other side of the piston is free to return through the spool valve to the pump. Therefore, the oil pressure pushing on one side of the piston helps it move in the desired direction. This provides the driver with power assistance in turning the sector shaft.

During a turn in the other direction, the oil pressure is applied to the opposite end of the piston. The amount of pressure applied depends on how much resistance to turning the wheel the driver encounters. If the resistance is high, such as during parking, the torsion bar and spool valve are turned more. This allows more high-pressure oil to the piston.

♦ 49-21 LINKAGE-TYPE POWER STEERING

In the linkage-type power-steering system, the power cylinder is not part of the steering gear. Instead, the power cylinder (or *booster* cylinder) is connected into the steering linkage. In addition, the control valve is included in the steering linkage. The control valve is either a separate assembly or integral with the power cylinder.

Figure 49-34 shows one linkage-type power-steering system in which the booster cylinder and control valve are separate units In operation, the steering gear works the same way as the mechanical types described in ♦49-14. However, the swinging end of the pitman arm is not directly connected to the steering linkage. Instead, it is connected to the control valve. As the end of the pitman arm swings when a turn is made, it actuates the control valve. Then, the control valve directs hydraulic oil pressure from the oil pump to the booster cylinder. Inside the booster cylinder, pressure is applied to one or the other side of a piston. Movement then takes place (actually, in this unit, the cylinder moves instead of the piston). This movement is transferred to the connecting rod in the steering linkage. Therefore, most of the force required to move the connecting rod and steer the car is furnished by the booster cylinder.

1. Control Valve Operation In many ways, the control valve is similar to that used in the in-line power-steering unit. When turning force passes through the steering gear and into the steering linkage, the force must pass through the control valve. This causes the valve to open and send high-pressure fluid to the power cylinder.

2. Booster Cylinder The cylinder is made up of two concentric shells. Oil flows between the two shells to enter the piston-rod end of the cylinder. The end of the piston rod is attached to the vehicle frame by a flexible connection. This allows some movement of the rod. The movement permits alignment of the rod with the cylinder as it moves back and forth during steering.

3. Integral Valve and Power Cylinder In this design, the linkage-type power-steering unit contains both the control valve and the power cylinder. The piston rod of the power cylinder is attached to the vehicle frame. The cylinder is linked to the steering linkage, forming a part of the linkage. This assembly is called the *power link*. It is a part of the linkage and, at the same time, supplies steering power.

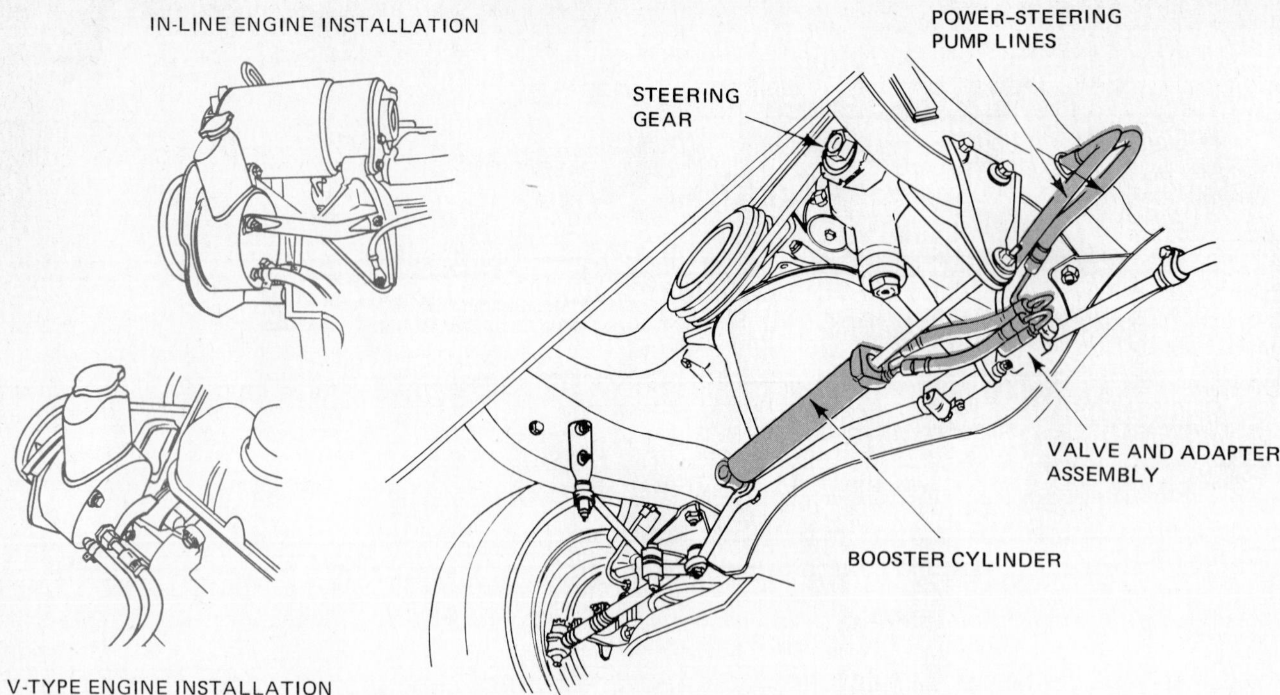

Fig. 49-34 Linkage-type power-steering system. *(Pontiac Motor Division of General Motors Corporation)*

♦ 49-22 RACK-AND-PINION POWER STEERING

A power rack-and-pinion steering gear is another design of integral power steering (Fig. 49-35). The rack functions as the power piston. The tie rods are attached between the rack and the spindle steering arms. The control valve is connected to the pinion gear.

Operation of the control valve is similar to that for the integral power-steering gear (♦49-20). When the steering wheel is turned, the resistance of wheels and the weight of

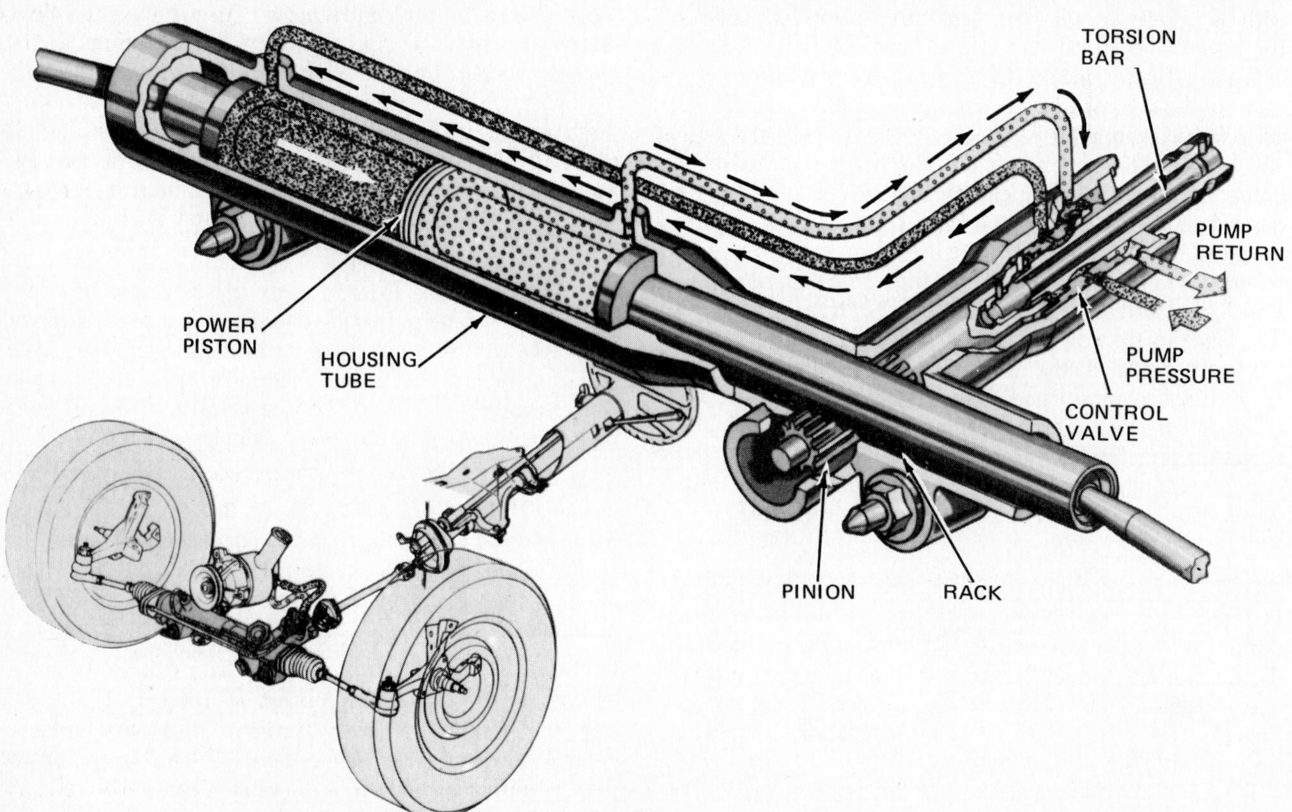

Fig. 49-35 Rack-and-pinion integral power-steering system. *(Ford Motor Company)*

the vehicle cause the torsion bar to twist. This twisting causes the rotary valve to move in its sleeve, aligning the fluid passages for the left, right, or neutral position. Oil pressure exerts force on the piston (Fig. 49-35) and helps move the rack to assist the turning effort. The piston is attached directly to the rack. The housing tube functions as the power cylinder.

The gear assembly is always filled with fluid, and all internal components are immersed in fluid. This makes periodic lubrication unnecessary, and also acts as a cushion to help absorb road shocks. On some rack-and-pinion power-steering gears, all fluid passages are internal except for the pressure and return hoses between the gear and pump.

STEERING WHEELS AND COLUMNS

♦ 49-23 TILT AND TELESCOPING STEERING WHEELS

Many cars have steering wheels that tilt up or down and also can be moved out of or into the steering column (Fig. 49-36). This makes it easier for the driver to get into or out of the car. The driver also can vary the position of the wheel to suit his or her build and can change the position during a long drive to vary driving posture.

Some automobiles have a steering column that can be moved inward toward the center of the car to make it easier for the driver to get in and out of the car. The pivot point on this arrangement is at the lower end of the steering column. There, a flex joint connects between the upper steering shaft and the worm shaft in the steering gear. This permits the steering shaft and steering column to be pivoted toward the center of the car.

A locking mechanism is connected to the transmission selector lever. The steering column is locked in the DRIVE position and in all selector-lever positions except PARK. To unlock the steering column, the selector must be moved to PARK. Also, if the steering column is moved out of the driving position, the selector lever is locked in PARK. This interlocking is a safety feature which prevents the steering column from being accidentally moved while the car is in operation.

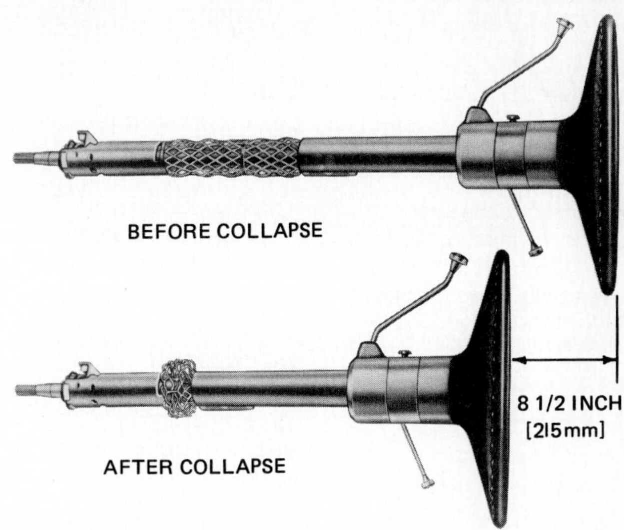

Fig. 49-37 Japanese lantern type of energy-absorbing steering column which collapses during impact. *(General Motors Corporation)*

♦ 49-24 COLLAPSIBLE STEERING COLUMN

The collapsible steering column (Figs. 49-37 to 49-39), used on modern cars as a protective device, will collapse on impact. If the car should become involved in a front-end collision that throws the driver forward, the steering column will absorb the energy of this forward movement and greatly reduce the possibility of injury. The steering shaft is made in two parts which are fitted together so that they can telescope as the steering column collapses.

The steering column shown in Fig. 49-37 is called the *Japanese lantern* design because, on impact, it folds up like a Japanese lantern. The type shown in Fig. 49-38 is a tube-and-

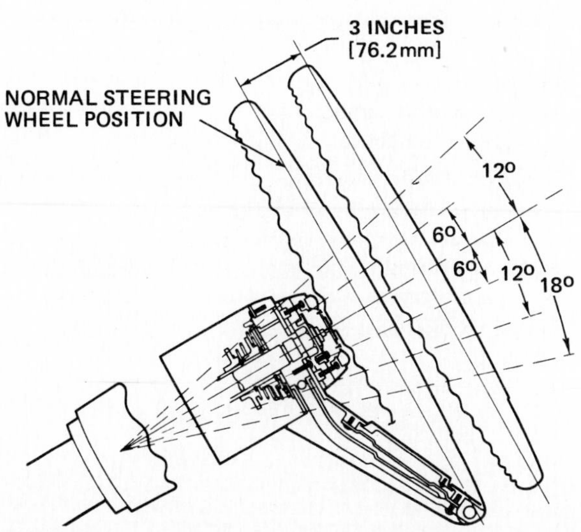

Fig. 49-36 Tilting and telescoping steering wheel. Lifting the release lever permits the wheel to be tilted to various positions. *(General Motors Corporation)*

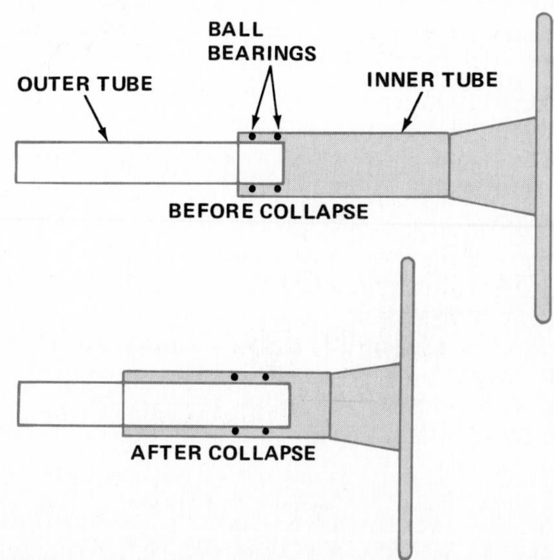

Fig. 49-38 Tube-and-ball type of energy-absorbing steering column. *(General Motors Corporation)*

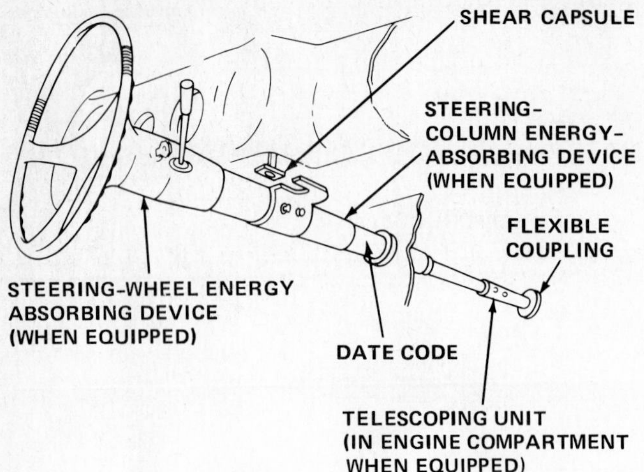

Fig. 49-39 Shear-capsule type of collapsible steering column.

ball design. In it, two tubes are placed one inside the other, with tight-fitting ball bearings between. On impact, the tubes are forced together, as shown in Fig. 49-38. The balls must plow furrows in the tubes to permit the relative motion. This absorbs the energy of the impact. The shear-capsule type (Fig. 49-39) absorbs shock by the cutting of the capsule, which then permits the column to collapse.

♦ 49-25 STEERING AND IGNITION LOCK

The steering-and-ignition lock assembly combines the ignition switch with the steering lock. The exposed part is the ignition switch, which is mounted in the right side of the steering column. The location is usually between the transmission selector lever (on cars with column shift) and the rim of the steering wheel. The key shown in Fig. 49-39 shows the typical location. The steering-and-ignition lock is described in ♦27-14 and illustrated in Fig. 27-18.

♦REVIEW QUESTIONS♦

Select the *one* correct, best, or most probable answer to each question. You can find the answers in the section indicated at the end of each question.

1. The purpose of the caster angle on an automobile is to (♦49-8)
   a. prevent tire wear
   b. bring the road contact of the tire under the point of load
   c. compensate for wear in the steering linkage
   d. maintain directional control
2. When turning a corner (♦49-10)
   a. the front wheels are toeing out
   b. the front wheels are turning on different angles
   c. the inside front wheel has a greater angle than the outside wheel
   d. all of the above
3. As viewed from the front of the car, the tilting of the front wheels away from the vertical is called (♦49-5)
   a. camber
   b. caster
   c. toe-in
   d. toe-out
4. The inward tilt of the center line of the ball joints is called (♦49-6)
   a. caster
   b. camber
   c. SAI
   d. included angle
5. Camber angle plus SAI angle is called the (♦49-7)
   a. caster
   b. included angle
   c. point of intersection
   d. toe-out
6. The point at which the center line of the wheel and the center line of the ball joints cross is called the (♦49-7)
   a. included angle
   b. point of departure
   c. point of intersection
   d. point of included angle
7. When the point of intersection is below the road surface, the front wheel will tend to (♦49-7)
   a. toe out
   b. toe in
   c. roll straight
   d. none of the above
8. The backward tilt of the center line of the ball joints from the vertical is called (♦49-8)
   a. positive caster
   b. negative caster
   c. positive camber
   d. negative camber
9. Positive caster tends to make front wheels (♦49-8)
   a. toe in
   b. toe out
   c. have neutral camber
   d. none of the above
10. In the pitman-arm steering gear, a gear sector or toothed roller is meshed with (♦49-14)
    a. a worm
    b. a ball bearing
    c. a roller bearing
    d. a steering wheel

CHAPTER 50
Steering and Suspension Service

After studying this chapter, and with proper instruction and equipment, you should be able to:

1. Diagnose troubles in manual- and power-steering systems.
2. Diagnose troubles in suspension systems.
3. Inspect and lubricate steering linkages.
4. Replace defective parts in the steering linkage.
5. Replace and adjust front-wheel bearings.
6. Inspect suspension systems and replace defective parts.
7. Perform a wheel alignment on vehicles with various types of suspension and steering systems.
8. Check rear-wheel alignment on a car with independent rear suspension.

♦ 50-1 STEERING AND SUSPENSION TROUBLE-DIAGNOSIS CHART

A variety of steering and suspension problems bring the driver to the mechanic. The skilled technician should be able to diagnose the cause of the problem by inspection and road testing. Sometimes, an apparent steering problem is actually in the suspension system.

♦ CAUTION ♦ Do not drive a car unless you have a valid driver's license. Do not make this test unless you have your instructor's permission. Then fasten your safety belt and conduct the test in the area designated by your instructor.

The chart that follows lists complaints and their possible causes.

STEERING AND SUSPENSION TROUBLE-DIAGNOSIS CHART*

Complaint	Possible Cause	Check or Correction
1. Excessive play in steering system (♦ 50-2)	a. Looseness in steering gear	Readjust, replace worn parts
	b. Looseness in linkage	Readjust, replace worn parts
	c. Worn ball joints or steering-knuckle parts	Replace worn parts
	d. Loose wheel bearing	Readjust
2. Hard steering (♦ 50-3)	a. Power steering inoperative	Refer to manufacturer's service manual
	b. Low or uneven tire pressure	Inflate to correct pressure
	c. Friction in steering gear	Lubricate, readjust, replace worn parts
	d. Friction in linkage	Lubricate, readjust, replace worn parts

*See ♦ 50-2 to 50-16 for detailed explanations of trouble causes and corrections listed.

STEERING AND SUSPENSION TROUBLE-DIAGNOSIS CHART *(Continued)*

Complaint	Possible Cause	Check or Correction
	e. Friction in ball joints	Lubricate, replace worn parts
	f. Incorrect alignment (caster, camber, toe, SAI)	Check alignment and readjust as necessary
	g. Frame misaligned	Straighten
	h. Front spring sagging	Replace or adjust
3. Car wander (♦ 50-4)	a. Low or uneven tire pressure	Inflate to correct pressure
	b. Linkage binding	Readjust, lubricate, replace worn parts
	c. Steering gear binding	Readjust, lubricate, replace worn parts
	d. Incorrect front alignment (caster, camber, toe, SAI)	Check alignment and readjust as necessary
	e. Looseness in linkage	Readjust, replace worn parts
	f. Looseness in steering gear	Readjust, replace worn parts
	g. Looseness in ball joints	Replace worn parts
	h. Loose rear springs	Tighten
	i. Unequal load in car	Readjust load
	j. Stabilizer bar ineffective	Tighten attachment, replace if damaged
4. Car pulls to one side during normal driving (♦ 50-5)	a. Uneven tire pressure	Inflate to correct pressure
	b. Uneven caster or camber	Check alignment, adjust as necessary
	c. Tight wheel bearing	Readjust, replace parts if damaged
	d. Uneven springs (sagging, broken, loose attachment)	Tighten, replace defective parts
	e. Wheels not tracking	Check tracking, straighten frame, tighten loose parts, replace defective parts
	f. Uneven torsion-bar adjustment	Adjust
	g. Brakes dragging	Repair
5. Car pulls to one side when braking (♦ 50-6)	a. Brakes grab	Readjust, replace brake lining, etc. (see Chap. 52)
	b. Uneven tire inflation	Inflate to correct pressure
	c. Incorrect or uneven caster	Readjust
	d. Causes listed under item 4	
6. Front-wheel shimmy at low speeds (♦ 50-7)	a. Uneven or low tire pressure	Inflate to correct pressure
	b. Loose linkage	Readjust, replace worn parts
	c. Loose ball joints	Replace worn parts
	d. Looseness in steering gear	Readjust, replace worn parts
	e. Front springs too flexible	Replace, tighten attachment
	f. Incorrect or unequal camber	Readjust
	g. Irregular tire tread	Replace worn tires, match treads
	h. Dynamic imbalance	Balance wheels
7. Front-wheel tramp (high-speed shimmy) (♦ 50-8)	a. Wheels out of balance	Rebalance
	b. Too much wheel runout	Balance, remount tire, straighten or replace wheel
	c. Defective shock absorbers	Repair or replace
	d. Causes listed under item 6	
8. Steering kickback (♦ 50-9)	a. Tire pressure low or uneven	Inflate to correct pressure
	b. Springs sagging	Replace; adjust torsion bars
	c. Shock absorbers defective	Repair or replace
	d. Looseness in linkage	Readjust, replace worn parts
	e. Looseness in steering gear	Readjust, replace worn parts
9. Tires squeal on turns (♦ 50-10)	a. Excessive speed	Take curves at slower speed
	b. Low or uneven tire pressure	Inflate to correct pressure
	c. Front alignment incorrect	Check and adjust
	d. Worn tires	Replace
10. Improper tire wear (♦ 50-11)	a. Wear at tread sides from underinflation	Inflate to correct pressure
	b. Wear at tread center from overinflation	Inflate to correct pressure
	c. Wear at one tread side from excessive camber	Adjust camber
	d. Featheredge wear from excessive toe-in or toe-out on turns	Correct toe-in or toe-out in turns

Complaint	Possible Cause	Check or Correction
	e. Cornering wear from excessive speeds on turns	Take turns at slower speeds
	f. Uneven or spotty wear from mechanical causes	Adjust brakes, align wheels, balance wheels, adjust linkage, etc.
	g. Rapid wear from speed	Drive more slowly for longer tire life
11. Hard or rough ride (♦ 50-12)	a. Excessive tire pressure	Reduce to correct pressure
	b. Defective shock absorbers	Repair or replace
	c. Excessive friction in spring suspension	Lubricate, realign parts
12. Sway on turns (♦ 50-13)	a. Loose stabilizer bar	Tighten
	b. Weak or sagging springs	Repair or replace
	c. Caster incorrect	Adjust
	d. Defective shock absorbers	Replace
13. Spring breakage (♦ 50-14)	a. Overloading	Avoid overloading
	b. Loose center or U bolts	Keep bolts tight
	c. Defective shock absorber	Repair or replace
	d. Tight spring shackle	Loosen, replace
14. Sagging springs (♦ 50-15)	a. Broken leaf	Replace
	b. Spring weak	Replace
	c. Coil spring short	Install shim
	d. Defective shock absorber	Repair or replace
15. Noises (♦ 50-16)	Could come from any loose, worn, or unlubricated part in the suspension or steering system.	

♦ 50-2 EXCESSIVE PLAY IN STEERING SYSTEM

Excessive looseness in the steering system means that there will be excessive free play of the steering wheel without corresponding movement of the front wheels. A small amount of free play makes steering easier. But when the play, or lash, becomes excessive, it may make steering harder. Many drivers complain about it. Excessive free play in the steering system reduces the ability of the driver to accurately steer and control the vehicle.

Excessive play may be due to wear or improper adjustment of the steering gear, worn parts or improper adjustments in the steering linkage, worn ball joints or steering-knuckle parts, or loose wheel bearings. In most cars with power steering, the steering-wheel rim should move 2 inches [51 mm] or less before the front wheels begin to move. On cars with manual steering, the maximum allowable free play is 3 inches [76 mm]. Figure 50-1 shows this measurement.

To check the amount of play in the steering system on vehicles with power steering, check the condition and tension of the drive belt for the power-steering pump. Then check the fluid level in the pump reservoir. Start the engine. Next, with the front wheels in the straight-ahead position, turn the steering wheel until the front wheels begin to move. Align a reference mark on the steering wheel with a mark on a ruler or scale (Fig. 50-1).

Now slowly turn the steering wheel in the opposite direction until the front wheels start to move again. The distance that the steering-wheel reference mark has moved along the ruler is the amount of free play in the steering system. If the steering-wheel rim moves too much before the front wheels begin to move, there is excessive play.

1. Steering-Linkage Check Steering linkage, including tie rods, can be checked for looseness. Raise the front of the car until the bottoms of the tires are slightly off the floor. Then grasp both front tires and push out on both at the same time (Fig. 50-2). Next pull in on both tires at the same time. Excessive movement means worn linkage parts.

Excessive movement in the steering linkage can cause wheel shimmy, vehicle wander, uneven braking, steering-control problems, and excessive tire wear. On vehicles with 16-inch [406 mm] diameter (or smaller) wheels, the maximum movement should be ¼ inch [6.35 mm] or less. Other steering and suspension check points for each car are shown in the manufacturer's service manual.

Note If the movement is excessive and the wheel bearings have not been adjusted, they may be loose. This will give a

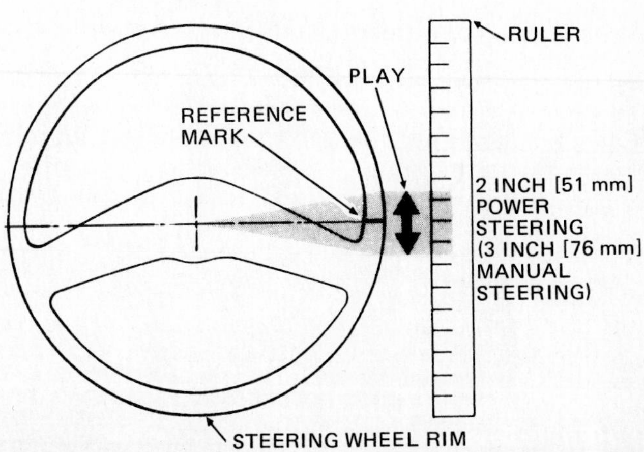

Fig. 50-1 Checking for play in the steering wheel. *(Motor Vehicle Manufacturers Association)*

Fig. 50-2 Checking looseness in steering linkage and tie rods. *(ATW)*

false reading by allowing excessive movement. To eliminate the effect of the wheel bearings, apply the brakes during the check. An assistant can do this, or you can use a portable brake depressor.

2. Inspecting Front-Wheel Bearings Loose wheel bearings can cause poor steering control, car wander, uneven front-brake action, and rapid tire wear. To check the front-wheel bearings, raise the car on a lift or use floor jacks, properly placed. The lift points differ according to the type of front end. If the spring is between the frame and the lower control arm, the car should be lifted at the frame cross member. Use this same lift point for torsion-bar suspension systems that have the torsion bar attached to the lower control arm. If the spring is above the upper control arm, lift the vehicle at the lower control arm, close to the ball joint. Use this same lift point if the torsion bar is attached to the upper control arm. In either case, the weight of the wheel should take up any play in the ball joints.

Now grasp the tire at the top and bottom (Fig. 50-3), and rock it in and out. Any movement is usually the result of loose front-wheel bearings. Look at the brake drum or disk and the backing plate or shield as you rock the wheel. If you see movement between the drum or disk and the plate or shield, the looseness is in the wheel bearing. Another check is to have an assistant apply the brakes as you try rocking the wheel. If this eliminates the free play, the wheel bearings are loose.

Fig. 50-3 Checking for wear in the ball joints and wheel bearings.

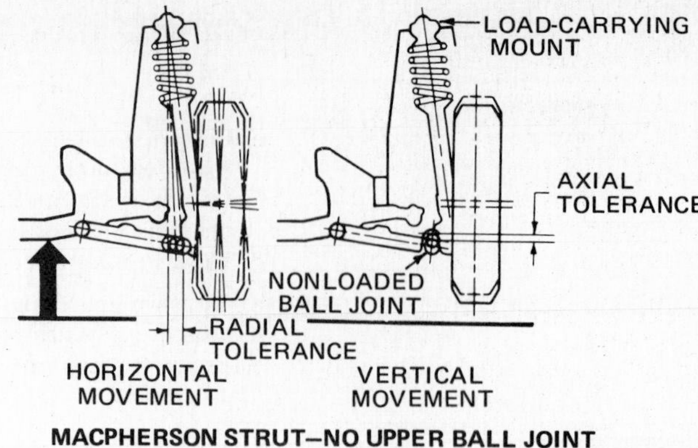

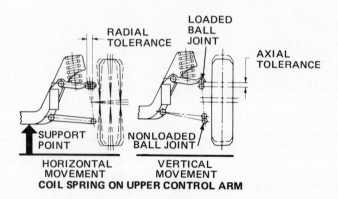

RADIAL TOLERANCE
LOADED BALL JOINT
AXIAL TOLERANCE
SUPPORT POINT
NONLOADED BALL JOINT
HORIZONTAL MOVEMENT
VERTICAL MOVEMENT
COIL SPRING ON UPPER CONTROL ARM

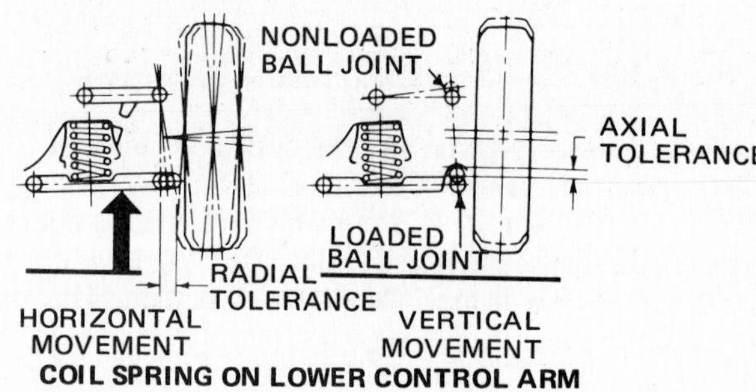

Fig. 50-4 Support points for checking ball joints in various front-suspension systems using coil springs. *(Motor Vehicle Manufacturers Association)*

In some inspection programs, the vehicle should be rejected if the wheel can be rocked more than ⅛ inch [3.2 mm], measured at the outer circumference of the tire. This amount of wheel wobble can make the vehicle unstable and hard to steer. The wheel bearings should be adjusted.

3. Inspecting Ball Joints Ball joints can be checked for wear while the wheel is supported as shown in Figs. 50-4 and 50-5. Axial play or tolerance is also called *vertical movement*. It is checked by moving the wheel straight up and down. Radial play or tolerance also is called *horizontal movement*. It is checked by rocking the wheel in and out at the top and bottom. Figure 50-3 shows this check.

The actual amount of play in a ball joint is measured with a dial indicator. In Fig. 50-6, the dial indicator is clamped to the lower control arm. The plunger tip rests against the steering-knuckle leg. With a pry bar, try to raise and lower the steering knuckle. As you do this, the play in the ball joint will show on the dial indicator. On the front-wheel-drive car

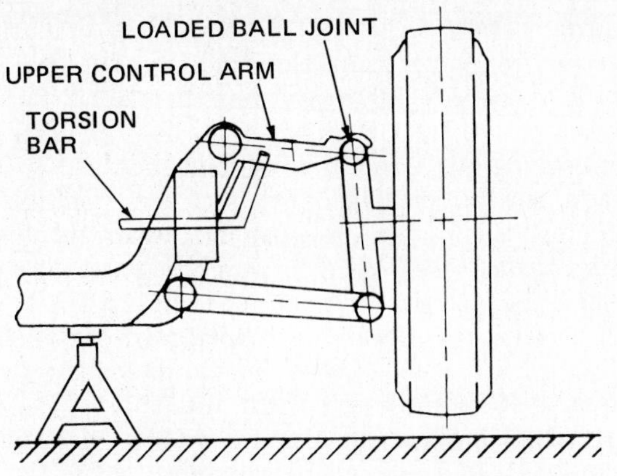

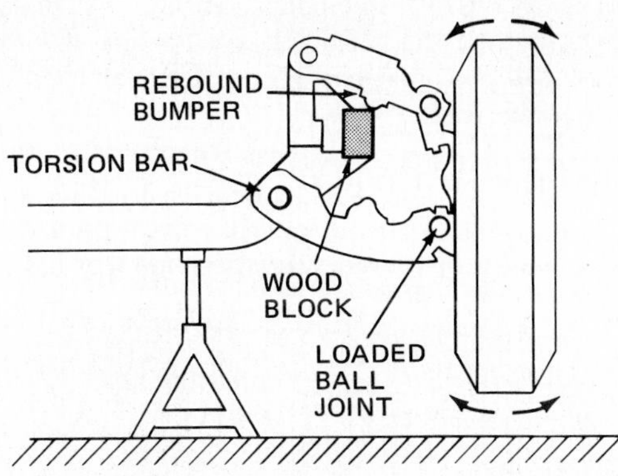

Fig. 50-5 Support points for checking ball joints in front-suspension systems using torsion bars.

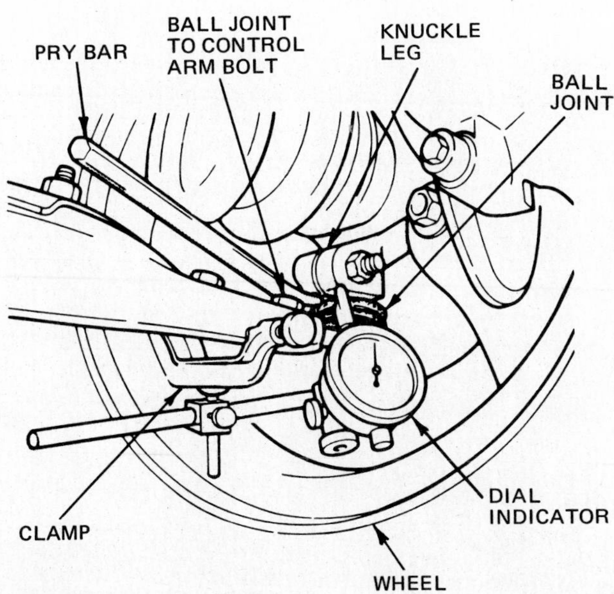

Fig. 50-6 A dial indicator can be used to measure the amount of play in a ball joint. *(Motor Vehicle Manufacturers Association)*

shown in Fig. 50-6, vertical movement of the ball joint must not exceed 0.050 inch [1.2 mm]. The axial (up-and-down) play in a typical ball joint should not exceed 1/16 inch [1.6 mm].

The ball joints on vehicles manufactured prior to 1973 must be inspected with the ball joints unloaded. This means that the weight of the car must be removed from them. Beginning with some 1973 vehicles, manufacturers used wear-indicating ball joints. It is much easier to check this type of ball joint for wear. A quick visual check is all that is necessary. The check is made with the ball joints loaded, or carrying the weight of the car.

Figures 50-4 and 50-5 identify the loaded and nonloaded ball joints for various suspension systems. The load is carried by the upper strut mount or by the ball joint for the control arm on which the spring acts. For example, when the spring is mounted on the upper control arm, the upper ball joint carries the load. When the spring is mounted on the lower control arm, the lower ball joint carries the load. Most wear occurs in the load-carrying ball joints. Nonloaded ball joints usually wear relatively little.

Figure 50-7 shows the use of a pry bar to check the ball joints. Pry under the front tire to see how much vertical movement the ball joints allow. Use only enough force to overcome the weight of the wheel assembly. If you use too much force, the ball joint may give you a false reading. You want to measure the movement of the wheel and ball joint as the joint is moved up to the LOAD position. Note the movement as indicated on the dial indicator.

Next, grasp the tire at top and bottom (Fig. 50-3), and try to wobble it. This is the test described earlier for inspecting front-wheel bearings. However, now we are assuming that the wheel bearings have been checked and either adjusted or found to be properly tightened. Therefore, we are now checking the horizontal movement of the ball joints. However, some manufacturers do not accept horizontal movement as an indicator of ball-joint wear.

The actual specifications for the allowable wear limits of the ball joints are listed in the manufacturer's service manual. Refer to the specifications for the car you are checking.

Note Some ball joints are preloaded with rubber or springs under compression. They should have very little

Fig. 50-7 Pry upward on the tire with a bar to check for loose ball joints. *(ATW)*

movement in a vertical direction. These ball joints are marked as PRELOADED in specification tables.

Any ball joint should be replaced if there is excessive play in it. On certain vehicles with the spring on the lower control arm, some manufacturers specify replacement of the lower ball joint whenever the upper ball joint is replaced. When checking ball joints on 1955 through 1970 Chevrolet cars, the lower ball joint should be replaced if the radial play exceeds 0.250 inch [6.35 mm]. Today this tolerance is less on most ball joints. The lower ball joint should also be replaced if axial play between the lower control arm and the spindle exceeds the tolerance specified in the manufacturer's service manual. This specification may vary from 0 to 0.150 inch [3.81 mm].

When the spring is on the upper control arm, the specifications for some cars do not allow any play in the lower ball joint. The lower ball joint should be replaced if it has any noticeable looseness. The upper ball joint should be replaced if the radial play exceeds 0.250 inch [6.35 mm].

The upper ball joint should also be replaced if the axial play between the upper control arm and the spindle exceeds the tolerance specified by the vehicle manufacturer. This specification may vary from 0 to 0.095 inch [2.41 mm] or more. For example, Ford specifies that no check for vertical movement, or axial play, is necessary when the spring is on the upper control arm. They recommend only that the radial play should be checked, using a dial indicator.

4. Inspecting Wear-Indicating Ball Joints Many cars have ball joints with wear indicators. The wear of this joint can be checked by visual inspection alone (Fig. 50-8). The amount of wear is indicated by the recession of the grease-fitting nipple into the ball-joint socket.

On a new ball joint, the nipple protrudes from the socket 0.050 inch [1.27 mm]. As the ball joint wears, the nipple recedes into the socket. When the wear has caused the nipple to recede 0.050 inch [1.27 mm] or more, the nipple will be level with or below the socket. Then replace the ball joint.

To check a wear-indicating ball joint, the vehicle should be supported on its wheels so that the ball joints are loaded. You may find it convenient to raise the car on a drive-on type of lift so that you can get to the ball joints easily.

Wipe the grease fitting to remove all dirt and grease. Then note the position of the grease-fitting nipple and compare it with Fig. 50-8. Use a steel scale to check the position of the nipple. However, the wear can also be checked with a screwdriver or even your fingernail. If the scale, screwdriver, or your fingernail passes over the nipple because it has recessed into the socket, then the ball joint is worn and should be replaced. Any ball joint that has cut, torn, or damaged seals should also be replaced.

5. Steering-Gear Check A quick check for looseness in the steering gear can be made by watching the pitman arm while an assistant turns the steering wheel one way and then

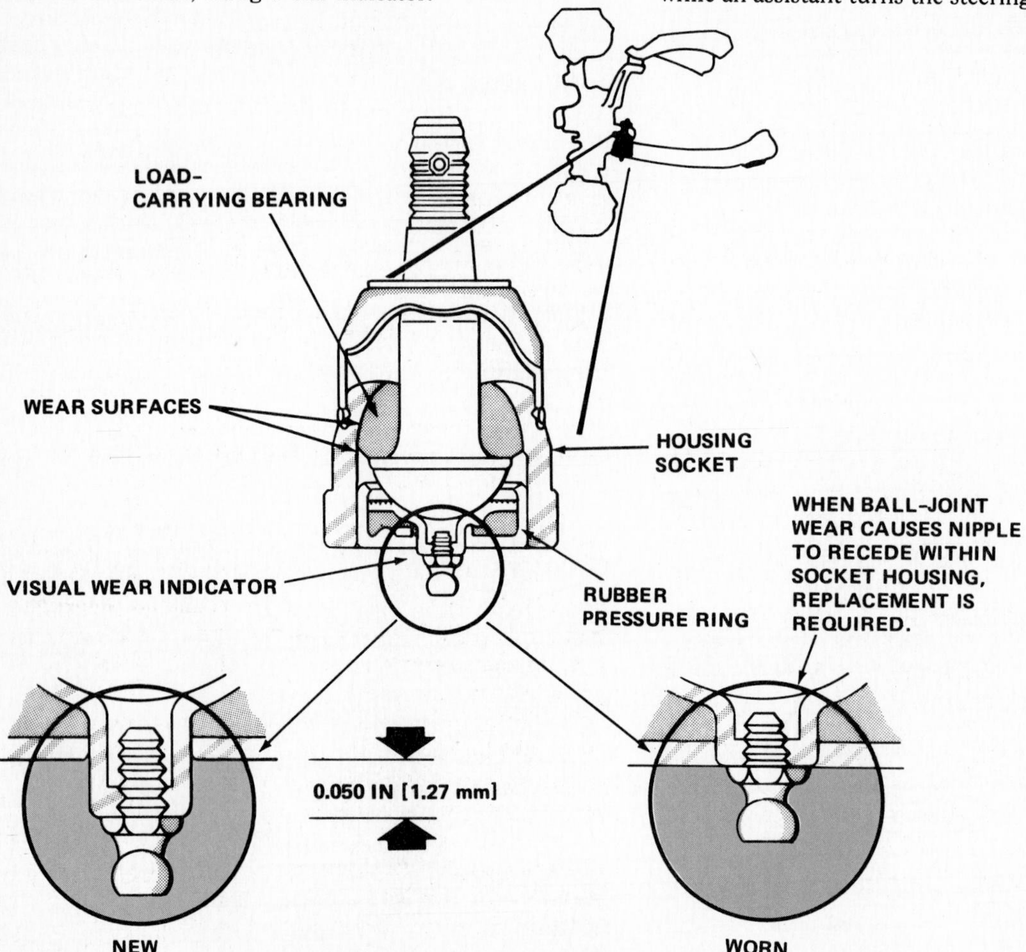

Fig. 50-8 How wear-indicating ball joints show that ball-joint replacement is necessary. In a worn ball joint, the grease-fitting nipple recedes into the socket, as shown to the right.

the other, with the front wheels on the floor. If, after reversal of steering-wheel rotation, excessive movement of the steering wheel is required to move the pitman arm, then the steering gear is worn or in need of adjustment. Steering-gear service is covered in the manufacturer's service manual.

♦ 50-3 HARD STEERING

If hard steering occurs, it is probably due to excessively tight adjustments in the steering gear or linkages. Hard steering can also be caused by low or uneven tire pressure; abnormal friction in the steering gear, in the linkage, or at the ball joints; or improper wheel or frame alignment.

If the car has power steering, its failure causes the steering system to revert to straight mechanical operation. A much greater steering force is then required from the driver. When this happens, the power-steering gear and the pump should be checked as outlined in the manufacturer's service manual.

The steering system may be checked for excessive friction by raising the front end of the car, turning the steering wheel, and checking the steering-system components to locate the source of excessive friction. Disconnect the linkage at the pitman arm. If this eliminates the frictional drag that makes the steering wheel hard to turn, then the friction is either in the linkage itself or at the steering knuckles. If the friction is not eliminated when the linkage is disconnected at the pitman arm, the steering gear is probably at fault.

If hard steering does not seem to be due to excessive friction in the steering system, the cause is probably incorrect front-wheel alignment, a misaligned frame, or sagging springs. Excessive caster, especially, causes hard steering. Wheel alignment is described later in the chapter.

♦ 50-4 CAR WANDER

Wander is the tendency of a car to veer away from a straight path without driver control. Frequent steering-wheel movements are necessary to prevent the car from weaving from one side of the road to the other. An example is when the driver must continually move the steering wheel back and forth to keep the car on the right side of the road or in the proper lane of traffic.

A variety of conditions can cause car wander. Low or uneven tire pressure, binding or excessive play in the linkage or steering gear, or improper front-wheel alignment will cause car wander. Any condition that causes tightness in the steering system will keep the wheels from automatically seeking the straight-ahead position. The driver has to correct the wheels constantly. This condition would probably cause hard steering, which is described in the previous section. Looseness or excessive play in the steering system might also cause car wander. These conditions tend to allow the wheels to waver slightly from their normal running position.

Several improper wheel-alignment angles may cause car wander. Excessively negative caster or uneven caster on the front wheels will tend to cause the wheels to swing away from the straight-ahead direction so that the driver must steer continually. An incorrect camber angle will do the same thing. Excessive toe-in may also cause the same condition.

♦ 50-5 CAR PULLS TO ONE SIDE DURING NORMAL DRIVING

Sometimes a car pulls to one side so that force must constantly be applied to the steering wheel to maintain straight-ahead travel. The cause could be uneven tire pressure, uneven caster or camber, a tight wheel bearing, uneven springs, uneven torsion-bar adjustment, or wheels not tracking. A lack of tracking means that the rear wheels are not following a path that is parallel to the path of the front wheels. Figure 50-9 shows this problem.

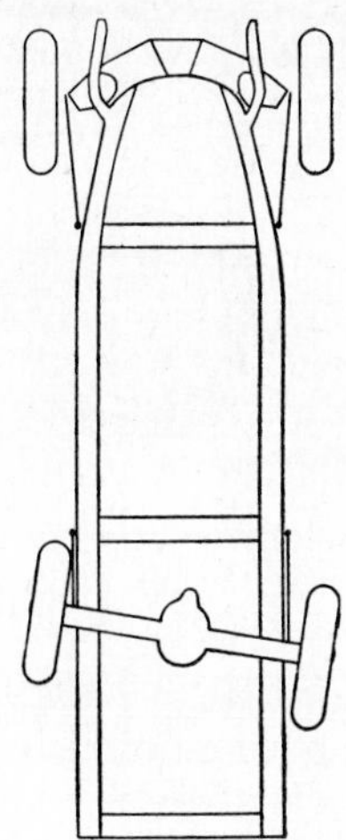

Fig. 50-9 If the frame is bent or the rear axle is swung back on one side, the car will not track properly. *(Applied Power, Inc.)*

Anything that makes one wheel drag or toe-in or toe-out more than the other causes the car to pull to that side. The methods used to check tracking and front-wheel alignment are described later.

♦ 50-6 CAR PULLS TO ONE SIDE WHEN BRAKING

The most likely cause of pulling to one side when braking is grabbing brakes. This happens when the brake lining on the shoes or pads becomes soaked with oil or brake fluid, when brake shoes are unevenly or improperly adjusted, or when a stuck wheel cylinder or caliper piston causes the shoes at one wheel to apply less braking force than the shoes at the wheel on the other side of the axle. Chapter 52 covers brake service. The conditions listed in ♦50-4 that cause car wander could also cause the car to pull to one side when braking. A pulling condition, from whatever cause, tends to become more noticeable as the car is braked to a stop.

♦ 50-7 LOW-SPEED FRONT-WHEEL SHIMMY

Front-wheel shimmy and front-wheel tramp are sometimes confused. Low-speed shimmy is the rapid oscillation of the wheel on the steering-knuckle support. The wheel tries to turn in and out alternately. This action causes the front end of the car to shake from side to side. Front-wheel tramp, or high-speed shimmy, is the tendency for the wheel-and-tire assembly to hop up and down and try to leave the pavement. Even when the tire does not leave the pavement, tramp can be observed as a rapid flexing-unflexing action of that part of the tire in contact with the pavement. The bottom of the tire first appears deflated as the wheel moves down, then appears inflated as the wheel moves up.

Low-speed shimmy can result from low or uneven tire pressure, excessive lateral runout, looseness in linkage, excessively soft springs, incorrect or unequal wheel camber, dynamic imbalance of the wheels, or tire-tread irregularities.

Fig. 50-10 Front-wheel tramp can be caused by an out-of-balance wheel and tire. *(Moog Automotive, Inc.)*

♦ 50-8 FRONT-WHEEL TRAMP

Front-wheel tramp is often called high-speed shimmy. This condition causes the front wheels to move up and down alternately. One of the most common causes of front-wheel tramp is unbalanced wheels, or wheels that have too much radial runout. An unbalanced wheel is heavy in one spot, like the wheel in Fig. 50-10. As it rotates, the heavy spot acts as an unbalanced rotating force. This tends to make the wheel hop up and down.

A similar action occurs if the wheel has too much radial runout. This is the amount that the wheel rotates out-of-round, instead of making a true circle as it turns. Defective shock absorbers, which fail to control spring oscillations, also cause wheel tramp. Any of the causes described in the previous section on front-wheel shimmy may also cause wheel tramp. Servicing of the wheel and tire is described in Chap. 53.

♦ 50-9 STEERING KICKBACK

Steering shock, or kickback, consists of sharp and rapid movements of the steering wheel that occur when the front wheels encounter obstructions in the road. Normally, some kickback to the steering wheel will always occur. When it becomes excessive, an investigation should be made. This condition could be the result of incorrect or uneven tire inflation, sagging springs, defective shock absorbers, or looseness in the linkage or steering gear. Any of these defects could permit road shock to carry excessively to the steering wheel.

♦ 50-10 TIRES SQUEAL ON TURNS

If the tires skid or squeal on turns, the cause may be excessive speed on the turns. If this is not the cause, it is probably low or uneven tire pressure, worn tires, or misalignment of the front wheels. Improper camber and toe settings may tend to cause tire squeal.

♦ 50-11 IMPROPER TIRE WEAR

Various types of abnormal wear occur on tires. The type of tire wear is often a good indication of a particular defect in the suspension or steering system, or improper operation or abuse. For example, if the tire is underinflated, the sides will bulge over, and the center of the tread will be lifted clear of the road (Fig. 50-11). The sides of the tread will take all the wear. The center will be barely worn.

Uneven tread wear shortens tire life. But, even more damaging is the excessive flexing of the tire sidewalls that takes place as the underinflated tire rolls on the pavement. The repeated flexing causes excessive heat, the fabric in the sidewalls to crack or break, and the plies to separate. This cracking and separation seriously weakens the sidewalls and may cause tire failure. In addition, the underinflated tire is unprotected against rim bruises. For example, when the tire hits a rut or stone on the road, or bumps a curb too hard, the tire will flex so much under the blow that it will actually be pinched on the rim. Pinching causes plies to break and leads to early tire failure.

Note The radial tire applies a much larger area of tread to the pavement, as explained in Chap. 53. Therefore, radial tires may appear to be running underinflated when compared with bias-ply tires.

Continuous high-speed driving on curves, both right and left, can produce tread wear that looks almost like underinflation wear. The side force on the tires as they round the curves causes the sides of the tread to wear. The only remedy is to reduce car speed on turns.

Overinflation causes the tire to ride on the center of its tread (Fig. 50-11), so that only the center of the tread wears. Uneven tread wear shortens tire life. But, equally damaging is the fact that the overinflated tire does not have normal "give" or flex when it meets a rut or bump in the road. Instead of flexing normally, the tire fabric takes the major shock. As a result, the fabric may crack or break so that the tire quickly fails.

Excessive toe-in or toe-out on turns causes the tire to be dragged sideways while it is moving forward. The tire on a front wheel that toes in 1 inch [25.4 mm] from straight ahead will be dragged sideways about 150 feet every mile [28.6 m every km]. This sideward drag scrapes off rubber. Characteristic of this type of wear are featheredges of rubber that appear on one side of the tread design (Fig. 50-12). If both front tires show this type of wear, the front end has improper toe.

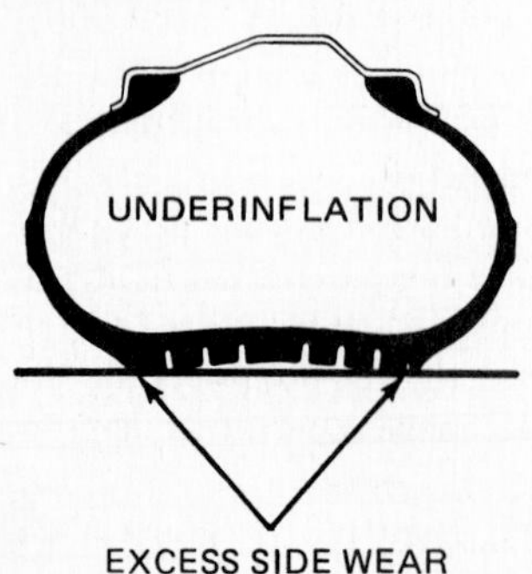

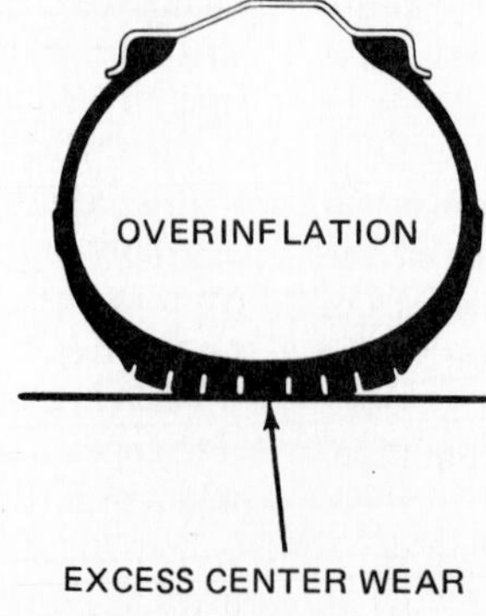

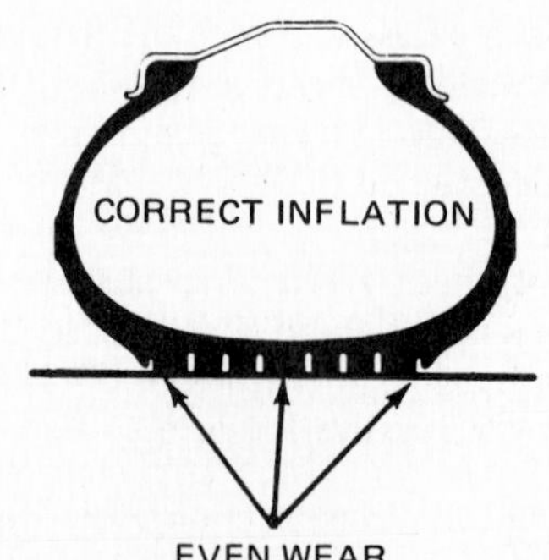

Fig. 50-11 Effects of overinflation and underinflation on tires. *(Moog Automotive, Inc.)*

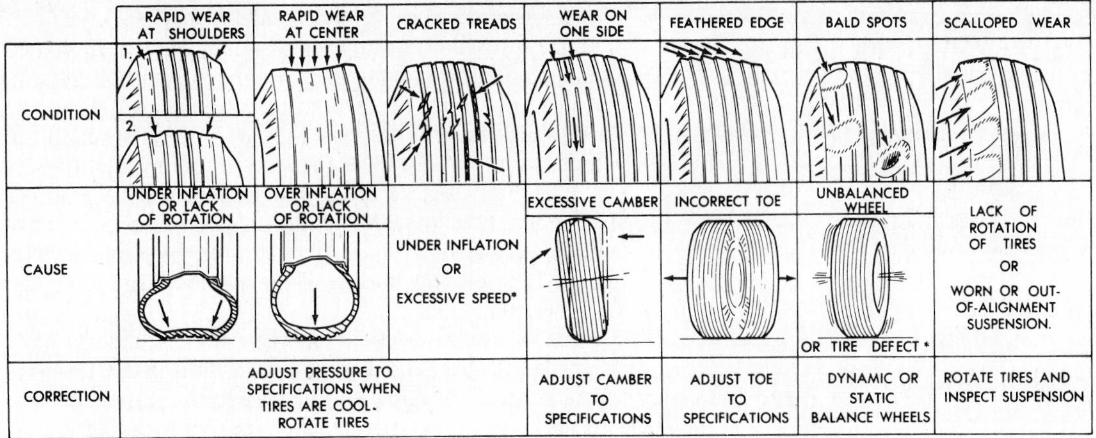

RAPID WEAR AT SHOULDERS
RAPID WEAR AT CENTER
CRACKED TREADS
WEAR ON ONE SIDE
FEATHERED EDGE
BALD SPOTS
SCALLOPED WEAR
CONDITION
1.
2.
CAUSE
UNDER INFLATION OR LACK OF ROTATION
OVER INFLATION OR LACK OF ROTATION
UNDER INFLATION OR EXCESSIVE SPEED*
EXCESSIVE CAMBER
INCORRECT TOE
UNBALANCED WHEEL
OR TIRE DEFECT *
LACK OF ROTATION OF TIRES OR WORN OR OUT-OF-ALIGNMENT SUSPENSION.
CORRECTION
ADJUST PRESSURE TO SPECIFICATIONS WHEN TIRES ARE COOL. ROTATE TIRES
ADJUST CAMBER TO SPECIFICATIONS
ADJUST TOE TO SPECIFICATIONS
DYNAMIC OR STATIC BALANCE WHEELS
ROTATE TIRES AND INSPECT SUSPENSION
*HAVE TIRE INSPECTED FOR FURTHER USE.

Fig. 50-12 Types of tire wear, their causes and corrections. *(Chrysler Corporation)*

But if only one tire shows this type of wear when both tires have been running in the same position on the car, a bent steering arm is indicated. This causes one wheel to toe-in more than the other.

Excessive camber of a wheel causes one side of the tire tread to wear more quickly than the other. This kind of wear is also shown in Fig. 50-12. If the camber is positive, the tire will tilt outward. If the camber is negative, the tire will tilt inward. Then heavy tread wear will appear on the inside.

Cornering wear, caused by taking curves at excessively high speeds, may be mistaken for camber wear or toe-in or toe-out wear. Cornering wear is due to centrifugal force acting on the car and causing the tires to roll and skid on the road. This produces a diagonal type of wear that rounds the outside shoulder of the tire and roughens the tread surface near the outside shoulder. In severe cornering wear, fins or sharp edges will be found along the inner edges of the tire treads. There is no adjustment that can be made to correct the steering system for this type of wear. The only solution is for the driver to slow down on curves.

Uneven tire wear, such as bald spots and scalloped wear, causes the tread to be unevenly or spottily worn (Fig. 50-12). This type of wear can result from several mechanical conditions. These include misaligned wheels, unequal or improperly adjusted brakes, unbalanced wheels, overinflated tires, out-of-round brake drums, and incorrect linkage adjustments.

High-speed operation causes much more rapid tire wear because of the higher temperature and greater amount of scuffing and rapid flexing to which the tires are subjected. The chart in Fig. 50-13 shows how tire wear increases with car speed. According to the chart, tires wear more than three times as fast at 70 mph [113 km/h] as they do at 30 mph [48 km/h]. Careful, slow driving with correct tire inflation pressures will greatly increase tire life.

♦ 50-12 HARD OR ROUGH RIDE

A hard or rough ride could be due to excessive tire pressure, improperly operating shock absorbers, or excessive friction in the spring suspension. The spring suspension can be checked for excessive friction in leaf-spring suspension systems. With a yardstick, measure from the floor to the lower edges of the car body, front and back. Then lift the front end of the car as high as possible by hand, and very slowly let it down. Carefully measure again and write down the distance.

Push down on the car bumper at the front end, and again slowly release the car. Remeasure the distance from the floor to the body. Note the difference in measurements. Repeat this action several times to obtain accurate measurements. The difference is caused by the friction in the suspension system and is called friction lag. After determining friction lag at the front end, check it at the rear of the car.

Excessive friction lag is corrected by lubricating the springs, shackles, and bushings (on types where lubrication is specified) and by loosening the shock-absorber mounts, shackle bolts, and U bolts. Then retighten the U bolts, shackle bolts, and shock-absorber mounts, in that order. This procedure permits realignment of parts that might have become misaligned and caused excessive friction.

A quick check of shock-absorber action on cars giving a hard or uneven ride may be made by bouncing each corner of the car in turn. Shock-absorber bounce testing is described in ♦48-22.

♦ 50-13 SWAY ON TURNS

Sway of the car body on turns or on rough roads may be due to a loose stabilizer bar or sway bar. Stabilizer-bar attachments to the frame, axle housing, or suspension arms should be checked. Weak or sagging springs could also cause exces-

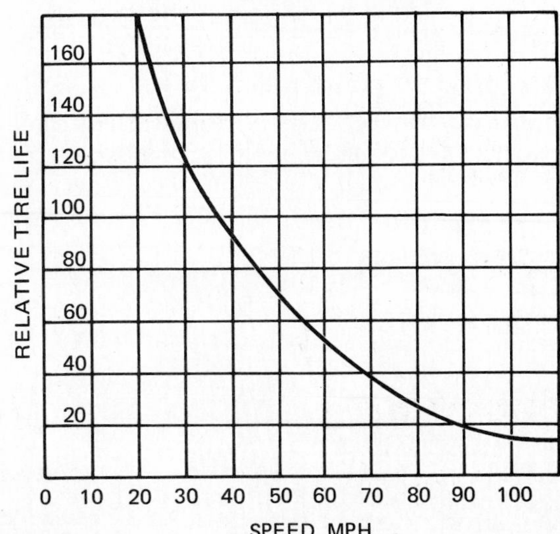

Fig. 50-13 Graph showing how tire wear increases with speed.

sive sway. If the shock absorbers are ineffective, they may permit excessive spring movement. This could cause strong body pitching and sway, particularly on rough roads. If the caster is excessively positive, it will cause the car to roll out, or lean out, on turns. Front-wheel alignment should be corrected.

♦ 50-14 SPRING BREAKAGE

Breakage of leaf springs can result from excessive overloading; loose U bolts that cause breakage near the center bolt, a loose center bolt that causes breakage at the center-bolt holes, an improperly operating shock absorber that causes breakage of the master leaf, or a tight spring shackle that causes breakage of the master leaf near or at the spring eye. Determining the point at which breakage has occurred usually indicates the cause.

♦ 50-15 SAGGING SPRINGS

For wheel alignment to be correct, the car must have the specified front-suspension height. Sagging springs allow the front-suspension height to drop. Springs will sag if they become weak, for example, from frequent overloading. When a coil-spring front-suspension system is overhauled, failure to return the shim to the coil-spring seat will make the spring seem shorter and give a sagging effect.

Not all coil springs require or use shims. On many cars, shimming under a spring is not recommended. If a torsion-bar suspension system sags, the torsion bars can be adjusted to restore normal car height. Defective shock absorbers may restrict spring action. This makes the springs appear to sag more than normal.

To check for sag of the coil spring, position the car on a smooth, level surface, bounce the front end several times, raise the front end, and allow it to settle. Then take the measurements shown in Fig. 50-14. The differences should be as noted in the specifications for the car being checked. Next, take the measurements on the other side. The difference between the two sides should be no greater than ½ inch [12.7 mm]. To make a correction, the coil springs must be replaced. On a car with torsion bars, a similar check may indicate that the torsion bars should be adjusted.

♦ 50-16 NOISES

The noises produced by defects in springs or shock absorbers will usually be either rattles or squeaks. Rattling noises can be produced by looseness of leaf-spring U bolts, metal spring covers, rebound clips, spring shackles, or shock-absorber mounts. These noises can usually be located by a careful examination of the various parts in the suspension system. Spring squeaks can result from a lack of lubrication in the spring shackles, at spring bushings, or in a leaf spring itself.

Squeaks in the shock absorber can result from tight or dry bushings. Rattles in the steering linkage may develop if linkage parts become loose. Sometimes squeaks during turns can develop because of a lack of lubrication in the joints or bearings of the steering linkage. This condition also produces hard steering.

Some of the connections between steering-linkage parts are made with ball sockets that can be lubricated. Others are permanently lubricated on original assembly. If these develop squeaks or excessive friction, they must be replaced.

WHEEL ALIGNMENT

♦ 50-17 SERVICING STEERING LINKAGES AND SUSPENSION

If any defects are found, causes must be determined and corrections made before attempting to align the wheels. Servicing steering and suspension includes removal, replacement, and adjustment of tie rods; removing and replacement of other linkage parts, such as the steering idler and upper and lower control arms; removal and replacement of springs; and removal and reinstallation of wheel hubs and drums or disks. In addition, the steering gear may require adjustment, or removal for service.

All of these services, if needed, must be performed before aligning the wheels. It does no good to do an alignment if the wheel bearings or other part is defective, worn out, or in need of adjustment. If service to any of the above components is required, refer to the manufacturer's service manual.

♦ 50-18 WHEEL ALIGNMENT

There are many types of wheel aligners. Some are mechanical types that attach to the wheel spindles (Fig. 50-15). Some have light beams that display the measurements on a screen in front of the car (Fig. 50-16). Others are electronic that indicate the measurements on meters, displays, or printouts.

When doing a front-wheel alignment, you check and adjust (if needed) caster, camber, and toe. You also measure SAI and turning radius. These are not adjustable. If they are out of specifications, it means parts are bent or damaged and must be replaced. However, before you make the alignment checks, the following prealignment inspection must first be made.

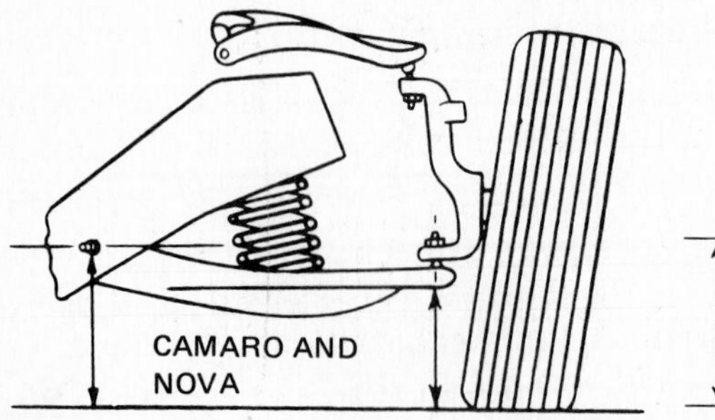

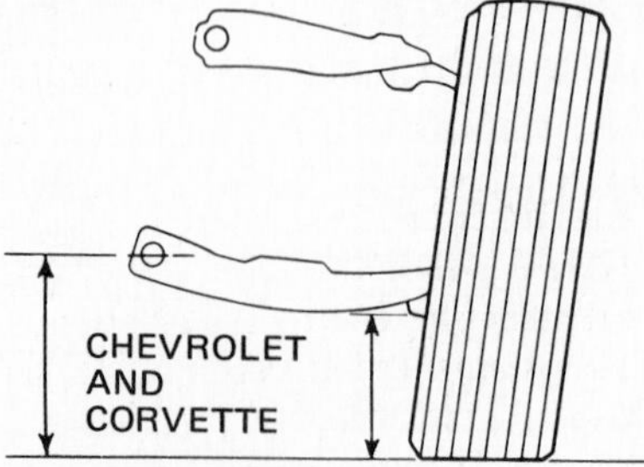

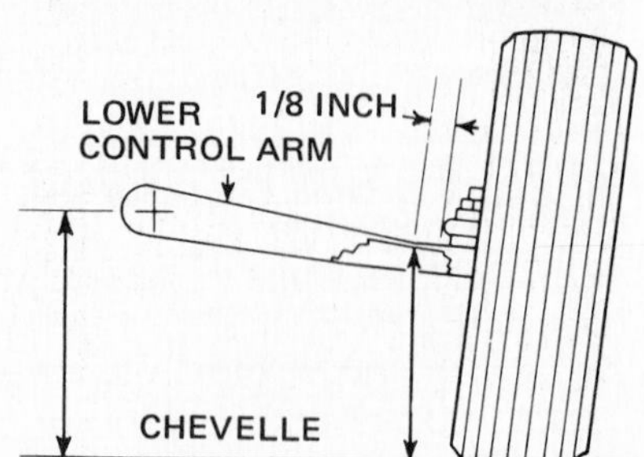

Fig. 50-14 Measurements to check for front-spring sag on various cars. *(Chevrolet Motor Division of General Motors Corporation)*

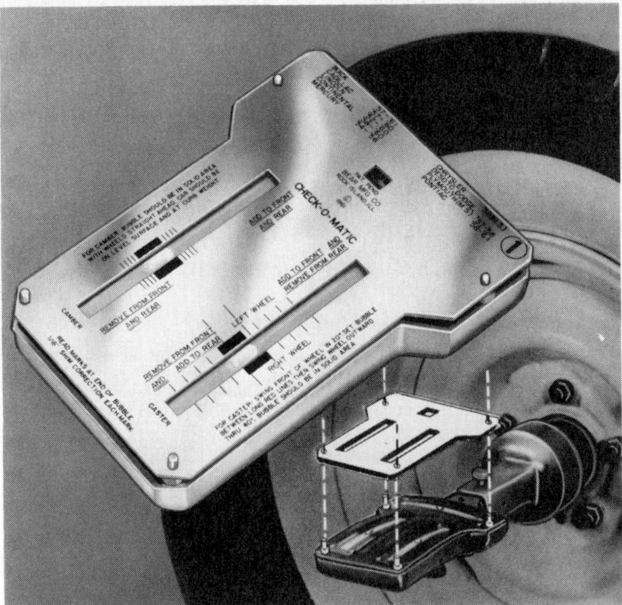

Fig. 50-15 A mechanical caster-camber gauge attached to the front-wheel spindle. *(Bear Manufacturing Company)*

- ♦ Check and correct tire pressure.
- ♦ Check and adjust wheel bearings.
- ♦ Check and adjust wheel run-out.
- ♦ Check ball joints and, if they are too loose, replace them.
- ♦ Check steering linkages, and make any corrections necessary.
- ♦ Check wheel balance, and correct it if necessary.
- ♦ Check rear leaf springs for cracks, broken leaves, and loose U bolts. Make any corrections necessary.
- ♦ Check front-suspension height.
- ♦ Check shock absorbers, and replace them if they are defective.

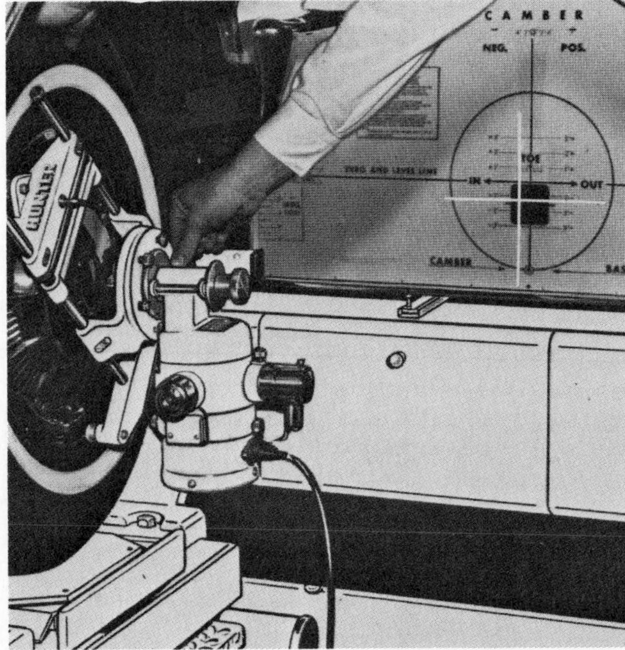

Fig. 50-16 Adjusting the wheel gauge to make a wheel-alignment check using a light-beam type of wheel aligner. Note the horizontal and vertical lines of light shining on the screen. *(Hunter Engineering Company)*

- ♦ Check wheel tracking. This means checking whether the rear wheels follow the front wheels or are off the track. If the wheels are off the track, it usually means a bent frame. The frame must be straightened before you can do a wheel alignment.

♦ 50-19 WHEEL BALANCE

The wheel may be checked for balance on or off the car. This is done by either of two methods: *static* or *dynamic.* In static balancing, the wheel is taken off the car and put on a "bubble" balancer to detect any imbalance (Fig. 50-17). A wheel that is out of balance is heavier in one section. This will cause the bubble in the center of the balancer to move off center. To balance the wheel, weights are added to the wheel rim until the bubble returns to center.

To dynamic-balance (or "spin-balance") a wheel, the wheel is spun either on or off the car. Figure 50-18 shows an electronic wheel balancer being used to balance a wheel on a car. Lack of balance shows up as a tendency for the wheel to move off center or out of line as it spins.

If the wheel is out-of-balance, one or more weights are installed on the wheel rim. In the shop, you will learn how to balance wheels.

♦ CAUTION ♦ To prevent injury from stones thrown out of the spinning tire, off-the-car wheel spinners should have a safety hood (Fig. 50-19). The hood fits around or over the tire while it is spinning to catch any stones that are thrown out.

♦ 50-20 ADJUSTING CAMBER AND CASTER

Several different ways to adjust camber and caster have been used. Some of the methods include removing and installing shims, turning a cam, shifting the inner control-arm shaft, and changing the length of the strut rod.

1. Adjustment by Installing or Removing Shims

The shims are located at the upper control-arm shafts. They are placed either inside or outside the frame bracket. Figure 50-20 shows the location of the shims in many General Mo-

Fig. 50-17 Bubble, or static, type of wheel balancer. *(Ammco Tools, Inc.)*

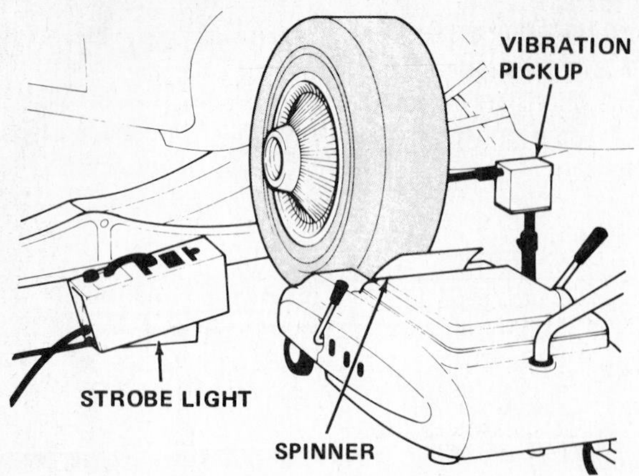

Fig. 50-18 An electronic type of dynamic wheel balancer. A magnet is attached to the brake backing plate. Any movement of the magnet is sensed through a short arm by a vibration pickup. This causes the strobe light to flash, indicating where to attach a wheel weight. *(Ford Motor Company)*

tors cars. The shims are inside the frame bracket. Figure 50-21 shows the location of the shims in many Ford cars. The shims are outside the frame bracket. When the shims are inside the frame bracket (Fig. 50-20), adding shims moves the upper control arm inward. This reduces positive camber.

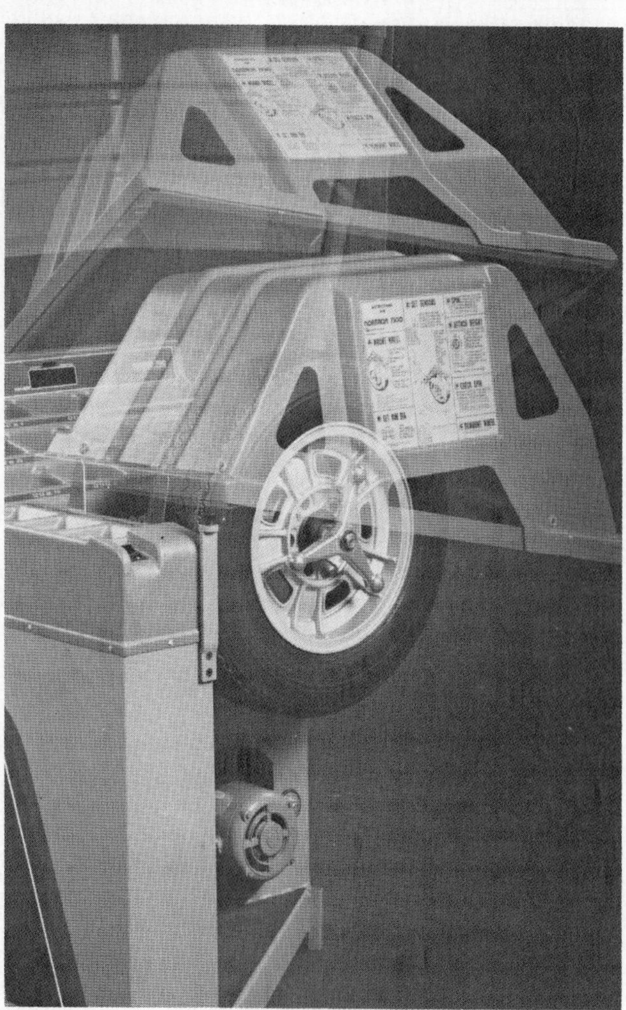

Fig. 50-19 Wheel balancer with safety hood installed over the tire. *(Hennessy Industries, Inc.)*

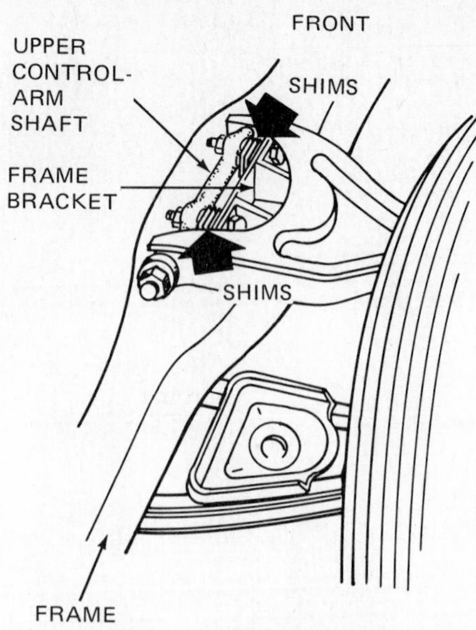

Fig. 50-20 Location of caster- and camber-adjusting shims (indicated by arrows). The shims and upper control-arm shaft are inside the frame bracket. *(Bear Manufacturing Company)*

When the shims and shaft are outside the frame bracket (Fig. 50-21), adding shims moves the upper control arm outward. This increases positive camber. If shims are added at one attachment bolt and removed from the other, the outer end of the upper control arm shifts one way or the other. This increases or decreases caster. Figure 50-22 shows these adjustments.

2. Adjustment by Turning a Cam There have been several variations of this method. Figure 50-23 shows the arrangement used on some Chrysler-built cars. The two bushings at the inner end of the upper control arm are attached to the frame brackets by two attachment bolts and cam assemblies. When the cam bolts are turned, the camber and caster

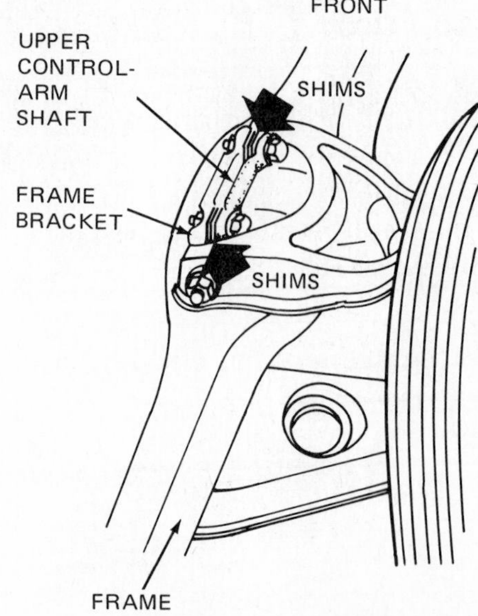

Fig. 50-21 Location of caster- and camber-adjusting shims (indicated by arrows). The shims and upper control-arm shaft are outside the frame bracket. *(Bear Manufacturing Company)*

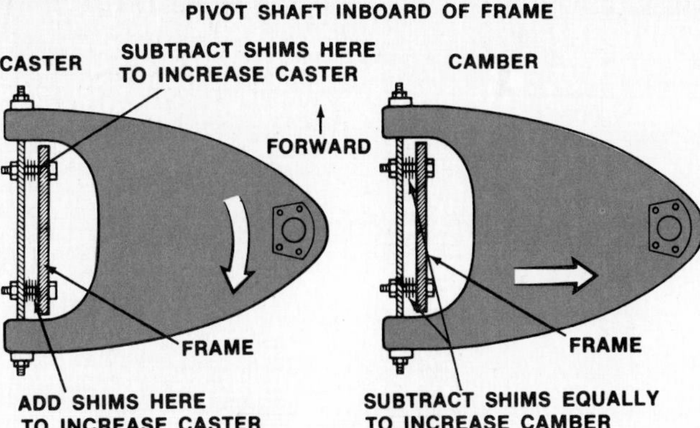

Fig. 50-22 Caster and camber adjustments on some cars using shims. *(Chevrolet Motor Division of General Motors Corporation)*

are changed. If both are turned the same amount and in the same direction, the camber is changed. If only one cam bolt is turned, or if the two are turned in opposite directions, the caster is changed.

3. Adjustment by Shifting Inner Shaft This system uses slots in the frame at the two points where the inner shaft is attached (Fig. 50-24). When the attaching bolts are loosened, the inner shaft can be shifted in or out to change camber. Only one end is shifted to change caster.

4. Adjustment by Changing Length of Strut Rod This type of adjustment is shown in Fig. 50-25.

♦ 50-21 ADJUSTING TOE

After correcting caster and camber, toe is adjusted (Fig. 50-26). Place the front wheels in the straight-ahead position. Then check the positions of the spokes in the steering wheel. If they are not centered, they can be properly positioned when toe is set. Toe is adjusted by turning the adjuster sleeves in the linkage. If the adjuster sleeves are turned to lengthen the tie rods, the toe-in is increased.

♦ 50-22 MACPHERSON-STRUT CAMBER AND CASTER ADJUSTMENTS

Some cars with MacPherson struts do not have any adjustment for camber or caster. Other cars with this type of suspension have a camber adjustment. On these cars, turn the cam bolt at the lower end of the strut to move the top of the wheel in or out to adjust the camber (Fig. 50-27).

Wrong alignment settings can cause excessive tire wear. Therefore, a caster-camber adjustment kit is available for some cars with nonadjustable MacPherson struts (Fig. 50-28). After the kit is installed, caster and camber can be adjusted by moving the top of the strut. The kit has a slotted plate that is installed between the strut and the inner fender. In the original design, the top of the strut is fixed in position because it is bolted through the holes in the inner fender.

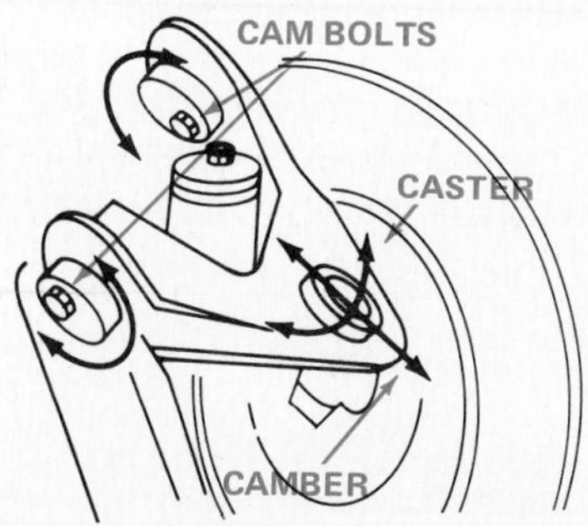

Fig. 50-23 Turning the cam bolts moves the upper control arm toward or away from the frame to adjust caster and camber. *(Ammco Tools, Inc.)*

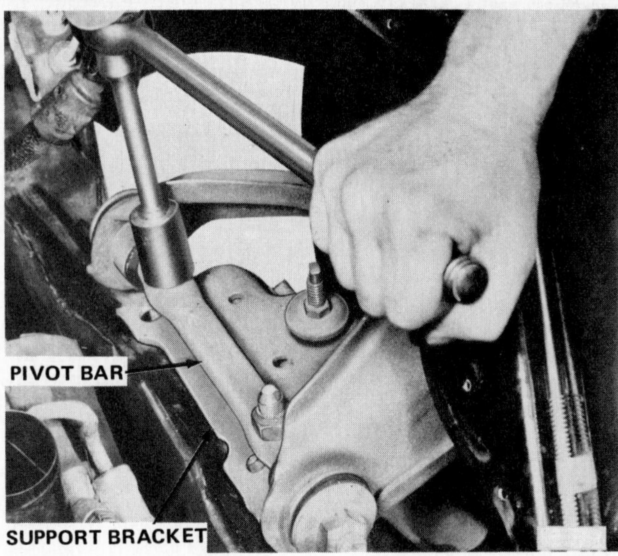

Fig. 50-24 Adjusting caster and camber by shifting the position of the inner shaft using slots in the frame. *(Chrysler Corporation)*

♦ 50-23 REAR-WHEEL ALIGNMENT

On cars with front-wheel drive and cars with independent rear suspension, such as the Chevrolet Corvette, the rear-wheel alignment can be checked. One method is to back the car onto the gauges used to align the front wheels. Camber will read in the normal manner. But toe-in will read as toe-out, and toe-out will read as toe-in. Caster is usually set at zero originally and needs no adjustment. Check that the strut rods are straight. If they are bent, replace them.

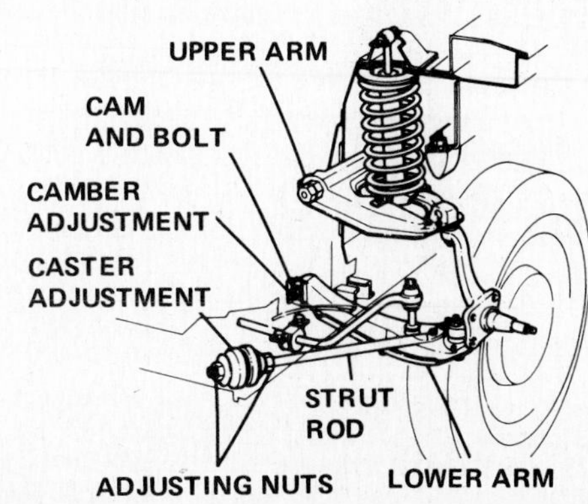

Fig. 50-25 Adjusting caster by changing the length of the strut rod. *(Ford Motor Company)*

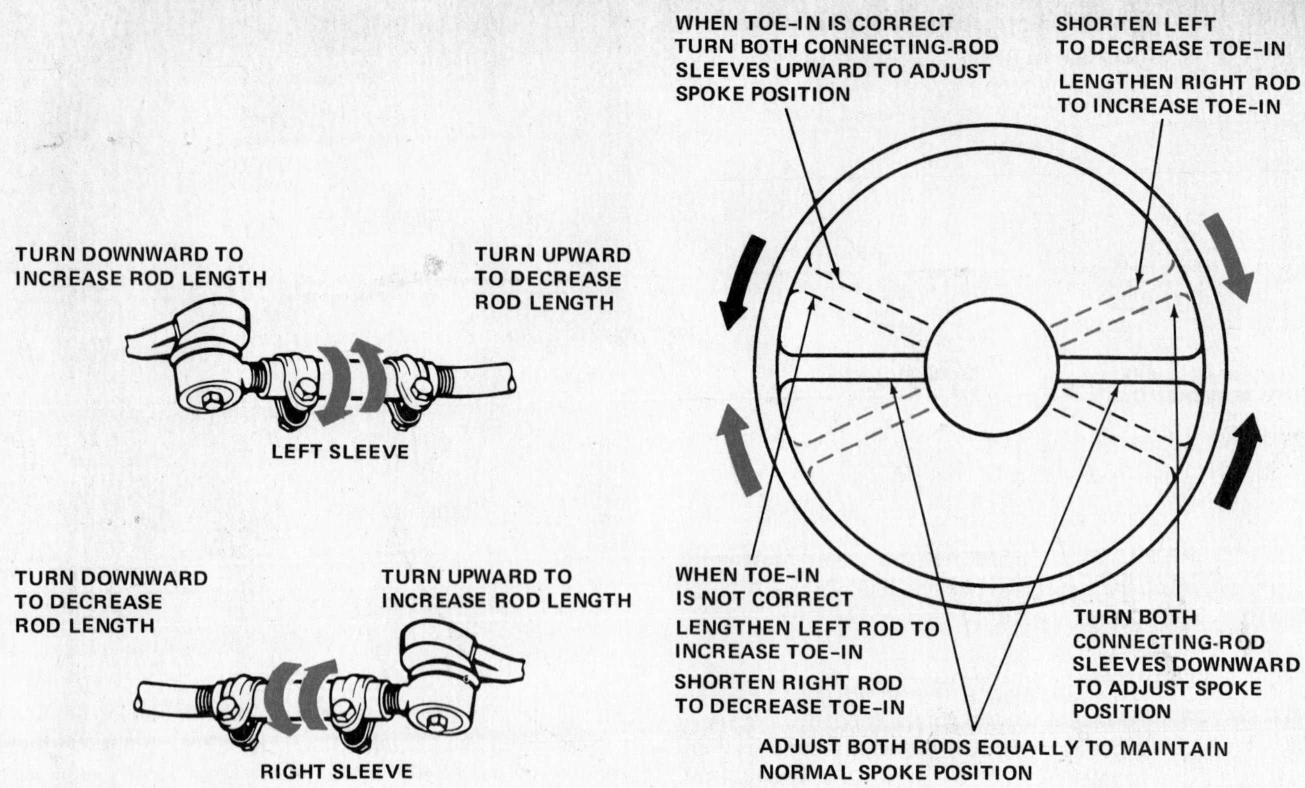
WHEN TOE-IN IS CORRECT TURN BOTH CONNECTING-ROD SLEEVES UPWARD TO ADJUST SPOKE POSITION
SHORTEN LEFT TO DECREASE TOE-IN
LENGTHEN RIGHT ROD TO INCREASE TOE-IN
TURN DOWNWARD TO INCREASE ROD LENGTH
TURN UPWARD TO DECREASE ROD LENGTH
LEFT SLEEVE
TURN DOWNWARD TO DECREASE ROD LENGTH
TURN UPWARD TO INCREASE ROD LENGTH
RIGHT SLEEVE
WHEN TOE-IN IS NOT CORRECT
LENGTHEN LEFT ROD TO INCREASE TOE-IN
SHORTEN RIGHT ROD TO DECREASE TOE-IN
TURN BOTH CONNECTING-ROD SLEEVES DOWNWARD TO ADJUST SPOKE POSITION
ADJUST BOTH RODS EQUALLY TO MAINTAIN NORMAL SPOKE POSITION

Fig. 50-26 Adjustments for toe-in and steering-wheel alignment. Left, spindle connecting-rod (tie-rod) adjustments. Right, adjustments to align steering wheel. *(Ford Motor Company)*

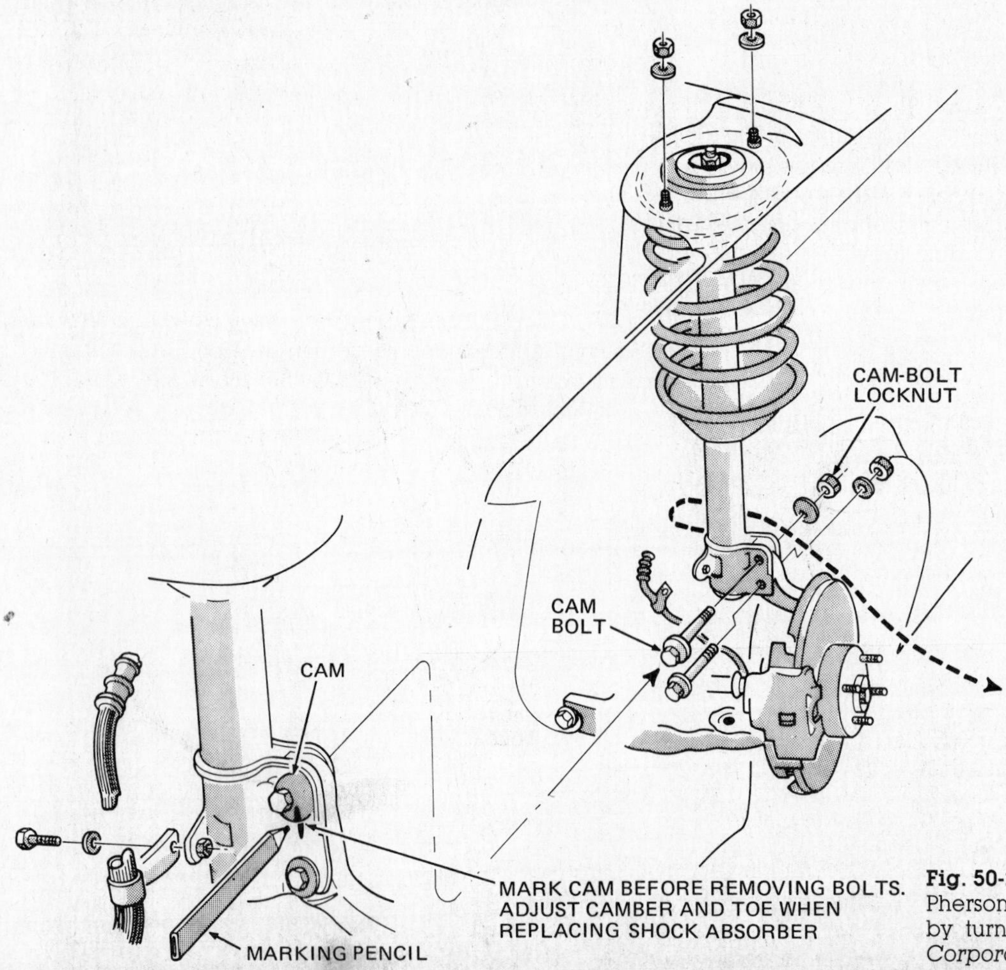

CAM-BOLT LOCKNUT

CAM BOLT

CAM

MARK CAM BEFORE REMOVING BOLTS. ADJUST CAMBER AND TOE WHEN REPLACING SHOCK ABSORBER

MARKING PENCIL

Fig. 50-27 On some cars with MacPherson struts, camber is adjusted by turning a cam bolt. *(Chrysler Corporation)*

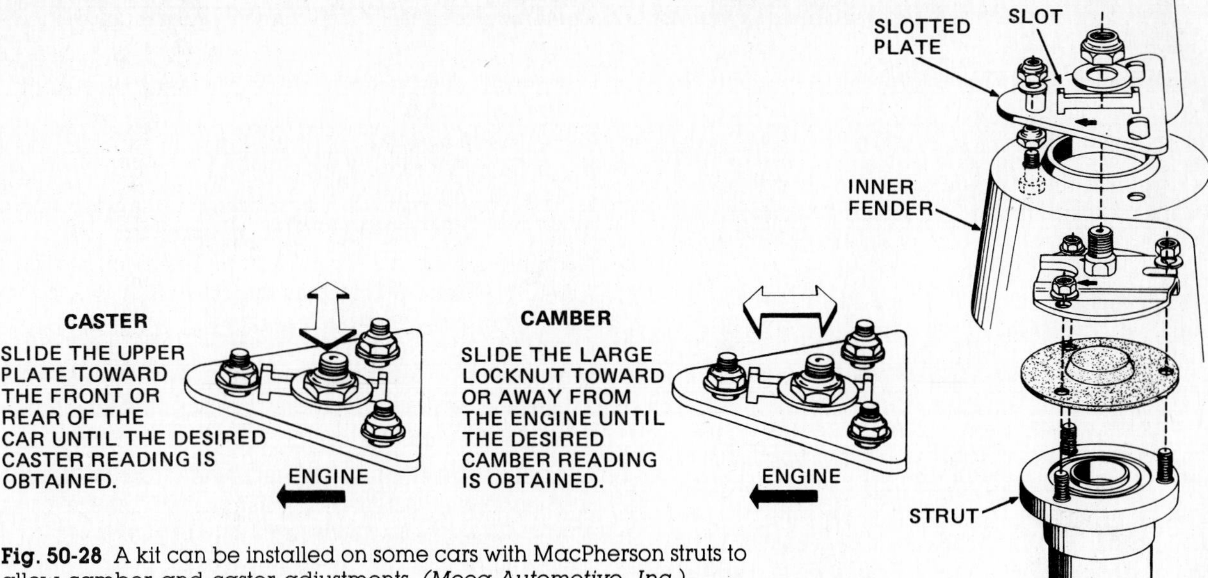
SLOTTED PLATE
SLOT
INNER FENDER
CASTER
SLIDE THE UPPER PLATE TOWARD THE FRONT OR REAR OF THE CAR UNTIL THE DESIRED CASTER READING IS OBTAINED.
ENGINE
CAMBER
SLIDE THE LARGE LOCKNUT TOWARD OR AWAY FROM THE ENGINE UNTIL THE DESIRED CAMBER READING IS OBTAINED.
ENGINE
STRUT

Fig. 50-28 A kit can be installed on some cars with MacPherson struts to allow camber and caster adjustments. *(Moog Automotive, Inc.)*

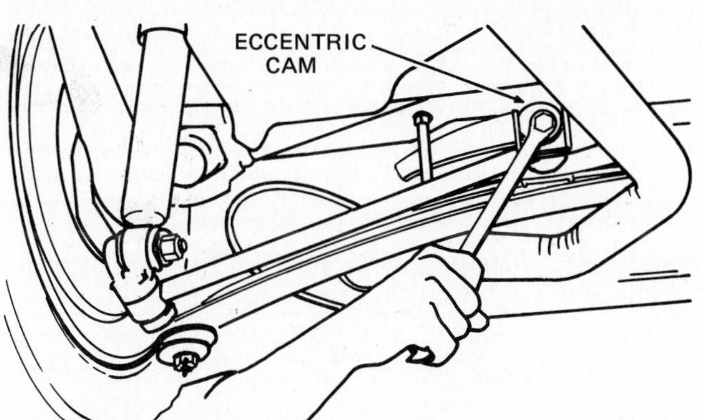
ECCENTRIC CAM

Fig. 50-29 Adjusting rear-wheel camber on a Chevrolet Corvette, which has independent rear suspension. *(Chevrolet Motor Division of General Motors Corporation)*

1. Camber Adjustment This is adjusted by turning the eccentric cam and bolt (Fig. 50-29). Loosen the cam-bolt nut and turn the assembly to adjust the camber. Then tighten the cam-bolt nut.

2. Toe-in Adjustment Adjust toe-in by inserting shims inside the frame side member on both sides of the torque-control-arm pivot bushing (Fig. 50-30).

♦ 50-24 SERVICING THE STEERING GEAR

Manual steering gears have two basic adjustments. One adjusts the worm-gear and steering-shaft end play. The other adjusts the backlash, or free play, between the worm and sector.

Other adjustments are required on power-steering gears. Refer to the manufacturer's service manual for the procedures.

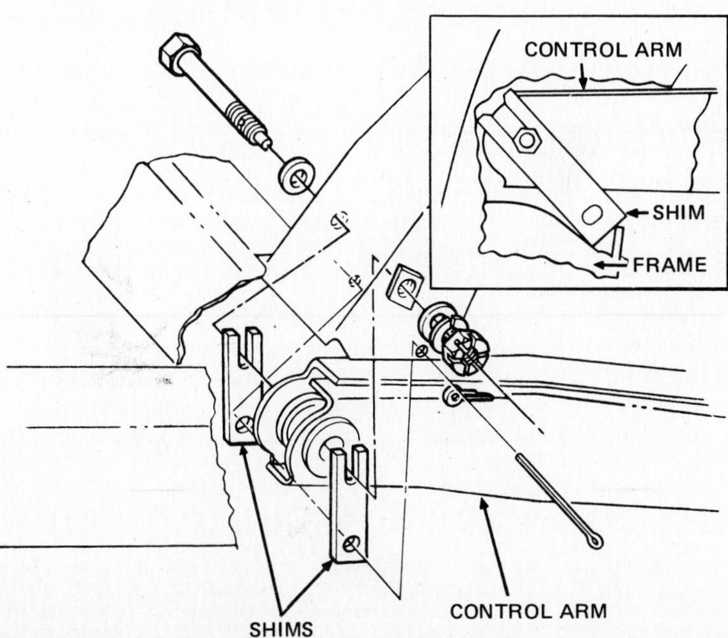
CONTROL ARM
SHIM
FRAME
SHIMS
CONTROL ARM

Fig. 50-30 Shim location to adjust rear-wheel toe on the Chevrolet Corvette. *(Chevrolet Motor Division of General Motors Corporation)*

♦ REVIEW QUESTIONS ♦

Select the *one* correct, best, or most probable answer to each question. You can find the answers in the section indicated at the end of each question.

1. A driver says that the front end of the car vibrates up and down while traveling at most road speeds. Mechanic A says that too much radial runout of the front tires could be the cause. Mechanic B says that static out-of-balance of the front tires could be the cause. Who is right? (♦50-8)
   a. A only
   b. B only
   c. either A or B
   d. neither A nor B
2. When a vehicle is road tested, the steering wheel shakes from side to side at higher speeds. Mechanic A says that this could be caused by front wheels not statically balanced. Mechanic B says that this could be caused by front wheels not dynamically balanced. Who is right? (♦50-7)
   a. A only
   b. B only
   c. either A or B
   d. neither A nor B
3. As a general rule, axial (up-and-down) play in a ball joint should not exceed (♦50-2)
   a. ¼ inch [6.35 mm]
   b. 1/16 inch [1.6 mm]
   c. ⅛ inch [3.2 mm]
   d. ½ inch [12.7 mm]
4. Mechanic A says that tire cupping is caused by wheel tramp. Mechanic B says that the cause is loose tie-rod ends. Who is right? (♦50-2)
   a. A only
   b. B only
   c. either A or B
   d. neither A nor B
5. Load-carrying ball joints with load indicators (♦50-3)
   a. should be unloaded for checking
   b. should be checked with ball joints loaded
   c. should be checked on a frame-contact hoist
   d. should be checked with a pry bar and ball-joint checking gauge

CHAPTER 51
Automotive Brakes

After studying this chapter, you should be able to:

1. Explain the use and types of friction in braking systems.
2. Describe the construction and operation of a dual braking system.
3. Discuss the operation of drum brakes.
4. Explain how disk brakes work.
5. Discuss the purpose and operation of parking brakes.
6. Describe the construction and operation of antilock brake systems.
7. Discuss power brakes and how they work.

♦ 51-1 AUTOMOTIVE BRAKES

This chapter describes the construction and operation of the various types of braking systems used in automotive vehicles. The braking system used most frequently operates hydraulically, by pressure applied through a liquid. These are the foot-operated brakes that the driver normally uses to slow or stop the car. They are called the *service brakes*. In addition, all cars have a parking-brake system which is mechanically operated by a separate foot or hand lever.

On some trucks and buses, the braking system is operated by air pressure (pneumatically). This type of brake is an *air brake*. Many boat and camping trailers have *electric brakes*. All of these braking systems depend upon friction between moving parts and stationary parts for their stopping force.

♦ 51-2 FRICTION

Friction is the resistance to motion between two objects in contact with each other. Three types of friction are dry, greasy, and viscous. Automotive brakes are examples of dry friction. Friction varies according to the force applied between the sliding surfaces, the roughness of the surfaces, and the material of which the surfaces are made. For example, suppose that a platform and its load weigh 110 pounds [50 kg]. Suppose that a force or pull of 55 pounds [25 kg] is required to move the platform along the floor (Fig. 51-1). Then reduce the load so that the platform and load weigh only 11 pounds [5 kg]. Now a force of only 5.5 pounds [2.5 kg] is required to move the platform along the floor. *Friction varies with the load.*

Now suppose we smoothed the floor and the sliding part of the platform with sandpaper. Then it would require less force to move the platform on the floor. *Friction varies with the roughness of the surfaces.*

Friction also varies with the type of material. For example, dragging a 110-pound [50-kg] bale of rubber across a con-

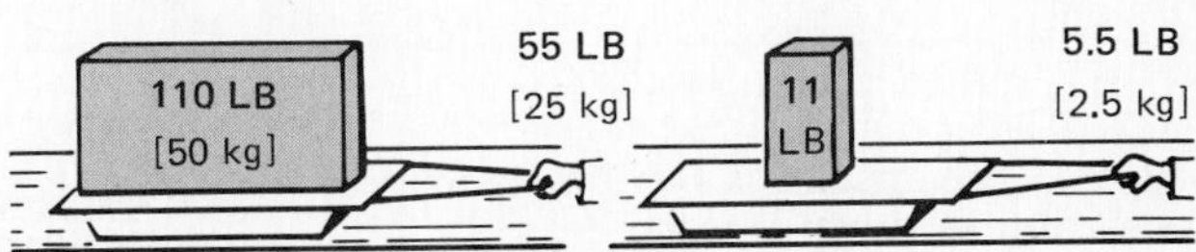

Fig. 51-1 Friction varies with the load applied between the sliding surfaces.

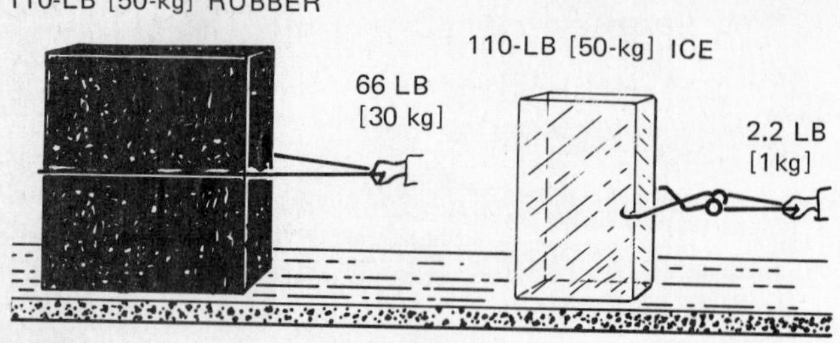

Fig. 51-2 Friction varies with the type of material.

crete floor might require a force of 66 pounds [30 kg] (Fig. 51-2). But to drag a 110-pound [50-kg] block of ice across the same floor might require a pull of only 2.2 pounds [1 kg].

♦ 51-3 FRICTION OF REST AND MOTION

More force is required to put an object into motion than is required to keep the object in motion (Fig. 51-3). In Fig. 51-3, it takes two people to start the object moving. After the object is moving, one person can keep it moving. Therefore, the friction of an object at rest is greater than the friction of an object in motion.

These two kinds of friction are called *static friction* and *kinetic friction*. The word *static* means at rest. The word *kinetic* means in motion, or moving. Therefore, static friction is friction of rest, and kinetic friction is friction of motion.

♦ 51-4 FRICTION IN CAR BRAKES

Friction is used in the car braking system. The friction between the brake drums or disks and brake shoes or pads slows or stops the car. This friction slows the rotation of the wheels.

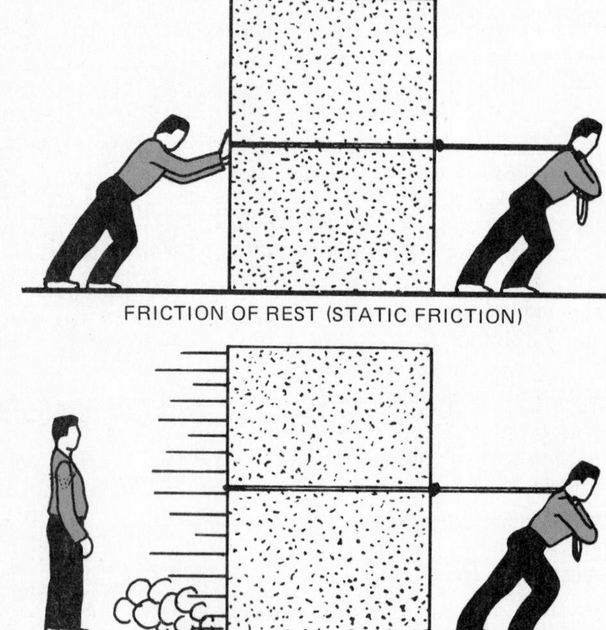

Fig. 51-3 Friction of rest is greater than friction of motion. In the example shown, it takes two people to overcome the friction of rest (static friction). But one person can keep the object moving by overcoming the friction of motion (kinetic friction).

Then friction between the tires and the road slows the motion of the car. Note that it is the friction between the tires and the road that stops the car. The car would *not* stop more quickly if the wheels were locked so that the tires skidded on the road.

If the brakes are applied so hard that the wheels lock, then the friction between the tires and the road is kinetic friction. This is the friction of motion as the tires skid on the road. When the brakes are applied a little less hard, the wheels are permitted to continue rotating. Then it is static friction that works between the tires and road. The tire surface is not skidding on the road but is rolling on it. The car stops more quickly if the brakes are applied just hard enough to get maximum static friction between the tires and road. If the brakes are applied harder than this, then the wheels will lock. The tires will skid, or slide, and the lesser kinetic friction will result.

HYDRAULICS AND PNEUMATICS

♦ 51-5 MEANING OF HYDRAULICS

Hydraulics is the use of a liquid under pressure to transfer force or motion, or to increase an applied force. Our special interest in hydraulics is related to the actions in automotive systems that result from pressure applied to a liquid. This is called *hydraulic pressure*. Hydraulic pressure is used in the brake system, in shock absorbers, and in power-steering systems. It is also used in automatic transmissions (control circuits and torque converter), engines (hydraulic valve lifters, oil pump, fuel pump, and water pump), and other parts.

♦ 51-6 INCOMPRESSIBILITY OF LIQUIDS

Increasing the pressure on a gas will compress the gas into a smaller volume (Fig. 51-4). However, increasing the pressure

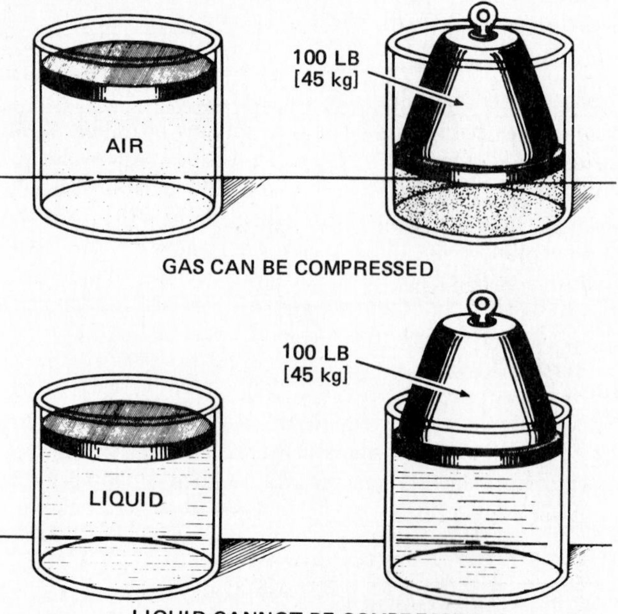

Fig. 51-4 Gas can be compressed when pressure is applied. However, a liquid cannot be compressed when pressure is applied.

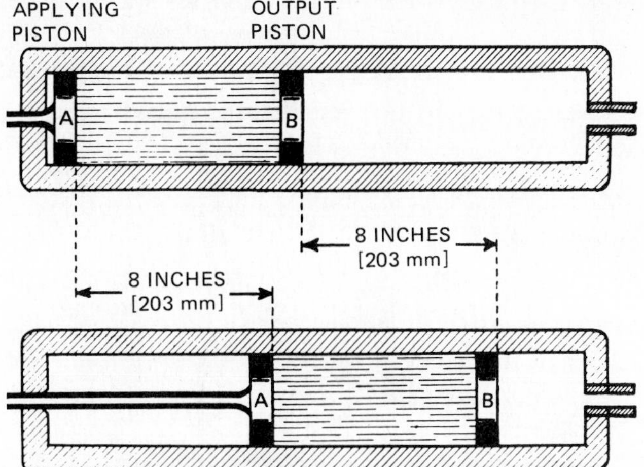

Fig. 51-5 Force and motion can be transmitted by a liquid. When the applying piston is moved 8 inches (203 mm) in the cylinder, the output piston is also moved 8 inches (203 mm).

on a liquid will not reduce its volume (Fig. 51-4). A liquid is incompressible. This is because the molecules of the liquid are relatively close together—as opposed to a gas, in which the molecules are relatively far apart.

Putting pressure on a gas can "squeeze out" some of the space between the molecules. But in a liquid, there is no extra space that can be squeezed out by the application of pressure. The molecules are already as close together as possible. Therefore, putting pressure on the molecules of a liquid cannot force them closer together.

♦ 51-7 TRANSMISSION OF MOTION BY LIQUID

Since liquid is not compressible, it can transmit motion. For example, Fig. 51-5 shows two pistons in a cylinder, with a liquid separating them. When the applying piston is pushed into the cylinder 8 inches [203 mm], the output piston will be pushed along in the cylinder for the same distance, or 8 inches [203 mm]. In Fig. 51-5, you could place a solid bar between piston A and piston B and get exactly the same effect.

Motion can also be transmitted from one cylinder to another by a tube or pipe. In Fig. 51-6, the applying piston is in cylinder A, and the output piston is in cylinder B. As the

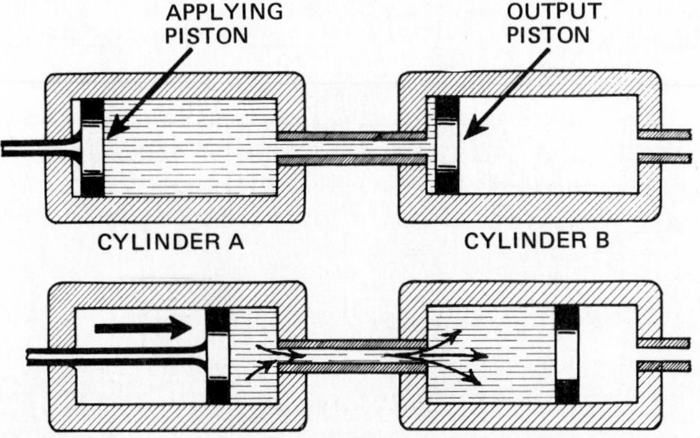

Fig. 51-6 Force and motion can be transmitted through a tube from one cylinder to another by a liquid, that is, by hydraulic pressure.

applying piston is moved into its cylinder, liquid is forced from cylinder A into cylinder B. This causes the output piston to be moved in its cylinder.

This is how the brake hydraulic system works (Fig. 51-7). The master cylinder has two pistons. As the brakes are applied, the pistons are moved to the left (in Fig. 51-7). This sends fluid under pressure to cylinders at the four wheels. Pistons in the wheel cylinders are forced to move out as shown by the arrows. This action applies the brakes, as explained later.

♦ 51-8 PNEUMATICS

Pneumatics means of or pertaining to air or other gases. Automobile tires are pneumatic tires because they are filled with air. Any tool that is run by compressed air is a pneumatic tool (Chap. 7).

Several types of pneumatic devices and systems have been used on cars and trucks. These include air suspension, air power steering, air brakes, and air shock absorbers. All of these are operated by compressed air.

DRUM BRAKES

♦ 51-9 BRAKE ACTION

A typical braking system (Figs. 51-7 and 51-8) includes two basic parts. These are the master cylinder with brake pedal and the wheel brake mechanism. The other parts are the connecting tubing, or brake lines, and the supporting arrangements.

Braking action starts at the brake pedal. When the pedal is pushed down, brake fluid is sent from the master cylinder to the wheels. At the wheels, the fluid pushes brake shoes, or pads, against revolving drums or disks. The friction between the stationary shoes or pads and the revolving drums or disks slows and stops them. This slows or stops the revolving wheels which, in turn, slow or stop the car.

Figure 51-7 shows the brake lines, or tubes, through which the fluid flows. There are two chambers and two pistons in the master cylinder. In the arrangement shown in Fig. 51-7 (called the *front-rear split*), one chamber is connected to the front-wheel brakes. The other chamber is connected to the rear-wheel brakes. This braking arrangement is called a *dual braking system*. Dual braking systems are used on modern cars. The purpose of splitting the system into two parts is that, if one part fails, the other will work and stop the car.

A second arrangement is called a *diagonal split*. One chamber in the master cylinder is connected to the right-front and left-rear wheel brakes. The other chamber is connected to the left-front and right-rear wheel brakes. Figure 51-9 compares the two arrangements.

In earlier braking systems, there was only one chamber in the master cylinder. It was connected to all four wheel brakes. If one part failed, the whole system failed. The dual braking system provides extra protection, because both sections seldom fail at once.

♦ 51-10 DUAL BRAKING SYSTEMS

In older cars, the master cylinder contained only one piston, as shown in Fig. 51-10. The dual braking system uses a two-piston master cylinder, as shown in Figs. 51-8 and 51-9. One brake line from the master cylinder goes to one set of wheel

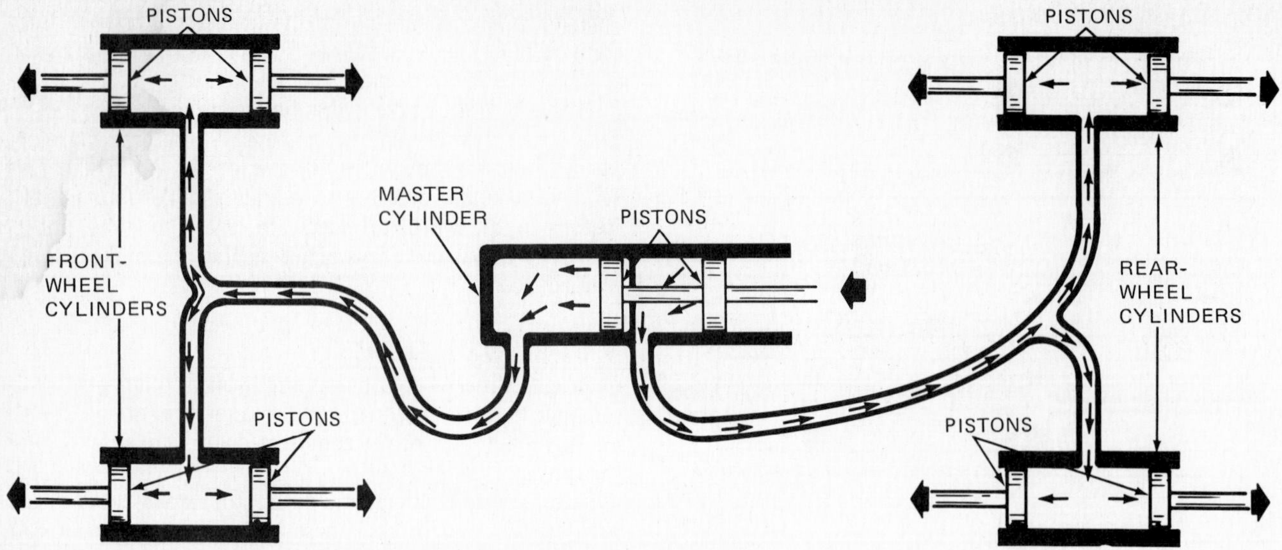

Fig. 51-7 Flow of brake fluid to the four wheel cylinders when the pistons are pushed into the master cylinder.

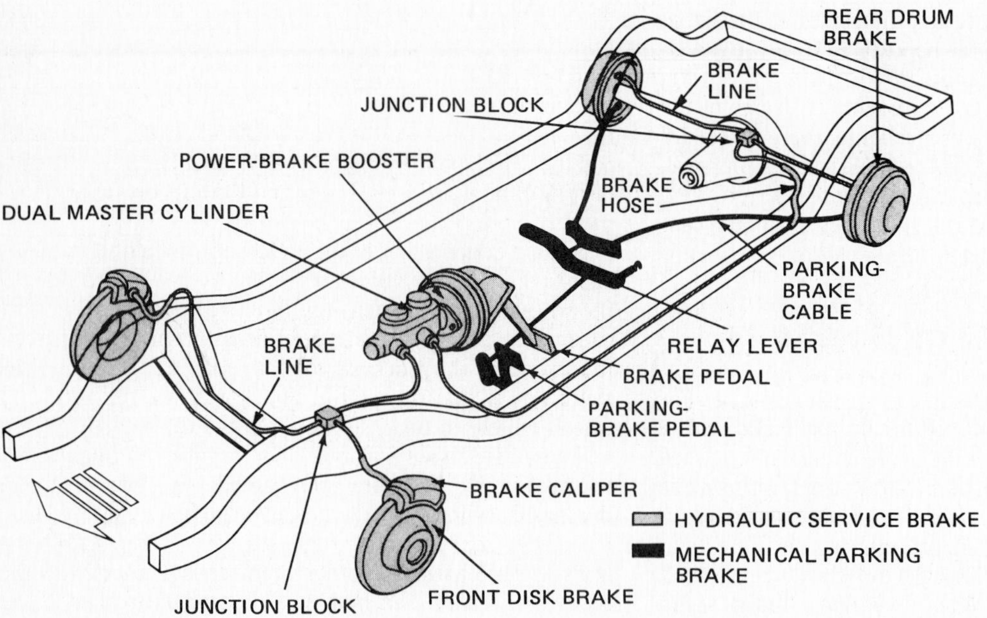

Fig. 51-8 Layout of the complete brake system on an automobile. *(Texaco, Inc.)*

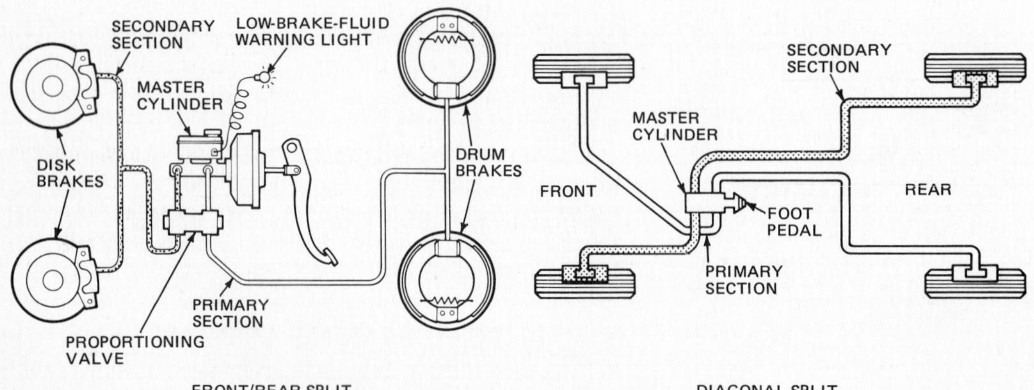

Fig. 51-9 Two basic types of the split braking system. *(Mazda Motors of America, Inc.; Ford Motor Company)*

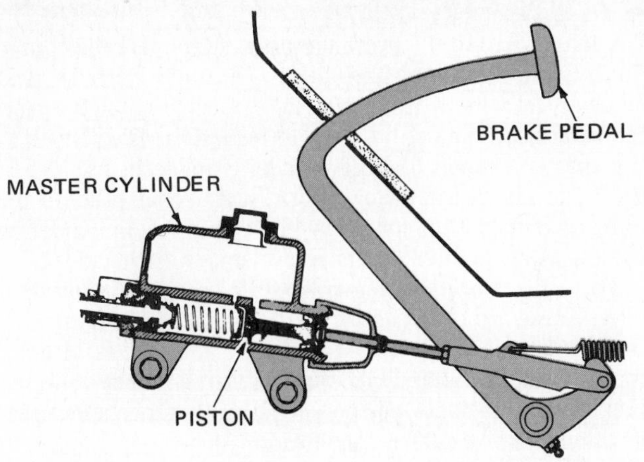

Fig. 51-10 Linkage between brake pedal and master cylinder. The master cylinder is the older type, using only one piston. *(General Motors Corporation)*

brakes. The other brake line from the master cylinder goes to the other set of wheel brakes. (Many brake systems include a power-brake booster, which is described later.)

The dual braking system includes a *pressure-differential valve*. The valve has switch contacts that are held open by equal pressures in both sections of the hydraulic system. If one section begins to leak, the pressure in that section will drop. Then the pressure-differential valve will connect the instrument-panel warning light to the battery. A red BRAKE light will come on to warn the driver of brake trouble, and that the brake system should be serviced at once. Although the other section of the dual braking system can provide emergency braking, the car is not safe to drive.

♦ 51-11 MASTER CYLINDERS

There are two types of master-cylinder construction. The older type has the reservoirs integrally cast with the cylinder to form a single-piece master-cylinder body (Fig. 51-11). It

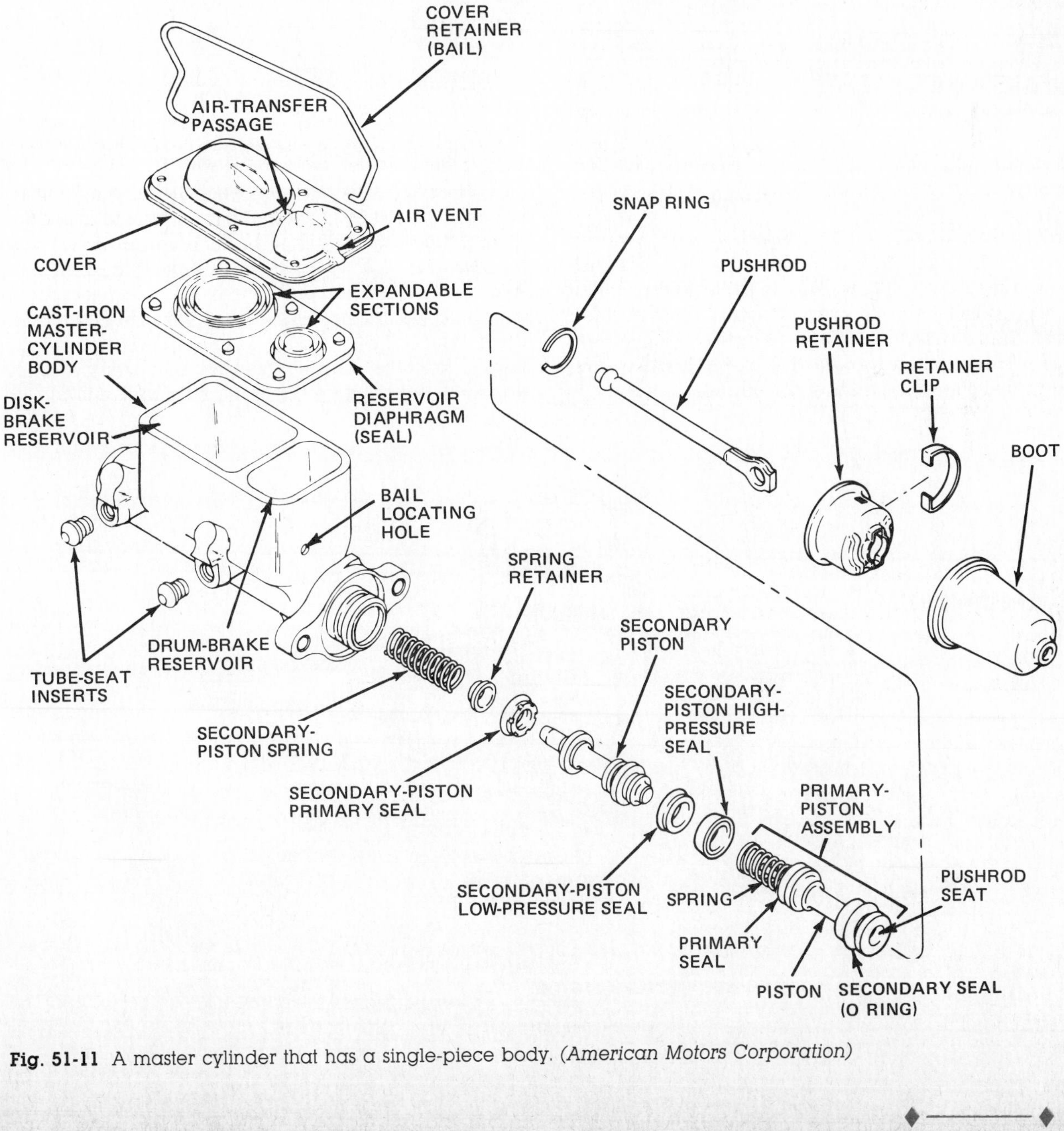

Fig. 51-11 A master cylinder that has a single-piece body. *(American Motors Corporation)*

usually is made of cast iron. The later type is sometimes called a *composite* master cylinder. It has reservoirs formed from sheet metal or plastic, then attached to the cylinder with rubber grommets or seals (Fig. 51-12). The separate cylinder to which the reservoirs are attached often is made of cast aluminum. These reservoirs may have translucent windows for visually checking the fluid level.

The two pistons in the master cylinder are called the *primary piston* and the *secondary piston* (Figs. 51-11 and 51-12). The primary piston is the piston that is directly actuated by the brake-pedal pushrod. When the master cylinder is mounted in the engine compartment against the fire wall, the primary piston is the rearward piston. This is the one closest to the fire wall. The secondary piston is at the forward end of the master cylinder.

There are two holes in the bottom of each reservoir. The small hole toward the front of each reservoir (in the direction of piston travel) is the *vent,* or *bypass, port*. The larger hole is the *replenishing,* or *compensating, port*.

♦ 51-12 BRAKE LINES

Brake fluid is carried by steel pipes called *brake lines* from the master cylinder to the various valves and then to the brakes at the wheels (Fig. 51-8). Brake lines usually pass under the floor pan of the car, where they are exposed to stones and other road damage. Because of this, brake lines often are wrapped with a wire "armor" to protect them against damage.

Steel tubing will crack if it vibrates or is flexed too much. Therefore, a short, flexible *brake hose* (or "flex hose") is used to connect the stationary brake lines to the brake assemblies that move up and down and swing with each front wheel for steering. Rear-wheel-drive cars use a junction block, with a single flexible hose at the rear axle (Fig. 51-8). This takes care of the up-and-down movement of the housing.

♦ 51-13 BRAKE FLUID

The liquid used in the hydraulic-brake system is called *brake fluid*. Brake fluid must have certain characteristics. It must be chemically inert; it must be little affected by high or low temperatures; it must provide lubrication for the pistons in the master cylinder, wheel cylinders, and calipers; and it must not attack the metal, plastic, and rubber parts in the braking system. Therefore, only brake fluid recommended by the car manufacturer (or its equivalent) must be added.

Three types of hydraulic-brake fluid are used in automobiles. These are classified by the Department of Transportation (DOT) as DOT 3, DOT 4, and DOT 5. This classification must appear on the brake-fluid container label. Silicone brake fluid is the only type now available that meets the specifications for the DOT 5 classification.

Careful Never put engine oil or any other type of mineral oil in the brake system. Mineral oil will cause rubber parts in the system, such as the piston cups and seals, to swell and break apart. This could cause complete brake failure. Use only the brake fluid recommended.

♦ 51-14 WHEEL-BRAKE MECHANISMS

Two different types of brakes may be used at the wheels, drum brakes and disk brakes (Fig. 51-13). In the drum brake, a drum is attached to and revolves with the wheel. Curved brake shoes fit inside the drum. When the car is moving and no braking is required, there are slight air gaps, or clearances between the shoes and the revolving drum. However, when the driver pushes down on the brake pedal, the action sends brake fluid to the wheel brake mechanism. A wheel cylinder in the brake pushes the shoes apart, as shown to the lower left in Fig. 51-13. The shoes are forced against the revolving drum. Friction between the stationary shoes and the revolving drum slows or stops the drum, the wheel, and the car.

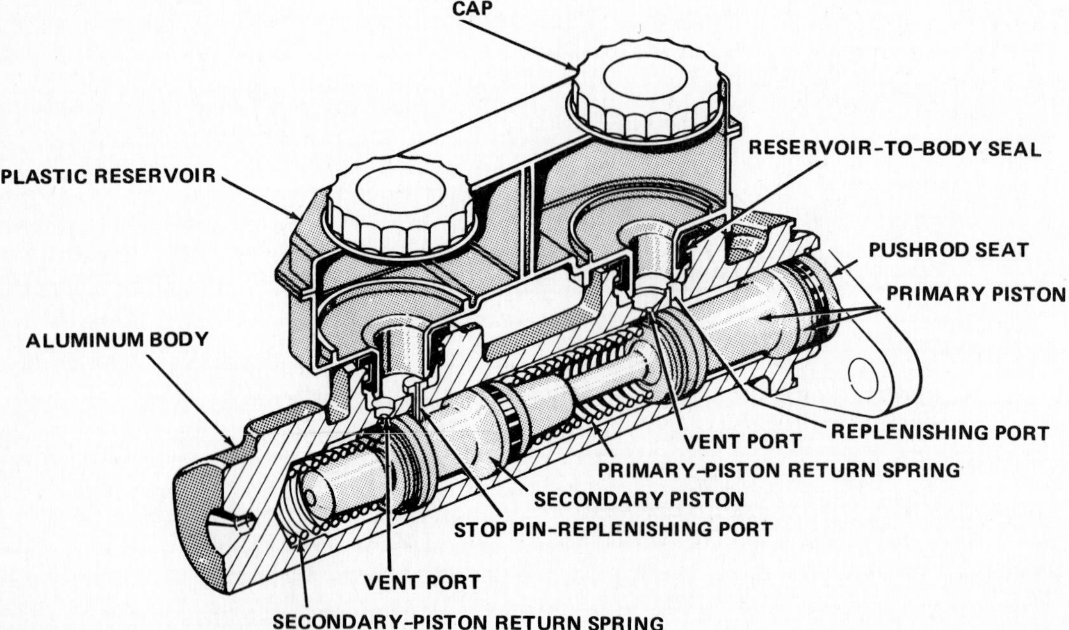

Fig. 51-12 A composite master cylinder, with separate plastic reservoirs attached to the body. *(Chrysler Corporation)*

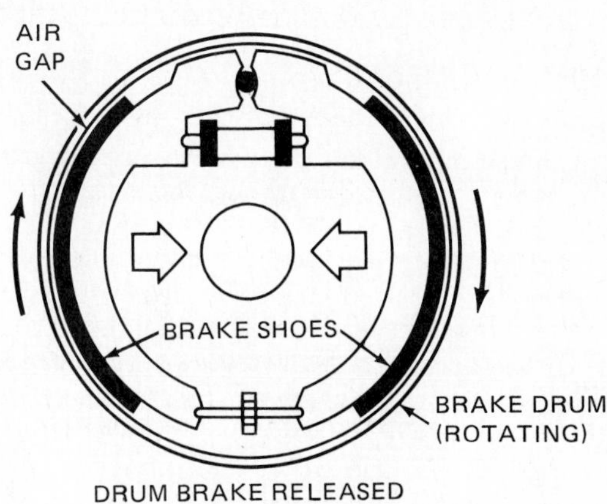

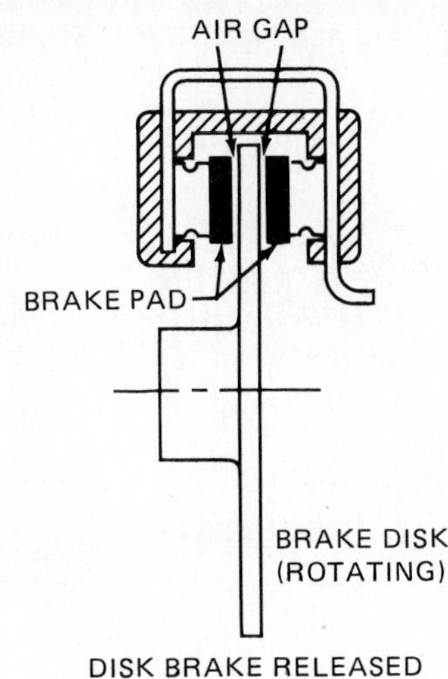

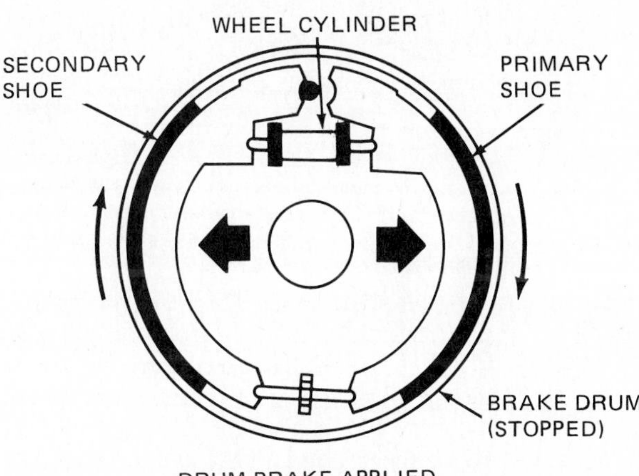

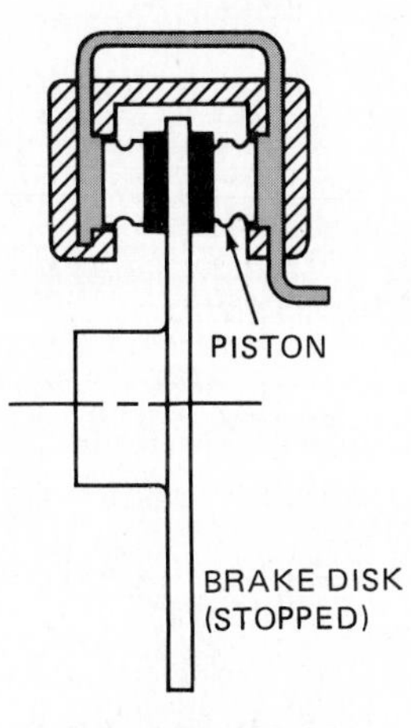

Fig. 51-13 The two types of wheel brakes.

In the disk brake, a disk (or *rotor*) is attached to the wheel and rotates with it. When the brakes are applied by the driver, the brake fluid forces a piston to push the two pads against the rotating disk (lower right in Fig. 51-13). Friction between the pads and the rotating disk slows or stops the disk, the wheel, and the car.

♦ 51-15 DRUM-BRAKE CONSTRUCTION

The drum brake has a steel or iron drum to which the wheel is bolted. The drum and wheel rotate together. Inside the drum, attached to the steering knuckle or rear axle, is the brake mechanism. The brake mechanism at the front wheels is attached to the steering knuckle. Figures 48-5 and 48-9 show the bolt holes through which the front-wheel brake is attached to the steering knuckle. Figure 48-22 shows the bolt holes through which the rear-wheel brake is attached to the axle housing. (This is a front-engine rear-wheel-drive car.)

Figure 51-14 shows how a drum brake is attached to the left rear wheel of a front-drive car. The brake is attached to the rear-suspension arm. The brake drum is mounted on the rear-wheel spindle by two bearings. The wheel is attached to the brake drum.

Figure 51-15 shows the active parts in a drum brake. There are two brake shoes at each wheel. The bottoms of the shoes are held apart by an adjusting screw, or *star wheel*. The tops of the shoes are held apart by actuating pins from the wheel cylinder. The shoes are made of metal. A facing of friction material, called *brake lining,* is riveted or cemented ("bonded") to each shoe (♦51-16). The linings are made of asbestos or similar friction material which can withstand the heat-producing braking action.

Both shoes in the drum brake are not the same. The shoe installed toward the front of the car is called the *primary shoe.* The shoe installed toward the rear is called the *secondary shoe.*

♦ 51-16 BRAKE LINING

Brake shoes are made of metal to which a lining of friction material is attached (Fig. 51-16). The lining has to be tough. During hard braking, the shoe may be pressed against the drum by a force of 1000 pounds [4448 N] or higher. Since friction increases as the applied force increases, a strong frictional drag is produced on the brake drum. This produces the braking effect at the wheel.

A large quantity of heat is produced by the frictional contact between the brake shoes and the drum. Under heavy braking conditions, drum-brake temperatures may reach 500°F [260°C]. Some heat flows through the brake linings to the shoes and backing plate. There the heat is carried away by

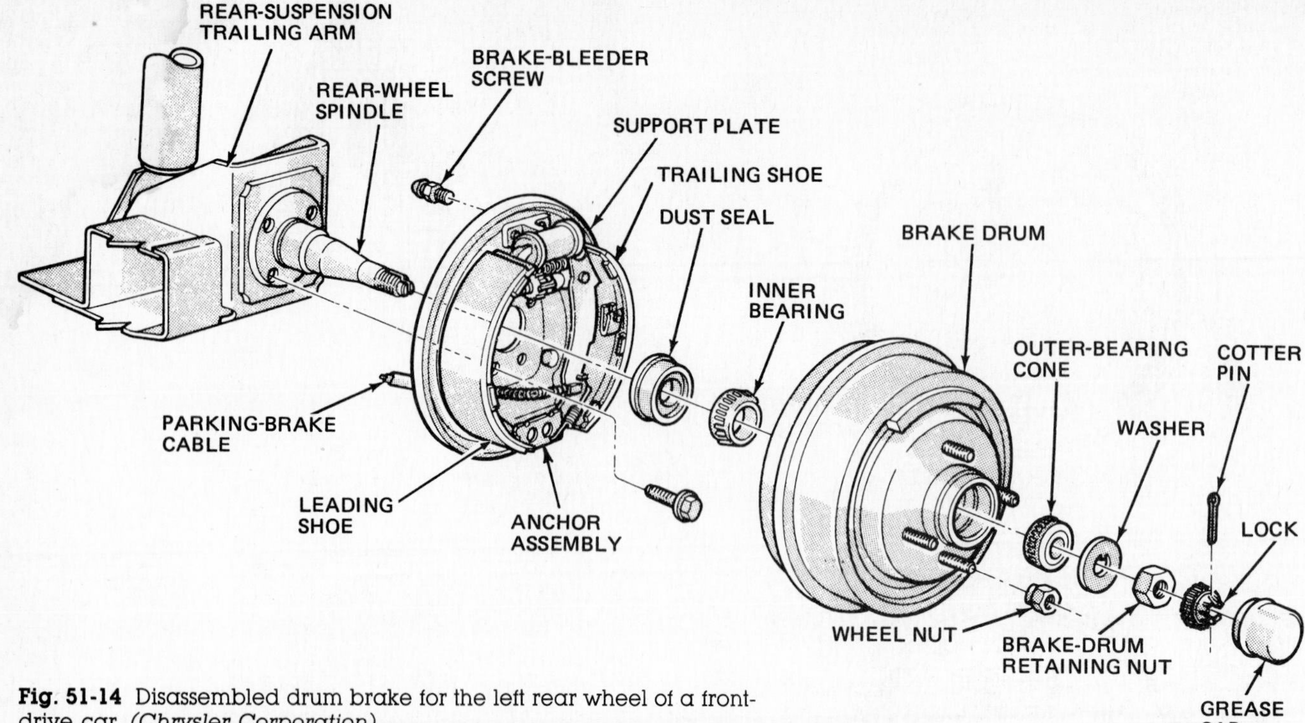

Fig. 51-14 Disassembled drum brake for the left rear wheel of a front-drive car. *(Chrysler Corporation)*

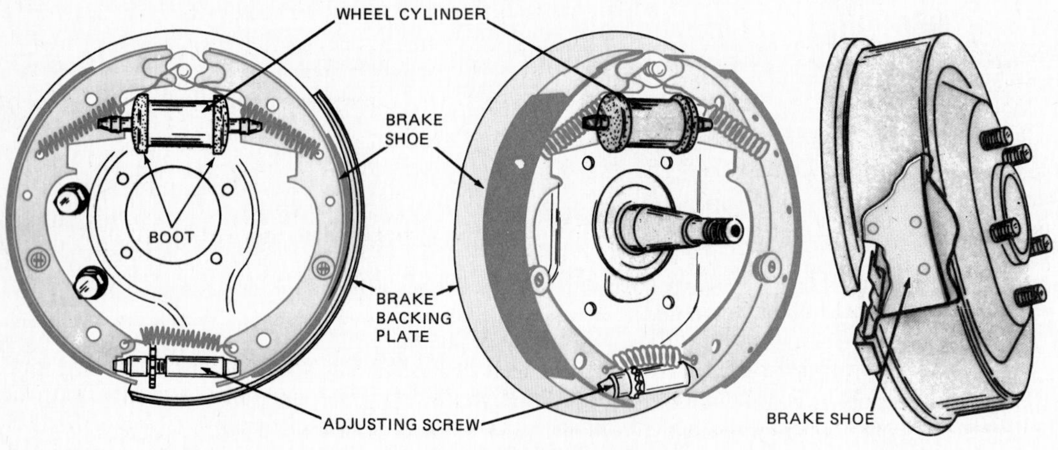

Fig. 51-15 Active parts of drum brakes for the rear (left) and front (center). To right, the brake drum has been partly cut away to show the shoe inside.

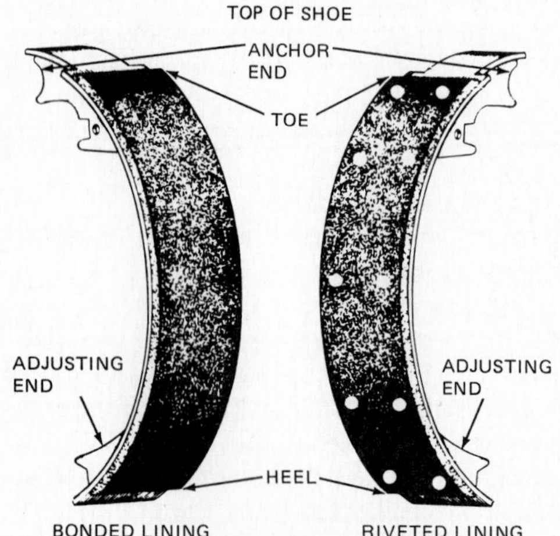

Fig. 51-16 Two methods of attaching the lining to the shoe. *(Bendix Corporation)*

the surrounding air. But most of the heat is absorbed by the brake drum. Some brake drums have cooling fins. They provide additional surface to get rid of the heat more quickly. To help provide drum cooling, some brake linings are grooved at the center.

Excessive temperature may damage brakes and burn the brake linings. When the linings and drums are excessively hot, the coefficient of friction changes. Then less effective braking action results. This is the reason that brakes "fade" after continuous use, as in driving down a mountain.

♦ 51-17 WHEEL CYLINDERS

Figures 51-17 and 51-18 show the construction of a drum-brake wheel cylinder. It has two pistons separated by a spring. Most automotive wheel cylinders are of this type. The pistons often are made of powdered metal with lubricant added to resist corrosion and sticking. Each piston is connected to a brake shoe by a solid connecting link that passes through

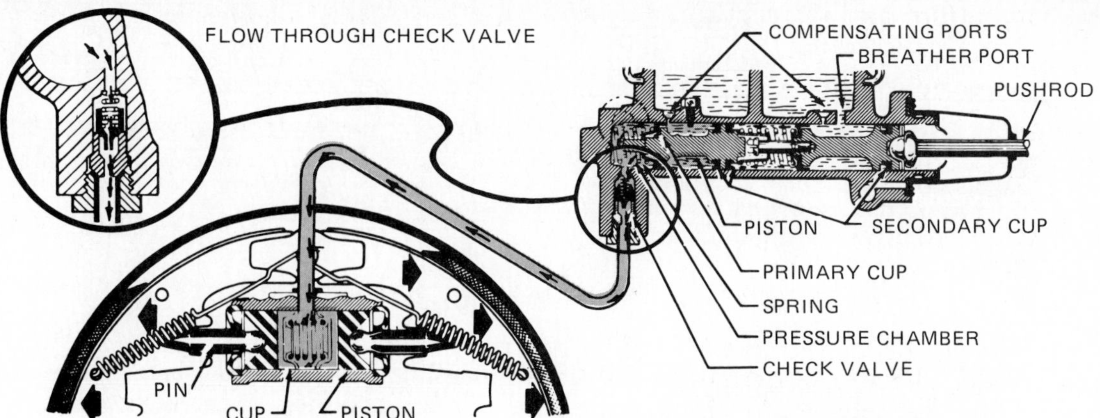

Fig. 51-17 Conditions in the drum-brake system with the brakes applied. Brake fluid flows from the master cylinder to the wheel cylinders. This causes the wheel-cylinder pistons to move outward and apply the brakes.

a rubber boot on the end of the cylinder. The boots may be either internal (Figs. 51-17 and 51-18) or external (Fig. 51-15).

A piston cup on the inner side of each piston provides a seal for the brake fluid. The piston cups are shaped so that the hydraulic pressure forces them tightly against the cylinder wall of the wheel cylinder. This produces an effective sealing action that holds the fluid in the cylinder.

The pistons in the wheel cylinders are usually larger at the front wheels. This is because, when the brakes are applied, the forward momentum of the car throws more of the weight on the front wheels. Therefore a greater braking force at the front wheels is necessary for balanced braking.

♦ 51-18 DRUM-BRAKE OPERATION

When the driver pushes the brake pedal down, brake fluid is forced from the master cylinder (Fig. 51-17). The fluid flows to the wheel cylinders. In each wheel cylinder, the additional fluid causes the hydraulic pressure against the pistons to increase. This forces the pistons apart, pushing the shoe-actuating pins out. Now the brake shoes are forced against the rotating drum. The resulting friction between the brake lining and the drum slows or stops the car.

When the driver's foot is lifted from the brake pedal, the brakes are released (Fig. 51-18). The shoe-return springs pull the shoes away from the drum. At the same time, return springs in the master cylinder push the pistons back toward the released position. All these actions force brake fluid out of the wheel cylinders and back into the master cylinder.

In some drum-brake systems, when the brakes are released a slight pressure is trapped in the brake lines and wheel cylinders. This pressure is called *residual line pressure*. It occurs when the spring-loaded check valve closes in the master cylinder (Fig. 51-17). The residual line pressure causes the piston cups in the wheel cylinders to be held tight against the wheel-cylinder wall, as shown by the arrows in Fig. 51-18. This prevents air leaks into the brake lines. It also prevents brake fluid leaks from the wheel cylinders.

♦ 51-19 DRUM-BRAKE SELF-ADJUSTERS

Most cars have self-adjusters that automatically adjust the brakes to compensate for brake-lining wear. Figure 51-19

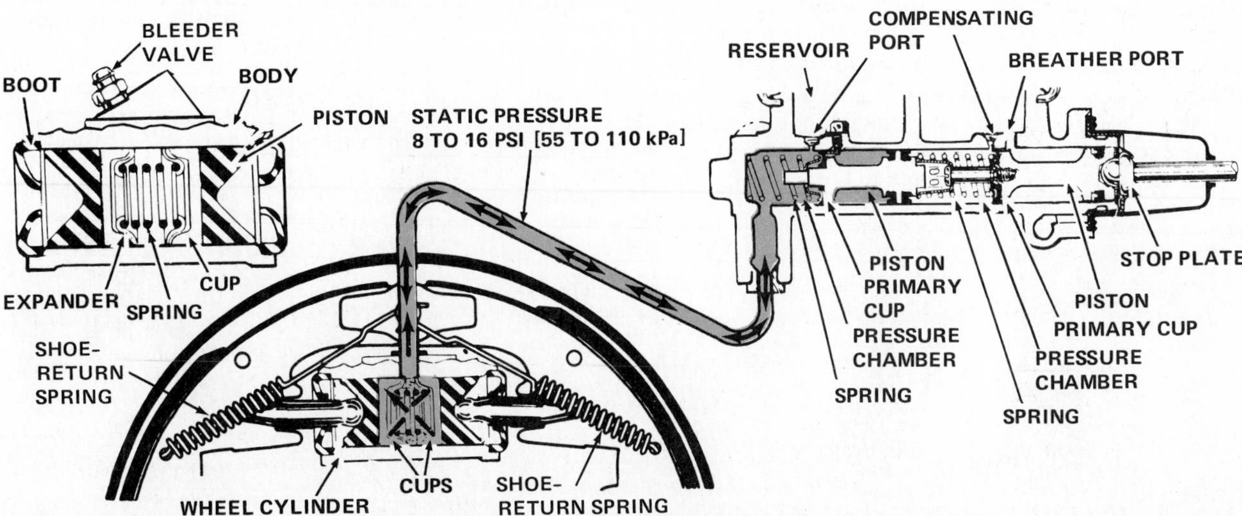

Fig. 51-18 Conditions in a drum-brake system when the brakes are released. Brake fluid returns to the master cylinder.

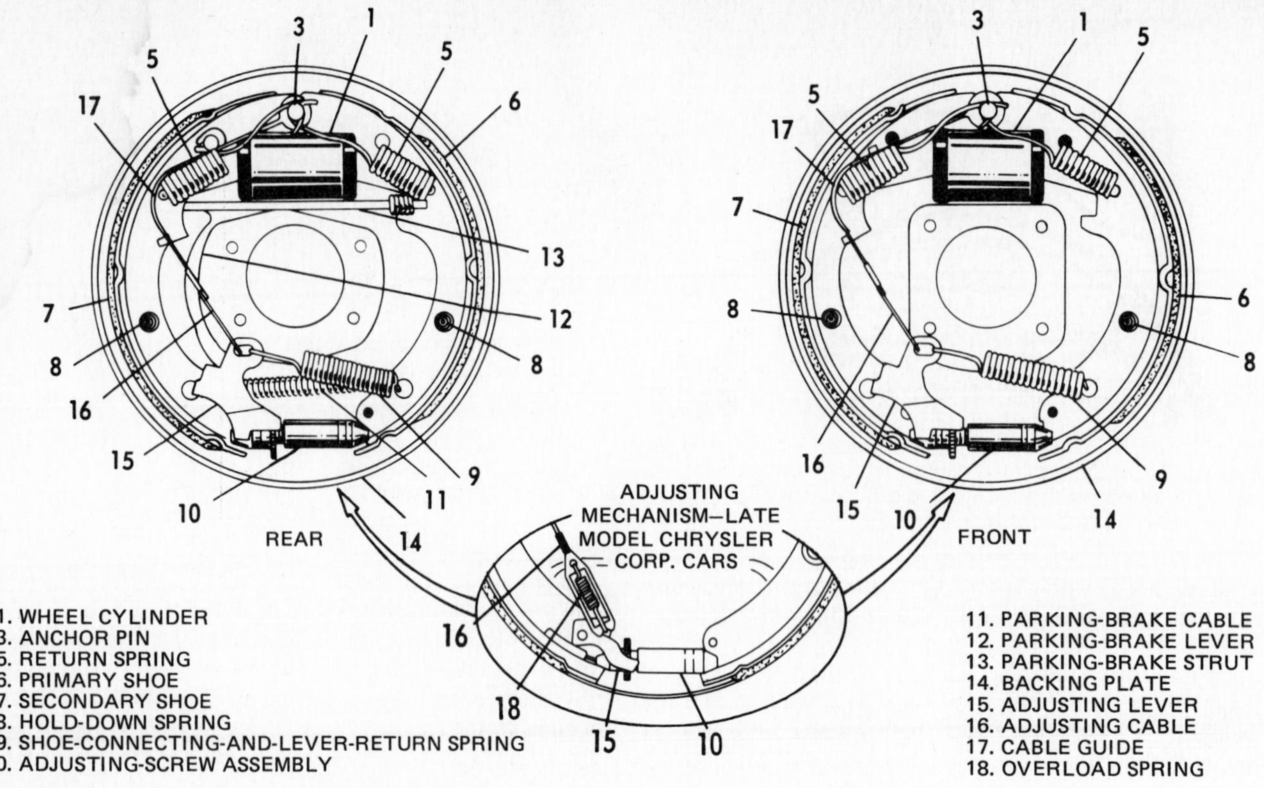

Fig. 51-19 Typical drum brakes for the front and rear wheels. Each brake assembly has a self-adjusting mechanism. *(Delco Moraine Division of General Motors Corporation)*

shows a typical arrangement. In most cars, the adjustment takes place, if it is necessary, when the car is moving backward and the brakes are applied.

As the brakes are applied with the car moving backward, friction between the primary shoe and the brake drum forces the primary shoe against the anchor pin. The wheel cylinder forces the upper end of the secondary shoe away from the anchor pin and downward (Fig. 51-19). This causes the adjuster lever to pivot on the secondary shoe so that the lower end of the lever is forced against the star wheel on the adjusting-screw assembly. If the brake linings have worn enough, the adjuster screw will be turned a full tooth. This spreads the lower ends of the brake shoes slightly to compensate for lining wear.

On some cars, the self-adjustment mechanism operates with the car moving forward when the brakes are applied. On other cars, adjustment takes place, if needed, whenever the brakes are applied. The car may be standing still or moving in either direction. Some cars have a special parking-brake rod-and-strut assembly that adjusts the brake shoes when the parking brake is applied.

DISK BRAKES

♦ 51-20 DISK BRAKES

The disk (also spelled disc) brake has a metal disk instead of a drum. Figure 51-20 shows a disk brake. It has a flat shoe, or pad, located on each side of the disk. To slow or stop the car, these two flat shoes are forced tightly against the rotating disk, or *rotor*. The shoes grip the disk as shown in Fig. 51-21. Fluid pressure from the master cylinder forces the pistons to move in. This action pushes the friction pads of the brake shoes tightly against the disk. The friction between the shoes and the disk slows and stops the disk.

There are three general types of disk brakes. They are floating-caliper, fixed-caliper, and sliding-caliper. Each type is described in the following sections. Power assist is often used with disk brakes. Either the vacuum brake booster (♦51-32) or the hydraulic brake booster (♦51-33) may be used.

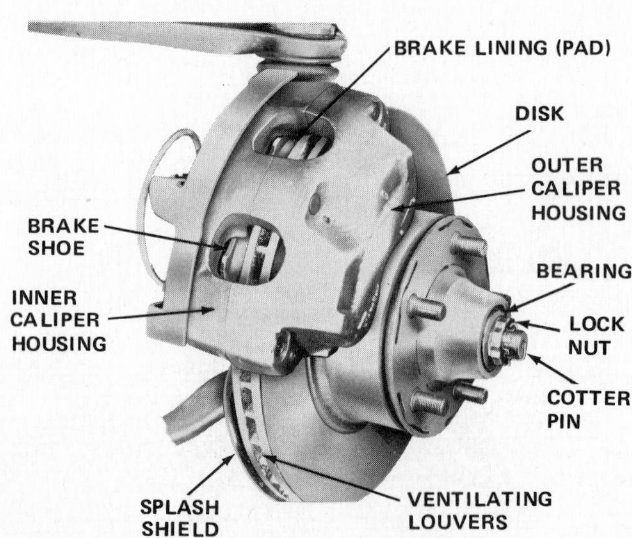

Fig. 51-20 A fixed-caliper disk-brake assembly. *(Chrysler Corporation)*

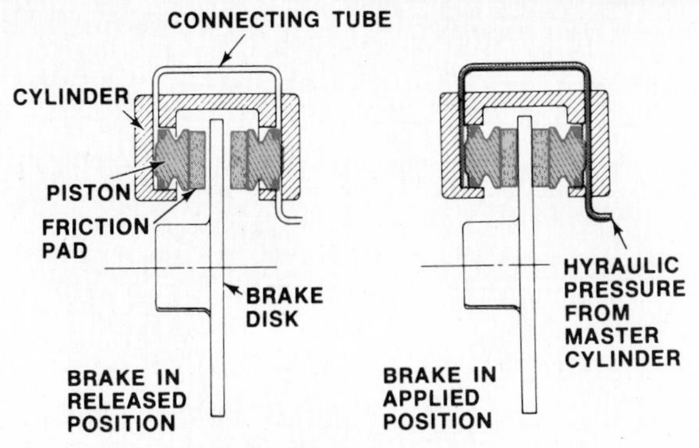

Fig. 51-21 In a disk brake, hydraulic pressure forces friction pads (the shoes and linings) inward against the brake disk to produce braking action.

♦ 51-21 FLOATING-CALIPER DISK BRAKE

Figure 51-22 shows a floating-caliper disk brake. Figure 51-23 shows a disassembled floating-caliper disk brake. The caliper is the part that holds the brake shoes on each side of the disk. In the floating-caliper brake, two steel guide pins are threaded into the steering-knuckle adapter. The caliper floats on four rubber bushings which fit on the inner and outer ends of the two guide pins. The bushings allow the caliper to swing in or out slightly when the brakes are applied.

When the brakes are applied, the brake fluid flows to the cylinder in the caliper and pushes the piston out. The piston then forces the shoe against the disk. At the same time, the pressure in the cylinder causes the caliper to pivot inward. This movement brings the other shoe into tight contact with the disk. As a result, the two shoes "pinch" the disk tightly to produce the braking action.

♦ 51-22 FIXED-CALIPER DISK BRAKE

This brake usually has four pistons, two on each side of the disk (Fig. 51-20). The reason for the name *fixed caliper* is that the caliper is bolted solidly to the steering knuckle. When the brakes are applied, the caliper cannot move. The four pistons are forced out of their caliper bores to push the inner and outer brake shoes in against the disk. Some brakes of this type have used only two pistons, one on each side of the disk.

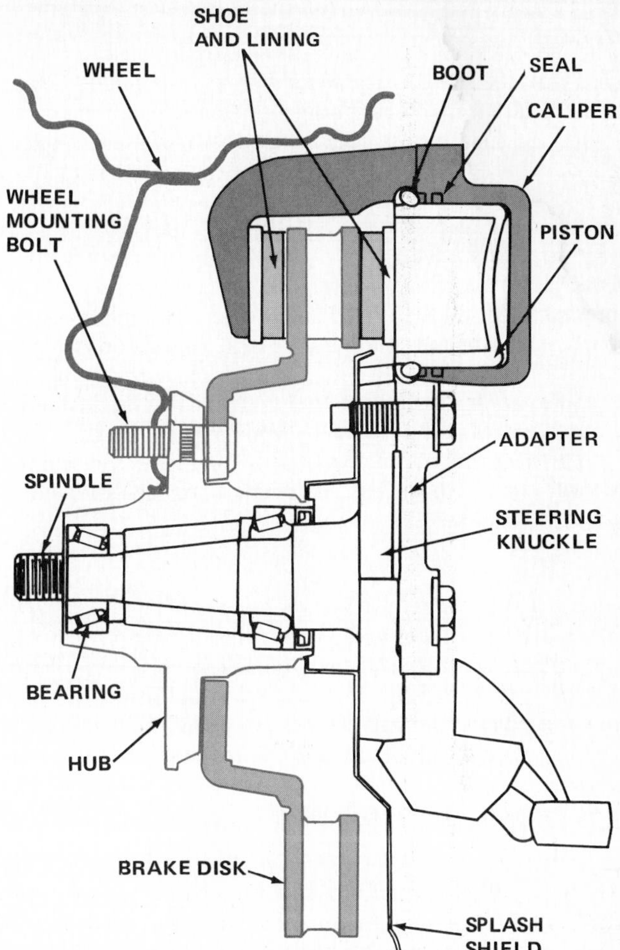

Fig. 51-22 A floating-caliper disk brake.

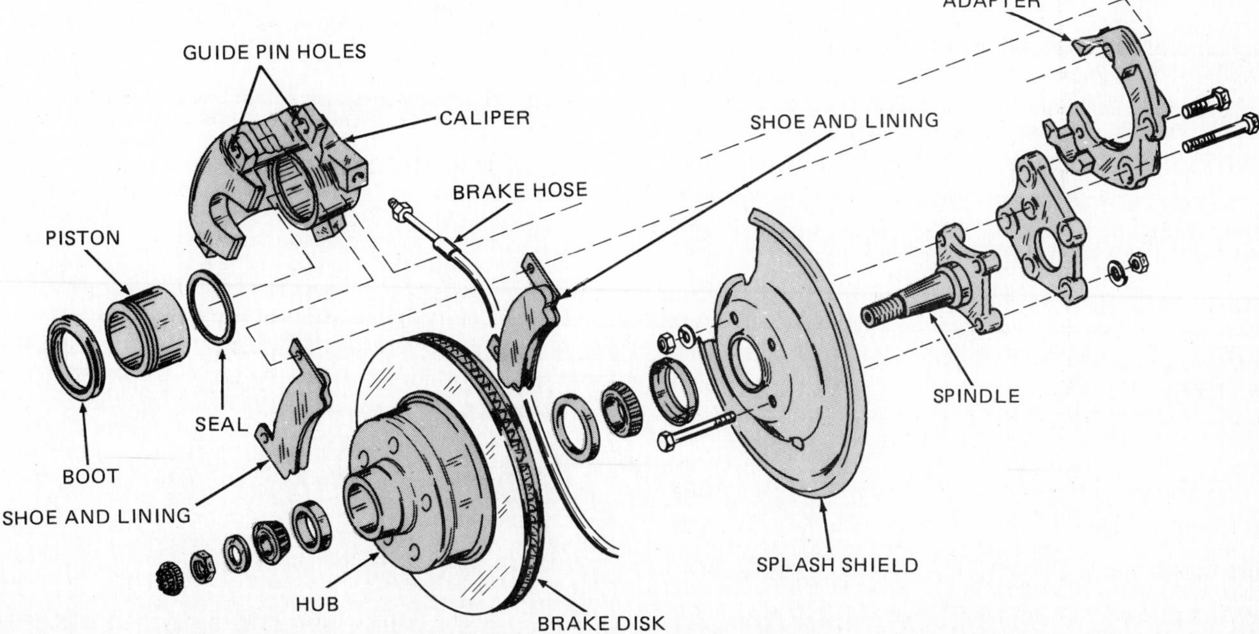

Fig. 51-23 Disassembled floating-caliper disk brake.

♦ 51-23 SLIDING-CALIPER DISK BRAKE

The sliding-caliper disk brake is similar to the floating-caliper disk brake (♦51-21). The difference is that the sliding caliper is suspended from rubber bushings on bolts. This permits the caliper to slide on the bolts when the brakes are applied.

♦ 51-24 SELF-ADJUSTMENT OF DISK BRAKES

Disk brakes are self-adjusting. Each piston has a seal on it to prevent fluid leakage (Fig. 51-24). When the brakes are applied, the piston moves toward the disk. This distorts the piston seal, as shown to the left in Fig. 51-24. When the brakes are released, the seal relaxes and returns to its original position. This pulls the piston away from the disk. As the brake linings wear, the piston "overtravels" and takes a new position in relation to the seal. This action provides self-adjustment of disk brakes.

♦ 51-25 WEAR SENSORS

Many disk-brake shoes have a wear indicator (Fig. 51-25). The purpose of the wear indicator is to make a noise. This warns the driver that the brake linings are thin and new linings are required. When the brake linings are worn, the tab touches the disk when the brakes are applied. This gives off a scraping noise.

♦ 51-26 METERING VALVE

Some cars equipped with front-disk and rear-drum brakes have a metering valve in the hydraulic line to the front brakes (Figs. 51-26 and 51-27). During light braking, the metering valve prevents the front disk brakes from applying until after the rear brake linings contact the drums. If the front brakes applied first, the rear wheels would lock. Then the tires would skid. The metering valve is sometimes called a *hold-off valve*.

♦ 51-27 PROPORTIONING VALVE

A proportioning valve is used in the rear brake line of some cars with front-disk and rear-drum brakes (Figs. 51-26 and 51-27). During hard braking, more of the car weight is transferred to the front wheels (Fig. 51-28). As a result, more braking is needed at the front wheels and less at the rear wheels. If normal braking continued, the rear wheels could lock. Then the tires would skid.

The proportioning valve reduces the pressure to the rear-wheel brakes when hard braking and high fluid pressures occur.

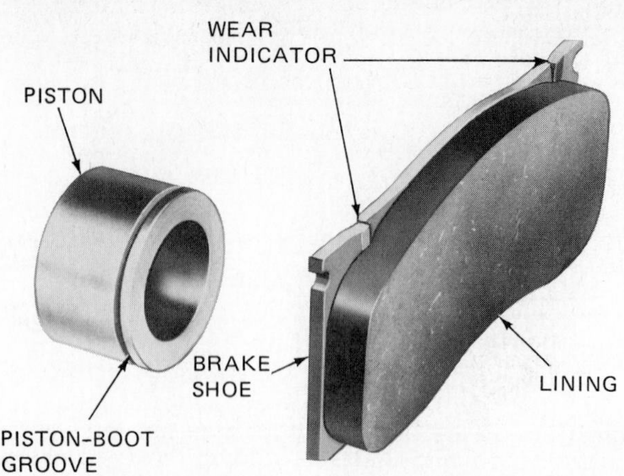

Fig. 51-25 Disk-brake piston and shoe which has a wear indicator attached. *(Chrysler Corporation)*

♦ 51-28 COMBINATION VALVE

Many cars equipped with front-disk and rear-drum brakes have a combination valve in the hydraulic system (Fig. 51-29). In the combination valve, the pressure-differential valve (♦51-10), metering valve (♦51-26), and proportioning valve (♦51-27) are combined into a single unit.

All combination valves include the failure warning switch, but not all contain the metering valve or proportioning valve. These valves are not used in all systems. However, all combination valves look the same from the outside. They all serve as the front junction block.

ANTILOCK BRAKE SYSTEMS

♦ 51-29 ANTILOCK BRAKE SYSTEMS

About 40 percent of all car accidents involve skidding. The most efficient braking takes place when the wheels are still

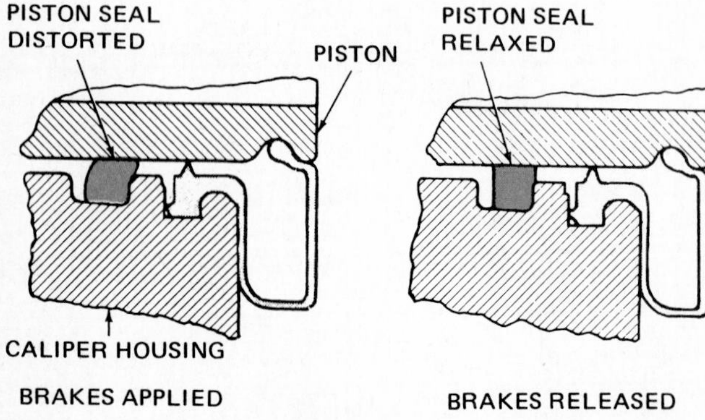

Fig. 51-24 Self-adjusting action of the piston seal when the brakes are applied and when they are released. *(Ford Motor Company)*

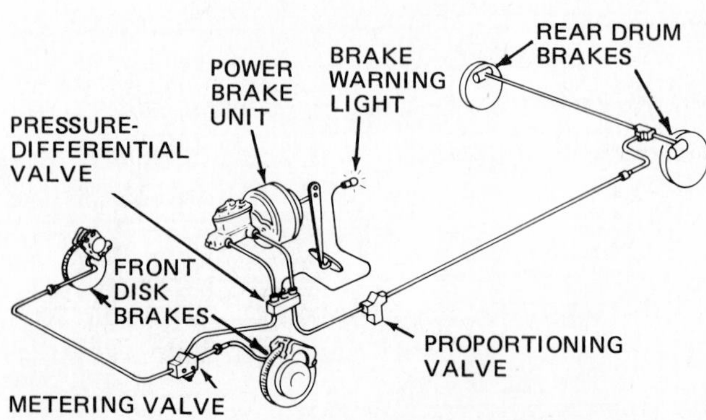

Fig. 51-26 Valves used in the hydraulic system of a car with front-disk and rear-drum brakes. *(Wagner Lockheed)*

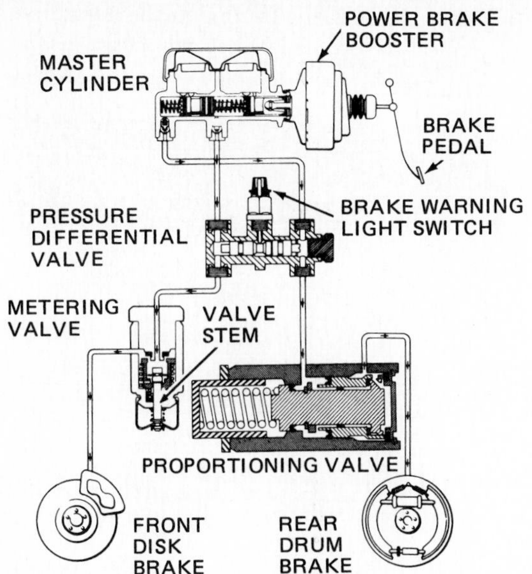

Fig. 51-27 Sectional views of the valves used in the hydraulic system of a car with front-disk and rear-drum brakes. *(Ford Motor Company)*

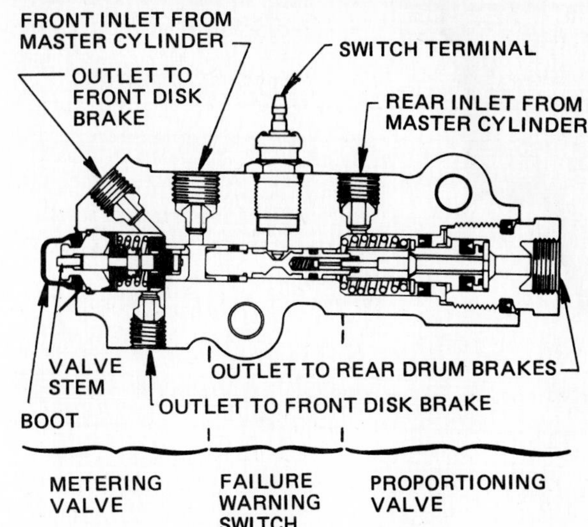

Fig. 51-29 Combination valve with warning-light switch, metering valve, and proportioning valve all in the same assembly. *(Delco Moraine Division of General Motors Corporation)*

revolving (♦51-4). If the brakes lock the wheels so that the tires skid, kinetic friction results, and braking is much less effective. To prevent skidding and provide maximum effective braking, several antilock devices have been developed. Some provide skid control at the rear wheels only. Others provide control at all four wheels.

Control means that as long as the wheels are rotating, the antilock device permits normal application of the brakes. But if the brakes are applied so hard that the wheels tend to stop turning and a skid starts to develop, the device comes into operation. It partly releases the brakes so that the wheels continue to rotate. However, braking continues, but it is held to just below the point where a skid would start. The result is maximum braking effect.

♦ 51-30 CHRYSLER SURE-BRAKE

The Chrysler antilock brake system is designed to prevent wheel lockup when the brakes are applied while the car is moving at above about 5 mph [8 km/h]. Figure 51-30 shows the antilock system layout on the car.

At the front wheel there is a magnetic wheel which acts as a speed sensor, attached to the brake disk. As the wheel and disk revolve, the magnetic wheel produces alternating current (ac) in the sensor. The sensor is a coil of wire, or a winding. A similar action takes place at the other wheels. These ac signals from the car wheels are fed into a logic control unit located in the trunk.

When the brakes are applied, the logic control compares the ac signals from the wheels. The frequency of the ac increases with speed. As long as the frequency of the ac from all wheels is about the same, normal braking will take place. However, if the ac from any wheel is slowing down too rapidly, it means that the wheel is also slowing down too rapidly. The tire is beginning to skid.

When the logic control unit senses a rapid drop in the frequency of the ac, it signals modulators at the front of the car. Figure 51-30 shows the locations of the front-wheel and rear-wheel modulators. The hydraulic pressure from the master cylinder to the wheel cylinders or calipers passes through

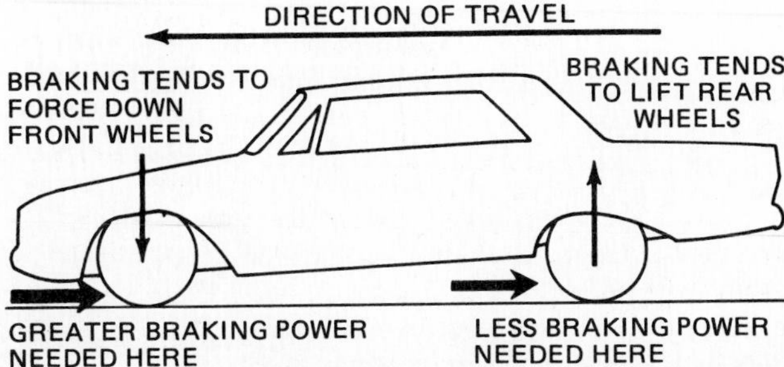

Fig. 51-28 During braking, the momentum of the car forces the front wheels down and the rear wheels up. This puts more of the braking effort on the front wheels. *(Ammco Tools, Inc.)*

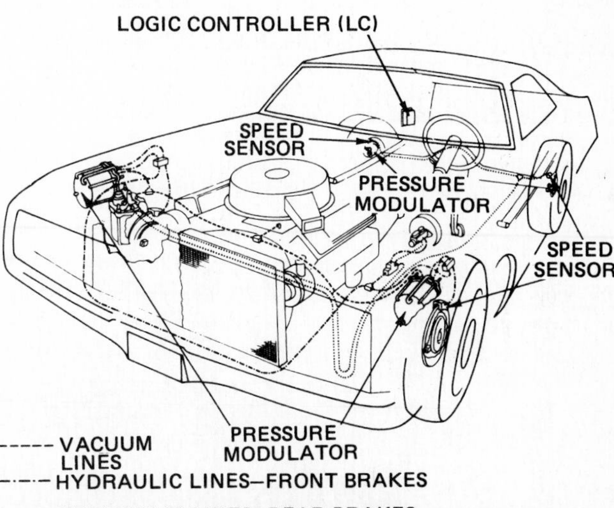

Fig. 51-30 Antilock brake system showing the locations of the components and their connections to the vacuum and hydraulic lines. *(Chrysler Corporation)*

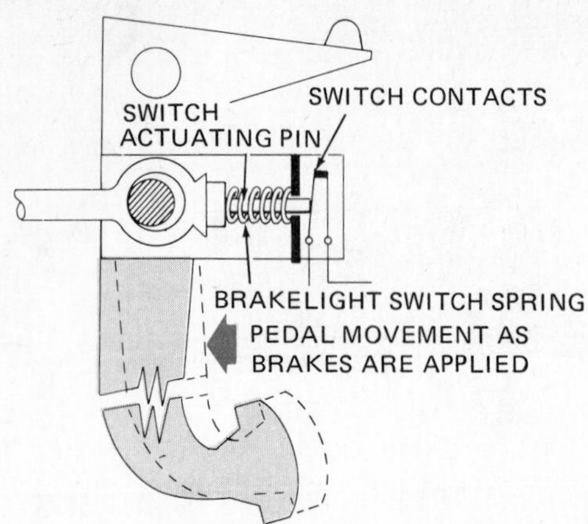

Fig. 51-31 Mechanical brakelight switch shown closed, with brakes applied. *(Ford Motor Company)*

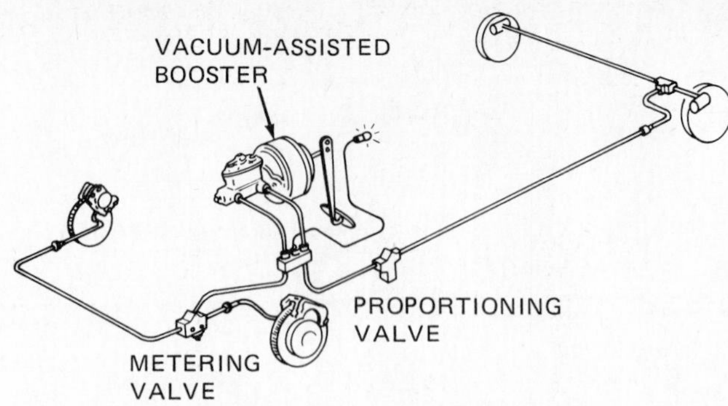

Fig. 51-33 A power-brake system using a vacuum-assisted brake booster. *(Wagner Lockheed)*

these modulators. When the logic control unit senses that a wheel is about to skid, it signals the modulator to reduce the hydraulic pressure to the brake for that wheel. When the pressure is reduced, the braking effect at that wheel is reduced, so the skid is prevented.

♦ 51-31 BRAKELIGHT SWITCH

Figure 51-31 shows the brakelight switch used in many cars. When the brake pedal is pushed down for braking, it causes the contacts in the switch to close. This connects the brakelights to the battery so that they come on. Figure 51-31 shows the action during braking.

POWER BRAKES

♦ 51-32 VACUUM BRAKE BOOSTER

Almost all American-built cars have power brakes. With power brakes, only a relatively light pedal force is required to brake the car. When the brake pedal is pushed down, a hydraulic or vacuum-operated booster takes over and furnishes most of the force for pushing the pistons into the master cylinder. The hydraulic booster is operated by oil pressure from the power-steering pump (♦51-33). The vacuum comes from the engine intake manifold. Figure 51-32 shows how the vacuum-booster works. The system includes a cylinder in which a tight-fitting piston can move. When vacuum is applied to one side of the piston, atmospheric pressure causes

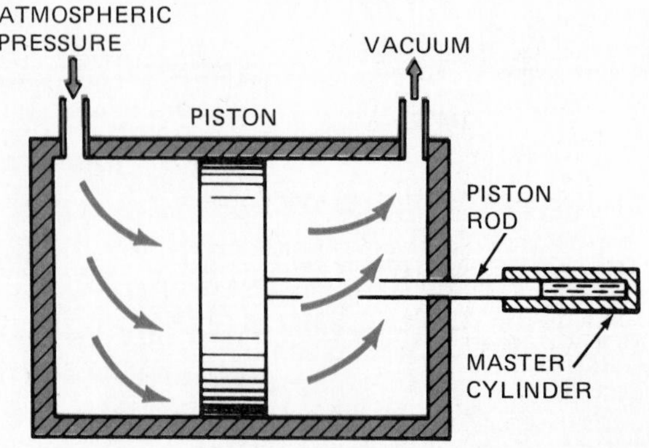

Fig. 51-32 If there is vacuum on one side of the piston and atmospheric pressure on the other side, the piston moves toward the vacuum side.

the piston to be pushed to the right as shown in Fig. 51-32. This movement pushes the piston rod into the master cylinder.

In the vacuum power-brake system, the brake pedal does not act directly on the master cylinder. Instead, brake-pedal movement operates a vacuum valve, which then admits vacuum to the power cylinder. Figure 51-33 shows the layout of a power-brake system using a vacuum-assist brake booster. Figure 51-34 is a sectional view of a vacuum brake booster.

Figure 51-35 shows what happens when the brake pedal is pushed down to apply the brakes. First, the pushrod moves the air valve away from the floating control valve. Now, atmospheric pressure can flow past the valves and into the space to the right of the plate and valve body, or the piston. The atmospheric pressure forces the piston to the left, so that the master-cylinder pistons are forced into the master cylinder. This action causes braking to take place.

The reaction disk, next to the air valve, gives the driver some braking "feel." A small proportion of the braking force being applied by the vacuum booster feeds back through the reaction disk. This feedback is felt by the driver through the pushrod and linkage to the brake pedal.

When the brake pedal is released, the air valve moves back to contact the floating control valve. This contact reseals the vacuum booster from atmospheric pressure so that the brakes are released. Some vacuum boosters have two diaphragms for additional braking force.

♦ 51-33 HYDRAULIC BRAKE BOOSTER

The hydraulic brake booster (Fig. 51-36) uses hydraulic pressure supplied by the power-steering pump (♦49-19) to assist in applying the brakes. Figure 51-37 is a disassembled hydraulic brake booster.

With the power-steering pump running, there is always pressure in the smaller cylinder in the brake booster. As the brakes are applied, the pressure through the input rod moves the lever. This forces the spool assembly (10 in Fig. 51-37) to move off center. The spool assembly now admits hydraulic pressure back of the large piston in the large cylinder. This pressure is applied to the output pushrod which is pressing against the pistons in the master cylinder. Then the master cylinder pistons move to send brake fluid to the wheel cylinders or calipers, producing braking.

Some medium and heavy trucks also use the hydraulic brake booster. These vehicles may have a separate pump to produce the hydraulic pressure, instead of the power-steering pump.

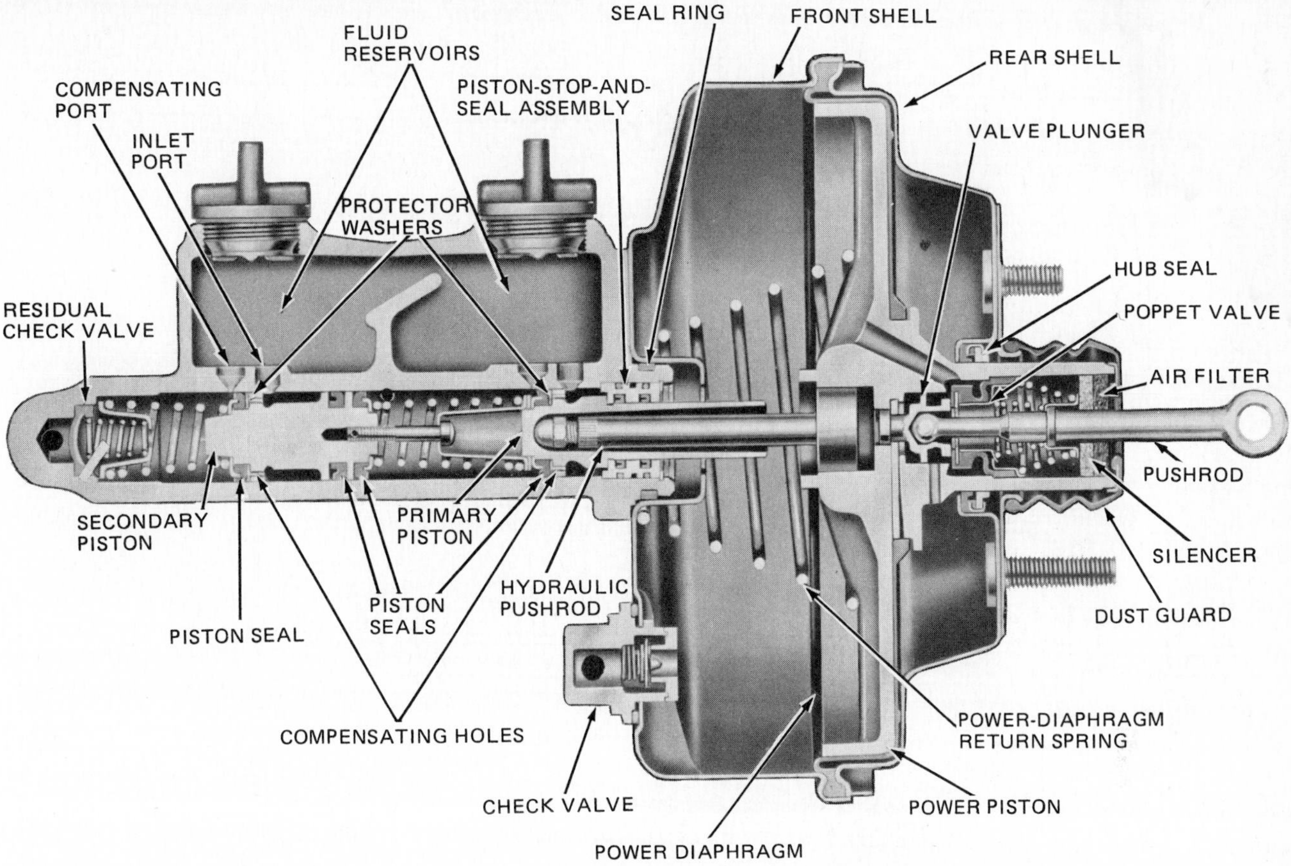

Fig. 51-34 Sectional view of a vacuum-assisted brake booster. *(General Motors Corporation)*

OTHER BRAKES

♦ 51-34 TRAILER BRAKES

Laws in many states specify that trailers weighing above 1000 pounds [454 kg] loaded must have brakes. There are two types, hydraulic and electric. The hydraulic-brake type is not connected to the towing vehicle. Instead, the hitch on the trailer has a master cylinder. When the towing vehicle brakes, the forward push of the trailer causes the master-cylinder tongue to push against the hitch ball. This activates the master cylinder so it sends hydraulic pressure to the trailer brakes.

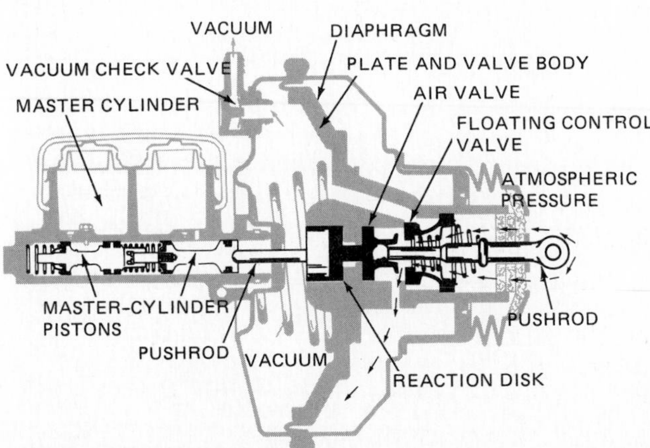

Fig. 51-35 Positions of the internal parts in the vacuum brake booster when the brakes are applied.

The electric trailer brake is wired to the brake pedal. When the brake pedal is depressed, electric current flows to the brake mechanisms at the trailer wheels. These mechanisms use electromagnets to force the brake shoes against the rotating brake drums.

♦ 51-35 PARKING BRAKES

Figure 51-8 shows the typical layout of the complete automobile braking system on the chassis. In addition to the hydraulic foot-operated service brake, the car has a separate mechanical *parking brake*. It may be operated by a foot pedal (Fig. 51-8) or by a hand lever (Fig. 51-38).

The parking brake holds the car stationary while it is

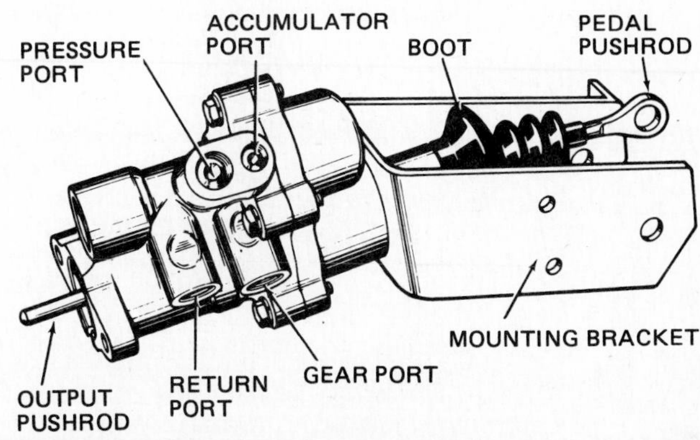

Fig. 51-36 A hydraulic-assisted brake booster. *(Chevrolet Motors Division of General Motors Corporation)*

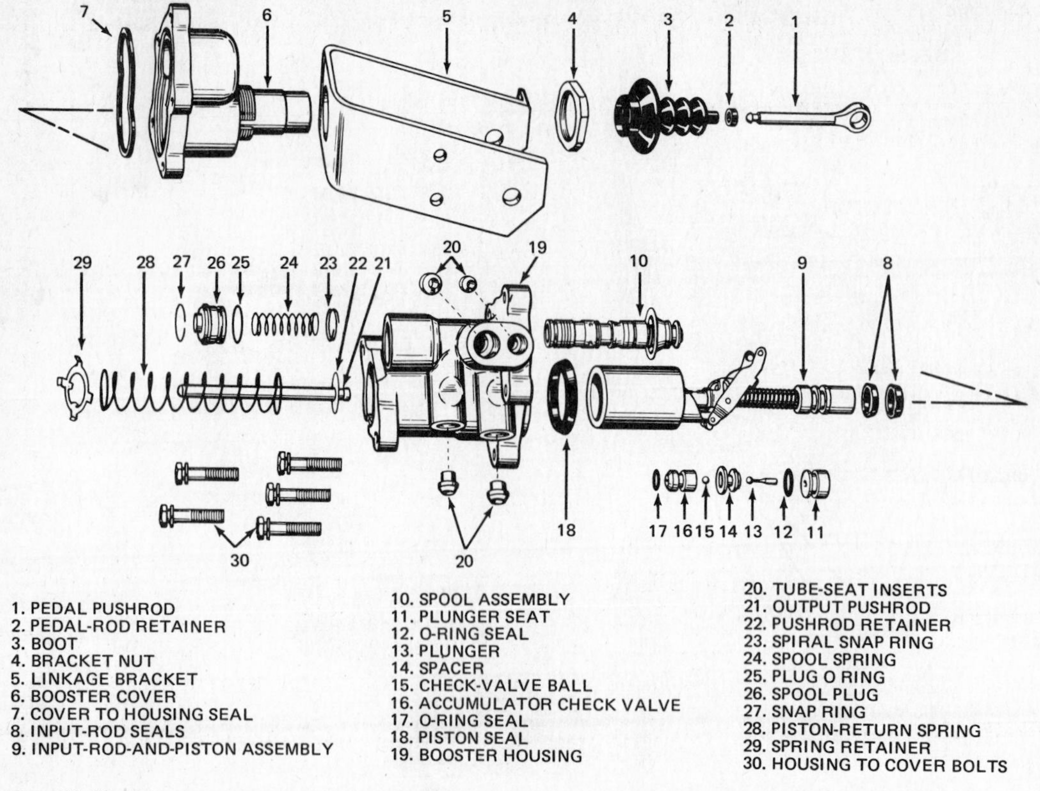

Fig. 51-37 Disassembled hydraulic brake booster. *(Chevrolet Motor Division of General Motors Corporation)*

parked. Since the parking brake is independent of the service brakes, it can be used as an emergency brake if the service brakes fail. When the parking brake is operated by a hand lever, some manufacturers call it the *hand brake*.

A parking brake is designed to hold one or more brakes continuously in an applied position. There are two basic types of parking brakes. These are known as *integral* and *independent*.

1. Integral Parking Brakes Integral parking brakes have parts which are common to both the parking-brake system and the service-brake system. For example, applying the parking brake may force the rear-wheel brake shoes or pads against the drum or rotor, just as the service brake does. However, the service brake operates the shoes or pads hydraulically, while the parking brake uses a separate mechanical system to transmit the braking force. Most cars built today have an integral parking brake.

2. Independent Parking Brakes These brakes are "independent" because they do not share any parts with the service brakes. One type has a pair of small curved brake shoes which fit into the hub of a disk-drum assembly. When the parking brake is applied, the shoes are forced against the drum in the hub. A second type, used on older cars and trucks, is located at the end of the transmission. It has a brake band that tightens on a drum, or internal curved shoes that move out against the drum.

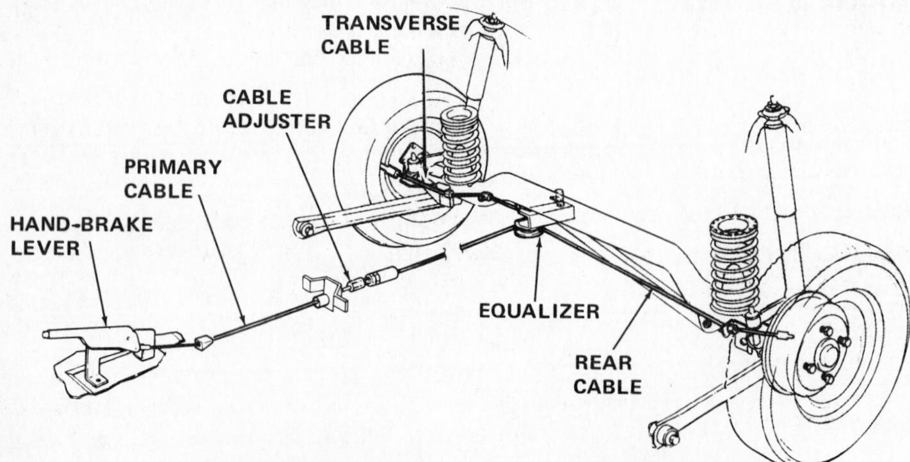

Fig. 51-38 A parking-brake system. Operation of the hand-brake lever pulls on the brake cables so that the rear brakes are mechanically applied. *(Ford Motor Company)*

◆ REVIEW QUESTIONS ◆

Select the *one* correct, best, or most probable answer to each question. You can find the answers in the section indicated at the end of each question.

1. The service brakes are operated (◆51-1)
   a. electrically
   b. hydraulically
   c. by air pressure
   d. by vacuum
2. During braking, the brake shoe is moved outward to force the lining against the (◆51-9)
   a. wheel piston or cylinder
   b. anchor pin
   c. brake drum
   d. wheel rim or axle
3. In most drum brakes, each wheel cylinder contains (◆51-17)
   a. one piston
   b. two pistons
   c. three pistons
   d. four pistons
4. When comparing the front and rear wheel-cylinder pistons, the pistons in the front wheel cylinders usually are (◆51-17)
   a. larger in diameter
   b. smaller in diameter
   c. the same size
   d. none of the above
5. In the dual-brake system, the master cylinder has (◆51-11)
   a. one piston
   b. two pistons
   c. three pistons
   d. four pistons
6. There are three types of disk brake (◆51-20)
   a. fixed-caliper, tab-action, and two-piston
   b. fixed-caliper, sliding-caliper, and floating-caliper
   c. floating-caliper, swinging-caliper, and proportioning
   d. all of the above
7. The antilock control system called Sure-Brake uses a vacuum-powered actuator, an electronic control module, and (◆51-30)
   a. wheel sensors
   b. a load-sensing valve
   c. a governor
   d. all of the above
8. When used, the purpose of the residual check valve is (◆51-18)
   a. to prevent tire skidding
   b. to prevent air from entering the hydraulic system
   c. to prevent wheel lockup by reducing the hydraulic pressure
   d. all of the above
9. The brake warning light warns the driver of (◆51-10)
   a. water in the master cylinder
   b. air in the hydraulic system
   c. failure of the primary or secondary section of the hydraulic system
   d. power-brake failure
10. Mechanic A says that most self-adjuster mechanisms are attached to the secondary brake shoe. Mechanic B says that self-adjuster mechanisms are attached to the primary brake shoe. Who is right? (◆51-19)
    a. A only
    b. B only
    c. both A and B
    d. neither A nor B
11. The brake warning light is turned on by the (◆51-10)
    a. metering valve
    b. proportioning valve
    c. pressure-differential valve
    d. none of the above
12. On a car with front-disk and rear-drum brakes, the front brakes grab when light pedal force is applied. This problem could be caused by a defective (◆51-26)
    a. proportioning valve
    b. pressure-differential valve
    c. metering valve
    d. check valve
13. On a car with front-disk and rear-drum brakes, the rear brakes lock up when the brake pedal is depressed normally. Which of these could cause this problem? (◆51-27)
    I. Leaking front-caliper seals
    II. Defective proportioning valve
    a. I only
    b. II only
    c. either I or II
    d. neither I nor II
14. In the dual master cylinder, the primary piston is the piston that is (◆51-11)
    a. directly actuated by the brake-pedal pushrod
    b. closest to the fire wall
    c. toward the rear of the car
    d. all of the above
15. To operate, a vacuum-assisted power-brake booster makes use of the pressure difference between intake-manifold vacuum and (◆51-32)
    a. venturi vacuum
    b. compressed air
    c. atmospheric pressure
    d. none of the above

CHAPTER 52
Brake Service

After studying this chapter, and with the proper instruction and equipment, you should be able to:

1. Diagnose troubles in drum-brake systems.
2. Diagnose troubles in disk-brake systems.
3. Test brakes on the equipment available in the shop and also on the road.
4. Adjust drum brakes.
5. Service drum and disk brakes, master cylinders, and brake lines.
6. Diagnose troubles in power-brake systems.
7. Overhaul a power-brake unit.

♦ CAUTION ♦ Breathing dust containing asbestos fibers can cause serious bodily harm. The United States Surgeon General has determined that ingesting asbestos fibers, either by breathing or through the mouth, can cause serious physical diseases. You should not sand or grind brake linings or clean brake parts with a dry brush or compressed air. Instead, wipe parts off with a cloth dampened with water. Wash your hands after working on brakes.

♦ 52-1 BRAKE TROUBLE DIAGNOSIS

The charts and the sections that follow them give you a means of tracing brake troubles to their causes. After you know the complaint, the charts provide you with the various possible causes. For each possible cause, the charts list the checks or corrections that you should make to eliminate the trouble. Following the trouble-diagnosis sections are sections on how to adjust and service automotive brakes.

Brake trouble diagnosis is divided into three parts. The first chart with the explanatory sections following it covers drum brakes. The second chart, covering disk brakes, follows. The third chart covers power brakes. However, some troubles can occur in any brake system, often from the same cause.

DRUM-BRAKE TROUBLE DIAGNOSIS

♦ 52-2 DRUM-BRAKE TROUBLE-DIAGNOSIS CHART

A variety of braking problems bring the driver to the mechanic. Few drivers will know exactly what is causing a trouble. The chart that follows lists possible troubles in drum-brake systems, their possible causes, and checks or corrections to be made. Following sections describe the troubles and causes or corrections in detail. The chart in ♦52-15 covers possible disk-brake troubles. The chart in ♦52-25 covers possible power-brake troubles.

DRUM-BRAKE TROUBLE-DIAGNOSIS CHART*

Complaint	Possible Cause	Check or Correction
1. Brake pedal goes to floorboard (♦ 52-3)	a. Linkage or shoes out of adjustment	Adjust
	b. Brake linings worn	Replace
	c. Lack of brake fluid	Add fluid; bleed system (see item 10 below)
	d. Air in system	Add fluid; bleed system (see item 9 below)
	e. Worn master cylinder	Repair
2. One brake drags (♦ 52-4)	a. Shoes out of adjustment	Adjust
	b. Clogged brake line	Clear or replace line
	c. Wheel cylinder defective	Repair or replace
	d. Weak or broken return spring	Replace
	e. Loose wheel bearing	Adjust bearing
3. All brakes drag (♦ 52-5)	a. Incorrect linkage adjustment	Adjust
	b. Trouble in master cylinder	Repair or replace
	c. Mineral oil in system	Replace damaged rubber parts; refill with recommended brake fluid
4. Car pulls to one side when braking (♦ 52-6)	a. Brake linings soaked with oil	Replace linings and oil seals; avoid overlubrication
	b. Brake linings soaked with brake fluid	Replace linings; repair or replace wheel cylinder
	c. Brake shoes out of adjustment	Adjust
	d. Tires not uniformly inflated	Inflate correctly
	e. Brake line clogged	Clear or replace line
	f. Defective wheel cylinder	Repair or replace
	g. Brake backing plate loose	Tighten
	h. Mismatched linings	Use same linings all around
5. Soft or spongy pedal (♦ 52-7)	a. Air in system	Add brake fluid; bleed system (see item 9 below)
	b. Brake shoes out of adjustment	Adjust
6. Poor braking action requiring excessive pedal force (♦ 52-8)	a. Brake linings soaked with water	Will be all right when dried out
	b. Shoes out of adjustment	Adjust
	c. Brake linings hot	Allow to cool
	d. Brake linings burned	Replace
	e. Brake drum glazed	Turn or grind drum
	f. Power-brake assembly not operating	Overhaul or replace
7. Brakes too sensitive or grab (♦ 52-9)	a. Shoes out of adjustment	Adjust
	b. Wrong linings	Install correct linings
	c. Brake linings greasy	Replace; check oil seals; avoid overlubrication
	d. Drums scored	Turn or grind drums
	e. Backing plates loose	Tighten
	f. Power-brake assembly malfunctioning	Overhaul or replace
	g. Brake linings soaked with oil	Replace linings and oil seals; avoid overlubrication
	h. Brake linings soaked with brake fluid	Replace linings; repair or replace wheel cylinders
8. Noisy brakes (♦ 52-10)	a. Linings worn	Replace
	b. Shoes warped	Replace
	c. Shoes rivets loose	Replace shoe or lining
	d. Drums worn or rough	Turn on grind drums
	e. Loose parts	Tighten
9. Air in system (♦ 52-11)	a. Defective master cylinder	Repair or replace
	b. Loose connections, damaged tube	Tighten connections; replace tube
	c. Brake fluid lost	See item 10 below
10. Loss of brake fluid (♦ 52-12)	a. Master cylinder leaks	Repair or replace
	b. Wheel cylinder leaks	Repair or replace
	c. Loose connections, damaged tube	Tighten connections; replace tube
Note: After repair, add brake fluid and bleed system.		
11. Brakes do not self-adjust (♦ 52-13)	a. Adjustment screw stuck	Free and clean up
	b. Adjustment lever does not engage star wheel	Repair; free up or replace adjuster
	c. Adjuster incorrectly installed	Install correctly

*See ♦ 52-3 to 52-14 for detailed explanation of the trouble causes and corrections listed in the chart.

DRUM-BRAKE TROUBLE-DIAGNOSIS CHART (*Continued*)

Complaint	Possible Cause	Check or Correction
12. Warning light comes on when braking (♦ 52-14)	a. One section has failed	Check both sections for braking action; repair defective section
	b. Pressure-differential valve defective	Replace

♦ 52-3 BRAKE PEDAL GOES TO FLOOR

When the brake pedal goes to the floor, there is no pedal reserve. Full pedal travel does not produce adequate braking. One section might fail, but it would be rare for both to fail at the same time. It is possible that the driver has continued to operate the car with one section out. (Either the driver ignored the warning light, or the light or the pressure-differential valve has failed.) Causes of failure could be linkage or brake shoes out of adjustment, linings worn, air in the system, lack of brake fluid, or a defective master cylinder.

♦ 52-4 ONE BRAKE DRAGS

If one brake drags, this means that the brake shoes are not moving away from the brake drum when the brakes are released. This trouble could be caused by incorrect shoe adjustment, or by a clogged brake line which does not release pressure from the wheel cylinder. It could also be due to sticking pistons in the wheel cylinder, to weak or broken brake-shoe return springs, or to loose wheel bearings. Loose wheel bearings could permit the wheel to wobble so that the brake drum comes in contact with the brake shoes even though they are retracted.

♦ 52-5 ALL BRAKES DRAG

When all brakes drag, the brake pedal may not have sufficient free travel. Then the pistons in the master cylinder do not fully retract. This would prevent the lip of the piston cup from clearing the compensating port, and hydraulic pressure would not be relieved as it should be. As a result, the wheel cylinders would not allow the shoes to retract.

A similar condition could result if engine oil was added to the system. Engine oil will cause the piston cups to swell. If they swelled enough, they would not clear the compensating ports even with the piston in the "fully retracted" position. A clogged compensating port would have the same result. Do not use a wire or drill to clear the port. This might produce a burr that would cut the piston cup. Instead, clear the port with alcohol and compressed air.

Clogging of the reservoir vent might also cause dragging brakes by pressurizing the fluid in the reservoir. This would prevent release of pressure on the fluid in the lines. A clogged vent could also cause leakage of air into the system (♦52-11).

♦ 52-6 CAR PULLS TO ONE SIDE

If the car pulls to one side when the brakes are applied, more braking force is being applied to one side than to the other. This happens if some of the brake linings have become soaked in oil or brake fluid, if brake shoes are unevenly or improperly adjusted, if tires are not evenly inflated, or if defective wheel cylinders or clogged brake lines are preventing uniform braking action at all wheels. A loose brake backing plate or the use of two different types of brake lining will cause the car to pull to one side when the brakes are applied. Misaligned front wheels or a broken spring could also cause this problem.

In a front-engine rear-wheel-drive car, linings will become soaked with oil if the lubricant level in the rear axle is too high. This may cause leakage past the oil seal (Fig. 52-1). At the front wheel, brake linings may get grease on them if the front-wheel bearings are over-lubricated or if the grease seal is defective or not properly installed. Wheel cylinders will leak brake fluid onto the brake linings if the cups are defective, if the cylinder bore is pitted, or if an actuating pin has been improperly installed (♦52-12). If the linings at a left wheel become soaked with brake fluid or oil, the car pulls to the left because the brakes are more effective on the left side. However, the direction of pull may depend on the type of friction material and the contaminant.

♦ 52-7 SOFT OR SPONGY PEDAL

If the pedal action is soft or spongy, there probably is air in the hydraulic system. Out-of-adjustment brake shoes could also cause this. Conditions that could allow air to enter the hydraulic system are described in ♦52-11.

♦ 52-8 POOR BRAKING ACTION REQUIRING EXCESSIVE PEDAL FORCE

A need for excessive pedal force could be caused by improper brake-shoe adjustment. The use of the wrong brake lining could cause the same trouble. Sometimes, brake linings that have become wet after a hard rain or after driving through water will not provide sufficient friction against the drums. However, normal braking action is usually restored after the brake linings have dried.

Another possible cause of poor braking action is excessive temperature. After the brakes have been applied for long periods, as in coming down a long hill, they begin to overheat. This overheating reduces braking effectiveness so that the brakes "fade." Often, if brakes are allowed to cool, braking efficiency is restored. However, excessively long periods of

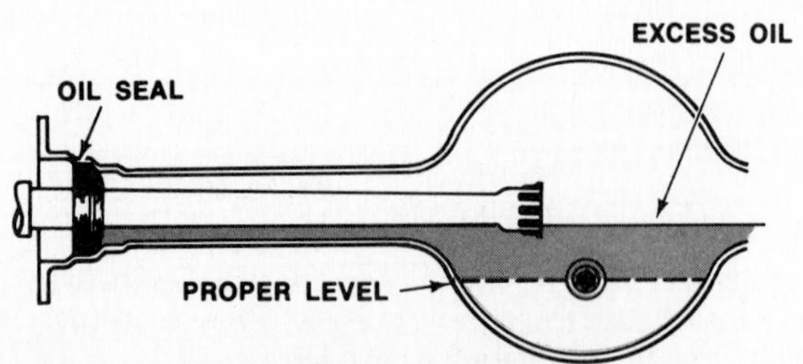

Fig. 52-1 In a front-engine rear-wheel-drive car, a high lubricant level in the rear axle may cause leakage past the oil seal. This would result in oil-soaked brake linings. *(Pontiac Motor Division of General Motors Corporation)*

braking at high temperature may char the brake linings so that they must be replaced. Further, this overheating may glaze the brake drum so that it becomes too smooth for effective braking action. Then the drum must be ground or turned to remove the glaze. Glazing can also take place even though the brakes are not overheated. Failure of the power-brake assembly will noticeably increase the force on the foot pedal required to produce braking.

♦ 52-9 BRAKES TOO SENSITIVE OR GRAB

If linings are greasy, or soaked with oil or brake fluid, the brakes tend to grab with slight pedal force. Then the linings must be replaced. If the brake shoes are out-of-adjustment, if the wrong linings are used, or if drums are scored or rough (Fig. 52-2), grabbing may result. A loose backing plate may cause the same condition. As the linings contact the drum, the backing plate shifts to give hard braking. A defective power-brake booster can also cause grabbing.

♦ 52-10 NOISY BRAKES

Brakes become noisy if the brake linings wear so much that the rivets contact the brake drum, if the shoes become warped so that the contact with the drum is not uniform, if shoe rivets become loose so that they contact the drum, or if the drum becomes rough or worn (Fig. 52-2). Any of these conditions may cause a squeak or squeal when the brakes are applied. Loose parts, such as the brake backing plate, also may rattle.

♦ 52-11 AIR IN SYSTEM

If air gets into the hydraulic system, poor braking and a spongy pedal will result. Air can get into the system if the air vent in the master-cylinder cover or cap becomes plugged (Fig. 52-3). This may tend to create a partial vacuum in the system on the return stroke of the piston. Air could then bypass the rear piston cup, as shown by the arrows in Fig. 52-3, and enter the system. Always check the vent and clean it when the cap or cover is removed.

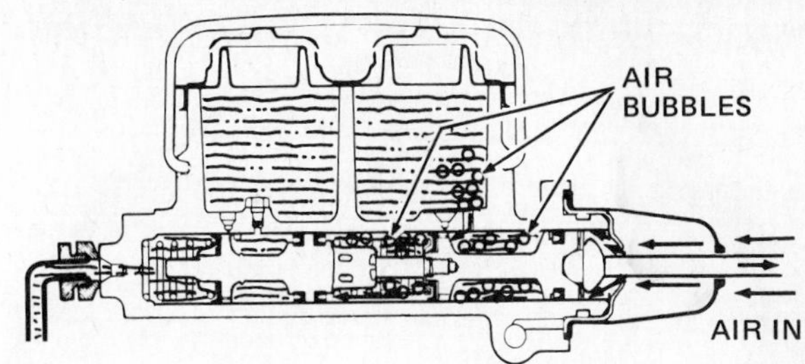

Fig. 52-3 If the air vent in the master-cylinder cover becomes clogged, air may be drawn into the system. The air is drawn past the low-pressure seal on the primary piston during the return stroke of the piston. *(General Motors Corporation)*

Air can also get into the hydraulic system if the master-cylinder residual check valve is leaky and does not maintain a slight pressure in the system. A leak could allow air to seep in around the wheel-cylinder piston cups. Without residual line pressure in the system, there would be no pressure holding the cups tight against the cylinder walls.

Probably the most common cause of air in the brake system is insufficient brake fluid in the master cylinder. If the brake fluid drops below the compensating port, the hydraulic system will draw air in as the piston moves forward when braking. Air in the system must be removed by adding brake fluid and bleeding the system. These procedures are described later.

♦ 52-12 LOSS OF BRAKE FLUID

Brake fluid can be lost if the master cylinder leaks, if the wheel cylinder leaks, if the line connections are loose, or if the line is damaged. One possible cause of wheel-cylinder leakage is incorrect installation of the actuating pin (Fig. 52-4). If the pin is cocked, the side thrust on the piston may permit leakage past the piston. Leakage from other causes at the master cylinder or wheel cylinder requires removal and repair or replacement of the defective parts.

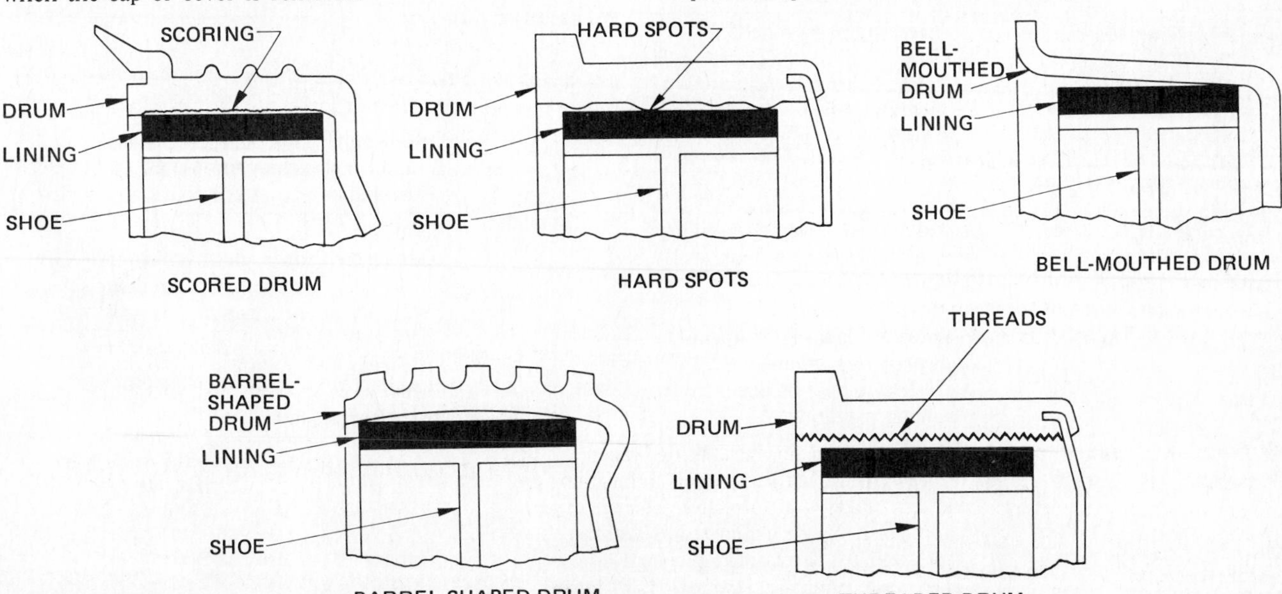

Fig. 52-2 Various types of brake-drum defects that require drum service. *(Bear Manufacturing Company)*

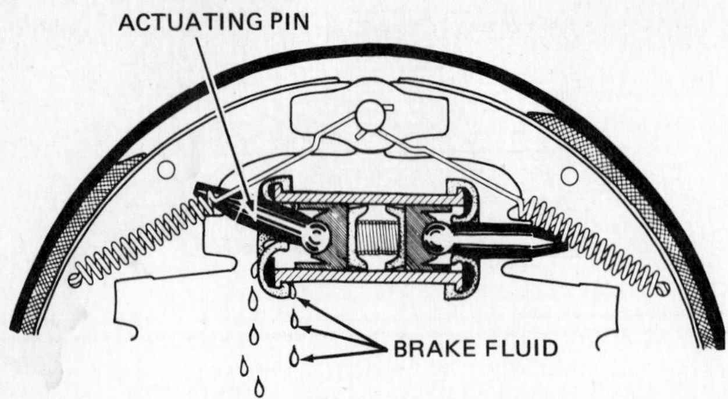

Fig. 52-4 Incorrect installation of the pin in the wheel cylinder will cause a side thrust on the piston which permits brake fluid to leak out past the cup. The pin must align with the notch in the brake shoe. *(Pontiac Motor Division of General Motors Corporation)*

♦ 52-13 BRAKES DO NOT SELF-ADJUST

Brakes do not self-adjust if the self-adjuster mechanism has been removed, the adjustment screw is stuck, the adjustment lever does not engage the star wheel, or the adjuster was incorrectly installed. Inspect the brake to find and correct the trouble.

♦ 52-14 WARNING LIGHT COMES ON WHEN BRAKING

If the warning light comes on when braking, it means there is low pressure in one section of the hydraulic system. One of the two braking sections has failed. Both sections should be checked so that the trouble can be found and eliminated. It is dangerous to drive with this condition. Even though the car slows, only half the wheels are being braked.

DISK-BRAKE TROUBLE DIAGNOSIS

♦ 52-15 DISK-BRAKE TROUBLE-DIAGNOSIS CHART

The chart that follows lists disk-brake troubles, their possible causes, and checks or corrections to be made. Following sections describe the troubles and causes or corrections to be made.

DISK-BRAKE TROUBLE-DIAGNOSIS CHART*

Complaint	Possible Cause	Check or Correction
1. Excessive pedal travel (♦ 52-16)	a. Excessive disk runout	Check runout; if excessive, install new disk
	b. Air leak, or insufficient fluid	Check system for leaks
	c. Improper brake fluid (low boiling point)	Drain and install correct fluid
	d. Warped or tapered shoe	Install new shoe
	e. Loose wheel-bearing adjustment	Readjust
	f. Damaged piston seal	Install new seal
	g. Power-brake malfunction	Check brake booster
2. Brake roughness or chatter (pedal pulsation) (♦ 52-17)	a. Excessive disk runout	Check runout; if excessive, install new disk
	b. Disk out of parallel	Check runout; if excessive, install new disk
	c. Loose wheel bearings	Readjust
3. Excessive pedal force, grabbing, or uneven braking (♦ 52-18)	a. Power-brake malfunction	Check brake booster
	b. Brake fluid, oil, or grease on linings	Install new linings
	c. Lining worn	Install new shoes and linings
	d. Incorrect lining	Install correct lining
	e. Frozen or seized pistons	Disassemble caliper and free pistons; install new caliper and pistons, if necessary.
4. Car pulls to one side (♦ 52-19)	a. Brake fluid or grease on linings	Install new linings
	b. Frozen or seized pistons	Disassemble caliper and free pistons
	c. Incorrect tire pressure	Inflate tires to recommended pressures
	d. Distorted brake shoes	Install new brake shoes
	e. Front end out of alignment	Check and align front end
	f. Broken rear spring	Install new rear spring
	g. Restricted hose or line	Check hoses and lines and correct as necessary
	h. Unmatched linings	Install correct lining
5. Noise (♦ 52-20): Groan	Brake noise when slowly releasing brakes (creep-groan). Not detrimental to function of disk brakes—no corrective action required. This noise may be eliminated by slightly increasing or decreasing brake-pedal force.	
Rattle	Brake noise or rattle at low speeds on rough roads may be due to excessive clearance between the shoe and the caliper. Install new shoe and lining assemblies to correct.	

*See ♦ 52-16 to 52-24 for detailed explanations of the trouble causes and corrections listed in the chart.

DRUM-BRAKE TROUBLE-DIAGNOSIS CHART (*Continued*)

Complaint	Possible Cause	Check or Correction
Scraping	a. Mounting bolts too long	Install mounting bolts of correct length
	b. Disk rubbing housing	Check for rust or mud buildup on caliper housing; check caliper mounting and bridge bolt tightness
	c. Loose wheel bearings	Readjust
	d. Linings worn, allowing wear indicator to scrape on disk	Replace pads
6. Brakes heat up during driving and fail to release (♦ 52-21)	a. Power-brake malfunction	Repair or replace brake booster
	b. Sticking pedal linkage	Free pedal linkage
	c. Driver riding brake pedal	Notify driver
	d. Frozen or seized piston	Disassemble caliper, clean cylinder bore, clean seal groove, and install new pistons, seals, and boots
	e. Residual check valve in master cylinder	Remove valve from master cylinder
	f. Incorrect linkage adjustment	Adjust linkage
7. Leaky caliper cylinder (♦ 52-22)	a. Damaged or worn piston seal	Install new seal
	b. Scores or corrosion on surface of piston or caliper bore	Disassemble caliper, clean cylinder bore; if necessary, install new pistons or replace caliper
8. Brake pedal can be depressed without braking action (♦ 52-23)	a. Piston pushed back in cylinder bores during servicing of caliper (and lining not properly positioned)	Reposition brake shoe and lining assemblies. Depress pedal a second time and if condition persists, look for the following causes:
	b. Leak in system or caliper	Check for leak, repair as required
	c. Damaged piston seal in one or more calipers	Disassemble caliper and replace piston seals as required
	d. Air in hydraulic system, or improper bleeding procedure	Bleed system
	e. Bleeder screw opens	Close bleeder screw and bleed entire system
	f. Leak past primary cup in master cylinder	Recondition master cylinder
9. Fluid level low in master cylinder (♦ 52-24)	a. Leaks in system or caliper	Check for leak, repair as required
	b. Worn brake-shoe linings	Replace shoes
10. Warning light comes on when braking (♦ 52-14)	a. One section has failed	Check both sections for braking action; repair defective system
	b. Pressure-differential valve defective	Replace

♦ 52-16 EXCESSIVE PEDAL TRAVEL

Anything that requires excessive movement of the caliper pistons will require excessive pedal travel. For example, if the disk has excessive runout, it will force the pistons farther back in their bores when the brakes are released. Therefore, additional pedal travel is required when the brakes are applied. Warped or tapered shoes, a damaged piston seal, or loose wheel bearings could cause the same problem. In addition, air in the brake lines, insufficient fluid in the system, or incorrect fluid (which has a low boiling point) will cause a spongy pedal and excessive pedal travel. If the power brake is defective, it could also cause greater than normal pedal travel.

♦ 52-17 BRAKE-PEDAL PULSATION

Brake-pedal pulsation is probably due to a disk with excessive runout or uneven thickness. The problem also may be caused by loose wheel bearings.

♦ 52-18 EXCESSIVE PEDAL FORCE, GRABBING, OR UNEVEN BRAKING

If excessive pedal force is required to obtain normal braking action, the power brake may be defective. In addition, if the linings are worn, hot, or water-soaked, they will not brake properly. Neither will a caliper that has a piston jammed in it. All these conditions require an excessively hard push on the brake pedal to provide braking action.

♦ 52-19 CAR PULLS TO ONE SIDE

Pulling to one side is due to uneven braking action. It could be caused by incorrect front-wheel alignment, uneven tire inflation, or a broken or weak suspension spring. Such things as brake fluid on the linings, unmatched linings, warped brake shoes, jammed pistons, floating or sliding caliper seized, or restrictions in the brake lines could cause the car to pull to one side when braking.

♦ 52-20 NOISE

The chart covers various noises and their causes. Refer to it for details.

♦ 52-21 BRAKES FAIL TO RELEASE

Brake-release failure could result from sticking pedal linkage, malfunctioning power brake, or pistons stuck in the calipers. It could also be due to the driver's riding the brake pedal or to failure of the master cylinder to release the pressure when the brakes are released.

♦ 52-22 LEAKY CALIPER CYLINDER

A leaky caliper cylinder could be due to a damaged or worn piston seal. The leak would also be caused by roughness on the surface of the piston as a result of scores, scratches, or corrosion.

♦ 52-23 BRAKE PEDAL CAN BE DEPRESSED WITHOUT BRAKING ACTION

If the brake calipers have been serviced, the pistons may be pushed back so far in their bores that a single full movement of the brake pedal will not produce braking. After any service on disk brakes, the brake pedal should be pumped several times. Then the master-cylinder reservoir should be filled to the proper level before the car is moved. Pumping the pedal moves the pistons into position so that normal brake-pedal application causes braking.

Other conditions can prevent braking action when the pedal is depressed. These include leaks or air in the hydraulic system. Leaks can occur at the piston seals, bleeder screws, or brake-line connections, or in the master cylinder. Also, the pressure-differential valve may be stuck, or the warning light may be burned out and both sections of the hydraulic system may have failed. This situation can occur, for example, if the warning-light bulb has burned out and the driver has been driving with one section defective. Then failure of the remaining section leaves the car with no braking action when the brake pedal is depressed.

♦ 52-24 FLUID LEVEL LOW IN MASTER CYLINDER

Brake-fluid level in the master cylinder should be ¼ inch [6 mm] below the top. Low fluid level could be due to leaks in the hydraulic system or caliper (♦52-23). Worn disk-brake-shoe linings also can cause this problem. As the linings wear, the fluid level lowers in the master cylinder.

POWER-BRAKE TROUBLE DIAGNOSIS

♦ 52-25 POWER-BRAKE TROUBLE-DIAGNOSIS CHART

The chart below relates various power-brake troubles to their possible causes and corrections. The chart and the sections that follow pertain to power-brake units only. Generally, the trouble-diagnosis charts in ♦52-2 and 52-15, which cover hydraulic brakes, also apply to power-brake systems. Therefore, the troubles listed in the charts, as well as the trouble corrections described earlier in the chapter, also apply to power brakes.

POWER-BRAKE TROUBLE-DIAGNOSIS CHART

Complaint	Possible Cause	Check or Correction
1. Excessive brake-pedal force (vacuum booster)	a. Defective vacuum check valve	Free or replace
	b. Hose collapsed	Replace
	c. Vacuum fitting plugged	Clear, replace
	d. Binding pedal linkage	Free
	e. Air inlet clogged	Clear
	f. Faulty piston seal	Replace
	g. Stuck piston	Clear, replace damaged parts
	h. Faulty diaphragm	Replace
	i. Causes listed under item 6 in chart in ♦ 52-2 or under item 3 in chart in ♦ 52-15	
2. Brakes grab	a. Reaction, or "brake-feel," mechanism damaged	Replace damaged parts
	b. Air-vacuum valve sticking	Free, replace damaged parts
	c. Causes listed under item 7 in chart in ♦ 52-2 or item 3 in chart in ♦ 52-15	
3. Pedal goes to floorboard	a. Hydraulic-plunger seal leaking	Replace
	b. Compensating valve not closing	Replace valve
	c. Causes listed under item 1 in chart in ♦ 52-2 or item 8 in chart in ♦ 52-15	
4. Brakes fail to release	a. Pedal linkage binding	Free
	b. Faulty check-valve action	Free, replace damaged parts
	c. Compensator port plugged	Clean port
	d. Hydraulic-plunger seal sticking	Replace seal
	e. Piston sticking	Lubricate, replace damaged parts

POWER-BRAKE TROUBLE-DIAGNOSIS CHART *(Continued)*

Complaint	Possible Cause	Check or Correction
	f. Broken return spring	Replace
	g. Causes listed under item 3 in chart in ♦ 52-2 or item 6 in chart in ♦ 52-15	
5. Loss of brake fluid	a. Worn or damaged seals in hydraulic section	Replace, fill, and bleed system
	b. Loose line connections	Tighten, replace seals
	c. Causes listed under item 10 in chart in ♦ 52-2 or items 7 and 9 in chart in ♦ 52-15	

BRAKE TESTING AND SERVICE

♦ 52-26 BRAKE TESTERS

There are two types of brake testers, the static and the dynamic. One type of static tester has four tread plates and registering columns (Fig. 52-5). To make the tests, the car is driven onto the tread plates at a specified speed and the brakes are applied hard. The stopping force at each wheel is registered on the four columns. If the readings are too low, or are unequal, brake service is needed.

The dynamic brake tester has rollers in the floor. The two wheels for which the brakes are to be tested are placed on the rollers (Fig. 52-6). If these are the drive wheels, the wheels are spun at the specified speed by the vehicle engine. For nondriving wheels, the rollers and wheels are spun by an electric motor. Then the throttle is released or the electric motor is turned off and the brakes are applied. The braking force at each wheel registers on meters. This shows if the brakes perform normally or if they need service.

♦ 52-27 BRAKE SERVICE

Any complaint of faulty braking action should be analyzed to determine the cause. Sometimes, all that is necessary in the earlier drum brakes is a brake-shoe adjustment to compensate for lining wear. On the later drum brakes with self-adjusters, the brakes automatically adjust to compensate for lining wear. However, the self-adjusters may fail. Other brake services include addition of brake fluid; bleeding the hydraulic system to remove air; repair or replacement of master cylinders, wheel cylinders, and calipers; replacement of brake linings; refinishing of brake drums or disks; and overhauling of power-brake units.

♦ 52-28 ADJUSTING DRUM BRAKES

Drum brakes without self-adjusters require periodic adjustment of the brake shoes. This compensates for brake-lining wear. Drum brakes are usually adjusted without removing the wheels and brake drums from the car. Self-adjusting brakes should require adjustment only after replacement of brake shoes, refinishing of the brake drums, or other service in which disassembly of the brakes was performed. This adjustment is called the *manual adjustment,* or *preliminary adjustment.* Adjustment procedures for drum brakes are given in the manufacturer's service manual.

♦ 52-29 REPLACING DRUM-BRAKE SHOES

When linings wear, the brake shoes must be replaced. To replace the shoes in a drum brake, the wheel and brake drum must be removed (Fig. 52-7). Then shoes with new linings are installed. Few shops today reline brake shoes, although this was common practice years ago.

Until a few years ago, most relined shoes were "arced." This meant the shoes were ground slightly to better fit the larger diameter of a used or refinished brake drum. However, grinding brake shoes is no longer universally recommended. This is because of the hazards resulting from the asbestos dust created during shoe grinding. See the CAUTION about asbestos dust at the beginning of this chapter.

♦ 52-30 SERVICING BRAKE DISKS AND DRUMS

Brake disks require replacement only if they become deeply scored or are warped out of line. Light scores and grooving

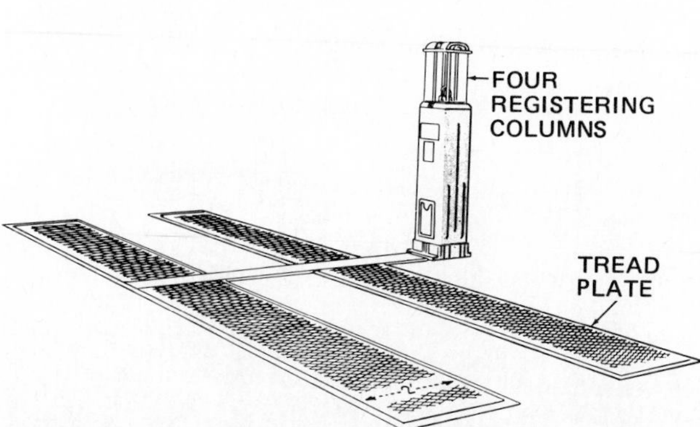

Fig. 52-5 Platform-type brake tester. *(Weaver Manufacturing Company)*

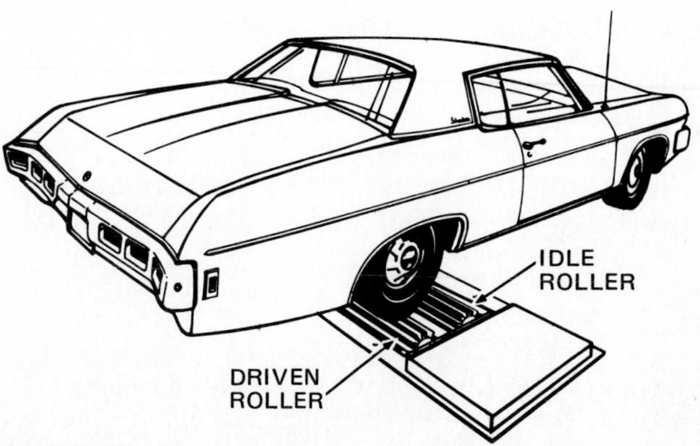

Fig. 52-6 Wheels on dynamic brake tester, ready for brake test. *(Clayton Manufacturing Company)*

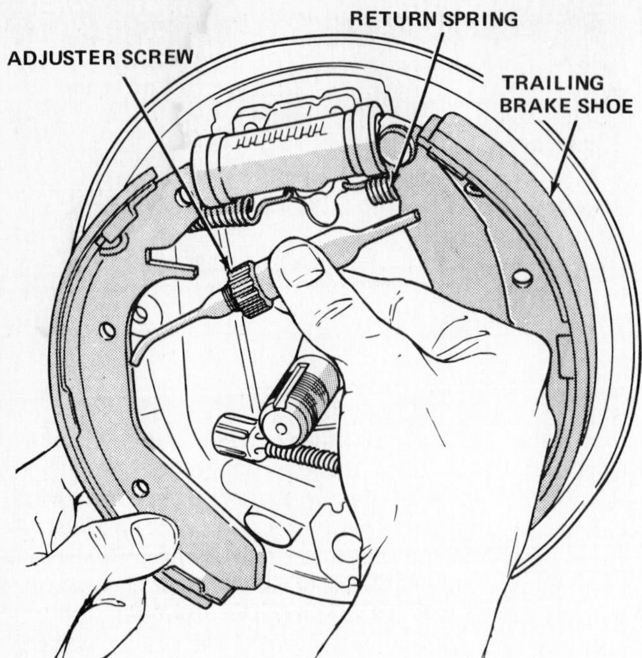

Fig. 52-7 Replacing brake shoes on a rear drum brake. *(Chrysler Corporation)*

are normal and will not affect braking. Some manufacturers recommend never grinding or refacing a scored brake disk. Instead, installation of a new disk is recommended. Refer to the manufacturer's service manual for the proper procedure to follow and tools to use.

Brake disks have a *discard dimension* (a number) cast into them. This dimension is the minimum thickness to which the disk can be refinished. If the disk must be refinished to a thinner diameter, discard it. The disk is too thin for safe use.

On drum brakes, the drum should be inspected for distortion, cracks, scores, roughness, or excessive glaze or smoothness. Glaze lowers friction and braking efficiency. Drums that are distorted or cracked should be discarded, and new drums installed. Light score marks can be removed with fine emery cloth. All traces of emery must be removed after smoothing the drum. Deeper scores, roughness, and glaze can be removed by turning or grinding the drum.

Many brake drums have their *discard diameter* cast into them. This dimension is the maximum allowable diameter. If it is necessary to turn or grind the drum to a larger diameter, discard it. The drum would be too thin for safe use. Brake drums should not be refinished to larger than the original diameter by more than 0.060 inch [1.5 mm]. This leaves 0.030 inch [0.76 mm] left for wear before the discard diameter is reached.

The diameters of the left and right drums on the same axle should be within 0.010 inch [0.25 mm] of each other. When the drum diameters on the same axle vary more than this, replace both drums.

♦ 52-31 WHEEL-CYLINDER SERVICE

Most wheel cylinders can be disassembled and rebuilt on the car. However, many manufacturers recommend that the wheel cylinder be removed from the backing plate and serviced on the bench. This makes it easier to properly and thoroughly clean, inspect, and reassemble the cylinder.

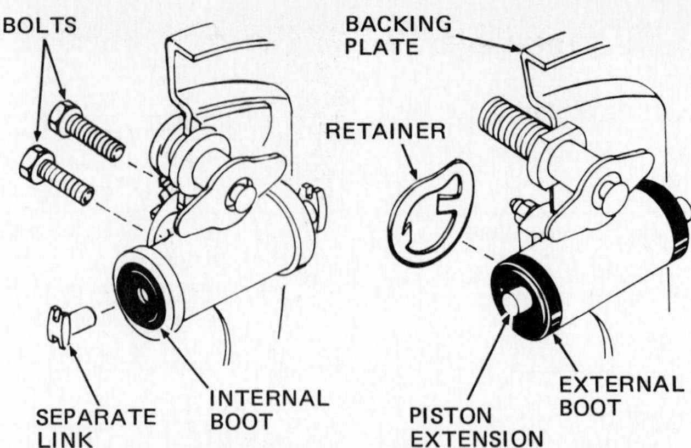

Fig. 52-8 Two methods of attaching the wheel cylinder to the backing plate. *(Delco Moraine Division of General Motors Corporation)*

To remove a wheel cylinder from the car, first remove the wheel and brake drum. Then disconnect the brake hose or tube from the wheel cylinder. Remove the wheel cylinder by taking out the attachment bolts or retainer (Fig. 52-8). Then tape the end of the hose or pipe shut to prevent any dirt from getting in.

Disassemble the wheel cylinder by first pulling off the boots. Then push out the pistons, cups, and springs. Clean all wheel-cylinder parts in clean brake fluid. Dry the parts with compressed air. Then place the dried parts on clean lint-free shop towels or paper. Check that all passages in the wheel cylinder and bleeder screw are clear by blowing through them with compressed air.

Inspect the cylinder bore for scoring and corrosion. Use crocus cloth to remove light corrosion and stains. Replace the wheel cylinder if crocus cloth does not remove the corrosion, or if the bore is pitted or scored. Some manufacturers permit the use of a brake-cylinder hone (Fig. 52-9) to remove scores and rust. However, the cylinder bore must not be honed more than 0.003 inch [0.08 mm] larger than its original diameter. If the scores do not clean out, replace the cylinder. The wheel cylinder also should be replaced if the clearance between the cylinder bore and the pistons is excessive.

When reassembling the wheel cylinder, lubricate all parts with clean brake fluid. Then assemble the wheel cylinder, using all parts in the repair kit. Install the bleeder screw, and torque it to specifications.

Careful Never allow any grease or oil to contact the rubber parts, or other internal parts, of the brake hydraulic system. Grease or oil will cause the rubber parts to swell, which may lead to brake failure.

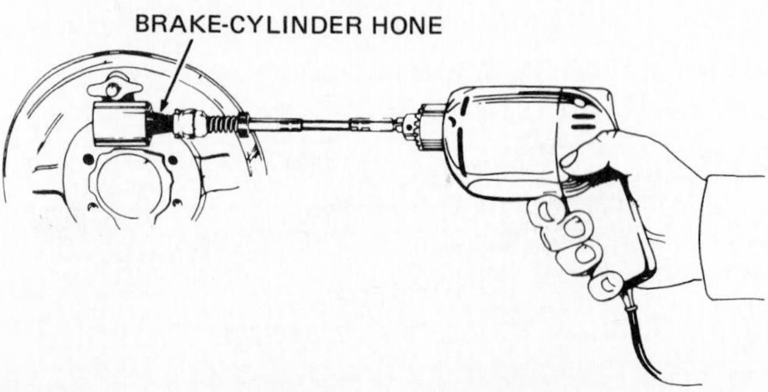

Fig. 52-9 Honing the bore in a wheel cylinder. *(Lisle Corporation)*

♦ 52-32 MASTER-CYLINDER SERVICE

Master cylinders may require disassembly for replacement of internal parts. However, some technicians prefer to install a new or rebuilt master cylinder. The service procedures for master cylinders used with disk brakes and drum brakes are very similar. One difference is that with disk brakes a larger brake-fluid reservoir is required. Figure 51-11 is a disassembled master cylinder used with a braking system that has front-wheel disk brakes and rear-wheel drum brakes. Note that the fluid reservoir for the front-disk brakes is larger than the other.

To service the master cylinder, clean the outside. Then remove the master cylinder from the car. Remove the cover and seal, and pour out any remaining brake fluid. Mount the master cylinder in a vise. If the car has a manual brake system, slide the boot to the rear, remove the retainer clip, and then remove the retainer, pushrod, and boot. Use the pushrod to force the primary piston inward, and remove the snap ring from the groove in the piston bore. Then remove the primary-piston assembly. The repair kit contains a complete new primary-piston assembly.

Remove the secondary-piston stop screw, if so equipped. Using the shop air hose, apply slight air pressure through the compensating port at the bottom of the reservoir. This will force out the secondary-piston assembly. Remove the piston seals from the secondary piston.

The outlet-tube seats, check valves, and springs must not be removed from some master cylinders. They are permanent parts of the master cylinder. However, these parts may be removed from other master cylinders. Follow the procedure in the manufacturers service manuals. Most disk-brake hydraulic systems do not use check valves.

Clean all parts in brake fluid or brake-cleaning solvent only. Blow dry with filtered compressed air. Blow out all passages and ports to be sure they are clear. If the master cylinder is scored, corroded, pitted, cracked, porous, or otherwise damaged, replace it. Some manufacturers permit slight honing of the master-cylinder bore. But if the pits or scoring are deep, a new master cylinder must be installed.

To assemble the master cylinder, dip all parts (except the body) into brake fluid. Insert the complete secondary-piston assembly, with return spring, into the master-cylinder bore. Install the secondary-piston stop screw, if so equipped. Put the primary-piston assembly into the bore. Depress the primary piston and install the snap ring in the bore groove. Install the pushrod, boot, and retainer on the pushrod, if so equipped. Install the pushrod assembly into the primary piston. Make sure the retainer is properly seated and holding the pushrod securely.

Careful Never add a stop screw to a master cylinder that does not have one, even though it has a threaded hole for the screw. If a stop screw is installed in a master cylinder of this type, the master cylinder may not function properly.

Position the inner end of the pushrod boot (if so equipped) in the retaining groove in the master cylinder. Put the seal into the cover, and install it on the master cylinder. Secure the cover with the retainer.

Note The master cylinder should be bled (♦52-36) before it is installed on the car. The procedure involves filling the master cylinder with brake fluid. Then the pistons are moved back and forth to get rid of any trapped air.

♦ 52-33 DISK-BRAKE SERVICE

On fixed-caliper brakes, the brake shoes can be replaced without removing the caliper assembly. With the car on a lift or safety stands, remove the wheel. Then use two pairs of pliers and pull on the tabs to pull each shoe out. Before installing new shoes, push the pistons in with slip-joint pliers. However, before you do this, you should remove some fluid from the master cylinder. Otherwise, pushing the pistons in will force enough fluid back into the master cylinder so it will overflow.

On the floating-caliper brake, remove both the wheel and caliper to replace the brake shoes. First, remove two-thirds of the fluid from the master-cylinder section for the disk brakes. Discard the fluid. Raise the car and remove the wheel cover and wheel. Use a C clamp as shown in Fig. 52-10 and tighten it to force the piston back into its cylinder. Remove the two mounting bolts and lift the caliper off. Support it with a wire hook so it does not hang from the brake hose. Remove the old shoes. Remove the sleeves and bushings from the four caliper ears.

On reinstallation, first install new sleeves and bushings and the shoes. Put each shoe back on the same side of the caliper from which it was removed. Make sure the piston is pushed back into its cylinder. Position the caliper over the disk and install the mounting bolts. Clinch the upper ears of the outboard shoe to hold it in place. The ears should be flat against the caliper. Add fresh brake fluid to the master cylinder. Pump the brake pedal several times to seat the linings against the disk and to get a firm pedal. Check and fill master cylinder as necessary.

♦ CAUTION ♦ Do not attempt to move the car until you feel a firm brake pedal.

If the caliper pistons or seals require replacement, the caliper assembly must be removed from the car. Then compressed air or a special piston remover can be used to remove the piston from the caliper. When installing a new seal on the piston, first dip the seal in clean brake fluid. Then install the

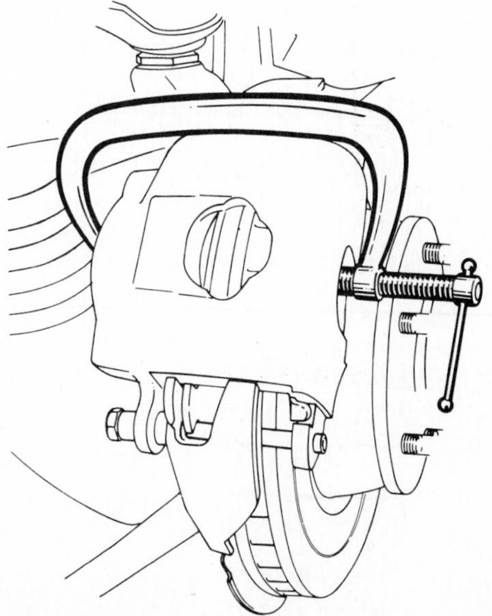

Fig. 52-10 Using a C clamp to force the piston into the caliper bore. *(Buick Motor Division of General Motors Corporation)*

piston in the bore, being careful not to unseat and twist the seal.

Slight roughness or corrosion in the caliber bores can be cleaned out with a hone. If the bore diameter must be increased more than 0.002 inch [0.05 mm], install a new caliper. If the piston is metal, and shows wear that has removed any of the chrome plating, install a new piston.

After assembly and installation of the caliper, the front-wheel bearings must be adjusted to specifications. Excessive play in the wheel bearings may affect disk-brake action.

♦ 52-34 HYDRAULIC-BRAKE TUBING REPAIR

Most hydraulic-brake tubing is made of double-walled, welded-steel tubing which is coated to resist rust. Only the tubing specified by the automotive manufacturer should be used. When replacing a tube, use the old tube as a pattern to form a new tube. Do not kink the tubing or make sharp bends.

Brake tubing must be cut off square with a special tube cutter. Do not use a jaw-type cutter or a hacksaw to cut brake tubing. Either of these can distort the tubing and leave heavy burrs that would prevent normal flaring of the tube. After the tube has been cut off, a flaring tool must be used to double-flare the end of the tube (Fig. 52-11).

♦ 52-35 FLUSHING HYDRAULIC SYSTEM

The process of removing all of the old brake fluid from the hydraulic system is called *flushing*. Some car manufacturers recommend flushing the hydraulic system when new parts are installed in it. The system *must* be flushed if there is any indication of brake-fluid contamination. Signs of contaminated brake fluid include corroded metal parts and soft or swollen rubber parts.

To flush the hydraulic system, install the pressure bleeder on the master cylinder (Fig. 52-12). If the car has a metering valve, it must be held in the open position. Place the brake-bleeder wrench on the bleeder valve in the wheel cylinder or caliper nearest the master cylinder. Install one end of a short bleeder hose on the bleeder valve. Place the other end in a transparent container.

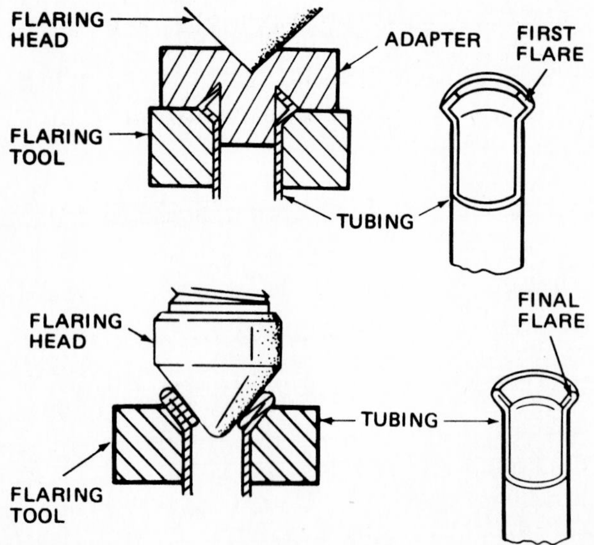

Fig. 52-11 Steps required to double-flare hydraulic-brake tubing.

Open the bleeder valve about 1½ turns, and let the fluid drain into the container. Close the bleeder valve when the fluid appears clean and clear. Then move on to the bleeder valve next closest to the master cylinder. Repeat the procedure at each wheel. When flushing is completed, check that the master cylinder is filled. About 1 quart [0.946 L] of clean, fresh brake fluid is needed to flush the hydraulic system.

If the hydraulic system is being flushed because of fluid contamination, replace all rubber parts in the brake system. This includes the rubber parts in the master cylinder, wheel cylinders and calipers, brake hoses, and combination valve. Then bleed the hydraulic system.

Some manufacturers recommend the use of a special flushing fluid. This fluid is used instead of new brake fluid during the flushing operation. Flushing is continued until all the old brake fluid has been flushed out. Then the flushing fluid is purged by applying clean, dry air through the master cylinder to blow the fluid out. Do no use too much air pressure. After all flushing fluid is out, fill the master-cylinder reservoir with new brake fluid. Then bleed the system as explained below.

♦ 52-36 FILLING AND BLEEDING HYDRAULIC SYSTEM

After flushing the hydraulic system (♦52-35), or at any time air may be in the hydraulic system, the hydraulic system must be filled and bled. For the brakes to operate properly, all air must be removed from the system.

The process of getting rid of any air trapped in a hydraulic brake line or component is called *bleeding*. In the bleeding process, brake fluid is forced through the brake line or component that has air in it. To bleed the brakes, install the pressure bleeder on the master cylinder (Fig. 52-12). If the vehicle has a metering valve, it must be held in the open position. Place the bleeder wrench on the bleeder valve in the wheel cylinder or caliper nearest the master cylinder. Install one end of the bleeder hose on the bleeder valve. The lower end of the bleeder hose is immersed in a clear container partly filled with fresh brake fluid. This allows you to see any air bubbles that come out of the line. It also prevents any air from being pulled back into the line. This can happen if the pressure on the brake fluid is released before the bleeder valve is closed.

Careful Clean away any dirt and grease from around the bleeder valve before opening it. This prevents any contamination from entering the line if the pressure on the fluid is released before the bleeder valve is closed.

If the master cylinder has a bleeder valve on it, bleed the master cylinder first. Open the bleeder valve about ¾ turn. Watch the flow of brake fluid from the end of the hose. As soon as the bubbles stop and brake fluid flows from the hose in a solid stream, close the bleeder valve. Repeat these steps at each wheel. Then disconnect the pressure bleeder from the master cylinder. Check that the master cylinder is filled with brake fluid. Wipe up any spilled brake fluid. Install the master-cylinder cover seal and cover. Pump the brake pedal several times. Be sure that a firm brake pedal is obtained before moving the vehicle.

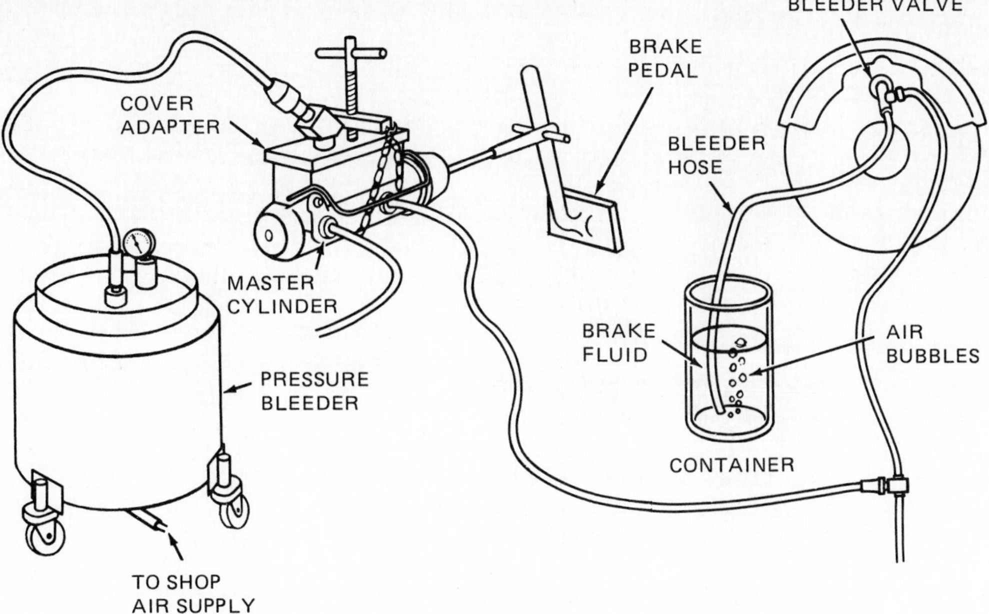

Fig. 52-12 Bleeding a hydraulic system with a pressure bleeder. *(Pontiac Motor Division of General Motors Corporation)*

♦ REVIEW QUESTIONS ♦

Select the *one* correct, best, or most probable answer to each question. You can find the answer in the section indicated at the end of each question.

1. The driver of a car with front-disk and rear-drum brakes complains that the brake pedal moves slowly to the floor while the pedal is depressed at a traffic light. This problem could be caused by (♦52-3)
   a. a leaking cup in the master cylinder
   b. a leaking power-brake booster
   c. a leaking residual check valve in the master cylinder
   d. an internal leak in the combination valve
2. The owner of a car with four-wheel disk brakes says that the brake pedal pulsates when the brakes are applied. Any of the following could be the cause *except* (♦52-17)
   a. a deeply scored or grooved rotor
   b. excessive rotor runout or wobble
   c. uneven rotor thickness
   d. a loose wheel bearing
3. The disk-brake reservoir on a dual master cylinder is low on fluid. Mechanic A says the brake pads may be worn. Mechanic B says the residual check valve may be faulty. Who is right? (♦52-24)
   a. A only
   b. B only
   c. both A and B
   d. neither A nor B
4. A car pulls to the right when the brakes are applied. Mechanic A says that the cause could be a defective master cylinder. Mechanic B says that the cause could be a defective metering or combination valve. Who is right? (♦52-6)
   a. A only
   b. B only
   c. either A or B
   d. neither A nor B
5. In the master cylinder, the brake-fluid level should be (♦52-24)
   a. not much more than ⅛ inch [3.2 mm] below the top
   b. level with the top
   c. ¼ inch [6 mm] below the top
   d. ½ to ¾ inch [12.7 to 19 mm] below the top

CHAPTER 53
Tires and Wheels: Construction and Service

After studying this chapter, and with the proper instruction and equipment, you should be able to:

1. Explain the difference between radial and bias-ply tires.
2. Explain each marking on the side of a tire.
3. Describe the Uniform Tire Quality Grading System.
4. Explain the differences between the various types of spare tires.
5. Diagnose tire wear.
6. Check tire pressure and add air.
7. Mount and inflate a tubeless tire.
8. Repair a flat tire.

♦ 53-1 PURPOSE OF TIRES

Tires have two functions. First, they are air-filled cushions that absorb most of the shocks caused by road irregularities. The tires flex, or give, as they meet these irregularities. Therefore they reduce the effect of the shocks on the passengers in the car. Second, the tires grip the road to provide good traction. Good traction enables the car to accelerate, brake, and make turns without skidding.

♦ 53-2 TIRE CONSTRUCTION

There are two general types of tires: those with inner tubes and those without tubes, called *tubeless* tires. On the inner-tube type, both the tube and the tire casing are mounted on the wheel rim. The tube is a hollow rubber doughnut. It is inflated with air after it is installed inside the tire and the tire is put on the wheel rim (Fig. 53-1). This inflation causes the tire to resist any change of shape.

Tubes are used in some truck tires and in motorcycle tires. Tubes are seldom used in passenger-car tires today. Cars use tubeless tires. The tubeless tire does not use an inner tube. Instead, the tubeless tire is mounted on the wheel rim so that the air is retained between the rim and the tire (Fig. 53-2).

The amount of air pressure used in the tire depends on the type of tire and the operation. Passenger tires are inflated to about 22 to 36 psi [155 to 248 kPa]. Heavy-duty tires on trucks or buses may be inflated to 100 psi [690 kPa].

Tire casings and tubeless tires are made in about the same way. Layers of cord, called *plies*, are shaped on a form and impregnated with rubber. The rubber sidewalls and treads are then applied (Fig. 53-3). They are vulcanized into place to form the completed tire. The term *vulcanizing* means heating the rubber under pressure. This process molds the rubber into the desired form and gives it the proper wear characteristics and flexibility. The number of layers of cord, or plies, varies according to the intended use of the tire. Passenger-car tires have 2, 4, or 6 plies. Heavy-duty truck and bus tires may have up to 14 plies. Tires for heavy-duty service, such as earthmoving machinery, may have up to 32 plies.

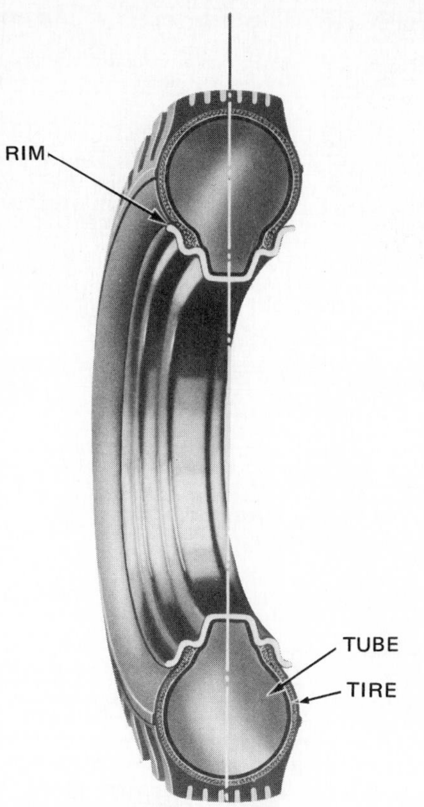

Fig. 53-1 Tire and tire rim cut away showing the tube.

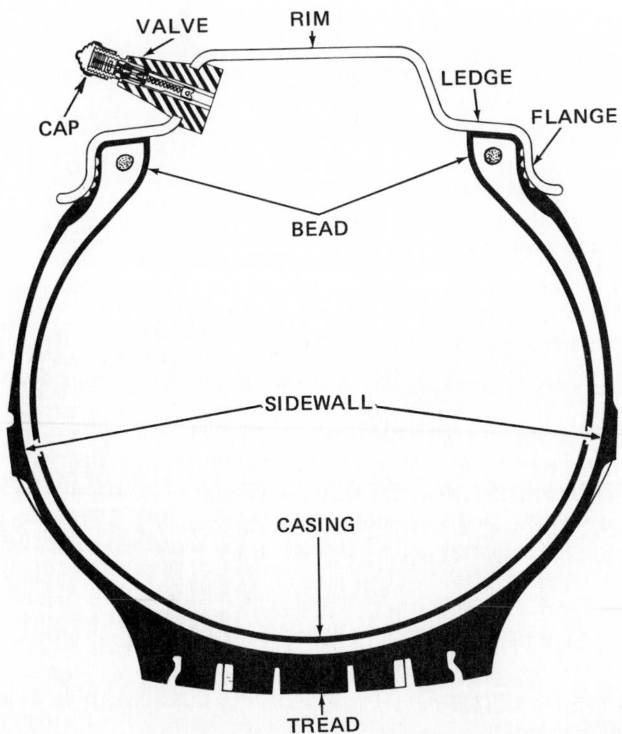

Fig. 53-2 In a tubeless tire, the tire bead rests between the ledges and flanges of the rim to produce an airtight seal. *(Pontiac Motor Division of General Motors Corporation)*

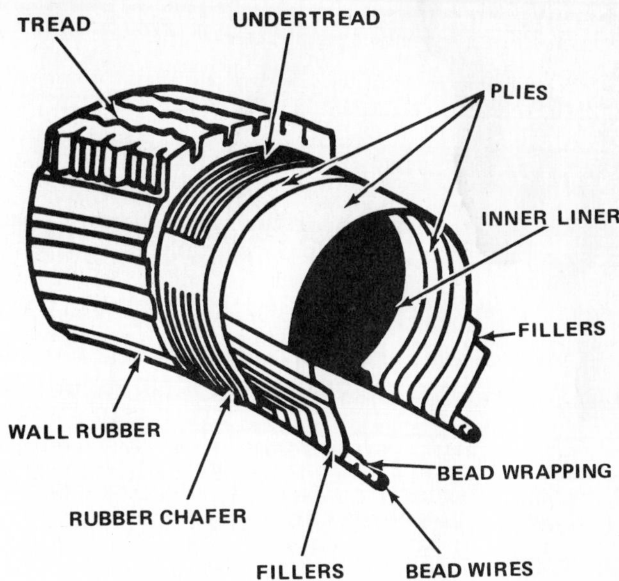

Fig. 53-3 Construction of a tubeless tire. *(Chevrolet Motor Division of General Motors Corporation)*

♦ 53-3 BIAS VS. RADIAL PLIES

There are two ways to apply the plies, on the bias and radially. For many years most tires were of the bias type, as shown to the left in Fig. 53-4. These tires have the plies criss-crossed. One layer runs diagonally one way, and the other layer runs diagonally the other way. This arrangement makes a carcass that is strong in all directions because of the overlapping plies. However, the plies tend to move against each other in bias tires. This movement generates heat, especially at high speed. Also, the tread tends to "squirm," or close up, as it meets the road. This increases tire wear.

Tires with radial plies, as shown to the right in Fig. 53-4 and in Fig. 53-5, were introduced to remedy these problems. In a radial tire, all plies run parallel to each other and vertical to the tire bead. Belts are applied on top of the plies to provide added strength parallel to the bead. Then the tread is vulcanized on top of the belts, which are made of rayon, nylon, glass fiber, or steel mesh. All radial tires work in the same way, regardless of the belt material.

Radial tires are installed on almost all cars built in the United States. The radial is more flexible than a bias-ply tire, so more of the tread stays on the pavement (Fig. 53-6). On a radial, the tread has less tendency to heel up when the car goes around a curve (Fig. 53-7). This keeps more rubber on the road and reduces the tendency of the tire to skid.

Also, better fuel economy is claimed for cars with belted tires. The tires roll on the road with less resistance and use up less power. Radial tires wear more slowly than bias-ply tires. This is because the radial-tire tread does not squirm as the tire meets the pavement. The bias-ply tire tread tends to squirm (Fig. 53-8). As the treads pinch together, they slide sideways. This causes tread wear. There is less heat buildup on the highway in the radial tire. This also slows radial-tire wear.

Bias-ply tires may also be belted (Fig. 53-9). However, even some manufacturers who make the belted-bias tire recommend the belted radial as the superior tire.

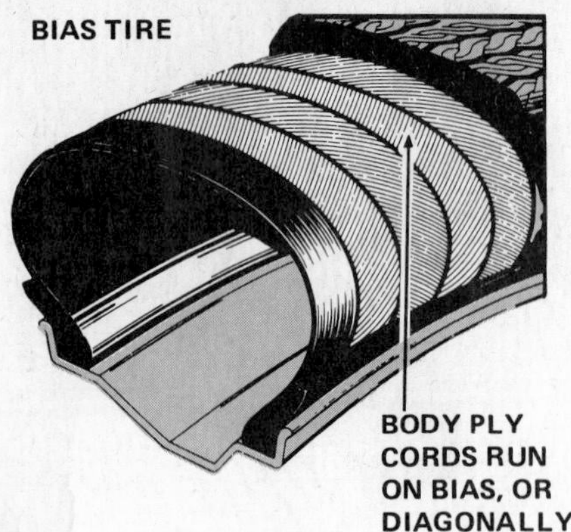

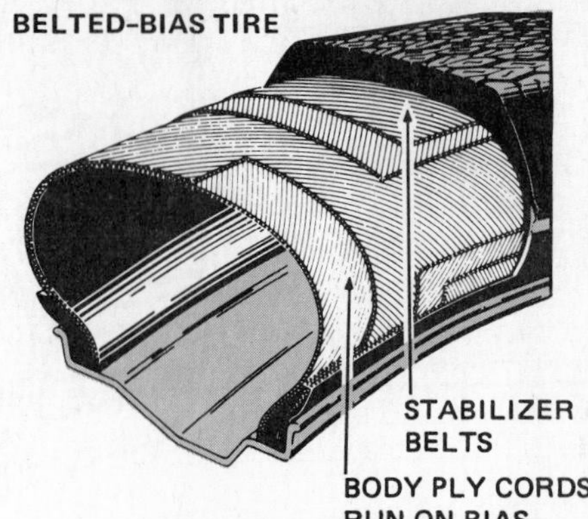

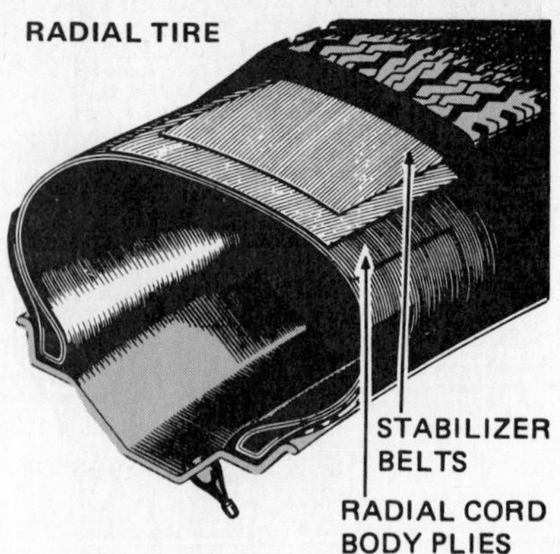

BODY PLY CORDS RUN ON BIAS FROM BEAD TO BEAD. BUILT WITH 2 TO 4 PLIES. CORD ANGLE REVERSED ON EACH PLY. TREAD IS BONDED DIRECTLY TO TOP PLY.

STABILIZER BELTS ARE APPLIED DIRECTLY BENEATH THE TREAD. BODY PLY CORDS RUN ON BIAS, SIMILAR TO BIAS TIRE CONSTRUCTION.

RADIAL PLY CORDS RUN STRAIGHT FROM BEAD TO BEAD WITH STABILIZER BELTS APPLIED DIRECTLY BENEATH THE TREAD.

Fig. 53-4 The three basic tire constructions. *(Firestone Rubber Company)*

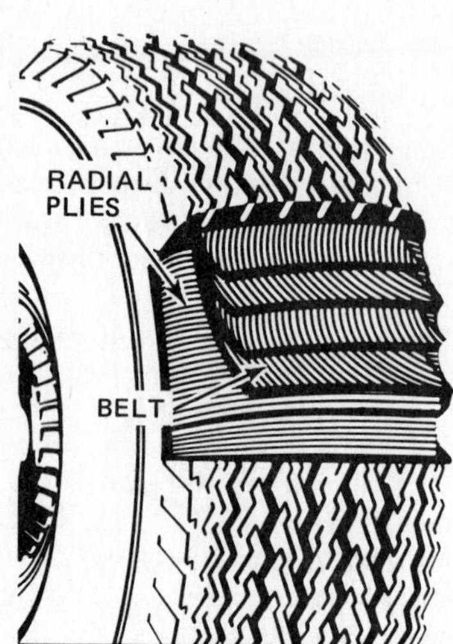

Fig. 53-5 Belted radial tire, partly cut away to show radial plies and belt. *(B. F. Goodrich Company)*

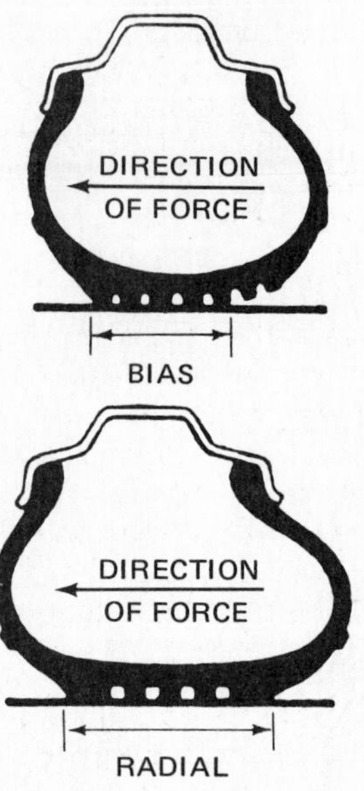

Fig. 53-7 The difference in the amount of tread a nonbelted bias-ply tire and a belted radial tire put on the pavement during a turn.

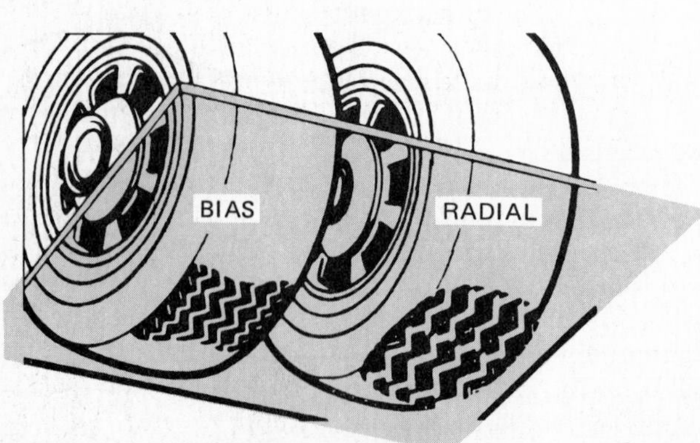

Fig. 53-6 Tread patterns ("footprints") of a nonbelted bias-ply tire and a belted radial tire on a flat surface. The radial tire puts more rubber on the road.

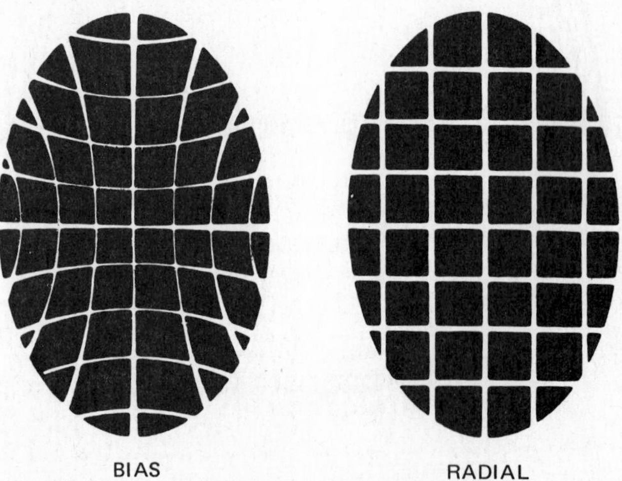

Fig. 53-8 Bias-ply tire tread tends to squirm as it meets the road. The radial tire tread tends to remain apart.

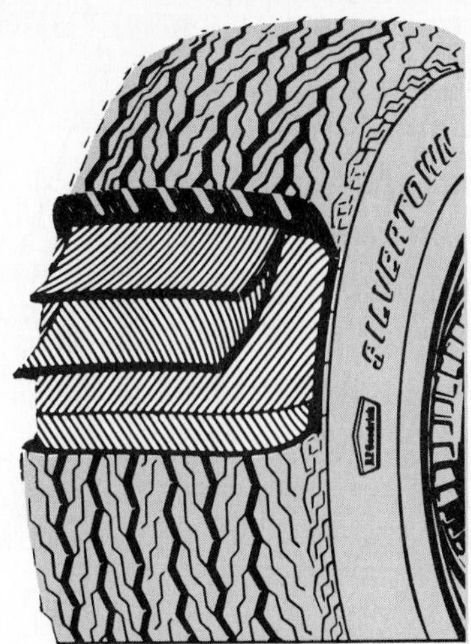

Fig. 53-9 Belted-bias tire, partly cut away to show bias plies and belt. *(B. F. Goodrich Company)*

Careful Never mix belted-radial and bias-ply tires, either belted or unbelted, on a car. Mixing the two types can cause poor car handling and increase the possibility of skidding. This is especially important with snow tires. Bias-ply snow tires on the rear and belted radials on the front can result in oversteer and spin-out on wet or icy roads.

♦ 53-4 TIRE TREAD

The tread is the part of the tire that rests on the road. There are many different tread designs. Figure 53-10 shows a few. Snow tires have large rubber cleats molded into the tread. The cleats cut through snow to improve traction.

Some tires have steel studs that stick out through the tread. Studs help the tire get better traction on ice and snow. However, some people claim studded tires shorten the life of the road surface. For this reason, studded tires are banned in some localities.

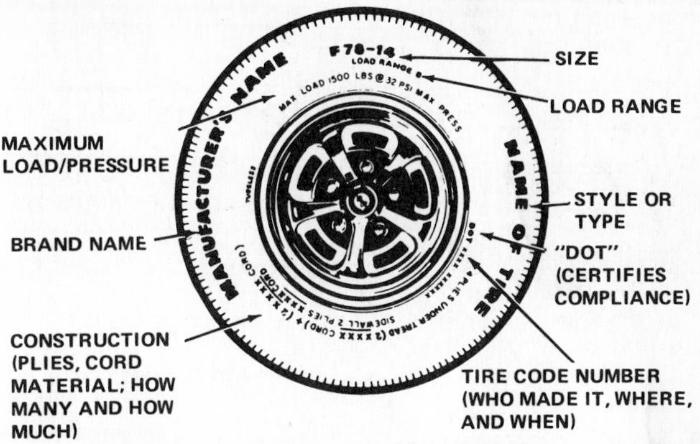

Fig. 53-11 Markings on the sidewall of a tire.

♦ 53-5 TIRE VALVE

Air is put into the tire, or into the inner tube, through a valve that opens when an air hose is applied to it (Fig. 53-2). The valve is sometimes called a *Schrader valve*. On a tubed tire, the valve is mounted in the inner tube and sticks out through a hole in the wheel rim. On the tubeless tire, the valve is mounted in the hole in the wheel rim. When the valve is closed, spring force and air pressure inside the tire or tube hold the valve on its seat. Most valves carry a valve cap. The cap is screwed down over the end of the valve. This protects the valve from dirt and acts as an added safeguard against air leaks.

♦ 53-6 TIRE SIZE AND MARKINGS

Tire size is marked on the sidewall of the tire. An older tire might be marked 7.75-14. This means that the tire fits on a wheel that is 14 inches [356 mm] in diameter at the rim where the tire bead rests. The 7.75 means the tire itself is about 7.75 inches [197 mm] wide when it is properly inflated.

Figure 53-11 shows a tire with an explanation of each mark

Fig. 53-10 Types of tire tread. *(B. F. Goodrich Company; Goodyear Tire and Rubber Company)*

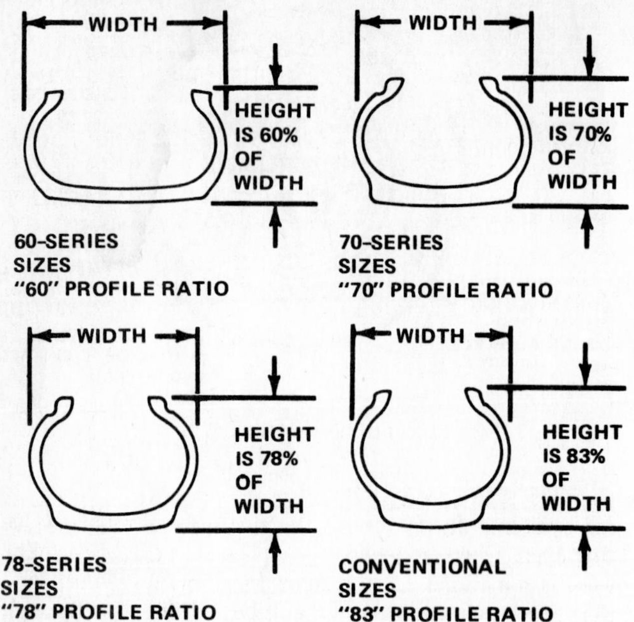

Fig. 53-12 Four aspect ratios of car tires. *(American Motors Corporation)*

on it. Tires carry several markings on the sidewall. The markings include a letter code to designate the type of car the tire is designed for. D means a lightweight car. F means intermediate. G means a standard car. H, J, and L are for large luxury cars and high-performance vehicles. For example, some cars use a G78-14 tire. The 14 means a rim diameter of 14 inches [356 mm]. The 78 indicates the ratio between the tire height and width (Fig. 53-12). This tire is 78 percent as high as it is wide. The ratio of the height to the width is called the *aspect ratio*, or the *profile ratio*. Four aspect ratios are 83, 78, 70, and 60. The lower the number, the wider the tire looks. A 60 tire is only 60 percent as high as it is wide.

The addition of an R to the sidewall marking, such as GR78-14, indicates that the tire is a radial. Also, if a tire is a radial, the word *radial* must be molded into the sidewall. Some radials are marked in the metric system. For example, a tire marked 175R13 is a radial tire which measures 175 mm [6.9 inches] wide. It mounts on a wheel with a rim diameter of 13 inches [330 mm]. Wheel diameter (for cars sold in the United States) is given in inches in all tire-sizing systems.

Some cars use metric-size tires. The meaning of each letter and number of a metric tire is shown in Fig. 53-13. This is the latest size designation for tires. Comparing the two tire-size labels, a tire formerly marked as an ER78-14 is now marked P195/75R14. An SL ("standard load") on the sidewall means a maximum inflation pressure of 35 psi [240 kPa]. When the tires are marked XL ("extra load"), maximum inflation pressure is 41 psi [280 kPa].

To identify the load that an older-type tire can safely carry, each tire is classified by load range. The load range indicates the allowable load for the tire as inflation pressure is increased. The tire in Fig. 53-11 is marked "Load Range B." Most passenger car tires are in load range B. There are three load ranges for passenger-car tires, B, C, and D. Under the oldest system, "ply rating" was used to indicate load range. The load range B tire has the same load-carrying capacity as a tire with 4-ply rating. Load range C equals a 6-ply-rating tire. Load range D equals an 8-ply-rating tire.

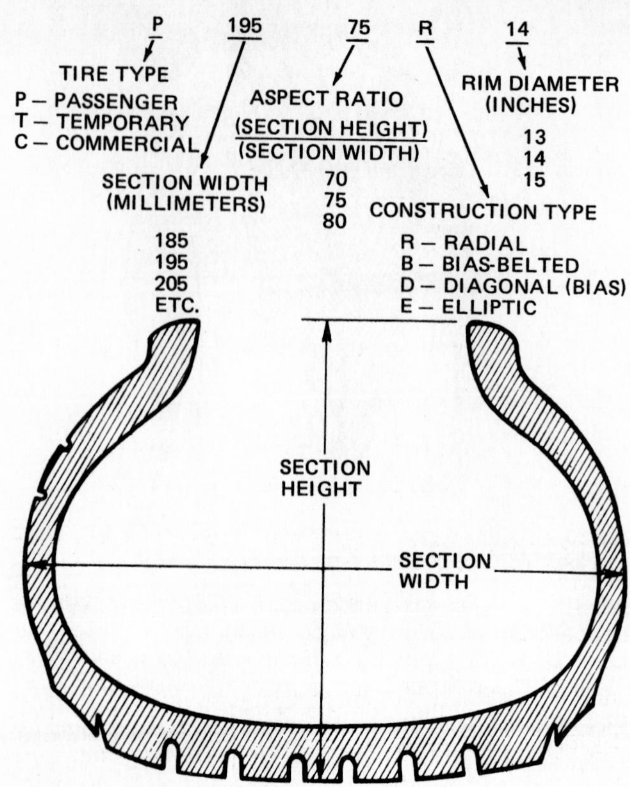

Fig. 53-13 Meanings of the tire size designations for a metric tire. *(Chevrolet Motor Division of General Motors Corporation)*

In addition, new tires comply with the Uniform Tire Quality Grading System (UTQGS) of the Department of Transportation (DOT). Under this system, tires are graded for tread wear, traction, and temperature resistance. Each tire is graded with a series of numbers and letters indicating comparative tread wear, traction, and temperature. The grades are molded on the tire sidewall (Fig. 53-14).

1. Tread Wear Tread wear is a number grade, such as 90 or 120. A tire with a tread-wear grade of 150 can be expected to last 50 percent longer on the government test course than a 100 tread-wear tire, if the tire is not abused. A tire with a tread-wear grade of 100 should last for about 30,000 miles [48,280 km]. The tire graded 150 should last about 45,000 miles [72,420 km].

2. Traction Traction is graded in three steps, A, B, and C. An A tire has better traction than a B tire. Grade C tires have poor traction on wet roads.

3. Temperature Resistance Temperature resistance is also graded in three steps. The temperature grades of A (the

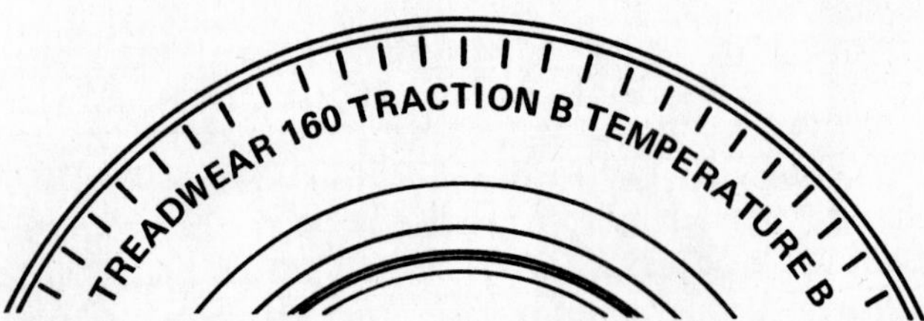

Fig. 53-14 Uniform Tire Quality Grading System markings on the tire sidewall. *(ATW)*

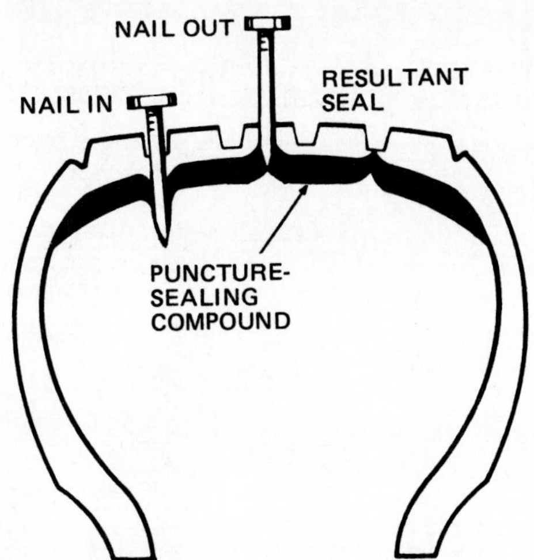

Fig. 53-15 Sealing action in a puncture-sealing tire. *(Pontiac Motor Division of General Motors Corporation)*

highest), B, and C represent the tire's resistance to the generation of heat when tested under laboratory conditions on a specified indoor test wheel. Sustained high temperature can cause the materials in a tire to deteriorate. This leads to reduced tire life, and to excessive temperatures which can lead to sudden tire failure. Grade C corresponds to the minimum standard that all passenger-car tires must meet. Grades B and A represent levels of performance on the laboratory test wheel higher than the minimum required by law.

♦ 53-7 PUNCTURE-SEALING TIRES

Some tubeless tires have a coating of plastic material in the inner surface. When the tire is punctured, this plastic material is forced by the internal air pressure into the hole left when the nail or other object is removed. This action is shown in Fig. 53-15. The plastic material then hardens, sealing the hole.

♦ 53-8 TUBES

Three types of rubber, one natural and two synthetic, have been used to make tubes. Today the most common tube material is butyl. A butyl tube can be identified by its blue stripe. The other synthetic rubber type (GR-S) has a red stripe. Natural rubber is not striped.

Three special types of inner tubes are described below.

1. Radial-Tire Inner Tube The construction of an inner tube for use in a radial tire differs from the tube used in a bias tire. A radial tire flexes in such a manner that it concentrates the flex action in one area—the edge of the belts in the shoulder of the tire. This concentration of stress may cause the splice of a conventional tube to fail. Therefore, conventional tubes must never be used in radial tires.

To overcome the problem, the radial tube is made from a special rubber compound that is designed to overcome this concentrated stress. Then the tube is spliced on a special machine which makes a stronger splice.

2. Puncture-Sealing Tubes Some tubes have a coating of the same plastic material used in the puncture-sealing tire (♦53-7). It flows into and seals any holes left by punctures. In some tubes, the plastic material coats the inside of the tube. In others the material is retained between an inner rubber diaphragm and the tube in a series of cells. This latter construction prevents the material from flowing as a result of centrifugal force and thereby building up in certain spots in the tube. If the material were allowed to build up, it would cause the tire-tube-and-wheel assembly to be unbalanced.

3. Safety Tube The safety tube is really two tubes in one, one smaller than the other and joined at the rim edge. When the tube is filled with air, the air flows first into the inside tube. From there it passes through an equalizing passage into the space between the two tubes. Therefore both tubes are filled with air. If a puncture or blowout occurs, air is lost from between the two tubes. However, the inside tube, which has not been damaged, retains its air pressure. It is sufficiently strong to support the weight of the car until the car can be slowed and stopped. Usually, the inside tube is reinforced with nylon fabric. The nylon takes the suddenly imposed weight of the car, without giving way, when a blowout occurs.

♦ 53-9 SPARE TIRES

Most cars carry a spare tire. It is installed if one of the four tires on the car goes flat. In many cars, the "spare" is a regular tire mounted on the standard wheel. In addition, three types of temporary spare tires are used (Fig. 53-16). These

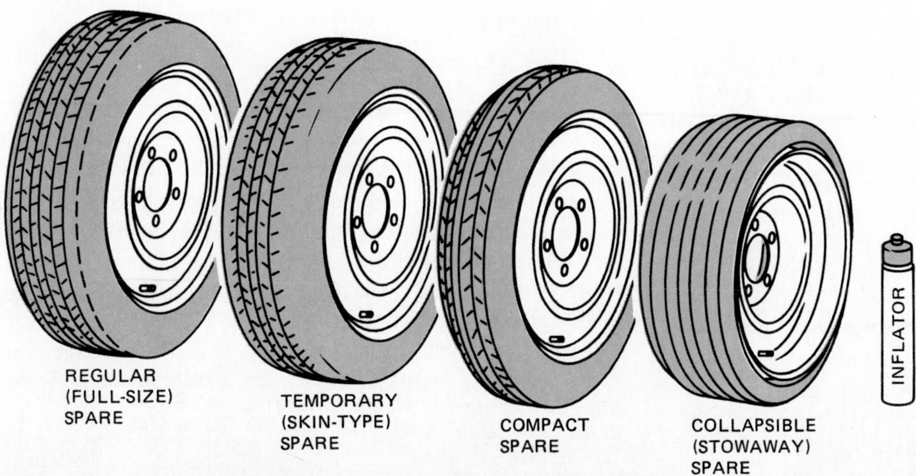

Fig. 53-16 Types of spare tires. *(General Motors Corporation)*

temporary spare tires are designed to reduce vehicle weight and to increase storage space. Also, their light weight and smaller size makes them easier to handle. All have bias-ply construction.

The temporary skin-type spare (Fig. 53-16) is a lightweight tire with full-size diameter. It is used as the spare in some cars that have a limited-slip differential. Another type of temporary spare is the collapsible spare tire. This tire (Fig. 53-16) is installed on the wheel in a deflated condition. The deflated tire is slightly larger than the rim. A pressure can of inflation propellant, called the inflator, is stored in the luggage compartment. Instructions on how to install and safely inflate collapsible spare tires are printed on the inflator can. The procedure is described in ♦53-26.

♦ CAUTION ♦ The collapsible spare tire must not be driven more than 150 miles [240 km] and at a speed of 50 mph [80 km/h] or less. To exceed these limits is to risk a blowout of the collapsible spare tire. Also, the tire must not be inflated from the air hose in the shop or in a service station. This can cause the tire to explode.

♦ 53-10 COMPACT SPARE TIRE

Another tire that saves space in the luggage compartment is the compact spare tire (Fig. 53-16). This spare is lighter and considerably smaller than the standard tire. The compact spare tire can be driven for the 1000- to 3000-mile [1609- to 4800-km] life of the tread. It is mounted on a narrow 51 × 4 wheel. The tire must not be mounted on any other wheel. No other tire, wheel cover, or trim ring should be installed on the special wheel. Also, the compact spare tire should not be used on the rear of a car equipped with a nonslip differential. The collapsible spare tire is smaller in diameter than the tire on the other side of the rear axle. Differential action would have to take place continuously. This may cause damage and failure of the differential.

The compact spare tire is for emergency use only. As soon as the standard tire has been repaired, reinstall it on the car. The compact spare carries an inflation pressure of 60 psi [415 kPa]. It gives a rough and noisy ride.

♦ 53-11 WHEELS

Most cars use a pressed steel or disk wheel. This type of wheel also is called a *safety-rim* wheel. Figure 53-17 shows how a disk wheel is made. The outer part, called the *rim*, is in one piece, and it is welded to the disk. This forms the seamless and airtight wheel that is needed to mount a tubeless tire. The center of the rim is smaller in diameter than the rest. This gives the rim the name *drop center*. The center well is necessary to permit removal and installation of the tire. The bead of the tire must be pushed off the bead seat and into the smaller diameter. Only then can the other side of the bead be worked up over the rim flange. Tire service is described later.

The 14-inch [356-mm] wheel is used on most cars today. However, some smaller cars have 12-inch [305-mm] or 13-inch [330-mm] wheels. Fourteen-inch [356-mm] wheels are available in three widths. The width used on a car depends on the tire specified for the car. The rim widths are 4.5 inches [114 mm], 5 inches [127 mm], and 6 inches [152 mm]. Optional larger tires usually require wider rims.

Most manufacturers recommend that a wheel be replaced if it is bent or leaks air. The new wheel must be exactly the

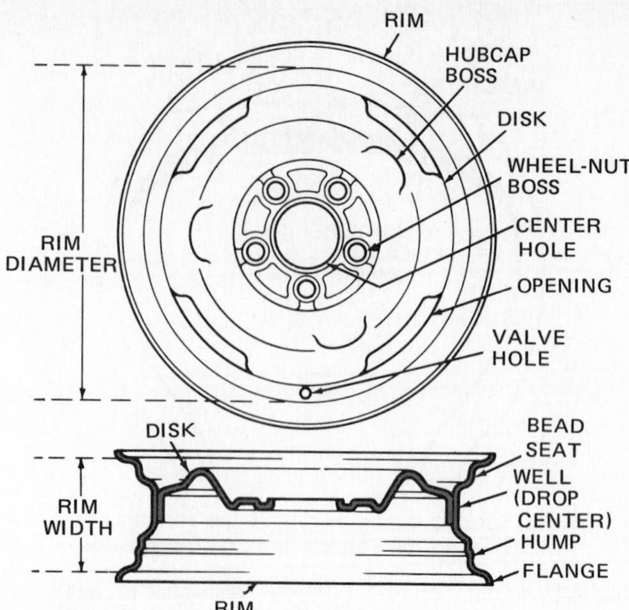

Fig. 53-17 Construction of a car wheel. *(American Motors Corporation)*

same as the old wheel. Installation of the wrong wheel could cause the wheel bearing to fail, the brakes to overheat, the speedometer to read inaccurately, and the tire to rub the body and frame.

♦ 53-12 SPECIAL WHEELS

Plain steel wheels, decorated with hub caps or wheel covers, are used on cars today. A variety of special wheels are available. These special wheels can be classified as styled steel or styled aluminum wheels. The "mag" wheel is very popular. It looks like the magnesium wheels used on some race cars. Magnesium metal is very light. However, for passenger cars, most mag wheels are made of aluminum. The term *mag wheel* can mean almost any chromed, aluminum-offset, or wide-rim wheel of spoke design.

Some aluminum wheels are lighter than the steel wheels they replace. Lighter wheels reduce unsprung weight. This improves handling and performance. Also, some aluminum wheels can improve brake and tire performance by allowing them to run cooler. Aluminum transmits and dissipates heat faster than steel.

TIRE SERVICE

♦ 53-13 TIRE SERVICE

Tire service includes periodic checking of the air pressure and addition of air as needed. Failure to maintain correct air pressure can cause rapid tire wear and early tire failure. Incorrect air pressure can also cause handling problems (♦53-14). Tire service also includes periodic inspection of the tire for abnormal wear, cuts, bruises, and other damage. In addition, tire service includes removal, repair, and replacement of tires.

♦ 53-14 TIRE INFLATION AND TIRE WEAR

The driver has more effect on tire life than anything else. Rapid tire wear results from quick starts and stops, heavy braking, high speed, taking corners too fast, and striking or rubbing curbs. Too little air in the tire can cause hard steering, front-wheel shimmy, steering kickback, and tire squeal on turns. A tire with too little air will wear on the shoulders and not in the center of the tread, as shown at the upper left in Fig. 53-18.

Also, the underinflated tire is subject to rim bruises. If the tire strikes a rut or stone, or bumps a curb too hard, it flexes so much that it is pinched against the rim. Any of these kinds of damage can lead to early tire failure.

With overinflation, the tire rides on the center of the tread. Then, only the center wears, as shown at the upper center in Fig. 53-18. In addition, the overinflated tire will not flex normally. As a result, the tire fabric can be weakened or even broken.

Other types of tire wear are discussed below.

1. Toe-in or Toe-out Tire Wear Excessive toe-in or toe-out on turns causes the tire to be dragged sideways as it moves forward. For example, a tire on a front wheel that toes in 1 inch [25.4 mm] from straight ahead will be dragged sideways about 150 feet [46 m] every mile [1.6 km]. This sideways drag scrapes off rubber, as shown at the upper right in Fig. 53-18. Note the feather edges of rubber that appear on one side of the tread. If both sides show this type of wear, the toe is incorrect. If only one tire shows this type of wear, a steering arm probably is bent. This causes one wheel to toe in or out more than the other.

2. Camber Wear If a wheel has excessive camber, the tire runs more on one shoulder than on the other. The tread wears excessively on that side, as shown at the lower left in Fig. 53-18.

3. Cornering Wear Cornering wear, shown at the lower center in Fig. 53-18, is caused by taking curves at excessive speeds. The tire skids and tends to roll, producing the diagonal type of wear shown. This is one of the more common causes of tire wear. Suggest that the driver slow down around curves.

4. Uneven Tire Wear Uneven tire wear, with the tread unevenly or spottily worn, is shown at the lower right in Fig. 53-18. It can result from several mechanical problems. These include misaligned wheels, unbalanced wheels, uneven or "grabby" brakes, overinflated tires, and out-of-round brake drums.

5. High-Speed Wear Tires wear more rapidly at high speed than at low speed. Tires driven consistently at 70 to 80 mph [113 to 129 km/h] will give less than half the life of tires driven at 30 mph [48 km/h].

Note See ♦50-11 for additional information on causes of improper tire wear.

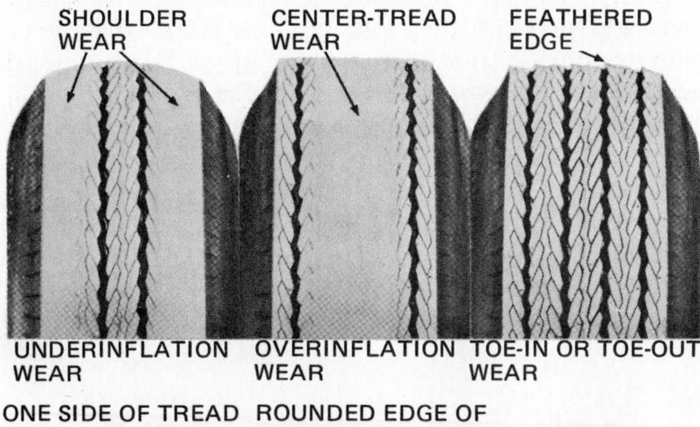

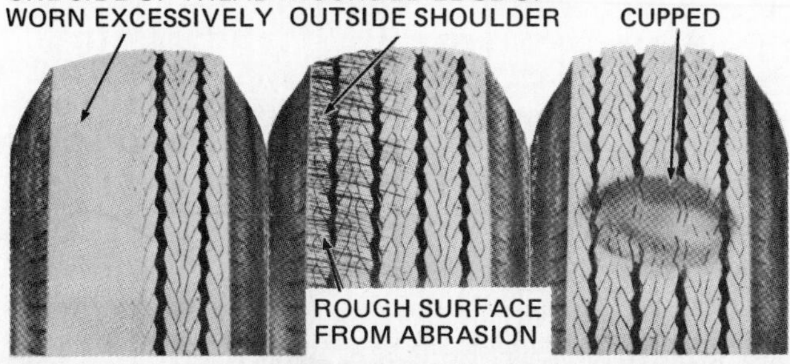

Fig. 53-18 Patterns of abnormal tire-tread wear. *(Buick Motor Division of General Motors Corporation)*

♦ 53-15 RADIAL-TIRE WADDLE

Waddle is side-to-side movement of the front or rear of the car as it moves forward. It is caused by the steel belt not being straight inside the tire. It is most noticeable at low speeds, 5 to 30 mph [8 to 48 km/h]. It may also be felt as ride roughness at 50 to 70 mph [80 to 113 km/h].

To determine where the faulty tire is on the car (front or rear), make a road test. If the faulty tire is on the rear, the rear end of the car will shake from side to side. From the driver's seat, it may feel as though someone were pushing on the side of the car.

If the faulty tire is on the front, the waddle is more visual. The front body sections will appear to move back and forth or from side to side.

♦ 53-16 CHECKING TIRE PRESSURE AND INFLATING TIRES

To check tire pressure and inflate the tires, you must know the correct tire pressure for the tire you are servicing. You find this spec on the tire sidewall (on many tires), in the shop manual, in the owner's manual, or on a label in the glovebox or on the car door (or at some similar place) on the car. Specifications are for cold tires. Tires that are hot from being driven or from sitting in the sun will have an increased air pressure. Air expands when hot. Tires that have just come off an interstate highway may show as much as a 5 to 7 psi [35 to 48 kPa] increase.

As a hot tire cools, it loses pressure. So, never bleed a hot tire to reduce the pressure. If you do this, then when the tire cools, its pressure could drop below the specified minimum.

There are times when the tire pressure should be on the high side. For example, one tire manufacturer recommends adding 4 psi [28 kPa] for turnpike speed, trailer pulling, or extra-heavy loads. But never exceed the maximum pressure specified on the tire sidewall.

♦ CAUTION ♦ Never stand over a tire while inflating it. If the tire should explode, you could be injured.

If the tire valve has a cap, always install the cap after checking pressure or adding air.

Inflation specifications are often given in kilopascals (kPa) instead of pounds per square inch (psi). The conversion table in Fig. 53-19 will help you convert from one to the other.

kPa	psi	kPa	psi
140	20	215	31
145	21	220	32
155	22	230	33
160	23	235	34
165	24	240	35
170	25	250	36
180	26	275	40
185	27	310	45
190	28	345	50
200	29	380	55
205	30	415	60

Conversion: 6.9 kPa = 1 psi

Fig. 53-19 Inflation pressure chart, showing conversion from kilopascals (kPa) to pounds per square inch (psi).

♦ 53-17 TIRE ROTATION

The amount of wear a tire gets depends on its location on the car. For example, on a car with rear-wheel drive, the right rear tire wears about twice as fast as the left front tire. This is because many roads are slightly crowned (higher in the center), and also because the right rear tire is driving. The crown causes the car to lean out a little, so that the right tires carry more weight. The combination of this and carrying power through the right rear tire causes it to wear faster. To equalize wear as much as possible, tires should be rotated any time uneven wear is noticed and at the distance specified by the car manufacturer. One manufacturer recommends rotating radial tires after the first 7500 miles [12,000 km] and then every 15,000 miles [24,000 km]. Bias tires should be rotated every 7500 miles [12,000 km].

Figure 53-20 shows the recommended rotation pattern for bias, bias-belted, and radial tires. Bias and bias-belted tires can be switched from one side of the car to the other. However, radial tires must not be switched from one side to the other. This would reverse their direction of rotation and cause handling and wear problems.

Use the four-wheel rotation pattern shown in Fig. 53-20 on cars that have a nonrotatable spare. Only a regular full-size spare tire should be rotated into service during a five-wheel rotation. If the car carries any type of temporary spare tire, leave it in place. Perform a four-wheel rotation.

Always check tire pressure after switching tires. Many cars require that the front tires carry different pressures from the rear tires. Therefore, when you switch tires from front to back, the pressure will require adjustment to meet the specifications.

Careful Studded tires should never be rotated. A studded tire should be put back on the wheel from which it was removed. Before you remove a studded tire, mark its location on the sidewall. A studded tire should be put back on the same wheel from which it was removed, mounted so that it rolls in the same direction. Reversing its rotation can result in serious handling problems.

♦ 53-18 TIRE INSPECTION

The purpose of inspecting the tires is to determine whether they are safe for further use. When an improper wear pattern is found, the technician must know the cause of the abnormal tread wear (Fig. 53-18). The technician must correct the cause or tell the driver what is wrong. When the tires are found to be in good condition, they can be rotated (Fig. 53-20). After the tires are cool, check and adjust inflation pressures.

When inspecting a tire, check for bulges in the sidewalls. A bulge is a danger signal. It can mean that the plies are separated or broken and that the tire is likely to go flat. A tire with a bulge should be removed. If the plies are broken or separated, the tire should be thrown away. To make a complete tire inspection, remove all stones from the tread. This is to make sure that no tire damage is hidden by the stones. Also, any time the tire is to be spin-balanced, remove all stones from the tread. This will ensure that no person is struck and injured by stones thrown from the tread as the tire rotates.

Many tires have tread-wear indicators, which are filled-in sections of the tread grooves. When the tread has worn down enough to show the indicators (Fig. 53-21), the tire should be replaced. There also are special tire-tread gauges that can be inserted into the tread grooves to measure how much tread remains. A quick way to check tread wear is with a Lincoln penny inserted in the tread grooves. If at any point you can see all of Lincoln's head, the tread is excessively worn. Some state laws require a tread depth of at least $\frac{1}{32}$ inch [0.79 mm] in any two adjacent grooves at any location on the tire.

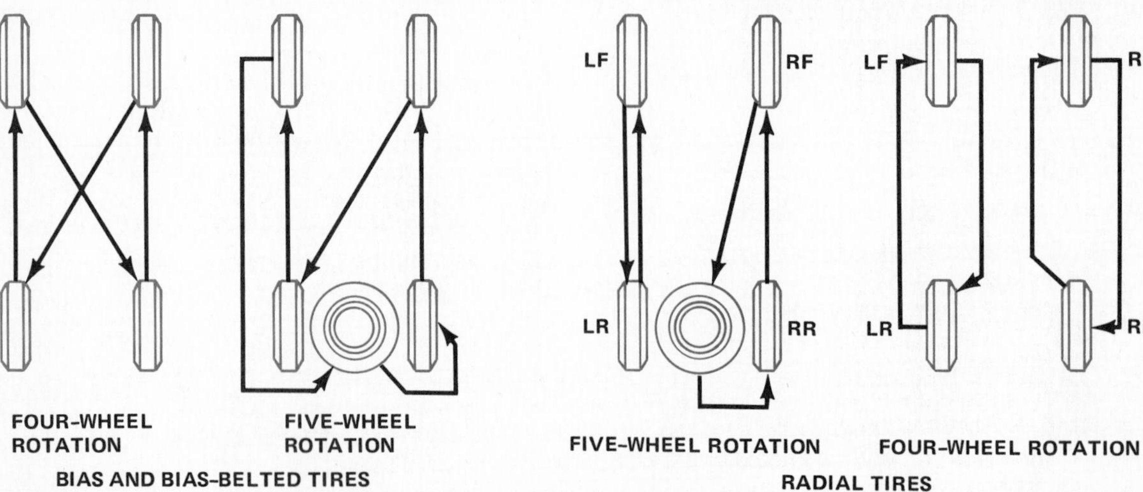

Fig. 53-20 Tire rotation patterns for tires with and without a rotatable spare tire. *(Chevrolet Motor Division of General Motors Corporation)*

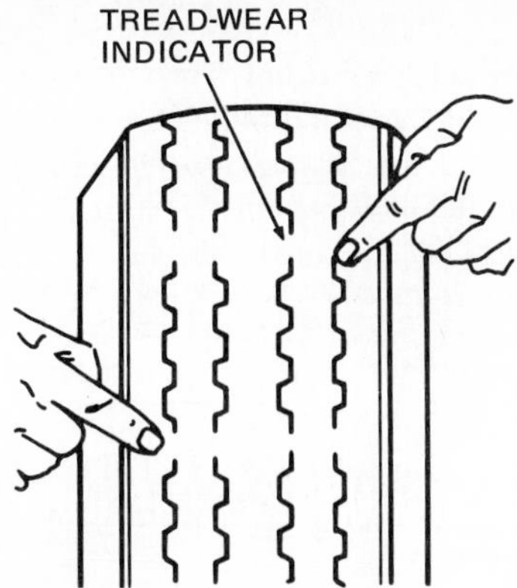

Fig. 53-21 A tire tread worn down so much that the tread-wear indicators show.

A tire can look okay from the outside and still have internal damage. To completely inspect a tire, remove it from the rim. Then examine it closely, inside and out.

♦ 53-19 TUBE INSPECTION

Tubes usually give little trouble if they are correctly installed. However, careless installation can cause trouble. For example, if the wheel rim is rough or rusty or if the tire bead is rough, the tube may wear through. Dirt in the casing can cause the same trouble. Another condition that can cause trouble is installing a tube that is too large in the tire. Sometimes an old tube (which may have stretched) is put in a new tire. When a tube that is too large is put into a casing, the tube can overlap at some point. The overlap will rub and wear and possibly cause early tube failure.

A radial tire that is used with a tube must have a special radial-tire inner tube in it. If regular tubes are used in radial tires, the tube splice may come apart.

Always check carefully around the valve stem when inspecting a tube. If the tube has been run flat or at low pressure, the valve stem may be broken or tearing away from the tube. Valve-stem trouble requires installation of a new tube.

♦ 53-20 REMOVING WHEEL FROM CAR

Radial tires must be removed from the car to be repaired (Fig. 53-22). Also, if the tire has been run flat, remove the tire for inspection. To repair a tire, first take off the hub cap or wheel cover. Then remove the wheel from the car. If you are using a lug wrench, loosen the lug nuts before raising the car. It is easier to loosen the lug nuts first, because the wheel will not turn if the car weight is on it. On some cars the lug nuts on the right side of the car have right-hand threads. The lug nuts on the left side have left-hand threads. The reason is that the forward rotation of the wheels tends to tighten the nuts, not loosen them.

♦ 53-21 DEMOUNTING A TIRE FROM DROP-CENTER RIM

With the wheel off the car, make a chalk mark across the tire and rim so that you can reinstall the tire in the same position. This preserves the balance of the wheel and tire. Next, release the air from the tire. This can be done by holding the tire valve open or removing the valve core.

♦ CAUTION ♦ The air coming out could shoot dirt particles into your eyes, so protect them with glasses or a shield.

The tire should then be removed from the rim, using a tire changer (Fig. 53-23). At one time, tire irons—flat strips of steel—were used to remove and install tires. They can damage the tire bead so that it will not seal, and this ruins the tire. Do not use tire irons!

♦ 53-22 USING SHOP TIRE CHANGERS

Today many shops have an air-powered tire changer (Fig. 53-23). After the bead is pushed off the rim (this is called *breaking the bead*), a tool is used with the tire changer to lift the bead up over the rim (Fig. 53-24). The powered tire changer also has a tool to remount the tire on the rim.

♦ 53-23 REMOUNTING TIRE ON RIM

To mount the tire on the rim, use the tire changer. Coat the rim and beads with rubber lubricant or a soap-and-water mixture. This will make the mounting procedure easier. Do not use a nondrying lubricant, such as antifreeze, silicone, grease, or oil. They will allow the tire to "walk around" the rim so that the tire balance is lost. Oil or grease will damage the rubber.

Fig. 53-22 Using an impact wrench to remove the lug nuts.

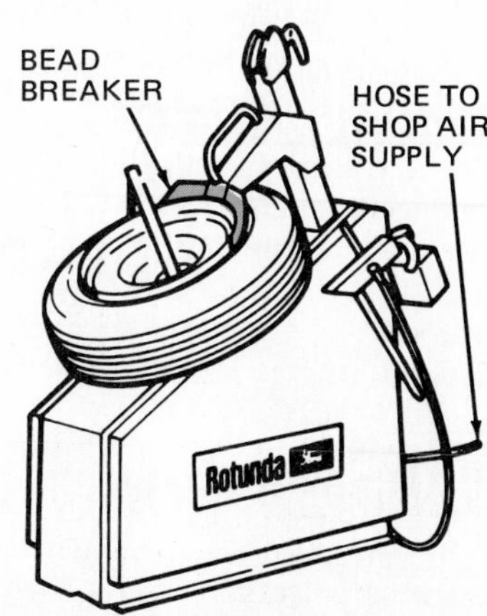

Fig. 53-23 An air-powered tire changer breaks the bead so that the tire can be removed from the wheel. *(Ford Motor Company)*

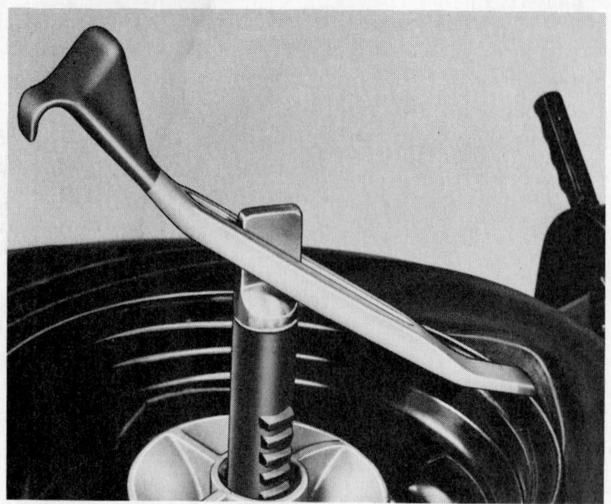

Fig. 53-24 Using the powered tire changer to lift the upper bead above the rim. The center post rotates, carrying the bead-lifting tool around with it.

When you are remounting the same tire that was removed from the rim, make sure the chalk marks on the tire and rim align. After the tire is on the rim, reposition the beads against the bead seat. Slowly inflate the tire (Fig. 53-25). If the beads do not hold air, use a tire-mounting band to spread the beads. You usually will hear a "pop" as the beads seat on the rim. Then install the valve core and inflate the tire to the recommended pressure.

♦ CAUTION ♦ Do not stand over the tire while inflating it. If the tire should explode, you could be injured.

♦ 53-24 CHECKING THE WHEEL

When the tire is off the wheel, check the rim for dents and roughness. Steel wool can be used to clean rust spots from standard steel wheels. Aluminum wheels should be cleaned only with mild soap and water. File off nicks or burrs. Then clean the rim to remove all filings and dirt. A wheel that has been bent should be discarded. A bent wheel may be weakened by heating, welding, or straightening so that it could fail on the highway.

Some wheels now have decorative plastic inserts. The plastic can be cleaned by using a sponge and soap and water.

♦ 53-25 TIRE-VALVE SERVICE

If the valve in the wheel requires replacement, remove the old valve and install a new one. There are two types: the snap-in type, and the type that is secured with a nut. To remove the snap-in type, cut off the base of the old valve. Lubricate the new valve with rubber lubricant. Then attach a tire-valve installing tool to the valve and pull the new valve into place.

On the clamp-in type that is secured with a nut, remove the nut to take the old valve out. Be sure to tighten the nut sufficiently when installing the new valve.

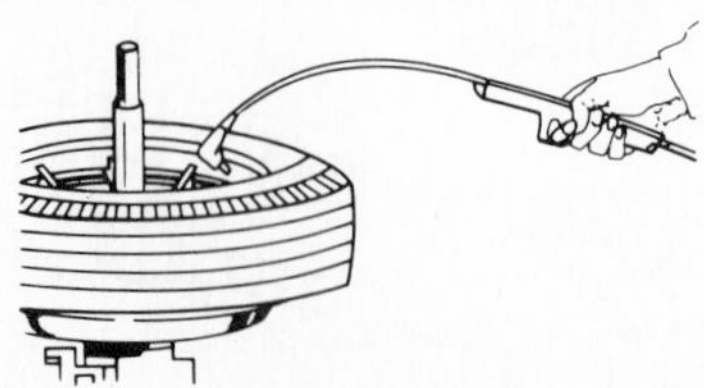

Fig. 53-25 Inflating the tire while seating the beads. Never exceed a pressure of 40 psi (276 kPa) in a passenger-car tire. *(Tire Industry Safety Council)*

♦ 53-26 SERVICING COLLAPSIBLE SPARE TIRE

The space-saving collapsible tire (♦53-9) is installed on the wheel deflated. The wheel must be installed on the car before the tire is inflated.

♦ CAUTION ♦ Do not inflate the tire before the wheel is mounted on the car. Follow the safety cautions listed on the inflator can.

The inflator can has detailed instructions on how to install and inflate the tire. Briefly, here is the inflation procedure.

1. If the temperature is 10°F [−12°C] or below, the inflator must be heated. Put the inflator over the defroster outlet of the car. Set the heater at DEFROST at the highest temperature. Then run the blower at the fastest speed for 10 minutes.
2. Do not inflate the tire off the car. First bolt the wheel to the car with the air valve at the bottom. Remove the plastic cap from the inflator and the cap from the tire valve.
3. Push the inflator onto the valve stem until you hear the sound of gas entering the tire.

♦ CAUTION ♦ Keep your hands off the metal parts of the inflator. They become extremely cold during discharge. You could freeze your fingers!

4. When the sound stops, wait 1 minute. Then remove the inflator and install the valve cap. The gas in the inflator, when completely used up, will properly fill the tire.

After you have inflated the tire and have removed the jack so that the tire rests on the pavement, the tire may look underinflated. This can happen especially in cold weather. Drive slowly for the first mile [1.6 km]. This will warm up the tire and increase the pressure.

5. If the inflator is the nonrefillable type, dispose of it in a safe waste receptacle. Do not burn or puncture the inflator. If it is the refillable type, it can be recharged with the proper equipment.
6. The collapsible spare tire must not be driven farther than necessary. The maximum distance is 150 miles [240 km]. As soon as the regular tire has been repaired and installed in place of the collapsible spare tire, remove the valve core from the collapsible spare tire. This will allow the gas to escape so that the tire collapses. It can then be stored, as before, in the luggage compartment.

Instead of a can of inflator, some manufacturers supply an electric air compressor to inflate the tire. Porsche, for example, supplies an electric air compressor. It is plugged into the cigarette-lighter socket and gets power from the car battery. The manufacturer warns that you should not use any other equipment to inflate the tire.

◆ CAUTION ◆ This tire must not be driven more than 150 miles (240 km) and at a speed of only 50 mph (80 km/h) or less. To exceed these limits is to risk a blowout of the collapsible spare tire. Also, the tire must not be inflated from the usual air hose at the shop or in a service station. This can cause the tire to explode.

TIRE AND TUBE REPAIR

◆ 53-27 TUBE REPAIR

If a tire has been punctured but has no other damage, it can be repaired with a patch. Remove the tube from the tire to find the leak. Inflate the tube and then submerge it in water. Bubbles will appear where there is a leak. Mark the spot. Then deflate the tube and dry it.

There are two ways to patch a tube leak. They are the cold-patch method and the hot-patch method. With the cold-patch method (also known as *chemical vulcanizing*), first make sure the rubber is clean, dry, and free of oil or grease. Buff, or roughen, the area around the leak. Then cover the area with vulcanizing cement. Let the cement dry until is is tacky. Press the patch into place. Roll it from the center out with a "stitching tool" or with the edge of a patch-kit can.

With the hot-patch method, prepare the tube in the same way as for the cold patch. Put the hot patch into place and clamp it. Then, with a match, light the fuel on the back of the patch. As the fuel burns, the heat vulcanizes the patch to the tube. After the patch has cooled, recheck the tube for leaks by submerging the tube in water.

Another kind of hot patch uses a vulcanizing hot plate. The hot plate supplies the heat required to bond the patch to the tube.

◆ 53-28 TIRE REPAIR

No attempt should be made to repair a tire that has been badly damaged. If the plies are torn or have holes in them, the tire should be thrown away. A puncture bigger than ¼ inch [6.35 mm] should not be patched. Instead, the tire should be replaced. Even though you might be able to patch the tire, it would be dangerous to use. The tire might blow out on the highway.

To repair small holes in a tubeless tire, first make sure that the object that caused the hole has been removed. Check the tire for other puncturing objects. Sometimes a tubeless tire can carry a nail for a long distance without losing air.

A radial tire should be removed from the wheel for repair. The repair plug should be of the head type and applied from inside the tire (♦53-30 and Fig. 53-26). Figure 53-27 shows the area of a tire in which a puncture can be repaired. Punctures outside this area require replacement of the tire.

Leaks from a tubeless tire are located in the same way as leaks from a tube. With the tire on the wheel and inflated, submerge the tire and wheel in water. Bubbles will show the location of any leaks. If a water tank is not available, coat the tire with soapy water. Soap bubbles will show the location of leaks.

If air leaks from around the spoke welds of the wheel, you can repair the leaks. Clean the area and apply two coats of cold-patch vulcanizing cement on the inside of the rim. Allow the first coat to dry before applying the second coat. Then cement a strip of rubber patching material over the area.

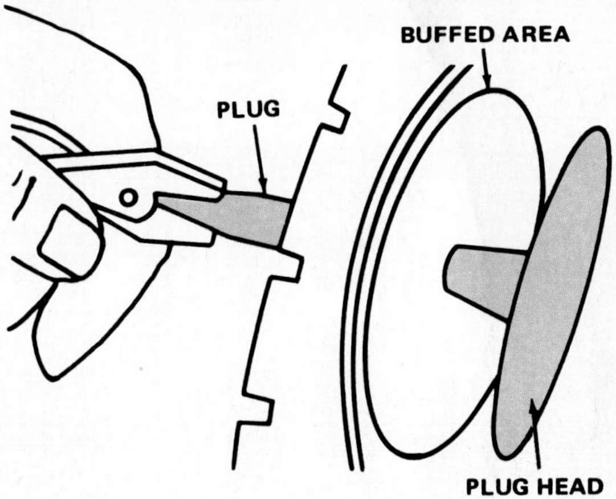

Fig. 53-26 Installing a head-type plug in a radial tubeless tire. *(Rubber Manufacturers Association)*

◆ 53-29 PLUGGING A MOUNTED TUBELESS TIRE

A temporary repair of a small puncture can be made with the tire still mounted on the rim. However, this repair is only a temporary fix. At the first opportunity, the tire must be removed from the rim and repaired from the inside, as explained in ♦53-30.

Remove the puncturing object, and clean the hole with a rasp. Apply the special vulcanizing fluid (supplied with the repair kit) to the outside of the hole. Push the snout of the vulcanizing-fluid can into the hole to get fluid inside the tire.

There are different kinds of rubber plugs. One kind is installed with a plug needle. To use this plug, first cover the hole with vulcanizing fluid. Then select a plug of the right size for the hole. The plug should be at least twice the diameter of the hole. Roll the small end of the plug into the eye of the needle. Dip the plug into vulcanizing fluid. Push the needle and plug through the hole (Fig. 53-28). Then pull the needle out. Trim off the plug ⅛ inch [3.2 mm] above the tire surface. Check for leakage. If there is no leakage, the tire can be inflated and installed on the car.

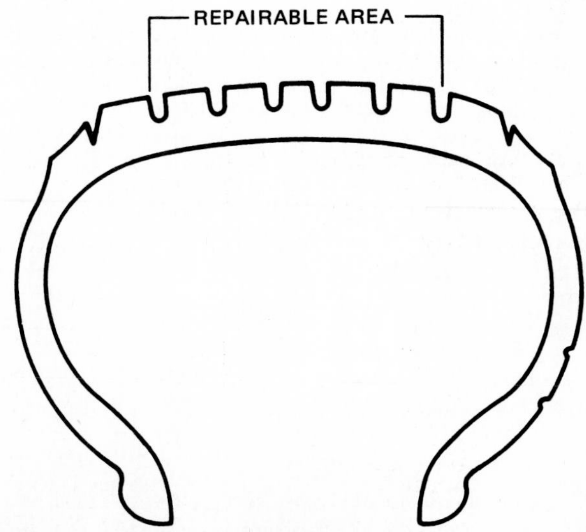

Fig. 53-27 The area of a tire in which a puncture can safely be patched. *(Chrysler Corporation)*

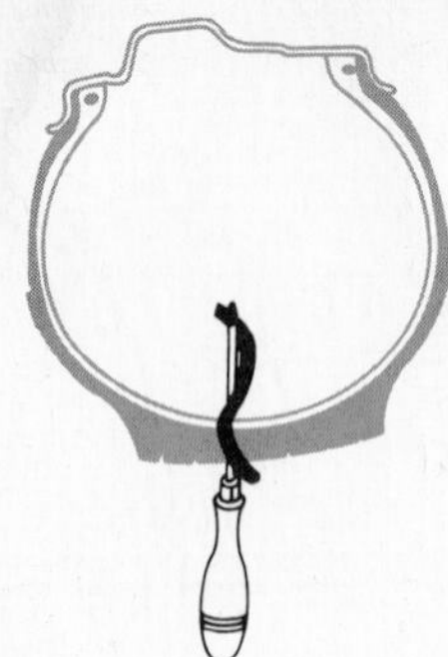

Fig. 53-28 A tire cut away to show a needle being used to insert the rubber plug into a hole in the tire. *(ATW)*

♦ 53-30 REPAIRING A DEMOUNTED TUBELESS TIRE

There are three methods of repairing holes in tires: the rubber-plug method, the cold-patch method, and the hot-patch method. Permanent repairs are made from inside the tire—with the tire off the rim.

1. Rubber-Plug Method Rubber plugs can be used in the same way as explained in ♦53-28. The repair is made from inside the tire and the inside area around the puncture is buffed and cleaned. Then the plug is installed from inside the tire (♦53-28).

2. Cold-Patch Method In the cold-patch method, first clean and buff the inside around the puncture. Then pour a small amount of vulcanizing fluid around the hole. Allow it to dry for 5 minutes. Next, remove the backing on the patch, and place the patch over the puncture. Stitch it down with the stitching tool. Start stitching at the center and work out, making sure to stitch down the edges.

Careful Dirt must not get on the fluid or patch during the repair job. Dirt could allow leakage.

3. Hot-Patch Method The hot-patch method is very similar to the cold-patch method. The difference is that heat is applied after the patch has been put into place over the area. This is done by lighting the patch with a match, or using with an electric hot plate, according to the type of patch being used.

After the repair job is done, mount the tire on the rim. Inflate the tire and test it for leakage (♦53-28).

♦ 53-31 REPAIRING DEMOUNTED TUBE-TYPE TIRE

If a tire that uses a tube has a small hole, clean out the hole. Repair the tube so it will hold air. No repair to the tire is necessary. However, if the hole is ¼ inch [6.3 mm] or larger, it should be repaired with a patch on the inside. This prevents dirt or water from working in between the tire and causing tube failure.

♦ 53-32 TIRE RETREADING

Tire retreading, or "recapping," is a specialized process that involves applying new tread material to the old casing and vulcanizing it into place. Only casings that are in good condition should be recapped. Recapping cannot repair a casing with broken or separated plies or other damage. Recapping requires special equipment.

The tire is cleaned, and the tread area is roughened by rasping it or buffing it on a wire wheel. Then a strip of new rubber tread, called *camelback*, is placed around the tread. The casing with the camelback then goes into the recapping machine. The machine is clamped shut, and heat is applied for the specified time. This vulcanizes the new tread onto the old casing.

♦ 53-33 BALANCING WHEEL-AND-TIRE ASSEMBLY

After a tire change or repair, the tire-and-wheel assembly should be checked for balance. Wheel balancing on and off the car is covered in ♦50-19.

♦ REVIEW QUESTIONS ♦

Select the *one* correct, best, or most probable answer to each question. You can find the answers in the section indicated at the end of each question.

1. One purpose of tires is to (♦53-1)
   a. grip the road and provide good traction
   b. substitute for springs
   c. act as brakes
   d. none of the above
2. Two general types of tires are (♦53-2)
   a. tube type and tubeless
   b. solid and tubeless
   c. air and pneumatic
   d. split-rim and drop-center
3. Vulcanizing means (♦53-2)
   a. heating rubber under pressure
   b. spraying with a special paint
   c. melting rubber while stirring it
   d. none of the above
4. Bias-ply tires have (♦53-2)
   a. all plies running parallel to one another
   b. belts of steel mesh in the tires
   c. one ply layer that runs diagonally one way and another layer runs diagonally the other way
   d. all of the above
5. In radial tires (♦53-3)
   a. one ply layer runs diagonally one way, and the other layer runs diagonally the other way
   b. all plies run parallel to one another and vertical to the tire bead
   c. inner tubes are always used
   d. none of the above

CHAPTER 54
Ventilation, Heating, and Air-Conditioning Systems

After studying this chapter, you should be able to:

1. Explain the need for ventilating the passenger compartment.
2. Describe the operation of the heater.
3. Name the basic parts and explain the operation of the refrigeration system.
4. Explain the difference between a thermostatic expansion valve and an orifice valve.
5. Describe the purpose and operation of the suction-throttling valve.
6. Explain the difference between manually controlled and automatic heater–air-conditioner systems.

♦ 54-1 VENTILATING THE PASSENGER COMPARTMENT

For health and comfort, fresh air must be allowed to pass through the passenger compartment, replacing stale and smoke-filled air inside the car. This process is called *ventilation*. There are two methods, uncontrolled and controlled.

Uncontrolled ventilation occurs when windows are opened. Controlled ventilation is of two types, ram-air and power. In the ram-air system (Fig. 54-1), ducts that can be opened admit air to the passenger compartment. Forward movement of the car forces, or *rams,* air through the ducts and into the car. This basic system is used on many vehicles, including those with heaters and air conditioners.

♦ 54-2 POWER VENTILATING SYSTEM

In this system, a blower motor is added to ensure adequate circulation of outside air through the car (Fig. 54-2). The

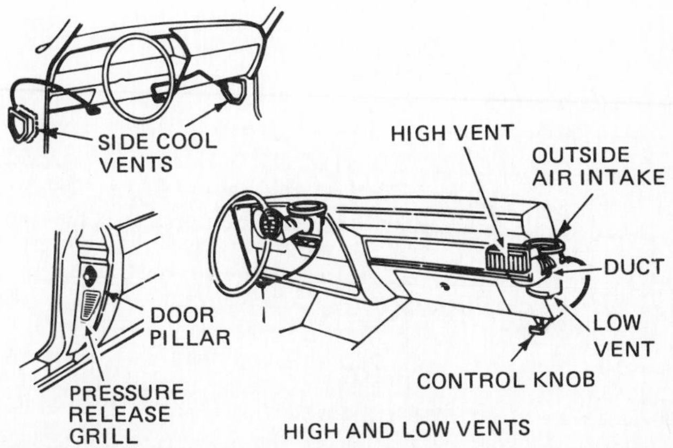

Fig. 54-1 Basic components in ram-air ventilating system. *(Ford Motor Company)*

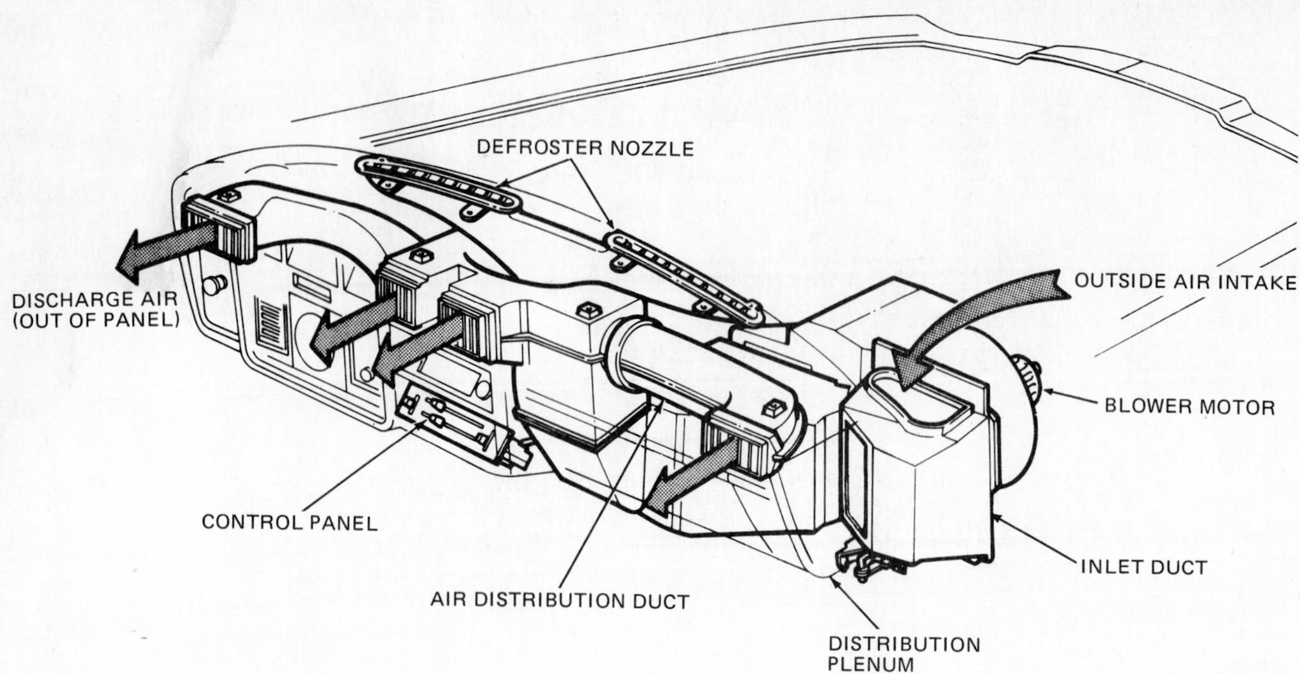

Fig. 54-2 A power ventilating system, which is usually combined with the heater and defroster. *(Ford Motor Company)*

blower motor can be operated at various speeds to suit the ventilating needs of the car occupants. Normally, the system includes a heater, and often an air-conditioner system.

♦ 54-3 HEATER

Figure 54-3 shows a car heater system. The *heater core* is a small radiator, much like the radiators in engine cooling systems (Chap. 21). Coolant from the engine cooling system flows through the heater core. This heats the heater core. Air from the outside or from the car interior flows through the heater core air passages. This warms the air. The system has three doors. The temperature door can be moved to allow more or less air to pass through the heater core. This determines how much air is heated. The air door can be moved to allow full air flow, or no air flow, or any position in between. The *defroster* door can be moved to direct heated air on the inside of the windshield, or to the heater outlet in the car.

The doors can be controlled by cables or by vacuum motors. Figure 54-4 shows a heater-control panel for a system using cables. The two levers are connected by cables to the heater-system doors. The heater-control panel for a vacuum-motor system is shown in Fig. 54-5. The levers control the operation of the doors by directing intake-manifold vacuum to the vacuum motors. This causes the motors to operate the doors to produce the desired results.

♦ 54-4 AIR-CONDITIONER FUNDAMENTALS

The air conditioner cools and dries the air in the passenger compartment. It uses a refrigerator which works in the same way as refrigerators in homes. The refrigerator works by evaporation. Evaporation is what happens when a liquid turns to gas, or vapor. Put a little water on your hand. Blow on your hand. Your hand feels cold because, as the water evaporates, it takes heat from your hand.

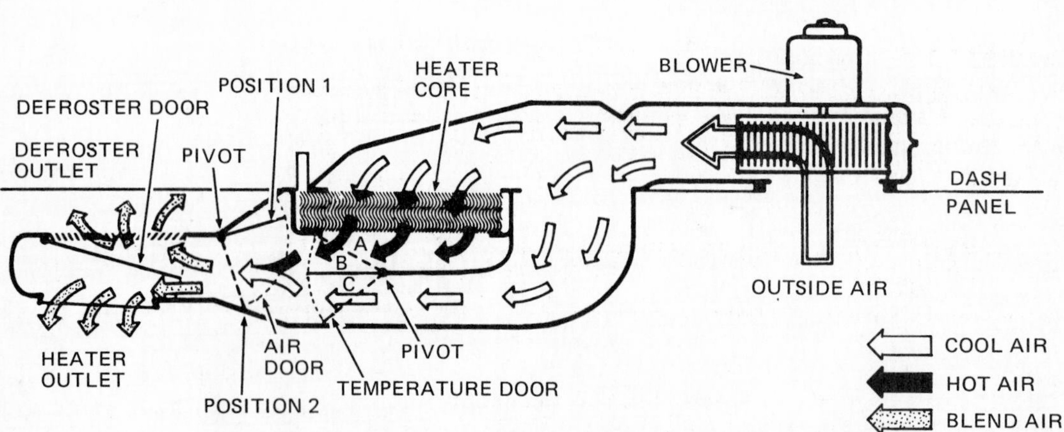

Fig. 54-3 Three doors that control the airflow through an automotive heater. *(Chevrolet Motor Division of General Motors Corporation)*

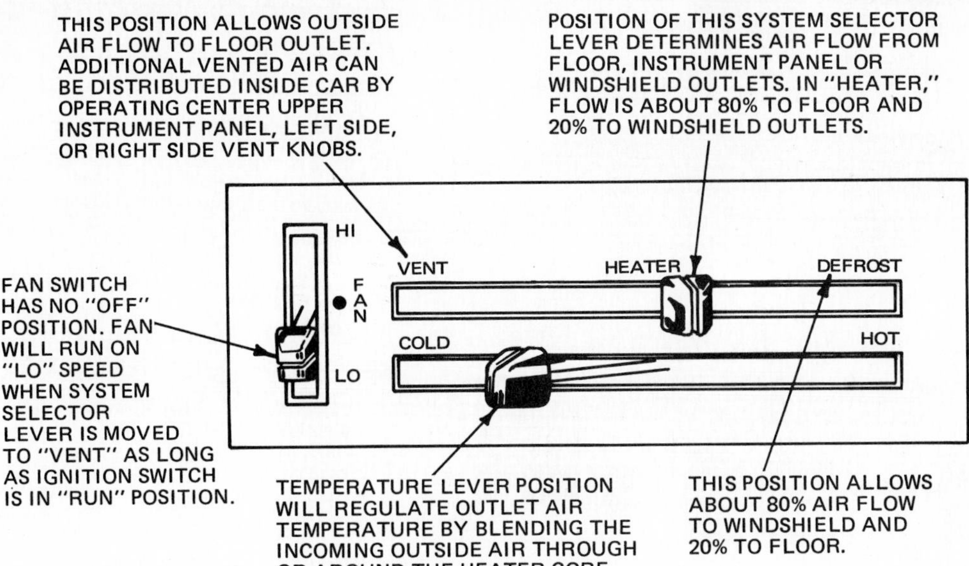
THIS POSITION ALLOWS OUTSIDE AIR FLOW TO FLOOR OUTLET. ADDITIONAL VENTED AIR CAN BE DISTRIBUTED INSIDE CAR BY OPERATING CENTER UPPER INSTRUMENT PANEL, LEFT SIDE, OR RIGHT SIDE VENT KNOBS.
POSITION OF THIS SYSTEM SELECTOR LEVER DETERMINES AIR FLOW FROM FLOOR, INSTRUMENT PANEL OR WINDSHIELD OUTLETS. IN "HEATER," FLOW IS ABOUT 80% TO FLOOR AND 20% TO WINDSHIELD OUTLETS.
HI
VENT
HEATER
DEFROST
FAN
COLD
HOT
LO
FAN SWITCH HAS NO "OFF" POSITION. FAN WILL RUN ON "LO" SPEED WHEN SYSTEM SELECTOR LEVER IS MOVED TO "VENT" AS LONG AS IGNITION SWITCH IS IN "RUN" POSITION.
TEMPERATURE LEVER POSITION WILL REGULATE OUTLET AIR TEMPERATURE BY BLENDING THE INCOMING OUTSIDE AIR THROUGH OR AROUND THE HEATER CORE.
THIS POSITION ALLOWS ABOUT 80% AIR FLOW TO WINDSHIELD AND 20% TO FLOOR.

Fig. 54-4 Heater control panel which is located on the car instrument panel. *(Chevrolet Motor Division of General Motors Corporation)*

The basic refrigeration system used in air conditioners is shown in Fig. 54-6. The compressor is driven by a belt from the engine crankshaft pulley. The compressor circulates a special fluid, called *refrigerant,* through the system. When the system is operating, the compressor draws low-pressure refrigerant vapor from the evaporator. Then the compressor compresses this vapor—puts it under high pressure—and sends it to the condenser.

The condenser is a heat exchanger, built much like the radiator in the engine cooling system (Chap. 21). The condenser does the same job: it gets rid of heat. When refrigerant vapor is compressed, it gets very hot.

The hot compressed vapor from the compressor enters the condenser. There, the vapor loses heat to the air passing through the air passages in the condenser. With this loss of heat, the vapor condenses or changes to a liquid. The liquid flows out of the condenser and into the receiver. The receiver is a storage place for liquid refrigerant. From there, it flows through the sight glass (if so equipped) and through the high-pressure line to the orifice tube. The sight glass is used on some cars. It allows the technician to determine if the liquid refrigerant flow is normal.

The orifice acts as a restriction to control the flow of refrigerant. The liquid refrigerant that does get through the orifice enters the evaporator. The evaporator is at low pressure, and this causes the refrigerant to evaporate. As it evaporates, it takes heat from air passing through the air passages in the evaporator. This air is from the car interior. It is cooled by passing through the evaporator.

Also, as the air is cooled, moisture in the air condenses on the outside of the evaporator. This is the same principle as moisture condensing on the outside of a glass of a cold drink. So the air is cooled and dried. Dry air feels cooler than moist air at the same temperature.

Figure 54-7 shows the air-conditioning system installed in a car. The condenser is up front, ahead of the engine-cooling-system radiator. Note that this system has a thermostatic expansion valve instead of an orifice as in Fig. 54-6.

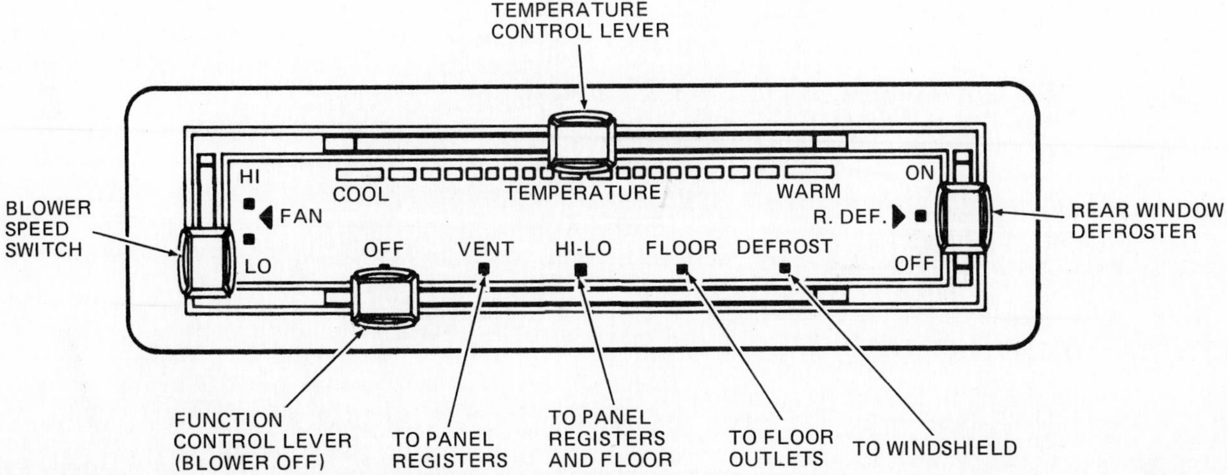
TEMPERATURE CONTROL LEVER
HI
COOL
TEMPERATURE
WARM
ON
FAN
R. DEF.
LO
OFF
VENT
HI-LO
FLOOR
DEFROST
OFF
BLOWER SPEED SWITCH
REAR WINDOW DEFROSTER
FUNCTION CONTROL LEVER (BLOWER OFF)
TO PANEL REGISTERS
TO PANEL REGISTERS AND FLOOR
TO FLOOR OUTLETS
TO WINDSHIELD

Fig. 54-5 Control panel for a heater system. *(Ford Motor Company)*

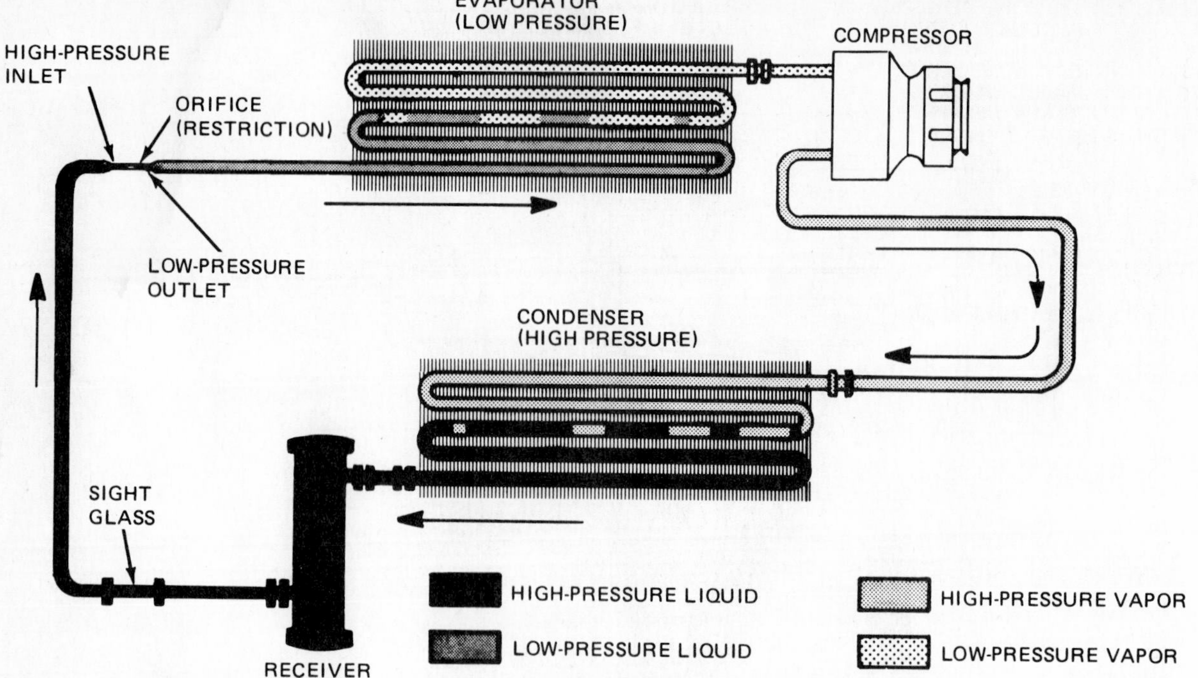

Fig. 54-6 Basic refrigeration system. Arrows show flow of refrigerant.

Figure 54-8 shows the basic refrigeration system using a thermostatic expansion valve. This valve serves the same basic purpose as the orifice shown in Fig. 54-6.

♦ 54-5 RESTRICTING REFRIGERANT FLOW

There must be a restriction between the condenser and the evaporator or else the system would not work. Without a restriction, the liquid refrigerant would flow freely and there would be no pressure difference between the condenser and evaporator. The pressure must be high in the condenser so the vapor will be hot. This allows the vapor to lose heat and condense. The pressure must be low in the evaporator so the liquid refrigerant can evaporate and remove heat from the air circulating between it and the car interior. The restriction is produced by either the orifice or the thermostatic expansion valve. Either one acts a flow-control valve.

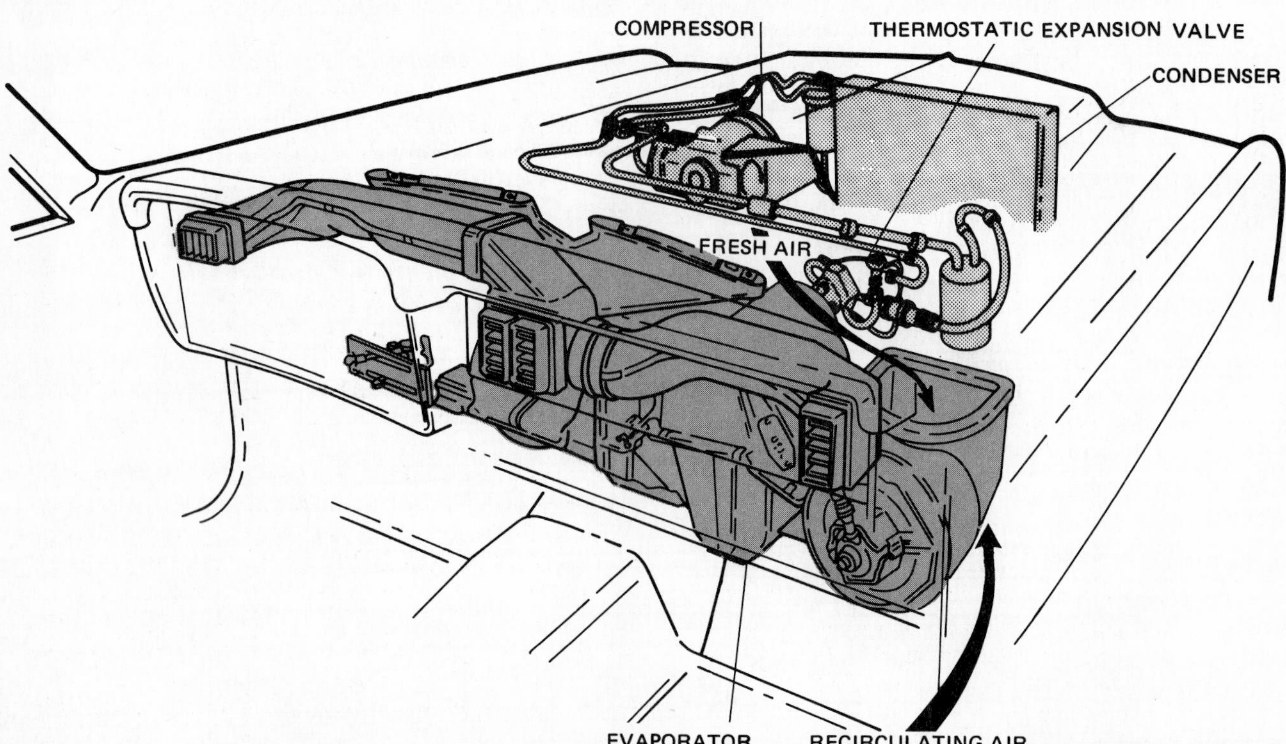

Fig. 54-7 Location of air-conditioning components in car. *(Ford Motor Company)*

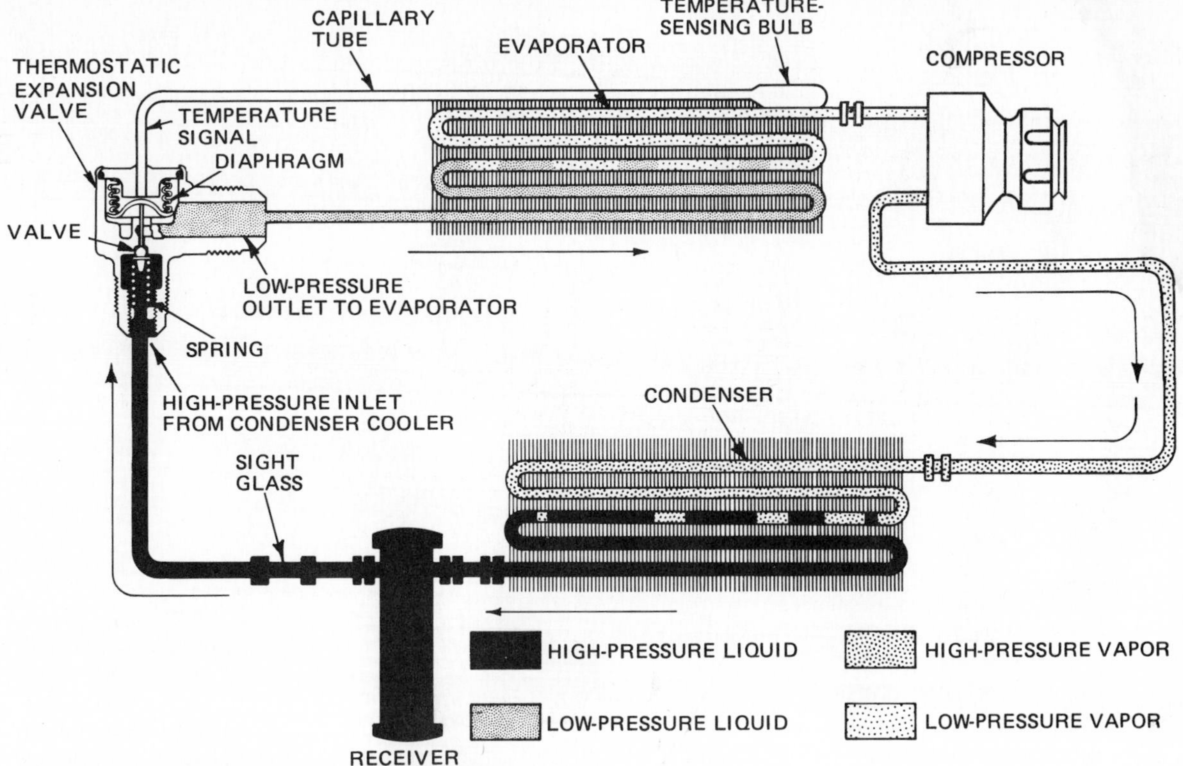

Fig. 54-8 Refrigeration system using a thermostatic expansion valve.

♦ 54-6 ORIFICE AND THERMOSTATIC EXPANSION VALVE

Both of these are types of flow-control valves. They restrict the flow of liquid refrigerant to the evaporator. The orifice is a small fixed-diameter opening through which the liquid refrigerant must flow.

The thermostatic expansion valve is a variable-orifice valve. It includes a temperature-sensing bulb which is placed near the outlet of the evaporator. The bulb senses the temperature at that point. If the temperature goes too high, it means there is too little cooling. The increased temperature causes the gas inside the temperature-sensing bulb to expand. This puts pressure on the temperature-signal diaphragm. The pressure causes the diaphragm to push the valve down. This opens the passage further, admitting more liquid refrigerant to the evaporator. As the liquid refrigerant enters, it evaporates, removing heat from the evaporator. When the temperature falls enough, the gas in the temperature-sensing bulb contracts so its pressure goes down. Then the valve spring pushes the valve closed. In operation, the valve assumes a position that admits just enough liquid refrigerant to maintain the evaporator at the proper low temperature.

♦ 54-7 ANTI-ICING CONTROLS

In addition to the restriction in the line between the condenser and evaporator, the system needs a second control. This is a control to prevent the formation of ice on the evaporator. Without such a control, the temperature of the evaporator could continue to fall until it was below the freezing temperature of water. If this happened, any moisture in the air passing through the air passages of the evaporator would condense on the evaporator coils and freeze. This would block the flow of air through the evaporator and prevent normal cooling.

There are two types of anti-icing control. In one, the control turns the compressor off if the pressure at the outlet of the evaporator falls too low. Unless the compressor is turned off, ice will start to form on the evaporator.

The second type of control uses a suction-throttling valve (Fig. 54-9). The suction-throttling valve is located in the system as shown in Fig. 54-10. It monitors the pressure of the vapor coming from the evaporator. If this pressure falls too low, the suction-throttling valve closes off the passage from the evaporator to the compressor.

In the suction-throttling valve, spring force and atmospheric pressure oppose the evaporator pressure. If the evaporator pressure rises too high, it means its temperature is going too high. The pressure overcomes the spring and atmospheric pressure and moves the piston so the valve opens. Now refrigerant vapor can flow from the evaporator to the compressor. This lowers the evaporator temperature. In operation, the piston is positioned at approximately the point which allows maximum cooling but without danger of ice forming on the evaporator.

The vacuum connection comes into operation when the driver moves the temperature-control lever to maximum cooling. This action admits engine vacuum into the chamber in back of the diaphragm. The vacuum causes the diaphragm to move (to the right in Fig. 54-10). As it moves, it carries the diaphragm rod with it. Now, the piston can move farther. This allows more refrigerant vapor to flow through the evaporator for additional cooling.

Figure 54-9 shows one type of suction-throttling valve. There are several other designs. All work in the same way. They prevent excessively low pressures and temperatures in the evaporator.

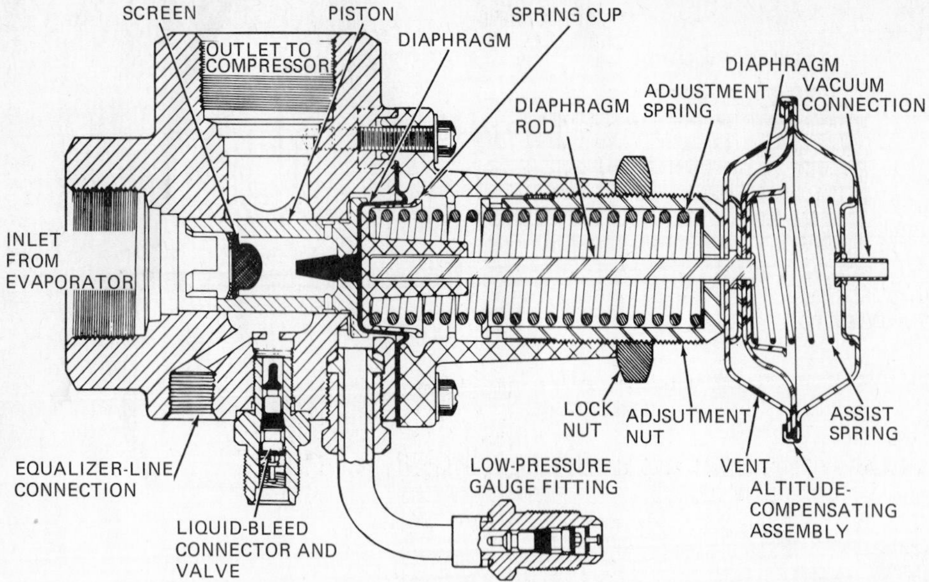

Fig. 54-9 Sectional view of a suction-throttling valve. *(Pontiac Motor Division of General Motors Corporation)*

♦ 54-8 ELECTRIC PRESSURE SWITCH

In many late-model automotive air-conditioning systems, the suction-throttling valve has been replaced with an electric pressure switch. This switch is used in systems having an orifice tube (Fig. 54-11). The switch has a diaphragm linked to a set of contact points. When the pressure applied to the diaphragm is low, the contact points are open. But when the pressure and temperature go up enough, the diaphragm closes the contact points. This connects the battery to the electric clutch on the compressor. The clutch then engages and drives the compressor. The compressor pulls refrigerant vapor from the evaporator and the evaporator pressure and temperature fall. The compressor cycles on and off to hold the evaporator temperature at the desired value.

♦ 54-9 THERMOSTATIC CYCLING SWITCH

Many systems using a cycling compressor have a thermostatic cycling switch (Fig. 54-12) instead of an electric pressure switch. The thermostatic cycling switch senses the tem-

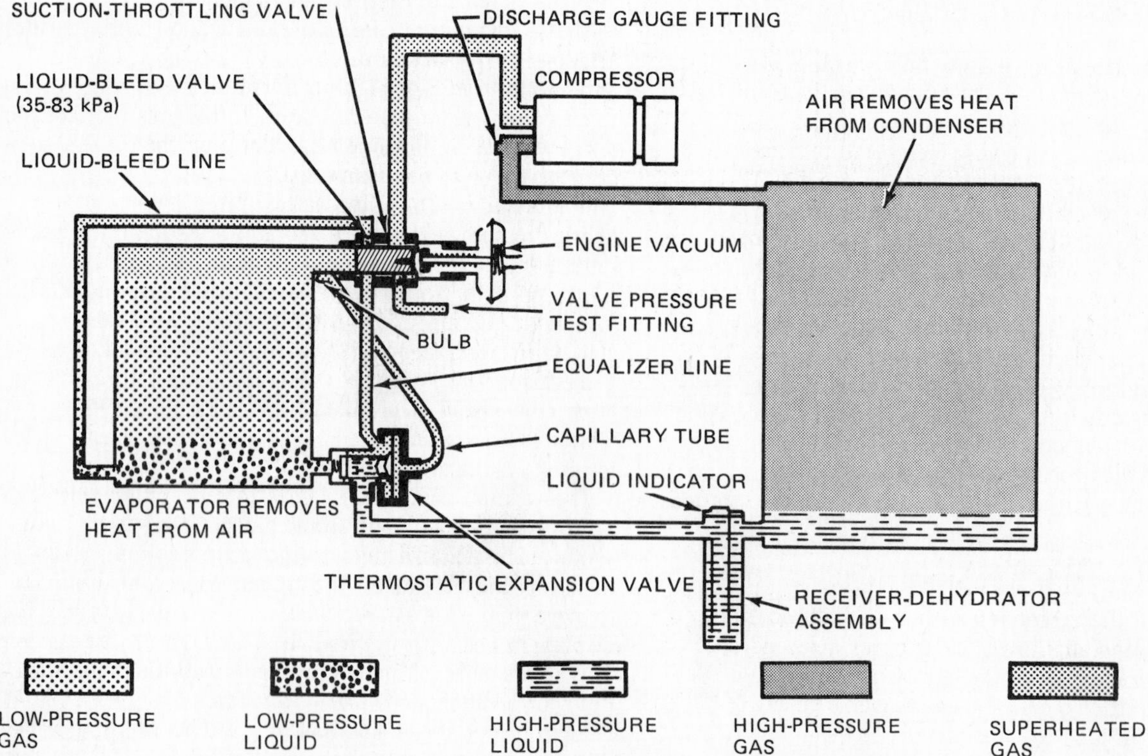

Fig. 54-10 Schematic of an air-conditioning system using a suction-throttling valve. *(Pontiac Motor Division of General Motors Corporation)*

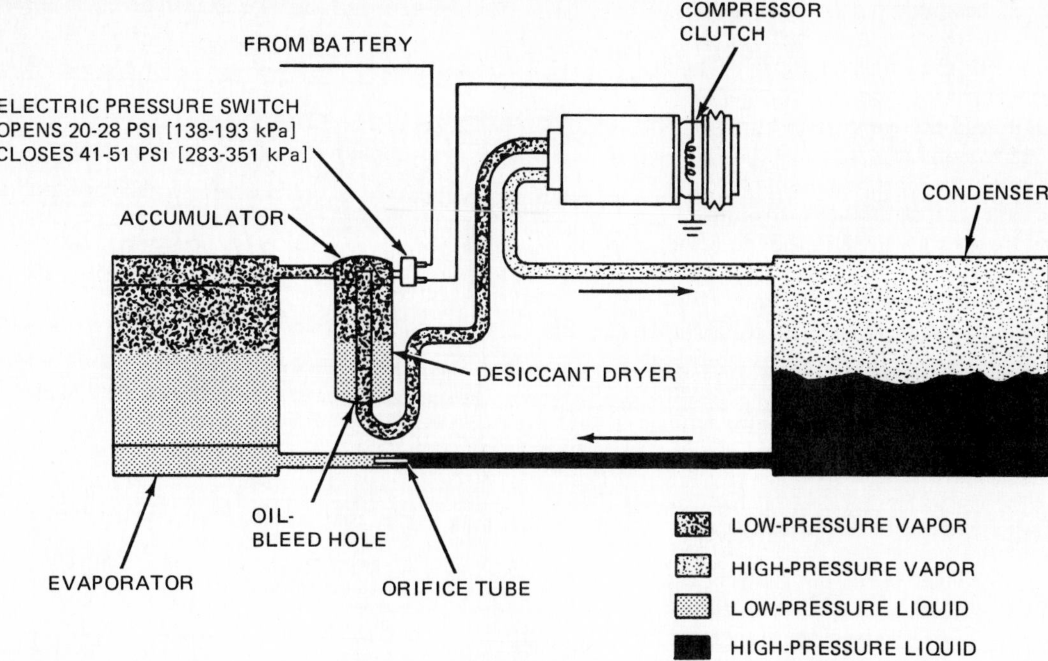

Fig. 54-11 Schematic of a cycling-clutch orifice tube (CCOT) system using an electric pressure switch to cycle the compressor clutch. *(Oldsmobile Division of General Motors Corporation)*

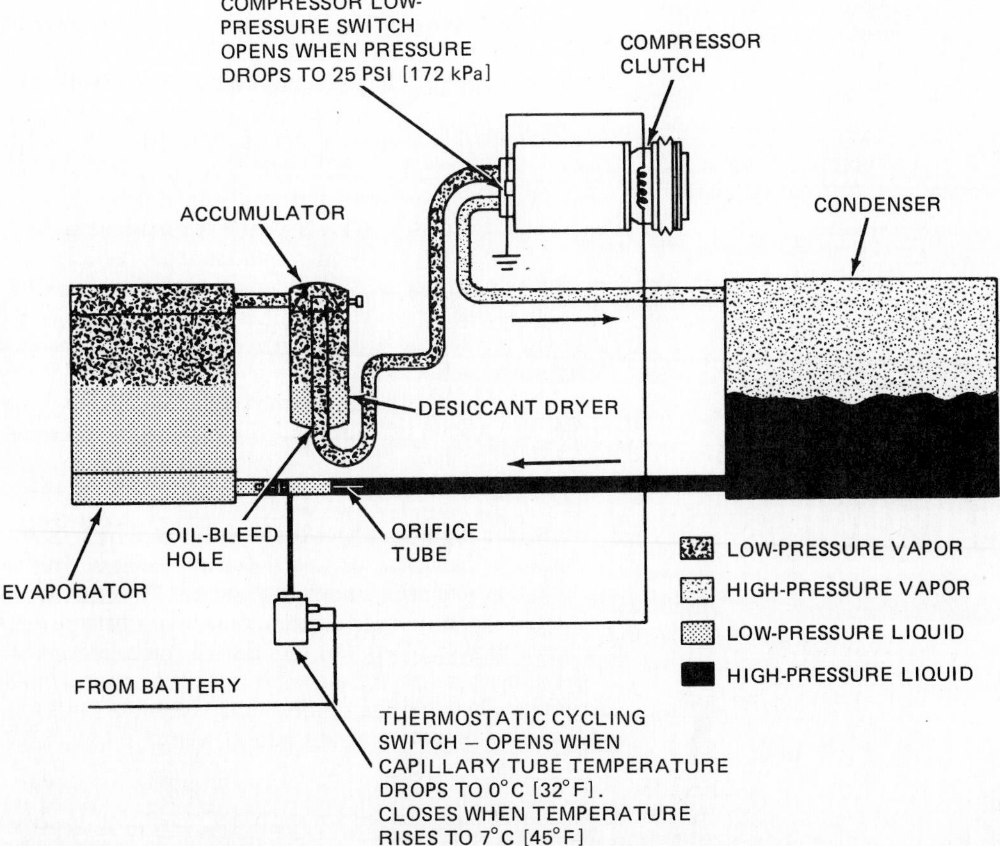

Fig. 54-12 Schematic of a CCOT system using a thermostatic cycling switch to cycle the compressor clutch. *(Oldsmobile Division of General Motors Corporation)*

perature of the refrigerant coming from the orifice tube. When the temperature falls to freezing, the switch opens the compressor-clutch circuit. This stops the compressor. When the temperature increases to several degrees above freezing, the switch closes its points and the compressor clutch engages. The compressor starts working again.

Figure 54-13 shows one type of thermostatic cycling switch. The bulb is filled with gas that expands with temperature. When the temperature goes up enough, the increased gas pressure closes the contacts. When the temperature falls, the gas pressure is reduced and the points open. Many thermostatic cycling switches do not have an external adjusting knob as shown in Fig. 54-13. Instead, they have an internal screw. The knob or screw adjusts the temperature at which the contacts open and close. Therefore, this controls the temperature range in which the evaporator operates.

♦ 54-10 CYCLING AND NONCYCLING COMPRESSORS

Some compressors run continuously unless the controls are in the OFF position. These systems are called *evaporator-pressure control systems* (Fig. 54-10). Icing of the evaporator is prevented by operation of a valve such as the suction-throttling valve.

Many automotive air conditioners use a cycling-clutch compressor (Figs. 54-11 and 54-12). Most cycling-clutch compressor systems have an orifice tube to control the flow of refrigerant from the condenser to the evaporator. When the pressure and temperature of the evaporator go up too much, the cycling switch connects the compressor clutch to the battery so that the compressor goes to work. There are two types of cycling switches. They are the electric pressure type (♦54-8) and the thermostatic cycling switch (♦54-9).

♦ 54-11 COMPRESSOR CLUTCH

Figure 54-14 shows how one type of compressor mounts on the side of the engine and is driven by a belt from the crank-

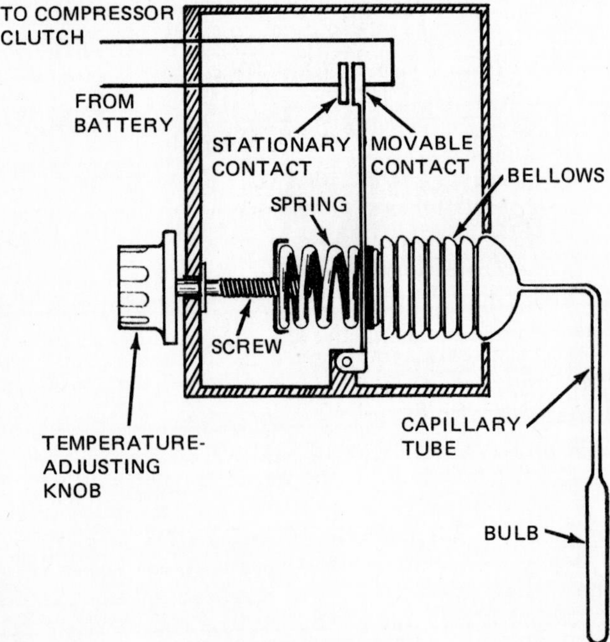

Fig. 54-13 Basic construction of a thermostatic cycling switch.

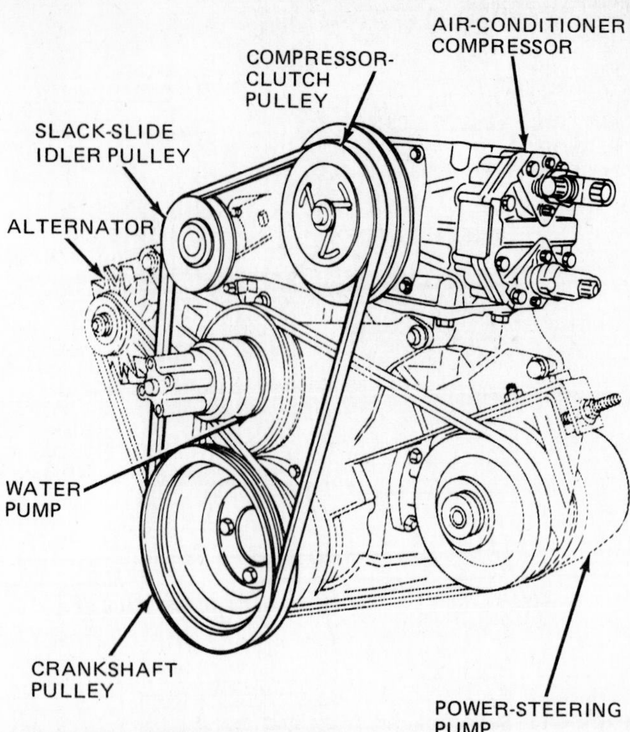

Fig. 54-14 Belt drive of the air-conditioner compressor and other belt-driven accessories. *(Ford Motor Company)*

shaft pulley. The compressor is a pump that compresses low-pressure refrigerant vapor into high-pressure refrigerant vapor and sends it to the condenser.

The compressor has a magnetic clutch (the compressor clutch). Figure 54-15 shows the magnetic clutch mounted on the compressor. The clutch includes a coil which is attached to the compressor housing. The compressor pulley is mounted on ball bearings so it can run freely when the clutch is off.

Figure 54-16 shows how the magnetic clutch works. In Fig. 54-16A, the clutch coil is disconnected from the battery. The pulley is turning freely on its ball bearings. The compressor shaft is not turning, so the compressor is not working.

In Fig. 54-16B, the clutch coil has been connected to the battery by the cycling switch. Current flow through the coil produces a magnetic field. This magnetically locks the pulley to the clutch-plate-and-hub assembly (Fig. 54-15). Now, the compressor shaft must turn and the compressor operates.

♦ 54-12 TYPES OF COMPRESSORS

A variety of compressors have been used (Fig. 54-17). Some have pistons that move up and down in cylinders. This is similar to the actions in automotive engines. The other types are rotary and they work in different ways. Some have pistons that move back and forth in cylinders. The cylinders are arranged either parallel to the compressor shaft or are set in a circle vertically to the shaft. Other rotary types are similar to some vane-type power-steering pumps.

♦ 54-13 SIGHT GLASS

A sight glass is used in some air-conditioning systems (Figs. 54-8 and 54-18). It enables the technician to observe the condition of the liquid refrigerant as it leaves the condenser. Chapter 55 explains how the sight glass is used.

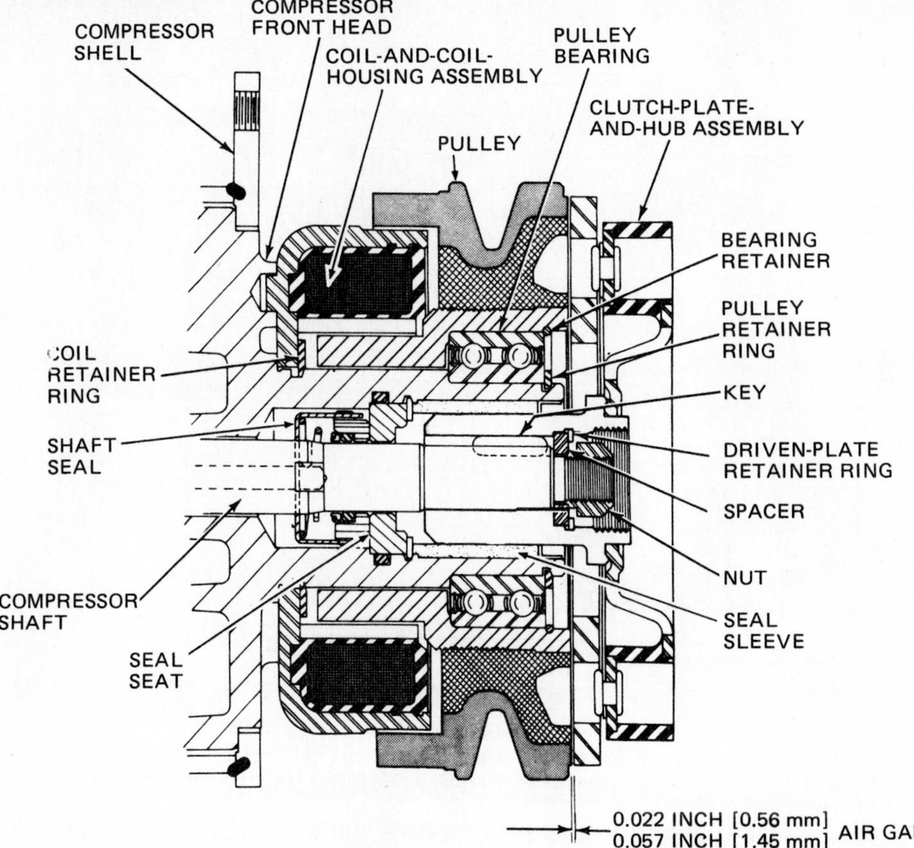

Fig. 54-15 Sectional view of magnetic clutch. *(General Motors Corporation)*

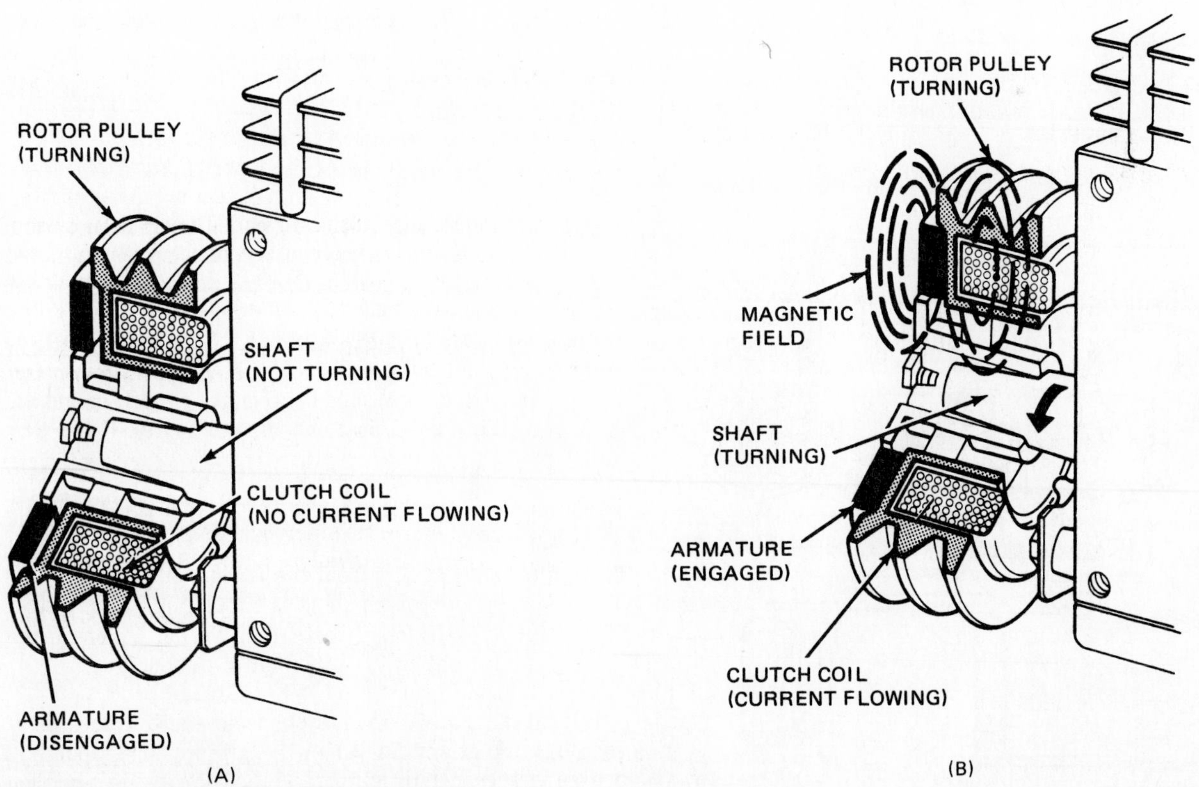

Fig. 54-16 Operation of magnetic clutch. (A) Cutaway view of magnetic clutch in the OFF position. (B) Magnetic clutch in the ON position. *(Warner Electric Brake and Clutch Company)*

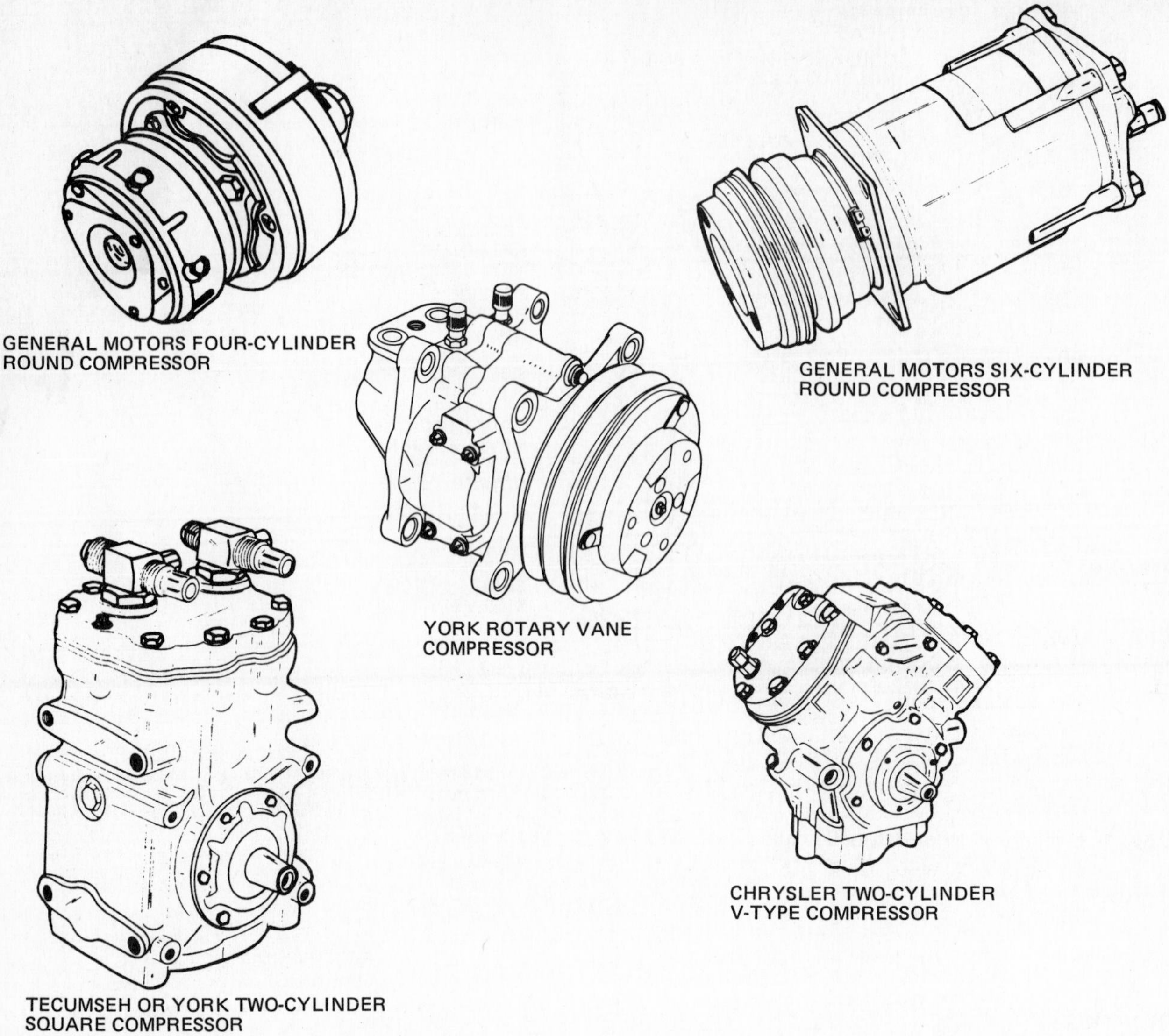

GENERAL MOTORS FOUR-CYLINDER
ROUND COMPRESSOR
GENERAL MOTORS SIX-CYLINDER
ROUND COMPRESSOR
YORK ROTARY VANE
COMPRESSOR
CHRYSLER TWO-CYLINDER
V-TYPE COMPRESSOR
TECUMSEH OR YORK TWO-CYLINDER
SQUARE COMPRESSOR

Fig. 54-17 Various types of air-conditioner compressors.

BUBBLES
SIGHT GLASS
FOAM
OIL STREAKS

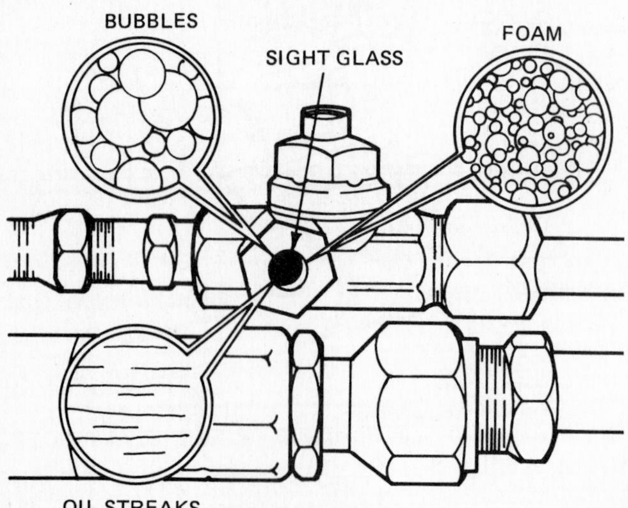

Fig. 54-18 A typical sight glass showing different conditions that can be observed. *(Ford Motor Company)*

Most late-model automotive air conditioners do not use a sight glass. On these, a gauge set is connected for diagnosis.

♦ 54-14 RECEIVERS AND ACCUMULATORS

Some refrigerant systems have a receiver-dehydrator (receiver-drier), which is located between the condenser and the evaporator (Figs. 54-6, 54-8, 54-10, and 54-19). It has four functions:

1. To receive liquid refrigerant and hold it in reserve for the evaporator
2. To filter out any dirt from the refrigerant
3. To adsorb any moisture that is circulating in the system
4. To trap any refrigerant vapor that did not liquefy in the condenser, holding the vapor until it condenses

Instead of a receiver, many systems now have an accumulator (Figs. 54-11 and 54-12) in the low-pressure vapor line between the evaporator outlet and the compressor inlet. The purpose of the accumulator (or *suction accumulator/drier*) is to prevent excessive liquid refrigerant from reaching the compressor. The accumulator allows any liquid refrigerant

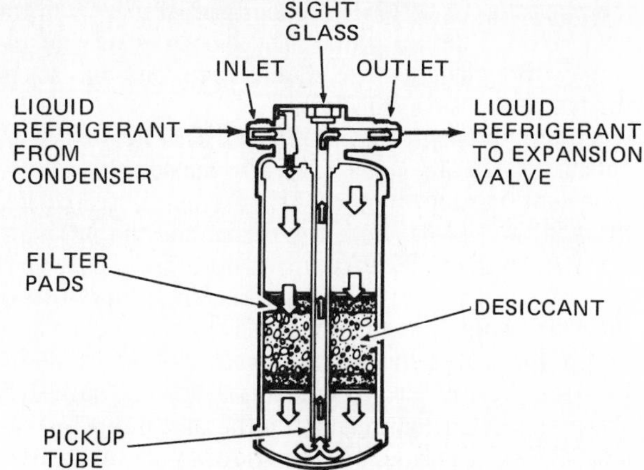

Fig. 54-19 Sectional view of a receiver-drier, which is also called a receiver-dehydrator. *(Ford Motor Company)*

leaving the evaporator to vaporize before reaching the compressor.

An oil-bleed hole near the bottom of the accumulator housing allows compressor oil and some liquid refrigerant to flow to the compressor at a controlled rate. A fine-mesh filter screen surrounds the bottom of the pickup tube. This prevents any contamination from plugging the bleed hole. The accumulator also has a desiccant bag to adsorb any moisture that is circulating in the system.

♦ 54-15 SAFETY DEVICES

Several safety devices are used in automotive air conditioners to protect the components if anything goes wrong. Some systems include all of these devices.

1. Ambient Switch This switch (Fig. 54-20) is located in the air-inlet duct. It senses the temperature of the air entering from outside the car. If this temperature falls below 32°F [0°C], it opens so the compressor clutch will not engage. Air conditioning is not required. Also, operation at low temperatures could damage the compressor seals.

2. Low-Pressure Cutoff Switch This switch senses evaporator pressure and disengages the compressor clutch if the pressure drops too low. A very low evaporator pressure usually means that the system has lost refrigerant. This means that refrigerant oil has also been lost. Oil loss could damage the compressor.

3. High-Pressure Relief Valve This valve opens if the pressure in the system goes too high. This might happen if the air flow through the condenser is blocked with leaves or other trash. Also, high pressure could occur if the system were overcharged during service.

4. Thermal Limiter and Superheat Switch This switch is open normally. If the system loses some or all refrigerant, the switch senses the low pressure and high refrigerant temperature. Its contacts close and the action overheats the fuse so that it blows. This stops the compressor.

♦ 54-16 REFRIGERANT

Refrigerant is a very special substance. It is a liquid at very low temperatures and a vapor at fairly low temperatures. The refrigerant used in automotive air conditioners boils at

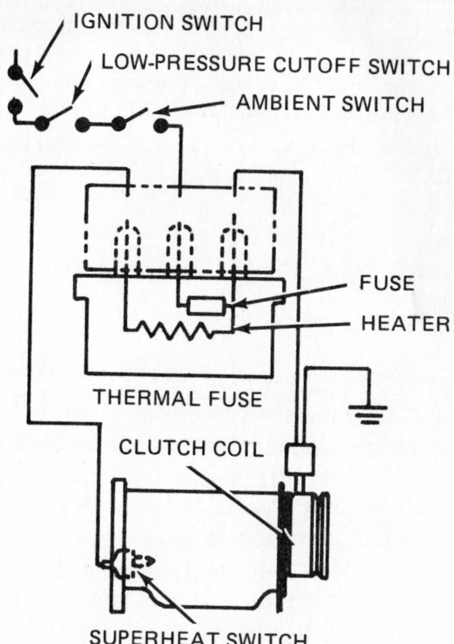

Fig. 54-20 Wiring system of the compressor clutch, showing the safety devices. *(General Motors Corporation)*

−22°F [−30°C] at atmospheric pressure. The ideal refrigerant must not be toxic or poisonous, flammable, or explosive. It must be reasonably inert and must not cause damage to any metal, rubber, or other substance it touches. The refrigerant that has all these desirable qualities is a special chemical. It is called dichlorodifluoromethane. It has the chemical formula CCl_2F_2. It is commonly known as Freon-12, Refrigerant-12, R-12, or simply Freon. However, Freon is actually the trademark of the manufacturer, Du Pont.

♦ CAUTION ♦ Always wear eye protection when handling refrigerant or servicing the automotive air conditioner. Although R-12 is safe if handled properly, it is dangerous when handled carelessly or improperly. If you get R-12 on your hand, it will freeze the flesh, be very painful, and may cause permanent injury. If you get R-12 in your eye, it can freeze the eye and cause loss of sight. Also, if R-12 comes in contact with a flame, it turns into poisonous phosgene gas. However, R-12 is a safe refrigerant if handled properly. Do not actually work with automotive air conditioners and refrigerant until you learn how to handle refrigerant. There is more on safety in the following chapter on automotive air-conditioning service.

♦ 54-17 REFRIGERANT OIL

The air-conditioning system needs oil to keep its moving parts and seals lubricated. Because there is no provision for adding oil to some compressors, it must be put in when the system is first assembled. Refrigerant oil is nonfoaming and highly refined with all impurities such as wax, moisture, and sulfur removed.

Some compressors have a sump in which a reserve of oil is kept. Oil is pumped from this sump to lubricate the compressor. Some of the oil mixes with the refrigerant and circulates in the system. Therefore, there is always oil circulating. When a component is replaced in the system, it is necessary to measure how much oil the old part has in it. You then put that same amount of new oil back into the system when the new component is installed.

Refrigerant oil is supplied in special airtight containers. The oil rapidly picks up moisture from the air if the container is left open. Therefore, the container should be opened only as long as is necessary to remove the amount of oil required. The following chapter explains how to add oil when a new component is installed in the system.

HEATER–AIR-CONDITIONER SYSTEMS

♦ 54-18 MANUALLY CONTROLLED AND AUTOMATIC SYSTEMS

Some heater–air-conditioner systems are manually controlled. The driver selects the desired mode by movement of the function-control lever. Desired temperature is selected by moving the temperature-control lever. In the automatic system, the driver can select automatic control and the desired temperature. Then the system will maintain that temperature, providing heat or cooling as required.

♦ 54-19 MANUAL CONTROL

Figure 54-21 shows the control panel for a manually controlled heater–air-conditioner system. Figure 52-22 shows the layout of the complete system with the air distribution patterns. This system has a fixed orifice and compressor-clutch-cycling electric pressure switch (♦54-8).

Figure 54-23 shows the system schematically. The temperature-blend door is controlled by a cable from the temperature-control lever (Fig. 54-21). If the temperature is not what the driver wants, the control lever must be moved manually. This repositions the temperature-blend door.

The other doors are controlled by vacuum motors. The vacuum motors are operated by the function-control lever (Fig. 54-21). When this lever is moved, it connects the different vacuum motors to the engine intake manifold. Manifold vacuum then causes the vacuum motors to move the appropriate doors. The operating modes are as follows:

1. In OFF, the outside-recirculate door is in the recirculating air position. No outside air enters. The panel doors are in the no-vacuum position. The floor-defrost door is in the floor (full-vacuum) position. The blower is off and the temperature-blend door maintains the position as set by the temperature-control lever.
2. When the function-control lever is moved to MAX A/C, the outside-recirculate door is closed to outside air (vacuum applied to vacuum motor). The car air is recirculated through the system. Both the partial and the full panel doors are in the panel (vacuum) position and the air flows out of the instrument-panel registers. The temperature of the recirculating air can be changed by changing the position of the temperature-control lever.
3. In NORM A/C, the outside-recirculate door is opened to admit outside air (no vacuum on the vacuum motor). All other doors are positioned the same as in the MAX A/C position. The temperature of the air is controlled in the same way, by the position of the temperature-control lever and the blend door.
4. In the VENT position, outside air is admitted (no vacuum on the outside-recirculate-door vacuum motor). The air passes through the system and exits from the instrument-panel registers. It is not cooled because the compressor is not operating. Movement of the function-control lever opens a switch that disconnects the compressor clutch from the battery.
5. In the HI-LO HEAT position, operation is the same as in the VENT position, except that the airflow is divided between the panel registers and the floor ducts. The partial panel door is in the panel position (vacuum). The full panel door (no vacuum) routes some of the air toward the floor-defrost door. This door is positioned by the vacuum on the vacuum motor to send the rest of the airflow toward the floor area.
6. In the FLOOR HEAT position, the airflow is directed to the floor outlets with a small amount directed upward through the defrost outlets to the windshield. The compressor is not operating, so there is no cooling.
7. In the DEFROST position, all air is directed to the defroster nozzles with a small air bleed to the floor. Maximum defrosting is obtained by moving the temperature-control lever to the highest temperature. The compressor operates to dehumidify the air and reduce windshield fogging.

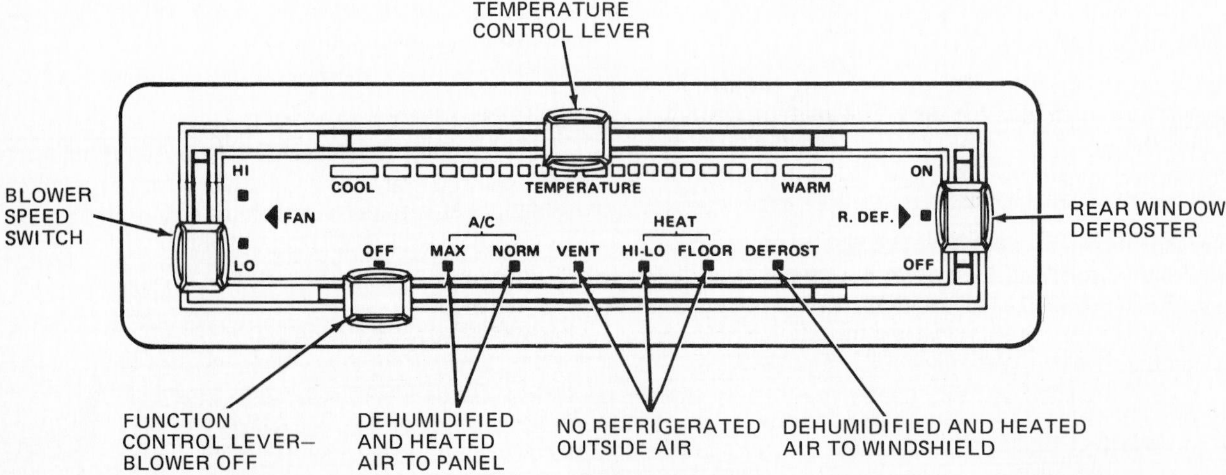

Fig. 54-21 Control panel for a manual air-conditioner–heater system. *(Ford Motor Company)*

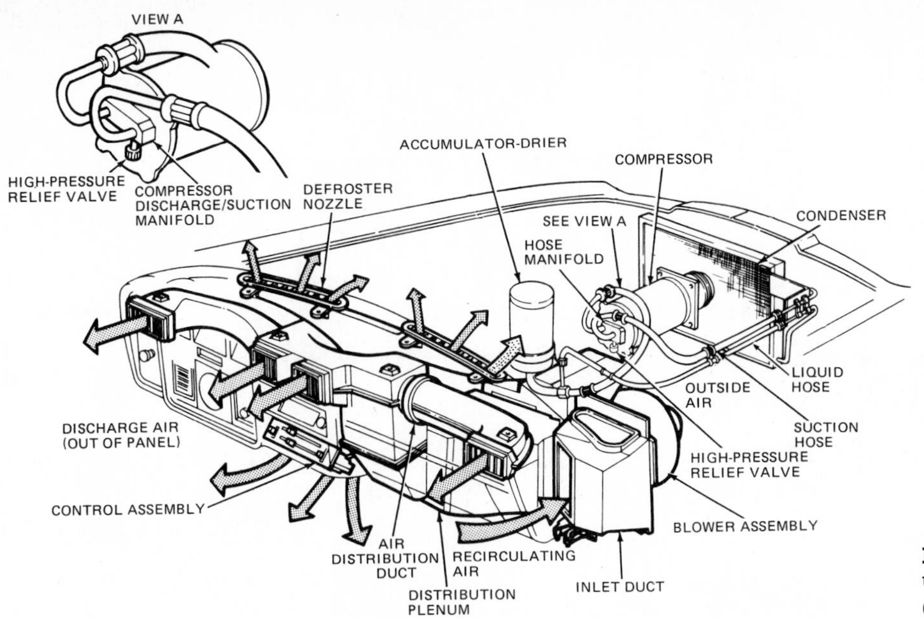

Fig. 54-22 Airflow patterns in manual air-conditioner–heater system. *(Ford Motor Company)*

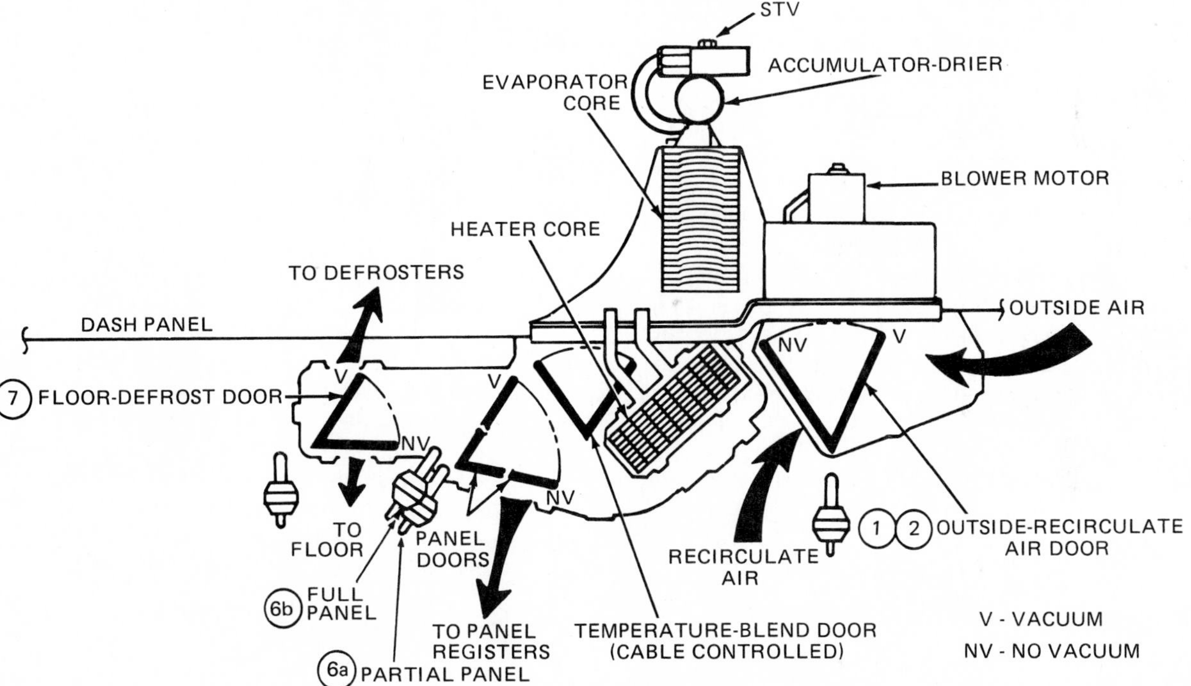

FORD AND MERCURY
MANUAL A/C — HEATER SYSTEM VACUUM MOTOR TEST CHART

FUNCTION CONTROL LEVER POSITION		VACUUM MOTORS APPLIED WITH VACUUM			
		OUTSIDE — RECIRC.	PANEL		FLOOR-DEFROST
			PARTIAL	FULL	
OFF		1-2	—	—	7
A/C	MAX *	1-2	6a	6b	7
	NORM *	—	6a	6b	7
VENT		—	6a	6b	7
HEAT	HI-LO	—	6a	—	7
	FLOOR	—	—	—	7
DEFROST *		—	—	—	—
VACUUM LINE COLOR CODE		WHITE	ORANGE	BLUE	YELLOW

— No Vacuum *Compressor clutch engaged when ambient cut-off switch is closed.

Fig. 54-23 Airflow in manual air-conditioner–heater systems in different operating modes. Chart shows which vacuum motors are applied. *(Ford Motor Company)*

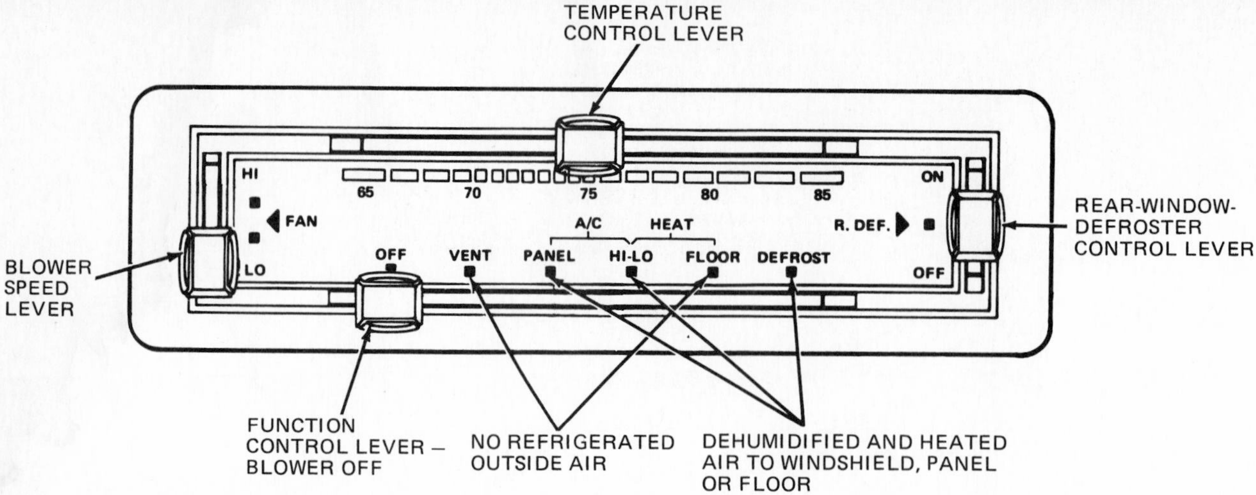

Fig. 54-24 Control panel for an automatically controlled air-conditioner–heater system. *(Ford Motor Company)*

♦ 54-20 AUTOMATIC CONTROL

The basic difference between the manually controlled system (♦54-19) and the automatic system is that the latter automatically maintains the preset temperature. In the automatic system (Figs. 54-24 to 54-26), the temperature-blend door is automatically positioned to produce the temperature preselected by the driver. Figure 54-24 shows the control panel. Figure 54-25 shows the system as it appears on the car. Figure 54-26 shows the system schematically.

All of the doors are positioned by vacuum motors except for the temperature-blend door. This door is positioned by a servomotor which uses modulated (varied) vacuum. The system has a temperature sensor which senses the inside air temperature. If it is not at the setting of the temperature-control lever, the sensor changes the amount of vacuum applied to the servomotor. If the temperature is too high, more vacuum is applied to the servomotor. This moves the temperature-blend door to pass more cold air from the evaporator core. The operating modes of the system shown in Figs. 54-24 to 54-26 are:

1. OFF In the OFF position, the outside-recirculating air door is in the recirculating-air position. The panels are in the no-vacuum position and the floor-defrost door is in the floor position.

2. VENT In the VENT position, outside air is directed through the panel registers. The compressor is not running.

3. PANEL In the PANEL position, with the car interior temperature considerably above the temperature setting, the modulated vacuum is high at the servomotor. This causes the

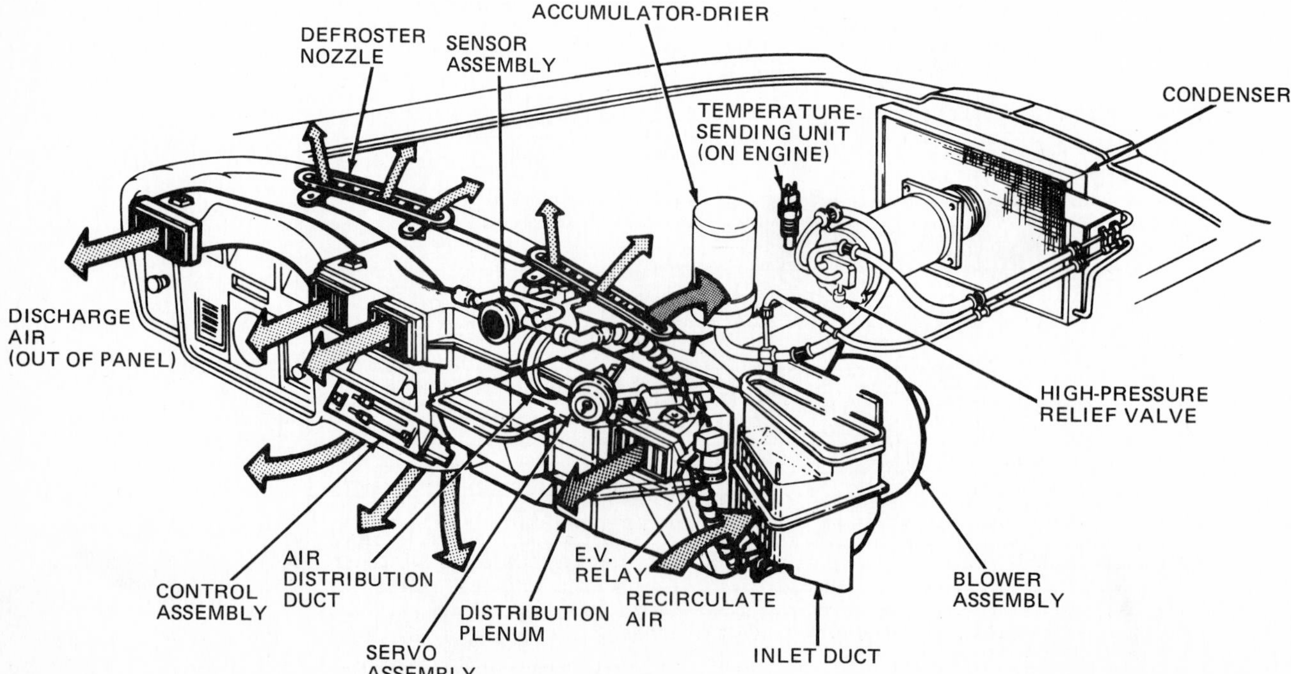

Fig. 54-25 Automatic temperature-control system, showing airflow in different operating modes. *(Ford Motor Company)*

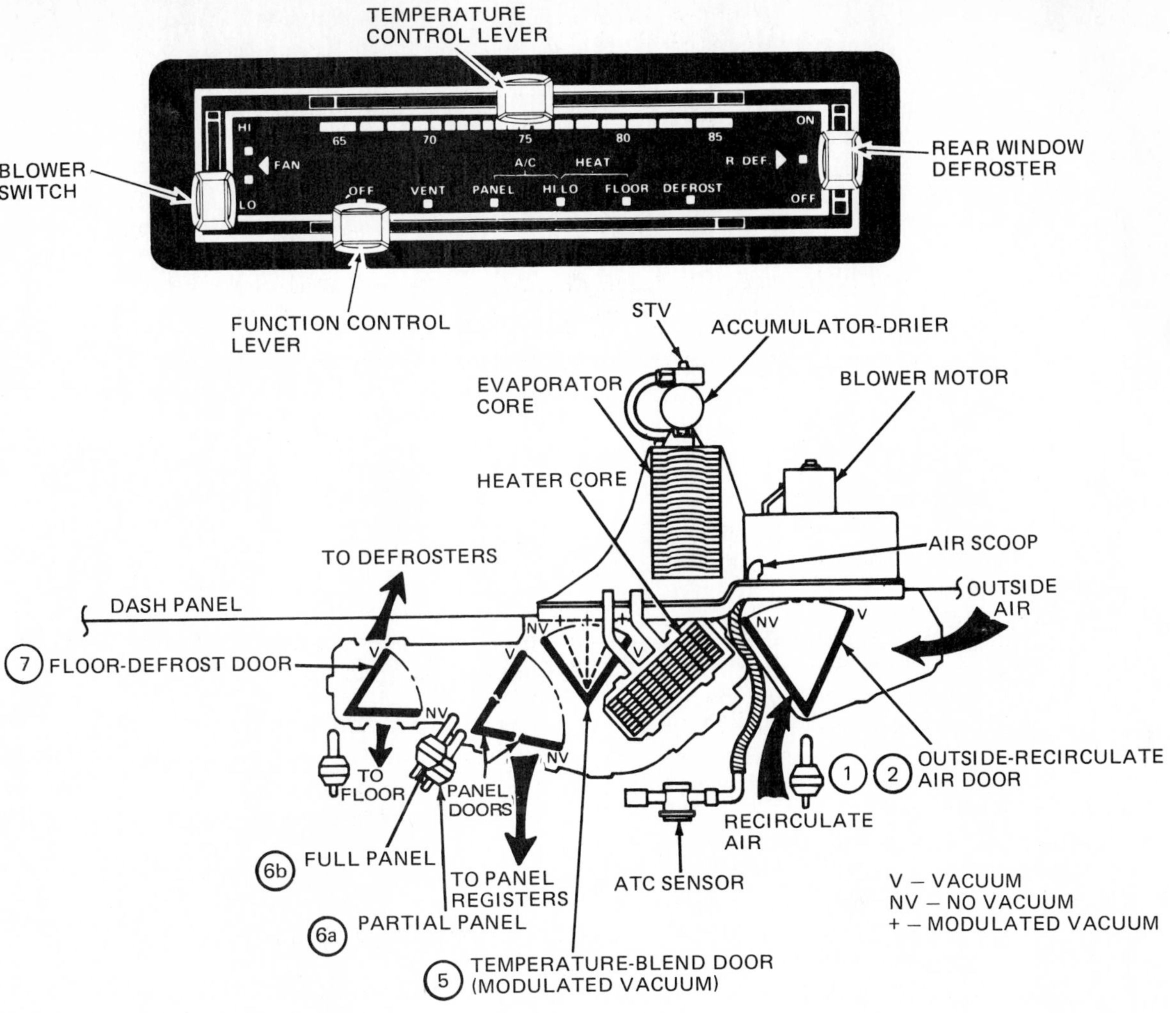

AUTOMATIC TEMPERATURE CONTROL SYSTEM VACUUM MOTOR TEST CHART

	FUNCTION CONTROL LEVER POSITION	ATC REQUIREMENT	VACUUM MOTORS APPLIED WITH VACUUM: OUTSIDE-RECIRC.	TEMP. BLEND +	PANEL: PARTIAL	PANEL: FULL	FLOOR-DEFROST
	OFF	–	1-2	+	–	–	7
	VENT	COOL	–	5	6a	6b	7
		MID	–	+	6a	6b	7
		WARM	–	–	6a	6b	7
A/C	PANEL*	COOL	1-2**	5	6a	6b	7
		MID	–	+	6a	6b	7
		WARM	–	–	6a	6b	7
	HI-LO*	COOL	–	5	6a	–	7
HEAT		MID	–	+	6a	–	7
		WARM	–	–	6a	–	7
	FLOOR	COOL	–	5	–	–	7
		MID	–	+	–	–	7
		WARM	1-2 †	–	–	–	7
	DEFROST*	COOL	–	5	–	–	–
		MID	–	+	–	–	–
		WARM	–	–	–	–	–
	VACUUM LINE COLOR CODE		WHITE	TAN	ORANGE	BLUE	YELLOW

* COMPRESSOR CLUTCH ENGAGED WHEN AMBIENT CUT OFF SWITCH IS CLOSED
† DELAYED BY ENGINE TEMP. SENDER AND E.V. RELAY UNTIL 120°F. ENGINE COOLANT TEMP.
** DEPENDENT ON HEATING/COOLING REQUIREMENT OF SERVO DIVERTER VALVE
+ MODULATED VACUUM
– NO VACUUM

Fig. 54-26 Automatic temperature-control system, showing operating conditions in different modes. *(Ford Motor Company)*

outside-recirculating door to move to the recirculating-air position. No outside air is taken in. The temperature-blend door moves to the maximum A/C position. All air bypasses the heater core. The car air is recirculated and cooled. The speed at which it circulates depends on the blower-speed-switch position.

As the interior temperature falls and nears the temperature-control-lever setting, the servomotor moves the temperature-blend door near a midposition. Now part of the cooled and dehydrated air is sent through the heater core. If the in-car temperature drops below the temperature setting, the blend door moves to admit more heated air and less cooled air. If the in-car temperature goes above the temperature setting, the blend door moves to admit more cooled air and less heated air. This is an automatic action to maintain the temperature at the preset level.

4. HI-LO In the HI-LO position, the operation is the same as in the PANEL position, except that the airflow is divided between the panel registers and the floor ducts. The system does not recirculate the interior air.

5. FLOOR In the FLOOR position, with the car interior temperature below the temperature setting, the temperature-blend door opens fully. Blower operation is delayed until the engine coolant reaches a temperature of about 120°F [49°C]. Then the blower comes on and all air circulates through the heater core. The compressor is not operating, so no cooling takes place. The air exits through the floor ducts. Some of the air goes up to the defroster vents.

When less heating is required to maintain the temperature setting, the temperature-blend door moves to a midposition. The outside-recirculating door opens to admit outside air. The blend of heated and outside air exits through the floor ducts and defroster vents, as before.

6. DEFROST In the DEFROST position, the panel doors and the floor-defrost door are in the full-defrost position (no vacuum). All treated air is directed to the defroster vents. A little bleeds through the floor ducts. The temperature-blend door moves in this position, but maximum heat to the windshield can be obtained by moving the temperature lever to the highest position. The compressor operates in this position to dehumidify the air and reduce windshield fogging.

Note The systems described above are those used on Ford-built vehicles. Other heater–air-conditioner systems operate in a similar manner. However, some systems have more advanced controls. For example, Electronic Touch Climate Control is used in some late-model General Motors' cars. This system uses a control panel (Fig. 54-27) which does not have any buttons, levers, or switches. Instead, the various operating modes, and temperature, are selected by touching the appropriate spots on the panel. This system replaces all vacuum lines with electric wiring. Small electric motor–driven actuators position the mode and temperature doors inside the air-conditioner unit.

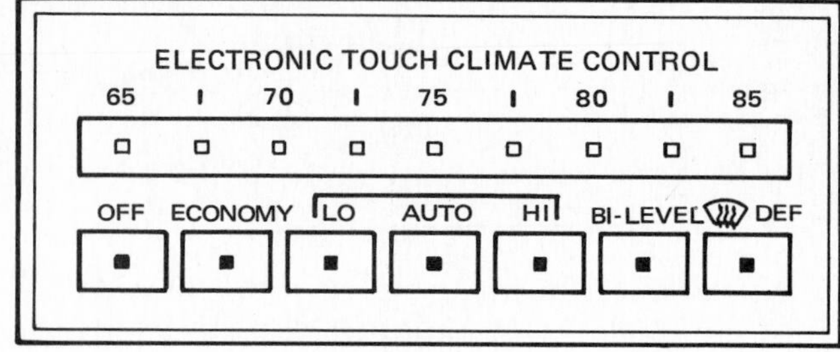

Fig. 54-27 Electronic touch climate-control panel.

♦ REVIEW QUESTIONS ♦

Select the *one* correct, best, or most probable answer to each question. You can find the answers in the section indicated at the end of each question.

1. The ram-air system of ventilation uses (♦54-1)
   a. no blower motor or fan
   b. one blower motor
   c. two blower motors
   d. the air-conditioner motor
2. The small radiator in the system which warms the air passing through is called the (♦54-3)
   a. evaporator
   b. heater core
   c. condenser
   d. pumper
3. The purpose of the defroster is to (♦54-3)
   a. wash dirt off the windshield
   b. warm the passenger compartment
   c. clear the windshield
   d. none of the above
4. The air-distribution system is controlled by (♦54-3)
   a. ducts and doors
   b. cables and vacuum motors
   c. shutters and nozzles
   d. both a and b
5. The air-conditioner system that turns the compressor on and off with signals from the passenger compartment is called (♦54-10)
   a. a defective system
   b. a cycling-clutch system
   c. an evaporator control system
   d. a condenser control system

CHAPTER 55
Heater and Air-Conditioner Service

After studying this chapter, and with the proper instruction and equipment, you should be able to:

1. Diagnose troubles in the heater or air-conditioner system.
2. Define *evacuating the system* and *recharging the system*.
3. Use the sight glass to check refrigerant in the system.
4. Locate refrigerant leaks.
5. Connect a gauge set and evacuate and recharge a system.
6. Remove and replace components of the heater–air-conditioner system.

♦ 55-1 HEATING-SYSTEM TROUBLE DIAGNOSIS

Heating problems are usually reported by the driver as (1) little or no heat, (2) failure of the blower to work, (3) coolant leaks, or (4) too much heat. This last complaint would occur only when the system cannot be turned down so it continues to deliver too much heat. The trouble-diagnosis chart that follows lists various complaints, their possible causes, and the checks or corrections to be made. Following sections describe air-conditioner tests and service procedures.

HEATER TROUBLE-DIAGNOSIS CHART

Complaint	Possible Cause	Check or Correction
1. Little or no heat	a. Air circulation insufficient	Blower motor or switch at fault, air leaks from heater housing, temperature door or cable out of adjustment, loose carpet obstructing airflow
	b. Coolant hose to heater blocked	Unkink hose, replace defective hose
	c. Air in heater core	Bleed air out
	d. Heater core clogged	Repair or replace core
	e. Coolant valve or vacuum motor malfunctioning (♦55-2)	Repair or replace to permit coolant circulation

HEATER TROUBLE-DIAGNOSIS CHART *(Continued)*

Complaint	Possible Cause	Check or Correction
	f. Coolant level low in engine cooling system	Add coolant, bleed air out of system, check system to locate and repair leaks
	g. Engine-cooling-system thermostat stuck open	Replace thermostat
2. Blower motor inoperative	a. Blown fuse, poor electrical connections	Check for cause of blown fuse and correct, tighten connections
	b. Motor defective	Replace
	c. Resistor open	Replace
3. Coolant leaks	Check hoses, hose connections, heater core, water valve	
4. Too much heat	a. Temperature-door cable out of adjustment	Readjust
	b. Engine-cooling system thermostat stuck closed	Replace thermostat
5. Insufficient defrosting	a. Defrost-door control cable out of adjustment	Readjust cable
	b. Defrost outlets blocked	Remove obstructions
	c. Any cause of little or no heat	See item 1
6. Vent door does not operate	Defective vacuum motor, leaky vacuum connections, or a defective control assembly	
7. Controls hard to operate	Loose or binding control cable or a sticky door	Repair
8. Odors from heater	Air leaks around blower case, coolant leaks around heater core.	Tighten bolts and see that seals and gaskets are in place; remove and repair heater core

♦ 55-2 TESTING THE VACUUM-CONTROL SYSTEM

To check the vacuum system, connect a vacuum gauge in the vacuum line between the check valve and the control assembly. Move the function-control lever to OFF. Start the engine, and let it run until there is a good vacuum reading on the gauge. Then turn the ignition key to the accessory position to see if vacuum holds with the engine not running. If vacuum is now lost, there is a leak in the check valve or line.

If vacuum is held, repeat the test with the function-control lever in the VENT position. In this position, vacuum is applied to the vent-door vacuum motor and the coolant-valve vacuum motor. If vacuum now leaks (with the ignition key in the accessory position), one or both vacuum motors are leaking and must be replaced.

If vacuum is held with the function-control lever on VENT, check the operation of the doors to make sure they work normally. If the vent door does not open when the control lever is on VENT, disconnect the vacuum hose from the vent-door vacuum motor. Connect the hose to the vacuum gauge (Fig. 55-1). Now start the engine and see if the gauge registers a vacuum. If it does not, the line is plugged or there is a defect in the control assembly. If there is a vacuum, then either the trouble is in the vacuum motor, or the door is jammed closed. The coolant-valve vacuum motor can be checked in the same way. Check that vacuum is available to the vacuum motor.

♦ 55-3 REPLACING HEATER-SYSTEM COMPONENTS

Figure 55-2 shows how the blower motor-and-wheel assembly is removed from one model of car. On some cars, an access opening must be cut in the inner fender skirt. When this is necessary, be careful to avoid damaging the case while cutting the opening.

The heater core is removed in different ways, depending on its location. As a first step, the engine cooling system must be drained. A typical control assembly, with cables, vacuum motors, and vacuum lines, is shown in Fig. 55-3. When you are working on any heater system, refer to the manufacturer's service manual for the car you are servicing.

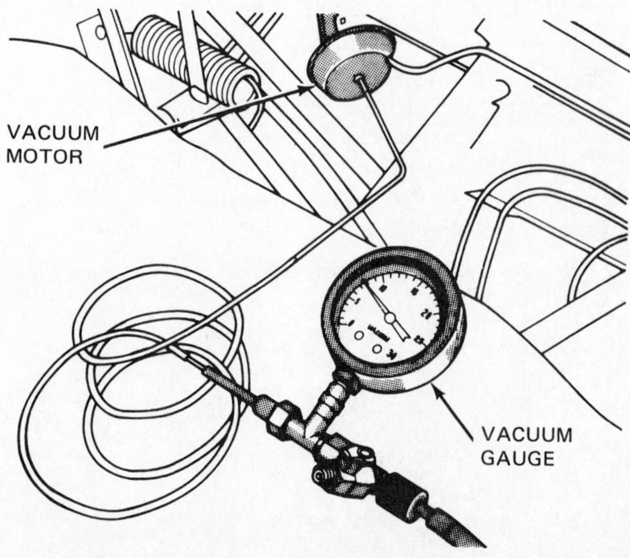

Fig. 55-1 Checking the operation of the vacuum motor with a vacuum gauge. *(Chrysler Corporation)*

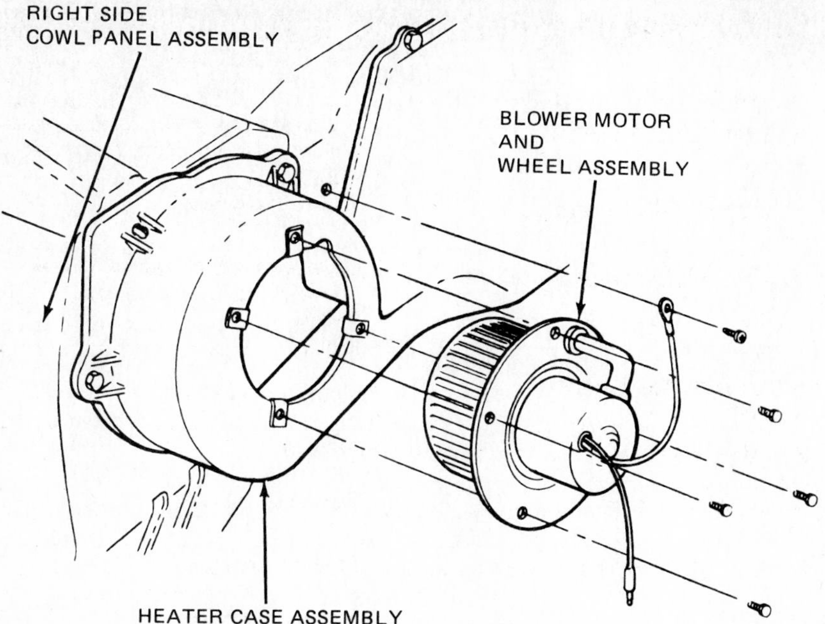

Fig. 55-2 Blower motor-and-wheel assembly installation. *(Ford Motor Company)*

AIR-CONDITIONER SERVICE

♦ **CAUTION** ♦ Before doing any work on air conditioners, reread the Caution in ♦54-16. Eye protection must always be worn. The safety rules must be followed when handling or working around refrigerant.

♦ 55-4 ENEMIES OF THE AIR CONDITIONER

The three worst enemies of the air conditioner are air, moisture, and dirt. They cannot enter a properly operating air conditioner in good condition. But they can enter if parts deteriorate or are damaged, as in an accident. Also, improper

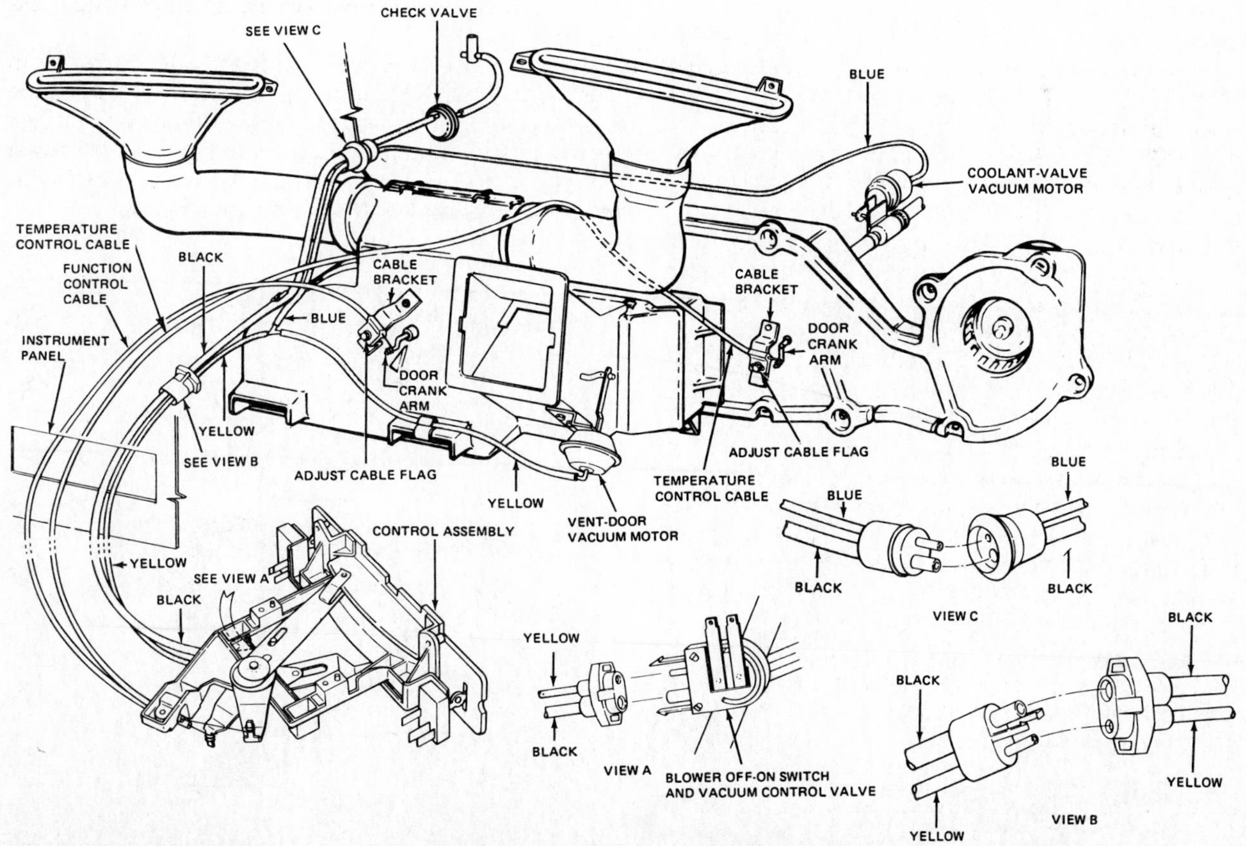

Fig. 55-3 Heater-system layout as viewed from back of the instrument panel. *(Ford Motor Company)*

servicing of the system can allow dirt, air, or moisture to enter the system. Figure 55-4 lists the contaminants that will damage the system.

In addition, high pressure and temperature can damage the system. One possible cause of high pressure in the condenser and compressor is a condenser clogged with leaves or other trash. This increases the condenser temperature and pressure so the compressor has to work harder. This could lead to compressor failure. Some systems have a high-pressure relief valve which lets some of the refrigerant escape if the pressure rises dangerously high.

♦ 55-5 AIR-CONDITIONER–HEATER TROUBLES

A variety of troubles may be reported by the driver. These include not enough heat, not enough cooling, mixed-up circulation (heat coming out of the air-conditioner ducts, for example), and the system quitting when the car is accelerated. The last condition may be normal on some cars.

The sight glass in some systems may be helpful in diagnosing the system (Fig. 55-5). A more accurate procedure is to use a gauge set (♦55-8). The gauges measure the refrigerant pressures in various parts of the system while it is operating. Variations from the specified pressures mean trouble. Another diagnostic tool is the leak detector. It pinpoints locations of refrigerant leaks (♦55-7).

Contaminant	Effects
Moisture	• Causes valves to freeze • Forms hydrochloric and hydrofluoric acid • Causes corrosion and rust
Air	• Causes high head pressure and high temperatures • Accelerates refrigerant instability • Oxidizes oil and causes varnish • Brings in moisture • Reduces cooling capacity
Dirt	• Causes clogged screens and orifices • Provides reactants to cause acids • Abrasive action • Additives hasten breakdown
Alcohol	• Attacks zinc and aluminum • Promotes copper plating • Hastens refrigerant breakdown • Serves no useful purpose
Dye	• Precipitates out and restricts valves • Results in oil overcharge • Breaks down • Detects only large leaks
Rubber	• Deteriorates and clogs system
Metal particles	• Clog screens and valves • Gall bearings • Break reed valves • Score moving parts
Incorrect oil	• Provides poor lubrication, contains wax, and forms sludge, all of which plug passages and cause valves to stick • Breaks down itself and causes breakdown of refrigerant • Contains additives that break down or causes refrigerant to break down, and some may form butyl rubber • Contains moisture

Fig. 55-4 Chart showing the various contaminants of the refrigeration system and the effects of each.

♦ 55-6 BASIC AIR-CONDITIONER SERVICE PROCEDURES

When the refrigeration system in the air conditioner is operating, the high-pressure side is hot and the low-pressure side is cold. The simplest checking procedure is to touch or feel the hoses and valves. For example, the hose from the condenser to the orifice tube should be warm. The hose from the orifice tube to the evaporator should be cold. The thermostatic expansion valve should be hot on the inlet side and cold on the outlet side. If a hose changes temperature along its length, it may have an internal restriction.

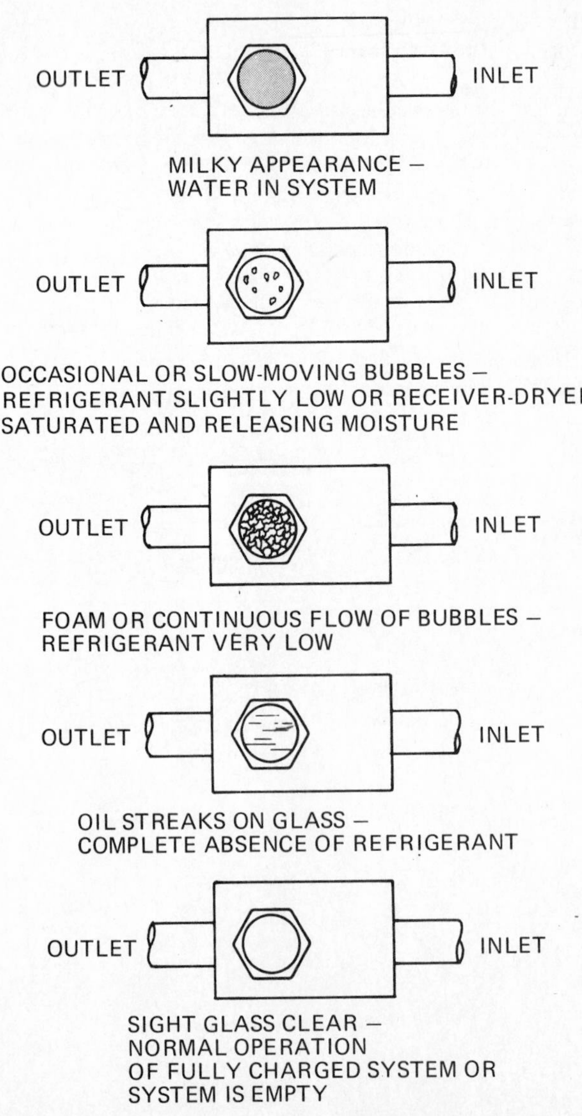

Fig. 55-5 Typical sight-glass indications for various conditions in the refrigeration system. *(Sun Electric Corporation)*

For a complete diagnosis, the sight glass (where present), a leak detector, and a gauge set should be used as explained in following sections.

♦ 55-7 LEAK DETECTOR

Two types of leak detectors are the flame type (butane or propane) and the electronic type. In addition, soapy water can be used. Also, refrigerant dye can be injected into the system. The dye will show up leaks. However, some manufacturers warn that the dye can stick valves and will show up only larger leaks. Soapy water brushed on connections can also show up only larger leaks. The flame and electronic leak detectors can locate very small leaks.

1. Flame Detector Figure 55-6 shows a flame leak detector and how the flame looks if there is no leak, a small leak, or a large leak. To use this leak detector, the gas-control valve is turned on and the gas is lit. It heats the reactor plate red hot. Then the end of the search hose is moved under all connections and along hoses to detect any refrigerant leak. The refrigerant is heavier than air and drops down toward the floor.

If the surrounding air is contaminated with refrigerant vapors, the leak detector will show the presence of refrigerant all the time. Therefore, the flame-type leak detector should be used only in well-ventilated areas.

♦ CAUTION ♦ Be careful when using a flame-type leak detector. Never breathe the fumes or black smoke that is given off when a leak has been located. They contain poisonous phosgene gas. Never use a flame-type leak detector in any area where combustible vapor, fuel, or dust is present. The flame could set off an explosion or a fire. Keep the torch stove and flame from contacting any surfaces which can be easily damaged by the heat. Use the torch only in a well-ventilated area.

2. Electronic Leak Detector The electronic leak detector (Fig. 55-7) is more sensitive than the flame-type detector. To use the electronic leak detector, place the sensitivity switch in the SEARCH position. Turn the control knob clockwise until a steady ticking sound is heard. Then move the probe tip slowly over and under the places suspected of leaking. These include connections, hoses that look damaged, valves, compressor seal points, receiver, condenser, and evaporator. Always position the probe tip *under* suspected leak areas. Refrigerant is heavier than air and drops downward toward the floor.

When the probe tip passes a leak, the ticking sound will increase. A large leak will increase the signal to a high-pitched squeal. To detect small leaks, move the sensitivity switch to the PINPOINT position (if the tester is so equipped). Then readjust the control knob until you get the steady ticking again. Now, pass the probe tip around the suspected area again. The PINPOINT setting changes the leak-detector sensitivity. Now the ticking will not increase until the probe tip is at the leak.

♦ 55-8 USING THE GAUGE SET

The gauge to the left (Fig. 55-8) is the low-pressure gauge. The dial is graduated in pounds per square inch (psi) of pressure from 0 to 100 or 150 (varies with different designs). It is also graduated on the vacuum side in inches [millimeters] of vacuum from 0 to 30 inches. This is the gauge used to check pressure on the low-pressure side of the system.

The gauge to the right in Fig. 55-8 is the high-pressure gauge. It is graduated from 0 to 300 or 500 psi (varies with

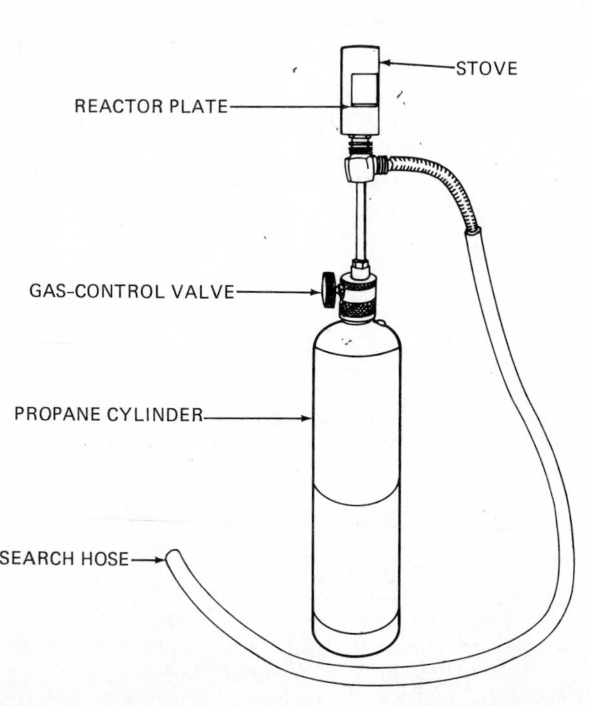

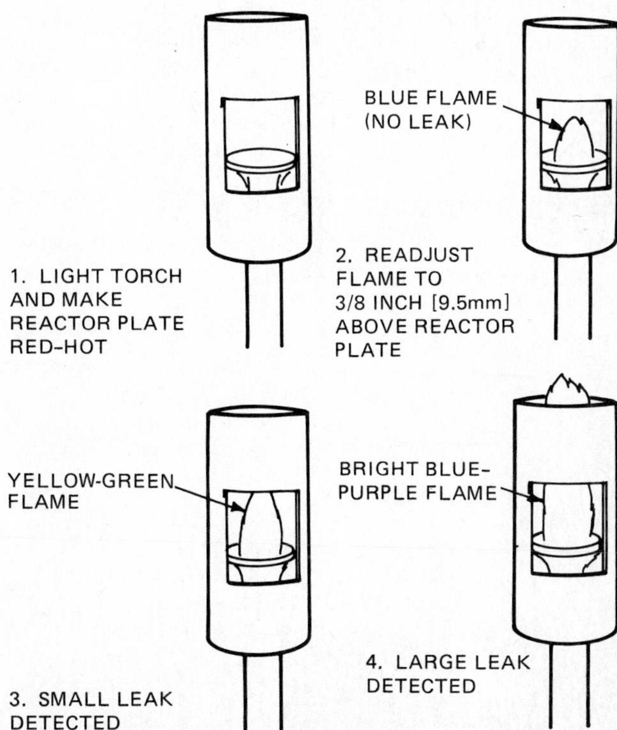

Fig. 55-6 A flame or torch type of leak detector. *(ATW)*

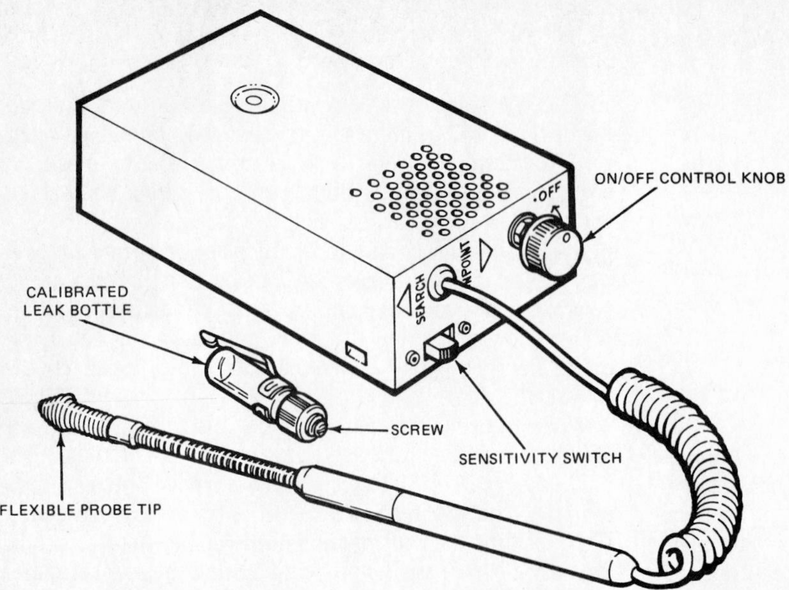

Fig. 55-7 Electronic leak detector. *(Ford Motor Company)*

different designs). This is the gauge used to check the high-pressure side of the system.

On some gauge sets, the low-pressure gauge housing and the low-pressure hose are blue. The high-pressure gauge housing and the high-pressure hose are red. This color coding is used to provide quick identification of the hoses and the connections to be made.

The center manifold fitting is common to both the low- and high-pressure sides. It is used for evacuating the system or adding refrigerant. When this fitting is not in use, it should be capped.

To use the gauge set, proceed as follows: (Fig. 55-9):

♦ CAUTION ♦ Always wear safety goggles! Never open the high-pressure gauge valve while charging the system. The high pressure will enter the refrigerant container and cause it to explode. Wear gloves or wrap a rag around the fitting when you make or break a connection in the refrigeration system.

1. With the engine stopped, remove the protective caps from the system service valves.
2. Make sure that all valves on the manifold-gauge set are closed (turned in to seat).

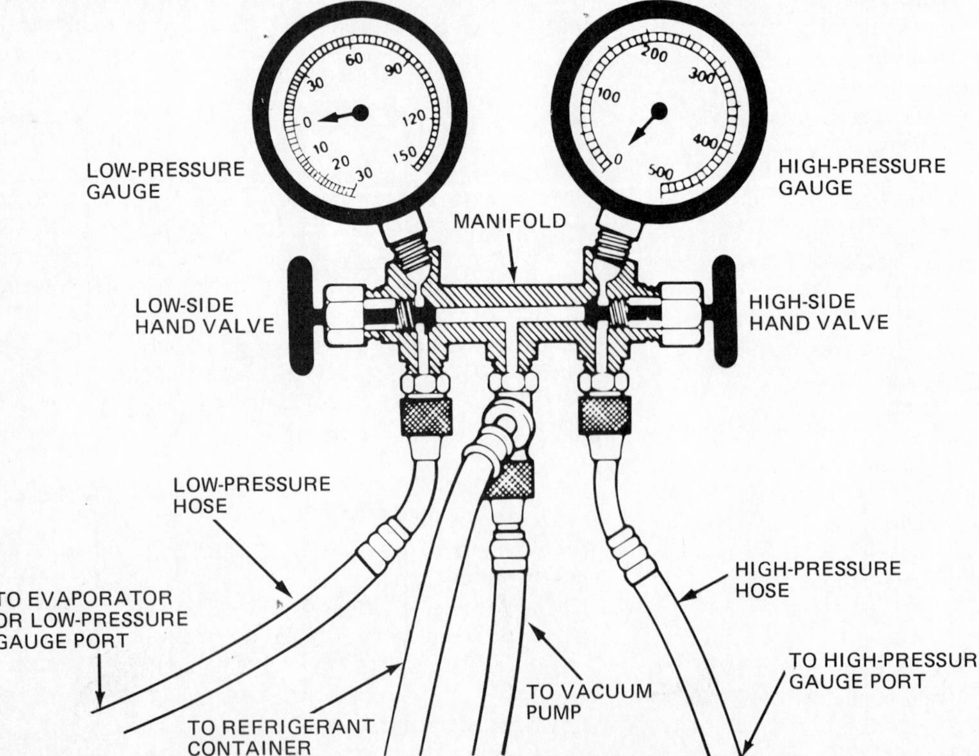

Fig. 55-8 Gauge set showing the passages through the manifold and how the valves open and close the passages to the attached hoses. The gauges read the system pressures even with the hand valves closed. *(Ford Motor Company)*

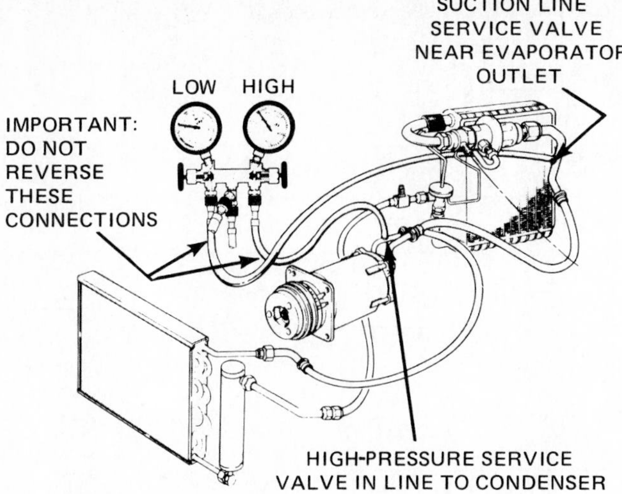

Fig. 55-9 Gauge-set connections for a performance test on one type of air-conditioning system. *(Ford Motor Company)*

3. Leave the center hose connection on the manifold-gauge set capped or connected to a refrigerant container through a hose.
4. Connect the high-pressure gauge hose to the valve fitting on the high side of the system.
5. Purge the test hoses by opening the high-pressure gauge valve one turn. Then crack open the low-pressure valve for 3 seconds. This allows refrigerant vapor to flow through the gauge set and hoses to purge them by forcing out all air and moisture.
6. Close both valves and connect the low-pressure hose to the valve fitting on the low side of the system. Purge the center manifold connection by opening either valve slightly and cracking the cap on the center fitting for 3 seconds. If the center fitting is connected by a hose to a refrigerant container, purge the hose to the container.
7. Stabilize the system before checking pressures. This is done as follows:
   a. Start the engine and adjust idle speed to 1500 to 2000 rpm.
   b. Turn on air conditioner and set for maximum cooling with blower fan on high speed.
   c. Close car doors and windows.
   d. Operate air conditioner for 5 minutes.
   e. If the system has a sight glass, check it (Fig. 55-5).
   f. Open the gauge valves and note the pressures. Compare these with the pressures specified for the system you are checking, as shown in the manufacturer's service manual. The amount the pressure varies from specifications indicates the condition in the system. The most common problem is low refrigerant. Some refrigerant loss can occur from one season to the next. This is normal.

Refrigerant loss occurs as a result of hose porosity, vibration, and the general construction of the components. For example, seals can allow some slight passage of refrigerant. Replacement of this lost refrigerant is the most common service required on air-conditioner systems.

Figure 55-10 relates various high and low pressures to the clutch cycling time. The chart shows which component might be causing an incorrect pressure reading.

Figure 55-11 shows test-pressure ranges indicating normal operation of various systems. Variations from these pressures indicate abnormal operation.

♦ 55-9 CHARGING STATION

Many shops have charging stations to service air conditioners (Fig. 55-12). A typical charging station is assembled on a two-wheel cart which can be rolled up to the car being serviced. It has the gauge set, a container of refrigerant, a vacuum pump, a leak detector, and two valves. These are in the line to the vacuum pump and to the refrigerant container.

♦ 55-10 EVACUATING THE SYSTEM

If the system has leaked, or damage has occurred, the system must be evacuated and recharged. The leak or damage must be repaired before recharging the system. Evacuating requires two steps. These are discharging the refrigerant and then using the vacuum pump to remove all traces of refrigerant, water vapor, and air.

♦ 55-11 DISCHARGING THE SYSTEM

1. Ford Method Connect the gauge set. Both gauge valves should be fully closed. Place the open end of the gauge-set center hose near a shop exhaust outlet. Slowly depressurize the system by opening the low-pressure gauge valve slightly. Slow discharge prevents loss of oil from the system.

2. General Motors and Chrysler Method This calls for fast discharging of the system. It causes a considerable loss of oil. The center hose from the gauge set should be placed in a container to catch the oil. Then, before recharging the system, the same amount of fresh oil should be added to the system.

♦ 55-12 USING THE VACUUM PUMP

With the gauge set connected and the center hose connected to the vacuum pump, start the pump. The high-pressure and low-pressure gauge valves and the vacuum-pump valve should be open. Operate the pump until the vacuum reading is the value specified in the manufacturer's service manual. If the vacuum reading is not up to specifications, the system is leaking. The leak must be found and corrected.

When evacuation is complete, turn off the gauge valves and prepare to recharge the system.

♦ 55-13 RECHARGING THE SYSTEM

With the gauge set connected as shown in Fig. 55-8, close the valve to the vacuum pump. Different recharging procedures are used, depending on the type of refrigerant container. A typical procedure using small cans of refrigerant is shown in Fig. 55-13. The refrigerant manifold can be connected to three small cans. Most systems take three cans for a full recharge. Some systems may take more.

To help the refrigerant vaporize so it can flow into the system, the cans are placed in a pan of hot water, as shown in Fig. 55-13. The water temperature must not exceed 125°F [52°C]. To purge any air from the charging hose, loosen the hose at the gauge set. Then slightly open one of the manifold

NOTE: Normal system conditional requirements must be maintained to properly evaluate refrigerant system pressures. Refer to chart applicable to system under test.

HIGH PRESSURE		LOW PRESSURE	CLUTCH CYCLE TIME			COMPONENT — CAUSES
@ LIQUID LINE	@ COMPRESSOR		RATE	ON	OFF	
HIGH	HIGH	HIGH	LOW	NORMAL TO LOW	NORMAL	ENGINE OVERHEATING
HIGH	HIGH	NORMAL TO HIGH	LOW	LONG	SHORT	CONDENSER — Inadequate Airflow
HIGH	HIGH	NORMAL				AIR IN REFRIGERANT REFRIGERANT OVERCHARGE (a)
NORMAL TO HIGH	NORMAL TO HIGH	NORMAL TO HIGH				HUMIDITY VERY HIGH (b)
NORMAL	NORMAL	HIGH	NORMAL TO LOW	NORMAL	NORMAL TO LOW	CLUTCH CYCLING SWITCH — High Setting
NORMAL	NORMAL	NORMAL	HIGH	SHORT	NORMAL	FIXED ORIFICE TUBE — Partially Restricted A/C LIQUID LINE — Partially Restricted
			LOW	LONG	SHORT	MOISTURE IN REFRIGERANT SYSTEM EXCESSIVE REFRIGERANT OIL
NORMAL TO LOW	NORMAL TO LOW	NORMAL	NORMAL TO LOW	LONG	SHORT	FIXED ORIFICE TUBE — Restricted
			HIGH	SHORT	NORMAL TO SHORT	EVAPORATOR CORE — Partially Restricted
			LOW	SHORT	LONG	EVAPORATOR — Low Airflow
			HIGH	SHORT	NORMAL TO SHORT	A/C SUCTION LINE — Partially Restricted
			HIGH	VERY SHORT	LONG	A/C LIQUID LINE — Restricted
			HIGH	SHORT TO VERY SHORT	SHORT TO VERY SHORT	LOW REFRIGERANT CHARGE
			CONTINUOUS RUN			CLUTCH CYCLING SWITCH — Sticking Closed
LOW	NORMAL	NORMAL	HIGH	SHORT	NORMAL	CONDENSER — Partially Restricted
LOW	LOW	HIGH	NORMAL TO LOW	LONG	SHORT	FIXED ORIFICE TUBE — O-Ring Leaking
			RAPID	VERY SHORT	VERY SHORT	CLUTCH CYCLING SWITCH — Low Setting
			LOW	LONG	SHORT	COMPRESSOR — Low Performance
			CONTINUOUS RUN			A/C SUCTION LINE — Restricted
NORMAL TO LOW	NORMAL TO LOW	NORMAL TO LOW	HIGH	VERY SHORT	LONG	CONDENSER — Restricted
LOW	LOW	NORMAL				EVAPORATOR CORE — Restricted
ERRATIC OPERATON OR COMPRESSOR NOT RUNNING			—	—	—	CLUTCH CYCLING SWITCH — Dirty Contacts or Sticking Open.

ADDITIONAL POSSIBLE CAUSE COMPONENTS
ASSOCIATED WITH INADEQUATE COMPRESSOR OPERATION

• COMPRESSOR CLUTCH Slipping • LOOSE DRIVE BELT
• CLUTCH COIL With Open, Blown Circuit, Blown Fuse or Loose Mounting
• CONTROL ASSEMBLY SWITCH
— Dirty Contacts or Sticking Open

ADDITIONAL POSSIBLE CAUSE COMPONENTS
ASSOCIATED WITH A DAMAGED COMPRESSOR

• CLUTCH CYCLING PRESSURE SWITCH — Sticking Closed or Compressor Clutch Seized
• SUCTION ACCUMULATOR/DRIER — Refrigerant Oil Bleed Hole Plugged
• REFRIGERANT LEAKS

(a) Compressor may make noise on initial run. This is a slugging condition caused by excessive liquid refrigerant.
(b) Compressor clutch may not cycle during very high humidity condition.

Fig. 55-10 Refrigerant system pressure evaluation chart. *(Ford Motor Company)*

Ambient (Outside Air) Temperature (°F)	At high-Pressure Test fitting* (psi)	AT LOW-PRESSURE TEST FITTING (psi)				
		STV, POA, or VIR Systems	Cycling-Clutch System With TXV and Rec-Dehyd**	Cycling-Clutch System with Orifice Tube and Accumulator (CCOT)**	CCOT System with Pressure-Cycling Switch	Chrysler Corp. with Evaporator-Pressure-Regulator Valve
60	120–170	28–31	7–15	—	—	—
70	150–125	28–31	7–15	24–31	24–31	22–30
80	180–275	28–31	7–15	24–31	24–31	22–37
90	200–310	28–31	7–15	24–32	24–31	25–37
100	230–330	28–35	10–30	24–32	24–36	—
110	270–360	28–38	10–35	24–32	—	—

*Pressures may be slightly higher on very humid days or lower on very dry days.
**Pressure just before clutch disengages (cycles off).

Fig. 55-11 Approximate test-pressure ranges for normal operation of various air-conditioning systems. *(AC-Delco Division of General Motors Corporation)*

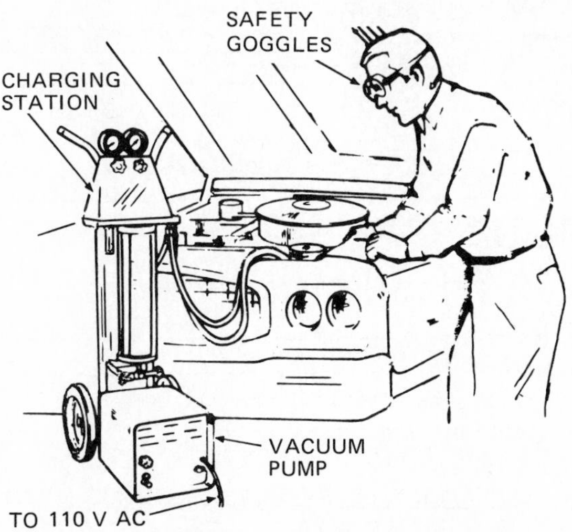

Fig. 55-12 Air-conditioner charging station. Note the technician is wearing safety goggles. *(Ford Motor Company)*

valves to allow refrigerant to flow through the hose. When refrigerant vapor starts to escape from the loose connection, tighten the connection.

Both gauge valves should be closed. Open all refrigerant-can valves. Start the engine and move the controls to the A/C low-blower position. Charge the system through the low-pressure side by slowly opening the low-pressure gauge valve. Adjust the valve to prevent excessive pressure (50 psi or 345 kPa). The high-pressure gauge valve must be closed.

Maintain fast-idle speed until all cans are empty. Close the gauge-set valves and the refrigerant manifold valves. Disconnect the gauge set. Check the system for normal operation.

REMOVING AND INSTALLING COMPONENTS

♦ 55-14 REPLACING HOSE

To replace a hose, the system must be discharged. Then it must be evacuated and recharged after the hose has been installed. Two types of fittings are used, flare fittings and O rings (Fig. 55-14). The fittings are inserted in the hose and secured with suitable clamps.

Careful Use only approved replacement refrigerant hose. Do not use heater hose.

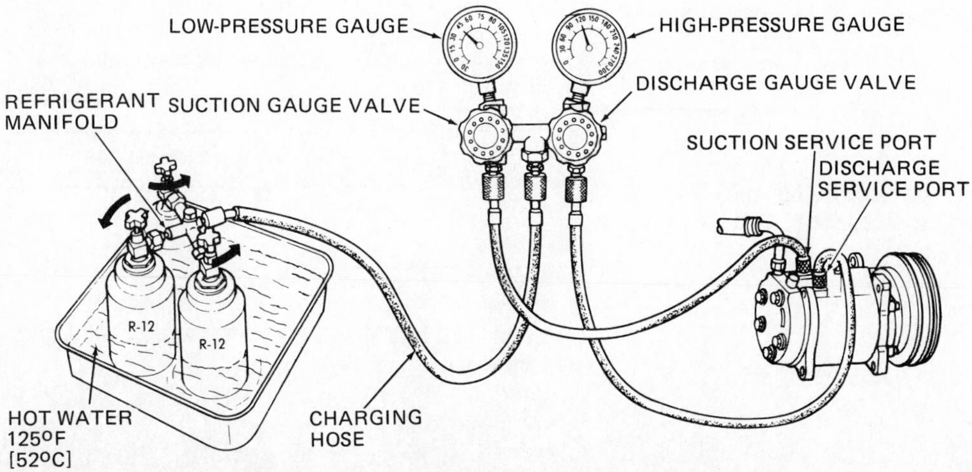

Fig. 55-13 Connections of gauges, refrigerant manifold, and refrigerant cans for recharging the system. *(Chrysler Corporation)*

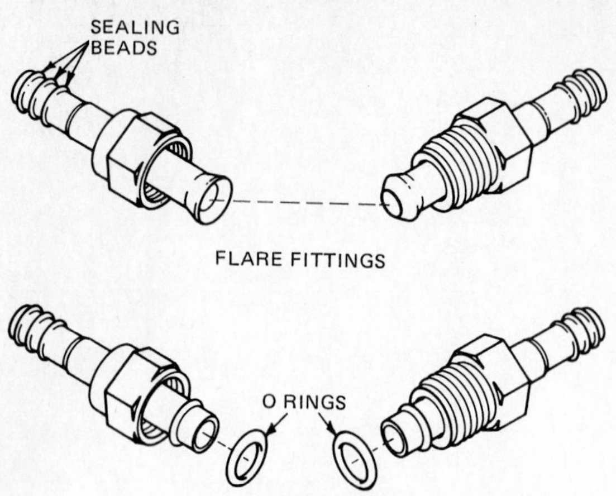

Fig. 55-14 Types of fittings used to make connections in the refrigerant system.

♦ 55-15 REMOVING AND INSTALLING OTHER COMPONENTS

When removing and installing air-conditioner components, make sure that the proper amount of oil is held in the system. When removing a compressor, for example, the procedure recommended by Ford for a six-cylinder (round) compressor is as follows.

Run the engine at fast idle for 10 minutes at maximum cooling and high blower speed. This distributes the oil through the system. Discharge the system and remove the compressor. Drain the oil from the compressor into a measuring cup. Discard this oil.

Replacement compressors have oil in them. Drain this oil. Pour the specified amount of clean new oil into the compressor. Install the drain plug and then install the compressor on the car.

Note If the compressor is being replaced (for example, because it is locked up), the new compressor will contain the specified amount of refrigerant oil.

For other components, manufacturers specify the amount of new oil that should be added before installation. For example, one specification is to add 3 ounces [90 cc] of oil to the evaporator and 1 ounce [30 cc] to the receiver.

♦ REVIEW QUESTIONS ♦

Select the *one* correct, best, or most probable answer to each question. You can find the answers in the section indicated at the end of each question.

1. When you feel a high-pressure pipe, it should be (♦55-6)
   a. warm
   b. cold
   c. the same temperature as the outside air
   d. the same temperature as the low-pressure pipe
2. When you feel a low-pressure pipe, it should be (♦55-6)
   a. warm
   b. cold
   c. the same temperature as the outside air
   d. the same temperature as the high-pressure pipe
3. If there is very little or no difference in temperature between the low-pressure and the high-pressure pipes the system is (♦55-6)
   a. full
   b. operating normally
   c. empty or nearly empty
   d. none of the above
4. When there is a noticeable temperature difference between the pipes at the compressor, the (♦55-6)
   a. refrigerant charge is probably okay
   b. refrigerant charge is probably low
   c. compressor is inoperative
   d. none of the above
5. Using a vacuum pump to pump out all refrigerant, air, and moisture is called (♦55-10)
   a. evacuating the system
   b. recharging the system
   c. purging the system
   d. emptying the system
6. Putting in a fresh charge of clean refrigerant is called (♦55-13)
   a. evacuating the system
   b. recharging the system
   c. purging the system
   d. emptying the system
7. One possible cause of excessive pressure in the air-conditioning system is (♦55-4)
   a. low outside temperature
   b. refrigerant leaks
   c. clogging of the air passages in the condenser
   d. clogging of the air passages in the compressor
8. The three major enemies of the air conditioner are (♦55-4)
   a. gasoline, oil, and electricity
   b. air, moisture, and dirt
   c. cold weather, humid days, and high altitude
   d. all of the above
9. A vacuum motor does not work. Mechanic A says the vacuum hose is disconnected or leaking. Mechanic B says the vacuum motor is defective. Who is right? (♦55-2)
   a. A only
   b. B only
   c. both A and B
   d. neither A nor B
10. Oil should be added after you have discharged and evacuated the system, and (♦55-15)
    a. any time a system component is replaced
    b. every 4 months
    c. every time the engine oil is changed
    d. at no other time

APPENDIX: CONVERSION TABLES

MILLIMETERS TO INCHES

mm	1	2	3	4	5	6	7	8	9	10	11	12	13
inches	0.0394	0.0787	0.1181	0.1575	0.1968	0.2362	0.2756	0.3150	0.3543	0.3937	0.4331	0.4724	0.5118
mm	14	15	16	17	18	19	20	21	22	23	24	25	26
inches	0.5512	0.5905	0.6299	0.6693	0.7087	0.7480	0.7874	0.8268	0.8661	0.9055	0.9449	0.9842	1.0236
mm	27	28	29	30	31	32	33	34	35	36	37	38	39
inches	1.0630	1.1024	1.1417	1.1811	1.2205	1.2598	1.2992	1.3386	1.3779	1.4173	1.4567	1.4961	1.5354
mm	40	41	42	43	44	45	46	47	48	49	50	51	52
inches	1.5748	1.6142	1.6535	1.6929	1.7323	1.7716	1.8110	1.8504	1.8898	1.9291	1.9685	2.0079	2.0472
mm	53	54	55	56	57	58	59	60	61	62	63	64	65
inches	2.0866	2.1260	2.1653	2.2047	2.2441	2.2835	2.3228	2.3622	2.4016	2.4409	2.4803	2.5197	2.5590
mm	66	67	68	69	70	71	72	73	74	75	76	77	78
inches	2.6984	2.6378	2.6772	2.7165	2.7559	2.7953	2.8346	2.8740	2.9134	2.9527	2.9921	3.0315	3.0709
mm	79	80	81	82	83	84	85	86	87	88	89	90	91
inches	3.1102	3.1496	3.1890	3.2283	3.2677	3.3071	3.3464	3.3858	3.4252	3.4646	3.5039	3.5433	3.5827
mm	92	93	94	95	96	97	98	99	100				
inches	3.6220	3.6614	3.7008	3.7401	3.7795	3.8189	3.8583	3.8976	3.9370				

INCHES TO MILLIMETERS

inches	1/64	1/32	3/64	1/16	5/64	3/32	7/64	1/8	9/64	5/32	11/64	3/16	13/64
mm	0.3969	0.7937	1.1906	1.5875	1.9844	2.3812	2.7781	3.1750	3.5719	3.9687	4.3656	4.7625	5.1594
inches	7/32	15/64	1/4	17/64	9/32	19/64	5/16	21/64	11/32	23/64	3/8	25/64	13/32
mm	5.5562	5.9531	6.3500	6.7469	7.1437	7.5406	7.9375	8.3344	8.7312	9.1281	9.5250	9.9219	10.3187
inches	27/64	7/16	29/64	15/32	31/64	1/2	33/64	17/32	35/64	9/16	37/64	19/32	39/64
mm	10.7156	11.1125	11.5094	11.9062	12.3031	12.7000	13.0969	13.4937	13.8906	14.2875	14.6844	15.0812	15.4781
inches	5/8	41/64	21/32	43/64	11/16	45/64	23/32	47/64	3/4	49/64	25/32	51/64	13/16
mm	15.8750	16.2719	16.6687	17.0656	17.4625	17.8594	18.2562	18.6531	19.0500	19.4469	19.8437	20.2406	20.6375
inches	53/64	27/32	55/64	7/8	57/64	29/32	59/64	15/16	61/64	31/32	63/64		
mm	21.0344	21.4312	21.8281	22.2250	22.6219	23.0187	23.4156	23.8125	24.2094	24.6062	25.0031		

FAHRENHEIT TO CELSIUS (Centigrade)

°F	−20	−15	−10	−5	0	1	2	3	4	5	10	15	20
°C	−28.9	−26.1	−23.3	−20.6	−17.8	−17.2	−16.7	−16.1	−15.6	−15.0	−12.2	−9.4	−6.7
°F	25	30	35	40	45	50	55	60	65	70	75	80	85
°C	−3.9	−1.1	1.7	4.4	7.2	10.0	12.8	15.6	18.3	21.1	23.9	26.7	29.4
°F	90	95	100	105	110	115	120	125	130	135	140	145	150
°C	32.2	35.0	37.8	40.6	43.3	46.1	48.9	51.7	54.4	57.2	60.0	62.8	65.6
°F	155	160	165	170	175	180	185	190	195	200	205	210	212
°C	68.3	71.1	73.9	76.7	79.4	82.2	85.0	87.8	90.6	93.8	96.1	98.9	100.0
°F	215	220	225	230	235	240	245	250	255	260	265		
°C	101.7	104.4	107.2	110.0	112.8	115.6	118.3	121.1	123.9	126.6	129.4		

FEET TO METERS

ft	0	1	2	3	4	5	6	7	8	9	ft
	m	m	m	m	m	m	m	m	m	m	
—		0.305	0.610	0.914	1.219	1.524	1.829	2.134	2.438	2.743	—
10	3.048	3.353	3.658	3.962	4.267	4.572	4.877	5.182	5.486	5.791	10
20	6.096	6.401	6.706	7.010	7.315	7.620	7.925	8.230	8.534	8.839	20
30	9.144	9.449	9.754	10.058	10.363	10.668	10.973	11.278	11.582	11.887	30
40	12.192	12.497	12.802	13.106	13.411	13.716	14.021	14.326	14.630	14.935	40
50	15.240	15.545	15.850	16.154	16.459	16.764	17.069	17.374	17.678	17.983	50
60	18.288	18.593	18.898	19.202	19.507	19.812	20.117	20.422	20.726	21.031	60
70	21.336	21.641	21.946	22.250	22.555	22.860	23.165	23.470	23.774	24.079	70
80	24.384	24.689	24.994	25.298	25.603	25.908	26.213	26.518	26.822	27.127	80
90	27.432	27.737	28.042	28.346	28.651	28.956	29.261	29.566	29.870	30.175	90
100	30.480	30.785	31.090	31.394	31.699	32.004	32.309	32.614	32.918	33.223	100

Source: Buick Motor Division of General Motors Corporation

MILES TO KILOMETERS

mile	0	1	2	3	4	5	6	7	8	9	mile
	km	km	km	km	km	km	km	km	km	km	
—		1.609	3.219	4.828	6.437	8.047	9.656	11.265	12.875	14.484	—
10	16.093	17.703	19.312	20.921	22.531	24.140	25.750	27.359	28.968	30.578	10
20	32.187	33.796	35.406	37.015	38.624	40.234	41.843	43.452	45.062	46.671	20
30	48.280	49.890	51.499	53.108	54.718	56.327	57.936	59.546	61.155	62.764	30
40	64.374	65.983	67.593	69.202	70.811	72.421	74.030	75.639	77.249	78.858	40
50	80.467	82.077	83.686	85.295	86.905	88.514	90.123	91.733	93.342	94.951	50
60	96.561	98.170	99.779	101.39	103.00	104.61	106.22	107.83	109.44	111.04	60
70	112.65	114.26	115.87	117.48	119.09	120.70	122.31	123.92	125.53	127.14	70
80	128.75	130.36	131.97	133.58	135.19	136.79	138.40	140.01	141.62	143.23	80
90	144.84	146.45	148.06	149.67	151.28	152.89	154.50	156.11	157.72	159.33	90
100	160.93	162.54	164.15	165.76	167.37	168.98	170.59	172.20	173.81	175.42	100

SQUARE INCHES TO SQUARE CENTIMETERS

in^2	0	1	2	3	4	5	6	7	8	9	in^2
	cm^2	cm^2	cm^2	cm^2	cm^2	cm^2	cm^2	cm^2	cm^2	cm^2	
—		6.452	12.903	19.355	25.806	32.258	38.710	45.161	51.613	58.064	—
10	64.516	70.968	77.419	83.871	90.322	96.774	103.226	109.677	116.129	122.580	10
20	129.032	135.484	141.935	148.387	154.838	161.290	167.742	174.193	180.645	187.096	20
30	193.548	200.000	206.451	212.903	219.354	225.806	232.258	238.709	245.161	251.612	30
40	258.064	264.516	270.967	277.419	283.870	290.322	296.774	303.225	309.677	316.128	40
50	322.580	329.032	335.483	341.935	348.386	354.838	361.290	367.741	374.193	380.644	50
60	387.096	393.548	399.999	406.451	412.902	419.354	425.806	432.257	438.709	445.160	60
70	451.612	458.064	464.515	470.967	477.418	483.870	490.322	496.773	503.225	509.676	70
80	516.128	522.580	529.031	535.483	541.934	548.386	554.838	561.289	567.741	574.192	80
90	580.644	587.096	593.547	599.999	606.450	612.902	619.354	625.805	632.257	638.708	90
100	645.160	651.612	658.063	664.515	670.966	677.418	683.870	690.321	696.773	703.224	100

CUBIC INCHES TO CUBIC CENTIMETERS

in^3	0	1	2	3	4	5	6	7	8	9	in^3
	cc	cc	cc	cc	cc	cc	cc	cc	cc	cc	
—		16.387	32.774	49.161	65.548	81.935	98.322	114.709	131.097	147.484	—
10	163.871	180.258	196.645	213.032	229.419	245.806	262.193	278.580	294.967	311.354	10
20	327.741	344.128	360.515	376.902	393.290	409.677	426.064	442.451	458.838	475.225	20
30	491.612	507.999	524.386	540.773	557.160	573.547	589.934	606.321	622.708	639.095	30
40	655.483	671.870	688.257	704.644	721.031	737.418	753.805	770.192	786.579	802.966	40
50	819.353	835.740	852.127	868.514	884.901	901.289	917.676	934.063	950.450	966.837	50
60	983.224	999.611	1015.998	1032.385	1048.772	1065.159	1081.546	1097.933	1114.320	1130.707	60
70	1147.094	1163.482	1179.869	1196.256	1212.643	1229.030	1245.417	1261.804	1278.191	1294.578	70
80	1310.965	1327.352	1343.739	1360.126	1376.513	1392.200	1409.288	1425.675	1442.062	1458.449	80
90	1474.836	1491.223	1507.610	1523.997	1540.384	1556.771	1573.158	1589.545	1605.932	1622.319	90
100	1638.706	1655.093	1671.481	1687.863	1704.255	1720.642	1737.029	1753.416	1769.803	1786.190	100

CUBIC FEET TO CUBIC METERS

ft^3	0	1	2	3	4	5	6	7	8	9	ft^3
	m^3	m^3	m^3	m^3	m^3	m^3	m^3	m^3	m^3	m^3	
—		0.0283	0.0566	0.0850	0.1133	0.1416	0.1699	0.1982	0.2265	0.2549	—
10	0.2832	0.3115	0.3398	0.3681	0.3964	0.4248	0.4531	0.4814	0.5097	0.5380	10
20	0.5663	0.5947	0.6230	0.6513	0.6796	0.7079	0.7362	0.7646	0.7929	0.8212	20
30	0.8495	0.8778	0.9061	0.9345	0.9628	0.9911	1.0194	1.0477	1.0760	1.1044	30
40	1.1327	1.1610	1.1893	1.2176	1.2459	1.2743	1.3026	1.3309	1.3592	1.3875	40
50	1.4159	1.4442	1.4725	1.5008	1.5291	1.5574	1.5858	1.6141	1.6424	1.6707	50
60	1.6990	1.7273	1.7557	1.7840	1.8123	1.8406	1.8689	1.8972	1.9256	1.9539	60
70	1.9822	2.0105	2.0388	2.0671	2.0955	2.1238	2.1521	2.1804	2.2087	2.2370	70
80	2.2654	2.2937	2.3220	2.3503	2.3786	2.4069	2.4353	2.4636	2.4919	2.5202	80
90	2.5485	2.5768	2.6052	2.6335	2.6618	2.6901	2.7184	2.7468	2.7751	2.8034	90
100	2.8317	2.6800	2.8884	2.9167	2.9450	2.9733	3.0016	3.0300	3.0583	3.0866	100

GALLONS (U.S.) TO LITERS

U.S. gal	0	1	2	3	4	5	6	7	8	9	U.S. gal
	l	l	l	l	l	l	l	l	l	l	
—		3.7854	7.5709	11.3563	15.1417	18.9271	22.7126	26.4980	30.2834	34.0638	—
10	37.8543	41.6397	45.4251	49.2105	52.9960	56.7814	60.5668	64.3523	68.1377	71.9231	10
20	75.7085	79.4940	83.2794	87.0648	90.8502	94.6357	98.4211	102.2065	105.9920	109.7774	20
30	113.5528	117.3482	121.1337	124.9191	128.7045	132.4899	136.2754	140.0608	143.8462	147.6316	30
40	151.4171	155.2025	158.9879	162.7734	166.5588	170.3442	174.1296	177.9151	181.7005	185.4859	40
50	189.2713	193.0568	196.8422	200.6276	204.4131	208.1985	211.9839	215.7693	219.5548	223.3402	50
60	227.1256	230.9110	234.6965	238.4819	242.2673	246.0527	249.8382	253.6236	257.4090	261.1945	60
70	264.9799	268.7653	272.5507	276.3362	280.1216	283.9070	287.6924	291.4779	295.2633	299.0487	70
80	302.8342	306.6196	310.4050	314.1904	317.9759	321.7613	325.5467	329.3321	333.1176	336.9030	80
90	340.6884	344.4738	348.2593	352.0447	355.8301	359.6156	363.4010	367.1864	370.9718	374.7573	90
100	378.5427	382.3281	386.1135	389.8990	393.6844	397.4698	401.2553	405.0407	408.8261	412.6115	100

POUNDS TO KILOGRAMS

lb	0	1	2	3	4	5	6	7	8	9	lb
	kg	kg	kg	kg	kg	kg	kg	kg	kg	kg	
—		0.454	0.907	1.361	1.814	2.268	2.722	3.175	3.629	4.082	—
10	4.536	4.990	5.443	5.897	6.350	6.804	7.257	7.711	8.165	8.618	10
20	9.072	9.525	9.979	10.433	10.886	11.340	11.793	12.247	12.701	13.154	20
30	13.608	14.061	14.515	14.969	15.422	15.876	16.329	16.783	17.237	17.690	30
40	18.144	18.597	19.051	19.504	19.958	20.412	20.865	21.319	21.772	22.226	40
50	22.680	23.133	23.587	24.040	24.494	24.948	25.401	25.855	26.308	26.762	50
60	27.216	27.669	28.123	28.576	29.030	29.484	29.937	30.391	30.844	31.298	60
70	31.751	32.205	32.659	33.112	33.566	34.019	34.473	34.927	35.380	35.834	70
80	36.287	36.741	37.195	37.648	38.102	38.555	39.009	39.463	39.916	40.370	80
90	40.823	41.277	41.730	42.184	42.638	43.092	43.545	43.998	44.453	44.906	90
100	45.359	45.813	46.266	46.720	47.174	47.627	48.081	48.534	48.988	49.442	100

POUNDS PER SQUARE INCH TO KILOPASCALS

psi	0	1	2	3	4	5	6	7	8	9	psi
	kPa	kPa	kPa	kPa	kPa	kPa	kPa	kPa	kPa	kPa	
—	0.0000	6.8948	13.7895	20.6843	27.5790	34.4738	41.3685	48.2663	55.1581	62.0528	—
10	68.9476	75.8423	82.7371	89.6318	96.5266	103.4214	110.3161	117.2109	124.1056	131.0004	10
20	137.8951	144.7899	151.6847	158.5794	165.4742	172.3689	179.2637	186.1584	193.0532	199.9480	20
30	206.8427	213.7375	220.6322	227.5270	234.4217	241.3165	248.2113	255.1060	262.0008	268.8955	30
40	275.7903	282.6850	289.5798	296.4746	303.3693	310.2641	317.1588	324.0536	330.9483	337.8431	40
50	344.7379	351.6326	358.5274	365.4221	372.3169	379.2116	386.1064	393.0012	399.8959	406.7907	50
60	412.6854	420.5802	427.4749	434.3697	441.2645	448.1592	455.0540	461.9487	468.8435	475.7382	60
70	482.6330	489.5278	496.4225	503.3173	510.2120	517.1068	524.0015	530.8963	537.7911	544.6858	70
80	551.5806	558.4753	565.3701	572.2648	579.1596	586.0544	592.9491	599.8439	606.7386	613.6334	80
90	620.5281	627.4229	634.3177	641.2124	648.1072	655.0019	661.8967	668.7914	675.6862	682.5810	90
100	689.4757	696.3705	703.2653	710.1601	717.0549	723.9497	730.8445	737.7393	744.6341	751.5289	100

POUND-FEET TO KILOGRAM-METERS

lb-ft	0	1	2	3	4	5	6	7	8	9	lb-ft
	kg-m	kg-m	kg-m	kg-m	kg-m	kg-m	kg-m	kg-m	kg-m	kg-m	
—		0.138	0.276	0.415	0.553	0.691	0.829	0.967	1.106	1.244	—
10	1.382	1.520	1.658	1.796	1.934	2.073	2.211	2.349	2.487	2.625	10
20	2.764	2.902	3.040	3.178	3.316	3.455	3.593	3.731	3.869	4.007	20
30	4.146	4.284	4.422	4.560	4.698	4.837	4.975	5.113	5.251	5.389	30
40	5.528	5.666	5.804	5.942	6.080	6.219	6.357	6.495	6.633	6.771	40
50	6.910	7.048	7.186	7.324	7.462	7.601	7.739	7.877	8.015	8.153	50
60	8.292	8.430	8.568	8.706	8.844	8.983	9.121	9.259	9.397	9.535	60
70	9.674	9.812	9.950	10.088	10.227	10.365	10.503	10.641	10.779	10.918	70
80	11.056	11.194	11.332	11.470	11.609	11.747	11.885	12.023	12.161	12.300	80
90	12.438	12.576	12.714	12.855	12.991	13.129	13.267	13.405	13.544	13.682	90
100	13.820	13.958	14.096	14.235	14.373	14.511	14.649	14.787	14.925	14.064	100

GLOSSARY

Accelerator pump In the carburetor, a pump (linked to the accelerator) which momentarily enriches the air-fuel mixture when the accelerator is depressed at low speed.

Accumulator A device used in automatic transmissions to cushion the shock of clutch and servo actions.

Advance The moving ahead of the ignition spark in relation to piston position; produced by centrifugal or vacuum devices in accordance with engine speed and intake-manifold vacuum, or electronically.

Aerobic gasket material A type of formed-in-place gasket, also known as self-curing, room-temperature-vulcanizing (RTV), and silicone rubber. A material that cures only in the presence of air, normally used on surfaces that flex or vibrate.

Air-aspirator system An air-injection system using a valve opened and closed by pulses in the exhaust system.

Air bleed An opening into a fuel passage through which air can pass, or bleed, into the fuel as it moves through the passage.

Air brake A braking system that uses compressed air to supply the force required to apply the brakes.

Air cleaner A device, mounted on or connected to the engine air intake, for filtering dirt and dust out of the air being drawn into the engine.

Air filter A filter that removes dirt and dust particles from air passing through it.

Air-fuel ratio The proportions of air and fuel (by weight) supplied to the engine cylinders for combustion.

Air-injection system An exhaust emission control system that injects air at low pressure into the exhaust manifold to help complete the combustion of unburned hydrocarbons and carbon monoxide in the exhaust gas.

Air-mass metering In some fuel-injection systems, fuel metering controlled primarily by engine speed and the amount of air actually entering the engine. Also called airflow metering.

Air pollution Contamination of the air by natural and manufactured pollutants.

Air spring In the suspension system, a flexible bag filled with compressed air, which compresses to absorb shock.

Alternating current (AC or ac) An electric current that flows first in one direction and then in the other.

Alternator In the vehicle electric system, a device that converts mechanical energy into electric energy for charging the battery and operating electrical accessories. Also known as an ac generator.

Ampere A unit of measure for current. One ampere corresponds to a flow of 6.28×10^{18} electrons per second.

Anaerobic sealant A material that cures only when subjected to pressure and the absence of air; the material hardens when squeezed tightly between two surfaces.

Antifreeze A chemical, usually ethylene glycol, that is added to the engine coolant to raise its boiling point and lower its freezing point.

Antilock brake system A system installed with the brakes to prevent wheel lockup during braking.

Arcing Name given to the spark that jumps an air gap between two electrical conductors; for example, the arcing of the distributor contact points.

Armature A part moved by magnetism, or a part moved through a magnetic field to produce current.

Aspect ratio The ratio of tire height to width. For example, a G78 tire is 78 percent as high as it is wide. The lower the number, the wider the tire.

Atom The smallest particle into which an element can be divided.

Atomization The spraying of a liquid through a nozzle so that the liquid is broken into a very fine mist.

Automatic choke A device that positions the choke valve automatically in accordance with engine temperature or time.

Automatic level control A suspension system that compensates for variations in load in the rear of the car; positions the rear at a predesigned level regardless of load.

Automatic transmission A transmission in which gear ratios are changed automatically, eliminating the necessity of hand-shifting gears.

Automatic-transmission fluid Any of several types of special oil used in automatic transmissions.

Axle A theoretical or actual crossbar supporting a vehicle on which one or more wheels turn.

Axle ratio The ratio between the rotational speed (rpm) of the drive shaft and that of the driven wheel; gear reduction in the final drive, determined by dividing the number of teeth on the ring gear by the number of teeth on the pinion gear.

Backlash In gearing, the clearance between the meshing teeth of two gears.

Ball joint A flexible joint consisting of a ball within a socket, used in front-suspension systems and valve-train rocker arms.

Ball-joint inclination See Steering-axis inclination.

Band In an automatic transmission, a hydraulically controlled brake band installed around a metal clutch drum; used to stop or permit drum rotation.

Battery An electrochemical device for storing energy in chemical form so that it can be released as electricity; a group of electric cells connected together.

Battery acid The electrolyte used in a battery; a mixture of sulfuric acid and water.

BDC See Bottom dead center.

Bead That part of the tire which is shaped to fit the rim; the bead is made of steel wires, wrapped and reinforced by the plies of the tire.

Bearing A part that transmits a load to a support and in so doing absorbs most of the friction and wear of the moving parts. A bearing is usually replaceable.

Belt In a tire, a flat strip of material—fiberglass, rayon, or woven steel—which underlies the tread, all around the circumference of the tire.

Bias-ply tire A tire in which the plies are laid diagonally, crisscrossing one another at an angle of about 30 to 40 degrees.

Bleeding A process by which air is removed from a hydraulic system (brake or power steering) by draining part of the fluid or operating the system to work out the air.

Blowby Leakage of compressed air-fuel mixture and burned gases (from combustion) past the piston rings into the crankcase.

Boost pressure The pressure in the intake manifold while the turbocharger is operating.

Bottom dead center (BDC) The piston position at the lower limit of its travel in the cylinder, when the cylinder volume is at its maximum.

Brake An energy-conversion device used to slow, stop, or hold a vehicle or mechanism; a device that changes the kinetic energy of motion into useless and wasted heat energy.

Brake drum A metal drum mounted on a car wheel to form the outer shell of the brake; the brake shoes press against the drum to slow or stop drum-and-wheel rotation for braking.

Brake fluid A special fluid used in hydraulic brake systems to transmit force through a closed system of tubing known as the brake lines.

Brake horsepower Power available from the engine crankshaft to do work, such as moving the vehicle; bhp = torque × rpm/5252.

Brakelights Lights at the rear of the car which indicate that the brakes are applied.

Brake lines The tubes and hoses connecting the master cylinder to the wheel cylinders or calipers in a hydraulic brake system.

Brake lining A high-friction material, often a form of asbestos, attached to the brake shoe by rivets or a bonding process. The lining takes the wear when the shoe is pressed against the brake drum, or rotor.

Brush A block of conducting substance, such as carbon, which rests against a rotating ring or commutator to form a continuous electric circuit.

BTDC Abbreviation for before top dead center; any position of the piston between bottom dead center and top dead center, on the upward stroke.

Bushing A one-piece sleeve placed in a bore to serve as a bearing surface.

Cables Stranded conductors, usually covered with insulating material, used for connections between electric devices.

Caliper In a disk brake, a housing for pistons and brake shoes, connected to the hydraulic system; holds the brake shoes so that they straddle the disk.

Cam A rotating lobe or eccentric that can be used with a cam follower to change rotary motion to reciprocating motion.

Camber The tilt of the top of the wheels from the vertical; when the tilt is outward, the camber is positive. Also, the angle which a front-wheel spindle makes with the horizontal.

Camshaft The shaft in the engine which has a series of cams for operating the valve mechanisms. It is driven by gears, or by sprockets and a toothed belt or chain from the crankshaft.

Carbon (C) A black deposit that is left on engine parts such as pistons, rings, and valves by the combustion of fuel and which inhibits their action.

Carbon dioxide (CO_2) A colorless, odorless gas that results from complete combustion; usually considered harmless. The gas absorbed from air by plants in photosynthesis; also used to carbonate beverages.

Carbon monoxide (CO) A colorless, odorless, tasteless, poisonous gas that results from incomplete combustion. A pollutant contained in engine exhaust gas.

Carburetor The device in an engine fuel system which mixes fuel with air and supplies the combustible mixture to the intake manifold.

Caster Tilting of the steering axis forward or backward to provide directional steering stability. Also, the angle which a front-wheel kingpin makes with the vertical.

Catalytic converter A mufflerlike device for use in an exhaust system. It converts harmful exhaust gases into harmless gases by promoting a chemical reaction between the catalysts and the pollutants.

Centrifugal advance A rotating-weight mechanism in the distributor that advances and retards ignition timing through the centrifugal force resulting from changes in the rotational speed of the distributor shaft.

Cetane number A measure of the ignition quality of diesel fuel, or how high a temperature is required to ignite it. The lower the cetane number, the higher the temperature required to ignite a diesel fuel.

Charcoal canister A container filled with activated charcoal; used to trap gasoline vapor from the fuel tank and carburetor while the engine is off.

Charging rate The amperage flowing from the alternator into the battery.

Chassis The assembly of mechanisms that makes up the major operating part of the vehicle; usually assumed to include everything except the car body.

Check valve A valve that opens to permit the passage of air or fluid in one direction only, or operates to prevent (check) some undesirable action.

Chip One or more integrated circuits manufactured as a very small package capable of performing many functions.

Choke In the carburetor, a device used when starting a cold engine. It "chokes off" the airflow through the air horn, producing a partial vacuum in the air horn for greater fuel delivery and a richer mixture. It operates automatically on most cars.

Circuit The complete path of an electric current, including the current source. When the path is continuous, the circuit is closed and current flows. When the path is broken, the circuit is open and no current flows. Also used to refer to fluid paths, as in refrigerant and hydraulic systems.

Circuit breaker A resettable protective device that opens an electric circuit to prevent damage when the circuit is overheated by excess current flow.

Clearance The space between two moving parts, or between a moving and a stationary part, such as a journal and a bearing. The bearing clearance is filled with lubricating oil when the mechanism is running.

Clutch A coupling that connects and disconnects a shaft from its drive while the drive mechanism is running. In an automobile power train, the device which engages and disengages the transmission from the engine. In an air-conditioning system, the device that engages and disengages the compressor shaft from its continuously rotating drive-belt pulley.

Clutch shaft The shaft on which the clutch is assembled, with the gear that drives the countershaft in the transmission on one end. On the clutch-gear end, it has external splines that can be used by a synchronizer drum to lock the clutch shaft to the main shaft for direct drive.

Clutch solenoid In automotive air conditioners, a solenoid that operates a clutch on the compressor drive pulley. When the clutch is engaged, the compressor is driven and cooling takes place.

Coil In an automobile ignition system, a transformer used to step up the battery voltage (by induction) to the high voltage required to fire the spark plugs.

Combination valve A brake-warning-lamp valve in combination with a proportioning and/or metering valve.

Combustion Burning; fire produced by the proper combination of fuel, heat, and oxygen. In the engine, the rapid burning of air and fuel in the combustion chamber.

Combustion chamber The space between the top of the piston and the cylinder head, in which the fuel is burned.

Commutator A series of copper bars at one end of a generator or starting-motor armature, electrically insulated from the armature shaft and insulated from one another by mica. The brushes rub against the bars of the commutator, which form a rotating connector between the armature windings and brushes.

Compression Reduction in the volume of a gas by squeezing it into a smaller space. Increasing the pressure reduces the volume and increases the density and temperature of the gas.

Compression-ignition engine An engine operating on the diesel cycle, in which the fuel is injected into the cylinders, where the heat of compression ignites it.

Compression ratio The volume of the cylinder and combustion chamber when the piston is at BDC, divided by the volume when the piston is at TDC.

Compression stroke The piston movement from BDC to TDC immediately following the intake stroke, during which both the intake and exhaust valves are closed while the air or air-fuel mixture in the cylinder is compressed.

Compressor The component of an air-conditioning system that compresses refrigerant vapor to increase its pressure and temperature.

Computer Command Control (CCC) system Electronic engine control system with a feedback carburetor or fuel-injection system, and a self-diagnostic capability, used by General Motors.

Condensation A change of state during which a gas turns to liquid, usually because of temperature or pressure changes. Also, moisture from the air, deposited on a cool surface.

Condenser In the ignition system, a capacitor connected across the contact points to reduce arcing by providing a storage place for electricity (electrons) as the contact points open. In an air-conditioning system, the radiatorlike heat exchanger in which refrigerant vapor loses heat and returns to the liquid state.

Conductor Any material or substance that allows current or heat to flow easily.

Connecting rod In the engine, the rod that connects the crank on the crankshaft with the piston.

Constant-velocity joint Two closely coupled universal joints arranged in such a way that their acceleration-deceleration effects cancel out. This results in an output drive-shaft speed that is always identical with the input drive-shaft speed, regardless of the angle of drive.

Contact points In the contact-point ignition system, the stationary and the movable points in the distributor which open and close the ignition primary circuit.

Contaminants Anything other than refrigerant and refrigerant oil in a refrigeration system; includes rust, dirt, moisture, and air.

Control arm A part of the suspension system designed to control wheel movement precisely.

Control system A system in which one or more outputs are forced to change as time progresses. The basic control system has three parts: (1) sensors (inputs) that measure various conditions and send this information to (2) the control unit (system decision maker) which decides how much change, if any, is needed and then signals one or more (3) actuators (outputs) to take the appropriate action.

Coolant The liquid mixture of about 50 percent antifreeze and 50 percent water used to carry heat out of the engine.

Cooling system The system that removes heat from the engine by the forced circulation of coolant and thereby prevents engine overheating. In a liquid-cooled engine, it includes the water jackets, water pump, radiator, and thermostat.

Countershaft The shaft in the transmission which is driven by the clutch gear; gears on the countershaft drive gears on the main shaft when the latter are engaged.

Coupling point In a torque converter, the speed at which the oil begins to strike the back faces of the stator vanes; occurs when the turbine and pump speeds reach a ratio of approximately 9:10.

Crankcase The lower part of the engine in which the crankshaft rotates, including the lower section of the cylinder block and the oil pan.

Crankcase emissions Pollutants emitted into the atmosphere from any portion of the engine crankcase ventilation or lubrication system.

Crankcase ventilation The circulation of air through the crankcase of a running engine to remove water, blowby, and other vapors; prevents oil dilution, contamination, sludge formation, and pressure buildup.

Crankshaft The main rotating member, or shaft, of the engine, with cranks to which the connecting rods are attached; converts up-and-down (reciprocating) motion into circular (rotary) motion.

Crankshaft gear A gear, or sprocket, mounted on the front of the crankshaft; used to drive the camshaft gear, chain, or toothed belt.

Cross-fire injection Type of gasoline fuel-injection system using two throttle-body injection units to supply air-fuel mixture through passages that cross over to feed cylinders on the opposite side of the engine.

Curb idle The normal hot-idle speed of an engine.

Current A flow of electrons, measured in amperes.

Cycle A series of events that repeat themselves. In the automotive engine, the four piston strokes that together produce the power.

Cycling-clutch orifice-tube system An air-conditioning system in which a small restriction (the orifice tube) in the refrigerant line acts as the flow-control valve and the compressor clutch is automatically engaged and disengaged (cycling clutch) to prevent evaporator icing.

Cylinder block The basic framework of the engine, in and on which the other engine parts are attached. It includes the engine cylinders and the upper part of the crankcase.

Cylinder head The part of the engine that covers and encloses the cylinders. It contains cooling fins or water jackets and the valves.

Cylinder liner See Cylinder sleeve.

Cylinder sleeve A replaceable sleeve, or liner, set into the cylinder block to form the cylinder bore.

Dashpot A device that controls the rate at which the throttle valve closes.

Defroster The part of the car heater system designed to melt frost or ice on the inside or outside of the windshield; includes the required duct work.

Desiccant A drying agent. In a refrigeration system, desiccant is placed in the receiver or accumulator to remove any moisture circulating in the system.

Detent A small depression in a shaft, rail, or rod into which a pawl or ball drops when the shaft, rail, or rod is moved; this provides a locking effect.

Detergent A chemical added to engine oil that helps keep internal parts of the engine clean by preventing the accumulation of deposits.

Detonation Commonly referred to as spark knock or ping. In the combustion chamber of a spark-ignition engine, an uncontrolled second explosion (after the spark occurs at the spark plug) with spontaneous combustion of the remaining compressed air-fuel mixture, resulting in a pinging sound.

Diagonal-brake system A dual-brake system with separate hydraulic circuits connecting diagonal wheels together (RF to LR and LF to RR).

Dial indicator A gauge that has a dial face and a needle to register movement; used to measure variations in dimensions and distances too small to be measured accurately by other means.

Diaphragm A thin dividing sheet or partition that separates an area into compartments; used in fuel pumps, modulator valves, vacuum-advance units, and other control devices.

Diesel cycle An engine operating cycle in which air is compressed and then fuel oil is injected into the compressed air at the end of the compression stroke. The heat produced by the compression ignites the fuel oil, eliminating the need for an electronic ignition system.

Diesel engine An engine operating on the diesel cycle and burning diesel fuel oil instead of gasoline.

Diesel fuel A light oil sprayed into the cylinders of a diesel engine near the end of the compression stroke.

Dieseling A condition in which a spark-ignition engine continues to run after the ignition is off; caused by carbon deposits or hot spots in the combustion chamber glowing sufficiently to furnish heat for combustion.

Differential A gear assembly between axles that permits one wheel to turn at a different speed from the other, while transmitting power from the drive shaft to the wheel axles.

Dimmer switch A two-position switch operated by the driver to select the high or low headlight beam.

Diode A solid-state electronic device that allows the passage of an electric current in one direction only. Used in the alternator to change alternating current to direct current for charging the battery.

Dipstick See Oil-level indicator.

Direct current (DC or dc) Electric current that flows in one direction only.

Direct drive Condition in a gearset when both the input shaft and the output shaft turn at the same speed, with a ratio of 1:1.

Directional signals A system on the car that flashes lights to indicate the direction in which the driver intends to turn.

Discharge To bleed some or all refrigerant from a system by opening a valve or connection and permitting the refrigerant to escape.

Disk brake A brake in which brake shoes, in a viselike caliper, grip a revolving disk to stop it.

Dispersant A chemical added to oil to prevent dirt and impurities from clinging together in lumps that could clog the engine lubricating system.

Displacement In an engine, the total volume of air or air-fuel mixture an engine is theoretically capable of drawing into all cylinders during one operating cycle. Also, the volume swept out by the piston in moving from one end of a stroke to the other.

Distributor Any device that distributes. In the ignition system, the rotary switch that directs high-voltage surges to the engine cylinders in the proper sequence. See Ignition distributor.

Distributorless ignition system An electronic ignition system without a separate ignition distributor. Sensors signal the position of the crankshaft to the ECM, which then electronically times and triggers the system and controls distribution of the secondary voltages.

Diverter valve In the air-injection system, a valve that diverts air-pump output into the air cleaner or the atmosphere during deceleration; this prevents backfiring and popping in the exhaust system.

Drivability The general operation of a vehicle, usually rated from good to poor; based on characteristics of concern to the average driver, such as smoothness of idle, even acceleration, ease of starting, quick warm-up, and not overheating.

Drive line The driving connection between the transmission and the differential; made up of one or more drive shafts.

Driven disk The friction disk in a clutch.

Drive pinion A rotating shaft with a small gear on one end that transmits torque to another gear; used in the differential. Also called the clutch shaft, in the manual transmission.

Drive shaft An assembly of one or two universal joints connected to a shaft or tube; used to transmit power from the transmission to the differential. Also called the propeller shaft.

Drum brake A brake in which curved brake shoes press against the inner circumference of a metal drum to produce the braking action.

Dry-charged battery A new battery that has been charged, and then stored with the electrolyte removed. Electrolyte must be added to activate the battery at the time of sale.

Dual-bed catalytic converter A catalytic converter which combines two bead-type converters (with different catalysts) in a single housing to control HC, CO, and NO_x.

Dwell In a contact-point distributor, the number of degrees of distributor-cam rotation that the points stay closed before they open again.

Dynamic balance The balance of an object when it is in motion (for example, the dynamic balance of a rotating wheel).

Eccentric A disk or offset section (of a shaft, for example) used to convert rotary to reciprocating motion. Sometimes called a cam.

Efficiency The ratio between the power of an effect and the power expended to produce the effect; the ratio between an actual result and the theoretically possible result.

Electric current A movement of electrons through a conductor such as a copper wire; measured in amperes.

Electric system In the automobile, the system that electrically cranks the engine for starting; furnishes high-voltage sparks to the engine cylinders to fire the compressed air-fuel charges; lights the lights; and powers the heater motor, radio, and other accessories. Consists, in part, of the starting motor, wiring, battery, alternator, regulator, ignition distributor, and ignition coil.

Electrode In a spark plug, the spark jumps between two electrodes. The wire passing through the insulator is the center electrode. The small piece of metal welded to the spark-plug shell (and to which the spark jumps) is the side, or ground, electrode.

Electrolyte The mixture of sulfuric acid and water used in lead-acid storage batteries. The acid enters into chemical reaction with active material in the plates to produce voltage and current.

Electromagnet A coil of wire (usually around an iron core) that produces magnetism as electric current passes through it.

Electromagnetic induction The characteristic of a magnetic field that causes an electric current to be created in a conductor if it passes through the field, or if the field builds and collapses around the conductor.

Electron A negatively charged particle that circles the nucleus of an atom. The movement of electrons is an electric current.

Electronic control module (ECM) See Electronic control unit.

Electronic control unit (ECU) The system computer that receives information from sensors and is programmed to operate various circuits and systems based on that information.

Electronic engine control system An engine control system that uses various sensors to send information (in the form of electrical signals) to the electronic control unit, which then computes a control (output) signal that is sent to various actuators.

Electronic fuel-injection system A system that injects fuel into the engine and includes an electronic control unit to time and meter the fuel flow.

Electronic ignition system An ignition system that uses transistors and other semiconductor devices as an electric switch to turn the primary current on and off.

Electronics Electrical assemblies, circuits, and systems that use electronic devices such as transistors and diodes.

Emission control Any device or modification added onto or designed into a motor vehicle for the purpose of reducing air-polluting emissions.

End play As applied to a crankshaft, the distance that the crankshaft can move forward and back in the cylinder block.

Energy The capacity or ability to do work. The most common forms are heat, mechanical, electrical, and chemical. Usually measured in work units of pound-feet [kilogram-meters], but also expressed in heat-energy units (Btu's [joules]).

Engine A machine that converts heat energy into mechanical energy. A device that burns fuel to produce mechanical power; sometimes referred to as a power plant.

Engine control system Any system (controlled by the driver, automatically, or electronically) that varies engine inputs to obtain the desired engine output.

Engine efficiency The ratio of the power actually delivered to the power that could be delivered if the engine operated without any power loss.

Engine mounts Flexible rubber insulators through which the engine is bolted to the car body or frame.

Evacuate To use a vacuum pump to pump any air and moisture out of an air-conditioner refrigerant system; required whenever any component in the refrigerant system has been replaced.

Evaporation The transformation of a liquid to the gaseous state.

Evaporative control system A system that prevents the escape of fuel vapors from the fuel tank or air cleaner while the engine is off. The vapors are stored in a charcoal canister or in the engine crankcase until the engine is started.

Evaporator The heat exchanger in an air conditioner in which refrigerant changes from a liquid to a gas (evaporates), taking heat from the surrounding air as it does so.

Exhaust emissions Pollutants emitted into the atmosphere through any opening downstream of the exhaust ports of an engine.

Exhaust-gas analyzer A device for sensing the amounts of air pollutants in the exhaust gas of a motor vehicle. The analyzers used in automotive shops check HC and CO; those used in testing laboratories can also check NO_x.

Exhaust-gas recirculation (EGR) system An NO_x control system that recycles a small part of the inert exhaust gas back through the intake manifold to lower the combustion temperature.

Exhaust manifold A device with several passages through which exhaust gases leave the engine combustion chambers and enter the exhaust piping system.

Exhaust pipe The pipe connecting the exhaust manifold to the next component in the exhaust system.

Exhaust stroke The piston stroke (from BDC to TDC) immediately following the power stroke, during which the exhaust valve opens so that the exhaust gases can escape from the cylinder to the exhaust manifold.

Exhaust system The system that collects the exhaust gases and discharges them into the air. Consists of the exhaust manifold, exhaust pipe, muffler, tail pipe, and resonator (if used).

Expansion plug A slightly dished plug that is used to seal core passages in the cylinder block and cylinder head. When driven into place, it is flattened and expanded to fit tightly.

Expansion tank A tank connected by a hose to the filler neck of an automobile radiator; the tank provides room for heated coolant to expand and to give off any air that may be trapped in the coolant. Also, a similar device used in some fuel tanks to prevent fuel from spilling out of the tank through expansion.

Expansion valve A flow-control valve located between the condenser and evaporator in an air conditioner; controls the amount of refrigerant sprayed into the evaporator.

Fan The bladed device in back of the radiator that rotates to draw cooling air through the radiator or around the engine cylinders; an air blower, such as the heater fan and the A/C blower.

Fan belt A belt (or belts), driven by the crankshaft, whose primary purpose is to drive the engine fan and water pump.

Feedback carburetor A carburetor used with an oxygen sensor and a control system to automatically adjust the air-fuel ratio for minimum exhaust emissions.

Field coil A coil, or winding, in a generator or starting motor which produces a magnetic field as current passes through it.

Filter A device through which air, gases, or liquids are passed to remove impurities.

Filter separator A combination fuel filter and vapor separator located between the fuel pump and carburetor on some cars.

Final drive The final speed-reduction gearing in the power train.

Firing order The order in which the engine cylinders fire, or deliver their power strokes, beginning with number 1 cylinder.

Fixed-caliper disk brake Disk brake using a caliper which is fixed in position and cannot move; the caliper usually has four pistons, two on each side of the disk.

Flasher An automatic-reset circuit breaker used in the directional-signal and hazard-warning circuits.

Flex plate On a vehicle with automatic transmission, a light plate bolted to the crankshaft to which the transmission torque converter is attached.

Floating-caliper disk brake Disk brake using a caliper mounted through rubber bushings which permit the caliper to float, or move, when the brakes are applied; there is one large piston in the caliper.

Float level The float position at which the needle valve closes the fuel inlet to the carburetor, to prevent further delivery of fuel.

Fluid Any liquid or gas.

Fluid coupling A device in the power train consisting of two rotating members; transmits power from the engine, through a fluid, to the transmission.

Flush In an air conditioner, to wash out the refrigerant passages with Refrigerant-12 to remove contaminants. In a brake system, to wash out the hydraulic system and the master and wheel cylinders, or calipers, with clean brake fluid to remove dirt or impurities that have gotten into the system.

Flywheel On a vehicle with manual transmission, a heavy metal wheel attached to the crankshaft which rotates with it; helps smooth out the power surges from the engine power strokes; also serves as part of the clutch and engine cranking system.

Flywheel ring gear A gear, fitted around the flywheel, that is engaged by teeth on the starting-motor drive to crank the engine.

Force Any push or pull exerted on an object; measured in pounds and ounces, or in newtons (N) in the metric system.

Four-stroke cycle The four piston strokes—intake, compression, power, and exhaust—that make up the complete cycle of events in the four-stroke-cycle engine. Also called four-cycle and four-stroke.

Four-wheel drive A vehicle with drive axles at both front and rear, so that all four wheels can be driven.

Frame The assembly of metal structural parts and channel sections that supports the car engine and body and is supported by the wheels.

Freon-12 Refrigerant used in automobile air conditioners. Also known as Refrigerant-12 and R-12.

Friction The resistance to motion between two bodies in contact with each other.

Friction disk In the clutch, a flat disk, faced on both sides with friction material and splined to the clutch shaft. It is positioned between the clutch pressure plate and the engine flywheel. Also called the clutch disk or driven disk.

Front-end geometry The angular relationship between the front wheels, wheel-attaching parts, and car frame. Includes camber, caster, steering-axis inclination, toe, and turning radius.

Front-wheel drive A vehicle having its drive wheels located on the front axle.

Fuel Any combustible substance. In a spark-ignition engine, the fuel (gasoline) is burned and the heat of combustion expands the resulting gases, which force the piston downward and rotate the crankshaft.

Fuel filter A device located in the fuel line that removes dirt and other contaminants from fuel passing through.

Fuel-induction system In the spark-ignition engine, the system that supplies the air-fuel mixture to the cylinders; includes either a carburetor or a fuel-injection system.

Fuel-injection system A system that delivers fuel under pressure into the combustion chamber or into the intake airflow.

Fuel line The pipe or tubes through which fuel flows from the fuel tank to the carburetor or fuel-injection system.

Fuel pump The electrical or mechanical device in the fuel system which forces fuel from the fuel tank to the carburetor or fuel-injection system.

Fuel system In an automobile, the system that delivers the fuel and air to the engine cylinders. Consists of the fuel tank and lines, gauge, fuel pump, carburetor or fuel-injection system, and intake manifold.

Fuse A device designed to open an electric circuit when the current is excessive, to protect equipment in the circuit.

Fuse block A boxlike unit that holds the fuses for the various electric circuits in an automobile.

Fusible link A type of circuit protector in which a special wire melts to open the circuit when the current is excessive.

Gap The air space between two electrodes, as the spark-plug gap or the contact-point gap.

Gas The state of matter in which the matter has neither a definite shape nor a definite volume; air is a mixture of several gases. In an automobile, the discharge from the tail pipe is called the exhaust gas. Also, gas is a slang expression for the liquid fuel gasoline.

Gasket A thin layer of soft material, such as paper, cork, rubber, or copper, placed between two flat surfaces to make a tight seal.

Gasohol An engine fuel made by mixing 10 percent ethyl alcohol with 90 percent unleaded gasoline.

Gasoline A liquid blend of hydrocarbons, obtained from crude oil; used as the fuel in most automobile engines.

Gear lubricant A type of grease or oil designed especially to lubricate gears.

Gear ratio The number of revolutions of a driving gear required to turn a driven gear through one complete revolution. For a pair of gears, the ratio is found by dividing the number of teeth on the driven gear by the number of teeth on the driving gear.

Gears Mechanical devices that transmit power or turning force from one shaft to another; gears contain teeth that mesh as the gears turn.

Gearshift A linkage-type mechanism by which the gears in an automobile transmission are engaged and disengaged.

Generator A device that converts mechanical energy into electrical energy; can produce either ac or dc electricity. In automotive usage, a dc generator (now seldom used).

Glow plug A small electric heater installed in the precombustion chamber of diesel engines to preheat the chamber for easier starting in cold weather.

Governor A device that controls, or governs, another device, usually on the basis of speed or load.

Governor pressure Pressure at one end of a shift valve, which is controlled by governor action. One of two signals that determine when an automatic transmission will shift.

Grease Lubricating oil to which thickening agents have been added.

Ground The return path for current in an electric circuit.

Halogen headlamp A sealed-beam headlamp with a small inner bulb filled with halogen which surrounds the tungsten filament.

Heat of compression Increase of temperature brought about by the compression of air or air-fuel mixture; the source of ignition in a diesel engine.

Heat-control valve In the engine, a thermostatically operated valve in the exhaust manifold; diverts heat to the intake manifold to warm it before the engine reaches normal operating temperature.

Heater core A small radiator mounted under the passenger side of the dash, through which hot coolant circulates; when heat is needed, air is circulated through the hot core.

Heat sink A device for absorbing heat from one medium and transferring it to another. The diodes in alternators are usually mounted in heat sinks which remove the heat from the diodes to prevent them from overheating.

High-energy ignition (HEI) system A General Motors electronic ignition system without contact points and with all ignition-system components contained in the distributor. Capable of producing 35,000 volts, or more.

Horn relay A relay connected between the battery and the horns. When the horn button is pressed, the relay is energized; it then connects the horns to the battery.

Horsepower A measure of mechanical power, or the rate at which work is done. One horsepower equals 33,000 ft-lb (foot-pounds) of work per minute, the power necessary to raise 33,000 pounds a distance of one foot in one minute.

Hub The center part of a wheel.

Hydraulic brake booster A brake booster that uses hydraulic pressure supplied by the power-steering pump to assist in applying the brakes.

Hydraulic brakes A braking system that uses hydraulic pressure to force the brake shoes against the brake drums, or rotors, as the brake pedal is depressed.

Hydraulics The use of a liquid under pressure to transfer force or motion, or to increase an applied force.

Hydraulic valve lifter A valve lifter that uses oil pressure from the engine lubricating system to keep the lifter in constant contact with the cam lobe and with the valve stem, pushrod, or rocker arm.

Hydrocarbon (HC) A compound containing only carbon and hydrogen atoms, usually derived from fossil fuels such as petroleum, natural gas, and coal; an agent in the formation of photochemical smog. Gasoline is a blend of liquid hydrocarbons refined from crude oil.

Hydrogen (H) A colorless, odorless, highly flammable gas whose combustion produces water; the simplest and lightest element.

Hydropneumatic suspension A suspension system using air, gas, fluid, or a combination of these.

Idle-mixture screw The adjustment screw (on some carburetors) that can be turned in or out to lean or enrich the idle mixture.

Idler arm In the steering system, a link that supports the tie rod and transmits steering motion to both wheels through the tie-rod ends.

Idle solenoid An electrically operated plunger used to provide a predetermined throttle setting at idle.

Idle speed The speed, or rpm, at which the engine runs when the accelerator pedal is fully released and there is no load on the engine.

Ignition The action of the spark in starting the burning of the compressed air-fuel mixture in the combustion chamber of a spark-ignition engine. In a diesel engine, the start of the burning of fuel after its temperature has been raised by the heat of compression.

Ignition coil The ignition-system component that acts as a transformer to step up (increase) the battery voltage to many thousands of volts. The high-voltage surge from the coil is transmitted to the spark plug to ignite the compressed air-fuel mixture.

Ignition distributor The unit in the ignition system which usually contains the mechanical or electronic switch that closes and opens the primary circuit to the coil at the proper time and then distributes the resulting high-voltage surges to the spark plugs.

Ignition lag In a diesel engine, the delay in time between the injection of fuel and the start of combustion.

Ignition module See Electronic control unit.

Ignition resistor A resistance connected into the ignition primary circuit to reduce the battery voltage to the coil during engine operation.

Ignition switch The switch in the ignition system (usually operated with a key) that opens and closes the ignition-coil primary circuit. May also be used to open and close other vehicle electric circuits.

Ignition system In the automobile, the system that furnishes high-voltage sparks to the engine cylinders to fire the compressed air-fuel mixture. Consists of the battery, ignition coil, ignition distributor, ignition switch, wiring, and spark plugs.

Ignition timing The delivery of the spark from the coil to the spark plug at the proper time for the power stroke, relative to the piston position.

Included angle In the front-suspension system, camber angle plus steering-axis-inclination angle.

Induction The action of producing a voltage in a conductor or coil by moving the conductor or coil through a magnetic field, or by moving the field past the conductor or coil.

Inertia The property of an object that causes it to resist any change in its speed or direction of travel.

Infrared analyzer A test instrument used to measure very small quantities of pollutants in exhaust gas. See Exhaust-gas analyzer.

In-line steering gear A type of integral power steering; uses a recirculating-ball steering gear to which is added a control valve and an actuating piston.

Instrument voltage regulator A thermostatic device that keeps the voltage to the instrument-panel gauges at about 5 volts.

Insulation Material that stops the travel of electricity (electrical insulation) or heat (heat insulation).

Intake manifold A casting with several passages through which air or air-fuel mixture flows from the air intake or carburetor to the ports in the cylinder head or cylinder block.

Intake stroke The piston stroke from TDC to BDC immediately following the exhaust stroke, during which the intake valve opens and the cylinder fills with air or air-fuel mixture from the intake manifold.

Integrated circuit Many very small solid-state devices capable of performing as a complete electronic circuit, with one or more integrated circuits manufactured as a "chip."

Interaxle differential A two-position differential located between two driving axles; used in the transfer case of some four-wheel-drive vehicles.

Internal gear A gear with teeth pointing inward, toward the hollow center of the gear.

Jet A calibrated passage in the carburetor through which fuel flows.

Journal The part of a rotating shaft which turns in a bearing.

Jump starting Starting a car that has a dead battery by connecting a charged battery to the starting system.

Kickdown In automatic transmissions, a system that produces a downshift when the accelerator is pushed down to the floorboard.

Knock A heavy metallic engine sound that varies with engine speed; usually caused by a loose or worn bearing; name also used for detonation, pinging, and spark knock. See Detonation.

Knuckle A steering knuckle; a front-suspension part that acts as a hinge to support a front wheel and permit it to be turned to steer the car. The knuckle pivots on ball joints or, in older cars, on kingpins.

Lash The amount of free motion in a gear train, between gears, or in a mechanical assembly, such as the lash in a valve train.

Lead (pronounced "leed") A cable or conductor that carries electric current.

Lead (pronounced "led") A heavy metal; used in lead-acid storage batteries.

Leak detector Any device used to locate an opening where refrigerant may escape. Common types are flame, electronic, dye, and soap bubbles.

Light A gas-filled bulb enclosing a wire that glows brightly when an electric current passes through it; a lamp. Also, any visible radiant energy.

Limited-slip differential A differential designed so that when one wheel is slipping, the most torque is supplied to the wheel with the best traction; also called a nonslip differential.

Linkage A hydraulic system, or an assembly of rods or links, used to transmit motion.

Locking hubs Hubs that can be disengaged so that the front axle can freewheel while the vehicle is in two-wheel drive.

Locking torque converter A torque converter in which the pump can be mechanically locked to the turbine, eliminating any loss through the fluid.

Lubricating system The system in the

engine that supplies engine parts with lubricating oil to prevent contact between any two moving metal surfaces.

MacPherson-strut suspension A suspension system which combines a coil spring and a shock absorber into a single strut assembly that requires only a lower control arm.

Magnetic Having the ability to attract iron. This ability may be permanent, or it may depend on a current flow, as in an electromagnet.

Magnetic clutch A magnetically operated clutch used to engage and disengage the air-conditioner compressor.

Magnetic field The space around a magnet which is filled by invisible lines of force.

Magnetic lines of force The imaginary lines by which a magnetic field may be visualized.

Magnetic switch A switch with a winding which, when energized by connection to a battery or alternator, causes the switch to open or close a circuit.

Magnetism The ability, either natural or produced by a flow of electric current, to attract iron.

Main bearings In the engine, the bearings that support the crankshaft.

Main jet The fuel nozzle, or jet, in the carburetor that supplies the fuel when the throttle is partially to fully open.

Manifold A device with several inlet or outlet passageways through which a gas or liquid is gathered or distributed. See Exhaust manifold, Intake manifold, and Manifold gauge set.

Manifold gauge set A high-pressure and a low-pressure gauge mounted together as a set; used for checking pressures in the air-conditioning system.

Manifold pressure The pressure in the intake manifold while the turbocharger is operating.

Manifold vacuum The vacuum in the intake manifold that develops as a result of the vacuum in the cylinders on their intake strokes.

Manual valve A spool valve in the valve body of an automatic transmission that is manually positioned by the driver through linkage.

Master cylinder The liquid-filled cylinder in the hydraulic braking system or clutch where hydraulic pressure is developed when the driver depresses a foot pedal.

Meshing The mating, or engaging, of the teeth of two gears.

Metering rod and jet A device consisting of a small, movable, cone- or step-shaped rod and a jet; used to increase or decrease fuel flow according to engine throttle opening, engine load, or a combination of both.

Metering valve A valve in the disk-brake system that prevents hydraulic pressure to the front brakes until after the rear brakes are applied.

Microprocessor The small, on-board solid-state electronic device that acts as the central processing unit. Sensors provide input information which the microprocessor uses to determine the desired response (if any) as an output signal.

Modulator A pressure-regulated governing device; used, for example, in automatic transmissions.

Molecule The smallest particle into which a substance can be divided and still retain the properties of that substance.

Motor A device that converts electrical energy into mechanical energy; for example, the starting motor.

Muffler In the engine exhaust system, a device through which the exhaust gases must pass and which reduces the exhaust noise. In an air-conditioning system, a device to minimize pumping sounds from the compressor.

Multiple-disk clutch A clutch with more than one friction disk; usually there are several driving disks and several driven disks, alternately placed.

Multiple-point injection A gasoline fuel-injection system in which only air enters the intake manifold. As the air approaches the intake valve, an injection valve opens in the valve port, spraying fuel into the airstream. Also called port injection.

Multiple-viscosity oil An engine oil that has a low viscosity when cold (for easier cranking) and a higher viscosity when hot (to provide adequate engine lubrication).

Mutual induction The condition in which a voltage is induced in one coil by a changing magnetic field caused by a changing current in another coil. The magnitude of the induced voltage depends on the number of turns in the two coils.

Needle valve A small, tapered, needle-pointed valve that can move into or out of a seat to close or open the passage through it. Used to control the fuel level in the carburetor float bowl.

Negative One of the two poles of a magnet, or one of the two terminals of an electrical device.

Negative terminal The terminal from which electrons flow in a complete electric circuit. On a battery, the negative terminal can be identified as the battery post with the smaller diameter; the minus sign (−) is often also used to identify the negative terminal.

Neutral In a transmission, the setting in which all gears are disengaged and the output shaft is disconnected from the drive wheels.

Neutral-start switch A switch wired into the ignition switch to prevent engine cranking unless the transmission selector lever is in NEUTRAL.

Nitrogen (N) A colorless, tasteless, odorless gas that constitutes 78 percent of the atmosphere by volume and is a part of all living tissues.

Nitrogen oxides (NO_x) Any chemical compound of nitrogen and oxygen; a basic air pollutant. Automotive exhaust emission levels of nitrogen oxides are limited by law.

North pole The pole from which the lines of force leave a magnet.

Nozzle The opening, or jet, through which fuel or air passes as it is discharged.

Octane rating A measure of the antiknock properties of a gasoline. The higher the octane rating, the more resistant the gasoline is to spark knock or detonation.

Ohm The unit of electrical resistance.

Oil A liquid lubricant usually made from crude oil and used for lubrication between moving parts. In a diesel engine, oil is used for fuel.

Oil clearance The space between the bearing and the shaft rotating within it.

Oil cooler A small radiator that lowers the temperature of oil flowing through it.

Oil filter A filter that removes impurities from the engine oil passing through it.

Oil-level indicator The dipstick that is removed and inspected to check the level of oil in the crankcase of an engine or compressor.

Oil pan The detachable lower part of the engine which encloses the crankcase and acts as an oil reservoir.

Oil-pressure indicator A gauge that indicates (to the driver) the oil pressure in the engine lubricating system, or a light that comes on if the oil pressure drops too low.

Oil pump In the lubricating system, the device that forces oil from the oil pan to the moving engine parts.

Open circuit In an electric circuit, a break or opening which prevents the passage of current.

Orifice A small opening, or hole, into a cavity.

Orifice tube A restriction in the refrigerant line of an air conditioner that acts as a flow-control valve.

O ring A type of sealing ring made of a special rubberlike material; in use, the O ring is compressed into a groove to provide the sealing action.

Oscilloscope A high-speed voltmeter that visually displays voltage variations on a television-type picture tube. Used to check engine ignition systems; also used to check charging systems and electronic fuel-injection systems.

Output shaft The main shaft of the transmission; the shaft that delivers torque from the transmission to the drive shaft.

Overdrive Transmission gearing that causes the output shaft to overdrive, or turn faster than the input shaft.

Overhead-camshaft (OHC) engine An engine in which the camshaft is mounted over the cylinder head, instead of inside the cylinder block.

Overhead-valve (OHV) engine An engine in which the valves are mounted in the cylinder head above the combustion chamber, instead of in the cylinder block. In this type of engine, the camshaft is mounted in the cylinder block, and the valves are actuated by pushrods.

Overrunning clutch A drive unit that transmits rotary motion in one direction only. In the other direction, the driving member overruns and does not pass the motion to the other member. Widely used as the drive mechanism for starting motors.

Oxygen (O) A colorless, tasteless, odorless, gaseous element that makes up about 21 percent of air. Capable of combining rapidly with all elements except the inert gases in the oxidation process called burning. Combines slowly with many metals in the oxidation process called rusting.

Oxygen sensor A device in the exhaust pipe which measures the amount of oxygen in the exhaust gas and sends this information as a varying voltage signal to the electronic control unit.

Parallel circuit The electric circuit formed when two or more electrical devices have their terminals connected, positive to positive and negative to negative, so that each may operate independently of the others from the same power source.

Parking brake Mechanically operated brake that is independent of the foot-operated service brakes on the vehicle; set when the vehicle is parked.

Particle A very small piece of metal, dirt, or other impurity which may be contained in the air, fuel, or lubricating oil used in an engine.

Passage A small hole or gallery in an assembly or casting, through which air, coolant, fuel, or oil flows.

Pawl An arm pivoted so that its free end can fit into a detent, slot, or groove at certain times to hold a part stationary.

PCV valve The valve that controls the flow of crankcase vapors in accordance with ventilation requirements for different engine speeds and loads.

Petroleum The crude oil from which gasoline, lubricating oil, and other such products are refined.

Photochemical smog Smog caused by hydrocarbons and nitrogen oxides reacting photochemically in the atmosphere. The reactions take place under low wind velocity, bright sunlight, and an inversion layer in which the air mass is trapped (as between the ocean and mountains in Los Angeles). Can cause eye and lung irritation.

Pickup coil In an electronic ignition system, the coil in which voltage is induced by the reluctor.

Pilot bearing A small bearing, in the center of the flywheel end of the crankshaft, which carries the forward end of the clutch shaft.

Ping Engine spark knock or detonation that occurs usually during acceleration. Caused by excessive advance of ignition timing or low-octane fuel.

Pinion gear The smaller of two meshing gears.

Piston A movable part, fitted into a cylinder, which can receive or transmit motion as a result of pressure changes in a fluid. In the engine, the round plug that slides up and down in the cylinder and which, through the connecting rod, forces the crankshaft to rotate.

Piston engine An internal-combustion engine using reciprocating pistons.

Piston pin The cylindrical or tubular metal piece that attaches the piston to the connecting rod. Also called the wrist pin.

Piston rings Rings fitted into grooves in the piston. There are two types: compression rings for sealing the compression pressure in the combustion chamber, and oil rings to scrape excessive oil off the cylinder wall.

Pitman arm In the steering system, the arm that is connected between the steering-gear sector shaft and the steering linkage, or tie rod. It swings back and forth for steering as the steering wheel is turned.

Pivot A pin or shaft upon which another part rests or turns.

Planetary-gear system A gear set consisting of a central sun gear surrounded by two or more planet pinions which are, in turn, meshed with a ring (or internal) gear; used in automatic transmissions and transfer cases.

Planet carrier In a planetary-gear system, the carrier or bracket that contains the shaft upon which the planet pinion turns.

Planet pinions In a planetary-gear system, the gears that mesh with, and revolve about, the sun gear; they also mesh with the ring (or internal) gear.

Plastigage A plastic material available in strips of various sizes; used to measure crankshaft main-bearing and connecting-rod-bearing clearances.

Plate In a battery, a rectangular sheet of spongy lead. Sulfuric acid in the electrolyte chemically reacts with the lead to produce an electric current.

Plies The layers of cord in a tire casing; each of these layers is a ply.

Polarity The quality of an electric component or circuit that determines the direction of current flow.

Pollutant Any substance that adds to the pollution of the atmosphere. In a vehicle, any such substance in the exhaust gas from the engine or escaping from the fuel tank or air cleaner.

Port In the engine, the passage to the cylinder opened and closed by a valve, and through which gases flow to enter and leave the cylinder.

Ported vacuum switch (PVS) A coolant-temperature-sensing vacuum-control valve used in distributor and EGR vacuum systems. Sometimes called the vacuum-control valve or coolant override valve.

Port injection See Multiple-point injection.

Positive crankcase ventilation (PCV) A crankcase ventilation system; uses intake-manifold vacuum to return the crankcase vapors and blowby gases from the crankcase to the intake manifold to be burned, thereby preventing their escape into the atmosphere.

Positive terminal The terminal to which electrons flow in a complete electric circuit. On a battery, the positive terminal can be identified as the battery post with the larger diameter; the plus sign (+) is often also used to identify the positive terminal.

Post A point at which a cable is connected to the battery.

Potential energy Energy stored in a body because of its position. A weight raised to a height has potential energy because it can do work coming down. Likewise, a tensed or compressed spring contains potential energy.

Power The rate at which work is done. A common power unit is the horsepower, which is equal to 33,000 ft-lb/min (foot-pounds per minute).

Power brakes A brake system that uses hydraulic or vacuum and atmospheric pressure to provide most of the force required for braking.

Power plant The engine or power source of a vehicle.

Power steering A steering system that uses hydraulic pressure from a pump to multiply the driver's steering force.

Power stroke The piston stroke from TDC to BDC immediately following the compression stroke, during which both valves are closed and the fuel burns, expanding the compressed gas, thereby forcing the piston down to transmit power to the crankshaft.

Power train The mechanisms that carry power from the engine crankshaft to the drive wheels; includes the clutch, transmission, drive shaft, differential, and axles.

Precombustion chamber In some engines, a separate small combustion chamber where combustion begins.

Preignition Ignition of the air-fuel mixture in the combustion chamber by some unwanted means, before the ignition spark occurs at the spark plug.

Preload In bearings, the amount of load placed on a bearing before actual operating loads are imposed. Proper preloading requires bearing adjustment and ensures alignment and minimum looseness in the system.

Pressure cap A radiator cap with valves which causes the cooling system to operate under pressure at a higher and more efficient temperature.

Pressure-differential valve The valve in a dual-brake system that turns on a warning light if the pressure drops in one part of the system.

Pressure plate That part of the clutch which exerts force against the friction disk; it is mounted on and rotates with the flywheel.

Pressure regulator A device that operates to prevent excessive pressure from developing. In hydraulic systems, a valve that opens to release oil from a line when the oil pressure reaches a specified maximum.

Pressure-relief valve A valve that opens to relieve excessive pressure.

Pressurize To apply more than atmospheric pressure to a gas or liquid.

Prevailing-torque fasteners Nuts and bolts designed to have a continuous resistance to turning.

Primary The low-voltage circuit of the ignition system.

Primary winding The outer winding of relatively heavy wire in an ignition coil.

Printed circuit An electric circuit made by applying a conductive material to an insulating board in a pattern that provides current paths between components mounted on or connected to the board.

Propane enrichment Procedure for setting the idle mixture on some cars with catalytic converters.

Proportioning valve A valve that reduces pressure to the rear wheels during hard braking.

Purge To remove, evacuate, or empty trapped substances from a space. In an air conditioner, to remove moisture and air from the refrigerant system by flushing with nitrogen or Refrigerant-12.

Purge valve A valve used on some charcoal canisters in evaporative emission control systems; limits the flow of vapor and air to the carburetor during idling.

Races The metal rings on which ball or roller bearings rotate.

Rack-and-pinion steering gear A steering gear in which a pinion on the end of the steering shaft meshes with a rack of gear teeth on the major cross member of the steering linkage.

Radial tire A tire in which the plies are placed radially, or perpendicular to the rim, with a circumferential belt on top of them. The rubber tread is vulcanized on top of the belt and plies.

Radiator In the cooling system, the device that removes heat from coolant passing through it; receives hot coolant from the engine and sends the coolant to the engine at a lower temperature.

Radio choke An electric coil used to prevent static in the radio caused by opening and closing of the contact points in the instrument voltage regulator.

Ratio Proportion; the relative amounts of two or more substances in a mixture. Usually expressed as a numerical relationship, as in 2:1.

Readout The visual delivery or display of information from an electronic device, circuit, or system.

Rear-end torque The reaction torque that acts on the rear-axle housing when torque is applied to the wheels; tends to turn the axle housing in a direction opposite to wheel rotation.

Receiver-dehydrator In a car air conditioner, a container for storing liquid refrigerant from the condenser. A sack of desiccant in this container removes small traces of moisture that may be left in the system after purging and evacuating.

Recirculation-ball-and-nut steering gear A type of steering gear in which a nut, meshing with a gear sector, is assembled on a worm gear; balls circulate between the nut and worm threads.

Rectifier A device that changes alternating current to direct current; in the alternator, a diode.

Refrigerant A substance used to transfer heat in an air conditioner, through a cycle of evaporation and condensation.

Refrigeration Cooling of an object or a substance by removal of heat through mechanical means.

Regulator In the charging system, a device that controls alternator output to prevent excessive voltage.

Relay An electrical device that opens or closes a circuit or circuits in response to a voltage signal.

Relief valve A valve that opens when a preset pressure is reached. This relieves or prevents excessive pressures.

Reluctor In an electronic ignition system, the metal rotor (with a series of tips) which replaces the cam used in a contact-point distributor.

Reserve capacity A battery rating; the number of minutes a battery can deliver a 25-ampere current before the cell voltages drop to 1.75 volts per cell.

Residual line pressure A slight pressure maintained in the hydraulic system for some drum-brake systems by a check valve in the master cylinder.

Resistance The opposition to a flow of current through a circuit or electrical device; measured in ohms. A voltage of one volt will cause one ampere to flow through a resistance of one ohm. This is known as Ohm's law, which can be written in three ways: amperes = volts/ohms; ohms = volts/amperes; and volts = amperes × ohms.

Resonator A device in the exhaust system, similar to a muffler, that reduces exhaust noise.

Retard Usually associated with the engine spark-timing mechanisms; the opposite of spark advance. To delay the occurrence of the spark in the combustion chamber.

Return spring A pull-back spring, often used in brake systems.

Ring gear A large gear carried by the differential case; meshes with and is driven by the drive pinion.

Ring ridge The ridge formed at the top of a cylinder as the cylinder wall below is worn away by piston-ring movement.

Road-draft tube A method of removing fumes and pressure from the engine crankcase; used prior to crankcase emission control systems. The tube, which was connected into the crankcase and suspended slightly above the ground, depended on venturi action to create a partial vacuum as the vehicle moved. The method was ineffective below about 20 mph [32 km/h].

Rocker arm In engines with the valves in the cylinder head, a device that rocks on a shaft (or pivots on a stud) as the cam moves the pushrod, causing a valve to open.

Roller tappet A valve lifter with a hardened steel roller on the end riding against the camshaft.

Room-temperature-vulcanizing sealant See Aerobic gasket material.

Rotary engine An engine, such as a gas turbine or a Wankel, in which the power is delivered to a spinning rotor (and *not* to a reciprocating piston as in the piston engine).

Rotor A revolving part of a machine, such as an alternator rotor, disk-brake rotor, distributor rotor, or Wankel-engine rotor.

RTV sealant See Aerobic gasket material.

Run-on See Dieseling.

Runout Wobble.

Schrader valve A spring-loaded valve through which a connection can be made to a refrigeration system; also used in tires.

Screens Pieces of fine mesh used to prevent solid particles from circulating through any liquid or vapor system and damaging vital moving parts. In an air conditioner, screens are located in the receiver-dehydrator, expansion valve, orifice tube, and compressor.

Seal A material, shaped around a shaft, used to close off the operating compartment of the shaft, preventing oil leakage.

Sealed-beam headlight A headlight that contains the filament, reflector, and lens in a single sealed unit.

Seat The surface upon which another part rests, as a valve seat. Also, to wear into a good fit.

Secondary air Air that is pumped to thermal reactors, catalytic converters, exhaust manifolds, or the cylinder-head exhaust ports to promote the chemical reactions that reduce exhaust-gas pollutants.

Secondary circuit The high-voltage circuit of the ignition system; consists of the coil, rotor, distributor cap, spark-plug cables, and spark plugs.

Sector A gear that is not a complete circle; specifically, the gear sector on the pitman shaft in many steering systems.

Segments The copper bars of a commutator.

Self-adjuster A mechanism used on drum brakes; compensates for shoe wear by automatically adjusting the shoe-to-drum clearance.

Self-diagnostic system Indicating devices on the car that alert the driver when something is wrong in the system.

Self-induction The inducing of a voltage in a current-carrying coil of wire because the current in that wire is changing.

Semiconductor A material that acts as an insulator under some conditions and as a conductor under other conditions.

Sensor Any device that receives and reacts to a signal, such as a change in voltage, temperature, or pressure.

Series circuit An electric circuit in which the devices are connected end to end, positive terminal to negative terminal. The same current flows through all the devices in the circuit.

Serpentine belt A single multigrooved belt used with multigrooved pulleys on some engines to drive all the engine-driven devices (except the camshaft).

Service brake The foot-operated brake used for retarding, stopping, and controlling the vehicle during normal driving conditions.

Servo A device in a hydraulic system that converts hydraulic pressure to mechanical movement. Consists of a piston that moves in a cylinder as hydraulic pressure acts on it.

Shackle The swinging support by which one end of a leaf spring is attached to the car frame.

Shift lever The lever used to change gears in a transmission. Also, the lever on the starting motor which moves the drive pinion into or out of mesh with the flywheel teeth.

Shift valve In an automatic transmission, a valve that moves to produce the shifts from one gear ratio to another.

Shimmy Rapid oscillation. In wheel shimmy, for example, the front wheel turns in and out alternately and rapidly; this causes the front end of the car to oscillate or shimmy.

Shock absorber A device placed at each vehicle wheel to regulate spring rebound and compression.

Shoe In the brake system, a metal plate that supports the brake lining and absorbs and transmits braking forces.

Short circuit A defect in an electric circuit which permits current to take a short path, or circuit, instead of following the desired path.

Shroud A hood placed around an engine fan to improve air flow.

Side clearance The clearance between the sides of moving parts when the sides do not serve as load-carrying surfaces.

Sight glass In a car air conditioner, a viewing glass or window set in the refrigerant line, usually in the top of the receiver-dehydrator; the sight glass allows a visual check of the refrigerant passing from the receiver to the evaporator.

Silicone rubber See Aerobic gasket material.

Single-point injection See Throttle-body injection.

Slip joint In the power train, a variable-length connection that permits the drive shaft to change its effective length.

Slip rings In an alternator, the rings that form a rotating connection between the armature windings and the brushes.

Smog A term coined from the words "smoke" and "fog." First applied to the foglike layer that hangs in the air under certain atmospheric conditions; now generally used to describe any condition of dirty air and/or fumes or smoke.

Solenoid An electromechanical device which, when connected to an electrical source such as a battery, produces a mechanical movement. This movement can be used to control a valve or to produce other movements.

Solenoid relay A relay that connects a solenoid to a current source when its contacts close; specifically, the starting-motor solenoid relay.

Solenoid switch A switch that is opened and closed electromagnetically, by the movement of a solenoid core. Usually, the core also causes a mechanical action, such as the movement of a drive pinion into mesh with flywheel teeth for cranking.

Solid-state device A device that has no moving parts except electrons. Diodes and transistors are examples.

Solvent A cold liquid cleaner used to wash away grease and dirt.

South pole The pole at which magnetic lines of force enter a magnet.

Spark advance See Advance.

Spark-ignition engine An engine operating on the Otto cycle, in which the fuel is ignited by the heat from an electric spark as it jumps the gap at the end of the spark plug.

Spark plug The assembly, which includes a pair of electrodes and an insulator, that provides a spark gap in the engine cylinder.

Spark-plug heat range The distance heat must travel from the center electrode to reach the outer shell of the spark plug and enter the cylinder head.

Spark test A quick check of the ignition system; made by holding the metal spark-plug end of a spark-plug cable about ¼ inch [6 mm] from the cylinder head, or block; cranking the engine; and checking for a spark.

Specific gravity The weight per unit volume of a substance as compared with the weight per unit volume of water.

Splayed crankpins The slight spreading apart of a crankpin in a V-type engine so that each rod has its own crankpin. This reduces vibration in some V-6 engines that have a 90 degree angle between the banks.

Spool valve A rod with indented sections; used to control oil flow in automatic transmissions.

Sprag clutch In an automatic transmission, a one-way clutch; a clutch that can transmit power in one direction but not in the other.

Spring A device that changes shape under stress or force but returns to its original shape when the stress or force is removed; the component of the automotive suspension system that absorbs road shocks by flexing and twisting.

Stabilizer bar An interconnecting shaft between the two lower suspension arms; reduces body roll on turns.

Starting motor The electric motor that cranks the engine, or turns the crankshaft, for starting.

Starting-motor drive The drive mechanism and gear on the end of the starting-motor armature shaft; used to couple the starting motor to, and disengage it from, the flywheel ring-gear teeth.

Static balance The balance of an object while it is not moving.

Static friction The friction between two bodies at rest.

Stator The stationary member of a machine, such as an electric motor or generator, in or about which a rotor revolves; in an electronic ignition system, a small magnet embedded in plastic (or a light-emitting diode) which when used with a reluctor replaces contact points. Also, the third member, in addition to the turbine and pump, in a torque converter.

Steering arm The arm attached to the steering knuckle that turns the knuckle and wheel for steering.

Steering axis The centerline of the ball joints in a front-suspension system.

Steering-axis inclination (SAI) The inward tilt of the steering axis from the vertical.

Steering gear That part of the steering system that is located at the lower end of the steering shaft; changes the rotary motion of the steering wheel into the linear motion of the front wheels for steering.

Steering kickback Sharp and rapid movements of the steering wheel as the front wheels encounter obstructions in the road; the shocks of these encounters "kick back" to the steering wheel.

Steering knuckle The front-wheel spindle that is supported by upper and lower ball joints and by the wheel; the part on which a front wheel is mounted and which is turned for steering.

Steering ratio The number of degrees that the steering wheel must be turned to pivot the front wheels 1 degree.

Steering shaft The shaft extending from the steering gear to the steering wheel.

Steering system The mechanism that enables the driver to turn the wheels for changing the direction of vehicle movement.

Stoichiometric ratio In a spark-ignition engine, the ideal air-fuel-mixture ratio of 14.7:1, which must be maintained on engines with dual-bed and three-way catalytic converters.

Stratified charge In a spark-ignition engine, an air-fuel charge with a small layer or pocket of rich air-fuel mixture. The rich mixture is ignited first; then ignition spreads to the leaner mixture filling the rest of the combustion chamber. The diesel engine is a stratified-charge engine.

Streamlining The shaping of a car body or truck cab so that it minimizes air resistance and can be moved through the air with less energy.

Stroke In an engine cylinder, the distance that the piston moves in traveling from BDC to TDC or from TDC to BDC.

Strut A bar that connects the lower control arm to the car frame; used when the lower control arm is attached to the frame at only one point. Also called a brake-reaction rod.

Suction pressure The pressure at the air-conditioner compressor inlet; the compressor intake pressure, as indicated by a gauge set.

Suction-throttling valve In an air conditioner, a valve located between the evaporator and the compressor; controls the temperature of the air flowing from the evaporator, to prevent freezing of moisture on the evaporator.

Sulfation The lead sulfate that forms on battery plates as a result of the battery action that produces electric current.

Sulfuric acid See Electrolyte.

Sulfur oxides (SO_x) Acids that can form in small amounts as the result of a reaction between hot exhaust gas and the catalyst in a catalytic converter.

Supercharger In the intake system of the engine, a pump that pressurizes the ingoing air or air-fuel mixture. This increases the amount of fuel that can be burned, increasing engine power. If the supercharger is driven by the engine exhaust gas, it is called a turbocharger.

Suspension The system of springs and other parts which supports the upper part of a vehicle on its axles and wheels.

Suspension arm In the front suspension, one of the arms pivoted on the frame at one end and on the steering-knuckle support at the other end.

Sway bar See Stabilizer bar.

Switch A device that opens and closes an electric circuit.

Synchronizer A device in the transmission that synchronizes gears about to be meshed, so that no gear clash will occur.

Synthetic oil An artificial oil that is manufactured; not a natural mineral oil made from petroleum.

Temperature indicator A gauge that indicates to the driver the temperature of the engine coolant, or a light that comes on if the coolant gets too hot.

Thermal Of or pertaining to heat.

Thermal efficiency Ratio of the energy output of an engine to the energy in the fuel required to produce that energy.

Thermistor A heat-sensing device with a negative temperature coefficient of resistance; as its temperature increases, its electrical resistance decreases. Used as the sensing device for engine-temperature indicating devices.

Thermometer An instrument that measures heat intensity (temperature) by the thermal expansion of a liquid.

Thermostat A device for the automatic regulation of temperature; usually contains a temperature-sensitive element that expands or contracts to open or close off the flow of air, a gas, or a liquid.

Thermostatic cycling switch In some cycling-clutch air-conditioning systems, a switch that turns the compressor clutch off if the evaporator temperature is below freezing to prevent icing of the evaporator; used instead of an electric pressure switch.

Thermostatic expansion valve Component of a refrigeration system that controls the rate of refrigerant flow to the evaporator. Commonly called the expansion valve.

Thermostatic gauge An indicating device (for fuel quantity, oil pressure, engine temperature) that contains a thermostatic blade or blades.

Thermostatic vacuum switch A temperature-sensing device extending into the coolant; connects full manifold vacuum to the distributor when coolant overheats. The resultant spark advance causes an increase in engine rpm, which lowers the coolant temperature.

Three-way catalytic converter A catalytic converter that uses rhodium and other catalysts to limit the amounts of the air pollutants HC, CO, and NO_x in the exhaust gas.

Throttle-body injection A type of gasoline fuel-injection system which sprays fuel under pressure into the intake air at only one place, usually the throttle body on the intake manifold. Also called single-point injection.

Throttle pressure Pressure at one end of the shift valve, which changes as engine intake-manifold vacuum changes. One of the two signals that determine when an automatic transmission will shift.

Throttle valve A disk valve in the carburetor base that pivots in response to accelerator-pedal position; allows the driver to regulate the volume of air or air-fuel mixture entering the intake manifold, thereby controlling the engine speed. Also called the throttle plate.

Throwout bearing In the clutch, the bearing that can be moved in to the release levers by clutch-pedal action, to disengage the engine crankshaft from the transmission.

Thrust bearing In the engine, the main bearing that has thrust faces to prevent excessive end play, or forward and backward movement, of the crankshaft.

Tie rods In the steering system, the rods that link the pitman arm to the steering-knuckle arms; small steel components that connect the front wheels to the steering mechanism.

Timing In an engine, delivery of the ignition spark or operation of the valves (in relation to the piston position) for the power stroke. See Ignition timing and Valve timing.

Tire The casing-and-tread assembly (with or without a tube) that is mounted on a car wheel to provide pneumatically cushioned contact and traction with the road.

Tire-wear angle Usually, a reference to camber because tilting the wheel puts more load on one side of the tire tread than on the other side.

Toe The amount, in inches, millimeters, or degrees, by which the front of a front wheel points inward (toe-in) or outward (toe-out).

Top dead center (TDC) The piston position when the piston has reached the upper limit of its travel in the cylinder and the center line of the connecting rod is parallel to the cylinder walls.

Torque converter In an automatic transmission, a fluid coupling that incorporates a stator to permit a torque increase.

Torsional vibration Rotary vibration that causes a twist-untwist action on a rotating shaft, so that a part of the shaft repeatedly moves ahead of, or lags behind, the remainder of the shaft; for example, the action of a crankshaft responding to the cylinder firing impulses.

Torsion-bar spring A long, straight bar that is fastened to the frame at one end and to a control arm at the other. Spring action is produced by a twisting of the bar.

Tracking Rear wheels following directly behind (in the tracks of) the front wheels.

Transaxle A power-transmission device that combines the functions of the transmission and the drive axle (final drive and differential) into a single assembly.

Transfer case An auxiliary transmission mounted behind the main transmission. Used to divide engine power and transfer it to both front and rear differentials, either full time or part time.

Transistor A solid-state electronic device that can be used as an electric switch or as an amplifier.

Transmission An assembly of gears that provides the different gear ratios, as well as neutral and reverse, through which engine power is transmitted to the final drive to rotate the drive wheels.

Tread The part of the tire that contacts the road. It is the thickest part of the tire and is cut with grooves to provide traction for driving and stopping.

Tuneup A procedure for inspecting, testing, and adjusting an engine, and replacing any worn parts, to restore engine performance.

Turbocharger A supercharger driven by the engine exhaust gas.

Turning radius The difference between the angles each of the front wheels makes with the car frame during turns, usually measured with the outside wheel turned 20 degrees. During a turn, the inner wheel toes out more. Also called toe-out on turns.

Two-stroke cycle The two piston strokes during which fuel intake, compression, combustion, and exhaust take place in a two-stroke-cycle engine.

Universal joint In the power train, a jointed connection in the drive shaft that permits the driving angle to change.

Vacuum Negative gauge pressure, or a pressure less than atmospheric pressure. Vacuum can be measured in pounds per square inch (psi) but is usually measured in inches or millimeters of mercury (Hg); a reading of 30 inches [762 mm] Hg would indicate a perfect vacuum.

Vacuum advance The advancing (or retarding) of ignition timing by changes in intake-manifold vacuum, reflecting throttle opening and engine load. Also, a mechanism on the ignition distributor that uses intake-manifold vacuum to advance the timing of the spark to the spark splugs.

Vacuum modulator In some automatic transmissions, a device that modulates, or changes, the main-line hydraulic pressure to meet changing engine loads.

Vacuum motor Pressure-operated device with spring and diaphragm which positions a lever in different locations as the pressure changes.

Vacuum pump A mechanical device used to evacuate a system.

Vacuum switch A switch that closes or opens its contacts in response to changing vacuum conditions.

Valve A device that can be opened or closed to allow or stop the flow of a liquid or gas.

Valve body A casting located in the oil pan, which contains most of the valves for the hydraulic control system of an automatic transmission.

Valve clearance The clearance between the rocker arm and the valve-stem tip in an overhead-valve engine; the clearance in the valve train when the valve is closed.

Valve float A condition in which the engine valves do not close completely or fail to close at the proper time.

Valve guide A cylindrical part or hole in the head in which a valve is assembled and in which it moves up and down.

Valve lifter A cylindrical part of the engine which rests on a cam of the camshaft and is lifted, by cam action, so that the valve is opened. Also called a lifter, tappet, valve tappet, or cam follower.

Valve overlap The number of degrees of crankshaft rotation during which the intake and exhaust valves are open together.

Valve rotator A device installed in place of the valve-spring retainer, which turns the valve slightly as it opens.

Valve seat The surface against which a valve comes to rest to provide a seal against leaking.

Valve-seat inserts Metal rings inserted in cylinder heads to act as valve seats (usually for exhaust valves).

Valve spring The spring in each valve assembly which has the job of closing the valve.

Valve tappet See Valve lifter.

Valve timing The timing of the opening and closing of the valves in relation to the piston position.

Valve train The valve-operating mechanism of an engine; includes all components from the camshaft to the valve.

Vane A flat, extended surface that is moved around an axis by or in a fluid. Part of the internal revolving portion of an air-supply pump.

Vaporization A change of state from liquid to vapor or gas, by evaporation or boiling; a general term including both evaporation and boiling.

Vapor-liquid separator A device in the evaporative emission control system; prevents liquid fuel from traveling to the engine through the charcoal-canister vapor line.

Vapor lock A condition in the fuel system in which gasoline vaporizes in the fuel line or fuel pump; bubbles of gasoline vapor restrict or prevent fuel delivery to the carburetor.

Vapor-recovery system An evaporative emission control system that recovers gasoline vapor escaping from the fuel tank and carburetor float bowl. See Evaporative control system.

Vapor-return line A line from the fuel pump to the fuel tank through which any vapor that has formed in the pump is returned to the tank.

Variable-speed fan An engine fan that will not exceed a predetermined speed or will rotate only as fast as required to prevent engine overheating.

Variable-venturi (VV) carburetor A carburetor in which the size of the venturi changes according to engine speed and load.

Vent An opening through which air can leave an enclosed chamber.

Ventilation The circulating of fresh air through any space to replace impure air.

Venturi In the carburetor, a narrowed passageway or restriction that increases the velocity of air moving through it; produces the vacuum responsible for the discharge of fuel from the fuel nozzle.

Venturi vacuum Vacuum in the venturi of a carburetor which increases with the speed of the airflow through it.

Vibration A rapid back-and-forth motion; an oscillation.

Vibration damper A device attached to the crankshaft of an engine to oppose crankshaft torsional vibration (the twist-untwist actions of the crankshaft caused by the cylinder firing impulses). Also called a harmonic balancer.

Viscosity The resistance to flow exhibited by a liquid. A thick oil has greater viscosity than a thin oil.

Voice alert system A type of indicating device in the car which can speak several words or phrases to the driver.

Voice command system A type of interactive control system in which the driver can give the car certain spoken commands and the system responds by performing the act or by providing the information desired.

Volatility A measure of the ease with which a liquid vaporizes. Volatility has a direct relationship to the flammability of a fuel.

Voltage The force which causes electrons to flow in a conductor. The difference in electrical pressure (or potential) between two points in a circuit.

Voltage regulator A device that prevents excessive alternator or generator voltage by alternately inserting and removing a resistance in the field circuit.

Volumetric efficiency A measure of how completely the engine cylinder fills up on the intake stroke of a spark-ignition engine.

Waddle Side-to-side movement of the front or rear of the car as it moves forward, caused by the steel belt not being straight inside a radial tire.

Wankel engine A rotary engine in which a three-lobe rotor turns eccentrically in an oval chamber to produce power.

Wastegate A control device on a turbocharger to limit boost pressure, thereby preventing engine and turbocharger damage.

Water jackets The spaces between the inner and outer shells of the cylinder block or head, through which coolant circulates.

Water pump In the cooling system, the device that circulates coolant between the engine water jackets and the radiator.

Wear sensor A tab on a brake shoe that causes a squealing sound when the brake pad is worn thin.

Wheel A disk or spokes with a hub at the center (which revolves around an axle) and a rim around the outside for mounting of the tire.

Wheel alignment A series of tests and adjustments to ensure that wheels and tires are properly positioned on the vehicle.

Wheel balancer A device that checks a wheel-and-tire assembly (statically, dynamically, or both) for balance.

Wheelbase The distance between the centerlines of the front and rear axles. For trucks with tandem rear axles, the rear centerline is considered to be midway between the two rear axles.

Wheel cylinders In a hydraulic braking system, hydraulic cylinders located in the brake mechanisms at the wheels. Hydraulic pressure from the master cylinder causes the wheel cylinders to move the brake shoes into contact with the brake drums for braking.

Wheel tramp Tendency for a wheel to move up and down so it repeatedly bears down hard, or "tramps," on the road. Sometimes called high-speed shimmy.

Wiring harness A group of individually insulated wires, wrapped together to form a neat, easily installed bundle.

INDEX